THE FAMILIES
AND GENERA
OF VASCULAR PLANTS

Edited by K. Kubitzki

Springer

Berlin
Heidelberg
New York
Hong Kong
London
Milan
Paris
Tokyo

Volumes published in this series:

Volume I Pteridophytes and Gymnosperms
Edited by K.U. Kramer and P.S. Green (1990)
Date of publication: 28.9.1990

Volume II Flowering Plants. Dicotyledons. Magnoliid, Hamamelid and Caryophyllid Families
Edited by K. Kubitzki, J.G. Rohwer, and V. Bittrich (1993)
Date of publication: 28.7.1993

Volume III Flowering Plants. Monocotyledons: Lilianae (except Orchidaceae)
Edited by K. Kubitzki (1998)
Date of publication: 27.8.1998

Volume IV Flowering Plants. Monocotyledons: Alismatanae and Commelinanae (except Gramineae)
Edited by K. Kubitzki (1998)
Date of publication: 27.8.1998

Volume V Flowering Plants. Dicotyledons: Malvales, Capparales and Non-betalain Caryophyllales
Edited by K. Kubitzki and C. Bayer (2003)
Date of publication: 12.9.2002

Volume VI Flowering Plants. Dicotyledons: Celastrales, Oxalidales, Rosales, Cornales, Ericales
Edited by K. Kubitzki (2004)
Date of publication: 21.1.2004

Volume VII Flowering Plants. Dicotyledons: Lamiales (except Acanthaceae including Avicenniaceae)
Edited by J.W. Kadereit (2004)

The Families and Genera of Vascular Plants

Edited by K. Kubitzki

VII *Flowering Plants · Dicotyledons*

Lamiales (except Acanthaceae including Avicenniaceae)

Volume Editor:
J.W. Kadereit

With 60 Figures, 8 in Color

Springer

Professor Dr. Klaus Kubitzki
Universität Hamburg
Biozentrum Klein-Flottbek und Botanischer Garten
Ohnhorststraße 18
22609 Hamburg
Germany

Professor Dr. Joachim W. Kadereit
Johannes Gutenberg-Universität Mainz
Institut für Spezielle Botanik und Botanischer Garten
55099 Mainz
Germany

ISBN 3-540-40593-3 Springer-Verlag Berlin Heidelberg New York

Library of Congress Cataloging-in-Publication Data

Flowering plants, dicotyledons : Lamiales (except Acanthaceae including Avicenniaceae) / volume editor, J.W. Kadereit.
p. cm. – (The families and genera of vascular plants ; 7)
Includes bibliographical references.
ISBN 3-540-40593-3 (alk. paper)
1. Lamiales – Classification. I. Kadereit, J.W. (Joachim W.), 1956– II. Series.
QK495.A12F57 2004
583′.96 – dc22 2003065479

Springer-Verlag is a part of Springer Science+Business Media

springeronline.com

Printed in Germany

Production: PRO EDIT GmbH, Heidelberg
Cover design: design & production GmbH, Heidelberg
Typesetting: SNP Best-set Typesetter Ltd., Hong Kong
Printed on acid free paper 5 4 3 2 1 0 – 31/3150Di

Preface

This volume, the seventh in this series, deals with the families of the Lamiales, a group of the asterids or Sympetalae, which was recognised by early workers such as Bartling and Lindley under the name Labiatiflorae or Personatae, and which has only moderately been modified in circumscription over time. Nevertheless, our considerably increased knowledge of the group and the results of molecular studies provide a more detailed basis for understanding relationships among and evolution within the families of the Lamiales, although, as J.W. Kadereit emphasises in his introduction, in this connection and also in relation to family delimitation, many questions remain currently unanswered.

My sincere thanks go to all authors for their scholarly treatment of families in this volume. In particular, the authors/organisers who handled the four large families with more than 100 genera, Bignoniaceae, Gesneriaceae, Labiatae and Scrophulariaceae, deserve special mention for their herculean work. The authors also emphasised the various themes of diversification, differing from group to group, that are so important for the understanding and interpretation of taxonomic diversity, such as phytochemical diversification in Labiatae and pollinator interactions in Gesneriaceae, to cite only two salient examples. In the circumscription of Scrophulariaceae, a pragmatic position has been taken; since this series aims primarily at providing baseline information for families and genera, the problem of a phylogenetically satisfactory classification of the Scrophulariaceae complex that avoids paraphyly at our present state of knowledge must be left undecided, and indeed is not the primary focus of this volume. Sincere thanks are also extended to Professor J.W. Kadereit for patiently bringing his work on the present volume to an end, in spite of all discouraging vicissitudes.

I am most grateful to all copyright holders for so generously authorising the use of their precious illustrative material. Finally, it is a pleasure to thank, also in the name of Professor Kadereit, the staff of Springer-Verlag and ProEdit, Heidelberg, for their dedicated work on the present volume. I would like to express special thanks to the copy editor, Monique Delafontaine, for the many improvements she made to the manuscript.

Hamburg, December 2003 K. Kubitzki

Contents

List of Contributors

Atkins, S.	Royal Botanic Gardens Kew, Richmond, Surrey TW9 3AB, UK
Barthlott, W.	Botanisches Institut und Botanischer Garten der Rheinischen Friedrich-Wilhelms-Universität, Meckenheimer Allee 170, 53115 Bonn, Germany
Budantsev, A.L.	Komarov Botanical Institute, Herbarium, Prof. Popov Street, St. Petersburg 197376, Russia
Cantino, P.D.	Department of Environmental and Plant Biology, Ohio University, Athens, OH 45701-2979, USA
Carolin, R.	Pulman's Cottage, 30 Pulman Street, Berry, N.S.W. 2535, Australia
Conn, B.J.	Royal Botanic Garden, Sydney, Australia
Conran, J.G.	Environmental Biology, School of Earth and Environmental Sciences, The University of Adelaide, SA 5005, Australia
Erbar, C.	Heidelberger Institut für Pflanzenwissenschaften (HIP) – Biodiversität und Pflanzensystematik, Im Neuenheimer Feld 345, 69120 Heidelberg, Germany
Fischer, E.	Institut für Biologie, Universität Koblenz/Landau, Universitätsstr. 1, 56070 Koblenz, Germany
Grayer, R.	Royal Botanic Gardens Kew, Richmond, Surrey TW9 3AB, UK
Green, P.S.	Royal Botanic Gardens Kew, Richmond, Surrey TW9 3AB, UK
Ihlenfeldt, H.-D.	Ostseestr. 40, 24369 Waabs, Germany
Harley, M.M.	Royal Botanic Gardens Kew, Richmond, Surrey TW9 3AB, UK
Harley, R.M.	Royal Botanic Gardens Kew, Richmond, Surrey TW9 3AB, UK
Jensen, S.R.	Department of Chemistry, The Technical University of Denmark, 2800 Lyngby, Denmark
Kadereit, J.W.	Institut für Spezielle Botanik und Botanischer Garten, Johannes Gutenberg-Universität Mainz, 55099 Mainz, Germany
Kok, R. de	Royal Botanic Gardens Kew, Richmond, Surrey TW9 3AB, UK
Kornhall, P.	Department of Systematic Botany, Evolutionary Biology Centre Uppsala University, Norbyv. 18D, 752 36 Uppsala, Sweden
Krestovskja, T.	Komarov Botanical Institute, Herbarium, Prof. Popov Street, St. Petersburg 197376, Russia
Kubitzki, K.	Institut für Allgemeine Botanik, Universität Hamburg, Ohnhorststr. 18, 22609 Hamburg, Germany

LEINS, P.	Heidelberger Institut für Pflanzenwissenschaften (HIP) – Biodiversität und Pflanzensystematik, Im Neuenheimer Feld 345, 69120 Heidelberg, Germany
LINDER, H.P.	Institute for Systematic Botany, University of Zürich, Zollikerstrasse 107, 8008 Zürich, Switzerland
LOHMANN, L.G.	Missouri Botanical Garden, P.O. Box 299, St. Louis, Missouri 64507, USA
MORALES, R.	Herbario, Real Jardin Botanico, Plaza de Murillo 2, Madrid 28014, Spain
NORMAN, E.M.	Department of Biology, Stetson University, Deland, FL 32723, USA
OXELMAN, B.	Department of Systematic Botany, Evolutionary Biology Centre Uppsala University, Norbyv. 18D, 752 36 Uppsala, Sweden
PATON, A.J.	Royal Botanic Gardens Kew, Richmond, Surrey TW9 3AB, UK
RYDING, O.	Botanical Museum, University of Copenhagen, Gothersgade 130, 1123 Copenhagen K, Denmark
SCHWARZBACH, A.E.	Department of Biological Sciences, Kent State University, Kent, OH 44242, USA
SEBSEBE DEMISSEW	The National Herbarium, Science Faculty, Addis Ababa University, P.O. Box 3434, Addis Ababa, Ethiopia
SEINE, R.	European Astronaut Centre, Linder Höhe, 51147 Köln, Germany
STRUWE, L.	Department of Ecology, Evolution, and Natural Resources, Rutgers University – Cook College, 237 Foran Hall, 59 Dudley Road, New Brunswick, NJ 08901-8551, USA
THEISEN, I.	Botanisches Institut und Botanischer Garten der Rheinischen Friedrich-Wilhelms-Universität, Meckenheimer Allee 170, 53115 Bonn, Germany
THIV, M.	Staatl. Museum für Naturkunde, Rosenstein 1, 70191 Stuttgart, Germany
UPSON, T.	Cambridge University Botanic Garden, Cory Lodge, Bateman Street, Cambridge CB2 1JF, UK
WAGENITZ, G.	A. v. Haller-Institut, Abteilung Systematische Botanik, Untere Karspüle 2, 37073 Göttingen, Germany
WAGSTAFF, S.J.	Allan Herbarium, Landcare Research, P.O. Box 69, Lincoln 8152, New Zealand
WEBER, A.	Institut für Botanik, Universität Wien, Rennweg 14, 1030 Wien, Austria

Lamiales: Introduction and Conspectus

J.W. Kadereit

Lamiales as presented here are a well-supported clade of the Lamiids (Bremer et al. 2002) found in all major, recent molecular studies of the Asterids (APG II 2003; see also Wagenitz 1992). The Lamiids, comprising approximately 23,600 species in 1140 genera, are herbaceous or less commonly woody plants with often opposite leaves. The flowers are sympetalous and mostly 5-merous, often representing zygomorphic lip flowers with a showy corolla and with a single whorl of stamens (the five stamens often reduced to four or two), and a bicarpellate ovary with the two carpels in median position (Endress 1994). Anthers very often have an outgrowth of the connective into the anther locule, known as pollen sac placentoid (Hartl 1964), glandular hairs with a head with only vertical cell walls are often present, and pollen tubes appear to lack callose (Prósperi and Cocucci 1979; Cocucci 1983). Both micropylar and chalazal endosperm haustoria are common (Yamazaki 1974), endosperm formation is cellular, and embryogeny normally is of the Onagrad type (Polygonad in Oleaceae; Albach et al. 2001). Protein bodies in the nuclei are widespread (Speta 1977, 1979; Bigazzi 1984, 1989a, b). Chemically, typical constituents of the order are 6- and/or 8-hydroxylated flavone glycosides (Grayer et al. 1999), verbascosides and cornoside as characteristic caffeic acid derivatives (Mølgaard and Ravn 1988; Jensen 1992), shikimate-derived anthraquinones (Jensen 1992), C11-decarboxylated iridoids (Jensen 1991, 1992), and stachyose or other oligosaccharides as carbon storage compounds (Hegnauer 1989). As will be discussed below, several of the characters listed above, including the zygomorphic lip flowers, originated only within the order (Stevens 2001).

In large part, Lamiales as circumscribed here correspond to the Tubiflorae (Wettstein 1935; Melchior 1964) or Personatae (Drude 1885–1887; Warming and Möbius 1929; Huber 1991) of earlier authors, and to Cronquist's (1981) Lamiales, Callitrichales, Plantaginales and Scrophulariales, to Thorne's (1992) Bignoniales, and to Takhtajan's (1997) Oleales, Scrophulariales, Lamiales, Callitrichales and Hippuridales. In comparison to the latter three systems, the Lamiales as circumscribed here differ by the exclusion of Boraginaceae (incl. Lennoaceae), treated as part of Lamiales by Cronquist (1981), and by the inclusion of Byblidaceae (Cronquist 1981: Rosales; Thorne 1992: Pittosporales; Takhtajan 1997: Aralianae-Byblidales), Carlemanniaceae (Cronquist 1981: Dipsacales; Thorne 1992: not recognized; Takhtajan 1997: Rubiales) and Plocospermataceae (Cronquist 1981: Gentianales; Thorne 1992: Gentianales; Takhtajan 1997: Gentianales), treated as belonging elsewhere by Cronquist (1981), Thorne (1992) and Takhtajan (1997). There is good molecular evidence for the inclusion of these three families in Lamiales (Byblidaceae: Hedrén et al. 1995; Bremer et al. 2001; Albach et al. 2001; Carlemanniaceae: Savolainen, Fay et al. 2000; Bremer et al. 2001; Plocospermataceae: Bremer et al. 2001), and other characters in support of this can be found. Thus, Plocospermataceae contain cornoside as a caffeic acid derivative commonly found in the order (Jensen 1996, 2000), the carnivorous Byblidaceae share embryological characters with the equally carnivorous Lentibulariaceae (Lang 1901; Conran 1996), and Carlemanniaceae have flowers with only two stamens which may prove to be morphological support of their sister-group relationship with Oleaceae.

The exact relationships of Lamiales within Lamiids are not finally resolved, but it is clear from molecular evidence (e.g. Downie and Palmer 1992; Olmstead et al. 1992, 1993; Backlund and Bremer 1997; Olmstead et al. 2000; Savolainen, Chase et al. 2000; Savolainen, Fay et al. 2000; Soltis et al. 2000; Albach et al. 2001; Bremer et al. 2001; Olmstead et al. 2001) that they are part of a larger clade containing Boraginaceae/Boraginales, Gentianales, Solanales, Vahliaceae. In some of these analyses (Olmstead et al. 2000; Savolainen, Chase et al. 2000) they are sister to Gentianales. Inclusion of Vahliaceae into Lamiales should be considered (Savolainen, Fay et al. 2000; Albach et al. 2001).

Although the Lamiales clearly represent a monophylum, their division into families in this treatment differs from that proposed by APG II (2003; see Table 1).

Table 1. Families recognized in this volume compared with the APG classification

APG II	FGVP
Acanthaceae	Acanthaceae
Bignoniaceae	Bignoniaceae
= Scrophulariaceae s.str.	Buddlejaceae
Byblidaceae	Byblidaceae
= Plantaginaceae s.l.	Callitrichaceae
Calceolariaceae	Srophulariaceae s.l.
Carlemanniaceae	Carlemanniaceae
Orobanchaceae	Cyclocheilaceae
Gesneriaceae	Gesneriaceae
= Plantaginaceae s.l.	Globulariaceae
= Plantaginaceae s.l.	Hippuridaceae
Labiatae	Labiatae
Lentibulariaceae	Lentibulariaceae
Martyniaceae	Martyniaceae
= Scrophulariaceae s.str.	Myoporaceae
Orobanchaceae	Nesogenaceae
Oleaceae	Oleaceae
Orobanchaceae	= Scrophulariaceae s.l.
Paulowniaceae	= Scrophulariaceae s.l.
Pedaliaceae	Pedaliaceae
Phrymaceae	Phrymaceae
Plantaginaceae s.l.	Plantaginaceae s.str.
Plocospermataceae	Plocospermataceae
Schlegeliaceae	= Scrophulariaceae s.l.
Scrophulariaceae	Scrophulariaceae
Stilbaceae	Stilbaceae
Tetrachondraceae	Tetrachondraceae
= Pedaliaceae	Trapellaceae
Verbenaceae	Verbenaceae

Apart from the inclusion of Avicenniaceae in Acanthaceae here (Schwarzbach and McDade 2002), in contrast to APG II, and the inclusion of Trapellaceae in Pedaliaceae by APG II, also in contrast to the present work, the major difference between APG II and this FGVP volume lies in the treatment of Scrophulariaceae s.l. and allies.

Following the molecular evidence last summarized and expanded by Olmstead et al. (2001), APG II divide Scrophulariaceae s.l. and allies into Orobanchaceae, Paulowniaceae, Plantaginaceae (incl. Callitrichaceae, Globulariaceae and Hippuridaceae; these are the Veronicaceae of Olmstead et al. 2001), Schlegeliaceae and Scrophulariaceae s.str. (incl. Buddlejaceae and Myoporaceae). Calceolariaceae are recognized as an additional family by Olmstead et al. (2001), and there is increasing evidence that Phrymaceae, containing *Phryma* and several genera mainly (but not only) of Scrophulariaceae-Mimuleae, should be recognized in an expanded circumscription (Beardsley and Olmstead 2002). In contrast, Scrophulariaceae in the present treatment include Calceolariaceae, Orobanchaceae, Paulowniaceae, and Schlegeliaceae. Buddlejaceae, Callitrichaceae, Globulariaceae, Hippuridaceae, Plantaginaceae s.str. as well as Myoporaceae are treated as separate families. Also, Phrymaceae here are treated as a monotypic family. Such treatment of Scrophulariaceae s.l. is non-phylogenetic in at least one respect. It is quite clear that Scrophulariaceae s.l. are paraphyletic when Buddlejaceae, Callitrichaceae, Globulariaceae, Hippuridaceae, Myoporaceae and Plantaginaceae, and probably also Phrymaceae are treated as separate families. Buddlejaceae and Myoporaceae group with one part of Scrophulariaceae s.l. with good support (= Scrophulariaceae s.str.: 78%; Olmstead et al. 2001), and Callitrichaceae, Globulariaceae and Hippuridaceae group with another part of Scrophulariaceae s.l., although with only low support (= Plantaginaceae = Veronicaceae: 48%; Olmstead et al. 2001). The recognition of a close relationship between most of the preceding families and Scrophulariaceae is by no means novel and had already been recognized (and partly formalized) by, for example, Hallier (1903) and Hartl (1965–1974). Following Olmstead et al. (2001), recognition of Scrophulariaceae s.l. including Buddlejaceae, Callitrichaceae, Globulariaceae, Hippuridaceae, Myoporaceae and Plantaginaceae s.str. is not significantly less parsimonious than dividing them into the above five families, provided that Calceolariaceae are excluded. This latter family groups quite distant from Scrophulariaceae s.l. Considering this and the comparatively small sample of genera analysed so far, treatment of Scrophulariaceae in a wide circumscription is justifiable, but not when Buddlejaceae, Callitrichaceae, Globulariaceae, Hippuridaceae, Myoporaceae and Plantaginaceae (and probably Phrymaceae) are excluded, and *Calceolaria* and allies are included. However, it is very likely that eventually Calceolariaceae, Orobanchaceae, Paulowniaceae, Plantaginaceae/Veronicaceae (incl. Callitrichaceae, Globulariaceae and Hippuridaceae), Schlegeliaceae, Scrophulariaceae s.str. (incl. Buddlejaceae and Myoporaceae) and an expanded Phrymaceae will have to be recognized because these lineages are, at least partly, more closely related to other families of Lamiales than to each other. Accordingly, the treatment of Scrophulariaceae s.l. (see Fischer, this Vol.) indicates the likely contents of the above lineages.

As regards the treatment of Labiatae/ Verbenaceae here and in APG II, in comparison with mainly Cronquist (1981) but partly also Thorne (1992) and Takhtajan (1997), major parts of Verbenaceae as traditionally circumscribed have been transferred to Labiatae, based on substantial morphological and molecular evidence (Cantino 1992a, b; Wagstaff and Olmstead 1997; Wagstaff et al. 1998), and Tetrachondraceae, treated as part of Labiatae by Cronquist (1981), and Cyclocheilaceae, Nesogenaceae, Phrymaceae and Stilbaceae, treated as part of Verbenaceae by Cronquist (1981; Cyclocheilaceae, Nesogenaceae not mentioned), are treated as separate families here. In terms of this transfer, Verbenaceae can be recognized mainly by their racemose inflorescence, their usually colporate or porate pollen grains and their stigmatic lobes which are apically thickened with stigmatic tissue, and Labiatae by their primarily cymose inflorescence (sometimes racemose through reduction), usually colpate pollen grains and slender stigmatic lobes. The ovules are attached to the sides of the false septa of the ovary in the Labiatae, whereas in the Verbenaceae they are attached to the carpel margins. Verbenaceae and Labiatae do not appear to be closely related to each other (Olmstead and Reeves 1995; Wagstaff and Olmstead 1997; Albach et al. 2001). Of the above five segregate families of Labiatae/Verbenaceae, the Tetrachondraceae, Cyclocheilaceae, Phrymaceae and Stilbaceae were analysed by Wagstaff and Olmstead (1997) and did not group with either Labiatae or Verbenaceae. The same applies to Nesogenaceae (Bremer et al. 2001).

Although family delimitation is reasonably unproblematic apart from Scrophulariaceae s.l., family interrelationships are still largely unclear. It seems likely that Plocospermataceae are sister to the remainder of the order (Oxelman et al. 1999; Savolainen, Fay et al. 2000), followed by Oleaceae plus Carlemanniaceae as a clade (Savolainen, Fay et al. 2000; Bremer et al. 2001), and Tetrachondraceae (Oxelman et al. 1999; Olmstead et al. 2000; Savolainen, Fay et al. 2000). There is evidence that, of the remaining and mostly lip-flowered families, Gesneriaceae are sister to the rest (Olmstead et al. 2001; Bremer et al. 2002). Beyond that, there is little statistical support for any relationship which led Savolainen, Fay et al. (2000) to suggest to combine all families (except Carlemanniaceae, Oleaceae and Plocospermataceae; Tetrachondraceae not included) in Scrophulariaceae s.l.

The distribution of the basal lineages Plocospermataceae (New World), Oleaceae (Old World and New World, with an exclusive occurrence of basal lineages in the Old World; Wallander and Albert 2000), Carlemanniaceae (Old World) and Tetrachondraceae (New World and Old World), in combination with the mostly very wide distribution of the remaining families, does not readily suggest where the order might have originated. The origin of the order dates back at least to the Eocene (Magallón and Sanderson 2001).

The above phylogenetic topology allows to trace the evolution of at least some characters in the order. Most notably, the zygomorphic lip flower with a variable number of petals forming the lower and upper lip (Donoghue et al. 1998; Endress 1999), and with a usually (albeit variably) reduced number of stamens (Endress 1999), appears to have originated only after the divergence of the three basal lineages (Plocospermataceae, Oleaceae + Carlemanniaceae, Tetrachondraceae). Somewhat zygomorphic flowers with only two stamens, however, are also found in Carlemanniaceae, and Oleaceae typically have two stamens only. Within the lip-flowered Lamiales, a reversal to actinomorphic flowers appears to have taken place in Byblidaceae at the family level, but many times within several families (Endress 1994; Donoghue et al. 1998; Endress 1999). Although Lamiales have the second-highest diversification rate among (investigated) angiosperm orders (Magallón and Sanderson 2001), this can not be explained by its lip flowers and associated flower biological diversification because high diversification rates are characteristic of all euasterids (Magallón and Sanderson 2001). Shikimate-derived anthraquinones (acetate-derived in Oleaceae; Jensen 1992), stachyose and other oligosaccharides as carbon storage compounds (mannitol in Oleaceae), 6- and/or 8-hydroxylated flavone glycosides (not present in Oleaceae; Grayer et al. 1999), as well as chalazal endosperm haustoria also appear to be synapomorphies of only the lip-flowered Lamiales (Stevens 2001). Iridoids of the IIa biosynthetic route sensu Jensen (1991) may have originated only after the divergence of Gesneriaceae. No iridoids have been found in Plocospermataceae (Jensen 1992) and Gesneriaceae, and Oleaceae have iridoids originating along a different biosynthetic pathway (Jensen 1991; Albach, Soltis and Soltis 2001). Finally, anatomical evidence (Carlquist 1992) suggests that while woodiness is primary in Oleaceae, it may be secondary at least in some representatives of several of the lip-flowered Lamiales (Acanthaceae, Gesneriaceae, Labiatae, Scrophulariaceae s.l., Stilbaceae, Verbenaceae).

The very close relationship among the different lineages of the lip-flowered Lamiales probably is also reflected in many parallel evolutionary trends in the different families. As summarized by Endress (1994), such trends are found, for example, in petal aestivation (ascending, descending, quincuncial, contort), stamen number (reduction from 4 to 2, reduction or loss of upper or lower pair; reversal to full complement of 5; Endress 1999), ovary structure (bi- or unilocular through variation of length of synascidiate zone as well as degree of protrusion of placentas), ovule number (reduction from many to few), anther structure and synorganization (monothecal, synthecal) and other characters often closely associated with pollination syndromes (Endress 1994).

The absence of well-supported hypotheses on the phylogenetic relationships among and, in most cases, also within the families of the lip-flowered Lamiales, the difficulties in family delimitation, particularly in Scrophulariaceae s.l., and the often very wide geographical distribution of the above taxa or assemblages make any attempt to explain the evolution of this group in terms of geographical and/or ecological diversification highly speculative. In addition, summary statements about geographical distribution and ecological properties, particularly of the very large families, are far from accurate.

However, a certain degree of geographical and ecological vicariance perhaps can be recognized when the large families or assemblages, i.e. Acanthaceae, Bignoniaceae, Gesneriaceae, Labiatae, Scrophulariaceae s.l. and Verbenaceae, are considered, and the geographical distribution of basal lineages, as far as identified, is accepted as at least an indication of their geographical origin. The possible sister-group relationship of constituent lineages of Scrophulariaceae s.l. to the remaining large families can not be taken into account in the following argument because such relationships are not yet clear.

Of the large families, Scrophulariaceae s.l. and Labiatae have their distribution centres mainly in extratropical zones (or additionally in tropical mountains in the case of Scrophulariaceae) and there often inhabit open habitats. If Viticoideae, Symphorematoideae and Prostantheroideae-Chloantheae of the Labiatae indeed should prove to be basal lineages of the family (Cantino et al. 1992; Wagstaff and Olmstead 1997; Harley et al., this volume), its origin might lie in the Old World tropics. This may also apply to Gesneriaceae, where the possibly basal lineages are distributed mainly in the Old World (Smith et al. 1997; Burtt 1998). In contrast, the geographical distribution of Verbenaceae suggests a New World origin (Atkins, this volume), and Bignoniaceae are distributed mainly in the New World but their origin in either the Old or New World is equivocal, considering the pantropical distribution of the presumably basal Tecomeae (Spangler and Olmstead 1999). The origin of Acanthaceae in the Old or New World is equally unclear because presumably basal lineages of this family (Nelsonioideae: Old and New World, Thunbergioideae: Old World, Mendoncioideae: Old World; Hedrén et al. 1995; Scotland et al. 1995; McDade and Moody 1999; Schwarzbach and McDade 2002) are distributed in both the Old and New World. By containing many lianas and trees, Bignoniaceae could be regarded as ecologically vicarious with the mostly herbaceous Acanthaceae and Gesneriaceae, but not in comparison with the often woody (although commonly shrubby) Verbenaceae. What is most difficult to understand is the wide sympatry of Acanthaceae and Gesneriaceae. Although Gesneriaceae contain a large number of epiphytic taxa in the Neotropics, and Acanthaceae, with their mostly active ejaculation of seeds, are specialized in terms of dispersal ecology, this is not likely to be explanation enough. Possibly their phytochemical differentiation (Gesneriaceae: lack of iridoids; Acanthaceae: calcium oxalate cystoliths), perhaps implying specializations in terms of herbivore and pathogene defence, allows the largely sympatric distribution of these two families. This hypothesis, however, has no observational basis.

When such considerations are extended to the smaller families of the order, Byblidaceae and Lentibulariaceae are distinguished by their carnivory, Stilbaceae by their largely ericoid habit and limitation to the Cape region of South Africa, Trapellaceae by being aquatic (a common condition also in parts of Scrophulariaceae s.l.), and Pedaliaceae and Martyniaceae by their occurrence in arid regions and their very specialized fruit morphology.

Conspectus of families as treated in this volume

1. Ovary with a single ovule
  2. Leaves verticillate, perianth a low collar, flowers with 1 stamen; aquatic, amphibious or rarely terrestrial herbs with drupaceous fruits. 1/1, temperate to boreal parts of both hemispheres **Hippuridaceae**
  2. Leaves alternate or opposite, perianth of calyx and corolla, flowers with 4 stamens
    3. Leaves alternate, flowers mostly in a capitulum, rarely in a spike; shrubs, subshrubs or perennial herbs with woody base, corolla zygomorphic, fruit a nut. 2/25, western Eurasia, Macaronesia, North Africa, Somalia and Sokotra **Globulariaceae**
    3. Leaves opposite, flowers in a raceme; perennial herbs, corolla zygomorphic, fruit a nut. 1/1, temperate eastern North America and eastern Asia **Phrymaceae**
1. Ovary with at least two ovules, sometimes only one developing into seed
  4. Mangrove trees or shrubs with erect pneumatophores, cosmopolitan **Avicenniaceae** (included in Acanthaceae; not treated in this volume)
  4. Not as above
    5. Plants carnivorous with various types of leaf traps; leaves alternate, fruits capsular
      6. Corolla more or less actinomorphic, nectary spur absent; placentation axile; evergreen shrubs or annual herbs. 1/6, Australia and New Guinea **Byblidaceae**
      6. Corolla zygomorphic, nectary spur present; placentation free central; plants herbaceous. 3/320, cosmopolitan **Lentibulariaceae**
    5. Plants not carnivorous
      7. Habit ericoid, leaves densely crowded, verticillate; shrubs or shrublets with actinomorphic or zygomorphic corolla and capsular fruits. 6/28, South Africa **Stilbaceae**
      7. Habit not ericoid
        8. Flowers unisexual, male flowers with a single stamen, or flowers functionally unisexual (anthers sterile in functional females, ovary without style and stigma in functional males), style of functionally female flowers with 4 branches
          9. Flowers unisexual, male flowers with 1 stamen; aquatic to terrestrial herbs with decussate leaves and schizocarpic fruits. 1/40–50, cosmopolitan **Callitrichaceae**
          9. Flowers functionally unisexual, functionally male flowers with 5 (or 6) stamens, style of functionally female flowers with 4 slender branches; terrestrial shrubs with actinomorphic corolla and capsular fruits. 1/1, northern Central America **Plocospermataceae**
        8. Flowers usually bisexual, when unisexual male flowers with 2 or more stamens or female flowers not with style of 4 branches
          10. Calyx reduced to rim or absent, flowers enclosed by 2 large bracteoles; subshrubs to shrubs with subopposite to opposite leaves, zygomorphic corolla and capsular or schizocarpic fruits. 2/4, Horn of Africa to Arabia **Cyclocheilaceae**
          10. Calyx present and not reduced to rim
            11. Corolla scarious; herbaceous or rarely woody plants with parallel-veined and usually alternate leaves and actinomorphic corolla, fruit a capsule or nut. 1/270, widespread in temperate regions and tropical mountains **Plantaginaceae**
            11. Corolla not scarious
              12. Flowers minute ($\ll$1 cm), actinomorphic, tetramerous, style gynobasic; small perennial herbs with opposite leaves, fruit a capsule or schizocarp. 2/3, America and New Zealand **Tetrachondraceae**
              12. Flowers usually larger, actinomorphic or zygomorphic, tetra- or mostly pentamerous; if style gynobasic, flowers not actinomorphic
                13. Stamens 2, no staminodes present, ovary inferior; terrestrial perennial herbs or shrubs with opposite leaves, corolla 4- or 5-merous, weakly zygomorphic to subregular, fruit a dry or fleshy capsule. 2/5, tropical Asia **Carlemanniaceae**
                13. Not as above
                  14. Plants with conspicuous mucilaginous trichomes on all aerial parts, pedicels usually with 2 to several extrafloral nectaries; woody to herbaceous plants with usually zygomorphic corolla, fruit dehiscent or not, usually with emergences such as spines, horns and wings. 13/70, arid parts of the Old World **Pedaliaceae**
                  14. Not as above
                    15. Corolla tetramerous, actinomorphic, plants usually woody
                      16. Stamens 2, ovules usually 2 per locule; trees, shrubs or climbers with mostly opposite leaves, fruit a drupe, capsule, berry or samara. 25/600, cosmopolitan **Oleaceae**
                      16. Stamens 4, ovules usually more than 2 per locule; trees, shrubs, climbers or suffrutescent herbs with mostly opposite leaves, fruit usually a capsule. 5/100, widely distributed in subtropical to tropical regions **Buddlejaceae**
                    15. Corolla usually pentamerous, zygomorphic, plants woody or herbaceous; if flowers tetramerous and actinomorphic, plants with 4 stamens and two ovules per locule

17. Plants aquatic, ovary inferior, abaxial carpel abortive and empty; rhizomatous perennial herbs with opposite leaves, corolla zygomorphic, fruit indehiscent. 1/1–2, eastern Asia **Trapellaceae**
17. Not as above
    18. Locules usually 1- or 2-ovulate, often divided by false septum
        19. Locule 1-ovulate, false septum absent, fruit dry, indehiscent, 2-seeded; plants herbaceous with opposite leaves, corolla zygomorphic. 1/9, southern Pacific and east coast of Africa **Nesogenaceae**
        19. Locule 2-ovulate, fruit most often drupaceous or a dry 4-parted schizocarp; trees, shrubs and herbs
            20. Stigma lobes thickened with conspicuous stigmatic tissue, inflorescence racemose, pollen usually colporate or porate, leaves usually opposite; herbaceous to woody plants, corolla usually zygomorphic and fruits drupaceous or schizocarpic. 34/1200, mainly New World **Verbenaceae**
            20. Stigma lobes slender, stigmatic tissue inconspicuous, inflorescence usually thyrsic but sometimes racemose, pollen usually colpate, leaves usually opposite; herbaceous to woody plants, corolla usually zygomorphic and fruits drupaceous or schizocarpic. 236/7173, cosmopolitan **Labiatae**
    18. Locules usually more than 2-ovulate, when 2-ovulate, not divided by false septum
        21. Fruit dehiscence usually explosive, seeds with enlarged funiculus serving seed dispersal, cells often with cystoliths; herbs, rarely climbers, shrubs or trees, leaves mostly opposite, corolla zygomorphic; mainly tropical **Acanthaceae** (not included in this volume)
        21. Fruit dehiscence usually not explosive, seeds without enlarged funiculus, cells usually without cystoliths
            22. Seeds usually flattened, winged, leaves mostly opposite, often compound; trees, shrubs or lianas, flowers zygomorphic, fruit usually a capsule. 104/860, mainly subtropics to tropics **Bignoniaceae**
            22. Seeds usually not flattened, not winged, plants woody to herbaceous
                23. Plants with secretory cavities visible as pellucid dots in the leaves; small trees or shrubs with mostly alternate leaves, corolla actinomorphic or zygomorphic, fruit a drupe, capsule or schizocarp. 4/330, temperate to tropical, mainly in southern hemisphere **Myoporaceae**
                23. Not as above
                    24. Fruits dehiscent or indehiscent, usually with two apical horns (sometimes indistinct), exocarp detaching at maturity; plants herbaceous with usually opposite leaves and conspicuous glandular trichomes, corolla zygomorphic. 5/16, mostly arid parts of the New World **Martyniaceae**
                    24. Not as above
                        25. Flowers usually in pair-flowered cymes in the axils of foliage leaves, rarely cymes reduced to single flowers and/or subtending leaves reduced to bracts; anthers often cohering in pairs or all coherent; ovary usually unilocular with parietal placentae; leaves usually opposite or whorled, rarely alternate; herbs, rarely shrubs or trees, corolla usually zygomorphic, fruit a capsule or berry. 140–150/3500, mainly subtropical and tropical **Gesneriaceae**
                        25. Flowers usually in conventional cymes in leaf or bract axils, cymes often reduced to single flowers; anthers usually not coherent; ovary usually bilocular with axile placentae; leaves often alternate (the basal ones sometimes opposite); herbs, rarely climbers, shrubs or trees, corolla usually zygomorphic, fruit usually a capsule, rarely a berry or drupe. 306/5850, cosmopolitan, mainly extratropical or in tropical mountains **Scrophulariaceae** (s.l.)

References

Albach, D.C., Soltis, P.S., Soltis, D.E. 2001. Patterns of embryological and biochemical evolution in the Asterids. Syst. Bot. 26: 242–262.

Albach, D.C., Soltis, P.S., Soltis, D.E., Olmstead, R.G. 2001. Phylogenetic analysis of Asterids based on sequences of four genes. Ann. Missouri Bot. Gard. 88: 163–212.

APG (Angiosperm Phylogeny Group) 2003. An update of the angiosperm phylogeny group classification for the orders and families of flowering plants: APG II. Bot. J. Linn. Soc. 141: 399–436.

Backlund, A., Bremer, B. 1997. Phylogeny of the Asteridae s.str. based on *rbc*L sequences, with particular reference to the Dipsacales. Plant Syst. Evol. 207: 225–254.

Beardsley, P.M., Olmstead, R.G. 2002. Redefining Phrymaceae: the placement of *Mimulus*, tribe Mimuleae, and *Phryma*. Am. J. Bot. 89: 1093–1102.

Bigazzi, M. 1984. The occurrence of intranuclear inclusions in the Labiatae, Verbenaceae and Scrophulariaceae. Caryologia 37: 269–292.

Bigazzi, M. 1989a. Ultrastructure of nuclear inclusions and the separation of Verbenaceae and Oleaceae (incl. *Nyctanthes*). Plant Syst. Evol. 163: 1–12.

Bigazzi, M. 1989b. Occurrence, ultrastructure and developmental features of nuclear inclusions in the tribe Antirrhineae (Scropulariaceae). I, II. Caryologia 42: 313–343.

Bremer, K., Backlund, A., Sennblad, B., Swenson, U., Andreasen, K., Hjertson, M., Lundberg, J., Backlund, M., Bremer, B. 2001. A phylogenetic analysis of 100+ genera and 50+ families of euasterids based on morphological and molecular data with notes on possible higher level morphological synapomorphies. Plant Syst. Evol. 229: 137–169.

Bremer, B., Bremer, K., Heidari, N., Olmstead, R.G., Anderberg, A.A., Källersjö, M., Barkhordarian, E. 2002. Phylogenetics of asterids based on 3 coding and 3 non-coding chloroplast DNA markers and the utility of non-coding DNA at higher taxonomic levels. Mol. Phylogenet. Evol. 24: 274–301.

Burtt, B.L. 1998. Climatic accommodation and phytogeography of the Gesneriaceae of the Old World. In: Mathew, P., Sivadasan, M. (eds.) Diversity and taxonomy of tropical flowering plants. Calicut: Mentor Books.

Cantino, P.D. 1992a. Evidence for a polyphyletic origin of the Labiatae. Ann. Missouri Bot. Gard. 79: 361–379.

Cantino, P.D. 1992b. Toward a phylogenetic classification of the Labiatae. In: Harley, R.M., Reynolds, T. (eds.) Advances in Labiate science. Kew: Royal Botanic Gardens, pp. 27–37.

Cantino, P.D., Harley, R.M., Wagstaff, S.J. 1992. Genera of Labiatae: status and classification. In: Harley, R.M., Reynolds, T. (eds.) Advances in Labiate science. Kew: Royal Botanic Gardens, pp. 511–522.

Carlquist, S. 1992. Wood anatomy of sympetalous dicotyledon families: a summary, with comments on systematic relationships and evolution of the woody habit. Ann. Missouri Bot. Gard. 79: 303–332.

Cocucci, A.E. 1983. New evidence from embryology in angiosperm classification. Nordic J. Bot. 3: 67–73.

Conran, J.G. 1996. The embryology and relationships of the Byblidaceae. Austral. Syst. Bot. 9: 243–254.

Cronquist, A. 1981. An integrated system of classification of flowering plants. New York: Columbia University Press.

Donoghue, M.J., Ree, R.H., Baum, D.A. 1998. Phylogeny and the evolution of flower symmetry in the Asteridae. Trends Plant Sci. 3: 311–317.

Downie, S.R., Palmer, J.D. 1992. Restriction site mapping of the chloroplast DNA inverted repeat: a molecular phylogeny of the Asteridae. Ann. Missouri Bot. Garden 79: 266–283.

Drude, O. 1885–1887. Die systematische und geographische Anordnung der Phanerogamen. In: Schenk, A. (ed.) Handbuch der Botanik III, 2: 175–496.

Endress, P.K. 1994. Diversity and evolutionary biology of tropical flowers. Cambridge: Cambridge University Press.

Endress, P.K. 1999. Symmetry in flowers: diversity and evolution. Int. J. Plant Sci. 160 (Suppl. 6): S3–S23.

Grayer, R.J., Chase, M.W., Simmonds, M.S.J. 1999. A comparison between chemical and molecular characters for the determination of phylogenetic relationships among plant families: an appreciation of Hegnauer's "Chemotaxonomie der Pflanzen". Biochem. Syst. Ecol. 27: 369–393.

Hallier, H. 1903. Über die Abgrenzung und Verwandtschaft der einzelnen Sippen bei den Scrophulariaceae. Bull. Herb. Boissier, 2ème sér. 3: 181–207.

Hartl, D. 1964. Das Placentoid der Pollensäcke, ein Merkmal der Tubifloren. Ber. Dtsch. Bot. Ges. 76: 70–72.

Hartl, D. 1965–1974. Scrophulariaceae. In: Hegi, G. (ed.) Illustrierte Flora von Mitteleuropa 6 (1), 2. Aufl. München: Lehmann.

Hedrén, M., Chase, M.W., Olmstead, R.G. 1995. Relationships in the Acanthaceae and related families as suggested by cladistic analysis of *rbc*L nucleotide sequences. Plant Syst. Evol. 194: 93–109.

Hegnauer, R. 1989. Chemotaxonomie der Pflanzen. Bd. 8. Basel: Birkhäuser.

Huber, H. 1991. Angiospermen. Leitfaden durch die Ordnungen und Familien der Bedecktsamer. Stuttgart: G. Fischer.

Jensen, S.R. 1991. Plant iridoids, their biosynthesis and distribution in angiosperms. In: Harborne, J.B., Tomas-Barberan, F.A. (eds.) Ecological chemistry and biochemistry of plant terpenoids. Oxford: Oxford University Press, pp. 133–158.

Jensen, S.R. 1992. Systematic implications of the distribution of iridoids and other chemical compounds in the Loganiaceae and other families of the Asteridae. Ann. Missouri Bot. Gard. 79: 284–302.

Jensen, S.R. 1996. Phenylethanoid glycosides in *Sanango racemosa* and in the family Gesneriaceae. Phytochemistry 43: 777–783.

Jensen, S.R. 2000. Chemical relationships of *Polypremum procumbens*, *Tetrachondra hamiltonii* and *Peltanthera floribunda*. Biochem. Syst. Ecol. 28: 45–51.

Lang, F.X. 1901. Untersuchungen über Morphologie, Anatomie und Samenentwicklung von *Polypomphylox* und *Byblis gigantea*. Flora 88: 149–206.

Magallón, S., Sanderson, M.J. 2001. Absolute diversification rates in angiosperm clades. Evolution 55: 1762–1780.

McDade, L.A., Moody, M.L. 1999. Phylogenetic relationships among Acanthaceae: evidence from non-coding *trnL-trnF* chloroplast DNA sequences. Am. J. Bot. 86: 70–80.

Melchior, H. 1964. Syllabus der Pflanzenfamilien II. Berlin: Bornträger.

Mølgaard, P., Ravn, H. 1988. Evolutionary aspects of caffeoyl ester distribution in dicotyledons. Phytochemistry 27: 2411–2421.

Olmstead, R.G., Reeves, P.A. 1995. Evidence for the polyphyly of the Scrophulariaceae based on chloroplast *rbc*L and *ndh*F sequences. Ann. Missouri Bot. Gard. 82: 176–193.

Olmstead, R.G., Michaels, H.J., Scott, K.M., Palmer, J.D. 1992. Monophyly of the Asteridae and identification of their

major lineages inferred from DNA sequences of *rbc*L. Ann. Missouri Bot. Gard. 79: 249–265.
Olmstead, R.G., Bremer, B., Scott, K., Palmer, J.D. 1993. A parsimony analysis of the Asteridae s.l. based on *rbc*L sequences. Ann. Missouri Bot. Gard. 80: 700–722.
Olmstead, R.G., Kim, K.-J., Jansen, R.K., Wagstaff, S.J. 2000. The phylogeny of the Asteridae sensu lato based on chloroplast *ndh*F gene sequences. Mol. Phylogenet. Evol. 16: 96–112.
Olmstead, R.G., DePamphilis, C.W., Wolf, A.D., Yound, N.D., Elisons, W.J. Reeves, P.A. 2001. Disintegration of the Scrophulariaceae. Am. J. Bot. 88: 348–361.
Oxelman, B., Backlund, M., Bremer, B. 1999. Relationships of the Buddlejaceae s.l. investigated using parsimony jackknife and branch support analysis of chloroplast *ndh*F and *rbc*L sequence data. Syst. Bot. 24: 164–182.
Prósperi, C.H., Cocucci, A.E. 1979. Importancia taxonómica de la calosa de los tubos polínicos en Tubiflorae. Kurtziana 12/13: 75–81.
Savolainen, V., Chase, M.W., Hoot, S.B., Morton, C.M., Soltis, D.E., Bayer, C., Fay, M.F., De Bruijn, A.Y., Sullivan, S. Qiu, Y.-L. 2000. Phylogenetics of flowering plants based on combined analysis of plastid *atp*B and *rbc*L gene sequences. Syst. Biol. 49: 306–362.
Savolainen, V., Fay, M.F., Albach, D.C., Backlund, A., van der Bank, M., Cameron, K.M., Johnson, S.A., Lledo, M.D., Pintaud, J.-C., Powell, M., Sheahan, M.C., Soltis, D.E., Soltis, P.S., Weston, P., Whitten, W.M., Wurdack, K.J., Chase, M.W. 2000. Phylogeny of the eudicots: a nearly complete familial analysis based on *rbc*L gene sequences. Kew Bull. 55: 357–309.
Schwarzbach, A.E., McDade, L.A. 2002. Phylogenetic relationships of the mangrove family Avicenniaceae based on chloroplast and nuclear ribosomal DNA sequences. Syst. Bot. 27: 84–98.
Scotland, R.W., Sweere, J.A., Reeves, P.A., Olmstead, R.G. 1995. Higher-level systematics of Acanthaceae determined by chloroplast DNA sequences. Am. J. Bot. 82: 266–275.
Smith, J.F., Wolfram, J.C., Brown, K.D., Carroll, C.L., Denton, D.S. 1997. Tribal relationships in the Gesneriaceae: evidence from DNA sequences of the chloroplast gene *ndh*F. Ann. Missouri Bot. Gard. 84: 50–66.
Soltis, D.E., Soltis, P.S., Chase, M.W., Mort, M.E., Albach, D.C., Zanis, M., Savolainen, V., Hahn, W.H., Hoot, S.B., Fay, M.F., Axtell, M., Swensen, S.M., Prince, L.M., Kress, W.J., Nixon, K.C., Farris, J.S. 2000. Angiosperm phylogeny inferred from 18S rDNA, *rbc*L and *atp*B sequences. Bot. J. Linn. Soc. 133: 381–461.
Spangler, R.E., Olmstead, R.G. 1999. Phylogenetic analysis of Bignoniaceae based on the cpDNA gene sequences *rbc*L and *ndh*F. Ann. Missouri Bot. Gard. 86: 33–46.
Speta, F. 1977. Proteinkörper in Zellkernen: neue Ergebnisse und deren Bedeutung für die Gefäßpflanzensystematik nebst einer Literaturübersicht für die Jahre 1966–1976. Candollea 32: 133–163.
Speta, F. 1979. Weitere Untersuchungen über Proteinkörper im Zellkern und ihre taxonomische Bedeutung. Plant Syst. Evol. 132: 1–26.
Stevens, P.F. 2001 onwards. Angiosperm Phylogeny website. Version 3, May 2002. http://www.mobot.org/MOBOT/research/APweb/.
Takhtajan, A. 1997. Diversity and classification of flowering plants. New York: Columbia University Press.
Thorne, R.F. 1992. An updated phylogenetic classification of flowering plants. Aliso 13: 365–389.
Wagenitz, G. 1992. The Asteridae: evolution of a concept and its present status. Ann. Missouri Bot. Gard. 79: 209–217.
Wagstaff, S.J., Olmstead, R.G. 1997. Phylogeny of Labiatae and Verbenaceae inferred from *rbc*L sequences. Syst. Bot. 22: 165–177.
Wagstaff, S.J., Hickerson, L., Spangler, R., Reeves, P.A., Olmstead, R.G. 1998. Phylogeny in Labiatae s.l., inferred from cpDNA sequences. Plant Syst. Evol. 209: 265–274.
Wallander, E., Albert, V.A. 2000. Phylogeny and classification of Oleaceae based on *rps16* and *trnL-F* sequence data. Am. J. Bot. 87: 1827–1841.
Warming, E., Möbius, M. 1929. Handbuch der systematischen Botanik. 4. Aufl. Berlin: Bornträger.
Wettstein, R. 1935. Handbuch der systematischen Botanik. 4. Aufl. Leipzig: F. Deuticke.
Yamazaki, T. 1974. A system of Gamopetalae based on the embryology. Fac. Sci. Univ. Tokyo Sect. 3 (Bot.) 11: 263–281.

Bignoniaceae

E. Fischer, I. Theisen and L.G. Lohmann

Bignoniaceae Durande, Notions Elém. Bot.: 256 (1782).

Usually trees or lianas, sometimes shrubs, rarely herbs; lianas mostly with unusual vascular structure. Leaves opposite, sometimes verticillate, rarely alternate, 2–3-foliolate, pinnate or palmate, less often simple, terminal pinna of liana species often modified into a tendril; axillary bud scales (pseudostipules) often present. Flowers in thyrses, racemes or solitary, terminal or axillary, usually conspicuous. Calyx with five sepals, sometimes bilobed or unlobed, rarely with calyptra. Corolla with five petals, often 2-lipped, rarely subrotate, imbricate or rarely valvate. Androecium attached to tube, stamens 4, didynamous in 2 pairs, the fifth (adaxial) stamen staminodial or absent, rarely all 5 stamens fertile or 2 fertile and 3 staminodial. Ovary with annular or cupular nectary disc, style with 2-lobed stigma, ovary superior, 2-carpellate, bilocular with a separating septum, sometimes unilocular or 4-locular, placentation axile or in unilocular ovary with 2 or 4 intruding parietal placentae, ovules numerous, anatropous or hemitropous. Fruit a bivalved capsule, often with a replum, rarely fleshy and indehiscent. Seeds usually flat, winged, less often with thicker, corky seed body and wings reduced or absent, sometimes embedded in pulp, endosperm usually absent.

A pantropical family with 104 genera and about 860 species, mainly distributed in the Neotropics.

Vegetative Morphology. The family contains predominantly lianas and shrubs in Bignonieae, and trees or shrubs and rarely lianas in Tecomeae, often reaching up to 30 m. In Tecomeae, exceptionally herbaceous genera are present (e.g. *Argylia*, *Incarvillea*). Crescentieae and Coleeae are trees and shrubs, Tourretieae and Eccremocarpeae herbaceous to wiry climbers. Twigs are usually solid, rarely hollow (e.g. *Stizophyllum*). Some genera (e.g. *Catophractes*, *Rhigozum* in Tecomeae, *Sphingiphila* in Bignonieae, *Parmentiera* in Crescentieae, Siegert and Wagner 1994) have distinct spines. Leaves are usually pinnately to tripinnately compound, trifoliate, or palmately compound, sometimes simple, opposite or occasionally verticillate, rarely alternate, the terminal pinna of lianas is usually modified into a simple, bifid, trifid or multifid tendril (Fig. 1A), in some genera with adhesive discs (e.g. *Distictis*, *Pithecoctenium*), or uncinate, cat-claw-like (e.g. *Macfadyena*, *Parabignonia*, *Melloa*, *Dolichandra*). Compound leaves often are 2–3-foliolate. *Phyllarthron* has leaves formed by a linear series of superimposed articules representing the winged rachis and petiole of a reduced compound leaf. Stipules are absent but leafy, stipule-like bud scales (pseudostipules) are often present in Bignonieae. Interpetiolar glandular fields and interpetiolar ridges are often present in Bignonieae.

Vegetative Anatomy. Leaf and stem anatomy show features of other Lamiales (Metcalfe and Chalk 1950). The lianas and shrubs of tribe Bignonieae are distinctive by showing anomalous secondary thickening resulting from the failure of certain segments of an otherwise normal cambium to produce secondary xylem while continuing the production of phloem. Thus, a secondary xylem interrupted by deep furrows or wedges of phloem is produced, resulting in a 4- or many-lobed (up to 32) or furrowed xylem cylinder with surrounding 4 or up to 32 phloem arms in cross section. This includes the development of wedge-shaped masses of phloem in the xylem, and the occurrence of secondary rings of xylem and phloem in the secondary cortex. All genera of tribe Bignonieae have anomalous wood structure, while this feature is absent in liana genera of Tecomeae. Temperate zone genera have ring-porous wood, the tropical genera do not. Wood rays vary from uniseriate to eight cells wide and from homogeneous to heterogeneous, parenchyma is paratracheal, aliform, confluent or in concentric bands (Gentry 1980). Wood fibres are sometimes septate and ripple marks are present or absent (Record and Hess 1940). The indumentum comprises various simple, forked, stellate, dendroid and thick-stellate trichomes (Gentry 1980). Lepidote scales and glands are of taxonomic importance.

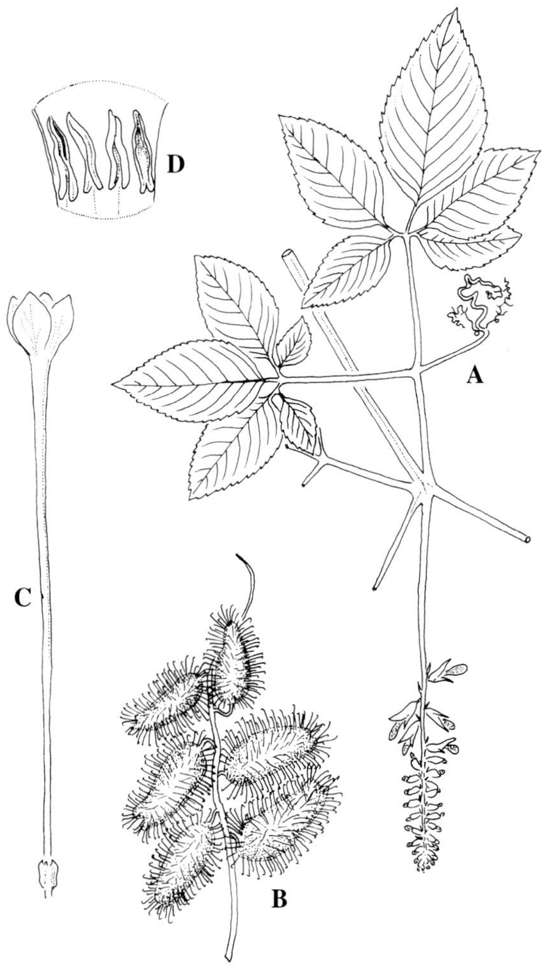

Fig. 1. Bignoniaceae. **A, B** *Tourretia lappacea*. **A** Habit. **B** Fruits. **C, D** *Sphingiphila tetramera*. **C** Flower. **D.** Stamens

Inflorescence Structure. Terminal to axillary inflorescences occur which are cymose, thyrsic or racemose (Figs. 1A, 2A). Both cymes and racemes can be reduced to a single flower. Ramiflorous to cauliflorous cymes or racemes occur especially in bat-pollinated species (in Coleeae and Crescentieae). Axillary inflorescences can include reduced juvenile leaves which suggest that the inflorescence is terminal on an immature branch. The lower branches of a fundamentally terminal inflorescences, however, are often in the axils of the upper leaves of a branch. Usually, thyrses or racemes are bracteate and only rarely frondose.

Floral Morphology. Flowers are usually large and showy. The calyx is usually 5-lobed, sometimes spathaceous (e.g. *Spathodea*, *Markhamia*, *Neojobertia*, *Gardnerodoxa*, *Dolichandra*, *Callichlamys*) or truncate, the lobes suppressed (Fig. 2A). *Lundia* is characterized by an evenly calyptrate calyx apex. Unusual double calyces with a thin, frilly submarginal rim and thick inner margin occur in *Amphilophium*, *Haplolophium* and *Glaziovia* (Gentry 1980). A double calyx with five irregular, submarginal outer teeth and an irregularly bilabiate inner calyx is found in *Delostoma*. The corolla is usually zygomorphic (Fig. 2A), sometimes strongly so, often bilabiate, rarely nearly actinomorphic, 5-lobed. Shape and colour of the corolla vary with pollination syndrome. The general structure of the corolla resembles that of other Lamiales. External pubescence is of systematic importance. Corolla lobes can be well developed or reduced (in Crescentieae). *Sphingiphila* is the only actinomorphic genus with only four lobes (Fig. 1C). Stamens are usually didynamous, 4 in pairs, with an additional staminode, usually reduced but sometimes longer than the fertile stamens (e.g. *Digomphia*, *Jacaranda*); occasionally only 2 stamens are fertile (*Catalpa*, *Pseudocatalpa*) and 3 are staminodial, rarely 5 fertile stamens (*Catophractes*, *Rhigozum*, *Oroxylum*) are present. Filaments are attached to the corolla tube and alternate with the lobes; in *Sphingiphila*, anthers are sessile on the corolla tube (Fig. 1D); anthers usually are dithecal, monothecal in some *Jacaranda*, in *Colea* and *Ophiocolea*, and opening by longitudinal slits. Thecae can be parallel, or divaricate, bent near the base and reflexed forward in *Cuspidaria* and in some *Arrabidaea*. An annular or cupular nectary disc usually is present around the base of the ovary, but absent e.g. in *Lundia*, *Cydista*, *Phryganocydia* and *Clytostoma*. The ovary is superior, 2-carpellate, bilocular with a septum, but sometimes unilocular, and placentation is axile or, in unilocular ovaries, with 2 or 4 intruding parietal placentae. A few genera have an incomplete partition perpendicular to the septum which gives the appearance of a 4-locular ovary (e.g. *Tourretia*). Ovules are numerous per cell, and uniseriate to multiseriate on the placenta. Rarely 6, 4, 2 or 1 pair of ovules per locule (e.g. *Arrabidaea*) are found. The stigma is bilobed to capitate, and in some genera (e.g. *Incarvillea*, *Bignonia*, *Tecoma*, *Catalpa*) the lobes are sensitive (Endress 1996).

Embryology. The endothecium of anthers often develops fibrous thickenings and the anther epidermis is persistent. Microsporogenesis is simultaneous and the initial microspore tetrads are tetrahedral. The tapetum is amoeboid. Ovules are orthotropous, anatropous or hemitropous, unitegmic and tenuinucellate. An endothelium is differentiated, and embryo sac development follows the Polygonum type. In the embryo sac, three persistent or ephemeral antipodal cells and two large synergids are developed. Endosperm formation is cellular, and chalazal or chalazal and micropylar haustoria are present (Shirke 1976; Fan et al. 1980; Ruwen 1981).

Pollen Morphology. The family shows high diversity in pollen features (Buurman 1978; Gentry and Tomb 1979; Gentry 1980). Apertures range from the most frequently occurring, 3-colpate type to monocolpate and inaperturate, and to 4–5-colpate, zonocolpate, pericolpate, spiroaperturate and syncolpate. The tectate, semitectate or intectate exine varies from psilate to reticulate, spinulose or areolate. Mostly monads but also tetrads (e.g. *Catalpa*, *Chilopsis*, *Cuspidaria*) or polyads are known. Although pollen morphology can be important for generic delimitation, it cannot be used as a diagnostic character for tribes. Gentry (1980) assumed that 3-colpate and finely reticulate pollen is ancestral.

Karyology. The basic chromosome number for almost all tribes (surveys in Goldblatt and Gentry 1979; Piazzano 1998) is x = 20; most species studied are diploids ($2n = 40$). The Old World tribe Oroxyleae has $2n = 28$ or 30. Further exceptions include $2n = 22$ (*Incarvillea*), $2n = 30$ (*Millingtonia*), $2n = 36$ (*Jacaranda*, *Tecoma*, *Mansoa*), $2n = 38$ (*Spathodea*, *Newbouldia*) and $2n = 80$ (*Macfadyena*). Gentry (1980) suggested an initial aneuploid loss of one chromosome from an ancestral hexaploid stock with x = 7.

Pollination and Reproductive Systems. Bignoniaceae show very diverse pollination syndromes, but only few detailed studies are available (e.g. Weber and Vogel 1986; Zyhra 1988; James and Knox 1993; Gussmann and Gottsberger 1996). Flowers are variable in colour, form, and time of anthesis. Most Bignoniaceae are bee-pollinated (e.g. *Arrabidaea*, *Tabebuia*, *Anemopaegma*), sometimes by *Xylocopa* species (e.g. *Pithecoctenium*, *Paragonia*), but some genera have corollas adapted to bird-pollination, e.g. *Pyrostegia*, *Campsis*, *Dolichandra*, *Martinella* in the New World, and *Spathodea*, *Lamiodendron*, *Fernandoa*, *Tecoma* in the Old World. They all have orange-red to red corollas with considerable nectar production. Other pollinators are butterflies (e.g. *Tynanthus*) or hawk moths (e.g. *Leucocalantha*, some species of *Arrabidaea*, *Ekmanianthe*, *Tanaecium* or *Sphingiphila* with fragrant, nocturnal flowers with long, salverform corolla tube; Fig. 1C). Chiropterophily occurs mainly in genera with cauliflorous (e.g. *Crescentia*) or flagelliflorous inflorescences (*Kigelia*) or stiff inflorescences overtopping the canopy (e.g. *Oroxylum*, *Fernandoa*). Pollination by lemurs has been reported for a Malagasy species, *Rhodocolea nycteriphila* (Coleeae, Zyhra 1988). Outcrossing is promoted by the sensitive stigma.

Seeds. Seeds are usually flattened, often winged (Fig. 3B). Only few genera with indehiscent fruits have vestigial seed wings or are wingless. The wings can be formed by fused capillary hairs (e.g. *Catalpa*, *Sparattospermum*) or usually by a thin membrane (Hesse and Morawetz 1980). In most genera seeds are bialate, but in taxa such as *Anemopaegma*, *Pithecoctenium* or *Jacaranda* the seed body is encircled by the wing.

Fruit and Dispersal. Fruits are mostly dehiscent (Fig. 2B) by two either septicidal or loculicidal valves and release winged, wind-dispersed seeds. Water-dispersed seeds with a thicker, floating, corky seed body and reduced or absent wings have evolved independently in many genera from wind-dispersed seeds (Gentry 1980). They are confined to swamps or seasonally inundated forests (e.g. *Anemopaegma* spp., *Arrabidaea* spp., *Adenocalymma* spp., *Memora* spp.). Tribes Crescentieae and Coleeae have large indehiscent fruits with wingless seeds embedded in their pulp. Seed dispersal here is by mammalian predators (e.g. elephants and baboons in *Kigelia*; Fig. 3A). Entirely fleshy berries occur in Malagasy Coleeae (*Colea*, *Ophiocolea*, *Phyllarthron*, *Phylloctenium*, *Rhodocolea*).

Phytochemistry. The reader is referred to Hegnauer (1964, 1989). Iridoid glycosides, iridoid alkaloids and flavones are usually present (von Poser et al. 2000). Lapacho, a yellow, skin-irritating powder from wood, is especially characteristic for the family, as are other naphthaquinones of the lapachol-type, and anthraquinones (tectochinon type). The quinonoid jacaranone has been extracted from *Jacaranda*. Glycosides, e.g. catalpin which is similar to aucubin, have been recorded. True tannins are replaced by derivates of complex esters and glycosides of cinnamic, benzoic and caffeic acids. Coloured pigments are mostly flavones rather than flavonols: anthocyanins (e.g. cyanidin, hyperin, chrysin-rutenosid) as well as various carotinoids. Seeds contain high percentages of oil (8–40%, fatty acids: oil, linol, linolen, palmitin acids).

Distribution and Habitats. Bignoniaceae are widely distributed in tropical and subtropical regions, with a few species in temperate climates (e.g. *Bignonia*, *Campsis*, *Catalpa* in North America, *Campsidium* in the Valdivian forest of Chile, *Incarvillea* in eastern Asia). Diversity centres are in northern South America. Bignonieae are predominantly neotropical with declining diversity towards Central America and North America (Gentry 1990, 1992). Tecomeae occur both in the New and Old World with highest diversity again in northern South America. Most of Bignoniaceae in Africa and Asia belong to this tribe (e.g. *Fernandoa*, *Stereospermum*, *Dinklageodoxa*, *Newbouldia*, *Dolichandrone*, *Lamiodendron*, *Radermachera*). Oroxyleae are exclusively tropical Asian, Coleeae are endemic to Madagascar with the exception of *Kigelia* (Gentry 1988). Tribes Tourretieae, Eccremocarpeae and Crescentieae are confined to the Neotropics. The preferred habitats are tropical rainforests or dry forests to savanna or cerrado habitats. Only few genera are found in Andean montane rainforests (*Tourretia*, *Eccremocarpus*), extratropical rainforests (e.g. *Campsidium*) or alpine meadows (e.g. *Incarvillea*).

Economic Uses. The reader is referred to Gentry (1980). Some species of *Tabebuia* and *Catalpa* are used for timber. *Jacaranda*, *Spathodea*, *Pyrostegia*, *Tecomaria*, *Tabebuia*, *Crescentia*, and *Campsis* provide ornamentals widely cultivated in subtropical and tropical regions. Several species have been reported to contain active compounds against major diseases such as cancer (bark of *Tabebuia*), diabetes, syphilis, malaria, hepatitis, rabies and leishmaniasis. *Crescentia alata* seeds and pulp are used in Nicaragua to produce a drink called "Semilla de Jicaro". The onion and garlic-smelling species (*Mansoa*) and the clove-smelling species (*Tynanthus*) are used as condiments throughout the Neotropics. *Tanaecium nocturnum* is used as love potion and hallucinogen. *Arrabidaea chica* produces a red dye which is used to dye basket fibres and for body paint by native pleo pleo in the Amazon. Fruits of *Crescentia* are used for "maraca" rattles.

Fossil Records. Fossil pollen which closely resembles pollen of *Jacaranda* has been recorded from the Oligocene of Puerto Rico (Graham and Jarzen 1969).

Subdivision and Relationships. Excluding *Paulownia* and tribe Schlegelieae, which were formerly considered to belong to Bignoniaceae (Armstrong 1985) and now are placed into "Paulowniaceae" and "Schlegeliaceae" (see Scrophulariaceae, this volume), Bignoniaceae are monophyletic (Spangler and Olmstead 1999). Within Lamiales they are close to Avicenniaceae, Lamiaceae, Lentibulariaceae, Phrymaceae, Schlegeliaceae, Verbenaceae and Acanthaceae (Bremer et al. 2001). Gentry (1980) divided the family into eight tribes, differing mainly in ovary structure and fruit characters (see key below): Tecomeae, Oroxyleae, Bignonieae, Eccremocarpeae, Tourretieae, Coleeae, Crescentieae and Schlegelieae. The latter tribe should be excluded and assigned to a separate family (Spangler and Olmstead 1999, see above). In a phylogenetic analysis of Bignoniaceae based on the analysis of the cpDNA sequences *rbc*L snd *ndh*F (Spangler and Olmstead 1999), tribes Bignonieae, Crescentieae, and Coleeae came out as monophyletic, whereas tribe Tecomeae formed a paraphyletic group. The Crescentieae, Bignonieae, Coleeae, Oroxyleae and Eccremocarpeae are each derived from within Tecomeae. The monogeneric Eccremocarpeae came out near the base of Bignoniaceae. The paraphyly of Tecomeae presents some taxonomic problems. The most basal Bignoniaceae, *Jacaranda* and *Podranea*, belong to Tecomeae (Spangler and Olmstead 1999). Other members of Tecomeae (e.g. *Cybistax*, *Tabebuia*, *Radermachera*), however, are highly derived within the family. Grose and Olmstead (2002) showed that a clade within Tecomeae contains *Tabebuia*, *Cybistax*, *Zeyheria*, *Godmania*, *Ekmanianthe* and tribe Crescentieae. Olmstead et al. (2002) showed that herbaceousness has evolved independently at high elevations in the Andes (*Argylia*) and the Himalayas (*Incarvillea*). Indehiscent fruits have evolved independently in Central America (Crescentieae), Madagascar (Coleeae) and Africa (*Kigelia*). In a study on the phylogeny of Bignonieae, Lohmann (2002) showed that several of the traditionally circumscribed genera (e.g. *Arrabidaea*, *Cuspidaria*, *Tanaecium*, *Haplolophium*) are not monophyletic. The monophyly of some other genera (e.g. *Anemopaegma*, *Lundia*, *Tynanthus*, *Pleonotoma*) is supported. A generic classification of Bignonieae that only recognizes monophyletic genera is being prepared (L. G. Lohmann, in prep.). Flower colour and fruit shape are poor generic characters. In contrast, tendril type, pseudostipule shape and stem anatomy represent synapomorphies of well-supported clades in Bignonieae. In the present treatment, the delimitation of tribes and genera follows Gentry (1980).

Key to the Tribes

1. Fruit dehiscent; trees, shrubs, herbs or lianas 2
- Fruit indehiscent; trees, shrubs or lianas 6
2. Placentation parietal, fruit without septum; wiry vines; above 3000 m altitude in the Andes **Eccremocarpeae 94. *Eccremocarpus*** (p. 34)
- Placentation axile, fruit with septum 3
3. Herbaceous, rather succulent vines; inflorescence subspicate, upper flowers mostly sterile; ovary 4-locular; capsule densely uncinate-spiny, the 4 valves not splitting to the base **Tourrettieae 95. *Tourrettia*** (p. 35)
- Woody vines, trees, shrubs or erect herbs; inflorescence never spicate, without sterile flowers; ovary bilocular; capsule spiny or not, the two valves dehiscent to base 4
4. Lianas with terminal pinna of compound leaves often modified into a tendril; stems with anomalous phloem arms; fruits dehiscing parallel to the septum (septicidal) (or occasionally 4-valved); predominantly New World **Bignonieae** (p. 23)
- Trees, shrubs, herbs or non-tendrillate lianas (*Perianthomega* has a twining petiole); stems without anomalous phloem arms; fruits dehiscing perpendicular to septum (loculicidal) (except in a few Old World species); Old and New World 5
5. Fruits dehiscing parallel to septum (septicidal) or 4-valved; tropical Asia **Oroxyleae** (p. 23)
- Fruits dehiscing perpendicular to septum (loculicidal); widespread **Tecomeae** (p. 13)
6. Leaves pinnately compound (or reduced to superimposed articules or rarely simple); Madagascar and Africa **Coleeae** (p. 35)
- Leaves palmately compound or simple; neotropics; flowers cauliflorous to terminal **Crescentieae** (p. 37)

1. Tribe Tecomeae Endl., Gen. Pl.: 711 (1839).

Trees, shrubs or lianas, interpetiolar glands and ridges usually absent (except e.g. *Campsis* with interpetiolar ridge), pseudostipules usually absent, rarely subulate or foliaceous. Leaves pinnate, palmate or simple. Calyx 5-dentate to irregularly lobed or spathaceous. Pollination various (bees, hawk moths, birds, bats). Stamens 4–5, anthers bithecal (mono-thecal in some *Jacaranda*), thecae straight, divaricate. Ovary bilocular, placentation axile. Fruit dehiscing perpendicular to septum (loculicidal). North to South America, Africa, Madagascar and tropical Asia.

Key to Genera of Tecomeae

1. Leaves pinnately or bipinnately (rarely 3–4-pinnately) compound; trees, shrubs, lianas or herbs 2
- Leaves palmately compound or simple or unifoliolate; erect trees, shrubs or herbs 28
2. Herbs of alpine habitats; Nepal, Himalaya **18. *Incarvillea***
- Trees, shrubs or lianas 3
3. Lianas 4
- Trees or shrubs 11
4. Leaves ternate, petiole twining; calyx more than 2.5 cm long, southern Brazil to Bolivia **28. *Perianthomega***
- Leaves pinnate, petiole not twining; calyx less than 2 cm long 5
5. Corolla distinctly curved, lower lip recurved, dark red to blackish red; Madagascar **29. *Perichlaena***
- Corolla not curved, lower lip not recurved, light to orange red or white to magenta; South America, Africa or Asia 6
6. Flowers white with purple throat or magenta; calyx either shallowly 2–3-lobed or membranaceous, cupular or 5-dentate 7
- Flowers red or red orange; calyx coriaceous and 5-dentate 10
7. Calyx 5-lobed to 5-dentate, more than 15 mm long; Africa and Asia 8
- Calyx shallowly 2–3-lobed, less than 10 mm long; flowers white with purple throat; Southeast Asia **25. *Pandorea***
8. Inflorescence a raceme; Southeast Asia **41. *Tecomanthe***
- Inflorescence thyrsic 9
9. Calyx deeply 5-dentate; flowers light magenta; southern Africa **30. *Podranea***
- Calyx 5-lobed, flowers white; West Africa **12. *Dinklageodoxa***
10. Corolla slightly constricted at apex, less than 8 mm wide at mouth of tube; calyx less than 12 mm long; Valdivian region in South America **3. *Campsidium***
- Corolla not constricted at apex, more than 14 mm wide at mouth of tube; calyx more than 12 mm long; North America, China **4. *Campsis***
11 (3). Leaves pinnate (rarely simple in *Digomphia*) 12
- Leaves 2–3(–4)-pinnate 27
12. Calyx spathaceous 13
- Calyx tubular, campanulate or tubular to campanulate, sometimes somewhat spathaceously split (*Fernandoa*) 17
13. Foliaceous pseudostipules present; corolla yellow or brownish red, campanulate, funnel-shaped, dilated above a very short, cylindrical tube; Africa, Asia **21. *Markhamia***
- Foliaceous pseudostipules absent; corolla white, lavender or bluish-purple to orange-red 14
14. Corolla broadly campanulate, orange-red with yellow border, more than 4.5 cm wide at mouth; calyx more than 4.5 cm long; Africa **36. *Spathodea***
- Corolla tubular to campanulate to funnel-shaped, white or lavender, less than 3.5 cm wide at mouth; calyx less than 3 cm long 15
15. Staminode present 16
- Staminode absent; corolla lavender, apex of calyx often slightly bifid; West Africa **23. *Newbouldia***
16. Staminode shorter than stamens, apex of staminode undivided; corolla white, funnel-shaped; Southeast Asia **13. *Dolichandrone***
- Staminode longer than stamens, strongly bifurcate, each branch 6–7 mm long; South America **11. *Digomphia* (p.p.)**
17 (12). Fertile stamens 5; corolla white, pink or yellow; leaves fascicled on branchlets; spiny shrubs; southern Africa **32. *Rhigozum***
- Fertile stamens 4, sometimes long or short staminodes present; leaves verticillate or decussate; spines absent 18
18. Leaves verticillate 19
- Leaves decussate 20
19. Pinnae in 3–4 pairs, subrotund or obtuse at apex, lower surface of pinnae and calyx adpressed stellate-tomentose **17. *Heterophragma***

- Pinnae in 4–10 pairs, acuminate at apex, glabrous on both sides, calyx with minute tuberculate to warty glands **34. *Santisukia***
20. Staminode longer than stamens, tip enlarged 21
- Staminode minute, much shorter than stamens or absent 22
21. Apex of staminode strongly bifurcate, each branch 6–7 mm long; calyx with 3–5 lanceolate to ovate lobes **11. *Digomphia* (p.p.)**
- Apex of staminode capitate or barely bifid; calyx cupular or deeply 5-lobed **19. *Jacaranda* (p.p.)**
22. Leaf rachis with a sharp, keel-like ridge above, pinnae in 8–12 pairs; capsules winged **24. *Pajanelia***
- Leaf rachis not keeled above, pinnae in less than 6 pairs; capsules not winged 23
23. Pinnae serrate, crenate or pinnatifid; extrafloral nectaries present; America, Africa **40. *Tecoma***
- Pinnae entire; Africa, Asia 24
24. Calyx regularly 5-lobed, rather thin, reticulately veined, shortly hairy; Southeast Asia **20. *Lamiodendron***
- Calyx lobes unequal, often less than 5, calyx not thin and reticulately veined 25
25. Pinnae underneath with hairy domatia; ovary densely stellate-tomentose; capsule flattened, with flat septum, if terete with 10 ribs **15. *Fernandoa***
- Pinnae underneath without domatia; ovary glabrous; capsule terete, with a terete septum, never with many ribs 26
26. Evergreen trees; calyx cupular, truncate; ovules in many rows in each locule, valves thin **31. *Radermachera***
- Deciduous trees; calyx cylindrical or campanulate; ovules in 2 rows in each locule, valves hard **38. *Stereospermum***

27 (11). Calyx open in bud; corolla tube curved; leaf rachis narrowly winged, pinnae subsessile; ovules in 1–2 rows per locule **27. *Pauldopia***
- Calyx closed in bud; corolla tube straight; leaf rachis usually not winged, pinnae stalked; ovules in many rows in each locule **31. *Radermachera***

28 (1). Leaves simple or unifoliolate, present at anthesis, sometimes alternate or in whorls 29
- Leaves palmately compound, sometimes deciduous at anthesis, always opposite unless herbs 41

29. Fertile stamens 5 (6–7); corolla subrotate with 5 (6–7) patent lobes; inflorescence a raceme; spiny shrubs; southern Africa **6. *Catophractes***
- Fertile stamens 4 or 2, staminodes usually present; corolla usually bilabiate; inflorescence a thyrse or raceme; spines absent; Asia or America 30
30. Plants from Asia (Arabia to Southeast Asia) 31
- Plants from America 32
31. Corolla yellow; stamens strongly exserted; capsule ellipsoid; Southeast Asia **10. *Deplanchea***
- Corolla orange; stamens slightly exserted; capsule linear; Arabia to India **42. *Tecomella***
32. Leaves linear, less than 1.4 cm wide 33
- Leaves narrowly oblong-elliptic to ovate or elliptic, more than 1.4 cm wide 34
33. Flowers yellow; leaves whorled; capsule fusiform, with cross-shaped septum; seeds hyaline-membranaceously winged **2. *Astianthus***
- Flowers magenta; leaves alternate; capsule linear, with flat septum; seed wing of tuft of trichomes at each end **7. *Chilopsis***
34. Fertile stamens 2; seeds pilose, wings of separate trichomes loosely fused basally; North America to Antilles **5. *Catalpa***
- Fertile stamens 4; seeds glabrous, wings membranaceous or absent; Central and South America 35
35. Staminode longer than stamens; flowers blue, lavender or purplish blue; capsule oblong-elliptic to suborbicular, flattened perpendicular to septum 36
- Staminode much shorter than stamens; flowers yellow, white, red or magenta; capsule linear to narrowly oblong, terete or flattened parallel to septum 37
36. Calyx cupular, coriaceous; staminode capitate; leaves rough **19. *Jacaranda* (p.p.)**
- Calyx split to base, membranaceous; staminode strongly bifid or 4-parted; leaves smooth **11. *Digomphia* (p.p.)**
37. Anthers strongly exserted (up to 2 cm); corolla greenish purple to yellowish brown; flowers single, mostly ramiflorous; capsule spirally twisted, thickened septum interrupted by a series of pits, fragmenting transversally; Cuba **37. *Spirotecoma* (p.p.)**
- Anthers included to slightly exserted; corolla yellow, red, white, lavender or magenta; flowers usually in terminal thyrses or racemes; capsule straight or only slightly curved, septum flat, not fragmenting 38
38. Flowers yellow; calyx evenly and sharply 5-dentate, with triangular teeth, never double with submarginal teeth; leaves in whorls of 4–5 or serrate 39
- Flowers red, white, lavender or magenta; calyx not evenly and sharply dentate, sometimes double with conspicuously submarginal teeth (*Delostoma*); leaves opposite, usually entire (serrate in *Delostoma*) 40
39. Capsule linear; leaves serrate, opposite; calyx cupular; corolla tubular to campanulate **40. *Tecoma* (p.p.)**
- Capsule oblong-ellipsoid; leaves entire, in whorls of 4–5; calyx campanulate; corolla broadly infundibuliform **33. *Romeroa***
40. Corolla tube glabrous outside; calyx never with conspicuous submarginal teeth; capsule linear, ± terete; leaves entire; Antilles and lowland South America **39. *Tabebuia* (p.p.)**
- Corolla tube pubescent outside; calyx often double with conspicuous submarginal teeth; capsule narrowly elliptic or elliptic-oblong, somewhat flattened perpendicular to septum; leaves conspicuously serrate or serrulate to subentire **9. *Delostoma***

41 (28). Herbs; southern Andes of Chile and Argentina north to lomas of South Peru **1. *Argylia***
- Trees, shrubs or lianas; Central to South America or Southeast Asia 42

42. Lianas; corolla campanulate or curved, glabrous or inside of lobes bearded; capsule narrowly oblong, stipitate; Southeast Asia **22. *Neosepicea***
- Trees or shrubs; northwest Argentina to Mexico and the Antilles 43
43. Capsule oblong or orbicular; calyx bilabiate to near base or membranaceous and deeply 5-dentate 44
- Capsule linear to oblong-ovoid; calyx cupular to tubular, mostly 5-denticulate to shallowly bilabiate or subspathaceous, rarely deeply 5-dentate when coriaceous 45
44. Capsule oblong, surface smooth except for prominent longitudinal ribs; pinnae cuneate, pubescent with simple hairs; flowers greenish; calyx 5-dentate **8. *Cybistax***
- Capsule orbicular, verrucose-surfaced, without ribs; leaves densely pubescent below with dendroid hairs; flowers brownish to yellowish or orange; calyx deeply 2-lobed **43. *Zeyheria***
45. Fertile stamens 2; corolla white and salverform, less than 5 cm long; capsule woody, fusiform linear **26. *Paratecoma***

- Fertile stamens 4; corolla various, white and salverform-tubular only when more than 15 cm long 46
46. Flowers mostly ramiflorous, single or in reduced few-flowered racemes; corolla yellow to magenta or yellowish brown; stamens long-exserted; capsule curved or spirally twisted, septum thickened, with pits or depressions in which seeds are impressed; Cuba and Hispaniola **37. *Spirotecoma***
- Flowers terminal or axillary, often in multiflowered racemes or thyrses; stamens usually included or sub-exserted, strongly exserted only when corolla white or red; capsule usually straight (spirally twisted in *Godmania*), septum thin and flat; Mexico and Antilles to Argentina 47
47. Capsule spirally twisted; anthers pilose, leaves 5–9-foliolate, pinnae often cuneate at base; corolla broadly campanulate from base, without narrow basal tube; calyx campanulate **16. *Godmania***
- Capsule ± straight; anthers glabrous; leaves (1–)3–9-foliolate, pinnae usually conspicuously petiolulate; corolla salverform to infundibuliform or campanulate, broadly campanulate only when tube base narrowly cylindrical; calyx cupular to tubular 48
48. Seeds with terminal trichomes instead of wings; flowers white or pale pink, with purplish stripes in throat; calyx membranaceous with a thinner irregularly rupturing apex **35. *Sparattosperma***
- Seed wing an entire membrane or reduced and corky; corolla colour various, when white the throat uniformly white or yellow; calyx coriaceous, apex often evenly denticulate 49
49. Calyx circumscissely caducous, falling with corolla; corolla white, the lobes irregularly laciniate-frilly; anthers subexserted, capsule longitudinally striate-costate, sometimes with conspicuously raised lenticels **14. *Ekmanianthe***
- Calyx falling after anthesis or persistent in fruit; corolla purple, yellow, red or white, lobes rarely lacinate-frilly; anthers usually included; capsule without conspicuously raised costae or lenticels **39. *Tabebuia* (p.p.)**

1. *Argylia* D. Don

Argylia D. Don, Edinburgh Philos. J. 9: 260 (1823); Gentry, Fl. Neotrop. 25, Bignoniaceae II: 6–11 (1992), rev.

Perennial rosulate herbs or chamaephytic sub-shrubs, stems sometimes terminating in spine, glandular-pubescent. Leaves alternate or subopposite, long-petiolate, palmately compound, pinnae often deeply pinnatisect. Inflorescence a terminal, contracted raceme. Calyx campanulate, deeply 5-dentate. Corolla yellow or yellow-orange, campanulate. Stamens 4, included, anthers glabrous or occasionally with a few trichomes, thecae divergent, staminode often absent. Ovary linear, lepidote to glandular-pubescent, with numerous ovules. Disc reduced. Capsule fusiform or linear, subtended by the persistent calyx, valves glabrous. Seeds thin, oblong, wingless or with a very reduced membranaceous margin. $2n = 30$. Twelve species, Peru, Chile, Argentina.

2. *Astianthus* D. Don

Astianthus D. Don, Edinburgh Philos. J. 9: 262 (1823); Gentry, Fl. Neotrop. 25, Bignoniaceae II: 11–16 (1992), rev.

Small tree or shrub, stems minutely lepidote or glabrous. Leaves verticillate, simple, linear, entire. Inflorescence a terminal thyrse. Calyx campanulate, 5-dentate. Corolla yellow, tubular to campanulate. Stamens 4, included, anthers glabrous, thecae divergent. Ovary oblong, glabrous, with numerous ovules. Disc cupular. Capsule terete, narrowly fusiform, not subtended by calyx. Valves glabrous, the septum cross-shaped in cross section, seeds tiny, flattened, the hyaline membranaceous wing completely surrounding the seed body. One rheophytic species, *A. viminalis* (H.B.K.) Baill., from Mexico to Nicaragua.

3. *Campsidium* Seem.

Campsidium Seem., Bonplandia 10: 147 (1862); Gentry, Fl. Neotrop. 25, Bignoniaceae II: 16–17 (1992), rev.

Liana, stems glabrous, except for a few, minute lepidote scales. Leaves opposite, sessile, pinnately 5–17-foliolate, the pinnae scattered-lepidote with reddish scales, plants otherwise glabrous. Inflorescence a terminal few-flowered bracteate raceme. Calyx campanulate, 5-dentate. Corolla bright red, tubular, constricted at apex. Stamens 4, included, anthers glabrous, thecae parallel, partially fused to filament apex. Ovary ovate, glandular-papillose, ovules multiseriate per locule. Disc reduced, cupular. Capsule narrowly ellipsoid stipitate, subtended by calyx, valves inconspicuously lepidote, septum thin and uniformly flattened. Seeds elliptic, with a small central body completely surrounded by a brownish-striate hyaline wing. One species, *C. valdivianum* (Phil.) Skottsberg in the Valdivian forest of Chile and Argentina.

4. *Campsis* Lour.

Campsis Lour., Fl. Cochinch.: 358, 377 (1790), nom. cons.; Gentry, Fl. Neotrop. 25, Bignoniaceae II: 17 (1992), rev.

Lianas, stems inconspicuously lepidote to variously pubescent. Leaves opposite, imparipinnately compound, without tendrils. Inflorescence a terminal, few-flowered bracteate thyrse. Calyx campanulate, 5-dentate. Corolla orange or red orange, tubular to campanulate. Stamens 4, included, anthers glabrous, staminode rudimentary. Ovary flattened ellipsoid or oblong-ellipsoid, lepidote, with multiseriate ovules per locule. Disc conical-annular. Capsule narrowly oblong, subtended by calyx, valves rather woody, glabrous except for lepidote scales and plate-shaped glands, seeds thin, bialate. $2n = 40$. Two species, *C. grandiflora* (Thunb.) Schumann in China, and *C. radicans* (L.) Bureau in eastern North America.

5. *Catalpa* Scop.

Catalpa Scop., Intr. Hist. Nat.: 170 (1777); Gentry, Fl. Neotrop. 25, Bignoniaceae II: 17–23 (1992), reg. rev.; Paclt, Candollea 13: 241–285 (1952), rev.

Small trees, stems lepidote to pubescent. Leaves opposite or verticillate, simple. Inflorescence an axillary, bracteate thyrse with usually many flowers. Calyx membranaceous, bilabiately split to near base. Corolla white to pink or light yellow, broadly campanulate, strongly bilabiate with 3 lower lobes longer than the upper lobes. Stamens 2, included, anthers glabrous, thecae divergent, staminodes 3, minute. Ovary oblong, lepidote, with many multiseriate ovules per locule. Capsule terete, sometimes subtended by calyx, valves glabrous. Seeds thin, winged, the wings formed by separate or basally partially fused hairs, the body also sometimes dorsally plumose. $2n = 40$. Ten species, four in eastern Asia, two in south-eastern North America and four in the Antilles.

6. *Catophractes* D. Don

Catophractes D. Don, Proc. Linn. Soc. Lond. 1: 4 (1839).

Shrub or small tree, stems spiny, with dense grey woolly hairs. Leaves opposite, simple, often fascicled. Inflorescence an axillary or terminal frondose raceme with usually few flowers. Calyx tubular, shortly splitting at one side with 5(6) filiform lobes. Corolla white, campanulate, slightly bilabiate with 5 (6–7) patent lobes. Stamens 5 (6–7), exserted, anthers glabrous, thecae connate in upper part, free and parallel below. Ovary oblong, with few ovules. Disc cupular. Capsule elliptic, flattened parallel to septum, subtended or not by calyx, valves thick, woody, warty. Seeds with papery wings. One species, *C. alexandri* D. Don from subtropical and southern Africa.

7. *Chilopsis* D. Don

Chilopsis D. Don, Edinburgh Philos. J. 9: 261 (1823); Gentry, Fl. Neotrop. 25, Bignoniaceae II: 23–24 (1992), rev.

Shrub or small tree. Leaves mostly alternate, simple. Inflorescence a terminal raceme. Calyx spathaceous, bilabiate to near base. Corolla lavender to magenta, throat with yellow ridges and purple nectar lines, broadly tubular to campanulate. Stamens 4, included, anthers glabrous, thecae divaricate, staminode short filiform. Ovary cylindrical-conical, lepidote. Disc patelliform-annular. Capsule linear, terete, not subtended by calyx, valves glabrous. Seeds with wings formed by separate or basally partially fused hairs. $2n = 40$. One species, *C. linearis* (Cav.) Sweet, from deserts of northern Mexico and south-western North America.

8. *Cybistax* Mart. ex Meisn.

Cybistax Mart. ex Meisn., Pl. Vasc. Gen. 1: 300; 2: 208 (1840); Gentry, Fl. Neotrop. 25, Bignoniaceae II: 24–30 (1992), rev.

Shrub or tree, stems glabrous to lepidote. Leaves alternate, long-petiolate, palmately 5–7-foliolate. Inflorescence a terminal, few-flowered thyrse. Calyx campanulate, 5-dentate. Corolla light green, tubular to campanulate. Stamens 4, included, anthers glabrous, thecae divaricate, staminode present. Ovary ovoid-oblong, minutely lepidote, with multiseriate ovules. Capsule oblong, not subtended by calyx, valves longitudinally 12-ribbed. Seeds with hyaline-membranaceous wing. One species, *C. antisyphilitica* (Mart.) Mart., tropical South America.

9. *Delostoma* D. Don

Delostoma D. Don, Edinburgh Philos. J. 9: 263 (1823); Gentry, Fl. Neotrop. 25, Bignoniaceae II: 30–37 (1992), rev.

Trees or shrubs, stems glabrous to pubescent. Leaves opposite, simple. Inflorescence a terminal raceme or few-flowered thyrse. Calyx cupular, 5-parted, sometimes with a double margin. Corolla red or magenta, tubular to campanulate or tubular, stamens 4, included or exserted, anthers glabrous, thecae parallel. Ovary ovoid, glandular-lepidote, with multiseriate ovules per locule. Capsule narrowly elliptic, not subtended by calyx, valves glabrous. Seeds thin, with hyaline wing. $2n = 42$. Four species, Andes from Colombia and Venezuela to Peru.

10. *Deplanchea* Vieill.

Deplanchea Vieill., Bull. Soc. Linn. Normandie 7: 96 (1862); Steenis, Fl. Malesiana I, 8: 135–141 (1977), reg. rev.
Diplanthera Banks & Sol. (1810).

Trees, stems glabrous to pubescent. Leaves in whorls of 3–4, simple. Inflorescence a terminal few-flowered thyrse. Calyx campanulate, 5-dentate. Corolla yellow, tubular to campanulate. Stamens 4, exserted, anthers glabrous, thecae divergent, staminode rarely present. Ovary ovoid-elliptic, glabrous, with numerous ovules. Disc annular. Capsule ellipsoid, not subtended by calyx, valves glabrous. Seeds with hyaline wing. Five species, Malesia, Australia, New Caledonia.

11. *Digomphia* Benth.

Digomphia Benth., Lond. J. Bot. 5: 364 (1846); Gentry, Fl. Neotrop. 25, Bignoniaceae II: 37–40 (1992), rev.

Trees or shrubs, stems glabrous to pubescent. Leaves opposite, imparipinnate or simple. Inflorescence a terminal few-flowered (often 1–2) thyrse. Calyx spathaceous, 3–5-lobed. Corolla lavender or bluish purple, tubular to campanulate. Stamens 4, included, anthers glabrous, thecae divergent, staminode present. Ovary elliptic, flattened, glabrous. Disc annular. Capsule oblong, not subtended by calyx, valves glabrous. Seeds thin, winged. Three species, Guyana Shield.

12. *Dinklageodoxa* Heine & Sandwith

Dinklageodoxa Heine & Sandwith, Kew Bull. 16: 223 (1962).

Liana, branchlets glabrous. Leaves opposite, imparipinnately compound, with 1–2 pairs of pinnae. Inflorescence a terminal 10–20-flowered bracteate thyrse. Calyx campanulate, 5-parted. Corolla white, outside lilac at base, inside with purplish lines and dots, tubular to campanulate. Stamens 4, included, anthers glabrous, thecae straight, staminode present. Ovary elliptic, glabrous, with numerous ovules. Disc ovoid. Fruit unknown. One species, *D. scandens* Heine & Sandwith, Liberia.

13. *Dolichandrone* (Fenzl.) Seem.

Dolichandrone (Fenzl.) Seem., Ann. Mag. Nat. Hist., ser. 3, 10: 31 (1862); Steenis, Fl. Malesiana I, 8: 141–144 (1977), reg. rev.

Trees, stems glabrous or pubescent. Leaves alternate, simple or 1-pinnate. Inflorescence a terminal few-flowered raceme or thyrse. Calyx spathaceous, 5-parted. Corolla white, fragrant, nocturnal (bat-pollinated), salver-shaped, tube narrowly cylindrical, lobes crisped. Stamens 4, included, anthers glabrous, thecae divergent, staminode present. Ovary subcylindrical, glabrous, with ovules in 4–6 rows on 2 placentae per locule. Disc annular. Capsule subcylindrical to flattened. Seeds hyaline-winged. $2n = 40$. Ten species, East Africa (1), Indomalesia (5), tropical Australia (3), New Caledonia (1).

14. *Ekmanianthe* Urb.

Ekmanianthe Urb., Repert. Spec. Nov. Regni Veg. 20: 308 (1924); Gentry, Fl. Neotrop. 25, Bignoniaceae II: 42–47 (1992), rev.

Trees, stems lepidote. Leaves opposite, long-petiolate, palmately compound. Inflorescence a terminal few-flowered raceme or thyrse. Calyx cupular, 5-parted. Corolla greenish-cream or white, tubular-salverform to campanulate, deeply laciniate lobes poorly defined. Stamens 4, exserted to subexserted, anthers glabrous, thecae straight, staminode present. Ovary linear, lepidote, with ovules in 2 series. Disc fleshy. Capsule linear, not subtended by calyx, valves longitudinally costate, lepidote. Seeds thin, bialate, wings hyaline. Two species, *E. longiflora* (Griseb.) Urban, *E. actinophylla* Urb., Cuba, Hispaniola.

15. *Fernandoa* Welw. ex Seem.

Fernandoa Welw. ex Seem., J. Bot. 3: 330 (1865); Perrier, Fl. Madagascar: 18–24 (1938), reg. rev.; Steenis, Blumea 23: 121–138 (1976), reg. rev.; Steenis, Fl. Malesiana I, 8: 159 (1977), reg. rev.; Bidgood, Kew Bull. 49: 381–390 (1994), reg. rev.
Kigelianthe Baill. (1891).
Haplophragma Dop (1926).

Shrubs or trees, stems glabrous. Leaves opposite, imparipinnate. Inflorescence a terminal few-flowered raceme or thyrse. Calyx campanulate, irregularly lobed. Corolla crimson, red-orange to yellow, broadly cup-shaped, limb bilabiate. Stamens 4, subexserted, anthers glabrous, thecae divaricate, staminode present. Ovary cylindrical, narrow, glabrous, with pluriseriate ovules per locule. Disc cupular. Capsule subcylindrical, not subtended by calyx, valves glabrous. Seeds winged. Fourteen species, Africa (4), Madagascar (3), southern China and Southeast Asia to Sumatra (7).

16. *Godmania* Hemsl.

Godmania Hemsl., Diagn. Pl. Nov.: 35 (1879); Gentry, Fl. Neotrop. 25, Bignoniaceae II: 47–51 (1992), rev.

Small to medium-sized trees, stems pubescent. Leaves opposite, long-petiolate, 5–9-foliolate.

Inflorescence a terminal thyrse. Calyx campanulate, 5-lobed. Corolla yellow and brown, broadly campanulate. Stamens 4, included, anthers pubescent, thecae straight, staminode present. Ovary linear-oblong or linear-conical, lepidote, with multiseriate ovules per locule. Disc annular. Capsule linear, not subtended by calyx, valves longitudinally striate-costate, twisted. Seeds bialate, wings irregular. Two species, *G. macrocarpa* (Benth.) Hemsl., *G. dardanoi* (J.C Gomes) A. Gentry, from Mexico to Brazil and Bolivia.

17. *Heterophragma* DC.

Heterophragma DC., Biblioth. Universelle Genève II, 17: 129 (1838); Santisuk, Fl. Thailand 5, 1: 56–57 (1987), reg. rev.

Trees, stems pubescent. Leaves verticillate, imparipinnate, with 3–4 pairs of pinnae, adpressed-stellate below. Inflorescence a terminal tomentose thyrse. Calyx tubular, irregularly lobed. Corolla cream-white to pale yellow, tubular, straight, lobes subequal. Stamens 4, included, anthers glabrous, thecae straight, staminode present. Ovary elliptic, glabrous, with numerous ovules per locule. Capsule subcylindrical, not subtended by calyx, valves woody. Seeds with broad hyaline wings. $2n = 40$. Two species, *H. sulfureum* Kurz, *H. quadriloculare* (Roxb.) K. Schum., India, Southeast Asia.

18. *Incarvillea* Juss.

Incarvillea Juss., Gen. Pl.: 138 (1789); Griers, Notes Roy. Bot. Gard. Edinburgh 23: 303–340 (1961), rev.

Annual, biennial or perennial herbs, rosulate, stems glabrous to pubescent. Leaves alternate or subopposite, long-petiolate, pinnately compound, with 1–11 pairs of pinnae. Inflorescence a terminal or axillary raceme. Calyx campanulate, 5-dentate. Corolla large, pink with yellow throat, tubular to campanulate, limb bilabiate. Stamens 4, included, anthers glabrous, thecae divergent, at base with spur-like appendage, staminode present. Ovary linear-oblong, sulcate, with 4-seriate ovules per locule. Disc cupular. Capsule linear, not subtended by calyx, valves coriaceous. Seeds orbicular, winged. $2n = 22$. About 14 species, Himalaya, Central and East Asia.

19. *Jacaranda* Juss.

Jacaranda Juss., Gen. Pl.: 138 (1789); Morawetz, Denkschr. Oesterr. Akad. Wiss., Mat.-Naturwiss. 123: 1–184 (1982), rev.; Gentry & Morawetz in Gentry, Fl. Neotrop. 25, Bignoniaceae II: 51–105 (1992), rev.

Trees, sometimes shrubs or xylopodial subshrubs, stems glabrous to pubescent. Leaves opposite, bipinnate or pinnate, rarely simple, opposite. Inflorescence a terminal or axillary few- to many-flowered thyrse. Calyx campanulate to tubular, variously shaped. Corolla blue or purple-blue to magenta, tubular to campanulate. Stamens 4, included, anthers glabrous, thecae unequal or anthers monothecous, staminode elongate, glandular, exceeding the stamens. Ovary flattened-cylindrical, variously pubescent, with ovules ca. 8-seriate per locule. Disc annular. Capsule elliptic or oblong, not subtended by calyx, valves variously pubescent. Seeds thin, with hyaline or brownish wings. $2n = 36$. About 50 species, from Guatemala and the Antilles to Argentina.

20. *Lamiodendron* Steenis

Lamiodendron Steenis, Nova Guinea II, 8: 379 (1957); Steenis, Fl. Malesiana I, 8: 161–162 (1977), rev.

Evergreen tree, stems glabrous. Leaves opposite, imparipinnate. Inflorescence an axillary, few-flowered raceme or thyrse, often with abortive flowers in leaf axils. Calyx campanulate, regularly 5-lobed. Corolla brilliant orange to apricot, broadly campanulate, subrotate. Stamens 4, subexserted, anthers glabrous, thecae divaricate, staminode present. Ovary cylindrical, 4-sulcate, each locule with two placentae, glabrous, with several ovules per locule. Disc cupular. Capsule linear, not subtended by calyx, valves glabrous. Seeds winged. One species, *L. magnificum* Steenis, New Guinea.

21. *Markhamia* Seem. ex Baill.

Markhamia Seem. ex Baill., Hist. Pl. 10: 47 (1888).

Shrubs or trees. Leaves opposite, imparipinnate, pseudostipules foliaceous or subulate. Inflorescence a terminal or axillary, few-flowered raceme or thyrse. Calyx spathaceous, cuspidate or uncinate at apex. Corolla yellow to brown-orange, tubular to campanulate, limb bilabiate. Stamens 4, included, anthers glabrous, thecae divaricate, staminode present. Ovary oblong, narrow, with numerous ovules, 4–6-seriate per locule. Disc annular to cup-shaped. Capsule long, not subtended by calyx, valves glabrous. Seeds winged. $2n = 40$. Ten species, tropical Africa (8) and Southeast Asia (2).

22. *Neosepicaea* Diels

Neosepicaea Diels, Bot. Jahrb. Syst. 57: 500 (1922); Steenis, Fl. Malesiana I, 8: 165–167 (1977), reg. rev.
Haussmannia F. Muell. (1864).

Lianas, stems glabrous to pubescent. Leaves opposite, long-petiolate, palmately 3–5-foliolate. Inflorescence a terminal or axillary many-flowered bracteate thyrse. Calyx cupular to campanulate, 5-parted. Corolla brown, red or orange to pink, shortly campanulate, often curved. Stamens 4, included, anthers glabrous, thecae divaricate, staminode present. Ovary oblong, lepidote, with numerous ovules per locule. Disc cupular. Capsule stipitate, lepidote, oblong-linear, terete, subtended by calyx or not, valves glabrous. Seeds thin-winged, rectangular. Four species, Moluccas, New Guinea, Queensland.

23. *Newbouldia* Seem. ex Bureau

Newbouldia Seem. ex Bureau, Monogr. Bignon. 17: 49 (1864).

Medium-sized tree, stems usually glabrous. Leaves opposite, imparipinnate, 3–6 pairs of pinnae, 3-verticillate. Inflorescence a terminal bracteate dense thyrse. Calyx spathaceous, split on one side, shortly bilobed at apex. Corolla magenta or pinkish with darker veins, tubular to campanulate. Stamens 4, included, anthers glabrous, thecae divergent, staminode present. Ovary linear, lepidote. Disc annular. Capsule linear, subcylindrical, not subtended by calyx, valves lepidote. Seeds bialate, wings hyaline. $2n = 40$ (38). One species, *N. laevis* (P. Beauv.) Seem. ex Bureau, tropical West Africa from Senegal to Congo.

24. *Pajanelia* DC.

Pajanelia DC., Biblioth. Universelle Genève II, 17: 130 (1838); Steenis, Fl. Malesiana I, 8: 162–164 (1977), reg. rev.

Tree, stems pubescent. Leaves opposite, 1-pinnate, rachis and petiole sharply keeled above. Inflorescence a terminal or axillary, few-flowered thyrsic fascicle. Calyx campanulate, irregularly lobed. Corolla pale yellowish, inside dull-purple, nocturnal, ventricose-campanulate, bilabiate. Stamens 4, included, anthers glabrous, thecae divergent, staminode present. Ovary oblong-cylindrical, lepidote, ovules numerous, multiseriate. Capsule flat, obovate-lanceolate, not subtended by calyx, valves broadly winged. Seeds flattened, with hyaline wings. $2n = 40$. One species, *P. longifolia* (Willd.) Schum., from Malabar through Southeast Asia.

25. *Pandorea* (Endl.) Spach.

Pandorea (Endl.) Spach., Hist. Nat. Veg. 9: 136 (1840); Steenis, Fl. Malesiana I, 8: 174–178 (1977), reg. rev.

Evergreen lianas, branchlets glabrous. Leaves opposite, imparipinnately compound, with (1–)2–4(–7) pairs of pinnae. Inflorescence an axillary bracteate thyrse, sometimes reduced to a raceme. Calyx cupular or campanulate, truncate or irregularly lobed. Corolla white with crimson throat to yellow with red spots, tubular to campanulate. Stamens 4, included, anthers glabrous, thecae divaricate, staminode present. Ovary linear to narrowly ovate, lepidote, with numerous ovules, multiseriate per locule. Disc annular. Capsule narrowly oblong, stipitate at base, terete or flattened, subtended by calyx, valves glabrous. Seeds with hyaline wings. $2n = 38$. Six species, eastern Malesia, Australia, New Caledonia.

26. *Paratecoma* Kuhlm.

Paratecoma Kuhlm., Bol. Serv. Florest. Brasil 4: 3 (1931); Gentry, Fl. Neotrop. 25, Bignoniaceae II: 106–109 (1992), rev.

Large tree, stems glabrous. Leaves alternate, palmately 5–7-foliolate. Inflorescence an axillary thyrsic fascicle. Calyx broadly campanulate, minutely 5-denticulate. Corolla white, salverform-tubular, bilabiate. Stamens 2, subexserted, anthers glabrous, thecae straight, staminodes 3. Ovary oblong-tetragonal, lepidote, ovules alternatingly 4-seriate (attached at different levels) per locule. Disc cupular. Capsule fusiform-linear, not subtended by calyx, valves thick and woody, lepidote. Seeds bialate. One species, *P. peroba* (Rec.) Kuhlm., coastal Brazil, threatened.

27. *Pauldopia* Steenis

Pauldopia Steenis, Acta Bot. Neerl. 18: 425 (1969).

Shrub or small tree, stems glabrous. Leaves opposite, bipinnate, with alate rachis. Inflorescence a terminal or axillary pendent thyrse. Calyx campanulate, truncate, minutely 5-lobed. Corolla pale yellow with reddish-brown nectar guides, tubular to campanulate, tube curved, limb 5-lobed. Stamens 4, included, anthers glabrous, thecae divergent, staminode present. Ovary linear, lepidote, with numerous ovules, 1–2-seriate per locule. Disc annular. Capsule subfalcate-linear, terete, not subtended by calyx. Seeds with narrow margin, not winged. One species, *P. ghorta* (Buch.-

Ham. ex G. Don) Steenis, Northeast India to Southeast Asia.

28. *Perianthomega* Bureau ex Baill.

Perianthomega Bureau ex Baill., Hist. Pl. 10: 33 (1888); Gentry, Fl. Neotrop. 25, Bignoniaceae II: 109–111 (1992), rev.

Liana, stems lepidote. Leaves opposite, triternate, petiole frequently becoming twining and woody. Inflorescence a terminal, few-flowered bracteate thyrse. Calyx broadly campanulate, 5-denticulate. Corolla white to cream, tubular to campanulate. Stamens 4, included, anthers glabrous, thecae divaricate, staminode present. Ovary linear-oblong, densely pubescent, with numerous ovules per locule. Disc broadly patelliform. Capsule linear, not subtended by calyx, valves glabrous. Seeds bialate, with hyaline tipped wings. One species, *P. vellozoi* Bureau, southern Brazil, northern Paraguay and southern Bolivia.

29. *Perichlaena* Baill.

Perichlaena Baill., Hist. Pl. 10: 50 (1888); Perrier, Fl. Madagascar: 16–18 (1938), rev.

Climbing shrub, branchlets glabrous. Leaves opposite, imparipinnate. Inflorescence an axillary, few-flowered (6–10) thyrse. Calyx cylindrical, 5-dentate. Corolla dark to blackish red, distinctly curved, tube cylindrical and straight below stamen insertion, limb bilabiate, upper lip erect, lower lip recurved. Stamens 4, exserted, anthers glabrous, staminode present. Ovary linear, lepidote, with numerous ovules, 1-seriate per locule. Disc annular. Capsule linear, flattened, not subtended by calyx, valves lepidote. Seeds with hyaline wings. One species, *P. richardi* Baill., Madagascar.

30. *Podranea* Sprague

Podranea Sprague in Thiselton-Dyer, Fl. Cap 4(2): 449 (1904); Gentry, Fl. Neotrop. 25, Bignoniaceae II: 111–113 (1992), reg. rev.

Liana or climbing shrub, branchlets glabrous. Leaves opposite, imparipinnate. Inflorescence a terminal bracteate thyrse. Calyx campanulate, 5-dentate. Corolla lavender with magenta markings, tubular to campanulate. Stamens 4, included, anthers glabrous, thecae straight, divaricate, staminode present. Ovary linear, tetragonal, glabrous, with numerous ovules, 6-seriate per locule. Disc annular. Capsule linear, subtended by calyx, valves leathery. Seeds winged. One species, *P. ricasoliana* (Tanf.) Sprague, tropical east and southern Africa.

31. *Radermachera* Zoll. & Moritzi

Radermachera Zoll. & Moritzi, in Zollinger, Syst. Verz. 3: 53 (1855); Steenis, Blumea 23: 121–138 (1976); Steenis, Fl. Malesiana I, 8: 149–157 (1977), reg. rev.
Mayodendron Kurz (1875).

Trees, rarely shrubs, stems pubescent. Leaves opposite, 1–3-pinnate, rachis sulcate. Inflorescence a terminal or axillary thyrse, rarely a raceme. Calyx spathaceous or irregularly lobed, rarely truncate. Corolla white, pink with yellow streaks or yellow to golden red, salverform or narrow funnel-shaped, lobes rounded. Stamens 4 (in one species 5), included, anthers glabrous, thecae divergent, staminode present. Ovary elongate, glabrous, with numerous ovules, multiseriate per locule. Disc annular. Capsule linear, terete, not subtended by calyx, valves smooth, membranaceous. Seeds flat, bialate with narrow hyaline wing. $2n = 40$. About 15 species, tropical Southeast Asia.

32. *Rhigozum* Burch.

Rhigozum Burch., Trav. S. Africa 1: 299 (1822).

Spiny shrubs or small trees, stems glabrous to pubescent. Leaves opposite, simple, with 1 pair of pinnae or imparipinnate, solitary or clustered on small lateral branchlets. Inflorescence an axillary solitary flower or few-flowered thyrsic fascicle on lateral short branches. Calyx campanulate, 5-denticulate. Corolla white, pink or bright yellow, campanulate with 5 spreading lobes. Stamens 5, subexserted, anthers glabrous, thecae divergent, staminode absent. Ovary oblong-elliptic, pubescent, with numerous ovules, 2-seriate per locule. Disc saucer-shaped. Capsule oblong to elliptic, flattened, not subtended by calyx, valves glabrous. Seeds winged. Seven species, mostly in tropical and South Africa (5), Ethiopia and Somalia (1), Madagascar (1).

33. *Romeroa* Dugand

Romeroa Dugand, Mutisia 8: 1 (1952); Gentry, Fl. Neotrop. 25, Bignoniaceae II: 113–115 (1992), rev.

Small to medium-sized tree, stems glabrous, with a few lepidote scales. Leaves opposite, simple, clustered near branch apices, in whorls of 4–5. Inflorescence a terminal subcorymbose raceme.

Calyx campanulate, deeply 3–4-lobed. Corolla yellow, broadly campanulate. Stamens 4, included, anthers glabrous, thecae divergent, staminode present. Ovary oblong-ellipsoid, lepidote, with numerous ovules, multiseriate per locule. Disc annular. Capsule oblong-ellipsoid, long-stiptate, subtended by calyx, valves lepidote. Seeds not winged. One species, *R. verticillata* Dugand, Colombia.

34. *Santisukia* Brummitt

Santisukia Brummitt, Kew Bull. 47: 436 (1992); Santisuk, Flora of Thailand 5: 57–60 (1987), rev.
Barnettia Santisuk (1973).

Small to medium-sized trees, stems puberulous. Leaves verticillate, pinnate, with 4–10 pairs of pinnae. Inflorescence a terminal thyrse. Calyx tubular to campanulate, irregularly lobed. Corolla white to pale lilac with greenish-yellow throat, tubular to campanulate, lobes subequal, crisped. Stamens 4, included, anthers glabrous, thecae divergent, staminode present. Ovary oblong-ellipsoid, lepidote, with numerous ovules per locule. Capsule oblong-ellipsoid, not subtended by calyx, valves with minute tuberculate glands, septum cruciform. Seeds bialate with hyaline-membranaceous wings. Two species, *S. pagetii* (Craib) Brummitt, *S. kerrii* (Barnett & Sandwith) Brummitt, Thailand.

35. *Sparattosperma* Mart. ex Meisn.

Sparattosperma Mart. ex Meisn., Pl. Vasc. Gen. 1: 300; 2: 208 (1840); Gentry, Fl. Neotrop. 25, Bignoniaceae II: 115–118 (1992), rev.

Medium-sized to large trees, branchlets glandular-lepidote, surface sticky. Leaves opposite, long-petiolate, palmately 3–5-foliolate. Inflorescence a terminal thyrse. Calyx 2–3-labiate. Corolla white or pale pink, broadly tubular to campanulate, margins of lobes irregularly laciniate and ciliate. Stamens 4, included, anthers glabrous, thecae divaricate, staminode present. Ovary narrowly oblong, glandular, with numerous ovules, multiseriate per locule. Disc cylindrical. Capsule linear-cylindrical, ± terete, not subtended by calyx, valves longitudinally costate. Seeds very long and narrow, hyaline-membranaceous wings fragmented into hair-like segments. Two species, *S. catingae* A.H. Gentry from Brazil and *S. leucanthum* (Vellozo) K. Schum. from Venezuela to Paraguay.

36. *Spathodea* P. Beauv.

Spathodea P. Beauv., Fl. Oware 1: 46 (1805); Gentry, Fl. Neotrop. 25, Bignoniaceae II: 118–120 (1992), reg. rev.

Tree, stems lepidote. Leaves opposite, imparipinnate, 9–15-foliolate. Inflorescence a terminal subumbellate raceme. Calyx spathaceous, narrowing to a recurved tip, 1-parted to base, water-filled in bud. Corolla red-orange to crimson, broadly campanulate above and cylindrical at base. Stamens 4, included, anthers glabrous, thecae divaricate, staminode present. Ovary narrowly oblong, papillose, with numerous ovules, multiseriate per locule. Disc annular. Capsule narrowly oblong-elliptic, subtended by calyx, valves keeled. Seeds flat, with broad hyaline-membranaceous wings. $2n = 26, 36, 38$. One species, *S. campanulata* P. Beauv., tropical Africa.

37. *Spirotecoma* Baill. ex Dalla Torre & Harms

Spirotecoma Baill. ex Dalla Torre & Harms, Gen. Siphon.: 467 (1904); Gentry, Fl. Neotrop. 25, Bignoniaceae II: 120–126 (1992), reg. rev.

Shrubs to large trees, stems glabrous or lepidote. Leaves opposite, 1-foliolate or palmately 3–7-foliolate. Inflorescence an axillary few-flowered raceme or 1–2 fascicled flowers. Calyx tubular to campanulate, irregularly dentate. Corolla dark reddish purple or yellowish brown, tubular-infundibuliform above a tubular base, 5-lobed, strongly bilabiate, curved and oblique. Stamens 4, exserted, thecae pendulous, staminode present. Ovary oblong, densely blackish lepidote, with numerous ovules. Disc annular. Capsule subterete, linear, curved or spirally twisted, not subtended by calyx, valves lepidote. Seeds bialate. Five species, Cuba, Hispaniola.

38. *Stereospermum* Cham.

Stereospermum Cham., Linnaea 7: 720 (1833); Steenis, Fl. Malesiana I, 8: 145–149 (1977), reg. rev.; Perrier, Fl. Madagascar: 4–16 (1938), reg. rev.
Siphocolea Baill. (1887).

Small to medium-sized trees or shrubs, stems pubescent. Leaves opposite, imparipinnate. Inflorescence a terminal or axillary thyrse. Calyx campanulate to tubular, irregularly 2–5-lobed. Corolla pink, slightly zygomorphic, tube campanulate. Stamens 4, included, anthers glabrous, thecae divergent, staminode present. Ovary linear-oblong. 4-angled, lepidote, with ovules 2-seriate per locule.

Disc annular. Capsule linear-cylindrical, not subtended by bracts, valves glabrous. Seeds winged. $2n = 40$. About 20 species, centred in tropical Asia, 3–4 species in tropical Africa.

39. *Tabebuia* Gomes ex DC.

Tabebuia Gomes ex DC., Biblioth. Universelle Genève, ser. 2, 17: 130 (1838); Gentry, Fl. Neotrop. 25, Bignoniaceae II: 127–273 (1992), rev.

Shrubs to large trees, stems glabrous to variously pubescent. Leaves opposite, simple, 1-foliolate or palmately 3–7(–9)-foliolate, opposite. Inflorescence a terminal thyrse, sometimes reduced to few-flowered raceme or solitary flower. Calyx cupular, campanulate to tubular, truncate to bilabiate. Corolla white, yellow, lavender, magenta or red, tubular to campanulate. Stamens 4, exserted or included, anthers glabrous, thecae divaricate, staminode present. Ovary linear-oblong, variously pubescent, with ovules multiseriate per locule. Disc annular. Capsule subterete, elongate-linear to short-oblong, sometimes subtended by calyx, valves smooth to verrucose-muricate. Seeds thin, bialate or wingless. $2n = 40$. About 100 species, from northern Mexico and the Antilles to Argentina, greatest diversity on Hispaniola and Cuba.

40. *Tecoma* Juss.

Tecoma Juss., Gen. Pl.: 139 (1789); Gentry, Fl. Neotrop. 25, Bignoniaceae II: 273–293 (1992), reg. rev.
Tecomaria Spach. (1840).

Shrubs or small trees, rarely subscandent in *T. capensis* Lindl., stems variously pubescent. Leaves opposite, simple, 3-foliolate or imparipinnately compound. Inflorescence a terminal raceme or thyrse. Calyx cupular, 5-parted. Corolla yellow or orange-red, tubular to campanulate to narrowly tubular-salverform. Stamens 4, included, anthers glabrous or pilose, thecae divaricate to divergent. Ovary narrowly cylindrical, lepidote, with ovules 2-seriate per locule. Disc cupular. Capsule linear, sometimes subtended by calyx, valves smooth. Seeds thin, bialate, wings hyaline-membranaceous. $2n = 36$. Fourteen species, tropical and South Africa (2 spp.), tropical America, especially Andes to Arizona.

41. *Tecomanthe* Baill.

Tecomanthe Baill., Hist. Pl. 10: 41 (1888); Steenis, Fl. Malesiana I, 8: 167–174 (1977), reg. rev.

Lianas, climbing or creeping, stems glabrous or pubescent. Leaves opposite, pinnate, with 1–7 pairs of pinnae. Inflorescence a cauliflorous or axillary few-flowered, pendent raceme. Calyx cupular, equally 5-lobed. Corolla pink, campanulate. Stamens 4, included, anthers glabrous, thecae divaricate, staminode present. Ovary cylindrical, lepidote, with ovules multiseriate on 2 placentae per locule. Disc annular. Capsule linear-terete, sometimes subtended by calyx, flattened, stipitate, valves coriaceous or woody. Seeds orbicular, winged. $2n = 36, 38$. Five species, Malesia, Australia, New Zealand (*T. speciosa* W. Oliver, only a single tree left which is propagated).

42. *Tecomella* Seem.

Tecomella Seem., J. Bot. 1: 18 (1863).

Shrub or small tree, stems variously pubescent. Leaves opposite, often fascicled, simple, lanceolate. Inflorescence a few-flowered thyrse, terminal on short branches. Calyx campanulate, 5-dentate. Corolla orange, tubular to campanulate, with wide throat. Stamens 4, exserted, anthers glabrous, thecae divergent, staminode present. Ovary oblong-lanceolate, lepidote, with numerous ovules per locule. Disc cupular. Capsule linear, not subtended by calyx, somewhat 4-costate, valves flat, coriaceous. Seeds elliptic, wings only marginally membranaceous. One species, *T. undulata* (Sm.) Seem., from Arabia to western India.

43. *Zeyheria* Mart.

Zeyheria Mart., Nov. Gen. Sp. Pl. 2: 65 (1826); Gentry, Fl. Neotrop. 25, Bignoniaceae II: 293–297 (1992), reg. rev.

Shrubs or trees, stems stellate pubescent. Leaves opposite, palmately (3–)5-foliolate. Inflorescence an axillary or terminal dichasial thyrse. Calyx cupular, bilabiately split to near base. Corolla brownish or tannish outside, yellowish to dark orange or purplish inside, not contracted to narrower tube near base. Stamens 4, exserted, anthers glabrous, thecae divergent to divaricate. Ovary globose-ellipsoid, sometimes stipitate, with multiseriate ovules per locule. Disc annular. Capsule round to broadly obovoid, subtended by calyx, slightly flattened, valves tomentose, verrucose with spine-like emanations. Seeds thin, with hyaline wing. Two species, *Z. montana* Mart., *Z. tuberculosa* Bureau, subamazonian Brazil to Bolivia.

2. Tribe Oroxyleae A.H. Gentry (1980).

Trees, shrubs or lianas, interpetiolar glands and ridges absent, pseudostipules absent. Leaves pinnate. Calyx cupular, truncate or campanulate. Pollination various (bees, hawk moths). Stamens 4–5, thecae parallel to divaricate, with one fertile theca, the sterile theca spur-like or both thecae fertile. Ovary bilocular, placentation axile. Fruit dehiscing parallel to septum (septicidal). Tropical Asia.

Key to Genera of Oroxyleae

1. Trees or shrubs 2
- Lianas 3
2. Flowers fleshy, very coarse; corolla wide, violet or reddish purple, 7–10 cm long; fertile stamens 5; capsule flat, sword-shaped, 60–120 cm long; leaves 1-pinnate **47. *Oroxylum***
- Flowers not fleshy, much smaller and thinner; corolla white; fertile stamens 4; capsule smaller and of different shape; leaves 3–4-pinnate **45. *Millingtonia***
3. Corolla salver-shaped, with narrowly cylindrical tube, 5–19 cm long, not geniculate; leaves 3-foliolate **46. *Nyctocalos***
- Corolla not salver-shaped, tubular, tube 3–4.5 cm long, geniculate above the basal tube; leaves 5–8-foliolate **44. *Hieris***

44. *Hieris* Steenis

Hieris Steenis, Bull. Jard. Bot. Buitenzorg III, 10: 278 (1928); Steenis, Fl. Malesiana I, 8: 127–128 (1977), rev.

Liana, stems glabrous. Leaves opposite, pinnate with (1–)2(–3) pairs of pinnae. Inflorescence a terminal or axillary dense raceme. Calyx cupular, truncate. Corolla yellow, with whitish to pale lilac lobes, tubular to campanulate, tube geniculate in lower third. Stamens 4, included, anthers glabrous, thecae divaricate, staminode present. Ovary ovate, subterete, with ovules in two rows of 3–4 per locule. Disc annular. Capsule unknown. One species, *H. curtisii* (Ridl.) Steenis, Penang (Malayan Peninsula).

45. *Millingtonia* L.f.

Millingtonia L.f., Suppl. Pl.: 45 (1782); Steenis, Fl. Malesiana I, 8: 133–134 (1977), rev.

Medium-sized tree, stems glabrous. Leaves opposite, 2–3-pinnate. Inflorescence an axillary, lax, many-flowered thyrse. Calyx truncate-campanulate, 5-lobed. Corolla white, fragrant, nocturnal, salver-shaped, with a very long, slender tube, limb bilabiate. Stamens 4, exserted, anthers glabrous, with one fertile theca, the sterile theca spur-like. Ovary linear-elliptic, lepidote, with numerous ovules. Disc cup-shaped. Capsule linear, not subtended by calyx, valves glabrous. Seeds winged. $2n = 30$. One species, *M. hortensis* L.f., Southeast Asia and Malesia.

46. *Nyctocalos* Teijsm. & Binn.

Nyctocalos Teijsm. & Binn., J. Bot. Neerl. 1: 366 (1861); Steenis, Fl. Malesiana I, 8: 123–127 (1977), reg. rev.

Lianas, stems glabrous. Leaves opposite, pinnate, 3(–5)-foliolate. Inflorescence an axillary or terminal short 8–12-flowered raceme. Calyx cupular, truncate. Corolla white, nocturnal, fragrant, almost actinomorphic, salver-shaped, tube long, narrow, dilated above, limb with 5 subequal lobes. Stamens 4 with staminode or 5, without staminode, subexserted, anthers glabrous, thecae versatile, divergent. Ovary cylindrical, lepidote, with ovules multiseriate per locule. Disc annular. Capsule flat, stalked, valves with central prominent rib. Seeds winged. Three species, Southeast Asia and West Malesia.

47. *Oroxylum* Vent.

Oroxylum Vent., Dec. Gen. Nov.: 8 (1808); Steenis, Fl. Malesiana I, 8: 128–131 (1977), reg. rev.

Pachycaul tree, stems glabrous. Leaves opposite, 2–3(–4)-pinnate, pinnae entire. Inflorescence a terminal or axillary dense raceme or thyrse. Calyx campanulate, 5-parted. Corolla funnel-shaped, with 5 subequal lobes, white, nocturnal, fetid, bat-pollinated. Stamens 5, included, anthers glabrous, thecae parallel, staminode absent. Ovary linear, with ovules multiseriate per locule. Disc annular. Capsule sword-shaped, linear, flat. Seeds large, winged. $2n = 28$. One species, *O. indicum* (L.) Kurz, from Sri Lanka to Celebes and Timor.

3. Tribe Bignonieae Dumort. (1829).

Lianas, rarely shrubs or trees, interpetiolar glands and ridges present or absent, pseudostipules foliaceous, subulate or absent. Leaves pinnate, rarely simple, terminal pinna often replaced by a tendril. Calyx 5-lobed, truncate, cupular or spathaceous. Pollination various (bees, hawkmoths, birds, bats and butterflies). Stamens 2–4, anthers dithecous, thecae usually straight, divaricate. Ovary bilocular, placentation axile. Fruit dehiscing parallel to septum (septicidal) or 4-valved. Neotropical, one sp. in North America.

Key to Genera of Bignonieae

1. Small trees; lateral branches thorn-tipped; corolla-limb with 4 ovate to rhombic lobes; stamens without filament **89. *Sphingiphila***
- Lianas or climbing shrubs; lateral branches not thorn-tipped; corolla-limb 5-lobed, stamens with distinct filaments 2
2. Fertile stamens 2, 3 staminodes present **84. *Pseudocatalpa***
- Fertile stamens 4, a single staminode present or absent 3
3. Tendrils trifid-uncinate, with claw-like apices; stem cross section with many, irregularly divided phloem arms; fruit 4-parted 4
- Tendrils simple, bifid or 3- to multifid, never uncinate or with claw-like apices; stem cross section usually with 4–32, radially symmetrical phloem arms 7
4. Flowers yellow 5
- Flowers red or magenta 6
5. Apex of sepals without a recurved apicle; inflorescence bracts minute; fruit linear, flattened **67. *Macfadyena***
- Apex of sepals with a recurved apicle; inflorescence bracts foliaceous; fruit elliptic-oblong, expanded **72. *Melloa***
6. Stems cylindrical; interpetiolar line present; corolla red; calyx 2-lobed **60. *Dolichandra***
- Stems tetragonal; interpetiolar line absent; corolla magenta; calyx 5-lobed **76. *Parabignonia***
7. Calyx spathaceous, split along its entire length 8
- Calyx cupular, tubular, or campanulate, truncate, or slightly 2–5-parted, split to less than half of its length 12
8. Flowers magenta, with wine-coloured nectar guides; without nectar disc; calyx with recurved tip forming a hood-like structure; pseudostipules foliaceous, well developed **79. *Phryganocydia***
- Flowers yellow or red, without nectar guides; with nectar disc; calyx without a recurved tip forming a hood-like structure; pseudostipules subulate, poorly developed 9
9. Tendrils simple; calyx thick and spongy; fruit oblong-elliptic **53. *Callichlamys***
- Tendrils trifid or multifid; calyx membranous; fruit linear 10
10. Stems cylindrical; leaves 3-foliolate **69. *Manaosella***
- Stems hexagonal; leaves pinnate or bipinnate 11
11. Interpetiolar line present; calyx 5-split down to base, glandular-pubescent, without plate-shaped glands; fruit glandular-pubescent **62. *Gardnerodoxa***
- Interpetiolar line absent; calyx 2-split down to base, glabrous, with plate-shaped glands; fruit glabrous **75. *Neojobertia***
12. Leaves biternate to ternately bipinnate or tripinnate 13
- Leaves 2- or 3- or 5-foliolate, biternate only in *Arrabidaea inaequalis*, rarely simple 14
13. Branches acutely tetragonal and winged; calyx without glands; corolla white or cream **82. *Pleonotoma***
- Branches terete, never winged; calyx with glands; corolla bright yellow **73. *Memora***
14. Anthers exserted; corolla orange to red, tubular to narrowly tubular 16
- Anthers included; corolla yellow, pink, magenta or white, tubular to campanulate 19
16. Stems hexagonal; tendrils trifid **85. *Pyrostegia***
- Stems cylindrical; tendrils multifid or simple 17
17. Tendrils multifid; corolla tube curved; calyx cupular; south-eastern North America **52. *Bignonia***
- Tendrils simple; corolla tube straight; calyx tubular; western Amazonia or Brazilian savannas 18
18. Pseudostipules foliaceous; corolla orange; calyx with glands; western Amazonia **68. *Macranthisiphon***
- Pseudostipules absent; corolla red; calyx without glands; Brazilian savannas **80. *Piriadacus***
19. Calyx with a "double margin", inner margin straight, outer margin wavy 20
- Calyx with a "simple margin" 22
20. Stems terete; inflorescence a raceme; fruit glabrous; seeds glabrous **63. *Glaziovia***
- Stems hexagonal; inflorescence a thyrse; fruit usually verrucose and tomentose, if smooth, then always tomentose; seeds pubescent 21
21. Corolla straight, bilabiate, split for about half of its length, closed at anthesis; calyx truncate, without glands **49. *Amphilophium***
- Corolla strongly curved (90° angle above base), not bilabiate, split for about a quarter of its length, open at anthesis; calyx 5-parted, with glands **64. *Haplolophium***
22. Pinnae pellucid-punctate; branchlets hollow **90. *Stizophyllum***
- Pinnae not pellucid-punctate; branchlets not hollow 23
23. Tendrils bifid; petiolar glandular fields present **77. *Paragonia***
- Tendrils simple or trifid; petiolar glandular fields absent 24
24. Corolla pink, magenta or red (sometimes white with purple markings in *Roentgenia*) 25
- Corolla yellow, cream or white 36
25. Corolla with nectar guides; nectar disc absent 26
- Corolla without nectar guides; nectar disc conspicuous 29
26. Inflorescence bracteate, bracts linear and elongate **86. *Roentgenia***
- Inflorescence ebracteate, or with poorly developed bracts 27
27. Ovary oblong, verrucose-muricate; pseudostipules a group of subulate scales, resembling a miniature bromeliad **55. *Clytostoma***
- Ovary linear, lepidote; pseudostipules not resembling a bromeliad, foliaceous or small and inconspicuous 28
28. Pseudostipules small and inconspicuous; plants usually smelling of garlic; capsule strongly tuberculate, with tubercles up to 5 mm long **70. *Mansoa***
- Pseudostipules foliaceous; plants never smelling of garlic; capsule smooth **57. *Cydista***
29. Calyx urceolated, dilated, with 5 conspicuous ridges on the sides **61. *Fridericia***
- Calyx neither urceolated nor dilated 30
30. Calyx >11 mm, long, bilobed to irregularly 3–4-lobed **71. *Martinella***
- Calyx <10 mm long, truncate to denticulate 31
31. Thecae reflexed forward by 90°; pollen in tetrads **56. *Cuspidaria***
- Thecae straight or only slightly curved; pollen simple 32
32. Branchlets hexagonal **59. *Distictis***
- Branchlets cylindrical 33
33. Pseudostipules foliaceous, well developed **87. *Saritaea***
- Pseudostipules absent or poorly developed 34
34. Interpetiolar glands absent; tendrils trifid **83. *Potamoganos***
- Interpetiolar glands present; tendrils simple 35
35. Inflorescence ebracteate or with minute bracts; capsule linear **51. *Arrabidaea***
- Inflorescence bracteate, with conspicuous linear bracts about as long as calyx; capsule elliptic **93. *Xylophragma***

36. Branchlets acutely tetragonal 74. *Mussatia*
- Branchlets terete 37
37. Capsule oblong or elliptic to orbicular 38
- Capsule linear 39
38. Calyx drying black; inflorescence an elongate raceme, with foliaceous bracts; ovary sessile; fruit oblong, sessile, covered by lenticels; seeds bialate or corky and wingless 48. *Adenocalymma*
- Calyx usually drying olive; inflorescence a contracted raceme, ebracteate or with minute bracts; ovary stipitate; fruit elliptic to orbicular, stipitate, smooth, without lenticels; seeds completely surrounded by a broad wing 50. *Anemopaegma*
39. Corolla narrowly tubular, flowers nocturnal 40
- Corolla campanulate, flowers diurnal 41
40. Tendrils simple; corolla > 10 cm long, smelling of almond (cyanide); fruits oblong-cylindrical 91. *Tanaecium*
- Tendrils trifid; corolla < 8 mm long, not smelling of almond; fruits linear, flattened 65. *Leucocalantha*
41. Anther thecae villous; nectar disc absent 66. *Lundia*
- Anther thecae glabrous; nectar disc present 42
42. Corolla tiny, <2 cm long, strongly bilabiate, upper 2 lobes fused, smelling of clove 92. *Tynanthus*
- Corolla large, >4 cm long, not bilabiate, upper 2 lobes free, not smelling of clove 43
43. Branchlets hexagonal; pinnae cordiform, cartaceous; tendrils multifid; calyx 5-parted; capsule densely spiny 81. *Pithecoctenium*
- Branchlets cylindrical; pinnae elliptic; tendrils trifid, coriaceous; calyx bilabiate to subspathaceous or truncate; capsule smooth 44
44. Calyx bilabiate to subspathaceous 88. *Spathicalyx*
- Calyx truncate 45
45. Interpetiolar glandular fields absent; pseudostipules foliaceous, never resembling a miniature bromeliad; capsule oblong 58. *Distictella*
- Interpetiolar glandular fields present; pseudostipules a group of subulate scales, resembling a miniature bromeliad; capsule elongated (subtetragonal or linear) 46
46. Corolla white; calyx tubular, coriaceous; fruit subtetragonal 54. *Ceratophytum*
- Corolla yellow; calyx cupular, membranous; fruit linear 78. *Periarrabidaea*

48. *Adenocalymma* Mart. ex Meisn.

Adenocalymma Mart. ex Meisn., Pl. Vasc. Gen. 1: 300, 2: 208 (1840); Gentry, Fl. Ecuador 7: 11–15 (1977), reg. rev.

Lianas, stems with 4 phloem arms in cross section, branchlets terete, glabrous to pubescent, without interpetiolar glands, with or without interpetiolar ridge, pseudostipules small and subulate. Leaves 3-foliolate, terminal pinna often replaced by a simple tendril. Inflorescence a narrow axillary or terminal bracteate raceme, bracts foliaceous. Calyx cupular to tubular, 5-denticulate, membranous, with glands. Corolla yellow, tubular to campanulate, straight. Stamens 4, included, anthers glabrous, thecae straight. Ovary narrowly cylindrical, lepidote or puberulent, ovules 2-seriate per locule. Disc annular. Capsule oblong, not flattened, woody, with distinct lenticels. Seeds bialate or almost wingless. About 60 species, tropical America.

49. *Amphilophium* Kunth

Amphilophium Kunth, J. Phys. Chim. Hist. Nat. Arts 87: 451 (1818); Gentry, Fl. Ecuador 7: 15–21 (1977), reg. rev.

Lianas, stems with multiples of 4 phloem arms in cross section, branchlets sharply hexagonal, variously pubescent, without interpetiolar glands, with interpetiolar ridge, pseudostipules foliaceous. Leaves 3-foliolate, terminal pinna often replaced by a trifid tendril. Inflorescence a terminal bracteate thyrse, usually on lateral branch. Calyx campanulate, truncate, with a double margin, inner margin straight, outer margin wavy, coriaceous, without glands. Corolla cream or purple, tubular, straight, thick, bilabiate, split about half of its length, closed at anthesis. Stamens 4, included, anthers glabrous, thecae straight. Ovary ovate-cylindrical, ovules multiseriate per locule. Disc annular. Capsule oblong-elliptic, flattened, woody, smooth to tuberculate. Seeds pubescent, winged. $2n = 40$. Six species from Mexico and West Indies to Argentina.

50. *Anemopaegma* Mart. ex Meisn.

Anemopaegma Mart. ex Meisn., Pl. Vasc. Gen. 1: 300; 2: 208 (1840); Gentry, Fl. Ecuador 7: 24–29 (1977), reg. rev.
Clytostomanthus Pichon (1946).

Shrubs or lianas, stems with multiples of 4 phloem arms in cross section, branchlets terete, glabrous to variously pubescent, with or without interpetiolar glands or ridges, pseudostipules foliaceous or absent. Leaves 3–(5)-foliolate, terminal pinna often replaced by a simple or trifid tendril. Inflorescence an axillary or terminal raceme, ebracteate or with minute bracts. Calyx cupular, usually truncate, sometimes 5-parted, membranous, with glands. Corolla bright to pale yellow, tubular to campanulate, straight. Stamens 4, included, anthers glabrous, thecae straight. Ovary ellipsoid, stipitate, lepidote or pubescent, ovules 2–6-seriate per locule. Disc annular. Capsule elliptic to orbicular, stipitate, valves smooth, without lenticels. Seeds flat, completely surrounded by a broad, hyaline-membranaceous wing. $2n = 40$. About 60 species from Mexico to Brazil and Argentina.

51. *Arrabidaea* DC.

Arrabidaea DC., Biblioth. Universelle Genève II, 17: 126 (1838); Gentry, Fl. Ecuador 7: 29–45 (1977), reg. rev.

Lianas, rarely small trees or shrubs, stems with 4 phloem arms in cross section, branchlets terete, glabrous to variously pubescent, with or without interpetiolar glands or ridge, pseudostipules small or rarely foliaceous. Leaves 3-foliolate, terminal pinna often replaced by a simple tendril, rarely pinnate or simple. Inflorescence an axillary or terminal bracteate and many-flowered thyrse. Calyx cupular, truncate or bilabiate, 5-denticulate, membranous, with or without glands. Corolla red or pale pink to purple, campanulate. Stamens 4, included, anthers glabrous, thecae straight or slightly curved. Ovary narrowly cylindrical to linear-tetragonal, usually minutely lepidote, ovules 2-seriate per locule. Disc cupular. Capsule linear, flattened, woody, valves usually smooth, rarely verrucose, with or without lenticels. Seeds bialate, wings usually hyaline. $2n = 40$. About 100 species in tropical America from Mexico and the West Indies to Argentina.

52. *Bignonia* L.

Bignonia L., Sp. Pl.: 622 (1753), nom. cons.

Liana, stems with multiples of 4 phloem arms in cross section, branchlets terete, glabrous to pubescent, without interpetiolar glands, with interpetiolar line, pseudostipules foliaceous. Leaves 3-foliolate, the terminal pinna replaced by a multifid tendril with adhesive discs. Inflorescence a short axillary bracteate cyme. Calyx cup-shaped and with irregular to nearly obsolete lobes, thick, with glands. Corolla dull red or orange outside, yellow within, tubular, curved, thick. Stamens 4, included, anthers glabrous, thecae straight. Ovary cylindrical, lepidote, ovules 2-seriate per locule. Disc annular. Capsule elongate, linear, flattened, woody, smooth, with lenticels. Seeds glabrous, with broad wings. $2n = 40$. One species, *B. capreolata* L., southeast North America.

53. *Callichlamys* Miq.

Callichlamys Miq., Linnaea 18: 254 (1845); Gentry, Fl. Ecuador 7: 45–48 (1977), reg. rev.

Liana, shrubby when young, stems with 4 phloem arms in cross section, branchlets terete, glabrous, without interpetiolar glands and ridge, pseudostipules subulate, poorly developed. Leaves 3-foliolate, terminal pinna often replaced by a simple tendril. Inflorescence a short axillary (rarely terminal) 1–12-flowered bracteate raceme. Calyx inflated-cupular, irregularly lobed, thick and spongy. Corolla campanulate, tube contracted above the ovary, chartaceous, straight. Stamens 4, included, anthers glabrous, thecae straight. Ovary ovate-cylindrical, lepidote, ovules 4–8-seriate per locule. Disc annular. Capsule oblong-elliptic, flattened, woody, smooth, with lenticels. Seeds thin, wings brown. $2n = 40$. One species, *C. latifolia* (L. Rich.) K. Schum. from Mexico to Brazil.

54. *Ceratophytum* Pittier

Ceratophytum Pittier, J. Wash. Acad. Sci. 18: 62 (1928); Gentry, Fl. Ecuador 7: 48–50 (1977), reg. rev.

Liana, stems with 4 phloem arms in cross section, branchlets terete, pubescent, with interpetiolar glands, without interpetiolar ridge, pseudostipules a group of subulate scales, resembling a miniature bromeliad. Leaves 3-foliolate, terminal pinna often replaced by a trifid tendril. Inflorescence a few-flowered, terminal bracteate thyrse. Calyx cupular, truncate, coriaceous, lepidote. Corolla cream to pale yellow, tubular to campanulate, straight, thick. Stamens 4, included, anthers glabrous, thecae straight. Ovary narrowly cylindrical, lepidote, ovules 6–8-seriate per locule. Disc cupular. Capsule elongated, subtetragonal, not flattened, woody, smooth, with lenticels. Seeds bialate, wings papery. One species, *C. tetragonolobum* (Jacq.) Sprague & Sandw., from Mexico to Trinidad and Guayana and south to Peru and Bolivia.

55. *Clytostoma* Miers ex Bureau

Clytostoma Miers ex Bureau, Adansonia 8: 353 (1868); Gentry, Fl. Ecuador 7: 50–54 (1977), reg. rev.

Lianas, stems with multiples of 4 phloem arms in cross section, branchlets tetragonal to subterete, glabrous to pubescent, without interpetiolar glands, with interpetiolar line, pseudostipules a group of subulate scales, resembling a miniature bromeliad. Leaves simple to 3-foliolate, terminal pinna often replaced by simple tendril. Inflorescence a few-flowered, ebracteate axillary cyme. Calyx cupular to campanulate, truncate, membranous, without glands. Corolla magenta to purple, with darker nectar guides, tubular to campanulate, straight, membranous. Stamens 4, included, anthers glabrous, thecae straight, divaricate, ovary oblong, verrucose-muricate, ovules 2(–4)-seriate per locule. Disc absent. Capsule oblong to suborbicular, expanded, woody, echinate, without lenticels. Seeds glabrous, thick, corky, wingless or winged on only one side. $2n = 40$. Nine species from Mexico to Uruguay, Argentina and Brazil.

56. *Cuspidaria* DC.

Cuspidaria DC., Biblioth. Universelle Genève II, 17: 125 (1838).
Saldhanhaea Bureau (1864).

Lianas, stems with 4 phloem arms in cross section, branchlets terete to subcylindrical, glabrous to pubescent, with or without interpetiolar glands and ridges, pseudostipules inconspicuous or absent. Leaves simple to 2–3-foliolate, the terminal pinna often replaced by a simple tendril. Inflorescence a terminal or axillary bracteate thyrse. Calyx cupular, 5-denticulate, membranous, without glands. Corolla cream to pale yellow or magenta, tubular to campanulate, straight, membranous. Stamens 4, included, anthers glabrous, thecae reflexed forward by 90°. Ovary ovoid-ellipsoidic, 4-sulcate, lepidote, ovules 2–4-seriate per locule. Disc annular. Capsule oblong-linear, slightly flattened, woody, often 4-winged, without lenticels. Seeds glabrous, bialate, wings papery. Nine species, tropical America from Panama, Amazonian Brazil to Bolivia and Argentina.

57. *Cydista* Miers

Cydista Miers, Proc. Roy. Hort. Soc. Lond. 3: 191 (1863); Gentry, Fl. Ecuador 7: 56–60 (1977), reg. rev.; Hauk, Ann. Missouri Bot. Gard. 84: 815–840 (1997), rev.
Levya Bureau ex Baill. (1888).

Lianas, stems with multiples of 4 phloem arms in cross section, branchlets terete, glabrous or pubescent, with or without interpetiolar glands or ridge, pseudostipules foliaceous. Leaves simple or 3-foliolate, the terminal pinna often replaced by a simple tendril. Inflorescence an axillary or terminal raceme or thyrse. Calyx cupular, truncate or irregularly bilabiate, membranous, with or without glands. Corolla white to magenta, with darker nectar guides, tubular to campanulate, straight, membranous. Stamens 4, included, anthers glabrous, thecae straight. Ovary cylindrical, lepidote or pubescent, ovules 2-seriate per locule. Disc absent. Capsule linear, flattened, woody, smooth, with lenticels. Seeds glabrous, bialate. $2n = 40$. Six species from Mexico to the West Indies, Brazil and Paraguay.

58. *Distictella* Kuntze

Distictella Kuntze, in Post & Kuntze, Lex. Gen. Pl.: 182 (1903); Gentry, Fl. Ecuador 7: 67–70 (1977), reg. rev.

Lianas, rarely shrubs, stems with multiples of 4 phloem arms in cross section, branchlets terete, glabrous to pubescent, without interpetiolar glands, with or without interpetiolar line, pseudostipules foliaceous. Leaves 3-foliolate, the terminal pinna often replaced by a trifid tendril. Inflorescence an axillary or terminal raceme or thyrse. Calyx cupular, truncate, thick, with groups of glands. Corolla white, tubular to campanulate, strongly curved, thick. Stamens 4, included, anthers glabrous, thecae straight. Ovary ovoid-oblong, lepidote, ovules 4–8-seriate per locule. Disc annular. Capsule oblong, convex, woody, smooth, with sparse lenticels. Seeds glabrous, with reduced wings. Ten species, from Costa Rica to Brazil.

59. *Distictis* Mart. ex Meisn.

Distictis Mart. ex Meisn., Pl. Vasc. Gen. 1: 300; 2: 208 (1840).
Phaedranthus Miers (1863), nom. cons.
Macrodiscus Bureau (1864).
Wunschmannia Urb. (1908).

Lianas, stems with multiples of 4 phloem arms in cross section, branchlets hexagonal, stellate-pubescent, without interpetiolar glands or ridges, pseudostipules present or absent. Leaves 3-foliolate, terminal pinna often replaced by a trifid tendril. Inflorescence a few-flowered terminal raceme. Calyx cupular, truncate, coriaceous, stellate-pubescent. Corolla white to lavender, tubular to campanulate above a narrow base. Stamens 4, included, anthers glabrous, thecae straight. Ovary oblong, ovules few-seriate per locule. Disc annular. Capsule elliptic-oblong, woody, curved. Seeds bialate, usually papillose or pubescent. $2n = 40$. Ten species, Mexico to Colombia and Peru.

60. *Dolichandra* Cham.

Dolichandra Cham., Linnaea 7: 657 (1832).

Lianas, stems with many, irregularly divided phloem arms in cross section, branchlets cylindrical, with interpetiolar glands and interpetiolar ridge, pseudostipules subulate. Leaves 3-foliolate, terminal pinna often replaced by a trifid, uncinate tendril with claw-like apices. Inflorescence a few-flowered axillary cyme. Calyx spathaceous, 2-lobed. Corolla red, tubular, basally constricted, straight, membranous. Stamens 4, exserted, anthers glabrous, thecae straight. Ovary oblong to ovoid, stipitate, lepidote, ovules multiseriate per locule. Disc annular. Capsule oblong-elliptic, coriaceous, 4-parted, flattened. Seeds glabrous, winged. One species *D. cynanchoides* Cham., Argentina, Paraguay, Uruguay, and southern Brazil.

61. ***Fridericia*** Mart.

Fridericia Mart. in Nees, Fridericia & Zollernia: 5 (1827); Sandwith & Hunt, Fl. Ilustrada Catarinense: 119–122 (1974), reg. rev.

Liana, stems with 4 phloem arms in cross section, branchlets terete, glabrous, without interpetiolar glands, with interpetiolar ridges, pseudostipules foliaceous. Leaves 3-foliolate, terminal pinna often replaced by a simple tendril. Inflorescence a terminal thyrse. Calyx urceolate, dilated, with 5 conspicuous ridges, 5-dentate and 5-winged, thick, without glands. Corolla rose to lilac, tubular, lobes subequal, straight, thick. Stamens 4, included, anthers glabrous, thecae straight. Ovary oblong, lepidote, ovules 2-seriate per locule. Disc cupular. Capsule oblong-linear, subtetragonal, flattened, thick, smooth, without lenticels. Seeds glabrous, with large membranaceous wings. One species, *F. speciosa* Mart., S Brazil.

62. ***Gardnerodoxa*** Sandwith

Gardnerodoxa Sandwith, Kew Bull. 1954: 611 (1955).

Lianas, stems with 4 phloem arms in cross section, branchlets hexagonal, glabrous to glandular-pubescent, without interpetiolar glands, with interpetiolar ridge, pseudostipules subulate. Leaves 3-pinnate, terminal pinna frequently a multifid tendril. Inflorescence a terminal cyme. Calyx cupular, 5-parted, membranous, without plate-shaped glands. Corolla yellow or red, tubular to campanulate, straight, membranous. Stamens 4, included, anthers glabrous, thecae straight. Ovary lepidote, ovules few, 2-seriate per locule. Disc annular. Capsule linear, cylindrical, flattened, woody, glandular-pubescent, without lenticels. Seeds glabrous, bialate. Three species, in dry forests of eastern Brazil.

63. ***Glaziovia*** Bureau

Glaziovia Bureau, Adansonia 8: 380 (1868).

Liana, stems with multiples of 4 phloem arms in cross section, branchlets cylindrical, pubescent, without interpetiolar glands and ridges, pseudostipules elliptic, foliaceous. Leaves 3-foliolate, terminal pinna often replaced by a trifid tendril with adhesive discs on tips. Inflorescence a terminal raceme. Calyx cupular, with a double margin, inner margin straight, outer margin undulate, coriaceous, without glands. Corolla yellowish or purple, tubular, bilabiate, curved. Stamens 4, included, anthers glabrous, thecae straight. Disc annular. Ovary oblong, smooth, ovules 2- to multiseriate per locule. Capsule elliptic, flattened, woody, smooth, without lenticels. Seeds glabrous, winged. One species, *G. bauhinioides* Bureau, Atlantic forest of eastern Brazil.

64. ***Haplolophium*** Cham.

Haplolophium Cham., Linnaea 7: 556 (1832), nom. cons.; Gentry, Novon 2: 163–165 (1992), rev.
Urbanolophium Melch. (1927).

Lianas, stems with multiples of 4 phloem arms in cross section, branchlets hexagonal, with stellate hairs, with interpetiolar glands, with interpetiolar ridge, pseudostipules foliaceous. Leaves 3-foliolate, the terminal pinna replaced by a trifid, disc-tipped tendril. Inflorescence an axillary, few-flowered thyrse. Calyx cupular, with a double margin, inner margin straight, outer margin wavy, 5-parted, thick, with glands. Corolla white to wine-red, tubular to campanulate, tube bent at 90° angle above base, lobes subequal, curved, thick. Stamens 4, included, anthers glabrous, thecae straight. Ovary subglobose, glabrous, ovules multiseriate per locule. Disc annular. Capsule ellipsoid, convex, woody, verrucose and tomentose, with lenticels. Seeds pubescent with wings. Four species, Brazil, Peru.

65. ***Leucocalantha*** Barb. Rodr.

Leucocalantha Barb. Rodr., Vellosia, ed. 2, 1: 46 (1891).

Liana, stems with 4 phloem arms in cross section, branchlets cylindrical to slightly angled, whitish-pubescent, with interpetiolar glands, without interpetiolar ridge, pseudostipules absent. Leaves 3-foliolate, terminal pinna often modified into trifid tendril. Inflorescence a terminal cyme. Calyx tubular, minutely 5-parted, glabrous, coriaceous, with sparse patelliform glands close to the margin. Corolla white, narrowly tubular, whitish-pubescent, nocturnal, straight, membranous. Stamens 4, included, anthers glabrous, thecae straight. Ovary elliptic, glandular-lepidote, ovules 1–3-seriate. Disc annular. Capsule linear, flattened, woody, smooth, without lenticels. Seeds glabrous, winged. One species, *L. aromatica* Barb. Rodr., Amazonian Brazil.

66. *Lundia* DC.

Lundia DC., Biblioth. Universelle Genève II, 17: 127 (1838); Gentry, Fl. Ecuador 7: 80–85 (1977), reg. rev.

Lianas, stems with 4 phloem arms in cross section, branchlets terete, glabrous to pubescent, with or without interpetiolar glands and ridge, pseudostipules inconspicuous or absent. Leaves 3-foliolate, the terminal pinna often replaced by a simple or trifid tendril. Inflorescence a terminal or axillary thyrse. Calyx calyptrate or bilabiately cupular and truncate, membranous, without glands. Corolla white to magenta, tubular to campanulate, straight, membranous to coriaceous. Stamens 4, included, anthers densely pubescent, thecae straight. Ovary oblong-conical, lepidote, ovules 2–6-seriate per locule. Disc absent. Capsule elongate-linear, flattened, woody, median nerve conspicuously raised, with lenticels. Seeds glabrous, bialate. $2n = 40$. Twelve species, from Guatemala to Brazil and Bolivia.

67. *Macfadyena* A. DC.

Macfadyena A. DC., Prodr. 9: 179 (1845); Gentry, Fl. Ecuador 7: 85–90 (1977), reg. rev.
Doxantha Miers (1863).

Lianas, stems with many, irregularly divided phloem arms in cross section, branchlets terete, glabrous, with or without interpetiolar glands, with interpetiolar ridge, pseudostipules small. Leaves 3-foliolate, the terminal pinna often replaced by a trifid tendril with claw-like apices. Inflorescence a (1–)few- to several-flowered axillary bracteate cyme. Calyx campanulate, truncate to subspathaceous, irregularly lobed, membranous, without glands. Corolla yellow, tubular to campanulate, straight, membranous. Stamens 4, included, anthers glabrous, thecae straight. Ovary linear-oblong, lepidote, ovules 2–4-seriate per locule. Disc annular. Capsule linear, narrow, flattened, 4-parted, smooth, with lenticels. Seeds glabrous, bialate, wings short and brown. $2n = 40$, 80. Three to four species, tropical America.

68. *Macranthisiphon* Bureau ex K. Schum.

Macranthisiphon Bureau ex K. Schum., in Engler & Prantl, Nat. Pflanzenfam. IV, 3b: 212, 219 (1894); Gentry, Fl. Ecuador 7: 90–92 (1977), reg. rev.

Liana or arching-branched shrub, stems with multiples of 4 phloem arms in cross section, branchlets terete, pubescent, without interpetiolar glands or ridge, pseudostipules foliaceous. Leaves simple or 3-foliolate, terminal pinna often replaced by a simple tendril. Inflorescence a few-flowered (2–3) terminal thyrse. Calyx cupular, truncate, with glands. Corolla orange or red-orange, narrowly thick-tubular, straight, membranous. Stamens 4, exserted, anthers glabrous, thecae straight. Ovary linear, subterete, lepidote, ovules 2-seriate per locule. Disc annular. Capsule linear, flattened, woody, smooth, without lenticels. Seeds glabrous, bialate, wings hyaline-membranaceous. One species, *M. longiflorus* (Cav.) K. Schum., western Amazonia.

69. *Manaosella* J.C. Gomes

Manaosella J.C. Gomes, Arch. Jard. Bot. Rio de Janeiro 9: 83 (1949).

Liana, stems with 4 phloem arms in cross section, branchlets pubescent, cylindrical but with ca. 8 ribs, without interpetiolar glands, with interpetiolar ridge, pseudostipules small or absent. Leaves 3-foliolate, terminal pinna often replaced by much-branched (to 22-fid) tendril. Inflorescence a few- to several-flowered axillary raceme. Calyx spathaceous, split to half of its length, membranous, without glands. Corolla yellow to magenta or greenish-white, tubular to campanulate. Stamens 4, included, anthers glabrous, thecae straight. Ovary linear, pubescent, ovules 2-seriate per locule. Disc annular. Capsule elongate-linear, flattened, woody, smooth, with lenticels, glandular-pubescent. Seeds glabrous, bialate, wings membranaceous. One species, *M. cordifolia* (A. DC.) A.H. Gentry, Venezuela, Brazil and Bolivia.

70. *Mansoa* DC.

Mansoa DC., Biblioth. Universelle Genève II, 17: 128 (1838); Gentry, Fl. Ecuador 7: 92–96 (1977), reg. rev.
Pachyptera DC. ex Meisn. (1840).
Pseudocalymma A. Samp. & Kuhlm. (1934).

Lianas, stems with multiples of 4 phloem arms in cross section, branchlets terete, glabrous to pubescent, with or without interpetiolar glands or ridge, pseudostipules small and inconspicuous. Leaves 3-foliolate, the terminal pinna often replaced by a trifid tendril. Inflorescence an axillary few-flowered thyrse. Calyx cupular to tubular-campanulate, 5-dentate, cartaceous, with sparse glands. Corolla magenta, tubular to campanulate, straight, membranous. Stamens 4, included, anthers glabrous, thecae straight. Ovary linear, lepidote, ovules 2-seriate per locule. Disc annular. Capsule linear-oblong, flattened, woody, strongly tuberculate with tubercles up to 5 mm long, without lenticels. Seeds glabrous, bialate. $2n$ = 36(–38). Fifteen species, southern Mexico to Brazil and Argentina.

71. *Martinella* Baill.

Martinella Baill., Hist. Pl. 10: 30 (1888); Gentry, Fl. Ecuador 7: 96–99 (1977), reg. rev.

Lianas, stems with 4 phloem arms in cross section, branchlets glabrous, terete, without interpetiolar glands, with interpetiolar ridge, pseudostipules absent. Leaves 2-foliolate, often with trifid tendril. Inflorescence an axillary raceme. Calyx tubular to campanulate, large (>11 cm long), bilobed to irregularly 3–4-lobed, membranous, without glands. Corolla deep wine-red, tubular to campanulate, straight, membranous. Stamens 4, included, anthers glabrous, thecae straight. Ovary linear-cylindrical, sparsely lepidote, ovules (2–)4-seriate per locule. Disc cupular. Capsule linear, strongly flattened, woody, smooth, with lenticels. Seeds glabrous, bialate. $2n$ = 40. Two species, *M. obovata* (Kunth) Bureau & Schum., *M. iquitosensis* A. Samp., southern Mexico to Brazil and Bolivia.

72. *Melloa* Bureau

Melloa Bureau, Adansonia 8: 379 (1868); Sandwith & Hunt, Fl. Ilustrada Catarinense: 113–115 (1974), reg. rev.

Liana, stems with many, irregularly divided phloem arms in cross section, branchlets terete, glabrous, without interpetiolar glands, with interpetiolar ridge, pseudostipules acute or subulate. Leaves 2-foliolate, terminal pinna often replaced by a trifid tendril with claw-like apices. Inflorescence an axillary, few-flowered bracteate thyrse. Calyx broadly campanulate, obliquely truncate to subspathaceous with small recurved apicule, membranous, without glands. Corolla yellow, tubular to campanulate, straight, membranous. Stamens 4, included, anthers glabrous, thecae straight. Ovary flattened-ovoid, ribbed, glabrous, ovules multiseriate per locule. Disc appearing double. Capsule elliptic-oblong, 4-parted, expanded, woody, with lenticels. Seeds glabrous, bialate, the wings hyaline-membranaceous. One species, *M. quadrivalvis* (Jacq.) A.H. Gentry, Mexico to Argentina.

73. *Memora* Miers

Memora Miers, Proc. Roy. Hort. Soc. Lond. 3: 185 (1863); Gentry, Fl. Ecuador 7: 99–100 (1977), reg. rev.

Lianas or shrubs, stems with 4 phloem arms in cross section, branchlets terete, glabrous to variously pubescent, without interpetiolar glands, with or without interpetiolar ridge, pseudostipules minutely foliaceous. Leaves pinnately or bipinnately (tripinnately) compound, sometimes 2-foliolate or with simple to trifid tendril. Inflorescence a few-flowered axillary raceme. Calyx cupular, bilabiate or truncate, membranous, with glands. Corolla bright yellow, tubular to campanulate, straight, membranous. Stamens 4, included, anthers glabrous, thecae straight. Ovary linear to oblong, glabrous or more or less lepidote, ovules 2-seriate per locule. Disc annular. Capsule usually linear and strongly flattened, sometimes oblong-cylindrical, woody, smooth, with lenticels. Seeds glabrous, bialate to corky or wingless. $2n$ = 40. Thirty-five species, tropical America.

74. *Mussatia* Bureau ex Baill.

Mussatia Bureau ex Baill., Hist. Pl. 10: 32 (1888); Gentry, Fl. Ecuador 7: 100–103 (1977), reg. rev.

Lianas, stems with multiples of 4 phloem arms in cross section, branchlets acutely tetragonal, glabrous, without interpetiolar glands, with interpetiolar ridge, pseudostipules foliaceous. Leaves 3-foliolate, terminal pinna often replaced by a simple tendril. Inflorescence a terminal bracteate thyrse. Calyx shallowly campanulate, irregularly lobed or split, membranous, with glands. Corolla yellow with purplish to brownish nectar guides, tubular to campanulate, straight, membranous. Stamens 4, included, anthers glabrous, thecae divergent. Ovary oblong, lepidote, ovules 4–6-seriate per locule. Disc cupular. Capsule oblong to narrowly oblong, slightly flattened, woody, rugulose-tuberculate or inconspicuously short-muricate, with lenticels. Seeds glabrous, bialate. Two species, *M. hyacinthina* (Standl.) Sandwith, *M. prieuri* (A. DC.) Bureau ex K. Schum., Mexico to Bolivia.

75. *Neojobertia* Baill.

Neojobertia Baill., Hist. Pl. 10: 35 (1888).

Liana, stems with 4 phloem arms in cross section, branchlets hexagonal, glabrous, without interpetiolar glands or ridge, pseudostipules subulate, poorly developed. Leaves 3-pinnate, terminal pinna frequently modified into a multifid tendril. Inflorescence a terminal cyme. Calyx spathaceous, split on one side down to base, membranous, with glands. Corolla yellow, tubular to campanulate, straight, membranous. Stamens 4, included, anthers glabrous, thecae straight. Ovary oblong, lepidote, ovules 2-seriate per locule. Disc annular. Capsule linear, flattened, woody, with lenticels. Seeds glabrous, winged. One species, *N. candolleana* (Mart. ex DC.) Bureau & K. Schum. in caatingas and savannas of Bahia up to Ceará in Brazil.

76. *Parabignonia* Bureau ex K. Schum.

Parabignonia Bureau ex K. Schum., in Engler & Prantl, Nat. Pflanzenfam. IV, 3b: 229 (1894).

Lianas, stems with many, irregularly divided phloem arms in cross section, branchlets tetragonal, pubescent, with or without interpetiolar glands, without interpetiolar ridge, pseudostipules absent. Leaves 3-foliolate, terminal pinna often replaced by a trifid tendril with claw-like apices. Inflorescence an axillary raceme, sometimes 2–3-flowered. Calyx campanulate, 5-lobed, glandular, membranous. Corolla magenta, tubular to campanulate, straight, membranous. Stamens 4, included, anthers glabrous, thecae straight. Ovary linear-oblong, lepidote to papillate, ovules irregularly 3–4-seriate per locule. Disc cupular. Capsule linear, flattened, 4-parted, woody, smooth, with lenticels. Seeds glabrous, bialate. Two species, *P. steyermarkii* Sandwith, *P. maximiliani* (Mart.) Bureau ex K. Schum., from Venezuela to Ecuador and Brazil.

77. *Paragonia* Bureau Fig. 2

Paragonia Bureau, Bull. Soc. Bot. France 19: 17 (1872); Hauk, Ann. Missouri Bot. Gard. 85: 460–474 (1998), rev.
Sanhilaria Baill. (1888).

Lianas, stems with 4 phloem arms in cross section, branchlets terete, glabrous, without interpetiolar glands, with interpetiolar ridge, pseudostipules of subulate scales appressed to twig or resembling a miniature bromeliad. Leaves 3-foliolate, terminal pinna often replaced by a minutely bifid tendril.

Fig. 2. Bignoniaceae. *Paragonia pyramidata*. **A** Habit. **B** Fruit

Inflorescence a many-flowered, terminal thyrse. Calyx cupular, truncate to lobed, thick, without glands. Corolla magenta, tubular to campanulate, straight, membranous. Stamens 4, included, anthers glabrous, thecae straight. Ovary narrowly cylindrical, lepidote, ovules 2-seriate per locule. Disc annular. Capsule elongate-linear, convex, woody, finely tuberculate, with lenticels. Seeds glabrous, bialate, the wings papery. $2n = 40$. Two species, *P. pyramidata* (Rich.) Bureau, *P. brasiliensis* (Baill.) A.H. Gentry, tropical America.

78. *Periarrabidaea* A. Samp.

Periarrabidaea A. Samp., Ann. Acad. Brasil. Sci. 6: 175 (1934).

Liana, stems with 4 phloem arms in cross section, branchlets cylindrical, glabrous, with interpetiolar glands, without interpetiolar ridge, pseudostipules a group of subulate scales, resembling a miniature bromeliad. Leaves 3-foliolate, terminal pinna often modified into a trifid tendril. Inflorescence an axillary dichasial thyrse. Calyx tubular to campanulate, truncate to very shortly divided, membranous, without glands. Corolla yellow, campanulate, straight, membranous. Stamens 4, included, anthers glabrous, thecae straight. Ovary cylindrical, lepidote, ovules 2–4-seriate. Disc annular. Capsule linear, coriaceous, flattened, smooth, with lenticels. Seeds glabrous, winged.

One species, *P. truncata* A. Samp., restricted to Amazonia, mainly Brazilian Amazonia to Guyana region.

79. *Phryganocydia* Mart. ex Bureau

Phryganocydia Mart. ex Bureau, Bull. Soc. Bot. France 19: 18 (1872); Gentry, Fl. Ecuador 7: 113–116 (1977), reg. rev.

Lianas, stems with multiples of 4 phloem arms in cross section, branchlets terete, glabrous to pubescent, without interpetiolar glands or ridges, pseudostipules foliaceous, well developed. Leaves simple or 2-foliolate, terminal pinna often replaced by a simple tendril. Inflorescence an axillary, few-flowered dichasial thyrse, sometimes reduced to single flower. Calyx spathaceously split down to base, tip recurved to hood-like, membranous, without glands. Corolla magenta with wine-coloured nectar guides, tubular to campanulate, straight, membranous. Stamens 4, included, anthers glabrous, thecae straight. Ovary rounded-conical to narrowly cylindrical, lepidote, ovules 2-seriate per locule. Disc absent. Capsule oblong-linear, flattened, thick, smooth, without lenticels. Seeds glabrous, bialate or wingless. $2n = 40$. Three species, Costa Rica to Brazil and Bolivia.

80. *Piriadacus* Pichon

Piriadacus Pichon, Bull. Soc. Bot. France 92: 225 (1946).

Liana, stems with 4 phloem arms in cross section, branchlets cylindrical, pubescent, with or without interpetiolar glands or ridges, pseudostipules absent. Leaves 3-foliolate, terminal pinna often replaced by a simple tendril. Inflorescence an axillary thyrse. Calyx tubular, 2–5-parted, without glands. Corolla red, tubular. Stamens 4, exserted, anthers glabrous, thecae slightly curved. Ovary ellipsoid, lepidote. Disc annular. Capsule linear, flattened, smooth. Seeds bialate, wings papery. One species, *P. erubescens* (DC.) Pichon, from Espirito Santo to Piaui, Brazil.

81. *Pithecoctenium* Mart. ex Meisn.

Pithecoctenium Mart. ex Meisn., Pl. Vasc. Gen. 1: 300; 2: 208 (1840); Gentry, Fl. Ecuador 7: 116–120 (1977), reg. rev.

Lianas, stems with multiples of 4 phloem arms in cross section, branchlets hexagonal, glabrous to pubescent, without interpetiolar glands, with interpetiolar ridge, pseudostipules spathulate. Leaves 3-foliolate, terminal pinna often replaced by a multifid tendril. Inflorescence an axillary, few-flowered thyrse or raceme. Calyx cupular, truncate, minutely 5-dentate, thick, with glands. Corolla white, tubular to campanulate, tube curved at right angle, thick. Stamens 4, included, anthers glabrous, thecae straight. Ovary ellipsoid-cylindrical, contracted above disc, densely pubescent, ovules multiseriate per locule. Disc annular. Capsule elliptic, flattened, densely spiny, without lenticels. Seeds glabrous, winged, wings papery. $2n = 40$. Three species, Mexico to Brazil and Argentina.

82. *Pleonotoma* Miers

Pleonotoma Miers, Proc. Roy. Hort. Soc. Lond. 3: 184 (1863); Gentry, Fl. Ecuador 7: 120–121 (1977), reg. rev.
Nestoria Urb. (1916).

Lianas, stems with 4 phloem arms in cross section, branchlets acutely tetragonal, winged, without interpetiolar glands, with interpetiolar ridge, pseudostipules foliaceous or absent. Leaves biternate or triternate or ternately bipinnate or tripinnate, terminal pinna often replaced by trifid tendril. Inflorescence a few-flowered, terminal or axillary raceme. Calyx cupular, truncate, minutely denticulate, thick, without glands. Corolla white to cream, narrowly campanulate to funnel-shaped, straight, membranous. Stamens 4, included, anthers glabrous, thecae straight. Ovary oblong, lepidote, ovules 2-seriate per locule. Disc cupular. Capsule elongate-linear to elongate-oblong, flattened, thick, smooth, with lenticels. Seeds glabrous, bialate. $2n = 40$. Fourteen species, Costa Rica to Argentina.

83. *Potamoganos* Sandwith

Potamoganos Sandwith, Recueil Trav. Bot. Neerl. 34: 220 (1937).

Liana, stems with multiples of 4 phloem arms in cross section, branchlets terete, glabrous, without interpetiolar glands or ridge, pseudostipules absent. Leaves 3-foliolate, terminal pinna often replaced by a trifid tendril. Inflorescence a terminal raceme. Calyx shortly cupular, minutely 5-dentate, thick, without glands. Corolla magenta, tubular to campanulate, straight, membranous. Stamens 4, included, anthers glabrous, thecae straight. Ovary elliptic, densely lepidote, ovules 4-seriate per locule. Disc annular. Fruit unknown. One species, *P. microcalyx* (G. Mey.) Sandwith, Guyana Shield in north-eastern South America.

84. *Pseudocatalpa* A.H. Gentry

Pseudocatalpa A.H. Gentry, Brittonia 25: 241 (1973).

Liana, stems with 4 phloem arms in cross section, branchlets several-angled, without interpetiolar glands, with interpetiolar ridge, pseudostipules short, not foliaceous. Leaves 3-foliolate, terminal pinna often replaced by a simple tendril. Inflorescence an axillary, short cyme. Calyx short, cupular, minutely 5-dentate, membranous, without glands. Corolla white, campanulate, straight, membranous. Stamens 2, included, anthers glabrous, thecae straight, staminodes 3. Ovary elliptic, glabrous, ovules 2-seriate per locule. Disc annular. Capsule long, flattened, smooth, without lenticels. Seeds glabrous, bialate. One species, *P. caudiculata* (Standl.) A.H. Gentry, northern Central America.

85. *Pyrostegia* C. Presl

Pyrostegia C. Presl, Abh. Koenigl. Boehm. Ges. Wiss. V, 3: 523 (1845); Gentry, Fl. Ecuador 7: 124–126 (1977), reg. rev.

Lianas, stems with multiples of 4 phloem arms in cross section, branchlets hexagonal, pubescent, with interpetiolar glands, without interpetiolar ridge, pseudostipules small and inconspicuous. Leaves 3-foliolate, the terminal pinna often replaced by a trifid tendril. Inflorescence a few-flowered, terminal or axillary thyrse. Calyx cupular, truncate, minutely 5-denticulate, membranous, with glands. Corolla orange-red, narrowly tubular to campanulate, straight, membranous. Stamens 4, exserted, anthers glabrous, thecae straight. Ovary linear-tetragonal, lepidote. Disc annular. Capsule linear, flattened, thick, smooth, with lenticels. Seeds glabrous, bialate. $2n = 40$ (60). Four species, tropical South America.

86. *Roentgenia* Urb.

Roentgenia Urb., Ber. Deutsch. Bot. Ges. 34: 747 (1916); Gentry, Fl. Ecuador 7: 126–127 (1977), reg. rev.

Lianas, stems with multiples of 4 phloem arms in cross section, branchlets terete to subtetragonal, pubescent, with interpetiolar glands, without interpetiolar ridge, pseudostipules minutely foliaceous or absent. Leaves 3-foliolate, the terminal pinna often replaced by a trifid tendril. Inflorescence an axillary or terminal, few-flowered bracteate thyrse. Calyx cupular, subtruncate to slightly lobed, membranous, without glands. Corolla white with purple markings to lavender, with darker nectar guides, tubular to campanulate, straight, membranous. Stamens 4, included, anthers glabrous, thecae straight. Ovary linear-oblong, lepidote, ovules 2-seriate per locule. Disc absent. Capsule linear, flattened, woody, smooth, with lenticels. Seeds glabrous, bialate. Two species, *R. bracteomana* (K. Schum. ex Sprague) Urb., *R. sordida* (Bureau & K. Schum.) Sprague & Sandwith, Venezuela to Amazonian Brazil.

87. *Saritaea* Dugand

Saritaea Dugand, Caldasia 3: 262 (1945); Gentry, Fl. Ecuador 7: 127–129 (1977), reg. rev.

Liana, stems with multiples of 4 phloem arms in cross section, branchlets terete, without interpetiolar glands or ridge, pseudostipules foliaceous, well developed. Leaves 3-foliolate, the terminal pinna often replaced by a simple tendril. Inflorescence a terminal, few-flowered thyrse. Calyx cupular, truncate, thick, with glands. Corolla magenta, tubular to campanulate, straight, membranous. Stamens 4, included, anthers glabrous, thecae straight. Ovary linear-cylindrical, lepidote, ovules 2-seriate per locule. Disc cupular. Capsule linear, flattened, woody, smooth, with lenticels. Seeds glabrous, bialate, wings hyaline. $2n = 40$. One species, *S. magnifica* (W. Bull) Dugand, Colombia and Ecuador.

88. *Spathicalyx* J.C. Gomes

Spathicalyx J.C. Gomes, Notul. Syst. (Paris) 15: 220 (1956); Gentry, Fl. Ecuador 7: 135–137 (1977), reg. rev.

Lianas, stems with 4 phloem arms in cross section, branchlets terete, glabrous, without interpetiolar glands and ridge, pseudostipules subulate. Leaves 3-foliolate, terminal pinna often replaced by a trifid tendril. Inflorescence an axillary or terminal thyrse. Calyx bilabiate or subspathaceous, 5-denticulate, membranous, with glands. Corolla white or orange-yellow, tubular to campanulate, straight, membranous. Stamens 4, included, anthers glabrous, thecae straight. Ovary oblong to linear-oblong, lepidote, ovules 4–8-seriate per locule. Disc annular. Capsule linear, flattened, woody, smooth, with lenticels. Seeds glabrous, bialate, the wings hyaline. Two species, *S. duckei* (A. Samp.) A.H. Gentry, *S. xanthophylla* J.C. Gomes, lower Amazonia to coastal Brazil.

89. *Sphingiphila* A.H. Gentry

Sphingiphila A.H. Gentry, Syst. Bot. 15: 277 (1990).

Shrub, stems without anomalous phloem growth, branchlets subterete, pubescent, without interpetiolar glands and ridge, short-shoot lateral branches thorn-tipped, pseudostipules absent. Leaves opposite or clustered in fascicles, mostly simple, in part 3-foliolate. Inflorescence a 1- or 2-flowered, axillary thyrsic fascicle. Calyx cupular, 4-denticulate, membranous, with glands. Corolla white, narrowly elongate-salverform, limb with 4 rhombic-obovate lobes, straight, membranous. Stamens 4, included, sessile, anthers glabrous, thecae parallel. Disc present. Ovary cylindrical, lepidote, ovules 2-seriate per locule. Capsule linear, flattened, smooth, with lenticels. Seeds glabrous, bialate, wings hyaline. One species, *S. tetramera* A.H. Gentry, Paraguay.

90. *Stizophyllum* Miers

Stizophyllum Miers, Proc. Roy. Hort. Soc. Lond. 3: 197 (1863); Gentry, Fl. Ecuador 7: 140–143 (1977), reg. rev.

Lianas, stems with 4 phloem arms in cross section, branchlets terete, hollow, pubescent, without interpetiolar glands and ridge, pseudostipules inconspicuous or spathulate. Leaves 3-foliolate, the terminal pinna often replaced by simple or trifid tendril. Inflorescence a few-flowered axillary raceme. Calyx campanulate, sometimes inflated, irregularly bilabiate to 5-lobed, membranous, without glands. Corolla white to magenta, tubular to campanulate, straight, membranous. Stamens 4, included, anthers glabrous or slightly pubescent, thecae straight. Ovary linear-tetragonal, lepidote, ovules 2-seriate per locule. Disc annular. Capsule narrow, linear, cylindrical, convex, pubescent, without lenticels. Seeds glabrous, bialate. $2n = 40$. Three species, southern Mexico to Bolivia.

91. *Tanaecium* Sw.

Tanaecium Sw., Prodr.: 6, 91 (1788).

Lianas, stems with multiples of 4 phloem arms in cross section, branchlets terete, glabrous, with interpetiolar glands, without interpetiolar ridge, pseudostipules inconspicuous or absent. Leaves 3-foliolate, terminal pinna often replaced by a simple tendril. Inflorescence an axillary raceme or thyrse. Calyx cupular, truncate, 5-denticulate, thick, with glands. Corolla white, narrowly tubular, long (>10 cm), straight, thick. Stamens 4, exserted, anthers glabrous, thecae straight or arcuate. Ovary oblong, lepidote, ovules multiseriate per locule. Disc cupular. Capsule oblong-cylindrical, convex, woody, smooth, without lenticels. Seeds glabrous, bialate or wingless. $2n = 40$. Six species, Costa Rica to Bolivia, *T. nocturnum* (Barb. Rodr.) Bureau & Schum. smelling of almonds (cyanide) and containing hallucinogenic compounds.

92. *Tynanthus* Miers

Tynanthus Miers, Proc. Roy. Hort. Soc. Lond. 3: 193 (1863); Gentry, Fl. Ecuador 7: 165–167 (1977), reg. rev.
Schizopsis Bureau (1864).

Lianas, stems with 4 phloem arms in cross section, branchlets terete to subtetragonal, glabrous to pubescent, without interpetiolar glands, with or without interpetiolar ridge, pseudostipules foliaceous or absent. Leaves 3-foliolate, the terminal pinna often replaced by simple or trifid tendril. Inflorescence an axillary, many-flowered thyrse. Calyx cupular, subtruncate to 5-denticulate, membranous, without glands. Corolla white, often tiny (<2 cm long), tubular to campanulate, bilabiate, upper 2 lobes fused, curved, membranous. Stamens 4, subexserted, anthers glabrous, thecae divaricate, bent at middle and reflexed towards corolla mouth. Ovary conical, densely pubescent, ovules 2–4-seriate per locule. Disc annular. Capsule linear, flattened, thick, smooth, without lenticels. Seeds glabrous, bialate, wings hyaline. Fourteen species, Mexico to Bolivia.

93. *Xylophragma* Sprague

Xylophragma Sprague, Hooker's Icon. Pl.: ad t. 2770 (1903).

Lianas, stems with 4 phloem arms in cross section, branchlets terete, pubescent, with interpetiolar glands, without interpetiolar ridge, pseudostipules absent. Leaves 3-foliolate, terminal pinna often replaced by a simple tendril. Inflorescence a bracteate axillary thyrse. Calyx tubular, truncate to 5-denticulate, thick, with glands. Corolla light magenta, tubular to campanulate, large, subrotate, 5-lobed, straight, membranous to thick. Stamens 4, included, anthers glabrous, thecae straight. Ovary flattened-ovoid, densely lepidote, ovules 6–8-seriate per locule. Disc cupular. Capsule elliptic, flattened, woody, smooth, with lenticels. Seeds glabrous, bialate, wings hyaline. $2n = 40$. Five species, Mexico to Bolivia and Paraguay.

4. Tribe Eccremocarpeae Hogg (1858).

Suffrutescent climber, interpetiolar glands and ridges absent, pseudostipules absent. Leaves bipinnatisect to tripinnatisect. Calyx campanulate. Flowers usually bee-pollinated. Stamens 4, anthers dithecous, thecae parallel. Ovary unilocular, placentation parietal. Fruit dehiscent, valves remaining apically united after dehiscence. South America.

94. *Eccremocarpus* Ruiz et Pav. Fig. 3

Eccremocarpus Ruiz et Pav., Fl. Peruv. Prodr.: 90 (1794); Gentry, Fl. Ecuador 7: 70–72 (1977), reg. rev.; D'Arcy, Ann. Missouri Bot. Gard. 84: 103–111 (1997), rev.

Suffrutescent climbers, stems without anomalous phloem growth, branchlets deeply ridged, pubescent. Leaves opposite, bipinnatisect to tripinnatisect, terminal pinna often modified into a multifid tendril. Inflorescence an axillary, few-flowered bracteate raceme. Calyx campanulate, 5-parted, membranous, without glands. Corolla red or orange to yellow with green tip, tubular, lobes small, not reflexed, straight, membranous. Stamens 4, subexserted, anthers glabrous, thecae straight. Ovary ovoid, contracted at base, lepidote, ovules multiseriate on two parietal placentae. Disc annular. Capsule ovoid-ellipsoid, flattened, woody, smooth, with lenticels. Seeds glabrous, winged. Three species from Colombia to Chile.

5. Tribe Tourrettieae K. Schum. (1894).

Herbaceous climber, interpetiolar glands and ridges absent, pseudostipules absent. Leaves biternate. Calyx campanulate, 2-parted. Flowers usually bee-pollinated. Stamens 4, anthers dithecous, thecae divaricate. Ovary with 4 locules, placentation axile. Fruit septicidally dehiscent, spiny. South America.

95. *Tourrettia* Foug.

Tourrettia Foug., Mem. Acad. Sci. (Paris) 1784: 205 (1787), nom. cons.; Gentry, Fl. Neotrop. 25, Bignoniaceae I: 110–114 (1980), rev.

Herbaceous annual climber, stems without anomalous phloem growth, branchlets tetragonal, pubescent. Leaves opposite, biternate with a subterminal, spirally coiling tendril. Inflorescence a bracteate raceme. Calyx campanulate, deeply 2-parted, hirsute, red. Corolla purplish to greenish, tubular, bilabiate, straight, membranous. Stamens 4, included, anthers glabrous, thecae straight. Ovary ovoid, densely short-tuberculate with thick trichomes, ovules uniseriate per locule. Disc cupular. Capsule ovoid, strongly echinate with long spines. Seeds glabrous, with brown wing. $2n$ = ca. 40. One species, *T. lappacea* (L'Herit.) Willd., Cordillera from Mexico to Argentina.

Fig. 3. Bignoniaceae. **A** *Kigelia africana*. Fruit. **B** *Eccremocarpus scaber*. Seed. SEM ×40. (Photo E. Fischer)

6. Tribe Coleeae Boj. (1837).

Trees or shrubs, interpetiolar glands and ridges absent, pseudostipules present or absent. Leaves pinnate or simple. Calyx cupular or campanulate, 5-dentate. Flowers usually bee-pollinated (bat-pollinated in *Kigelia*, lemur-pollinated in *Rhodocolea*). Stamens 4, anthers monothecous (*Colea*, *Ophiocolea*) or dithecous, thecae divaricate or not (*Kigelia*). Ovary unilocular with 2 parietal placentae (*Colea*) or bilocular, placentation axile. Fruit indehiscent, fleshy or rarely fibrous with hard cortex (*Kigelia*). Africa, Madagascar.

Key to Genera of Coleeae

1. Anthers monothecous; leaves nearly always verticillate; pseudostipules verticillate, conspicuous 2
- Anthers dithecous; leaves rarely verticillate, mostly opposite; pseudostipules minute or absent 3
2. Ovary elongated, cylindrical, smooth, bilocular; fruit cylindrical and very long (20–50 cm); Madagascar **98. *Ophiocolea***
- Ovary ovate-conical and short, mostly covered with glands, warts or costules, unilocular with 2 parietal placentae at least in upper half; fruit short, ovate, rounded, subreniform or rarely oblong, covered by rough warts, verrucose or with large and irregular wings; Madagascar, Seychelles, Mauritius **96. *Colea***
3. Inflorescence with pendent flagellae, up to 1 m long; corolla large, 6–12 cm long; fruit sausage-shaped, up to 1 m long; leaves imparipinnate; Africa **97. *Kigelia***
- Inflorescence axillary or cauliflorous, less than 50 cm long; corolla less than 4–5 cm long; fruit less than 30 cm long; leaves imparipinnate, simple or composed of articules; Madagascar, Comoro Islands 4
4. Leaves imparipinnate, always opposite **101. *Rhodocolea***
- Leaves simple or composed of superposed articules, opposite, verticillate or fasciculate 5
5. Spiny shrubs; persistent petioles sclerified to spines and opposite; leaves deciduous, simple, membranaceous, nearly always entire, often 2–4 pairs in pseudofascicles in axils of spines **100. *Phylloctenium***
- Trees or shrubs without spines, leaves persistent, composed of superposed articules (articulate rachis), rarely reduced to one articule **99. *Phyllarthron***

96. *Colea* Bojer ex Meisn.

Colea Bojer ex Meisn., Pl. Vasc. Gen. 1: 301; 2: 210 (1840), nom. cons.; Perrier, Fl. Madagascar: 32–54 (1938), rev.

Small, often pachycaul trees or shrubs, pseudostipules verticillate. Leaves verticillate, rarely opposite, imparipinnately compound. Inflorescence a few-flowered, usually cauliflorous thyrse. Calyx cupular to 5-dentate. Corolla white, yellow, rose, mauve with yellowish throat, dark red, tubular to campanulate, straight, thick. Stamens 4, included, anthers monothecous, with rudimentary sterile theca, staminode short. Ovary ovate-conical, covered with glands, warts or costules, ovules multiseriate. Disc cupular. Fruit a berry, short, ovate, rounded, subreniform or rarely oblong, covered by rough warts, verrucose or with large and irregular wings. Seeds orbicular, smooth. Twenty-one species, Madagascar, Mauritius (1), Seychelles (1).

97. *Kigelia* DC. Fig. 3

Kigelia DC., Biblioth. Universelle Genève II, 17: 135 (1838).

Medium-sized trees, pseudostipules absent. Leaves opposite, imparipinnately compound, 7–9-foliolate. Inflorescence a pendulous thyrse, hanging flagelliflorously below lower branches. Calyx large, cupular, irregularly 5-dentate. Corolla brownish-maroon, nocturnal, tubular to campanulate, straight, thick. Stamens 4, included, anthers glabrous, thecae thick, not divaricate. Ovary cylindrical, lepidote, ovules ca. 10-seriate per locule. Disc annular. Fruit very large (up to 50 cm long), sausage-shaped, fibrous and woody. Seeds glabrous, wingless. $2n = 40$. One or two variable species, *K. africana* (Lam.) Benth., *K. moosa* Sprague, tropical Africa.

98. *Ophiocolea* H. Perrier

Ophiocolea H. Perrier, Ann. Inst. Bot. Geol. Colon. Marseille V, 6: 31 (1938); Perrier, Fl. Madagascar: 25–32 (1938), rev.

Small trees or shrubs, pseudostipules verticillate. Leaves verticillate to 3–6, imparipinnately compound. Inflorescence a cauliflorous, few-flowered thyrsic fascicle. Calyx cupular to 5-dentate. Corolla white, yellow to pinkish, tubular to campanulate, straight, membranous. Stamens 4, included, anthers monothecous, with rudimentary sterile theca, staminode short. Ovary linear, elongated, cylindrical, smooth, ovules 4-seriate per locule. Disc cupular. Fruit a berry, cylindrical and very long (20–50 cm). Seeds glabrous, wingless. Five species, Madagascar, Comoro Islands (1).

99. *Phyllarthron* DC.

Phyllarthron DC., Ann. Sci. Nat., Bot. II, 11: 296 (1839); Perrier, Fl. Madagascar: 68–84 (1938), rev.
Zaa Baill. (1889).

Trees or shrubs, pseudostipules absent. Leaves opposite or verticillate, persistent, consisting of linear series of articulated segments. Inflorescence an axillary or terminal thyrse. Calyx campanulate, 5-dentate. Corolla white to rose, tubular to campanulate, straight, membranous. Stamens 4, included, anthers dithecous, thecae divaricate.

Ovary linear to ovate, elongated, glandular pubescent, ovules multiseriate per locule. Disc cupular. Fruit a berry, variable in shape, smooth. Seeds orbicular, glabrous, wingless. $2n = 40$. About 15 species, Madagascar and Comoro Islands (1).

100. *Phylloctenium* Baill.

Phylloctenium Baill., Bull. Mens. Soc. Linn. Paris: 692 (1887); Perrier, Fl. Madagascar: 65–68 (1938), rev.
Paracolea Baill. (1887).

Spiny shrubs, pseudostipules minute, verticillate, scale-like. Leaves opposite, simple, membranaceous, nearly always entire, often to 2–4 pairs in pseudofascicles in axils of spines, persistent petioles sclerified to spines. Inflorescence a few-flowered axillary thyrse, often reduced to solitary flower. Calyx cupular, truncate or minutely 5-dentate. Corolla white to yellowish with purple or violet-red nectar guides, tubular to campanulate, straight, membranous. Stamens 4, included, anthers dithecous, thecae divaricate. Ovary linear, elongated, cylindrical, smooth, ovules multiseriate per locule. Disc cupular. Fruit a berry, cylindrical, elongate, smooth. Seeds obovate-cuneiform, wingless. Two species, *P. bernieri* Baill., *P. decaryanum* H. Perrier, Madagascar.

101. *Rhodocolea* Baill.

Rhodocolea Baill., Bull. Mens. Soc. Linn. Paris: 693 (1887); Perrier, Fl. Madagascar: 54–65 (1938), rev.

Large trees or shrubs, pseudostipules absent. Leaves opposite, imparipinnately compound, rachis articulate. Inflorescence an axillary, terminal or cauliflorous thyrse. Calyx campanulate, 5-dentate. Corolla white to rose, tubular to campanulate. Stamens 4, included, anthers dithecous, thecae divaricate. Ovary elongated, cylindrical, smooth, ovules multiseriate per locule. Disc cupular. Fruit a berry, cylindrical, elongate, smooth or with 6 costules. Seeds orbicular, wingless. Seven species, Madagascar.

7. Tribe Crescentieae DC. (1845).

Trees, interpetiolar glands and ridges absent, pseudostipules absent. Leaves palmate or simple. Calyx irregularly rupturing apically to spathaceous. Flowers bat-pollinated. Stamens 4, anthers dithecous, thecae thick, divergent. Ovary incompletely bilocular (*Amphitecna*) or unilocular, with parietal placentae. Fruit indehiscent, with a fleshy cortex and fibrous core. Central and South America.

Key to Genera of Crescentieae

1. Leaves opposite, 3–7-foliolate, petiole wingless or narrowly winged, calyx spathaceously split; fruit not hard **104. *Parmentiera***
– Leaves alternate or fasciculate, usually simple (if compound, the winged petiole as wide as the pinnae), calyx usually bilabiate, rarely spathaceous, fruit hard (calabash) 2
2. Leaves alternate, not fasciculate; corolla lobes reduced and more or less fused; placentation axile at base of ovary and parietal on two placentae above; seeds 13 × 16 mm **102. *Amphitecna***
– Leaves fasciculate; corolla lobes triangular; placentation parietal on four placentae; seeds 8 × 6 mm **103. *Crescentia***

102. *Amphitecna* Miers

Amphitecna Miers, Trans. Linn. Soc. Lond. 26: 163 (1868); Gentry, Fl. Neotrop. 25, Bignoniaceae I: 50–82 (1980), rev.
Dendrosicus Raf. (1838), nom. rej.
Enallagma (Miers) Baill. (1888), nom. cons.
Neotuerkheimia Donn. (1909).

Small to medium-sized trees. Leaves alternate, simple. Inflorescence a solitary flower or few-flowered thyrsic fascicle, terminal or cauliflorous on short shoots from older twigs. Calyx large, bilabiate, sometimes subtruncate, rarely spathaceous. Corolla greenish-white, tubular to campanulate, usually with inconspicuous transverse fold in throat, straight, thick. Stamens 4, subexserted, anthers glabrous, thecae thick, divergent. Ovary ovoid-elliptic, lepidote, ovules ca. 4-seriate in each locule. Fruit (pepo or calabash) spherical to ellipsoid, indehiscent, with relatively fragile, woody excocarp and soft pulp. Seeds thick, glabrous, not winged. $2n = 40$. Eighteen species from Mexico and West Indies to Venezuela and Ecuador.

103. *Crescentia* L.

Crescentia L., Sp. Pl.: 626 (1753), nom. cons.; Gentry, Fl. Neotrop. 25, Bignoniaceae I: 82–96 (1980), rev.

Small to medium-sized trees. Leaves simple or 3-foliolate on thick twigs in alternate fascicles. Inflorescence a cauliflorous, 1- or 2-flowered thyrsic fascicle from nodes on trunk and older twigs. Calyx large, bilabiately split. Corolla off-tan usually with maroon pencilling, tubular to campanulate, lobes deltoid, acuminate, straight, thick. Stamens 4, subexserted, anthers glabrous, thecae thick, divergent. Ovary ovoid-elliptic, lepidote, ovules multiseriate on 4 parietal placentae. Fruit (pepo or calabash) large, spherical to ellipsoidal, indehiscent, with hard woody exocarp and soft pulp. Seeds glabrous, not winged. $2n = 40$. Six species in tropical America from Mexico and the West Indies to Amazonian Brazil.

104. *Parmentiera* DC.

Parmentiera DC., Biblioth. Universelle Genève II, 17: 135 (1838); Gentry, Fl. Neotrop. 25, Bignoniaceae I: 97–110 (1980), rev.

Shrubs to large trees, often with spine below each node. Leaves opposite to subopposite, palmately 3–5-foliolate. Inflorescence a solitary flower or few-flowered thyrsic fascicle, terminal or usually cauliflorous on trunk. Calyx spathaceous, split ventrally to base. Corolla white or greenish-white, broadly campanulate, with a transverse fold in the throat, straight, thick. Stamens subexserted, anthers glabrous, thecae thick, divergent. Ovary cylindrical, ovules multiseriate. Disc annular. Fruit (pepo or calabash) elongate-oblong, indehiscent, with a fleshy exocarp and fibrous core. Seeds glabrous, with vestigial wing. $2n = 40$. Nine species, Mexico to North West Colombia.

Selected Bibliography

Armstrong, J.E. 1985. The delimitation of Bignoniaceae and Scrophulariaceae based on floral anatomy, and the placement of problem genera. Am. J. Bot. 72: 755–766.

Bremer, K., Backlund, A., Sennblad, B., Swenson, U., Andreasen, K., Hjertson, M., Lundberg, J., Backlund, M., Bremer, B. 2001. A phylogenetic analysis of 100+ and 50+ families of euasterids based on morphological and molecular data with notes on possible higher level morphological synapomorphies. Plant Syst. Evol. 229: 137–169.

Buurman, J. 1978. Contribution to the pollen morphology of the Bignoniaceae, with special reference to the tricolpate type. Pollen Spores 19: 447–519.

Endress, P.K. 1996. Diversity and evolutionary biology of tropical flowers. Cambridge University Press.

Fan, R., Wu, Q., Zou, H. 1980. Embryological studies of *Catalpa fargesii* f. *duclouxii* (Dode) Gilmour. J. Navjing Tech. Coll. Forest Prod. 2: 67–73.

Gentry, A.H. 1980. Bignoniaceae part 1 – Tribes Crescentieae and Tourrettieae. Flora Neotrop. 25 (1): 1–131. New York Botanical Garden.

Gentry, A.H. 1988. Distribution and evolution of the Madagascar Bignoniaceae. In: Goldblatt, P., Lowry, P.P. (eds.) Modern systematic studies in African botany. Monogr. Syst. Bot. Missouri Bot. Gard. 25: 175–185.

Gentry, A.H. 1990. Evolutionary patterns in neotropical Bignoniaceae. Mem. New York Bot. Gard. 55: 118–129.

Gentry, A.H. 1992. Distributional patterns of Central American and West Indian Bignoniaceae. Tulane Stud. Zool. Bot. Suppl. 1: 111–125.

Gentry, A.H., Tomb, A.S. 1979. Taxonomic implications of Bignoniaceae palynology. Ann. Missouri Bot. Gard. 66: 756–777.

Goldblatt, P., Gentry, A.H. 1979. Cytology of Bignoniaceae. Bot. Notiser 132: 475–482.

Graham, A., Jarzen, D.M. 1969. Studies in neotropical palaeobotany 1. The oligocene communities of Puerto Rico. Ann. Missouri Bot. Gard. 56: 308–357.

Grose, S.O., Olmstead, R.G. 2002. Patterns of evolution in Bignoniaceae: the *Tabebuia/Crescentia conundrum*. Am. J. Bot. 89, Suppl. Abstracts: 126.

Gussman, A.B., Gottsberger, G. 1996. Differences in floral morphology, floral nectar constituents, carotenoids, and flavonoids in petals of orange and yellow *Pyrostegia venusta* (Bignoniaceae) flowers. Phyton (Austria) 36: 161–171.

Hegnauer, R. 1964. Bignoniaceae. Chemotaxonomie der Pflanzen 3: 268–281.

Hegnauer, R. 1989. Bignoniaceae. Chemotaxonomie der Pflanzen 8: 128–138.

Hesse, M., Morawetz, W. 1980. Skulptur und systematischer Wert der Samenoberfläche bei *Jacaranda* und anderen Bignoniaceae. Plant Syst. Evol. 135: 1–10.

James, E.A., Knox, R.B. 1993. Reproductive biology of the Australian species of the genus *Pandorea* (Bignoniaceae). Austral. J. Bot. 41: 611–626.

Lohmann, L.G. 2002. Phylogeny and morphological evolution of Bignonieae (Bignoniaceae). Am. J. Bot. 89, Suppl. Abstracts: 134.

Metcalfe, C.R., Chalk, L. 1950. Anatomy of the Dicotyledons. Oxford: Clarendon Press, pp. 1002–1013.

Olmstead, R.G., Zyhra, M.L., Grose, S.O., Eckert, A.J., Spangler, R.E. 2002. Molecular phylogeny, evolution, and classification of the Bignoniaceae. Am. J. Bot. 89, Suppl. Abstracts: 240.

Piazzano, M. 1998. Numeros cromosomicos en Bignoniaceae de Argentina. Kurtziana 26: 179–189.

Record, S., Hess, R. 1940. American timbers of the family Bignoniaceae. Trop. Woods 63: 9–38.

Ruwen, F. 1981. A study on the embryological development of *Catalpa ovata*. J. Nanjing Tech. Coll. Forest Prod. 4: 64–74.

Shirke, D.R. 1976. Contributions to the life-history of *Spathodea campanulata* Beauv. Indian Sci. Congr. Assoc. Proc. 63: 88.

Siegert, A., Wagner, F. 1994. Zum Bau einiger Bignoniaceen-Dornen unter besonderer Berücksichtigung der Blatt-Dornen von *Parmentiera aculeata* (H.B.K.) Seem. Beitr. Biol. Pfl. 67: 387–437.

Spangler, R.E., Olmstead, R.G. 1999. Phylogenetic analysis of Bignoniaceae based on the cpDNA gene sequences *rbc*L and *ndh*F. Ann. Missouri Bot. Gard. 86: 33–46.

Von Poser, G.L., Schripsema, J., Henriques, A.T., Jensen, S.R. 2000. The distribution of iridoids in Bignoniaceae. Biochem. Syst. Ecol. 28: 351–366.

Weber, A., Vogel, S. 1986. The pollination syndrome of *Deplanchea tetraphylla* (Bignoniaceae). Plant Syst. Evol. 154: 237–250.

Zyhra, M.L. 1988. Phylogenetics, biogeography, and pollination ecology of endemic Malagasy Coleeae (Bignoniaceae). Ph.D. Dissertation. University of Wisconsin-Madison.

Buddlejaceae[1]

B. Oxelman, P. Kornhall and E.M. Norman

Buddlejaceae K. Wilh., Samenpflanzen: 90 (1910) ("Buddleiaceae"), nom. cons.

Shrubs or trees, rarely lianas or suffrutescent herbs. Leaves opposite, rarely alternate, petiolate to sessile, usually with interpetiolar lines, sometimes connate, simple. Lamina pinnately veined; margins entire, serrate, crenate, or dentate. Indumentum of stellate and glandular multicellular hairs. Plants hermaphrodite or dioecious (most American *Buddleja* species). Flowers in cymes of various kinds, usually in globose heads, panicles, or verticils; tetramerous, actinomorphic, except for the often unequal calyx lobes. Calyx gamosepalous, campanulate, cup-shaped, or tubular, often unequal but not bilabiate, persistent. Corolla gamopetalous, imbricate or valvate, tubular, funnelform, salverform, or campanulate. Stamens 4, often inserted on the upper half of the corolla tube, all equal, free, alternating with the corolla lobes. Anthers bilocular, tetrasporangiate, dehiscing with longitudinal slits, introrse. Pollen grains 3–4(5)-colporate, 2-celled. Ovary superior, bilocular, rarely 4-celled, with anatropous ovules, with a single style and a bilobed stigma. Fruit usually a capsule, rarely a berry, 4- to more than 100-seeded. Seeds endospermic; sometimes winged.

Five genera with ca. 100 species in subtropical and tropical America, Africa, and Asia.

Vegetative Morphology. All members of the Buddlejaceae are woody. Plants growing in arid areas are usually small shrubs whereas larger shrubs, sometimes climbing, and trees (up to 30 m high) are found in more mesic regions. Most recent authors concur with Hasselberg (1937) who argued that the interpetiolar lines seen in Buddlejaceae are formed by the leaves, and not homologous to the stipules in Loganiaceae.

Vegetative Anatomy. Growth rings are more or less conspicuous. Perforation plates are simple and lateral wall pitting consists of alternate pits with the pit cavity circular to oval in outline. Helical thickenings of vessel walls are usually present. Vascular tracheids (Carlquist 1997) are present in species with xeromorphic wood, and libriform fibres are always present. Wood parenchyma is scanty, paratracheal. Both uniseriate and multiseriate rays occur.

Most wood characteristics can be found elsewhere in the Asteridae, and the variation in the family appears to be correlated with levels of xeromorphy (Carlquist 1997). In contrast to Loganiaceae, but in accordance with most members of the Lamiales, there is no intraxylary phloem (Solereder 1895). The cork is pericyclic (Solereder 1895). Stomata are found on the lower epidermis and are either paracytic, anomocytic, or anisocytic, and these types are often mixed on one plant. Glandular and stellate trichomes are found on most aboveground parts of the plant, except older and woody stems, the inner corolla surface, the upper part of the style, and the stamens. The inner surface of the corolla is usually covered by unicellular, short, blunt hairs. Location and density of these often are of diagnostic value at the species level. Otherwise, the degree of hairiness varies considerably, and many xeric taxa have a very thick indumentum of stellate hairs which, in conjunction with the glands, give the plant a canescent appearance. The upper surfaces of the leaves of some high-altitude species are glabrous. A small group of South American species are characterised by peculiar unicellular trichomes on their placentae (Norman 2000).

Inflorescence. Flowers are arranged in compound cymes of various kinds, usually forming thyrsoid synflorescenses (Fig. 4A, B). Norman (2000) considers an inflorescence of paired cymes in axillary position along terminal or subterminal portions of the axes as basic for the family. From this type, numerous independent reductions of pedicels and/or secondary axes are hypothesised to have occurred, leading to verticillate, spicate, or globose heads. The inflorescence of *Gomphostigma* has been termed a raceme (Leeuwenberg 1977) but is better interpreted as thyrsoid, with single-flowered individual inflorescences.

[1] See also Scrophulariaceae.

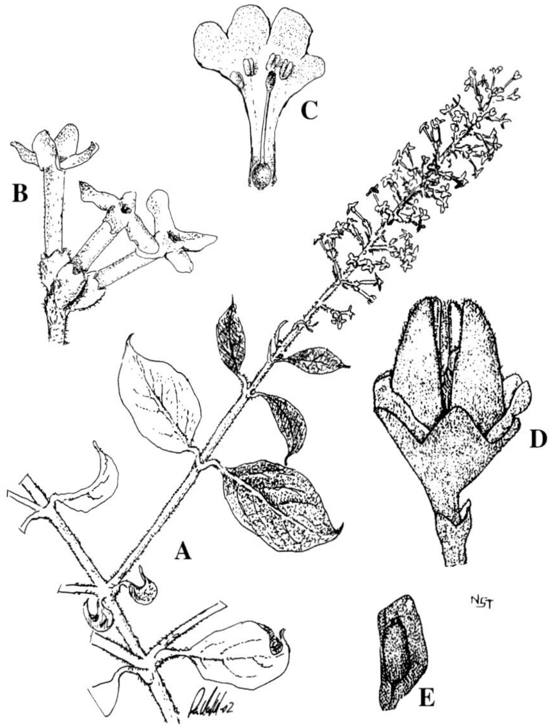

Fig. 4. Buddlejaceae. *Buddleja tucumanensis*. **A** Habit. **B** Flower. **C** Open flower. *Buddleja thyrsoides*. **D** Fruit. **E** Seed. (**A–C** P. Kornhall; **D, E** Fabris 1965)

Flower Structure. The flowers are actinomorphic and usually tetramerous (Fig. 4B, C). They are usually rather small, but vary between 3 and 40 mm. The synsepalous calyx is tubular or campanulate, and covered with a mixture of stellate tomentum and glandular trichomes. The corolla is white, yellow, orange, pinkish, or purplish. Aestivation of the corolla is often imbricate but sometimes both valvate and imbricate aestivation can be found on the same plant. The outside surface of the corolla is covered with stellate and glandular trichomes. The four stamens are inserted on the corolla and alternate with the corolla lobes. The anthers are introrse, tetrasporangiate, and dehisce by longitudinal slits. Anther wall formation follows the Dicotyledonous type. Mature pollen grains are 2-celled.

In dioecious taxa the flowers contain sterile, slightly reduced organs of the opposite sex.

The gynoecium is superior and consists of two fused carpels. The ovary is usually 2-celled, or 4-celled (some *Nicodemia*). There is a single, apical, persistent style, attenuate from the ovary, with a slightly bilobed, papillate stigma.

Embryology. There is a rich literature on embryo development in Buddlejaceae (reviewed by Hakki 1980; Norman 2000). Placentation is axile, with more than 20 pendulous or ascending ovules per locule. The ovules are hemianatropous, unitegmic, and tenuinucellate, with a Polygonum-type embryo sac. Endosperm formation is cellular, and both chalazal and micropylar haustoria are present. The chalazal haustorium is associated with a cone-like region rich in proteins, a hypostase. The mature endosperm is rich in protein and oils, and ruminate due to the inward expansion of endothelial cells (Maldonado de Magnano 1986). Its development conforms to the Verbascum type. Embryo development follows the Veronica variation of the Onagrad type (Maldonado de Magnano 1987).

Pollen Morphology. Pollen grains are usually rather small, ranging from c. 10 to 50 μm, subprolate to sphaeroidal, and usually three-colporate. Keenan (1969b) demonstrated an association between ploidy level and the number of apertures. All diploids are tri-colporate, whereas most polyploids are tetra- (rarely penta-) colporate and on the average larger. The sexine is thicker than the nexine, the tectum is smooth and columellae are frequently indistinct. Punt (1980) recognised 19 characters with variable states within Buddlejeae of the Loganiaceae. Despite the more restrictive taxonomic inclusiveness employed here, it seems that almost all the variation described by Punt is retained, reinforcing the great diversity in pollen morphology.

Karyology. Chromosomes in Buddlejaceae are usually small and similar in morphology (Moore 1947). Most chromosome numbers can be inferred to be based on $x = 19$. Gadella (1980) hypothesised that this may have been derived from an initial hybridisation between two basic stocks having $x = 7$ and $x = 12$, the latter being a common base number in taxa closely related to Buddlejaceae. If this is true, *Gomphostigma* ($x = 7$) may represent one of these lineages. Particularly suggestive is the fact that *Nuxia* ($x = 19$) and *Retzia* ($x = 12$) strongly group with Stilbaceae ($x = 12$) in phylogenetic trees based on cpDNA sequences (Oxelman et al. 1999). In light of their morphological similarity, it may be worth exploring whether *Nuxia* and Buddlejaceae (excluding *Gomphostigma*) share a common parental stock. Within the $x = 19$ group, ploidy levels up to 24 (*Buddleja colvilei*; Keenan 1969a) have been recorded. About 40% of the chromosome counts made have revealed polyploid taxa. It appears that polyploidy is especially common in Asian taxa. Speta (1977) has reported protein bodies in the cell nuclei similar to those found in *Veronica*, but expressed some reservations about this similarity. However, the drawing of protein bodies of *Scrophularia calliantha* in Speta (1979) is similar to that of *Buddleja* in Speta (1977).

Pollination and Reproductive Systems. Norman (2000) reported small bees, honey bees, flies, syrphids, bumble bees, wasps, butterflies, and hummingbirds as visitors of neotropical *Buddleja*. The butterfly bush, *Buddleja davidii*, received its English name because of the ability to attract butterflies.

Most American species of *Buddleja* are functionally dioecious (Norman 2000 and references therein). Some taxa are polygamous, i.e. hermaphroditic plants exist together with the unisexual ones. Staminate flowers often have more conspicuous stamens and a corolla with a more open throat than pistillate flowers, often accompanied by a longer corolla tube and wider lobes (Norman 2000).

Hermaphroditic flowers of the Buddlejaceae are usually herkogamous. In the American *Buddleja* sect. *Stachyoides*, the stigma is raised above the anthers, which is typical for hummingbird-pollinated flowers. In Old World flowers, the situation is reverse, with the stigma located below the anthers. The latter situation is common among flowers with narrow tubes that are visited by butterflies (Norman 2000). Pollen samples from mosses in the subalpine forest belt in Guatemala contained up to 1% *Buddleja* pollen, despite the entomophilous nature of the flowers and the patchy distribution of the trees (Islebe and Hooghiemstra 1995).

Fruit, Seed, and Dispersal. Most Buddlejaceae taxa have ovoid, oblong, or subglobular capsules which dehisce septicidally (Fig. 4D), but indehiscent fruits occur in *Buddleja scordioides*. *Nicodemia* is characterised by having berries. Seeds of American *Buddleja* species are usually very small, often less than 1 mm in diameter. The testa is reticulate and sometimes winged (Fig. 4E). The wings probably serve for wind dispersal. Some taxa have air pockets between the exotesta and the endothelium, possibly enhancing water dispersal along washes and streams (Norman 2000). The berries of *Nicodemia* and the indehiscent fruits of *Buddleja scordioides* are probably attractive to birds or mammals.

Phytochemistry. Jensen (2000b) gives a thorough review of the secondary compounds found in the family. The iridoids found are from the same biosynthetic pathway as in other taxa of the Lamiales, but significantly different from those in the Gentianales. Caffeic acid esters of phenylethanoid glycosides (CPGs), which are never found in Gentianales, are almost consistently present and constitute a phytochemical synapomorphy for the Lamiales. The distribution of these compounds, and of flavonoids (Tomas-Barberan et al. 1988), is of potential systematic value. For example, the iridoid unedoside is only found in *Nuxia* and Stilbaceae (incl. *Retzia*), thus supporting the chloroplast DNA data. In contrast, all Buddlejaceae species investigated have been found to contain the iridoids aucubin and catalpol instead. The lack of iridoids together with the presence of sorbitol, cornoside, and CPGs in *Polypremum* and *Tetrachondra* (Jensen 2000a) also corroborates the results from chloroplast DNA sequences (Oxelman et al. 1999), and supports grouping *Polypremum* in the Tetrachondraceae instead of in the Buddlejaceae.

Subdivisions and Relationships Within the Family. The phylogenetic status of the included genera has not been thoroughly tested and a study using molecular data is needed. It is probable that *Buddleja*, as circumscribed here, is paraphyletic. This is indicated by the chloroplast DNA sequence data of Oxelman et al. (1999), although neither sampling nor resolution is sufficient in that study. Later work by Wallick et al. (2000) and P. Kornhall (more cpDNA sequences; unpubl.) strengthens this view. The other genera, however, are probably monophyletic and contain only few similar species and are diagnosed by characters unique in the family. Norman (2000) has pointed out a close relationship between *Emorya* and *Buddleja* ser. *Stachyoides*. *Chilianthus* is enigmatic because of its great morphological similarity with *Nuxia*. Unpublished chloroplast sequences of *Chilianthus saligna* (Oxelman et al., in prep.) strongly suggest a membership of *Chilianthus* in Buddlejaceae, rather than a relationship with Stilbaceae.

Affinities. Buddlejaceae has long been classified within the Loganiaceae, despite numerous properties which are shared with scrophulariaceous taxa (Oxelman et al. 1999 and references therein). Recent molecular work (Oxelman et al. 1999), often well corroborated by biochemical data (Jensen 2000a, 2000b), has narrowed the traditional view of Buddlejaceae further. *Desfontainia* and *Plocosperma* have affinities outside the Lamiales. *Sanango* (Norman 1994; Dickison 1994; Jensen 1994; Wiehler 1994, Oxelman et al. 1999) and *Pelthanthera* (Oxelman et al. 1999, Jensen 2000a) are related to the Gesneriaceae. *Polypremum* is closely related to *Tetrachondra* (Oxelman et al. 1999, Jensen 2000a), which together are the sister group of the entire Lamiales except Oleaceae. *Nuxia* and *Retzia* are related to Stilbaceae, although the morphology and chromosome numbers of *Nuxia* are puzzling. The remaining Buddlejaceae belong to the "Scroph I" clade of Olmstead and Reeves (1995). This clade includes Scophulariaceae tribe Hemimerideae as well as the Myoporaceae (to which the formerly buddlejaceous *Androya* belongs; Oxelman et al. 1999). These two groups branch off early in the Scroph I clade with moderate support. A strongly supported subclade includes Buddlejaceae, Scophulariaceae tribes Manuleae, Verbasceae, and a few other taxa (Oxelman et al., in prep.).

Distribution and Habitats. The family is mainly found in tropical and subtropical and tropical montane regions in America, Africa, and Asia. Because South Africa contains the greatest diversity within the family, it has been suggested to be the centre of origin (Moore 1947). Most plants grow in sunny habitats under various soil conditions. Some species are weedy, for example, the South American *B. stachyoides* which is adventitious in Australia (Leeuwenberg 1979).

Parasites. Closely related species of *Peronospora* have been found to infect *Buddleja* species and *Verbascum*, *Scrophularia*, and *Digitalis* (Mohrbutter 1937). Black spots in ring formations on the leaves of *Buddleja incana*, *B. bullata*, and *B. skutchii* are formed by the ascomycete *Dermatodothis buddleyae*, and *Cladosporium* sp. causes black, powdery infections on fruiting branches of *B. scordioides* (Norman 2000). The dipteran leaf miner *Dizyomyza verbasci* attacks *Scrophularia* and *Verbascum*, as well as *B. davidii* (Mohrbutter 1937). Another dipteran, *Asphodylia buddleia*, causes abnormal enlargement of flower buds in *B. racemosa*. Beetles, homopterans, and many lepidopterans are also known to feed on *Buddleja* species (reviewed by Norman 2000). Many of them are known to feed also on *Verbascum* and *Scrophularia*.

Economic Importance. *Buddleja* species have been extensively used for various medical purposes, reviewed by Bisset (1980) and Norman (2000). Over a dozen species are cultivated as ornamentals, the most widespread being *B. davidii*, which is hardy also in cool temperate climates. Grain yields of wheat have been reported to increase when soils are amended with *Buddleja coriacea* foliage (Mahboubi et al. 1997).

Conservation. Norman (2000) lists 17 America taxa, all rather narrow endemics, many from SE Brazil, for which there are only five or fewer known collections. Another four of the Old World taxa treated by Leeuwenberg (1979) are of similar conservation status using the same criteria.

Key to the Genera

1. Fruit a berry — **5. *Nicodemia***
– Fruit a capsule — 2
2. Synflorescence resembling a raceme, with single cup-shaped flowers in leaf axils (southern Africa) — **4. *Gomphostigma***
– Synflorescence paniculate, individual inflorescenses usually many-flowered; if single-flowered, then corolla tubular — 3
3. Stamens short, included or scarcely exserted from corolla — **1. *Buddleja***
– Stamens distinctly exserted from corolla — 4
4. Corolla tubular (Mexico and southern USA) — **3. *Emorya***
– Corolla cup-shaped (southern Africa) — **2. *Chilianthus***

1. *Buddleja* L. Fig. 4

Buddleja L., Sp. Pl.: 112 (1753); Leeuwenberg, Meded. Landbouwhoogeschool Wageningen 79: 1–163 (1979), rev. of Old World spp.; Norman, Fl. Neotrop. Monogr. 81: 1–225 (2000), rev. of New World spp.

Shrubs or trees, mostly with a grey indumentum of stellate and glandular hairs. Leaves opposite, rarely alternate or whorled, entire, crenate or dentate. Synflorescence thyrsoid, variously densely packed, from globose to loosely paniculate. Corolla cup-shaped, campanulate, funnel-shaped, tubular, or salverform, white, yellow, purple, or pinkish, often with a dark orange throat, aestivation valvate or imbricate. Anthers usually included in corolla. Fruit a two-celled capsule. $2n = 30$, 38, 76, 114, 152, 228, and 304. About 90 species. Throughout the range of the family.

2. *Chilianthus*

Chilianthus Burchell, Trav. S. Africa 1: 94 (1822); Leeuwenberg, Meded. Landbouwhoogeschool Wageningen 79: 1–163 (1979), rev. as *Buddleja*.

Shrubs or small trees, mostly with a grey indumentum of stellate and glandular hairs. Leaves opposite, entire, crenate or dentate. Synflorescence thyrsoid, large and laxly paniculate. Corolla cup-shaped, white, creamy, yellow, or greenish, less than 5 mm long, aestivation imbricate. Anthers exserted from corolla. Fruit a 2-celled capsule. $2n = 38$. Three species. Southern Africa.

3. *Emorya*

Emorya Torr., Bot. Mex. Bound.: 121 (1859); Norman, Fl. Neotrop. Monogr. 81: 1–225 (2000), rev.

Shrubs, mostly with a grey indumentum of stellate and glandular hairs. Leaves opposite, sinuate-dentate. Synflorescence thyrsoid, with rather few 1–15-flowered cymes. Corolla tubular, 25–35 mm long. Anthers exserted from corolla. Fruit a 2-celled capsule. $2n = 38$. Two species. Southern USA and Mexico.

4. *Gomphostigma*

Gomphostigma Turcz., Bull. Soc. Imp. Nat. Moscou 16: 53 (1843); Leeuwenberg, Meded. Landbouwhoogeschool Wageningen 77: 15–31 (1977), rev.

Small shrubs or suffrutescent herbs, glabrous or with stellate hairs. Leaves opposite, entire or subentire. Synflorescence thyrsoid, with single-flowered inflorescences. Corolla cup-shaped, white, less than 5 mm long, aestivation imbricate. Anthers exserted from corolla. Fruit a two-celled capsule. $2n = 28$. Two species. Southern Africa.

5. *Nicodemia*

Nicodemia Tenore, Cat. Orto Bot. Napoli: 88 (1845); Leeuwenberg, Meded. Landbouwhoogeschool Wageningen 79: 1–163 (1979), rev. as *Buddleja*.

Shrubs or small trees, mostly with a grey indumentum of stellate and glandular hairs. Leaves opposite, entire, crenate or dentate. Synflorescence thyrsoid, large and laxly paniculate. Corolla tubular, yellow, orange, or greenish, up to 12 mm long, aestivation imbricate. Anthers included or almost included in corolla. Fruit a berry, 2- or 4-celled. $2n = 38$. Eight species. Madagascar, southern Africa.

Selected Bibliography

Bisset, N.G. 1980. Phytochemistry. In: Engler & Prantl, Die natürlichen Pflanzenfamilien, ed. 2, 28b I. Berlin: Duncker & Humblot.

Carlquist, S. 1997. Wood anatomy of Buddlejaceae. Aliso 15: 41–56.

Dickison, W.C. 1994. A re-examination of *Sanango racemosum*. 2. Vegetative and floral anatomy. Taxon 43: 601–618.

Fabris, H.A. 1965. Loganiaceae. In: Cabrera, A.L. (ed.) Flora de la provincia de Buenos Aires 5: 21–26.

Gadella, T.W.J. 1980. Cytology. In: Engler & Prantl, Die natürlichen Pflanzenfamilien, ed. 2, 28b I. Berlin: Duncker & Humblot.

Hakki, M.I. 1980. Embryology. In: Engler & Prantl, Die natürlichen Pflanzenfamilien, ed. 2, 28b I. Berlin: Duncker & Humblot.

Hasselberg, G.B.E. 1937. Zur Morphologie des vegetativen Sprosses der Loganiaceen. Uppsala: Almqvist & Wiksells Boktryckeri.

Islebe, G.A., Hooghiemstra, H. 1995. Recent pollen spectra of Highland Guatemala. J. Biogeogr. 22: 1091–1099.

Jensen, S.R. 1994. A re-examination of *Sanango racemosum*. 3. Chemotaxonomy. Taxon 43: 619–623.

Jensen, S.R. 2000a. Chemical relationships of *Polypremum procumbens*, *Tetrachondra hamiltonii* and *Peltanthera floribunda*. Biochem. Syst. Ecol. 28: 45–51.

Jensen, S.R. 2000b. Chemistry of the Buddlejaceae. In: Norman, E. (ed.) Buddlejaceae. Fl. Neotrop. Monogr. 81.

Keenan, J. 1969a. Notes on *Buddleja* I. Polyploid pattern in *B. colvilei*. Notes Roy. Bot. Gard. Edinburgh 29: 197–198.

Keenan, J. 1969b. Notes on *Buddleja* II. Pollen. Notes Roy. Bot. Gard. Edinburgh 29: 199–202.

Leeuwenberg, A.J.M. 1977. The Loganiaceae of Africa XVI. *Gomphostigma* Turcz. Meded. Landbouwhoogeschool Wageningen 77: 15–30.

Leeuwenberg, A.J.M. 1979. The Loganiaceae of Africa XVIII. *Buddleja* L. II. Revision of the African and Asiatic species. Meded. Landbouwhoogeschool Wageningen 79: 1–163.

Leeuwenberg, A.J.M., Leenhouts, P.W. 1980. Taxonomy. In: Engler & Prantl, Die natürlichen Pflanzenfamilien, ed. 2, 28b I. Berlin: Duncker & Humblot.

Mahboubi, P., Gordon, A.M., Stoskopf, N., Voroney, R.P. 1997. Agroforestry in the Bolivian Altiplano: evaluation of tree species and greenhouse growth of wheat with soils treated with tree leaves. Agrofor. Syst. 37: 59–77.

Maldonado de Magnano, S. 1986. Estudios embriologicos en *Buddleja* (Buddlejaceae) I: endosperma y episperma. Darwiniana 27: 225–236.

Maldonado de Magnano, S. 1987. Estudios embriologicos en *Buddleja* (Buddlejaceae) II: embriogenesis. Darwiniana 28: 391–395.

Mohrbutter, C. 1937. Embryologische Studien an Loganiaceen. Planta 26: 64–80.

Moore, R.J. 1947. Cytotaxonomic studies in the Loganiaceae I. Chromosome numbers and phylogeny in the Loganiaceae. Am. J. Bot. 34: 527–538.

Norman, E. 1994. A re-examination of *Sanango racemosum*. 1. Morphology and distribution. Taxon 43: 591–600.

Norman, E. 2000. Buddlejaceae. Fl. Neotrop. Monogr. 81: 1–225.

Olmstead, R.G., Reeves, P.A. 1995. Evidence for the polyphyly of the Scrophulariaceae based on chloroplast *rbc*L and *ndh*F sequences. Ann. Missouri Bot. Gard. 82: 176–193.

Oxelman, B., Backlund, M., Bremer, B. 1999. Relationships of the Buddlejaceae s. l. investigated using parsimony jackknife and branch support analysis of chloroplast *ndh*F and *rbc*L sequence data. Syst. Bot. 24: 164–182.

Oxelman, B., Kornhall, P., Olmstead, D., Bremer, B. in prep. A chloroplast phylogeny of Lamiales with particular reference to Scrophulariaceae s.str.

Punt, W. 1980. Pollen morphology. In: Engler & Prantl, Die natürlichen Pflanzenfamilien, ed. 2, 28b I. Berlin: Duncker & Humblot.

Solereder, H. 1895. Loganiaceae. In: Engler & Prantl, Die natürlichen Pflanzenfamilien IV, 2. Leipzig: Engelmann.

Speta, F. 1977. Proteinkörper in Zellkernen: neue Ergebnisse und deren Bedeutung für die Gefässpflanzensystematik nebst einer Literaturübersicht für die Jahre 1966–1976. Candollea 32: 133–163.

Speta, F. 1979. Weitere Untersuchungen über Proteinkörper in Zellkernen und ihre taxonomische Bedeutung. Plant Syst. Evol. 132: 1–26.

Tomas-Barberan, F.A., Grayer-Barkmeijer, R.J., Gil, N.I., Harborne, J.B. 1988. Distribution of 6-hydroxy-, 6-methoxy- and 8-hydroxyflavone glycosides in the Labiatae, the Scrophulariaceae and related families. Phytochemistry 27: 2631–2645.

Wallick, K.P., Molvray M., Kores P.J., Elisens, W.J. 2000. A phylogenetic analysis of Buddlejaceae s.s. inferred from the *trn*L region of cpDNA. Poster Abstract Botany 2000. Am. J. Bot. Suppl. 87 (6): 183.

Wiehler, H. 1994. A re-examination of *Sanango racemosum*. 4. Its new systematic position in Gesneriaceae. Taxon 43: 619–623.

Byblidaceae

J.G. Conran and R. Carolin

Byblidaceae (Engl. & Gilg.) K. Domin, Acta Bot. Bohemica 1: 3–4 (1922), nom. cons.

Perennial carnivorous evergreen shrubs with secondary growth, or annual or ephemeral herbs. Rootstock a woody caudex or indistinct, roots fibrous. Stems perennial and branched, short-lived and elongating at flowering from branched subterranean shoots, erect, scrambling or floating. Leaves cauline, alternate, estipulate, subulate-linear, with insect-trapping, mucilage-secreting glandular hairs and a small, knob-like apical swelling. Flowers solitary in upper leaf axils, hermaphrodite, medium-sized and showy, calyx and corolla well developed; perianth actinomorphic, 5-merous, calyx connate basally, imbricate, persistent, corolla connate basally, appearing free, rotate, cerise to pale lilac, rarely white, underside pale lilac, white or yellow; limb 5-lobed; lobes imbricate in bud; stamens five, opposite the sepals, inserted at the base of the tube; filaments free, twisted to make the anthers face abaxially and the flower zygomorphic; anthers basifixed, 2-thecate, 4-sporangiate, introrse, opening by pore-like, short apical slits. Gynoecium of two united carpels. Ovary superior, 2-locular with axile placentation; ovules many, anatropous, unitegmic, tenuinucellate; style terminal, filiform, sigmoid, stigma minutely capitate or bilobed, papillate, twisted to face the same direction as the anther pores. Fruit a smooth, loculicidal, cartilaginous capsule. Seeds small, angular, black, prominently sculptured with smooth or denticulate longitudinal ridges, with or without transverse ridges; endosperm hard, copious, starchy; embryo small, linear.

One genus, *Byblis* with six species native to Australia, one extending to New Guinea. All species are cultivated as ornamental curiosities.

Vegetative Morphology. Roots are fibrous, consisting of a short-lived taproot and numerous adventitious roots, especially in the annual species, with thicker roots in the perennial species. The rootstock is a short, branched subterranean caudex in the perennial species, otherwise absent, and the single or branched aerial stems are woody or herbaceous.

Flowering stems are erect, leafy, unbranched, consisting of the upper shoots, elongating from a more or less basal rosette in one perennial species. The leaves are non-sheathing, linear, widest where they join the stem and tapering, persistent, unifacial or slightly bifacial, circular or reniform in cross section (Fig. 5). The midrib is either grooved or inconspicuous. The entire leaf surface is covered with both tentacular and sessile glandular-secretory hairs. Vernation is flat in *B. gigantea* and *B. lamellata* whereas the remainder have inversely circinnate vernation.

Vegetative Anatomy. Vessels occur in the roots and stems and have predominantly simple perforations, although there are scattered scalariform and multiperforate vessel elements in the woody stems of *Byblis gigantea* as well as unusual, narrow, tracheid-like vessels. In *B. gigantea* there are secondary thickenings resembling annual rings, and there are tracheids and lignified ray cells with bordered pits (Carlquist 1976). Sieve-tube plastids are of the dicotyledonous Ss type (Behnke 1991). Laticifers are absent.

The roots are reported to be triarch (Lang 1901) or diarch (Lloyd 1942). Solereder (1908) described the stems (scapes) of *Byblis* as consisting of a sclerenchymatous cylinder of fibres adjoining a ring of collateral or concentric vascular bundles with central phloem. Diels (1930) noted that there were nine vascular bundles and an internally sclerenchymatous endodermis (starch sheath) in the stem (rhizome). The sclerenchymatous pith consists of large pitted cells (Metcalfe and Chalk 1930).

The leaves are rounded, reniform or basally more or less triangular in cross section. There is a well-developed and vascularised apical hydathode (Lang 1901). The epidermis is thick-walled and consists of elongate, straight-walled cells which mature into a series of parallel ridges and furrows running along the long axis of the leaves. Stomata are located in longitudinal grooves along the leaf and are of the paracytic (rubiaceous) type (Metcalfe and Chalk 1979).

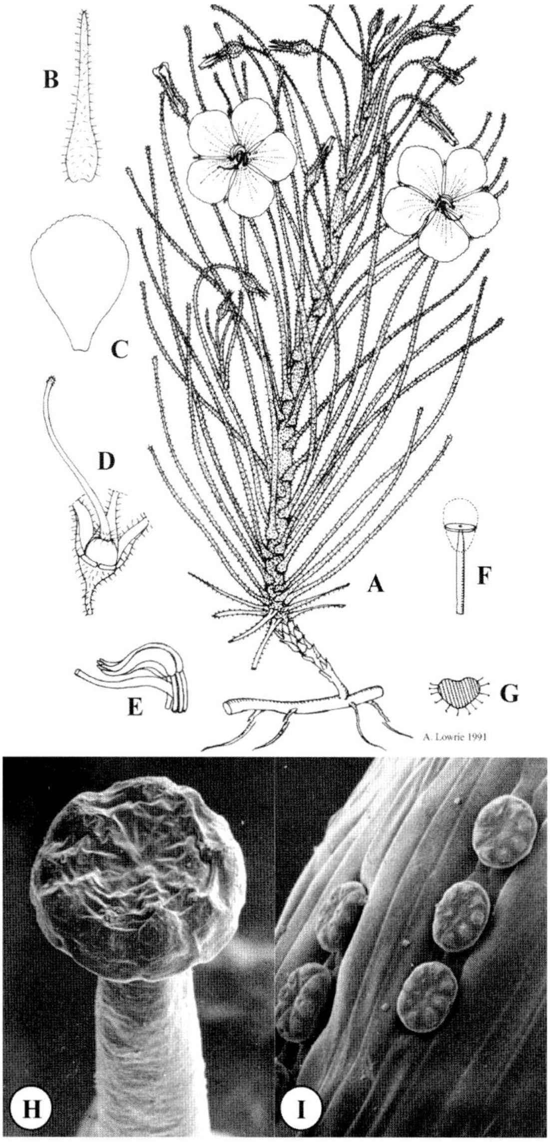

Fig. 5. Byblidaceae. **A–G** *Byblis gigantea*. **A** Habit. **B** Sepal. **C** Petal. **D** Ovary. **E** Stamens enclosing ovary and style base. **F** Stalked gland. **G** Leaf in transverse section. **H–I** *B. filifolia*. **H** Stalked gland. **I** Sessile gland. (**A–G** A. Lowrie, in Lowrie 1998; **H–I** Photo G. Jauzdems)

There are numerous epidermal glands on all plant surfaces except the corolla. Stalked glands with 2–8 basal cells, an elongate stalk cell with spiral primary thickenings, and 8 to usually 32 radiating capital cells with adaxial pores arranged near the centre of the head at the junction of cell pairs (Fig. 5) secrete large quantities of mucilage. The glands closely resemble those of *Pinguicula* in the Lentibulariaceae. In the epidermal grooves, there are numerous sessile glands arranged in single-file groups between the stomata. These 'sessile' glands consist of a pair of basal cells, a very short stalk cell, and eight radiating capital cells, and were differentiated by Lloyd (1942) into digestive hairs and short mucilage (= stalked) hairs. Mesophyll cells have an undifferentiated chlorenchyma and the central mesophyll is achlorophyllous. There are three vascular bundles with associated pillars of sclerenchymatous support tissue.

Inflorescence Structure. The flowers are solitary in the upper leaf axils on long pedicels; prophylls are absent.

Floral Morphology and Anatomy. Flowers in Byblidaceae are hermaphrodite. The calyx consists of 5 small, green sepals which are shortly fused at the base. The rotate, actinomorphic corolla consists of 5 broadly obcuneate petals. These are very shortly fused at the base and appear free. They are contorted in bud. The petal apices are conspicuously dentate.

The pedicels have ten vascular traces from which five diverge to supply the sepals, followed by the petals and then the stamens, each with a single trace. The carpels are 3-traced and the style is vascularised by dorsal bundles (Chrtek et al. 1989; Rani 1994).

The 5 stamens are inserted opposite the sepals and are attached to the base of the corolla tube. The stamens are more or less unequal and contorted adaxially to the main stem axis. The anthers are basifixed, dehiscing apically by short, pore-like slits derived from two lateral slits in the upper anther half and confluent apically. There is a thick connective, a ventral anther furrow and protruding papillate dorsal apex. The apically restricted, 1-layered endothecium is discontinuous over the connective and a pollen sac placentoid is present (Endress and Stumpf 1991).

Embryology. The anther wall formation is of the Dicotyledonous type and the middle layers are ephemeral with apical fibrous thickenings. The tapetum is glandular-secretory and binucleate. Microsporogenesis is simultaneous and the tetrads are tetrahedral and the pollen binucleate. The ovule is anatropous, unitegmic and tenuinucellate, with a massive integument and well-developed endothecium (Lang 1901). Megasporogenesis produces a linear tetrad and the micropylar cells degenerate. Embryo sac development is of the Polygonum type, with persistent antipodal cells, pyriform synergids, and the ellipsoidal embryo sac contains abundant starch grains (Conran 1996).

Endosperm formation is cellular, and there are prominent micropylar and chalazal haustoria. The endosperm is non-ruminate. Embryogenesis appears to be of the Onagrad type, and the nucellus degenerates early in seed development (Conran 1996).

Pollen Morphology. Pollen grains are oblate-sphaeroidal and tricolpate (sometimes tetracolpate), 35–41 µm diameter (Erdtman 1952), with a thick punctitegillate sexine and an equally thick nexine.

Karyology. Chromosomes in the Byblidaceae have been reported for *B. gigantea* (s. lat. and including *B. lamellata*) with conflicting counts of $2n = 18$ (Kress 1970) and $2n = 14$ (Kondo 1973). Chromosome numbers for *B. liniflora* are also contentious, with Kondo (1973) reporting $2n = 24$, but Peng and Kenton (1982) finding $2n = 32$. Although *Byblis liniflora* was confirmed as $2n = 32$ by Conran et al. (2002a), the remainder of the annual tropical species were found to have counts of $2n = 16$.

Pollination. The flowers of the Byblidaceae are generally considered to be buzz-pollinated, with the abaxially oriented stamens and apically porate anthers releasing pollen by the vibration of an approaching pollinator insect's wings (Lowrie 1998). The flowers are visited by small native bees and hover flies (Conran, pers. observ.). Juniper et al. (1989) claimed that the flowers of *B. gigantea* s. lat. are unusual in placing the flowers amongst the traps but, as they are borne on long petioles and held away from the leaves, this reduces the risk to visiting pollinators. De Buhr (1975) postulated mimicry by *Byblis gigantea* s. lat. of *Thysanotus multiflorus* (Lilianae: Laxmanniaceae), with which it co-occurs and co-flowers. There are no studies of the relative rewards offered by either species (both are buzz-pollinated), and it is unknown whether their pollinators are conspecific. The northern Australian *Byblis* species are also possible mimics, sharing morphology and habitat with the carnivorous *Drosera indica* (Droseraceae).

Fruit and Seed Morphology. The fruit is an apiculate ovoid or 2-lobed loculicidal capsule. The fruits are glandular hairy in early development, becoming eglandular and smooth to ridged with reticulate venation at maturity.

The seeds are angular, black and prominently ridged and sculptured, the sculpturing characteristic at the species level for some taxa (Lowrie and Conran 1998). The micropyle is located close to the chalazal scar. There is a collapsed and sculptured exotesta, sclerenchymatous mesotesta and unthickened endotesta (Conran 1996). The endosperm is starchy and there are occasional deposits of calcium oxalate (Netolitzky 1926).

Germination in the family is epigeal phanerocotylar. The seedlings have a well-developed taproot and the hypocotyl is glandular hairy. The cotyledons are linear and leaf-like. The first leaf pair is opposite while later leaves are alternate, the first leaves resemble the later ones and both these and the cotyledons are glandular-hairy.

Ecology. The Byblidaceae are considered to be carnivorous, with the stalked glands functioning as trapping glands (Lloyd 1942), whereas the sessile glands show evidence of being both digestive and absorptive (Fenner 1904; Bruce 1905), although this latter point was challenged by Hartmeyer (1997, 1998). The family shares with the Droseraceae a relationship with predatory bugs of the genus *Setocoris* (Hemiptera: Miridae), where captured prey are killed and fed upon by the bugs (Hamilton 1903; Lowrie 1998). Although there are no studies for the Australian species, investigation of the relationship between the South African proto-carnivorous *Roridula* (Roridulaceae) and its associated Mirid, where the insects' faeces are deposited on the leaves providing efficient and rapid turnover of nutrients (Ellis and Midgley 1996; Midgley and Stock 1998), suggests that a similar relationship may well exist for the Byblidaceae.

The northern, annual species live in ephemeral soaks and seasonally inundated areas, whereas the perennial SW Western Australian species *B. gigantea* is found on seasonally waterlogged *Leptospermum*/Restionaceae low scrub (Speck and Baird 1984), and *B. lamellata* on deep sands in fire-prone dry heathlands. Following fire, *B. gigantea* survives by regenerating from the deep, woody perennial rootstock (Baird 1984), whereas *B. lamellata* is a taller branched shrub which can resprout from aboveground stems after fire as well as from the rhizomes. In all species, germination is triggered as a response to smoke from fires, and can be stimulated with a 100 ppm solution of gibberellic acid (Lowrie 1996), with the northern Australian species germinating at hot, wet temperatures above 25 °C, whereas the SW Western Australian species tend to require cool, moist conditions, usually below 10 °C.

PHYTOCHEMISTRY. Iridoids are thought to be present in the family (Gibbs 1974). It is unknown to which iridoid type they belong (Jensen et al. 1975). Naphthoquinones are absent (Zenk et al. 1969), as are proanthocyanins and cyanogens (Cronquist 1981).

SUBDIVISIONS AND RELATIONSHIPS WITHIN THE FAMILY. Byblidaceae are monogeneric, but there are two species groups representing the annual northern taxa and southern perennial species (Conran and Lowrie 1993; Lowrie 1998; Lowrie and Conran 1998; Conran et al. 2002b; Conran 2003).

AFFINITIES. Byblidaceae have been traditionally associated with the Droseraceae (Bentham and Hooker 1865) and/or Roridulaceae, the latter sometimes included within Byblidaceae (Cronquist 1981, 1988) and related more broadly to the Cornales and other Rosidae. Recent studies involving serology (Grund and Jensen 1981) and molecular sequences of the chloroplast gene *rbc*L (Albert et al. 1992) and ribosomal 18S rRNA (Conran and Dowd 1993) found instead that the family's affinities were with the Asteridae, with the *rbc*L study placing them sister to the Lentibulariaceae, supporting the embryological studies of Lang (1901) and Conran (1996).

DISTRIBUTION AND HABITATS. The family is predominantly Australian, with *B. liniflora* extending to southern Papua New Guinea (van Steenis 1971). The species grow in seasonally wet, open savannahs, heathlands or, in the case of *B. lamellata*, deep sands in open heathlands.

ECONOMIC IMPORTANCE. Byblidaceae are of interest to specialist carnivorous plant collectors. Although the family have been removed from CITES Appendix II, *B. gigantea* s. stricto is currently listed as on the Priority Flora list for Western Australia as DRF (declared rare flora), and has been recommended as priority 3 and IUCN:VUDN1.

Only one genus:

Byblis Salisb. Fig. 5

Byblis Salisb., Parad. Lond. 2: t. 95 (1808).

Description as for family.

Selected Bibliography

Albert, V.A., Williams, S.E., Chase, M.W. 1992. Carnivorous plants: phylogeny and structural evolution. Science 257: 1491–1495.

Baird, A.M. 1984. Observations on regeneration after fire in the Yule Brook Reserve near Perth, Western Australia. J. Roy. Soc. W.A. 67: 1–13.

Behnke, H.-D. 1991. Distribution and evolution of forms and types of sieve-element plastids in the dicotyledons. Aliso 13: 167–182.

Bentham, G., Hooker, J.D. 1865. Genera Plantarum vol. I. London: L. Reeve.

Bruce, A.N. 1905. On the activity of the glands of *Byblis gigantea*. Notes Roy. Bot. Gard. Edinburgh 16: 9–14.

Carlquist, S. 1976. Wood anatomy of Byblidaceae. Bot. Gaz. 137: 35–38.

Chrtek, J., Slavíková, Z., Studnička, M. 1989. Beitrag zur Leitbündelanordnung in den Kronblättern von ausgewählten Arten der fleischfressenden Pflanzen. Preslia 61: 107–124.

Conran, J.G. 1996. The embryology and relationships of the Byblidaceae. Austral. Syst. Bot. 9: 243–254.

Conran, J.G. 2003. Byblidaceae. In: Orchard, A. (ed.) Flora of Australia vol. 10. CSIRO, Melbourne (in press).

Conran, J.G., Dowd, J.M. 1993. The phylogenetic relationships of the *Byblis-Roridula* (Byblidaceae-Roridulaceae) complex inferred from 18S rRNA partial sequences. Plant Syst. Evol. 188: 73–86.

Conran, J.G., Lowrie, A. 1993. *Byblis liniflora* subsp. *occidentalis* (Byblidaceae). A new subspecies of from north-western Australia. Austral. Syst. Bot. 6: 175–179.

Conran, J.G., Houben, A., Lowrie, A. 2002a. Chromosome numbers in Byblidaceae. Austral. J. Bot. 50: 583–586.

Conran, J.G., Lowrie, A., Moyle-Croft, J. 2002b. A revision of *Byblis* (Byblidaceae) in south-western Australia. Nuytsia 15: 11–20.

Cronquist, A. 1981. An integrated system of classification of flowering plants. New York: Columbia University Press.

Cronquist, A. 1988. The evolution and classification of flowering plants. New York: New York Botanic Gardens.

De Buhr, L.E. 1975. Observations on *Byblis gigantea* in Western Australia. Carniv. Plants Newslett. 3: 60–61.

Diels, L. 1930. Byblidaceae. In: Engler & Prantl, Die natürlichen Pflanzenfamilien, ed. 2, 13a. Leipzig: Engelmann.

Ellis, A.G., Midgley, J.J. 1996. A new plant-animal mutualism involving a plant with sticky leaves and a resident Hemipteran insect. Oecologia 106: 478–481.

Endress, P.K., Stumpf, S. 1991. The diversity of stamen structure in 'Lower' Rosidae (Rosales, Fabales, Proteales, Sapindales). Bot. J. Linn. Soc. 107: 217–293.

Erdtman, G. 1952. Pollen morphology and plant taxonomy: Angiosperms. New York: Hafner.

Fenner, C.A. 1904. Beiträge zur Kenntnis der Anatomie, Entwickelungsgeschichte und Biologie der Laublätter und Drüsen einiger Insectivoren. Flora 93: 335–434.

Gibbs, D. 1974. Chemotaxonomy of flowering plants. Montreal: McGill Queens University Press.

Grund, C., Jensen, U. 1981. Systematic relationships of the Saxifragales revealed by serological characteristics of seed proteins. Plant Syst. Evol. 137: 1–22.

Hamilton, A.G. 1903. Notes on *Byblis gigantea*. Proc. Linn. Soc. N.S.W. 28: 680–684.

Hartmeyer, S. 1997. Carnivory of *Byblis* revisited: a simple method for enzyme testing on carnivorous plants. Carniv. Plant Newslett. 26: 39–45.

Hartmeyer, S. 1998. Carnivory in *Byblis* revisited II: the phenomenon of symbiosis on insect trapping plants. Carniv. Plant Newslett. 27: 110–113.

Jensen, S.R., Nielsen, B.J., Dahlgren, R. 1975. Iridoid compounds, their occurrence and systematic importance in the Angiosperms. Bot. Notiser 128: 148–180.

Juniper, B.E., Robins, R.J., Joel, D.M. 1989. Carnivorous plants. London: Academic Press.

Kondo, K. 1973. The karyotypes of the species of *Byblis*. Bull. Torrey Bot. Club 100: 367–369.

Kress, A. 1970. Zytotaxonomische Untersuchungen an einigen Insektenfängern (Droseraceae, Byblidaceae, Cephalotaceae, Roridulaceae, Sarraceniaceae). Ber. Deutsch. Bot. Ges. 83: 55–62.

Lang, F.X. 1901. Untersuchungen über Morphologie, Anatomie und Samenentwicklung von *Polypomphylox* und *Byblis gigantea*. Flora 88: 149–206.

Lloyd, F.M. 1942. The carnivorous plants, 2nd edn. Waltham, Mass.: Chronica Botanica Co.

Lowrie, A. 1996. An easy method to smoke treat carnivorous plant seed. Bull. Austral. Carniv. Plant Soc. 15: 3–5.

Lowrie, A. 1998. Carnivorous plants of Australia, vol. 3. Perth: University of Western Australia Press.

Lowrie, A., Conran, J.G. 1998. A taxonomic revision the genus *Byblis* (Byblidaceae) in northern Australia. Nuytsia 12: 59–74.

Metcalfe, C.R., Chalk, L. 1930. Anatomy of the Dicotyledons, vol. I. Oxford: Clarendon Press.

Metcalfe, C.R., Chalk, L. 1979. Anatomy of the Dicotyledons, 2nd edn., vol. I. Oxford: Clarendon Press.

Midgley, J.J., Stock, W.D. 1998. Natural abundance of Delta N-15 confirms insectivorous habit of *Roridula gorgonias*, despite it having no proteolytic enzymes. Ann. Bot. 82: 387–388.

Netolitzky, F. 1926. Anatomie der Angiospermen - Samen. In: Linsbauer, K. (ed.) Handbuch der Pflanzenanatomie, 10. Berlin: Borntraeger, pp. 1–364.

Peng, C.-I., Kenton, A. 1982. Chromosome numbers of *Byblis liniflora* Salisb. (Byblidaceae). Ann. Missouri Bot. Gard. 69: 417–419.

Rani, R.S. 1994. Floral anatomy and the affinities of Byblidaceae. Rheedea 4: 144–150.

Solereder, H. 1908. Systematic Anatomy of the Dicotyledons, vol. 1. Introduction, Polypetalae and Gamopetalae. Oxford: Clarendon Press.

Speck, N.H., Baird, A.M. 1984. Vegetation of the Yule Brook Reserve near Perth, Western Australia. J. Roy. Soc. W.A. 66: 147–162.

van Steenis, C.G.G.J. 1971. Byblidaceae. Flor. Males. Bull. Suppl. ser. I 7: 135–137.

Zenk, M.H., Fürbringer, M., Steglich, W. 1969. Occurrence and distribution of 7-Methyljulone and Plumbagin in the Droseraceae. Phytochem. 8: 2199–2200.

Callitrichaceae[1]

C. Erbar and P. Leins

Callitrichaceae Link, Enumeratio hort. berol. I: 7 (1821).

Slender annual or perennial, much-branched herbs, aquatic, amphibious or terrestrial. Leaves decussate, in rosettes at floating branch tips, small, entire, estipulate. Submersed leaves usually linear and often bifid at the apex. Floating or aerial leaves linear, elliptic, oblong or spathulate. Glandular, fan-like scale hairs in the leaf axils, in terrestrial and amphibious species additionally peltate, disc-like hairs. Plants monoecious. Unisexual flowers minute, often subtended by a pair of horn-shaped prophylls, usually solitary in the leaf axil or, rarely, one male and one female in the same leaf axil. Staminate flower consisting of a single stamen with a slender filament and a reniform anther with two locules. Anther opening by one slit across the top of the anther. Pistillate flower consisting of a single bicarpellate ovary with two, usually elongate stylodia. Each primary locule in transversal position divided into two chambers. Each chamber with a single pendulous and axile ovule. Ovule with one massive integument, tenuinucellate, (hemi-)anatropous and epitropous. Fruit separating into 4 drupe-like mericarps. Mericarps rounded, keeled or winged.

A monogeneric family of about (17–)40–50 species of almost cosmopolitan distribution (widespread and centered in temperate zones, but also occurring in tropical regions).

Vegetative Morphology. The growth form of *Callitriche* is terrestrial, amphibious or obligately submersed. Analysis of cpDNA suggests that the terrestrial growth form may be derived from the amphibious growth form in the family (Philbrick and Jansen 1991).

Individuals of terrestrial species produce small, prostrate mats ranging from 2 to 7 cm in diameter and only aerial flowers. The habit of amphibious *Callitriche* is variable, but most individuals grow in clumps and are much branched. Plants may grow terrestrially on moist soils, or aquatically. Aquatic individuals can either be entirely submersed, or the principal stems can reach the water surface and support a rosette of floating leaves (Fig. 6A). Depending on environmental conditions, amphibious species can produce only aerial flowers, both aerial and submersed flowers on the same individual, or only submersed flowers. Annual forms of obligately submersed species are relatively small plants of one to few, upright stems. The perennial submersed forms are larger and consist of multiple stems that branch abundantly.

The stems are usually erect above, and creeping and rooting below. When submersed they are elongate and when emersed contracted. The leaves are decussate, often rosette-forming at the tip of floating stems. The floating or aerial leaves are linear, elliptic, oblong or spathulate and 3-nerved. The submersed leaves are usually linear, 1-nerved and often have a bifid apex.

Vegetative Anatomy. Phloem plastids are of the S-type (Behnke 1991). Like in many aquatic plants, the vascular system is strongly reduced and assembled in a weak central cord enclosing a very slender pith (only 2 or 3 cells in cross section), with few vessels in the internodes turning into short nodal tracheids (Hegelmaier 1864).

Plants that grow submersed have a well-developed cuticle but lack stomata (Hegelmaier 1864). Stomata are numerous on both surfaces of the leaves and on the stems in terrestrial species. They are absent from the stems but occur sporadically on the leaves in submersed forms. Stomata are confined to the upper surface of floating leaves and are ephemeral in submersed plant parts.

All *Callitriche* species have glandular, scale-like hairs in the leaf axils. The mucilage produced beneath the cuticle in these hairs protects the young leaves in bud. The glandular scales have a unicellular foot and a multicellular head in which the cells are arranged fan-like. In the submersed species (sect. *Pseudocallitriche*) each head cell is replaced by a series of 2–4 cells (Fig. 6H). Peltate (stellate) hairs on stems, leaves, prophylls and pedicels of the female flowers are recorded in terrestrial and amphibious species (sect. *Eucallitriche*). The peltate hairs consist of a stalk cell and

[1] See also Scrophulariaceae.

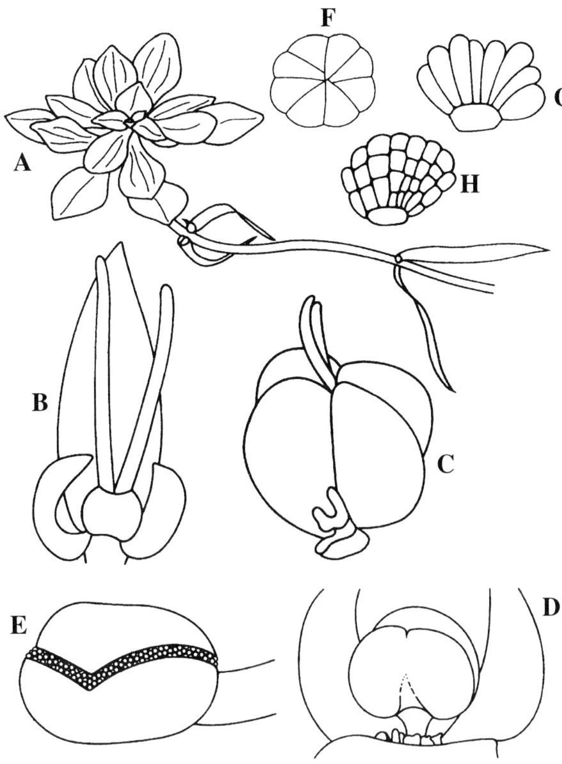

Fig. 6. Callitrichaceae. **A–F** *Callitriche obtusangula*. **A** Habit. **B** Female flower with two prophylls in the axil of a bract. **C** Fruit. **D** Male flower. **E** Reniform anther with two locules opening by one slit across the top of the anther. **F** Peltate hair of stems, leaves, prophylls and pedicels in surface view. **G** *C. verna*. Peltate hair in longisection. **H** *C. autumnalis*. Fan-like scale hair found in the leaf axils in longisection. (**A–F** Orig., **G–H** After Hegelmaier 1864)

a varying number of disc cells (ranging from 4 to 20; Hegelmaier 1864; Fig. 6F, G). Shape and number of cells in the disc are of diagnostic importance.

Flower Structure. The flowers are minute and unisexual. Usually they are solitary in a leaf axil. An accessory bud – vegetative or floral – is sometimes borne in the leaf axil below the principal bud. When both the principal and the accessory bud are floral, the upper usually develops into a male, and the lower into a female flower. As the lower flower may lack prophylls, these two flowers can have the appearance of a single bisexual flower.

Staminate and pistillate flowers occur in various combinations in the axils of opposite leaves. Usually, female flowers develop in the axils of the lower leaves, and male flowers are formed in the upper leaf axils, but both staminate and pistillate flowers can also develop at the same node.

Although all flowers lack a perianth, those of several species are subtended by a pair of delicate, gas-inflated, horn-like or sack-like prophylls in transversal position (Fig. 6B, D). Prophylls are absent in submersed and terrestrial species as well as in terrestrial and submersed forms of some of the amphibious species. Prophylls, however, always occur below flowers situated in the axil of floating leaves. The inflated, gas-filled prophylls may aid in holding the flowers above the water surface (flotation) to facilitate aerial pollination.

The male flower (Fig. 6D) consists of one stamen with a slender filament and a reniform anther with two locules opening by one slit (Fig. 6E; adaxial sporangia primarily synthecous, abaxial sporangia secondarily synthecous). Young developmental stages of the anther show an adaxial, kidney-shaped protuberance and two broad, longitudinal protuberances at the abaxial side. The outer shape is in accordance with the histological differentiation. Adaxially, archesporial cells are formed continuously in the whole transversal bulge, whereas the archespor is separated by sterile tissue in the abaxial half of the stamen primordium. Later on, the sterile tissue dissolves (*C. obtusangula*, Leins and Erbar 1988; Leins 2000).

The female flower consists of a (sometimes) stalked, somewhat compressed, superior, bicarpellate ovary crowned by a pair of erect or recurved stylodia (Fig. 6B). The stylodia are papillate and stigmatic over their entire length. They are usually long and slender but short in the terrestrial and submersed forms of amphibious species with internal geitonogamy (see Pollination and Reproductive Systems). The base of the stylodia is surrounded by the elevated, upper parts of the ovary.

The two carpels arise in transversal (lateral) position. Each ovary locule becomes later divided into two by false septa formed from the dorsal part of the carpel. There is a single pendulous (hemi-) anatropous ovule with the micropyle orientated upwards in each of the four locules.

Amphibious species that have internal geitonogamy (see Pollination and Reproductive Systems) show some variation in flower structure. In forms with floating rosettes, anthers reach full development and produce a large number of pollen grains (about 1200–3000 per anther). These flowers also develop full-size ovaries and stylodia. In submersed and terrestrial forms, the anthers are typically only poorly developed and produce a small number of pollen grains (as low as four per anther). Although these flowers form a full-size ovary, they lack well-developed stylodia. When these are present as rudimentary structures, characteristic papillae are typically absent (Osborne and Philbrick 1994).

Embryology. The adaxial sporangia of neighbouring thecae are fused primarily and the abaxial ones become fused secondarily during ontogeny. The anther dehisces by one slit which runs across the apex of the stamen (*C. obtusangula*, Leins and Erbar 1988, 1990; Leins 2000).

Prior to dehiscence, the wall of the anther is composed of epidermis and fibrous endothecium, but the latter without thickenings in *C. autumnalis* and the entirely submersed forms (e.g. *C. hermaphroditica*; Davis 1966; Philbrick and Osborn 1994).

The tapetum is secretory and its cells become multinucleate. Cell wall formation after meiosis is simultaneous and the pollen grains are 3-celled when shed.

The ovule is (hemi-)anatropous, unitegmic and tenuinucellate and is in an epitropous position (raphe ventral and turned-up micropyle). The nucellus is very reduced. The edges of the very large integument of partially thick-walled cells form a fairly long and very narrow micropyle. Frequently, the micropylar canal is difficult to trace over its entire length.

The unicellular archespor functions directly as the megaspore mother cell, in which meiosis takes place. The chalazal megaspore of a linear tetrad develops into a Polygonum-type embryo sac. The synergids are pyriform, the polar nuclei fuse before fertilization, and the three small antipodes are ephemeral. A well-developed endothelium of tall, plasma-rich cells is differentiated.

Endosperm formation is cellular from the beginning, with the development of terminal haustoria. The aggressive micropylar haustorium is formed of two or four, long, twisted cells which penetrate the micropyle and pass through the integument to the funicle. The chalazal haustorium is considerably smaller and its four cells form a short hook at the base of the endosperm.

Embryogeny follows the Onagrad type and a long filamentous suspensor is developed (Jørgensen 1923; Davis 1966).

Pollen Morphology. In *Callitriche* several pollen types are found. Pollen morphology is related to mode of pollination.

The ovoid or ± spheroidal or irregular pollen grains are 3(4)-colporate or have small, ± irregular thin areas or rifts or are inaperturate (Erdtman 1966). A recent study (Cooper et al. 2000) reports the lack of well-defined apertures and describes only aperture-like areas, characterized by a reduction in the otherwise thickened basal layer.

An intectate two-layered exine is present in aerially flowering species. The exine has a reticulate, semi-reticulate or irregularly echinate surface. The sexine is thicker than the granular nexine. Pollen of these terrestrial species, however, has a thicker nexine (30% of total exine) than the amphibious species (10–20% of total exine). In amphibious taxa with internal geitonogamy (see Pollination), such as *C. heterophylla* and *C. trochlearis*, exine thickness is reduced and aperture-like areas may facilitate pollen tube germination. Amphibious taxa without internal geitonogamy (e.g. *C. cophocarpa*) have inaperturate pollen with a thinner exine. Pollen of the hypohydrophilous *C. hermaphroditica* is exineless in perennial forms of the species, and the exine is only rudimentary in annual forms. In the annual forms the exine forms either a thin, homogeneous to granular band that is generally less than 0.1 μm thick, or the exine is only present as small, irregularly shaped granules. The intine of the pollen grains in this species is well-developed. The poorly developed exine in *C. hermaphroditica* is the likely cause for the irregular shape and smooth external appearance of the pollen grains (Martinsson 1993; Philbrick and Osborn 1994; Cooper et al. 2000).

In general, terrestrial taxa have well-defined exines, amphibious taxa have thinner exines, and obligately submersed taxa have exineless pollen.

Precocious germination is reported (see Pollination).

Karyology. Chromosome numbers are rather variable in *Callitriche*, and diploid numbers range from $2n = 6$ to $2n = 40$. Based on $x = 5$, Philbrick (1994) assumed in North American species a euploid series ($2n = 10, 20, 49$) and an aneuploid reduction series ($2n = 8, 6$). For European species, $2n = 12$, $2n = 28$ and $2n = 38$ have been reported (see Federov 1969).

Pollination and Reproductive Systems. In most species, plants are proterogynous. When both staminate and pistillate flowers develop at one node, pistillate flowers become receptive prior to anther dehiscence.

A remarkable array of pollination systems has been claimed to exist in *Callitriche*: anemophily, epihydrophily (pollen transfer at the water surface), and hypohydrophily (underwater pollination; see Schotsman 1982; Philbrick and Anderson 1992). Some species of *Callitriche* are reported to have more than one pollination system (Schotsman 1982; Jäggi and Cook 1998).

In terrestrial and amphibious species with aerial flowers, geitonogamy appears to be more frequent than xenogamy (Philbrick and Anderson 1992; Jäggi and Cook 1998). Geitonogamy apparently takes place most readily when both staminate and pistillate flowers co-occur at one node, either by pollen falling from a dehiscent anther to the stylodia, or by the elongate stylodia coming into direct contact with the open anther (Osborn and Philbrick 1994).

In amphibious species, besides typical geitonogamy, a special reproductive mechanism termed internal geitonogamy occurs (e.g. *Callitriche heterophylla*, *C. verna*; Philbrick 1984). This involves pollen tube growth from within anthers of staminate flowers, traversing large amounts of vegetative tissue into the base of pistillate flowers (Philbrick and Les 2000). This is a unique self-fertilization system known only from *Callitriche* (documented in seven species, e.g. *C. heterophylla*, *C. verna*). After germination within the indehisced anther, the pollen tubes grow from the anther through the filament and then follow one of three routes: (1) they grow to and enter the base of an ovary in the same leaf axil; (2) they traverse the node and enter the base of an ovary in the opposite leaf axil; or (3) they grow within the stem and enter the base of an ovary at an adjacent node (Philbrick and Anderson 1992).

Internal geitonogamy expands the scope of ecological conditions in which seed production can occur and allows fertilization of aerial and submersed flowers. In amphibious species, only those with internal geitonogamy produce seeds in submersed flowers. Internal geitonogamy is correlated not only with the amphibious growth habit but also with polyploidy, although not all polyploid amphibious species of *Callitriche* have internal geitonogamy. This breeding system is functionally analogous to cleistogamy.

Epihydrophily has also been claimed for various amphibious species. This mode of pollination may be correlated with thinner exines and absence of apertures by decreasing the likelihood of water uptake and thereby enabling the pollen to be more buoyant for surface dispersal. Epihydrophily has been investigated in detail in *Callitriche cophocarpa* (Jäggi and Cook 1998). Anthers are lifted about 0.5 cm over the water surface and pollen grains remain stuck to the anther flaps unless they are mechanically removed by wind or water. Wind and waves are also needed to transfer the pollen over the floating leaves to the surface of the water. The unwettable pollen is mostly dispersed in clumps on the water surface and may float for three days. Around the individual leaves the surface tension of the water is lowered because their adaxial leaf surface is unwettable; this may ease the passage of pollen to receptive stigmas. Anemophily and hyphydrophily may also take place in *C. cophocarpa* (Jäggi and Cook 1998).

Hypohydrophily has been ascribed to the obligately submersed *Callitriche hermaphroditica* and *C. truncata* and also to several amphibious species. In *C. hermaphroditica*, pollen germination takes place within the locule of the dehisced anther, resulting in a tangled mass of pollen tubes. Geitonogamy seems possible, as the elongate stylodia of pistillate flowers bend to the contiguous node and thus become positioned close to staminate flowers (Osborn and Philbrick 1994). Recently, molecular data provided the first direct evidence for xenogamic hypohydrophily in *Callitriche hermaphroditica* (Philbrick 1993). However, the mechanism of pollen transfer in cross-pollination remains unclear. Confirmation of hypohydrophily in this species is in accordance with the lack of an exine and the reduction of endothecial thickenings, well known for underwater cross-pollination. The efficiency of successful pollination seems to be influenced by the geometry of the pollen grain. The more efficient, filiform shape and greater dimensions are achieved here by the precocious outgrowth of the pollen tube and the resultant, tangled mass of pollen grains.

Callitriche is the only known genus in which both anemophily and hydrophily are reported to co-occur. Anemophily is thought to be primitive within Callitrichaceae (Osborn et al. 2001).

In *Callitriche obtusangula*, it often can be observed that flies run over carpets of floating leaf rosettes. By this, they may touch emergent anthers and stigmas and contribute to pollination accidentally but effectively (Leins 2000).

Fruit and Seed. The fruit of *Callitriche* is a quadripartite schizocarp (Fig. 6C). Fruits are ca. 1–2 mm broad, round to oblong-oval in outline, and the mericarps have rounded, keeled or winged margins.

The one-seeded mericarps are drupe-like, with a very thin, fleshy outer part (which dries up quickly) and a hard inner part. Three different layers can be distinguished in the pericarp. The exocarp is formed by one layer of large, thin-walled epidermal cells. The mesocarp consists of 2–3(–5) layers of parenchymatous cells. The innermost of these layers has sclerified thickenings. The sclerified part of the mesocarp extends into the wing, when present. The sclerified cells within the wing are generally larger and more irregular than the others. The endocarp is composed of two layers of sclereids orientated crosswise to each other. The sclerified cell layer of the mesocarp and the endocarp stick together and form the stone wall of the drupe-like mericarp (Hegelmaier 1864; Jørgensen 1923; Martinsson 1991).

The reniform seeds are minute and contain an oily endosperm in which a straight or slightly curved embryo with two short cotyledons is embedded. The originally very thick integument becomes much stretched by the growth of the endosperm, resulting in a very thin testa of thin-walled, colourless, parenchymatous cells.

Dispersal. The drupe-like mericarps, whose fleshy parts dry up after maturation, show no adaptations to water dispersal. The diaspores are not appreciably lighter than water but are transported by currents due to their smallness (van der Pijl 1982). Also water-fowl contribute to their dispersal (Hegi 1925). Plants of riverbanks or ponds could easily be dispersed by becoming embedded in mud on feet or other parts of birds.

The seeds lack dormancy.

The plants can reproduce vegetatively by the principal stem dying off and the side branches developing independently; each node of the stem is able to form adventive roots.

Phytochemistry. *Callitriche* contains carbocyclic iridoids: aucubin is accompanied by traces of catalpol (Wieffering 1966; Fikenscher et al. 1969). Phenolic compounds like flavones and caffeic acid have been reported. Carbohydrates are stored as sucrose (Hegnauer 1967).

Subdivision and Relationships Within the Family. Hegelmaier (1864) subdivided the genus into two sections: sect. *Eucallitriche* containing terrestrial and amphibious taxa, and sect. *Pseudocallitriche* with obligately submersed species. Fassett (1951) proposed a third section, *Microcallitriche*, which is a subset of Hegelmaier's sect. *Eucallitriche* that contains the terrestrial species. Recent cladistic analysis of cpDNA restriction fragment variation, combined with several morphological characters, largely supports these classifications (Philbrick and Jansen 1991). Phylogenetic analyses using morphological and *rbc*L sequence characters support a two-section classification (*Callitriche*, *Pseudocallitriche*; Philbrick and Les 2000).

Affinities. In the past, the family has been considered to be related to Haloragaceae (De Candolle 1828; Hegelmaier 1864), Caryophyllaceae (Clarke 1865) or Euphorbiales (Baillon 1858). Based on embryological and phytochemical characters, most modern systems associate Callitrichaceae with the Lamiales-Scrophulariales (Takhtajan 1987; Thorne 1992). Cronquist (1981) united the Callitrichaceae, Hippuridaceae and Hydrostachyaceae into one order Callitrichales near Plantaginales and Scrophulariales. A position within the Lamiales-Scrophulariales (with a closer relationship to the Scrophulariaceae) is confirmed by the study of DNA sequences (Olmstead et al. 1992). Another molecular study (Olmstead and Reeves 1995) resolves the Callitrichaceae (together with the Hippuridaceae) as an aquatic clade within Scrophulariaceae.

Since molecular studies provide evidence that Scrophulariaceae, as circumscribed in most taxonomic treatments, are polyphyletic, this family has been subdivided into five families (Olmstead et al. 2001). The new Veronicaceae (Antirrhinaceae in Savolainen et al. 2000, see also Reveal et al. 1999), besides including traditional members of the Scrophulariaceae like *Antirrhinum*, *Digitalis*, and *Veronica*, contain the small families Plantaginaceae, Globulariaceae, Hippuridaceae and Callitrichaceae. However, support for Veronicaceae is weak, whereas support for the clade that includes *Antirrhinum* with Globulariaceae, Plantaginaceae, Hippuridaceae, and Callitrichaceae is very strong (Olmstead et al. 2001). *Callitriche* and *Hippuris* form a smaller clade within Veronicaceae; *Callitriche* is the closest relative of *Hippuris*. Whereas Veronicaceae at present lack a morphological family description, *Callitriche*/*Hippuris* share distinctive characters such as aquatic habit and very reduced floral morphology due to either wind pollination or water pollination (Olmstead and Reeves 1995; Reeves and Olmstead 1998; Olmstead et al. 2001).

Distribution and Habitats. Terrestrial species of *Callitriche* occur in seasonally wet or moist habitats. Amphibious species can grow completely submersed, stranded on moist soil at the water's edge, or as intermediate aquatic forms in which the stems extend to the water surface and develop a floating rosette of leaves. They occur in fresh or brackish water, in slow-moving streams and rivers, roadside ditches, and vernal pools. Aquatic species occur in flowing water and lakes.

The family is almost cosmopolitan in distribution (found even on Greenland and in the Antarctic), with most species indigenous to temperate zones of both hemispheres, but also in many tropical regions.

Economic Importance. *Callitriche* has no economic use.

The occurrence of some species can be used to predict the presence of particular pollutants (analyses mainly in S and SW Germany). *Callitriche obtusangula*, for example, starts to grow in primarily oligotrophic, base-rich, slow-moving water as soon as pollution takes place. The species prefers the ammonium- and phosphate-eutrophied areas of the water (Kohler et al. 1973). Growth of *Callitriche platycarpa* is probably promoted by greater amounts of nitrate (Wiegleb 1978). *Callitriche hamulata*, on the other hand, is hampered by too high nitrate- and ammonium load (Monschau-Dudenhausen 1982).

One genus:

Callitriche L. Fig. 6

Callitriche L., Syst. ed. VI: 82 (1748), Sp. Pl. 2: 969 (1753).

Characters as for the family.
(17–)40–50 ssp., cosmopolitan.

Selected Bibliography

Baillon, H. 1858. Recherches sur l'organogénie du *Callitriche* et sur ses rapports naturels. Bull. Soc. Bot. France 5: 337–341.

Behnke, H.D. 1991. Distribution and evolution of forms and types of sieve-element plastids in the dicotyledons. Aliso 13: 167–182.

Clarke, B. 1865. On the structure and affinities of Callitrichaceae. J. Bot. 3: 36–39.

Cooper, R.L., Osborn, J.M., Philbrick, C.T. 2000. Comparative pollen morphology and ultrastructure of the Callitrichaceae. Am. J. Bot. 87: 161–175.

Cronquist, A. 1981. An integrated system of classification of flowering plants. New York: Columbia Univ. Press.

Davis, G.L. 1966. Systematic embryology of the angiosperms. New York: Wiley.

De Candolle, A.P. 1828. Prodromus Systematis Naturalis Regni Vegetabilis III. Paris: Treuttel & Würtz, pp. 65–72.

Erdtman, G. 1966. Pollen morphology and plant taxonomy. New York: Hafner Publ. Co.

Fassett, N.C. 1951. *Callitriche* in the New World. Rhodora 53: 137–155, 161–182, 185–194, 209–222.

Federov, Al.A. (ed.) 1969. Chromosome numbers of flowering plants. Leningrad: Nauka.

Fikenscher, L.H., Hegnauer, R., Ruijgrock, H.W.L. 1969. Iridoide Pflanzenstoffe (Pseudoindikane) als systematische Merkmale. Pharm. Weekblad 104: 561–566.

Hegelmaier, F. 1864. Monographie der Gattung *Callitriche*. Stuttgart: Ebner & Seubert.

Hegi, G. 1925. Illustrierte Flora von Mitteleuropa V. 1. München: Lehmanns.

Hegnauer, R. 1967. Chemical characters in plant taxonomy: some possibilities and limitations. Pure Appl. Chem. 14: 173–187.

Jäggi, M., Cook, C.D.K. 1998. Reproductive biology of *Callitriche cophocarpa* Sendner (Callitrichaceae). Candollea 53: 101–115.

Jørgensen, C.A. 1923. Studies on Callitrichaceae. Bot. Tidsskr. 38: 81–126.

Kohler, A., Wonneberger, R., Zeltner, G. 1973. Die Bedeutung chemischer und pflanzlicher "Verschmutzungsindikatoren" im Fließgewässersystem Moosach (Münchener Ebene). Arch. Hydrobiol. 72: 533–549.

Leins, P. 2000. Blüte und Frucht. Aspekte der Morphologie, Entwicklungsgeschichte, Phylogenie, Funktion und Ökologie. Stuttgart: Schweizerbart'sche Verlagsbuchhandlung.

Leins, P., Erbar, C. 1988. Einige Bemerkungen zur Blütenentwicklung und systematischen Stellung der Wasserpflanzen *Callitriche*, *Hippuris* und *Hydrostachys*. Beitr. Biol. Pflanzen 63: 157–178.

Leins, P., Erbar, C. 1990. The possible relationship of Hydrostachyaceae based on comparative ontgenetical flower studies. Mitt. Inst. Allg. Bot. Hamburg 23b: 723–729.

Martinsson, K. 1991. Geographical variation in fruit morphology in Swedish *Callitriche hermaphroditica* (Callitrichaceae). Nord. J. Bot. 11: 497–512.

Martinsson, K. 1993. The pollen of Swedish *Callitriche* (Callitrichaceae) – trends towards submergence. Grana 32: 198–209.

Monschau-Dudenhausen, K. 1982. Wasserpflanzen als Belastungsindikatoren in Fließgewässern dargestellt am Beispiel der Schwarzwaldflüsse Nagold und Alb. Beih. Veröff. Naturschutz Landschaftspflege Bad.-Württ. 28: 1–118.

Olmstead, R.G., Reeves, P.A. 1995. Evidence for the polyphyly of the Scrophulariaceae based on chloroplast *rbc*L and *ndh*F sequences. Ann. Missouri Bot. Gard. 82: 176–193.

Olmstead, R.G., Michaels, H.J., Scott, K.M., Palmer, J.D. 1992. Monophyly of the Asteridae and identification of their major lineages inferred from DNA sequences of *rbc*L. Ann. Missouri Bot. Gard. 79: 249–265.

Olmstead, R.G., DePamphilis, C.W., Wolfe, A.D., Young, N.D., Elisons, W.J., Reeves, P.A. 2001. Disintegration of the Scrophulariaceae. Am. J. Bot. 88: 348–361.

Osborn, J.M., Philbrick, C.T. 1994. Comparative pollen structure and pollination biology in the Callitrichaceae. Acta Bot. Gallica 141: 257–266.

Osborn, J.M., El-Ghazaly, G., Cooper, R.L. 2001. Development of the exineless pollen wall in *Callitriche truncata* (Callitrichaceae) and the evolution of underwater pollination. Plant Syst. Evol. 228: 81–87.

Philbrick, C.T. 1984. Pollen tube growth within vegetative tissues of *Callitriche* (Callitrichaceae). Am. J. Bot. 71: 882–886.

Philbrick, C.T. 1993. Underwater cross-pollination in *Callitriche hermaphroditica* (Callitrichaceae): evidence from random amplified polymorphic DNA markers. Am. J. Bot. 80: 391–394.

Philbrick, C.T. 1994. Chromosome counts for *Callitriche* (Callitrichaceae) in North America. Rhodora 96: 383–386.

Philbrick, C.T., Anderson, G.J. 1992. Pollination biology in the Callitrichaceae. Syst. Bot. 17: 282–292.

Philbrick, C.T., Jansen, R.K. 1991. Phylogenetic studies of North American *Callitriche* (Callitrichaceae) using chloroplast DNA restriction fragment analysis. Syst. Bot. 16: 476–491.

Philbrick, C.T., Les, D.H. 2000. Phylogenetic studies in *Callitriche*: implications for interpretation of ecological, karyological and pollination system evolution. Aquat. Bot. 68: 123–141.

Philbrick, C.T., Osborn, J.M. 1994. Exine reduction in underwater flowering *Callitriche* (Callitrichaceae): implications for the evolution of hypohydrophily. Rhodora 96: 370–381.

Pijl, L. van der 1982. Principles of dispersal in higher plants. 3rd ed. Berlin, Heidelberg, New York: Springer.

Reeves, P.A., Olmstead, R.G. 1998. Evolution of novel morphological and reproductive traits in a clade containing *Antirrhinum majus* (Scrophulariaceae). Am. J. Bot. 85: 1047–1056.

Reveal, J.L., Judd, W.S., Olmstead, R.G. 1999. Proposal to conserve the name Antirrhinaceae against Plantaginaceae (Magnoliophyta). Taxon 48: 182.

Savolainen, V., Fay, M.F., Albach, D.C., Backlund, A., van der Bank, M., Cameron, K.M., Johnson, S.A., Lledo, M.D., Pintaud, J.-C., Powell, M., Sheahan, M.C., Soltis, D.E., Soltis, P.S., Weston, P., Whitten, W.M., Wurdack, K.J., Chase, M.W. 2000. Phylogeny of the eudicots: a nearly complete familial analysis based on *rbc*L gene sequences. Kew Bull. 55: 257–309.

Schotsman, H.D. 1982. Biologie florale des *Callitriche*: étude sur quelques espèces d'Espagne méridionale. Adansonia II, 3–4: 111–160.

Takhtajan, A. 1987. Systema Magnoliophytorum (in Russian). Leningrad: Nauka.

Thorne, R.F. 1992. Classification and geography of the flowering plants. Bot. Rev. 58: 225–348.

Wieffering, J.H. 1966. Aucubinartige Glucoside (Pseudoindikane) und verwandte Heteroside als systematische Merkmale. Phytochemistry 5: 1053–1064.

Wiegleb, G. 1978. Untersuchungen über den Zusammenhang zwischen hydrochemischen Umweltfaktoren und Makrophytenvegetation. Arch. Hydrobiol. 83: 443–484.

Carlemanniaceae

M. Thiv

Carlemanniaceae Airy Shaw, Kew Bull. 19: 511 (1965).

Perennial herbs or shrubs. Leaves opposite, petiolate, simple, elliptic to ovate, acuminate, pinnately nerved, dentate or crenate-serrate, estipulate. Flowers in axillary or terminal cymes, pedicellate or (sub)sessile, hermaphrodite, zygomorphic to subregular, sometimes heterostylous. Calyx tube adnate to the ovary; lobes 4 or 5, unequal, persistent, linear to spathulate. Corolla sympetalous, tubular, funnelform or campanulate, 4- or 5-merous, sometimes with hairs inside the tube; lobes imbricate or induplicate-valvate, ovate, rounded. Stamens two, inserted at the middle of corolla tube, alternating with the petals; filaments short, anthers oblong-linear, opening laterally by slits, connivent around the style, latrorse, dorsifixed; pollen 3- or 5–6-colpate. Disc conspicuous, cylindrical or conical, epigynous. Ovary inferior, bicarpellate, syncarpous, bilocular, style filiform, stigma clavate, ovules numerous, on subglobose axile placentas at the middle of the septum or on placentas ascending from the inner base of each loculus; ovules anatropous, unitegmic, tenuinucellate. Fruit a bilocular, dry or fleshy capsule, 2- or 5-valvate. Seeds numerous, ovoid, smooth, with endosperm.

A family of two genera and five species in tropical Asia.

Vegetative Anatomy. Glandular hairs occur on all external surfaces and parts of the corolla. Glands on vegetative and floral parts, however, differ from each other. Whereas vegetative glands consist of a short unicellular stalk and a mostly 4-celled head, glandular corolla hairs have a long unicellular stalk and a multicellular head (Solereder 1893; Tange 1998).

Anatomical studies of *Carlemannia* were conducted by Solereder (1893). This genus is characterised by star-shaped crystals in the palisade parenchyma, simple vessel perforations, starch-containing wood prosenchyma, and isolated groups of bast fibres between the primary cortex and the bast. Solereder (1893) observed that the outer cell layer of the primary cortex of *Carlemannia* becomes the phellogen. In contrast to Rubiaceae, typically having paracytic stomata, Solereder (1893) described diacytic stomata for *Silvianthus.*

Pollen Morphology. The palynology of Carlemanniaceae was studied by Majumdar and Chanda (1978). Both genera have oblate pollen grains with a broadly reticulate, intectate exine structure. They mainly differ in the number of apertures, *Silvianthus* having tricolpate and *Carlemannia* 5–6-colpate pollen grains.

Reproductive Systems and Pollination. Data on the reproductive systems of *Silvianthus tonkinensis* were published by Tange (1998). Within the family, heterostyly has only been reported for this species. It is self-incompatible and has distylous flowers with different pollen sizes. Tange (1998) found that only few flowers of an inflorescence open at the same time. Anthesis starts in the early morning, and lasts for one day. During anthesis, copious nectar is produced by the disc. These observations, together with the presence of a yellow, tubular-funnelform corolla and heterostyly, may indicate entomophily (bees) for this species.

Fruit, Seed and Dispersal. The fruit morphology of the two genera differs remarkably. *Silvianthus* has a white, fleshy capsule which opens by 5 irregular, longitudinal valves, each terminated by a calyx lobe. As a result, the placentas with the mature seeds are presented (Tange 1998). Fruits of *Carlemannia* are dry, 2-valvate capsules with loculicidal dehiscence (Fig. 7C). The seed morphology of *Carlemannia* and *Silvianthus* was described by Fedotova (1996) and Tange (1998), respectively. Seeds of Carlemanniaceae are ovoid and have a small embryo, which is located centrally or towards one end. The copious endosperm stores large quantities of oil. In *Carlemannia* the presence of proteins has also been reported. Main differences between the seed morphology of the two genera concern the shape of exotesta cells, being isodiametric-polygonal with all walls thick-

Fig. 7. Carlemanniaceae. **A–C** *Carlemannia tetragona*. **A** Stem with leaves and inflorescences. **B** Stem with leaves and infructescences. **C** Fruit. **D, E** *Silvianthus bracteatus*. **D** Stem with leaves and inflorescence. **E** Stem with leaves and infructescence. (Wu and Chen 1991)

ened in *Carlemannia*, and narrow and thickened radial walls in *Silvianthus*. Furthermore, seeds of *Silvianthus* differ from those of *Carlemannia* by having additional layers of parenchymatic cells below the exotesta and a row of flattened cells between ridges bordering the endosperm, which represent the former endothelium cells. According to Tange (1998), the seeds of *Silvianthus* are probably dispersed by birds which are attracted by the colour contrast between the black seeds and the white fruit.

Phytochemistry. According to Bremer et al. (2001), cephalarosides, aucubins and alkaloids could not be detected in Carlemanniaceae.

Affinities. Originally placed as a morphological aberrant group into Rubiaceae by Bentham (1853), this opinion was followed by, for example, Hooker (1873) and Schumann (1897). Solereder (1893) concluded from his anatomical studies that Carlemanniaceae could be related to Caprifoliaceae. Also Bremekamp (1939), Kern and van Steenis (1951) and Hutchinson (1967) subscribed to this view. Hallier (1910a, 1910b), however, suggested that Carlemanniaceae are closely related to Gesneriaceae or Saxifragaceae. Bremekamp (1939) suggested a closer affinity to Valerianaceae, based on the shared presence of a reduced number of stamens. Moreover, Airy Shaw (1965) discussed a relationship of Carlemanniaceae to Verbenaceae. APG (1998) placed the Carlemanniaceae unassigned to order in the Euasterids II clade. The molecular and morphological analyses by Savolainen et al. (2000) and Bremer et al. (2001) indicated an inclusion of Carlemanniaceae into Lamiales of the Euasterids I, unresolved at the base of this order, or as weakly supported sister group to Oleaceae. Such a relationship would be supported by the presence of only 2 stamens in these two families.

Distribution and Habitats. The distribution area of Carlemanniaceae covers the tropical part of SE Asia from Assam to Sumatra. The species mainly grow in montane forests or along streams in evergreen hillside forests (Kern and van Steenis 1951; Tange 1998).

Key to the Genera

1. Herbs; flowers tetramerous, in lax corymbose inflorescences; pedicels 0.5–2 cm long, corolla lobes imbricate; fruit a 2-valvate dry capsule **1. *Carlemannia***
– Shrubs; flowers pentamerous, in dense inflorescences, (sub)sessile; corolla lobes induplicate-valvate; fruit a 5-valvate fleshy capsule **2. *Silvianthus***

1. *Carlemannia* Benth. Fig. 7A–C

Carlemannia Benth. in Hook. J. Bot. 5: 307 (1853).

Perennial herbs. Leaves crenate-serrate; flowers tetramerous, in lax corymbose inflorescences (Fig. 7A); pedicels 0.5–2 cm long, corolla lobes imbricate; pollen grains 5–6-colpate; placentas ascending from the inner base of each loculus, fruit a 2-valvate dry capsule. Three spp., in eastern Himalayas, Assam, Burma, Yunnan, Sumatra.

2. *Silvianthus* Hook. f. Fig. 7D, E

Silvianthus Hook. f. in Hook. Ic. Pl. 11: 36, t. 1048 (1868).

Shrubs. Leaves dentate; flowers pentamerous, in dense inflorescences, (sub)sessile (Fig. 7D); corolla lobes induplicate-valvate; pollen grains tricolpate;

placentas at the middle of the septum, fruit a 5-valvate fleshy capsule. Two spp., in Assam, Burma, Yunnan, Laos, Thailand, Vietnam.

Selected Bibliography

Airy Shaw, H.K. 1965. A new species of the genus *Silvianthus* Hook. f., and on the family Carlemanniaceae. Kew Bull. 19: 507–512.

APG 1998. An ordinal classification for the families of flowering plants. Ann. Missouri Bot. Gard. 85: 531–553.

Bentham, G. 1853. On three new genera connected with the Indian Flora. J. Bot. (Hooker) 5: 304–309.

Bremekamp, C.E.B. 1939. On the position of the genera *Carlemannia* Benth. and *Silvianthus* Hook. f. Recueil Trav. Bot. Neerl. 36: 372.

Bremer, K., Backlund, A., Sennblad, B., Swenson, U., Andreasen, K., Hjertson, M., Lundberg, J., Backlund, M, Bremer, B. 2001. A phylogenetic analysis of 100+ genera and 50+ families of euasterids based on morphological and molecular data with notes on possible higher level morphological synapomorphies. Plant Syst. Evol. 229: 137–169.

Fedotova, T.A. 1996. Morphology of the fruit and seed of *Carlemannia* species (Carlemanniaceae). Bot. Zhurn. (Moscow) 81: 24–34.

Hallier, H. 1910a. Über *Juliania*, eine Terebinthaceengattung mit Cupula, und die wahren Stammeltern der Katzenblütler. Just's Bot. Jahresber. 36, 3: 217–227.

Hallier, H. 1910b. Über Phanerogamen von unsicherer oder unrichtiger Stellung. Meded. Rijks-Herb. 1: 1–41.

Hooker, J.D. 1873. Rubiaceae. In: Bentham, G., Hooker, J.D. Genera Plantarum vol. 2. London: Reeve, pp. 7–151.

Hutchinson, J. 1967. The genera of flowering plants, vol. 2. Oxford: Clarendon, pp. 86.

Kern, J.H., Steenis, C.G.G.J. van 1951. Caprifoliaceae. In: Steenis, C.G.G.J. van (ed.) Flora Malesiana I, 4. Djakarta: Noordhoff-Kolff, pp. 175–194.

Majumdar, S., Chanda, S. 1978. Pollen morphology and taxonomy of *Carlemannia* and *Silvianthus* of the family Rubiaceae s.l. Trans. Bose Res. Inst. Calcutta 41: 99–105.

Savolainen, V., Fay, M.F., Albach, D.C., Backlund, A., Bank, M. van der, Cameron, K.M., Johnson, S.A., Lledó, M.D., Pintaud, J.-C., Powell, M., Sheahan, M.C., Soltis, D.E., Soltis, P.S., Weston, P., Whitten, W.M., Wurdack K.J., Chase, M.W. 2000. Phylogeny of the eudicots: a nearly complete familial analysis based on *rbc*L gene sequences. Kew Bull. 55: 257–309.

Schumann, K. 1897. Rubiaceae. In: Engler & Prantl, Die natürlichen Pflanzenfamilien IV, 4, pp. 1–156.

Solereder, H. 1893. Ein Beitrag zur anatomischen Charakteristik und zur Systematik der Rubiaceae. Bull. Herb. Boissier 1: 167–183.

Tange, C. 1998. *Silvianthus* (Carlemanniaceae) a genus and family new to Thailand. Thai For. Bull. Bot. 26: 59–65.

Wu, C.Y., Chen, J. 1991. Flora Yunnanica, vol. 5. Beijing: Science Press.

Cyclocheilaceae

Sebsebe Demissew

Cyclocheilaceae Marais, Kew Bull. 35: 805 (1981).

Subshrubs to shrubs with simple and glandular hairs. Leaves simple, entire, opposite, subopposite, fasciculate on short shoots; stipules absent. Flowers bisexual, solitary, on short shoots from axils of stem leaves, sessile, but with a pedicel-like axis below bracts. Bracts and prophylls present, prophylls membranous, conspicuous, ± free or joined. Calyx absent or reduced to disc-like rim. Corolla gamopetalous, tubular, infundibuliform, 5-lobed. Stamens 4, didynamous, inserted on corolla tube, filaments pilose, anther thecae 2, opening longitudinally. Style simple, curved, stigma lingulate or spathulate. Ovary superior, compressed, 1–2-locular, ovules 2–10 on long funiculi. Placentation axile or parietal. Fruit capsular or schizocarpic, compressed, discoid. Seeds without endosperm.

The family contains two genera and four species mainly from the horn of Africa. *Asepalum* extends into wider parts of Africa.

Flower Structure. The calyx of the flower is usually interpreted as being missing or reduced to a small, swollen rim at the base of the corolla (Fig. 8E), rather than treating the prophylls as a pair of sepals. A study of floral vascularization could possibly resolve this question. According to Junell (1934), the ovary is composed of two median-placed carpels with one to several ovules in each. These are basal or superposed and anatropous to epitropous, with elongate funicles. Junell (1934) pointed out the similarity in gynoecial structure between Cyclocheilaceae and Nesogenaceae, and of both to Stilbaceae.

Inflorescence. Various interpretations can be offered for the structure of the flower and the short shoot on which it is borne. The shoot could be interpreted as a reduced, single-flowered cyme, in which the stem leaf would then be the bract, and with two pairs of bracts, the lower pair small and inconspicuous and the upper pair (prophylls) larger. However, there is no direct evidence of a cymose origin, in which case it is more appropriate to treat the structure simply as a short shoot bearing a terminal flower with a lower pair of small bracts and an upper larger pair.

Pollen Morphology. According to Raj (1985), the pollen grains are usually tricolpate, rarely tetracolpate, their diameters (equatorial) vary in the range 25–34 μm, their length (polar) being 31–39 μm. They are prolate spheroidal to prolate, occasionally syncolpate at one pole. Membranes are thin, densely granular, irregularly ruptured. Exine tectate perforate to microreticulate. The sculpture consists of 5–8 elements which are triangular, angular or rounded, and are either loosely or compactly arranged on the tectum around a perforation. The pollen grains show strong resemblance to the Australian genus *Nesogenes* in Nesogenaceae, but not to Acanthaceae, Scrophulariaceae, Verbenaceae nor Labiatae, as previously assumed.

Fruit, Seed and Dispersal. Fruits are either loculicidal capsules or schizocarps. The prophylls are enlarged during the fruiting stage and enclose the fruits (Fig. 8B). This entire structure falls off easily and acts as the dispersal unit. The enlargement of the prophylls increases the surface area of the dispersal unit which helps dispersal by wind, which is prevalent in the dry open habitats where these plants grow. The wind-dispersed diaspores of the various species of the family are comparable to dispersal mechanisms found in, for example, *Lippia carviodora* (Verbenaceae) and the genus *Capitanya* (Lamiaceae) which also are adapted to similar habitats (Sebsebe Demissew and Puff 1989). The seeds lack endosperm.

Distribution and Habitat. The family is restricted to the Somali-Masai regional centre of endemism in eastern and north-eastern Africa, extending to Arabia. The three species of *Cyclocheilon* (*C. kelleri*, *C. physocalyx* and *C. somalense*) are known from the northern parts of Somalia, and *C. kelleri* also occurs in southern and eastern Ethiopia. They mainly grow in open woodland or bushland on red, sandy limestone soils at

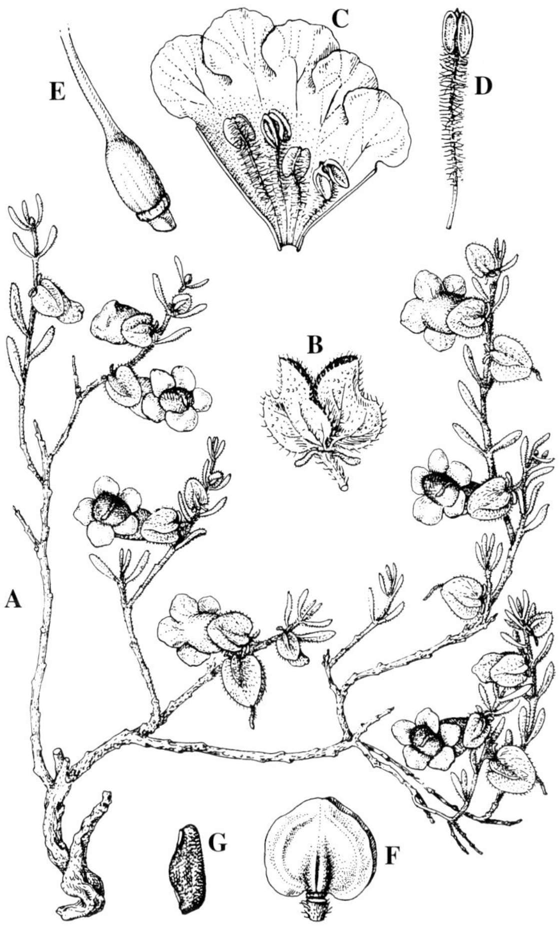

Fig. 8. Cyclocheilaceae. *Cyclocheilon physocalyx.* **A** Flowering branch. **B** Prophylls. **C** Corolla opened. **D** Stamen. **E** Ovary enlarged, showing calyx rim. **F** Fruit. **G** Seed. (Marais 1981)

250–1875 m. The monotypic *Asepalum* with *A. eriantherum* occurs in *Acacia-Commiphora* bushland and *Combretum-Terminalia* woodland at 800–1650 m. It covers wider areas (compared to the three species in *Cyclocheilon*) in eastern and north-eastern Africa, extending from northern Tanzania, Kenya, eastern Uganda, Ethiopia, Somalia to Yemen.

Affinities. *Cyclocheilon* was originally described by Oliver (1895, 1896) in the Scrophulariceae-Gerardieae. The genus *Asepalum* was originally described as a species of *Tinnea* (Labiatae: Scutellarioideae, see this volume) by Vatke (1882), before being transferred to *Cyclocheilon* by Engler (1897). Stapf (1901) transferred *Cyclocheilon* to Verbenaceae and later, Airy Shaw (1966) removed it to Dicrastylidaceae, a segregate of Chloanthaceae, which is now Labiatae subfam. Prostantheroideae tribe Chloantheae (see this volume). Junell (1934) suggested a relationship to Acanthaceae rather than Verbenaceae, but also pointed out a close similarity in gynoecial structure with Nesogenaceae and Stilbaceae. Marais (1981) also believed *Cyclocheilon* and *Asepalum* to be related to Acanthaceae, based on overall morphology, but created a new family, Cyclocheilaceae, for them. Recently, *rbc*L sequencing has been carried out on *Cyclocheilon* and, according to Olmstead (pers. comm.), the genus seems to fit best within Orobanchaceae (see R.G. Olmstead and B. Bremer, in prep.). Recent molecular studies on Scrophulariaceae and the parasitic Orobanchaceae suggest that a major reorganisation of these families is needed (Olmstead and Reeves 1995; Young et al. 1999; Olmstead et al. 2001). At present there exists no evidence that Cyclocheilaceae are parasitic, as are other members of the Orobanchaceae.

Uses. According to Marais (1981), the roots of *Cyclocheilon physocalyx* and *Asepalum eriantherum* have been used to dye leather red.

Key to the Genera

1. Ovary and fruit 1-locular or incompletely 2-locular; fruit loculicidal; bracts inserted at the same level as the prophylls; prophylls hairy **1. *Cyclocheilon***
– Ovary and fruit 2-locular; fruit breaking into two mericarps at maturity; bracts inserted below the prophylls; prophylls glabrous **2. *Asepalum***

1. *Cyclocheilon* Oliv. Fig. 8

Cyclocheilon Oliv., Bull. Misc. Inf. Kew 1895: 222 (1895).

Small subshrubs, branching from base; stems whitish to purplish, pubescent; leaves covered with scabrid hairs, estipulate; bracts 2, inserted ± at the same level as, and opposite to conspicuous prophylls, pubescent; prophylls 2, rotund, much larger than bracts, pubescent, free or fused towards the often slightly keeled base, accrescent in fruit. Ovary ± ovoid, laterally compressed, 1-locular or incompletely 2-locular with 1–5 ovules in each locule. Style curved, stigma spathulate, c. 1/3 of style length. Fruit capsular, rounded, loculicidal, 4-seeded. Seeds curved, bean-shaped, reticulate. Three species, *C. kelleri* Engl., *C. physocalyx* Chiov. and *C. somalensis* Oliv. In northern Somalia and eastern Ethiopia.

2. *Asepalum* Marais

Asepalum Marais, Kew Bull. 35: 809 (1981).

Branching shrub with brittle stems, stems whitish or pale greyish, young parts with strigose hairs, older parts glabrescent; leaves scabrid, readily caducous; flowers sessile, appearing fasciculate on young flowering shoots, shoots elongating during fruiting; bracts 2, small, in the upper half of the pedicel below the prophylls, opposite or alternate; prophylls 2, conspicuous, glabrous, deeply cordate, fused at the base, accrescent. Ovary 2-locular, with one basal ovule in each loculus on a long sigmoid funicle. Stigma lingulate, as long as style. Fruit a schizocarp, obcordate, with remains of the style at the apex, breaking into 2 mericarps, each mericarp with ridge where seeds develop. Seeds obovoid, glabrous, with pronounced shoulder on one side where mericarps break up. One species, *A. eriantherum* (Vatke) Marais, known from northeastern and eastern parts of Africa (Ethiopia, Somalia, Kenya, Tanzania and Uganda) and Arabia (Yemen).

Selected Bibliography

Airy Shaw, K. 1966. In: Willis, J.C. (ed.) Dictionary of flowering plants and ferns, 7th edn. Cambridge: Cambridge University Press.

Engler, A. 1897. Contribuzioni alla conoscenza della flora dell'Africa orientali, VIV. Annuar. Istit. Bot. Roma 7: 13–31.

Junell, S. 1934. Zur Gynäceummorphologie und Systematik der Verbenaceen und Labiaten nebst Bemerkungen über ihre Samenentwicklung. Symb. Bot. Ups. 4: 1–219.

Marais, W. 1981. Two new gamopetalous families, Cyclocheilaceae and Nesogenaceae, for extra-Australian Dicrastylidaceae. Kew Bull. 35: 791–812.

Marais, W. 1984. Cyclocheilaceae. In: Polhill, R.M. (ed.) Flora Trop. East. Afr. Rotterdam: A.A. Balkema.

Oliver, D. 1895. Diagnoses Africanae, VII. Bull. Misc. Inf. Kew: 222–223.

Oliver, D. 1896. Hooker's Icones Pl., t. 2439.

Olmstead, R.G., Reeves, P.A. 1995. Evidence for the polyphyly of the Scrophulariaceae based on chloroplast *rbc*L and *ndh*F sequences. Ann. Missouri Bot. Gard. 82: 176–193.

Olmstead, R.G., dePamphilis, C.W., Wolfe, A.D., Young, N.D., Elisons, W.J., Reeves, P.A. 2001. Disintegration of the Scrophulariaceae. Am. J. Bot. 88: 348–361.

Raj, B. 1985. A contribution to the pollen morphology of Nesogenaceae and Cyclocheilaceae. Pollen Spores 27: 295–306.

Sebsebe Demissew, Puff, C. 1989. Observations on *Lippia carviodora*, a dioecious Verbenaceae. Beitr. Biol. Pflanzen 64: 221–230.

Stapf, O. 1901. Hooker's Icones Pl., t. 2685.

Vatke, W. 1882. Plantae in itinere africano ab J.M. Hildebrandt collectae. Oesterr. Bot. Zeitschr. 43: 507–541.

Young, N.E.D., Steiner, K.E., dePamphilis, C.W. 1999. The evolution of parasitism in the Scrophulariaceae: plastid gene sequences refute an evolutionary transition series. Ann. Missouri Bot. Gard. 86: 876–893.

Gesneriaceae

A. Weber

Gesneriaceae [Rich. & Juss. ex] DC., Essai propr. méd. pl., ed. 2: 192 (1816) ("Gessnerieae"), nom. cons.

Perennial or rarely annual herbs, subshrubs, shrubs or rarely small trees; perennial herbs with fibrous roots or with rooting above- or underground stems, rootstocks, rhizomes, scaly rhizomes, or tubers; terrestrial, epiphytic or climbing. Stem erect, ascending, decumbent, creeping, pendulous, or ± absent. Leaves opposite, sometimes in whorls of three or four, or in near-distichous or spiral-alternate arrangement; usually petiolate; stipules absent; lamina usually undivided, rarely lobed or pinnately dissected. Number of leaf pairs sometimes reduced to the cotyledonary pair, with one of the two cotyledons growing up to a large, foliar organ. Indumentum of stem and leaves of glandular and eglandular hairs, rarely absent. Inflorescences a foliose or (rarely) bracteose indeterminate thyrse with axillary pair-flowered cymes; cymes sometimes reduced to solitary flowers; bracteolate or rarely ebracteolate. Flowers usually showy, zoophilous, rarely auto- or cleistogamous, 5- (rarely 4-)merous. Sepals free or connate to a variable extent, equal or unequal and then calyx two- or three-lipped. Corolla sympetalous, zygomorphic with bilabiate limb, rarely actinomorphic; shape variable, long-tubular, infundibuliform, funnel-shaped, campanulate to flat-rotate; white, blue, violet, red, orange, yellow, greenish, brownish, or various colours combined, with or without markings. Stamens mostly 4 or 2, rarely (in the actinomorphic flowers) 5; 1 or 3 staminodes usually present. Filaments always adnate to the corolla, inserted at the base, the middle or near the mouth of the corolla, either short or long and slender, in the tetrandrous species often didynamous; anthers of variable shape, basi- or (rarely) dorsifixed, dithecal, opening by longitudinal slits or by apical (rarely basal) pores, often cohering at apex or face to face, in the tetrandrous-didynamous species the anthers cohering in pairs or all four, included in the corolla, less commonly exserted. Nectary either adnate to the ovary base or free; annular or one-sided or divided into separate glands, sometimes absent. Ovary syncarpous, composed of two median carpels; mostly unilocular with lateral, T-shaped placentae; rarely bilocular with axile placentae, rarely one carpel sterile; ovary superior, semi-inferior or inferior; style usually well developed; stigma capitate, stomatomorphic, bilabiate or variously bifid; ovules numerous, anatropous, unitegmic, tenuinucellate. Fruit dehiscent (a dry or fleshy capsule) or indehiscent; dry capsules elongate, ovoid or globose, dehiscing septicidally or loculicidally by two or septicidally plus loculicidally by four valves or by a single dorsal slit, rarely irregularly or circumscissile; 2-valved capsules twisted in some genera; fleshy capsules often showy and exposing a central cone of seeds, placental tissue and/or funicles (display capsules); indehiscent fruits either hard or fleshy (berry); berries sometimes vividly coloured. Seeds numerous, minute, usually oval or polygonal in outline; dispersed by wind, water (rain), birds, or ants. Testa frequently striate or reticulate, often with various additional ornamentations, rarely with short or hair-like axial appendages. Endosperm single- or several-layered. Embryo with two equal cotyledons.

A family comprising between 140 and 150 genera and more than 3500 species, distributed mainly in the tropics and subtropics both of the Old and the New World, with transgressions to the north (Europe: Pyrenees, Balkan Peninsula; Asia: Himalayas, China incl. N China) and the south (SE Australia, New Zealand, S Chile).

Seedling Structure. As in other dicotyledons, the embryo consists of two equal cotyledons, hypocotyl, and radicle. Germination is phanerocotylar, the two unfolding cotyledons being the first photosynthetic organs. In the Coronantheroid (southern hemisphere) and Gesnerioid (neotropical) Gesneriaceae, seedling development corresponds to the usual pattern with equally developing and short-lived cotyledons (isocotyly) and with a normal shoot apical meristem between. In the palaeotropical Epithematoid and Didymocarpoid Gesneriaceae, the two cotyledons show unequal growth (see below), the development of a shoot apical meristem is often delayed, and in

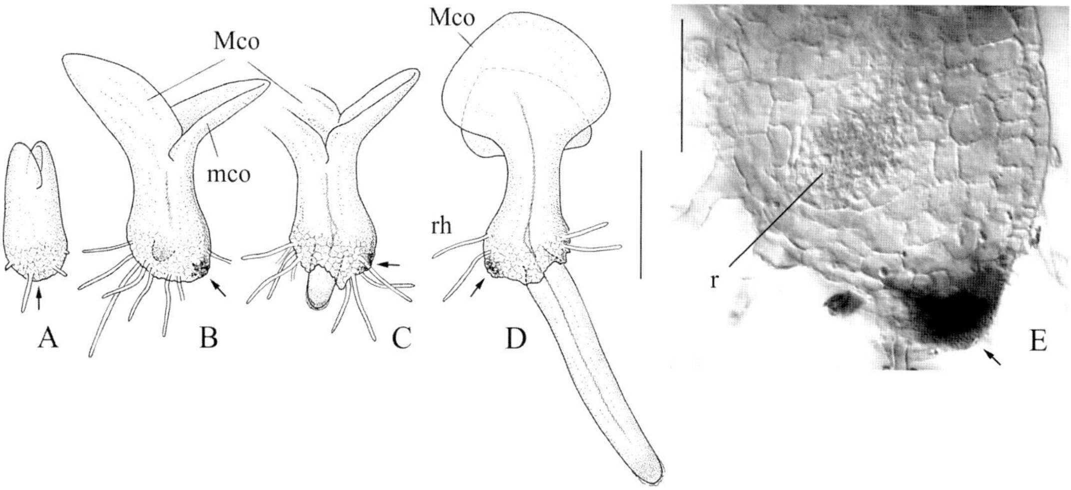

Fig. 9. Gesneriaceae. Seedling development of *Rhynchoglossum gardneri*. **A** Embryo after release from the testa; note equal cotyledons and first rhizoids (*rh*) emerging from the hypocotyl base. **B–D** Differentiation of macro- (*Mco*) and microcotyledon (*mco*), emergence of the first root (note endogenous origin). **E** Phase-contrast photo of hypocotyl base with endogenous first root. *Arrows* Site of former suspensor attachment. *Bar* 0.5 mm (**A–D**), 0.1 mm (**E**). (Weber 1978a)

some taxa an ordinary shoot apical meristem in the form of a meristematic dome does not develop at all. Prior to radicle development, the epidermal cells of the hypocotyl base grow out to hair-like or tubular rhizoids ("myzotriches", Haccius and Troll 1961). These may form a distinct collar between the hypocotyl base and the radicle (*Aeschynanthus*), or may be evenly distributed. They serve as anchor and adsorption organs for the young seedling.

In a number of palaeotropical taxa there are remarkable modifications in seedling development. In *Streptocarpus* a successive delay of the radicle development can be observed (Jong 1970). In *Rhynchoglossum gardneri* (Weber 1978a) and *Monophyllaea singularis* (Imaichi et al. 2001) the radicle does not develop into a primary root any more: the first root is an endogenous lateral root breaking through the hypocotyl base (Fig. 9).

As first observed in *Streptocarpus* (Caspary 1858; Crocker 1860), and later in many more palaeotropical Gesneriaceae (Fritsch 1904; Hill 1938), the two cotyledons show unequal development after germination: one, the macrocotyledon, grows faster and often develops into a foliar organ (Figs. 9B–D, 14A–C). In *Monophyllaea* (most species), *Acanthonema*, *Trachystigma* and in the "unifoliate" species of *Streptocarpus* (and sometimes facultatively in species of *Epithema*, *Chirita* sect. *Microchirita*, etc.), the macrocotyledon is the only foliar organ of the plant and shows continuous growth (Fig. 12). In contrast, the smaller cotyledon (microcotyledon) stops growth very early, withers and is without function (Fig. 14B, C). The phenomenon of "anisocotyly" (Fritsch 1920) is characteristic of palaeotropical Gesneriaceae and is the basis for recognition of subfamily Cyrtandroideae (Burtt 1963). The question whether the macrocotyledon is genetically determined, or is induced by gravity, light or other factors has been clarified recently by experiments with *Streptocarpus* and *Chirita* seedlings (Saueregger and Weber, in press): at (and a short time after) germination both cotyledons have the potential to develop into a macrocotyledon. Determination is by light: the cotyledon receiving more light in early stages is the one that develops into the macrocotyledon.

In *Streptocarpus* (Schenk 1943; Jong 1970, 1978; Jong and Burtt 1975) and *Monophyllaea singularis* (Weber 1987, 1990; Imaichi et al. 2001), the two cotyledons are separated by the intercalation of a "mesocotyl", by which the macrocotyledon is elevated above the microcotyledon. In *Streptocarpus* the meristem situated at the base of the macrocotyledon ("groove meristem") elongates, later producing inflorescences and (in the plurifoliate species) "phyllomorphs" of the same structure as the seedling (but without the equivalent of a microcotyledon). In *Monophyllaea singularis*, however, the "groove meristem" loses its meristematic properties and the inflorescences are produced adventitiously (see Exceptional Morphologies).

Stem, Leaves and Roots. The stem may be well developed, bearing leaf pairs or alternate leaves at marked intervals (caulescent habit, e.g. Fig. 14F) or short to acaulescent with contracted internodes and ± densely aggregated leaves. In the latter group, a distinction can be made between proper rosette plants, with the (usually spirally arranged) leaves forming a dense, plane, basal rosette (Fig. 14E), and "rosulate" plants with rather loosely tufted leaves. The stem may be thick or thin, erect or creeping, branched or unbranched. Production of side branches is from axillary buds; adventitious (serial) branches may also occur. In some taxa, axillary stolons with elongated internodes are produced (see Growth Patterns).

The leaves are estipulate, but in *Rhytidophyllum tomentosum* and *R. auriculatum* stipule-like "auricles" are formed at the petiole base (Weber 1973a, Skog 1976). A petiole is usually present; sessile leaves are rare (e.g. the solitary leaves of *Monophyllaea* and unifoliate *Streptocarpus*). The lamina is usually simple, the form ranging from linear (e.g. *Aeschynanthus stenophyllus*) through lanceolate, elliptic, ovate, and cordate to almost orbicular. The margin can be entire, crenate, dentate or serrate. Sinuate and deeply lobed leaves are found in some species of *Ridleyandra* (Weber and Burtt 1998c) and *Chirita* (*C. juliae*, *C. pinnatifida*). In extreme cases the incisions reach the midrib: the pectinate leaves of *Henckelia pectinata* and the pinnately lobed leaves of *Chirita pinnata* (Wang et al. 1992) and *Cremosperma filicifolium* (Kvist and Skog 1988). Peltation occurs occasionally (e.g. *Drymonia peltata*, *Sinningia tuberosa*, *Cyrtandra peltata*, *Metapetrocosmea peltata*).

The lamina surface may be plane and smooth, or mamillate to micromamillate, the mamillae appearing as invaginations when seen from below. In some taxa (e.g. *Episcia*, *Cyrtandra splendens*) the mamillae are persistent, in others they are flattened by diffuse laminal growth (e.g. *Petrocosmea heterophylla*).

Opposite (decussate) leaf arrangement prevails and is the primary condition in the family. Whorls of three sometimes occur as an abnormal condition or can be a consistent character (e.g. *Cyrtandra* sect. *Verticillatae*, *Didissandra ternata*, *Gasteranthus ternatus*, several species of *Achimenes*, *Kohleria*, *Sinningia* and *Chirita*. (Pseudo)whorls of four or more leaves are rare; examples include *Sinningia leucotricha*, *Columnea fritschii*, *C. ultraviolacea*, *Aeschynanthus speciosus* and *A. angustifolius*. In the latter species, leaf number increases gradually along the successive stem nodes, combined with a change of leaf form from ovate to linear (heterophylly). Especially in rosette plants, a slight change of the divergence angle as well as asymmetrical growth of the leaves result in avoidance of an overlap of successive leaf pairs. As a result, rosettes and tufts often look as if composed of spirally arranged leaves, but the leaves are in fact opposite (e.g. *Ridleyandra*, *Henckelia* p.p.).

Unequal size of opposite leaves (anisophylly) is frequent (Fig. 11A–D). Pronounced anisophylly is combined with leaf asymmetry and changes in the divergence angles ["*Goldfussia*" (= *Strobilanthes*) type of anisophylly; Goebel 1928; Troll 1937]. In strongly anisophyllous plants the smaller leaves are often reduced to small, stipule-like appendages placed in two adjacent ranks on the upper side of the shoot axis, while the larger and strongly asymmetrical leaves are arranged in two nearly opposite ranks. The latter usually turn their laminas into a single plane so that the whole shoot looks like a pinnate leaf with alternately arranged leaflets. Strongly anisophyllous taxa are found in all major groups of Gesneriaceae; striking examples include *Dalbergaria* = *Columnea* sect. *Collandra*, *Monopyle*, *Lembocarpus*, *Paradrymonia*, *Whytockia*, *Loxonia* and *Stauranthera*. In *Rhynchoglossum* the smaller leaves are usually completely aborted, and the remaining, strongly asymmetric larger leaves approach a distichous arrangement ("alterniphylly"; Fig. 11E; Weber et al. 1992). The degree of anisophylly may change within individual shoots. Some species of *Cyrtandra* start with about two pairs of opposite and ± symmetrical leaves and switch to strong aniso- or alterniphylly, others from weak to strong anisophylly. Leaf development, different expressions of anisophylly depending on light level, and vascular architecture of anisophyllous neotropical Gesneriaceae have been studied by Dengler and Sánchez-Burgos (1988), Morgan and Dengler (1988) and Sánchez-Burgos and Dengler (1988).

Leaves can also be arranged spirally. This pattern is rare in caulescent plants; in the New World Gesneriaceae it is found in some genera of Beslerieae and in nearly all species of Gesnerieae, and in Asia in four of the five sections of *Henckelia*, in *Boeica* and others. An alternate phyllotaxis is often found in rosette plants: all European (*Ramonda*, Fig. 14E, *Haberlea*, *Jancaea*), numerous Chinese and Himalayan genera (e.g. *Corallodiscus*, *Petrocosmea*, with *P. forrestii* having a dense, flat rosette similar to *Aeonium tabuliforme*), the African *Saintpaulia* and others show this pattern. For the rosulate pattern of *Streptocarpus*, see Exceptional Morphologies.

The spiral pattern can be (at least in some caulescent plants) derived from the decussate pattern by "node disintegration" (intercalation of internodes between the paired leaves of the nodes; Haccius 1939, 1950). In *Henckelia* sect. *Glossadenia*, the characteristic alternate-spiral phyllotaxis may have a different origin. The section includes one species (*Henckelia leucantha*) which has a decussate-isophyllous habit, and another (*H. descendens*) with a strongly anisophyllous habit. It seems possible that the smaller leaves have been completely reduced in the remaining species of the section (as in *Rhynchoglossum*), and that the resulting alterniphyllous arrangement switched to a spiral one (Weber and Burtt 1998d).

Branching of the shoot is by the production of axillary shoots. In some genera a displacement of the axillary shoots (especially the inflorescences) onto the leaf petioles can be observed (recaulescence). By the formation of distinct shoot units, usually composed of a single leaf pair and a bracteose "terminal" inflorescence, and repetition of such units, a sympodial shoot architecture is achieved in some Epithematoid Gesneriaceae (*Loxonia*, *Stauranthera*, *Rhynchoglossum*, see Inflorescence Structure) and in the African *Schizoboea* (Weber 1980).

The roots of Gesneriaceae have received little attention so far. In principle, the root system consists of the primary root developing from the radicle, and lateral roots emerging endogenously from the basal (usually underground) parts of the stem. However, the primary root never seems to occupy a dominant position and is of subordinate functional importance. Developmental delay and complete suppression of the primary root is observed in some Old World Gesneriaceae (see Seedling Structure). Adventitious roots are produced from the aboveground stems in many genera, in some cases assisting the attachment of climbing or epiphytic plants (e.g. *Agalymla*, *Paradrymonia*, *Drymonia*, etc.).

Growth Patterns. The spectrum of life-forms in the family ranges from small annuals to trees reaching 13 m. The bulk of species is represented by perennial herbs. For illustration of some characteristic growth patterns of neotropical and palaeotropical Gesneriaceae, see Figs. 13 and 14.

I. Annuals. True annuals are rare and are only found in genera which also include perennial species: *Streptocarpus* (e.g. *S. nobilis*, *S. elongatus*), *Rhynchoglossum* (*R. obliquum*, *R. papuae*), *Monophyllaea* (*M. glabra*), *Chirita* (species of sect. *Microchirita*). The plants occur in (sub)tropical areas with a pronounced seasonal climate. The annual habit may be combined with autogamy, for example, in the partially cleistogamous *Streptocarpus nobilis*. Annual habit is almost unknown among the New World gesneriads, a possible exception being *Sinningia nordestina* (Chautems 2000; Chautems et al. 2000).

II. Perennial herbs. Among perennials, monocarpic and polycarpic forms can be distinguished. These forms (flowering and fruiting once vs. periodically) are adaptations to a seasonal climate. In the aseasonal (ever-wet) parts of the tropics, however, this distinction becomes blurred as flowering and fruiting is often continuous, lasting over many years without interruption. This pattern is here referred to as "prolonged monocarpy".

One of the most striking examples of monocarpy is the curious perennation mode of the "unifoliate" species of *Streptocarpus* (see Exceptional Morphologies).

Prolonged monocarpy can be observed in tropical areas with constant climatic conditions throughout the year. It is found, for instance, in many species of *Monophyllaea*. These soft herbaceous plants enter the reproductive stage soon after germination, and this stage lasts without interruption several to many years until the plants die. Death is often by external causes such as damage, desiccation, etc., while undisturbed plants appear almost immortal. In contrast, species of *Monophyllaea* growing in seasonal areas (*M. glabra* in the NW of the Malay Peninsula) are annual-monocarpic: the plants die when the dry season commences and only the seeds survive the unfavourable dry period.

Much commoner are polycarpic perennials. They represent the typical growth form in subtropical and warm-temperate areas of both the Old and New World, and almost all European, Indo-Chinese and North American taxa belong to this category. Polycarpic perennials are also found in tropical regions with a slightly to pronounced seasonal climate. Perennation serves to survive a period with adverse conditions, either a dry or a cold (winter) season (the European and Chinese "alpine" gesneriads).

A striking array of different perennation strategies can be observed in the family. These include survival by (1) ground stems, (2) stolons, (3) rhizomes and rootstocks, (4) scaly rhizomes, and (5) tubers. Additional strategies such as anabiosis and poikilohydry are discussed in Habitat and Special Adaptations.

1. Ground stems. These are distinguished from proper rhizomes by their unspecialised habit: the basal parts of the stems lie on the ground or somewhat buried in the ground, producing roots for anchorage and water uptake, but lack nutrients stored in the stem or in cataphylls. Such ground stems are found in most ground herbs of ever-wet rainforests (Burtt 1977a). The shoot axes grow and produce new leaves continuously. In erect-growing plants an average height is maintained in that the lower part of the stem becomes decumbent as apical growth proceeds. The lower stem parts produce roots and get buried in the course of time. They often become lignified and do not accumulate food-reserves. The development of lateral shoots on the prostrate stem may lead to multiplication and to the patchy habit observed in many species. The main growing parts of the shoots and plants are always aerial and directed upwards. Flowering and fruiting are continuous, the younger stem parts bear the flowers, the older ones the fruits. Sometimes, e.g. in species of *Cyrtandra*, the flowers and fruits do not develop until the respective stem part has become prostrate and leafless. The fruits are then placed directly on the ground.
2. Stolons. These are axillary side branches with markedly elongated internodes, creeping over the ground and producing adventitious roots. They serve for perennation, migration and multiplication. Conspicuous examples include *Alsobia* (with one stolon per node and monochasial branching), *Episcia* (with two stolons per node and dichasial branching, Fig. 13A), *Henckelia stolonifera*, *Boeica stolonifera* and *Chirita stolonifera*.
3. Rhizomes and rootstocks. Typical rhizomes (= distinctly thickened horizontal underground shoots) are found in some Gesneriaceae of seasonal subtropical or temperate zones. Their function is manifold: (a) perennation = survival of unfavourable periods; the aerial plant parts die back at the beginning of the unfavourable season, and new aerial shoots are produced from the rhizome apex, from axillary buds, or from both on the return of favourable conditions, (b) storage of nutrients (and to some extent water); these provide the energy for producing new aerial shoots, (c) migration: by the elongation of the rhizome the plant migrates into areas hitherto not exploited by the roots, (d) multiplication: branching of the rhizomes leads to multiplication of the aerial shoots and thus vegetative propagation, (e) anchorage: with the aid of the rhizome-borne roots, the aerial shoots are mechanically anchored in the ground. Remarkable examples of rhizomatous gesneriads are *Raphiocarpus sesquifolius* (S China) and *Platystemma violoides* (Himalayas): these plants possess a horizontal, thickened underground rhizome with apparent storage function (in *Platystemma* the rhizome is a string of small, tuberous swellings). From the rhizome apex aerial shoots are produced periodically. These consist of an elongate stem (in *Platystemma* only a few centimetres long) and a single foliage leaf. The aerial shoots thus strongly resemble a *Monophyllaea*, but the overall morphology is completely different.

Rootstocks are shorter, much contracted and grow ± vertically, but a clear demarcation from rhizomes is not possible. Rootstocks and transitional forms occur in most rosette plants. The perennation mode of *Didymocarpus* (in the restricted sense of Weber and Burtt 1998d) is exceptional. After fruiting, the herbaceous stems die back at the beginning of the unfavourable season. Already at this time, new leafy shoots are formed from the rootstock. They rise just above the soil surface and persist in that condition until favourable conditions return. Then the shoots grow up, flower and fruit. It is remarkable that the few species of *Didymocarpus* s.str. that grow in the ever-wet tropics of the Malay Peninsula (one also in Sumatra) have essentially retained this habit. The duration of the shoots, however, is not strictly annual and, in the absence of a dry season, there is no resting period.

Another remarkable plant is *Kaisupeea* (Burtt 2001a). As in *Didymocarpus*, the stem dies down to the ground in the dry season, but no new leaves are produced until the return of favourable conditions. Then light green cataphylls appear, corresponding to expanded leaf bases that presumably have covered the new bud underground. This bud then breaks through and produces an aerial, flowering stem.

4. Scaly rhizomes. This special type of rhizome ("Zwiebelsprosse" or "Kätzchen" in the German literature) is found only in

some neotropical Gesneriaceae (apparently a synapomorphy for tribe Gloxinieae, Zimmer et al. 2002). They represent underground appendages of the shoot axis which serve as survival and propagation structures. They are one to several cm long and consist of a thin central axis and closely packed pairs or whorls of small, fleshy leaf scales. They survive in the ground when the aboveground plant parts die back in the dry period, and sprout when favourable conditions return. As one plant produces usually several scaly rhizomes, multiplication also takes place. In *Achimenes* sometimes also aerial scaly rhizomes are found. Intermediate forms between stolons and scaly rhizomes are the so-called "stringy rhizomes" produced above ground in some species of *Gloxinia* (Skog, pers. comm.).

5. Tubers. Like scaly rhizomes, tubers are confined to the New World Gesneriaceae and are characteristic of *Sinningia* (Fig. 13B), *Chrysothemis*, *Lembocarpus*, *Rhoogeton* and some species of *Nautilocalyx*. Unlike scaly rhizomes, tubers have apparently arisen independently at least twice in the New World, in tribes Episcieae and Sinningieae (Zimmer et al. 2002). Tuber formation starts with the thickening of the hypocotyl in the seedling (H. Weber 1936, under *Corytholoma*) and epicotylar nodes may become integrated into the tuber in the course of development. The tuber may be buried in the ground but (in *Sinningia*) is more often exposed, with only the base and the roots penetrating the ground. The form of the tuber may be globose, pear-shaped or flat. The size depends on the species and the age of the plant. The largest tubers have been found in *Sinningia macrostachya*, reaching a diameter of up to 1 m (Gilmer and Hofacker 2001). The seasonal shoots arise from the upthrust or depressed central area of the tuber. The shoots usually produce both leaves and flowers, but a differentiation into flowering and vegetative shoots may occur (*Sinningia tuberosa*, see Exceptional Morphologies).

III. Subshrubs and shrubs. The terms woody herb, subshrub (suffrutex) and shrub (frutex) have been ambiguously used to characterise species and alliances of Gesneriaceae, but a clear categorisation is not possible. Lignification of stem bases is sometimes observed in caulescent species, with the shoots dying back to the lignified zone in unfavourable seasons. Slight lignification may extend to the entire shoots which nonetheless die after fruiting (*Didymocarpus antirrhinoides*, Weber and Burtt 1983, 1985). Lignification may be so slight that the shoots remain soft and flexible (e.g. *Solenophora*). In such cases, it is difficult to speak of a proper shrub.

Typical shrubs are branched, preferably at the base (species of *Cyrtandra*, *Besleria*, etc.). Often, however, a plant has a distinctly woody stem which remains unbranched and bears a tuft of leaves at the apex. Such single-stemmed (monaxial) shrubs are found in *Henckelia* sect. *Heteroboea*, *Hexatheca*, *Cyrtandra* spp., *Rhytidophyllum* and many others. Secondary growth is usual minimal, even in "shrubby" species, but significant secondary growth and lignification are found in some species with persistent stems, e.g. *Gesneria* and *Rhytidophyllum*.

IV. Arborescent shrubs and trees. *Cyrtandra* species may reach a height of 4 or 5 m and thus may be termed arborescent shrubs or small understorey trees. The largest trees are found in two genera of Coronantheroid Gesneriaceae, reaching a height of 10 m and more: *Depanthus* (New Caledonia) and *Negria* (Lord Howe Island, SW Pacific). According to Green (1973: t. 659), the latter has a strong trunk and spreading branches, but the wood is light and brittle as in the other (small) trees and shrubs of Gesneriaceae.

V. Climbers. Especially in the neotropical Gesneriaceae, there is no sharp boundary between epiphytes, climbers and lianas. In most cases germination takes place high on a host tree. Many species of *Drymonia* (e.g. *D. serrulata*, *D. alloplectoides*), however, root in the ground and their wiry stems climb trees, often producing elongate, horizontal lateral branches (Skog, pers. comm.). In the palaeotropics, the only genus consisting (with few exceptions) of true climbers rooting in the ground is the Malesian genus *Agalmyla* (Fig. 14G). It produces long climbing stems which cling to the trunk of a host tree by numerous short roots produced along the internodes. The disadvantage of the opposite leaf arrangement is eliminated by strong anisophylly: one of the two leaves of a node is much diminished in size, often even reduced to a small scale. The remaining large leaves then form two ranks, spreading from the trunk of the host tree. The usually epiphytic species of *Aeschynanthus* may germinate on the ground occasionally, but climb by normal nodal roots.

VI. Epiphytes. In the neotropical Gesneriaceae, epiphytes are found mainly in tribe Episcieae (rarely in other tribes: *Capanea*, Gloxinieae, and a few species of *Sinningia*, Sinningieae). Essentially epiphytic genera are *Drymonia*, *Paradrymonia*, *Rufodorsia*, *Oerstedina*, *Codonanthe* (Fig. 13C), *Codonanthopsis*, *Nematanthus*, and *Columnea* s.l. Within the latter genus, with nearly 300 species, a wide range of epiphytic growth patterns can be observed. One group has (sub)erect shrubby stems and branches with opposite isophyllous leaves (Fig. 13D), another group has spreading stems with strongly anisophyllous leaves arranged in a plane (Fig. 13E), and a third group has pendulous stems with the ± isophyllous leaves often arranged in a single plane (Fig. 13F). Variations are found in many details and have not been studied properly so far. The phenomorphology of several *Columnea* species was studied by Freiberg (1994, 1996a, 1996b) within the context of the microclimate of habitats. In general, epiphytism seems to be correlated with ornithochory: the genera listed above produce either display capsules or fleshy, often coloured berries attractive to birds. In the palaeotropical Gesneriaceae the main epiphytic genus is *Aeschynanthus* (Fig. 14H) in which almost all of the ca. 160 species are epiphytic. Epiphytic species are also found in *Lysionotus* and *Loxostigma*. In all these genera the seeds have distinct, often long, filiform appendages that facilitate dispersal by wind and anchorage on tree trunks and branches (see Mendum et al. 2001).

Exceptional Morphologies. A number of palaeotropical Gesneriaceae exhibit very unusual morphologies, in particular (1) *Streptocarpus* and (2) the Epithematoid Gesneriaceae. The Neotropical Gesneriaceae have a more conventional structure, although phenomena such as anisophylly, anisoclady, etc. are not uncommon. As examples for rather peculiar morphologies, (3) *Sinningia tuberosa* and *Lembocarpus amoenus* are described.

1. *Streptocarpus*. The strange morphology of *Streptocarpus* has been noted and studied since its introduction into cultivation in the early 19th century. Significant studies include Caspary (1858), Crocker (1860), Hielscher (1883), Pischinger (1903), Schenk (1943), Burtt (1970), Hilliard and Burtt (1971), Noel and van Staden (1975), Jong (1970, 1973, 1978) and Jong and Burtt (1975).

The genus is traditionally divided into two subgenera, subg. *Streptocarpella* and subg. *Streptocarpus*. Superficially, subg. *Streptocarpella* does not exhibit any peculiarities: the plants produce stems with decussate leaf pairs from which the axillary inflorescences arise. The seedling, however, is unusual by the strong anisocotyly and often the intercalation of an internode between the macro- and the microcotyledon (mesocotyl, Schenk 1943).

In subg. *Streptocarpus* a wide array of unusual morphologies can be observed. The most conspicuous is that in which only a single foliar organ is present in the form of a giant and ever-growing macrocotyledon (unifoliate habit; Fig. 12A, B). A mesocotyl is developed, but the hypocotyl and the mesocotyl remain short and develop into a stout stalk bearing the macrocotyledon at its top. From the base of the midrib, a series of inflorescences (richly and dichasially branched pair-flowered cymes) emerges. By their structure and orientation these can be identified as homologous to the axillary inflorescences of subg. *Streptocarpella*, but are increased in number by accessory shoot formation. The macrocotyledon and the stalk (hypo- and mesocotyl) form an integrated structure which has been termed "phyllomorph" (Jong 1970, 1973, 1978; Jong and Burtt 1975). Growth is by a trinity of meristems: (1) the "basal meristem" (situated at the lamina base and responsible for the continuous growth of the foliar structure), (2) the "petiolode meristem" (an intercalary meristem located in the upper part of the stalk = "petiolode"), and (3) the "groove meristem" (situated at the junction of the petiolode and the lamina). The most remarkable structure is the petiolode. It forms a continuous transition between the axial mesocotyl and the lamina base and functionally represents a petiole. Jong and Burtt (1975), therefore, regard it as a mixture of leaf- and stem-like properties.

The unifoliate species of *Streptocarpus*, consisting of the cotyledonary phyllomorph only, are monocarpic, and die after producing inflorescences, flowers and fruits. Maturity is reached in two or several years. In those species living in a markedly seasonal climate, a unique mode of survival of the unfavourable dry period has evolved: the lamina sheds a large distal part through abscission (60% in *S. molweniensis*). On the return of favourable conditions, growth is resumed from the remaining basal part of the lamina. Abscission is not along a predetermined abscission zone, and its initiation is probably mediated by changes in the levels of endogenous cytokinins (van Staden 1973).

The phyllomorphic organisation is also characteristic of the "plurifoliate" and "rosulate" species. Their plant body can be understood as a succession of phyllomorphs. In the plurifoliate species (e.g. *S. polyanthus*) two or few phyllomorphs are produced, each repeating the structure of the cotyledonary phyllomorph. Perhaps the most remarkable plurifoliate species is *S. fanniniae* (Jong 1970; Jong and Burtt 1975). The plant produces long petiolodes, so that a long-creeping, climbing and trailing habit is achieved. From the petiolodes vegetative buds and new branches are produced.

Rosulate species such as *S. baudertii*, *S. kentaniensis*, *S. rexii* and *S. gardenii* have a close resemblance to familiar rosette plants. However, the rosette is not made up of simple leaves but of phyllomorphs with a distinct petiolode at the lamina base. Among the rosulate species at least two distinct patterns can be recognised: the centric and the excentric pattern. In the first the phyllomorphs are arranged in a spiral phyllotactic sequence on a condensed vertical axis. In the latter, the phyllomorphs are arranged in two ranks on the upper surface of a horizontal rhizomatous axis. Here the axis is composed of an aggregation of petiolode bases. These are often pigmented, and have stomata and roots. The resemblance to a conventional rhizome is superficial. Both in the centric and excentric pattern the phyllomorphs produce vegetative buds on the petiolodes, and from these buds lateral rosettes or "branch rhizomes" develop. Roots arise regularly from the base of the petiolodes, so that each phyllomorph is provided with its own root system. Each individual phyllomorph of the "rosette" is monocarpic and perishes after flowering and fruiting.

Noteworthy are the experiments with growth regulators carried out by Dubuc-Lebreux (1976, 1978), Dubuc-Lebreux and Vieth (1976), Rosenblum (1981) and Rosenblum and Basile (1984). They showed that unifoliate species of *Streptocarpus* can be induced to produce additional phyllomorph-like structures.

2. Epithematoid Gesneriaceae. The morphology of this group has been studied by Oehlkers (1923), Hallé and Delmotte (1973), Weber (1975a, 1976b, 1976c, 1977b, 1978a, 1978b, 1982a, 1987, 1988, 1990), Burtt (1978), Feuillet (1979), Kohyama and Hotta (1986), Tsukaya (1997), Y.Z. Wang (2000) and Imaichi et al. (2001).

In appearance, the most peculiar genus of the alliance is *Monophyllaea*. Plants of this Malesian genus consist of a single, huge foliar organ (macrocotyledon), and a long stalk (hypocotyl) (Fig. 12C, D). The plant thus comprises the same elements as the unifoliate species of *Streptocarpus* but, inter alia, there are strong differences in the formation of the inflorescences. While in *Streptocarpus* the inflorescences are produced in a descending series in the axil of the macrocotyledon, those of *Monophyllaea* (long-pedunculate cincinni with much-condensed flower pairs) emerge in an alternate sequence, forming two ranks in an almost distichous arrangement. The cymes are associated with tiny bracts at their base.

Perhaps the most peculiar species of *Monophyllaea* is *M. singularis* from Borneo. In this species the flowers are arranged in numerous tiny groups along one side of the stalk (sometimes also along the midrib of the macrocotyledon). Here a mesocotyl is formed, as in *Streptocarpus*. The flower groups are produced in a most peculiar manner: mature epi- and hypodermal cells divide secondarily and form meristematic spots from which the flower groups emerge (Weber 1987, 1990; Imaichi et al. 2001). The groups correspond to the inflorescence aggregates of other *Monophyllaea* species, but comprise only few and epedunculate cymes with few flower pairs.

Also noteworthy are some *Monophyllaea* species which produce several leaves and exhibit a phyllomorphic structure (Burtt 1978).

The near-distichous position of the cymes within the inflorescence aggregates can be explained by anisophylly: of the four ranks of leaves/bracts (in decussate gesneriads), only two produce axillary cymes (Weber 1975a). This interpretation implies an ancestor with strong anisophylly and cymes produced exclusively in the axils of the large leaves. Such plants are found in the small Chinese genus *Whytockia*. Floral characters and molecular data (Mayer et al. 2003) confirm the close relationship between the two disjunct genera. It thus can be concluded that *Whytockia* is a relic of the stock from which *Monophyllaea* evolved (Weber 1976b).

It must be noted that in both genera (as well in the other members of Epithematoid Gesneriaceae), the apical meristem is not present in the conventional form of a meristematic dome, and growth is markedly rhythmical. This led Y.Z. Wang (2000: 231) to the erroneous interpretation that the shoot of *Whytockia* would be "a lateral branch system formed by sprouting of lateral buds in axils of small leaves substituting for the thoroughly restrained phyllogens".

Similar in habit to *Whytockia* are *Loxonia* and *Stauranthera*. In these genera, however, the inflorescences do not emerge from the axils of the large leaves but are opposite, seemingly emerging from the axils of the small leaves. Closer analysis shows that they represent terminal and complex structures, comprising two to several pair-flowered cymes in alternate position (alternicladic thyrses). The whole flowering region of *Loxonia* and *Stauranthera* is composed of sympodial units, each comprising a strongly anisophyllous leaf pair and the terminal complex inflorescence. By the growth of the subsequent units, the terminal inflorescences are bent to the opposite side of the large leaves.

In *Rhynchoglossum*, anisophylly has proceeded to its extreme: the small leaves are completely reduced and only the strongly asymmetrical large leaves remain, being placed in two almost distichous ranks. As in *Loxonia* and *Stauranthera*, the inflorescences are terminal but the cymes are reduced to single flowers. This results in unilateral racemes with two flower rows, often (especially in the small-flowered annual species) with highly increased flower number.

The genus *Epithema* is unique in several respects. Above the strongly unequal cotyledons (placed at the very base of the plant and decaying soon), a single foliar leaf is produced, recalling an additional macrocotyledon. It is followed by one to several opposite pairs of ± equal foliar leaves. The stem and the axillary shoots are terminated by an inflorescence consisting of a globose "head" (an extremely contracted pair-flowered cyme) and a large, cucullate bract. This type of inflorescence can be formally derived from a more elaborate thyrse, with several bract/inflorescence units in alternate position. Such thyrses are known from certain species of *Monophyllaea*. In *Epithema* the thyrse is reduced to a single bract/inflorescence unit. Inflorescences with two bracts/axillary inflorescences are found occasionally in *E. saxatile* (Weber 1988), thus giving some empirical evidence for the hypothesis.

The ecology, germination and growth interruption during the year was studied in *Epithema tenue* in West Africa (Hallé and Delmotte 1973). Here, in an area with two wet seasons, the situation is certainly different from tropical Asia and Malesia, where the plants seem to grow straight through to flowering.

The last and ill-known Chinese genus *Gyrogyne* (monotypic with *G. subaequifolia*) is not discussed here. By the plicate calyx it is said to be closely related to *Stauranthera*, the subequal leaves exhibiting an ancestral condition.

3. *Sinningia tuberosa* and *Lembocarpus amoenus* belong to the morphologically most remarkable species of neotropical Gesneriaceae. Both plants possess a storage organ in the form of a tuber and produce usually only one leaf per season. In *Sinningia tuberosa* the tuber is perennial and produces one (rarely two or more) tiny vegetative shoot(s) at its top. This shoot produces first a few pairs of tiny cataphylls and then a strongly anisophyllous leaf pair, only one of which develops into a large foliar leaf. The flowers are produced on separate short shoots. These form exclusively cataphylls from the axils of which the flowers emerge (Chautems and Weber 1999). In *Lembocarpus amoenus* the tuber is annual and a new tuber is formed each year on top of the old one. The tuber produces also a tiny shoot with an extremely anisophyllous leaf pair but, in contrast to *Sinningia*, no separate floral shoots are developed and the inflorescence is produced in the axil of the foliar leaf (Boggan 1991).

Vegetative Anatomy. Summaries of early anatomical studies were given by Solereder (1899, 1908) and later by Metcalfe and Chalk (1950, 1979). Some of the more recent papers devoted to leaf and stem anatomy, the indumentum, and anatomical features of taxonomic/phylogenetic interest include Bokhari and Burtt (1970), Burtt and Bokhari (1973): sclereids in *Cyrtandra*; Carlquist and Hoekman (1986): wood anatomy; Herat and Theobald (1979): anatomy of Sri Lanka taxa; Pongračić (1931): leaf anatomy, trichomes; Rechinger (1899), trichomes; Rosser and Burtt (1969): stem and leaf anatomy of Trichosporeae; Sachs (1915), leaf anatomy; Sahasrabudhe and Stace (1974): trichomes, stomata; Wiehler (1983): anatomy of neotropical Gesneriaceae.

I. Leaf anatomy. The leaves of Gesneriaceae are dorsiventral. The upper leaf epidermis consists of usually cubic cells and is covered by a cuticle of variable thickness. Particularly in epiphytes such as *Aeschynanthus*, *Codonanthe* and *Columnea*, a one- to several-layered, large-celled and thin-walled hypodermis is present. It originates from the upper protodermis, and serves as a water storage tissue. The mesophyll consists of a single layer of cylindrical palisade cells and a spongy parenchyma with cubic or elongated cells in loose arrangement. Here the vascular bundles are embedded. In some species of *Aeschynanthus* the midrib bundle is surrounded by an endodermis with Casparian strips, which can be regarded as a special adaptation to water-stress. Also embedded in the mesophyll are often cubic or elongated, uniformly thickened or hippocrepiform sclereids ("spicular cells") with strongly pitted walls. In *Cyrtandra* many forms of star-like astro- and branched dendrosclereids are found. The lower epidermis is similar to the upper one, but additionally comprises the stomata. Occasionally it is coloured by anthocyans which are stored in the epidermis cells and the lower spongy parenchyma cells.

In general, the stomata are anisocytic (anomocytic in young cotyledons). They are surrounded by 2 guard-cells and 3–4 unequal subsidiary cells. They are situated on the lower leaf surface, petiole and stem, in *Ramonda* they occur also on the upper leaf side (amphistomatic leaves). Distribution is either even or sometimes in distinct groups (islands) (e.g. *Gasteranthus*, *Gesneria*, *Napeanthus*). Stomatal turrets (domes) with groups of stomata are observed, inter alia, in *Cyrtandra*, *Napeanthus* and in the *Gesneria* alliance (Skog 1976). Domes with individual stomata are found in most members of tribe Gloxinieae (Wiehler 1983).

The ontogenetic types of stomata distinguished by Fryns-Claessens and van Cotthem (1973) seem to have some taxonomic relevance: the aniso-mesogenous type is found predominantly in the Gesnerioid Gesneriaceae, the anomo-mesoperigenous (or aperigenous) type in most Coronantheroid Gesneriaceae (Wiehler 1983). Data for palaeotropical Gesneriaceae are largely lacking.

The trichomes of Gesneriaceae are usually simple and uniseriate (rarely multiseriate: *Rhynchoglossum*), eglandular or glandular. Simple trichomes are widespread in the family, branched eglandular hairs occur occasionally, e.g. in *Paraboea* and *Trisepalum*, forming a characteristic arachnoid-interwoven indumentum. The hairs usually end in a single terminal cell, which is sometimes bent to form a small hook (*Epithema*, *Stauranthera*, *Monopyle* and others; Weber 1976c, 1977b). The "rod-like" hairs found in many species of *Henckelia* sect. *Glossadenia* are peculiar: they are made up of a single elongate cell terminated by a short, apiculus-like apical cell (Weber and Burtt, in prep.). Terminal eglandular cells in the form of head-like, star-like, twin-head or peltate structures are not known in the family. The glandular trichomes often have a multicellular head. Peltate glandular hairs are present in *Monophyllaea* (secreting calcium carbonate) and in *Aeschynanthus* sect. *Microtrichium*. *Didymocarpus* and *Streptocarpus* (and others) possess pigmented glands of yellowish brown to blackish-red colour (caused by the naphtaquinone "dunnione" in *Streptocarpus dunnii* and allies, Price and Robinson 1938, 1940; for the biosynthesis of naphtaquinones and anthraquinones in *Streptocarpus dunnii*, see Stöckigt et al. 1973 and Inoue et al. 1984). In *Didymocarpus* spp. the gland exudates have been identified as aurentiacin, benzoquinoid chalcones and flavokawin B (also a chalcone) (Wollenweber et al. 1981). These pigmented glands have a characteristic one- or several-celled head. A variety of forms is illustrated in Weber and Burtt (1983). Sometimes they give off a strong odour (e.g. *Didymocarpus aromaticus*).

The terminal cells of glandular hairs are sometimes thickened (calcified or silicified), with the lumina occasionally completely filled with cell wall material. The cells of the stalk are sometimes filled with red or blue liquids (probably anthocyanins).

The kind and composition of the indumentum is often used for species definition, occasionally also for defining genera (e.g. the separation of *Boea* and *Paraboea* is essentially based on the type of indumentum, Burtt 1984).

II. Stem anatomy has been studied by Rosser (in Rosser and Burtt 1969) in *Aeschynanthus* and *Agalmyla*. She reports epidermal cells with usually rectangular and parallel orientation to the stem axis, and with a thick and smooth, striate or thin cuticle. The cortex, often with crystals and crystal druses in the cells, consists of 2 layers. The outer cortex of old stems is made up of closely packed, collenchymatous cells, while the inner cortex comprises loosely packed, thin-walled cells. Hippocrepiform sclereids are frequent in *Aeschynanthus*, and the usual presence of a cortical sheath with a layer of hippocrepiform sclereids is noteworthy. In addition, an endodermis with uniformly thickened and suberised cells and Casparian strips on radial walls is present.

The stele of Gesneriaceae is generally a hollow cylinder of continuous phloem and xylem of which the latter may contain fibres and/or libriform tracheids. In the centre of the cylinder is a pith of large parenchymatic cells. In the pith crystals, starch grains and/or nests of sclereids can be found. A peculiar feature is the presence of vascular bundles in the pith of *Rhynchoglossum* and *Monophyllaea* ("medullary bundles", Hollstein 1878; Wonisch 1909a, 1909b; see also IV below).

Nodal anatomy has been studied mainly in the neotropical Gesneriaceae and provides useful characters for defining tribes (cf. Wiehler 1983). The nodes are mainly uni- or trilacunar, rarely 4- or 5-lacunar, and split off collateral vascular bundles. Split laterals occur in *Chirita lavandulacea* and probably other species of sect. *Microchirita* (Burtt, pers. comm.). Differences of node anatomy under natural and cultivation conditions have been reported for *Columnea ambigua* (1-lacunar vs. 3-lacunar; Howard 1970, as *Alloplectus ambiguus*).

Wood anatomy has been studied by Carlquist and Hoekman (1986). Growth rings are absent or nearly so. The vessel elements usually have simple perforation plates and alternating circular or oval pits of various sizes on the vessel walls. The vessels are usually grouped in radial chains. Imperforate tracheary elements are libriform fibres, rarely fibre tracheids. In most species they are septate. Often only multiseriate rays occur or rays are completely absent. *Cyrtandra* wood, also with multiseriate rays, is ray-less in early ontogeny. The absence of rays seems to indicate secondary woodiness in this and other genera. All features of the wood reflect the mesic habitats characteristic of Gesneriaceae, but moderate degrees of xeromorphy are found in some species with narrow vessels, grouped vessels, and grooves in vessel walls.

III. Cell inclusions are often observed, commonly in the form of calcium-oxalate crystals. Rhomboidal crystals are present in ray cells of *Cyrtandra* and *Rhytidophyllum*. Raphides often occur in association with the leaf vascular bundles (e.g. *Napeanthus*, *Aeschynanthus*). Cystoliths are found in *Rhynchoglossum* (Fritsch 1908). Starches have been surveyed by Reichert (1913).

IV. Some Epithematoid Gesneriaceae possess schizogenous secretory ducts. These are formed next to the xylem and run through the stem, the leaves, the inflorescence axes, pedicels and eventually into the floral organs (sepals, ovary). In *Monophyllaea* and *Rhynchoglossum* they are also found in the pith, in association with the medullary bundles. Resin or oily substances are thought to be the main components of the contents.

Inflorescence Structure. Full recognition of the special nature of the gesneriaceous inflorescence has been obtained only recently. It became apparent that the great diversity observed in the family can be referred to a single basic type: the floral region of the plant represents an indeterminate thyrse made up of axillary inflorescences in the form of "pair-flowered cymes" (in the synflorescence concept of Troll (1964), the inflorescence can be denoted as a polytelic system, the "florescences" being indeterminate thyrses and the "partial florescences" being pair-flowered cymes).

The "pair-flowered cyme" is a special type of cyme, found only in Gesneriaceae (Weber 1973b, 1982b, 1995; Wiehler 1983), a few genera of Scrophulariaceae (*Penstemon*, *Tetranema*, *Russelia*, *Calceolaria* and allies, Weber 1973b), and *Sanango* (Wiehler 1994). In this cyme, the terminal flower (T) of each cyme unit is associated with an additional flower in frontal position (front-flower, F; Fig. 10). Thus, each cyme unit seems to end in a flower pair. In general, the front-flower has no subtending bracteole, while the lateral flowers are usually subtended by bracteoles (± small prophylls of the successive cyme units). However, in a few taxa, the front-flower is also subtended by a bracteole, thus demonstrating that the extra-axillary origin of the front-flower is a secondary condition, due to suppression/abortion of the bracteole. In ontogeny, the front-flowers are initiated later than the lateral flowers, but develop more quickly and open earlier than the lateral flowers, just after the terminal flower (Weber 1995). The lateral flowers are usually subtended by bracteoles (= small-sized prophylls of the successive cyme units).

Phylogenetically, the pair-flowered cyme probably originated from a paniculate branching system and represents a stage intermediate to an ordinary cyme, the front-flowers being remnant flowers (Weber 1973b, 1982b). Empirical support for this view is that sometimes the front-flower is developed as a branch.

The most obvious variation of the pair-flowered cyme relates to flower number. Although a formal reduction series [many → several → few → two (= T1, F1) → one (T1) flower(s); Fig. 10A–E] can be easily established, it is usually difficult to recognise phylogenetic directions within a given alliance. By reduction of the pair-flowered cymes to solitary flowers and reduction of the subtend-

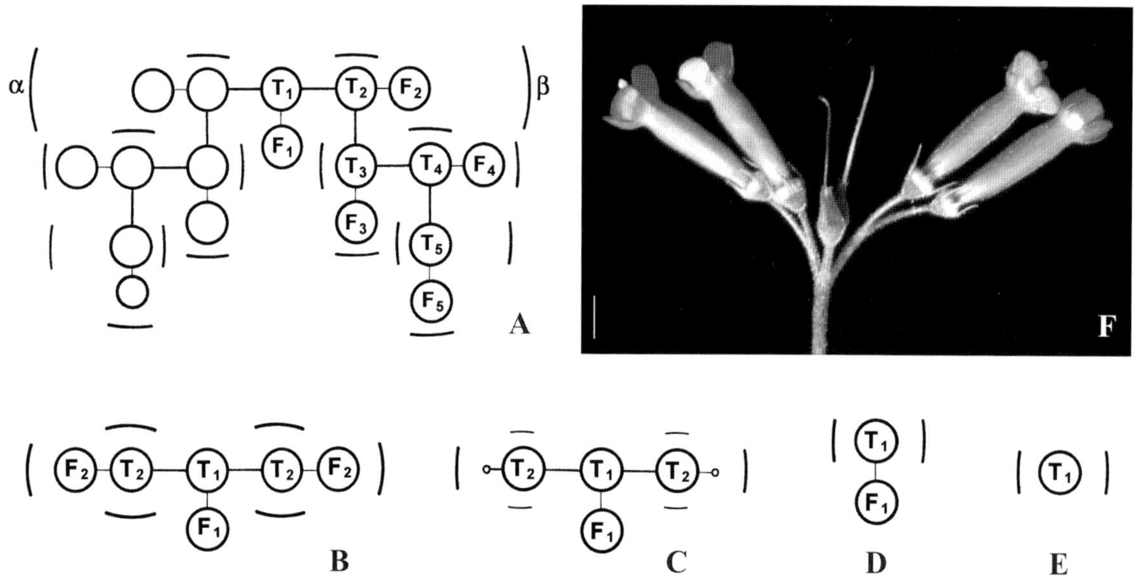

Fig. 10. Gesneriaceae. **A–E** Diagrams of pair-flowered cymes, arranged in a reduction series to a single flower (E). *T* Terminal flower of cyme unit, *F* front flower, α, β prophylls (first bracteoles). **F** Six-flowered cyme of *Sinningia lineata* (bracteoles are lacking in this species). **A–E** from Weber (1978c), **F** photo A. Weber. *Bar* 1 cm

ing leaves to bracts, the thyrse turns into a bracteose raceme (e.g. *Diastema*, *Smithiantha*, *Koellikeria*). Occasionally a spike is reached by extreme shortening of the pedicels (e.g. *Sinningia allagophylla*).

In general, the front-flowers are very stable elements in the cyme, but there are also a few cases in which they are subject to reduction, either irregularly (e.g. *Chrysothemis friedrichsthaliana*, Weber 1978c) or regularly. In the latter case, the cyme becomes a conventional cyme (*Loxonia hirsuta*, *Stauranthera caerulea*; Weber 1977b).

Additional variation relates to the symmetry of branching (compound pair-flowered dichasium, double cincinnus, cincinnus = scorpioid cyme), length of peduncle, length of the subsequent hypopodia, length of pedicels, presence (including form, size and coloration) of bracteoles, and eventual displacement of bracteoles. By regular displacement of one of the two bracteoles, a branching pattern as found in Solanaceae–Solaneae results, with two perpendicular bracteoles of unequal size at each node (*Drymonia*, Weber 1982b). By the enlargement and fusion of the first bracteole pair, a conspicuous cupule embracing the flowers is formed in some species of *Cyrtandra* (e.g. *C. cupulata*, *C. burbidgei*).

Pair-flowered cincinni with high flower number may proceed to a pseudomonopodial detachment of flower primordia (*Epithema*, *Monophyllaea*; probably also *Tylopsacas* and *Cremosperma*; Weber 1995).

By serial repetition, often more than one cyme (or solitary flower) can be present in a leaf axil. Particularly remarkable are the "crested" inflorescences of some species of *Chirita* sect. *Microchirita*: here the pair-flowered cymes are reduced to flower pairs (T1, F1) and such pairs are produced in considerable number in a line on the petiole of the subtending leaf (Weber 1975b). A similar pattern, including epipetiolar displacement, is found in some species of *Henckelia* sect. *Heteroboea* (Weber and Burtt 1998d). Examples from the New World Gesneriaceae include *Reldia alternifolia* (Kvist and Skog 1989) and *Gloxinia* spp. (Boggan, pers. comm.).

The outer appearance of the inflorescence, its visibility and attraction for pollinators depend essentially on the size of and distance between the leaves. In rosette plants the cymes are usually long pedunculate (scapose) and the flowers thus fully exposed. In caulescent plants the subtending leaves may be of equal size and form as in the vegetative region (frondose inflorescences), or they may become gradually reduced (frondo-bracteose inflorescences), or they may be ± abruptly reduced to small bracts (bracteose inflorescences).

In *Schizoboea* the bracteose inflorescence is reduced to a single pair of bracts and cymes, and sharply set off from the foliage leaves. The flowering region is composed of sympodial units, each with an isophyllous leaf pair and the condensed cyme pair at the top. As each foliage leaf is capable of producing a successive unit, a dichasial growth pattern results (Weber 1980).

The arrangement of the subtending leaves/bracts of the cymes (or solitary flowers) usually continues the arrangement of the vegetative leaves. Exceptions are found in *Smithiantha* and *Koellikeria*, where the foliage leaves are decussate, while the bracts in the inflorescence (indeterminate raceme) switch abruptly to an alternate arrangement.

Strong anisophylly is usually associated with alterniclady where only the large leaves of the pairs produce axillary cymes. In case of reduction of the anisophyllous leaves to (isophyllous) bracts in the inflorescence, only the bracts corresponding to the large leaves ("plus-bracts") produce an axillary cyme; therefore, the cymes are arranged in two rows (*Loxonia*, *Stauranthera*). In *Rhynchoglossum*, the cymes are reduced to single flowers, and the thyrse thus switches to a unilateral raceme (often misinterpreted as a scorpioid cyme). In these genera the flowering region is of sympodial structure (see also Exceptional Morphologies).

Flower Structure and Development. The flowers of Gesneriaceae are tetracyclic, sympetalous, primarily pentamerous and primarily zygomorphic.

The calyx consists of five sepals which are either free to the base or fused to a variable extent. Aestivation is imbricate or valvate (but even here the oblique edges of the sepals may overlap in a quincuncial manner). In *Stauranthera* and *Gyrogyne* the calyx is plicate. Normally, the sepals are of equal size. Bilabiate calyces result from a different extent of connation and/or different size of the three upper and two lower sepal lobes. In *Trisepalum*, *Cyrtandra stonei* and others, the upper lobes are almost completely connate whereas the two lower lobes are nearly free, so that the calyx appears to be trimerous. The shape of the sepals is usually simple (ovate, triangular, linear, etc.), the margin is entire or sometimes dentate or serrate; peculiar pectinate sepals are found in some species of *Columnea* s.l. (e.g. *C. incredibilis*, Kvist and Skog 1988). The calyx is usually green but may be vividly coloured, especially in ornithophilous flowers, either being of the same colour as the corolla (e.g. the bright red calyces of *Aeschynanthus* spp.) or making a conspicuous contrast with the corolla (e.g. species of *Columnea*, *Drymonia*, *Paradrymonia*, etc.). Remarkable are the tubular "water calyces" of *Aeschynanthus* spp. in the Old World tropics and *Chrysothemis* in the neotropics (Skutch 1992, as *Tussacia*).

The corolla is sympetalous and primarily pentamerous (rarely tetramerous: *Ramonda nathaliae*, *Bournea sinensis*, *Championia reticulata*). It exhibits a wide range of forms resulting from the different proportions and shapes of the tube and the limb. There is practically a continuum between flowers with a tube more than 10 cm long and such with almost none (sometimes even within a single genus or section: *Ridleyandra*, Weber and Burtt 1998c, *Henckelia* sect. *Glossadenia*, Weber and Burtt, in prep.). The tube may be narrow, gently or abruptly widening, pouched on the ventral or dorsal side, straight or curved. The very base is often gibbous for holding the nectar. Some taxa have a sac or spur at the base, either dorsally (e.g. *Gasteranthus*, *Sinningia*, *Nautilocalyx*, *Episcia*, *Achimenes* spp.) or ventrally (*Stauranthera*). The spur may hold nectar, but may also represent a structure giving strength and balance to a long-tubed flower. The limb is primarily bilabiate, with two upper and three lower lobes. There are also cases in which four lobes form the upper lip and the fifth the lower lip (*Calcareoboea*, *Allocheilos*). Resupination by twisting of the pedicel occurs in *Senyumia* and in some (mostly Andean) species of *Alloplectus*. Also the flowers of some *Nematanthus* species are presented in an upside-down position.

A regular limb can be sometimes found in tubular flowers (e.g. some species of *Cyrtandra*, *Kohleria* and *Sinningia*), but more often in short-tubed, campanulate or flat-faced ("rotate") flowers. Actinomorphic corollas often have five stamens (see below).

The corolla lobes are usually rounded, less commonly pointed or fimbriate. Some species of *Trichantha* (=Correct *Columnea* sect. *Ortholoma*) bear conspicuous rod-like or filiform appendages at the sinuses (e.g. *C. moorei*) or on the dorsal side of the corolla lobes. In *C. incredibilis* the appendages are up to 38 mm long and completely replace the 3–4 mm long corolla lobes in function (Kvist and Skog 1988).

Coloration of the corolla is extremely varied, with colours ranging from white through yellow, red, scarlet, crimson, violet, greenish, brownish to nearly black. Additional markings, stripes, dots, etc., are common. In many species of palaeotropical Gesneriaceae, two (often raised) yellow lines lead into the corolla mouth. Similar nectar guides are found in neotropical Gesneriaceae with whitish flowers, e.g. *Drymonia*, *Codonanthe*, *Achimenes*, *Nautilocalyx*.

The androecium is composed of 4 (rarely 5) or 2 fertile stamens, the other 1–3 stamens being reduced to sterile staminodia, or occasionally absent entirely. In case of diandry, the lateral staminodes are usually longer and more elaborate than the median (dorsal) staminode which is usually a tiny rudiment or even completely aborted. The presence of five stamens is associated with actinomorphy of the corolla (*Ramonda*, *Thamnocharis*, *Bournea*, *Tengia*, *Conandron*, *Bellonia*, *Napeanthus* p.p.). These flowers are mostly flat-faced (rotate), more rarely campanulate (*Bournea*) or suburceolate (*Tengia*). The actinomorphic flowers were formerly often regarded as primitive, but recognised as derived by Burtt (1970). They (or at least the flat-faced forms) exhibit the general syndrome of oligandric pollen flowers (see Vogel 1978), and one of the reasons why actinomorphic flowers have evolved secondarily in the Gesneriaceae is apparently the switch to a different pollination mode (buzz-pollination).

Reduction of the stamen number to 2 is only known in the palaeotropical Gesneriaceae (over 30 genera) and in a single genus of the Coronantheroid Gesneriaceae (*Sarmienta*), but not in the neotropical Gesneriaceae. It is usually the posterior (upper) pair that is reduced, but there are also some genera in which the anterior stamens are reduced (*Opithandra*, *Epithema*, *Sarmienta*, *Sepikea*?).

The fertile stamens are divided into a filament and a dithecal anther with four pollen chambers. In correlation with the corolla type, filament length varies considerably: in the short-tubed and flat flowers, in which the anthers are exposed (forming a cone), the filaments are usually short, in the long-tubed flowers they are long, sometimes slightly contorted. Insertion of the filaments is mostly at the base, less commonly in the middle and rarely at the top (near the mouth) of the corolla tube. Filament curving (for correct positioning of the anthers) and coiling (after pollen is shed) is often found when the anthers are coherent. The flat-flowered species of *Henckelia* sect. *Loxocarpus* have conspicuous filament knees that bear a yellow blotch and apparently act as anther/pollen dummies. In some palaeotropical Gesneriaceae there are sometimes spur-like appendages below the junction with the anther. In some species of *Ornithoboea* these appendages are globose knobs of yellowish colour apparently acting as anther dummies as well. In many species of Episcieae the filament bases are broad, sheath-like and fused to the dorsally split tube. The tube may provide protection for the superior ovary and channels the proboscis of a pollinator to the nectar, which is siphoned up from the nectary into the space between the corolla and the filament tube.

Anthers have a wide range of forms: the thecae may be parallel, divergent or transversal, insertion on the filament may be basi- or dorsifixed, the thecae may remain separate or be confluent at the top. The connective may be narrow or massive and separating the thecae strongly from each other (e.g. *Codonanthe*). Dehiscence is by longitudinal slits or less frequently by an apical or basal pore. Poricidal dehiscence and dry, powdery pollen is indicative of pollen flowers.

The anthers are often connate or coherent, either side by side, face to face, or at the apices. This is achieved by interlocking and fusion of the epidermal cells. Ultrastructural details of anther connation have been studied in *Sinningia cardinalis* (Lamond and Vieth 1972). When four stamens are present, all anthers may cohere (forming a cross-like structure) or, when the androeceum is didynamous, they may cohere pair-wise. The coherence is of functional importance, especially in flowers with exserted stamens: the cross- or pair-wise fusion holds the anthers in a stable position during pollination (Trapp 1956).

Perhaps the most specialised anther configuration is found in *Drymonia*, in which a kind of "salt-shaker mechanism" is present (Wiehler 1983): in the bud the four anthers are grouped coherently around the style, with their thecae facing inwards. As the anthers mature, they become connate along the length of the thecal margins and at their upper end, thus forming a rectangular or columnar unit. The individual thecae open by a short basal slit or pore. Just before anthesis, by lengthening of the corolla and the filaments, the curvature and the differential length of the filament pairs cause the anthers to turn upside-down, so that the pores become located apically. When entering the flower, the (insect) visitor touches the bottom of the anther unit which then tips over briefly like a salt-shaker and dusts the visitor with dry pollen. Not all the pollen is shed at once, but several visits are required to empty the anther unit.

After the pollen has been shed, the anthers are often removed from the corolla entrance by various filament movements (coiling, retraction, bending down- or upwards).

Normally, a nectary is developed between stamens and ovary. It can be assumed that the flowers of Gesneriaceae are primarily nectar flowers and that the (rather rare) lack of a nectary is a phylogenetically secondary condition. As floral ontogeny (Weber 1976a, 1976c, 1978a), floral anatomy (Wilson 1974a, 1974b), and the "adnate" nectary of the Coronantheroid Gesneriaceae (see below) demonstrate, the nectary is an enation of the ovary base and not of receptacular origin or a transformed inner stamen whorl.

There is a great variety of nectary forms. In the complete form it is cylindrical, cup- or ring-shaped, occasionally with shallow lobes at the rim. In the palaeotropical Gesneriaceae this is by far the prevailing form and other types (e.g. reduction to a ventral scale: *Henckelia* sect. *Glossadenia*) are rare. In the Coronantheroid Gesneriaceae the nectary is also complete, but adnate to (embedded in) the ovary base. In the neotropical Gesneriaceae only a dozen of the genera have a ring-shaped nectary; in the others it is strongly modified. Two evolutionary lines can be observed: (1) Beslerieae: complete ring → dorsally thickened ring → semi-annular structure → dorsal gland; (2) Gloxinieae, Sinningieae and Episcieae: ring → 5-lobed ring → 5 separate glands of equal size → 5 glands of unequal size (the two dorsal ones enlarged) → connation of the dorsal and stepwise reduction of the lateral and ventral glands → single dorsal gland. The nectary shape is of great importance in defining genera (Wiehler 1983).

Both in the palaeo- and neotropical Gesneriaceae, there are taxa in which the nectary is rudimentary (non-functional) or completely reduced (in pollen-flowers, deceptive pollen flowers and perfume-flowers; see Pollination).

The gynoecium is syncarpous and composed of two carpels in median position. It consists of the ovary, style and stigma.

The stigma may be inconspicuous (with a small area of papillae at the style tip), distinctly capitate, stomatomorphic (mouth-shaped) or bilobed. Under the latter term quite a number of different types must be distinguished (Burtt 1954). The position of the lobes may be median or lateral, with both or only the anterior carpel forming the two lobes.

A style seems to be always well developed. It may be straight or gently curved. In flowers with a distinct tube it is usually held immediately below the corolla roof. In some species of *Chirita* and in the Asiatic "*Streptocarpus*" *orientalis* and "*S*". *burmanicus*, it is held in a stable position between two flaps of the corolla roof. Position may be central or, in the case of enantiostyly, lateral (see below).

The outer shape of the ovary varies from (rarely) disc-like through globose, ovoid, elongated to long-cylindrical. In the long forms sometimes a narrow, tapering "stipe" (basal ovary part with sterile placentae) is present (species of *Didymocarpus* and *Aeschynanthus*).

Usually, the two carpels are developed equally, but there are exceptions in some E Asian taxa, in which the anterior carpel is much reduced and sterile: *Hemiboea*, *Briggsiopsis*, *Deinostigma*, some species of *Chirita*, *Metabriggsia* [this is said to have a single parietal placenta by Wang et al. (1990, 1992), but this is probably due to misinterpretation (wrong orientation of the ovary in which one of the carpels is completely reduced, thus representing a perfectly pseudomonomerous gynoecium)].

Ovary position may be roughly classified into superior, semi-inferior and inferior. Distinctly semi-inferior and completely inferior ovaries (including many transitional forms) occur only in the neotropical Gesneriaceae; slight sinkage into the receptacle is found in the other groups as well. Genera with superior, semi-inferior and inferior ovaries are found side by side in the Gloxinieae and in the Sinningieae, almost superior to fully inferior ovaries even in one and the same genus (*Sinningia*).

Position and shape of placentae is of great taxonomic importance. In fact, the distinction between uni- and bilocular (in terms of placentation: parietal versus axile) is the classical criterion separating Gesneriaceae from Scrophulariaceae. This, however, needs to be regarded critically in two respects: (1) there is principal congruence in the gynoecium architecture of the two families; in both families a completely bilocular portion (synascidiate zone) is found in the lower part and a unilocular portion (with incomplete septum; symplicate zone) in the upper part. Both portions may have placentae and be fertile; it is the different extent and proportion of the two portions that makes the difference (Weber 1971). (2) It is, therefore, not too surprising that converted ovary types occur in both families: unilocular ovaries in Scrophulariaceae (e.g. *Diplacus*, *Rehmannia*, *Dintera*, *Melampyrum*, *Lathraea*; see Hartl 1956), and ± completely bilocular ovaries, with axile placentae, in Gesneriaceae (*Monophyllaea*, *Whytockia*, *Allostigma*, *Sarmienta*, *Hemiboea*, *Deinostigma*, *Briggsiopsis*, some species of *Chirita*, the four latter genera with only one carpel fertile; Weber 1971; Wilson 1974a, 1974b). In general, however, in the Gesneriaceae, the bilocular part is short, sometimes without placentae, and in a few instances completely reduced.

The fused flanks of the two carpels (bearing the placentae at the margins) protrude into the ovary locule to a different extent. In some instances they touch in the centre, thus forming an incomplete septum, in others they are short or ± lacking, and then the placentae occupy a strictly parietal position. In ovaries with a complete septum (or in the lower, septate portion of the ovary), the placentae may remain separate or may form a unitary mushroom-shaped or pillow-like body in central (axile) position (Weber 1971; Wilson 1974a, 1974b).

The form of the individual placentae (in cross section) can be roughly classified into two types. In the Didymocarpoid Gesneriaceae they are rather thin, of equal thickness throughout their length (lamelliform), and curved back in the distal part. Especially in *Aeschynanthus* and allies, lamellae are long and markedly revolute. The production of ovules is usually confined to the distal, reflexed parts, sometimes even to the very ends. The sterile "inner" surfaces often touch along the median line of the locule. In some species of *Chirita* sect. *Microchirita* the epidermis of the touching, sterile portions of the placentae produces conspicuous folds that are covered by a thick mantle of mucilage at anthesis; in the reflexed, fertile portion the epidermal cells grow into short trichomes that fill the spaces between the funicles and again produce a thick film of mucilage. The mucilage is probably important for the growth of pollen tubes (Weber 1971).

The second type of placenta, occurring in the rest of the family (Epithematoid, Coronantheroid and Gesnerioid Gesneriaceae), is perhaps not uniform but it seems justified to keep it under a single heading. The prevailing form is that of a (roughly) right-angled triangle, with ovules being produced on all sides (except the short lamella of attachment to the carpel margin). Variation relates to the shape of the placenta (e.g. more rounded) and to the restriction of ovule production to the inner (e.g. species of *Columnea*, *Chrysothemis*) or outer (e.g. *Achimenes candida*) surface of the placentae.

Floral development has been studied in some members of Didymocarpoid Gesneriaceae (*Streptocarpus*, *Saintpaulia* and *Haberlea*: Oehlkers 1962; Harrison et al. 1999; Citerne et al. 2000) and Epithematoid Gesneriaceae (*Monophyllaea*, *Epithema*, *Rhynchoglossum*, *Whytockia*: Weber 1976a, 1976c, 1978a; Y.Z Wang et al. 1997; Y.Z. Wang, in prep.). Succession of floral organs follows the common pattern in Scrophulariales: sepals → petals (alternating with sepals) → stamens (alternating with petals) → carpels. Zygomorphy is expressed in different stages of floral development, either late (the young floral bud being actinomorphic) or early (with a dorsiventral gradient obvious already in the emergence of the sepal primordia: *Monophyllaea*, Weber 1976a). Staminode primordia seem to be always present in the young flower buds: they emerge either at the same time as the stamen primordia (but later showing successive delay) or slightly afterwards. The fusion of the corolla lobes to a sympetalous corolla is late (late sympetaly, Erbar 1991), taking place by meristem incorporation and -fusion across the backs of the anther primordia. The two carpel primordia arise ± simultaneously in median position. The form of the corolla is produced in the latest stages of floral development (well documented in Scrophulariaceae s.l.: Ritterbusch 1976, 1980; Wunderlin 1992).

The flowers of Gesneriaceae (especially *Streptocarpus*) have since long been the subject of genetic investigations (e.g. Lawrence et al. 1939; Lawrence and Sturgess 1957; Lawrence 1957, 1958), recently also combined with molecular genetics and systematics (Harrison et al. 1999; Citerne et al. 2000). A focus of interest is presently the genetic and developmental basis of floral actinomorphy (both in natural species and in peloric forms like the popular "Florist's Gloxinias") with the involvement of *cycloidea*-like genes (Cronk and Möller 1997; Möller et al. 1999; Citerne and Cronk 1999, Citerne et al. 2000).

Floral Anatomy. The floral anatomy of Gesneriaceae was studied by Wilson (1974a, b), paying special attention to the vascularisation of the nectary. As far as the gynoecium is concerned, some anatomical data were also provided by Weber (1971).

In Cyrtandroideae the nectary (when present) is not always vascularised. In the Didymocarpoid Gesneriaceae it is supplied by a branch from each of the stamen traces. The vascularisation of the gynoecium consists of the two carpel dorsal bundles, fused lateral bundles and (usually) fused ventral bundles. The carpel ventrals and laterals arise from two ventral-lateral mother strands, one on either side of the position of the loculus.

In the Epithematoid Gesneriaceae the vascularisation of the nectary is different: in *Rhynchoglossum* it is supplied by branching bundles arising from the central cylinder and carpel ventrals. These bundles supply also the gynoecial wall which – in contrast to all other Cyrtandroideae where the gynoecial wall is devoid of vascular tissue – is additionally richly vascularised. In *Epithema* the supply to the nectary arises from the central cylinder, and in *Monophyllaea* from the carpel dorsals and ventrals. The manner of formation of the carpel ventrals in *Epithema* is unique in the family: after producing traces to the sepals and petals, the vascular tissue condenses into a solid central cylinder; this breaks up into two large, median bundles and four smaller ones supplying the stamens and staminodia; the large bundles give rise to a carpel dorsal and then divide; each half swings around and fuses with the corresponding half of the other bundle, resulting in two large carpel ventrals. In the Epithematoid Gesneriaceae carpel laterals are generally small or lacking.

In the Coronantheroid and Gesnerioid Gesneriaceae the nectariferous tissue (when present) is nearly always vascularised, with varied nature and source of the vascular supply. With few exceptions, the gynoecium wall contains numerous vascular bundles in addition to the usual main bundles. The bundles completing the ring originate largely from either stamen traces in the receptacle or stamen bundles in the floral tube. The nectary is mostly vascularised from this ring. Vascularisation supports the view that the nectary is an enation and not an alternation of some pre-existing cycle of floral parts.

Like the vegetative organs, the floral parts (sepals, corolla, ovary, etc.) may be covered in various ways by epidermal trichomes in the form of eglandular or glandular hairs or sessile glands. Special cases include *Monophyllaea* and *Whytockia*, with chalk-secreting glands on the inner side of the sepals (Weber 1976a, b), *Henckelia geitleri*, with dense yellow glands on the style making it an anther dummy (Weber 1989a, as *Didymocarpus geitleri*), and *Emarhendia*, with patches of densely packed glands on the upper lip (Kiew et al. 1998) which secrete fatty oil (S. Vogel, pers. comm.). Occasionally, distinct rings or patches of hairs are found inside the corolla tube (e.g. *Hemiboea*, *Agalmyla* sect. *Agalmyla*, Hilliard and Burtt 2002). The hairs may be of functional significance in pollination. Boggan (pers. comm.) observed that many species of Sinningieae and Gloxinieae have flowers with glandular trichomes of various kinds and configurations inside the corolla tube, especially above the anthers, and that this character is strongly correlated with insect pollination. In contrast, hummingbird-pollinated flowers are nearly always glabrous inside. Studies at a broader scale are needed.

Embryology. Sporogenesis and embryology are still little known. Relevant papers include Balicka-Iwanowska (1899), Schnarf (1921, 1931), Laurent (1923), Oehlkers (1923), Glišić (1928, 1934), Thathachar (1942), Crété (1942, 1949, 1955a, 1955b), Arekal (1961), Swamy and Padmanabhan (1961), Holmqvist (1964), Anders (1966), Padmanabhan (1961, 1966), Adatia et al. (1971), Viemont and Astie (1974), Viemont (1976), Alimova and Yakovlev (1982), Johri et al. (1992), Svoma and Kiehn (1993), and Luegmayr (1994).

As far as is known, the anther anatomy is of the common dicotyledonous type, with four layers: epidermis, endothecium, middle (transitory) layer, and tapetum (Huber 1953; Weber 1976a). During meiosis of the pollen mother cells, the tapetal cells undergo one or several mitoses without cell wall formation, so that the tapetal cells become 2–5-nucleate. Fusion of nuclei results in polyploid tapetal cells. The developmental stages and manifold activities and functions of the tapetal cells have been studied in *Monophyllaea horsfieldii* by Luegmayr (1994), who also studied sporogenesis and the development of the male gametophyte by means of TEM. The development follows the regular pattern of angiosperms: pollen mother cell (larger than tapetum cells, large nucleus) → meiosis → tetrad stage (the four microspores arranged in a tetrahedral, occasionally decussate pattern; separation by callose; the highly active dictyosomes secrete numerous vesicles that form the primexine; at the same time secretion of Pro-Ubisch bodies by the tapetal cells) → free microspores (formation of the ectexine with tectum, columellae and foot layer and the lamellated endexine; between the columellae the primexine matrix is conspicuous) → vacuolisation of microspores → first mitosis → accumulation of storage substances → differentiation of the generative cells → mature pollen grain. The mature pollen grains are 2-cellular.

The ovules of Gesneriaceae are anatropous, unitegmic and tenuinucellate. The nucellus is (always?) single-layered and does not cover the chalazal part of the embryo sac. Embryo sac development conforms to the *Polygonum*-type. That of *Streptocarpus rexii*, described by Anders (1966), may serve as an example: at meiosis the diploid embryo mother cell measures 30–40 μm. The nucleus (10–12 μm) lies in the micropylar region, the chalazal region narrows down to 5 μm. After meiosis 3 of the 4 haploid nuclei (4–5 μm in diameter) degenerate, and the remaining embryo sac nucleus is the somewhat longer and thinner chalazal nucleus. After three mitoses the embryo sac becomes 8-celled. Fusion of the two polar nuclei results in a diploid primary endosperm nucleus. The three antipodals degenerate (but not in other taxa such as *Platystemma violoides*, Adatia et al. 1971), so that the mature embryo sac (now 140–180 µm long) comprises four cells: the egg cell, two synergids and the secondary endosperm nucleus. When the pollen tube intrudes into the embryo sac, one of the synergids degenerates. Fertilisation takes place in the usual double manner.

An overview of endosperm development has been provided by Johri et al. (1992). In *Platystemma violoides* the fertilised endosperm nucleus, lying in the centre of the embryo sac, divides earlier than the fertilised egg cell. Parallel transversal cell walls arise and divide the space into a micropylar and chalazal chamber. The micropylar chamber undergoes many transversal divisions; the uppermost pair of cells forms a micropylar haustorium, the other cells form a cellular endosperm of the *Brunella*-type (Schnarf 1931). In the chalazal chamber only one division takes place, which results in a 2-nucleate chalazal haustorium.

The micropylar chamber divides either transversally to form a linear row of three cells, or vertically. The micropylar haustorium is usually 2-celled and uninucleate, but 4-celled haustoria have been also reported.

The chalazal haustorium is either uninucleate, binucleate, or two-celled.

Endosperm presence/absence in the seeds is an old distinction criterion between the neotropical and palaeotropical Gesneriaceae (Brown 1839). However, also in the latter at least a single cell layer of endosperm is (sometimes? always?) present. A detailed survey is needed.

The embryo development (embryogeny) can be referred to the Onagrad-type. In the 8-cell stage the peripheral cells give rise to the dermatogen, whereas the inner ones function as the initials of the periblem and plerome and eventually take part in the formation of the hypocotyl. The cotyledons originate from a terminal cell group. At the opposite end some inner cells become the initials of the central cylinder of the primary root. Elongate and multinucleate suspensor cells have been reported for *Ramonda myconi*. The embryo of *Rhytidophyllum crenulatum* is said to lack cotyledons and tissue differentiation (Cook 1907), but confirmation is needed.

Pollen Morphology. Apart from older studies by LM (Erdtman 1952 and earlier authors), the pollen of Gesneriaceae has been recently studied in more detail by means of SEM and TEM (Gesnerioideae: Melhelm and Mauro 1973; Williams 1978; Fritze and Williams 1988; Cyrtandroideae, esp. SE Asian and Hawaiian taxa: Luegmayr 1993a, b). A detailed TEM study of the development of the generative cell (GC) of *Cyrtandra pendula* was carried out by Luegmayr (1993c).

The pollen grains are generally small- to medium-sized, 3(rarely 4)-colp(or)ate, isopolar monads with a perforate or reticulate exine. The form is usually spheroidal, rarely prolate, suboblate or oblate; in polar view the outline is circular, rarely subangular or triangular. The diameter varies in the range 9–24 µm in the palaeotropical Gesneriaceae, whereas in the neotropical members the grains are larger, ranging from 25–51 µm in diameter. Acetolysis seems to have no effect on pollen shape and size. The colp(or)ate apertures are usually constant in length in the palaeotropical Gesneriaceae, but vary in the New World Gesneriaceae from long to short. The pollen wall is composed of a tectate ectexine with distinct columellae, a foot layer, endexine and intine. The ectexine is mostly perforate or reticulate. Microreticulate or rugulate patterns (*Epithema*) are rather exceptional.

Decrease of the brochus size towards the apocolpium of reticulate exines (heterobrochate exine) is frequently observed in the neotropical Gesneriaceae, while in the palaeotropical ones the exine is usually homobrochate. Rod-like luminal processes are found, for instance, in *Stauranthera grandiflora* and *Streptocarpus silvaticus*. In *Chirita* and *Cyrtandra* spp. conical supratectate sculptural elements are present. Columellae and foot layer are always present, but vary considerably in length and thickness respectively. The endexine is lamellated, forming a sculptured (granular) aperture membrane (lacking in *Epithema*). The intine is 2-layered with the outer layer (exintine) being thickened in the apertural region and having channel-like structures.

Pollenkitt is often present on the wall surface.

A general assessment of the taxonomic value of pollen characters cannot be made. Large genera are often eurypalynous and as heterogeneous as in their macroscopical characters. Only few genera, such as *Aeschynanthus* and *Epithema*, proved to be stenopalynous.

Reproductive Systems and Pollination. With few exceptions, the flowers of Gesneriaceae are cross-pollinated and zoophilous, with insects (mainly bees), birds and bats acting as pollen vectors. The speciation and evolutionary diversification of the family is closely linked with the mode of pollination. A survey of various flower forms, colorations and pollination syndromes of neo- and palaeotropical Gesneriaceae is presented in Figs. 15 and 16. Field studies on pollination are still scarce (Vogel 1966; Jones and Rich 1972; Roelofs 1980; Snow and Snow 1980; Steiner 1985; Kato et al. 1993; Cory 1994; Sazima et al. 1995; San Martin-Gajardo and Freitas 1999; San Martin-Gajardo and Sazima 2002a, 2002b) and many of the considerations below are based on the interpretation of flower characters and pollination syndromes.

Except for a few chiropterophilous species of *Gesneria* with protogynous flowers, cross-pollination is promoted or enforced by protandry. Another device enhancing cross-pollination is enantiostyly (left-right-position of style). In Gesneriaceae, enantiostyly seems to be always of the monoecious type (both morphs occurring in one plant or inflorescence). In *Saintpaulia* it is only the style that changes position; in *Didymocarpus* and *Henckelia* the style and the anthers occupy reciprocal positions.

In entomophilous flowers, five types relating to the reward offered can be distinguished.

1. Nectar flowers (Fig. 15A–D): corolla with a distinct (often long) tube, coloration ± conspicuous, limb zygomorphic, nectar-secreting disc present, nectar often accumulating in distinct bulges at the corolla base; pollen sticky.
2. Deceptive nectar flowers. The flowers exhibit the general habit of nectar flowers but the nectary, if present at all, is non-functional. The flowers of some species of *Ornithoboea* may belong to this type: by the spreading calyx and the corolla shape, they strongly resemble orchid flowers; they may indeed mimic orchid flowers (*Calanthe*?) but at least bear the general habit of nectar flowers (Burtt 1958a; Weber 1979).
3. Pollen flowers. These exhibit the general character syndrome of oligandric pollen flowers (see Vogel 1978): nodding flower position; corolla short-tubed, campanulate or flat-faced, with (sub)regular limb; stamens sometimes five, anthers ± exposed, large, sometimes forming a cone or coherent, often bright yellow or orange, dehiscing by apical pores or transversal slits, pollen non-sticky, powdery; nectary non-functional or reduced. Flower exploitation is apparently by buzzing. Striking examples include *Ramonda* (Fig. 14E), *Conandron*, *Didymocarpus cordatus*, *Bellonia* (Fig. 15E), *Niphaea*, *Phinea* and *Napeanthus*.
4. Partially and fully deceptive pollen flowers. Simulation of copious pollen is by (a) yellow anther walls (e.g. *Saintpaulia*, Fig. 15F: by the robust, yellow wall the anthers are attractive for a much longer time than pollen is available, Vogel 1978, 1993), (b) a yellow blotch on the corolla above the anthers (e.g. *Petrocosmea kerrii*, Fig. 15G), (c) yellow filament knees (*Henckelia caerulea*: exserted filaments with yellow knees, the true anthers are of cryptic colour; Weber and Kiew 1983; Vogel 1993; Fig. 15H), (d) yellow blotches on the corolla near filament insertion (e.g. *Henckelia puncticulata*; Weber 1989a), (e) a patch of yellow, moniliform hairs at the palate (plastic imitation of pollen grains?: *Stauranthera*; Weber, in prep.), or (e) an exserted yellow style (*Henckelia geitleri*: style studded with yellow glands; Weber 1989a; Fig. 15I).
5. Perfume flowers. This flower type is known from *Gloxinia perennis* (Vogel 1966), other species of *Gloxinia* sect. *Gloxinia* and *Monopyle* (Wiehler 1983). The corolla is rather short- and broad-tubed, and an osmophore is present near the base, replacing the nectary (Fig. 15K). The osmophore consists of palisade-like, purple-brown epidermal cells that secrete tiny droplets of terpenes.

As far as can be inferred from the floral characters (pollination syndromes), insects involved in pollination include bees (Apidae, including the Euglossini in the neotropics, Anthophoridae, ?Xylocopinae and others), butterflies (rare) and moths (very rare). Wasps, dipterans, gnats, etc. are not definitely known as pollinators so far; if at all, they probably play an insignificant role. Unfortunately, observations of actual pollination are scanty. Field studies are imperative to confirm or refute the conclusions drawn from the pollination syndromes. This applies also to the presumably bird- and bat-pollinated plants discussed below.

Bee flowers providing nectar are usually broad-tubed, infundibuliform or (zygomorphic-) campanulate, with a wide-open mouth (gullet or bell flowers) and of light colour (white, blue, violet, yellow in various shades and combinations). Personate flowers (mouth closed by a vault of the palate) are rare, examples include *Rhynchoglossum* (Fig. 15D), *Didymocarpus antirrhinoides* and *D. corchorifolius*. Pollen deposition is mostly nototribic. In *Drymonia serrulata* glandular trichomes inside the corolla secrete an oil which is deposited on the anthers and transferred to the thorax of visiting bees (*Epicharis* spp., Anthophoridae). The oil facilitates pollen adhesion (Steiner 1985). The remarkable "salt-shaker"-mechanism of the coherent anthers of *Drymonia* is described under Floral Structure.

With regard to Euglossine bees, Wiehler (1983) distinguished two pollination syndromes: (1) Gynandro-euglossophily: both male and female Euglossines visit the flowers in search of nectar. The corolla is relatively large, broad-tubed, the limb is subregular, with rounded lobes, colour is light (white, light yellow or rose), and a nectary is present. The so-called episcioid flower (which apparently evolved independently in several neotropical genera such as *Episcia, Drymonia, Nautliocalyx, Paradrymonia*, etc.) represents a special subtype: colour white or whitish, corolla lobes conspicuously dentate or fimbriate, corolla tube with a dorsal sac or blunt spur at the base (Fig. 15J). Gynandro-euglossophily is a special, but not yet satisfactorily confirmed form of melittophily. Bees other than Euglossini may be legitimate pollinators, too, and therefore this pollination syndrome is here referred to as melitto-/euglossophily. (2) Andro-euglossophily: male Euglossine bees collect fragrance compounds, produced by the osmophore (see above under "perfume flowers"). The perfume is used by the male bees to mark the swarm areas and to attract female bees (Vogel 1966). Outside Gesneriaceae this pollination type is known from numerous orchids, some Solanaceae, Araceae and Euphorbiaceae.

Pollen and fully deceptive pollen flowers are devoid of nectar and are exclusively addressed to bees. For details see above.

Probably psychophilous flowers (nectariferous; corolla hypocrateriform, with a long, narrow tube; limb flat, red or blue) are known from some species of *Achimenes*, Sinningia (Fig. 15L) and *Episcia*. *Sinningia tubiflora* is a striking example exhibiting a sphingophilous syndrome (corolla long- and narrow-tubed, white, strongly fragrant; Perret et al. 2001; Fig. 15M).

The "key-hole flowers" of some *Streptocarpus* species (tube narrow, limb flat, tube entrance a small, laterally compressed hole) are possibly visited by long-proboscid flies.

Bird pollination (ornithophily) occurs in the palaeotropical and neotropical Gesneriaceae (Fig. 16A–I), as well as in the southern hemisphere Gesneriaceae (*Rhabdothamnus, Mitraria, Sarmienta, Asteranthera*). It is rather rare in the palaeotropical Gesneriaceae, where only few genera (*Aeschynanthus, Agalmyla*, Fig. 16A–C) and particular species of species-rich (essentially melittophilous) genera (*Henckelia, Streptocarpus, Cyrtandra*) exhibit an ornithophilous pollination syndrome (flowers bright red, tubular or salverform, with abundant nectar). Almost all species of *Agalmyla* and *Aeschynanthus* are climbers and epiphytes, and display their bright red flowers (often in dense bunches) high above the ground. In some Malayan species of *Aeschynanthus* the nectar sugar composition has been studied by Freeman et al. (1991). They found a low sucrose content, as is typical of ornithophilous flowers, and a roughly 1:1 ratio (45:53%) of fructose to glucose. Observations or special studies of pollination are largely lacking. The probable bird pollinators in Africa and Asia are Sunbirds (Nectariniidae) and, in the area from Sulawesi to New Zealand, Honey-eaters (Meliphagidae).

In contrast, bird pollination is apparently very frequent in the neotropical Gesneriaceae (perhaps 60% of the species). Here the pollen vectors are essentially hummingbirds (Trochilidae). It is remarkable that the distribution centre of neotropical Gesneriaceae and that of hummingbirds coincides (Colombia and Ecuador). There is a strong correlation of ornithophily and an epiphytic habit of the plants, but also many terrestrial species seem to be pollinated by hummingbirds.

The hummingbird flowers are typically red, orange or bright yellow. In the ornithophilous taxa of Episcieae, the corolla shape can be roughly classified into four types (Wiehler 1983): (1) "tubular": tube cylindrical, straight or slightly curved, limb subregular [e.g. species of *Achimenes, Moussonia, Kohleria, Sinningia* (Fig. 16D), in the majority of species of *Dalbergaria* = *Columnea* sect. *Collandra, Pentadenia* = *Columnea* sects. *Pentadenia* and *Stygnanthe* and *Trichantha* = *Columnea* sect. *Ortholoma*]; (2) "columneoid": tube narrow at base, expanding apically, limb strongly zygomorphic, divided into a prominent galea (consisting of the two enlarged dorsal lobes and two triangular lateral lobes) and a long and narrow, usually deflexed ventral lobe (e.g. *Columnea* s.str. = *Columnea* sect. *Columnea*; Fig. 16E, F); (3) "hypocyrtoid" (urceolate): tube ± strongly inflated, mouth much constricted ("target flowers"; e.g. *Pearcea hypocyrtiflora*, many species of *Gasteranthus* and *Nematanthus*, Fig. 16H, I); and (4) "converted melitto-/euglossine flower": corolla shape as in bee flowers, but coloration deep orange or red (instead of white or light yellow). This type apparently represents a recent switch from melitto-/euglossophily to ornithophily (e.g.

species of *Episcia*, *Nautilocalyx*, *Drymonia* and *Kohleria*, Fig. 16G). An additional type, combining characters of (1) and (2) (flowers tubular, limb strongly bilabiate, the large upper lip projecting forwards), is found in *Pheidonocarpa* and many species of *Sinningia* and *Gesneria*.

Apart from flower characters, also extrafloral cues may play a significant role in the attraction of hummingbirds: large, coloured bracteoles in the inflorescences (e.g. *Drymonia* spp.) or red blotches, margins or translucent windows on the foliage leaves (*Dalbergaria* = *Columnea* sect. *Collandra*) (Figs. 11A, 13E, 16I). The flowers are, in contrast, often rather inconspicuous and hidden in the foliage. As has been observed in *Columnea florida*, young and inexperienced hummingbirds (*Phaethornis longuemareus* and *Heliodoxa jacula*) first aim at the red spots on the leaves and then spend several seconds inspecting various portions of the plant before finding the flowers. Experienced birds hover briefly above the red spot and then dip quickly under the leaves and approach directly the axillary flowers (Jones and Rich 1972).

Bat pollination (chiropterophily) is known (or presumed) only in some neotropical Gesneriaceae and obviously has evolved independently within several genera (*Capanea*, *Drymonia*, *Kohleria*, *Gesneria*, *Rhytidophyllum*, *Sinningia* and *Paliavana*; Fig. 16J–L). The flowers are short- and broad-tubed or distinctly campanulate, the limb is ± zygomorphic, coloration is yellowish or greenish with brown or dirty violet spots; nectar is produced in abundance. The chiropterophilous pollination syndrome of some neotropical taxa has been described in detail by Vogel (1958, 1968, 1969a, b) and Skog (1976); for general aspects and a compilation of Gesneriaceae species and references, see Dobat and Peikert-Holle (1985). Pollination of *Sinningia brasiliensis* and *Paliavana prasinata* by Phyllostomidae-Glossophaginae has been recently documented by San Martin-Gajardo and Sazima (2002a, b). Both species exhibit a typical chiropterophilous pollination syndrome that includes nocturnal anthesis, unpleasant scent and increase of nectar secretion at night.

Some genera, especially of the neotropical Gesneriaceae, exhibit a wide range of pollination syndromes, e.g. *Sinningia* (melitto-/euglosso-, psycho-, sphingo-, ornitho-, chiropterophily), *Achimenes* (melitto-/euglosso-, psycho-, ornithophily), *Gloxinia* s.l. (melitto-/euglosso-, andro-euglosso-, ornithophily), *Kohleria* (melitto-/euglosso-, ornitho-, chiropterophily), *Gesneria* (ornitho-, chiropterophily) (Wiehler 1983; Perret et al. 2001, 2003; Vogel, pers. comm.).

In the Sinningieae, nectar sugar composition in relation to pollination syndromes was studied by Perret et al. (2001). In the hummingbird and bee flowers (representing 95% of the species), nectar was found to be sucrose-dominant, with an average sugar concentration of 23.9 ± 10.6 and 28.7 ± 10.6% respectively. The nectar of the sphingophilous and chiropterophilous flowers differed significantly in the hexose (fructose + glucose) dominance, with a low sugar concentration (7.1 ± 3.4%) in bat flowers.

Attraction of pollinators is primarily by visual cues. The majority of Gesneriaceae has flowers that are devoid of any floral scent. This particularly holds true for the ornithophilous flowers. However, for a number of species (from ca. 15 genera, essentially neotropical) fragrance has been reported (Boggan 1996). The range of odours is surprisingly wide: sweet honey-like, lemony, musky, clove-, cabbage-, carrion-like and others. Scent seems to be addressed to special pollinators, e.g. *Sinningia tubiflora*, with powerful lemony fragrance, to moths, or *Capanea* cf. *grandiflora*, with cabbage-like scent, to bats.

Self-pollination (facultative or obligatory autogamy) is known or suspected for a rather small number of Gesneriaceae (e.g. species of *Epithema*, *Rhynchoglossum*, *Monophyllaea*, *Sinningia pusilla*?). Their flowers are mostly small and inconspicuous, often produced in large numbers, and fruit set is almost 100%. In a few cases the flowers are cleistogamous (e.g. *Streptocarpus nobilis*, *Codonanthopsis dissimulata*, Moore 1973).

Fruit and Seed Structure, Seed Dispersal. Fruit structure is diverse within the family, and significant both in ecological and taxonomical respects. A special study surveying the whole family was carried out by Ivanina (1965, 1967), and important complements relating to the neotropical Gesneriaceae were provided by Wiehler (1983). A variety of fruits of palaeo- and neotropical Gesneriaceae is illustrated in Figs. 17 and 18.

The traditional classification of the gesneriaceous fruits into "capsules" and "berries" is better replaced by "dehiscent" and "indehiscent fruits", as the gesneriaceous berries often do not conform with the type which is commonly associated with that term. Dry capsular fruits are found in all major groups and there is little doubt that they represent the more primitive type.

Dry capsular fruits prevail by far in the palaeotropical Gesneriaceae (Fig. 18A–F). In many Didymocarpoid Gesneriaceae the capsule is elongated and pod-like (in some species of *Aeschy-*

nanthus and *Agalmyla* more than 40 cm long; Fig. 18A). Dehiscence is mostly by longitudinal slits, so that the fruit opens with valves, either septicidally (rare, e.g. *Ramonda, Jancaea, Haberlea, Corallodiscus*), loculicidally (the most frequent type; e.g. *Didymocarpus, Chirita*) or both (four-valved capsule; e.g. *Leptoboea, Boeica, Championia, Tengia*). In *Schizoboea, Lysionotus* and others, the fruit regularly disintegrates into six parts: four valves and the sclerified lateral bundles. In *Tetraphyllum* the sclerified bundles remain united at the style base, while the four valves are shed. In the tardily dehiscent fruits of *Didissandra*, the valves break into lamellae along the sclerified lateral and parietal bundles.

Regarding seed dispersal, these simple capsular fruits are unspecialised, the seeds are shaken out by wind, rain or passing animals, and simply fall to the ground.

In a number of genera (*Streptocarpus, Boea, Paraboea* p.p., *Trisepalum, Ornithoboea, Senyumia, Spelaeanthus*), the (loculicidally dehiscent) fruits are twisted, the two valves being helically wound (Fig. 18B). Twisting does not take place in the ripe, dry condition, but results from a special mode of growth starting after anthesis. The ecological significance seems to lie in the fact that seed release (by wind or water) is retarded and prolonged.

Another characteristic type is the "follicular" fruit, in which dehiscence takes place in the upper carpel only (loculicidally along the median carpel bundle). This (usually long) fruit is held ± horizontally, making an angle with the pedicel. In *Henckelia* sect. *Didymanthus* this is achieved by unequal growth of the receptacle, in sect. *Loxocarpus* mostly by unequal growth of the carpel base. This type of fruit is found in plants of ever-wet rainforests, and represents a special form of a rain-wash or rain-splash capsule: when the fruit opens dorsally, the seeds lie exposed and are washed away by the rain. Examples are *Ridleyandra* (Fig. 18C), *Henckelia* (Fig. 18D), *Paraboea* p.p., *Hovanella* and *Trachystigma*. By progressive shortening of the fruit, *Henckelia browniana* and allies approach a typical bowl-shaped rain-splash cup (Fig. 18E).

Short, ovoid or globose capsules open in the ways described above or by a circumscissile line (*Epithema*: a perfect rain-splash cup embedded in a funnel-shaped calyx; Weber 1976c; Fig. 18F), or the upper part of the fruit wall breaks irregularly into pieces (*Loxonia burttiana*; Weber 1977a).

In the Coronantheroid and Gesnerioid Gesneriaceae there are fewer genera with dry capsular fruits. Septicidal dehiscence is very rare (*Depanthus, Anetanthus*), loculicidal dehiscence is common (e.g. throughout Gloxinieae and Gesnerieae), and four-valved fruits are represented too (*Coronanthera, Negria, Rhabdothamnus, Napeanthus* spp.). According to J.K. Boggan (pers. comm.), the dry capsular fruits found in the Gloxinieae, Sinningieae, *Pheidonocarpa* and *Bellonia* are remarkably uniform: ovoid, rostrate with an upturned apex, the two valves opening only slightly, and only splitting the superior portion of the ovary; the insides of the valves with a row of stiff trichomes, releasing the seeds like a salt-shaker (Fig. 17A). The capsules of *Lembocarpus, Rhoogeton* and *Cremersia* (all of uncertain tribal affiliation) are quite different: here the two short valves open to 180°, the valves being pushed apart by pads of swelling tissue situated at the base of the valves (Boggan 1991, pers. comm.; Feuillet and Skog 2003). Horizontal, follicular fruits occur in particular species of *Diastema, Monopyle, Gloxinia lindeniana* (Fig. 17C) and *Kohleria amabilis*, suggesting dispersal by rain. An example of a bowl-shaped rain-splash cup is *Gesneria humilis*. Irregular capsule dehiscence has been reported for *Cremosperma, Cremospermopsis* and *Tylopsacas*.

The mode of capsule dehiscence is rather independent of ovary position. In fruits resulting from (semi-)inferior ovaries, either the apical, free ovary parts open or the inferior parts split into halves or on one (the upper) side (*Monopyle* spp.).

In the neotropical Gesneriaceae numerous forms of fleshy capsules are found (Fig. 17D–K). They are common in the Episcieae, developing from a superior ovary, but are also found in some species of Gloxinieae and Sinningieae, occurring here in conjunction with inferior ovaries, in which case the valves frequently split the ovary to the base, tearing open the hypanthium. In the Episcieae the fruit valves are often coloured and the seeds are exposed in a conspicuous manner (display fruits). Dispersal is apparently by birds; possibly the fruit is also eaten by mammals such as bats or monkeys (Wiehler 1983). At least two kinds of display capsules can be distinguished: (a) (semi-)fleshy capsules, in which the opening convex valves form a cup which presents the seeds with their prominent funicles (*Episcia, Nautilocalyx, Chrysothemis, Paradrymonia*, etc.; Fig. 17E, J), (b) fleshy capsules in which the valves curve back and become separated from the placentae, the latter forming a central cone covered by the glistening mass of funicles and seeds (e.g. *Alloplectus, Drymonia* and *Nematanthus*; Fig. 17G–I). Curious are the fruits of *Solenophora calycosa*: the fleshy capsules (resulting from an inferior ovary crowned by a persistent calyx) split in an irregular manner, exposing a central red globe of placentae and seeds (Fig. 17K).

Indehiscent fruits are common in the Gesnerioid and Coronantheroid Gesneriaceae. True (fleshy) berries, either white or coloured and of variable shape (globose, ovoid, depressed, etc.) are found in *Columnea* s.l. (Fig. 17L), *Corytoplectus*, *Rufodorsia*, *Neomortonia*, *Codonanthe*, *Drymonia* p.p., *Besleria*, *Asteranthera*, *Sarmienta* and *Mitraria*. Significant differences can be found in the nature of the pulp: in the episcioid genera it consists of plump and swollen funicular tissue, in *Besleria* of enlarged placental tissue. In a few *Besleria* species from Venezuela, French Guiana and Panama the berries peel, either irregularly or loculicidally plus septicidally. The bright-coloured, paper-thin pericarp becomes reflexed or coiled above the hanging remainder of the fruit which consists of attractively coloured placental tissue and seeds.

In the palaeotropical Gesneriaceae, indehiscent fruits are found only in two or three genera: *Cyrtandra*, *Sepikea* (if regarded as distinct from *Cyrtandra*) and *Rhynchotechum*. By far the largest genus is *Cyrtandra*, with ovoid or somewhat elongate and often very hard (sclerocarpous) fruits (Fig. 18G, H). It is still unknown how these fruits are dispersed. More typical (fleshy) berries are found in the Pacific (e.g. Hawaiian) species of *Cyrtandra* (Fig. 18I). These are eaten with some probability by birds.

Ovules and seeds are generally produced in great numbers. The seeds are always small, usually less than 1 mm long and 0.5 mm broad. Shape is variable, to some degree even in a single capsule. The ellipsoidal form dominates, less common are globose or spindle-shaped seeds. The seeds of neotropical Gesneriaceae tend to be larger, as they have generally a larger amount of endosperm. Here, the smallest seeds are found in the Gloxinieae, Gesnerieae and Beslerieae (ca. 0.5 mm), larger seeds (ca. 1 mm) occur in the Sinningieae, and the largest ones (usually over 1 mm) in the Episcieae (Boggan, pers. comm.). The largest seeds are found in *Codonanthe* (ca. 3 mm). In the species living as ant-nest epiphytes the capsules contain only ca. 35 seeds (Wiehler 1983).

Seed anatomy seems to be fairly uniform. The embryo is always surrounded by a layer of medium-sized cells, which is followed by a single- (palaeotropical Gesneriaceae; in older treatments these are wrongly described as being devoid of endosperm) or several-celled layer of endosperm. The endosperm is followed by another cell layer, in which the cell lumen is (always?) filled with tannins. The two outermost cell layers, originating from the integument of the unitegmic ovules, are the endothelium and the testa. The endothelium cells are much larger than the testa cells, their shape is quadrangular, rectangular, hexagonal or irregular. The endothelium cells are arranged in longitudinal rows, and invaginate to some degree the endosperm. From the outside, the seeds often look slightly furrowed, and can be designated as aulacospermous, as found in many Scrophulariaceae (Hartl 1959). Bothrospermous seeds are not known in Gesneriaceae.

The testa cells are usually polygonal, either isodiametric (rare) or elongated in longitudinal direction. The contiguous cell walls are usually well visible. The testa cells form straight or spirally arranged cell rows. Distinct spiral cell orientation is found in the Epithematoid Gesneriaceae as well as in most neotropical Gesneriaceae. The colour of the testa is usually light or dark brown, sometimes blackish. In the neotropical Gesneriaceae having display fruits, the seed surface is sometimes pink or reddish.

Seed coat structure (surface sculpture) was surveyed in the family by means of SEM by Beaufort-Murphy (1983). She worked out a detailed terminology and discussed the taxonomic significance of seed surface characters in various systematic groups. The studies of Sontag and Weber (1998) and Mühlbauer (in Weber and Burtt 1998d) demonstrated the potential for a better delimitation of genera such as *Didissandra*, *Ridleyandra*, *Didymocarpus*, *Henckelia*, *Hovanella*, etc.

Distinct ornamentation of the testa cells seems to be restricted to the Didymocarpoid Gesneriaceae. A "verruculate" testa is characteristic of African taxa (*Schizoboea*, *Hovanella*, *Saintpaulia*, *Linnaeopsis*, *Streptocarpus* subg. *Streptocarpella*): each testa cell produces a dome-shaped elevation in the centre. A great diversity is found in *Henckelia*: moniliform knots along the cell borders, evenly distributed knobs or knobs filling the cell centre. The nature of the knobs has been analysed in *Chirita* (Buchner et al., unpubl. data): they represent concretions within the outer cell wall. The wall consists of an inner and an outer lamella (with cuticula) and a matrix between the two. The sculpture elements are embedded in the matrix, rooting in the inner lamella. The form and arrangement of the sculpturing elements is subject to considerable variation.

Distinct seed appendages in the form of triangular or (more commonly) long filiform structures are produced in the genera hitherto accommodated in the (probably artificial) tribe Trichosporeae. The apical end of the seed bears always a single appendage which arises from an elongation of the ovule tip. A first step in the formation of a seed appendage may be seen in the apically pointed seeds of *Anna*. A consecutive step is the rather short, triangular appendage of *Aeschynanthus* sect. *Microtrichium*. In the other sections of *Aeschynanthus*, the apical appendage is hair-like, sometimes reaching a length of 4–5 cm. The hilar end of the seed may bear a single hair as well. This corresponds to the funicle in *Lysionotus*, but represents a separate outgrowth in *Agalmyla* and *Aeschynanthus*. In the latter genus there are also species with two, several or many hairs at the hilar end of the seed. The number of hairs has been used already by Bentham (1876) for distinguishing sections. Testa and seed appendage morphology in *Aeschynanthus* and its phytogeographical and taxonomic implications have been recently studied by Mendum et al. (2001). Seed ontogeny was examined by Christie and Mendum (2001). Dispersal of appendaged seeds is obviously by wind. A slight current of air is sufficient to blow the seeds away. In these epiphytic plants the hairs (especially when getting wet) also may help to attach the seeds to the bark of branches or tree trunks. In the species with a coma of hairs, the hairs spread in dry weather and form a parachute. The great effectiveness of a hair coma in wind dispersal explains the wide distribution of *Aeschynanthus* sect. *Polytrichium* in Asia and Malesia (Mendum et al. 2001).

The enlarged, fleshy, white funicles of the neotropical *Chrysothemis friedrichsthaliana* are rich in lipids and serve as a kind of elaiosome in ant dispersal (Lu and Mesler 1981). Also the seeds of *Codonanthe* bear a many-celled, hyaline and long funicle. In this genus the unusually large (ca. 3 mm long) seeds are clad by an aril-like cupule in the lower half. Some *Codonanthe* species are typical ant-garden plants. Seed transport by ants has been observed by Kleinfeldt (1978).

In summary, seed dispersal may be (a) in an unspecialised manner (autochory; seed release by wind, water or passing animals), (b) by wind (anemochory; by the small size and weight the grains may drift over some distance; distinct adaptations to wind dispersal are the hair-like appendages of *Aeschynanthus* and allies), (c) by water (ombrohydrochory; rain-wash and rain-splash capsules), (d) by animals (zoochory; birds, perhaps also small mammals; berries and display capsules), and (e) by ants (myrmecochory).

Karyology. The chromosome numbers known up to 1984 have been summarized by Skog (1984). Since then relatively few chromosomal studies have been published. Recent papers, all devoted to palaeotropical Gesneriaceae, include Kiehn et al. (1998), Jong and Möller (2000), Kiehn and Weber (1998, 2001), and Rashid et al. (2001).

At present, no basic number can be identified for the family as a whole. In general, the basic numbers are moderately high, ranging between $n = 8$ and 17. The lowest number found so far is $n = 4$ (*Chirita pumila*), the highest is $n = 64$ (*Streptocarpus hildebrandtii*).

In neotropical Gesneriaceae, chromosome numbers are largely consistent within genera. The apparent uniformity of numbers and the combination with breeding data have provided important arguments for the definition of genera: for the separation of *Eucodonia* ($n = 12$) from *Achimenes* ($n = 11$) the discordant chromosome number was regarded the most significant feature (Wiehler 1983); for the inclusion of *Rechsteineria* and *Lietzia* into *Sinningia* (both $n = 13$), or *Hypocyrta* into *Nematanthus* (both $n = 8$), hybridisation experiments played an essential role (Moore 1973; Wiehler 1983; Wiehler and Chautems 1995). If dysploidy is taken into account, the chromosome numbers are even consistent within tribes: in Gloxinieae most genera have $x = n = 13$, a few $n =$ 12, 11 or 10, in Episcieae the number is $n = 9$ almost throughout (two genera have $n = 8$), in Sinningieae $n = 13$, in Gesnerieae $n = 14$ (see Wiehler 1983; Burtt and Wiehler 1995). Based on a combination of molecular and karyological data, two base number are assumed by Zimmer et al. (2002) in the neotropical Gesneriaceae: $n = 16$ (Beslerieae and Napeantheae) and $n = 13$. The latter number has been retained in Gloxinieae (with rare reductions to 12, 11 or 10) and Sinningieae, and has been increased to $n = 14$ in Gesnerieae) and reduced to $n = 9$ in Episcieae (with occasional reduction to $n = 8$).

While the neotropical Gesneriaceae are relatively well known chromosomally, fewer data are available for the palaeotropical Gesneriaceae, especially for the Chinese and Indo-Chinese genera. The largest genus, *Cyrtandra*, with 450 to 600 species, is remarkably uniform. With the doubtful exception of three unidentified species from New Guinea ($n = 16$), all species examined so far (ca. 50) have $n = 17$. Other genera with a single number include *Agalmyla* ($n = 16$), *Ridleyandra* ($n = 17$) and *Petrocosmea* ($n = 17$), but relatively few species have been counted in these genera.

In a number of palaeotropical genera two or several basic numbers are found in one genus, evidently due to ascending or descending dysploidy (e.g. *Streptocarpus*: $n = 15, 16$; *Aeschynanthus*: $n = 14, 15, 16$; *Didymocarpus*: $n = 11, 12, 14, 16, 18$). Ascending dysploidy may contribute to explain speciation in *Monophyllaea*: the series of apparent morphological advancement, represented by *M. horsfieldii* → *M. hirticalyx* → *M. hendersonii*, is paralleled by increasing dysploidy: $n = 10, 11, 12$ (Kiehn and Weber 1998). The number $n = 9$ of the ancestral and sister genus *Whytockia* (Weber 1976b; Mayer et al. 2003) can be interpreted as the original basic number in a continuous series of ascending dysploidy. Descending dysploidy has likely led to different basic numbers in *Didymocarpus*. In the Malayan *D. cordatus* ($x = 14$), an accession with $x = 13$, and in *D. antirrhinoides* an accession with $x = 10$ has been found. These discordant numbers are apparently the result of fusion of two pairs of chromosomes, as two larger chromosomes are detectable in each diploid cell (Kiehn et al. 1998). As only decreasing numbers (14 → 13 and 11 → 10) have been discovered so far, decreasing dysploidy starting from $x = 16$ or 18 is the probable mechanism of chromosomal evolution in *Didymocarpus*.

Polyploidy occurs to a different extent in the major groups of Gesneriaceae. It is remarkably rare in the neotropical Gesneriaceae. Even large genera such as *Columnea* s.l., *Sinningia* and *Nautilocalyx* are highly uniform and contain at most a few tetraploid species (other examples for tetraploidy are some species or populations of *Achimenes* and *Eucodonia*). This uniformity both in number and diploid level strongly contrasts with the Coronantheroid Gesnerioideae, in which generally high numbers are found (*Mitraria*, *Sarmienta*, *Rhabdothamnus*: $n = \pm 37$, *Fieldia*: $n = \pm 40$, *Negria*: $n = \pm 45$. When seen in connection with the hypothesis of representing a remnant of an ancient gesneriad flora (see Origin and Phytogeographical History), these numbers could be interpreted as palaeopolyploid numbers.

In the palaeotropical Gesneriaceae there is a fair number of genera in which polyploidy has significantly contributed to speciation. In the large African and Madagascan genus *Streptocarpus* (ca. 135 spp., basic numbers $x = 15$ and 16), polyploid species (tetra- and octoploid, $n = 30, 32, 48, 64$) are only found in Madagascar and the Comoro Islands, while the species on the African mainland remained at the diploid level (Jong and Möller 2000).

Polyploidisation has played a major role in the evolutionary diversification of the SE Asian-Malesian genus *Aeschynanthus* (ca. 160 spp., $n = 14, 15, 16$, see Rashid et al. 2001). In this genus, not only is polyploidy (tetra- to octoploidy) characteristic of particular species but also different ploidy levels occur even within individual species, e.g. *A. parvifolius*, *A. ellipicus*. In the latter species even three different ploidy levels (2x, 4x, 6x) have been reported. As far as is known, in these species the diploids and polyploids cannot be distinguished morphologically, and polyploidisation may be the first step of speciation. In the large genus *Henckelia*, most species are diploid ($x = n = 9$) but polyploidy occurs in the *H. hispida* alliance (3x, 4x) and in *H. parviflora* (4x) (Kiehn et al. 1998).

The great variety of chromosome numbers found in *Chirita* ($n = 4, 9, 10, 14, 16, 17, 18$) is presumably due to the taxonomic heterogeneity of the genus and cannot be simply interpreted in terms of dysploidy and polyploidy. The only clearly natural alliance seems to be sect. *Microchirita*, in which three haploid numbers occur: $n = 9, 17, 18$. All diploid (and blue-flowered) species occur in the Malay Peninsula (some extending into S Thailand), while the (white and yellow-flowered) species with $n = 17$ have a more northern distribution (C and N Thailand, Vietnam). The latter are probably descending-dysploids from tetraploids with $n = 18$. From that group presently only one species (*C. lavandulacea*, blue-flowered, Vietnam) is known, for which both $n = 18$ and $n = 17$ have been reported (see Kiehn et al. 1998).

In general, the chromosomes of Gesneriaceae are very small. They measure usually less than 1 μm in mitotic metaphase. Larger chromosomes have been found only in *Agalmyla*, *Monophyllaea* and *Whytockia*, measuring up to 3 μm (Ratter 1975; Kiehn and Weber 1998; Y.Z. Wang et al. 1998).

Studies in meiotic chromosomes have been carried out by Eberle (1956, 1957). Noteworthy is the occurrence of a 4-merous ring and a heteromorphic bivalent in "*Episcia tesselata*" (= *Nautilocalyx bullatus*).

Phytochemistry. A general survey on chemical characters of Gesneriaceae has been worked out by Kvist (1984). Regarding flavonoids, anthocyans are found throughout the family (Harborne 1966, 1967; Lowry 1972). The widespread anthocyanins pelargonidin and malvidin (in the palaeotropical Gesneriaceae also cyanidin and delphinidin) and the glycosidic patterns 3-rutinoside and 5-glucoside-3-rutinoside are common. The apparently most primitive types of anthocyans (cyanidin- and pelargonidin-3-glucoside) have been detected in *Rhabdothamnus solandri*, a Coronantheroid gesneriad from New Zealand (Lowry 1973). While the listed anthocyanidins have a hydroxy-group at the 3-position, this group is mostly lacking in *Sarmienta* and in most neotropical Gesneriaceae. The resulting 3-desoxyanthocyanins, otherwise known only from ferns, the moss *Bryum*, *Sorghum* (Poaceae), *Chiranthodendron* (Sterculiaceae) and *Arrabidea* (Bignoniaceae), include apigeninidin, luteolinidin, and columnidin (6- or 8-hydroxy-luteolinidin). These pigments cause the bright red or orange flower colours found in ornithophilous genera and are also present in the leaves. In the palaeotropical Gesneriaceae 3-desoxyanthocyanidins are not known. Here, other flavonoid pigments, such as chalcones and aurones (present as the glucosides isosalipurposide and cernuoside, yellow flower colour), have been reported as well (Harborne 1966). In *Didymocarpus* spp. benzoquinoid chalcones have been identified as gland exudates (Wollenweber et al. 1981). Flavonols seem to be absent in Gesneriaceae, while flavones (luteolin or diosmetin) could be detected in most species examined (Harborne 1967).

Iridoids, although common in Scrophulariaceae and related families, are not known in the Gesneriaceae. Instead, cornosids, which are often vicarious to iridoids, occur in several genera (Jensen 1992). Caffeic acid is found in the leaves of several genera (e.g. *Aeschynanthus*, *Cyrtandra*, *Gloxinia*) (Hegnauer 1966: 203). Verbascoside, an ester of caffeic acid, has been found in *Conandron* (Mølgård and Ravn 1988) and *Nautilocalyx* (Jensen 1996). Kvist and Pedersen (1986) report a complete absence of verbascoside and conandroside in Epithemateae, Napeantheae, *Agalmyla* and *Saintpaulia*. In all other tribes both glycosides occur regularly. 3,4,6-trichydroxydopamine is frequently observed in East Asian genera, but not in the European genera (Kvist and Pedersen 1986). Quinones such as naphthaquinone (Price and Robinson 1938, 1940), hydroquinone (Kvist and Pedersen 1986) and anthraquinones (Lu et al. 1998a, b) have been reported in Old World Gesneriaceae, but have not been found in neotropical genera. Volatile oils are so far only known from leaves of *Didymocarpus pedicellatus* (didymocarpene, didymocarpenol and didymocarpol; Hegnauer 1966: 205 f.).

As far as can be concluded from the (still insufficient) data available, phytochemistry is in fair accordance with the major groups of Gesneriaceae. The Old World Gesneriaceae differ from the New World ones by the absence of 3-desoxyanthocyanins and the presence of chalcones and aurones. *Rhabdothamnus*, a member of the Coronantheroid Gesneriaceae, seems to possess the most primitive hydrolated anthocyans and lacks chalcones. Within palaeotropical Gesneriaceae, the Epithematoid Gesneriaceae are distinct by the absence of dihydroxy phenolics which are widespread in the Didymocarpoid Gesneriaceae. The same applies to tribe Napeantheae in the neotropical Gesneriaceae (Kvist and Pedersen 1986).

Affinities and Recognition of Gesneriaceae. All available molecular data are in agreement that Gesneriaceae represent a monophyletic family. Traditionally, this has been regarded to be closely related to Scrophulariaceae (in its conventional sense). However, no subgroup of that family could be identified as being particularly closely allied. Recently, Gesneriaceae have been included in several molecular analyses of Lamiales and higher groups (Olmstead and Reeves 1995; Wagstaff and Olmstead 1997; Oxelman et al. 1999a, b; Spangler and Olmstead 1999; Backlund et al. 2000; Albach et al. 2001; Olmstead et al. 2001; Bremer et al. 2001, 2002). These studies indicate that Gesneriaceae occupy a rather basal position and are sister to most of the remainder of Lamiales. Only Oleaceae, Tetrachondraceae, *Calceolaria* (separated from Scrophulariaceae as a distinct family, Calceolariaceae, by Olmstead et al. 2001), *Sanango* and *Peltanthera* (formerly included in Loganiaceae) have been found to diverge earlier.

It is unfortunate, however, that in most of these studies the Gesneriaceae are represented by one or two species only. Moreover, there are also papers showing Gesneriaceae to be nested higher up within Lamiales (Young et al. 1999; Wagstaff et al. 2000; Savolainen et al. 2000; Wikström et al. 2001), and these analyses also differ markedly in the placement of several other taxa. Such discrepancies undermine confidence in molecular phylogenies, and thus the conclusion that Gesneriaceae are near-basal in Lamiales must be considered very tentative.

In general, it is not difficult to distinguish a gesneriad from a member of Scrophulariaceae s.l. There are two key characters for discrimination: (1) unilocular (vs. bilocular) ovary, (2) pair-flowered (vs. conventional) axillary cymes. Additional characters, relating to special groups, or "negative" characters may be helpful as well.

1. Unilocular means that the fused carpel flanks that protrude into the ovary are not fused at the tips and do not form a complete septum. However, in the lower part of the ovary, there is usually a portion with a complete septum and it is only the expanded upper part of the ovary that has a unilocular structure. Secondly, there are some Gesneriaceae in which the septate part makes up the whole ovary (e.g. *Whytockia*, *Monophyllaea*, *Asteranthera*) and in which placentation is axile, as in Scrophulariaceae (Weber 1971; Wilson 1974a, 1974b). However, none of the genera showing this bilocular ovary type are particularly close or transitional to Scrophulariaceae. In Scrophulariaceae, which typically have a bilocular ovary, also genera with unilocular ovaries occur (*Melampyrum*, *Dintera*, *Diplacus*, *Lathraea* and others, Hartl 1959). Therefore, ovary structure does not provide an absolute criterion for distinction of the two families.
2. The presence of pair-flowered cymes is largely (but again not absolutely) distinctive for Gesneriaceae. There are some cases in which difficulties may arise: (a) some genera in Scrophulariaceae also have pair-flowered cymes (Weber 1973b): *Penstemon*, *Russelia*, *Tetranema*, *Calceolaria*, *Jovellana*, *Stemotria* (the three latter genera have recently been suggested to represent a family of their own: Calceolariaceae; Olmstead and Reeves 1995, Olmstead et al. 2001), (b) there are a few members of Gesneriaceae in which the "front-flowers" within the pair-flowered cymes are subject to partial or complete reduction: *Chrysothemis*, *Loxonia hirsuta*, *Stauranthera caerulea* (Weber 1977b, 1978c); (c) in both families the reduction of the (pair-flowered or conventional) cymes to single flowers is possible (common in Scrophulariaceae, rather rare in Gesneriaceae, e.g. *Diastema*, *Koellikeria*, *Smithiantha*, *Rhynchoglossum*); the thyrses thus switch to racemes, the different origin of which is no longer apparent.

Additional diagnostic or "negative" characters include the seeds (in Gesneriaceae always minute, elliptic, produced in large number, always aulacospermous, sometimes appendaged), the seed funicles (never forming conspicuous "collars", see Weber 1989b), and the lack of iridoids (making herbarium specimens turn black). Palaeotropical Gesneriaceae are well characterised by anisocotylous seedlings, and some neotropical alliances by a semi- or fully inferior ovary which is otherwise rare in the Lamiales.

Subdivision of the Family, Arrangement of the Genera and Generic Concepts. Gesneriaceae were one of the last major families to take on its present form. Originally two separate families were recognised: the neotropical Gesneriaceae, formally established by DeCandolle (1816) (until recently attributed to Dumortier, Commentat. Bot. 57, 1822), and the palaeotropical Didymocarpaceae (D. Don 1822) = Cyrtandraceae (Jack 1823). G. Don (1838) and Brown (1839) united the two families and they were followed by Bentham (1876), who was the first to publish an overall account of Gesneriaceae in its wide sense. Fritsch (1893–1894), the next and last author revising the family as a whole (the important treatment of Clarke 1883 was devoted to the paleotropical "Cyrtandreae" only), worked out a very detailed infrafamilial classification. He distinguished two subfamilies (Cyrtandroideae and Gesnerioideae) based on the position of the ovary, and divided each subfamily into numerous tribes and subtribes. In a number of (sub)tribes, both neo- and palaeotropical genera were placed side by side. Burtt (1963) reverted essentially to the original geographical separation in that he clearly recognised that the palaeotropical Gesneriaceae share a conspicuous and uncommon (synapomorphic) feature – the anisocotylous seedling. Burtt (1963) defined this group as subfamily Cyrtandroideae and distinguished 5 tribes: Cyrtandreae, Trichosporeae, Didymocarpeae, Klugieae, and Loxonieae. Additional tribes (Saintpaulieae, Rhynchotecheae, Ramondeae, Titanotricheae) were then suggested by Ivanina (1965, 1967) and Wang et al. (1990, 1998). The Klugieae and Loxonieae were later united on grounds of morphological data linking genera across the two tribes (Weber 1975a, 1976b, 1978a), and presently figure as tribe Epithemateae (Burtt 1997). Not included in the palaeotropical subfamily Cyrtandroideae were the Australian and SW Pacific genera *Fieldia*, *Lenbrassia*, *Negria*, *Coronanthera* and *Depanthus*. These were placed in tribe Coronanthereae and, together with the temperate South American tribe Mitrarieae (*Mitraria*, *Sarmienta*, *Asteranthera*), referred to the neotropical subfamily Gesnerioideae. In 1983, Wiehler united Coronanthereae and Mitrarieae and raised them to a subfamily of its own, Coronantheroideae. The morphological character linking these genera is that the nectary is adnate to the ovary base. Thus, Gesneriaceae are presently subdivided into three subfamilies (Burtt and Wiehler 1995): Coronantheroideae, Gesnerioideae and Cyrtandroideae (now to be called Didymocarpoideae for priority reasons). With regard to the Gesnerioideae, Wiehler (1983) proposed a subdivision into five tribes: Gloxinieae, Episcieae, Beslerieae, Napeantheae, and Gesnerieae.

The advent of cladistics (e.g. Boggan 1991; Smith 1997) and molecular systematics (e.g. Smith and Carroll 1997; Samuel et al. 1997a, 1997b; Smith and Atkinson 1998; Smith et al. 1997b, 1998; Smith 2000a, 2000b; Möller and Cronk 1997a, b, 1999, 2001a, b, in prep.; Atkins et al. 2001; Zimmer et al. 2002; Weber et al. 2002; Mayer et al. 2003; Perret et al. 2003; Roalson et al., in prep.; Pfosser et al., in prep.) provided a new set of data, and it became obvious that the former morphological classifications needed reconsideration and modification.

1. Number and content of subfamilies. The molecular data presently available (*ndh*F, *atp*B/*rbc*L-spacer, *trn*L-F) suggest that the original bipartition seems more appropriate than a tripartition. The Coronantheroideae appear associated with Gesnerioideae, either sister to them or nested within. The two together are sister to the palaeotropical Gesneriaceae. In the latter group two clear-cut groups can be recognised: the Epithemateae and the remainder of palaeotropical Gesneriaceae (Smith 2000c; Weber et al. 2002; Mayer et al. 2003; Pfosser et al., in prep.).

The primary division of the family is thus into two groups and each group can be divided in a similar manner into two, very unequal groups. The resulting four groups (here informally referred to as Coronantheroid, Gesnerioid, Epithematoid and Didymocarpoid Gesneriaceae) are convenient entities for phytogeographical discussion (Burtt 1998a), and the question arises whether they should be ranked as subfamilies.

The author's opinion is that this course should not be followed: (a) a subdivision into two subfamilies (Gesnerioideae and Didymocarpoideae (= Cyrtandroideae)), as suggested by the molecular data, is consistent with earlier classifications. (b) By the establishment of four subfamilies (by raising Epithemateae to a fourth subfamily, considered as a possibility by Burtt 1977b), important information on relationships, character distribution and phytogeography would become blurred: with regard to the Coronanthereae s.l., the link with the neotropical Gesneriaceae would not be obvious, and with regard to the Epithemateae the connection with the other palaeotropical Gesneriaceae would become non-apparent. Epithemateae are clearly a palaeotropical group, with their distribution centre in (sub)tropical Asia, and there exists a significant morphological link (synapomorphy) with the other palaeotropical Gesneriaceae: the anisocotyly of the seedlings.

Nonetheless, a final decision whether two, three or four subfamilies should be distinguished should await further studies.

2. Tribes. Wiehler's (1983) tribes of Gesnerioideae (Gloxinieae, Gesnerieae, Episcieae, Beslerieae, Napeantheae) appear well supported by the molecular data and, in addition, a sixth tribe was suggested – Sinningieae, comprising the three genera *Sinningia*, *Paliavana* and *Vanhouttea*, formerly placed in the Gloxinieae (Smith et al. 1997b; Smith 2000c; Zimmer et al. 2002; Perret et al. 2003). Moreover, it is important to note that Beslerieae plus Napeantheae are sharply separated from the remainder of Gesnerioid Gesneriaceae. This remarkable demarcation should also be reflected in a future classification. However, Beslerieae is probably the most poorly defined tribe, this being ± entirely based on the absence of bracteoles in the inflorescences; the discovery of bracteoles in a species of *Resia* places doubt on this definition of the tribe (or this placement of the genus) (Skog and De Jesus 1997). Several other putative members of the tribe (*Anetanthus*, *Tylopsacas*, ?*Cubitanthus*) have not been included in molecular analyses and should be investigated more thoroughly. The relationships within the tribes and of some odd genera were studied by Smith and Atkinson (1998), Smith (2000a, 2000b, 2001), and Roalson et al. (in prep.). It must be noted that some genera still cannot be clearly placed within one of the existing tribes (in the present account they are put into a separate section). This applies also to particular species (*Gloxinia sarmentiana*, Zimmer et al. 2002).

If Coronantheroideae are maintained as a subfamily, the original two small tribes Coronanthereae and Mitrarieae can be used for subdivision, as molecular data confirm their monophyly.

In palaeotropical Gesneriaceae, the Epithemateae form a distinct unit, in which the generic relationships have become reasonably clear by the morphological studies of Weber (1975–1978) and the molecular studies of Mayer et al. (2003). In contrast, the large remainder forms a still badly resolved and ill-understood assemblage. Above all, a tripartition into the tribes Didymocarpeae, Trichosporeae and Cyrtandreae is by no means supported by the molecular data. The genera of the traditional tribes Trichosporeae and Cyrtandreae appear nested high up within the Asiatic genera, and both groups seem to be polyphyletic and artificial. On the other hand, some geographical groups come out as monophyletic units: the three European genera (*Haberlea*, *Ramonda*, *Jancaea*) and all African and Madagascan genera investigated so far, with *Streptocarpus* paraphyletic to the other genera examined (*Saintpaulia*, *Linnaeopsis*, *Schizoboea*, *Colpogyne*, *Hovanella*). Both groups are nested (rather basally) within the large assemblage of Asiatic genera.

For a phylogenetic grouping of the large amount of Asiatic genera, much more information is needed than is presently available. The molecular data of Pfosser et al. (in prep.) suggest that *Corallodiscus* (rosette plants with septicidally dehiscent capsules) is basal. This genus is followed by a clade of caulescent, ± shrubby Asiatic genera with septi- plus loculicidally opening capsules. The European and the African genera form successive clades and thus appear embedded between the "basal" and the "advanced" Asiatic genera. Within the advanced Asiatics, the numerous genera with twisted (exceptionally also straight) fruits seem to form a large monophyletic group. Within the large remainder of straight-fruited genera the relationships are rather uncertain. However, the data seem to give valuable hints on unexpected, but morphologically conceivable relationships (e.g. the capsule-fruited *Boeica* and the berry-fruited *Rhynchotechum*, formerly placed in different tribes), or confirm doubts whether genera such as *Chirita* are monophyletic.

3. Consequences. On the one hand, it seems clear that (a) the available formal taxonomic categories are not sufficient to reflect fully the relationships of the groups gaining shape, and (b) that especially in the large group of Asiatic Gesneriaceae, the available information is not sufficient to give a reliable picture of relationships, and a subdivision into tribes (or whatever formal categories) is presently not possible.

For these reasons a mixed classification of Gesneriaceae is presented here which includes both formal and informal "categories". For subdivision of the Gesnerioid Gesneriaceae tribal names are used. For the Coronantheroid and the Epithematoid Gesneriaceae no formal subdivision seems necessary. The Didymocarpoid Gesneriaceae are not formally subdivided, but some geographical or morphological groups are given and commented on. This informal grouping is open to changes produced by future investigations. It is hoped that in due course the way will be paved for a more natural classification and formal taxonomic subdivision of the Didymocarpoid Gesneriaceae. The Conspectus (p. 104) summarises the informal and formal names of the groups referred to in the present account.

Generic concepts have changed in the course of time and, not surprisingly, are uneven in different parts of the family. Regarding the Neotropical Gesneriaceae, drastic changes in the generic definitions have been made by Wiehler (1983 and previous papers). Wiehler was the first gesneriologist to dispense consistently with the common use of similar flower shapes as indicators of relationships (for general considerations, see also Burtt 1974). Under the motto "pollination syndromes do not constitute genera", he remodelled a large number of neotropical genera to approach a more natural delimitation and resurrected several previously synonymised genera. Subsequent workers followed him in principle, but not always in detail. Cases of disagreement include the large genus *Columnea*, which was split by Wiehler (1973, 1978, 1983) into five genera, but retained in its original broad sense by Skog (1979), Kvist and Skog (1993), Smith (1994), Smith and Sytsma (1994a–c) and others. Conversely, Skog (1976) distinguished three genera in the Gesnerieae (*Gesneria*, *Rhytidophyllum* and the new *Pheidonocarpa*), while Wiehler (1983) suggested to lump them under *Gesneria*.

Strong differences in the generic concepts are apparent in the Asiatic Gesneriaceae. When revising the Chinese Gesneriaceae, W.T. Wang and collaborators used partly very narrow generic concepts. This resulted in the establishment of many new and often monotypic genera. Some (*Boeicopsis*, *Deltocheilos*, *Schistolobos*, *Buxiphyllum*, *Tumidinodus*) proved very short-lived and have been sunk into synonymy already. Except for *Didissandra* (in the delimitation of Weber and Burtt 1998a, strictly confined to Malesia) and *Raphiocarpus* (the Sino-Vietnamese species described under *Didissandra*), the Chinese genera are treated here as they are represented in the "Flora of China" (Wang et al. 1998).

Still problematic are some taxonomically old, large and ill-defined genera. One of them was *Didymocarpus*, in which species from Africa, Madagascar, mainland Asia, Malesia and Australia had been lumped. Recent studies (Weber and Burtt

1998d) lead to a split into three genera with clear phytogeographical relations: *Didymocarpus* s.str., an essentially Sino-Himalayan genus, *Henckelia*, a South Indian-Malesian genus, and *Hovanella*, a small Madagascan genus of African affinities. Another candidate for remodelling is *Chirita* which is essentially defined by the stigma form only. In the molecular studies of Pfosser et al. (in prep.), the species examined turn up in at least four different clades, and there is little doubt that the genus is artificial.

Distribution. The general distribution of Gesneriaceae is pantropical with adjoining subtropical and warm-temperate regions. For a more detailed discussion, the four major groups of the family must be considered separately (see also Burtt 1998a).

1. Coronantheroid Gesneriaceae. This small group (with 7–9 genera and ca. 20 species) shows an exceptional distribution pattern in Australasia (New Caledonia, E Australia, Lord Howe Island and New Zealand) and (with 3 monotypic genera) temperate South America (mainly Chile). In terms of species number, the Coronantheroid Gesnerioideae are a primarily Australasian and tropical group.
2. Gesnerioid Gesneriaceae. This large alliance, comprising over 50 genera with ca. 1500 species, is restricted to the neotropics, ranging from C and S Mexico through Central America to Uruguay and northern Argentina. Within that range, some alliances and genera show characteristic distribution patterns: e.g. Mexico and Central America: *Achimenes*, *Smithiantha*, *Eucodonia*, *Moussonia*, *Solenophora*; Caribbean area: Gesnerieae; Andes: *Capanea* and *Heppiella*; SE Brazil: *Paliavana*, *Vanhouttea* and most species of *Sinningia*. Large genera such as *Besleria* or the epiphytic *Columnea* s.l. show a very wide distribution (but often with distinct centres of diversity), while others have a narrow distribution and are locally endemic. The Andean regions of western Colombia, Ecuador, and northern Peru hold the largest number of species of Gesneriaceae in the New World. Also remarkable is the high endemism of the Guayanan shield, at both the generic and specific levels (e.g. *Tylopsacas*, *Lembocarpus*, *Rhoogeton*, *Lampadaria*, *Cremersia*; Skog and Feuillet, Flora of the Guianas, in prep.).
3. Epithematoid Gesneriaceae. This is a small group of 7 genera and over 80 species. The core distribution is in SE Asia, with two genera (*Loxonia*, *Monophyllaea*) being strictly Malesian, but the others transgressing the Tropic of Cancer in N India, Nepal, Bhutan and S China. *Epithema* ranges from SE Asia and Malesia (incl. Solomon Islands) to N India and Sri Lanka, with a wide disjunction to Central and West Africa. *Rhynchoglossum* has a wide distribution in Asia and Malesia (from India to New Guinea), but there is also one species (*R. azureum*) in tropical America (Mexico to Peru). This disjunction has been enigmatic for a long time, but molecular data (Mayer et al. 2003) suggest that the species is closely related to South Indian species and, therefore, seems to be a rather recent introduction (see also below).
4. Didymocarpoid Gesneriaceae. With over 70 genera and ca. 1900 species, this is the largest group of Gesneriaceae. The general distribution includes Europe (3 genera), Africa (9 genera), mainland Asia, Malesia and the Pacific. Outside Africa, two major phytogeographical groups are distinguished by Burtt (1998a): the Eurasian group, and the Sundaland group. The former includes the many and usually small genera of N India, the Himalayas and/or China (e.g. *Platystemma*, *Corallodiscus*, *Thamnocharis*, *Tengia*, *Bournea*, *Opithandra*, etc.), as well as genera having also (or only) species somewhat further south (Burma, Thailand, Laos, Vietnam, etc.). An example for the latter group is *Didymocarpus* (in the restricted sense of Weber and Burtt 1998d), which has its greatest species concentration in the Himalayas and in S China but spreads southwards and, with a few species, even reaches the Malay Peninsula and thus the everwet tropics. Other examples include *Petrocosmea*, *Raphiocarpus*, *Ornithoboea*, *Chirita* sect. *Chirita* and sect. *Gibbosaccus*. The genera of the Sundaic group have their core distribution in Malesia (Malay archipelago), sometimes with extensions to the west (S India), north and/or east. Strictly Malesian genera are *Hexatheca*, *Didissandra* (sensu Weber and Burtt 1998a, 1998b), *Ridleyandra*, *Agalmyla* and some monotypic genera of the Malay Peninsula. The large genus *Henckelia* also has species in S India and Sri Lanka, *Paraboea* reaches S China with some species, and *Cyrtandra*, the largest genus of Gesneriaceae, reaches with many species to the Hawaiian Islands in the east and with a single species to S Japan in the north.

Mainland Asia (excl. Malay Peninsula) harbours far (perhaps three times) more genera than the

Sundaic group. However, in terms of species numbers, the situation is reverse: there are many more species in Malesia and the Pacific than in Eurasia.

There are no endemic genera in the eastern part of Malesia. Some genera described for New Guinea and adjacent areas have been reduced, and the last surviving genus, *Sepikea*, "is probably only a species of *Cyrtandra*, as the description that the two upper stamens, not the two lower ones, are fertile is almost certainly a misobservation" (Burtt 1998a: 13).

There can be little doubt that the division of the Didymocarpoid Gesneriaceae into the three groups presented above is an oversimplification.

Regarding the geographical origin of the family, one hypothesis is that the family is of Chinese ("Cathaysian") origin. This possibility is mentioned in a broad study of spermatophytic genera endemic to China (H.S. Wang 1989, see also Ying et al. 1993). This may appear plausible at first sight because of the many (31) gesneriaceous genera endemic here, and the up to three times higher number of genera in mainland Asia as compared to Malesia. However, this opinion does not take into account that (a) the generic concepts of many Chinese genera are very narrow, (b) that the centre of evolutionary diversification is uncritically equalled with the centre of origin, and (c) that it is based solely on the consideration of Asiatic Gesneriaceae and ignores other parts of the family such as the neotropical Gesneriaceae. When considering these other parts, it is almost inevitable to link the present distribution with continental drift and plate tectonics. Recently, Burtt (1998a) has proposed the following "highly speculative" hypothesis. The essential points are:

1. Gesneriaceae is a family of Gondwana origin.
2. Coronantheroid Gesneriaceae are a relict group. Its Australasian members have survived on the Australian plate.
3. Coronantheroid Gesneriaceae invaded the Americas via the Antarctic and southern South America and gave rise to Gesnerioid Gesneriaceae. While Coronantheroid Gesneriaceae became nearly extinct (the three temperate South American genera being the last survivors), Gesnerioid Gesneriaceae evolved explosively in the American tropics.
4. Australasian Coronantheroid Gesneriaceae gave rise to the present palaeotropical Gesneriaceae. These moved northwards with the Indian plate and split very early into the Epithematoid and Didymocarpoid Gesneriaceae. Before this split, the vital mutation to anisocotyly must have taken place.
5. The presently small group of Epithematoid Gesneriaceae is a relict group that was once much larger and had a much wider distribution in Asia and Africa. *Epithema tenue* can be considered as the last remnant of Epithematoid Gesneriaceae on the African continent. *Rhynchoglossum* reached America (where it is now represented by *R. azureum*) via Africa from where it now has completely disappeared.
6. On the way north, a part of Didymocarpoid Gesneriaceae spread to Madagascar from where it colonised mainland Africa.
7. The Indian plate carried Didymocarpoid Gesneriaceae finally to the Asiatic continent. Here, a division took place between the plants of northern India and the south, probably as a result of desiccation. The northern group became established in the Sino-Himalayan area and spread from here, under active evolutionary diversification, east- and south-eastwards to China and adjacent areas, as well as westwards to Europe. The southern group spread from S India and Sri Lanka into Sundaland and moved eastwards. Strong support for the W~E migration can be seen in the fact that no endemic genera (the state of the monotypic *Sepikea* is doubtful) are found east of Wallace's line. Time was apparently too short for new genera to evolve in that area.

Although Burtt's (1998a) hypothesis appears very plausible from a geographical point of view, it is hard to conceive that the family is old enough to match the geological events of Gondwana break-up. It should be noted, however, that some molecular analyses of the Lamiales suggest that Gesneriaceae are basal to most of the Lamiales and may represent one of the oldest families in the order (see above under Affinities and Recognition of Gesneriaceae). Nonetheless, regarding *Epithema* and *Rhynchoglossum*, it seems more plausible to explain the disjunct distribution of individual species by rather recent introduction events than by continental drift. The close relationship of *Rhynchoglossum azureum* with South Indian species (Mayer et al. 2003) suggests that the species has reached America rather recently, probably via transpacific trips or migrations of early Polynesians. This is also suggested by the current rare localities which are usually near former population centres of ancient dwellers along the Pacific coast (L.E. Skog, pers. comm.).

Habitat and Special Adaptations. Gesneriaceae occupy an extremely wide range of habitats. The altitudinal range is from sea level to nearly 5000 m (*Corallodiscus kingianus*). Therefore, it is not easy to identify special and limiting ecophysiological factors for the family. One factor seems important for almost all Gesneriaceae, irrespective of habitat type – high humidity, necessary either throughout the year or at least in distinctive growing periods. At low altitudes, gesneriads are predominantly bound to forest habitats, growing in damp, shady places on slopes or moist rocks, often near streams or waterfalls. In the equatorial rainforests of Asia there are relatively few gesneriads in the lowlands (e.g. *Cyrtandra*, some species of *Henckelia*, and rock plants such as *Monophyllaea*, *Epithema*, *Paraboea*, etc.), but they abound in the montane forests with their cool and ever-moist conditions. This distribution is mirrored in the neotropics, with few species in the lowlands, e.g. Amazon basin, but abounding in the Andean uplands. More open habitats are rocky, damp and humid gorges, the home of some *Streptocarpus* species in South Africa. Also, in alpine habitats, as in the Himalayas and South China, the plants grow often in open habitats but seem to be generally restricted to shady humid places such as rock crevices, overhanging rocks, shallow caves and cave entrances.

Terrestrial plants growing under constantly warm and moist conditions are not in need of special morphological and/or ecophysiological devices. Plants living in areas and habitats with periodically changing climate (especially amount of precipitation), however, have had to develop special adaptations and survival strategies, restricting growth and flowering to a short period of favourable conditions. Morphologically, these adaptations include annual habit, development of seasonal shoots, underground storage organs such as rhizomes, scaly rhizomes, tubers, etc., and leaf abscission (see above under Growth Patterns and under Exceptional Morphologies). Physiologically, the most remarkable feature of Gesneriaceae to survive periodical desiccation is poikilohydry. These plants pass into a state of anabiosis in the dry period: the leaves shrivel and the plants look dead. However, when moist conditions return, the plants recover fully. Some species have been named after this ability: the Chinese *Boea hygrometrica*, the Australian *Boea hygroscopica* and the Madagascan *Streptocarpus revivescens*. The European gesneriads *Ramonda*, *Haberlea* and *Jancaea* also are resurrection plants, and most probably species of *Corallodiscus*, *Paraboea*, *Trisepalum*, *Henckelia* sect. *Henckelia* and others can be added (Burtt 1998a). Ecophysiological studies have been devoted to the European Gesneriaceae, e.g. by Contandriopoulos and Quezel (1967), Crawford (1989), Stevanovic and Glišić (1997) and Markovska et al. (1989, 1994, 1997). The last-named authors showed that a switch to CAM takes place in the drying and the revival stages of the plants. The Australian *Boea hygroscopica* was studied by Gaff (1981).

Ethnobotany. Ethnobotanical use of Gesneriaceae is mainly known from C and S America (Wiehler 1995). A major medicinal importance seems to be for snake bite treatments. Kvist (1986) and Kvist and Holm-Nielsen (1987) recorded species of *Columnea* s.l., *Drymonia*, and *Chrysothemis* being used by three Indian tribes in the coastal region of Ecuador. A hybrid of *Kohleria* (possibly *K. amabilis* × *K. hirsuta*) is cultivated and traded in Ecuador as a snake bite remedy. Two Indian tribes in Panama use *Chrysothemis friedrichsthaliana* to treat snake bites (Duke 1970).

A drink of the boiled scaly rhizomes of *Kohleria hirsuta* and *K. spicata* is used as a remedy for kidney afflictions in Colombia and Venezuela. In Colombia, *Kohleria* species are used to regulate kidney and liver functions. In Guatemala, *K. spicata* is said to have astringent properties. Less commonly, *K. spicata* is taken for venereal disease treatment (Kvist and Skog 1992). An Amazonian Indian tribe smokes the leaves of *Columnea picta* for stimulation (Kvist and Holm-Nielsen 1987). *Alloplectus sprucei*, *Drymonia serrulata*, *Columnea rubriacuta* are used for curing eczemas, *C. parviflora* for eczema of the scalp, and *Columnea* sp. is applied to burns (Kvist 1989). A decoction of *Napeanthus robustus* is used as a pain killer by Ecuadorean coastal Indians, and the shamans use gesneriads as magical plants in their rites (Kvist 1985). Vickers and Plowman (1984) found *Codonanthopsis dissimulata* and *Drymonia coriacea* to be used for toothache treatment by Amazonas Indians. They also treat boils with cultivated *Gloxinia perennis*.

Much less is known about the ethnobotany of palaeotropical Gesneriaceae. A compilation, listing also the chemical compounds, is under preparation by Eberwein and Weber (in prep.).

Uses. In the Western World, Gesneriaceae are widely grown for ornamentals. The most important genera are *Saintpaulia*, *Streptocarpus*, *Aeschynanthus*, *Sinningia*, *Achimenes*, *Episcia*, *Columnea* and *Nematanthus*. "African Violets" (species and hybrids of *Saintpaulia*) and "Florist's Gloxinias" (usually peloric forms, with *Sinningia speciosa* being the parental species) are among the most popular indoor ornamentals in America and Europe. According to Hubert and Hubert (1991), there exist probably more than 40000 named cultivars of *Saintpaulia*, of which some 7500 are registered by the "African Violet Society of America". Books on ornamental Gesneriaceae and their cultivation include Moore (1957), Fogg (1968), Elbert and Elbert (1984), Arthurs (1988), Clements (1988), Robey (1988), Wall (1990), and Erhardt and Erhardt (1993). There are several societies of gesneriad lovers and growers all over the world.

Legends to Figures 11–18 (p. 96–103)

Fig. 11. Gesneriaceae. Anisophylly and alterniphylly. **A, B** *Columnea sanguinea*. Shoot seen from below and above in different magnification. **C, D** *Cyrtandra dispar*. **C** Juvenile plant, note pseudodistichous leaf arrangement. **D** Close-up of stem, large leaves in pseudodistichous arrangement, small leaves reduced to triangular scales. **E** *Rhynchoglossum gardneri*. Shoot with alterniphyllous leaves (small leaves completely reduced). Photos A. Weber. *Bar* 10 cm (**A, C**), 5 cm (**B, E**), 1 cm (**D**)

Fig. 12. Gesneriaceae. Unifoliate taxa. **A** *Streptocarpus grandis*. **B** *Streptocarpus wendlandii*. **C** *Monophyllaea horsfieldii*. **D** *Monophyllaea hendersonii*. *Bar* 10 cm. Photos A. Weber

Fig. 13. Gesneriaceae. Growth patterns of neotropical taxa. **A** *Episcia lilacina*. Stoloniferous, mat-forming ground herb. **B** *Sinningia leucotricha*. Terrestrial herb with tuber. **C** *Codonanthe crassifolia*. Epiphytic climber. **D** *Columnea tulae*. Epiphyte with erect, isophyllous shoots. **E** *Columnea (Dalbergaria) florida*. Epiphyte with spreading, strongly anisophyllous shoots. **F** *Columnea flaccida*. Epiphyte with pendulous shoots. Photos A. Weber (**A**), W. Barthlott (**B**), A. Weissenhofer (**C, F**), W. Huber (**D, E**). *Bar* 5 cm (**A–D, F**), 10 cm (**F**)

Fig. 14. Gesneriaceae. Anisocotyly and growth patterns of palaeotropical taxa. **A–D** *Chirita lavandulacea*. **A** Seedlings a few days after sowing, mostly with equal cotyledons. **B** Older seedlings, cotyledons strongly unequal. **C** Young plant with first flower emerging from the axil of the macrocotyledon (*Mco*); *mco* microcotyledon; the pair of cotyledons is followed by an isophyllous leaf pair. **D** Portion of flowering plant with isophyllous leaf pairs. **E** *Ramonda myconi*. Terrestrial rosette plant with alternate leaves and scapose axillary inflorescences (note also radial symmetry of flowers). **F** *Henckelia malayana*. Caulescent plant with opposite leaves and axillary cymes. **G** *Agalmyla parasitica*. Climber clinging to bark of tree; leaves strongly anisophyllous, only the large leaves visible. **H** *Aeschynanthus parvifolius*. Epiphyte with pendent stems, leaves opposite, arranged in one plane. Photos A. Weber. *Bar* 0.5 cm (**A**), 1 cm (**B**), 5 cm (**C–H**)

Fig. 15. Gesneriaceae. Flower shapes and entomophilous pollination syndromes of palaeotropical (**A, B, D, F–I**) and neotropical (**C, E, J–M**) taxa. **A–D** Melittophilous nectar flowers. **A** *Henckelia hispida*. Corolla tubular-infundibuliform, lower lip with nectar guides. **B** *Ridleyandra morganii*. Corolla broad-tubular, mauve, with nectar guides. **C** *Paliavana tenuiflora*. Corolla large, broad-tubed. **D** *Rhynchoglossum gardneri*. Personate corolla. **E–I** Melittophilous nectarless pollen flowers and deceptive pollen flowers. **E** *Niphaea oblonga*. Corolla flat-faced, rotate, white, with central cone of yellow anthers. **F** *Saintpaulia ionantha*. Corolla slightly zygomorphic, with bright yellow, long-lasting anthers (partially deceptive pollen flower). **G** *Petrocosmea kerrii*. Corolla with large yellow blotch on corolla above anthers. **H** *Henckelia caerulea*. Stamens with yellow filament knee. **I** *Henckelia geitleri*. Corolla with yellow blotches behind anthers, style long exserted and yellow (three-dimensional anther dummy). **J, K** Gynandro- and andro-euglossophilous flowers. **J** *Alsobia dianthiflora*. Corolla with fimbriate limb, gynandro-euglossophilous. **K** *Gloxinia perennis*. Corolla campanulate, nectary replaced by reddish-brown osmophore, andro-euglossophilous. **L, M** Psycho- and sphingophilous flowers. **L** *Sinningia pusilla*. Corolla with narrow tube and narrow tube entrance. **M** *Sinningia tubiflora*. Corolla long-tubular, white. Photos A. Weber (**A, B–K**), A. Chautems (**C**), B. Renaud (**L, M**). *Bar* 1 cm

Fig. 16. Gesneriaceae. Flower shapes and vertebrate pollination syndromes of palaeotropical (**A–C**) and neotropical (**D–L**) species. **A–I** Ornithophilous flowers. **A** *Aeschynanthus speciosus*. **B** *Agalmyla chorisepala*. **C** *Agalmyla parasitica*. **D** *Sinningia macropoda*. **E** *Columnea verecunda*. **F** *Columnea gloriosa*. **G** *Kohleria bogotensis*. **H** *Nematanthus strigillosus*. Target flower. **I** *Nematanthus fritschii*. Pendulous, resupinate flower. **J–L** Chiropterophilous flowers. **J** *Kohleria allenii*. Probably transitional between ornithophily and chiropterophily. **K** *Sinningia brasiliensis*. **L** *Paliavana prasinata*. Photos A. Weber (**A–H, J, K**), S. Vogel (**I, L**). *Bar* 2 cm

Fig. 17. Gesneriaceae. Fruits of neotropical taxa. **A–C** Capsules with probably wind- or rain-dispersed seeds. **A** *Sinningia elatior*. Dry bivalved capsule, seeds (wind-dispersal). **B** *Phinaea multiflora*. Semi-fleshy, bivalved capsule exposing the seeds to rain. **C** *Gloxinia lindeniana*. Capsule resulting from an inferior ovary, splitting along upper side and exposing the seeds (rain-disperal). **D–K** Display capsules, seeds animal(esp. bird)-dispersed. **D, E** *Nematanthus wettsteinii*. Closed and open fleshy capsule. **F** *Nematanthus fornix*. White fleshy capsule before opening. **G** *Nematanthus hirtellus*. Capsule exposing a mass of seeds and funicles. **H** *Drymonia serrulata*. Display capsule with contrast colours. **I** *Drymonia strigosa*. **J** *Nematanthus australis*. Seeds exposed on long, fleshy funicles. **K** *Solenophora calycosa*. Irregularly opening fleshy capsule, exposing a globe of placentae with seeds on surface. **L** *Columnea* cf. *oerstediana*. Indehiscent, fleshy berry. Photos M. Peixoto (**A**), A. Weber (**B, C, I**), A. Chautems (**D–H, J, L**), A. Weissenhofer (**K**). *Bar* 1 cm

Fig. 18. Gesneriaceae. Fruits of palaeotropical taxa. **A–F** Capsules with wind- (**A**) or rain-dispersed (**C–H**) seeds. **A** *Aeschynanthus* cf. *hians*. Long, slender, bivalved capsule, seeds with filiform appendages (wind-dispersal). **B** *Streptocarpus orientalis*. Twisted capsule. **C** *Ridleyandra porphyrantha*. Elongated capsule, opening on the upper side only. **D** *Henckelia puncticulata*. Boat-shaped follicular capsule. **E** *Henckelia browniana* (*Loxocarpus incanus*). Bowl-shaped splash-cup. **F** *Epithema saxatile*. Cup-shaped splash-cup. **G–I** Indehiscent fruits. **G** *Cyrtandra wallichii*. Elongated, sclerocarpous "berry". **H** *Cyrtandra patula*. Barrel-shaped "berry". **I** *Cyrtandra pogonantha*. Fleshy berry. Photos A. Weber (**A–H**), M. Kiehn (**I**). *Bar* 1 cm

Fig. 11

Fig. 12

Fig. 13

Fig. 14

Fig. 15

Fig. 16

Fig. 17

Fig. 18

Conspectus of Gesneriaceae (with informal and formal names used in the text)

I. Coronantheroid Gesneriaceae (Southern hemisphere Gesneriaceae)
Genera 1–9
II. Gesnerioid Gesneriaceae (Neotropical Gesneriaceae)
1. Tribe Beslerieae Bartl.
Genera 10–16
2. Tribe Napeantheae Wiehler
Genus 17
3. Tribe Gloxinieae (G. Don) G. Don
Genera 18–34
4. Tribe Gesnerieae
Genera) 35–37
5. Tribe Sinningieae
Genera 38–40
6. Tribe Episcieae
Genera 41–56
7. Genera of uncertain tribal affiliation
Genera 57–62
Palaeotropical Gesneriaceae, Old World Gesneriaceae
III. Epithematoid Gesneriaceae
Genera 63–69
IV. Didymocarpoid Gesneriaceae
1. The basal Asiatic genera
Genera 70–76
2. The European genera
Genera 77–79
3. The African and Madagascan genera
Genera 80–88
4. The advanced Asiatic genera
Genera 89–147

1. Key to the Major Systematic and Geographical Groups

1. Nectary adnate to the ovary; E Australia, SW Pacific or temperate S America
I. Coronantheroid Gesneriaceae Key 2 (p. 104)
- Nectary, if present, not adnate to the ovary; plants of Europe, Africa, Asia, Malay archipelago, NE Australia and the Pacific, or the neotropics 2
2. Plants of the neotropics, seedlings with equal cotyledons (if with unequal cotyledons and two-ranked, strongly asymmentrical leaves, and with terminal, unilateral racemes of strongly bilabiate, personate flowers: *Rhynchoglossum azureum*, see Epithematoid Gesneriaceae, Key 9); nectary, if present, often consisting of separate glands; ovary often semi- or completely inferior; ovary and fruit globose or ovoid, rarely elongate, fruit a dry or fleshy capsule or a berry, capsule never long cylindrical and never twisted **II. Gesnerioid Gesneriaceae** Key 3 (p. 104)
- Plants of the Old World (except *Rhynchoglossum azureum*), seedlings with unequal cotyledons; nectary, if present, usually cup-shaped, occasionally lobed, but never consisting of separate glands, rarely unilateral; ovary always superior (at most slightly sunk into the receptacle); fruit usually a dry capsule, rarely a hard or fleshy berry (*Rhynchotechum*, *Cyrtandra*), capsule cylindrical or less commonly globose, sometimes twisted 3
3. Placentae triangular in cross section; ovary and fruit ± globose, not twisted, style sharply set off; seeds spirally striate-reticulate; plants usually fleshy-succulent, sometimes with strongly asymmetrical leaves
III. Epithematoid Gesneriaceae Key 9 (p. 107)
- Placentae lamelliform in cross section, recurved to revolute; ovary and fruit usually elongate, rarely ovoid or globose, fruit valves straight or twisted; style usually not well set off; seeds straight-reticulate or with various ornamentations; plants rarely fleshy-succulent
IV. Didymocarpoid Gesneriaceae 4
4. Plants of Africa and Madagascar Key 10 (p. 107)
- Plants of Europe Key 11 (p. 107)
- Plants of mainland Asia (excl. S India and Malay Peninsula incl. S Thailand) Key 12 (p. 107)
- Plants of S India, Sri Lanka, Malay archipelago, the Pacific and Australia Key 13 (p. 109)

2. Key to Genera of Coronantheroid Gesneriaceae

1. Plants from Australia and the SW Pacific (1. Tribe Coronanthereae) 2
- Plants from temperate South America (2. Tribe Mitrarieae) 7
2. Fruit a capsule 3
- Fruit a berry 6
3. Corolla actinomorphic, stamens 5; capsule dehiscing septicidally by 2 valves **2. *Depanthus***
- Corolla zygomorphic, stamens 4; capsule dehiscing septicidally plus loculicidally by 4 valves 4
4. Inflorescence epedunculate, 1–2 flowers per leaf axil **6. *Rhabdothamnus***
- Inflorescence with peduncle 5
5. Anthers free; tree with verticillate leaves; corolla large, yellow **5. *Negria***
- Anthers coherent; shrubs and trees with decussate, leathery leaves; corolla small **1. *Coronanthera***
6. Trees up to 13 m tall **4. *Lenbrassia***
- Small shrubs or woody creepers **3. *Fieldia***
7. Stamens 2 **8. *Sarmienta***
- Stamens 4 8
8. Flowers with small bracteoles; corolla with a prominent galea and 3 lobes, each 2 cm long; anthers all coherent; woody creepers or vines **9. *Asteranthera***
- Flowers with bracts embracing the calyx; corolla with short, subequal lobes, 7 × 7 mm; anthers free or coherent into pairs; small shrubs **7. *Mitraria***

3. Key to Tribes and Odd Genera of Gesnerioid Gesneriaceae (for genera not associated with a tribe, see section Genera of Uncertain Tribal Affiliation)

1. Ovary superior 2
- Ovary semi-inferior or inferior 14
2. Plants with a tuber 3
- Plants without a tuber 7
3. Fruit a fleshy capsule; funicles thick and fleshy; sepals connate to form a tubular, funnel-shaped or urceolate calyx of vivid yellow-green, orange or red colour (*Chrysothemis*) **6. Tribe Episcieae** p.p. Key 8 (p. 106)
- Fruit a dry or fleshy capsule, funicles thin; sepals free or only basally connate, usually green 4
4. Capsule valves opening to 180°, pushed apart by swelling pads of tissue at the bases of the valves 5
- Capsule not as above 6
5. Corolla with a dorsal sac or blunt spur at the base; nectary a single dorsal emarginate gland; tuber properties not known; leaves one to few **59. *Rhoogeton***
- Corolla without a dorsal sac or spur; nectary absent; tuber annual, hairy; plants producing a single leaf per season **58. *Lembocarpus***

6. Tuber small and poorly developed; corolla usually oblique in the calyx, with a dorsal sac or spur; filaments basally connate; capsule fleshy (*Nautilocalyx* p.p.) **6. Tribe Episcieae** p.p. Key 8 (p. 106)
- Tuber large and conspicuous; corolla usually erect in the calyx and usually without a sac or spur; filaments free; capsule usually dry and rostrate (*Sinningia* p.p.) **5. Tribe Sinningieae** Key 7 (p.106)
7. Cymes essentially ebracteolate; mostly terrestrial forest plants growing in deep shade; flowers often small **1. Tribe Beslerieae** Key 4 (p. 105)
- Cymes bracteolate 8
8. Fruit a fleshy capsule or berry; seeds with prominent fleshy funicles; nectary consisting of 2 dorsal connate or 5 separate glands; plants often epiphytic **6. Tribe Episcieae** Key 8 (p. 106)
- Fruit a dry capsule, if somewhat fleshy, then nectary consisting of a dorsal and ventral gland (*Lampadaria*), or nectary absent; plants terrestrial 9
9. Plants with thin, stiff, wiry stems and scaly rhizomes (*Goyazia*) **3. Tribe Gloxinieae** p.p. Key 5 (p. 105)
- Plants not as above 10
10. Nectary absent, corolla actino- or zygomorphic 11
- Nectary present, corolla zygomorphic 12
11. Low herbaceous plants, with leaves in a basal tuft or rosette; stomata clustered on mounds; fruit enclosed in the persistent calyx; anthers opening by longitudinal slits **2. Tribe Napeantheae** (only **17.** ***Napeanthus***)
- Woody, often spiny shrubs; stomata scattered; fruit not enclosed; anthers opening by an apical pore **57.** ***Bellonia***
12. Capsule somewhat fleshy; nectary consisting of a large separate dorsal and ventral gland **60.** ***Lampadaria***
- Capsule dry; nectary not as above 13
13. Sepals free, equal; corolla limb without glandular hairs; Fruit valves opening to 180° **61.** ***Cremersia***
- Sepals connate, unequal, calyx zygomorphic; corolla limb with glandular hairs; fruit dehiscing irregularly and disintegrating in the persistent calyx **62.** ***Cremospermopsis***
14. Ovary semi-inferior 15
- Ovary inferior, if semi-inferior then capsule with curved rostrum (*Pheidonocarpa*) 16
15. Plants with ± large tuber, if without tuber, then basal part of stem swollen and succulent (*Sinningia* p.p.) or with perennial stems and confined to SE Brazil (*Paliavana*, *Vanhouttea*), without scaly rhizomes **5. Tribe Sinningieae** Key 7 (p. 106)
- Plants with scaly rhizomes, rarely without and then plants herbaceous and not with perennial stems **3. Tribe Gloxinieae** p.p. Key 5 (p. 105)
16. Plants with opposite leaves and usually with scaly rhizomes (*Achimenes* p.p., *Gloxinia*, *Capanea* p.p.), if without scaly rhizomes, then nectary consisting of a large dorsal double gland or separate glands (*Solenophora*) **3. Tribe Gloxinieae** p.p. Key 5 (p. 105)
- Plants with alternate leaves (except *Pheidonocarpa*, opposite leaves with petiole bases joined across the nodes, and a few spp. of *Gesneria*), never with scaly rhizomes, nectary ring-shaped, occasionally lobed **4. Tribe Gesnerieae** Key 6 (p. 106)

4. Key to Genera of Tribe Beslerieae

1. Fruit a globose, fleshy berry (with pulp of placental tissue) **10.** ***Besleria***
- Fruit a fleshy or dry capsule 2
2. Capsule fleshy or semi-fleshy **11.** ***Gasteranthus***
- Capsule dry, valves often membranous 3
3. Leaves opposite (rarely whorled at nodes), iso- or anisophyllous 4
- Leaves alternate, forming a ± dense tuft 5
4. Nectary annular or semi-annular; capsule subglobose, dehiscing irregularly; seeds not as above **12.** ***Cremosperma***
- Nectary a lobed sheath surrounding the ovary; capsule ovoid-elongate, dehiscing septicidally (plus shortly loculicidally from the tips); seeds discoid, flat **13.** ***Anetanthus***
5. Leaves in a basal tuft; inflorescences many-flowered unilateral cymes (cincinni) with involute apex; nectary annular with 2 dorsal enlarged lobes **16.** ***Tylopsacas***
- Plants with distinct stem, leaves crowded at top, inflorescences and nectary not as above 6
6. Nectary ring-shaped; flowers in a dense head on long peduncle **14.** ***Resia***
- Nectary a single dorsal bilobed gland; flowers in leaf axil **15.** ***Reldia***

5. Key to Genera of Tribe Gloxinieae

1. Nectary annular 2
- Nectary composed of 1–5 individual glands or absent 9
2. Small herbs with wiry stems; leaves leathery, lateral veins reaching the margin and forming a marginal vein, ovary superior **34.** ***Goyazia***
- Plants without wiry stems; leaves not leathery, lateral veins ending before reaching margin; ovary mostly semi-inferior or inferior 3
3. Inflorescence a terminal raceme, flower-bearing bracts alternate 4
- Inflorescence with flower-bearing leaves or bracts opposite 5
4. Small herbs with leaves in a basal tuft; upper leaf surface dotted white; corolla ca. 1 cm long **20.** ***Koellikeria***
- Herbs 0.25 to 1 m tall, with plain green or red, marbled leaves; corolla ca. 3 cm long or longer **28.** ***Smithiantha***
5. Anthers free **33.** ***Heppiella***
- Anthers coherent 6
6. Subshrubs or shrubs, without rhizomes **30.** ***Moussonia***
- Herbs with scaly rhizomes 7
7. Stems and leaves densely lanate-villous **27.** ***Eucodonia***
- Stems and leaves without woolly indumentum 8
8. Plants from Central America, mainly Mexico **26.** ***Achimenes***
- Plants from South America and Panama to Canal Zone **18.** ***Gloxinia*** p.p.
9. Nectary composed of individual glands 10
- Nectary absent or non-functional 14
10. Ovary inferior; nectary a large, double-connate, dorsal gland **31.** ***Solenophora***
- Ovary semi-inferior; nectary of 2 or more glands 11
11. Inflorescence usually a terminal bracteose raceme, the lateral flowers emerging from the axils of bracts **22.** ***Diastema***
- Inflorescence not as above 12
12. Capsule fleshy, 2-valved **24.** ***Pearcea***
- Capsule dry 13
13. Stigma capitate; anthers and stigma exserted; plants frequently epiphytic or climbing **25.** ***Capanea***
- Stigma bilobed; anthers and stigma included; plants always terrestrial **23.** ***Kohleria***
14. Corolla campanulate, zygomorphic, often blue or white 15
- Corolla rotate or subrotate, always white 17

15. Leaves strongly anisophyllous **21. *Monopyle***
- Leaves isophyllous or nearly so 16

16. Anthers coherent; ovary almost completely inferior; corolla more than 2 cm long **18. *Gloxinia*** p.p. (*G.* sect. *Gloxinia*)
- Anthers free; ovary semi-inferior; corolla less than 1 cm long **19. *Anodiscus***

17. Filaments as long as or shorter than the anthers, straight **29. *Niphaea***
- Filaments longer than the anthers, those of the anterior stamens curved inwards **32. *Phinaea***

6. Key to Genera of Tribe Gesnerieae

1. Capsule rostrate, rostrum curved, as long or longer than floral tube; leaves opposite; petiole bases joined across the stem **35. *Pheidonocarpa***
- Capsule not rostrate, the free portion at most somewhat raised; leaves alternate (exceptionally opposite, then petiole bases not joined across the stem) 2

2. Capsule with costae, or verrucose, or glabrous, not villous; filaments inserted at very base of corolla tube; leaves glabrous, scabrous or pilose, smooth, rarely bullate **36. *Gesneria***
- Capsule without costae, usually villous; filaments inserted distinctly above base of corolla tube; leaves areolate, tomentose-hirsute, when glabrescent, then corolla tubular and red **37. *Rhytidophyllum***

7. Key to Genera of Sinningieae

1. Herbs, rarely shrubs, mostly with tubers, if without tubers, then stem fleshy and flowers white **38. *Sinningia***
- Shrubs without tubers 2

2. Corolla campanulate or funnel-shaped, variously coloured but not red **39. *Paliavana***
- Corolla tubular, cylindrical, red **40. *Vanhouttea***

8. Key to Genera of Tribe Episcieae

1. Thecae elongate, sagittate, dehiscing by basal pores, the 4 anthers coherent by their inward-facing thecae, forming a column-like "salt-shaker", with pores upside; pollen dry, powdery **45. *Drymonia***
- Thecae not sagittate, coherent into flat rectangles, squares, or into pairs; pollen usually sticky 2

2. Anther connective broad, thecae widely separated, rounded; plants usually associated with ant nests **55. *Codonanthe***
- Anther connective narrow, thecae oblong, ± touching 3

3. Fruit a capsule 4
- Fruit a berry 12

4. Plants with stolons 5
- Plants without stolons 6

5. Plants terrestrial, with 2 stolons per node **49. *Episcia***
- Plants epiphytic, with 1 stolon per node **50. *Alsobia***

6. Fruit a display capsule, with 2 fleshy valves reflexed, exposing a cone-shaped mass of fleshy funicles and seeds 7
- Fruit a cup-shaped capsule, with 2 valves opening to about 45°, with the seeds and funicles inside the cup 9

7. Leaves extremely anisophyllous, the smaller of the pair usually stipule-like and early caducous; corolla funnel-shaped, white, sometimes speckled wine-red; ant-nest epiphytes **56. *Codonanthopsis***
- Leaves isophyllous or weakly anisophyllous; corolla often with a pouch below the constricted entrance, or the entrance wide but laterally compressed; plants terrestrial or epiphytic 8

8. Leaf lamina with a hypodermis; plant with succulent leaves; restricted to the Atlantic coastal forest in Brazil **54. *Nematanthus***
- Leaf lamina without hypodermis; mostly terrestrial herbs, subshrubs or vines with non-succulent leaves; distributed throughout northwestern South America **44. *Alloplectus***

9. Sepals connate for at least 3/4 of their length; plants with succulent stems and underground tubers **42. *Chrysothemis***
- Sepals free; plants without tubers (except some spp. of *Nautilocalyx*) 10

10. Calyx actinomorphic, star-shaped **53. *Cobananthus***
- Calyx zygomorphic 11

11. Terrestrial herbs, some spp. with tubers; erect or creeping and rooting at nodes; stems succulent **41. *Nautilocalyx***
- Epiphytic herbs (in the form of tank bromeliads) or vines, more rarely terrestrial plants, without tubers; stems usually firm **43. *Paradrymonia***

12. Plants terrestrial **47. *Corytoplectus***
- Plants epiphytic 13

13. Berry bright orange **48. *Neomortonia***
- Berry white, pink, red, purple, or lavender 14

14. Corolla 8–20 mm long, campanulate, with a broad limb 15
- Corolla longer than 20 mm, if shorter, with a narrow limb (**46. *Columnea*** s.l.) 16

15. Corolla white, the back and upper lobes reddish; berry globose or depressed **51. *Rufodorsia***
- Corolla plain white or reddish; berry pointed **52. *Oerstedina***

16. Nectary composed of 5 separate glands, the 2 dorsal ones sometimes enlarged or connate **46c. *Pentadenia*** (= ***Columnea* sects. *Pentadenia*** and ***Stygnanthe***)
- Nectary a large, double-connate, dorsal gland (rarely an additional small ventral gland present) 17

17. Berry lavender, subglobose-depressed **46e. *Bucinellina*** (= ***Columnea* sect. *Buccinellina***)
- Berry white, pink, red, or purple, globose or ovoid 18

18. Berry globose, white, rarely pink or blotched with red; corolla always with a narrow, gradually widening (never inflated) tube, with a distinct galea, and a long, narrow, reflexed lower lobe; stem thin and pendent, or thicker and ascending **46a. *Columnea*** s.str. (= ***Columnea* sect. *Columnea***)
- Berry ovoid or pointed, yellow, orange or red, rarely white; corolla usually tubular or somewhat inflated, with a narrow, subregular limb, or sometimes bilabiate and *Columnea*-like, but then the tube inflated; stems stout, short, ascending or spreading, or thin or pendent, never vining, usually branching only at the base, conspicuous by the arrangement of the extremely anisophyllous leaves into a plane; the large leaves often with red patterns **46b. *Dalbergaria*** (= ***Columnea* sect. *Collandra***)
- Berry globose or ovoid, usually reddish, rarely white; corolla tubular or somewhat inflated, with a subregular limb, a galea present only in combination with a strongly inflated tube; stems usually slender, pendent, scandent, or ascending, branching freely, the opposite leaves of equal or strongly unequal size (only in a few species with slender, scandent, vining stems bare except at the tip) **46d. *Trichantha*** (= ***Columnea* sect. *Ortholoma***)

9. Key to Genera of Epithematoid Gesneriaceae

1. Sepals ovate to suborbicular, aestivation imbricate, with chalk-secreting glands on inner sides; ovary bilocular 2
- Sepals triangular, pointed, aestivation valvate; without glands; ovary unilocular 3
2. Plant with several, strongly anisophyllous leaf pairs; inflorescences from the axils of the large leaves **68. *Whytockia***
- Plant usually with a single leaf (macrocotyledon), rarely with several leaves of the same shape as the macrocotyledon; inflorescences from the axils of tiny bracts situated at the base of the leaf or from the stem **69. *Monophyllaea***
3. Leaves of upper nodes of equal size and shape, or slightly unequal 4
- Leaves of all nodes strongly unequal, the small leaf reduced to a cordate scale or stipule-like appendage, or leaves in alternate-distichous arrangement 5
4. All cauline leaves in pairs; calyx plicate; flowers few near apex of plant; fertile stamens 4 **64. *Gyrogyne***
- Lower cauline leaf solitary, upper ones in 1–2 pairs; inflorescence a many-flowered, dense and (falsely) terminal head, flowers in 4 rows; sepals not plicate, spreading in fruit; fertile stamens 2 **67. *Epithema***
5. Leaves alternate, near-distichous, strongly asymmetrical; inflorescence a long, terminal, unilateral raceme with flowers in two rows; corolla strongly bilabiate, upper lip small, reflexed; lower lip broad, tongue-shaped; fruit opening with valves **63. *Rhynchoglossum***
- Leaves opposite, strongly unequal in size; inflorescence a branched thyrse, upper part of fruit disintegrating irregularly 6
6. Leaves ± densely hairy with hooked hairs, grey-green; inflorescences small, opposite the large leaves; flowers small; calyx not plicate; corolla greenish-white, with a distinctly bilabiate limb, not spurred **66. *Loxonia***
- Leaves sparsely hairy; inflorescence large, terminal; flowers large; calyx plicate; corolla subrotate, evenly blue, or with an orange-yellow hairy blotch on the palate, with or without a thick conical spur at the base **65. *Stauranthera***

10. Key to Didymocarpoid Gesneriaceae of Africa and Madagascar

1. Fruit twisted **80. *Streptocarpus***
- Fruit not twisted 2
2. Plant unifoliate 3
- Plant with several to many leaves 4
3. Fruit short, ovoid; filaments flattened and toothed; disc unilateral **83. *Acanthonema***
- Fruit elongate, slightly curving near tip, dehiscing dorsally; filaments not flattened and toothed **84. *Trachystigma***
4. Fruit very short, subglobose **85. *Nodonema***
- Fruit ovoid-conical to long-linear 5
5. Corolla with short tube and wide, flat limb; the two cohering yellow anthers fully exposed **81. *Saintpaulia***
- Corolla and stamens not as above 6
6. Plant acaulescent, leaves in a rosette; corolla small (6–7 mm long), actinomorphic, broadly tubular-campanulate **88. *Colpogyne***
- Plant caulescent 7
7. Stem erect; fruit opening dorsally **87. *Hovanella***
- Stem creeping or subascending; fruit dehiscing by 2 or 4 valves 8
8. Leaves opposite; inflorescence a terminal flower cluster; stem dichasially branched below inflorescence; fruit splitting into 4 valves and 2 lateral ribs **86. *Schizoboea***
- Leaves alternate; inflorescence axillary; stem monopodially branched; fruit bivalved **82. *Linnaeopsis***

11. Key to Didymopcarpoid Gesneriaceae of Europe

1. Corolla tubular, limb 5-lobed, distinctly zygomorphic; stamens 4, didynamous, anthers cohering in pairs; disc present, annular **77. *Haberlea***
- Corolla flat-rotate or slightly campanulate; corolla lobes and stamens 4 or 5; stamens of equal length, anthers free; disc reduced 2
2. Corolla flat-rotate or shallowly campanulate, 5- or 4-merous; corolla lobes large; stamens as many as corolla lobes; leaves green above, rusty brown below **78. *Ramonda***
- Corolla campanulate, with distinct tube, 5-merous; lobes rather short; leaves white-woolly **79. *Jancaea***

12. Key to Didymocarpoid Gesneriaceae of Mainland Asia (China and adjacent areas in the SW and S: NW, N & E India, Nepal, Bhutan, Bangladesh, Burma, N & C Thailand, Cambodia, Laos, Vietnam, Taiwan, excl. S India and Malay Peninsula)

1. Fruit indehiscent, hard or fleshy 2
- Fruit a dehiscent capsule 3
2. Fertile stamens 4, corolla tubular-campanulate, tube short, at most 6 mm long **63. *Rhynchotechum***
- Fertile stamens 2, corolla tubular-infundibuliform, tube longer **146. *Cyrtandra***
3. Seeds with a long tapering apex or a distinct apical appendage 4
- Seeds not as above 7
4. Seeds with a long, tapering brown apex **112. *Anna***
- Seeds with a distinct hyaline or white apical appendage and one to many appendages at the hilar end 5
5. Plants epiphytic; corolla usually bright red, tubular to arcuate **132. *Aeschynanthus***
- Plants not as above 6
6. Leaves thin, those of a pair very unequal, not whorled; corolla ventricose, spotted inside; fertile stamens 4; anthers without hooks at connective **111. *Loxostigma***
- Leaves fleshy, ± equal, often whorled; corolla (always?) with yellow, raised lines in the throat; fertile stamens 2, anthers with hooked connective **113. *Lysionotus***
7. Corolla actinomorphic, 5- or 4-merous, stamen number equalling the number of corolla lobes 8
- Corolla zygomorphic 11
8. Corolla flat-faced (rotate) 9
- Corolla campanulate or suburceolate 10
9. Anthers free, connectives without terminal projections; disc present **104. *Thamnocharis***
- Anthers connate, forming a tube around the style; connectives with terminal projections; disc absent **107. *Conandron***
10. Corolla suburceolate, 2.5× longer than calyx, tube longer than limb; cyme lax, several-flowered, bracts small, opposite **106. *Tengia***
- Corolla campanulate, as long as calyx, tube shorter than limb; cyme umbelliform, many-flowered, bracts 6–9, forming an involucre **105. *Bournea***

11. Fertile stamens 4 — 12
- Fertile stamens 2 — 31
12. Plants stemless, leaves in a basal rosette or tuft — 15
- Plants with a distinct stem, leaves spaced along stem — 25
13. Calyx zygomorphic-bilabiate, upper lip 1-lobed, lower lip 4-lobed — **114. *Cathayanthe***
- Calyx actinomorphic — 14
14. Upper corolla lip 4-lobed — **92. *Paraisometrum***
- Upper corolla lip 2-lobed — 15
15. Stigma undivided — 16
- Stigma bilobed (with upper and lower lobe) — 19
16. Anthers all coherent; capsule opening along upper side — **115. *Beccarinda*** (p.p.)
- Anthers cohering in pairs or free; capsule 2- or 4-valved — 17
17. Anthers cohering in pairs, palate of corolla bearded — **70. *Corallodiscus***
- Anthers free, palate not bearded — 18
18. Ovary oblong; rosette plant — **89. *Oreocharis*** (p.p.)
- Ovary ovoid or conical; plant usually caulescent, rarely stemless — **72. *Boeica*** (p.p.)
19. Upper corolla lip entire or emarginate — 20
- Upper lip bilobed — 21
20. Anthers free, all or upper ones exserted, thecae confluent at apex — **90. *Tremacron***
- Anthers coherent in pairs, usually included, thecae not confluent — **93. *Ancylostemon*** (p.p.)
21. Ovary bilocular, only one locule with placentae — **105. *Briggsiopsis***
- Ovary unilocular, each carpel with placentae — 22
22. Lower lip with 3 rounded lobes; corolla ventricose — **108. *Briggsia*** (p.p.)
- Lower lip usually with 3 acute (rarely rounded) lobes; corolla variously shaped — 23
23. Anthers longitudinally dehiscent, thecae not confluent at apex — **91. *Isometrum***
- Anthers longitudinally dehiscent, thecae confluent at apex — 24
24. Anthers free — **89. *Oreocharis*** (p.p.)
- Anthers coherent in pairs — **93. *Ancylostemon*** (p.p.)
25. Stigma bilobed (with upper and lower lobe) — 26
- Stigma undivided — 28
26. Corolla tubular, gradually widening, not ventricose — **110. *Raphiocarpus***
- Corolla broad tubular to campanulate, ventricose — 27
27. Ovary unilocular, with two parietal placentae — **108. *Briggsia***
- Ovary bilocular, with one locule sterile — **105. *Briggsiopsis***
28. Small rhizomatous herb, with 1(–2) leaves at apex of aerial stem — **75. *Platystemma***
- Habit not as above — 29
29. All 4 anthers coherent, capsule making an angle with pedicel, opening along upper side — **115. *Beccarinda*** (p.p.)
- Anthers free; capsule in straight continuation of pedicel, capsule 2- or 4-valved — 30
30. Leaves opposite; thecae not confluent; corolla yellowish — **71. *Leptoboea***
- Leaves alternate; thecae confluent at apex; corolla white, pink, blue or purple — **72. *Boeica***
31. Plants stemless, leaves in a rosette or basal tuft — 32
- Plants with a distinct stem, leaves spaced along stem — 49
32. The two posterior stamens fertile — **103. *Opithandra***
- The two anterior stamens fertile — 33
33. Upper lip of corolla 4-lobed, lower lobe undivided — 34
- Upper lip of corolla 2-lobed or undivided, lower lip 3-lobed — 35
34. Cyme umbelliform, flowers forming a dense cluster; corolla much (4–5×) longer than limb; stamens inserted above middle of tube — **101. *Calcareoboea***
- Cyme lax; corolla tube shorter than limb; stamens inserted near corolla base — **102. *Allocheilos***
35. Upper lip of corolla undivided — 36
- Upper lip of corolla 2-lobed — 37
36. Anthers basifixed, thecae parallel, not confluent, lamina base broadly cuneate to rounded — **96. *Deinocheilos***
- Anthers dorsifixed, divaricate, apically confluent; lamina base cordate — **121. *Gyrocheilos***
37. Indumentum (especially on lower leaf surface) arachnoid-tomentose, hairs interwoven, often branched — 38
- Indumentum not as above — 39
38. Calyx regular, 5-partite, capsule straight or twisted — **138. *Paraboea***
- Calyx bilabiate, 3-partite (the 3 upper sepals combined in a lip, the lower ones free); capsule twisted — **139. *Trisepalum***
39. Capsule twisted — **137. *Boea***
- Capsule straight — 40
40. Anthers basifixed, thecae parallel — **94. *Petrocosmea***
- Anthers dorsifixed — 41
41. Stamens slightly exserted; parietal placentae not divided — **99. *Dayaoshania***
- Stamens included; parietal placentae bipartite — 42
42. Anthers free, thecae not confluent; capsule nearly globose; leaves peltate, lamina base rounded — **95. *Metapetrocosmea***
- Anthers apically coherent; capsule ovoid to linear; leaves not or rarely auriculate-peltate — 43
43. Stigma bilobed, corolla lobes acute — **97. *Lagarosolen***
- Stigma undivided; corolla lobes rounded or acute — 44
44. Corolla lobes acute — **100. *Dolicholoma***
- Corolla lobes rounded — 45
45. Capsule ovoid or ellipsoid, shorter than calyx or of equal length — 46
- Capsule elongate, much longer than calyx — 47
46. Corolla ± salverform, tube cylindric — **118. *Primulina***
- Corolla tubular — **117. *Chiritopsis***
47. Corolla suburceolate-tubular, 1–2.5 mm in diameter, limb subregular; stamens inserted near corolla base — **98. *Petrocodon***
- Corolla campanulate, tubular or infundibuliform; limb bilabiate; stamens inserted near or above middle of corolla tube — 48
48. Stigma capitate; capsule often stipitate — **120. *Didymocarpus***
- Only lower lobe of stigma developed, this usually 2-lobed; capsule not stipitate — **116. *Chirita***
49. Capsule twisted — 50
- Capsule straight — 55
50. Indumentum (esp. on lower leaf surface) arachnoid-tomentose, hairs interwoven and often branched — 51
- Indumentum not as above — 52
51. Calyx regular, with 5 acute lobes; fruits straight or twisted — **138. *Paraboea*** (p.p.)
- Calyx bilabiate, three upper lobes combined into an upper lip, lower lobes free; capsule twisted — **139. *Trisepalum***
52. Plant with annual, herbaceous, monocarpic aerial stem, dying after fruiting; resting bud protected by cataphylls — **142. *Kaisupeea***
- Plant not as above — 59
53. Flowers solitary from leaf axils; branched shrub — **140. *Rhabdothamnopsis***
- Flower in few- to many-flowered cymes — 60

54. Palate of corolla with a beard of club-shaped hairs and a line of hairs forming a circlet around the corolla mouth **141. *Ornithoboea***
- Palate of corolla without beard and circlet **137. *Boea***
55. Stem erect, with a tetramerous whorl or a pair of leaves at the top; stem with pairs of scale leaves, with axillary inflorescences in some species; fruit a 4-valved capsule, the main ribs of the carpels remaining united with the persistent style base **74. *Tetraphyllum***
- Habit and fruit not as above 56
56. Indumentum (esp. on lower leaf surface) arachnoid-tomentose, hairs interwoven and often branched **138. *Paraboea*** (p.p.)
- Indumentum not as above 57
57. Bracts of cymes conspicuous, at least 1 cm long, rounded, forming an involucre 58
- Bracts of cymes inconspicuous 61
58. Calyx bilabiate, the 2 upper lobes fused halfway, the 3 lower ones free **126. *Phylloboea***
- Calyx more or less regular 59
59. Ovary 1-locular, placentae parietal; corolla glabrous inside **119. *Metabriggsia***
- Ovary bilocular 60
60. Each locule with axile placentae; capsule not known, probably 2-valved; corolla glabrous inside **127. *Hemiboeopsis***
- Only one (upper) locule of ovary with placentae; capsule dehiscing along upper side; corolla with hairs inside **128. *Hemiboea***
61. Stigma capitate; capsule often stipitate **120. *Didymocarpus***
- Stigma not capitate; capsule not stipitate 62
62. Leaves alternate, capsule curved, opening along upper side; ovary bilocular, with only one fertile locule **123. *Deinostigma***
- Leaves opposite, capsule 2(–4)-valved, ovary mostly with two fertile locules (a few spp. of *Chirita* with one fertile locule) 63
63. Stigma with equal upper and lower lobe, stamens inserted below corolla mouth **122. *Didymostigma***
- Stigma not as above; stamens inserted at middle of corolla tube 64
64. Stigma lobes unequal, the upper one smaller, truncate or triangular, the lower one narrowly obtrapeziform, emarginate **124. *Pseudochirita***
- Stigma consisting of lower lobe only, lobe bifid, rarely undivided **116. *Chirita***

13. Key to Didymocarpoid Gesneriaceae of South India, Malesia, Australia and the Pacific (S India, Sri Lanka, Sumatra, Malay Peninsula incl. S Thailand, Borneo, Java, Lesser Sunda Islands, Sulawesi, Philippines, New Guinea, Solomon Islands, NE Australia, Pacific Islands)

1. Fruit indehiscent, a hard-walled or fleshy berry 2
- Fruit dehiscent, a dry (rarely somewhat fleshy) capsule 4
2. Fertile stamens 4 **73. *Rhynchotechum***
- Fertile stamens 2 3
3. Posterior stamen pair fertile (?) **147. *Sepikea***
- Anterior stamen pair fertile **146. *Cyrtandra***
4. Seeds with an apical (triangular or filiform) appendage and one to many appendages at the hilar end 5
- Seeds not appendaged 6
5. Leaves herbaceous, lateral veins visible; fertile stamens 4 (rarely 2); plants usually climbers (rarely terrestrial), rooting in the ground **134. *Agalmyla***
- Leaves leathery-fleshy, veins almost invisible; fertile stamens 2, plants usually epiphytic **132. *Aeschynanthus***
6. Fertile stamens 4 7
- Fertile stamens 2 12
7. Calyx 5-merous; corolla 4-merous, rotate **76. *Championia***
- Flowers not as above 8
8. Anthers of the two lower stamens dithecous, those of the upper pair monothecous; fruit cylindrical, subcarnose, opening irregularly **131. *Hexatheca***
- Anthers and fruit not as above 9
9. Anthers free; corolla flat-faced or slightly campanulate; leaves alternate **72. *Boeica***
- Anthers connate in pairs or all anthers cohering; corolla campanulate, tubular or infundibuliform; leaves opposite 10
10. All anthers cohering, thecae dehiscing by an apical pore **115. *Beccarinda***
- Anthers cohering in pairs (occasionally all coherent), thecae dehiscing by longitudinal slits 11
11. Fruit a slender, terete, tardily dehiscent capsule, with longitudinal ribs when dry; corolla infundibuliform **129. *Didissandra*** (p.p.)
- Fruit a usually thickish, somewhat fleshy capsule, without ribs, opening along the upper side; corolla mostly broadly tubular, rarely campanulate or flat-rotate **130. *Ridleyandra***
12. Indumentum (esp. on lower leaf surface) arachnoid-tomentose, hairs interwoven and often branched 13
- Indumentum not as above 14
13. Calyx regular, sepals free or basally connate; stigma usually not linguiform; fruits straight or twisted **138. *Paraboea***
- Calyx bilabiate, the 3 upper lobes united into a 3-lobed tip, the 2 lower ones free to base; stigma linguiform; fruits twisted **139. *Trisepalum***
14. Fruit globose-ovoid, small, enclosed in calyx **135. *Orchadocarpa***
- Fruit elongated 15
15. Fruit not twisted 16
- Fruit spirally twisted 20
16. Fruit a slender, terete, tardily dehiscent capsule, with longitudinal ribs when ripe **129. *Didissandra*** (p.p.)
- Fruit splitting early, not ribbed 17
17. Fruit 2-valved 18
- Fruit opening on upper side only 19
18. Stigma capitate **120. *Didymocarpus***
- Stigma without upper lobe, lower lobe enlarged, usually bifid **116. *Chirita***
19. Lobes of upper lip with densely packed glands with multicellular head; calcicole, delicate plants of cave entrances **145. *Emarhendia***
- Lobes of upper lip without glands; robust terrestrial plants, rarely saxicole on acid rocks **136. *Henckelia***
20. Twisting of fruit slight, 1 turn over entire length **144. *Spelaeanthus***
- Twisting of fruit strong, several turns over entire length 21
21. Corolla flat-faced, resupinate **143. *Senyumia***
- Corolla with distinct tube or campanulate to flat-faced, not resupinate 22

22. Corolla with a distinct, broad tube and spreading limb; palate with a beard of club-shaped hairs; a line of hairs forming a circlet around the corolla mouth **141. *Ornithoboea***
- Corolla campanulate or short-tubed with flat-faced limb; palate without a beard and without a circlet **137. *Boea***

Groups and genera of Gesneriaceae

I. Coronantheroid Gesneriaceae ("Southern Hemisphere Gesneriaceae")

Subfamily Coronantheroideae Wiehler (1983).
Subfamily Gesnerioideae [Link 1829], tribes Coronanthereae Fritsch (1893) and Mitrarieae (Hanst.) B.L. Burtt (1963), sensu Burtt (1963).

Subshrubs, shrubs or trees with isocotylous seedlings. Leaves opposite, sometimes whorled, with stomatal complexes predominantly of the anomo-mesoperigenous type. Axillary pair-flowered cymes sometimes reduced to single flowers. Flowers with 4 (rarely 5 or 2) fertile stamens, these of equal length or didynamous, staminode of fifth stamen present. Nectary, if present, embedded in the basal part of the ovary ("adnate nectary"). Ovary superior, ovoid, unilocular, rarely bilocular, placentae triangular in cross section, or axile. Seeds without surface ornamentation of testa cells. $n = 37$ to 45. Southern hemisphere.

1. *Coronanthera* C.B. Clarke

Coronanthera [Vieill. ex] C.B. Clarke, in A. & C. DC., Monogr. phan. 5/1: 170 (1883); Wiehler, Selbyana 6: 159–(1983), annot. list of spp.

Woody shrubs or trees to 15 m tall. Leaves iso- or somewhat anisophyllous, elliptic-ovate. Cymes pedunculate, few- to several-flowered, bracteoles small, lanceolate or ovate. Sepals connate at base, narrowly triangular, persistent. Corolla greenish-yellow, small to medium-sized, urceolate or tubular and somewhat narrowed at the mouth; limb bilabiate, lobes small, subequal, rounded. Stamens inserted near corolla base; anthers divaricate, cohering all or in pairs. Nectary present or reduced. Fruit an ovoid dry capsule with acute to subrostrate apex, dehiscence septicidal, or septi- plus loculicidal, valves sometimes apically coherent. Eleven spp., 10 in New Caledonia, one in the Solomon Islands.

2. *Depanthus* S. Moore

Depanthus S. Moore, J. Linn. Soc. Bot. 45: 373 (1921).

Small trees to 10 m or more tall. Leaves often anisophyllous. Cymes pedunculate, few-flowered. Sepals connate at base, triangular, persistent. Corolla short- and broad-tubed-campanulate to urceolate, limb almost regular, lobes small, recurved. Stamens 5, included, filaments of equal length, inserted near corolla base, tomentose in lower part, anthers free, locules apically not confluent. Nectary lacking? Ovary unilocular; stigma bilobed. Fruit an ovoid, shortly rostrate capsule, septicidally dehiscent. Two spp., New Caledonia. Possibly congeneric with *Coronanthera* (Burtt 1998a), differing by the actinomorphic campanulate flowers with 5 equal stamens.

3. *Fieldia* Cunn.

Fieldia Cunn., in Field, Geogr. Mem. New South Wales 363, t. 2 (1825), non Fieldia Gaudich. (1825).

Epiphytic, straggling-climbing subshrubs. Stem and branches slender, woody. Leaves distinctly anisophyllous, shortly petiolate, ovate to ovate-lanceolate, coarsely serrate. Flowers solitary in leaf axils, pendulous, long-pedunculate, bracts spathaceous, pedicel short and hidden by the bracteoles. Sepals free, linear-lanceolate. Corolla pale yellowish-green, tube cylindrical, limb actinomorphic, lobes small, rounded, spreading. Stamens inserted at the very base of the corolla, of nearly equal length; anthers small, free. Fruit an ovoid, spongy-subcarnose berry. $n = \pm 40$. One sp., *F. australis* Cunn., SE Australia.

4. *Lenbrassia* G.W. Gillett

Lenbrassia G.W. Gillett, J. Arnold Arbor. 55: 431 (1974); Morley, J. Adelaide Bot. Gard. 1: 187–192 (1978), notes, new var.; Burtt, Edinburgh J. Bot. 56: 371–379 (1999), incl. into *Fieldia*.

Small branched tree to 13 m tall, with hard and strong wood. Leaves slightly anisophyllous, (ob)lanceolate, unequal at base, sharply serrate, with scabrous pubescence. Cymes axillary, one- to three-flowered, peduncle short, bracteoles linear. Sepals connate in lower half, triangular, scabrous, persistent. Corolla tangerine orange, tube cylindrical, limb slightly bilabiate, lobes rounded, spreading. Stamens inserted distinctly above corolla base, twisted and curved; anthers subglobose, apically coherent. Nectary a swollen ring. Fruit a fleshy, light green, shiny, pubescent berry. One sp., *L. australiana* (C.T. White) G.W. Gillett, NE Australia. Although very different in habit, included into *Fieldia* by Burtt (1999).

5. *Negria* F. Muell.

Negria F. Muell., Fragm. Phytogr. Austral. 7: 151 (1871); Greene, Curtis' Bot. Mag. 179: 659 (1973), illustr. & notes.

Small tree to 9 m tall, wood light and brittle, branches thickish, spreading. Leaves in whorls of 3 or 4, clustered towards the branch ends, ovate or elliptic, softly pubescent below, with rough pubescence above, finely and shallowly denticulate. Cymes usually 3-flowered. Sepals connate in lower part to form a broad-tubed cup, lobes narrow. Corolla yellow, with orange-red spots on lobes and in throat; tube broad, slightly curved; limb oblique, bilabiate, lobes reflexed. Stamens of equal length, inserted in the middle of the corolla tube; anthers broadly ovoid, free, exserted. Nectary a broad ring. Fruit an ovoid, beaked capsule, dehiscing by four valves. $n = \pm 45$. One sp., *N. rhabdothamnoides* F. Muell., SW Pacific (Lord Howe Island).

6. *Rhabdothamnus* Cunn.

Rhabdothamnus Cunn., Ann. Nat. Hist. 1: 460 (1838); Allan, Fl. New Zeal. 1: 953–954 (1961), rev.

Much-branched, scabrous shrub to 2 m tall. Leaves small, broadly ovate to suborbicular, coarsely serrate-dentate, greyish-green by the dense indumentum. Flowers solitary in leaf axils. Sepals equal, ovate or triangular-acuminate, persistent. Corolla subcampanulate, brick-red or orange with red veins; tube broad, limb slightly bilabiate, lobes spreading, ciliolate. Stamens inserted at the corolla tube, long, slender, incurved above; anthers crosswise coherent. Nectary annular, rather obscure. Ovary ovoid; style long, curved; stigma inconspicuous, obscurely bilobed. Fruit an ovoid capsule, dehiscence by 4 valves. $n = 37$. One sp., *R. solandri* Cunn., New Zealand (North Island).

7. *Mitraria* Cav.

Mitraria Cav., An. Ci. Nat. 3: 230, t. 31 (1801), nom. cons., non *Mitraria* J.F. Gmel. (1791).

Straggling or epiphytic subshrub. Stem and branches softly woody, obscurely tetragonous. Leaves small, opposite or ternate, ovate-acute, coarsely crenate-serrate, nearly glabrous. Flowers axillary, solitary, pendulous; long-pedunculate, bracteoles large, ovate, connate on one side, embracing the calyx, pedicel short and hidden within bracteoles. Sepals free, the upper one larger. Corolla bright scarlet; tube curved, constricted at base, ventricose above, mouth slightly contracted; limb obscurely bilabiate, lobes subequal, obtuse, spreading. Stamens slightly exserted; anthers free. Nectary large, cupular. Ovary with long, slightly exserted style; stigma inconspicuous. Fruit a fleshy berry. $n = \pm 37$. One sp., *M. coccinea* Cav., S Chile and adjacent Argentina.

8. *Sarmienta* Ruiz & Pav.

Sarmienta Ruiz & Pav., Prodr. 4 (1794), nom. cons.
Urceolaria Huth (1758), nom. rejic.

Suffrutescent, usually epiphytic creeper or climber. Stems slender, flexuous, sparingly branched, rooting at the basal nodes. Leaves small, opposite, subsessile, ovate to suborbicular, rather fleshy, entire or obscurely crenate, glabrous, nerves obscure. Flowers axillary, solitary, pendulous, long-pedunculate with small bracteoles, the pedicel much shorter than the peduncle. Sepals free, linear to lanceolate. Corolla scarlet; tube elongate, strongly constricted at base, inflated in the middle and again narrowed below the oblique, subregular limb, lobes spreading. Fertile stamens 2, the posterior ones; inserted near corolla base, long exserted; anthers free, the sterile anterior stamens with long filaments and clavate tips or minute anthers, included; fifth stamen a short staminode. Nectary ring-shaped. Fruit a fleshy berry. $n = \pm 37$. One sp., *S. scandens* (J.D. Brandis) Pers., S Chile and Chiloe Island.

9. *Asteranthera* Hanst.

Asteranthera [Klotzsch & Hanst. ex] Hanst., Linnaea 26: 211 (1854).

Terrestrial or epiphytic creeping-climbing shrubs. Stems thin, woody, rooting at the basal nodes. Leaves small, elliptic, membranous-coriaceous, hairy above, glabrous and whitish below. Flowers solitary (rarely 2) in the leaf axils, erect; peduncle short, with tiny lanceolate bracteoles, pedicel longer than peduncle. Sepals almost free, narrowly triangular, dentate. Corolla red, with whitish lines on tube, lobes mottled; tube slightly constricted above the base and then gradually widening, limb strongly bilabiate, upper lip a long galea, projecting forwards, lower lip spreading or slightly reflexed. Stamens inserted above the corolla base, held below the galea, anthers crosswise coherent. Nectary a dorsally thickened ring. Ovary ovoid-oblong, bilocular; style long, slender; stigma small. Fruit a fleshy berry. One sp., *A. ovata* Hanst., S Chile and adjacent Argentina.

II. Gesnerioid Gesneriaceae ("Neotropical Gesneriaceae")

Subfamily Gesnerioideae Link (1829) ("Gesneriaceae"), sensu Wiehler (1983) and Burtt & Wiehler (1995).

Perennial herbs, subshrubs, shrubs, small trees, vines, climbers or epiphytes. Seedlings isocotylous. Leaves opposite or in whorls, less commonly alternate; stomata scattered or in groups, stomatal complexes predominatly of the aniso-mesogenous type. Pair-flowered cymes axillary, pedunculate or epedunculate, bracteolate or ebracteolate, one- to many-flowered. Flowers zygomorphic, rarely actinomorphic, of various shapes. Stamens 4, rarely 5, never 2, staminode usually present, anthers free or cohering in pairs or all together. Nectary annular or consisting of 2 to 5 separate glands, free from the ovary wall, rarely reduced and non-functional, or completely lacking. Ovary superior, semi- or completely inferior, usually ovoid or globose, unilocular, placentae triangular in cross section. Fruit a dry or fleshy 2- or 4-valved capsule, sometimes exposing the placentae and seeds as a central cone (display capsule), or an indehiscent berry. Seeds usually without surface ornamentation of testa cells. Chromosome numbers $n = 8$ to 16. Native to the neotropics.

1. Tribe Beslerieae Bartl. (1830).

Wiehler, Selbyana 6: 195–198 (1983), emend.

Perennial herbs, shrubs or small trees, without subterranean storage organs, often growing in humid forest and/or in deep shade. Nodes unilacunar. Leaves opposite or alternate, rarely whorled, often leathery. Epidermal cells of lower leaf surface with straight walls; stomata irregularly scattered or in groups. Petiolar vascular trace (always?) a deep crescent. Cymes axillary, usually ebracteolate. Stamens 4. Nectary annular, semi-annular, lobed or sheathed, or reduced to a single dorsal gland. Ovary superior. Fruit a berry or a dry or fleshy capsule with loculicidal, septicidal or loculicidal plus septicidal dehiscence. Chromosome numbers known only for a few *Besleria* species ($n = 16$). The tribe is weakly defined (essentially by the ebracteolate cymes) and heterogeneous. It seems to represent one of the earlier lineages of the neotropical Gesneriaceae (Smith 1996, 2000b, Smith et al. 1997b). Together with the monotypic Napeantheae, it forms a clade sister to the remaining Gesnerioid Gesneriaceae.

10. *Besleria* L.

Besleria [Plum. ex] L., Sp. pl. 2: 619 (1753); Wiehler, Selbyana 6: 195 (1983), notes; Morton, Contr. U.S. Natl. Herb. 26: 395–474 (1939), rev.; Skog, Ann. Missouri Bot. Gard. 65: 802–821 (1979), reg. rev.; Chautems, Rev. Brasil. Bot. 14: 51–59 (1991), reg. rev.; Feuillet & Steyermark in Berry et al., Fl. Venezuel. Guayana 5 (1999), reg. rev.; Skog, Monogr. Syst. Bot. Missouri Bot. Gard. 85: 1117–1119 (2001), reg. rev.; Chautems in Wanderley et al., Fl. Fanerogâm. São Paulo, reg. rev. (in press).
Pterobesleria C.V. Morton (1953).

Perennial herbs, shrubs or small trees to 5 m. Stem terete or quadrangular. Leaves opposite, rarely whorled, membranous to coriaceous. Stomata not in groups. Cymes fasciculate or subumbellate, rarely flowers solitary. Sepals connate (at least at base), calyx campanulate, urceolate or cyclindrical, lobes imbricate, rounded or acuminate, entire or serrulate. Corolla yellow, orange, red, or white; tube usually cylindrical, sometimes spurred or gibbous at base, or ventricose and contracted at throat; limb (sub)regular or bilabiate. Stamens with broad, flattened filaments, adnate to corolla base; anthers coherent at first, becoming free; locules confluent at apex. Nectary annular or semi-annular. Fruit a fleshy, globose, white, red or orange berry; pericarp thick, smooth or verrucose, pulp consisting of placental tissue. $n = 16$. Over 200 spp., throughout the neotropics, centres of diversity in the Andes of Colombia and Ecuador, some species endemic in SE Brazil.

11. *Gasteranthus* Benth.

Gasteranthus Benth., Pl. Hartw.: 233 (1846); Wiehler, Selbyana 1: 154–156 (1975), reestabl.; Skog, Ann. Missouri Bot. Gard. 65: 923–929 (1979), reg. rev.; Freiberg, Phyton 36: 303–309 (1996), 38: 167–173 (1998), Brittonia 52: 203–209, reg. rev.; Skog & Kvist, Syst. Bot. Monogr. 59: 1–118 (2000), rev.
Besleria L. subg. *Gasteranthus* (Benth.) Hanst. (1865).
Besleria L. sect. *Neobesleria* Morton (1939).

Terrestrial herbs or low subshrubs. Stems succulent or woody, erect, terete. Leaves opposite, sometimes whorled, membranous to subcoriaceous, stomata in groups (white dots on lower leaf surface). Cymes usually pedunculate, one- to many-flowered. Corolla usually spurred, funnelform with a wide limb, broadly tubular, or strongly urceolate with an inflated ventral poach and a small throat and limb ("hypocyrtoid"); colour whitish, pale or bright yellow, orange to red, often spotted. Stamens included, adnate to corolla base. Nectary a dorsally thickened ring, semi-annular or reduced to dorsal glands. Fruit a 2- or 4-valved, laterally and apically compressed, fleshy capsule. About 35 spp., from Guatemala and nearby Mexico to western South America south to Bolivia, centre of diversity in western Ecuador.

12. *Cremosperma* Benth.

Cremosperma Benth., Pl. Hartw. 234 (1846); Morton, J. Wash. Acad. Sci. 25: 283–291 (1935), rev.; Skog, Ann. Missouri Bot. Gard 65: 888–891 (1979), reg. rev.; Kvist & Skog, Nord. J. Bot. 8: 259–269 (1988), reg. rev.

Small terrestrial or epiphytic herbs. Stems softly woody, erect, ascending or creeping. Leaves opposite, rarely whorled, iso- to strongly anisophyllous, entire or incised. Cymes few, emerging from the upper leaf axils, usually long-pedunculate, umbellate or capitate (cincinni). Flowers small, usually with short pedicels. Calyx with an often 10-costate tube, lobes short. Corolla small, yellow, white or reddish, occasionally spotted; tube ± erect in the calyx, cylindrical to funnelform; limb with spreading to patent lobes. Stamens inserted below or at the middle of the corolla tube. Nectary annular or semi-annular. Fruit a subglobose, membranous capsule, dehiscing irregularly. Between 25 and 30 spp., from the Andes of Peru to Panama.

13. *Anetanthus* Benth.

Anetanthus [Hiern ex] Benth., in Benth. & Hook. f., Gen. pl. 2: 1025 (1876); Howard, J. Arnold Arbor. 56: 364–368 (1975), rev.; Wiehler, Selbyana 1: 374–404 (1976), rev.; Skog, Selbyana 7: 94–99 (1982), new sp.

Terrestrial or saxicolous herbs. Stem decumbent or pendent, slender. Leaves opposite, isophyllous, membranous, ovate, elliptic or lanceolate-acute. Cymes pedunculate, several-flowered. Sepals shortly connate. Corolla white, bluish, violet or red, tube subcylindrical, limb with round, slightly spreading lobes. Stamens adnate to corolla base, anthers coherent in pairs. Nectary cup-shaped, shallowly lobed. Fruit an elongate ovoid capsule, 2-valved, dehiscence by septicidal plus short loculicidal slits. Seeds suborbicular, flattened, narrowly winged. Two spp., Peru, Bolivia, C Colombia and SE Brazil.

14. *Resia* H.E. Moore

Resia H.E. Moore, Bot. Mus. Leafl. Harvard Univ. 20: 87 (1962); Skog & De Jesus, BioLlania, ed. espec. 6: 515–525 (1997), new subsp.

Terrestrial herbs or subshrubs. Stem erect (eventually becoming pendent), unbranched. Leaves crowded at apex, opposite (?) or alternate, subsessile, membranous or chartaceous, obovate to oblanceolate, side veins numerous. Cymes long-pedunculate, sometimes with bracteoles, flowers in a dense head. Sepals connate at base. Corolla white, orange or yellow; tube ventricose, limb with unequal spreading lobes, the dorsal ones triangular, the lateral ones obliquely obovate. Stamens included; filaments flattened and connate at base, inserted near corolla base; anthers reniform, all coherent; staminode filiform. Nectary annular. Fruit an ovoid or subglobose, laterally compressed dry capsule. Two spp., Colombia and Venezuela.

15. *Reldia* Wiehler

Reldia Wiehler, Selbyana 2: 124 (1977); Kvist & Skog, Nord. J. Bot. 8: 601–611 (1989), rev.

Terrestrial herbs with creeping rhizomes. Stem terete. Leaves alternate, often tufted near stem apex. Cymes epedunculate (?), often several in a leaf axil (cristate, epipetiolar inflorescences in *R. alternifolia* Wiehler). Sepals free nearly to base. Corolla small, funnel-shaped, usually white, often yellow in the throat; tube dorsally spurred; limb unequal. Stamens slightly exserted, inserted at corolla base; anthers coherent. Nectary usually reduced to a dorsal, bilobed gland. Ovary ovoid, stigma capitate. Fruit a dry, bivalved capsule, surrounded by the persistent calyx. Five spp., from Panama to northern Peru.

16. *Tylopsacas* Leeuwenb.

Tylopsacas Leeuwenb., Taxon 9: 220 (1960); Feuillet & Steyermark in Berry & al., Fl. Venezuel. Guayana 5 (1999), reg. rev.
Tylosperma Leeuwenb. (1958), nom. illeg., non Botsch. (1952).

Terrestrial herbs. Stem ± absent. Leaves rosulate, apparently alternate, petiolate, oblanceolate, apex acute, base decurrent. Cymes unilateral (cincinni), with numerous flower pairs, shorter than the leaves. Flowers small, pedicels thin. Sepals shortly connate at base, linear-lanceolate. Corolla white, tube slightly inflated in the middle, limb slightly bilabiate, with elliptic, rounded lobes. Stamens inserted near corolla base; anthers reniform, coherent in pairs. Nectary ring-shaped, with two enlarged dorsal lobes. Fruit a 2(–4?)-valved capsule, valves thin and translucent. Seeds subglobose, somewhat flattened. One sp., *T. cuneata* (Gleason) Leeuwenb., in the Guayana Highlands.

2. TRIBE NAPEANTHEAE Wiehler (1983).

Terrestrial herbs, without subterranean storage organs. Nodes unilacunar. Leaves opposite, usually leathery; epidermal cells of lower leaf surface with straight side walls; stomata in groups. Cymes axillary, bracteolate. Flowers usually rotate or campanulate, with regular or slightly bilabiate limb. Stamens 4, anthers free. Nectary absent. Ovary superior, ovoid or globose. Fruit a dry capsule with loculicidal or loculicidal plus septicidal dehiscence. Chromosome number unknown. The tribe is monotypic with *Napeanthus*. As shown by recent molecular data (Smith 2000b; Zimmer et al. 2002), it is closely related to Beslerieae. The two tribes form a clade which is sister to the rest of the Gesnerioid Gesneriaceae.

17. *Napeanthus* Gardn.

Napeanthus Gardn., Lond. J. Bot. 2: 13 (1843); Leeuwenb., Acta Bot. Neerl. 7: 340–354 (1958), rev; Skog, Ann. Missouri Bot. Gard. 65: 951–954 (1979), reg. rev.; Skog, Monogr. Syst. Bot. Missouri Bot. Gard. 85: 1127 (2001), reg. rev.
Hatschbachia L.B. Smith (1953).

Herbs with short decumbent or ± absent stem. Leaves in a rosette or tufted, (sub)isophyllous, petiole short or lacking, sessile leaves joined by a line across the stem; lamina membranous to papyraceous, linear-lanceolate to oblong-spathulate, often falcate; margin (sub)entire or crenate-serrate towards the apex, somewhat sinuate. Cymes one- to many-flowered, pedunculate, sub-umbellate. Sepals connate at base to up to middle, calyx campanulate. Corolla white to blue, pink or lilac; rotate or campanulate, sometimes ventricose or tubular; limb (sub)regular, rarely distinctly bilabiate, lobes spreading, usually emarginate at apex. Stamens 5, equal, or 4 and a staminode, inserted at corolla base; anthers free, orbicular or oblong, thecae apically confluent, dehiscing longicidally. Fruit a dry capsule, included in the persistent calyx, oblong, ovoid or globose, 2- or 4-valved, the valves not persistent after dehiscence. Over 20 spp., more to be expected; throughout the neotropics.

3. TRIBE GLOXINIEAE (G. Don) G. Don (1839).

Wiehler, Selbyana 6: 144–146 (1983), emend.

Usually terrestrial, perennial herbs, subshrubs or rarely trees, herbs often with scaly rhizomes. Nodes unilacunar. Leaves opposite, sometimes whorled, usually isophyllous; lamina thin, usually velvety; epidermal cells of lower leaf surface usually with sinuate side walls; stomata scattered. Petiolar vascular trace usually a shallow crescent. Cymes axillary, pedunculate or epedunculate, usually bracteolate, or reduced to single flowers and combined in a terminal raceme. Stamens 4. Nectary ring-shaped, or of 5 to 2 separate glands, or absent. Ovary semi-inferior or inferior, rarely almost superior. Fruit usually a dry, ovoid, rostrate, bivalved capsule with loculicidal dehiscence (for details see Fruit and Seed Structure), less common are fleshy, non-rostrate capsules of various forms. Chromosome number $n = 13$, rarely 12, 11, or 10. The tribe is here conceived as to exclude *Sinningia*, *Paliavana* and *Vanhouttea*, which are placed in the separate tribe Sinningieae. According to Smith (2001) and Smith and Atkinson (1998), the tribe is paraphyletic, while according to Zimmer et al. (2002) it is strongly supported as monophyletic, if *Bellonia*, *Lembocarpus* (both running here under Genera of Uncertain Tribal Affiliation) and *Gloxinia sarmentiana* are excluded. According to Zimmer et al. (2002) and Roalson et al. (in prep.), three alliances can be readily recognised: (a) the *Gloxinia* alliance, with *Gloxinia* s.l., *Anodiscus* and *Kollikeria* (nested within *Gloxinia* s.l.), *Monopyle* and *Diastema* (genera 18–22; scaly rhizomes present, nectary primarily annular, but 5 separate glands in *Diastema*, absent in *Gloxinia* p.p., *Monopyle* and *Anodiscus*; fruit a dry rostrate capsule, or modified in various ways; chromosome number $n = 13$), (b) the *Kohleria* alliance, with *Kohleria*, *Pearcea* and *Capanea* (23–25; scaly or "stringy" rhizomes present, but lacking in *Pearcea* or with fleshy, widely spaced scales; fruit a dry capsule in most *Kohleria* spp., but otherwise specialised; chromosome number $n = 13$ or 11), (c) the Central American alliance, with *Achimenes*, *Eucodonia*, *Smithiantha*, *Niphaea*, *Moussonia*, *Solenophora* (26–31; with or without scaly rhizomes; nectary annular, fruit mostly a dry rostrate capsule, except the odd *Solenophora*; chromosome number $n = 12$, 11 or 10). The remaining genera *Phinaea*, *Heppiella* and *Goyazia* (32–34) seem to occupy a rather isolated position.

18. *Gloxinia* L'Hér.

Gloxinia L'Hér. in Aiton, Hort. Keco. 2: 331 (1789); Wiehler, Selbyana 1: 380–388 (1976), emend. & rev.; Hoehne, Bot. Estado Sao Paulo 3: 315–335 (1964), reg. rev.; Skog, Ann. Missouri Bot. Gard. 65: 929–932 (1979), reg. rev.
Seemannia Regel (1855).

Perennial herbs with scaly rhizomes, occasionally aerial "stringy rhizomes" present. Stem usually erect. Leaves ± isophyllous, petiolate. Flowers usually solitary and ebracteolate in leaf or bract axils, showy, purple, orange, red, white, rarely white or yellow. Sepals free. Corolla tubular, funnelform or campanulate, erect or oblique in the calyx, limb with subequal lobes. Stamens included; filaments often coiling after anthesis; anthers apically coherent. Nectary annular or reduced, rarely absent. Ovary semi-inferior to inferior. Capsule usually dry, bivalved, rostrate, rarely fleshy. $n = 13$. Fifteen spp., C America and tropical South America, with greatest diversity in Bolivia and C Brazil. Various pollination syndromes present. Heterogeneous and apparently polyphyletic (Zimmer et al. 2002; Roalson et al., in prep.).

19. *Anodiscus* Benth.

Anodiscus Benth., in Benth. & Hook. f., Gen. pl. 2: 998 (1876); Wiehler, Selbyana 6: 174 (1983), notes.

Terrestrial, perennial herbs, without scaly rhizomes, 0.5–2 m tall. Leaves frequently whorled, lanceolate to ovate, shortly petiolate. Flowering stem an elongate raceme, with leaves reduced to bracts and solitary axillary flowers; lateral racemes usually present at base of terminal raceme. Flowers small, corolla white, with broad tube and slightly bilabiate limb of subequal lobes. Nectary lacking. Capsule dry, bivalved, rostrate. $n = 13$ (according to Wiehler 1983, but undocumented). One sp., *A. xanthophyllus* (Poepp. & Endl.) Mansf., in the Andes of Ecuador and Peru.

20. *Koellikeria* Regel

Koellikeria Regel, Ind. Sem. Hort. Bot. Turic. 4 (1847), Flora 31: 249 (1848); Wiehler, Selbyana 6: 174 (1983), notes; Skog, Ann. Missouri Bot. Gard. 65: 932–935 (1979), reg. rev.; Feuillet & Steyermark in Berry et al., Fl. Venezuel. Guayana 5 (1999), reg. rev.; Skog, Monogr. Syst. Bot. Missouri Bot. Gard. 85: 1125 (2001), reg. rev.

Small perennial herbs with scaly rhizomes. Leaves ± congested into a rosette, opposite, isophyllous; shortly petiolate. Flowering stem a terminal elongate raceme, with leaves reduced to minute bracts (in alternate arrangement) and solitary axillary flowers; often lateral racemes present as well. Flowers small, with short pedicels. Sepals free, equalling the corolla tube. Corolla shortly tubular, nearly erect in the calyx, zygomorphic, the upper lobes short, the lower ones much longer and broader, toothed. Stamens adnate to the corolla base, anthers coherent by their sides and apices. Nectary annular. Ovary semi-inferior, stigma bilobed. Capsule dry, bivalved, rostrate. $n = 13$. One sp., *K. erinoides* (DC.) Mansf., from Nicaragua to tropical South America.

21. *Monopyle* Benth.

Monopyle [Moritz ex] Benth., in Benth. & Hook. f., Gen. pl. 2: 997 (1876); Wiehler, Selbyana 6: 166–167 (1983); Skog, Ann. Missouri Bot. Gard. 65: 942–946 (1979), reg. rev.

Perennial herbs with scaly rhizomes. Stem erect or decumbent, terete, slender, usually branched, usually with uncinate trichomes. Leaves anisophyllous, sometimes appearing alternate, joined across the stem in a stipular line; petiolate; lamina membranous, the base usually strongly oblique, often bullate or areolate above. Flowers in a terminal thyrse or raceme, emerging from small bracts. Corolla somewhat zygomorphic; tube open-campanulate, base rounded, narrowed, or slightly saccate; limb 5-lobed, flaring, white or white with blue, sometimes yellow in the throat, rarely plain yellow. Stamens included; filaments slender, curved inwardly, adnate to corolla base; anthers coherent, thecae divergent, connective thick. Nectary usually lacking and replaced by an osmophore (if nectary present, annular, but much reduced). Ovary inferior, narrowly turbinate; stigma stomatomorphic. Capsule (always?) fleshy, linear-oblong, topped by the persistent calyx, opening by 1 or 2 longitudinal slits along its entire length. $n = 13$. Over 20 spp., the Western Cordillera from Bolivia to Guatemala.

22. *Diastema* Benth.

Diastema Benth., Bot. Voyage Sulphur 132 (1845); Wiehler, Selbyana 6: 170 (1983), notes; Skog, Ann. Missouri Bot. Gard. 65: 891–896 (1979), reg. rev.; Feuillet & Steyermark in Berry et al., Fl. Venezuel. Guayana 5 (1999), reg. rev.; Skog, Monogr. Syst. Bot. Missouri Bot. Gard. 85: 1123 (2001), reg. rev.

Slender perennial herbs with scaly rhizomes. Stem short, unbranched, hirsute or villous. Leaves ± isophyllous, usually long-petiolate, membranous, crenate, dentate, or serrate. Cymes usually reduced to solitary flowers and combined in a terminal, bracteose raceme. Sepals free, often spreading. Corolla white, sometimes with purple markings on the lobes, rarely red, tubular or funnelform, only slightly broader towards the limb, neither spurred nor ventricose; limb regular or slightly zygomorphic, lobes rounded. Stamens adnate to the corolla base, anthers orbicular, at first apically coherent, later free, thecae slender, dehiscent by slits above. Nectary of 5 glands of equal length, sometimes connate into 2–3 glands. Ovary semi-inferior, stigma bilobed. Capsule obovoid, apex convex, membranous, bivalved, usually splitting along the dorsal side to the base of the ovary. n = 13. Over 20 spp., Bolivia to Mexico, Venezuela, centre of diversity in Andean Colombia, Ecuador and Peru.

23. *Kohleria* Regel

Kohleria Regel, Ind. Sem. Hort. Bot. Turic. 4 (1847), Flora 31: 250 (1848); Wiehler, Selbyana 6: 168 (1983), notes; Kvist & Skog, Smiths. Contr. Bot. 79: 1–80 (1992), rev.; Morton, Baileya 15: 61–78 (1967), reg. rev.; Skog, Ann. Missouri Bot. Gard. 65: 935–942 (1979), reg. rev.; Feuillet & Steyermark in Berry et al., Fl. Venezuel. Guayana 5 (1999), reg. rev.; Skog, Monogr. Syst. Bot. Missouri Bot. Gard. 85: 1126 (2001), reg. rev.
Isoloma (Benth.) Decne. (1848), non J. Smith (1841).
Tydaea Decne. (1848).

Perennial herbs or subshrubs with scaly rhizomes. Stem usually erect, sometimes decumbent. Leaves opposite or in whorls of 3–4, ± isophyllous, pilose or hirsute, petiolate. Cymes few-flowered, pedunculate or epedunculate, fasciculate to umbellate, from leaf or bract axils, or flowers solitary. Sepals erect or recurved, as long as or longer than the corolla tube. Corolla red or orange, rarely pink, yellow, purple or greenish; tube usually somewhat inflated or ventricose, or campanulate above a narrow base, hirsute outside; limb bilabiate or subregular, usually spotted or striped. Stamens adnate to base of corolla tube; anthers at first coherent, later often becoming free; longicidally dehiscent. Nectary of five separate glands, rarely with two or three or all of the glands united at the base. Ovary more than half inferior; stigma bilobed. Fruit a dry, rarely fleshy, bivalved, rostrate capsule. n = 13. Seventeen spp., Western Cordillera from Peru to Mexico, Venezuela, Trinidad, the Guianas; centre of diversity in Colombia. Various pollination syndromes present. Some species are fine greenhouse ornamentals with spectacular flowers.

24. *Pearcea* Regel

Pearcea Regel, Gartenflora 16: 388 (1867); Kvist & Skog, Smiths. Contr. Bot. 84: 1–47 (1996), emend. & rev.
Parakohleria Wiehler (1978).

Perennial herbs, usually with conspicuous indumentum, occasionally with subterranean rhizomes with the scales widely spaced. Stem erect. Leaves of a pair (sub)equal, petiolate, usually (ovate-)lanceolate, often pilose or hirsute. Cymes pedunculate (rarely without peduncle), flowers congested. Sepals basally connate. Corolla bent downwards at base, narrowly to broadly funnelform, subventricose or ventricose, more rarely urceolate, throat often constricted and limb usually subregular, rarely bilabiate, usually villous or pilose outside; red, often basally and ventrally yellow, rarely entirely yellow. Stamens didynamous, usually included; filaments coiling in the female stage, anthers cohering apically and laterally. Nectary usually of five free glands, two sometimes connate. Ovary semi-inferior; stigma stomatomorphic or slightly bilobed. Capsule fleshy, bivalved. n = 11 (according to Wiehler also n = 13, but undocumented). Seventeen spp., eastern Andean slopes and adjacent lowlands from N Colombia through Ecuador and Peru to NW Bolivia, centres of diversity the N and C Ecuadorian Amazon region and C Peru.

25. *Capanea* Planch.

Capanea [Decne. ex] Planch., Fl. Serres Jard. Eur. 5: t. 499–500 (1849); Skog, Ann. Missouri Bot. Gard. 65: 821–825 (1979), reg. rev.
Campanea Decne. (1849), nom. inval.

Epiphytic or terrestrial herbs, shrubs or vines, with subterranean rhizomes with widely spaced, membranous (not fleshy) scales. Stem erect, scandent or pendent, densely pubescent. Leaves ± isophyllous, (sometimes very shortly) petiolate; membranous, asymmetric at base, usually tomentose. Cymes one- to several-flowered, with elongate peduncules and pedicels, pendent. Flowers large and showy, zygomorphic. Sepals shortly connate. Corolla erect in the calyx, tubular or campanulate, tube broad, ventricose, sometimes constricted at throat, limb wide, with subequal lobes, corolla tube purple or greenish, limb usually greenish with lines or spots. Stamens ± exserted; filaments adnate to corolla base, sometimes coiled; anthers divergent, apically coherent. Nectary of 5 large, usually separate glands. Ovary half inferior; style exserted; stigma stomatomorphic. Capsule almost globose, with persistent calyx, opening by 4 valves. n = 13. Eleven spp. have been described, but probably they can be reduced to two (Skog, pers. comm.). Northern South America to Guatemala.

26. *Achimenes* Pers.

Achimenes Pers., Syn. pl. 2: 164 (1806), nom. cons., non Achimenes P. Browne (1756) nec Achimenes Vahl (1791); Wiehler, Selbyana 1: 374–404 (1976), emend.; Morton, Baileya 10: 147–154 (1962), sect. rev.; Skog, Ann. Missouri Bot. Gard. 65: 789–793 (1979), reg. rev.; Skog, Monogr. Syst. Bot. Missouri Bot. Gard. 85: 1116–1117 (2001), reg. rev.

Perennial herbs with scaly rhizomes; stems erect or decumbent, unbranched. Leaves opposite or whorled, iso- or anisophyllous. Cymes one- to several-flowered. Sepals free. Corolla usually showy, red, purple, pink, orange, yellow or white, erect or oblique in the calyx, either tubular with a flat spreading limb, or obliquely funnelform, sometimes with a basal sac or spur on the upper side; limb subregular to bilabiate, with the three lower lobes larger, lobes broad, entire or fringed. Stamens didynamous, included; filaments slender, adnate to corolla base; anthers apically coherent. Nectary annular or cup-like. Ovary semi- to completely inferior, style curved, stigma bilobed, stomatomorphic or cupulate. Capsule dry, bivalved, rostrate. $n = 11, 22$. About 24 spp., Mexico and C America, the Caribbean, Colombia, Ecuador. Various pollination syndromes present. Some species are popular greenhouse ornamentals, or used as bedding plants in the southern USA.

27. *Eucodonia* Hanst.

Eucodonia Hanst., Linnaea 26: 201 (1854), Wiehler, Selbyana 1: 374–404 (1976), reestabl., emend. & rev.

Small terrestrial or saxicolous herbs with scaly rhizomes. Stem succulent. Leaves opposite, pseudoverticillate. Indumentum on stem and underside of leaves densely lanate-villous. Flowers solitary in the leaf axils, corolla obliquely campanulate, purple or lavender, with long pedicel. Capsule dry, bivalved, rostrate. $n = 12$. Two spp., C & S Mexico. Formerly included in *Achimenes*, but chromosome number different and hybrids sterile.

28. *Smithiantha* Kuntze

Smithiantha Kuntze, Revis. gen. pl. 2: 977 (1891); Wiehler, Phytologia 73: 220–241 (1992), notes; Wiehler, Gesneriana 1: 29–97 (1995), new spp.

Perennial herbs or subshrubs with scaly rhizomes. Stem erect, usually unbranched. Leaves opposite, petiolate, mostly cordate, soft-hairy, often mottled with purple and/or white. Flowering stem a terminal raceme, with leaves reduced to alternate bracts and solitary axillary flowers, flowers without bracteoles. Sepals short and small. Corolla red, orange, with darker spots, sometimes yellow inside, lavender, purple or white with darker stripes; tubular to campanulate, mouth not constricted, limb with small or prominent, subequal, rounded lobes. Stamens included, adnate to corolla base; anthers coherent. Nectary ring-shaped, sometimes shallowly lobed. Ovary semi-inferior. Capsule dry, bivalved, rostrate. $n = 12$. About 7 spp., C & N Mexico to Guatemala. Beautiful ornamentals.

29. *Niphaea* Lindl.

Niphaea Lindl., Bot. Reg. 27, Misc. 80 (1841).

Perennial herbs with scaly rhizomes. Stem low, erect, herbaceous, terete. Leaves opposite, isophyllous, petiolate, ovate, serrate, main and side veins sometimes silvery-white, softly hairy. Cymes epedunculate, bracteolate, one- to few-flowered, flowers long-stalked, ± composing a terminal corymb. Sepals free nearly to base. Corolla rotate, shallowly campanulate, white. Stamens adnate to corolla base; filaments short (at most as long as anthers), straight; anthers yellow, free, cells apically confluent, longicidally dehiscent. Nectary absent. Ovary semi-inferior, stigma stomatomorphic. Capsule dry, bivalved, rostrate. $n = 11$. Five spp. have been described, but reduction to two or three seems adequate. Mexico, Guatemala, Venezuela, Peru.

30. *Moussonia* Regel

Moussonia Regel, Ind. Sem. Hort. Bot. Turic. 4 (1847), Flora 31: 248. (1848); Wiehler, Selbyana 1: 22–31 (1975), reestabl. & rev.; Skog, Ann. Missouri Bot. Gard. 65: 947–951 (1979), reg. rev.; Skog, Monogr. Syst. Bot. Missouri Bot. Gard. 85: 1126–1127 (2001), reg. rev.
Kohleria sect. *Moussonia* (Regel) Fritsch (1894).

Perennial herbs or subshrubs without scaly rhizomes. Stem erect, branched. Leaves of the pairs equal. Flowers in (usually pedunculate) axillary cymes or solitary. Corolla red, orange to yellow, tube long, broadest at the middle, base and throat narrow, limb with subregular lobes. Stamens included, adnate to the corolla base. Nectary annular. Ovary semi-inferior to almost completely inferior; stigma capitate-stomatomorphic. Nectary annular. Capsule dry, bivalved, rostrate. $n = 11$. Eleven spp., from Mexico to Panama.

31. *Solenophora* Benth.

Solenophora Benth., Pl. Hartw. 68 (1840); Skog, Ann. Missouri Bot. Gard. 65: 987–990 (1979), reg. rev.; Weigend & Förther, Harvard Papers Bot. 7(1): 37–78 (2002), rev.

(Sub)shrubs or weak trees, scaly rhizomes absent. Stem usually elongate, ascending to erect, woody, terete to subquadrangular, pith usually prominent, brown to grey at maturity. Leaves petiolate, ± anisophyllous, often large, membranous, ovate to elliptic, base asymmetrical, margin dentate or serrate. Cymes one- to few-flowered, usually pedunculate and bracteolate. Sepals connate, fused part exceeding the ovary, lobes short or long, rarely calyx subtruncate. Corolla large and showy, tube elongate, not spurred, yellow to orange or reddish orange, throat ampliate, limb with broad, rounded lobes, often spotted. Stamens adnate to the corolla base, included or rarely exserted; filaments long, curved, broad at the base; anthers dehiscing longicidally. Nectary usually a single, bilobed gland (rarely of 2, 3 or 5 glands?). Ovary inferior, apex flattened or convex; style elongate, flattened, broader towards the apex; stigma stomatomorphic. Capsule usually globose, crowned by the persistent calyx, dehiscing irregularly and exposing the placentae and seeds. $n = 10$. Sixteen spp., C America.

32. *Phinaea* Benth.

Phinaea Benth. in Benth & Hook.f., Gen. pl. 2: 997 (1876); Skog, Ann. Missouri Bot. Gard. 65: 973–976 (1979), reg. rev.; Skog, Monogr. Syst. Bot. Missouri Bot. Gard. 85: 1128 (2001), reg. rev.

Small perennial herbs with scaly rhizomes. Stem short, erect. Leaves opposite, usually (sub)isophyllous, often congested at the stem apex, membranous, soft. Cymes epedunculate, bracteolate, one- to many-flowered, congested. Calyx campanulate, with spreading equal or subequal lobes. Corolla white (rarely pink), subrotate or campanulate, tube short, limb with spreading or erect, broad, subequal lobes. Stamens adnate to base of corolla tube; filaments slender, longer than the anthers, those of the anterior stamen pair distinctly longer and curved, anthers coherent or sometimes free, obovate to subglobose, with distinct (not confluent) cells, dehiscing by a short slit. Nectary absent, or annular and strongly reduced. Ovary semi-inferior, style curved; stigma stomatomorphic. Capsule globose, somewhat fleshy or dry, bivalved. $n = 11, 13, \pm 26$. Ten spp. described, more expected, Mexico to Peru, Venezuela, Cuba, Haiti. Flowers similar to *Niphaea*, but differing by the long, curved filaments of the anterior stamens.

33. *Heppiella* Regel

Heppiella Regel, Gartenflora 2: 353, t. 70 (1853); Kvist, Syst. Bot. 15: 720–735 (1990), rev.

Epipetric or (less commonly) epiphytic, perennial, usually with scaly rhizomes, ± suffruticose herbs. Leaves opposite or in whorls of three, membranous, glandular hairy. Cymes with or without peduncle, usually bracteolate, sometimes composing a terminal, corymboid thyrse, or reduced to single flowers. Sepals usually short, ± equal. Corolla red; tubular, straight, limb with small (sub)equal lobes. Stamens included or slightly exserted; adnate to corolla base; anthers free. Nectary ring-shaped. Ovary semi-inferior; stigma stomatomorphic. Capsule bivalved, dry, rostrate. $n = 13, 26$. Four spp., one very variable, Peru, Ecuador, Colombia, and western Venezuela.

34. *Goyazia* Taub.

Goyazia Taub., Bot. Jahrb. Syst. 21: 451 (1896); Wiehler, Selbyana 1: 395 (1976), emend. & rev.

Small saxicolous herbs with scaly rhizomes. Stem thin, stiff, wiry. Leaves of a pair somewhat unequal; petiole short, swollen; lamina thick-leathery, venation pericraspedodromous (side veins reaching the margins, joining the superadjacent secondaries and surrounding the total lamina at the margin). Flowers solitary in the leaf axils, pedicels short. Corolla with straight, subcampanulate tube, slightly widening upwards, limb oblique, lobes subequal. Stamens adnate to corolla base; filaments slender; anthers subreniform, coherent. Nectary annular, five-lobed. Ovary semi-inferior (approaching superior), ovoid-globose; stigma bilobed. Fruit a dry bivalved capsule. Two spp., C Brazil (Planalto).

4. Tribe Gesnerieae Dumort. (1829).

Terrestrial herbs, shrubs or rarely small trees, without subterranean storage organs. Nodes unilacunar. Leaves alternate, rarely opposite, often leathery. Epidermal cells of lower leaf surface with straight or sinuate side walls; stomata grouped into islands or irregularly scattered. Petiolar vascular trace usually a complete ring or nearly so. Cymes usually pedunculate and bracteolate. Nectary ring-shaped, sometimes lobed. Ovary inferior or semi-inferior. Fruit a dry capsule with loculicidal or loculicidal plus septicidal dehiscence. $n = 14$. The molecular data of Smith and Atkinson (1998) and Zimmer et al. (2002) suggest that the tribe is sister to the Gloxinieae. To some degree, *Pheidonocarpa* links the two tribes, its chromosome number ($n = 14$) agreeing with other Gesnerieae, but its opposite leaves, semi-inferior ovary, and rostrate capsule being more typical of Gloxinieae. Wiehler (1983) stressed the weak delimitation of the three genera and suggested to lump them under *Gesneria*, but no formal inclusions were made.

35. *Pheidonocarpa* L.E. Skog

Pheidonocarpa L.E. Skog, Smiths. Contr. Bot. 29: 40 (1976).

Perennial, suffruticose, fibrous-rooted plants. Stems woody, erect or decumbent. Leaves opposite; petioles short, the bases joined across the stem; lamina rigid, scabrous. Cymes many-flowered compound dichasia; peduncles much exceeding the subtending leaves. Sepals shortly connate. Corolla zygomorphic, pilose; tube ventricose, limb bilabiate. Stamens 4, filaments adnate to base of corolla tube; anthers coherent or free, exserted. Ovary semi-inferior. Nectary annular, 5-lobed. Fruit a dry capsule with curved rostrate apex, pilose, bivalved. $n = 14$. One sp., *P. corymbosa* (Swartz) L.E. Skog, with two subspp. in Cuba and Jamaica, respectively.

36. *Gesneria* L. s.str. (Skog 1976).

Gesneria L.; Sp. pl., ed. 1: 612 (1753); Skog, Smiths. Contr. Bot. 29: 43–182 (1976), rev.
Pentarhaphia Lindl. (1827).

Suffruticose herbs, shrubs, or small trees to 10 m tall. Stem erect, decumbent, or pendent, branched or not. Leaves alternate, exceptionally opposite, often crowed at branch tips; petiole sometimes very short, vascular trace a thick ring or U in cross section; lamina membranous or coriaceous, usually with adaxial multiple epidermis containing crystal cells. Cymes compound or few-flowered, or reduced to solitary flowers. Sepals slightly connate at base. Corolla tubular, ventricose or infundibuliform-campanulate; white, green red, orange, yellow or brown, limb regular or bilabiate, lobes spreading or reflexed. Filaments curved or geniculate at base, inserted at corolla base; anthers cohering pair-wise at their apices or in a row side by side. Nectary annular or lobed. Ovary inferior; stigma stomatomorphic, clavate, capitate or bilabiate. Capsule usually dry, dehiscing into 2 or 4 valves, occasionally forming a splash cup. $n = 7, 14, 28$. Over 60 spp., in the Caribbean region. Nine sections are recognised by Skog (1976): (1) sect. *Pentaraphia* (Lindl.) Fritsch, (2) sect. *Stenochonanthe* L.E. Skog, (3) sect. *Lachnoblaste* L.E. Skog, (4) sect. *Myrmekianthe* L.E. Skog, (5) sect. *Gesneria*, (6) sect. *Physcophyllon* L.E. Skog, (7) sect. *Dittanthera* (G. Don) L.E. Skog, (8) sect. *Duchartrea* (Decne.) Fritsch, and (9) sect. *Chorisanthera* (G. Don) L.E. Skog.

37. *Rhytidophyllum* Mart.

Rhytidophyllum Mart., Nov. gen. sp. pl. 3: 38 (1829).

Shrubs or trees, usually single-stemmed. Leaves alternate, mostly in a tuft on stem apex; petiolate to subsessile, sometimes with stipule-like "auricules" at the base; lamina lanceolate. Axillary cymes long-pedunculate, of several to many flowers in a double cincinnus or compound dichasium. Flowers rather short-tubed, obliquely campanulate, with wide mouth, limb spreading, yellow to greenish or brownish, rarely red or orange, with darker dots. Stamens inserted distinctly (to 6 mm) above the corolla base. Ovary inferior. Capsule globose to ovoid, dry, bivalved. $n = 14$. About 20 spp., in the Caribbean region. Weakly distinct from *Gesneria*.

5. Tribe Sinningieae Fritsch (1893).

Usually saxicolous, rarely terrestrial or epiphytic plants, perennial through tubers producing annual herbaceous shoots, in some cases tuber replaced by the basal stem portion somewhat swollen and succulent, or plants woody shrubs without specialised storage organ. Nodes unilacunar. Leaves opposite, sometimes whorled or rosulate; membranous, usually velvety; stomata scattered. Petiolar vascular trace usually a shallow crescent. Cymes bracteolate. Corolla usually erect in the calyx (even when strongly gibbous or spurred), size and shape very variable. Nectary of 5 to 2 separate glands, rarely (secondarily) annular. Ovary semi-inferior, rarely inferior or almost superior Fruit a dry, ovoid, rostrate (rarely somewhat fleshy and non-rostrate) bivalved capsule with loculicidal dehiscence. $n = 13$. Main distribution in SE Brazil. The reestablishment of the tribe for the three following genera (formerly placed in tribe Gloxinieae) was suggested by the molecular data of Smith et al. (1997b), and later confirmed by Zimmer et al. (2002) and Perret et al. (2003). Moreover, both the morphological-cladistic analysis of Boggan (1991) and the molecular data of Perret et al. (2003) suggest an inclusion of *Paliavana* and *Vanhouttea* into *Sinningia*.

38. *Sinningia* Nees

Sinningia Nees, Ann. Sci. Nat., Ser. 1, 6: 297 (1825); Wiehler, Selbyana 6: 171 (1983), notes; Skog, Ann. Missouri Bot. Gard. 65: 983–987 (1979), reg. rev.; Chautems, Candollea 45: 381–388 (1990), 46: 411–426 (1991), 52: 159–169 (1997), Gesneriana 1: 8–14 (1995), rev. in progr.; Wiehler & Chautems, Gesneriana 1: 5–7 (1995), inclusion of *Lietzia*; Feuillet & Steyermark, in Berry et al., Fl. Venezuel. Guayana 5 (1999), reg. rev; Skog, Monogr. Syst. Bot. Missouri Bot. Gard. 85: 1128 (2001), reg. rev.; Chautems in Wanderley et al., Fl. Fanerogâm. São Paulo (in press), reg. rev.
Corytholoma (Benth.) Decne. (1848).
Rechsteineria Regel (1848).
Lietzia Regel (1890).

Terrestrial, often saxicolous, rarely epiphytic perennial herbs or subshrubs, with hard-fleshy tubers (to 1 m in diameter), rarely stoloniferous with the stolons producing tubers. Annual-monocarpic stems erect, occasionally decumbent, mostly unbranched, rarely stemless. Leaves opposite or in whorls, sometimes congested on stem top, (sub)isophyllous, petiolate or sessile. Cymes axillary, few- to many-flowered, or flowers in terminal, bracteose racemes or spikes. Calyx with campanulate tube, lobes linear, triangular or lanceolate. Corolla red, orange, white, greenish-yellowish, blue or purple, tube broadly tubular, cylindrical or obliquely campanulate, often gibbous at base, or ventricose above, limb nearly regular or (more often) bilabiate. Stamens didynamous; filaments adnate to corolla base; anthers coherent, dehiscing longicidally. Nectary of 1 to 5 glands, sometimes two enlarged dorsally, free or fused, rarely 5 connate in a ring. Ovary semi-inferior, inferior, or almost superior; stigma stomatomorphic. Capsule dry, ovoid, rostrate or rarely fleshy and non-rostrate, bivalved. $n = 13$. Over 60 spp., from C America to N Argentina, but predominantly E & S Brazil. Various pollination syndromes present (Perret et al. 2001). Subdivision into sections is envisaged by Chautems (in prep.). *Sinningia speciosa* (Lodd.) Hiern is the parent plant of the popular "Florist's Gloxinias", with flowers in many colour variants and peloric forms.

39. *Paliavana* Vand.

Paliavana [Vell. ex] Vand., Fl. lusit. bras. spec. 40, t. 3, f. 17 (1788); Wiehler, Selbyana 6: 175–180 (1983), notes; Chautems, Candollea 56: 261–279 (2002), new sp., key.

Terrestrial shrubs or subshrubs with perennial stems and fibrous roots (no tubers or rhizomes). Leaves opposite. Corolla campanulate or funnel-shaped, large, green with maroon dots, greenish yellow, blue-purple, or cream-white with a green tinge. Nectary of five separate, large, rounded glands. Ovary semi-inferior. Capsule dry, bivalved, rostrate. $n = 13$. Six spp., SE and E Brazil.

40. *Vanhouttea* Lem.

Vanhouttea Lem., Bull. Soc. Hort. Orleans 1: 346 (1845); Wiehler, Selbyana 6: 180 (1983); notes; Chautems, Candollea 56: 261–279 (2002), new spp., key.

Terrestrial shrubs or subshrubs, with fibrous roots (no tubers or rhizomes). Leaves opposite. Corolla tubular, cylindrical, red. Nectary of 5 individual glands (sometimes connected at the base by a narrow ring). Ovary semi-inferior. Capsule dry, bivalved, rostrate. Eight spp., SE Brazil.

6. Tribe Episcieae Endl. (1839).

Wiehler, Selbyana 6: 181–195 (1983), notes.

Epiphytic or less frequently terrestrial herbs, shrubs or vines; exceptionally with tubers. Nodes trilacunar with split lateral bundles. Leaves opposite, sometimes whorled, iso- or anisophyllous; often succulent, or leathery; epidermal cells of lower leaf surface usually with straight side walls; stomata scattered. Petiolar vascular trace usually a shallow crescent. Cymes bracteolate. Flowers usually obliquely inserted in the calyx (especially when spurred). Stamens 4, filaments often shortly connate at base; anthers usually coherent and dehiscing longicidally. Nectary usually of a single dorsal, bilobed gland, rarely of 5 separate glands or other configurations. Ovary superior, globose or ovoid; stigma stomatomorphic, bilabiate or capitate. Fruit a soft berry or a fleshy, bivalved capsule with loculicidal dehiscence. Seeds usually with a fleshy funicle. Chromosome number $n = 9$, rarely 8. As confirmed by molecular data (Smith and Carroll 1997; Smith 2000a, b; Zimmer et al. 2002; Roalson et al., in prep.), the tribe represents a homogeneous group. Several alliances can be recognised: (a) the *Nautilocalyx* alliance, with *Nautilocalyx*, *Paradrymonia* s.l. and *Chrysothemis* (genera 37–39; probably basal in the tribe; mainly terrestrial, sometimes with tubers; fruit a fleshy, bivalved capsule), (b) *Alloplectus* and *Drymonia* (40–41; basal to (c) or nested within; terrestrial to epiphytic; fruit a fleshy capsule), (c) the *Columnea* alliance, with *Columnea* s.l., *Corytoplectus* and possibly *Neomortonia* (42–44; usually epiphytic, fruit a berry), (d) *Episcia* and *Alsobia* [45–46; stoloniferous herbs; affinity uncertain, *Episcia* may also be related to (a), *Alsobia* to (e)], (e) the Central American alliance with *Rufodorsia*, *Oerstedina* and *Cobananthus* (47–49; mostly epiphytes; fruit a berry or thin-walled capsule), and (f) *Nematanthus*, *Codonanthe* and *Codonanthopsis* (50–52; epiphytic shrubs, creepers or lianas; fruit a fleshy capsule, the seeds often presented on top of fleshy funicles). Some genera placed under Genera of Uncertain Tribal Affiliation may also belong to the Episcieae.

41. *Nautilocalyx* Hanst.

Nautilocalyx [Linden ex] Hanst., Linnaea 26: 207 (1854); Linden, Cat. p. 12 (1851) (Linden's catalogue has not yet been rediscovered); Wiehler, Selbyana 5: 29–45 (1978), emend. & rev.; Skog, Ann. Missouri Bot. Gard. 65: 954–958 (1979), reg. rev.; Feuillet & Steyermark in Berry et al., Fl. Venezuel. Guayana 5 (1999), reg. rev.

Terrestrial perennial herbs, sometimes with small tubers. Stem erect, ascending or decumbent, ± succulent. Leaves in distant pairs or rarely tufted. Cymes fascicled, or flowers solitary, bracteoles sometimes leafy. Sepals unequal, the upper one forced back around the corolla spur, green or coloured. Corolla white to yellow, usually with purple spots or lines, oblique in the calyx, tube laterally broadened, with a short basal sac or spur, limb with nearly equal lobes, entire, their margins toothed or fimbriate. Stamens included; filaments basally connate and adnate to the corolla base; anthers cohering in two pairs by their apices, dehiscing longicidally. Nectary of a single or two dorsal glands. Ovary superior; stigma bilobed. Fruit a bivalved capsule. n = 9, 18. Over 70 spp., throughout tropical America except SE Brazil.

42. *Chrysothemis* Decne.

Chrysothemis Decne., Rev. Hort. 21, ser. 3, 3: 242 (1849); Leeuwenb., Acta Bot. Neerl. 7: 329–340 (1958), rev.; Skog, Ann. Missouri Bot. Gard. 65: 825–830 (1979), reg. rev.; Feuillet & Steyermark in Berry et al., Fl. Venezuel. Guayana 5 (1999), reg. rev.; Skog, Monogr. Syst. Bot. Missouri Bot. Gard. 85: 1119 (2001), reg. rev.
Tussacia Benth. (1846).

Terrestrial or epiphytic perennial herbs with tubers. Stems succulent, erect or rarely decumbent, subquadrangular. Leaves (sub)isophyllous, membranous, short-petiolate, blade usually decurrent. Cymes one- to 9-flowered, often umbellate. Sepals connate, forming an inflated or campanulate to urceolate, sometimes winged calyx, yellow-green, orange or red. Corolla tube cylindrical, gibbous, internally with a ring of hairs at the stamen insertion, mouth oblique; limb flaring, lobes rounded, ± entire, yellow to orange with darker lines. Stamens didynamous, included; filaments flattened, basally connate and inserted shortly above the corolla base; anthers orbicular, sometimes coherent. Nectary a bilobed dorsal gland, rarely of 4 separate glands. Ovary globose, conical or ovoid, pilose; stigma bilobed. Fruit a fleshy, globose to ovoid, bivalved capsule, included in the persistent calyx. n = 9. Seven spp., Ecuador to Guatemala, C Brazil, the Guianas, Venezuela and Lesser Antilles. Some spp. are grown as ornamentals for their attractive red or orange calyces and bright yellow flowers.

43. *Paradrymonia* Hanst.

Paradrymonia Hanst., Linnaea 26: 180, 207 (1854); Wiehler, Selbyana 5: 45–59 (1978), emend. & rev.; Skog, Ann. Missouri Bot. Gard. 65: 964–973 (1979), reg. rev.; Feuillet & Steyermark in Berry et al., Fl. Venezuel. Guayana 5 (1999), reg. rev.; Skog, Monogr. Syst. Bot. Missouri Bot. Gard. 85: 1127 (2001), reg. rev.

Terrestrial herbs to subshrubs or epiphytic suffrutescent vines. Stems erect, ascending, or scrambling, often succulent. Leaves iso- to strongly anisophyllous (sometimes in a single colony of a species), in the terrestrial species often tufted, usually large and surpassing the short stem, lamina usually oblanceolate, membranous to fleshy. Cymes usually congested, many-flowered, rarely exceeding the petiole, flowers often in a compact cluster at the ground or hidden by the leaves. Sepals free, linear or lanceolate. Corolla white or yellow with red or purple spots or lines, funnelform or trumpet-shaped, spurred at the base. Stamens included; anthers coherent. Nectary of one or two glands. Fruit a bivalved capsule. n = 9. Over 70 spp., throughout the neotropics except SE Brasil.

44. *Alloplectus* Mart.

Alloplectus Mart., Nov. gen. sp. pl. 3: 53, t. 223 (1829), nom. cons.; Wiehler, Phytologia 27: 309–328 (1973), emend. & rev.; Skog, Ann. Missouri Bot. Gard. 65: 793–802 (1979), reg. rev.; Feuillet & Steyermark in Berry et al., Fl. Venezuel. Guayana 5 (1999), reg. rev.; Skog, Monogr. Syst. Bot. Missouri Bot. Gard. 85: 1117 (2001), reg. rev.; J.L. Clark, rev. in prep.
Crantzia Scop. (1777).

Terrestrial or less frequently epiphytic shrubs or coarse herbs with fibrous roots. Stems erect, more rarely scandent or vining, quadrangular. Leaves ± isophyllous, often long petiolate. Cymes one- to few-flowered, flowers densely crowded, peduncle ±

absent. Sepals free, usually coloured, entire to strongly serrate. Corolla oblique in the calyx, tubular, yellow or red, sometimes laterally compressed and constricted at the throat, or base gibbous with a spur, often ventricose below the mouth, limb of usually small lobes. Stamens included; filaments adnate to corolla base; anthers often transversally oblong, coherent or becoming free, dehiscing longicidally. Nectary a single, sometimes bilobed, large gland. Fruit a fleshy bivalved capsule surrounded by the persistent calyx. $n = 9$. Over 75 spp., throughout the neotropics except SE Brazil. Apparently polyphyletic, with a group of species with resupinate flowers representing *Alloplectus* s.str. (Clark and Zimmer 2003).

45. *Drymonia* Mart.

Drymonia Mart., Nov. gen. spec. pl. 3: 57, t. 224 (1829); Wiehler, Phytologia 27: 324 (1973); emend.; Skog, Ann. Missouri Bot. Gard. 65: 896–918 (1979), reg. rev.; Feuillet & Steyermark in Berry et al., Fl. Venezuel. Guayana 5 (1999), reg. rev.; Skog, Monogr. Syst. Bot. Missouri Bot. Gard. 85: 1123–1125 (2001), reg. rev.

Terrestrial or epiphytic (sub)shrubs or lianas. Stems quadrangular or terete, in climbers up to 5 m long, often with adventitious roots. Leaves ± isophyllous, petiolate, membranous or rarely coriaceous. Cymes one- to several-flowered, condensed; bracteoles often large, leaf-like and coloured, caducous. Sepals often large and coloured, free or shortly connate, unequal with the uppermost shortest. Corolla oblique in the calyx, usually funnelform and broader towards the mouth, spurred or saccate at the base, throat broad; limb with usually spreading, rounded, often fimbriate lobes, the lower lobes usually larger. Stamens included, didynamous; forming a "saltshaker"-like structure; filaments adnate to the corolla base, occasionally contorted; anthers sagittate, before anthesis coherent by the sides and faces, with the spurred bases up, dehiscing by 2 to 4 pores, after anthesis the filaments coiling and the anthers separating. Nectary a single dorsal gland. Fruit a fleshy capsule with brightly orange- or purple-coloured reflexed valves, displaying a cone-shaped mass of seeds and pulpy funicles. $n = 9$. Over 140 spp., throughout the neotropics, with distribution centres in Colombia and Ecuador.

46. *Columnea* L., s.lat.

Columnea L., Sp. pl., ed. 1: 638 (1753); Stearn, Bull. Brit. Mus. Nat. Hist. (Bot.) 4: 181–236 (1969), reg. rev.; Morton, Phytologia 21: 165–195 (1971), reg. rev.; Skog, Ann. Missouri Bot. Gard. 65: 838–888 (1979), reg. rev.; Kvist & Skog, Allertonia 6: 327–400 (1993), reg. rev.; Smith, Syst. Bot. Monogr. 44: 1–89 (1994), sect. rev.; Feuillet & Steyermark in Berry et al., Fl. Venezuel. Guayana 5 (1999), reg. rev.; Skog, Monogr. Syst. Bot. Missouri Bot. Gard. 85: 1120–1122 (2001), reg. rev.
Dalbergaria Tussac (1811).
Trichantha Hook.f. (1844).
Pentadenia (Planch.) Hanst. (1854).
Stygnanthe Hanst. (1854).
Bucinellina (Wiehler) Wiehler (1981).

Epiphytic, rarely epipetric or terrestrial herbs or subshrubs, shoots often conspicuously dorsiventral. Leaves isophyllous or strongly anisophyllous, lower leaf surfaces sometimes with a conspicuous pattern of red or purple towards the apex and/or along the margin and veins, or with translucent red "windows". Cymes in the anisophyllous species in the axils of the larger leaves only, epedunculate, one- to 10-flowered, bracteoles scale-like, less commonly large and hiding the calyces, frequently coloured. Sepals almost free, (sub)equal, entire or dissected. Corolla often red, less commonly yellow, cream or greenish, tube frequently ventricose, basally often dorsally gibbous, usually glandular hairy in the throat, limb subregular to strongly bilabiate ("galeate"), lobes 5 or (by the fusion of the dorsal lobes) 4, usually rounded. Stamens included or exserted; filaments usually basally fused and adnate to the corolla tube; anthers usually coherent. Nectary a single bilobed gland, less commonly 4 or 5, rarely 2 or 3 free glands. Ovary superior; stigma stomatomorphic or bilobed. Fruit a globose or ovoid, white or coloured berry. $n = 9$. Over 270 spp., from Mexico south to Ecuador and Bolivia and east to Brazil, centres of diversity in Colombia, Ecuador and Panama. Essentially hummingbird-pollinated. An ample discussion whether to split *Columnea* into separate genera or into sections is found in Wiehler (1983) and Kvist and Skog (1993). As almost any berry-fruited species fits into the vague concept of *Columnea* s.lat., Wiehler pleaded for splitting the genus into the following five "columneoid genera".

46a. *Columnea* L., s.str.

Columnea L., Sp. pl., ed. 1: 638 (1753); Wiehler, Phytologia 27: 309–328 (1973); Wiehler, Selbyana 6: 187 (1983), notes.
Columnea L. sect. *Columnea*

Epiphytic herbs or subshrubs. Stems ascending, spreading or pendent. Leaves iso-, rarely strongly anisophyllous. Cymes one- to few-flowered, sometimes borne below the foliage. Sepals ± equal. Corolla mostly "columneoid" (tube basally narrow, gradually widening, never inflated, upper and lateral lobes joined into a "galea", lowermost lobe narrow and reflexed); colour red or red with yellow. Stamens with cohering anthers. Nectary of a single dorsal gland. Berry globose or subglobose, usually white. *n* = 9. Over 75 spp., C America, the Caribbean islands, Colombia, Ecuador, Venezuela, Guayana Highlands and NE Brazil, the centre of diversity is Panama and Costa Rica.

46b. *Dalbergaria* Tussac

Dalbergaria Tussac, Fl. Antill. 1: 141, pl. 19 (1808); Wiehler, Selbyana 6: 189 (1983), notes.
Collandra Lem. (1847).
Columnea L. sect. *Collandra* (Lem.) Benth. (1876).

Epiphytic subshrubs with strongly anisophyllous shoots. Leaves subsessile, in close-set, nearly distichous pairs; the large leaf oblanceolate or elliptic, the small leaf stipule-like; lamina often marked with red patterns on the lower surface, or with translucent red windows. Cymes few-flowered, hidden from above. Corolla tubular, with a narrow subregular limb or somewhat inflated and with a bilabiate limb, rarely "columneoid". Nectary of two large, connate dorsal glands, sometimes with an additional small ventral gland. Berry ovoid, usually coloured. *n* = 9. Over 90 spp., Bolivia to the Guayana Highlands, north to Mexico, and Greater Antilles.

46c. *Pentadenia* (Planch.) Hanst.

Pentadenia (Planch.) Hanst., Linnaea 26: 187, 211 (1854); Wiehler, Phytologia 27: 314 (1973), emend. & rev.
Stygnanthe Hanst. (1854).
Columnea L. sect. *Pentadenia* (Planch.) Hanst. (1876).
Columnea L. sect. *Stygnanthe* (Hanst.) Benth. (1876).

Epiphytic subshrubs, lianas or thin-stemmed trailers. Stem erect, ascending, scrambling, or pendent. Leaves opposite, (sub)equal. Corolla tubular, limb subregular with small subequal lobes; or corolla tube slightly (rarely strongly) inflated and the limb bilabiate, rarely with the upper lobes fused into a galea. Nectary of five separate glands, the two dorsal ones sometimes connate, or reduced to the dorsal pair (variable even within a single individual; Smith 1991). Berry globose or subglobose, white or lavender. *n* = 9. Over 35 spp., from Bolivia along the Western Cordillera to Mexico (centre of diversity Ecuador), one sp. in the coastal range of Venezuela.

46d. *Trichantha* Hook.f.

Trichantha Hook.f., Icon. pl. 3: tab. 666 (1844); Morton, Contr. U.S. Natl. Herb. 38: 1–27 (1963), rev.; Wiehler, Selbyana 1: 33 (1975), notes.
Columnea L. sect. *Ortholoma* Benth. (1846).
Ortholoma (Benth.) Hanst. (1854).
Columnea L. sect. *Trichantha* (Hook.) Kuntze (1904).

Epiphytic, mostly vining, spreading or scandent subshrubs. Stem erect, decumbent or pendent. Leaves usually anisophyllous, somewhat fleshy, coriaceous or chartaceous. Cymes one- to several-flowered. Sepals laciniate, serrate or entire. Corolla tubular, straight, sometimes inflated, limb of subequal lobes or strongly bilabiate or dorsal lobes connate to form a galea (then tube centrally much inflated and distally constricted); sometimes with ± conspicuous appendages, flat, long and narrow emerging either from the sinuses or from the dorsal side of the lobes. Stamen included. Nectary of a single, dorsal, connate gland. Berry globose or ovoid, pink, red or purple, rarely white. *n* = 9. Over 70 spp., along the Western Cordillera from Bolivia to Guatemala, coastal Venezuela, Greater Antilles, the Guayana Highlands, and eastern Brazil; the centre of diversity is Colombia and Panama.

46e. *Bucinellina* (Wiehler) Wiehler

Bucinellina (Wiehler) Wiehler, Selbyana 5: 381 (1981).
Bucinella Wiehler (1977), nom. illeg.
Columnea sect. *Bucinellina* (Wiehler) Kvist & Skog (1993).

Epiphytic subshrubs, set apart from the other columneoid taxa by the flattened depressed berry, the oblate shape of the pollen grains and the corolla shape (limb with two dorsal patent lobes and three lateral and ventral reflexed lobes). *n* = 9. Two spp., SW Colombia and NW Ecuador.

47. *Corytoplectus* Oerst.

Corytoplectus Oerst., Centralamer. Gesner. 45 (1858); Wiehler, Phytologia 27: 309–328 (1973), reestabl., emend. & rev.; Feuillet & Steyermark in Berry et al., Fl. Venezuel. Guayana 5 (1999), reg. rev.

Terrestrial herbs. Stem suffrutescent or fleshy. Leaves ± isophyllous. Cymes erect in the leaf-axils, several- to many-flowered, umbel-like or congested. Sepals equal, of variable shape. Corolla erect in the calyx, tubular with some ventral inflation, constricted in the throat; limb narrow, with short, (sub)equal lobes. Stamens included. Nectary of 4, 2 or rarely a single dorsal, bilobed gland. Fruit a globose, shiny black or translucent berry with black seeds. $n = 9$. Fifteen spp., at higher elevations of the Western Cordillera (Bolivia to Panama and coastal Venezuela) and in the Guayana Highlands.

48. *Neomortonia* Wiehler

Neomortonia Wiehler, Selbyana 1: 17 (1975); Skog, Ann. Missouri Bot. Gard. 65: 958–961 (1979), reg. rev.; Wiehler, Selbyana 6: 194 (1983), notes.

Epiphytic or saxicolous herbs. Stems pendent, repent, or scrambling, to 1 m long, slender, 1–2 mm in diameter, frequently branched. Leaves opposite or rarely in whorls of three, ± isophyllous, petiole very short, lamina small (0.5–1.5 cm), membranous, ovate or elliptic, crenulate or serrulate. Flowers solitary in the leaf axils, showy, oblique in the calyx. Sepals free, entire or serrulate. Corolla white, eventually reddish on upper side, tube spurred at base, funnel-shaped, limb with spreading, fringed lobes; if conspicuous appendages missing, then with densely sericeous, villous, or lanate indumentum (the corolla colour actually is the colour of the indumentum). Stamens included; filaments curved, adnate to the corolla base, shortly connate in a tube around the ovary; anthers coherent in pairs, dehiscing longicidally. Nectary a dorsal bilobed gland, white. Ovary superior; stigma stomatomorphic. Fruit an ovoid, laterally compressed berry, orange; seeds surrounded by an aril on lower half, striate, yellow to brown. $n = 9$. Three spp., W Colombia and C America. Apparently polyphyletic (J.L. Clark, pers. comm.).

49. *Episcia* Mart.

Episcia Mart., Nov. gen. sp. pl. 3: 39 (1829); Wiehler, Selbyana 5: 25–28 (1978), emend. & rev.; Skog, Ann. Missouri Bot. Gard. 65: 918–923 (1979), reg. rev. (incl. *Alsobia*); Feuillet & Steyermark in Berry et al., Fl. Venezuel. Guayana 5 (1999), reg. rev.; Skog, Monogr. Syst. Bot. Missouri Bot. Gard. 85: 1125 (2001), reg. rev.

Stoloniferous, terrestrial or saxicolous herbs. Stem creeping, rooting at nodes, dichasially branched by the regular production of stolons. The lower leaf pair with axillary stolons, upper leaf pairs crowded, bearing the flowers; petiole short; dark green or with various patterns of variegation above, often coloured below. Cymes one- to 6-flowered, peduncles short or absent. Sepals free or shortly connate, upper sepal forced back around the corolla spur, green or coloured. Corolla white, yellow, blue, purple or red, salverform to campanulate, conspicuously spurred, contracted above the spur and at the throat, rarely ventricose, limb oblique, lobes spreading, rounded, minutely toothed or fimbriate. Stamens didynamous, included; filaments nearly straight, inserted at corolla base, depressed or coiling after anthesis; anthers coherent in pairs or in a square, becoming free, dehiscing longicidally. Nectary a single large, dorsal gland. Fruit an ovoid bivalved fleshy capsule. $n = 9$. Nine spp., from tropical South America to Nicaragua and the Guianas.

50. *Alsobia* Hanst.

Alsobia Hanst., Linnaea 26: 207 (1854); Wiehler, Selbyana 5: 28 (1978), reestabl. & rev.
Episcia sect. *Alsobia* (Hanst.) Benth. (1876).

Stoloniferous epiphytic herbs. Stem creeping or pendent, with one stolon per node in the axils of alternating leaves, the subsequent stolons giving the appearance of a single pendent stem with numerous plantlets (= stolon apices with opposite, crowded leaves) in a series. Leaves with a several-layered hypodermis on the adaxial side. Inflorescences epedunculate; flower and fruit as in *Episcia*, corolla white or white-punctate. $n = 9$. Two spp., Mexico, Guatemala and Costa Rica. Formerly included in *Episcia*.

51. *Rufodorsia* Wiehler

Rufodorsia Wiehler, Selbyana 1: 139 (1975); Skog, Ann. Missouri Bot. Gard. 65: 979–983 (1979), reg. rev.; Skog, Monogr. Syst. Bot. Missouri Bot. Gard. 85: 1128 (2001), reg. rev.

Epiphytic herbs or subshrubs. Stem erect, ascending or decumbent, with few branches. Leaves membranous, succulent or subcoriaceous, petiolate. Cymes few- to several-flowered, flowers much congested. Corolla infundibuliform, limb somewhat bilabiate; usually white, lobes and tube reddish on upper side. Stamens included, filaments adnate to corolla base; anthers apically coherent in two pairs. Nectary a bilobed dorsal gland. Fruit a white berry. $n = 9$. Four spp., Panama and Costa Rica.

52. *Oerstedina* Wiehler

Oerstedina Wiehler, Selbyana 2: 115 (1977); Skog, Ann. Missouri Bot. Gard. 65: 961–964 (1979), reg. rev.

Epiphytic perennial herbs or subshrubs. Stems erect or ascending, internodes often elongate. Leaves (sub)isophyllous, petiolate, ± elliptic, often coloured below. Cymes densely clustered in the upper leaf axils; peduncles short or lacking. Flowers inconspicuous, pedicels short. Sepals subequal, almost free to base, often reddish. Corolla tubular, white with reddish spots, or pink; tube somewhat inflated, gibbous at base, barely exceeding the sepals; limb with ± spreading lobes. Stamens included, filaments adnate to base of corolla tube. Nectary a single bilobed dorsal gland, or with an accessory ventral gland. Fruit an acute-acuminate berry. $n = 9$. Three spp., C America.

53. *Cobananthus* Wiehler

Cobananthus Wiehler, Selbyana 2: 94 (1977).

Terrestrial or facultatively epiphytic, suffrutescent herbs. Leaves slightly anisophyllous. Cymes of 1–3 flowers, bracteolate. Calyx orange or red, with a short pentagonal-campanulate tube, star-shaped. Corolla yellow, tubular, slightly gibbous, mouth not constricted; limb of small, rounded lobes. Stamens included; anthers suborbicular. Nectary a single bipartite dorsal gland. Fruit a subglobose, papyraceous, bivalved capsule. $n = 9$. One sp., *C. calochlamys* (Donn. Sm.) Wiehler, endemic to the region of Cobán, Alta Verapaz, Guatemala. Similar to *Alloplectus* and *Columnea*, differing from both by its thin-walled, subglobose capsule.

54. *Nematanthus* Schrad.

Nematanthus Schrad., Gött. Gel. Anz. 1: 718 (1821), nom. cons.; Wiehler; Selbyana 6: 190 (1983), notes; Chautems, Taxon 36: 656 (1987); Chautems, Diss. Bot. 112: 1–226 (1988), rev.
Orobanchia Vand. (1788), nom. rejic.
Hypocyrta Mart. (1788), p.p.

Epiphytic or saxicolous subshrubs or herbs. Stem ascendent, climbing, creeping or pendent, ± woody at base. Leaves often anisophyllous, subsessile or petiolate, elliptic to obovate, hard-fleshy, sometimes purple on the lower surface. Cymes 1- to 8-flowered, peduncle short or long and pendent. Sepals shortly connate, often coloured. Corolla yellow, orange, red, rose to dark purple, often resupinate; tube cylindrical or gradually widening, sometimes sharply bent at the base, usually strongly ventricose and making an angle of 20 to 90° with the pedicel, limb with short spreading or reflexed lobes, mouth constricted. Stamens didynamous, included; filaments adnate to the corolla base; anthers all cohering. Nectary a dorsal, bilobed gland. Fruit a fleshy, white or variously coloured capsule, included in the persistent calyx, exposing the coloured placentae and prominent funicles of the seeds. $n = 8$. Thirty spp., S & SE Brazil. Morphologically diverse and comprising four distinct flower types related to different pollination modes: (a) hypocyrtoid-non-resupinate, (b) hypocyrtoid-resupinate, (c) pendent-resupinate, with corolla laterally compressed and wide-open mouth, (d) large campanulate, white.

55. *Codonanthe* (Mart.) Hanst.

Codonanthe (Mart.) Hanst., Linnaea 26: 209 (1854), nom. cons.; Skog, Ann. Missouri Bot. Gard. 65: 830–838 (1979), reg. rev.; Wiehler, Selbyana 6: 192 (1983), notes; Chautems, Candollea 52: 159–169 (1997), new sp.; Feuillet & Steyermark in Berry et al., Fl. Venezuel. Guayana 5 (1999), reg. rev.; Skog, Monogr. Syst. Bot. Missouri Bot. Gard. 85: 1119–1120 (2001), reg. rev.

Epiphytic lignescent herbs, shrubs or lianas. Stems pendent, repent or erect, to 2 m tall, the nodes producing adventitious roots and/or extrafloral nectaries. Leaves iso- or slightly anisophyllous, shortly petiolate, fleshy, becoming coriaceous when dry, often with reddish extrafloral nectaries on the lower surface. Cymes one- to few-flowered, ± epedunculate, ebracteolate. Sepals subequal to very unequal, the dorsal one often recurved. Corolla white, pink, lilac, yellow, or deep purplish, often with reddish lines or spots; tube funnelform to subcampanulate, base saccate, spurred or gradually to abruptly broader towards the throat, or somewhat compressed or grooved, limb with spreading, rounded lobes. Stamens included; filaments adnate to the corolla base, shortly connate at base, recoiled or lowered after pollen shed; anthers oblong, with broad connectives, coherent in pairs by the tips or sides, later separating, dehiscing by pores. Nectary a single dorsal gland. Ovary superior, stigma bilobed or stomatomorphic. Fruit a fleshy, globose, ovoid or compressed berry or berry-like capsule, tardily dehiscent or dehiscing by two valves, red, pink, orange, or yellow green. Seeds red, pink or yellow, fusiform or ellipsoid, obliquely striate; funicle often fleshy or the seed surrounded by an aril. $n = 8$. Over 20 spp., throughout the neotropics, with species concentration in the Amazon river basin and S & E Brazil. Some species plants form a constituent element of the flora of ant nests.

56. *Codonanthopsis* Mansf.

Codonanthopsis Mansf., Repert. Spec. Nov. Regni Veg. 36: 120 (1934); Wiehler, Selbyana 6: 193 (1983), emend.; Feuillet & Steyermark in Berry et al., Fl. Venezuel. Guayana 5 (1999), reg. rev.

Epiphytic lignescent herbs or subshrubs. Stem terete, thickish, stiff. Leaves strongly anisophyllous, the smaller leaf stipule-like and early caducous, shortly petiolate, lamina fleshy or papyraceous, oblanceolate-acuminate. Cymes few-flowered, fasciculate. Flowers small, pedicels slender. Sepals ± free, the ventral lobes longer than the dorsal ones, the latter curved and appressed to the corolla spur. Corolla white with purple, or lilac, tube cylindrical, with a dorsal spur at the base, limb with short lobes. Stamens adnate to the corolla base, flattened and basally connate; anthers with narrow connective, dehiscing by an apical slit. Nectary a single dorsal gland. Fruit a fleshy capsule with loculicidal dehiscence, valves reflexed, displaying the placenta, seeds and funicles. $n = 8, 9$. Five spp., distributed along the river systems of the Amazon from Peru to Brazil. Some species are occasionally found in ant nests.

7. Genera of Uncertain Tribal Affiliation

57. *Bellonia* L.

Bellonia L., Sp. pl., ed. 1: 172 (1753); Xu & Skog, Acta Sci. Nat. Univ. Sunyatseni Suppl. 9: 95–107 (1990), rev.

Shrubs 0.4 to 1.2 m high. Stems woody, arching, axillary shoots modified to thorns in *B. spinosa* Swartz. Leaves (sub)equal in the pairs, thin, sharply toothed to scalloped and almost holly-like. Inflorescences axillary cymes or single flowers. Flowers 5- or often 6-merous. Sepals narrow, spreading. Corolla white, flat-faced, actinomorphic, with elliptical, rounded lobes. Stamens 5, exserted; filaments very short; anthers large, yellow, forming a cone in the flower centre, thecae dehiscing by apical pores. Nectary absent. Ovary semi-inferior. Fruit a dry, bivalved, rostrate capsule. $n = 13$. Two spp., Cuba and Hispaniola. Remarkable by the shrubby-woody habit, the presence of thorns (*B. spinosa*), and the white *Solanum*-like flowers exhibiting the syndrome of oligandrous pollen flowers (Vogel 1978). *Bellonia* has been placed in tribe Gloxinieae by Wiehler (1983). Unpublished molecular data (J.F. Smith, pers. comm.) suggest that the genus should be included in tribe Gesnerieae. Confirmation should be awaited, as the morphology is very different.

58. *Lembocarpus* Leeuwenb.

Lembocarpus Leeuwenb., Meded. Bot. Mus. Herb. Rijks Univ. Utrecht 146: 318 (VI. 1958), reimpr. Acta Bot. Neerl. 7: 318 (X. 1958); Boggan, Gloxinian 40: 20–22 (1990), notes.

Terrestrial, perennial herbs with a small, annual, hairy, subterranean tuber, from which a single leaf emerges per season. Long-pedunculate, bracteolate cyme emerging from the axil of the leaf, few-flowered or reduced to a single flower. Sepals free. Corolla broadly tubular, limb slightly oblique, with subequal spreading lobes; pale blue or white with lilac or purple. Stamina didynamous; anthers cohering in pairs. Nectary absent. Ovary superior; stigma bilobed. Fruit a bivalved loculicidal capsule, the valves opening to 180°. One sp., *L. amoenus* Leeuwenb., Surinam and French Guiana. *Lembocarpus* was placed by Wiehler (1983) near *Sinningia* in tribe Gloxinieae. Boggan (1990, 1991) pleaded for an inclusion into Episcieae, a placement also favoured by Feuillet and Skog (2003). The *ndh*F data of Smith (2001) confirm that the genus is not related to *Sinningia* (now tribe Sinningieae), but should be returned to Gloxinieae. Recent data of Smith (pers. comm.) are in favour of a position in Episcieae.

59. *Rhoogeton* Leeuwenb.

Rhoogeton Leeuwenb., Meded. Bot. Mus. Herb. Rijks Univ. Utrecht 146: 321 (VI. 1958), reimp. Acta Bot. Neerl. 7: 321 (X. 1958), reg. rev.; Feuillet & Steyermark in Berry et al., Fl. Venezuel. Guayana 5 (1999), reg. rev.

Terrestrial stemless plants with a small, fibrous-rooted tuber. Leaves one or few, petiolate, lamina rounded to oblong-ovate, penninerved. Cymes long-pedunculate, bracteolate, several-flowered, flowers condensed. Sepals free. Corolla "episcioid", tubular to trumpet-shaped, oblique in the calyx, tube with a dorsal sac or blunt spur at the base, limb with spreading lobes. Stamens didynamous, included; filaments connate at base and inserted near corolla base; anthers coherent in pairs, dehiscing by longitudinal slits. Nectary a single dorsal emarginate gland. Ovary superior; stigma bilobed. Fruit a bivalvate loculicidal capsule, the valves opening to 180°. Four spp., Guyana. Despite the episcioid flowers, the genus is regarded to be closely allied to *Lembocarpus* (Leeuwenberg 1958; Boggan 1991). Recent findings are controversial: morphological data and *ndh*F sequences place *Rhoogeton* in Episciceae, while ITS sequences imply its position outside of that tribe and possibly within Gloxinieae (Smith 2000b). Feuillet and Skog (2003) again suggest a possible position in Episcieae.

60. *Lampadaria* Feuillet & L.E. Skog

Lampadaria Feuillet & L.E. Skog, Brittonia 54: 344 (2003).

Terrestrial, caulescent herbs, with short internodes. Leaves opposite, petiolate, ovate-elliptic, venation pinnate, stomata scattered. Cymes long-pedunculate, bracteolate, flowers short-pedicellate. Sepals 5, free. Corolla campanulate, limb slightly bilabiate. Stamens 4, included, filaments adnate to the corolla base, anthers free, dehiscing longicidally. Nectary consisting of 2 large glands, a ventral and a dorsal one. Ovary superior, style curved, stigma capitate. Fruit a somewhat fleshy capsule, bivalved, dehiscing loculicidally. One sp., *L. rupestris* Feuillet & L.E. Skog, Guyana. Defined by an unusual combination of characters (inflorescence form, biglandular nectary, superior ovary, somewhat fleshy capsule). Placement in the Episcieae seems a possibility.

61. *Cremersia* Feuillet & L.E. Skog

Cremersia Feuillet & L.E. Skog, Brittonia 54: 347 (2003).

Terrestrial, caulescent herbs. Leaves opposite, petiolate, ovate-elliptic, venation pinnate, stomata scattered. Cymes pedunculate, bracteolate, flowers pedicellate. Sepals 5, free. Corolla salverform, limb bilabiate, 5-lobed. Stamens 4, included, filaments adnate to the corolla base, anthers cohering in a tetrad, thecae dehiscing longicidally. Nectary a dorsal gland. Ovary superior, style bent at apex, stigma capitate. Fruit a dry, bivalved capsule, dehiscing loculicidally, opening to 180°, seeds with short, thick funicles. One sp., *C. platula* Feuillet & L.E. Skog, French Guiana. Tribal placement is difficult, the best placement is perhaps in Episcieae. Several characters are shared with *Rhoogeton*, a genus also difficult to place.

62. *Cremospermopsis* L.E. Skog & L.P. Kvist

Cremospermopsis L.E. Skog & L.P. Kvist, Novon 9: 262–269 (2002).

Perennial herbs to subshrubs. Stem usually erect. Leaves opposite, ± equal, petiolate, mostly ovate. Cymes pedunculate, usually congested, mostly with 4 lanceolate to ovate bracteoles. Sepals 1/2 to 2/3 connate, often strongly unequal (esp. in the fruiting stage) with 3 to 4 usually lanceolate lobes and 1 or 2 much wider and ovate. Corolla small, white or yellow, funnelform to tubular, outside glandular hairy, esp. on the ± bilabiate limb. Stamens 4, adnate to the corolla tube at least 2/3 of their length, glabrous, anthers coherent, thecae confluent. Nectary annular or higher on both sides of ovary and low or absent in between. Ovary superior, stigma capitate. Fruit a dry capsule, compressed and with two grooves, dehiscing irregularly and disintegrating in the persistent calyx which may function as a splash-cup. Seeds more or less regularly spherical, very small, surface papillate. Two spp., Colombia (Antioquia and adjacent Bolivar). Defined by the unusual character combination (zygomorphic calyx, corolla limb with glandular hairs, and spherical papillate seeds). The greatest similarity is with *Cremosperma* in tribe Beslerieae.

III. Epithematoid Gesneriaceae

Subfam. Didymocarpoideae (Cyrtandroideae) tribe Epithemateae (Meisn.) C.B. Clarke (1874) (Burtt 1997).
Subfam. Didymocarpoideae (Cyrtandroideae) tribe Klugieae Fritsch (1893), sensu Burtt & Wiehler (1995).
Subfam. Didymocarpoideae (Cyrtandroideae) tribes Klugieae Fritsch (1893) and Loxonieae (A.DC.) B.L. Burtt (1963).

Perennial, poly- or monocarpic, rarely annual, caulescent herbs with at least a single, elongated internode (hypocotyl). Stems sometimes with medullary bundles and secretory ducts. Leaves one or many, often strongly anisophyllous, texture usually thin and delicate. Cymes pair-flowered, emerging from the axils of the foliage leaves, or subtending leaves reduced to bracts and cymes combined into terminal thyrses or (by reduction of the cymes to single flowers) racemes; bracteoles within the cymes lacking. Flowers zygomorphic. Fertile stamens 4, rarely 2; 1 (or 3) staminode(s) present. Nectary ring- or cup-shaped. Ovary globose or ovoid, shorter than calyx; unilocular with 4 triangular placentae, or bilocular with 4 or 2 axile placentae; style slender, stigma inconspicuous, capitate or slightly bilobed. Fruit an ovoid or globose capsule, shorter than the calyx, dehiscence in various ways. South and SE Asia to New Guinea, one sp. in W Africa, one in Central America. According to the molecular studies of Mayer et al. (2003), *Rhynchoglossum* is sister to the remaining genera. Strongly supported is the affinity of *Whytockia* and *Monophyllaea* (as predicted by Weber 1976b) and (less firmly) of *Stauranthera* and *Loxonia* (Weber 1977b), to which also *Epithema* is allied. Due to lack of material, *Gyrogyne* was not included in the molecular analysis, but the plicate calyx suggests a position close (ancestral?) to *Stauranthera*.

63. *Rhynchoglossum* Blume

Rhynchoglossum Blume, Bijdr. fl. Ned. Ind. 741 (1826) ("Rhinchoglossum"); Burtt, Notes Roy. Bot. Gard. Edinburgh 24: 167–171 (1962), prov. rev.; Wang et al. in Fl. China 18: 399–400 (1998), reg. rev.
Klugia Schltdl. (1833).

Perennial or annual herbs, glabrous or with sparse indumentum of many-celled, often branched, eglandular hairs. Stems terete, fleshy-succulent, with medullary bundles. Leaves alternate, nearly distichous, shortly petiolate, lamina obliquely ovate-cordate, base strongly asymmetrical. Inflorescences terminal, unilateral racemes with two flower rows, subtending bracts small, linear, mostly displaced onto the pedicels, a few sterile bracts often present on the flowerless side of the inflorescence axis. Sepals connate in the lower half, sometimes with wings at the lines of fusion, lobes narrowly triangular-acute. Corolla strongly zygomorphic; tube cylindrical, white; upper lip of two short, upright or reclined lobes, lower lip a large, roundish or elongate, entire or slightly threepartite tongue of azure colour. Fertile stamens 4 (the former genus *Klugia*) or 2; filaments inserted near middle of corolla, anthers coherent in pairs, thecae parallel or divaricate, apically confluent, dehiscing longicidally. Ovary unilocular. Fruit a globose or ovoid capsule included in the calyx, dehiscing loculicidally. $n = 10, 18, 21, 27$. About 10 spp., from India and S China to New Guinea, one sp. (or more?) in C America (Mexico to Peru).

64. *Gyrogyne* W.T. Wang

Gyrogyne W.T. Wang, Bull. Bot., Res., Harbin. 1: 41, 51, fig. 7 (1981), transl. & reimp. in Contr. New York Bot. Gard. 16: 21, 45, 55 fig. 6 (1986); Wang et al. in Fl. China 18: 396 (1998), rev.

Perennial, shortly rhizomatous herbs. Stem erect, with two or three nodes, unbranched. Leaves opposite, petiolate, slightly unequal in a pair, ovate, coarsely serrate. Cyme (pseudo)terminal, few-flowered. Sepals connate in lower half, plicate. Corolla white; tube short, broad, saccate at base; limb slightly shorter than tube, bilabiate, lobes rounded. Nectary annular. Stamens 4, included, filaments adnate to corolla base; subulate; thecae slightly divergent, not confluent, dehiscing longicidally, connective projecting apically. Ovary depressed globose, glabrous, unilocular; style slender, ca. 4 times longer than ovary; stigma depressed globose. Fruit unknown. One sp., *G. subaequifolia* W.T. Wang, S China (W Guangxi).

65. *Stauranthera* Benth.

Stauranthera Benth., Scroph. Ind. 57 (1835); Burtt, J. Arnold Arbor. 65: 129–133, new sp., notes; Wang et al. in Fl. China 18: 396 (1998), reg. rev.; Weber (in prep.), rev.
Miquelia Blume (VI. 1838), non C.F.W. Meissn. (IX. 1838).

Perennial (sometimes annual?) herbs with fleshy-succulent stems. Leaves opposite, strongly anisophyllous, large leaves obliquely elliptic or ovate, base very unequal, apex acute-acuminate or blunt; small leaves reduced to a tiny cordi- or reniform auricle. Inflorescence terminal, an alternicladic thyrse with pair- or single-flowered cymes or solitary flowers emerging from small bracts. Sepals with conspicuous plications at the sinuses. Corolla either rotate and widely campanulate, ecalcarate, or distinctly zygomorphic and spurred at the base, blue or white, sometimes with a yellow blotch at the palate. Stamens 4, filaments adnate to corolla base, anthers coherent and forming a cross-like figure, thecae slightly divergent, dehiscing longicidally. Nectary absent. Ovary unilocular. Capsule globose, falling irregularly into pieces in the upper half. $n = 18–20$. Five (or more) spp., from NE India and S China throughout Malesia to New Guinea.

66. *Loxonia* Jack

Loxonia Jack, Trans. Linn. Soc. 14: 40 (1823); Weber, Plant Syst. Evol. 127: 201–216 (1977), rev.

Perennial, rhizomatous herbs with indumentum of hooked hairs. Leaves opposite, strongly anisophyllous; large leaves obliquely ovate; small leaf stipule-like, suborbicular or reniform. Flowering region of sympodial architecture, axis flexuous. Inflorescences opposite to the large leaves (seemingly emerging from the axils of the small leaves), representing thyrses with one to several cymes in alternate-distichous arrangement. Cymes pair-flowered or (by loss of front-flowers) of conventional structure. Sepals free or obliquely connate to form a two-lipped calyx, lobes narrow-triangular. Corolla greenish-white, eventually with a purple dot on each side of the throat; tube short, limb bilabiate; mouth hairy. Stamens 4, didynamous, anthers coherent in pairs. Nectary a low, ventrally somewhat thickened ring. Ovary unilocular. Fruit an (always?) 4-valved capsule. $n = 22$. Three spp., W Malesia (Sumatra, Malay Peninsula, Borneo, Java).

67. *Epithema* Blume

Epithema Blume, Bijdr. fl. Nederl. Ind. 737 (1826); Hilliard & Burtt, Edinburgh J. Bot. 54: 111–113 (1997), new spp.; Wang et al. in Fl. China 18: 400 (1998), reg. rev.; Hilliard & Burtt, rev. in prep.

Monocarpic herbs with succulent stems. Cotyledons basal, the macrocotyledon developing into a large petiolate leaf that falls off early. Lowermost cauline leaf solitary, with long petiole and broadly ovate-cordate lamina, upper leaves opposite, in one or two pairs, subsessile, lamina ovate. Inflorescences pseudoterminal, densely contracted pair-flowered cymes, usually embraced by their cucullate bract. Sepals connate in lower half, lobes triangular, spreading after anthesis and the calyx forming a campanulate rain splash cup. Corolla white or bluish, tube cylindrical, limb bilabiate. Nectary cup-shaped. Fertile stamens 2, the posterior pair, filaments inserted above middle of corolla tube, anthers apically coherent, thecae divaricate, apically confluent; staminodes two, their filaments longer than those of the fertile stamens, distal parts upcurved. Ovary unilocular. Capsule globose, membranous, dehiscence circumscissile. $n = 8, 9, 12$. Over 20 spp., from N India and S China over SE Asia to New Guinea, one sp. in W Africa.

68. *Whytockia* W.W. Sm.

Whytockia W.W. Sm., Trans. Bot. Soc. Edinburgh 27: 338 (1919); Weber, Notes Roy. Bot. Gard. Edinburgh 40: 359–367 (1982), rev.; Y.Z. Wang, Acta Phytotax. Sin. 33: 1–5 (1995), ibid. 33: 296–301 (1995), new spp.; Y.Z. Wang & Li, Acta Phytotax. Sin. 35: 67–69, new sp.; Y.Z. Wang et al., Acta Phytotax. Sin. 36: 28–35 (1998); W.T. Wang et al. in Fl. China 18: 396–399 (1998), rev.
Oshimella Masamune & Suzuki (1934), nom. illeg.

Perennial herbs with decumbent or creeping stems, rooting from the nodes. Flowering stems ascending or erect; rarely branched or branches from the creeping part. Leaves in strongly unequal pairs, thin, membranous, pubescent, with scattered, sessile chalk glands on lower surface. Large leaves sessile or with a short petiole, ovate, strongly asymmetrical at base, serrate. Pair-flowered cymes emerging from the axils of the large leaves only. Sepals connate at base, aestivation imbricate, slightly striate (by the presence of secretory canals), chalk glands on inner sides. Corolla white or pink to light violet-purple, tubular-infundibuliform, limb bilabiate, lobes rounded. Stamens 4, didynamous; filaments adnate to corolla tube near base; anthers coherent, thecae divaricate. Ovary bilocular, with four axile placentae. Capsule subglobose, 2-valved or irregularly dehiscent. $n = 9$. About 8 spp., S China and Taiwan.

69. *Monophyllaea* R. Br.

Monophyllaea R. Br., Cyrtandreae 121 (1839), reimp. in Benn. & R. Br., Pl. jav. rar. 121 (1840); Burtt, Notes Roy. Bot. Gard. Edinburgh 37: 20 (1978), rev.; Weber, Beitr. Biol. Pflanzen 70: 405–406 (1998), new spp.
Horsfieldia Chifflot (1909), non Willd. (1805).
Moultonia J.B. Balf. & W.W. Sm. (1915).

Perennial monocarpic or annual herbs. Stem (hypocotyl) fleshy, with medullary vascular bundles, bearing a single large, oblong or suborbicluar, ever-growing leaf (macrocotyledon), a few species caulescent with 3–4 leaves. Inflorescences helicoid pair-flowered cymes, usually long-pedunculate, arising at or below the leaf base, occasionally also scattered along the midrib, with small subtending bracts at their base. Sepals 5, rarely 4, aestivation descending-imbricate, streaked with secretory canals and studded with chalk glands on the inner side. Corolla 5–15 mm long, tube broad, usually bearded inside. Stamens 4, the anterior ones strongly arcuate or curved back to bring anthers into contact, anthers with divergent thecae, confluent, usually all four cohering by their edges to form a down-facing plate, less commonly didynamous or all free. Nectary cupular, thick, undulate or lobulate. Ovary bilocular, with axile placentae; style persistent and swelling after anthesis. Fruit capsular, usually thin-walled, breaking up into 4 valves or the swollen style falling and leaving a large pore. $n = 10, 11, 12$. Over 30 spp., Malesia, from Sumatra to New Guinea and from S Thailand and Luzon to Java. Two subgenera: (1) subg. *Monophyllaea*, throughout Malesia, and (2) subg. *Moultonia* (Balf.f. & W.W. Sm.) B.L. Burtt, mainly Borneo.

IV. Didymocarpoid Gesneriaceae

Subfam. Didymocarpoideae (Cyrtandroideae) tribe Trichosporeae Nees, Flora 8: 143 (1825).
Subfam. Didymocarpoideae (Cyrtandroideae) tribe Cyrtandreae (Jack) Bartl., Ord. Nat. Pl. 185 (1830).
Subfam. Didymocarpoideae (Cyrtandroideae) tribe Didymocarpeae (D. Don) Endl., Gen. Pl. 716 (1839).

Perennial polycarpic or rarely annual, caulescent or acaulescent herbs. Stems without medullary bundles and secretory canals. Leaves many, opposite or less commonly alternate, texture variable.

Inflorescences pair-flowered cymes, rarely reduced to solitary flowers, nearly always emerging from the axils of foliage leaves; bracteoles nearly always present. Flowers zygomorphic, rarely actinomorphic. Fertile stamens 4 or 2; staminodes usually present. Ovary (with few exceptions) elongate, often long-cylindrical; usually unilocular, with lamelliform, curved to revolute placentae; the lower carpel sometimes sterile. Fruit a usually elongate to long, dry (very rarely somewhat fleshy) capsule, dehiscing septicidally (rare), septi- plus loculicidally, loculicidally (very common), by a single split along the midrib of the upper carpel ("follicular capsule"), irregularly, or fruit an indehiscent, hard or fleshy berry (*Cyrtandra*). Seeds often with ornamented testa cells, in a few genera with a single or several appendages (the former tribe Trichosporeae). Europe, Africa, Asia, Malesia, Australia and the Pacific. The Didymocarpoid Gesneriaceae are an amalgamation of the traditional tribes Didymocarpeae (incl. Ramondeae, Rhynchotecheae, Saintpaulieae, etc.) Trichosporeae and Cyrtandreae. The two latter tribes proved polyphyletic (first recognised by Smith et al. 1997b and confirmed by later studies) and are not retained here. The available molecular data show that the Didymocarpoid Gesneriaceae cannot be subdivided neatly into distinct alliances, but represent a partly confusing cascade of small clades and individual genera (Weber et al. 2002; Pfosser et al., in prep.). *Corallodiscus* comes out as the most basal genus, and this is followed by a clade comprising several genera around *Leptoboea*. Both are here subsumed under Basal Asiatic Didymocarpoids. This "group" is followed by well-supported clades comprising the European and the African genera. The large remainder, the Advanced Asiatic and Malesian Didymocarpoid Gesneriaceae, are still poorly understood (see below).

IV.1. The Basal Asiatic Genera

This "group" comprises two clades: *Corallodiscus* and the *Leptoboea*-alliance. *Corallodiscus* is a small genus of Chinese rosette plants, which indeed exhibits some primitive features: tetrandrous flowers and septicidally dehiscing capsules. It is probably close to the European genus *Haberlea* with similar characters. *Leptoboea* (for which molecular data are lacking, but differing from *Boeica* only by the opposite leaf arrangement), *Boeica*, *Rhynchotechum* (with indehiscent fruits and formerly placed in tribe Cyrtandreae along with *Cyrtandra*) and *Tetraphyllum* represent caulescent, mostly somewhat shrubby plants with 4-valved fruits (septicidal + loculicidal dehiscence). The same applies to *Championia* from Sri Lanka, for which molecular data are lacking and the inclusion of which is a mere guess. In habit very different is the tiny rhizomatous *Platystemma* which, however, has also 4-valved fruits. Fritsch (1893) placed all these genera (along with two or three misplaced ones) in his new tribe Championieae.

70. *Corallodiscus* Batalin

Corallodiscus Batalin, Trudy Imp. S.-Petersburgsk. Bot. Sada (Acta Hort. Petrop.) 12: 176 (1892); Burtt, Gard. Chron. ser. 3, 122: 212 (1947), reestabl., rev.; Wang et al., Fl. Reip. Pop. Sin. 69: 233–245 (1990); Wang et al. in Fl. China 18: 283–284 (1998), reg. rev.
Didissandra auct. non C.B. Clarke: Craib (1919).

Perennial rosette herbs. Leaves many, glabrous to densely woolly, broadly lanceolate, (ob)ovate, rhombic, elliptical, or suborbicular, base broadly or narrowly cuneate, venation conspicuous. Cymes one- to many-flowered, lax or rarely subumbellate, bracteoles apparently absent. Sepals free or fused halfway. Corolla blue to purple, rarely yellow or white, tubular, limb bilabiate; upper lip shorter than lower lip, lobes subequal, rounded or obtuse; lower lip with dense beard inside of tube. Stamens 4, inserted above corolla base (rarely above middle), included or slightly exserted; anthers dorsifixed, coherent in pairs; thecae divaricate, confluent, longicidally dehiscent. Nectary annular. Ovary oblong, stigma capitate. Capsule oblong to linear, much longer than the calyx, dehiscing loculicidally, sometimes also septicidally. $n = 10$, 20. Three to five spp., Bhutan, China, N and NE India, Nepal, Thailand. While Wang et al. (1990) recognised 9 spp. in China, Wang et al. (1998) reduced the number to three, due to a much broader circumscription of *C. lanuginosus*.

71. *Leptoboea* Benth.

Leptoboea Benth., in Benth. & Hook.f., Gen. pl. 2: 1025 (1876); Wang et al. in Fl. China 18: 286–287 (1998), reg. rev.; Hilliard in Grierson & Long, Fl. Bhutan 2(3): 1323–1324 (2001), reg. rev.
Championia C.B. Clarke (1874), non Gardner (1846).

Terrestrial shrubs with branched stems. Branches spongy. Leaves opposite, those of a pair equal or subequal, usually clustered on branches; petiole short, lamina ovate, apex acute, base broadly cuneate. Cymes with filiform peduncles and pedicels, 3–5-flowered, umbellate. Sepals free, equal. Corolla shallowly campanulate, tube slightly longer than limb, limb obscurely bilabiate, lobes (sub)equal, rounded. Stamens 4, inserted at corolla base, included, anthers free, basifixed, thecae subparallel, apically confluent, dehiscing poricidally. Nectary absent. Ovary ovoid, stigma capitate. Capsule linear, much longer than the calyx, dehiscence septicidal, eventually also loculicidal. Three spp., more to be described. Bhutan, N and NE India, S China, Myanmar, Thailand.

72. *Boeica* C.B. Clarke

Boeica C.B. Clarke, Commelyn. Cyrtandr. bengal. 118, tt. 85–87 (1874); Burtt, Notes Roy. Bot. Gard. Edinburgh 35: 369–374 (1976), notes, new sp.; Wang & Pan, Bull. Bot. Res. 2: 122 (1982), new sp.; Wang et al. in Fl. China 18: 287–289 (1998), reg. rev.
Boeicopsis H.W. Li (1982).

(Sub)shrubs or perennial rhizomatous herbs. Leaves alternate, spaced along stem, tufted at apex or forming a basal rosette, petiolate, lamina (ob)ovate to elliptical, base cuneate or attenuate, glabrous to densely woolly. Cymes lax or dense, one- to many-flowered. Sepals free or fused halfway. Corolla campanulate or flat-faced, limb slightly zygomorphic, lobes (sub)equal, rounded or acute. Stamens 4, inserted near corolla base, filaments short, anthers large, free, basifixed, thecae subparallel, apically confluent; dehiscence variable. Nectary ring-like or lacking. Ovary ovoid or conical, stigma capitate. Capsule short to linear, loculicidally and then septicidally dehiscent, style persistent. $n = 12$. About 12 spp., Bhutan, S China, N and NE India, Myanmar, N Vietnam, NW Malaya. Apparently heterogeneous and in need of revision. Wang et al. (1990) refer the seven Chinese spp. to two sections: (1) sect. *Boeica*, (2) sect. *Boeicopsis* (H.W. Li) K.Y. Pan.

73. *Rhynchotechum* Blume

Rhynchotechum Blume, Bijdr. 775 (1826); Burtt, Notes Roy. Bot. Gard. Edinburgh 24: 35–39 (1962), list of spp.; Wang, Guihaia 4: 183–189 (1984), reg. rev.; Wang et al. in Wu & Raven, Fl. China 18: 393–395 (1998), reg. rev.; J.C Wang & C.C Wang, Taiwania 45: 355–365 (2000), reg. rev.; Burtt, rev. in prep.
Isanthera Nees (1834).
Corysanthera [Wall. ex] Endl. (1839).

Shrubs or subshrubs, stem erect, woody, unbranched or branched. Leaves opposite (sometimes alternate?), (sub)equal in a pair, surface woolly, hirsute or glabrescent. Cymes (few- to) many-flowered, lax. Sepals 5, free, equal. Corolla tubular-campanulate, limb slightly bilabiate, lobes rounded. Stamens 4, included, inserted near the base or in the middle of the corolla, anthers free, thecae parallel, apically confluent, dehiscing longicidally. Nectary ring-shaped or absent. Ovary ovoid, stigma globose. Fruit indeshiscent, an ovoid or globose berry. $n = 10$. About 13–15 spp. (18 described), NE India, Nepal, Bhutan, SW and S China, SE Asia and Malesia to New Guinea. By the berry-like fruit the genus was formerly thought to be related with *Cyrtandra*, but new morphological (Burtt, in prep.) and molecular data (Pfosser et al., in prep.) suggest a close relationship with *Boeica*. Remarkable is the presence of anthrachinones (Lu et al. 1998a, 1998b).

74. *Tetraphyllum* C.B. Clarke

Tetraphyllum [Griff. ex] C.B. Clarke, in A. & C. DC. Monogr. phan 5/1: 136 (1883); Burtt, rev. in prep.

Perennial herbs with (lignescent?) monocarpic stem(s). Stem erect, caulescent, tawny villose, bearing a tetramerous (pseudo)whorl or a single leaf pair at the top and pairs of scale leaves (in some species with axillary cymes) along the stem. Leaves petiolate or (sub)sessile, ovate or lanceolate, partly unequal-sided. Cymes condensed, flowers nearly sessile; bracteoles small. Sepals 5, almost free to base. Corolla widely funnel-shaped, limb bilabiate or subregular, lobes obtuse. Stamens 4, didynamous, or 2, filaments adnate to corolla base, anthers coherent at the tips, opening by longitudinal slits. Nectary much reduced. Ovary ovoid, stigma obscurely bilobed. Capsule splitting loculicidally and septicidally into 4 valves, leaving the two main ribs united to the persistent style base. About 6 spp., NE India, Bangladesh, Myanmar, Thailand.

75. *Platystemma* Wall.

Platystemma Wall., Pl. Asiat. rar. 2: 42 (1831); Wang et al. in Fl. China 18: 284 (1998), reg. rev.

Small slender herbs with a horizontal rhizome (a series of tuberous swellings). Aerial shoots with a single, sessile, orbicular-cordate, membranous, hairy leaf. Cyme emerging from the leaf axil, peduncle slender, bearing one to few flowers, bracteoles very small, sublinear. Sepals free nearly to base, forming a wide, campanulate calyx. Corolla violet, tube short, limb wide, deeply bilabiate, lobes oblong, obtuse. Stamens 4, filaments short, curved; anthers apically coherent; thecae divaricate, apically confluent, dehiscing by a single short slit. Nectary annular. Ovary ovoid, style much longer than ovary, stigma inconspicuous. Capsule ovoid-oblong, splitting loculicidally (+ sepicidally?), partially enclosed in the persisting calyx. $n = 20$. One sp., *P. violoides* Wall., E & W Himalayas (Nepal, Bhutan, N India), SW China.

76. *Championia* Gardn.

Championia Gardn., Calcutta J. Nat. Hist. 6: 485 (1846); Theobald & Grupe in Dassanyake, Rev. Handb. Fl. Ceylon 3: 96–98 (1981), rev.

Low, erect, caulescent, sometimes branched, perennial herbs. Leaves opposite, petiolate. Cymes compound dichasia, flowers solitary or rarely in pairs, lax; peduncles slender, shorter than the leaves. Sepals 5, nearly free, linear. Corolla 4-merous (upper lip undivided), regular, flat-faced, tube very short. Stamens 4, equal, inserted on corolla tube; filaments short; anthers connivent, ovate-oblong, 4-locular with 2 locules smaller. Nectary absent. Ovary oblong-conical, stigma small, subcapitate. Capsule oblong, pointed, dehiscing loculicidally and then septicidally. One sp., *C. reticulata* Gardn., Sri Lanka. Unusual by the 5-merous calyx and the regular, rotate, 4-merous corolla.

IV.2. The European Genera

The three genera prove clearly related, with *Haberlea* apparently being sister to *Ramonda* and *Jancaea*. The fruits open septicidally as in *Corallodiscus*. These four genera (plus *Petrocosmea*) formed tribe Ramondeae of Fritsch (1893–1894).

77. *Haberlea* Friv.

Haberlea Friv., Magyar. Tud. Társ. Evkon. 2: 249, t. 1 (1835); Rix & Webb in Tutin et al. (eds.), Fl. Europaea 3: 284–285 (1972), rev.

Perennial rosette plants. Leaves alternate, shortly petiolate, spathulate, coarsely dentate-serrate, rough-hairy. Cymes long-pedunculate, flowers 2–10, in pairs, bracteoles small, narrow. Sepals lanceolate, unequally connate and forming a somewhat zygomorphic calyx. Corolla blue (rarely white), with yellow spots in the mouth; tube widening above, limb bilabiate, lobes ovate-elliptical. Stamens 4, didynamous, included, filaments inserted at corolla base, anthers cohering in pairs, thecae divaricate, apically confluent. Nectary annular. Ovary ovoid-conical, style slender, stigma inconspicuous. Capsule bivalved, septicidally dehiscent. $n = 19, 22, 24$. Two spp., Balkan Peninsula. Like *Ramonda* and *Jancaea*, a relict of the Tertiary flora.

78. *Ramonda* Rich.

Ramonda Rich., in Persoon, Syn. pl. 1: 216 (1805), nom. cons.; Rix & Webb. in Tutin et al. (eds.), Fl. Europaea 3: 284–285 (1972), rev.

Perennial rosette plants. Leaves alternate, radially spreading, (sub)sessile or with short winged petiole, lamina ovate-spathulate to elliptical, obtuse, coarsely crenate-dentate, rugose, with ferrugineous hairs on lower surface. Cymes long-pedunculate, flowers few, bracteoles absent. Flowers 5- or 4-merous. Sepals free, oblong-ovate. Corolla flat-faced, rotate, lobes elliptical-obovate, obtuse, lilac-blue, mouth whitish and with a conspicuous orange-yellow ring. Stamens equalling the number of corolla lobes; filaments short, anthers large, with parallel thecae, not coherent, forming roughly a cone, opening by an apical pore. Nectary absent. Ovary ovoid-conical, style much longer than the anther cone; stigma small, capitate. Capsule ellipsoidal-oblong, dehiscing septicidally. $n = \pm 18, 24, \pm 36$. Three spp., Spain (Pyrenees and adjoining mountains) and Balkan Peninsula.

79. *Jancaea* Boiss.

Jancaea Boiss., Fl. Orient. 1: 4 (Feb. 1875); Rix & Webb in Tutin et al. (eds.), Fl. Europaea 3: 284–285 (1972), rev.; Skog, Taxon 32: 491–492 (1983), orthogr.
Jankaea Boiss. (Sept.–Oct. 1875).

Perennial rosette plants. Leaves small, alternate, subsessile, elliptical or ovate-acute, with a dense, woolly silver-white hair cover. Cymes long-pedunculate, flowers few, 4(rarely 5)-merous. Sepals free, oblong. Corolla light lavender, tubular-campanulate, limb slightly bilabiate, lobes ovate, rounded. Stamens 4(5), included. Ovary ovoid, pubescent, style and anthers blackish. Capsule ellipsoidal, septicidally dehiscent. $n = 28$. One sp., *J. heldreichii* (Boiss.) Boiss., Greece (Thessalia: Mt. Olympus). The species has been sometimes included in *Ramonda* which is supported by molecular data (Möller et al. 1999).

IV.3. The African and Madagascan Genera

Despite the morphological heterogeneity (e.g. caulescent, rosulate, unifoliate habit, straight and twisted fruits), the available molecular data (Möller and Cronk 1997a, 1997b, 2001a, 2001b) suggest that the African Didymocarpoid Gesneriaceae form a monophyletic group, with all the smaller genera examined so far (*Saintpaulia*, *Linnaeopsis*, *Schizoboea*, *Colpogyne* and *Hovanella*) nested in *Streptocarpus*.

80. *Streptocarpus* Lindl.

Streptocarpus Lindl., Bot. Reg. 14, t. 1173 (1828); Hilliard & Burtt, *Streptocarpus*. An African plant study (1971), Pietermaritzburg: Univ. Natal Press, rev.; Hilliard & Burtt, Notes Roy. Bot. Gard. Edinburgh 43: 229–232 (1986), 46: 321–322 (1990), new spp.; Hilliard, Edinburgh J. Bot. 49: 75 (1992), new spp.; La Croix, Kew Mag. 11: 82–88 (1994), reg. rev.

Annual or perennial herbs of caulescent, acaulescent-rosulate, plurifoliate or unifoliate habit. Leaves opposite, petiolate and with determinate growth in the caulescent species, otherwise alternate or solitary and showing continued growth. Cymes pedunculate; eventually displaced to the lamina base or emerging from the leaf-stalk, many-, few- or (rarely) one-flowered. Sepals free to base, rarely connate at base. Corolla tubular or infundibuliform, limb bilabiate. Stamens 2, the anterior pair; anthers usually cohering face to face, thecae usually divaricate and apically confluent; lateral staminodes usually present. Nectary annular or shortly cupular. Ovary ovoid to cylindrical, ovules restricted to the recurved tips of placentae. Fruit a ± cylindrical capsule, twisting spirally during postanthetical development, dehiscing loculicidally, eventually splitting into 4 valves when old. Seeds reticulate or verruculose. n = 15, 16, 30, 32, 48, 64. About 140 spp. Two subgenera: (1) subg. *Streptocarpella* Engl., caulescent plants with opposite and petiolate leaves, n = usually 15, tropical Africa, Madagascar and Comoro Islands, and (2) subg. *Streptocarpus*, plants acaulescent and of phyllomorphic organisation, n = usually 16 and multiples thereof, mainly south-eastern Africa and Madagascar. Some species (and hybrids) of subg. *Streptocarpus* are popular ornamentals. The four Asiatic species still running under *Streptocarpus* are misplaced in this genus.

81. *Saintpaulia* H. Wendl.

Saintpaulia H. Wendl., Gartenflora 42: 321, t. 1391 (1893); Burtt, Notes Roy. Bot. Gard. Edinburgh 22: 547–567 (1958), rev.; ibid. 25: 191–195 (1964), new spp.

Perennial acaulescent or procumbent-caulescent herbs. Leaves opposite or alternate, petiolate, ± fleshy, suborbicular to elliptic. Cymes pedunculate, one- to several-flowered. Sepals almost free to base, linear or lanceolate. Corolla violet-blue or white; tube very short, limb spreading at right angles to it, bilabiate, lobes rounded. Stamens 2, the anterior ones; filaments stout, somewhat flattened and twisted; anthers large, robust, reniform, bright yellow, fully exposed, cohering face to face, dehiscing by arcuate confluent slits. Staminodes 2–3, abnormally 2 or all fertile (*cycloidea* mutant cultivars). Ovary short, conical; style exserted stiffly to the left or right of the corolla centre (enantiostyly). Capsule ovoid to linear-cylindrical, tardily dehiscent. Seeds verruculose. n = 14 (doubtful), 15 (30, only in cultivars). About 22 spp., E Africa (SE Kenya, N Tanzania). Some species, especially *S. ionantha* and allies, are the source of numerous cultivars and hybrids, grown as popular ornamentals.

82. *Linnaeopsis* Engl.

Linnaeopsis Engl., Bot. Jahrb. Syst. 28: 482 (1900); Burtt, Notes Roy. Bot. Gard. Edinburgh 22: 581–582 (1958), notes, new sp.; Weigend, Flora 195: 45–50 (2000), rev.

Perennial, subrosulate, creeping or shrubby-subscandent herbs. Leaves alternate, petiolate, lamina oblong, shallowly ovate, cordate or suborbicular, membranous or subcoriaceous, crenate or serrate, hairy. Cymes pedunculate, flowers few to ca. 10, in pairs, or solitary. Sepals nearly free to base, linear-lanceolate. Corolla white, obliquely campanulate with broad and short tube and wide-open mouth, limb ± spreading, slightly bilabiate, lobes small, rounded. Stamens 2, inserted near the corolla base; filaments short, curved; anthers cohering face to face. Nectary cup-shaped. Ovary conical, subglabrous or hairy. Capsule conical to cylindrical, straight or twisted. Seeds verruculose. Three spp., E Africa (N Tanzania).

83. *Acanthonema* Hook. f., nom. cons.

Acanthonema Hook. f., Bot. Mag. t. 5339 (1862); Burtt, in Hutch. & Dalz., Fl. West Trop. Africa, ed. 2, 2: 383 (1963), rev.; Burtt, Bull. Mus. Natl. Hist. Nat. B., Adansonia 3: 415–417 (1982), inclusion of *Carolofritschia*; Burtt, Fl. Cameroun 27: 16–21 (1984), reg. rev.; Burtt, Fl. Gabon 27: 10–15 (1985), reg. rev.
Carolofritschia Engl. (1899).

Perennial (?) monocarpic unifoliate herbs. Stem short, with fibrous roots. Leaf (macrocotyledon) of variable (usually oblong-cordate) shape and size, hairy. Cymes several- to many-flowered, shortly pedunculate, emerging in a series from the leaf base. Sepals free, linear-oblong. Corolla white with bluish lobes, infundibuliform; tube curved, slightly inflated towards the mouth; limb subregular, lobes short, rounded. Stamens 4, included, usually only 2 (the anterior pair) fertile; filaments adnate to the middle of corolla tube, somewhat thickened and flattened in the middle, those of the anterior pair with a tooth-like process at the apex; anthers cohering face to face, thecae divaricate and apically confluent. Nectary unilateral, bipartite, oblique, nearly half as long as ovary. Ovary conical, bilocular, with axile placentae, the recurved lamellae ovuliferous all over. Capsule ovoid, not twisted, dehiscence unknown (simply breaking with age?). Seeds reticulate. Two spp., tropical W Africa.

84. ***Trachystigma*** C.B. Clarke

Trachystigma C.B. Clarke, in A. & C. DC., Monogr. phan. 5/1: 131 (1883); Burtt, Fl. Cameroun 27: 24 (1984), reg. rev.; Burtt, Fl. Gabon 27: 15–16 (1985), reg. rev.

Herbs of the habit of *Acanthonema* or a unifoliate *Streptocarpus*. Leaf (macrocotyledon) cordate-ovate, sessile. Cymes emerging in a series from the leaf base, pedunculate, several- to many-flowered. Sepals nearly free to base, narrow-lanceolate. Corolla rather short-tubed, widening towards the mouth, limb slightly bilabiate, lobes rounded. Fertile stamens 2, scarcely exserted; filaments with an apical tooth; anthers with divergent thecae, coherent. Nectary unilateral, oblique. Ovary cylindrical, unilocular; style as long as ovary; stigma bilobed. Fruit a slender, cylindrical capsule, opening along the median line of the upper carpel only. One species, *T. mannii* C.B. Clarke, tropical W Africa (Gabon, perhaps also S Cameroun).

85. ***Nodonema*** B.L. Burtt

Nodonema B.L. Burtt, Bull. Mus. Natl. Hist. Nat. Paris B, Adansonia 4: 415 (1982); Burtt, Fl. Cameroun 27: 21–23 (1984), reg. rev.; Burtt, Fl. Gabon 27: 10–15 (1985), reg. rev.

Small, rhizomatous herbs. Stem (rhizome) short, with contracted internodes, growing horizontally, bearing few leaves. Leaves alternate (?), petiolate, lamina elliptic, ovate to suborbicular, margin serrate-dentate, hairy. Cymes few-flowered, with short peduncle, bracteoles very small. Sepals nearly free to base, narrow-triangular. Corolla white, with purple lines in the throat, infundibuliform, tube slightly inflated on ventral side, limb bilabiate, lobes rounded. Fertile stamens 2, filments adnate to corolla base, thecae slightly divergent. Nectary lacking. Ovary subglobose, style sharply set off. Capsule subglobose, pilose; mode of dehiscence unknown. One sp., *N. lineatum* B.L. Burtt, tropical W Africa (S Nigeria, Cameroun, Gabon).

86. ***Schizoboea*** (Fritsch) B.L. Burtt

Schizoboea (Fritsch) B.L. Burtt, Notes Roy. Bot. Gard. Edinburgh 33: 266 (1974); Burtt, Fl. Cameroun 27: 15–16 (1984), reg. rev.
Roettlera Vahl sect. *Schizoboea* Fritsch (1887).

Weak creeping herbs, rooting at the lower nodes. Stem dichasially branched, lateral branches produced just below the terminal inflorescence, branches comprising a single leaf pair and the inflorescence. Leaves opposite, (sub)isophyllous, petiolate, lamina ovate. Inflorescence "terminal", consisting of two lateral, condensed few-flowered cymes emerging from the axils of small bracts (Weber 1980). Sepals free to base. Corolla white, tube short and broad, slightly ventricose, limb subregular, with short, rounded lobes. Fertile stamens 2, the anterior ones, filaments inserted below middle of corolla tube, thecae slightly divergent. Nectary small, annular, shallowly 5-lobed. Ovary cylindrical-conical, style short, stigma capitate. Capsule cylindrical, sometimes slightly falcate, dehiscing into 4 valves and the lateral, sclerified vascular bundles. Seeds verruculose. One variable species, *S. kamerunensis* (Engl.) B.L. Burtt, tropical W, C and E Africa.

87. ***Hovanella*** A. Weber & B.L. Burtt

Hovanella A. Weber & B.L. Burtt, Beitr. Biol. Pflanzen 70 ("1997"): 333 (1998).
Didymocarpus sect. *Hova* C.B. Clarke (1883).

Perennial weak caulescent herbs. Stem erect, with few nodes, base decumbent, with fibrous roots. Leaves opposite, ± isophyllous, petiolate; lamina ovate, margin subentire or coarsely crenate. Cymes pedunculate, few-flowered. Sepals free to base. Corolla small, bluish to violet; tube short, broad, ventricose; limb slightly bilabiate, lobes small, rounded. Stamens 2, the anterior ones, inserted near the corolla base, anthers coherent face to face. Nectary a low, lobed ring. Ovary cylindrical, or channelled on the dorsal side [*H. madagascarica* (C.B. Clarke) A. Weber & B.L. Burtt], style short, stigma capitate. Capsule held horizontally and making a ± right angle to the pedicel, dehiscence along the midrib of the upper carpel. Seeds verruculose. n = 14. Two or three spp., E & C Madagascar.

88. *Colpogyne* B.L. Burtt

Colpogyne B.L. Burtt in Humbert & Leroy, Fl. Madagascar et Comores, fam. 180: 150 (1971).

Perennial rosette herbs with short rhizome. Leaves apparently alternate, sessile or shortly petiolate, oblong-lanceolate, crenate-serrate, woolly-hairy on both sides. Cymes long-pedunculate, few- to several-flowered, flowers not regularly in pairs, bracteoles lacking. Sepals nearly free to base, oval-elliptical. Corolla white or with purple margins, short-tubed, campanulate, limb actinomorphic, lobes small, reflexed. Stamens 2, inserted near corolla base, ± of same length as corolla tube; thecae divergent. Nectary annular, somewhat lobed. Ovary short, ovoid, somewhat compressed, with dorsal and ventral invaginations; style short; stigma capitate. Fruit a short, straight capsule. Seeds verruculose. One species, *C. betsiliensis* (Humbert) B.L. Burtt, C Madagascar.

IV.4. The Advanced Asiatic and Malesian Genera

This is a very heterogeneous assemblage of genera, the relationships of which are still very poorly understood. This results from the only slight differences of the genes analysed so far, from the lack of molecular and precise morphological data of many genera, and from the heterogeneity (para- or polyphyly) of particular genera. The arrangement of the genera is essentially according to conspicuous characters of habit, flower, fruit and distribution (with a rough distinction of (mainland-)"Asiatic" and "Malesian" distribution, according to the "Eurasian" and "Sundaic" group of Burtt 1998a). Some of the groups seem to represent natural alliances but others certainly do not, due to convergences and parallel evolution of characters (e.g. diandrous flowers may have originated from tetrandrous ones several times). Unless stated otherwise, the flowers are zygomorphic, the fruits straight (not twisted), opening loculicidally by two valves, and the seeds are unappendaged. Although no clear distinction can be made between rosette and tufted habit, the rosette plants typically possess a flat, compact rosette of leaves placed on the ground and long-pedunculate (scapose) axillary inflorescences, while in the tufted plants the leaves are arranged more loosely, they are often elevated above the ground and usually there are species in the alliance with ± distinct caulescent habit. (a) Asiatic (mainly S Chinese) genera with rosette habit and tetrandrous flowers (genera 89–93), (b) Asiatic (mainly S Chinese) genera with rosette habit and diandrous flowers, the anterior stamen pair fertile (94–102), (c) Asiatic (mainly S Chinese) genus with rosette habit and diandrous flowers, the posterior stamen pair fertile (103), (d) Asiatic (mainly S Chinese) genera with rosette habit and actinomorphic flowers (104–107; included in an expanded tribe Ramondieae Fritsch by Wang et al. 1990, 1992, 1998); only *Conandron* has so far been included in molecular analyses, showing no relation to the European Ramondieae); (e) Asiatic caulescent and tufted plants with tetrandrous flowers (108–110); (f) *Briggsia* allies with pointed or appendaged seeds (111–113, part of the former tribe Trichosporeae sensu Wang et al. 1990, 1998 or, with the exclusion of *Anna*, Burtt 1963); (g) Asiatic tetrandrous genera with follicular fruits (114–115, rather odd genera with fruits opening by a dorsal slit); (h) Asiatic tufted and caulescent genera with diandrous flowers (116–128); (i) Malesian genera with tetrandrous flowers (129–131); (j) (Asiatic-)Malesian genera with primarily tetrandrous flowers and appendaged seeds (132–134; formerly constituting the core genera of tribe Trichosporeae); (k) Malesian genera with diandrous flowers and straight fruits (135–136); (m) Malesian(-Asiatic) genera with diandrous flowers and mostly twisted fruits (137–145; this alliance, which includes also some genera with straight fruits, seems to represent a natural group and is well supported by molecular data, Pfosser et al., in prep.); (n) (Asiatic-)Malesian-Pacific genera with indehiscent fruits (146–147).

89. *Oreocharis* Benth., nom. cons.

Oreocharis Benth. in Benth. & Hook.f., Gen. pl. 2: 1021 (1876), non Lindl. (1847); Pan, Acta Phytotax. Sin. 25: 264–293 (1987), reg. rev. (in Chinese); Wang et al. in Fl. China 18: 251–261 (1998), reg. rev.
Perantha Craib (1918).
Dasydesmus Craib (1920).

Perennial acaulescent or subacaulescent-rosulate plants, often with rufescent-lanuginose indumentum. Leaves petiolate, ovate or elliptical-acute, reticulate below. Cymes pedunculate-scapose, flowers in corymbose or subumbellate arrangement, bracteoles small. Sepals nearly free to base, small. Corolla blue, purple, tubular to funnel-shaped, subbilabiate, lobes rotundate or lanceolate. Stamens 4, didynamous, equalling the corolla tube; filaments inserted at corolla base; anthers free. Nectary cylindrical, usually glabrous; stigma obscurely bilobed. Capsule bivalved, loculicidally dehiscent. $n = 17$. Twenty-eight spp., C & S China

(27 spp.), N Vietnam and Thailand. Four sections (Wang et al. 1990): (1) sect. *Stomactin* (C.B. Clarke) Fritsch, (2) sect. *Orthoanthera* K.Y. Pan, (3) sect. *Oreocharis*, and (4) sect. *Platyanthera* K.Y. Pan.

90. ***Tremacron*** Craib

Tremacron Craib, Notes Roy. Bot. Gard. Edinburgh 10: 217 (1918); Wang et al. in Fl. China 18: 261–265 (1998), rev.

Perennial acaulescent herbs. Leaves radial, shortly petiolate, lamina elliptic, ovate, cordate-ovate or cordate-orbicular, crenate or dentate, both surfaces pubescent. Cymes long-pedunculate, with several flowers, bracteoles linear. Sepals free to base, narrowly lanceolate. Corolla yellow, white or red, subglabrous, tubular-campanulate, limb bilabiate, upper lip short, emarginate, lower lip with 3 rounded lobes. Stamens 4, of nearly equal length, rarely didynamous, exserted; filaments inserted near corolla base; anthers globose, divaricate, thecae apically confluent. Nectary annular or cyathiform. Ovary oblong or cylindrical; stigma bilobed. Capsule oblong or cylindrical, bivalved. Seven spp., S China (Yunnan, Sichuan).

91. ***Isometrum*** Craib

Isometrum Craib, Notes Roy. Bot. Gard. Edinburgh 11: 250 (1919); Burtt, Notes Roy. Bot. Gard. Edinburgh 33: 91–93 (1960), rev.; Pan, Acta Bot. Yunnan. 8: 23–36 (1986), Acta Phytotax. Sin. 26: 432 (1988), rev.; Wang et al. in Fl. China 18: 263–267 (1998), rev.

Perennial acaulescent herbs with short vertical rhizome. Leaves sessile or petiolate, entire or pinnately lobed, usually hairy on both sides. Cymes with few to several flowers, pedunculate, peduncle and pedicels brown villose and glandular pubescent, rarely without glandular hairs; bracteoles 2, small. Sepals free to base, equal. Corolla tubular or campanulate, pale purple, sanguineous or blue-purple, limb bilabiate, upper lip 2-lobed, lower lip 3-lobed, lobes rounded. Stamens 4, didynamous, included; filaments inserted near corolla base; anthers coherent, thecae divergent. Nectary annular, entire or 5-lobed. Ovary short-cylindrical; style equal or shorter than ovary; stigma bilobed. Capsule cylindrical, glabrous or pubescent, bivalved. So far 14 spp. are known, C China (Sichuan, Gansu, Hubei, Shaanxi).

92. ***Paraisometrum*** W.T. Wang

Paraisometrum W.T. Wang in Weitzman et al., Novon 7: 431 (1997); Wang et al. in Fl. China 18: 249–250 (1998).

Perennial rhizomatous rosette herbs. Leaves many, blade pubescent to woolly, base cuneate to broadly cuneate. Cymes lax, few-flowered; bracteoles 2, opposite. Sepals free to base, equal. Corolla purple, zygomorphic, tube tubular-funnelform, longer than the limb; limb bilabiate; upper lip 4-lobed, over 2 times longer than the undivided, acute or rounded lower lip. Stamens 4, included, adnate to corolla tube above the middle; anthers basifixed, coherent in pairs by their adaxial surfaces, thecae parallel, not confluent, connective not projecting. Nectary annular. Ovary linear, stigma undivided, depressed capitate. Fruit and seeds unknown. One sp., *P. mileense* W.T. Wang, S China (SE Yunnan).

93. ***Ancylostemon*** Craib

Ancylostemon Craib, Notes Roy. Bot. Gard. Edinburgh 11: 233, 257 (1919); Burtt & Davidson, Notes Roy. Bot. Gard. Edinburgh 21: 215–216 (1954), rev.; Wang et al. in Fl. China 18: 268–272 (1998), rev.

Perennial acaulescent herbs. Leaves with long, pubescent petiole, lamina ovate, elliptic, elliptic-ovate or lanceolate, coarsely dentate, crenate or serrulate, pubescent on both surfaces. Cymes umbelliform, sometimes reduced to solitary flowers; bracteoles linear, villose. Sepals free or fused halfway, puberulous outside, linear-lanceolate or triangular. Corolla tubular, purple, reddish-orange or yellow-whitish, bilabiate; upper lip 2-lobed or undivided, lower lip 3- or rarely 5-lobed. Stamens 4, didynamous, glabrous, inserted at or above the middle of corolla tube; anthers coherent in pairs. Nectary annular. Ovary cylindrical, glabrous or puberulous; stigma bifid. Capsule slender, cylindrical, glabrous, bivalved. $n = 17$. Twelve spp., C & S China.

94. ***Petrocosmea*** Oliv.

Petrocosmea Oliv., Hook. Icon. pl. 18: t. 1716 (1887); Craib, Notes Roy. Bot. Gard. Edinburgh 11: 269–275 (1919), rev.; Wang, Acta Bot. Yunnan. 7: 49–68 (1985), reg. rev.; Wang et al. in Fl. China 18: 302–308 (1998), reg. rev.

Perennial acaulescent herbs. Leaves few to many, in a basal rosette, petiolate, ovate to orbicular, apex acute, obtuse or rounded, base cuneate to cordate, rarely peltate. Cymes pedunculate, few- to several-flowered, lax, rarely subumbellate; bracteoles present. Sepals free, equal or fused to a 3-partite calyx. Corolla short-tubed, flat-faced, blue, purple or white; tube broad, shorter than limb, limb bilabiate, lobes rounded. Stamens 2, filaments short, inserted near corolla base; anthers basifixed, coherent at apices, thecae parallel, apically confluent or not, longicidally dehiscent. Nectary lacking. Ovary ovoid or conical, stigma capitate or globose. Capsule ovoid or oblong, dehiscing loculicidally. $n = 16, 17$. About 27 spp., NE India, S China (24 spp.), Myanmar, Thailand, S Vietnam. Three sections (Wang 1985): (1) sect. *Petrocosmea*, (2) sect. *Anisochilus* Hemsl., (3) sect. *Deinanthera* W.T. Wang.

95. *Metapetrocosmea* W.T. Wang

Metapetrocosmea W.T. Wang, Bull. Bot. Res., Harbin. 1: 38 (1981); Wang et al. in Fl. China 18: 309 (1998), rev.

Perennial acaulescent herbs. Leaves with long petiole, elliptical, entire or repand. Cymes pedunculate, 1–3-flowered, bracteoles small. Sepals free, narrowly triangular. Corolla white, small, with short and broad tube, limb bilabiate, lobes rounded. Stamens 2, included, inserted near corolla base, filiform, apically thickened; anthers dorsifixed, white villose, thecae divaricate, apically not confluent, longicidally dehiscent. Nectary absent. Ovary broadly ovoid; stigma small, capitate. Capsule globose, smooth. One sp., *M. peltata* (Merr. & Chun) W.T. Wang, S China (Hainan Island). Differing from *Petrocosmea* in the longer corolla tube and in villous anthers with widely divaricate thecae.

96. *Deinocheilos* W.T. Wang

Deinocheilos W.T. Wang, Guihaia 6: 1 (1986); Wang et al. in Fl. China 18: 309–310 (1998), rev.

Perennial acaulescent herbs. Leaves many, alternate, petiolate, elliptical, dentate. Cymes pedunculate, few- to several-flowered, bracteoles linear. Sepals free, linear. Corolla white or pale purple, tubular to funnel-shaped, bilabiate, upper lip triangular or semi-orbicular, undivided, lower lip 3-lobed. Stamens 2, exserted; filaments long, slender, inserted above middle of corolla tube; anthers basifixed, apically coherent or not; staminodes sometimes absent. Nectary cyathiform. Ovary slender, cylindrical; style shorter than ovary; stigma small, complanate-capitate. Capsule slender cylindrical, bivalved. Two spp., S China (W Sichuan, S Jiangxi). Outwardly similar to the tetrandrous *Tremacron*.

97. *Lagarosolen* W.T. Wang

Lagarosolen W.T. Wang, Acta Bot. Yunnan. 6: 11, fig. 1 (1984); Fang et al., Acta Phytotax. Sin. 31: 463–471 (1993), new. sp.; Wang et al. in Fl. China 18: 310 (1998), rev.

Perennial acaulescent herbs. Leaves alternate, petiolate, broadly elliptic to ovate, ovate-orbicular, crenate, white strigose on both surfaces. Cymes shorter than leaves, dichasially branched, bracteoles large, narrow elliptical. Sepals free to base, linear-lanceolate. Corolla purple, with narrow, slightly widening tube, limb bilabiate, lobes narrowly triangular, acute. Stamens 2, included; filaments short, inserted at or above middle of corolla tube, straight; anthers oblong, apically coherent, thecae apically confluent. Nectary cyathiform. Ovary slender, cylindrical, puberulous or glandular pubescent; stigma bilobed. Fruit 4-valved. Two spp., S China.

98. *Petrocodon* Hance

Petrocodon Hance, J. Bot. 21: 167 (1883); Wang et al. in Fl. China 18: 348 (1998), rev.

Perennial acaulescent herbs with stout vertical rootstock. Leaves petiolate, oblong to lanceolate, entire or dentate, strigose. Cymes long-pedunculate, several- to many-flowered, subumbellate; bracteoles linear. Sepals free nearly to base, linear-lanceolate. Corolla white, funnel-shaped or urceolate-campanulate, with 5 nearly equal, triangular lobes. Stamens 2, inserted near corolla base; anthers coherent. Nectary annular. Ovary slender cylindrical; stigma inconspicuous. Capsule slender, cylindrical, loculicidally dehiscent. One sp., *P. dealbatus* Hance (with 2 varieties), China.

99. *Dayaoshania* W.T. Wang

Dayaoshania W.T. Wang, Acta Phytotax. Sin. 21: 319 (1983); Wang et al. in Fl. China 18: 291–292 (1998), rev.

Perennial acaulescent herbs with stout vertical rhizome. Leaves alternate, petiolate, lamina broadly elliptic, orbicular-ovate or suborbicular, margin entire or inconspicuously blunt-serrulate, both surfaces white puberulous. Cymes 2–4, long-pedunculate, 1–2-flowered; bracteoles 2, opposite. Sepals free, narrowly triangular. Corolla white or pale purple, campanulate, tube short, widening, limb bilabiate, lobes broadly triangular, acute. Stamens 2 (sometimes 1), adnate to corolla base; anthers dorsifixed, dark purple, longicidally dehiscent. Nectary annular. Ovary slender, cylindrical, densely puberulous; stigma bilobed. Capsule slender, cylindrical, puberulous bivalved. One sp., *D. cotinifolia* W.T. Wang; S China (E Guangxi).

100. *Dolicholoma* D. Fang & W.T. Wang

Dolicholoma D. Fang & W.T. Wang, Bot. Res. Acad. Sin. 1: 18 (1983); Wang et al. in Fl. China 18: 360–361 (1998), rev.

Perennial acaulescent herbs. Leaves many, long-petiolate, narrowly elliptic or ovate, white pubescent, lower surface sometimes purple, margin ciliate. Cymes with slender peduncle, one- to few-flowered; bracteoles linear, pedicels white pubescent. Sepals free, linear. Corolla rose pink, villous outside, puberulous inside; tube narrowly cylindrical, limb bilabiate, lobes long, narrowly triangular, acute, ciliate. Stamens 2, inserted below throat; anthers dorsifixed, apically coherent, thecae apically confluent. Nectary annular. Ovary narrowly ovoid, puberulous; style slender; stigma disciform. Capsule ellipsoid, 4-valved. One sp., *D. jasminiflorum* D. Fang & W.T. Wang; S China (SW Guangxi). The long- and narrow-tubed flowers with long, narrow, spreading corolla lobes suggest pollination by butterflies or moths.

101. *Calcareoboea* [C.Y. Wu ex] H.W. Li

Calcareoboea C.Y. Wu ex H.W. Li, Acta Bot. Yunnan. 4: 241 (1982); Wang et al. in Fl. China 18: 361 (1998), rev.

Perennial acaulescent herbs with a stout, elongate rhizome. Leaves subcoriaceous, serrate. Cymes long-pedunculate, umbellate; bracteoles leafy, forming an involucre. Pedicels nearly free to base, linear-lanceolate. Corolla vermilion, narrowly funnel-shaped, subbilabiate, upper lip of four short teeth; lower lip tongue-like of a single, patent lobe. Stamens 2, filaments complanate, curved, inserted in the upper part of corolla tube; anthers 2-locular, divaricate. Nectary cupular, rim crenate. Ovary slender cylindrical; stigma minute, capitate. Capsule slender, cylindrical, bivalved. One sp., *C. coccinea* [C.Y. Wu ex] H.W. Li, S China (SE Yunnan, W Guangxi), N Vietnam. Form and colour of the flowers suggest ornithophily. The inclusion of two or three odd species of *Didymocarpus* has been suggested by Burtt (2001b), but is controversial.

102. *Allocheilos* W.T. Wang

Allocheilos W.T. Wang, Acta Phytotax. Sin. 21: 321 (1983); Wang et al. in Fl. China 18: 361 (1998), rev.

Perennial acaulescent herbs with slender rhizome. Leaves radial, petiole long, complanate, villose lamina suborbicular or cordate-orbicular, crenate, upper surface appressed-villose, lower surface densely brown pubescent. Cymes long-pedunculate, 2–5-flowered, sparsely villose; bracteoles opposite, linear, puberulous; pedicels glandular pubescent. Sepals free, linear. Corolla funnel-shaped, bilabiate, upper lip undivided, lower lip 4-lobed (?), lobes triangular, acute. Stamens 2, filaments curved, narrowly linear, inserted above corolla base; anthers coherent, apices pubescent. Ovary ovoid to short-cylindrical, stigma subcapitate. Capsule slender, cylindrical, bivalved. Two spp., S China (Guizhou, E Yunnan).

103. *Opithandra* B.L. Burtt

Opithandra B.L. Burtt, Baileya 4: 162 (1956); Burtt, Notes Roy. Bot. Gard. Edinburgh 22: 303 (1958), prov. rev.; Wang et al. in Fl. China 18: 289–291 (1998), reg. rev.
Schistolobos W.T. Wang (1992).

Perennial subacaulescent herbs. Leaves alternate, tufted or in a dense rosette, petiolate, pilose or pubescent. Cymes with peduncles longer than leaves, one- to several-flowered, bracteoles small. Sepals free, lanceolate. Corolla pink to purple, tube of variable length, limb subregular or bilabiate, lobes porrect or patent. Fertile stamens 2, the posterior pair; filaments straight or slightly curved; anthers free or apically coherent, oblong or subglobose, thecae not confluent. Nectary low, cup-shaped. Ovary slender, tapering into the style; stigma deeply bilobed or small, pointed or puncticulate. Capsule cylindrical, dehiscing loculicidally. $n = 17$. Ten spp., S China, Japan. Four sections (Wang et al. 1990): (1) sect. *Opithandra*, (2) sect. *Microstigma* W.T. Wang, (3) sect. *Stenosiphon* W.T. Wang, and (4) sect. *Briggsioides* W.T. Wang.

104. *Thamnocharis* W.T. Wang

Thamnocharis W.T. Wang, Acta Phytotax. Sin. 19: 485 (1981); Wang et al. in Fl. China 18: 249 (1998), rev.

Perennial acaulescent herbs with short vertical rhizome. Leaves petiolate, elliptic, lanceolate or narrowly obovate, base cuneate, margin obscurely serrate, densely white puberulous on both surfaces. Cymes long-pedunculate, with two subulate bracteoles and 5 to 9 small actinomorphic flowers. Calyx, corolla and androecium 4- or 5-merous. Sepals free, narrowly triangular. Corolla purple or blue, with very short tube, lobes lanceolate-oblong, of equal size and shape. Stamens of unequal length; filaments slender; anthers basifixed, thecae parallel, apically not confluent. Nectary annular. Ovary ovoid, puberulous; style slender, glabrous; stigma truncate. Capsule cylindrical, sparsely strigose, loculicidally dehiscent. One sp., *T. esquirolii* (H. Lév.) W.T. Wang, S China (SW Guizhou).

105. *Bournea* Oliv.

Bournea Oliv., in Hook. f., Icon. pl., ser. 4, 23, tab. 2254 (1893); Wang et al. in Fl. China 18: 249–250 (1998), rev.

Perennial rosette herbs. Leaves petiolate; lamina ovate to oblong-ovate, base truncate or cordate, margin entire or crenate-sinuate, densely strigose. Cymes umbelliform, long-pedunculate, densely strigose; bracteoles 6 to 9 (to numerous), fairly long, linear-lanceolate, involucrate. Flowers actinomorphic, 4- or 5-merous. Sepals narrowly triangular, puberulous outside. Corolla campanulate, equalling the calyx in length, purple, lobes equal, rounded. Stamens slightly exserted, inserted near or above middle of corolla tube; anthers dorsifixed, thecae parallel, apically not confluent. Nectary annular. Ovary slender, cylindrical; style short; stigma bilobed or undivided. Capsule linear-oblanceolate, bivalved, dehiscence loculicidal. Two spp., SE China (Guangdong, SW Fujian).

106. *Tengia* Chun

Tengia Chun, Sunyatsenia 6: 279 (1946); Wang et al. in Fl. China 18: 250 (1998), rev.

Perennial rosette herbs with vertical rhizome. Leaves elliptic, narrowly elliptic or narrowly obovate, denticulate, upper surface strigose, lower surface grey pubescent along veins. Cymes with long, strigose peduncle, bearing 8 to 10 actinomorphic flowers. Sepals free, linear-lanceolate, ferrugineous. Corolla white, tinged rose pink, actinomorphic, suburceolate, tube longer than limb, slightly constricted at throat, lobes of equal size, triangular. Stamens included, inserted at corolla base; anthers globose, thecae divergent, apically confluent. Nectary annular. Ovary slender cylindrical, densely strigose; style slightly exserted, puberulous; stigma capitate or slightly bilobed. Capsule slender cylindrical, 4-valved. One sp., *T. scopulorum* Chun, with two varieties. S China (S Guizhou, Yunnan). Probably closely related to *Bournea* and the diandrous *Petrocodon* (Burtt 1958b, 1970).

107. *Conandron* Sieb. & Zucc.

Conandron Sieb. & Zucc., Abh. Math.-Phys. Cl. Königl. Bayer. Akad. Wiss. 3: 729, t. 5 (1843).

Perennial (sub)acaulescent herbs. Leaves few, (sub)glabrous, elliptical with cuneate, truncate or cordate base. Cymes pedunculate, few- to many-flowered, umbel-like. Sepals free nearly to base. Corolla actinomorphic, glabrous, tube ca. half as long as limb, limb 5-lobed, lobes equal, acute or obtuse. Stamens 5, inserted near corolla base, exserted; anthers dorsifixed, connate into a tube surrounding the style; thecae parallel, apically not confluent, longicidally dehiscent; connectives with long apical projections. Nectary lacking. Ovary ovoid, with parietal placentae; stigma globose. Capsule straight, narrowly ovoid, ca. 2.5× longer than the calyx, bivalved, loculicidally dehiscent. $n = 16$. One sp., *C. ramondioides* Sieb. & Zucc., E China, Taiwan and S Japan.

108. *Briggsia* Craib

Briggsia Craib, Notes Roy. Bot. Gard. Edinburgh 11: 236 (1919); Burtt, Notes Roy. Bot. Gard. Edinburgh 21: 195 (1954), ibid. 34: 101–106 (1975), notes; Pan, Acta Phytotax. Sin. 26(6): 450–457 (1988), new spp.; Wang et al. in Fl. China 18: 272–280 (1998), reg. rev.

Perennial acaulescent or caulescent herbs. Leaves in a basal rosette or spaced along the stem or tufted at stem top, opposite (rarely alternate?), those of a pair equal or unequal, petiolate, ovate or elliptic, base cuneate, attenuate or cordate, rarely peltate. Cymes one- to many-flowered, lax or rarely subumbellate. Sepals free, rarely fused halfway. Corolla blue, purple, red, orange or white, usually spotted within; tube broad, 2–3 times longer than limb, ventrally pouched; limb bilabiate, upper lip much or slightly shorter than lower lip, lobes subequal, rounded, truncate or acute. Stamens 4, inserted at corolla base, rarely near or above middle, included; anthers basifixed, coherent in pairs at the tips, thecae divergent, apically confluent or not, longicidally dehiscent. Nectary ring-like. Ovary oblong to linear; stigma 2-lobed, lobes equal, undivided. Capsule much longer than the calyx, loculicidally dehiscent. $n = 17, 34$. About 22 spp., C & S China (21 spp.), Bhutan, India, Myanmar, Vietnam. The delimitation of *Briggsia* and *Loxostigma* is still problematic (Burtt 1975; Wang et al. 1990: 272).

109. *Briggsiopsis* K.Y. Pan

Briggsiopsis K.Y. Pan, Acta Phytotax. Sin. 23: 216 (1985); Wang et al. in Fl. China 18: 280–281 (1998), rev.

Perennial rosette herbs. Leaves petiolate, ovate to suborbicular, pubescent, entire or repand. Cymes long-pedunculate, one- to few-flowered. Sepals nearly free to base, linear-lanceolate. Corolla infundibuliform, white, with purple streaks inside, limb bilabiate, lobes rounded. Stamens 4, didynamous, inserted at middle of corolla tube; anthers cohering in pairs at the apex. Nectary annular, deeply 5-lobed. Ovary oblong, bilocular, only one locule fertile, with axile placentae, the other sterile, stigma bilobed. Capsule oblong, oblique, loculicidally dehiscent. One sp., *B. delavayi* (Franch.) K.Y. Pan, S China (C & S Sichuan, NE Yunnan, Guizhou).

110. *Raphiocarpus* Chun

Raphiocarpus Chun, Sunyatsenia 6: 273, pl. 44 & fig. 32 (1946); Weber & Burtt, Beitr. Biol. Pflanzen 70: 153–177 (1998), reestabl., emend.
Didissandra sect. *Chiritoides* K. Fritsch (1894).

Shrubs, subshrubs or perennial herbs, in *R. sesquifolius* with a distinct underground rhizome and annual shoots with a single, strongly anisophyllous leaf pair. Leaves opposite, those of a pair equal or unequal, the pairs well-spaced or tufted at the stem apex, petiole present or absent, lamina ovate or broadly lanceolate, base cuneate to cordate. Cymes one- to several-flowered. Sepals 5, free or connate. Corolla usually large, tubular to funnel-shaped and widening above, limb bilabiate, lobes rounded, coloration variable. Stamens 4, inserted near or above middle of the corolla tube, included; anthers (always?) coherent, thecae divergent, dehiscing longicidally. Nectary ring-shaped. Ovary slender, stigma entire or bilobed. Capsule elongate, dehiscing loculicidally. Eleven spp., some more to be described, S China, N & C Vietnam. *Raphiocarpus* includes all species from China and Vietnam formerly placed in *Didissandra* (Weber and Burtt 1998a). It may prove artificial, when the species (and adjacent genera) become better known.

111. *Loxostigma* C.B. Clarke

Loxostigma C.B. Clarke in A. & C. DC., Monogr. phan. 5/1: 59 (1883); Burtt, Notes Roy. Bot. Gard. Edinburgh 34: 101–106 (1975), notes; Wang et al. in Fl. China 18: 372–374 (1998), reg.rev.

Terrestrial or epiphytic herbs, with underground rhizome or the lower part of the stem creeping and rooting. Leaves membranous, usually opposite, strongly anisophyllous; internodes sometimes condensed, giving irregular pseudowhorls of leaves. Cymes pedunculate, flowers 1–2 to many. Sepals free or almost free to base. Corolla tubular, ventricose in the middle, spotted within (similarly to *Briggsia*). Stamens 4, slightly didynamous; filaments inserted at base or near middle of corolla tube, curved, anthers equal, cohering by their tips either in pairs or all together. Nectary cupular. Ovary sessile, elongate; stigma obliquely bilamellate. Fruit sessile, somewhat compressed laterally, dehiscing loculicidally. Seeds with or without an appendage at each end. Seven (or more) spp., NE India, Bhutan, Sikkim, Nepal, Myanmar, S China, N Vietnam.

112. *Anna* Pellegr.

Anna Pellegr., Bull. Soc. Bot. France 77: 46 (1930). Burtt & Davidson, Notes Roy. Bot. Gard. Edinburgh 21: 232 (1955), new sp.; Wang & Pan, Bull. Bot. Res. Harbin 2(2): 144 (1982), reg. rev.; Wang et al. in Fl. China 18: 287–289 (1998), reg. rev.
Tumidinodus H.W. Li (1982).

Terrestrial subshrubs. Stem and branches sometimes angled. Leaves opposite, those of a pair unequal, petiolate, lamina ovate, base cuneate, apex acute, margin entire or serrate. Cymes several-flowered, lax. Sepals free, equal. Corolla white or yellowish; tube broad, with a ventral pouch, limb bilabiate, lobes obtuse or rounded. Stamens 4, inserted near the middle of corolla tube, included; anthers basifixed, coherent in pairs, thecae slightly divaricate, apically confluent, dehiscing longicidally. Nectary ring-like. Ovary linear, stigma disc-like, undivided. Capsule linear, much longer than the calyx, loculicidally dehiscent. Seed apex drawn out to a point and brown, forming not a distinct appendage as in the following genera. Three spp., S China, N Vietnam. Perhaps the most primitive genus in the alliance.

113. *Lysionotus* D. Don

Lysionotus D. Don, Edinburgh Phil. J. 7: 85 (1822); Wang, Guihaia 3: 249–284 (1983), reg. rev. (in Chinese); Hilliard & Burtt, Edinburgh J. Bot. 52: 215–224, notes, transfers; Wang et al. in Fl. China 18: 385–393 (1998), reg. rev.

Epiphytic herbs or subshrubs. Stem branched, branches ± woody, glabrous. Leaves opposite or in whorls of three, isophyllous or unequal in size, herbaceous or coriaceous, petiole short, lamina lanceolate or elliptical-acuminate, margin crenate-serrate. Cymes pedunculate, few- to many-flowered, bracteoles inconspicuous. Sepals nearly free to base, lanceolate. Corolla white or purplish, infundibuliform, tube straight, inflated in upper half, limb slightly bilabiate, mouth (always?) with two yellow folds, lobes rounded. Stamens 2, included, filaments flattened and curved; anthers cohering face to face, connective sometimes with an appendage, thecae subparallel. Nectary annular. Ovary (sub)stipitate, cylindrical, stigma slightly bilamellate. Capsule cylindrical, splitting loculicidally (later also septicidally). Seeds with a hair-like appendage at the apical end and a capillary funicle at the hilar end. $n = 15, 16$. About 30 spp., from N India and Nepal eastwards through S China, N Thailand, N Vietnam to S Japan.

114. *Cathayanthe* Chun

Cathayanthe Chun, Sunyatsenia 6: 282 (1946); Wang et al. in Fl. China 18: 284–285 (1998), rev.

Perennial acaulescent herbs with slender horizontal rhizome. Leaves emerging in a series (two rows?) from the rhizome, long-petiolate, lamina membranaceous, narrowly (ob)ovate-lanceolate. Cymes long-pedunculate, with a single pair of nodding flowers; bracteoles absent. Sepals connate, forming a bilabiate calyx, upper lip undivided, lower lip 4-lobed, lobes lanceolate. Corolla purple, tubular-campanulate, bilabiate, upper lip 2-lobed, lower lip 3-lobed, lobes elliptic, of equal size. Stamens 4, included, inserted at middle of corolla tube, curved; anthers coherent, thecae apically confluent, connective apiculate; staminodes absent. Nectary tubular. Ovary narrowly ellipsoidal, glandular pubescent; style slender; stigma semi-orbicular, depressed. Capsule narrowly ellipsoid, compressed, opening by a single slit. One sp., *C. biflora* Chun, S China (Hainan Island).

115. *Beccarinda* Kuntze

Beccarinda Kuntze, Rev. gen. 2: 470 (1891); Burtt, Notes Roy. Bot. Gard. Edinburgh 22: 61–64 (1955), prov. rev.; Wang et al. in Fl. China 18: 285–286 (1998), reg. rev.; Burtt, Beitr. Biol. Pflanzen 70: 377–382 (1998), new sp.
Slackia Griff. (1854), non Griff. (1848) nec Griff. (1851).
Petrodoxa Anthony (1934).

Perennial herbs, rarely subshrubs. Stemless with basal leaves or with leafy unbranched stem. Indumentum of white, fawn, brown or reddish-purple hairs. Leaves petiolate, elliptic or ovate to ovate-orbicular, base cordate. Cymes pedunculate, umbelliform or pseudocapitate. Sepals 5, free to base. Corolla purple-blue or red, campanulate, bilabiate. Stamens 4, inserted near corolla base, included, anthers all coherent, thecae subparallel, apically confluent, dehiscing poricidally. Nectary ring-like or inconspicuous. Ovary ovoid, stigma capitate. Capsule making an angle with the pedicel, straight, ovoid to oblong, opening on the upper side. $n = 10$. Around 8 spp., NE India, Myanmar, S China incl. Hainan, Vietnam (Tonkin), one sp. in Sumatra. The genus is not yet thoroughly understood. The discovery of a species in Sumatra (Burtt 1998c) reinforces Burtt's suggestion (1956) that the genus is of tropical affinity.

116. *Chirita* D. Don

Chirita [Buch.-Ham. ex] D. Don, Edinburgh Phil. J. 7: 83 (1822); Wood, Notes Roy. Bot. Gard. Edinburgh 33: 123–205 (1974), rev.; Wang, Bull. Bot. Res. 5(2): 1–97, 5(3): 37–86 (1985), reg. rev.; Wang et al. in Fl. China 18: 311–345 (1998), reg. rev.
Ceratoscyphus Chun (1946).
Deltocheilos W.T. Wang (1981).

Annual or perennial, acaulescent, rosulate or caulescent, erect or creeping herbs, sometimes slightly woody. Indumentum of simple, white or reddish hairs, or hairs hooked or glandular. Leaves opposite, sometimes ternate or whorled; petiole long or short, winged or not; lamina ranging from nearly linear to suborbicular; often ovate, sometimes lobed, pinnatifid or pinnate. Cymes axillary, sometimes displaced onto the petiole, one- to many-flowered, bracteoles free or connate, large, small or lacking. Sepals free or connate, linear-lanceolate to suborbicular, sometimes fimbriate. Corolla narrowly tubular to broadly funnel-shaped; limb slightly to distinctly bilabiate. Stamens 2, fused either apically or face to face or joined by a ligature, rarely free; thecae divergent, rarely parallel. Nectary cup-shaped, rim entire, cleft or lobed. Ovary very rarely stipitate, rarely one carpel sterile; stigma usually with the upper lobe absent and the lower lobe expanded into a bifid lamina. $n = 4, 9, 10, 14, 16, 17, 18$. Between 80 and 140 spp., from the Western Himalayas and E & S China to S India, Sri Lanka and W Malesia (Greater Sunda Islands). Despite the revisions of Wood (1974) and Wang (1985, Chinese species), the genus is still very problematic with regard to delimitation and subdivision. Wood (1974) recognised three sections: (1) sect. *Gibbosaccus* C.B. Clarke, (2) sect. *Chirita*, (3) sect. *Microchirita* C.B. Clarke. Highly heterogeneous, only sect. *Microchirita* seems to be a monophyletic alliance.

117. *Chiritopsis* W.T. Wang

Chiritopsis W.T. Wang, Bull. Bot. Res., Harbin 1: 21 (1981); Wang et al. in Fl. China 18: 345–348 (1998), rev.

Perennial acaulescent herbs with stout rhizome. Leaves with long petiole, elliptic, rhomboid-ovate, rhomboid or suborbicular, undivided or bipinnatifid, puberulous. Cymes pedunculate, several-flowered, bracteoles narrowly lanceolate. Sepals free, lanceolate. Corolla pale yellow or white, tubular, bilabiate, lobes rounded. Stamens 2, glabrous, inserted near corolla base, slightly curved; anthers coherent, apically confluent. Nectary annular or interrupted. Ovary ovoid; stigma bilobed, with enlarged and sometimes bilobed lower lip. Capsule ovoid, loculicidally dehiscent. Nine spp., S China. Differing from *Chirita* mainly in the small flowers and the short, ovoid capsule.

118. *Primulina* Hance

Primulina Hance, J. Bot. 21: 169 (1883); Hook. f., Bot. Mag. t. 7117 (1890); Wang et al. in Fl. China 18: 310–311 (1998), rev.

Perennial acaulescent herbs. Leaves with undulate-winged petiole, lamina ovate-suborbicular, lobed at margins, upper surface puberulous, lower surface also with glandular hairs. Cymes with peduncle as long as or shorter than leaves, umbellate, 3–7-flowered; bracteoles narrowly ovate. Sepals free, lanceolate. Corolla purple, salverform, lobes ovate-orbicular, rounded. Stamens 2, inserted above corolla base, anthers connivent; staminodes absent. Nectary with two subquadrangular lobes. Ovary ovoid, style short, stigma bifid. Capsule ellipsoid, loculicidally dehiscent to base. One sp., *P. tabacum* Hance, S China (N Guangdong). The glandular leaves are said to give off a strong smell of tobacco.

119. *Metabriggsia* W.T. Wang

Metabriggsia W.T. Wang, Guihaia 3: 1 (1983); Wang et al. in Fl. China 18: 293–294 (1998), rev.

Perennial caulescent herbs. Stem erect, villose. Leaves opposite, petiolate, ovate or elliptic. Cymes pedunculate, few-flowered, bracteoles 2, forming a globose involucre. Sepals free, linear-lanceolate. Corolla white, funnel-shaped, bilabiate, upper lip short, shallowly 2-lobed, lower lip with 3 rounded lobes. Stamens 2, inserted at middle of corolla tube, filiform, straight; anthers basifixed, apically coherent, not confluent. Nectary annular. Ovary slender, cylindrical, puberulous, only one carpel fertile, style slender, longer than ovary; stigma small, capitate. Capsule slender, cylindrical. Two spp., S China (NW Guangxi). Perhaps allied to *Briggsia*, but diandrous and ovary with only one carpel fertile.

120. *Didymocarpus* Wall.

Didymocarpus Wall., Edinburgh Philos. J. 1: 378 (1819), nom. cons.; Wang et al. in Fl. China 18: 349–358 (1998), reg. rev.; Weber & Burtt, Beitr. Biol. Pflanzen 70: 293–363 (1998), emend.; Weber et al., Ann. Naturhist. Mus. Wien 102 B: 441–475 (2000), materials for rev.

Perennial terrestrial or saxicolous, caulescent herbs. Stems seasonal, herbaceous, erect, with few nodes; renewal buds are produced in the humid season, are dormant the dry season and develop to flowering shoots in the next humid season. Leaves decussate, usually two (rarely several) pairs born on top of stem, in the tropical (Malayan) species leaves in several distant pairs. Cymes pedunculate, few- to many-flowered. Sepals free or connate halfway, cartilaginous, surface polished. Corolla usually claret-coloured or violet, in the Malayan species also white or yellow; usually straight tubular or trumpet-shaped, widening towards the mouth; limb slightly to strongly bilabiate, rarely personate, lobes rounded; Stamens 2, included; filaments slender; anthers cohering. Nectary cup-shaped, rim sometimes lobed. Ovary cylindrical, stigma capitate. Capsule usually stipitate, bivalved, dehiscing loculicidally. $n = 11, 12, 14, 16, 18, 22, 27, 28$. Over 70 spp., from N & NE India, Nepal and S China southwards to the Malay Peninsula and N Sumatra. The genus was redefined by Weber and Burtt (1998d) and restricted to an essentially Sino-Himalayan alliance extending southwards and reaching with a few spp. the ever-wet tropics. According to Wang et al. (1998), 31 spp. occur in China, but some have to be removed from the genus (Weber et al. 2000).

121. *Gyrocheilos* W.T. Wang

Gyrocheilos W.T. Wang, Bull. Bot. Res., Harbin 1: 28 (1981); Wang et al. in Fl. China 18: 359–360 (1998), rev.

Perennial acaulescent herbs with stout rhizome. Leaves long petiolate, coriaceous or papery, cordate or kidney-shaped, coarsely doubly dentate, pubescent or strigose, palmately nerved. Cymes pedunculate, several- to many-flowered; bracteoles ovate. Sepals connate, forming a campanulate calyx, lobes linear or oblong, blunt.

Corolla purple or rose pink, tube short, widening, limb bilabiate, upper lip semi-orbicular, undivided, lower lip with 3 rounded lobes. Stamens 2, included, filaments inserted above corolla base; anthers ellipsoid, coherent, thecae apically confluent. Nectary annular. Ovary slender, cylindrical, glabrous; style exserted; stigma capitate. Capsule slender, cylindrical, glabrous, loculicidally dehiscent. Four spp., S China.

122. ***Didymostigma*** W.T. Wang

Didymostigma W.T. Wang, Acta Phytotax. Sin. 22: 188 (1984); Feng & Lu, Acta Phytotax. Sin. 32: 565 (1994), new sp.; Wang et al. in Fl. China 18: 292 (1998), rev.

Annual caulescent herbs with square-angled stems. Leaves opposite, petiolate, ovate, pubescent, Cymes pedunculate; bracteoles large, similar in shape to leaves; pedicels pubescent. Sepals free, linear-lanceolate. Corolla pale purple, tubular-infundibuliform, limb bilabiate, lobes rounded. Stamens 2, inserted above middle of corolla tube; filaments slightly curved, linear; anthers ellipsoid, basifixed, apically coherent, thecae parallel, not confluent. Nectary cyathiform. Ovary slender, cylindrical, sparsely pubescent; stigma bilobed. Capsule slender, cylindrical, puberulous, loculicidally dehiscent. Two spp., SE China.

123. ***Deinostigma*** W.T. Wang & Z.Y. Li

Deinostigma W.T. Wang & Z.Y. Li, Acta Phytotax. Sin. 30(4): 356 (1992).

Small perennial caulescent herbs. Stem lignescent at base. Leaves alternate, petiolate, ovate or elliptical, with rounded or peltate base, margin repand-crenate. Cymes pedunculate, 1–2-flowered; bracteoles ovate. Sepals free, linear-lanceolate. Corolla white or lilac, tubular-infundibuliform, limb bilabiate, lobes rounded. Stamens 2, inserted near middle of corolla tube, thickened in upper part; anthers velutinous, coherent, thecae divaricate, apically confluent. Nectary absent. Ovary short-cylindrical, bilocular, with only one locule fertile; stigma with two subequal lobes. Capsule curved, opening only along the dorsal suture. One sp., *D. poilanei* (Pellegr.) W.T. Wang & Z.Y. Li, Vietnam (Annam). Affinities obscure (Wang et al. 1992: 357).

124. ***Pseudochirita*** W.T. Wang

Pseudochirita W.T. Wang, Bot. Res. Acad. Sin. 1: 21 (1983); Wang et al. in Fl. China 18: 293 (1998), reg. rev.

Perennial herbs up to 1 m tall, sericeous all over. Leaves opposite, slightly anisophyllous, petiolate, ovate or elliptic, serrate. Cymes pedunculate, few- to several-flowered; bracteoles oval to suborbicular. Sepals connate, forming a campanulate calyx, lobes rounded. Corolla white, limb bilabiate, lobes rounded. Stamens 2, inserted at middle of corolla tube, filiform, slightly curved; anthers basifixed, coherent, thecae parallel, apically not confluent, longicidally dehiscent. Nectary annular. Ovary slender, cylindrical; stigma unequally bilobed. Capsule slender, cylindrical, loculicidally dehiscent. One sp., *P. guangxiensis* (S.Z. Huang) W.T. Wang, S China (C & W Guangxi), Vietnam.

125. ***Allostigma*** W.T. Wang

Allostigma W.T. Wang, Acta Phytotax. Sin. 22: 185 (1984); Wang et al. in Fl. China 18: 293 (1998), rev.

Perennial caulescent herbs. Stems 4-angled, densely pubescent. Leaves opposite, petiolate, ovate to elliptic, serrate, lower surface densely yellow glandular-punctate, both surfaces puberulous. Cymes pedunculate, pubescent; bracteoles linear. Sepals 5, connate, lobes linear-lanceolate. Corolla lavender, funnel-shaped, bilabiate; upper lip with acute lobes, lower lip with rounded lobes. Stamens 2, inserted at middle of corolla tube; filaments slightly curved, wider at middle; anthers elliptic-globose, basifixed, connate at apex. Nectary cup-shaped. Ovary linear, puberulous; stigma bilobed, upper lobe small, triangular, lower lobe much larger, subtruncate. Capsule slender, cylindrical, puberulous. One sp., *A. guangxiense* W.T. Wang, S China (Guangxi). Very similar to *Pseudochirita*, but ovary with axile placentae.

126. ***Phylloboea*** Benth.

Phylloboea Benth., in Benth. & Hook. f., Gen. pl. 2: 1020 (1876); Burtt, Notes Roy. Bot. Gard. Edinburgh 23: 89–90 (1960); notes.

Lignescent caulescent herbs. Leaves large, opposite; petiole long, winged, lamina ovate. Cymes pedunculate, bracteoles conspicuous, rounded, persistent. Sepals connate, the upper ones ca. halfway, the lower ones only basally, calyx thus distinctly bilabiate. Corolla white, rose-coloured in the mouth, wide-campanulate, limb subregular to subbilabiate, lobes ovate, obtuse. Stamens 2; filaments curved; anthers reniform, connivent, thecae apically confluent, dehiscing by a single slit. Nectary unknown (lacking?). Ovary conical. Capsule narrow, longer than calyx, loculicidally dehiscent. One sp., *P. glandulosa* B.L. Burtt, Myanmar (Moulmein).

127. *Hemiboeopsis* W.T. Wang

Hemiboeopsis W.T. Wang, Acta Bot. Yunnan. 6: 397 (1984); Wang et al. in Fl. China 18: 301–302 (1998), reg. rev.

Caulescent subshrubs, to 0.8 m tall. Leaves opposite, petiolate, oblong or oblong-lanceolate, margin entire or repand-crenate. Cymes with short peduncle, 3- to 7-flowered, bracteoles large, suborbicular or ovate-orbicular. Sepals free, spathulate-linear. Corolla white or pale purple, funnel-shaped, slightly bilabiate, lobes rounded. Stamens 2, included; filaments slender, inserted below middle of corolla tube; anthers ellipsoid, dorsifixed, ventrally coherent, thecae apically confluent, connective with a triangular appendage at apex. Nectary annular. Ovary slender, cylindrical, bilocular, placentation axile; style with black glands near base; stigma bilobed. Capsule slender, cylindrical, glabrous, verrucose, probably bivalved. One sp., *H. longisepala* (H.W. Li) W.T. Wang, S China (SE Yunnan), Laos. Differing from *Hemiboea* by both carpels being fertile.

128. *Hemiboea* C.B. Clarke

Hemiboea C.B. Clarke, in Hook., Icon. pl. 18: tab. 1798 (1888); Li, Acta Phytotax. Sin. 25: 81–92, 220–230 (1987), notes; Wang et al. in Fl. China 18: 294–301 (1998), reg. rev.

Perennial caulescent herbs with erect stem. Leaves opposite, petiolate, ovate or lanceolate. Cymes in the axils of the uppermost leaves, pedunculate, several- to many-flowered, flowers enclosed by large, connate bracteoles forming a boat-shaped, semi-globose or urceolate involucre. Sepals connate in lower part. Corolla white, pale yellow or rose-coloured, funnel-shaped, limb bilabiate, lobes rounded. Stamens 2, filaments slender or filiform, inserted above base or below middle of corolla tube; anthers apically coherent, thecae parallel, not confluent. Nectary annular. Ovary slender, cylindrical, bilocular, only upper (adaxial) locule fertile; stigma truncate or capitate. Capsule cylindrical, straight or oblique in relation to pedicel, loculicidally dehiscent. $n = 16, 18$. About 23 spp., C & S China, Taiwan, N Vietnam, Japan (Ryukyu Islands). Two sections (Wang et al. 1990): (1) sect. *Subcapitatae* C.B. Clarke, (2) sect. *Hemiboea*.

129. *Didissandra* C.B. Clarke

Didissandra C.B. Clarke in A. & C. DC., Monogr. phan. 5/1: 65 (1883), nom. cons., p.p.; Weber & Burtt, Beitr. Biol. Pflanzen 70: 153–177, 191–223 (1998), emend. & rev.; Vitek et al., Beitr. Biol. Pflanzen 70: 275–291 (1998), checklist.
Ellobum Blume (1826), nom. rejic.

Perennial, lignescent herbs; caulescent, erect, decumbent-ascending or creeping, slightly to prominently woody. Leaves decussate or ternate, ± isophyllous, petiolate, ovate, usually hairy. Cymes pedunculate, unilateral (pair-flowered cincinni) or epedunculate, with condensed, dichasially arranged flowers. Sepals free, narrowly triangular, persistent. Corolla white, eventually with a violet or purplish tinge, mostly two yellow bars at tube entrance, tube cylindrical, limb bilabiate, sometimes the lower lip distinctly elongated. Stamens 4 or 2, didynamous, sometimes the posterior pair distinctly smaller. Nectary cup-shaped. Capsule slender, cylindrical, ribbed, dehiscing tardily, at first loculicidally, then shredding into several lamellae associated with the sclerenchymatous sheaths embracing the vascular bundles. $n = 10$. Eight spp., W Malesia. A segregate of *Didissandra* with two sections (Weber and Burtt 1998a): (1) sect. *Cyrtandroides* (2 spp., Malay Peninsula, Borneo), (2) sect. *Didissandra* (6 spp., Sumatra, Borneo, Java).

130. *Ridleyandra* A. Weber & B.L. Burtt

Ridleyandra A. Weber & B.L. Burtt, Beitr. Biol. Pflanzen 70: 171 (1998); ibid. 225–273 (1998), rev.

Perennial lignescent herbs. Stem usually woody, ± erect, hairy. Leaves usually clustered at stem apex, petiolate, obovate, elliptical or lanceolate, usually glabrous above, sometimes with minute sessile 2-celled glands; margin entire, dentate, serrate, deeply saw-toothed or wavy. Cymes pedunculate, one- to several-flowered, bracteoles broad-lanceolate to linear. Sepals free. Corolla long- and rather broad-tubed to very short-tubed with nearly flat limb; limb bilabiate or subrotate, lobes rounded. Stamens 4, didynamous, filaments sometimes with spur-like outgrowths; anthers cohering in pairs (occasionally all four) at the apex. Nectary cup-shaped or a low ring or lacking. Ovary ovoid to elongate; stigma (always?) formed by the lower carpel. Capsule elongate or short, thick-walled, ± fleshy or dry, dehiscing on the upper side. Surface cells of seeds with a longitudinal ridge in the middle. $n = 17$. Over 20 spp. Two sections: (1) sect. *Ridleyandra* (ca. 18 spp., Malay

Peninsula), (2) sect. *Stilpnothrix* (C.B. Clarke) A. Weber & B.L. Burtt (5 spp., Borneo).

131. *Hexatheca* C.B. Clarke

Hexatheca C.B. Clarke, in A. & C.DC, Monogr. phan. 5/1: 193 (1883); Burtt, Notes Roy. Bot. Gard. Edinburgh 46: 53–57 (1989), notes, new sp.

Shrubs or lignescent herbs. Stem erect, woody, usually unbranched. Leaves in a dense or loose tuft at stem apex, opposite or in whorls of three, radially spreading, petiole long, short or ill-defined, lamina lanceolate to broadly elliptical-acuminate. Cymes short- or long-pedunculate, several-flowered, bracteoles narrow-oblong to linear. Sepals free nearly to base, linear. Corolla white, with yellow mouth, infundibuliform-bilabiate, lobes rounded. Stamens 4, didynamous, inserted at corolla base; anthers of the two lower stamens bithecous, those of the upper pair monothecous. Nectary absent. Ovary conical, pilose-villose; style short; stigma subbilobed. Capsule cylindrical, subcarnose, dehiscing irregularly. $n = 17$. Four spp., Borneo.

132. *Aeschynanthus* Jack

Aeschynanthus Jack, Trans. Linn. Soc. 14: 42 (1823), nom. cons.; Burtt, Notes Roy. Bot. Gard. Edinburgh 28: 219–225 (1968), notes; Burtt & Woods, Notes Roy. Bot. Gard. Edinburgh 33: 471–489 (1975), notes, new spp., sect. rev.; Wang, Bull. Bot. Lab. N.E. Forestry Inst. 4: 26–30 (1984), new sect.; Woods, Kew Mag. 8: 115–23 (1991), notes, new spp.; Royen, Alpine Fl. New Guinea 4 (1983), inclusion of *Oxychlamys*; Wang et al. in Fl. China 18: 375–385 (1998), reg. rev.
Trichosporum D. Don (1822), nom. rejic.
Euthamnus Schltr. (1923).
Oxychlamys Schltr. (1923).

Epiphytic (rarely terrestrial) (sub)shrubs. Stem terete, rarely (sub)quadrangular, branched or not, erect, spreading or pendulous; clinging to the substrate by nodal roots. Leaves opposite, often arranged in a single plane, rarely whorled; petiole short, lamina (linear to) lanceolate, ovate, or cordate (to suborbicular), margin entire, rarely obscurely denticulate; fleshy or leathery, with a water-storing hypodermis below the upper epidermis; nerves usually obscure. Flowers in few-flowered axillary cymes or solitary, in leaf axils or emerging from scaly bracts and forming a pseudoterminal cluster; bracteoles ovate or scaly, rarely large and coloured, usually deciduous. Sepals either free and linear, or basally fused or entirely connate to form a cup-shaped or cylindrical, often showy (coloured) calyx with small rounded lobes. Corolla usually bright red, or rarely orange, yellow, greenish yellow or greenish; tube usually long, arcuate, limb bilabiate, often with dark spots or stripes, lobes rounded. Stamens 4, didynamous, included or exserted, anthers coherent in pairs at the tips. Nectary ring- or cup-shaped. Ovary stipitate, linear, stigma capitate or peltate. Capsule linear, in some spp. up to 50 cm long, loculicidally dehiscent. Seeds with a triangular appendage at both ends or a filiform appendage at the apical and one to many filiform appendages at the hilar end. $n = 14, 15, 16, 24, 48$. About 160 spp., from India and S China throughout Malesia to New Guinea and the Solomon Islands. Essentially ornithophilous. Seven sections are recognised, mainly based on the number and morphology of the seed appendages (Mendum et al. 2001): (1) sect. *Haplotrichium* Benth., (2) sect. *Microtrichium* C.B. Clarke, (3) sect. *Aeschynanthus*, (4) sect. *Diplotrichium* Benth., (5) sect. *Polytrichium* Benth., (6) sect. *Xantanthos* W.T. Wang, and (7) sect. "X" (not yet described). The sections fall into two groups (Denduangboripant and Cronk 2000; Mendum et al. 2001; Denduangboripant et al. 2001).

133. *Micraeschynanthus* Ridl. (doubtful genus)

Micraeschynanthus Ridl., Fl. Malay Penins. 5, suppl.: 324 (1925); Burtt, Notes Roy. Bot. Gard. Edinburgh 28: 222 (1968), notes.

Ridley (1925: 324): "Epiphyte, creeping. Leaves opposite, subsessile, fleshy, glabrous, round, ovate, acute. Flowers very small, 3 or 4 in very short axillary inflorescences. Sepals 5, lanceolate acuminate, glandular hairy. Corolla shorter, tube very short, lobes 5, rounded. Stamens 4, filaments very short, attached to tube; anthers large, elliptic. Ovary cylindrical, narrowed slightly above; stigma discoid with a transverse furrow. Fruit unknown". One doubtful sp., *M. dischidioides* Ridl., Malay Peninsula. Only known from a single specimen, perhaps an *Aeschynanthus* with an abnormally developed flower (Burtt 1968).

134. *Agalmyla* Blume

Agalmyla Blume, Bijdr. fl. Ned. Ind. 766 (1826); Hilliard & Burtt, Edinburgh J. Bot. 59(1): 1–210 (2002), rev.
Dichrotrichum [Reinw. ex] De Vriese (1856).
Tetradema Schltr. (1920).

Perennial climbers with short roots produced along the stem, a few spp. terrestrial shrubs. Leaves opposite, of herbaceous texture, usually strongly anisophyllous, isophyllous in the terrestrial species, large leaf variable in shape and indumentum, small leaf with short petiole and blade or scale-like and caducous. Cymes subsessile to long-pedunculate; bracteoles variable in size and conspicuousness; flowers much congested. Sepals usually connate. Corolla usually with a distinctly arcuate tube; limb weakly to strongly bilabiate, the upper lip of 2 or 4 lobes, the lower lip of 3 or 1 lobes, mouth open or laterally compressed; colour red, with or without black lines on the lobes, sometimes yellow in the throat; inside of the tube towards the base with or without an annulus or five patches of hairs. Fertile stamens 4, didynamous, rarely 2, exserted, inserted at middle of corolla tube; anthers synthecous, coherent in pairs by their apices, thecae parallel, dehiscing longicidally. Nectary annular to cupular. Ovary slender, cylindrical, often stipitate; stigma with two large lateral lobes. Capsule elongate, cylindrical, loculi- and then septicidally dehiscent. Seeds with a filiform appendage at each end, one brown, the other (hilar) hyaline. $n = 16$. Over 90 spp., Malesia, from Sumatra to New Guinea. Essentially ornithophilous. Three sections (Hilliard and Burtt 2002): (1) sect. *Agalmyla*, (2) sect. *Exannularia* Hilliard & Burtt, (3) sect. *Dichrotrichum* (De Vriese) Hilliard & Burtt. The first occurs west, the two others east of Wallace's line.

135. *Orchadocarpa* Ridl.

Orchadocarpa Ridl., J. Straits Branch Roy. Asiat. Soc. 44: 78 (1905).

Subacaulescent herbs. Leaves opposite, long-petiolate, elliptical-ovate, serrate, hairy. Cymes long-pedunculate, flowers several. Sepals free almost to base, hairy. Corolla lilac-blue to violet; with dark blue and yellow patches in the throat, tube short and broad, limb bilabiate, lower lip much longer than the upper, lobes rounded. Stamens 2, shortly exserted; anthers reniform, coherent. Nectary unknown (lacking?). Ovary subglobose-ovoid, short. Capsule straight, very short (shorter than the persisting sepals), slightly beaked. One sp., *O. lilacina* Ridl., Malay Peninsula (Main Range). Relationships obscure.

136. *Henckelia* Spreng.

Henckelia Spreng., Anleit. 2: 405 (1817); Weber & Burtt, Beitr. Biol. Pflanzen 70: 293–363 (1998), reestabl.; Vitek et al., Ann. Naturhist. Mus. Wien 102 B: 477–530, (2000), list & nomencl. of spp.
Loxocarpus R. Br. (1839).
Codonoboea Ridl. (1923).
Platyadenia B.L. Burtt (1971).

Perennial herbs, lignescent subshrubs or woody shrubs. Stem erect, ascending or creeping. Leaves opposite or alternate, short- to long-petiolate, lamina ranging from ribbon-like to suborbicular, often ovate or lanceolate, usually hairy. Cymes pedunculate, one- to many-flowered. Sepals 5, free or basally connate. Corolla frequently infundibuliform, more rarely campanulate or flat-faced. Stamens 2, usually included; filaments long, often with an apical spur, anthers cohering. Nectary cylindrical or cup-shaped, sometimes lacking. Ovary usually slender, conical or ovoid. Capsule usually long, held horizontally (making an angle with the pedicel) and opening on the upper side. $n = 9, 18$. Around 180 spp., many not described. From S India and Sri Lanka throughout Malesia to New Guinea. The genus was recently reestablished, with the inclusion of most of the Malesian species originally described under *Didymocarpus* (Weber and Burtt 1998d). Five sections: (1) sect. *Henckelia*, (2) sect. *Didymanthus* (C.B. Clarke) A. Weber & B.L. Burtt, (3) sect. *Heteroboea* (Benth.) A. Weber & B.L. Burtt, (4) sect. *Loxocarpus* (R. Br.) A. Weber & B.L. Burtt, (5) sect. *Glossadenia* A. Weber & B.L. Burtt; (1) in S India and Sri Lanka, (2)–(5) in Malesia.

137. *Boea* Lam.

Boea [Comm. ex] Lam., Encycl. Méth. 1: 401, fr. Béole (1785), éd. nouv. Padua 1: 396 ("1784"); Burtt, Notes Roy. Bot. Gard. Edinburgh 41: 401–452 (1984), emend. & rev.; Zhou & Guo, Acta Phytotax. Sin. 29: 477–478 (1991), new spp.; Wang et al. in Fl. China 18: 367–368 (1998), reg. rev.
Dorcoceras Bunge (1833).

Perennial, (sub)acaulescent or caulescent herbs, sometimes woody at base. Leaves opposite, alternate in some rosulate species, with indumentum of long, eglandular, simple hairs. Cymes few- to many-flowered, di- or monochasially branched; bracteoles small. Sepals free. Corolla white, bluish or violet, obliquely campanulate or flat-faced, limb subregular or distinctly bilabiate with the upper lip bilobed and paler than the darker violet, shortly 3-lobed lower lip. Stamens 2; filaments curved, especially when exserted, thickened and bright yellow; anthers with widely divergent thecae, cohering face to face, usually cream or pale brownish. Ovary cylindrical or conical-cylindrical; stigma small, capitate. Capsule loculicidally dehiscent, valves spirally twisted. $n = 8, 9, 16, 17$. About 15 spp., S China, NE India, N Thailand, Vietnam,

Philippines, Sulawesi, Sumbawa, Flores, New Guinea, Solomon Islands, Bismarck and Louisiade archipelago, Australia. Differing from *Paraboea* essentially in the type of indumentum.

138. ***Paraboea*** (C.B. Clarke) Ridl.

Paraboea (C.B. Clarke) Ridl., J. Straits Branch Roy. Asiatic Soc. 44: 63 (1905); Burtt, Notes Roy. Bot. Gard. Edinburgh 41: 401–452 (1984), emend., list of spp.; Xu & Burtt, Edinburgh J. Bot. 48: 1–18 (1991), new spp.; Xu, Ann. Missouri Bot. Gard. 82: 570–580 (1995), new spp.; Weitzmann et al., Novon 7: 423–435 (1997). new sp.; Wang et al. in Fl. China 18: 362–367 (1998), reg. rev.; Wong et al., Sandakania 13: 23–29, new sp.
Didymocarpus sect. *Paraboea* C.B. Clarke (1883).
Chlamydoboea Stapf (1913).
Buxiphyllum W.T. Wang & C.Z. Gao (1981).
Boea auct. mult. saltem p.p.

Perennial, acaulescent or caulescent herbs, often woody at the base, sometimes with straggling woody branches, occasionally monocarpic; plants usually woolly, at least the young stems. Leaves opposite, or spiral in some rosulate species; upper surface usually thinly woolly, often glabrescent or with scabrous hairs beneath the wool; lower surface usually densely arachnoid-hairy from interwoven and usually branched hairs, more rarely indumentum consisting of spaced or clustered, branched, stellate or dendroid hairs. Cymes pedunculate, few- to many-flowered, in some species (esp. the monocarpic ones) confined to the upper stem part and combined to form a terminal thyrse. Sepals almost free. Corolla obliquely campanulate or with short-tubed and flat-faced limb, slightly bilabiate, the two lips concolorous, white, blue or violet. Stamens 2, adnate to the corolla tube; filaments usually thick, not bright yellow when exserted. Ovary ± cylindrical, stigma small, capitate, very rarely sublinguiform. Capsule loculicidally dehiscent and spirally twisted, or straight, held horizontally and splitting on the upper side. $n = 9, 16, 17, 18$. Nearly 90 spp., from Sumatra to New Guinea and from S China and the Philippines to Borneo. The main diagnostic character is the arachnoid, interwoven indumentum.

139. ***Trisepalum*** C.B. Clarke

Trisepalum C.B. Clarke in A. & C.DC., Monogr. phan. 5/1: 138 (1883); Burtt, Notes Roy. Bot. Gard. Edinburgh 41: 401–452 (1984), rev.; Wang et al. in Fl. China 18: 370–371 (1998), reg. rev.
Dichiloboea Stapf (1913).

Perennial, acaulescent or caulescent herbs, occasionally monocarpic, often woody in the lower part, sometimes with stout straggling branches, densely woolly, at least the young stems. Leaves opposite (alternate in some rosulate species?), with arachnoid indumentum, hairs usually branched. Cymes pedunculate, di- or monochasially branched, sometimes combined to form a terminal thyrse; pedicels of a flower pair usually of unequal length; bracteoles inconspicuous to large and then overlapping to form a strobilus-like cone. Sepals unequally connate, the three upper ones forming a shortly 3-dentate or undivided lip, the lower ones free or shortly connate. Corolla white or light violet, obliquely campanulate. Stamens 2, included; anthers with divergent thecae. Ovary conical-cylindrical; stigma linguiform, downcurved over the corolla mouth. Capsule usually twisted and dehiscent along both carpels, occasionally along only one, occasionally not twisted and laterally compressed. $n = 18$. About 13 spp., from S China to the north of the Malay Peninsula.

140. ***Rhabdothamnopsis*** Hemsl.

Rhabdothamnopsis Hemsl., J. Linn. Soc. Bot. 35: 517 (1903); Wang et al. in Fl. China 18: 371 (1998), rev.

Small shrubs to 0.5 m tall, branched at base, stems subquadrangular, slightly woody. Leaves opposite, shortly petiolate or subsessile, lamina obliquely elliptic to (ob)ovate, denticulate or crenate, upper surface sparsely puberulous, lower surface glabrous. Flowers solitary in leaf axils; peduncle slender. Sepals free, linear-lanceolate. Corolla purple, tubular to funnel-shaped, bilabiate, lobes ovate-orbicular. Stamens 2, included, inserted below the middle of the corolla tube; anthers coherent, thecae apically confluent. Nectary annular. Ovary slender, cylindrical, glandular-pubescent; style glandular-puberulous; stigma capitate. Capsule cylindrical, spirally twisted. One sp., *R. sinensis* Hemsl., S China.

141. ***Ornithoboea*** C.B. Clarke

Ornithoboea [Parish ex] C.B. Clarke, in A. & C. DC., Monogr. phan. 5/1: 147 (1883); Burtt, Notes Roy. Bot. Gard. Edinburgh 22: 287–299 (1958), rev.; Wang et al. in Fl. China 18: 361–370 (1998), reg. rev.
Lepadanthus Ridl. (1909).
Brachiostemon Hand.-Mazz. (1934).
Sinoboea Chun (1946).

Perennial caulescent herbs. Leaves opposite, often slightly unequal, petiolate, lamina ± ovate, obliquely subcordate at base, margin crenate or serrate-dentate. Indumentum variable, in some species densely woolly or arachnoid. Cymes sometimes unilateral (pair-flowered cincinni). Sepals free, persistent, often reflexed. Corolla tube funnel-shaped or distinctly inflated in the upper part; limb bilabiate, the upper lip of two short blunt lobes, the lower lip much larger, 3-lobed; palate with a beard of conspicuous unicellular club-shaped hairs; a line of hairs forming a circlet around the corolla mouth. Stamens 2, included; filaments straight or geniculate and with a distinct knob at the knee; anthers cohering face to face, reniform. Ovary conical, pubescent or pilose. Capsule spirally twisted. n = (16?), 17. Eleven spp., from S China to the northern part of the Malay Peninsula. Some species, especially *O. arachnoidea* (Diels) Craib, have flowers very similar to orchid flowers (floral mimicry?, Burtt 1958a; Weber 1979).

142. *Kaisupeea* B.L. Burtt

Kaisupeea B.L. Burtt, Nord. J. Bot. 21(2): 116 (2001).

Perennial rhizomatous herbs, stems annual-monocarpic, often with bladeless cataphylls at base. Indumentum of spreading gland-tipped hairs and sessile globular glands, eglandular hairs intermixed. Leaves opposite, petiole ill-defined, lamina obovate or obliquely elliptic-lanceolate, with short, appressed, acute eglandular hairs and globular glands. Cymes pedunculate, several-flowered. Sepals almost free. Corolla obliquely campanulate, held ± horizontally. Stamens 2, filaments adnate to corolla base, short, thick and curved, anthers divaricate, back of connective hairy. Nectary annular, very small. Ovary ellipsoid-conical, studded with globular glands, or with gland-tipped hairs intermixed; stigma capitate. Capsules pendent on decurved pedicels, loculicidally dehiscent, valves straight or twisted. n = 18. Three spp., Myanmar, Thailand, S Laos.

143. *Senyumia* Kiew, A. Weber & B.L. Burtt

Senyumia Kiew, A. Weber & B.L. Burtt, Beitr. Biol. Pflanzen 70: 400 (1998).

Small saxicolous herbs. Stem decumbent, slightly woody, woolly-hairy. Leaves opposite, tufted, petiole rather short, lamina thin-membranous, broadly lanceolate to ovate, coarsely serrate, teeth blunt, long hairy on both sides. Cymes with slender peduncle and several flower pairs in dichasial arrangement. Flowers resupinate. Sepals 5, linear-lanceolate, free. Corolla pure white, with short tube and large, bilabiate limb, lobes reflexed. Stamens 2, filaments short, anthers large, yellow, cohering. Nectary absent. Ovary ovoid. Capsule spirally twisted, loculicidally dehiscent. n = 9. One sp., *S. minutiflora* (Ridl.) Kiew, A. Weber & B.L. Burtt, Malay Peninsula (Gunung Senyum), in damp limestone caves.

144. *Spelaeanthus* Kiew, A. Weber & B.L. Burtt

Spelaeanthus Kiew, A. Weber & B.L. Burtt, Beitr. Biol. Pflanzen 70: 401 (1998).

Small saxicolous herbs. Stem thin, to 10 cm long, ascending, with long sticky hairs. Leaves opposite, loosely tufted, petiolate, ovate, margin serrate with blunt teeth, grey-green, with sticky hairs. Cymes with slender peduncle and few to many flower pairs. Sepals free, linear to ovate. Corolla small, pure white, shaped like a shallow bowl, limb bilabiate, with rounded lobes, mouth wide open. Stamens 2, included. Ovary conical. Nectary absent. Capsule straight or slightly twisted, thin-walled, loculicidally dehiscent. n = 8. One sp., *S. chinii* Kiew, A. Weber & B.L. Burtt, Malay Peninsula. In and at the entrance of limestone caves.

145. *Emarhendia* Kiew, A. Weber & B.L. Burtt

Emarhendia Kiew, A. Weber & B.L. Burtt, Beitr. Biol. Pflanzen 70: 398 (1998).

Small saxicolous herbs. Stem ascending, with long sticky hairs. Leaves opposite, tufted, thin-membranous, viscid-hairy, petiolate, ovate, with (often obliquely) cordate base, margin deeply and coarsely (partly doubly) serrate. Cymes with long slender peduncle and several flower pairs. Sepals 5, free. Corolla small, tubular-campanulate, bluish-white, limb bilabiate, the two lobes of upper lip with a patch of short, densely packed glands, lower lip 3-lobed. Stamens 2, included; anthers cohering. Nectary absent. Ovary conical. Capsule straight, held horizontally, opening along upper side. n = 8. One sp., *E. bettiana* (M.R. Hend.) Kiew, A. Weber & B.L. Burtt, Malay Peninsula (Pahang: Bkt. Charas and vicinity), on damp limestone rocks and cave entrances.

146. *Cyrtandra* J.R. & G. Forst.

Cyrtandra J.R. & G. Forst., Char. gen. 3 (1775); C.B. Clarke in A. & C. DC., Monogr. phan. 5/1: 201–287 (1883), rev.; Burtt, Edinburgh J. Bot. 47: 201–233 (1990), 49: 285–296 ((1992) 1993), new spp.; Backer & Bakh.f., Fl. Java 2: 543 (1965), reg. rev.; Gillett, Contr. U.S. Natl. Herb. 37: 107–159 (1967), reg. rev.; Gillett, J. Arnold Arbor. 54: 105–110 (1973), reg. rev.;

Gillett, Univ. Calif. Publ. Bot. 66: 1–59 (1973), reg. rev.; St. John, Bernice P. Bishop Mus. Bull. 229: 1–465 (1988), reg. rev.; Wagner, Herbst & Sohmer, Manual of Flowering Plants of Hawaii 1: 735–781 (1990), reg. rev.; Wang et al. in Fl. China 18: 395 (1998); Burtt, Phytomorphol. Gold. Jubilee Iss. 2001: 393–404, notes.
Whitia Blume (1823).
Cyrtandropsis Lauterb. (1910).
Protocyrtandra Hosok. (1934).
Cyrtandroidea F. Br. (1935).

Lignescent herbs, shrubs or small trees. Stem terete or angular, erect or ascending, branched or unbranched. Leaves opposite, iso- to strongly anisophyllous, petiolate, lamina ranging from narrowly lanceolate to suborbicular, margin entire or dentate, venation pinnate. Cymes short- to long-pedunculate, placed in upper or lower (leafless) part of stem, sometimes growing underground; bracteoles free or connate and forming a conspicuous cup; flowers in dense clusters or in open cymes or solitary. Sepals 5, free or connate, calyx sometimes bilabiate, persistent or caducous. Corolla variously (usually light-)coloured, funnelform, campanulate or tubular, limb bilabiate, rarely (sub)regular, lobes rounded. Stamens usually 2, rarely 4, included or (rarely) exserted; anthers free or coherent, thecae subparallel, sometimes apically confluent, dehiscing longicidally. Nectary ring- or cup-shaped. Ovary ovoid to oblong, stigma capitate or bilobed. Fruit indehiscent, oblong, ovoid or globose, hard-walled, leathery or fleshy. n = (16?), 17. Around 600 spp., from the Nicobar Islands, S China and S Japan throughout Malesia and the Pacific to the Hawaiian Islands. *Cyrtandra* is the largest genus of Gesneriaceae and has the widest distribution in the Old World. Centres of diversity are Borneo, New Guinea and the Pacific Islands.

147. *Sepikea* Schltr.

Sepikea Schltr., Bot. Jahrb. Syst. 58: 306 (1923).

Large terrestrial herbs. Stem unbranched, terete. Leaves opposite, with short petiole, lanceolate-elliptical. Indumentum of soft, brownish-yellowish hairs. Cymes long-pedunculate, few-flowered. Sepals fused halfway and forming a cylindrical-tubular calyx, lobes equal, narrow. Corolla white, tubular-infundibuliform, tube ca. 2 times as long as calyx limb with subequal, patent, oval lobes. Fertile stamens 2, the posterior pair (probably an incorrect observation: Burtt 1998a, 2002) included, filaments inserted at middle of corolla tube. Nectary annular, shallowly lobed. Ovary cylindrical-conical, stigma bifid. Fruit short-cylindrical, indehiscent. One sp., *S. cylindrocarpa* Schltr., New Guinea (Sepik area). Probably congeneric with *Cyrtandra*.

Genera of Uncertain Familial Affiliation

Sanango Bunting & Duke

Sanango Bunting & Duke, Ann. Missouri Bot. Gard. 48: 270 (1961).
Gomaranthus S. Rauschert (1982).

Small trees, ca. 15 m tall, with very hard wood. Leaves opposite, petiolate, the margins joined across the shoot axis, lamina elliptical to obovate. Inflorescence a terminal, bracteose thyrse with opposite pair-flowered cymes. Sepals connate in the lower part, narrowly triangular, ± equal. Corolla tubular, tube curved and slightly inflated in the lower part, limb subregular, lobes rounded, with cochlear aestivation in bud. Stamens 4, included, adnate to the corolla base; filaments dilated; anthers subglobose, introrse, versatile, dehiscent by a single, horseshoe-shaped line. Staminode present, small. Nectary cup-shaped, ± as high as the ovary. Ovary superior, globose, slightly 4-partite, incompletely divided in the upper part, completely divided (with axile placentae) in the lower part; style very short; stigma capitate-bilobed, the lobes laterally joined with the style and directing downwards. Fruit a capsule, depressed at the apex, style persisting for a long time, dehiscence septicidal, then loculicidal. $n = 16$ (M. Kiehn, unpubl. data). One sp., *S. racemosum* (Ruiz & Pav.) Barringer, Peru. Previously placed in the Loganiaceae, Buddlejaceae and Scrophulariaceae, but considered to belong to Gesneriaceae by Wiehler (1994). The molecular confirmation of Smith et al. (1997a) proved erroneous; current data place *Sanango* outside or at the base of Gesneriaceae (J. Smith, pers. comm., Pfosser et al., in prep.).

Cubitanthus Barringer

Cubitanthus Barringer, J. Arnold Arbor. 65: 145 (1984); Chautems, Rev. Brasil. Bot. 14: 51–59 (1991), reg. rev.

Perennial herbs, with indumentum of long multicellular hairs. Stem decumbent, 4-winged, with adventitious roots at the nodes. Roots fibrous. Leaves opposite; petiole slightly winged; blade serrate, venation pinnate. Flowers solitary in leaf axils; stalk slightly winged at base, ebracteolate. Sepals free, lanceolate-acute. Corolla cylindrical, limb bilabiate, upper lip subentire, not reflexed, lower lip 3-lobed, densely villous at base. Stamens 4, geniculate, adnate to corolla base; anthers bilocular, coherent in pairs. Nectary entire, annular. Ovary superior, with parietal placentae, style elongate, stigma entire, slightly enlarged. Capsule septicidal, 2-valved, included within persistent calyx. Seeds ovoid, slightly striate. One sp., *C. alatus* (Cham. & Schltdl.) Barringer, Brazil (Bahia). The species was originally assigned to Scrophulariaceae, which position is also indicated by molecular data (Perret et al., in prep.; Pfosser et al., in prep.).

Jerdonia Wight

Jerdonia Wight, Icon. Pl. Ind. orient. 4(2): 10, t. 1352 (1848).

(Sub)acaulescent perrennialherbs. Leaves long-petiolate, elliptic. Cymes long-pedunculate, few-flowered, bracts small. Sepals free, lanceolate. Corolla pale lilac with red lines; tube elongate, inflated above; limb oblique, bilabiate, lobes ovate, obtuse. Stamens 4, filaments short, dilated upwards, two with distinct spurs; anthers connivent and coherent at the apices; thecae divergent, apically confluent. Nectary annular. Ovary ovoid, unilocular, with 4 parietal placentae; style slender; stigma small, obscurely bilobed. Fruit a broadly ovoid capsule, loculicidally dehiscent. Seeds ellipsoid, testa striate-reticulate, with longitudinal furrows (slightly star-shaped in cross section). One sp., *J. indica* Wight, SW India (Nilghiri and Anamally Hills). The genus has a number of anomalous features (see Appendix in Weber 1989b). Its affinity is obscure and the inclusion in Gesneriaceae has been doubted. Burtt (1963, 1977b) pleaded for an accommodation in Scrophulariaceae.

Excluded Genera

Brookea Benth.

Brookea Benth., in Benth. & Hook. f., Gen. pl. 2: 939 (1876).

= Scrophulariaceae, this volume, gen. 42.

Charadrophila Marloth

Charadrophila Marloth, Bot. Jahrb. Syst. 26: 358 (1899).

= Scrophulariaceae.

Although often included in Gesneriaceae, the seedling, inflorescence, flower, gynoecium, fruit, and seed characters place the genus unequivocally in Scrophulariaceae (s.l.), where it was originally described (Weber 1989b).

Cyrtandromoea Zoll.

Cyrtandromoea Zoll., Syst. Verz. Ind. Arch. 3: 55, 58 (1858); Burtt, Bull. Bot. Surv. India 7: 73–88 (1965), prov. rev.
Busea Miq. (1856).

= Scrophulariaceae.

The genus was transferred by Burtt (1965) to Scrophulariaceae. The molecular data of Smith et al. (1997a) suggest a position in Gesneriaceae, but this needs confirmation. Apart from morphological characters, also the presence of iridoids (otherwise unknown in Gesneriaceae) is in favour of a position in Scrophulariaceae (s. l.).

Rehmannia Fisch. & C. Mey.

Rehmannia Libosch. ex Fisch. & C. Mey., Index Sem. Hortus Petrop. 1: 36 (1835), nom. cons.

= Scrophulariaceae.

Occasionally included in Gesneriaceae, but possibly belonging to Scrophulariaceae (s. l.).

Titanotrichum Soler.

Titanotrichum Soler., Ber. Deutsch. Bot. Ges. 27: 393 (1909).
= Scrophulariaceae

The genus was originally described under Scrophulariaceae, but was later repeatedly regarded to belong to Gesneriaceae. Wang et al. (1990, 1992, 1998) placed it in the latter family as a tribe of its own (Titanotricheae). Detailed morphological and molecular studies are currently being carried out (C.N. Wang, in prep.).

Acknowledgments

The "Fonds zur Förderung der wissenschaftlichen Forschung in Osterreich" (FWF) is thanked for financial support (proj. no. P 13107-B10) and for enabling colour reproduction of Figs. 11–18.

Selected Bibliography

Adatia, R.D., Sharma, Y.B., Vijayraghavan, M.R. 1971. Studies in the Gesneriaceae I. Morphology and embryology of *Platystemma violoides* Wall. Bot. Not. 124: 25–38.
Albach, D.C., Soltis, P.S., Soltis, D.E., Olmstead, R.G. 2001. Phylogenetic analysis of asterids based on sequences of four genes. Ann. Missouri Bot. Gard. 88: 163–212.

Alimova, G.K., Yakovlev, M.S. 1982. A contribution to the embryology of *Streptocarpus rexii* (Gesneriaceae) (in Russian). Bot. Zh. 67: 470–479.

Anders, O. 1966. Entwicklung und Bau des Embryosacks von *Streptocarpus rexii*. Flora, Abt. B. 156: 446–451.

Arekal, G.D. 1961. Embryology of *Klugia notoniana*. Bot. Gaz. 123: 144–150.

Arthurs, K.L. 1988. How to grow African Violets. Menlo Park: Lane Publ.

Atkins, H., Preston, J., Cronk, Q.C.B. 2001. A molecular test of Huxley's line: *Cyrtandra* (Gesneriaceae) in Borneo and the Philippines. Biol. J. Linn. Soc. 72: 143–159.

Backlund, M., Oxelman, B., Bremer, B. 2000. Phylogenetic relationships within the Gentianales based on *ndh*F and *rbc*L sequences, with particular reference to the Loganiaceae. Am. J. Bot. 87: 1029–1043.

Balicka-Iwanowska, G. 1899. Contribution à l'étude du sac embryonnaire chez certains Gamopetales. Flora 86: 47–71.

Beaufort-Murphy, H.T. 1983. The seed-surface morphology of the Gesneriaceae, utilizing the scanning electric microscope and a new system for diagnosing seed morphology. Selbyana 6: 220–422.

Bentham, G. 1876. Gesneriaceae. In: Bentham, G., Hooker, J.D. Genera plantarum 2(2): 990–1025. London: Lovell Reeve & Co.

Boggan, J.K. 1990. *Lembocarpus amoenus*. Gloxinian 40: 20–22.

Boggan, J.K. 1991. A morphological study and cladistic analysis of *Sinningia* and associated genera, with particular reference to *Lembocarpus, Lietzia, Paliavana*, and *Vanhouttea* (Gesneriaceae: Gloxinieae). M.Sc. Thesis. New York, Ithaca: Cornell University.

Boggan, J.K. 1996. Fragrant flowers in Gesneriaceae. Gloxinian 46: 17–27.

Bokhari M.H., Burtt, B.L. 1970. Studies in the Gesneriaceae of the Old World. XXXII. Foliar sclereids in *Cyrtandra*. Notes Roy. Bot. Gard. Edinburgh 30: 11–22.

Bremer, K., Backlund, A., Sennblad, B., Swenson, U., Andreasen, K., Hjertson, M., Lundberg, J., Backlund, M., Bremer, B. 2001. A phylogenetic analysis of 100+ genera and 50+ families of euasterids based on morphological and molecular data with notes on possible higher level morphological synapomorphies. Plant Syst. Evol. 229: 137–169.

Bremer, B., Bremer, K., Heidari, N., Erixon, P., Anderberg, A.A., Olmstead, R.G., Källersjö, M., Barkhordarian, E. 2002. Phylogenetics of asterids based on 3 coding and 3 non-coding chloroplast DNA markers and the utility of non-coding DNA at higher taxonomic levels. Molec. Phylogen. Evol. 24: 274-301.

Brown, R. 1839. On Cyrtandreae. (Part reprint from J.J. Bennett, Plantae Javanicae Rariores, part 1, mostly preprint from part 2: 105–122. See Mabberley 1986). London: R. & J.E. Taylor.

Burtt, B.L. 1954. Studies in the Gesneriaceae of the Old World. I. General introduction. Notes Roy. Bot. Gard. Edinburgh 21: 185–192.

Burtt, B.L. 1956. Studies in the Gesneriaceae of the Old World. X. The genus *Beccarinda*. Notes Roy. Bot. Gard. Edinburgh 2: 61–64.

Burtt, B.L. 1958a. Studies in the Gesneriaceae of the Old World. XI. The genus *Ornithoboea*. Notes Roy. Bot. Gard. Edinburgh 22: 287–299.

Burtt, B.L. 1958b. Studies in the Gesneriaceae of the Old World. XIII. Miscellaneous transfers and reductions. Notes Roy. Bot. Gard. Edinburgh 22: 309–314.

Burtt, B.L. 1963. Studies in the Gesneriaceae of the Old World. XXIV. Tentative keys to the tribes and genera. Notes Roy. Bot. Gard. Edinburgh 24: 205–220.

Burtt, B.L. 1965. The transfer of *Cyrtandromoea* from Gesneriaceae to Scrophulariaceae, with notes on the classification of that family. Bull. Bot. Surv. Ind. 7: 73–88.

Burtt, B.L. 1968. Studies in the Gesneriaceae of the Old World. XXIX. A reconsideration of generic limits in tribe Trichosporeae. Notes Roy. Bot. Gard. Edinburgh 28: 219–225.

Burtt, B.L. 1970. Studies in the Gesneriaceae of the Old World. XXXI. Some aspects of functional evolution. Notes Roy. Bot. Gard. Edinburgh 30: 1–10.

Burtt, B.L. 1974. Patterns of structural change in the flowering plants. Trans. Bot. Soc. Edinburgh 42: 133–142.

Burtt, B.L. 1975. Studies in the Gesneriaceae of the Old World. XL. The genus *Loxostigma*. Notes Roy. Bot. Gard. Edinburgh 34: 101–105.

Burtt, B.L. 1977a. Notes on rain forest herbs. In: Mabberley, D.J., Chang, K.L. (eds.) Tropical botany. Gard. Bull. Singapore 29: 73–80.

Burtt, B.L. 1977b. Classification above the genus, as exemplified by Gesneriaceae, with parallels from other groups. Plant Syst. Evol., Suppl. I: 97–109.

Burtt, B.L. 1978. Studies in the Gesneriaceae of the Old World. XLV. A preliminary revision of *Monophyllaea*. Notes Roy. Bot. Gard. Edinburgh 37: 1–60.

Burtt, B.L. 1984. Studies in the Gesneriaceae of the Old World. XLVII. Revised generic concepts for *Boea* and its allies. Notes Roy. Bot. Gard. Edinburgh 41: 401–452.

Burtt, B.L. 1997. Old World Gesneriaceae. V. Suprageneric names. Edinburgh J. Bot. 54: 85–90.

Burtt, B.L. 1998a. Climatic accommodation and phytogeography of the Gesneriaceae of the Old World. In: Mathew, P., Sivadasan, M. (eds.). Diversity and taxonomy of tropical flowering plants. Calicut: Mentor Books.

Burtt, B.L. 1998b. Taxonomic history of *Didymocarpus* and *Henckelia* (Gesneriaceae). Beitr. Biol. Pflanzen 70: 365–375.

Burtt, B.L. 1998c. New species of phytogeographical interest in *Beccarinda* and *Henckelia* (Gesneriaceae). Beitr. Biol. Pflanzen 70: 377–382.

Burtt, B.L. 1999. Old World Gesneriaceae. VI. Six miscellaneous notes. Edinburgh J. Bot. 56: 371–379.

Burtt, B.L. 2001a. *Kaisupeea*: a new genus of Gesneriaceae centred in Thailand. Nord. J. Bot. 21: 115–119.

Burtt, B.L. 2001b. Flora of Thailand: annotated checklist of Gesneriaceae. Thai Forest Bull. 29: 81–102.

Burtt, B.L. 2002. A survey of the genus *Cyrtandra* (Gesneriaceae). Phytomorphology Golden Jubilee Issue 2001: 393–401.

Burtt, B.L., Bokhari, M. 1973. Studies in the Gesneriaceae of the Old World. XXXVI. Foliar sclereids in New Guinea and Pacific *Cyrtandra*. Notes Roy. Bot. Gard. Edinburgh 32: 397–402.

Burtt, B.L., Wiehler, H. 1995. Classification of the family Gesneriaceae. Gesneriana 1: 1–4.

Carlquist, S., Hoekman, D.A. 1986. Wood anatomy of Gesneriaceae. Aliso 11: 279–297.

Caspary, R. 1858. (Title unknown). Verh. Nat.-Hist. Vereins Preuss. Rheinl. & Westf. 15, Sitzungsber. LXXIV (paper not seen; for an anonymous report on Caspary's observations regarding the unequal cotyledons of *Streptocarpus polyanthus* and *S. rexii*, see Flora 1859: 120).

Chautems, A. 2000. Introducing *Sinningia nordestina*: an unusual annual new species from northeastern Brazil. Gloxinian 50(2): 18–21.

Chautems, A., Weber, A. 1999. Shoot and inflorescence architecture of the neotropical genus *Sinningia* (Gesneriaceae). In: Kurmann, M.H., Hemsley, A.R. (eds.) The evolution of plant architecture. Kew: Royal Botanic Gardens, pp. 305–322.

Chautems, A., Baracho, G.S., Sigueira Filho, J.A. 2000. A new species of *Sinningia* (Gesneriaceae) from northeastern Brazil. Brittonia 52: 49–53.

Christie, F., Mendum, M. 2001. The ontogeny of *Aeschynanthus* seeds – a comparative study using scanning electron microscopy. Bot. J. Linn. Soc. 138: 197–207.

Citerne, H., Cronk, Q.C.B. 1999. The origin of the peloric *Sinningia*. New Plantsman 6: 219–222.

Citerne, H.L., Möller M., Cronk, Q.C.B. 2000. Diversity of *cycloidea*-like genes in Gesneriaceae in relation to floral symmetry. Ann. Bot. 86: 167–176.

Clark, J.L., Zimmer, E., 2003. A preliminary phylogeny of *Alloplectus* (Gesneriaceae): implications for the evolution of flower resupination. Syst. Bot. 28: 365–375.

Clarke, C.B. 1883. Cyrtandreae. In: A. & C. De Candolle: Monographiae Phanerogamarum 5/1. Paris: Masson.

Clements, T. 1988. African Violets. London: David & Charles.

Contandriopoulos, J., Quezel, P. 1967. A propos des phénomènes de reviviscence chez *Ramonda nathaliae*. Compt. Rend. 91. Congrès National Soc. Savantes, Rennes 1996.

Cook, M.T. 1907. The embryology of *Rhytidophyllum*. Bull. Torrey Bot. Club 34: 179–185, pl. 10.

Cory, C.S. 1994. The effects of plant neighborhoods on the reproduction of a hummingbird pollinated Costa Rican native plant, *Columnea raymondii* Morton (Gesneriaceae). Thesis. University of California.

Crawford, R.M.M. 1989. Studies in plant survival. Oxford: Blackwell.

Crété, P. 1942. Recherches histologiques et physiologiques sur l'embryologie des Labiatiflores. Contribution à l'étude des formations haustoriales. D.Sci. Thesis. Université de Paris.

Crété, P. 1949. Recherches embryologiques sur les Gesnériacées. Développement de l'albumen et de l'embryon chez *Chirita lavandulacea* Stapf. Bull. Soc. Bot. France 96: 234–235.

Crété, P. 1955a. L'application de certaines données embryologique à la systématique des Orobanchacées et de quelques familles voisines. Phytomorphology 5: 422–435.

Crété, P. 1955b. L'origine du sac embryonnaire et le développement de l'albumen chez l'*Alloplectus sanguineus* Mart. (Gesneriacées). Bull. Soc. Bot. France 102: 5–6.

Crocker, C.W. 1860. Notes on the germination of certain species of Cyrtandreae. J. Proc. Linn. Soc. Bot. 5: 65–66.

Cronk, Q.C.B., Möller, M. 1997. Genetics of floral symmetry revealed. Trends Ecol. Evol. 12: 85–86.

De Candolle, A.P. 1816. Essai sur les propriétés médicales des plantes, ed. 2. Paris: Crochard.

Denduangboripant, J., Cronk, Q.C.B. 2000. High intraindividual variation in internal transcribed spacer sequences in *Aeschynanthus* (Gesneriaceae): implications for phylogenetics. Proc. Roy. Soc. Lond. B 267: 1407–1415.

Denduangboripant, J., Mendum, M., Cronk, Q.C.B. 2001. Evolution in *Aeschynanthus* (Gesneriaceae) inferred from ITS sequences. Plant Syst. Evol. 228: 181–197.

Dengler, N.G., Sánchez-Burgos, A.A. 1988. Effect of light level on the expression of anisophylly in *Paradrymonia ciliosa* (Gesneriaceae). Bot. Gaz. (Crawfordsville) 14: 158–165.

Dobat, K., Peikert-Holle, T. 1985. Blüten und Fledermäuse. Bestäubung durch Fledermäuse und Flughunde (Chiropterophilie). Frankfurt/Main: Kramer.

Don, D. 1822. Descriptions of two new genera of Nepaul plants. Edinburgh Phil. J. 7: 82–86.

Don, G. 1838. A general history of dichlamydaeous plants, vol. 4. London: Rington et al.

Dubuc-Lebreux, M.-A. 1976. Effets de quelques régulateurs de croissance sur le phyllomorphe cotylédonaire de *Streptocarpus wendlandii* Spreng. (Gesneriaceae). Ann. Sci. Nat. Bot., sér. 12, 17: 259–276.

Dubuc-Lebreux, M.-A. 1978. Modification of unifoliate habit of *Streptocarpus wendlandii* and *Streptocarpus michelmorei* by some growth regulators. Phytomorphology 28: 224–238.

Dubuc-Lebreux, M.-A., Vieth, J. 1976. Effets de quelques régulateurs de croissance sur la morphologie florale et inflorescentielle de *Streptocarpus rexii* (Hook.) Lindl. Bull. Soc. Bot. France 123: 273–291.

Duke, J.A. 1970. Ethnobotanical observations on the Chocó Indians. Econ. Bot. 24: 344–366.

Eberle, P. 1956. Cytologische Untersuchungen an Gesneriaceen. I. Mitteilung. Die Struktur der Pachytänchromosomen sowie eine Reihe neu bestimmter Chromosomenzahlen. Chromosoma 8: 285–316.

Eberle, P. 1957. Über das Vorkommen eines Viererringes und eines heteromorphen Bivalents in der Meiosis von *Episcia tesselata* Linden. Z. Indukt. Abstammungs-Vererbungsl. 88: 184–188.

Eberwein, R., Weber, A. in prep. Medicinal use and chemical compounds of Gesneriaceae.

Elbert, V.F., Elbert, G.A. 1984. The Miracle Houseplants. African Violets and other easy-to-bloom plants in the Gesneriad family. New York: Crown Publishers.

Erbar, C. 1991. Sympetaly – a systematic character? Bot. Jahrb. Syst. 112: 417–451.

Erdtman, G. 1952. Pollen morphology and plant taxonomy. Angiosperms (An introduction to Palynology I). Stockholm: Almqvist & Wiksell.

Erhardt, A., Erhardt, W. 1993. Schöne Usambaraveilchen und andere Gesnerien. Stuttgart: Ulmer.

Feuillet, C. 1979. Contribution à l'étude morphologique et architecturale des Gesneriacees. Thèse. Paris: Université Pierre et Marie Curie, pp. 1–91, figs. 1–94.

Feuillet, C., Skog, L.E. 2003. Novae Gesneriaceae Neotropicarum. IX. New genera and species from the Guianas. Brittonia 54: 344–351.

Fogg, H.G.W. 1968. Begonias, Sinningias and African Violets. London: Gifford.

Freeman, C.E., Worthington, R.D., Jackson, M.S. 1991. Floral nectar sugar compositions of some South and Southeast Asian species. Biotropica 23(4b): 568–574.

Freiberg, M. 1994. Phänomorphologie epiphytischer Gesneriaceen unter besonderer Berücksichtigung des Mikroklimas. Diss. Fak. Naturwiss., Universität Ulm.

Freiberg, M. 1996a. Phenotype expression of epiphytic Gesneriaceae under different microclimatic conditions in Costa Rica. Ecotropica 2: 49–57.

Freiberg, M. 1996b. Spatial distribution of vascular epiphytes on three emergent canopy trees in French Guiana. Biotropica 28: 345–355.

Fritsch, K. 1893–1894. Gesneriaceae. In: Engler, A., Prantl, K. Die natürlichen Pflanzenfamilien IV, 3b: 133–144 (1893), 145–185 (1894). Leipzig: W. Engelmann.

Fritsch, K. 1904. Die Keimpflanzen der Gesneriaceen mit besonderer Berücksichtigung von *Streptocarpus*, nebst vergleichender Studien über die Morphologie dieser Familie. Jena: Fischer.

Fritsch, K. 1908. Über das Vorkommen von Cystolithen bei *Klugia zeylanica*. In: Linsbauer, L. (ed.) Wiesner Festschrift. Wien: Konegen, pp. 412–416.

Fritsch, K. 1920. Über den Begriff der Anisokotylie. Ber. Deutsch. Bot. Ges. 38: 69–73.

Fritze, K.J., Williams, N.H. 1988. The taxonomic significance of pollen morphology in the *Columnea* alliance (Gesneriaceae: Gesnerioideae). Ann. Missouri Bot. Gard. 75: 168–191.

Fryns-Claessens, W., van Cotthem, W. 1973. A new classification of the ontogenetic types of stomata. Bot. Rev. 39: 71–138.

Gaff, D. 1981. The biology of resurrection plants. In: Pate, J.S., McComb, A.J. (eds.) The biology of Australian plants. Perth: Univ. West Austral. Press.

Gilmer, K., Hofacker, A. 2001. *Sinningia macrostachya* (Lindl.) Chautems. Kakteen Sukk. 52: 155–158.

Glisic, L.M. 1928. Development of the female gametophyte and endosperm in *Haberlea rhodopensis* Friv. Bull. Inst. Jard. Bot. Univ. Beograd 1: 1–13.

Glisic, L.M. 1934. Zur Kenntnis der Samenentwicklung der Gesneriaceen. Über die Endosperm- und Haustorienbildung von *Roettlera*. Bull. Inst. Jard. Bot. Univ. Beograd 3: 94–111.

Goebel, K. 1928. Organographie der Pflanzen. Vol. 3/1. Jena: Fischer.

Green, P.S. 1973. *Negria rhabdothamnoides*. Curtis's Bot. Mag., n.s., t. 659.

Haccius, B. 1939. Untersuchungen über die Bedeutung der Distichie für das Verständnis der zerstreuten Blattstellung bei den Dikotylen. Bot. Archiv 40: 58–150.

Haccius, B. 1950. Weitere Untersuchungen zum Verständnis der zerstreuten Blattstellung bei den Dikotylen. Sitzungsber. Heidelberg. Akad. Wiss., Math.-Naturwiss. Kl., Heidelberg 6: 1–51 & 289–337.

Haccius, B., Troll, W. 1961. Über die sogenannten Wurzelhaare an den Keimpflanzen von *Drosera*- und *Cuscuta*-Arten. Beitr. Biol. Pflanzen 36: 139–157.

Hallé F., Delmotte, A. 1973. Croissance et floraison de la Gésneriacée africaine *Epithema tenue* C.B.Clarke. Adansonia II, 13: 273–287.

Hanstein, J. 1865. Die Gesneraceen des Königlichen Herbariums und der Gärten zu Berlin, nebst Beobachtungen über die Familie im Ganzen. II. Abschnitt, Drittes Stück. Linnaea 34: 225–462.

Harborne, J.B. 1966. Comparative biochemistry of the flavonoids. II. 3-desoxyanthocyanins and their systematic distribution in ferns and gesnerads. Phytochemistry 5: 589–600.

Harborne, J.B. 1967. Comparative biochemistry of the flavonoids. VI. Flavonoid patterns in the Bignoniaceae and the Gesneriaceae. Phytochemistry 6: 1643–1651.

Harrison C.J., Möller, M., Cronk, Q.C.B. 1999. Evolution and development of floral diversity in *Streptocarpus*. Ann. Bot. 84: 49–60.

Hartl, D. 1956. Morphologische Studien am Pistill der Scrophulariaceen. Österr. Bot. Z. 103: 185–242.

Hartl, D. 1959. Das alveolierte Endosperm bei Scrophulariaceen, seine Entstehung, Anatomie und taxonomische Bedeutung. Beitr. Biol. Pflanzen 35: 95–110.

Hegnauer, R. 1966. Chemotaxonomie der Pflanzen, Bd. 4. Stuttgart: Birkhäuser, pp. 20–206.

Herat, K.M., Theobald, W.L. 1979. Comparative studies of vegetative anatomy and morphology of the Gesneriaceae of Sri Lanka. Bot. J. Lin. Soc. 78: 285–298.

Hielscher, T. 1883. Anatomie und Biologie der Gattung *Streptocarpus*. Beitr. Biol. Pflanzen 3: 1–24.

Hill, A.W. 1938. The monocotylous seedlings of certain dicotyledons, with special reference to the Gesneriaceae. Ann. Bot. (London) 2: 127–143.

Hilliard, O.M., Burtt, B.L. 1971. *Streptocarpus*. An African plant study. Pietermaritzburg: Univ. Natal Press.

Hilliard, O.M., Burtt, B.L. 2002. The genus *Agalmyla* (Gesneriaceae-Cyrtandroideae). Edinburgh J. Bot. 59: 1–210.

Hollstein, R.T.H. 1878. Ueber den Gefässbündelverlauf im Stamme der Gesneraceen. Diss. Universität Halle, p. 1–40.

Holmqvist, C. 1964. Some facts about the embryology of Gesneriaceae. In: 10th Int. Bot. Congr., Edinburgh. Abstract no. 221, p. 96.

Howard, R.A. 1970. Some observations on the nodes of woody plants with special reference to the problem of 'split-lateral' versus the 'common gap'. In: Robson, N.K.B., Cutler, D.F., Gregory, M. (eds.) New research in plant anatomy. Bot. J. Linn. Soc. suppl. 1: 195–214.

Huber, E. 1953. Beitrag zur anatomischen Untersuchung der Antheren von *Saintpaulia*. Sitzungsber. Akad. Wiss. Wien, Math.-Naturwiss. Kl., Abt. I, 162: 227–234.

Hubert, N., Hubert, H. 1991. Blommor besvarad kärlak. Wiken.

Imaichi, R., Inokuchi, S., Kato, M. 2001. Developmental morphology of the one-leaf plant *Monophyllaea singularis* (Gesneriaceae). Plant Syst. Evol. 229: 171–185.

Inoue, K., Ueda, S., Nayeshiro, H., Moritome, N., Inouye, H. 1984. Biosynthesis of naphtaquinones and anthraquinones in *Streptocarpus dunnii* cell cultures. Phytochemistry 23: 313–318.

Ivanina, L.I. 1965. Application of the carpological method to the taxonomy of Gesneriaceae. Notes Roy. Bot. Gard. Edinburgh 26: 383–402.

Ivanina, L.I. 1967. The family Gesneriaceae (The Carpological Review). Leningrad: Komarov Bot. Inst.

Jack, W. 1823. On Cyrtandraceae, a new natural order of plants. Trans. Linn. Soc. 14: 23–44.

Jensen, S.R. 1992. Systematic implications of the distribution of iridoids and other chemical compounds in the Asteridae. Ann. Missouri. Bot. Gard. 79: 284–302.

Jensen, S.R. 1996. Caffeolyl phenylethanoid glycosides in *Sanango racemosum* and in the Gesneriaceae. Phytochemistry 43: 777–783.

Johri, B.M., Ambegaokar, K.B., Srivastava, P.S. 1992. Comparative embryology of angiosperms. Berlin, Heidelberg, New York: Springer.

Jones, C.E., Rich, P.V. 1972. Ornithophily and extrafloral color patterns in *Columnea florida* Morton (Gesneriaceae). Bull. South. Calif. Acad. Sci. 71: 113–116.

Jong, K. 1970. Developmental aspects of vegetative morphology in *Streptocarpus*. Ph.D. Thesis. University of Edinburgh.

Jong, K. 1973. *Streptocarpus* (Gesneriaceae) and the phyllomorph concept. Acta Bot. Neerl. 22: 244–245.

Jong, K. 1978. Phyllomorphic organisation in rosulate *Streptocarpus*. Notes Roy. Bot. Gard. Edinburgh 36: 369–396.

Jong, K., Burtt, B.L. 1975. The evolution of morphological novelty exemplified in the growth patterns of some Gesneriaceae. New Phytol. 75: 297–311.

Jong, K., Möller, M. 2000. New chromosome counts in *Streptocarpus* (Gesneriaceae) from Madagascar and the Comoro Islands and their taxonomic significance. Plant Syst. Evol. 224: 173–182.

Kato, M., Itino, T., Nagamitsu, T. 1993. Melittophily and ornithophily of long-tubed flowers in Zingiberaceae and Gesneriaceae in West Sumatra. Tropics 2: 129–142.

Kiehn, M., Weber, A. 1998. Chromosome numbers of Malayan and other palaeotropical Gesneriaceae. II. Tribes Trichosporeae, Cyrtandreae and Epithemateae. Beitr. Biol. Pflanzen 70: 445–470.

Kiehn, M., Weber, A. 2001. Karyological differentiation and speciation in Malesian Gesneriaceae. In: Saw, L.G., Chua, L.S.L., Khoo, K.C. (eds.) Taxonomy: the cornerstone of biodiversity. Proc. 4th Int. Flora Malesiana Symp. 1998. Kuala Lumpur: Forest Res. Inst. Malaysia (FRIM), pp. 139–142.

Kiehn, M., Hellmayr, E., Weber, A. 1998. Chromosome numbers of Malayan and other palaeotropical Gesneriaceae. I. Tribe Didymocarpeae. Beitr. Biol. Pflanzen 70: 407–444.

Kiew, R., Weber, A., Burtt, L. 1998. Three new genera of Gesneriaceae from limestone of Peninsular Malaysia. Beitr. Biol. Pflanzen 70: 383–403.

Kleinfeldt, S.E. 1978. Ant-gardens: the interaction of *Codonanthe crassifolia* (Gesneriaceae) and *Crematogaster longispina* (Formicidae). Ecology 59: 449–456.

Kohyama, T., Hotta, M. 1986. Growth analysis of Sumatran *Monophyllaea*, possessing only one leaf throughout perennial life. Pl. Sp. Biol. 1: 117–125.

Kvist, L.P. 1984. Gesneriaceae's taxonomi med specielt henblik påkemiske karakterer. (The taxonomy of Gesneriaceae with special reference to chemical characters). Thesis. Aarhus University, pp. 1–190.

Kvist, L.P. 1985. Ethnobotanical collecting in Ecuador. Gloxinian 35: 14–15.

Kvist, L.P. 1986. Gesneriads and snake bite. Gloxinian 36: 8–13.

Kvist, L.P. 1989. Popular names and medical uses of *Columnea*. Gloxinian 39: 21–25.

Kvist, L.P., Holm-Nielsen, L.B. 1987. Ethnobotanical aspects of lowland Ecuador. Opera Bot. 92: 83–107.

Kvist, L.P., Pedersen, J.A. 1986. Distribution and taxonomic implications of some phenolics in the family Gesneriaceae determined by EPR Spectroscopy. Biochem. Syst. Ecol. 14: 385–405.

Kvist, L.P., Skog, L.E. 1988. *Columnea incredibilis* and *Cremosperma filicifolium*, two remarkable new Gesneriaceae from western Colombia. Nord. J. Bot. 8: 259–269.

Kvist, L.P., Skog, L.E. 1989. Revision of *Reldia* (Gesneriaceae). Nord. J. Bot. 601–611.

Kvist, L.P., Skog, L.E. 1992. Revision of *Kohleria* (Gesneriaceae). Smiths. Contr. Bot. 79: 1–83.

Kvist, L.P., Skog, L.E. 1993. The genus *Columnea* (Gesneriaceae) in Ecuador. Allertonia 6: 327–400.

Lamond, M., Vieth, J. 1972. L'androcée synanthéré du *Rechsteineria cardinalis* (Gesnériacées). Une contribution au problème des fusions. Can. J. Bot. 50: 1633–1637.

Laurent, V. 1923. Zur Entwicklungsgeschichte von *Corytoloma cyclophyllum* Dus. n. sp. ined. Svensk Bot. Tidskr. 17 (2): 165–174.

Lawrence, W.J.C. 1957. Studies on *Streptocarpus*. IV. Genetics of flower colour patterns. Heredity 11: 337–357.

Lawrence, W.J.C. 1958. Studies on *Streptocarpus* Lindl. V. Speciation and gene systems. Heredity 12: 333–356.

Lawrence, W.J.C., Sturgess, V.C. 1957. Studies on *Streptocarpus* III. Genetics and chemistry of flower colour in the garden forms, species and hybrids. Heredity 11: 303–336.

Lawrence, W.J.C., Scott-Moncrieff, R., Sturgess, V.C. 1939. Studies on *Streptocarpus*. I. Genetics and chemistry of flower colour in the garden strains. J. Genet. 38: 299–306.

Lowry, J.B. 1972. Anthocyanins of some Malaysian members of the Gesneriaceae. Phytochemistry 11: 3267–3269.

Lowry, J.B. 1973. *Rhabdothamnus solandri*: some phytochemical results. New Zeal. J. Bot. 11: 555–560.

Lu, K.L., Mesler, M.R. 1981. Ant dispersal in a neotropical forest floor Gesneriad. Biotropica 13: 159–160.

Lu, Y., Xu, P.-J., Chen, Z.-N., Liu, G.M. 1998a. The anthraquinones of *Rhynchotechum vestitum*. Phytochemistry 47: 315–317.

Lu, Y., Xu, P.-J., Chen, Z.-N., Liu, G.M. 1998b. Anthraquinone glycosides from *Rhynchotechum vestitum*. Phytochemistry 49: 1135–1137.

Luegmayr, E. 1993a. Pollen characters of Old World Gesneriaceae (Cyrtandroideae) with special reference to SE Asian taxa. Grana 32: 221–232.

Luegmayr, E. 1993b. Pollen of Hawaiian *Cyrtandra* (Gesneriaceae), including notes on Southeast Asian taxa. Blumea 38: 25–38.

Luegmayr, E. 1993c. The generative cell and its close association with the endoplasmatic reticulum of the vegetative cell in pollen of *Cyrtandra pendula* (Gesneriaceae). Protoplasma 177: 73–81.

Luegmayr, E. 1994. Pollenmorphologische und pollenontogenetische Untersuchungen an ausgewählten Gesneriaceae. Diss. Formal- & Naturwiss. Fak. Univ. Wien.

Markovska, Y.K., Kimenov, G.P., Tsonev, T.D. 1989. Presence of Crassulacean Acid Metabolism in higher poikilohydric plants *Haberlea rhodopensis* Friv. and *Ramonda serbica* Panc. Photosynthetica 23: 364–367.

Markovska, Y.K., Tsonev, T.D., Kimenov, G.P., Tutekova, A. 1994. Physiological changes in higher poikilophydric plants – *Haberlea rhodopensis* Friv. and *Ramonda serbica* Panc. during drought and rewatering at different light regimes. J. Plant Physiol. 144: 100–108.

Markovska, Y., Tsonev, T., Kimenov, G. 1997. Regulation of CA and respiratory recycling by water supply in higher poikilohydric plants *Haberlea rhodopensis* Friv. and *Ramonda serbica* Panc. at transition from biosis to anabiosis and vice versa. Bot. Acta 110: 18–24.

Mayer, V., Möller, M., Perret, M., Weber, A. 2003. Phylogenetic position and generic differentiation of Epithemateae (Gesneriaceae) inferred from cpDNA sequence data. Am. J. Bot. 90(2): 321–329.

Melhelm, T., Mauro, C. 1973. Pollen morphological studies in Gesneriaceae. Hoehnea 3: 13–27, fig. 1–32.

Mendum, M., Laßnig, P., Weber, A., Christie, F. 2001. Testa and seed appendage morphology in *Aeschynanthus* (Gesneriaceae): phytogeographical patterns and taxonomic implications. Bot. J. Linn. Soc. 135: 195–213.

Mctcalfe, C.R., Chalk, L. 1950. Anatomy of the dicotyledons. Vol. II. Oxford: Clarendon Press.

Metcalfe, C.R., Chalk, L. 1979. Anatomy of the dicotyledons. Oxford: Clarendon Press.

Mølgård, P., Ravn, H. 1988. Evolutionary aspects of caffeoyl ester distribution in dicotyledons. Phytochemistry 27: 2411–2421.

Möller, M., Cronk, Q.C.B. 1997a. Origin and relationships of *Saintpaulia* H.Wendl. (Gesneriaceae) based on ribosomal DNA internal transcribed spacer (ITS) sequences. Am. J. Bot. 84: 956–965.

Möller, M., Cronk, Q.C.B. 1997b. Phylogeny and disjunct distribution: evolution of *Saintpaulia* (Gesneriaceae). Proc. Roy. Soc. Lond. Ser. B, 264: 1827–1836.

Möller, M., Cronk, Q.C.B. 1999. New approaches to the taxonomy of *Saintpaulia* and *Streptocarpus*. In: Andrews, S., Leslie, A.C., Alexander, C. (eds.) Taxonomy of cultivated plants. Proc. 3rd Int. Symp. Kew: Royal Botanical Gardens, pp. 253–264.

Möller, M., Cronk, Q.C.B. 2001a. Evolution of morphological novelty: a phylogenetic analysis of growth patterns in *Streptocarpus* (Gesneriaceae). Evolution 55: 918–929.

Möller, M., Cronk, Q.C.B. 2001b. Phylogenetic studies in *Streptocarpus* (Gesneriaceae): reconstruction of biogeographic history and distribution patterns. Syst. Geogr. Pl. 71: 545–555.

Möller, M., Clokie, M., Cubas, P., Cronk, Q.C.B. 1999. Integrating molecular phylogenies and developmental genetics: a Gesneriaceae case study. In: Hollingsworth, P.M., Bateman,

R.M., Gornall, R.J. (eds.) Molecular systematics and plant evolution. London: Taylor & Francis.
Moore, H.E. 1957. African Violets and their relatives. New York: Macmillan.
Moore, H.E. 1973. A synopsis of the genus *Codonanthe* (Gesneriaceae). Baileya 19: 4–33.
Morgan, M.T., Dengler, N.G. 1988. Vascular architecture in isophyllous and facultatively anisophyllous species of *Pentadenia* (Gesneriaceae). Am. J. Bot. 75: 1485–1494.
Noel, A.R.A., van Staden, J. 1975. Phyllomorph senescence in *Streptocarpus molweniensis*. Ann. Bot. 39: 921–929.
Oehlkers, F. 1923. Entwicklungsgeschichte von *Monophyllaea horsfieldii*. Beih. Bot. Centralbl., Abt. A 39: 128–151.
Oehlkers, F. 1962. Entwicklungsgeschichtliche Untersuchungen. II. Die Blütenentwicklung von *Streptocarpus wendlandii*. Z. Bot. 50: 217–236.
Olmstead, R.G., Reeves, P.A. 1995. Evidence for polyphyly of the Scrophulariaceae based on chloroplast *rbc*L and *ndh*F sequences. Ann. Missouri Bot. Gard. 82: 176–193.
Olmstead, R.G., DePamphilis, C.W., Wolfe, A.D., Young, N.D., Elisons, W.J., Reeves, P.A. 2001. Disintegration of the Scrophulariaceae. Am. J. Bot. 88: 348–361.
Oxelman, B., Backlund, M., Bremer, K. 1999a. Relationships of the Buddlejaceae s.l. investigated using parsimony, jackknife and branch support analysis of chloroplast *ndh*F and *rbc*L sequence data. Syst. Bot. 24: 14–182.
Oxelman, B., Kornhall, P., Bremer, B., Olmstead, R.G. 1999b. A new circumscription of Scrophulariaceae. Abstr. XVI Int. Bot. Congr., St. Louis, 1999.
Padmanabhan, D. 1961. A contribution to the embryology of *Epithema carnosum*. J. Madras. Univ. 31B(1): 37–46.
Padmanabhan, D. 1966. The octant embryo of *Epithema carnosum* Benth. Proc. Ind. Acad. Sci. B, 64: 6.
Perret, M., Chautems, A., Spichiger, R., Peixoto, M., Savolainen, V. 2001. Nectar sugar concentration in relation to pollination syndromes in Sinningieae (Gesneriaceae). Ann. Bot. 87: 267–273.
Perret, M., Chautems, A., Spichiger, R., Kite, G., Savolainen, V. 2003. Systematics and evolution of tribe Sinningieae (Gesneriaceae): evidence from phylogenetic analyses of six plastid DNA regions and nuclear *ncpGS*. Am. J. Bot. 90: 445–460.
Pischinger, F. 1903. Über Bau und Regeneration des Assimilationsapparates von *Streptocarpus* und *Monophyllaea*. Sitzungsber. Akad. Wiss. Wien, Math.-Naturwiss. Kl., 111: 278–302.
Pongracic, O. 1931. Beiträge zur Anatomie der Gesneriaceen. Sitzungsber. Akad. Wiss. Wien, Math.-Naturwiss. Kl., Abt. I, 140: 183–219.
Price, J.R., Robinson, R. 1938. A new natural colouring matter of the naphthalene group. Nature 143: 147.
Price, J.R., Robinson, R. 1940. Dunnione. J. Chem. Indust. 1940: 1493.
Rashid, M.H., Jong, K., Mendum, M. 2001. Cytotaxonomic observations in the genus *Aeschynanthus* (Gesneriaceae). Edinburgh J. Bot. 58: 31–43.
Ratter, J.A. 1975. A survey of chromosome numbers in the Gesneriaceae of the Old World. Notes Roy. Bot. Gard. Edinburgh 33: 527–543.
Rechinger, K. 1899. Vergleichende Untersuchungen über die Trichome der Gesneriaceen. Österr. Bot. Z. 49: 89–92, 142–146, 180–183, 207–213.
Reichert, E.T. 1913. Starches of Gesneriaceae. In: The differentiation and specificity of starches in relation to genera, species, etc. Part II. Carnegie Institution of Washington, pp. 886–890.
Ritterbusch, A. 1976. Die Organopoiese von *Calceolaria tripartita* R. et P. (Scrophulariaceae). Bot. Jahrb. Syst. 95: 267–320.
Ritterbusch, A. 1980. The modelling of growth and development, a transformational approach to floral organogenesis. Flora 169: 498–509.
Roalson, E.H., Boggan, J.K., Skog, L.E., Zimmer, E.A. in prep. Generic boundaries in the Gloxinieae (Gesneriaceae): tracking diversification and convergence in floral form with nuclear, chloroplast, and morphological cladistic data sets.
Robey, M.J. 1988. African Violets - gifts from nature. London: Cornwall Books.
Roelofs, F.M. 1980. The reproductive biology of *Cyrtandra grandiflora* (Gesneriaceae) on Oahu. Pacific Sci. 33: 223–231.
Rosenblum, I.M. 1981. An approach toward understanding some of the morphogenetic bases of phylogeny of *Streptocarpus* (Gesneriaceae). Ph.D. Thesis. New York: City University of New York.
Rosenblum, I.M., Basile, D.V. 1984. Hormonal regulation of morphogenesis in *Streptocarpus* and its relevance to evolutionary history of the Gesneriaceae. Am. J. Bot. 71: 52–64.
Rosser, E.M., Burtt, B.L. 1969. Studies in the Gesneriaceae of the Old World. XXX. Anatomical characters in the tribe Trichosporeae. Notes Roy. Bot. Gard. Edinburgh 29: 39–58.
Sachs, M. 1915. Anatomisch-systematische Untersuchungen über die Blattstruktur bei den Gesnerioideen und einigen Triben der Cyrtandroideen. Diss. Univ. Erlangen, pp. 1–54.
Sahasrabudhe, S., Stace, C.A. 1974. Developmental and structural variation in the trichomes and stomata of some Gesneriaceae. New Bot. 1: 46–62.
Samuel, R., Kiehn, M., Pinsker, W. 1997a. Relationships among six genera of Gesneriaceae-Cyrtandroideae inferred from cpDNA, *atp*B/*rbc*L spacer sequences. In: Hollingsworth, P.M., Bateman, R.M., Gornall, R.J. (eds.) Advances in plant molecular systematics. Glasgow: University of Glasgow.
Samuel, R., Pinsker, W., Kiehn, M. 1997b. Phylogeny of some species of *Cyrtandra* (Gesneriaceae) inferred from the *atp*B/*rbc*L cpDNA intergene region. Bot. Acta 110: 503–510.
Sánchez-Burgos, A.A., Dengler, N.G. 1988. Leaf development in isophyllous and facultatively anisophyllous species of *Pentadenia* (Gesneriaceae). Am. J. Bot. 75: 1472–1484.
San Martin-Gajardo, I., Freitas, L. 1999. Hummingbird pollination in *Besleria longimucronata* Hoehne (Gesneriaceae) in southeastern Brazil. Biociências (Porto Alegre) 7: 13–24.
San Martin-Gajardo, I., Sazima, M. 2002a. Floral biology and pollination of Sinningieae species in southeastern Brazil. In: Möller, M., Mendum, M., Cronk, Q.C.B. (eds.) Gesneriaceae Worksh., Talk Abstr. Royal Botanic Garden Edinburgh, September 2002.
San Martin-Gajardo, I., Sazima, M. 2002b. Pollination biology of two *Paliavana* species (Sinningieae) in southeastern Brazil. In: Möller, M., Mendum, M., Cronk, Q.C.B. (eds.) Gesneriaceae Worksh., Poster Abstr. Royal Botanic Garden Edinburgh, September 2002.
Saueregger, J., Weber, A. in press. Determining factors of anisocotyly in Gesneriaceae. Edinburgh J. Bot.
Savolainen, V., Fay, M.F., Albach, D.C., Backlund, A., van der Bank, M., Cameron, K.M., Johnson, S.A., Lledó, M.D., Pintaud, J.-C., Powell, M., Sheahan, M.C., Soltis, D.E., Soltis, P.S., Weston, P., Whitten, W.M., Wurdack, K.J., Chase, M.W. 2000. Phylogeny of the eudicots: a nearly complete familial analysis based on *rbc*L gene sequences. Kew Bull. 55: 257–309.

Sazima, I., Buzato, S., Sazima, M. 1995. An assemblage of hummingbird-pollinated flowers in a montane forest in southeastern Brazil. Bot. Acta 109: 149–160.

Schenk, W. 1943. Morphologisch-anatomische Untersuchungen an der Gattung *Streptocarpus* Lindl. Bot. Arch. 44: 217–284

Schnarf, K. 1921. Kleine Beiträge zur Entwicklungsgeschichte der Angiospermen. II. *Klugia zeylanica* (R. Brown) Gardn. Österr. Bot. Z. 70: 255–261.

Schnarf, K. 1931. Vergleichende Embryologie der Angiospermen. Berlin Heidelberg New York: Springer.

Skog, L.E. 1976. A study of the tribe Gesnerieae, with a revision of *Gesneria* (Gesneriaceae: Gesnerioideae). Smiths. Contr. Bot. 29: 1–182.

Skog, L.E. 1979. Gesneriaceae. In: Flora of Panama, Family 175. Gesneriaceae. Ann. Missouri Bot. Gard. 65: 783–996.

Skog, L.E. 1984. A review of the chromosome numbers in the Gesneriaceae. Selbyana 7: 252–273.

Skog, L.E. 2001. Gesneriaceae. In: Stevens, W.D. et al. (eds.) Flora of Nicaragua. Vol. II. Monogr. Syst. Bot. Missouri Bot. Gard. 85: 1114–1128.

Skog, L.E., de Jesus, F.F. 1997. A review of *Resia* (Gesneriaceae). BioLlania (ed. esp. 6): 515–525.

Skutch, A.F. 1992. *Tussacia friedrichsthaliana*, a terrestrial herb with aquatic flowers. Brenesia 37: 151–156.

Smith, J.F. 1991. The evolution and systematics of *Columnea* sections *Pentadenia* and *Stygnanthe* (Gesneriaceae). Ph.D. Thesis. University of Wisconsin, USA.

Smith, J.F. 1994. Systematics of *Columnea* section *Pentadenia* and section *Stygnanthe* (Gesneriaceae). Syst. Bot. Monogr. 44: 1–89.

Smith, J.F. 1997 ("1996"). Tribal relationships within the Gesneriaceae: a cladistic analysis of morphological data. Syst. Bot. 21: 497–513.

Smith, J.F. 2000a. Phylogenetic resolution within the tribe Episcieae (Gesneriaceae): congruence of ITS and *ndh*F sequences from parsimony and maximum-likelihood analyses. Am. J. Bot. 87: 883–897.

Smith, J.F. 2000b. A phylogenetic analysis of tribes Beslerieae and Napeantheae (Gesneriaceae): parsimony and maximum-likelihood analyses of *ndh*F sequences. Syst. Bot. 25: 71–80.

Smith, J.F. 2000c. Phylogenetic signal common to three data sets: combining data which initially appear heterogeneous. Plant Syst. Evol. 221: 179–198.

Smith, J.F. 2001. The phylogenetic relationships of *Lembocarpus* and *Goyazia* (Gesneriaceae) based on *ndh*F sequences. Ann. Missouri Bot. Gard. 88: 135–201.

Smith, J.F., Atkinson, S. 1998. Phylogenetic analysis of the tribes Gloxinieae and Gesnerieae (Gesneriaceae): data from *ndh*F sequences. Selbyana 19: 122–131.

Smith, J.F., Carroll, C.L. 1997. Phylogenetic relationships of the Episcieae (Gesneriaceae) based on *ndh*F sequences. Syst. Bot. 22: 713–724.

Smith, J.F., Sytsma, K.J. 1994a. Evolution in the Andean epiphytic genus *Columnea* (Gesneriaceae): part I. Morphological variation. Syst. Bot. 19: 220–235.

Smith, J.F., Sytsma, K.J. 1994b. Evolution in the Andean epiphytic genus *Columnea* (Gesneriaceae): part II. Chloroplast DNA restriction site variation. Syst. Bot. 19: 317–336.

Smith, J.F., Sytsma, K.J. 1994c. Molecules and morphology: congruence of data in *Columnea* (Gesneriaceae). Plant Syst. Evol. 194: 37–52.

Smith, J.F., Burke, C.C., Wagner, W.L. 1996. Interspecific hybridization in natural populations of *Cyrtandra* (Gesneriaceae) on the Hawaiian Islands: evidence from RAPD markers. Plant Syst. Evol 200: 61–77.

Smith, J.F., Brown, K.D., Carroll, C.L., Denton, D.S. 1997a. Familial placement of *Cyrtandromoea*, *Titanotrichum* and *Sanango*, three problematic genera of the Lamiales. Taxon 46: 65–74.

Smith, J.F., Wolfram, J.C., Brown, K.D., Carroll, C.L., Denton, D.S. 1997b. Tribal relationships in the Gesneriaceae: evidence from DNA sequences of the chloroplast gene *ndh*F. Ann. Missouri Bot. Gard. 84: 50–66.

Smith J.F., Kresge, M.E., Möller, M., Cronk Q.C.B. 1998. A cladistic analysis of *ndh*F sequences from representative species of *Saintpaulia* and *Streptocarpus* sections *Streptocarpus* and *Streptocarpella* (Gesneriaceae). Edinburgh J. Bot. 55: 1–11.

Snow, D.W., Snow, B.K. 1980. Relationships between hummingbirds and flowers in the Andes of Colombia. Bull. Brit. Mus. Nat. Hist. Zool. 38: 105–139.

Solereder, H. 1899. Systematische Anatomie der Dicotyledonen. Stuttgart: Enke.

Solereder, H. 1908. Ergänzungsband zu Systematische Anatomie der Dicotyledonen. Stuttgart: Enke.

Sontag, S., Weber, A. 1998. Seed coat structure of *Didissandra*, *Ridleyandra* and *Raphiocarpus* (Gesneriaceae). Beitr. Biol. Pflanzen 70: 179–190.

Spangler, R.E., Olmstead, R.G. 1999. Phylogenetic analysis of the Bignoniaceae based on the cpDNA gene sequences *rbc*L and *ndh*F. Ann. Missouri Bot. Gard. 86: 33–46.

Steiner, K.E. 1985. The role of nectar and oil in the pollination of *Drymonia serrulata* by *Epicharis* bees (Anthophoridae) in Panama. Biotropica 17: 217–229.

Stevanovic, B., Glisic, O. 1997. Eco-anatomical differences between Balkan endemo-relict species of Gesneriaceae. Bocconea 5: 661–665.

Stöckigt, J., Srocka, U., Zenk, M.H. 1973. Structure and biosynthesis of a new anthraquinone from *Streptocarpus dunnii*. Phytochemistry 12: 2389–2391.

Svoma, E., Kiehn, M. 1993. Beiträge zur Embryologie der Gattung *Cyrtandra* (Gesneriaceae). In: Heiselmayer, P. (ed.) Österr. Botanikertreffen, Kurzf. Beitr., 50. Univ. Salzburg.

Swamy, B.G.L., Padmanabhan, D. 1961. The quadrant proembryo of *Epithema carnosum*. Proc. Ind. Acad. Sci. B, 63: 4.

Thathachar, T. 1942. Studies in Gesneriaceae, gametogenesis and embryogeny of *Didymocarpus tomentosa*. J. Ind. Bot. Soc. 21: 185–193.

Trapp, A. 1956. Zur Morphologie und Entwicklungsgeschichte der Staubblätter sympetaler Blüten. Bot. Studien, Heft 5.

Troll, W. 1937. Vergleichende Morphologie der Höheren Pflanzen. Bd. 1, Teil 1. Berlin: Borntraeger.

Troll, W. 1964. Die Infloreszenzen. Typologie und Stellung im Aufbau des Vegegationskörpers. Bd. 1/1. Berlin: Borntraeger.

Tsukaya, H. 1997. Determination of the unequal fate of cotyledons of a one-leaf plant, *Monophyllaea*. Development 124: 1275–1280.

Van Staden, J. 1973. Changes in endogenous cytokinin levels during abscission and senescence of *Streptocarpus* leaves. J. Exp. Bot. 24: 667–673.

Vickers, W.T., Plowman, T. 1984. Useful plants of the Siona and Secoya Indians of Eastern Ecuador. Fieldiana Bot. N.S. 15: 1–63.

Viemont, J.-D. 1976. Ontogenèse de l'appareil végétatif chez le *Streptocarpus rexii* Lindl. Bull. Soc. Bot. France 123: 9.

Viemont, J.-D., Astie, M. 1974. Quelques particularités du développement des deux *Streptocarpus*. Bull. Soc. Bot. France 121: 11–22.

Vogel, S. 1958. Fledermausblumen in Südamerika. Österr. Bot. Z. 104: 491-530.
Vogel, S. 1966. Parfümsammelnde Bienen als Bestäuber von Orchidaceen und *Gloxinia*. Österr. Bot. Z. 113: 302-361.
Vogel, S. 1968. Chiropterophilie in der neotropischen Flora. Neue Mitteilungen I. Flora B 157: 562–602.
Vogel, S. 1969a. Chiropterophilie in der neotropischen Flora. Neue Mitteilungen II. Flora B 158: 185–222.
Vogel, S. 1969b. Chiropterophilie in der neotropischen Flora. Neue Mitteilungen III. Flora B 158: 289–323.
Vogel, S. 1978. Evolutionary shifts from reward to deception in pollen flowers. In: Richards, A.J. (ed.) The pollination of flowers by insects. London: Academic Press, pp. 89–96.
Vogel, S. 1993. Betrug bei Pflanzen: Die Täuschblumen. Abhandl. Akad. Wiss. u. Lit. Mainz, Math.-Naturwiss. Kl., Heft 1: 1–48.
Wagstaff, S.J., Olmstead, R.G. 1997. Phylogeny of Labiatae and Verbenaceae inferred from *rbc*L sequences. Syst. Bot. 22: 165–179.
Wagstaff, S.J., Martinson, K., Swenson, U. 2000. Divergence estimates of *Tetrachondra hamiltonii* and *T. patagonica* (Tetrachondraceae) and their implications for austral biogeography. New Zeal. J. Bot. 39: 595–606.
Wall, B. 1990. African Violets and related plants. London: Cassell.
Wang, H.S. 1989. A study of the origin of spermatophytoc genera endemic to China. Acta Bot. Yunnan. 11: 1–16.
Wang, W.T. 1985. The second revision of the genus *Petrocosmea* (Gesneriaceae). Acta Bot. Yunnan. 7: 49–68.
Wang, Y.Z. 2000. The growth pattern of shoots in *Whytockia* (Gesneriaceae) with phylogenetic implications. Acta Phytotax. Sin. 38: 231–235.
Wang, Y.Z. in prep. Flower morphology and development of *Whytockia*.
Wang, W.T., Pan, K.Y., Zhang, Z.Y., Li, Z.Y. 1990. Gesneriaceae. In: Wang, W.T. (ed.) Flora Reipublicae popularis Sinicae. Vol. 69. Beijing: Science Press, pp. 125–581.
Wang, W.T., Pan, K.Y., Li, Z.Y. 1992. Keys to the Gesneriaceae of China. Edinburgh J. Bot. 49: 5–74.
Wang, Y.Z., Gao, Z., Liang, H., Wu, Z. 1997. Floral morphogenesis of *Rhynchoglossum omeiense* (Gesneriaceae) and its phylogenetic implication. Acta Bot. Yunnan. 19: 265–270.
Wang, W.T., Pan, K.Y., Li, Z.Y., Weitzmann, A., Skog, L.E. 1998. Gesneriaceae. In: Wu, Z.Y., Raven, P.H. (eds.) Flora of China. 18. Scrophulariaceae through Gesneriaceae. Beijing: Science Press & St. Louis: Missouri Bot. Gard. Press.
Wang, Y.Z., Gu, Z.J., Hong, D.Y. 1998. Karyotypes of *Whytockia* (Gesneriaceae). Acta Phytotax. Sin. 36: 28–35.
Weber, A. 1971. Zur Morphologie des Gynoeceums der Gesneriaceen. Österr. Bot. Z. 119: 234–305.
Weber, A. 1973a. Über die "Stipeln" von *Rhytidophyllum* Mart. (Gesneriaceae). Österr. Bot. Z. 121: 107–109.
Weber, A. 1973b. Die Struktur der paarblütigen Partialfloreszenzen der Gesneriaceen und bestimmter Scrophulariaceen. Beitr. Biol. Pflanzen 49: 429–460.
Weber, A. 1975a. Beiträge zur Morphologie und Systematik der Klugieae und Loxonieae (Gesneriaceae). I. Die Sproß- und Infloreszenzorganisation von *Monophyllaea* R.Br. Bot. Jahrb. Syst. 95: 174–207.
Weber, A. 1975b. The cristate inflorescence of *Chirita* sect. *Microchirita* (Gesneriaceae). Notes Roy. Bot. Gard. Edinburgh 34: 221–230.
Weber, A. 1976a. Beiträge zur Morphologie und Systematik der Klugieae und Loxonieae (Gesneriaceae). II. Morphologie, Anatomie und Ontogenese der Blüte von *Monophyllaea* R.Br. Bot. Jahrb. Syst. 95: 435–454.
Weber, A. 1976b. Beiträge zur Morphologie und Systematik der Klugieae und Loxonieae (Gesneriaceae). III. *Whytockia* als morphologische und phylogenetische Ausgangsform von *Monophyllaea*. Beitr. Biol. Pflanzen 52: 183–205.
Weber, A. 1976c. Beiträge zur Morphologie und Systematik der Klugieae und Loxonieae (Gesneriaceae). IV. Wuchsform, Infloreszenz- und Blütenmorphologie von *Epithema*. Plant Syst. Evol. 126: 287–322.
Weber, A. 1977a. Beiträge zur Morphologie und Systematik der Klugieae und Loxonieae (Gesneriaceae). V. Revision der Gattung *Loxonia* (Gesneriaceae). Plant Syst. Evol. 127: 201–216.
Weber, A. 1977b. Beiträge zur Morphologie und Systematik der Klugieae und Loxonieae (Gesneriaceae). VI. Morphologie und Verwandtschaftsbeziehungen von *Loxonia* und *Stauranthera*. Flora 166: 153–175.
Weber, A. 1978a. Beiträge zur Morphologie und Systematik der Klugieae und Loxonieae (Gesneriaceae). VII. Sproß-, Infloreszenz- und Blütenbau von *Rhynchoglossum*. Bot. Jahrb. Syst. 99: 1–47.
Weber, A. 1978b. Beiträge zur Morphologie und Systematik der Klugieae und Loxonieae (Gesneriaceae). VIII. Ein typologischer Vergleich zwischen *Rhynchoglossum klugioides* und *Loxonia*. Linzer Biol. Beitr. 10: 217–228.
Weber, A. 1978c. Transitions from pair-flowered to normal cymes in Gesneriaceae. Notes Roy. Bot. Gard. Edinburgh 36: 355–368.
Weber, A. 1979. *Ornithoboea arachnoidea* – an "orchid-flowered" gesneriad. Gloxinian 29: 9–12.
Weber, A. 1980. Die Gattungsmerkmale von *Schizoboea* (Gesneriaceae-Didymocarpeae). Plant Syst. Evol. 134: 183–192.
Weber, A. 1982a. Contributions to the morphology and systematics of Klugieae and Loxonieae (Gesneriaceae). IX. The genus *Whytockia*. Notes Roy. Bot. Gard. Edinburgh 40: 113–121.
Weber, A. 1982b. Evolution and radiation of the pair-flowered cyme in Gesneriaceae. Austral. Syst. Bot. Soc. Newslett. 30: 23–41.
Weber, A. 1987. Organogenesis and morphological organization of *Monophyllaea singularis* (Gesneriaceae). In: Greuter, W. et al. Abstr. Vol. XIV Int. Bot. Congr. Berlin 1987, p. 219.
Weber, A. 1988. Contributions to the morphology and systematics of Klugieae-Loxonieae (Gesneriaceae). X. Development and interpretation of the inflorescence of *Epithema*. Beitr. Biol. Pflanzen 63: 431–451.
Weber, A. 1989a. *Didymocarpus geitleri*, a remarkable new species of Gesneriaceae with deceptive pollen flowers. Plant Syst. Evol. 165: 95–100.
Weber, A. 1989b. Family position and conjectural affinities of *Charadrophila capensis* Marloth. Bot. Jahrb. Syst. 111: 87–119.
Weber, A. 1990. *Monophyllaea singularis* – architecture and development of a singular plant. Abstr. Gesneriaceae Worksh., Montréal 1990. Montréal: Jard. Bot.
Weber, A. 1995. Developmental aspects of the pair-flowered cyme of Gesneriaceae. Gesneriana 1: 18–28.
Weber, H. 1936. Vergleichend-morphologische Studien über die sproßbürtige Bewurzelung. Nova Acta Leopold. 4(21): 230–298.
Weber, A., Burtt, B.L. 1983. *Didymocarpus corchorifolius* and its allies (Gesneriaceae). Blumea 28: 291–309.
Weber, A., Burtt, B.L. 1985. Supplementary notes on "*Didymocarpus corchorifolius* and its allies (Gesneriaceae)". Blumea 31: 155–159.

Weber, A., Burtt, B.L. 1998a. *Didissandra*: redefinition and partition of an artificial genus of Gesneriaceae. Beitr. Biol. Pflanzen 70: 153–177.
Weber, A., Burtt, B.L. 1998b. Revision of the genus *Didissandra* (Gesneriaceae). Beitr. Biol. Pflanzen 70: 191–223.
Weber, A., Burtt, B.L. 1998c. Revision of the genus *Ridleyandra* (Gesneriaceae). Beitr. Biol. Pflanzen 70: 225–273.
Weber, A., Burtt, B.L. 1998d. Remodelling of *Didymocarpus* and associated genera. Beitr. Biol. Pflanzen 70: 293–363.
Weber, A., Burtt, B.L. in prep. Revision of *Henckelia* sect. *Glossadenia* (Gesneriaceae).
Weber, A., Kiew, R. 1983. Gesneriads of Peninsular Malaysia. Nature Malaysiana 8(3): 24–31.
Weber, A., Till, S., Eberwein, R. 1992. Komplexe Blütenstände bei Acanthaceen: *Crabbea velutina* und *Dicliptera laxata*. Flora 187: 227–257.
Weber, A., Burtt, B.L., Vitek, E. 2000. Materials for a revision of *Didymocarpus* (Gesneriaceae). Ann. Naturhist. Mus. Wien 102B: 441–475.
Weber, A., Jang, Ch.G., Pfosser, M., Mayer, V., Kiehn, M., Perret, M., Möller, M. 2002. Classification of Old World Gesneriaceae: solved and open problems. In: Möller, M., Mendum, M., Cronk, Q.C.B. (eds.) Gesneriaceae Worksh., Talk Abstr., Royal Botanic Garden Edinburgh, 25–28 September 2002.
Wiehler, H. 1973. One hundred transfers from *Alloplectus* and *Columnea* (Gesneriaceae). Phytologia 27: 309–328.
Wiehler, H. 1978. Miscellaneous transfers and new species in neotropical Gesneriaceae. Selbyana 5: 61–93.
Wiehler, H. 1983. A synopsis of Neotropical Gesneriaceae. Selbyana 6: 1–219.
Wiehler, H. 1994. A re-examination of *Sanango racemosum*. 4. Its new systematic position in Gesneriaceae. Taxon 43: 625–632.
Wiehler, H. 1995. Medicinal Gesneriads. 122 species of the rain forest plant family Gesneriaceae used medicinally in the neotropics. Gesneriana 1: 98–120.
Wiehler, H., Chautems, A. 1995. A reduction of *Lietzia* to *Sinningia*. Gesneriana 1: 5–7.
Wikström, N., Savolainen, V., Chase, M.W. 2001. Evolution of the angiosperms: calibrating the family tree. Proc. Roy. Soc. Lond. B 268: 2211–2220, appendices A–B, pp. 1–15.
Williams, N.H. 1978. Pollen structure and the systematics of the neotropical Gesneriaceae. Selbyana 2: 310–322.
Wilson, C.L. 1974a. Floral anatomy in Gesneriaceae. I. Cyrtandroideae. Bot. Gaz. (Crawfordsville) 135: 247–256.
Wilson, C.L. 1974b. Floral anatomy in Gesneriaceae. II. Gesnerioideae. Bot. Gaz. (Crawfordsville) 135: 256–268.
Wollenweber, E., Rehse, C., Dietz, V.H. 1981. The occurrence of aurentiacin and flavokawin B in *Pityrogramma triangularis* var. *pallida* and *Didymocarpus* species. Phytochemistry 20: 1167–1168.
Wonisch, F. 1909a. Die Sekretgänge von *Monophyllaea*, *Klugia* and *Rhynchoglossum* (Gesneriaceae). Österr. Bot. Z. 59: 209–215, pl. 1.
Wonisch, F. 1909b. Über den Gefässbündelverlauf bei den Cyrtandroideen. Sitzungsber. Kaiserl. Akad. Wiss., Math.-Naturwiss. Cl., Abt. 1, 118: 453–486.
Wunderlin, U. 1992. Untersuchungen zur vergleichenden Entwicklungsgeschichte von Scrophulariaceen-Blüten. Die Entwicklung der Blüten von *Digitalis lanata*, *Digitalis lutea*, *Digitalis isabelliana*, *Erinus alpinus*, *Wulfenia baldaccii*, *Alonsoa warscewiczii* und *Nemesia capensis*. Diss. Bot. 188: 1–313.
Ying, T.S., Zhang, Y.L., Boufford, D.E. 1993. The endemic genera of seed plants of China. Beijing: Science Press.
Young, N.D., Steiner, K.E., dePamphilis, C.W. 1999. The evolution of parasitism in Scrophulariaceae/Orobanchaceae: plastid gene sequences refute an evolutionary transition series. Ann. Missouri Bot. Gard. 86: 876–893.
Zimmer, E.A., Roalson, E.H., Skog., L.E., Boggan, J.K., Idnurm, A. 2002. Phylogenetic relationships in the Gesnerioideae (Gesneriaceae) based on nrDNA ITS and cpDNA *trn*L-F and *trn*E-T spacer region sequences. Am. J. Bot. 89: 296–311.

Globulariaceae[1]

G. Wagenitz

Globulariaceae DC. in Lam. et DC., Fl. Franç., ed. 3. 3: 427 (1805) ("Globulariae"), nom. cons.
Scrophulariaceae trib. Globularieae (DC.) Barringer, Novon 3: 17 (1993).

Shrubs, subshrubs or perennial herbs with woody base. Leaves simple, alternate, coriaceous, without stipules. Inflorescence usually a capitulum with an involucre, rarely a spike (*Poskea*). Flowers subtended by bracts, zygomorphic or nearly actinomorphic, bisexual, hypogynous; calyx with five acute lobes, corolla sympetalous, blue, usually bilabiate; stamens 4, inserted in the upper part of the corolla tube, often two of them higher up, alternate with the lobes; anthers reniform, with two confluent, bisporangiate loculi, opening by a single slit; ovary superior, usually surrounded by a disk or several glands, one-locular with one, mostly shortly bifid style; ovule solitary, anatropous, pendulous. Fruit small, indehiscent, remaining in the persistent calyx. Seed with a straight long embryo surrounded by scanty endosperm.

Two genera and ca. 25 species mainly in the Mediterranean region and S Europe (*Globularia*), and Somalia and Socotra (*Poskea*).

Vegetative Morphology. Globulariaceae are generally perennial plants with at least a woody base, ranging from small shrubs to subshrubs or rosette plants (Fig. 19A). When the basal innovation takes place with very short branches, small cushion shrubs (*G. dumulosa*) result, when long branches are appressed to the ground, espalier dwarf shrubs are formed (*G. cordifolia*). The stems can have adventitious roots. In *G. nudicaulis*, large-leaved rosettes are formed while other species are half-rosette plants. Leaves are always coriaceous and persisting for several years.

Vegetative Anatomy. The anatomy of the leaves has been studied by Kupffender (1891) and Heckel (1894). The structure is variable in *Globularia*. The lamina is dorsiventral or centric, and the chlorenchyma in the latter type can consist of palisade or isodiametric cells. Stomata are present on both surfaces and usually ranunculaceous, although other types are sometimes intermixed. The epidermis frequently contains chloroplasts. Small glandular hairs with a short stalk cell and a head of two or four cells in one plane are characteristic. In some species, they secrete calcareous material. Uniseriate hairs occur, too, these are glandular in *Poskea*. Branched sclerenchymatous idioblasts are known from the mesophyll of *G. orientalis*. The structure of the principal vein has been used as a systematic character. The fascicle is either connected to the epidermis on both sides by collenchyma, or it is surrounded by mesophyll. In the cortex of the root of several species, idioblasts with partly thickened walls are found (Luhan 1954).

The anatomy of the wood of *G. salicina* has more recently been studied by Carlquist (1992). Significant is the presence of tracheids (instead of fibre-tracheids or libriform fibres) associated with a lack of vessel grouping. Diffuse axial parenchyma is present in addition to scanty vasicentric parenchyma. The rays are of the Heterogeneous Type IIA. Tracheids and diffuse axial parenchyma are remarkably primitive characters not known in the Scrophulariaceae.

Inflorescence Structure. Inflorescences are polytelic (without terminal flower) and either spikes (*Poskea*) or capitula (*Globularia*; Fig. 19A). Bracts are present. In *Globularia* several sterile bracts form an involucre at the base of the capitulum. The inflorescences are terminating the primary stem and the branches in *Poskea* and most species of *Globularia*. In the Macaronesian species of *Globularia* sect. *Lytanthus*, the capitula are only formed on short lateral branches.

Flower Structure. The flowers of *Globularia* are remarkably variable, especially in their lower lip. Variation between and within species has been shown by Schwarz (1938; Fig. 19B–E).

Flower Anatomy. Development and anatomy of the flower have been studied by Eckardt (1937). He

[1] See also Scrophulariaceae.

Fig. 19. Globulariaceae. **A** *Globularia vulgaris*, habit and flower. **B** *Poskea socotrana*. Corolla. **C** *Globularia salicina*. Corolla. **D** *Globularia repens*. Corolla. **E** *Globularia nudicaulis*. Corolla. (**A** Folch i Guillèn 1981, **B–E** Schwarz 1938)

could show that the ovary is pseudomonomerous with an adaxial solid carpel (rarely with a cavity) and an abaxial fertile one.

Embryology. The embryological literature is scanty, and only *G. vulgaris* has been studied in detail by Rosén (1940) and Crété (1943). The single ovule is pendulous (micropyle pointing upwards), anatropous, unitegmic and tenuinucellate. The chalazal megaspore develops into an embryo sac of the Polygonum type, surrounded by an endothelium. Endosperm development is of the cellular type with micropylar and chalazal haustoria. The large chalazal haustorium can break through the integument. Somatic apospory has been noted by Rosén (1940). The straight embryo is lying centrally in the endosperm.

Pollen Morphology. Pollen morphology is well known mainly through the studies by Praglowski and Gyllander (1970), Punt and Marks (1991) and Argue (1993). *Globularia* is a stenopalynous genus, and the pollen of *Poskea socotrana* is similar to that of *Globularia*. The main features are: pollen grains tricolporate, usually prolate spheroidal to subprolate, pertectate (subtectate in *Poskea*), infrabaculate, scabrate (smooth in *Poskea*). The colpi are long and sunken with thin, usually somewhat irregular margins and a granular or spinulose membrane, the endoaperture is lolongate and not very distinct. The exine is 1–1.5 μm thick, and the sexine has a baculate zone and a tectum with very fine scabrae and minute perforations difficult to see with the light microscope.

The pollen morphology of the Scrophulariaceae–Selagineae is superficially similar to that of Globulariaceae but differs by its reticulate sexine and the nearly smooth colpus membrane.

Karyology. Chromosome numbers are known for 17 species of *Globularia*. The base number is $x = 8$ with $2n = 16$ and 32. Other numbers (probably miscounts) have been reported only rarely. Tetraploidy occurs mainly within species.

Hybridisation. Hybridisation apparently is rare in *Globularia*. *Globularia fuxeensis* seems to be a rare stabilized hybrid between *G. repens* and *G. nudicaulis* (Contandriopoulos and Cauwet 1968). Two other hybrids are reported from the Canary Islands, and from the Pyrenees.

Pollination. No detailed study is available. According to older literature reporting observations mainly in the Alps, butterflies are the most important pollinators. Other insects with a sufficiently long proboscis can also feed on the nectar secreted by the disc. Other flower visitors probably only collect pollen.

Fruit, Seed and Dispersal. The small indehiscent fruits with a membranous pericarp have no recognizable adaptations for dispersal. A study from the Canary Islands (Gonzalez Martin et al.

1994) showed low seed set combined with a high percentage of viable seeds (with embryo).

PHYTOCHEMISTRY. There has been relatively little interest in the chemistry of the family, as it contains no commercially used plants and scarcely any medicinal ones. Only recently especially the iridoid substances have been studied more intensively. Aucubin has been known to occur as early as 1934, and is probably responsible for the widespread tendency of the leaves to turn black when dried. Further iridioids present are globularin, catalpol, globularicisin, globularidin, globularimin, globularinin, globularifolin, asperulin, and lytanthosalin (cf. Hegnauer 1989).

AFFINITIES. A close affinity to Scrophulariaceae s.l. is undisputed but must be reconsidered in the light of recent molecular systematic studies of this and allied families. Only one study is available including *Globularia* (Oxelman et al. 1999). From this it appears that *Globularia* splits off early in the "Scrophulariaeae II" (Plantaginaceae) clade. The similarities between *Selago* and *Globularia*, especially the unilocular ovary, are apparently due to parallelism. The wood anatomy (Carlquist 1992) is of special interest in this context. Several characters considered as primitive occur in *Globularia* but not in the Scrophulariaceae s.l.

DISTRIBUTION AND HABITATS. The Globulariaceae have their centre of distribution in the Mediterranean region and S Europe, with a considerable number of locally endemic species of *Globularia* in the Cape Verde and Canary Islands, the Atlas, the Baleares, Italy, Greece and Turkey. *Globularia vulgaris* is peculiarly disjunct by occurring in Sweden (Öland and Gotland), France and Spain. The genus *Poskea* is confined to Somalia and Socotra. The species are mostly found in open, dry and often rocky habitats.

PARASITES. No galls on *Globularia* are known. The smut *Puccinia globulariae* occurs only on *Globularia* and has an isolated position within *Puccinia*. Other fungi on *Globularia* are *Trichodothella blumeri*, *Septoria globulariae* and two species of *Phyllosticta*.

CONSERVATION. Several species of *Globularia* have restricted distribution areas and are endangered as, e.g., *G. stygia* in Greece and *G. neapolitana* in Italy.

KEY TO THE GENERA

1. Inflorescence a dense spike. Corolla nearly actinomorphic **1. *Poskea***
– Inflorescence a capitulum with an involucre. Corolla distinctly zygomorphic (bilabiate) **2. *Globularia***

1. *Poskea* Vatke Fig. 19B

Poskea Vatke, Linnaea 43: 321 (1882); Taylor, J. Bot. 71: 310–312 (1933) rev.; Braggio Morucchio, Webbia 24: 624–634 (1970), rev.
Cockburnia Balf.f. (1883).

Low shrubs with coriaceous undivided leaves. Flowers small, in dense spikes at the end of the branches. Calyx with five lanceolate teeth, corolla divided up to 1/4 or 1/3 into five equal or slightly unequal lobes, four stamens inserted near the middle of the corolla or above, ovary ovoid with persistent style with slightly capitate stigma. Three species in northern Somalia and Socotra, two of these are only known from a single collection.

2. *Globularia* L. Fig. 19A, C–E

Globularia L., Spec. Pl. 1: 95 (1753); Schwarz, Bot. Jahrb. Syst. 69: 318–373 (1938), rev.; Schwarz, Drudea 3: 5–16 (1963), cytol. evol.; Holländer & Jäger, Flora 189: 223–254 (1994), morph., chorol.; Holländer & Jäger, Hercynia N.F. 31: 143–171 (1998), *G. bisnagarica* life-history.
Carradoria DC. (1848).
Lytanthus Wettst. (1895).

Low shrubs, subshrubs or perennials with woody base. Leaves coriaceous, simple, lanceolate to suborbicular or spathulate, sometimes emarginate or with few teeth in the upper part. Inflorescence a capitulum with bracts and an involucre at the end of the stem or on short lateral branches with the end of the stem remaining vegetative. Flowers zygomorphic, calyx with five acute lobes as long as or longer than the tube, corolla bilabiate, the larger upper lip with three short teeth or deeply laciniate, the lower lip consisting of two linear lobes 2/3 as long as the upper lip or much shorter, sometimes completely lacking or only one lobe present; stamens four, inserted in the tube; ovary ovoid, style simple and shortly bifid at the apex. Fruit indehiscent included in the persisting calyx. A genus of 22 to 23 species mainly in the Mediterranean area extending to Macaronesia, northern Europe and north-eastern Africa.

Selected Bibliography

Argue, C.L. 1993. Pollen morphology in the Selagineae, Manuleae (Scrophulariaceae) and selected Globulariaceae, and its taxonomic significance. Am. J. Bot. 80: 723–733.

Carlquist, S. 1992. Wood anatomy of sympetalous dicotyledon families: a summary, with comments on systematic relationships and evolution of the woody habit. Ann. Missouri Bot. Gard. 79: 303–332.

Contandriopoulos, J., Cauwet, A.-M. 1968. A propos du *Globularia fuxeensis* Gir. des Pyrénées. Nat. Monspel. Sér. Bot. 19: 29–35.

Crété, P. 1943. Recherches histologiques et physiologiques sur l'embryologie des Labiatiflores. Embryologie du *Globularia vulgaris* L. Bull. Soc. Bot. France 90: 29–32, 36–39.

Eckardt, T. 1937. Untersuchungen über Morphologie, Entwicklungsgeschichte und systematische Bedeutung des pseudomonomeren Gynoeceums. Nova Acta Leopold. N.F. 5(26): 2–112.

Folch i Guillèn, R. 1981. La vegetació dels països catalans. Barcelona: Inst. Catalan. Hist. Nat. Mem. 10.

Gonzalez Martin, M., Cabrera Perez, M.A., Gonzalez Artiles, F.J. 1994. Germination of Canarian species of *Globularia* L. Inv. Agrar. Producc. Protecc. Veg. 9: 29–34.

Heckel, E. 1894. Étude monographique de la famille des Globulariacées au point de vue botanique, chimique et thérapeutique. Paris: Ann. Fac. Sci. Marseille Suppl. 3.

Hegnauer, R. 1989. Chemotaxonomie der Pflanzen. Vol. 8 (Globulariaceae, pp. 520–522). Basel: Birkhäuser.

Kupffender, H. 1891: Beiträge zur Anatomie der Globulariaceen und Selaginaceen und zur Kenntnis des Blattcambiums. Diss. Phil. Kiel: Univ. Erlangen.

Luhan, M. 1954. Über das Vorkommen von Sklerenchym-Idioblasten bei *Globularia*-Arten. Ber. Deutsch. Bot. Ges. 67: 346–355.

Oxelman, B., Backlund, M., Bremer, B. 1999. Relationships of the Buddlejaceae s.l. investigated using parsimony jackknife and branch support analysis of chloroplast *ndhF* and *rbcL* sequence data. Syst. Bot. 24: 164–182.

Praglowski, J., Gyllander, K. 1970. Globulariaceae. In: World Pollen Flora 4: 1–21.

Punt, W., Marks, A. 1991. The northwest European pollen flora, 49. Globulariaceae. Rev. Palaeobot. Palynol. 69: 109–112.

Rosén, W. 1940. Notes on the embryology of *Globularia vulgaris* L. Bot. Notis. 1940: 253–261.

Schwarz, O. 1938. Die Gattung *Globularia*. Bot. Jahrb. Syst. 69: 318–373.

Hippuridaceae[1]

P. LEINS and C. ERBAR

Hippuridaceae Vest, Anleit. Stud. Bot.: 265, 278 (1818), nom. cons.

Aquatic or amphibious, rarely terrestrial herbs with creeping rhizome. Stems usually erect and emergent or sometimes floating. Leaves verticillate (4–16 per whorl), linear, estipulate. Stems and leaves with peltate glandular hairs. Flowers small, bisexual or rarely unisexual, solitary in the axils of the upper aerial leaves. Perianth a low collar. Stamen solitary, inserted on abaxial side of flower, with slender filament and tetrasporangiate, introrse anther. Gynoecium inferior, monocarpellate (or pseudomonomerous), style elongate. Ovule one, pendulous, anatropous, unitegmic, tenuinucellate. Fruit a drupe. Fruit wall consisting of cup-shaped sclerified pericarp topped by a plug formed by sclerified funicle and upper part of integument.

A single genus and species, widely distributed in temperate and boreal regions of the Northern Hemisphere, and in Australia and southern S America, with ecological races in the Arctic and Baltic.

VEGETATIVE MORPHOLOGY. The perennial, aquatic, amphibious or rarely terrestrial herbs have a stout, submersed, creeping, sympodial rhizome, from which many erect branches (30–60 cm, rarely 150 cm high) arise. The shoots bear closely set whorls of leaves (Fig. 20A). Other shoots may be floating and more than 2 m long. The erect shoots bear flowers when emerging above the water surface; the uppermost part of the shoot, however, is sterile. After flowering they die down to the rhizome. The non-flowering submersed shoots usually remain green throughout the winter. The submersed leaves are long (5–8 cm), linear, pale green and flaccid. The emergent leaves are shorter (ca. 20 mm), obovate to linear, dark green and rigid (Schindler 1904).

The species is very variable and there are some ecologically specialized races in the Arctic and the Baltic Sea. One form (described as *H. tetraphylla*) has only four very short, obovate leaves per whorl. Leaf number and shape, however, are to some extent under environmental control and can be influenced by high salinity or low temperature.

VEGETATIVE ANATOMY. Phloem plastids are of the S-type (Behnke 1991).

The anatomical structure of the plant is that of a hygrophyt. The vascular system is reduced to an axile strand of thin-walled tissue, the outer part consisting of a narrow zone of phloem and the inner part of a broader region of xylem. The vessels are small in diameter, with spiral or reticulate thickenings. The erect stems have a well-defined endodermis. The broad cortex, except for two subepidermal cell layers, is permeated by an anastomosing system of vertically elongated, intercellular spaces.

In the leaves, the outer part of the mesophyll, with only poorly developed palisade tissue, is interspersed by numerous air cavities which are especially well developed below the stomata. The palisade tissue is not clearly differentiated from the spongy parenchyma of the centre of the mesophyll. Vascular bundles are poorly developed, embedded in the mesophyll, and surrounded by a single-layered sheath of parenchymatous cells. Stomata occur on both surfaces (Schindler 1904; Metcalfe and Chalk 1957).

Peltate hairs are almost entirely confined to the upper side of the leaves and the shoot. The peltate hairs consist of a unicellular foot cell and a multicellular, disc-like head (Fig. 20C, D). In the leaf axil, the stalk cell supports several cells in a fan-like arrangement (Fig. 20E). The peltate hairs are glandular and the mucilage protects young leaves. On the aerial part of the plant the hairs are deciduous, the head breaking off from the unicellular stalk. The latter is partly sunk in the epidermis, and its wall is strongly cutinised, whereas the cell walls of the disc cells are not cutinised (Solereder 1899; Uphof 1962).

FLOWER STRUCTURE. The small and inconspicuous flowers (green with dark red anther) are borne solitary in the axil of leaves of the aerial part of the

[1] See also Scrophulariaceae.

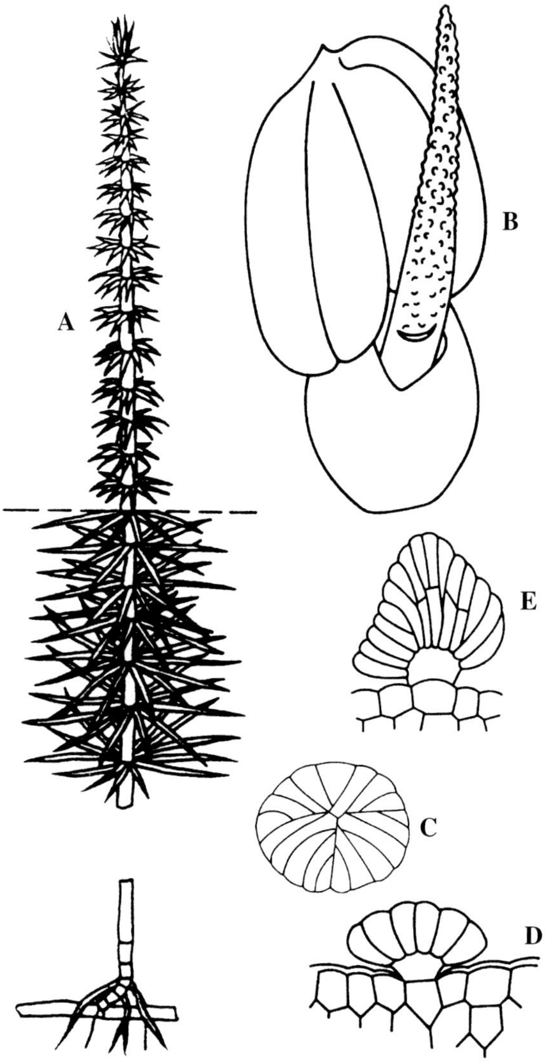

Fig. 20. Hippuridaceae. *Hippuris vulgaris*. **A** Habit. **B** Flower. **C** Peltate hair in surface-view. **D** Peltate hair in longisection. **E** Fan-shaped peltate hair. (**A** After Rothmaler 1987. **B, C** original. **D, E** after Solereder 1899)

stem. The upper flowers are sessile, the lower ones usually shortly stalked. The flowers are generally bisexual (Fig. 20B), but sometimes unisexual by failure of stamen or pistil development. The perianth is reduced to a low collar that surrounds the superior parts of the flower. The single stamen is inserted on the abaxial side of the flower and has a massive, tetrasporangiate, introrse anther which opens by two longitudinal slits. The inferior 1-locular ovary contains a single ana- and epitropous ovule pendulous from the adaxial ovary wall. The ring-like gynoecial primordium passes through a sloping, cup-like stage with a transversal slit. The abaxial part overtops the adaxial one and develops into the style with a basal transversal slit (Leins and Erbar 1988; Leins 2000). The style is long, slender and lies in the groove between the thecae of the anther. It has stigmatic papillae throughout its length.

The inferior gynoecium can be either regarded as monocarpellate or interpreted as pseudomonomerous (Leins and Erbar 1988, 1990).

Embryology. The anther is tetrasporangiate. Cells of the secretory tapetum become 2-nucleate. The pollen grains are 3-celled when shed.

The ovule is anatropous, unitegmic and tenuinucellate. The apical cells of the nucellus undergo periclinal divisions and form a nucellar cup. The rest of the nucellus is only few-celled. The large integument forms a long and narrow micropylar canal which is difficult to trace through its entire length. At the base of the funicle, a tissue plug forming a lid on the micropyle develops.

The archesporial cell functions directly as the megaspore mother cell. Cytokinesis in the megaspore mother cell accompanies meiosis. In its further development, the chalazal megaspore of a linear tetrad follows the Polygonum type. The synergids are filiform, the polar nuclei fuse early, and the three antipodes persist until after fertilization.

The funicular lid closes the micropyle, and the pollen tube grows along the surface of the funicle and the integument. Shortly after reaching the integument, the pollen tube enters the ovule by growing through the integument. At the upper part of the embryo sac, it bends and penetrates the embryo sac laterally (Juel 1911).

Endosperm formation is ab initio cellular. Embryogeny follows the Onagrad type, and a large haustorial suspensor is formed (Juel 1910, 1911; Davis 1966).

Pollen Morphology. The pollen grains are 4–6-colpoidate, suboblate and have a thin sexine (Erdtman 1966).

Karyology. The most frequently reported chromosome number is $2n = 32$ (e.g. Federov 1969). There are also counts of $2n = 16$ (see Moore 1973) and $2n = 30$ (Harada 1952).

Pollination and Reproductive Systems. The flowers are proterogynous and wind-pollinated. The filament elongates only after the style has dried up.

Fertilization seems to occur only rarely; mature fruits develop abundantly in some years only.

Thus, vegetative propagation is of high importance (Schindler 1904).

FRUIT AND SEED. The fruit of *Hippuris* is a drupe. The 2–3 mm large fruits are ovoid and smooth. The pericarp consists of a thin, fleshy exocarp and a hard endocarp. The pericarp below the persistent base of the style has no thickened cell walls, so that the sclerified tissue forms a cup. The opening of this cup is closed by a plug formed by the funicle and the integument around the micropyle which become sclerified. The other parts of the testa are only thin and membranous. The plug is opened by the radicle during germination (Juel 1910, 1911).

DISPERSAL. The mature drupes fall off and sink into the mud where they pass the winter. The fleshy parts decay.

Adaptations to water dispersal are wanting but the diaspores may be transported over short distances by currents or wind. Long-distance dispersal may be effected by waterfowl carrying the diaspores embedded in mud on their feet. Occasionally, the fruits may be frozen in the ice sheet and dispersed by ice during melting (Hegi 1926).

PHYTOCHEMISTRY. *Hippuris* produces the carbocyclic iridoids aucubin and catalpol (Wieffering 1966). The phenolic compounds caffeic and ferulic acids, kaempferol and scopoletin are accumulated. Carbohydrates are stored as stachyose (Hegnauer 1967).

AFFINITIES. Previously, Hippuridaceae have been associated with Haloragaceae (e.g. De Candolle 1828). Primarily on the basis of embryological and phytochemical evidence, the family has been accepted as part of the Asteridae (around the Lamiales-Scrophulariales; e.g. Takhtajan 1987; Thorne 1992). Cronquist (1981) united Hippuridaceae, Callitrichaceae and Hydrostachyaceae into one order Callitrichales near Plantaginales and Scrophulariales. A position near Cornales (Thorne 1968) or Cornales-Dipsacales (Leins and Erbar 1990) has also been suggested on the basis of gynoecial morphology.

Results from DNA sequences reveal that Hippuridaceae (together with Callitrichaceae) are an aquatic clade within Scrophulariaceae (Olmstead and Reeves 1995). Since molecular studies provide evidence that Scrophulariaceae, as circumscribed in most taxonomic treatments, are polyphyletic, the family has been disintegrated into 5 families (Olmstead et al. 2001). The new Veronicaceae (Antirrhinaceae in Savolainen et al. 2000, see also Reveal et al. 1999) include (besides traditional members of Scrophulariaceae such as *Antirrhinum*, *Digitalis*, and *Veronica*) Plantaginaceae, Globulariaceae, Callitrichaceae and Hippuridaceae. Nevertheless, support for Veronicaceae is weak, whereas support for the clade that includes *Antirrhinum* with Globulariaceae, Plantaginaceae, Callitrichaceae, and Hippuridaceae is very strong (Olmstead et al. 2001). In this clade, *Hippuris* and *Callitriche* form a subclade; *Hippuris* is the closest relative of *Callitriche*. Whereas Veronicaceae at present lack a morphological family circumscription, the *Hippuris*/*Callitriche* clade is associated with distinctive characters such as aquatic habit and very reduced floral morphology resulting from either wind or water pollination (Olmstead and Reeves 1995; Reeves and Olmstead 1998; Olmstead et al. 2001).

DISTRIBUTION AND HABITATS. The usually amphibious, rarely submersed herbs grow in still or slow-flowing, preferably base-rich waters (up to 2.5 m depth) like streams, ponds, shallow lakes, and mossy banks. The terrestrial form is found in wet meadows and on mud that can dry out.

The (broadly defined) species is widely distributed throughout the temperate and cold regions of the Northern Hemisphere (Europe to northern Iceland and northern Norway; Asia; N America; Greenland) and occurs also in Antarctic America and Australia.

PARASITES. *Hippuris vulgaris* is the host of *Uromyces lineatus* which produces aecidia (Hegnauer 1989).

ECONOMIC IMPORTANCE. The submersed shoots remain green in winter and are an important winter food for many animals. The Eskimos gather and eat young shoots.

One monotypic genus:

Hippuris L. Fig. 20

Hippuris L., Syst. ed. I (1735), Sp. Pl. 1: 4 (1753).

Characters as for the family.

Monotypic, *H. vulgaris* L., northern hemisphere, Antarctic, S Australia.

Selected Bibliography

Behnke, H.D. 1991. Distribution and evolution of forms and types of sieve-element plastids in the dicotyledons. Aliso 13: 167–182.

Cronquist, A. 1981. An integrated system of classification of flowering plants. New York: Columbia University Press.

Davis, G.L. 1966. Systematic embryology of the angiosperms. New York: Wiley.

De Candolle, A.P. 1828. Prodromus Systematis Naturalis Regni Vegetabilis III. Paris: Treuttel & Würtz, pp. 65–72.

Erdtman, G. 1966. Pollen morphology and plant taxonomy. New York: Hafner Publ. Co.

Federov, Al.A. (ed.) 1969. Chromosome numbers of flowering plants. Leningrad: Nauka.

Harada, I. 1952. Chromosome studies of some dicotyledonous water plants. Jap. J. Genet. 27: 117–120.

Hegi, G. 1926: Illustrierte Flora von Mitteleuropa vol. 2. München: Lehmann.

Hegnauer, R. 1967. Chemical characters in plant taxonomy: some possibilities and limitations. Pure Appl. Chem. 14: 173–187.

Hegnauer, R. 1989. Chemotaxonomie der Pflanzen VIII. Basel: Birkhäuser.

Juel, O. 1910. *Cynomorium* und *Hippuris*. Svensk Bot. Tidskr. 4: 151–159.

Juel, O. 1911. Studien über die Entwicklung von *Hippuris vulgaris*. Nova Acta Regiae Soc. Sci. Upsal. IV, 2, N. 11: 1–26.

Leins, P. 2000. Blüte und Frucht. Aspekte der Morphologie, Entwicklungsgeschichte, Phylogenie, Funktion und Ökologie. Stuttgart: Schweizerbart'sche Verlagsbuchhandlung.

Leins, P., Erbar, C. 1988. Einige Bemerkungen zur Blütenentwicklung und systematische Stellung der Wasserpflanzen *Callitriche*, *Hippuris* und *Hydrostachys*. Beitr. Biol. Pflanzen 63: 157–178.

Leins, P., Erbar, C. 1990. The possible relationship of Hydrostachyaceae based on comparative ontogenetical flower studies. Mitt. Inst. Allg. Bot. Hamburg 23b: 723–729.

Metcalfe, C.R., Chalk, L. 1957. Anatomy of the dicotyledons II. Oxford: Clarendon Press.

Moore, R.J. (ed.) 1973. Index to plant chromosome numbers 1967–1971. Regnum Vegetabile 90.

Olmstead, R.G., Reeves, P.A. 1995. Evidence for the polyphyly of the Scrophulariaceae based on chloroplast *rbc*L and *ndh*F sequences. Ann. Missouri Bot. Gard. 82: 176–193.

Olmstead, R.G., DePamphilis, C.W., Wolfe, A.D., Young, N.D., Elisons, W.J., Reeves, P.A. 2001. Disintegration of the Scrophulariaceae. Am. J. Bot. 88: 348–361.

Reeves, P.A., Olmstead, R.G. 1998. Evolution of novel morphological and reproductive traits in a clade containing *Antirrhinum majus* (Scrophulariaceae). Am. J. Bot. 85: 1047–1056.

Reveal, J.L., Judd, W.S., Olmstead, R.G. 1999. Proposal to conserve the name Antirrhinaceae against Plantaginaceae (Magnoliophyta). Taxon 48: 182.

Rothmaler, W. 1987. Exkursionsflora. Band 3: Atlas der Gefäßpflanzen, 6. Aufl. Berlin: Volk und Wissen.

Savolainen, V., Fay, M.F., Albach, D.C., Backlund, A., van der Bank, M., Cameron, K.M., Johnson, S.A., Lledo, M.D., Pintaud, J.-C., Powell, M., Sheahan, M.C., Soltis, D.E., Soltis, P.S., Weston, P., Whitten, W.M., Wurdack, K.J., Chase, M.W. 2000. Phylogeny of the eudicots: a nearly complete familial analysis based on *rbc*L gene sequences. Kew Bull. 55: 257–309.

Schindler, A.K. 1904. Die Abtrennung der Hippuridaceen von den Halorrhagaceen. Bot. Jahrb. Syst. 34, Beibl. 77: 1–77.

Solereder, H. 1899. Systematische Anatomie der Dicotyledonen. Stuttgart: Enke.

Takhtajan, A. 1987. Systema Magnoliophytorum (in Russian). Leningrad: Nauka.

Thorne, R.F. 1968. Synopsis of a putatively phylogenetic classification of the flowering plants. Aliso 6: 57–66.

Thorne, R.F. 1992. Classification and geography of the flowering plants. Bot. Rev. 58: 225–348.

Uphof, J.C.Th. 1962. Plant hairs. In: Zimmermann, W., Ozenda, P.G. (eds.) Encyclopedia of plant anatomy IV, 5, 2nd edn. Berlin-Nikolassee: Bornträger.

Wieffering, J.H. 1966. Aucubinartige Glucoside (Pseudoindikane) und verwandte Heteroside als systematische Merkmale. Phytochemistry 5: 1053–1064.

Labiatae

R.M. Harley,[1] S. Atkins, A.L. Budantsev, P.D. Cantino, B.J. Conn, R. Grayer, M.M. Harley, R. de Kok, T. Krestovskaja, R. Morales, A.J. Paton, O. Ryding and T. Upson

Labiatae Adans., Fam. Pl. 2: 180 (1763).
Lamiaceae Martynov (1820), nom. alt.

Trees, shrubs, subshrubs or perennial or annual herbs, rarely climbers, aromatic or not. Roots rarely tuberous. Stems often quadrangular, erect to prostrate, sometimes forming stolons or large or slender rhizomes. Indumentum usually present, of glandular and non-glandular trichomes, often hair-like, rarely scale-like, usually multicellular-uniseriate, simple, branched, dendroid or stellate, sometimes gland-tipped, large-headed subsessile glands rarely absent.

Leaves opposite, often decussate, sometimes whorled, very rarely alternate, simple, entire, toothed or lobed, sometimes compound and then digitate or pinnate, petiolate or sessile, rarely forming a basal rosette, estipulate. Inflorescence often bracteate, bracts sometimes conspicuous, persistent or deciduous, rarely (*Lavandula* and *Scutellaria*) spirally arranged, composed of cymes, bracteolate or not, and often arranged in a terminal, lax or congested, indeterminate thyrse which may be paniculate, raceme-like with cymes often 1-flowered, or spike-like, or rarely congested into a capitulum, with or without a distinct involucre of bracteoles. Flowers hypogynous, usually bisexual, or less often unisexual due to gynodioecy or gynomonoecy, very rarely due to dioecy. Perianth biseriate, sepals 4–5(–9), connate, actinomorphic to zygomorphic, sometimes 2-lipped, lobes 2–many, often 5, equal or unequal, rarely obsolete, some lobes often fused, or lips entire, calyx-tube (5–)10–15-nerved, straight or curved, throat hairy or glabrous, calyx often accrescent, rarely inflated or fleshy in fruit. Petals (4–)5(–16) connate, actinomorphic to more often slightly to strongly zygomorphic, often 2-lipped, rarely 1-lipped, lobes (2–)4–5(–16), equal or unequal, porrect to patent, one or other lip often concave to galeate, corolla-tube short to elongate, rarely spurred, often with annulus of hairs or appendaged within, rarely corolla resupinate. Stamens attached within corolla-tube, usually 4 or 2 by abortion and then staminodes often present, or stamens 5–8(–16), when 4 often didynamous (rarely a fifth, posterior vestigial staminode present), free or rarely monodelphic, filaments short or often elongate, usually exserted from corolla-tube and sometimes long-exserted from corolla; parallel, divergent or ascending and sometimes included within or lying under the posterior corolla-lip, or declinate and then sometimes included within the anterior corolla-lip, anthers usually dithecous, tetrasporangiate or monothecous by abortion, thecae parallel or divergent, occasionally widely separated by an elongate connective, or apically confluent or syn thecous, opening by longitudinal slits or rarely by pores. Disc at base of ovary often present, usually fleshy, entire or irregularly or often 4-lobed, anterior lobe sometimes longer than others, nectariferous. Gynoecium 2-carpellate, often 4-locular by intrusion of carpel wall forming "false septum", or rarely imperfectly 2-locular and free towards apex, ovary usually 4-ovuled, 2-locular ovaries generally with loculi 2-ovuled, and 4-locular ovaries with 1 ovule per loculus, ovary entire or lobed, with terminal style, or more often deeply 4-lobed, the loculi often separated and with style gynobasic, style usually with 2 equal or unequal stigma-lobes, rarely entire with 1 stigma-lobe vestigial, or stigma capitate or very rarely 4-lobed. Ovules anatropous to hemianatropous, usually basal or sub-basal, erect, rarely orthotropous, apical, pendulous, borne laterally or submarginally on the placenta, unitegmic, tenuinucellate. Fruit drupaceous, often with pyrenes, or dry, indehiscent, or separating into two 2-seeded or frequently, four 1-seeded mericarps, sometimes fewer by abortion. Mericarps (nutlets) often with sculptured, tuberculate, hairy or rarely winged pericarp, mucilage cells often present. Seeds albuminous or exalbuminous, epigeal. Embryo straight or bent, investing or spatulate.

The family contains 236 genera and about 7173 species, almost cosmopolitan, but absent from the coldest regions of high latitude or altitude.

[1] For details of authorship see p. 275.
[2] In the generic descriptions, calyx and corolla-lobe numbers of zygomorphic flowers are often represented by a formula, with the number of lobes in the posterior lip followed by that of the anterior lip, e.g. 3/2.

Vegetative Morphology. Labiatae are composed of annual, biennial and perennial herbs, subshrubs, shrubs or trees, including large forest forms (e.g. *Gmelina*, *Vitex*, *Tectona* spp.), and even woody climbers, especially Symphorematoideae, as well as some Viticoideae, Ajugoideae and *Petraeovitex*. The range in habit includes herbs with prostrate to ascending stems, sometimes with specialised stolons or rhizomes or stem tubers (especially in some species of *Plectranthus*, *Stachys*), and occasionally species with leaves crowded into a basal rosette (e.g. *Horminum pyrenaicum*, *Hyptis pedalipes*, *Salvia taraxacoides*). Subshrubs may possess a stout, woody, subterranean caudex which can give rise to many aerial stems (e.g. *Ziziphora clinopodioides*, *Eriope obovata*). Stem internodes are often quadrangular in cross section, the angles sometimes winged, and rarely taking over the function of photosynthesis, with leaves much reduced (e.g. some *Rhabdocaulon* spp.). Some genera have species with fistulose internodes, these rarely much inflated and covered in a waxy bloom in species of *Eriope*, *Hypenia* (Harley 1991; see Pollination and Reproductive Biology below). Almost all Labiatae bear non-glandular and glandular trichomes (see Vegetative Anatomy below) and many, especially Nepetoideae, are aromatic, due to the presence of terpenoids. The leaves, which lack stipules, are almost invariably opposite (Fig. 22A) and often decussate or sometimes whorled, although alternate or subopposite leaves have been regularly observed in species of *Amasonia* and the related *Monochilus*, as well as in *Aeollanthus* spp. (Ryding 1986). The lamina is usually simple (Fig. 22A), toothed, lobed or entire, and expanded, often petioled. Venation is usually camptodromous to semicraspedodromous, occasionally brochidodromous (Hickey 1973), although there is much variation. Pinnate, trifoliolate or digitate leaves have been recorded regularly in some genera, notably *Vitex* (Fig. 23A; Viticoideae), *Cedronella*, *Perrierastrum* (Nepetoideae), *Petraeovitex*, and *Peronema* (Incertae sedis), more sporadically in others (e.g. *Salvia*, *Hyptis*). Especially in species adapted to more xeric conditions, the lamina may be reduced and often coriaceous, sometimes with a revolute margin and sometimes developing an ericoid or cupressoid appearance. Some Old World genera, especially *Plectranthus* and *Aeollanthus*, contain species which have fleshy or even succulent leaves, presumably as a response to environmental conditions. Spinose leaves (e.g. *Acanthomintha*, *Chamaesphacos*, *Salvia* spp.) or bracteoles (e.g. *Galeopsis*, *Notochaete*) or axillary spines (e.g. *Lagochilus*) occur in some genera. Scale-leaves occur on the rhizomes of some species (e.g. *Mentha*, *Kurzamra*). Roots are usually fibrous and can be adventicious on prostrate stems. Tuberous roots occur in a number of unrelated genera.

Vegetative Anatomy. Unless otherwise stated, much of this information has been extracted from Metcalfe and Chalk (1950), while Carlquist (1992) should be consulted for a substantial review of wood anatomy, in relation to ecology, of Labiatae as traditionally delimited. Stems are frequently quadrangular in cross section, with a subepidermal layer of collenchyma, often restricted to angles, or sometimes along the sides (e.g. *Pycnostachys*) and forming a closed ring in some genera of Nepetoideae (*Aeollanthus*, *Plectranthus*, *Anisochilus*) and Lamioideae (*Chelonopsis*, *Colquhounia*, *Pogostemon*). A collenchymatous hypodermis has been reported in *Clerodendrum inerme* (Mullan 1933) and, in *Bystropogon*, there are fibres associated with it. In *Tectona*, stone cells are associated with the collenchymatous cortex. Collenchyma is apparently absent from the cortex of various genera of Westringieae and also from *Cuminia*. The cortex of a number of species, some of semi-aquatic habitats, have large intercellular spaces (e.g. in *Mentha*, *Stachys*, *Pogostemon* and *Prunella*). An endodermis is reported in *Holmskioldia* and *Scutellaria*, and also in many genera of Lamioideae and Nepetoideae, where it is said to be often suberized. However, Metcalfe and Chalk (1950) do not make mention of the endodermis in any genera of the more woody subfamilies: Symphorematoideae, Viticoideae, Ajugoideae or Prostantheroideae. Cork is produced, either superficially or in deeper layers of the stem in many genera. The pericycle may contain isolated bundles of sclerenchymatous elements or may form a continuous ring in *Ajuga*, *Cymaria*, *Marrubium*, *Tectona*, *Clerodendrum* and *Sphenodesma*, while in the last two and in *Oxera*, stone cells are associated with the fibres. However, the sclerenchymatous elements seem to be less well developed in the pericycle of many Lamioideae and Nepetoideae, and are reported as absent from *Lamium*, *Galeopsis*, *Comanthosphace* (Lamioideae) and from *Mentha*, *Thymus*, *Micromeria*, *Monarda*, *Lallemantia*, *Horminum* and *Prunella* (Nepetoideae: Mentheae). Primary vascular bundles usually include four principal ones, opposite the stem angles (in cross section). Secondary phloem includes fibres in various groups, and associated stone cells are also reported in Symphorematoideae and in *Tectona* and *Callicarpa* (Incertae sedis). A similar situation is found in various genera of Prostantheroideae: tribe Chloantheae: *Chloanthes*, *Cyanostegia* and *Pityrodia*, while in tribe Westringieae, *Hemiandra* was found to have the phloem fibres forming a continuous cylinder. Phloem fibres are also reported in species of *Scutellaria* and *Tinnea* (Scutellarioideae) and in *Meriandra*, *Cunila* and *Hoslundia*. Xylem vessels usually have simple perforations. Xylem develops into a continuous cylinder at a very early stage in most groups, although exceptions are reported in *Horminum* (Nepetoideae) and *Physostegia* (Lamioideae), and the cylinder is often traversed by narrow rays. Broader, lignified rays occur in more woody species of *Vitex* (Viticoideae) and *Tectona* (Incertae sedis). Medullary bundles are reported in *Tectona* and in *Teijsmanniodendron* (Viticoideae). Wood tends to be semi-ring-porous or ring-porous, or semi-ring-porous to diffuse in Hyptidinae (Nepetoideae) as reported by Rudall (1981a), with vessels typically arranged in small groups in a tangential pattern, with simple perforations (except rarely in some *Vitex* species), occasionally scalariform in some Hyptidinae (Rudall 1981a), although vessels are said to be mostly solitary in *Hoslundia* (Nepetoideae). Spiral thickening is reported in a range of genera. Vessel element dimensions are often minute to medium-sized but can cover a wide range; this is probably related to habit and xeromorphy (Rudall 1981a). In Hyptidinae, intervascular pitting is alternate, with small bordered pits. In several Viticoid genera, tyloses with large, simple pits have been recorded. Rays are often heterocellular and vary from 1 to 12 cells wide. Rudall (1985) reported perforated ray cells in *Hyptis hagei* (Nepetoideae). Pits to rays are simple and small or, in Hyptidinae, elongated (Rudall 1981a), pits to fibres being small and, in some genera, septate. Parenchyma is paratracheal and rather sparse in most genera, although more abundant and aliform or confluent in some species of *Aegiphila* (Ajugoideae), *Gmelina* (Viticoideae), *Callicarpa* and *Peronema* (Incertae sedis).

Regarding leaves of Labiatae, Rudall (1980) pointed out that, at least in the xeromorphic subtribe Hyptidinae, leaf anatomy shows much variation and is taxonomically useful at species level. Leaves of Labiatae are usually dorsiventral, but many examples of isobilateral leaf types are reported. Leaves may be persistent or deciduous, especially in dry habitats. Rarely, leaves may be deciduous at onset of flowering, as in *Madlabium* and some species of *Plectranthus*. Cantino (1990) examined leaf trichomes and stomata of Labiatae and Verbenaceae. Multicellular glandular trichomes are widespread in Labiatae and usually more frequent on the abaxial surface of the lamina. Variation in the subsessile glandular trichomes, which possess a very small stalk obscured by the much larger head, appears taxonomically significant in some cases. Cantino (1990) recognised eleven types based on head cell number (one to many), and presence or absence of tangential walls and partial or tertiary radial walls. The most complex type, with a many-celled head, described as broad and scale-like, was found only in species of Symphorematoideae, Viticoideae and Ajugoideae. Variation in the capitate glandular trichomes, with a relatively smaller head and obvious stalk, appear to have less taxonomic importance. *Zhumeria* and *Meriandra* (Nepetoideae: Salviinae) appear to be unique in the family, with glandular dendroid trichomes (Bokhari and Hedge 1976). Non-glandular hairs may be unicellular or multicellular, and either simple and uniseriate, or branched in various ways. Unicellular hairs are widespread throughout the family, and are nearly always accompanied by multicellular hairs, which distinguishes Labiatae from most Verbenaceae (which are characterised by the possession of unicellular hairs only, in almost all genera). In Prostantheroideae, however, species of tribe Westringieae have only unicellular trichomes except in very few cases, while tribe Chloantheae is characterised by the possession of branched multicellular hairs, often in great profusion. Branched, multicellular hairs are also a feature of many Symphorematoideae, especially on the involucral bracteoles where they appear stellate. Several genera of Lamioideae, *Comanthosphace*, *Rostrinucula*, *Leucosceptrum*, *Gomphostemma*, *Phlomis* and *Marrubium*, possess branched hairs. Two species of *Scutellaria* (Scutellarioideae) possess such trichomes, but elsewhere in this subfamily they appear to be absent. In the Nepetoideae, the presence of branched hairs is often a valuable diagnostic character at species level. They are present in tribe Mentheae, especially in subtribe Salviinae: *Zhumeria*, *Meriandra*, *Perovskia*, *Rosmarinus*, *Lepechinia*, *Chaunostoma* and several species of *Salvia*. In subtribe Nepetinae, branched hairs are found in one section of *Nepeta*, and in Menthinae and tribe Ocimeae they occur sporadically in a number of genera, while in Ajugoideae they appear to be restricted to a few species of *Teucrium*. Extrafloral nectaries on the leaves have been reported in *Clerodendrum*, *Amasonia*, *Monochilus* (Ajugoideae) and in *Callicarpa*

(Incertae sedis). In some species of *Ocimum* (Nepetoideae: Ocimeae), they can develop in place of the abscission-scar of deciduous bracts. Hydathodes occur at the leaf-margins of some species of Hyptidinae (Rudall 1980). Ten types of stomata were reported by Cantino (1990), from which this summary is based. Anomocytic and diacytic stomata are most frequently encountered. Diallelocytic stomata with three subsidiary cells also occur widely and were found in nearly all species of subfamily Nepetoideae examined. However, they appeared to be absent from Prostantheroideae, where anisocytic stomata are particularly common. Paracytic stomata are very rare in Nepetoideae, but characteristic of *Prostanthera* (Prostantheroideae), *Trichostema*, *Oxera* (Ajugoideae), members of Symphorematoideae and *Petraeovitex* (Incertae sedis). Other stomatal types tend to occur with lesser frequency, or show little correlation with taxonomy. The distribution of stomata on the leaf can vary greatly, with the abaxial surface often more densely covered. In some species, the stomata are restricted to the abaxial surface. Rudall (1979, 1980) noted much variation in distribution and also presentation of stomata in both *Eriope* and other Hyptidinae (Nepetoideae), which includes many xeromorphic species and a wide range of habit. Stomata could be sunk in pits or in grooves, sometimes surrounded by hairs or glands, or in hairy species may be raised. The leaf cuticle can be thickened, especially in species from more xeric habitats. Rudall (1980) noted characteristic striation patterns on leaf cuticles of some species of Hyptidinae, and variation in cuticular pattern between different populations of *Eriope hypenoides* Mart. ex Benth. (Nepetoideae: Ocimeae) (Rudall 1979). Cells of the epidermis may be smooth to papillose (*Bystropogon*) or bullate in some genera, and can have wavy to straight anticlinal walls (Rudall 1980). A hypodermis of large cells is reported below the adaxial epidermis of leaves of *Rosmarinus* (Nepetoideae). Presence or absence of a hypodermis helped to characterise two sections of *Hyptis*, some species of which showed many macromorphological similarities (Rudall 1985). In *Vitex* (Viticoideae), the mesophyll is composed of a compact palisade tissue, a condition apparently similar to that found in some *Salvia* (Nepetoideae) species. In some species of Hyptidinae (Rudall 1980) and in *Amasonia* and the related *Monochilus* (Ajugoideae), a multilayered palisade mesophyll has been reported and, also in these genera, the vascular tissue is without sclerenchyma. A similar situation apparently occurs in some Nepetoideae and Lamioideae, where the vascular bundles are accompanied by a parenchymatous sheath, although no further details are given. Rudall (1980) reported bundle sheath extensions both with and without lignification in different species of Hyptidinae.

The petiole in cross section most frequently exhibits an arc of widely spaced vascular bundles. This may be deeply crescentic in *Callicarpa* (Incertae sedis) and *Vitex* (Viticoideae) and in *Physostegia* (Lamioideae) and various genera of Nepetoideae, or shallow in some species of *Plectranthus*. In some cases this median arc has incurved ends, e.g. in species of *Caryopteris* and *Clerodendrum* (Ajugoideae) *Ballota* (Lamioideae) and *Hyptis* (Nepetoideae), where the vascular tissue is also continuous, as it is in *Melittis* (Lamioideae) and various other genera of Nepetoideae. Elsewhere, the vascular arc may tend towards a cylinder, as in *Leonurus*, *Phlomis* and *Stachys* (Lamioideae) and in some species of *Hyptis* (Nepetoideae). In *Phlomis tuberosa* (Lamioideae) the cylinder consists of numerous separate bundles, while in *Oxera* (Ajugoideae) the numerous bundles are collateral. Accessory bundles occur frequently, especially in the petiolar wings, and indeed were found in all genera of Lamioideae and Nepetoideae examined. In *Congea* and *Symphorema* (Symphorematoideae), 1–2 additional bundles were also reported on the adaxial side of the petiole. Large sac-like secretory cells have been reported in the mesophyll of *Brazoria* and *Physostegia* (Lamioideae). Crystalline inclusions, such as needles, rods or octohedra, are said to be often present; large, solitary crystal inclusions have been reported in parenchymatous tissues of *Tectona* (Incertae sedis), while Abu-Asab and Cantino (1987) report sac-like idioblasts in the leaf mesophyll of three related genera, *Brazoria*, *Warnockia* and *Physostegia* (Lamioideae).

Inflorescence Structure. Inflorescences of Labiatae are usually bracteate (the bracts sometimes conspicuous in *Lavandula*, *Hemizygia* and *Salvia viridis* (Nepetoideae), and often also bracteolate. Throughout this text, "bracts" refers to the foliar structures, often reduced or modified, that subtend the basic inflorescence unit, which is typically cymose. The term bracteole is used for those usually much reduced foliar structures which are within, or form an integral part of, the cyme, e.g. forming an involucre around a capitulum which is itself subtended by a bract. Bracts and bracteoles are normally opposite, so that the flowers which they subtend are also, and may be persistent or deciduous at anthesis. Rarely, as in *Lavandula* (Nepetoideae), the bracts and flowers are spirally arranged. Many species have sessile flowers, but in *Catoferia* (Nepetoideae) the 3-flowered cymes have the pedicels fused to the inflorescence axis, so that the sessile flowers arise directly on the axis, some way above the bracts. The cymose nature of the inflorescence of Labiatae distinguishes it from those of Verbenaceae, which are racemose. In a number of genera, such as *Nepeta*, *Isodon* and *Hyptidendron* (Nepetoideae), the cymose nature of the inflorescence is clearly apparent. In the latter, some species have dichasial and some species monochasial cymes. The cymes are frequently aggregated towards the end of an indeterminate flowering stem, to form a thyrse. The nature of the inflorescence in Labiatae is controlled by three basic processes: (1) reduction, whereby the number of flowers in a cyme may be reduced; (2) contraction, whereby the axis either of the flowering stem, or of the peduncle subtending the cyme, or of the subordinate branches within the cyme, including the pedicel, is reduced in size or obsolete; (3) elongation, where the opposite process is applied to these axes. Cymes are often congested by reduction of the peduncles, so that the flowers are fasciculate. An opposite pair of such cymose fascicles is referred to as a verticillaster. Sometimes the fascicles of flowers are many-flowered and held in the axils of leaf-like bracts (see, for example, *Mentha arvensis* and many Lamioideae). Sometimes the bracts and bracteoles are greatly reduced and the cymes form a terminal raceme-like thyrse, although the individual units retain their cymose structure. Sometimes the cymes are reduced to single flowers, so that the thyrse appears truly raceme-like as, for example, in *Scutellaria* (Scutellarioideae) and in *Eriope* and *Hypenia* (Nepetoideae). In these cases, the bracteoles are usually present at the base of the flowers, indicating the cymose origin of the unit. In other examples, the bracteoles may be absent, as in *Aeollanthus* and *Endostemon tenuiflorus*, where the raceme-like inflorescences can only be indirectly inferred to have a cymose origin due to their systematic position in Labiatae. Sometimes, the cymes are highly diffuse, and in some prostrate, herbaceous species (e.g. *Mentha cunninghamii*, *Clinopodium brownei*), leaf-like bracts subtend single flowers on long, erect pedicels or, less often, 3-flowered pedunculate cymes, which occur sporadically along the stem. The contraction of the stem axis can result in a dense, spike-like inflorescence, as in some *Stachys* species, or a terminal, head-like inflorescence, as in *Mentha aquatica*. In *Monarda* and especially *Monardella*, the thyrse itself, together with the sessile or subsessile flowers, may be congested to form a terminal head, usually surrounded by an involucre of conspicuous bracts, the cymose nature of the inflorescence being lost. In some cases, the cyme itself may be structurally complex, and various types of cymose branching can occur. In *Plectranthus hereroensis*, from South Africa, the first branch of the cyme is dichasial, with the two resultant branches being cincinnate. A similar pattern occurs in *Hyptis pectinata*, where the cymes of subsessile flowers are usually congested to form a terminal spike-like thyrse. Where the flowers are sessile or subsessile, various types of inflorescence can occur, as is well displayed in Hyptidinae (Nepetoideae). In *Eriope*, for example, the cymes are one-flowered, with the thyrse raceme-like, each flower on an elongate peduncle which is easily mistaken for the pedicel. However, each flower has two minute, paired bracteoles at the base of the calyx, so that the true pedicel is obsolete. Other examples have both the peduncle and the pedicel reduced. In *Hyptis* sect. *Hyptis*, the sessile flowers are arranged in a capitulum, due to the contraction of all the branches within the cyme, the bracteoles being displaced to form an involucre, with the direct association between the flower and its bracteole obscured. These capitula may often then be arranged in various types of synflorescence. A similar but simpler situation occurs in the inflorescences of the Symphorematoideae, where the bracteoles form wing-like structures around small heads of flowers. The cymose nature of the *Hyptis* capitulum is demonstrated by the irregular pattern of anthesis within it, as opposed to that of some Verbenaceae, such as *Lantana*, where the centripetal pattern demonstrates its racemose structure. Interestingly, comparative studies on the capitula of *Hyptis* indicate that they are derived from a series of cymes where the lowest branch is dichasial, and the later branches cincinnate, as in *Plectranthus hereroensis*. The unusual alternate flower arrangement in the inflorescences of some *Aeollanthus* species may be also due to a cincinnate cymose pattern. The complex inflorescence terminology used by Weberling (1992), based on earlier papers by Troll, has not been employed here, although many of the variations that occur in the Labiatae inflorescence were described in Weberling's detailed work.

Floral Morphology and Anatomy. The hypogynous flower of Labiatae (Fig. 24C) has a biseriate perianth, with sepals 4–8(–9), but usually

5, connate, actinomorphic to zygomorphic, often 2-lipped, with 2–5 or rarely up to 15 lobes, lobes rarely obsolete, calyx-tube (5–)10–15-nerved, cylindrical, campanulate or infundibuliform, rarely urceolate, sometimes gibbous, straight or curved, with throat straight or oblique, often accrescent, rarely inflated, in fruit. Petals (4–)5, rarely to 16 in *Symphorema*, connate. Corolla-lobe number is usually 4–5, often reduced to 2 or 3, especially in some 2-lipped Scutellarioideae, Lamioideae and Nepetoideae, or may be more in some Symphorematoideae, in some Prostantheroideae tribe Chloantheae (e.g. *Lachnostachys*) and in *Tectona* (Incertae sedis). The corolla may be actinomorphic but is usually weakly to strongly zygomorphic (Fig. 23C), (2–)4–5(–16)-lobed and 2-lipped (1-lipped in *Teucrium*), the lips often unequal, the upper often 2-lobed and the lower 3-lobed, porrect to patent, upper or lower lip sometimes concave, hooded or galeate, corolla-tube short to elongate, straight, curved or geniculate, sometimes constricted to gibbous, saccate, or rarely spurred. In some species with cleistogamous flowers, the corolla may be greatly reduced (see Pollination and Reproductive Biology). Stamens (Fig. 24C) epipetalous, attached within tube, usually equal in number to, or less than the number of corolla-lobes, typically 4 in most groups (with the posterior stamen absent or rarely reduced), especially Ajugoideae (*Aegiphila* sometimes with 5 stamens), Prostantheroideae tribe Westringieae, Scutellarioideae, Lamioideae and Nepetoideae, and the stamens may be didynamous. Some genera or species in these latter groups have the four stamens reduced to two. Rarely, it is the anterior pair that is staminodal or absent, as in *Hypogomphia* and in a few species of *Nepeta*, but more usually it is the posterior pair. In *Oxera* (Ajugoideae), the transition from 4 to 2 stamens has occurred by reduction of either the posterior or anterior pairs. Both conditions are found in *O. robusta* (de Kok and Mabberley 1999a), while both 4 and 2 stamens occur in *O. robusta* and *O. coriacea* (de Kok and Mabberley 1999b). Higher stamen numbers occur most frequently in Symphorematoideae, Viticoideae and in some Prostantheroideae tribe Chloantheae, groups formerly included in Verbenaceae. In *Symphorema polyandra*, for example, with actinomorphic flowers, there are normally 16 corolla-lobes and adnate stamens, while in one aberrant individual, there were consistently and, in addition, eight free stamens arising proximal to the gynoecium – a situation apparently not previously recorded from the Lamiales. It is perhaps possible that this situation arose due to a mutant peloric form, arising from a typically zygomorphic flower, becoming genetically fixed, although such events must be extremely rare. Peloria in Labiatae, and other families, are well reviewed by Rudall and Bateman (2003). Filaments usually free, but in one group of *Plectranthus* (Nepetoideae), the stamen-filaments have become post-genitally united at the base, a character at one time used to define the now obsolete genus *Coleus*. They may be hairy or glabrous, rarely appendaged, as in *Prunella*, short or often elongate, usually exserted from corolla-tube and sometimes long-exserted from corolla, parallel, divergent or ascending and sometimes included within or lying under the posterior corolla-lip or, in Ocimeae, declinate and then sometimes included within the anterior corolla-lip. Stamen position can also be modified by resupination (see Pollination and Reproductive Biology). Anthers are usually dithecous or monothecous by abortion, with thecae parallel or divergent to divaricate, distinct and occasionally widely separated by an elongate or thickened connective. This is particularly well developed in *Salvia*. In this genus, the flowers possess only two stamens, each normally with a single fertile theca and a very elongate connective, with the lower theca forming a sterile appendage that usually acts as a lever during pollination. In many genera, the thecae may be apically confluent or synthecous (as in Nepetoideae: Ocimeae), sometimes appendaged or hairy, ellipsoid to elongate, opening by longitudinal slits or rarely by pores. In some species of *Plectranthus* (Nepetoideae), the anthers are usually ± orbicular, and rarely with the outer edges of the thecae fused in the lower half to form a pouch, which opens apically. In a few Lamioideae (*Leucosceptrum*, *Comanthosphace* and *Rostrinucula*), the anthers are wider than long, thus appearing transversally dehiscent. In *Horminum*, the anthers are adherent in pairs, while in some two-stamened genera, e.g. *Ziziphora*, *Rhododon*, *Salvia* subg. *Calosphace*, the anthers or connectives may be united in different ways. A disc at the base of the ovary is often present, usually fleshy, entire or irregularly or often 4-lobed, anterior lobe sometimes longer than others, nectariferous. The gynoecium of the Labiatae is 2-carpellate with usually 4 (or 2) ovules, but often becomes 4-locular, with one ovule per loculus. In Lamioideae and Nepetoideae, the ovary is deeply 4-lobed and the style gynobasic, while in the other subfamilies the stylar position is usually terminal or may be intermediate, with the ovary unlobed, or lobed less deeply. Harley (1976) used the presence or absence of a persistent "stylopodium" in the gynobasic flowers of some Hyptidinae as a diagnostic marker for certain genera and some sections of *Hyptis*. The stylopodium is the persistent, basal part of the style, separated from the longer, distal portion, which is deciduous, by a small joint. Rudall (1981b), however, found that the fold of tissue, which forms the abscission point, is present in many Labiatae with a gynobasic style, but in most groups this fold is found below the level of the top of the gynoecium, so that the small, persistent stylopodium, although present, is inconspicuous and hidden from view. In other groups within the Hyptidinae, notably *Eriope*, *Hyptidendron* and some sections of *Hyptis*, where the stylopodium is said to be present, the point of abscission is well above the ovary, thereby making the structure conspicuous. In *Marsypianthes*, however, the gynoecium is unusual, with the style appearing to be terminal. The ovary-lobes are fused to the style on their ventral face, and abscission of the style takes place at the top of the ovary. The stylopodium therefore remains attached to the developing nutlets until maturity. Floral vasculature in Labiatae has been described by Hillson (1959) for some genera of subfamilies Ajugoideae, Prostantheroideae: Westringieae, Scutellarioideae, Lamioideae and all four tribes of Nepetoideae. Similarly, Rao (1952) examined a number of genera, formerly Verbenaceae but today included in Labiatae, in subfamilies Viticoideae, Ajugoideae and Scutellarioideae. For further details, reference should also be made to these works and to Stauffer (1937). In a study of the floral anatomy of the Nepetoideae subtribe Hyptidinae, Rudall (1981b) noted that the vasculature in this group is fairly uniform and as described for other members of Ocimeae (cited as Ocimoideae). The most variable feature appeared to be the number of calyx traces, from 5–15, which initially arise from the vascular cylinder. In Viticoideae, Rao (1952) however, showed that *Gmelina* is unusual, the number of traces being greatly increased by traces branching and uniting, while in hexamerous flowers of *Tectona* (now placed in Incertae sedis), there are 12 traces present. Secretory hairs, acting as hydathodes, have been reported on the inner surface of calyces of *Clerodendron* flowers (Koorders 1897). The resultant "water calyx" appears to have a protective function for the developing flower. Many Labiatae display a fleshy disc at the base of the gynoecium, which may be variously lobed. Frei (1955) demonstrated its nectarial function. In *Scutellaria* (Scutellarioideae), the disc is modified so that the nutlets sit on a peg-like gynophore. The structure of the disc was investigated by Rudall (1981b), who found it to be highly vascularized in Hyptidinae (also indicating a nectarial function), with anomocytic stomata around the tops of the disc-lobes. The lobes in these species were supplied by dorsal carpellary traces composed entirely of phloem. The disc-lobes alternate with the carpels except in *Lavandula* and in *Hyptidendron asperrima*. Some groups apparently still await detailed anatomical investigation of their floral structure.

Early studies by Van Tieghem (1907) demonstrated the very similar placentation found in Labiatae, including Viticoideae (at that time included in Verbenaceae), as opposed to that of Verbenoideae, which today represents Verbenaceae sensu stricto. Junell (1934) confirmed much of this work, and carried out a detailed survey of placentation and gynoecial structure within the two families. The position of the ovules in relation to the carpel margins clearly distinguishes the two groups. Labiatae, including those groups formerly included in Verbenaceae, such as Viticoideae and Chloanthoideae (sic), bear the ovules on the inner face of the carpel wall – i.e. lateral, while the ovules of Verbenaceae are attached to the carpel margin – i.e. marginal. Junell (1934) did not find any intermediate forms between these groups but noted, however, that in some "Stachyoideae", which today includes genera mainly from subfamilies Lamioideae and Nepetoideae, the carpel margin has become reduced, so that the lateral position of ovules is less easy to see. Data on variation in ovule position, and other minor modifications to placentation, can be found on reference to this classic paper. Each ovule is usually initiated at the base of the locule, becoming anatropous or hemianatropous and erect, with a single, rather thick integument. In Symphorematoideae, however, the ovary is imperfectly 2-locular, being free towards the apex, in the region where the ovules, which are orthotropous and pendulous, are borne. This appears to be quite different to the situation in other Labiatae. Junell (1934), however, demonstrated that the structure in Symphorematoideae is actually very similar to that found in many Viticoideae, citing the example of *Teijsmanniodendron* (Viticoideae) as being in an intermediate position between the Symphorematoideae on the one hand, and other members of the Viticoideae on the other. Wunderlich (1967) made detailed studies of pollen, embryo sac and endosperm development within Labiatae, but unfortunately did not examine those genera traditionally placed in Verbenaceae, as had Junell (1937) before her. However, her studies, as those of Billings (1909), indicated a need for a taxonomic reappraisal of the family in the light of their findings.

Embryology. Wall development of the tetrasporangiate anther is of the Dicotyledonous type, that is, with suppression of periclinal divisions in the inner secondary parietal layer, which develops directly into the tapetum, the outer layer giving rise to both the endothecium and to the single middle layer which is ephemeral (Davis 1966). The endothecium develops fibrous thickenings, of the Baseplate type, sometimes disrupted (Manning 1996), although these fail to differentiate in the pollen grains of cleistogamous flowers of *Lamium amplexicaule*, which germinate in situ. The cells of the glandular tapetum become multinucleate. Simultaneous cytokinesis in the pollen mother cell follows meiosis and the pollen tetrads are tetrahedral or decussate. Mature pollen of Labiatae is tricolpate and 2-celled in six of the seven subfamilies of Labiatae. The exception is Nepetoideae, in which the mature pollen is hexacolpate and 3-celled.

In a recent paper on the megagametophyte of Labiatae, Rudall and Clark (1992) examined a wide range of material, including not only genera traditionally placed in Labiatae, but also *Clerodendrum* (Ajugoideae) and a number of Verbenaceae sensu stricto, as well as reviewing the data from a large number of other genera by previous authors. The functional megaspore is derived from the cell at the chalazal cell of a linear tetrad, and divides to form a Polygonum-type embryo sac. Rudall and Clark (1992) recognised a number of different embryo sac shapes, suggesting that this character, as well as the pattern of early endosperm formation, may have taxonomic importance. The Ocimeae, for instance, generally had a more linear embryo sac. Carlson and Stuart (1936) showed a remarkable variation in the embryo sac of different species of *Salvia* (Nepetoideae), which could have taxonomic significance at an infrageneric level. Endosperm is cellular, occupying the chalazal region, and its presence or absence in the mature seed is usually characteristic of different subfamilies of Labiatae. The ovules are anatropous to hemianatropous, usually basal or sub-basal, erect, rarely orthotropous, apical, pendulous, borne laterally or submarginally on the placenta, unitegmic and tenuinucellate. Agamospermy is reported in *Pycnanthemum* (Chambers 1961).

Pollen Morphology. There are many references to the pollen morphology of Labiatae that treat just one or two taxa, a tribe or a subtribe. However, there are also numerous accounts which provide pollen data for a wider range of taxa. These include Erdtman (1945), Risch (1956), Wunderlich (1967), Raj (1974, 1983), Vij and Kashyap (1975), Nabli (1976), Gill and Chinnappa (1982), Pozhidaev (1989), Abu Asab (1990), Trudel and Morton (1992), and Abu-Asab and Cantino (1994).

Pollen grains of Labiatae are single, isopolar and, usually, 3- (Fig. 21A, B) or 6-colpate (Fig. 21C). Colpus number has a striking association with the subfamilies: in I. Symphorematoideae, II. Viticoideae, III. Ajugoideae, IV. Prostantheroideae, V. Scutellarioideae, and VI. Lamioideae – the pollen are generally tricolpate, and in VII. Nepetoideae – hexacolpate (Erdtman 1945). Less frequently 4- (Fig. 21E) or 5-colpate (Fig. 21F) grains occur in subfamilies I–VI. Rarely, in *Tripora divaricata* (Ajugoideae) the pollen grains are triporate (Fig. 21D). Aperture number within species is fairly stable in 3-colpate (or triporate) pollen, but samples of hexacolpate pollen sometimes include grains with 8, or possibly more, colpi, while taxa with predominantly 4-colpate pollen, for example, *Sideritis* spp. (Lamioideae), may include some 5-colpate grains, and taxa with predominantly 5-colpate grains, for example, *Cornutia* spp. (Viticoideae), may include the odd 4-colpate grain. The equatorial outline is usually rounded triangular (Fig. 21B) or circular in tricolpate grains, ellipsoid (Fig. 21C) or circular (Fig. 21J) in hexacolpate grains, more or less circular in 5-colpate grains (Fig. 21F) or square ("maltese cross"; Fig. 21E) in 4-colpate grains.

Pollen shape is frequently adversely affected by standard acetolysis procedure which destroys the colpus membranes (Fig. 21A, E, F) and effectively collapses the natural shape of the grain (Demissew and Harley 1992). This leads to pollen grains sometimes being erroneously described as "prolate" in the literature when natural shape is more likely to range from oblate spheroidal to subprolate. The polar length varies in the range ca. 12–103 µm, and the equatorial width in the range ca. 12–117 µm. Very large pollen grains are found, for example, in *Trichostema* (Ajugoideae), *Orthosiphon* and *Syncolostemon* (Nepetoideae), while very small pollen grains (<18 µm polar length and/or equatorial width) occur in a number of lamioid genera, for example, *Craniotome*, *Paralamium*, and *Pogostemon*, and also in *Scutellaria* (Scutellarioideae).

The colpi are usually long (Fig. 21N, O, R) with acute apices (Fig. 21A–C, J, M, N). The diameter of the apocolpial area between the apices of the colpi also varies among taxa. In some taxa the colpus

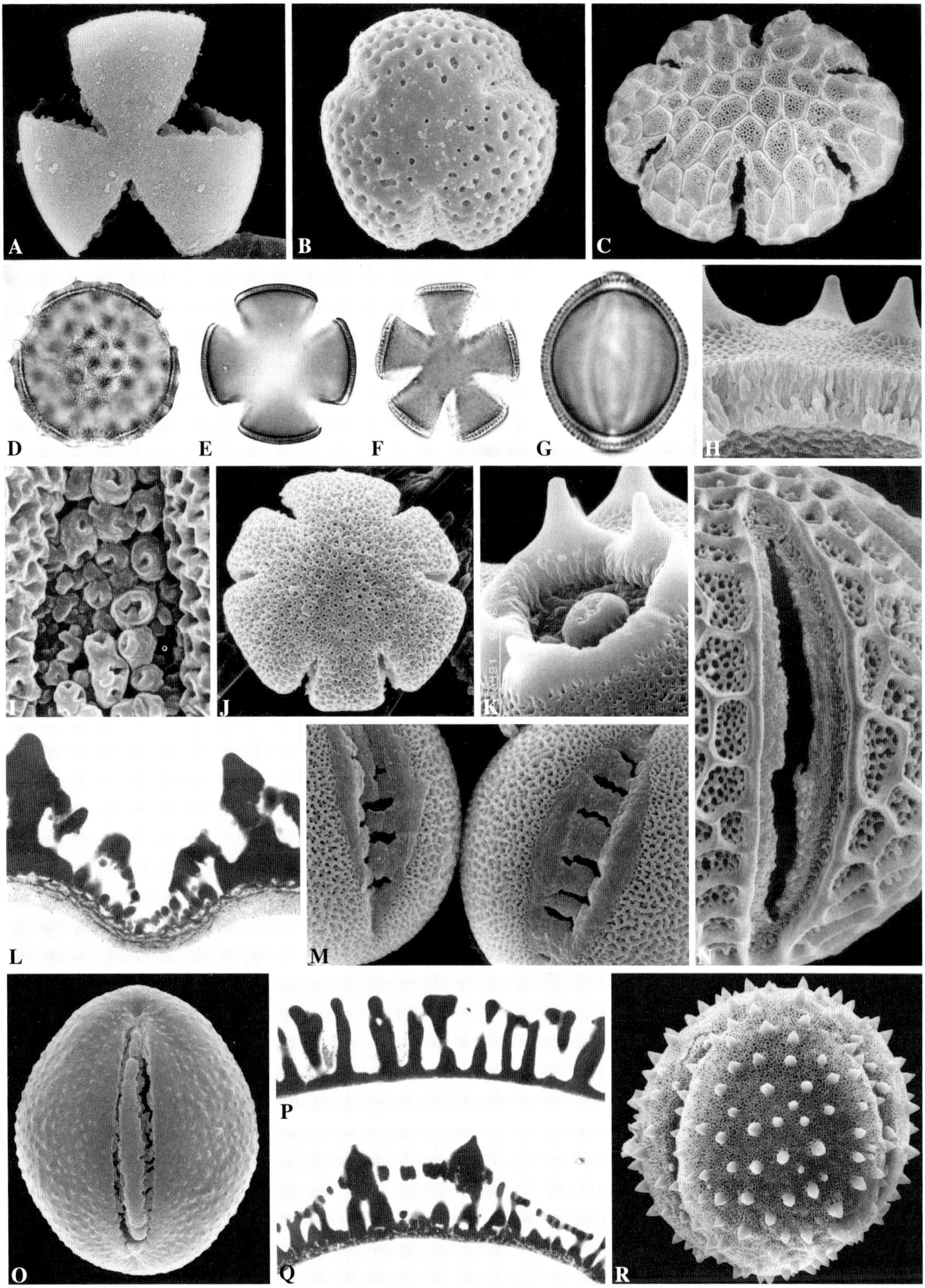

Fig. 21. Labiatae pollen grains. **A** *Lamium maculatum*. Polar view (note fragments of colpus membrane on aperture margins). SEM ×1650. **B** *Marrubium supinum*. Polar view. SEM ×1740. **C** *Ocimum urticifolium*. Polar view. SEM ×1080. **D** *Tripora divaricata*. Polar view. LM × ca. 300. **E** *Sideritis candicans*. Polar view. LM × ca. 600. **F** *Cornutia latifolia*. Polar view. LM × ca. 400. **G** *Stachys hypoleuca*. Polar view. LM ×1000. **H** *Tripora divaricata*. Fracture through mesocolpial wall. SEM ×3250. **I** *Microcorys exserta*. Equatorial view, colpus. SEM ×4800. **J** *Endostemon tenuiflorus*. Polar view, SEM ×920. **K** *Tripora divaricata*. Equatorial view, operculate pore. SEM ×5200. **L** *Ocimum transamazonicum*. Transverse ultrathin section through colpus. TEM ×10000. **M** *Leonurus cardiaca*. Equatorial view, colpus. SEM ×3000. **N** *Becium dhoforense*. Equatorial view, colpus (note colpus membrane, in position but ruptured). SEM ×1600. **O** *Teucrium canadense*. Equatorial view, operculate colpus. SEM ×1600. **P** *Endostemon tomentosus*. Transverse ultrathin section through mesocolpium. TEM ×8000. **Q** *Ocimum basilicum*. Transverse ultrathin section through mesocolpium. TEM ×5000. **R** *Trichostema setaceum*. Equatorial view. SEM ×900. Photos M.M. Harley

margo is very well demarcated (Fig. 21N). The colpus membrane is usually fragile (Fig. 21A, N – see also comments above relating to shape), and the surface smooth to slightly scabrate (Fig. 21M, N) and, in some tricolpate pollen, has a characteristic tendency to develop splits, horizontally to the aperture margins, for example, in *Leonurus cardiaca* (Fig. 21M). In a few genera the aperture membrane is covered by small ectexinous elements bearing some structural resemblance to the main ectexine, for example, in *Microcorys exserta* (Fig. 21I). In *Teucrium* spp. (Fig. 21O) the colpi are operculate. In some taxa with hexacolpate pollen grains, the mesocolpial width is not constant and may alternate between wide and narrow, as in *Nepeta amoena* (Jamzad et al. 2000), and *Endostemon* (Fig. 21J, and Paton et al. 1994).

Wall thickness varies in the range 1–7 µm. In some taxa the pollen grain wall in the apocolpial region is thicker than in the mesocolpial region (Fig. 21G). In general thicker walls tend to be associated with larger pollen grain size. The ectexine (Fig. 21H, P, Q) usually comprises a somewhat irregular tectum, and a well-defined infratectum, while the foot layer is frequently irregular and often quite narrow and granular at the exine/intine interface (Fig. 21Q). The infratectum is frequently much wider than the tectum or foot layer and usually simple columellate (Fig. 21P, Q); the columellae may be widely spaced, or densely packed. Rarely, in *Tripora* (Fig. 21H) and *Teucrium* (Ajugoideae), the infratectum comprises fine, densely packed, branching columellae (Abu-Asab 1990). In a few other genera of Ajugoideae (Abu-Asab 1990), a granular infratectum is observed: *Amethystea*, *Rubiteucris* and *Clerodendrum*. In species of Labiatae which have been thin-sectioned, an acetolysis-resistant endexine is more or less absent in apocolpial and mesocolpial regions of the grain, and present but often not very thick at the aperture margins where it is frequently lamellated with ectexine (Fig. 21L).

The topology of the tectum in Labiatae is highly variable. A high percentage of taxa in both subfamilies have a bi-reticulate exine (Fig. 21C, N), and there is a wide range of variants of the bi-reticulate exine within the family, perhaps nowhere more strikingly so than in the pollen of taxa within subtribe Ocimineae (Nepetoideae) (M.M. Harley 1992). Other exine types include spinulose, for example, *Teucrium canadense* (Fig. 21O), spinose, for example, *Tripora divaricata* and *Trichostema setaceum* (Fig. 21H, R), finely or coarsely perforate, as in many species of *Marrubium* (Fig. 21B) and *Sideritis*, psilate-punctate or microperforate is typical for *Lavandula* species, finely granular, for example, *Lamium maculatum* (Fig. 21A) or simple reticulate, for example, in some of the genera in tribe Prostantheroideae, tribe Westringeae: *Hemigenia*, *Microcorys*, and *Westringia*.

Karyology. Chromosomes of Labiatae are generally small, ranging from about 0.75 to about 6 µm. Most chromosome studies on Labiatae have either been general surveys, usually within a geographical area, such as Morton (1962) for West Africa, or more detailed studies within a particular genus, e.g. Chambers (1961) for *Pycnanthemum*, Harley and Brighton (1977) for *Mentha*, Irving (1976) for *Hedeoma*, and Marrero (1992) for *Sideritis* subg. *Marrubiastrum*, but there is very little in the way of a general review of the chromosomes of Labiatae. Nevertheless, it is clear from many accounts that chromosome variation and behaviour has played an important role in the evolution of the group (but see comments by Irving 1976 for *Hedeoma*). Numbers range from about $2n = 10$ in *Stachys arvensis*, to $2n = 240$ in *Salvia ombrophila*. Polyploidy is extremely common throughout the family, with auto- and allopolyploidy occurring but the latter most frequently, and with segmental allopolyploidy reported in *Mentha spicata* (Harley and Brighton 1977). Singh (1995), with counts taken from a wide range of sources, has estimated that polyploidy occurs in just over 58% of Labiatae, while Morton (1973) has estimated that polyploidy occurs in 68.8% of the British Labiate flora. Similar, but rather lower values were obtained for Western Himalayan Labiatae: 42.35% (Gill 1984) and for Canada: 51.4% (Gill 1981). In all these floras, the range in chromosome number is very great, with many genera with dysploid series, and with lower base numbers in the range 5–10 or even 11 in many genera. Gill (1981) has made a rather subjective estimate of base numbers in the family, suggesting that $x = 7$ is the ancestral form, but numbers based on 8, 9 and 10 are also frequent. Chromosome change in number has not only been achieved by polyploidy, however, but also by structural rearrangement, such as Robertsonian change, as in *Sideritis* (see Marrero 1992), and perhaps by loss of duplicated pairs in polyploid species in Hyptidinae (see Harley and Heywood 1992) and in *Hedeoma* (Irving 1976). Such changes can result in the establishment of new base numbers. Higher, secondary base numbers are often assumed to be due to polyploidy, perhaps following hybridisation between members of a dysploid series, and base numbers such as x = 13, 15, 17, 19 are by no means unusual – see, for example, *Congea* (Symphorematoideae), *Marrubium*, *Physostegia*, *Lagochilus* (Lamioideae), *Rhododon*, *Acanthomintha*, *Marsypianthes*, *Pycnostachys* (Nepetoideae), etc. Reduction of base number has apparently occurred in a few weedy annual species, such as *Stachys arvensis* (as reported above) and *Salvia hispanica* with $2n = 12$. Reduction in base number to x = 6 also occurs in the diploid, weedy species *Hypenia salzmannii*, where other members of the genus, from undisturbed savannah, have x = 10. Some groups with a wide range of basic numbers – *Scutellaria*, *Sideritis*, *Salvia* – probably have a long history of chromosome change and hybridisation, while in *Pycnanthemum*, another genus with a range of dysploid numbers, agamospermy is reported (Chambers 1961). Polytypic genera with a single, constant diploid number, such as *Eriope* with $2n = 20$, are apparently rare. In the New World, for *Salvia* species of subgen. *Calosphace*, a range of base numbers occurs, with most species having x = 11. A geographical pattern has been noted (Harley and Heywood 1992), which may throw light on the origin of the group. In Central America, where most active speciation seems to be occurring, most species are diploids, often with associated B chromosomes. In South America, Andean species so far counted are tetraploids or hexaploids, while the more isolated eastern South American species appear to be high polyploids whose base number is uncertain. Similarly, in the large genus *Hyptis*, which is primarily South American, diploids ($2n = 16$) are only known from a single group of mainly Andean species which appears to be still undergoing speciation, whereas only tetraploids and higher polyploids are recorded from the other groups within the genus. Many groups, such as the Australian Prostantheroideae, await karyological investigation.

Pollination and Reproductive Biology. Huck (1992) has published an overview of pollination biology in the Labiatae, from which some of what follows has been taken. Cross-pollination in Labiatae is effected by animals, of which by far the most diverse group are the insects. No cases of wind- or water-pollination have been recorded. There appears to have been a long history of co-evolution between Labiatae and its pollinators, with visual attraction of the flower or inflorescence playing a major role, although scent may be important in some cases (see below). Species with small, massed flowers, such as *Mentha aquatica* or *Lycopus* spp. (Nepetoideae), of the less specialised brush blossom type, regularly have a large and varied number of insect visitors, including not only bee species but also wasps (Vespidae), Diptera, especially hoverflies (Syrphidae), Lepidoptera and even Coleoptera. Those species with two-lipped flowers, in which access to the nectar is less accessible, tend to rely on fewer pollinator species. While bees constitute the majority of pollinators of Labiatae, Byragi Reddy and Subba Reddi (1994) found that wasps were at least as important as bees in the pollination of *Vitex negundo* (Viticoideae). Also, Potgieter et al. (1999) have demonstrated that long-tongued beeflies (Diptera: Nemestrinidae and Acroceridae) are the principal pollinators of certain Southern African *Plectranthus* species (Nepetoideae) with elongate corolla-tubes. Some species are regularly visited by Lepidoptera or Diptera, and are adapted for pollination by them. One such example is *Clerodendrum* (Ajugoideae), where several species with perfumed flowers are adapted to hawk-moth-pollination (Tanaka 1973; Primack et al. 1981; Endress 1994). Some *Origanum* species (Nepetoideae), such as *O. amanum*, are possibly butterfly-pollinated. In the large tropical genus *Hyptis* (Nepetoideae), most species are bee-pollinated, and do not appear to have perfumed flowers. However, a few, e.g. *Hyptis martiusii*, have weakly perfumed flowers with a very slender, more elongate corolla-tube and are frequently visited by butterflies.

Perfumed flowers have been recorded sporadically in the Labiatae, including such bee-pollinated species as *Melissa officinalis* (Nepetoideae), where the major component of scent is beta-caryophyllene, produced by osmophores on the inner surface of the adaxial corolla-lip (Schultze et al. 1992). Mattern and Vogel (1994) investigated 142 species of Labiatae (excluding subfamilies Viticoideae and Symphorematoideae) and demonstrated that in 16% of those examined, the calyx fragrances, produced from glandular trichomes, were substantially different from those found in the leaves. There appeared to be little systematic significance in the results, although these may indicate that flower fragrance plays a greater role in attracting pollinators of Labiatae than previously thought.

The large majority of Labiatae, however, are adapted to pollination by Apidae, both pollen and nectar being the principal rewards, and many have often developed sophisticated pollination mechanisms. In many cases, pollen-collection is much more specific in terms of host plant than is nectar-collection, which may be from the same or different plant species. Attraction in these cases is probably largely visual. Flag blossoms with conspicuously coloured bracts near the apex of the inflorescence include *Lavandula stoechas*, *Hemizygia* spp. and *Salvia viridis*. Other species have markedly contrasting colours between the calyx and corolla, as in *Clerodendrum*, *Amasonia* (Ajugoideae), *Holmskioldia* (Scutellarioideae) and *Marsypianthes burchellii* (Nepetoideae). Many species have corollas with coloured honey-guides, especially on the lower lip. Some of the adaptations are less obvious and involve close pollinator/flower interrelationships. There are often close associations between individual bee and labiate species. In a study of Labiatae pollination in a Greek Mediterranean phryganic ecosystem, Petanidou and Vokou (1993) found that, of the ten plant species examined, solitary bees formed 47.3% of all pollinator species and a significant number of these were monotrophic, that is, entirely restricted to pollinating members of the family. Petanidou (1996) has demonstrated the interdependency of the bee species and of the plant species, mainly Labiatae, that provide them with nectar and pollen, in the phrygana. Harley (1988a), in field studies of a range of *Eriope* species (Nepetoideae: Ocimeae), found that these were each exclusively visited by very few (one to three) solitary bee species, which were normally also visiting one other non-Labiatae species in the vicinity. In this case, corolla size and tube length appeared to play a role in pollinator selection. The introduction of the Old World hive bee, *Apis mellifera*, has often caused a breakdown in the delicate interrelations between species and their pollinators, due to the far more generalist foraging behaviour. Harley (1986a) reported hybrids between two locally endemic species of *Hyptis* sect. Pachyphyllae in Brazil. While native bee species were faithful to only one or other of these, *Apis* regularly made visits to both. Another example of breakdown of pollination barriers is reported in *Monarda* (Nepetoideae) by Whitten (1981). Here, two intercompatible species, one adapted to hummingbird- and the other to bumblebee- (*Bombus* spp.) pollination, produced hybrids when growing in proximity. There still appear to be few observations on pollination of forest tree species of Labiatae, but it is likely that these are mainly bee-pollinated. Regarding other groups, Subba Reddi et al. (1996) record *Xylocopa* bees on *Gmelina asiatica* (Viticoideae), while smaller *Ceratina* bees are the principal pollinators of teak, *Tectona grandis* (Incertae sedis), according to Tangmitcharven and Owens (1997).

In a number of genera there has been a shift towards bird-pollination, especially in such groups as the New World *Salvia* subgen. *Calosphace* (Nepetoideae), many of whose red- and large-flowered species are pollinated by hummingbirds (Trochilidae). Honeycreepers are said to pollinate some species of *Stenogyne* (Lamioideae) in Hawaii. In the Brazilian *Hypenia* (Nepetoideae), many of which have elongate, tubular red corollas, one group with resupinate flowers has become adapted to bird-pollination, while those with normal flowers are usually visited by bees (Atkinson 1998). In the Old World, the pollinators of southern African *Salvia africana-lutea* are recorded to be sunbirds (Nectariinidae), as are the usually orange flowers of *Leonotis* spp. (Vos 1995), while the flowers of *Prostanthera* sect. *Klanderia* (Prostantheroideae) in Australia are pollinated by honeyeaters (Meliphagidae; Conn 1986). In the former example, it is interesting that the only annual, weedy species of *Leonotis* is autogamous. In the New World, a number of cosmopolitan genera (e.g. *Stachys*, *Scutellaria* and *Clinopodium*) have evolved adaptations to hummingbird-pollination in some species, while Chávez (1974) has indicated *Catoferia* as a possible bat-pollinated plant, although the appearance and delicate nature of its corollas would make hawk-moth-pollination seem more likely.

Most Labiatae have 2-lipped flowers, with the stamens directed towards the posterior corolla-lip (e.g. *Lamium*, *Stachys*, *Salvia*, *Monarda*), where nototribic pollination can occur.

In some groups with resupinate flowers, an important shift in pollinator species must have occurred. Resupination is found in a number of genera of Ajugoideae: in *Monochilus* and many species of *Clerodendrum*, while in some species of *Trichostema*, it occurs only after anthesis. Resupinate flowers are also found in Nepetoideae, in Nepetinae such as *Lophanthus* and *Hymenocrater* and occasionally in *Marmoritis*. In Menthinae it is recorded sporadically in *Salvia* and *Micromeria*, while in *Cyclotrichium* the corolla is resupinate by twisting of its tube. In Ocimeae, it is recorded only in some species of *Hypenia*, mentioned above. In the Ocimeae (Nepetoideae), however, the stamens are regularly directed towards the anterior lip, and pollination is sternotribic, thus favouring bees with different foraging behaviour. A few genera or species have developed a triggered pollination system. In *Trichostema* (Ajugoideae), Howe (1986) reported that the flowers were triggered by the pollinating insect's weight, which causes the stamens to curve down and come into contact with its body. In *Aeollanthus* sect. *Cucullilabium* (Nepetoideae subtribe Plectranthinae), the stamens are held under tension until released from the lower lip (Hedge 1972; Ryding 1986). This explosive mechanism is best developed in genera of subtribe Hyptidinae. Here, the declinate stamens are held under tension within the hinged, concave anterior corolla-lip, whose margins enclose them (Burkhart 1939; Harley 1971). In the larger-flowered genera, such as *Eriope* and *Hypenia*, pollination is often by traplining bees, where very few flowers are open at any one time. Here, an elaborate adaptation, known as "the greasy pole syndrome", has been reported (Harley 1991). This is a suite of characters including waxy, swollen, virgate inflorescence stems, which deter ants from stealing the nectar and "illegitimately" triggering the flowers. One of the best-known and most elaborate types of pollination mechanism within Labiatae is that of *Salvia* (Nepetoideae). The flowers possess only two stamens, normally with a single fertile theca and a very elongate connective which articulates with the often shorter filament. The lower theca forms a sterile appendage that usually acts as a lever, tapping the pollen directly onto the back of the pollinator. There is considerable variation on this theme within the genus, which includes both insect and bird pollinators. In some species (e.g. *S. verticillata*), there is no articulation between the filament and connective, but there is a "transfer of function" between the filament and connective in that the posterior lip of the corolla is basally narrowed and readily folds back exposing the stamens (Hedge, pers. comm.). Some genera show modifications of the anther or anther-connective, which may be extended into a spur. In *Dicerandra* (Nepetoideae), the anther-thecae bear an apical spur which has been shown to trigger pollen discharge (Huck 1987). The many other adaptations within Labiatae, involving the form of corolla and stamens, are reviewed by Huck (1992), but further detailed field studies are needed.

Owens and Ubera-Jiménez (1992) have reviewed the breeding systems of Labiatae. Self-compatibility is widespread in the family and, indeed, up to the present, no convincing evidence of self-incompatibility in Labiatae has been presented. There is, however, a wide range of adaptations, apart from those dealt with above, which promote outbreeding. Both dichogamy and herkogamy occur widely: both temporal separation, such as protandry, and spatial separation, such as the rapid growth of stigmas just prior to maturation, or the lateral curving of the stamens after pollen is shed, in *Nepeta*, for example, thereby avoiding contact with the centrally placed stigmas. However, a few species, often weeds such as *Salvia verbenaca* and *Lamium amplexicaule*, can regularly produce cleistogamous flowers that are obligately selfed. These have also been reported in *Schnabelia* (Ajugoideae). Another means of promoting outbreeding is gynodioecy which is widespread in the family, especially in Nepetoideae tribe Mentheae (e.g. *Origanum*, *Mentha*, *Ziziphora*, *Glechoma*, etc.) and in some Lamioideae (e.g. *Ballota*, *Marrubium*, *Stachys*, etc.). Gynodioecy has also been reported in *Scutellaria* (Scutellarioideae) by Loew (1894). In the Ocimeae, however, gynodioecy is recorded only in *Lavandula* (Knuth 1909). Temporal gynodioecy is reported in *Rosmarinus* (Ubera-Jiménez and Hidalgo-Fernández 1992) and temporal gynomonoecy occurs frequently in *Mentha* and *Lycopus* (Webber and Ball 1980). Dioecious Labiatae are rare, being reported only in *Callicarpa* (Incertae sedis) by Kawakubo (1998), in which pollen from female flowers is non-aperturate and unable to germinate, *Aegiphila* (Ajugoideae), *Colebrookea* (Lamioideae), *Lepechinia* sect. *Parviflorae*, *Nepeta* sect. *Oxynepeta* and *Tetradenia* (Nepetoideae). There is some evidence that heterostyly may occur in *Perovskia* (Nepetoideae), although this has not been reported previously (see generic account in this volume), and needs confirmation. The only other example reported for the family is in *Salvia brandegeei* (Nepetoideae) by Barrett et al. (2000). It is of interest that both are members of subtribe Salviinae. Agamospermy is at present known only in *Pycnanthemum* (Chambers 1961), in which a range of diploid chromosome numbers from $2n$ = 36–ca. 108 is reported, as well as intraspecific hybridisation. Some of the species with higher ploidy levels frequently have male-sterile individuals that are apomictic.

Fruit and Seed Structure. The fruit wall and seed of Labiatae are intimately connected, the fruit being either indehiscent and usually dispersed as a unit, or breaking into usually two or four, rarely more, mericarps (Fig. 24D). Building on the work of earlier morphologists, notably Wagner (1914) and Wojciechowska (1966), Ryding, in a series of papers, has made a detailed study of pericarp structure in Labiatae. These are summarised at length, with phylogenetic conclusions, in Ryding (1995), from which the following account is largely taken. The ontogenetic relationships of the various tissues found in the fruit wall of Labiatae cannot be easily accommodated in the simple categories exocarp, mesocarp and endocarp but, nevertheless, all taxonomic groups of Labiatae appear to demonstrate variations from a single base pattern. Ryding recognised four main tissue regions, which he terms exocarp, mesocarp, the sclerenchyma region and the endocarp. The exocarp is usually thin, and mostly composed of a single epidermal layer. In Nepetoideae, but in no other subfamily, the majority of genera and species possess characteristic mucilaginous cells which, on contact with water, disrupt to extrude hygroscopic spiral fibrillae. These have also been examined by Hedge (1970) in *Salvia* and by Budantsev and Lobova (1997) in Nepetinae (Nepetoideae). Hedge (1970), who found much variation in pericarp structure among the species that he studied, commented on the various adaptive functions of the mucilage cells that had been proposed, but concluded that mucilage probably plays an important role in anchoring the nutlet to the soil. Ryding (2002) and some earlier authors have suggested that the mucilage may also serve as a means of dispersal by epizoochory. Ryding (1995) noted that variations in the pattern of distribution of mucilaginous and non-mucilaginous cells, of the exocarp, could have taxonomic significance. Fleshy fruits, found in many genera in subfamilies Viticoideae and Ajugoideae, are usually due to modification of the mesocarp, but in *Prasium* (Lamioideae) and in *Clerodendrum cephalanthum* (Ajugoideae) the exocarp cells are vertically very elongate, forming a fleshy layer. In *Amasonia*, *Spartothamnella* (Ajugoideae), *Gmelina* (Viticoideae) and *Tectona* (Incertae sedis), the exocarp is sclerenchymatous, while in *Lepechinia* and the closely related *Chaunostoma* (Nepetoideae) these cells have long, parallel pit canals. Cell shape, presence of glandular or non-glandular trichomes and other structures, such as tubercles in *Aeollanthus* spp. (Nepetoideae), are all formed from exocarp cells, but many apparently superficial features of the nutlet are the result of variation in structures formed by the mesocarp or sclerenchyma layer. The mesocarp is usually composed of parenchymatous cells, and may be differentiated into a number of layers. Stone cells, fibres or collenchymatous cells are occasionally found. In *Hyptis* sect. *Minthidium*, some species such as *H. fasciculata* exhibit a pair of conspicuous air-sacs on the nutlet, which are formed from collenchyma cells. In the myrmecochorous nutlets of *Lamium* (Lamioideae) and *Ajuga* (Ajugoideae), there is a layer of fatty tissue – the elaiosome – on the abcission scar, which is attractive to ants (Lüönd and Lüönd 1981; Packham 1983). Crystal inclusions in the inner mesocarp appear to be a feature of most Salviinae (Nepetoideae), while many Plectranthinae have crystals in the second innermost layer. Papillae, formed by parenchymatous mesocarp cells and which are externally apparent, are a feature of several Scutellarioideae. In *Tinnea*, these can form a wing. The origin of the sclerenchyma layer, which forms the pyrene in many Labiatae, is often obscure and it may be derived from mesocarp or endocarp. It can display a wide range of cell types, often including a layer of vertically arranged, elongate bone cells, and stone cells are also often present. The sclereids may be elongate, star-shaped or more irregular. In some cases, the outer layer of the pyrene may be heavily sculptured, and this can account for some of the visible surface sculpturing on the nutlets of many groups, such as the Nepetoideae and Ajugoideae. In a few genera the sclerenchyma layer seems to be lacking and may be replaced by elongate cells or fibres, often with large pits or scalariform thickening, as in *Comanthosphace*, where these cells may be of vascular origin. The endocarp is usually composed of a thin, single-celled layer. Ryding's (1995) cladogram, based on pericarp characters, largely supports the findings of Cantino (1992a), based on a range of other morphological characters. For further details, reference should be made to this paper (Ryding 1995).

A brief summary of some of the seed characters, found in Labiatae, is provided by Cantino and Sanders (1986). The presence or absence of endosperm is of importance at the higher levels of classification within the family, but this character must be treated with caution. As pointed out by Wunderlich (1967), the seeds of Labiatae generally contain only small amounts of, or no endosperm. The one exception appears to be Prostantheroideae, where the seeds of *Prostanthera* were described by Bentham (1876) as albuminous. This is also true of the only other tribe of this subfamily: Chloantheae, formerly included in Verbenaceae. Guignard (1893) was the first to carry out a detailed survey of seed anatomy of Labiatae, and he recognised two groups, based on abundance of endosperm. The first group, which today is represented by Nepetoideae, possesses only a single layer of endosperm persisting in the mature seed, and can be considered exalbuminous. His second group, which he termed albuminous, has several persistent layers of endosperm, although microscopical examination is often necessary to observe them. This latter group included many genera now placed in subfamily Lamioideae: genera erroneously considered to be exalbuminous by

Bentham (1876), as well as *Teucrium* and *Ajuga* (Ajugoideae) and perhaps *Scutellaria*, where only two endosperm layers were observed. For further details and the few exceptions, see Cantino and Sanders (1986). There appear to be few if any reliable records of endosperm, however, for the many genera, now in Viticoideae, Ajugoideae or Scutellarioideae, which have been transferred from Verbenaceae. Most authors (Bentham 1876; Lam 1919) have stated that these groups are exalbuminous but, for the reasons stated above, they need reinvestigation, as these findings may prove erroneous. Embryo shape was investigated for a wide range of Angiosperm seeds by Martin (1946), and Cantino and Sanders (1986) have summarised the results for Labiatae. Three embryo types occur: spatulate, bent and investing. The investing type, in which the straight embryo has cotyledons that enclose it, was found only in the Nepetoideae, while most Lamioideae had the spatulate type, where the embryo is of this form. The bent embryo type is recorded only in *Catoferia* (Nepetoideae) and in *Scutellaria* (Scutellarioideae). The embryos of *Ajuga* and *Teucrium* (Ajugoideae) are said to be slightly curved (Wojciechowska 1966). Other subfamilies require further study.

Dispersal. Diaspores of most Labiatae, such as Lamioideae and Nepetoideae, are typically mericarps. The dispersal of these groups, with fruits which break into four mericarps or nutlets, has been reviewed by Bouman and Meeuse (1992). Species of these subfamilies are frequently members of open savannah communities, subject to a seasonal climate. In many examples, dispersal seems to be without specialised adaptations, the dry mericarps being released from the calyx during the dry season and being subject to the vagaries of such habitats, where air movements or rain wash or sudden flooding can carry seeds, mixed with other debris, some distance from the parent plant. However, many species appear to have special adaptations to aid dispersal. Winged nutlets occur in *Tinnea* (Ajugoideae), *Scutellaria scutellarioides* (Scutellarioideae) and in *Hyptidendron arboreum* and *Eriope latifolia* (Nepetoideae). Some genera have species with nutlets with apical hair or scales, which can presumably assist dispersal by wind. These include *Achyrospermum* and *Eremostachys* (Lamioideae), while nutlet pericarps of *Thymus* (Nepetoideae) are reported to have gas-filled cavities. Dispersal by water is probably frequent among species of aquatic or semi-aquatic habitats, and some, e.g. *Scutellaria galericulata* (Bouman and Meeuse 1992), possess nutlets which can float for long periods without losing their viability. In *Mentha aquatica* (Nepetoideae), the pericarp appears to be composed of loose cells which enable the nutlet to float, while in species adapted to seasonally drier conditions, such as *M. suaveolens*, the nutlets have a darker, more compact surface which causes them to sink when wetted. In *Hyptis fasciculata* (Nepetoideae), a plant of wet habitats, the nutlets have a pair of "air-sacs" which probably assist dispersal when water levels are high. Myxospermy, which is frequent in this subfamily, may have functions related to dispersal, both by fixing the mericarps to the soil or by epizoochory (see preceding section). Animal dispersal is frequent in the family. Among the genera with four mericarps, a few have a fleshy pericarp, especially in Lamioideae. Bouman and Meeuse (1992) cite *Prasium*, *Stenogyne* and *Phyllostegia* as examples. Field observations on means of dispersal, however, appear to be lacking. The hooked mericarps in *Rostrinucula* (Lamioideae) may adhere to passing animals (Ryding, pers. comm.). Myrmecochory appears to be rather uncommon, only being reported in *Lamium* (as *Lamiastrum* – Lamioideae) and *Ajuga* (Ajugoideae). In both of these, there is a layer of fatty tissue – the elaiosome – on the abcission scar, which is attractive to ants who strip it off, before leaving the light-sensitive seeds, still surrounded by pericarp, to germinate (Lüönd and Lüönd 1981; Packham 1983). In very many cases, the calyx or other parts of the plant play an important role in dispersal. In some cases this may involve the entire plant, such as dispersal by wind of desert tumbleweeds such as occur in *Teucrium* and *Salvia* (Zohary 1937). In *Rhaphiodon* (Nepetoideae), the compact, spherical infructescence is the dispersal unit. It becomes detached at maturity as a spiny "bur", each fruiting-calyx being furnished with several long, rigid spines. The nutlets appear to be released when the tough calyx finally disintegrates. More frequently, the calyx, with the nutlets still inside, becomes the dispersal unit. *Scutellaria mexicana* and the Asiatic *S. utriculata* (Scutellarioideae) have large inflated calyces that are wind-dispersed (Paton 1992), and this also seems to occur in the N African *Saccocalyx* (Nepetoideae) (Boumann and Meeuse 1992), in *Tinnea* sect. *Membranaceae* (Scutellarioideae) and in some steppe species of *Salvia* (Nepetoideae; Hedge, pers. comm.). The fruiting-calyx of *Colebrookea* (Lamioideae), which can easily become detached from the plant, develops long and narrow, plumose lobes and may be blown by the wind (Ryding, pers. comm.). This is probably also true of *Kurzamra pulchella*, from the

Atacama desert of Chile, which possesses similarly plumose calyx-lobes. In *Hoslundia opposita* (Nepetoideae), the calyx becomes fleshy and brightly coloured, suggesting dispersal by birds. In many cases, the calyx takes on the function of a capsular fruit, by not permitting the mericarps to escape until conditions are suitable. This may be brought about by hygroscopic movements of the calyx-lobes, as in *Eriope stricta* (Nepetoideae). In *Marsypianthes* (Nepetoideae), the calyx-lobes are connivent until the rather large mericarps are mature, when they become widely spreading. The mericarps, which have involute, fimbriate wings, are probably then dispersed by wind. Very many species, e.g. *Eriope* spp., *Plectranthus barbatus*, *Clinopodium* spp. (Nepetoideae), have a dense tuft of hairs in the calyx-throat, which helps to retain the nutlets within the calyx. Probably in this case, the nutlets are retained until the calyx becomes dry and the hairs wither and become very brittle, when any movements will assist in dislodging them, so that the nutlets may not be released until towards the end of the dry season. Spinose calyx-lobes, especially when hooked at the tip, as in *Oncinocalyx* (Ajugoideae), *Marrubium vulgare*, *Notochaete* sp. (Lamioideae) and *Hyptis uncinata* (Nepetoideae), can provide an effective means of animal dispersal (Munir 1991). *Marrubium vulgare* is now distributed almost worldwide, dispersed in sheep's wool. Paton (1992) made a detailed study of dispersal mechanisms in species of *Scutellaria*. In sect. *Scutellaria*, the mature nutlets are loose within the closed calyx, whose upper lip is furnished with a projecting scutellum, and is deciduous. Any slight touch dislodges the upper lip, and no doubt consequently the nutlets. In *Scutellaria* and many other genera, such as *Ocimum* (Nepetoideae), the fruiting pedicels become rigid and springy, and often strongly curved, so that passing animals, which brush past, may cause the nutlets to be catapulted to some distance from the calyx. In *Aeollanthus* (Nepetoideae), Ryding (1986) describes various dispersal mechanisms with a rather similar function.

The success of seeds after dispersal is beyond the scope of this paper. The influence of mature plants over seed germination, due to exudation of inhibiting substances – allelopathy – is reported by Heisey and Delwiche (1984) in *Trichostema* (Ajugoideae).

Apart from the above, dispersal in some Labiatae is achieved by vegetative means. Many, especially temperate, genera have extensive rhizome systems. In *Mentha*, these are frequently carried by water, along riversides or elsewhere, and some sterile clones have very natural distributions. Some species, e.g. *Plectranthus luteus* and *P. melleri*, have axillary stem tubers, with modified, swollen internodes that drop off and no doubt aid dispersal, as do the subterranean tubers of such species as the Chinese artichoke, *Stachys affinis* (Lamioideae) and *Plectranthus esculentus* (Nepetoideae). These last two species are widely cultivated and their distribution, as that of many others, is much assisted by anthropogenic factors. Among the indehiscent-fruited Labiatae, there seems to be less information available about their dispersal mechanisms. Those with a fleshy pericarp, such as species of *Vitex* (Viticoideae), *Clerodendrum* (Ajugoideae), *Callicarpa* and *Tectona* (Incertae sedis), are probably often distributed by birds. The red fruits of *Aegiphila* (Ajugoideae) are taken by Yellow-bellied Elaenias, *Elaenia flavogaster* (Tyrannidae) in NE Brazil (pers. obs.).

PHYTOCHEMISTRY. Species belonging to the Labiatae produce a great variety of secondary compounds, but are especially renowned for their essential oils, which are produced in glandular hairs on the surfaces of leaves and inflorescences. Numerous papers and books have been written on the essential oils in species of the family, and good reviews are those by Hegnauer (1966, 1989), Lawrence (1992) and Başer (2000). Abstracts of papers on essential oils in the Labiatae written between 1973 and 1993 can be found in Cousins (1994). However, essential oil production is not a feature of the family as a whole, as it is only common in the subfamily Nepetoideae (Hegnauer 1989). The volatile oils in this subfamily mainly consist of monoterpenoids, e.g. menthone in *Mentha* and thymol in *Thymus*, sesquiterpenoids such as caryophyllene, and sometimes volatile phenylpropanoids such as eugenol and methyl chavicol, e.g. in *Ocimum* species (Grayer et al. 1996). There is often a considerable infraspecific variation in essential oils. For example, species of *Mentha* and *Ocimum* comprise many different essential oil chemotypes. Genera rich in essential oils outside the Nepetoideae include *Pogostemon* (patchouli, subfam. Lamioideae), *Sideritis* (subfam. Lamioideae) and *Prostanthera* (subfam. Prostantheroideae) (Hegnauer 1966, 1989). However, instead of essential oils, most species of the Lamioideae, Viticoideae, Ajugoideae and Scutellarioideae produce iridoids, which are monoterpenoid lactones, to which frequently sugars are attached (iridoid glycosides). Surveys of the iridoid distribution in the Labiatae have been carried out by Kooiman (1972), and of genera previously belonging to the Verbenaceae by Rimpler and co-workers (see Hegnauer 1990). Iridoids are not completely lacking from the Nepetoideae, but are very rare. In the genus *Nepeta*, non-glycosidic volatile iridoids such as nepetalactone are present in the essential oils of many species, and some species of this genus also contain iridoid glycosides. The presence of iridoids is probably a plesiomorphic character in the Labiatae, and essential oil production an apomorphic feature, because iridoids are generally present in closely related, "more primitive" families such as the Verbenaceae and Scrophulariaceae, whereas essential oils are rare in these. The distribution of iridoids and essential oils within the Labiatae is highly correlated with pollen features and thus with the classification by Erdtman (1945), who divided the family into two subfamilies, the Nepetoideae and the Lamioideae sensu lato, on the basis of the two main pollen types.

A class of terpenoid constituents characteristic of Labiatae as a whole, but uncommon in related families, are diterpenoids. These have been found in genera belonging to subfamilies Viticoideae, Ajugoideae, Prostantheroideae, Scutellarioideae, Lamioideae and Nepetoideae, and show a great variety of skeletal types. Representatives of the subfamily Symphorematoideae do not yet seem to have been studied for these constituents. The most common diterpenoids in Labiatae are labdanes, neoclerodanes, abietanes, primaranes and *ent*-kauranes (Hegnauer 1989; Vestri Alvarenga et al. 2001). Labdanes are characteristic of many genera in the subfamily Lamioideae, e.g. *Ballota*, *Galeopsis*, *Lagochilus*, *Leonotis*, *Leonurus*, *Leucas*, *Marrubium* and *Roylea*. However, *ent*-kaurane diterpenoids are more common in *Sideritis*, whereas species of *Stachys* produce mainly neoclerodane diterpenoids. Some neoclerodanes have also been reported from another Lamioideae genus, *Leonurus*. However, the great majority of neoclerodanes in the Labiatae have been found in the subfamilies Scutellarioideae and Ajugoideae, especially in the genera *Scutellaria* (98 different types), *Ajuga* (82 types) and *Teucrium* (196 types) (Vestri Alvarenga et al. 2001).

Neoclerodanes also have been reported from genera in the Ajugoideae that used to belong to Verbenaceae, such as *Clerodendrum*, *Peronema* and *Caryopteris*. Furthermore, neoclerodanes are the only type of diterpenoid that has been reported from Prostantheroideae (from *Cyanostegia* and *Pityrodia*), but presumably few species of this subfamily have been surveyed for diterpenoids. Abietane diterpenoids are characteristic of subfamily Nepetoideae and are found in genera belonging to most tribes and subtribes, e.g. *Salvia*, *Lepechinia*, *Meriandra* (tribe Mentheae, subtribe Salviinae), *Horminium* (Mentheae, Menthinae), *Nepeta* (Mentheae, Nepetinae), *Hyptis* (Ocimeae, Hyptidinae), *Fuerstia*, *Hoslundia*, *Orthosiphon*, *Tetradenia* (Ocimeae, Ociminae), *Plectranthus*, *Aeollanthus*, *Dauphinea* (Ocimeae, Plectranthinae), and *Isodon* (Ocimeae, subtribe *Hanceolinae*). The largest number of different abietanes has been found in *Salvia* (275 types) and *Plectranthus* (incl. *Coleus* and *Solenostemon*, 135 types). *Salvia* species also produce a variety of diterpenoids of different skeletal types such as neoclerodanes and primaranes. Species in the genus *Isodon* do not produce abietanes, but mainly *ent*-kauranes and fujinanes. *Ent*-kauranes have also been reported from *Plectranthus* and *Salvia*, but fujinanes are exclusively found in *Isodon*. In Viticoideae, only few but a wide variety of diterpenoid types have been found, which occur in all the other subfamilies (labdanes, neoclerodanes, abietanes, primaranes and *ent*-kauranes), supporting the theory that Viticoideae is a polyphyletic group, basal to the remaining subfamilies of Labiatae. Many diterpenoids found in the Labiatae have biological activity, such as antibacterial, antifungal and insect antifeedant activities, and therefore may have an ecological function in the plants (Cole 1992; Simmonds et al. 1996). Higher terpenoids such as triterpenoids and triterpene saponins are also common in Labiatae. According to Hegnauer (1989), pentacyclic triterpene acids probably occur in nearly every species of the family.

Apart from producing many different terpenoids, species of Labiatae also produce a large variety of phenolic constituents, which are often accumulated in the plants in large concentrations. The most common groups of phenolics in this family are caffeic acid esters and flavonoids. Caffeic acid esters provide important chemosystematic markers in Labiatae at the subfamily level. The two major types are (1) rosmarinic acid and (2) phenylethyl caffeoylglycosides, and their occurrence is again strongly correlated with pollen type (Pedersen 2000). The presence of rosmarinic acid, which is an ester of caffeic acid with α-hydroxydihydrocaffeic acid, is almost entirely restricted to Nepetoideae, and the compound has been found in a large percentage of the species and genera investigated for this subfamily. However, in a few species of Nepetoideae it appears to be replaced by a different caffeic acid ester, chlorogenic acid (Pedersen 2000). Rosmarinic acid, when present, may constitute more than 3% of the dried weight of Nepetoideae species. It is a powerful antioxidant, but this means that it is easily oxidised when the plant material is dried, causing herbarium material to turn black. However, dark discoloration of herbarium material can also be caused by disintegration of iridoids in plants, so that black herbarium specimens are not confined to Nepetoideae. The second major group of caffeic acid conjugates in Labiatae, phenylethyl caffeoylglycosides, e.g. acteoside or verbascoside, are absent from Nepetoideae, but characteristic of most of the other subfamilies. Pederson (2000) found this type of compound in most of the species investigated for Ajugoideae, Lamioideae and Prostantheroideae when fresh material was used, but much less frequently in herbarium material, which suggests that these compounds also disintegrate on drying. In Scutellarioideae, acteoside-like compounds were only found in 2 out of 10 species investigated. In the genus *Scutellaria*, esters of phenolic acids with the iridoid glycoside catalpol are common (Cole 1992), so perhaps these esters partly replace the phenylethyl caffeoylglycosides in Scutellarioideae. In Viticoideae, phenylethyl caffeoylglycosides are generally present in species of *Callicarpa*, *Cornutia* and *Geunsia*, and in some species of *Gmelina* and *Premna*. However, they are absent from the genera *Hymenopyramis*, *Vitex*, and *Teijsmannidendron*. They are replaced by rosmarinic acid in *Hymenopyramis*, and by chlorogenic acid in *Teijsmannidendron* and in many species of *Vitex* (Pederson 2000), which are compounds associated with Nepetoideae. Therefore, like the distribution of diterpenoid types, the distribution of caffeic acid esters supports the theory that Viticoideae are a polyphyletic group, which is basal to the other subfamilies of Labiatae, including Nepetoideae.

The two main types of flavonoids present in Labiatae are (1) "internal" flavonoid glycosides and (2) "external" flavonoid aglycones. Internal flavonoid glycosides are polar compounds because they have sugars attached, and they are thought to occur in the cell sap of vacuoles, mainly in green tissues (leaves, stems and fruits). Flavonoid glycosides are present in nearly every species of the angiosperms, gymnosperms, ferns and liverworts and in some mosses. The most common type in Labiatae are the flavone O-glycosides, frequently hydroxylated in the 6-position. Flavone glycosides hydroxylated in the 8-position also occur, but are much rarer. The latter are useful taxonomic markers below the genus level, especially at the sectional level, e.g. in *Sideritis*, *Teucrium* (Tomás-Barberán and Gil 1992) and *Lavandula* (Upson et al. 2000). Apart from flavone O-glycosides, many species also produce flavone C-glycosides, especially the compound vicenin-2, and flavonol O-glycosides. Flavanone O-glycosides are quite common in subfamily Nepetoideae, tribe Mentheae, subtribe Menthinae (Tomás-Barberán and Gil 1992). External flavonoid aglycones do not contain sugars and are often highly methylated and therefore non-polar (insoluble in water). Their distribution is restricted to certain plant families, and in Labiatae they tend to occur in species producing essential oils and/or diterpenoids in glandular hairs on the plant surfaces. For that reason they are also called "surface" flavonoids. Surface flavonoids with a 5-hydroxy-6,7-dimethoxy substitution pattern of the A-ring, e.g. cirsimaritin, are typical for the family as a whole (Tomás-Barberán and Gil 1992). Additional hydroxylation/methylation patterns of surface flavonoids are sometimes characteristic at the tribal level. For example, flavones with a 5,6-dihydroxy-7,8-dimethoxy substitution pattern of the A-ring, e.g. thymusin, are found in Nepetoideae tribe Mentheae, subtribe Menthinae, but not in other groups of Labiatae (Tomás-Barberán and Gil 1992), whereas a 5,7-dihydroxy-6,8-dimethoxy substitution pattern (e.g. in nevadensin) is characteristic of the tribe Ocimeae, subtribe Ociminae (Grayer et al. 2001). The genus *Scutellaria* produces a large number of very unusual external flavonoids, which often possess an unsubstituted B-ring or 2′-oxygenation. Many of these *Scutellaria* flavonoids are unique in the plant kingdom.

The fatty acid compositions of the seed oils of Labiatae provide further chemical characters at various levels of classification, e.g. at subfamilial and generic level (Hegnauer 1989). The seed oils in species of subfamily Nepetoideae are usually very rich in linolenic acid, and the ratio linolenic acid/linoleic acid was found to vary from 1.5 to 6.6 (Marin et al. 1991). In Ajugoideae, this ratio varied from 0.5 (species of *Ajuga*) to 1.0 (species of *Teucrium*), whereas in Scutellarioideae this ratio was as low as 0.02. The linolenic acid/linoleic acid ratio in species of the subfamily Lamioideae is also very low (Hegnauer 1989). However, the seed oils in this subfamily are characterised by the occurrence of allenic fatty acids such as laballenic acid, which is present in high concentrations in some species, e.g. 25.1% in *Phlomis tuberosa* (Aitzetmüller and Tsevegüren 1998). Another allenic fatty acid, lamenallenic acid, seems to be restricted in occurrence to the genus *Lamium*.

Some nitrogen-containing compounds, such as alkaloids and cyanogenic glycosides, have been reported from Labiatae, but they are not prominent (Hegnauer 1989). Betaines are generally present in the family, in relatively high concentrations in Lamioideae and Ajugoideae, and low concentrations in Nepetoideae (Blunden et al. 1996).

Distribution and Habitats. Hedge (1992) has made a review of the biogeography of Labiatae, excluding those genera formerly included in Verbenaceae. The following is largely taken from his account, together with additional data, where these are lacking. Hedge recognised six regions of high Labiate diversity: (1) Mediterranean and SW Central Asia; (2) Africa south of the Sahel and Madagascar; (3) China; (4) Australia; (5) South America; (6) Northern America and Mexico. To this should be added (7) Indomalesian region (SE Asia), which is the centre of diversity of Symphorematoideae and many Viticoideae. The remaining subfamilies are primarily adapted to more open, savannah and upland habitats, and are characteristic of the other regions, cited by Hedge (1992). Several species-rich genera have very wide natural distributions, often with several centres of diversity. These fall generally into two groups, based on their distribution: those primarily tropical in origin, including *Vitex* (Viticoideae), *Clerodendrum* (Ajugoideae), *Callicarpa* (Incertae sedis), *Ocimum*, *Hyptis* (Nepetoideae), and those which probably had a temperate origin, but often extend into the montane tropics, such as *Ajuga*, (Ajugoideae), *Scutellaria* (Scutellarioideae), *Stachys* (Lamioideae), *Salvia*, *Clinopodium*, *Mentha*, (Nepetoideae). *Teucrium* (Ajugoideae) is unusual, in that it has a distinct distribution in the southern hemisphere, and yet is most species-rich in the north, spreading down to the mountains of tropical Africa. At species level, *Prunella vulgaris* (Nepetoideae) must have one of the widest, seemingly natural distributions in temperate regions and in tropical and subtropical mountains, although it is probably introduced and completely naturalized in many areas.

1. The Mediterranean and SW Central Asiatic region. This is one of the greatest centres of diversity of many genera, e.g. *Teucrium* (Ajugoideae), *Scutellaria* (Scutellarioideae) *Phlomis*, *Stachys*, *Lamium*, *Sideritis*, (Lamioideae), *Salvia*, *Satureja*, *Thymus*, *Origanum*, *Nepeta* (Nepetoideae), and many others. Genera endemic to the area include *Cedronella*, *Horminum* and *Cleonia* (Nepetoideae) and *Melittis*, *Prasium*, *Moluccella*, *Isoleucas* (Lamioideae), but many other genera centred here extend eastwards as far as NW China. A few genera are circum-boreal, extending also into North America (see below).

2. Africa south of the Sahel and Madagascar. The majority of genera to be found in this mainly tropical region belong to Nepetoideae tribe Ocimeae, and two subtribes of this, that are almost entirely restricted to the Old World: Plectranthinae and Ociminae. Endemic genera of these groups include *Tetradenia*, *Aeollanthus*, *Pycnostachys*, *Holostylon*, *Hoslundia*, *Syncolostemon*, *Haumaniastrum*, *Thorncroftia*. Other endemic genera include *Tinnea* (Scutellarioideae) and *Acrotome* (Lamioideae). *Leonotis* is composed of many species endemic to Africa, but has one weedy species extending to other continents. *Achyrospermum* (Lamioideae), although not endemic, is of interest, being one of the few forest genera from this subfamily. Many other genera, such as *Leucas*, *Leonotis*, *Stachys* (Lamioideae), *Salvia* and *Micromeria* (Nepetoideae tribe Mentheae), extend elsewhere, especially into Arabia and other parts of subtropical or tropical Asia. Southern Africa, which includes the temperate Cape region, is a secondary centre of diversity of *Stachys*, with 41 spp., and *Salvia* (26 spp.) (see Codd 1985). Madagascar (see Hedge 1998), with 90% endemism among Labiatae at the species level, has, at present, only three small endemic genera: *Capitanopsis*, *Madlabium* and *Dauphinea* (Nepetoideae tribe Ocimeae). These are related genera, falling within the same clade as *Plectranthus*, and their status requires further evaluation.

3. China has a very large and diverse Labiate flora, perhaps as many as 100 genera (Li Xi-wen and Hedge 1994; Chen Shou-liang and Gilbert 1994), reflecting its area and diversity of climate and habitat. Genera with a substantial number of Chinese species include *Callicarpa* (Incertae sedis), *Premna* (Viticoideae), *Clerodendrum*, *Ajuga* (Ajugoideae), *Gomphostemma*, *Chelonopsis*, *Phlomis*, *Paraphlomis* (Lamioideae), *Scutellaria* (Scutellarioideae), *Elsholtzia*, *Salvia*, *Nepeta*, *Dracocephalum*, *Isodon* (Nepetoideae). Endemics include *Schnabelia* (Ajugoideae), *Wenchengia* (Scutellarioideae), *Bostrychanthera*, *Loxocalyx* (Lamioideae), *Rostrinucula*, *Heterolamium* and *Hanceola* (Nepetoideae), *Holocheila* and *Ombrocharis* (Incertae sedis). The correct systematic position and status of the two latter genera is still in doubt. Between

China and the Indomalesian region (see 7 below) rises the high Himalayan massif, which has a number of genera endemic or typical of it. These include *Colquhounia*, *Alajja*, *Eriophyton*, *Roylea*, *Notochaete* (Lamioideae) and *Marmoritis* (Nepetoideae).

4. Australia. The Australian region is remarkable in having an endemic subfamily of Labiatae: Prostantheroideae. This has two tribes: Chloantheae and Westringieae, the former at one time included within the Verbenaceae. The largest of the 16 genera is *Prostanthera*, with about 100 species. Other genera, occurring in the area, include *Viticipremna*, and *Gmelina* (Viticoideae) and *Callicarpa* (Incertae sedis), which extend from the Indomalesian region into the more tropical parts of Australia. Other endemic genera include *Spartothamnella*, *Oncinocalyx* and *Huxleya* (Ajugoideae). A number of other widespread genera, all of which have their centres of diversity elsewhere, include *Teucrium*, *Ajuga* (Ajugoideae), *Scutellaria* (Scutellarioideae) and the following Nepetoideae: *Lycopus*, *Mentha*, *Salvia* (Mentheae) and *Plectranthus* (Ocimeae).

5. South America. The South American Labiate flora, which contains two very large genera, *Salvia* and *Hyptis*, but rather few genera, falls naturally into three regions: (A) Andean; (B) Guianan and Brazilian shields; (C) temperate South America.

   A. The Andean Labiatae are primarily derived from north temperate genera, and the various groups have many links to Central America, including Mexico. The largest group is *Salvia* subg. *Calosphace* with its Andean centre of diversity linked to a much larger one in Mexico. *Scutellaria*, *Stachys* and *Clinopodium* also appear to have been derived from a southern migration from Central America, but have radiated to produce endemic groups, in some cases associated with adaptations to humming-bird-pollination. Other Andean Nepetoid genera include the monotypic *Obtegomeria* from Colombia, related to *Clinopodium*, *Minthostachys*, which appears most closely related to the Macaronesian *Bystropogon*, and *Lepechinia*, which extends to Mexico with one Hawaiian extension. The small neotropical genus *Catoferia* (Nepetoideae, Ocimeae) also links Central and S America, but its closest relatives would seem to be eastern Asiatic members of the Old World subtribe Ociminae, perhaps the genus *Orthosiphon*. The only other member of the subtribe native to the Americas is *Ocimum*, which extends from Mexico to South America. One species extends into lowland areas of the Andes.

   B. Guianan and Brazilian shields. This region, including Amazonian and eastern South America, mostly overlying much older geological formations, has a very different flora, dominated by the Nepetoid Ocimeae, subtribe Hyptidinae. This is composed of the large genus *Hyptis* with about 300 species, and a number of smaller endemic, satellite genera, such as *Peltodon*, *Rhaphiodon*, *Hypenia*, *Hyptidendron* and *Eriope*. The primarily Old World genus *Ocimum* (Nepetoideae: Ocimeae) is represented by several species endemic to the area. It seems probable that *Ocimum* originally had an Old World tropical origin, perhaps reaching the New World via W Africa in the early Tertiary. A few species of *Salvia* (Mentheae) reach the mountains of eastern Brazil and extend south into temperate S America. Other endemic genera to be found in this region include *Cornutia* (Viticoideae), *Amasonia* and *Monochilus* (Ajugoideae), and more widespread genera such as *Vitex* and *Clerodendrum* (Viticoideae) and *Aegiphila* (Ajugoideae). The latter, with over 100 species, extends throughout tropical America.

   C. Temperate South America. A number of endemic genera occur in this region, belonging to Nepetoideae, tribe Mentheae: *Kurzamra*, *Rhabdocaulon*, *Glechon* and *Hoehnea*, while *Cunila* and *Hesperozygis* are trans-equatorial, being also represented in Mexico.

6. Northern America and Mexico. This region has quite a large number of endemic genera, both in the west, especially California, as well as in the east. Most of these belong to tribe Mentheae (Nepetoideae) and include *Monarda*, *Monardella*, *Pycnanthemum*, *Rhododon*, *Stachydeoma*, *Pogogyne*, *Acanthomintha*, *Neoeplingia* and *Blephilia*. Endemic Lamioideae include *Warnockia*, *Physostegia*, *Macbridea*, *Brazoria* and *Synandra*. *Trichostema* (Ajugoideae) is a near-endemic,

extending into the Caribbean. More widespread, species-rich genera represented in the area include *Scutellaria* (Scutellarioideae), *Stachys* (Lamioideae), *Salvia*, *Lycopus*, *Clinopodium* and *Hedeoma* (Nepetoideae), the last a New World genus extending into Andean and temperate South America. Of particular interest is *Lycopus* (Mentheae), with a circumboreal distribution but well represented in North America, while *Mentha* and *Dracocephalum*, both primarily Old World genera, each have one species native in North America. One or two members of the species-rich genus *Clerodendrum* (Ajugoideae), primarily of the Old World tropics, occur in southern areas. The primarily tropical Viticoideae is represented by *Vitex*, which in the New World is confined to the Caribbean and Central America from Mexico southwards, but there are also two genera endemic to the Caribbean: *Petitia* and *Pseudocarpidium*. Mention should also be made here of *Callicarpa* (Incertae sedis). This is a mainly Asiatic genus, with a few New World species, especially *C. americana*, in the S United States and Mexico, and others in the Caribbean.

7. Indomalesian region (SE Asia). This region is the centre of diversity of Symphorematoideae and many Viticoideae, both of which include many forest trees and lianas, such as *Sphenodesma* (Symphorematoideae), *Gmelina* and *Teijsmanniodendron* (Viticoideae). The last two, together with *Tsoongia* and *Paravitex*, are endemic to the region. Other speciose viticoid genera, such as *Premna*, *Vitex* and *Callicarpa* (the last now treated as Incertae sedis), extend much more widely through the tropics, some even into tropical America. Subfamily Ajugoideae includes the monotypic *Hosea*, endemic to Indomalesia, and several other genera, such as *Clerodendrum*. However, this, and a number of smaller genera which extend into, or have their centre of diversity in tropical Africa, Madagascar, Polynesia or Australia, are temperate or subtropical. The remarkable genus *Oxera*, with 21 species, is almost entirely restricted to New Caledonia (de Kok 1997). *Gomphostemma*, one of the few genera of Lamioideae adapted to true forest habitats, is also primarily Indomalesian, as is *Platostoma* (Nepetoideae). Finally, mention must be made of two other woody genera, endemic to the region: *Hymenopyramis* and *Tectona* (Incertae sedis), the latter genus including commercially important timber trees.

A number of genera merit special mention: *Lycopus* and *Mentha* are species of wet habitats with a mainly north temperate distribution – Europe, Asia, N America, with species in Australia, and *Mentha* also in Africa; *Agastache* is primarily in N America to Mexico, with one species in eastern Asia; *Collinsonia* (including *Keiskea*) and *Meehania* are found in China and Japan and the eastern United States; *Cunila* and *Hedeoma* occur in temperate N and S America (and *H. mandonianum* in the high Bolivian Andes); *Otostegia* and *Meriandra* both have disjunctions between NE Africa to Arabia and in C Asia, from Afghanistan to India. *Lavandula* has centres of diversity both in the Mediterranean and in NE Africa to Arabia, and extensions to India and Macaronesia. A number of endemic island genera occur: *Teucridium* (New Zealand), *Haplostachys*, *Stenogyne* and *Phyllostegia* (Hawaii); *Bystropogon* (Macaronesia); *Cuminia* (Juan Fernandez I.); *Oxera* (New Caledonia and Pacific); *Petitia* and *Pseudocarpidium* (Bahamas and Greater Antilles).

Direct evidence for the origin of the family is sparse, as Cretaceous fossil remains are almost nonexistent or at best equivocal (see below), although the diversity and present-day distribution points to an origin during this period. Molecular studies (see below) indicate that the earliest ancestors must have been Viticoid- or Symphorematoid-like plants, suggesting a likely Gondwanan origin, probably with fairly rapid radiation into other areas during Late Cretaceous and early Tertiary times, with subsequent diversification. An Asiatic origin also seems likely for the Nepetoideae. Much of the present-day species diversity of the family, however, is no doubt due to speciation during Neogene times.

Palaeobotany. Labiatae are poorly represented in the fossil record, and many of the records must be considered dubious owing to the difficulty of determination. Perhaps the earliest record we have is that cited by Collinson et al. (1993) of fossil leaves of *Premnophyllum trigonum* from the Upper Cretaceous of Bohemia. All other records are from the Tertiary. Taylor (1990) cites a record of fossil leaves of *Clerodendrum* (Ajugoideae) from the Middle Eocene of N America and, in the same work, various records of flowers of *Holmskioldia* (Scutellarioideae) of Oligocene age, but the identity of these should be viewed with caution. Nutlets described as *Ajuginucula smithii* by Reid and Chandler (1926) from the Bembridge Oligocene flora, Isle of Wight, indicate the presence of Ajugoideae, and Mai (1985) has recorded

Lycopus (Nepetoideae) fruits from the Oligocene of Europe. Pollen records of Labiatae have been mostly of the hexacolpate Nepetoideae, no doubt due to greater ease of recognition, and have been reviewed by Muller (1981). The earliest of these is of Early Eocene age, from Rajasthan, NW India (Kar 1996). The material has been identified as pollen of *Ocimum* (Nepetoideae: tribe Ocimeae). While it appears to be Nepetoid pollen, possibly Ocimeae, the grains do not match any extant species of *Ocimum*, but show more features in common with other species, belonging to genera such as *Orthosiphon*, *Plectranthus* or *Aeollanthus* (M.M. Harley, pers. comm.). Owing to the very large number of closely related genera in this group, which must have radiated rapidly from the Miocene onwards, it may be unwise to try to identify the pollen with any extant genus. Emboden (1964) identified *Salvia* pollen from Upper Miocene deposits from Alaska, and Labiatae pollen has also been reported of about the same age from Europe (Van Campo 1976). Of greater interest, perhaps, is the discovery of Nepetoid pollen in the Messel Oil shales of early Middle Eocene age from Germany (M.M. Harley et al., in prep.). The occurrence of Nepetoideae pollen, relatively early in the fossil record, lends credence to the idea that the group arose early from ancestors with tricolpate pollen. Bees have probably played an important role as pollinators of Labiatae from the earliest times. The success of the Labiatae, and the co-evolution of their flowers with the Apidae, are doubtless related events. The Apidae had arisen at least by the Late Cretaceous, and probably much earlier (Michener and Grimaldi 1988).

Affinities, Subdivision and Relationships Within the Family. The earliest full taxonomic treatment of the Labiatae was that of Bentham (1832–1836), who created order out of what Lindley (1829) had earlier described as "the disgrace of Botany!". Bentham's accounts of the Labiatae (modified in 1876) were based on very detailed studies, especially of floral structure, which are still valid today. His classification was followed by that of Briquet (1895–1897), who made various modifications to Bentham's system, by no means all of which were improvements. This system, with modifications by Melchior (1964), is still used by many people today.

Labiatae have long been considered closely allied to Verbenaceae, sharing many characters in common. Cronquist (1981) also included Boraginaceae and Lennoaceae within his small order Lamiales. Cantino (1982b), in a cladistic analysis using the characters that unite Labiatae with either Scrophulariaceae or Boraginaceae, demonstrated that the Labiatae have a much closer affinity to the former, rather than the latter family whose relationships lie elsewhere. However, both van Tieghem (1907) and Junell (1934), using micromorphological characters of the gynoecium, such as placentation, found that the genera of Verbenaceae with cymose inflorescences showed a number of characters which differentiated them from subfamily Verbenoideae, characterised by racemose inflorescences. Junell also concluded that the placentation of the cymose genera was similar to that of Labiatae, which also possess cymose inflorescences, and that the two groups could be united into a single family. In a study of pollen morphology, Erdtman (1945) showed that Labiatae could be separated into two groups – Nepetoideae, with hexacolpate pollen which is shed at the three-celled stage, and the Lamioideae, which mostly have tricolpate pollen shed at the two-celled stage, a character which they share with many of Verbenaceae. This division was further investigated by Cantino and Sanders (1986), using a much broader range of characters including myxospermy, presence or absence of endosperm, embryo shape, as well as a number of chemical characters. They concluded that while Nepetoideae showed a number of synapomorphies and appeared to be monophyletic, the same was not true of Lamioideae. Cantino (1992a, 1992b) then went on to show, in a detailed morphological analysis of Verbenaceae and Labiatae, that the latter family was polyphyletic, with several clades arising independently from within Verbenaceae. A more natural classification would result if the cymose genera of subfamilies Caryopteridoideae, Chloanthoideae, Viticoideae and tribe Monochileae were transferred from Verbenaceae to Labiatae, leaving only subfam. Verbenoideae, with its characteristic racemose inflorescences and tricolporate pollen, in the reconstituted Verbenaceae. This is very similar to the changes proposed earlier by Junell. A provisional list of accepted genera of Labiatae, with a breakdown into subfamilies, tribes and subtribes, was published by Cantino et al. (1992) and largely adopted by Thorne (1992). Since then, there has been much support for the proposed changes based on analyses of cpDNA sequences. These studies, while providing welcome corroboration for conclusions based on non-molecular data, have also helped to clarify the relationship between Labiatae in its new delimitation, and other groups, including Verbenaceae. The most recent findings of Olmstead et al. (2001) suggest that the closest relatives of Verbenaceae sensu stricto are Martyniaceae and Bignoniaceae, rather than Labiatae, although bootstrap support for this is weak. In the same analysis, sister groups of Labiatae appear to be *Paulownia* with *Mimulus* and Orobanchaceae. Firm morphological data to support these conclusions, which were based on molecular data only, are lacking but see Bremer et al. (2001) in which molecular, chemical and morphological data were combined, in their analysis of the Euasterids. There is now, however, a clearer picture of phylogenetic relationships within the family, with the arrangement of subfamilies and genera of Nepetoideae presented here being largely in line with some of these findings (Wagstaff et al. 1995, 1998). Some of the major changes to the traditional classification include the recognition, within Labiatae, of subfams. Symphorematoideae (occasionally considered as a distinct family: Symphoremataceae) and Viticoideae, formerly part of Verbenaceae; the addition of a number of genera, formerly Verbenaceae, in subfam. Ajugoideae; the inclusion of tribe Chloantheae (= Chloanthoideae, = Chloanthaceae) within subfam. Prostantheroideae; the inclusion of *Holmskioldia*, formerly Verbenaceae) within subfam. Scutellarioideae; the reorganisation of Nepetoideae into three tribes: Elsholtzieae, Mentheae and Ocimeae; the division of the Mentheae into three subtribes: Salviinae, Menthinae and Nepetinae; the repositioning of *Lavandula*, sole member of Lavandulinae, within tribe Ocimeae, together with four other subtribes: Hanceolinae, Hyptidinae, Ociminae and Plectranthinae. The first of these, which has only recently been recognised, includes the large, primarily Asiatic genus *Isodon*, which has often been linked in the past with *Plectranthus*.

Of the seven subfamilies, the least satisfactory, as circumscribed by Cantino et al. (1992), is Viticoideae, which is clearly paraphyletic, or possibly polyphyletic. Recent molecular findings (Olmstead, unpubl. data) indicate that *Hymenopyramis* is sister to *Petraeovitex* and that the two are not close to other genera of Viticoideae. According to the same study, both *Callicarpa* and *Tectona* come out in relatively basal positions, *Callicarpa* being weakly supported as sister group to Prostantheroideae, while *Tectona* is weakly supported as sister group to most of the family. The removal of these four genera to Incertae sedis would leave Viticoideae more homogeneous. At present, different kinds of data (molecular or morphological) are giving conflicting results, and it is clear that the whole group is in need of much more detailed studies. With molecular and morphological analyses continuing, we can expect further changes to some groups in the future.

One of the most distinctive groups, which is consistently monophyletic in most studies, is Nepetoideae, partly characterised by its hexacolpate pollen. A molecular study by Wagstaff et al. (1998) suggests that the subfamily may have been derived early in the history of the family from Symphorematoid or Viticoid ancestors, and the gynobasic style may have evolved independently in Lamioideae. The origin of Nepetoid Labiatae remains something of a mystery. They are more readily identified in the fossil record than other Labiatae, due to their distinctive hexacolpate pollen which first appears in the Eocene, and an Asiatic origin seems likely. Three tribes have been recognised within Nepetoideae: Elsholtzieae, Mentheae and Ocimeae. Two genera have not been assigned to tribe. Cladistic studies have shown *Melissa* to be basal to a clade that includes both Elsholtzieae and Mentheae, and therefore not readily assignable to any tribe. This emphasises the more isolated position of Ocimeae and could be resolved by combining the other two tribes to include *Melissa* also. Recent molecular data on Ocimeae (Suddee 2001) indicate that the subtribes Hyptidinae, Plectranthinae and Ociminae each form a clade, and that *Isodon* (Hanceolinae) is the sister group to the remaining Ocimeae. *Heterolamium*, for which good material is lacking, has in the past been assigned to Ocimeae, but on closer examination appears more probably to be a member of Mentheae subtribe Nepetinae. It is at present included in Incertae sedis.

Several genera cannot be assigned with confidence to a subfamily. The distinctive inflorescence structure of three of them, *Acrymia*, *Cymaria*, and *Garrettia*, suggests that they are closely related. The latter two also have similar pollen (Abu-Asab and Cantino 1993). DNA sequence data (R.G. Olmstead, unpubl. data) suggest that *Cymaria* is the sister group of Lamioideae, but *Acrymia* and *Garrettia* were not included in that study. Some aspects of the morphology of *Holocheila* suggest a relationship to Scutellarioideae, although it lacks the characteristic pericarp outgrowths (tubercles and hair-like structures) of that subfamily. The pollen sculpturing of *Holocheila* is very similar to *Scutellaria*, *Renschia*, and *Tinnea* (Abu-Asab and Cantino 1993). The attachment of the ovary-lobes to a vertical extension of the disc is reminiscent of the gynophore of *Scutellaria*. It is unclear whether the seemingly unithecal anther of *Holocheila* is derived through complete fusion of two thecae, as in *Pogostemon*, or loss of one theca, as in *Scutellaria*, but the lateral attachment of the anther suggests the latter (Cantino, pers. comm.). Although morphology suggests that *Holocheila* is related to Scutellarioideae, DNA sequence data (R.G. Olmstead, unpubl. data) place it within Lamioideae. Further studies are needed to determine the relationships of *Ombrocharis*. Pollen studies on inadequate material (Cantino and Sanders 1986) suggest it may belong to Nepetoideae, although the morphological characters are more typical of Lamioideae. Other genera, assigned to Incertae sedis, have been discussed above, in relation to Viticoideae.

Of the 236 genera that comprise the Labiatae, 50% of the known species are restricted to the 10 largest genera, namely *Salvia* (900 spp.), *Clerodendrum* (500 spp.), *Scutellaria* (360 spp.), *Stachys* and *Plectranthus* (each with 300 spp.), *Hyptis* (280 spp.), *Teucrium* and *Vitex* (each with 250 spp.), *Thymus* (220 spp.) and either *Premna* or *Nepeta* (each with 200 spp.). On the other hand, there are 143 genera with fewer than ten species, of which 66 genera, representing 28% of the total number of genera in the family, have only a single species. This number seems very high, and may represent a lack of knowledge of the true relationships of some of these genera, especially in the large subfamily Lamioideae, where over 30% of the genera are monotypic. The classification presented here represents our present knowledge but, with active research still proceeding, we may expect further changes in the future.

Economic Importance. Labiatae have a very wide range of uses, and the literature is extensive (for recent reviews, see, for Old World: Rivera Nuñez and Obón de Castro 1992, for New World: Heinrich 1992). Many species of the subfamily Viticoideae are forest trees, and some of these, such as various species of *Gmelina* and *Vitex*, are important for their timber which is of considerable commercial value. Also included in this category is *Tectona grandis* (Incertae sedis), commercially known as teak, and one of the foremost timber trees. Traditionally, however, the family is known for the many aromatic plants which have been widely used since antiquity, especially in the Mediterranean region and in many parts of Asia. In some cases, these plants are economically very significant. Many of them are members of subfamily Nepetoideae, which are noted for their essential oils, held in special glands, especially on the foliage. The use of such leaves, from earliest times, especially for culinary purposes, is best documented for the Mediterranean region, where many examples originated. These include species of *Salvia* (sage), *Mentha* (peppermint and spearmint), *Thymus* (thyme), *Origanum* (oregano and marjoram), *Satureja* (savory), *Rosmarinus* (rosemary) and *Melissa* (lemon balm). From the tropics come *Ocimum basilicum*, *O. tenuiflorum* (basil) and *Plectranthus amboinicus*. Many other aromatic species are used in perfumes or toilet preparations, such as *Lavandula*, which was formerly employed to perfume laundry, and *Pogostemon* (patchouli – a member of Lamioideae). *Nepeta* species are found to be strongly attractive to cats. Many of the aforementioned are used to flavour drinks or refreshing teas: *Salvia*, *Mentha*, *Sideritis* (Greek mountain tea – Lamioideae). The dried leaves of many of the above may also be used to make potpourri. The mucilaginous nutlets of *Salvia hispanica* are used to make a refreshing drink in Mexico, while the oily nutlets of *Perilla frutescens* are crushed to extract oil for use in making paint. The medicinal use of Labiatae is widespread especially in folk medicine, not only those genera mentioned above but also many others. Heinrich (1992) points out that the family is of outstanding importance in its use in indigenous medical systems of North American Indian cultures, where it is ranked third in ethnobotanical importance. Many local native species, including not only aromatic Nepetoideae but also Lamioideae, Scutellarioideae and various other groups, are used to cure a very wide range of diseases, some of which can be shown to contain biologically active compounds. References to a range of literature are given in the papers already cited. Aromatherapy uses essential oils of many Labiatae. Recently, interest has centred on biologically active diterpenes in *Salvia* (Rodríguez-Hahn et al. 1992), which have been used in the treatment of cardiac problems. Another field that has attracted attention is that of specific antifeedant compounds, especially in *Ajuga* and *Teucrium* spp., which can reduce the predations of crop pests (Simmonds and Blaney 1992). Some species, such as *Hoslundia opposita* (Nepetoideae), are claimed to be poisonous to cattle (Verdcourt and Trump 1969), while field observations indicate that some *Mentha* species are distasteful to grazing animals. A number of Lamioideae, such as *Leonurus cardiaca*, have been used for cardiac problems (Shiu-ying Hu 1974), and some Lamioideae have been shown to have deleterious effects if consumed in large quantities. Another valuable attribute of many Labiatae is as a rich source of nectar for the honey industry. They play an important role, especially in the production of many Mediterranean honeys, produced from natural vegetation rich in Labiatae. Many members of the Labiatae have attractive flowers and are of major horticultural importance. Over 50 genera are cultivated in European gardens alone, while many more are found in other regions of the world. These include *Congea* (Symphorematoideae), *Vitex*, *Callicarpa* (Viticoideae), *Ajuga*, *Clerodendrum*, *Caryopteris* (Ajugoideae), *Prostanthera* (Prostantheroideae), *Scutellaria*, *Holmskioldia* (Scutellarioideae), *Lamium*, *Stachys*, *Phlomis*, *Physostegia* (Lamioideae), *Salvia*, *Rosmarinus*, *Thymus*, *Monarda*, *Agastache*, *Nepeta* (Nepetoideae).

Classification of the Labiatae (Lamiaceae)

I. Subfam. Symphorematoideae Briq. in Engler & Prantl (1895).
(Gen. 1–3).
II. Subfam. Viticoideae Briq. in Engler & Prantl (1895).
(Gen. 4–13).
III. Subfam. Ajugoideae Kostel. (1834).
(Gen. 14–37).
IV. Subfam. Prostantheroideae Luerss. (1882).
1. Tribe Chloantheae Benth. & Hook. f. (1876).
(Gen. 38–47).
2. Tribe Westringieae Bartl. (1830).
(Gen. 48–53).
VI. Subfam. Scutellarioideae (Dumort.) Caruel (1884).
(Gen. 54–58).
VII. Subfam. Lamioideae Harley
(Gen. 59–121).
VII. Subfam. Nepetoideae (Dumort.) Luerss. (1882).
1. Tribe Elsholtzieae (Burnett) Sanders & Cantino (1984).
(Gen. 122–126).
2. Tribe Mentheae Dumort. (1827).
a. Subtribe Salviinae (Dumort.) Endl. (1838).
(Gen. 127–134).
b. Subtribe Menthinae (Dumort.) Endl. (1838).
(Gen. 135–177).
c. Subtribe Nepetinae (Dumort.) Coss. & Germ. (1845).
(Gen. 178–189).
d. Mentheae: Incertae sedis:
(Gen. 190–191: *Melissa* L., *Heterolamium* C.Y. Wu).
3. Tribe Ocimeae Dumort. (1829).
a. Subtribe: Lavandulinae Endl. (1838).
(Gen. 192).
b. Subtribe: Hanceolinae (C.Y. Wu) A.J. Paton, Ryding & Harley (2002).
(Gen. 193–195).
c. Subtribe: Hyptidinae Endl. (1838).
(Gen. 196–203).
d. Subtribe: Ociminae (Dumort.) Schmidt in Mart. (1858).
(Gen. 204–215).
e. Subtribe: Plectranthinae Endl. (1838).
(Gen. 216–226).
VIII. Labiatae: Incertae sedis:
(Gen. 227–236: *Tectona* L.f., *Callicarpa* L., *Hymenopyramis* Wall. ex Griff., *Petraeovitex* Oliv., *Peronema* Jack, *Garrettia* H.R. Fletcher, *Cymaria* Benth., *Acrymia* Prain, *Holocheila* (Kudo) S. Chow), *Ombrocharis* Hand.-Maz.).

Key to the Subfamilies

1. Style terminal or subterminal. If rarely appearing gynobasic, then calyx 2-lipped with lips entire, rounded **3**
– Style gynobasic **2**
2. Pollen usually hexacolpate, 3-celled, plants often strongly aromatic with volatile terpenoids, rosmarinic acid present, embryo-investing **Nepetoideae** (p. 228)
– Pollen usually tricolpate, 2-celled, plants rarely aromatic, terpenoids and rosmarinic acid usually very poorly represented, iridoid glycosides often frequent, embryo spatulate **Lamioideae** (p. 211)
3. Lianas. Flowers in 3–7-flowered capitate cymes with an involucre of bracts. Ovary imperfectly 2-locular, free towards apex, ovules pendulous, apically attached, orthotropous **Symphorematoideae** (p. 191)
– Trees, shrubs or herbs, rarely lianas. Flowers in various types of inflorescence. Ovary 2–4(or rarely more)-locular, ovules anatropous to hemianatropous, erect **4**
4. Calyx usually 2-lipped with lips entire, or shallowly 5-lobed with rounded lobes. Corolla zygomorphic, usually 2-lipped. Fruit schizocarpic, with 4 mericarps, pericarp usually tuberculate or plumose. Anther slits often ciliate **Scutellarioideae** (p. 209)
– Calyx and corolla actinomorphic to 2-lipped. Fruit schizocarpic or indehiscent and often drupaceous. If breaking into 4 mericarps, then pericarp smooth to verrucose or reticulate, not as above. Anther slits not as above **5**
5. Seeds albuminous. Ovary lobed or not, if ovary unlobed, then indumentum usually of stellate or branched hairs or scales. Australia **Prostantheroideae** (p. 204)
– Seeds usually exalbuminous. Ovary lobed or unlobed, indumentum usually of simple hairs **6**
6. Corolla rarely 2-lipped, ± actinomorphic to 1-lipped. Pollen often with supratectal spines, spinules or verrucae, columellae usually branched **Ajugoideae** (p. 196)
– Corolla often 2-lipped, sometimes weakly so. Pollen not as above, columellae unbranched **Viticoideae** (p. 193)

I. Subfam. Symphorematoideae Briq. in Engler & Prantl (1895).

Woody climbers with simple or stellate hairs. Leaves opposite, entire or toothed, petiolate. Inflorescence of 3–7-flowered, capitate cymes (Fig. 22), pedunculate from the axils of leaf-like, sometimes reduced, often caducous bracts and surrounded by an involucre of 3–4, or often 6, usually conspicuous and accrescent, often coloured bracteoles, cymes often arranged in panicles or axillary thyrses. Flowers sessile, often perfumed. Calyx usually actinomorphic, 5–8-lobed, lobes sometimes bifid, sometimes with an epicalyx at base of lobes, opposite or alternate with calyx-lobes. Corolla actinomorphic, 5–16-lobed, with stamen and corolla-lobe numbers equal, or (*Congea*) 2-lipped, 5-merous with 4 didynamous stamens; stamens 4–18, inserted near corolla-throat, anthers 2-thecous, thecae parallel, separate. Style terminal, stigma-lobes 2. Disc absent. Ovary

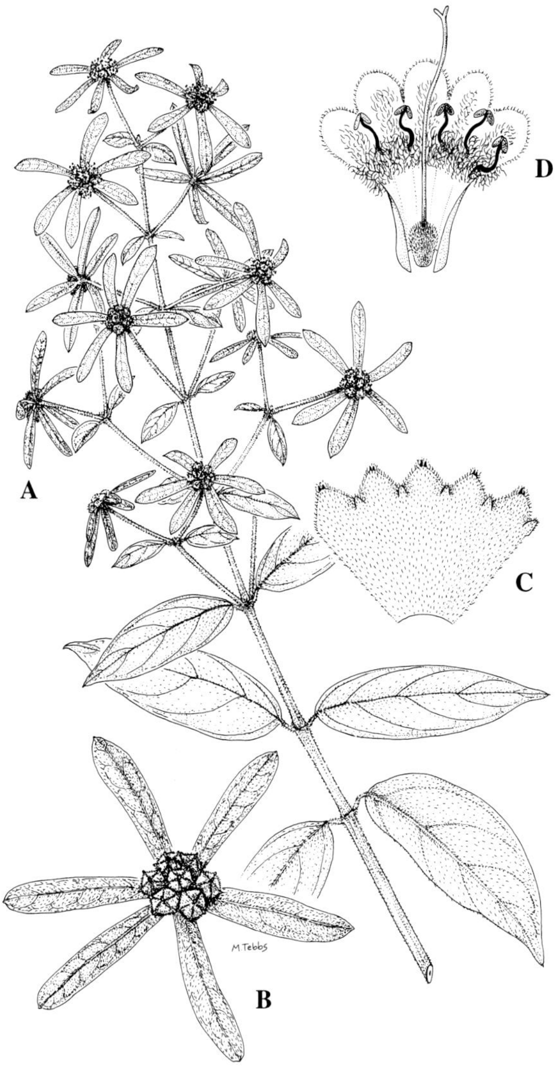

Fig. 22. Labiatae subfam. Symphorematoideae. *Sphenodesme mollis*. **A** Habit. **B** Fruiting cyme. **C** Calyx, external view. **D** Corolla opened out to show stamens and gynoecium. Drawing by M. Tebbs

imperfectly 2-locular, free towards apex, locules 2-ovulate, with incomplete septum, ovules pendulous, apically attached, orthotropous. Fruit weakly drupaceous or dry, indehiscent, 1(–2)-seeded by abortion, seeds ? exalbuminous.

1. Involucral bracteoles 6, flowers with actinomorphic or sub-actinomorphic corolla, stamens 5–16 2
– Involucral bracteoles 3–4, flowers with 2-lipped corolla, stamens 4 **3. *Congea***
2. Flowers 6–16-merous, stamen-filaments elongate, anthers exserted from corolla-tube, style elongate **2. *Symphorema***
– Flowers 5–6(–7)-merous, if 6-merous, then stamens with very short filaments, included in corolla-tube and style very short **1. *Sphenodesme***

1. *Sphenodesme* Jack Fig. 22

Sphenodesme Jack, Malayan Misc. 1, 1: 19 (1820); Munir, Gard. Bull. Singapore 21: 315–378 (1966), rev.

Climbing shrubs; leaves entire; inflorescence often paniculate, terminal and axillary, usually with elongate, thyrsoid branches from axils of reduced leaves, cymes 3-, 5- or 7-flowered, pedunculate, from axils of small, leaf-like, caducous bracts, bracteoles of involucre 6, accrescent, usually conspicuous in fruit, often greenish; calyx narrowly infundibular, 5–10-nerved, ± accrescent, often inflated in fruit, usually actinomorphic, 5–6(–7)-lobed, rarely irregularly lobed, 5-merous species mostly with bifid or emarginate lobes, often inflexed and appearing 10-lobed, sometimes with 5 accessory, smaller lobes ("epicalyx"), erect to reflexed, in sinuses of larger lobes (or rarely opposite and external to them), or calyx rarely 2-lipped, lips minutely toothed, calyx-tube 5–10-nerved; corolla white, pale yellow or bluish to dark purple, longer than calyx, actinomorphic or slightly irregular, 5–6(–7)-lobed, lobes usually subequal, linear, acute to obovate-oblong and rounded, corolla-tube slightly exserted from calyx, straight, hairy-annulate near insertion of stamens; stamens 5 or 6(–7), 5-merous species usually with stamens, exserted, isomerous with corolla-lobes and alternating with them, 6-merous flowers with short-filamented stamens included within corolla-tube, thecae elliptic-ovoid, parallel, separate; style filiform, long and exserted from corolla or very short, stigma-lobes 2, subequal; ovary usually hairy or glandular towards apex. About 14 species, India, China, Burma and Thailand to Malaysia. Munir (1966) recognised two sections: sect. *Sphenodesme*: stamens exserted from corolla-tube, filaments elongate, style elongate, epicalyx often present, flowers 5-merous; sect. *Brachynema* (Griff.) Clarke: stamens included in corolla-tube, filaments very short, style very short, epicalyx absent, flowers usually 6-merous.

2. *Symphorema* Roxb.

Symphorema Roxb., Pl. Coromandel 2: 46, t. 186 (1805); Munir, Gard. Bull. Singapore 22: 153–171 (1967), rev.

Climbing shrubs; leaves entire or toothed, often sinuate-dentate; inflorescence terminal or axillary, often with short, thyrsoid branches from axils of caducous leaves, cymes capitate, 7-flowered, pedunculate, from axils of reduced, leaf-like, usually caducous bracts, bracteoles of involucre 6, accrescent and conspicuous in fruit, often white-

or pink-tinged, becoming wing-like and sometimes scarious, rarely toothed; flowers sessile or shortly pedicellate, sometimes fragrant; calyx narrowly infundibular to cylindrical, accrescent in fruit, actinomorphic, 3–8-lobed, lobes equal to subequal, triangular, erect, tube indistinctly nerved, glabrous within, longer than lobes; corolla white, pinkish or lilac, actinomorphic, 6–16-lobed, slightly exserted from calyx, lobes equal, ± ligulate, erect or curled, delicate, glabrous, corolla-tube cylindrical, subequalling lobes, glabrous within; stamens 6–18, equal, usually as many as corolla-lobes and alternating with them, slightly shorter than corolla (rarely with additional 8 free stamens from base of ovary in *S. polyandra*), thecae oblong, parallel, separate; style terminal, elongate, included in corolla, stigma-lobes 2, elongate, truncate or subulate at apex; ovary sometimes emarginate at apex. $2n = 24, 28, 36$. Three species from India and Ceylon to Burma, Thailand, Cambodia, Vietnam and the Philippines, monsoon forests.

3. *Congea* Roxb.

Congea Roxb., Pl. Coromandel 3: 90, t. 293 (1820); Munir, Gard. Bull. Singapore 21: 259–314 (1966), rev.

Climbing shrubs; leaves entire; inflorescence terminal or axillary, of 3–9-flowered capitate cymes, pedunculate, involucre of 3–4 elongate, spreading outer bracteoles (often referred to as bracts), rarely with 2 inner bracteoles, inconspicuous, subulate, outer bracteoles spathulate to ovate-elliptic, rounded at apex, rarely bilobed, very showy, purple or white-tomentose on adaxial surface and sometimes connate at base to form a small cup-shaped structure; flowers sessile or shortly pedicellate; calyx actinomorphic, weakly accrescent in fruit, narrowly infundibuliform, 5-lobed, lobes equal, triangular, calyx-tube longer than lobes, 10-nerved, densely hairy within, especially in throat; corolla lilac or pink, 2-lipped, 5-lobed (2/3), longer or shorter than calyx, posterior lip with 2 rounded lobes, anterior lip much shorter with 3 lobes rounded, slightly spreading, corolla-tube cylindrical, straight, slightly expanded around ovary near base, inserted or exserted from calyx, densely annulate within near throat; stamens 4 (posterior stamen absent), didynamous, divergent, exserted or rarely included, filaments sometimes very long, becoming curled, thecae rounded, parallel, separate; style terminal, exserted, stigma-lobes 2, very short. $2n = 34$. Seven to ten species, India (Assam), Bangladesh, Burma, Thailand, Cambodia, Laos, Vietnam, Malaya, Sumatra, and SW China. *Congea tomentosa* Roxb. is widely cultivated.

II. Subfam. Viticoideae Briq. in Engler & Prantl (1895).

Trees, shrubs, lianas, rarely herbs (*Premna herbacea*), non-aromatic with volatile terpenoids and rosmarinic acid usually absent, and iridoid glycosides and acteosides present. Leaves simple or compound (Fig. 23), decussate or whorled. Inflorescence cymose, cymes simple, corymbose,

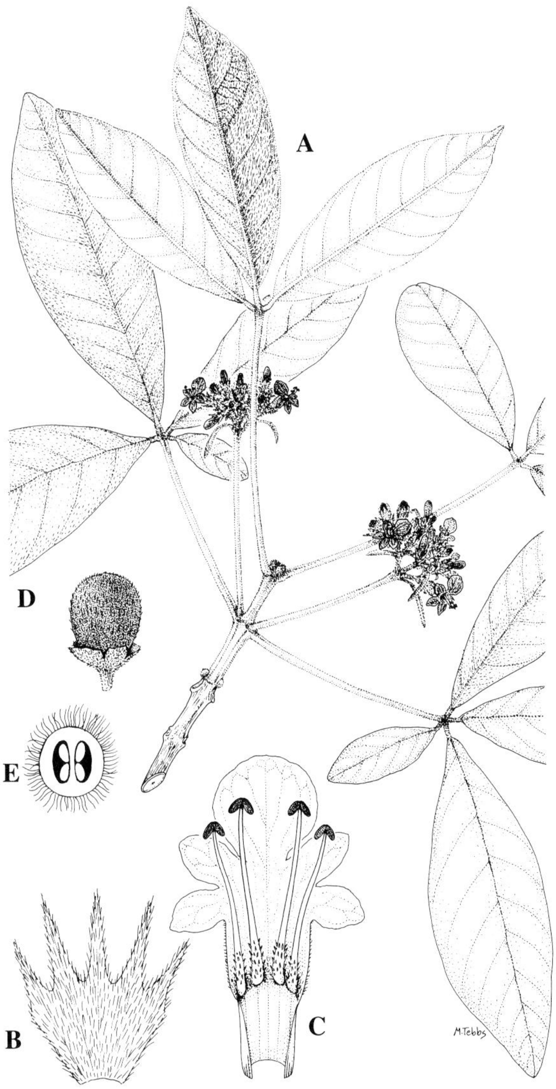

Fig. 23. Labiatae subfam. Viticoideae. *Vitex polygama*. **A** Habit. **B** Calyx opened out, external surface. **C** Corolla opened out to show stamens. **D** Fruit with persistent calyx, side view. **E** Fruit, transverse section. Drawing by M. Tebbs

panicled, umbellate, or occasionally reduced to single flower; bracts usually present, sometimes leaf-like. Calyx actinomorphic to weakly zygomorphic, campanulate or tubular, rim truncate or 4–5-lobed, sometimes 2-lipped. Corolla actinomorphic or zygomorphic, limb 4–5-lobed, often 2-lipped, tube straight or funnel-shaped, lobes spreading. Stamens 4(5), rarely reduced to 2 (*Cornutia*), sometimes with 1 or more staminodes present, equal or didynamous, slightly exserted, anther-thecae parallel or divergent. Pollen mostly 3-colpate, rarely 4–5-colpate (*Cornutia*), with exine structure microreticulate to tectate, or finely rugulose to microreticulate, or tectate-perforate and without spinal sculpturing. Ovary unlobed to shallowly 4-lobed, 2- or 4-locular, with 1 or 2 ovules in each locule, ovules attached laterally, usually at top of locule. Style terminal or subterminal, stigma 2-lobed. Fruit drupaceous with one pyrene with 4 locules (but often with 2 or more locules aborted), or 2 pyrenes each with 2 locules.

1. Fertile stamens 4(5) — 2
- Fertile stamens 2 plus 2 staminodes — **5. *Cornutia***
2. Calyx and corolla regularly 4-lobed, actinomorphic; stamens inserted at top of tube, filaments very short — **4. *Petitia***
- Calyx and corolla 4- or 5-lobed, lobes unequal, zygomorphic; stamens inserted in lower middle or lower part of tube, filaments long — 3
3. Calyx and corolla 4-lobed — 4
- Calyx and corolla 5-lobed — 6
4. Inflorescence axillary — **8. *Tsoongia***
- Inflorescence terminal — 5
5. Leaves simple — **6. *Premna***
- Leaves digitately compound (leaflets 3–6(7)) — **7. *Viticipremna***
6. Corolla-tube widely funnel-shaped, greatly enlarged at throat; style with stigma-lobes unequal (posterior branch much shorter); fruit a drupe with very fleshy mesocarp — **12. *Gmelina***
- Corolla-tube narrowly funnel-shaped, or cylindrical, not greatly enlarged at throat; style with stigma-lobes more or less equal; fruit a drupe with dry or only somewhat fleshy mesocarp — 7
7. Inflorescence axillary only; fruit flattened, 4-lobed, with 2 pyrenes, each with 2 locules — **13. *Pseudocarpidium***
- Inflorescence terminal and axillary; fruit not flattened or lobed, with 1 pyrene with 4 locules — 8
8. Petiole with well-developed and conspicuously swollen articulations of the petioles and petiolules; fruit a capsule — **11. *Teijsmanniodendron***
- Petiole without articulations; fruit a drupe — 9
9. Leaves simple — **9. *Paravitex***
- Leaves compound — **10. *Vitex***

4. *Petitia* Jacq.

Petitia Jacq. in Enum. Pl. Carib. 1 (1760); Moldenke, Rep. Spec. Nov. 42: 229–251 (1937), rev.

Trees or shrubs; leaves simple; inflorescence axillary, cymose, paniculate, dense; flowers scented, pale yellow to white; calyx campanulate, subtruncate or regularly 4-lobed; corolla infundibular to hypocrateriform, actinomorphic, 4-lobed; stamens 4, included or slightly exserted, equal in length, inserted at top of tube with filaments very short, thecae parallel; ovary 2-locular, each locule with 2 ovules, style terminal, with stigma bilobed, lobes unequal; fruit a small drupe containing a single 2–4-seeded pyrene. Two species, Bahamas and Greater Antilles.

5. *Cornutia* L. Fig. 21F

Cornutia L., Sp. Pl. 2: 623 (1753); Moldenke, Rep. Spec. Nov. 40: 153–205 (1936), rev.

Trees or shrubs; leaves simple, usually aromatic; inflorescence terminal, dense to lax, pyramidal-thyrsoid cymes, or axillary simple cymes; flowers not scented, blue violet or purple; calyx campanulate, truncate or minutely 4–(5)-dentate; corolla hypocrateriform, zygomorphic, 4-lobed, 2-lipped, stamens 2 plus 2 staminodes, inserted at or just above the middle of tube, only a little exserted, didynamous, thecae widely divergent at base, connective thickened; ovary 2-carpellate, each carpel 2-locular, ovules 4; style terminal, with stigma-lobes short, unequal; fruit drupaceous with fleshy exocarp, pyrene hard, 4-seeded (or 3–1 by abortion). Twelve species. Tropical America.

6. *Premna* L.

Premna L., Mant. ii, 154, 252 (1771); Munir, J. Adelaide Bot. Gard. 6: 1–43 (1984), reg. rev.
Pygmaeopremna Merr. (1910).

Trees, shrubs or lianas, rarely small herbs; leaves simple; inflorescence terminal, cymose, compound, lax to dense, flowers purplish or white; calyx tubular or campanulate, truncate or toothed, often obscurely 2-lipped; corolla infundibular, 4-lobed, 2-lipped (occasionally ± actinomorphic), stamens 4, inserted in middle to lower part of tube, slightly exserted, thecae parallel or divergent; ovary bicarpellate, syncarpous, 4-locular with one ovule in each locule, style terminal with two short stigmatic lobes; fruit drupaceous, with hard pyrene with four seeds. Numbers of species vary from 50 (Mabberley 1997) to 200 (Verdcourt 1992). Tropical and subtropical Asia, Africa, Australia and Pacific Islands.

7. *Viticipremna* H.J. Lam

Viticipremna H.J. Lam, Verbenac. Malay. Archip. 162 (1919); Munir, J. Adelaide Bot. Gard. 7: 181–200 (1985), rev.

Trees or shrubs; leaves digitately compound, leaflets 3–6(–7), usually petiolulate; inflorescence terminal, cymose, compound, much-branched, flower white to pale yellow; calyx truncate or 4-lobed, persistent; corolla hypocrateriform, 4-lobed, ±2-lipped, lobes subequal; stamens 4, inserted at base of tube, slightly exserted, thecae divergent; ovary bicarpellate, 4-locular, 1 ovule in each locule; style terminal, with 2 short, unequal stigmatic lobes; fruit a small drupe, pyrene hard, 4-celled. Five species, Philippines to Celebes, New Guinea, Australia and Fiji.

8. *Tsoongia* Merrill

Tsoongia Merrill in Philipp. J. Sci. 23: 264 (1923).

Tree or shrub; leaves simple, lobed or palmately trifoliolate; inflorescence axillary, cymose, lax, few-flowered, flower yellow; calyx campanulate, 4-lobed, 2-lipped; corolla infundibular, zygomorphic, 4-lobed, lobes subequal, weakly 2-lipped; stamens 4, equal in length, inserted at middle of tube, exserted; ovary 2-locular, 1–2 ovules per locule; style terminal with stigma-lobes short, bifid. Fruit drupaceous, globose, remaining undivided. One species, *Tsoongia axilliflora* Merr. China to SE Asia.

9. *Paravitex* Fletcher

Paravitex Fletcher in Bull. Misc. Inf. 1937: 74–75 (1937).

Shrub or liana; leaves simple; inflorescence terminal, in dichotomous or trichotomous cymes, flower white with yellow blotches on anterior lip; calyx campanulate, entire or sinuate-lobate with lobes subequal; corolla 5-lobed, 2-lipped; stamens 4, didynamous, inserted near base of tube, little exserted; ovary 4-locular, each locule with 1 ovule, but with only one locule developing; fruit a drupe with one pyrene, 1-seeded by abortion. One species, *Paravitex siamica* Fletcher, Thailand.

10. *Vitex* L. Fig. 23

Vitex L., Sp. Pl. 2: 638 (1753); Munir, J. Adelaide Bot. Gard. 10: 31–80 (1987), reg. rev.
Neorapinia Moldenke (1955).

Trees or shrubs; leaves usually digitately compound with 3–8 leaflets but sometimes unifoliolate, sometimes in whorls; inflorescence axillary or terminal, a dichasial cyme, sometimes compound into a thyrse or panicle, flower blue, yellow, white to purple; calyx campanulate or tubular, 3–5-lobed, or truncate, actinomorphic to weakly 2-lipped; corolla infundibular, zygomorphic, 5-lobed, 2-lipped, tube straight or curved; stamens 4, didynamous, attached near base of tube; usually exserted, thecae dorsifixed, divergent; pollen 3-colpate; ovary bicarpellate, carpels 2-locular and 2-seeded; style terminal with stigma-lobes short, equal; fruit drupaceous, with hard pyrene, 4-seeded. $2n = 12, 16$. Approximately 250 species, in Old and New World tropics, few in temperate regions.

11. *Teijsmanniodendron* Koorders

Teijsmanniodendron Koorders in Ann. Jard. Buitenz. 19: 19 (1904); Kostermans, Reinwardtia 1: 75–106 (1951), rev.

Trees; leaves simple or palmately compound, petioles and petiolules with conspicuously swollen articulation; inflorescence terminal and/or axillary, a lax panicle of distant, abbreviated cymes; calyx campanulate, 5-lobed; corolla 5-lobed, 2-lipped; stamens 4 (rarely 5), didynamous, inserted towards base of tube, slightly exserted; ovary 4-locular, locules 1-ovulate; fruit a non-dehiscent capsular fruit with a single seed (by abortion of the other 3 ovules). About 14 species, SE Asia and Malaysia.

12. *Gmelina* L.

Gmelina L. Sp. Pl. 2: 626 (1753); Munir, J. Adelaide Bot. Gard. 7: 91–116 (1984), reg. rev.

Trees, or shrubs, sometimes climbing or scrambling; stems sometimes armed with spines; leaves simple; inflorescence terminal, cymose, compound, branched, or seldom reduced to single flower, blue or purple-violet, white, yellow, brownish, mauve or purple pink; calyx tubular or campanulate, truncate or 4–5-lobed, often with large glands on surface; corolla infundibular, (4–)5-lobed, ±2-lipped, tube widely funnel-shaped, greatly enlarged at throat; stamens 4, inserted at bottom of tube, didynamous, only slightly exserted, thecae parallel or divergent; ovary bicarpellate, syncarpous, 4-locular, with one ovule in each locule; style terminal with two unequal stigmatic lobes, posterior branch much shorter;

fruit a fleshy, succulent drupe with hard pyrene, 4-seeded (rarely 2-seeded). $2n = 36, 38, 40$. About 33 species, Asia, Australia, New Guinea and New Caledonia.

13. *Pseudocarpidium* Millsp.

Pseudocarpidium Millsp., Publ. Field. Columb. Mus. Bot. Ser. 2: 181 (1906); León & Alain, Flora of Cuba 4: 314–316 (1957), reg. rev.

Small trees or shrubs; leaves simple; inflorescence axillary, cymose, lax, flower mauve or blue; calyx campanulate, actinomorphic, 5-lobed, lobes equal; corolla tubular-infundibular to hypocrateriform, 5-lobed, 2-lipped; stamens 4, inserted at base of tube, hardly exserted, thecae parallel; ovary 4-locular, style terminal, with stigma shortly 2-lobed; fruit a flattened, rounded schizocarp, dry or slightly fleshy, composed of 2 pyrenes, each with 2 locules. Eight species, Hispaniola, Cuba and the Bahamas.

III. Subfam. Ajugoideae Kostel. (1834).

Subfam. *Teucrioideae* (Dumort.) Caruel (1884).

Plants woody or herbaceous, usually perennial, sometimes aromatic or ill-scented, but volatile terpenoids and rosmarinic acid usually absent, and iridoid glycosides and acteosides present, hairs usually simple. Leaves opposite, alternate, or whorled, mostly simple, rarely dissected, lobed, or compound. Inflorescence cymose, cymes sometimes reduced to single flowers (either solitary or racemose), bracts usually present (absent in *Caryopteris*), sometimes brightly coloured. Calyx usually actinomorphic, usually exannulate, 4–5(–6)-lobed. Corolla actinomorphic to weakly zygomorphic or 1-lipped (infrequently 2-lipped), limb 4–5-lobed, tube usually cylindrical to infundibular, rarely annulate, corolla in bud often asymmetrical (expanding abruptly on anterior side). Stamens 4, rarely 2 or 5, equal or didynamous (anterior pair longer), usually exserted (often greatly so; Fig. 24), thecae parallel to divaricate, separate or confluent at dehiscence. Pollen mostly tricolpate, tectate-perforate (rarely microreticulate), usually with supratectal spines, spinules or verrucae (infrequently psilate, granulate, or suprareticulate), columellae usually branched (rarely simple, granular, or undifferentiated). Ovary unlobed to shallowly 4-lobed, placentation axile. Style terminal or subterminal, usually

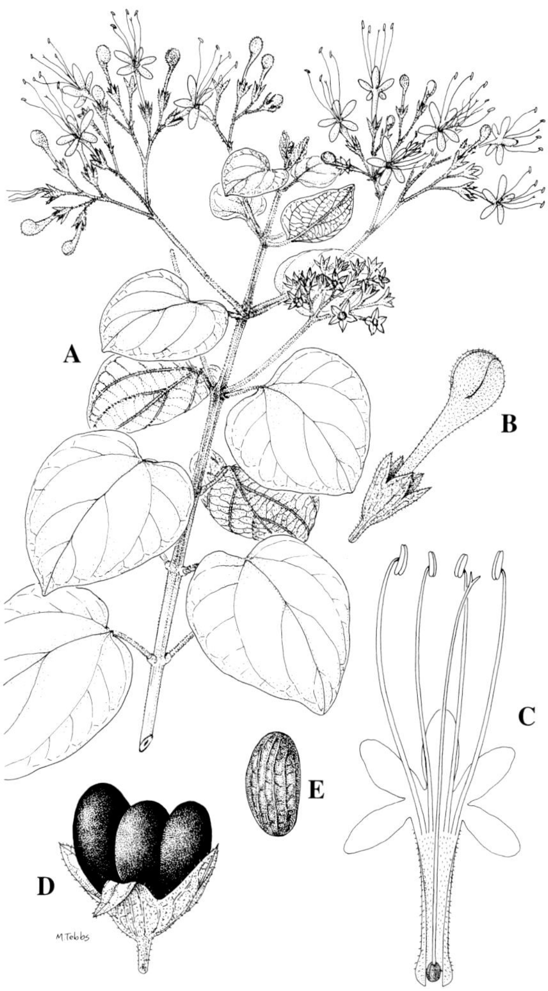

Fig. 24. Labiatae subfam. Ajugoideae. *Clerodendrum umbellatum*. **A** Habit. **B** Flower in bud, side view. **C** Corolla opened out to show stamens and gynoecium. **D** Fruit, side view, example with only 3 loculi. **E** Pyrene. Drawing by M. Tebbs

deciduous. Disc usually poorly developed to absent. Fruit drupaceous with four 1-seeded pyrenes, a schizocarp with 4 dry or fleshy mericarps, or rarely dry, indehiscent and 4-seeded, surface of dry fruits usually reticulate-ridged. Endosperm usually scanty or absent. Embryo spatulate.

1. Fertile stamens two, the posterior (or rarely anterior) pair reduced to staminodes 2
– Fertile stamens four or five 3
2. Annual herb with deeply lobed leaves **35. *Amethystea***
– Woody plants with entire leaves **23. *Oxera***
3. Calyx broadly infundibular to patelliform, greatly accrescent; corolla 1-lipped (lower lip only); fruit dry **15. *Karomia***
– Calyx not as above; corolla and fruit various 4

4. Stamens as many as corolla-lobes 5
- Stamens fewer than corolla-lobes 7
5. Ovary unlobed (style terminal); fruit drupaceous **21. *Aegiphila****
- Ovary shallowly 4-lobed in flower (style subterminal), becoming deeply 4-lobed during fruit development; fruit a fleshy or dry schizocarp 6
6. Corolla actinomorphic or nearly so **22. *Faradaya***
- Corolla zygomorphic **24. *Hosea*****
7. Anther-thecae distinct at dehiscence, parallel 8
- Anther-thecae confluent at dehiscence, usually divergent to divaricate 15
8. Fruit a schizocarp of dry, basally winged mericarps; lianas and scandent shrubs with a terminal, corymboid thyrse **14. *Glossocarya***
- Mericarps not basally winged or fruit not a schizocarp 9
9. Bracts brightly coloured, showy; leaves usually alternate **17. *Amasonia***
- Either bracts not showy or leaves opposite or whorled 10
10. Leaves alternate, or sometimes subopposite or subverticillate; inflorescence a raceme, bracts much smaller than foliage leaves; Brazilian herbs and subshrubs **18. *Monochilus***
- Leaves usually opposite or whorled, inflorescence cymose or flowers solitary in axils of normal foliage leaves; mostly woody plants 11
11. Corolla actinomorphic, stamens inserted near top of corolla-tube and much shorter than corolla-lobes; Australian subshrubs with linear leaves **19. *Huxleya***
- Corolla zygomorphic or stamens not as above 12
12. Fruit a dry schizocarp (4 nutlets, often narrowly winged); corolla zygomorphic but not 1-lipped, the anterior lobe much larger than other four; Asiatic shrubs 13
- Fruit usually drupaceous (intact or a schizocarp); if fruit dry, then plant a non-Asiatic herb or subshrub; corolla usually either 1-lipped or subactinomorphic 14
13. Stamens spreading, straight; bracts usually absent; anterior corolla-lobe dentate to fimbriate-laciniate; corolla-tube annulate **37. *Caryopteris***
- Stamens parallel, arched; bracts present; anterior corolla-lobe entire; corolla-tube exannulate **30. *Discretitheca***
14. Flower-bud markedly asymmetrical, corolla expanding abruptly on lower side only; anterior corolla-lobe frequently much larger than the other four; anthers usually basifixed (occasionally approaching versatile); stigma-lobes frequently unequal **16. *Rotheca***
- Flower-bud symmetrical or, if asymmetrical, usually expanding abruptly on upper side due to resupination; anterior corolla-lobe only slightly (if at all) larger than the others; anthers versatile; stigma-lobes equal **20. *Clerodendrum***
15. Corolla marcescent, persisting as sheath around fruit **29. *Ajuga***
- Corolla not marcescent 16
16. Leaf-margin entire or lobed but not toothed 17
- Leaf-margin toothed (or both toothed and lobed) 21
17. Calyx-lobes conspicuously hooked **26. *Oncinocalyx***
- Calyx-lobes not hooked 18
18. Fruit drupaceous; Australian shrubs **25. *Spartothamnella***
- Fruit a schizocarp of 4 nutlets 19
19. Corolla not 1-lipped; pollen colpi not operculate; N. America **36. *Trichostema***
- Corolla 1-lipped; pollen colpi operculate; widespread 20
20. Ovary unlobed or lobed up to a quarter of its length; nutlets coherent but separating with slight pressure; calyx actinomorphic; flowers tiny; New Zealand **27. *Teucridium***
- Ovary usually lobed from a quarter to half its length; nutlets separating freely; calyx zygomorphic or actinomorphic; flowers usually larger **28. *Teucrium***
21. Pollen colpi operculate; corolla usually 1-lipped (anterior lip only); ovary usually lobed from a quarter to half its length **28. *Teucrium***
- Pollen colpi not operculate; corolla usually 2-lipped or unlipped; ovary unlobed or lobed up to a quarter of its length 22
22. Calyx strongly 2-lipped; nutlets subglobose **34. *Rubiteucris***
- Calyx ± actinomorphic; nutlets obovoid or boat-shaped 23
23. Tall, spreading shrubs; pollen verrucate; nutlets or pyrenes boat-shaped, arcuate in cross section, attachment-scar as long as nutlet; calyx-lobes 5–6 **31. *Pseudocaryopteris***
- Herbs or low shrubs; pollen spinulose to spinose; nutlets obovoid, not arcuate in cross section, attachment-scar 0.5–0.9× as long as nutlet; calyx-lobes 4–5 24
24. Flowers in lax, many-flowered axillary cymes, grading into a terminal, paniculiform thyrse; pollen triporate; tall herb **32. *Tripora***
- Flowers solitary or in 3–5-flowered axillary cymes; pollen tricolpate; low herbs or shrubs **33. *Schnabelia***

* Rare individuals of *Clerodendrum* with 5 stamens will also key out here. They differ in having hermaphrodite flowers, whereas in most species of *Aegiphila* the flowers appear to be functionally unisexual with either the stamens or pistils much reduced in size.

** Rare individuals of *Oxera* with four stamens will also key out here. They differ in having a well-developed nectar-disc (absent in *Hosea*) and fleshy mericarps that are rounded at the apex (dry and apically acute in *Hosea*).

14. *Glossocarya* Wall. ex Griff.

Glossocarya Wall. ex Griff., Calcutta J. Nat. Hist. 3: 366 (1843); Munir, J. Adelaide Bot. Gard. 13: 17–34 (1990), reg. rev.

Lianas and scandent shrubs; leaves opposite, petiolate, broadly ovate to obovate or subrotund, entire; inflorescence a lax, terminal, corymboid thyrse; calyx actinomorphic, accrescent, shallowly 5-lobed, nearly truncate in fruit; corolla white to pale yellowish, actinomorphic to weakly zygomorphic, limb 5-lobed, spreading, lobes similar or anterior lobe slightly larger, corolla in bud usually slightly asymmetrical; stamens 4, didynamous to subequal, long-exserted, filaments straight, anther-thecae parallel, separate at dehiscence; pollen with supratectal spinules; ovary unlobed; stigma-lobes equal; disc absent; nutlets clavate, smooth, pubescent, basally winged, attachment-scar lateral, as long as nutlet; endosperm absent. Nine to 13 species of forests and thickets in SE Asia, Papua New Guinea and Australia (Queensland).

15. *Karomia* Dop

Karomia Dop, Bull. Mus. Hist. Nat. (Paris) II, 4: 1052 (1932); Fernandes, Garcia de Orta, Ser. Bot., Lisboa 7: 33–46 (1985), reg. rev.

Trees and shrubs, at least some aromatic; leaves opposite, petiolate, ovate to obovate or suborbicular, entire or crenate; inflorescence a lax axillary cyme or terminal thyrse; calyx actinomorphic, greatly accrescent, broadly infundibular in fruit, 5-lobed, lobes ovate to dome-shaped or obsolescent; corolla purple or whitish, 1-lipped with only anterior lip present, 5-lobed, anterior lobe largest, concave, sometimes fimbriate, tube straight or curved, sometimes gibbous, corolla in bud usually asymmetrical; stamens 4, subequal, long-exserted, filaments arched, anther-thecae parallel or divaricate, confluent or separate at dehiscence; pollen psilate or bearing blunt, irregularly shaped supratectal protuberances; ovary unlobed; stigma-lobes equal or unequal; disc poorly developed to absent; fruit dry, obconical and apically truncate or with four horn-like projections, or globose, usually pubescent, 4-seeded, remaining intact or splitting into four mericarps, attachment-scar as long as mericarp; endosperm absent. Nine species, one in Vietnam, the others in Madagascar and eastern and southern Africa.

16. *Rotheca* Raf.

Rotheca Raf., Fl. Tell. 4: 69 (1838); Steane and Mabberley, Novon 8: 204–206 (1998), gen. conspectus.
Cyclonema Hochst. (1842).
Clerodendrum L. subg. *Cyclonema* (Hochst.) Guerke (1895).
Clerodendrum L. subg. *Clerodendrum* sect. *Konocalyx* Verdc. (1992).

Shrubs, subshrubs and perennial herbs (rarely lianas or small trees), ill-scented; leaves opposite or whorled, usually sessile or short-petiolate, simple, entire or toothed; inflorescence a terminal thyrse (usually lax, paniculiform) or axillary cyme, occasionally reduced to solitary, axillary flowers; calyx actinomorphic or nearly so, not accrescent, 5-lobed or truncate; corolla blue, purple or white (rarely yellow), subactinomorphic to 1-lipped and then with only anterior lip present, 5-lobed, lobes subequal or anterior lobe largest and often concave and different in colour, tube straight or curved, often gibbous, corolla in bud asymmetrical; stamens 4, didynamous to subequal, long-exserted, filaments arched or straight, anther-thecae parallel, separate at dehiscence; pollen with supratectal spinules; ovary unlobed in flower but becoming lobed during fruit development; stigma-lobes equal or frequently unequal; disc poorly to well developed; fruit drupaceous, 4-lobed, the lobes sometimes elongating greatly during development, eventually separating into fleshy 1-seeded mericarps; endosperm absent. $2n$ = 48, 52, 184. Fifty to 60 species, mostly of sub-Saharan Africa but a few in Madagascar, the Indian Ocean islands, southern Asia, and Indonesia. Although *Rotheca* has been included within *Clerodendrum* by most authors, there is strong molecular evidence that its inclusion renders *Clerodendrum* polyphyletic (Steane et al. 1997, 1999). The two genera also differ in their iridoid compounds (Stenzel et al. 1988).

17. *Amasonia* L. f.

Amasonia L. f., Suppl. Pl. 48: 294 (1781), nom. cons.; Moldenke, Feddes Rep. Spec. Nov. Regn. Veg. 46: 193–228 (1939), rev.
Taligalea Aubl. (1775), nom. rejic.

Shrubs, subshrubs and perennial herbs; leaves usually alternate (occasionally opposite), sessile or short-petiolate, oblanceolate to obovate or elliptical, variously toothed to repand (rarely entire); inflorescence a terminal thyrse or raceme, bracts brightly coloured, red to purple or yellow; calyx bright red or purple, actinomorphic, accrescent or not, 5-lobed; corolla yellow to white, subactinomorphic to weakly zygomorphic, 5-lobed, anterior lobe usually largest; stamens 4, didynamous, exserted, filaments straight or upcurved (probably varying with stage of anthesis), anthers versatile, thecae parallel, separate at dehiscence; pollen with supratectal spinules; ovary unlobed; stigma-lobes subequal to unequal; disc absent; fruit drupaceous with four 1-seeded pyrenes; endosperm absent. Eight species of northern and central South America.

18. *Monochilus* Fisch. & C.A. Mey.

Monochilus Fisch. & C.A. Mey., Index Sem. Hort. Petrop. 1: 34 (1835).

Perennial herbs and subshrubs; leaves alternate, subopposite or subverticillate, petiolate or sessile, simple; inflorescence racemose, terminal and axillary; calyx actinomorphic to 1-lipped, accrescent, 5-lobed; corolla white, 1-lipped with only anterior lip present, but facing upwards due to resupination, 5-lobed with median lobe largest, tube deeply cleft between posterior (lower) lobes; stamens 4, didynamous to subequal, slightly exserted from corolla-tube, filaments straight or arched (probably varying with stage of anthesis), thecae parallel, separate at dehiscence; pollen operculate, psilate except for scattered, supratectal granules; ovary unlobed; stigma-lobes very short, equal to subequal or essentially absent; disc absent; fruit drupaceous with four 1-seeded pyrenes; endosperm absent. Two species, Brazil.

19. *Huxleya* Ewart

Huxleya Ewart in Ewart & B. Rees, Proc. Roy. Soc. Victoria 25: 109 (1912); Munir, J. Adelaide Bot. Gard. 13: 35–41 (1990), rev.; de Kok, Grayer & Kite, Austr. Syst. Bot. 13: 425–428 (2000), gen. redescription, relationships.

Subshrub; leaves opposite, sessile, linear, entire; flowers axillary, usually solitary (occasionally in 3-flowered cymes); calyx actinomorphic, 5-lobed, greatly accrescent; corolla white, actinomorphic, 5-lobed, lobes spreading, tube narrowly cylindrical; stamens 4, equal, slightly exserted, filaments short, straight, inserted near rim of corolla-tube, anthers versatile, thecae parallel, separate at dehiscence; pollen 6-pantocolpate, operculate, with supratectal spinules; ovary unlobed or very shallowly 4-lobed; stigma-lobes equal to subequal; disc absent; fruit a schizocarp with four fleshy, 1-seeded mericarps. One species, *Huxleya linifolia* Ewart & B. Rees, northern Australia.

20. *Clerodendrum* L. Fig. 24

Clerodendrum L., Sp. Pl. 2: 637 (1753); L., Gen. Pl. ed. 5: 285 (1754); Thomas, Bot. Jahrb. Syst. 68: 1–98 (1936), reg. rev.
Siphonanthus L. (1754).
Volkameria L. (1754).
Tetraclea A. Gray (1853).
Kalaharia Baill. (1891).

Trees, lianas, shrubs, subshrubs and rarely perennial or annual herbs, often ill-scented; leaves opposite (rarely alternate or whorled), usually petiolate, simple, entire or toothed, rarely lobed; inflorescence a terminal thyrse (usually lax, paniculiform or corymboid, less often capitate) or axillary cyme (occasionally reduced to solitary flower); flowers often resupinate; calyx usually actinomorphic, usually accrescent, often showy, 5-lobed (rarely 2–4-lobed or truncate); corolla variable in colour, actinomorphic to subactinomorphic, usually hypocrateriform, or 1-lipped with anterior lip present, but usually facing upwards due to resupination, 5-lobed, anterior lobe sometimes slightly larger than others, corolla in bud symmetrical or asymmetrical; stamens 4, subequal to didynamous, exserted, filaments straight or arched (usually upwards), anthers versatile, thecae parallel, separate at dehiscence; pollen tricolpate (rarely 6-pantocolpate), with supratectal spinules, columellae branched to granular; ovary unlobed to shallowly 4-lobed; stigma-lobes equal or subequal; disc usually poorly developed to absent; fruit drupaceous with four 1-seeded pyrenes or a schizocarp with four fleshy (rarely dry), 1-seeded mericarps (or rarely two 2-seeded mericarps); endosperm absent. $2n = 22$–108 (most frequently 48 or 52; $x = 12, 13$). Four hundred to 500 species, mostly palaeotropical but some neotropical and a few temperate; many used as ornamentals.

21. *Aegiphila* Jacq.

Aegiphila Jacq., Observ. Bot. 2: 3 (1767); Moldenke, Brittonia 1: 245–477 (1934), rev.

Trees, shrubs, and woody vines, frequently dioecious; leaves usually opposite (occasionally alternate or subverticillate), petiolate or sessile, simple, usually entire; cymes axillary, paniculiform to capitate (rarely reduced to solitary flowers), often merging to form a terminal thyrse; calyx actinomorphic, often coriaceous, accrescent, shallowly 4–5-lobed or truncate, sometimes torn irregularly into 2–4 deeper lobes by corolla elongation; corolla usually white to yellow or greenish (occasionally red to purple), actinomorphic, hypocrateriform or infundibular, 4–5-lobed; stamens 4–5 (usually as many as corolla-lobes), equal, usually exserted when fertile, included when sterile (i.e. in functionally pistillate flowers), filaments straight, thecae parallel, separate at dehiscence; pollen with supratectal spinules; ovary unlobed, stigma-lobes equal; disc poorly developed or absent; fruit drupaceous, oblate to globose, with one to four 1-seeded pyrenes; endosperm absent. One hundred and sixteen species, mostly neotropical.

22. *Faradaya* F. Muell.

Faradaya F. Muell., Fragm. 5: 21 (1865); de Kok & Mabberley, Blumea 44: 321–342 (1999), rev.
Schizopremna Baill. (1891).

Lianas and small trees; leaves opposite (rarely whorled), petiolate, ovate to obovate, entire, with flat, orbicular glands near base of abaxial surface; inflorescence lax, cymose, axillary (sometimes cauliflorous) or terminal, often paniculiform; calyx actinomorphic, leathery, not accrescent, truncate or 4-lobed in bud, usually remaining closed until torn irregularly into 2–4 deep lobes by corolla elongation (less frequently calyx opening normally, remaining shallowly 4-lobed or truncate); corolla white or creamy, fleshy or stiff, actinomorphic or nearly so, 4-lobed, lobes spreading, entire or notched; stamens 4, equal or subequal, exserted, filaments straight, thecae parallel, separate at dehiscence; pollen with supratectal spinules; ovary shallowly 4-lobed, stigma-lobes very short, equal; disc absent to well developed; fruit drupaceous, the lobes elongating greatly, eventually separating into 4 fleshy, 1-seeded mericarps (often fewer by abortion) with apex rounded, mericarp attachment-scar basal; endosperm absent. Three species of Eastern Malaysia, Papua New Guinea, Indonesia, Melanesia, Polynesia, and north-eastern Australia.

23. *Oxera* Labill.

Oxera Labill., Sertum austro-caledonicum 23 (1824); de Kok & Mabberley, Kew Bull. 54: 257–264, 265–300 (1999), rev.

Lianas, shrubs, and small trees; leaves opposite, occasionally alternate or whorled, petiolate or sessile, simple, entire or rarely sinuate; inflorescence a lax, axillary or cauliflorous cyme or paniculiform or racemoid thyrse; calyx actinomorphic, not accrescent, 4–5-lobed (rarely truncate) in bud, remaining closed in some species until torn irregularly into 1–5 lobes by corolla elongation, opening normally in others; corolla white to yellow or orange, weakly zygomorphic to 2-lipped, 4–5-lobed, lobes highly variable in shape, anterior lobe often widened distally and emarginate, fimbriate or galeate; stamens usually 2 (the posterior or rarely the anterior pair reduced to staminodes), rarely 4, usually exserted, filaments straight or curved, thecae parallel, separate at dehiscence; pollen with supratectal spinules; ovary shallowly 4-lobed; stigma-lobes very short, equal; disc well developed; fruit drupaceous, the lobes elongating greatly, eventually separating into 4 fleshy, 1-seeded mericarps (often fewer by abortion) with apex rounded, mericarp attachment-scar basal; endosperm absent. Twenty-one species of South Pacific islands (mostly New Caledonia).

24. *Hosea* Ridl.

Hosea Ridl., J. Straits Branch Roy. Asiat. Soc. 50: 124, 125 (1908).
Hoseanthus Merr. (1917).

Lianas; leaves opposite, petiolate, ovate, entire; cymes lax, axillary, subtended by showy orange bracts; calyx 4-lobed, basically actinomorphic but often forming two 2-lobed lips due to incomplete separation of lobes, not accrescent; corolla orange, zygomorphic, 4-lobed, lobes erect, anterior and lateral lobes similar, obovate, posterior lobe longer, oblong; stamens 4, didynamous, long-exserted, filaments arched, thecae parallel, separate at dehiscence; pollen with supratectal spinules; ovary shallowly 4-lobed; stigma-lobes very short, equal; disc absent; fruit dry (or very slightly fleshy?), the lobes elongating greatly, eventually separating into potentially 4 (but usually fewer due to abortion) large, fusiform, 1-seeded mericarps with apex acute, mericarp attachment-scar basal; endosperm absent. One species, *Hosea lobbiana* (Clarke) Ridl., Sarawak and Brunei.

25. *Spartothamnella* Briq.

Spartothamnella Briq. in Engler & Prantl, Nat. Pflanzenfam. IV, 3a: 161 (1895); Munir, J. Adelaide Bot. Gard. 1: 3–25 (1976), rev.
Spartothamnus A. Cunn. ex Walp. (1847), non *Spartothamnus* Webb & Berthel. ex C. Presl (1845).

Shrubs; stems sharply quadrangular; leaves opposite, sessile or subsessile, elliptic to oblong or linear-lanceolate, entire, often caducous; cymes axillary, tight, 1–3-flowered; calyx actinomorphic, not accrescent, 5-lobed; corolla white, zygomorphic, not clearly lipped or 1-lipped and then only anterior lip present, 5-lobed, anterior lobe largest; stamens 4, didynamous, included to slightly exserted, filaments arched, thecae divaricate, confluent at dehiscence; pollen operculate, with supratectal verrucae; ovary shallowly 4-lobed; stigma-lobes equal; disc absent; fruit drupaceous, globose, orange-red, pyrenes 4, 1-seeded, reticulately ridged; endosperm absent. Three species, Australia. Previously placed within "Chloanthoideae" (Prostantheroideae tribe Chloantheae), but probably closely related to *Teucrium*, *Teucridium*, and *Oncinocalyx*, with which it shares a distinctive pollen morphology (Abu-Asab and Cantino 1993).

26. *Oncinocalyx* F. Muell.

Oncinocalyx F. Muell., S. Sci. Rec. 3: 69 (1883); Munir, J. Adelaide Bot. Gard. 14: 77–84 (1991), rev.

Perennial herb or low shrub; stems sharply quadrangular; leaves opposite, sessile, linear to narrowly elliptical, entire, strongly revolute; inflorescence a tight axillary cyme of (1–)3–5 flowers; calyx actinomorphic, not accrescent,

5-lobed, lobes subulate, apically recurved-uncinate; corolla white, zygomorphic, 5-lobed, anterior lobe largest; stamens 4, equal or didynamous, included to slightly exserted, filaments straight, thecae divaricate, confluent at dehiscence; pollen operculate, with supratectal verrucae; ovary shallowly 4-lobed; stigma-lobes equal; disc poorly developed; nutlets obovoid-trigonal, reticulately ridged, pubescent, glandular, attachment-scar lateral, 0.6× as long as nutlet; endosperm present. One species, *Oncinocalyx betchei* F. Muell., eastern Australia (Queensland, New South Wales).

27. *Teucridium* Hook. f.

Teucridium Hook. f., Fl. Nov.-zel. 1: 203 (1853).

Shrub, non-aromatic; stems sharply quadrangular; leaves opposite, petiolate, broadly ovate to subrotund, entire or irregularly lobed; flowers small, axillary, usually solitary (occasionally in tight, few-flowered cymes); calyx actinomorphic, not accrescent, 5-lobed; corolla white to pale blue, very small, 1-lipped with anterior lip present, 5-lobed, anterior lobe largest; stamens 4, didynamous, exserted, filaments arched, thecae divaricate, confluent at dehiscence; pollen operculate, with supratectal verrucae; ovary unlobed or very shallowly 4-lobed; stigma-lobes subequal; disc absent; nutlets obovoid, rugose, pubescent, glandular, attachment-scar lateral, 0.7× as long as nutlet; endosperm absent. $2n = 64$. One species, *Teucridium parvifolium* Hook. f., New Zealand.

28. *Teucrium* L. Fig. 21 (Buchstaby Michael Zahl)

Teucrium L., Sp. Pl. 2: 562 (1753); Gen. Pl. ed. 5: 247 (1754).
Melosmon Raf. (1837).
Kinostemon Kudo (1929).

Shrubs, subshrubs, or perennial herbs (rarely annual or biennial), often aromatic or ill-scented; leaves opposite, sessile or petiolate, simple, margin entire, variously toothed or pinnately lobed; inflorescence an axillary cyme or paniculiform to spiciform thyrse, cymes often reduced to single flowers located in leaf-axils or in bracteate racemes or heads; calyx actinomorphic to 2-lipped, accrescent or not, 5-lobed, lobes all similar or posterior lobe different from other 4, or posterior, lateral and anterior lobes differentiated as distinct types, calyx-tube often ± gibbous; corolla highly variable in colour, usually 1-lipped with anterior lip present, rarely 2-lipped, 5-lobed, anterior lobe much larger than others, usually concave; stamens 4, didynamous, exserted, filaments arched or straight, thecae divaricate, usually confluent at dehiscence; pollen operculate, usually with supratectal verrucae (rarely suprareticulate or partially psilate); ovary shallowly 4-lobed; stigma-lobes equal or unequal; disc absent to well developed; nutlets obovoid to oblate or subglobose, usually reticulately ridged to alveolate, pubescent or glabrous, attachment-scar lateral, 0.3–0.8× as long as nutlet; endosperm usually present. $2n = 10–104$ (most frequently 26 or 32). About 250 species, subcosmopolitan but mostly extratropical, with centre of diversity in Mediterranean region. A number of sections have been recognised (Briquet 1895–1897; Kästner 1989; Navarro and El Oualidi 1997).

29. *Ajuga* L.

Ajuga L., Sp. Pl. 2: 561 (1753); Gen. Pl. ed. 5: 246 (1754).
Chamaepitys Hill (1756).

Perennial, annual, and biennial herbs (rarely subshrubs); inflorescence of dense, verticillastrate cymes, sometimes reduced to solitary flowers or merging to form a terminal, spiciform or subcapitate thyrse; calyx actinomorphic or weakly zygomorphic (posterior lobe reduced), not accrescent, 5-lobed; corolla highly variable in colour, 2-lipped or 1-lipped and then with only anterior lip present, marcescent, persisting as a sheath around fruit, posterior lip short or obsolescent, usually 2-lobed or retuse, anterior lip 3-lobed, median lobe very large, broadening distally and usually emarginate or bipartite; stamens 4, didynamous, usually slightly exserted, thecae divergent to divaricate, confluent at dehiscence; pollen suprareticulate, columellae simple or branched; ovary shallowly 4-lobed; stigma-lobes equal to unequal; disc poorly to well developed, usually asymmetrical; fruit usually a schizocarp of 4 nutlets (rarely drupaceous, 4-lobed), nutlets oblate to obovoid, reticulately ridged to foveolate, usually glabrous, attachment-scar lateral, 0.55–0.85× as long as nutlet; endosperm present. $2n = 32$ in most species (16, 24, 28, 30, 62, 64, c. 86). Forty to 50 species, palaeotemperate, mostly Eurasian. *Ajuga reptans* L. is used as an ornamental groundcover.

30. *Discretitheca* P.D. Cantino

Discretitheca P.D. Cantino, Syst. Bot. 23: 379 (1999).

Shrub; leaves opposite, petiolate, ovate, dentate; cymes lax, axillary, grading into terminal, paniculiform thyrse; calyx actinomorphic, slightly accrescent, 5-lobed; corolla pink, zygomorphic, 5-lobed (4/1), posterior lobes ± similar, anterior lobe larger, concave, corolla in bud strongly asymmetrical, corolla-tube slightly upturned near base; stamens 4, didynamous, long-exserted, arched, thecae parallel, separate at dehiscence; pollen 6-pantocolpate, with supratectal spinules; ovary unlobed; stigma-lobes unequal; disc poorly developed to absent; fruit a dry schizocarp, requiring slight pressure to split into nutlets, nutlets reticulately ridged, minutely pubescent and glandular, obovoid, obscurely trigonal, attachment-scar as long as nutlet; endosperm absent. One species, *Discretitheca nepalensis* (Moldenke) P.D. Cantino, Nepal.

31. ***Pseudocaryopteris*** P.D. Cantino

Pseudocaryopteris P.D. Cantino, Syst. Bot. 23: 380 (1999).
Caryopteris sect. *Pseudocaryopteris* Briq. (1895).

Tall shrubs, often scrambling or spreading; leaves opposite, petiolate, ovate to lanceolate, variously toothed or rarely subentire; inflorescence of axillary cymes or axillary or terminal paniculiform thyrses; calyx ± actinomorphic, accrescent or not, 5–6-lobed; corolla white to deep red, purple, or blue, 5-lobed (4/1), 4 posterior lobes ± similar, anterior lobe larger, corolla in bud ± asymmetrical; stamens 4, equal or didynamous, exserted, arched, thecae divergent to divaricate, confluent at dehiscence; pollen with supratectal verrucae, columellae sparingly branched; ovary unlobed; stigma-lobes equal; disc poorly developed; fruit a dry or slightly fleshy schizocarp, sometimes requiring pressure to split into nutlets, nutlets boat-shaped (arcuate in cross section), smooth or slightly ridged, glabrous or pubescent, attachment-scar as long as nutlet; endosperm absent. $2n = 40, 60$. Three species, S China, India, Nepal, Bhutan, Thailand, Myanmar, and Pakistan.

32. ***Tripora*** P.D. Cantino Fig. 21D, H, K

Tripora P.D. Cantino, Syst. Bot. 23: 380 (1999).

Perennial herb, ill-scented; leaves opposite, petiolate, ovate, dentate to crenate; cymes lax, axillary, grading into terminal, paniculiform thyrse; calyx actinomorphic, slightly accrescent, 5-lobed; corolla purple with basal part of anterior lobe white, zygomorphic, 5-lobed (4/1), 4 posterior lobes ± similar, anterior lobe larger, emarginate or erose, slightly concave, corolla in bud strongly asymmetrical; stamens 4, didynamous, long-exserted, filaments arched, thecae divaricate, confluent at dehiscence; pollen triporate, with supratectal spines; ovary shallowly 4-lobed; stigma-lobes usually unequal; disc moderately well developed; nutlets obovoid, reticulately ridged, puberulent, attachment-scar lateral, 0.5–0.7× as long as nutlet; endosperm scanty. One species, *Tripora divaricata* (Maxim.) P.D. Cantino, China, Japan, Korea.

33. ***Schnabelia*** Hand.-Mazz.

Schnabelia Hand.-Mazz., Kaiserl. Akad. Wiss. Wien, Math.-Naturwiss. Kl., Anz. 58: 92 (1921); Cantino, Wagstaff & Olmstead, Syst. Bot. 23: 369–386 (1999), gen. recircumscription, redescription.
Chienodoxa Y.Z. Sun (1951).

Low shrubs and perennial herbs, sometimes prostrate, decumbent or scandent, slightly aromatic; stems sometimes winged; leaves opposite, petiolate, simple or occasionally 3-lobed, ovate to triangular or suborbicular, dentate to crenate, sometimes caducous; inflorescence of axillary cymes, 1–5-flowered; cleistogamous [cl.] flowers sometimes present, minute, with corolla and stamens shorter than calyx; calyx (in both cl. and chasmogamous [ch.] flowers) actinomorphic or occasionally with 1 lobe smaller than others, 4–5-lobed, lobes broadly ovate to linear-lanceolate; corolla of ch. flowers purple, pink or white, zygomorphic, unlipped or 1-lipped with anterior lip present, 5-lobed, 4 posterior lobes similar, anterior lobe larger, often (always?) concave, corolla in bud asymmetrical; stamens of ch. flowers 4, didynamous, exserted, filaments arched or ± straight, thecae parallel to divaricate, confluent at dehiscence; pollen with supratectal spinules, columellae branched or granular; ovary unlobed to very shallowly 4-lobed; stigma-lobes equal to subequal; disc poorly developed or absent; nutlets obovoid, reticulately ridged, pubescent, attachment-scar lateral, 0.5–0.9× as long as nutlet. Five species of forests and thickets in China.

34. ***Rubiteucris*** Kudô

Rubiteucris Kudô, Mem. Fac. Sci. Taihoku Imp. Univ. 2(2): 297 (1929); Cantino, Wagstaff & Olmstead, Syst. Bot. 23: 369–386 (1999), gen. recircumscription, redescription.
Cardioteucris C.Y. Wu (1962).

Rhizomatous perennial herb, normally unbranched; leaves opposite, long-petiolate, broadly ovate, trilobed or trifoliolate, coarsely crenate to dentate; inflorescence a terminal thyrse; calyx 2-lipped (3/2), broadly campanulate, gaping, accrescent, in fruit gibbous, lips shallowly lobed; corolla white, 2-lipped (2/3) or 1-lipped (0/5), anterior lobe largest, corolla in bud asymmetrical; stamens 4, didynamous, long-exserted, filaments arched, thecae divergent to divaricate, confluent at dehiscence; pollen with supratectal spinules, columellae branched or granular; ovary very shallowly 4-lobed to unlobed; stigma-lobes equal; disc absent; nutlets broadly obovoid to subglobose, smooth to faintly ridged, glabrous, attachment-scar lateral, 0.7–1.0× as long as nutlet; endosperm scanty. Two species of forests and open slopes in China, India, and Myanmar.

35. *Amethystea* L.

Amethystea L., Sp. Pl. 1: 21 (1753); Gen. Pl. ed. 5: 13 (1754).

Annual herb, slightly aromatic; leaves opposite, petiolate, deeply 3-lobed, serrate; inflorescence a lax, terminal paniculiform thyrse; flowers minute; calyx actinomorphic, accrescent, 5-lobed; corolla pale lavender, zygomorphic, 5 lobed, anterior lobe largest, concave; fertile stamens 2, posterior pair reduced to staminodes, filaments arched, thecae divaricate, confluent at dehiscence; pollen with supratectal spinules; ovary unlobed; stigma-lobes unequal; disc absent; nutlets obovoid, reticulately ridged, glabrous, attachment-scar lateral, 0.7–0.8× as long as nutlet; endosperm present. $2n = 20, 26$. One species, *Amethystea caerulea* L., widespread in open, disturbed habitats in temperate Asia.

36. *Trichostema* L. Fig. 21R

Trichostema L., Sp. Pl. 2: 598 (1753); L., Gen. Pl. ed. 5: 260 (1754); Lewis, Brittonia 5: 276–303 (1945), rev.; Lewis & Rzedowski, Madroño 25: 151–154 (1978), reg. rev.
Isanthus Michaux (1803).
Eplingia L.O. Williams (1973).

Shrubs, subshrubs, and annuals, strongly aromatic; leaves opposite, petiolate or sessile, elliptic to linear or ovate, usually entire; inflorescence of axillary cymes, sometimes reduced to solitary flowers or grading into paniculiform terminal thyrse; calyx actinomorphic or zygomorphic, accrescent, 5-lobed (3/2), posterior lobes in zygomorphic calyces forming long upper lip (appearing to be lower lip in fruit due to post-anthesis resupination), anterior lobes shorter; corolla violet to pink, zygomorphic, 5-lobed, four posterior lobes usually similar, entire, anterior-most lobe larger, entire or erose, tube straight or distally upturned, corolla in bud strongly asymmetrical to nearly symmetrical; stamens 4, usually didynamous, usually long-exserted, filaments strongly arched to nearly straight, thecae divergent to nearly parallel, confluent at dehiscence; pollen with supratectal spines or spinules, columellate stratum of exine undifferentiated or composed of branched columellae; ovary shallowly 4-lobed; stigma-lobes equal or unequal; disc well developed to absent; nutlets obovoid, reticulately ridged, pubescent or glabrous, attachment-scar lateral, 0.2–0.5× as long as nutlet; endosperm scanty or absent. $2n = 14, 20, 28, 38$. Seventeen species of dry habitats, ranging from southern Canada to Mexico, Florida, and the Bahamas. Five sections are recognised. Calyx strongly zygomorphic, inverted in fruit: sect. *Trichostema*. Annual herbs; calyx actinomorphic: sect. *Orthopodium* Benth. Inflorescence densely lanate; leaves linear, revolute: sect. *Chromocephalum* F.H. Lewis. Not as above; corolla bright pink; stamens nearly straight: sect. *Rhodanthum* F.H. Lewis. Not as above; corolla violet; stamens arched: sect. *Paniculatum* F.H. Lewis.

37. *Caryopteris* Bunge

Caryopteris Bunge, Plantarum Mongolico-Chinensium Decas Prima 27 (1835); Cantino, Wagstaff & Olmstead, Syst. Bot. 23: 369–386 (1999), gen. recircumscription, redescription, relationships.

Shrubs and subshrubs, strongly aromatic; leaves opposite, petiolate, simple, toothed or entire; inflorescence of axillary and terminal cymes, corymboid, ebracteate; calyx actinomorphic, accrescent, 5-lobed; corolla bluish purple or pale green to yellowish, zygomorphic, 5-lobed (4/1), posterior lobes similar, ± entire, anterior lobe much larger, concave, dentate to fimbriate-laciniate, corolla in bud ± symmetrical; stamens 4, equal to didynamous, exserted, filaments straight, divergent, thecae parallel, separate at dehiscence; pollen suprareticulate or psilate, columellae sparingly branched; ovary unlobed; stigma-lobes equal; disc absent; fruit a schizocarp of 4 nutlets, nutlets boat-shaped (arcuate in cross section) or nearly flat, laterally winged, usually smooth, glabrous or pubescent, attachment-scar as long as nutlet; endosperm scanty. $2n = 26, 52, 40?$ Seven species, China, Mongolia, Tibet, Taiwan, Korea, and Japan. The artificial hybrid *C.* × *clandonensis* A. Simmonds ex Rehder is widely grown as an ornamental.

IV. Subfam. Prostantheroideae Luerssen (1882).

Dicrastylidaceae Drummond ex Harvey (1855), nom. inval.
Verbenaceae tribe Viticeae subtribe Chloanthinae Benth. (1870).
Verbenaceae tribe Chloantheae Benth. (1876).
Subfam. Chloanthoideae Briq. (1895).
Subfam. Chloanthoideae Briq. (1895).
Stilbaceae subfam. Chloanthoideae Briq.: sensu Moldenke (1959).

Subshrubs, shrubs or small trees, non-aromatic or occasionally with distal branches and leaves aromatic. Indumentum usually of simple, stellate or branched hairs (occasionally of scales), sparse to dense, or indumentum lacking. Leaves simple, entire or toothed. Inflorescence cymose, usually thyrsoid, often appearing racemoid with cymes 1-flowered. Flowers usually zygomorphic, rarely almost actinomorphic, calyx actinomorphic or rarely 2-lipped, corolla 4–8- or often 5-lobed, stamens 4–8, all fertile or thecae variously reduced (*Hemiandra*, *Hemigenia*), or 2 staminodes present (*Hemiphora*, *Microcorys*, *Westringia*, *Wrixonia*). Pollen 3-, 6- and/or 8-colpate. Ovary unlobed (Fig. 25), or shallowly to deeply 4-lobed (sometimes unequally lobed), style terminal, often persistent in young fruit. Disc poorly to well developed. Fruit dry, entire, lobed or deeply divided. Seeds albuminous.

Key to the Tribes

1. Style arising from unlobed ovary; stamens 4–8, or 2 fertile and 2 staminodes (in *Hemiphora*); indumentum usually of stellate or branched hairs, or scales (some *Pityrodia*), or indumentum lacking and then fruiting-calyx enlarged and toothed (*Cyanostegia*) Tribe **Chloantheae**
- Style arising from between 4-lobed ovary; stamens 4, or 2 fertile and 2 staminodes; indumentum of simple hairs, very rarely of branched hairs, or indumentum lacking Tribe **Westringieae** (p. 207)

IV. 1. Tribe Chloantheae Benth. & Hook. f. (1876).

Shrubs or subshrubs, non-aromatic, indumentum often lanate with stellate or variously branched hairs (but glabrous, glutinous in *Cyanostegia*) or indumentum of peltate scales in some *Pityrodia*. Stamens 4–8, or reduced to 2 fertile stamens and 2 staminodes (*Hemiphora*), anthers 2-thecous, thecae distinct, connective lacking basal appendage, ovary ± unlobed or sometimes 2-lobed in *Pityrodia*, style terminal. Fruit ± entire, lobed or separating into 2 or 4 mericarps (nutlets).

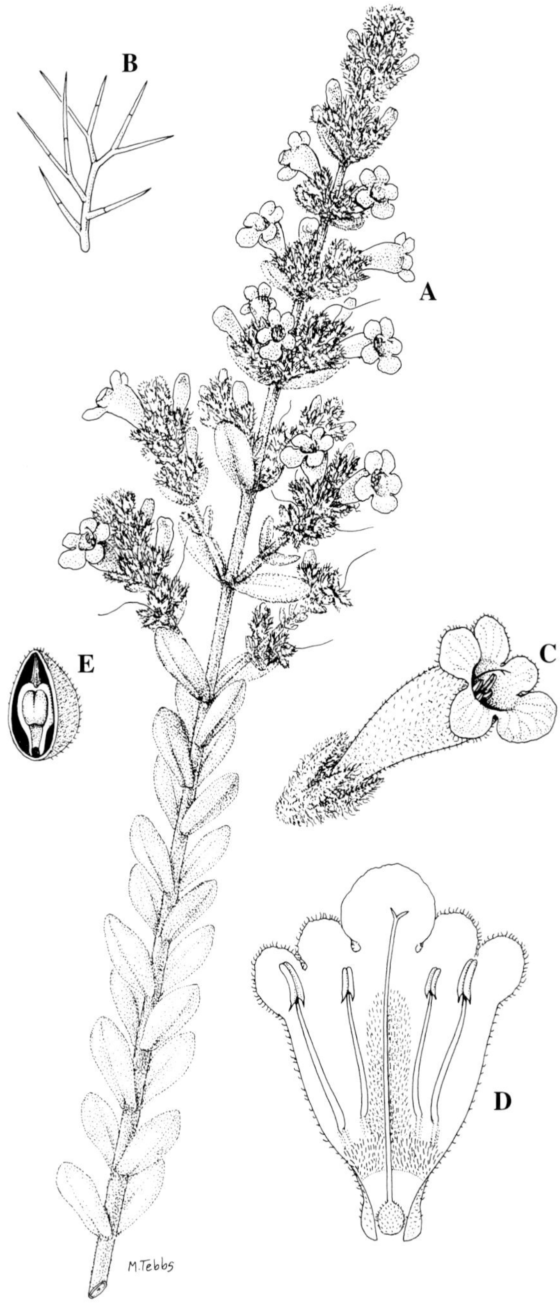

Fig. 25. Labiatae subfam. Prostantheroideae. *Pityrodia terminalis*. **A** Habit. **B** Dendroid trichome. **C** Flower, side view. **D** Corolla opened to show stamens and gynoecium. **E** Mericarp showing inner surface. Drawing by M. Tebbs

Key to the Genera

1. Flowers ± actinomorphic, isomerous; corolla regular or nearly so, 4–8-lobed; stamens 4–8 2
- Flowers zygomorphic, heteromerous; corolla mostly ±2-lipped or unequally 5-lobed; stamens 4 or only 2 fertile 6

2. Perianth 4-lobed 3
- Perianth 5–8-lobed 4
3. Flowers in dense spikes; stamens included; style entire (not lobed) **43. *Physopsis***
- Flowers in ± capitate clusters, heads solitary or in corymbose panicles; stamens shortly exserted; style shortly 2-lobed **42. *Mallophora***
4. Style deeply 2-branched; corolla usually 5-lobed **41. *Dicrastylis***
- Style entire or very minutely notched distally; corolla (5–)6–8-lobed 5
5. Corolla distinctly lobed; stamens inserted between lobes, included or exserted **47. *Newcastelia***
- Corolla-tube truncated, not lobed; stamens inserted on rim of corolla-tube, exserted **45. *Lachnostachys***
6. Leaves decurrent **38. *Chloanthes***
- Leaves not decurrent 7
7. Fertile stamens 2, anterior pair reduced to staminodes **39. *Hemiphora***
- Fertile stamens 4; staminodes lacking 8
8. Fruiting-calyx distinctly enlarged, toothed; filaments dilated in distal third **40. *Cyanostegia***
- Fruiting-calyx ± unchanged, untoothed; filaments of uniform width throughout 9
9. Leaves and branches with branched hairs; corolla tubular, with inner surface hairy; stamens inserted near middle of corolla-tube; anthers with basal appendage; ovary hairy; mericarps 2 **46. *Pityrodia***
- Leaves and branches with stellate hairs; corolla widely spreading, with inner surface glabrous; stamens inserted at base of corolla; anthers lacking a basal appendage; ovary glabrous; mericarps 4 **47. *Brachysola***

38. *Chloanthes* R. Br.

Chloanthes R. Br., Prodr. 1: 513 (1810); Munir, J. Adelaide Bot. Gard. 1: 83–105 (1977), rev.; Conn, Fl. New South Wales 3: 619 (1992), reg. rev.

Shrubs or subshrubs; stems ± round in cross section; indumentum lanate with hairs branched; leaves sessile, decurrent, decussate or ternate, bullate-rugose; inflorescence cymose, with flowers on leafy branches (hence appearing axillary) and part-inflorescences usually 1-flowered; bracteoles 2; flowers zygomorphic; calyx 5-lobed; corolla pale mauve, greenish blue or greenish yellow, distinctly 2-lipped, 5-lobed (2/3), or occasionally unequally 5-lobed in distal half; anterior lip with median lobe rarely larger than others; corolla-tube expanded in throat; stamens 4, inserted below middle of corolla-tube, just exserted, anthers with shortly divergent basal lobes; ovary ± unlobed; style with stigma-lobes present; fruit drupaceous, usually separating into two 2-locular mericarps. Four species, all endemic to Australia.

39. *Hemiphora* F. Muell.

Hemiphora F. Muell., Syst. Cens. Austr. Pl. 1: 103 (1882); Munir, J. Adelaide Bot. Gard. 1: 161–166 (1978), rev.

Shrubs or subshrubs; branches ± round in cross section; indumentum densely lanate with hairs branched; leaves sessile, decussate or ternate, bullate-rugose; inflorescence terminal, leafy and hence appearing axillary, cymose, unbranched, with part-inflorescences 1-flowered; bracteoles 2; flowers zygomorphic, sessile or shortly pedicellate; calyx deeply 5-lobed, densely woolly; corolla reddish purple, 2-lipped, 5-lobed (2/3), anterior lip shorter than posterior, more deeply lobed, tube expanded distally; fertile stamens 2, anterior pair reduced to staminodes, slightly exserted, anthers dorsifixed, thecae free in lower half, staminodes 2, with anthers greatly reduced, included; ovary ± unlobed; style with stigma-lobes minute; fruit dry, ± globose, indehiscent, 4-seeded. One species, *H. elderi* (F. Muell.) F. Muell., endemic to Australia.

40. *Cyanostegia* Turcz.

Cyanostegia Turcz., Bull. Soc. Nat. Mosc. 22(2): 35 (1849); Munir, Brunonia 1: 46–67 (1978).
Bunnya F. Muell. (1865), rev.

Shrubs or subshrubs; branches ± round in cross section, mostly glutinous; leaves sessile or petiolate, decussate; inflorescence terminal, cymose, openly paniculate, or with flowers on leafy branches (hence appearing axillary), part-inflorescences 1- to many-flowered; bracteoles 2; calyx broadly campanulate, 5-lobed or distally sinuately toothed; corolla deep blue to dark purple, zygomorphic, 5-lobed, anterior lobe larger than other 4, tube broadly campanulate; stamens 4, inserted near base of corolla-tube, exserted; anthers with thecae free in lower half; ovary ± unlobed; style with stigma-lobes minute and often unequal; fruit dry, hard, indehiscent, usually ± oblique, surrounded by enlarged calyx, 1- or 2-seeded. Five species, all endemic in Australia.

41. *Dicrastylis* J. Drumm. ex Harv.

Dicrastylis J. Drumm. ex Harv., Hook. J. Bot. Kew Misc. 7: 56 (1855); Munir, J. Adelaide Bot. Gard. 14: 85–92 (1991), rev.

Shrubs or subshrubs; branches ± round in cross section; indumentum densely lanate with hairs branched; leaves sessile or petiolate, decussate or whorled, sometimes almost alternate; inflorescence terminal, cymose, with part-inflorescences pedunculate or sessile, pyramidal or spike-like panicles, or the cymes sessile, congested to form decussate or whorled clusters, usually many-flowered; bracteoles 2; flowers ± actinomorphic, sessile or pedicellate; calyx 5-lobed; corolla white, pale mauve, purplish blue, mostly 5-lobed in distal half, with lobes equal, tube expanded in throat; stamens mostly 5, inserted in corolla-tube, exserted or scarcely so; anthers dorsifixed, free and divergent in lower half; ovary ± unlobed; style with stigma-lobes well developed; fruit dry, globose, indehiscent, 1- or 2-seeded. About 26 species, all endemic to Australia.

42. *Mallophora* Endl.

Mallophora Endl., Ann. Wein. Mus. 2: 206 (1838); Munir, Brunonia 1: 407–689 (1978), rev.
Lachnocephalus Turcz. (1849).

Shrubs; branches ± round in cross section; indumentum densely lanate with hairs branched; leaves sessile, decussate; inflorescence terminal, thyrsoid, subglobose and capitate or both capitate and spike-like (elongate), solitary or in corymbose panicles, with part-inflorescences shortly pedunculate or sessile, congested to form decussate clusters, many-flowered; bracteoles 2; flowers ± actinomorphic, subsessile; calyx deeply 4-lobed, densely hairy on outer surface; corolla white, 4-lobed in distal half, lobes almost equal, tube cylindrical; stamens 4, inserted in corolla-tube, scarcely exserted, anthers dorsifixed, free and divergent in lower half; ovary ± unlobed; style with stigma-lobes short; fruit dry, ± globose, indehiscent, usually 1-seeded. Two species, endemic to Australia.

43. *Physopsis* Turcz.

Physopsis Turcz., Bull. Soc. Nat. Mosc. 22(2): 34 (1849); Munir, Brunonia 1: 407–689 (1978), rev.

Shrubs; branches ± round in cross section; indumentum densely lanate with hairs branched; leaves sessile, decussate; inflorescence terminal, thyrsoid, elongate, dense and spike-like, with part-inflorescences shortly pedunculate or sessile, congested to form decussate clusters, 1- to few-flowered; bracteoles 2; flowers ± actinomorphic, sessile or shortly pedicellate; calyx 4-lobed, densely hairy on outer surface; corolla ochraceous-yellow to yellow, 4-lobed in distal half, lobes equal; tube cylindrical; stamens 4, included within corolla-tube, anthers dorsifixed, thecae free in lower half; ovary ± unlobed; style often becoming eccentric (especially when only one ovule develops), with stigma entire or stigma-lobes short; fruit dry, ± globose, indehiscent, often enclosed by calyx, usually 1-seeded. Two species, both endemic to Australia.

44. *Newcastelia* F. Muell.

Newcastelia F. Muell., Hook. J. Bot. Kew Misc. [8: 326 (1856), nom. nud.]; 9: 22 (1857); Munir, Brunonia 1: 407–689 (1978), rev.

Shrubs or subshrubs; branches ± round in cross section; indumentum densely lanate with hairs branched; leaves sessile or petiolate, decussate or in whorls of 3 or rarely 4; inflorescence a terminal thyrse, branched or less frequently unbranched, with part-inflorescences pedunculate, spike-like to almost head-like, many-flowered; bracteoles 2; flowers ± actinomorphic, sessile or subsessile; calyx 5- or 6-lobed, rarely more than 6-lobed, densely woolly; corolla ± purple to blue, or yellow to whitish yellow, 5- or 6-lobed in distal half, rarely 4- or more than 6-lobed, lobes equal, tube expanded in throat; stamens 5 or 6, rarely more, inserted in corolla-tube, exserted or included, anthers dorsifixed, free and divergent in lower half; ovary ± unlobed; style with stigma-lobes minute or stigma almost entire; fruit dry, globose, indehiscent, 1- or 2-seeded. Twelve species, all endemic to Australia.

45. *Lachnostachys* Hook.

Lachnostachys Hook., Icon. Pl. N.S. 1: 415, t. 414 (1842); Munir, Brunonia 1: 643–679 (1978), rev.
Walcottia F. Muell. (1859).
Pycnolachne Turcz. (1863).

Shrubs or subshrubs; branches ± round in cross section; indumentum densely lanate with hairs branched; leaves sessile, decussate; inflorescence a terminal thyrse, often leafy and hence appearing axillary, ± branched or unbranched, with part-inflorescences pedunculate, elongate and spike-like, or ± lax, corymbose or pyramidal-paniculate, many-flowered; bracteoles 2; flowers ± actinomorphic, sessile or subsessile; calyx 5–8(–9)-lobed, densely woolly; corolla ± as large as calyx, ± purple, lobes absent or very minutely 5–8-lobed, rarely more, tube campanulate; stamens 5–8, rarely more, inserted on rim of corolla-tube, exserted, anthers dorsifixed, free and divergent in lower half; ovary ± unlobed; style with stigma-lobes minute or stigma entire; fruit dry, globose, indehiscent, 1-seeded. Six species, all endemic to Australia.

46. *Pityrodia* R. Br. Fig. 25

Pityrodia R. Br., Prodr. 1: 513 (1810); Munir, J. Adelaide Bot. Gard. 2: 1–138 (1979), rev.
Premna auct. non L., sensu Spreng. (1825).
Quoya Gaudich. (1828).
Dasymalla Endl. (1839).
Dennisonia F. Muell. (1859).
Depremesnilla F. Muell. (1876).

Shrubs or subshrubs; branches ± round or quadrangular in cross section, indumentum of branched hairs or lepidote scales; leaves opposite or in whorls of 3–4; inflorescence thyrsoid, with flowers either on leafy branches (hence appearing axillary) and part-inflorescences usually few-flowered or terminal, leafy to bracteate, botryoidal or paniculate; bracteoles 2; flowers zygomorphic; calyx 5-lobed; corolla white, pale mauve, red, deep pink, purple-blue, occasionally pale yellow, tubular, distinctly 2-lipped, 5-lobed (2/3), posterior lip erect to extended forwards, 2-lobed, or unequally 5-lobed in distal half, anterior lip usually spreading, 3-lobed, tube expanded in throat, hairy within; stamens 4, inserted near mid-point of corolla-tube, included or shortly exserted; anthers dorsifixed, thecae free in lower half; ovary hairy, 2-lobed or ± unlobed; style rarely displaced laterally with stigma-lobes distinct; fruit separating into two 2-locular mericarps. About 45 species, all endemic to Australia. *Pityrodia* appears to be polyphyletic and parts of it may need to be united with *Chloanthes* R. Br. and *Hemiphora* F. Muell.

47. *Brachysola* Rye

Brachysola Rye, Nuytsia 13: 331–338 (2000).
Chloanthes sect. *Brachysolenia* F. Muell. (1876).
Pityrodia sect. *Brachysolenia* (F. Muell.) E. Pritz. (1904).

Shrubs; branches ± round; indumentum of non-glandular stellate hairs and simple glandular hairs; leaves opposite or almost whorled; inflorescence a terminal dichasium, 1–7-flowered; flowers zygomorphic; calyx 5-lobed; corolla bluish purple with base more deeply coloured surrounding a yellow viscid centre, 2-lipped with lips widely spreading, 5-lobed (2/3), posterior lip 2-lobed, anterior lip 3-lobed, tube broadly expanded in throat, glabrous within; stamens 4, inserted near base of corolla, included, anthers dorsifixed near base; ovary ± unlobed; style terminal, with base becoming exceeded by developing fruit; stigma-lobes distinct; fruit separating into four 1-locular mericarps. Two species, both endemic to Australia.

IV. 2. Tribe Westringieae Bartl. (1830).

Tribe *Prostanthereae* Benth. (1834).

Shrubs or subshrubs, usually non-aromatic, except in *Prostanthera*, indumentum usually of simple, glandular and/or non-glandular hairs, less frequently with branched hairs, never lanate, or less frequently indumentum lacking. Fertile stamens 4 or 2, anthers 2-thecous or often 1-thecous, connective lacking basal appendage except in *Prostanthera*; ovary 4-lobed, style terminal but inserted between ovary-lobes, stigma-lobes present. Fruit with 4 mericarps (nutlets).

1. Stamens 4 (all fertile) 2
– Stamens 2 (fertile); staminodes 2 4
2. Calyx 2-lipped; anthers 2-thecous; connective not extending beyond insertion of the filament but often with 1 or 2 appendages; leaves and branchlets frequently aromatic when crushed **51. *Prostanthera***
– Calyx 5-lobed, sometimes almost 2-lipped; anthers 1-thecous; connective (including the other ± reduced, mostly sterile theca) elongated beyond insertion of the filament, appendages lacking; leaves and branchlets non-aromatic 3
2. Anther-connective terminating in a minute lobe (sterile theca), not readily visible **48. *Hemiandra***
– Anther-connective elongated and distinct, posterior staminal pair terminating in a short, dilated, crested end and anterior staminal pair attenuated and glabrous or with an imperfect theca at the end **49. *Hemigenia***
4. Anterior stamens fertile; anthers 2-thecous, appendages lacking; posterior stamens sterile (staminodes), terminating in very small, head-like ends **52. *Wrixonia***
– Anterior stamens sterile (staminodes), connective white, reduced to 2 linear or linear-clavate branches; posterior anthers fertile; anthers 1-thecous 5
5. Connective of fertile stamens elongated, extending beyond the insertion of the filament into a short, dilated and bearded end **50. *Microcorys***
– Connective of fertile stamens not or only minutely extended beyond the insertion of the filament **53. *Westringia***

48. *Hemiandra* R. Br.

Hemiandra R. Br., Prodr. 1: 502 (1810).

Shrubs or subshrubs, erect to ± prostrate, non-aromatic; branches mostly ± round in cross section (except distal branchlets usually quadrangular in cross section); leaves decussate, pungent; inflorescence thyrsoid, with flowers on leafy branches (hence appearing axillary) and part-inflorescences usually few-flowered; bracteoles 2; calyx ± distinctly 2-lipped, 5-lobed; corolla white, pale mauve, red, often pinkish, blue-green, occasionally yellow, 2-lipped, 5-lobed (2/3), posterior lip erect to extended forwards, anterior lip usually spreading, with middle lobe often 2-fid, tube expanded in throat; stamens 4, anthers 1-thecous, sterile theca reduced to tooth-like appendage, connective lacking a basal appendage, elongated beyond insertion of the filament; fruit of 4 mericarps, enclosed or not enclosed by calyx-lobes. About 14 species, all endemic to Australia.

49. *Hemigenia* R. Br.

Hemigenia R. Br., Prodr. 1: 502 (1810).

Shrubs or subshrubs, non-aromatic or ± aromatic when crushed; branches mostly ± round in cross section (except distal branchlets usually quadrangular in cross section), leaves decussate, or in whorls of 3–4; inflorescence thyrsoid, with flowers on leafy branches (hence appearing axillary) and part-inflorescences cymose, usually few-flowered; bracteoles 2; calyx often appearing 2-lipped, 5-lobed (sometimes unequally); corolla pale to dark mauve, often pinkish, occasionally white, distinctly 2-lipped, 5-lobed (2/3), posterior lip erect to extended forwards, anterior lip usually spreading, tube expanded in throat; stamens 4, posterior pair usually 2-thecous, but one theca greatly reduced and sometimes absent, anterior stamens with anthers 1-thecous, other theca sterile and often terminated by trichomes, all anther-connectives lacking a basal appendage, elongated beyond their insertion on the filament; fruit of 4 mericarps, enclosed or not enclosed by calyx-lobes. About 50 species, all endemic to Australia.

50. *Microcorys* R. Br. Fig. 21I

Microcorys R. Br., Prodr. 1: 502 (1810).

Shrubs or subshrubs, non-aromatic; branches mostly ± round in cross section (except distal branchlets usually quadrangular in cross section); leaves decussate or whorled; inflorescence thyrsoid, with flowers on leafy branches (hence appearing axillary) and part-inflorescences usually few-flowered; bracteoles 2; calyx often apparently 2-lipped, 5-lobed; corolla pale to dark mauve, pinkish or red, occasionally almost white, ±2-lipped, 5-lobed (2/3), posterior lip erect to extended forwards, anterior lip usually spreading, tube expanded in throat; fertile stamens 2, anterior pair reduced to staminodes, inserted on posterior side of corolla-throat, anthers 1-thecous, other theca sterile, often terminated by trichomes, connective lacking a basal appendage, elongated beyond its insertion on the filament into a short, dilated and bearded end, staminodes inserted on anterior side of corolla-throat, with thecae reduced to 2 sterile linear-clavate branches, connective lacking a basal appendage; fruit of 4 mericarps, usually not enclosed by calyx-lobes. About 20 species all endemic to Australia.

51. *Prostanthera* Labill.

Prostanthera Labill., Nov. Holl. Pl. Sp. 2: 18 (1806); Conn, J. Adelaide Bot. Gard. 6: 207–348 (1984), rev.; Fl. S. Australia 3: 1209–1219 (1986), reg. rev.; Nuytsia 6: 351–411 (1988), reg. rev.; Fl. New South Wales 3: 646–662 (1992), reg. rev.; Fl. Victoria 4: 445–455 (1999), reg. rev.
Cryphia R. Br. (1810).
Klanderia F. Muell. (1852).
Eichlerago J. Carrick (1977).

Shrubs or subshrubs, rarely trees; branches and leaves usually ± strongly aromatic when crushed; branches mostly ± round in cross section (except distal branchlets usually quadrangular in cross section); leaves decussate, rarely ternate; inflorescence thyrsoid, with flowers either on leafy branches (hence appearing axillary) and part-inflorescences usually few-flowered or inflorescence terminal, leafy to bracteate, botryoidal or paniculate; bracteoles 2; calyx 2-lipped; corolla white, pale mauve, olivaceous, red, often pinkish, mid (metallic) blue-green, occasionally yellow, distinctly 2-lipped, 5-lobed (2/3), posterior lip erect to extended forwards, anterior lip usually spreading, tube expanded in throat; stamens 4, anthers 2-thecous, with connective often elongated into a basal appendage; fruit of 4 mericarps, enclosed by calyx-lobes, often with calyx closed by infolded anterior calyx-lobe. About 100 species, all endemic to Australia. Two sections have been recognised (Bentham 1870). Calyx-lobes unequal in length; corolla-tube short, broadly expanded distally; anterior lobe longer and more spreading than the erect posterior lobes; fruit enclosed by inwardly folded anterior calyx-lobe, posterior calyx-lobe usually recurved: section *Prostanthera*. Calyx-lobes ± equal in length; corolla-tube long, slightly expanded distally, corolla-lobes ± equal; fruit not enclosed by the calyx-lobes, anterior lobe recurved, posterior lobe usually directed forwards: section *Klanderia* (F. Muell.) Benth. The species in sect. *Prostanthera* are pollinated by insects, whereas those of sect. *Klanderia* are bird-pollinated.

52. *Wrixonia* F. Muell.

Wrixonia F. Muell., Fragm. 10: 18 (1876); Carrick, J. Adelaide Bot. Gard. 1: 27–34 (1976), rev.

Shrubs or undershrubs, non-aromatic; branches ± round in cross section; leaves decussate; inflorescence frondose-bracteate, thyrsoid, terminal, few-flowered; bracteoles 2; calyx 2-lipped; corolla white, distinctly 2-lipped, 5-lobed (2/3), posterior lip with lobes directed outwards, anterior lip with median lobe directed outwards and forwards to recurved, lateral lobes spreading, tube expanded in throat; stamens 2, posterior pair reduced to staminodes, inserted on anterior side of corolla-throat, anthers 2-thecous, staminodes inserted on posterior side of corolla-throat, staminodes ± club-shaped, white; fruit of 4 mericarps, enclosed by inward-folded anterior calyx-lobe, posterior calyx-lobe usually recurved. Two species, endemic to Australia.

53. *Westringia* Sm.

Westringia Sm., Vetensk. Akad. Handl. Stockh. 171 (1797); Conn, Fl. S. Australia 3: 1230–1232 (1986), reg. rev.; Nuytsia 6: 335–349 (1988), rev.; Fl. New South Wales 3: 641–645 (1992), reg. rev.; Fl. Victoria 4: 441–444 (1999), reg. rev.

Shrubs or undershrubs, non-aromatic; branches often triangular or quadrangular in cross section; leaves in whorls of 3–6, less frequently decussate; inflorescence thyrsoid on leafy branches, part-inflorescences 1-flowered, bracteoles 2; calyx 5-lobed; corolla white to pale mauve, distinctly 2-lipped, 5-lobed (2/3), posterior lip with lobes directed outwards, anterior lip with median lobe directed outwards and forwards to recurved, lateral lobes spreading, tube expanded in throat; stamens 2 (anterior pair reduced to staminodes), inserted on posterior side of corolla-throat, anthers 1-thecous, dorsifixed, appearing basifixed, staminodes inserted on anterior side of corolla-throat, thecae reduced to 2 lobes, white; fruit of 4 mericarps, not enclosed by calyx-lobes. About 25 species, endemic to Australia.

V. Subfam. Scutellarioideae (Dumort.) Caruel (1884).

Shrubs and herbs, usually perennial, rarely aromatic with volatile terpenoids and rosmarinic acid absent, iridoid glycosides present and acteosides sometimes present. Leaves opposite, alternate, or whorled, petiolate to subsessile, simple. Inflorescence of axillary cymes, often reduced to single flowers (solitary or forming racemes). Bracts present, often reduced, bracteoles present or absent. Calyx usually 2-lipped (Fig. 26), ± accrescent, usually enclosing fruit at maturity, shallowly 5-lobed or lips entire, lips and lobes usually rounded, venation usually obscure. Corolla zygomorphic, usually 2-lipped, 4–5-lobed, tube cylindrical to infundibular, often curved. Stamens 4, anterior pair longer (rarely subequal), included or exserted, filaments usually pubescent, often apically geniculate, thecae parallel to divaricate, separate at dehiscence, anther slits often ciliate. Pollen tricolpate, tectate-perforate to microreticulate, usually suprareticulate (rarely psilate), col-

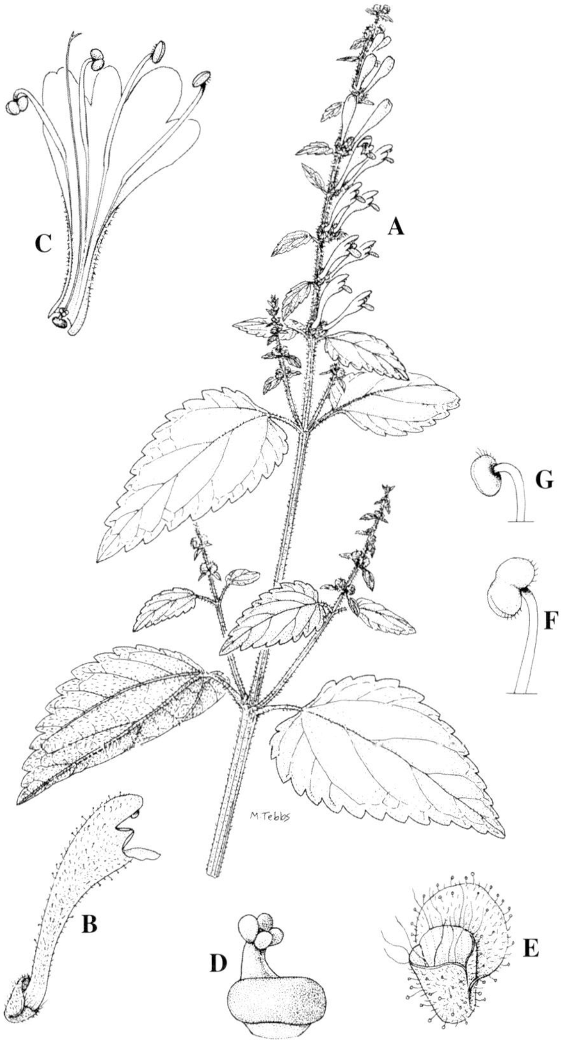

Fig. 26. Labiatae subfam. Scutellarioideae. *Scutellaria altissima*. **A** Habit. **B** Flower, side view. **C** Corolla opened out. **D** Disc and young gynoecium showing gynophore. **E** Calyx. **F** Anterior stamen. **G** Posterior stamen. Drawing by M. Tebbs

umellae simple. Ovary shallowly to deeply 4-lobed (rarely lobed on sides only). Style terminal to subterminal (rarely gynobasic), deciduous, its apex unequally 2-lobed. Disc poorly to well developed. Fruit a schizocarp. Nutlets 4, dry or rarely fleshy, surface frequently tuberculate or bearing long, plumose, hair-like processes, attachment-scar usually lateral. Seeds albuminous or not. Embryo spatulate or bent.

1. Calyx actinomorphic, patelliform, subtruncate, nutlets somewhat fleshy **54. *Holmskioldia***
- Calyx bilabiate; nutlets dry 2
2. Calyx with 3-lobed upper lip and 2-lobed lower lip; nutlets attached to receptacle by slender stalks; leaves alternate **55. *Wenchengia***
- Calyx-lips unlobed; nutlets not stalked; leaves usually opposite or whorled 3
3. Anterior stamens with one theca aborted; nutlet attachment-scar minute; upper lip of calyx usually bearing a conspicuous projection (scutellum) **58. *Scutellaria***
- Anther-thecae all fertile; attachment-scar 1/4 to 3/4 as long as nutlet; scutellum absent 4
4. Nutlet attachment-scar nearly as long as nutlet, bearing a vertical membranaceous outgrowth on one side; pericarp tuberculate but lacking hair-like outgrowths **56. *Renschia***
- Nutlet attachment-scar no more than half as long as nutlet, lacking membranaceous outgrowth; pericarp not tuberculate but bearing long, usually plumose, hair-like outgrowths, which may coalesce to form a wing **57. *Tinnea***

54. *Holmskioldia* Retz.

Holmskioldia Retz., Observ. Bot. 6: 31 (1791); Fernandes, Garcia de Orta, Ser. Bot., Lisboa 7: 33–46 (1985), gen. recircumscription.

Tall shrub with long, arching branches; leaves opposite, petiolate, broadly ovate to deltoid, finely toothed; inflorescence of lax cymes, grading into a terminal thyrse; bracts much reduced, paired bracteoles usually present; calyx actinomorphic, strongly accrescent, 5-lobed, lobes broadly rounded, tube patelliform, subtruncate; corolla bright red, zygomorphic, 5-lobed, lobes with margins recurved, posterior pair smallest, anterior lobe largest, tube curved; stamens slightly exserted, thecae parallel; ovary lobed on sides only, glandular-verrucose; disc poorly developed; nutlets slightly fleshy, tuberculate, glabrous, conspicuously glandular, attachment-scar lateral, 0.5–0.6× as long as nutlet; endosperm present; embryo spatulate. $2n = 32, 36$. One sp., *Holmskioldia sanguinea* Retz., native to the subtropical Himalayan region but widely grown as an ornamental in warm climates and often naturalized.

55. *Wenchengia* C.Y. Wu & S. Chow

Wenchengia C.Y. Wu & S. Chow, Acta Phytotax. Sin. 10: 250 (1965).

Low shrub; stem sympodially branched, distally leafy; leaves alternate, petiolate, oblanceolate, crenate to repand; inflorescence a terminal raceme; bracts present, reduced; calyx 2-lipped, 5-lobed (3/2), somewhat accrescent, enclosing nutlets at maturity (?), posterior lip shallowly 3-lobed, anterior lip much larger, nearly truncate, emarginate, tube broadly campanulate, gaping; corolla pink, 2-lipped (2/3), tube downcurved; stamens subequal, included, thecae slightly divergent; ovary shallowly lobed; disc poorly developed; nutlets attached to receptacle by slender stalks, dry, obovoid, laterally compressed, longitudinally ridged, apically tuberculate, each tubercle with a hair-like projection, attachment-scar lateral, 0.3× nutlet; endosperm present; embryo spatulate. One sp., *Wenchengia alternifolia* C.Y. Wu & S. Chow, southern China (Hainan), forest understorey; possibly extinct, known only from two collections made in the 1930s.

56. *Renschia* Vatke

Renschia Vatke, Linnaea 43: 94 (1881); Vollesen, Bot. Tidsskr. 70: 1–63 (1975), rev.

Low shrub; leaves opposite, petiolate, broadly ovate to orbicular, sinuate to sparsely crenate; inflorescence a terminal raceme; bracts present, reduced, bracteoles present; calyx 2-lipped, slightly accrescent, campanulate in flower, urceolate in fruit, lips entire, rounded, becoming appressed at maturity and enclosing nutlets, eventually dehiscent; corolla bluish purple, 2-lipped (2/3), tube straight; stamens barely exserted, thecae parallel, anther slit minutely ciliate; ovary shallowly lobed; disc well developed, symmetrical; nutlets dry, clavate, curved, glabrous, dorsal surface tuberculate, attachment-scar lateral, 0.8× nutlet, with a vertical membranaceous outgrowth along one side; endosperm present; embryo spatulate. One sp., *Renschia heterotypica* (S. Moore) Vatke, northern Somalia.

57. *Tinnea* Kotschy ex Hook. f.

Tinnea Kotschy ex Hook. f., Bot. Mag.: tab. 5637 (1967); Vollesen, Bot. Tidsskr. 70: 1–63 (1975), rev.

Shrubs, subshrubs, and perennial herbs; leaves opposite, alternate or whorled, usually entire or few-toothed; inflorescence of cymes sometimes reduced to solitary flowers or merging to form a terminal thyrse or raceme; bracts present, reduced or not, bracteoles present; calyx 2-lipped, inflated-accrescent, lips entire, rounded, becoming appressed at maturity and enclosing nutlets, eventually dehiscent or not; corolla reddish brown, orange, yellow, or violet, bilabiate (2/3), tube straight or upcurved, anterior lip 2–3× the posterior, median lobe much larger than others, usually strongly recurved laterally; stamens usually slightly exserted, filaments apically geniculate and often swollen, basally bearded and sometimes prolonged downwards as appendage, anterior anthers larger than posterior, thecae parallel to divaricate, anther slit usually minutely ciliate; ovary-lobes separate both above and below the style attachment (1/4–1/2 way to base of ovary); disc well developed, symmetrical; nutlets dry, clavate, pubescent or glabrous, the dorsal side bearing long, hair-like structures, usually plumose and coalescent to form a peltate, wing-like flying device, attachment-scar lateral, 0.25–0.5× nutlet; endosperm present or absent; embryo spatulate. $2n = 32$. Nineteen spp. of tropical and southern Africa, mainly in fire-prone grassland, woodland, and scrub. Vollesen (1975) distinguished subgenus *Tinnea*, with dehiscent fruiting-calyx and winged nutlets (the wings composed of coalescent, plumose, hair-like outgrowths), from subgenus *Membranacea* (Robyns & Lebrun) Vollesen, with indehiscent calyx and simple hair-like outgrowths on the nutlets.

58. *Scutellaria* L. Fig. 26

Scutellaria L., Sp. Pl. 2: 598 (1753); L., Gen. Pl. ed. 5: 260 (1754); Paton, Kew Bull. 45: 399–450 (1990), rev.
Perilomia Kunth (1818).
Salazaria Torrey (1858).
Harlanlewisia Epling (1955).

Perennial (rarely annual) herbs, subshrubs or shrubs, usually non-aromatic (a few species reported to be aromatic); leaves opposite, petiolate, entire, toothed or rarely pinnatifid; inflorescence of decussately or helically arranged flowers in axillary or terminal, often secund racemes or solitary in axils of foliage leaves; bracts present, reduced or not, bracteoles present or absent; calyx 2-lipped, accrescent, lips entire, rounded, becoming appressed at maturity, enclosing nutlets until posterior lip detaches to release them (or rarely the calyx indehiscent, inflated), posterior lip usually folded to produce a scutellum; corolla white, pink, blue, yellow or red, zygomorphic, tube straight or curved, often basally geniculate, usually 2-lipped, 4-lobed (3/1), with galeate posterior lip and unlobed anterior lip (less often weakly zygomorphic with 4 subequal, ± spreading lobes); stamens usually included, thecae parallel to divergent, anther slit usually ciliate, anterior anthers dimidiate due to abortion of outer theca; pollen suprareticulate or rarely psilate; ovary-lobes separate both above and below the style attachment (1/3–2/3 way to base of ovary) or rarely the style gynobasic; disc well developed, often with anterior side swollen at the base of a peg-like gynophore; nutlets dry, usually ellipsoid to obovoid or subglobose, pubescent or glabrous, variously sculptured (often tuberculate), attachment-scar lateral or rarely basal; endosperm absent; embryo usually bent with an incumbent radicle, more rarely bent at right angles or slightly curved. $2n = 12–88$ (most frequently 20, 22, 24, 32, 34). About 360 spp., subcosmopolitan but poorly represented in moist tropical lowlands. Paton (1990) distinguished subgenus *Scutellaria*, with one-sided inflorescences in which flowers are subtended by leaves or leaf-like bracts, from subgenus *Apeltanthus* Juz, with four-sided inflorescences and cucullate bracts.

VI. Subfam. Lamioideae Harley (2003).

Subfam. *Stachyoideae* (Dumort.) Luerss. (1882); Briq. in Engl. & Prantl (1895) et auct. mult., p.p., typ. incl.
Subfam. *Prasioideae* (Benth.) Thomé (1889).
Subfam. *Pogostemonoideae* P.D. Cantino, Harley & Wagstaff in Harley & Reynolds (1992).

Herbs or shrubs or rarely small trees, mostly non-aromatic or only slightly aromatic, rarely with spines in leaf-axils (*Lagochilus*, *Otostegia*), iridoid glycosides often present. Leaves opposite or sometimes whorled. Inflorescence thyrsoid or raceme-like, cymes 1- to many-flowered; lower bracts mostly large and similar to foliage leaves, or sometimes all bracts distinctly smaller than leaves, bracteoles present or sometimes absent, usually small, sometimes spiny. Calyx actinomorphic to 2-lipped, (3–)5(–30)-lobed. Corolla usually 2-lipped, mostly 4- or 5-lobed, 2/3 or 1/3 (the latter due to fusion of upper 2 lobes), rarely 3/1 (*Pogostemon*), posterior lip often hooded, entire to 2-lobed (3-lobed in *Pogostemon*), anterior lip usually flat and 3-lobed with mid-lobe usually larger than lateral lobes or narrow and entire (in *Pogostemon*), corolla-tube annulate or not. Stamens 4, exserted or included in tube, ascending to posterior lip of corolla, or sometimes spreading, anterior pair inserted closer to proximal part of corolla-tube or

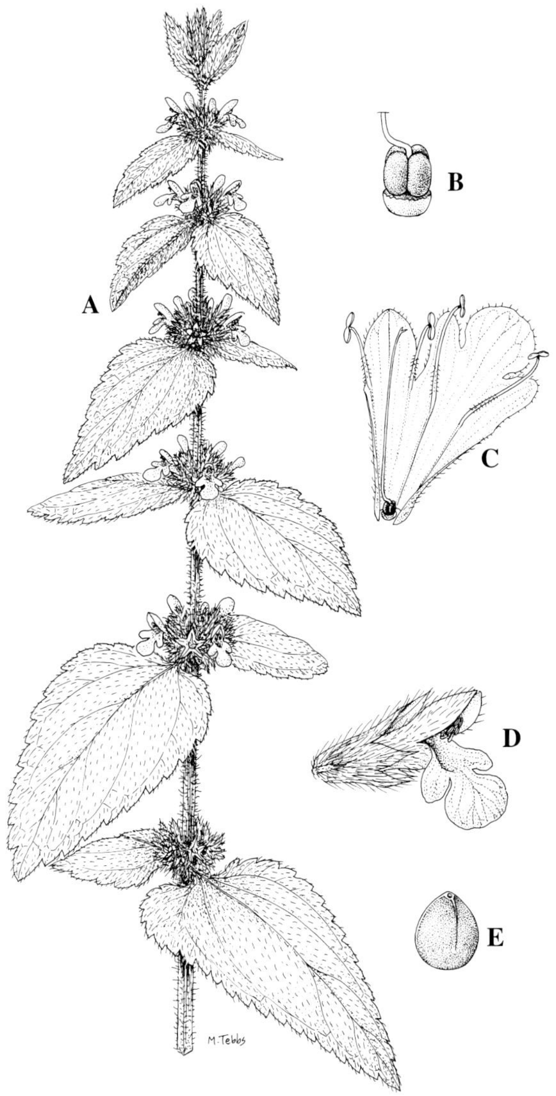

Fig. 27. Labiatae subfam. Lamioideae. *Stachys alpina*. **A** Habit. **B** Disc and gynoecium. **C** Corolla opened out. **D** Flower, side view. **E** Mericarp. Drawing by M. Tebbs

at same level as posterior pair, and usually longer, fertile or rarely anterior pair sterile (in *Hypogomphia*), or gynodioecious (*Colebrookea*), anther-thecae 2, dehiscing longitudinally, rarely by valves (*Galeopsis*), weakly delimited or confluent, sometimes well delimited, divaricate, narrowly divergent or parallel. Pollen 3-colpate (pantoporate in some *Sideritis* spp.). Disc well developed (Fig. 27), symmetric or with a larger lobe at abaxial side. Style gynobasic, with stigma-lobes usually equal or with anterior lobe longer, sometimes apparently entire due to reduction of upper lobe (*Acrotome*). Ovary 4-locular (by development of wall in initially 2-locular ovary), deeply 4-lobed, locules 1-ovulate. Fruit schizocarpic, mericarps (nutlets) 4, or due to abortion fewer, usually dry, sometimes fleshy, not mucilaginous. Seeds albuminous. Embryo spatulate. Some genera are not well delimited and requiring further study. Further subdivision of the Lamioideae has not been attempted here.

1. Corolla (3/1) weakly 2-lipped; stamens spreading or subdeclinate, exserted from the corolla, filaments usually with moniliform hairs along the centre **59. *Pogostemon***
- Corolla (1–2/3)2-lipped; stamens directed to upper lip of corolla, exserted or not exserted from the corolla, filaments without moniliform hairs 2
2. Plants with at least some branched hairs 3
- Plants without branched hairs 18
3. Nutlets rostrate and hooked at apex **69. *Rostrinucula***
- Nutlets neither rostrate nor hooked at apex 4
4. Nutlets winged at apex **72. *Colquhounia***
- Nutlets not winged at apex 5
5. Anthers transversally dehiscent; stamens long-exserted from the corolla 6
- Anthers not transversally dehiscent; not or only shortly exserted from corolla 7
6. Corolla-tube annulate **68. *Comanthosphace***
- Corolla-tube exannulate **67. *Leucosceptrum***
7. Nutlets densely pubescent at the apex **113. *Pseuderemostachys***
- Nutlets glabrous at the apex 8
8. Stamens included in the corolla-tube; posterior lip of the corolla flat or only slightly hooded 9
- Stamens exserted beyond the mouth of the corolla-tube; posterior lip of the corolla hooded or sometimes flat 12
9. Leaves sagittate **112. *Metastachydium***
- Leaves variously shaped, not sagittate 10
10. Corolla weakly 2-lipped; NE Iran to NW India **101. *Phlomidoschema***
- Corolla strongly 2-lipped 11
11. Both pairs of stamens with thecae equal and fertile; calyx 5–30-lobed; posterior lip of corolla 2-lobed; nutlets subtruncate or rounded at apex **111. *Marrubium***
- Anterior stamens with thecae dimidiate or sterile; calyx 5-lobed; posterior lip of corolla entire or 2-lobed; nutlets rounded at apex **103. *Sideritis***
12. Calyx 10–20-lobed, tube ± dilated into a limb **114. *Ballota***
- Calyx 5-lobed, tube not dilated or dilated into a very narrow limb 13
13. Calyx-lobes uncinate **95. *Notochaete***
- Calyx-lobes straight 14
14. Nutlets ± fleshy, rounded at apex; posterior lip of corolla often short, shallowly hooded (S & E Asia) **63. *Gomphostemma***
- Nutlets dry, truncate or rounded at apex; posterior lip of corolla elongate, shallowly or deeply hooded 15
15. Stigma-lobes strongly unequal or rarely equal; calyx-lobes mostly broad at base and abruptly narrowed to a narrow apex, often expanded at margins; leaves pinnatisect or unlobed 16
- Stigma-lobes equal or slightly unequal; calyx-lobes variously shaped, not as above; leaves unlobed 17
16. Nutlets densely bearded at apex; leaves pinnatisect or unlobed **93. *Eremostachys***
- Nutlets glabrous to hairy, hardly densely bearded at apex; leaves unlobed **92. *Phlomis***

17. Calyx subequally or slightly unequally 5-lobed, lobes subulate or triangular; verticillasters 2- to many-flowered **100. *Stachys***
- Calyx unequally 5-lobed with 3 posterior lobes much larger than lower 2; verticillasters 2-flowered. Yemen **118. *Isoleucas***
18. Nutlets fleshy 19
- Nutlets not fleshy 22
19. Posterior lip of corolla deeply hooded **102. *Prasium***
- Posterior lip of corolla shallowly hooded or almost flat 20
20. Cymes long-pedunculate; flowers over 30 mm long. China **65. *Bostrychanthera***
- Cymes sessile or shortly pedunculate; flowers variable in length. Hawaii 21
21. Inflorescences not terminal; posterior lip of corolla usually longer than anterior **106. *Stenogyne***
- Inflorescences mostly terminal; posterior corolla-lip shorter than anterior **105. *Phyllostegia***
22. Nutlets winged or densely scaly at apex 23
- Nutlets neither winged nor scaly at apex 26
23. Nutlets scaly at apex **66. *Achyrospermum***
- Nutlets winged, but not scaly 24
24. Inflorescence with lax, pedunculate cymes; corolla over 15 mm long **64. *Chelonopsis***
- Inflorescence with sessile or subsessile cymes; corolla variable in length 25
25. Leaves spinose-aristate **107. *Chamaesphacos***
- Leaves not spinose **72. *Colquhounia***
26. Posterior lip of corolla bearded along margin 27
- Posterior lip of corolla not bearded 32
27. Calyx 5-lobed, lobes mostly broad at base and abruptly narrowed to narrow apex, often expanded at margins; leaves pinnatisect or undivided 28
- Calyx 5–10(–12)-lobed, lobes variously shaped, not as above; leaves not divided 30
28. Nutlets glabrous to hairy not densely bearded at apex; leaves unlobed **92. *Phlomis***
- Nutlets densely bearded at apex; leaves pinnatisect or unlobed 29
29. Anther-thecae confluent, posterior lip of corolla short, weakly hooded **94. *Pseudomarrubium***
- Anther-thecae not confluent, posterior lip of corolla longer than lower lip, hooded **93. *Eremostachys***
30. Calyx very widely funnel-shaped to expanded and flattened in distal part **117. *Otostegia***
- Calyx tubular to narrowly funnel-shaped up to near throat, sometimes oblique with recurved lower lip 31
31. Corolla 5–25 mm long, with whitish hairs, anterior lip longer or shorter than posterior lip, not withering **119. *Leucas***
- Corolla 17–50 mm long, with reddish or yellowish hairs, anterior lip much shorter than posterior lip and soon withering **120. *Leonotis***
32. Leaves ± palmately divided; stamens included or shortly exserted from corolla-tube 33
- Leaves crenate, dentate, entire or rarely pinnately divided; stamens included or long-exserted from corolla-tube 36
33. Bracteoles spiny, usually longer than calyces; spines usually present in axils of foliage leaves; posterior lip of corolla usually long and narrow, 2-lobed **108. *Lagochilus***
- Bracteoles spiny or not, hardly longer than calyces; spines absent from axils of foliage leaves; posterior lip of corolla short, entire or shallowly 2-lobed 34
34. Corolla-tube annulate or rarely exannulate; nutlets apically truncate **98. *Leonurus***
- Corolla-tube exannulate; nutlets apically rounded 35
35. Corolla-tube exserted from calyx **97. *Panzerina***
- Corolla-tube included in calyx **96. *Lagopsis***
36. Style entire, included in corolla-tube. Africa **121. *Acrotome***
- Style bifid, included in or exserted from corolla-tube 37
37. Annual and biennial herbs 38
- Perennial herbs, subshrubs or shrubs 51
38. Anterior lip of corolla with two conical protuberances near base **88. *Galeopsis***
- Anterior lip of corolla without protuberances 39
39. Calyx closing after anthesis. North America 40
- Calyx remaining open after anthesis 41
40. Anterior calyx-lobes dentate **78. *Brazoria***
- Anterior calyx-lobes entire **79. *Warnockia***
41. Calyx 4-lobed **76. *Synandra***
- Calyx 5-lobed 42
42. Nutlets over 2.2 times as long as broad, rounded at apex; inflorescence racemoid **74. *Thuspeinanta***
- Nutlets up to 2.3 times as long as broad, rounded or truncate at apex; inflorescence thyrsoid or racemoid 43
43. Anterior pair of stamens sterile, posterior pair fertile **75. *Hypogomphia***
- All stamens fertile 44
44. Stamens included in corolla-tube **103. *Sideritis***
- Stamens exserted from mouth of corolla-tube 45
45. Cymes distinctly pedunculate **89. *Microtoena***
- Cymes sessile or subsessile 46
46. Calyx-tube widely campanulate, often with large, widely funnel-shaped limb, 5–10-lobed; plants subglabrous or very sparsely hairy **116. *Moluccella***
- Calyx not widely campanulate, 5-lobed; plants ± hairy 47
47. Anthers hairy; corolla-tube strongly broadened, often saccate near throat **87. *Lamium***
- Anthers glabrous; corolla-tube slightly broadened towards throat 48
48. Nutlets apically rounded or almost so, glabrous **100. *Stachys***
- Nutlets apically truncate at apex, hairy or glabrous 49
49. Corolla-tube exannulate **99. *Chaiturus***
- Corolla-tube annulate 50
50. Leaves shallowly crenate **86. *Matsumurella***
- Leaves deeply crenate **98. *Leonurus***
51. Nutlets connate to calyx; lobes of fruiting-calyx long, linear and plumose; shrubs **60. *Colebrookea***
- Nutlets free from calyx; lobes of fruiting-calyces variously shaped, not both long and linear; herbs or shrubs 52
52. Nutlets irregularly reticulate, rounded at apex; corolla at least 25 mm long. Europe and Turkey **81. *Melittis***
- Nutlets not or minutely and indistinctly reticulate, rounded or truncate at apex; corolla variable in length 53
53. Leaves incised to lobed **108. *Lagochilus***
- Leaves not incised or lobed 54
54. Calyx-tube expanded to widely funnel-shaped or almost flattened limb or, if tube not so expanded, calyx-lobes never 5 55
- Calyx-tube not expanded to widely funnel-shaped to almost flat limb, calyx-lobes 5 60
55. Calyx 3-lobed; nutlets longitudinally ridged; inflorescence racemoid **77. *Macbridea***
- Calyx 4–12-lobed; nutlets not distinctly ridged; inflorescence thyrsoid or racemoid 56
56. Calyx 4-lobed **108. *Lagochilus***
- Calyx 5–12-lobed 57
57. Calyx-limb not circular, ± 2-lipped **117. *Otostegia***

- Calyx-limb circular or subcircular 58
58. Calyx expanded to very broad limb, limb over 20 mm in diameter, lobes minute **116. *Moluccella***
- Calyx slightly expanded, less than 20 mm in diameter, lobes long and narrow 59
59. Leaves spinulose at apex, blade over 3 times as long as broad, glabrous or almost so. Pakistan **115. *Sulaimania***
- Leaves not spinulose at apex, blade less than 3.5 times as long as broad, moderately hairy. Europe and W Asia **114. *Ballota***
60. Bracteoles long and spinose **108. *Lagochilus***
- Bracteoles short or not spinose 61
61. Stamens included in corolla-tube **103. *Sideritis***
- Stamens exserted from corolla-tube 62
62. Calyx-lobes unequal (1/2/2); posterior lobe largest, lateral smallest and triangular, anterior lobes lanceolate-triangular. E Asia **73. *Paralamium***
- Calyx 2-lipped or not, if 2-lipped anterior lip scarcely longer, lobes subequal or unequal (3/2 or 1/4) 63
63. Calyx 2-lipped with anterior lip longer; lobes (3/2) unequal; plants rhizomatous. E Asia **84. *Loxocalyx***
- Calyx-lobes subequally 5-lobed or (1/4) unequal; plants rhizomatous or not 64
64. Inflorescence racemoid; all cymes 1-flowered 65
- Inflorescence thyrsoid; at least some cymes 2- to many-flowered 69
65. Plants creeping. Taiwan and Japan 66
- Plants not creeping 67
66. Inflorescence dense, bracts ± obscuring flowers **85. *Ajugoides***
- Inflorescence lax, bracts not obscuring flowers **71. *Suzukia***
67. Calyx ± closed in fruit; posterior lip of corolla undulate, flat; nutlets hairy. Hawaii **104. *Haplostachys***
- Calyx open in fruit; posterior lip of corolla hooded, sometimes ± flat; nutlets glabrous or hairy 68
68. Posterior lip of corolla flat or slightly hooded; plants glabrous or sparsely hairy below inflorescence. N America **80. *Physostegia***
- Posterior lip of corolla hooded; plants sparsely to densely hairy 69
69. Inflorescence with lax, prominently pedunculate cymes. S and E Asia 70
- Inflorescence with ± dense, very shortly pedunculate cymes 71
70. Stigma-lobes subequal; corolla weakly 2-lipped, less than 6 mm long **62. *Craniotome***
- Stigma-lobes unequal; corolla strongly 2-lipped, mostly over 6 mm long **89. *Microtoena***
71. Anterior lip of corolla with lateral lobes acute or broad with apex mucronate; plants rhizomatous **87. *Lamium***
- Anterior lip of corolla with lateral lobes rounded or emarginate at apex; plants rhizomatous or not 72
72. Anthers hairy; upper leaves orbicular or broadly rhombic, mostly crowded; flowers over 20 mm long. Alpine plants, C and E Asia 73
- Anthers glabrous (sometimes hairy in *Stachyopsis*); leaves variously shaped; flowers mostly less than 20 mm long 74
73. Corolla 30–40 mm long; filaments of posterior anthers not widened at base **82. *Alajja***
- Corolla 21–28 mm long; filaments of posterior anthers widened at base **83. *Eriophyton***
74. Corolla-tube saccate, swollen and ventrally abruptly widened **61. *Eurysolen***
- Corolla-tube neither saccate, swollen nor ventrally abruptly widened 75
75. Calyx-lobes oblong, widened above point of insertion; shrub **109. *Roylea***
- Calyx-lobes narrowed towards apex 76
76. Leaves in basal rosette; stem short or almost absent **91. *Lamiophlomis***
- All leaves arranged along stem 77
77. Posterior pair of stamens longer than anterior pair, with hairy filaments, with 1-thecous or 2-thecous anthers **66. *Anisomeles***
- Posterior pair of stamens shorter than anterior pair, with glabrous or sometimes hairy filaments, with normal 2-thecous anthers (but thecae often confluent) 78
78. Plant stoloniferous and with short, erect flowering stems; inflorescence short and dense with flowers obscured by large leaf-like bracts. Japan **85. *Ajugoides***
- Plant not stoloniferous; inflorescence variable, not as above 79
79. Nutlets truncate at apex 80
- Nutlets rounded at apex 83
80. Corolla less than 10 mm long, tube exannulate; nutlets hairy at apex **99. *Chaiturus***
- Corolla over 10 mm long, tube annulate or not, nutlets glabrous or hairy at apex 81
81. Calyx-lobes spinescent; nutlets obliquely truncate at apex; anthers glabrous or hairy **110. *Stachyopsis***
- Calyx-lobes not spinescent; nutlets not or only slightly oblique at apex; anthers glabrous 82
82. Calyx-lobes narrowly triangular to lanceolate **86. *Matsumurella***
- Calyx-lobes broadly triangular at base and mucronate at apex, obsolete or sometimes triangular to lanceolate **90. *Paraphlomis***
83. Calyx-lobes broadly triangular at base and mucronate at apex, obsolete or sometimes triangular to lanceolate, not spinose. S and E Asia **90. *Paraphlomis***
- Calyx-lobes triangular, lanceolate or subulate, spiny or not 84
84. Calyx-tube with 10 strongly raised veins, basal part of tube with veins as broad as or broader than intervening spaces **114. *Ballota* (*nigra*)**
- Calyx-tube without distinct veins or with few to many slightly to moderately raised veins, if 10, basal part of tube with veins narrower than intervening spaces **100. *Stachys***

59. *Pogostemon* Desf.

Pogostemon Desf., Mem. Mus. Hist. Nat. Paris 2: 154 (1815); Bhatti & Ingrouille, Bull. Nat. Hist. Mus. Lond. (Bot.) 27, 2: 77–148 (1997), rev.; Ingrouille & Bhatti, Bot. J. Linn. Soc. 128: 159–183 (1998), intragen. relationships.
Dysophylla Blume (1826).
Eusteralis Raf. (1837).
Anuragia Raizada (1976).

Annual or perennial herbs or shrubs, with simple and/or branched hairs; leaves opposite or in whorls of 3–12; inflorescence thyrsoid; bracts and bracteoles small; calyx 5-lobed (3/2), lobes subequal or anterior longer; corolla weakly 2-lipped, 4-lobed (3/1), purple or white, not hooded, anterior lip narrow, entire; stamens equal or posterior pair slightly longer, spreading or subdeclinate, exserted from the corolla, filaments usually bearded along centre with moniliform hairs, anthers about as long as broad, thecae confluent; stigma-lobes equal to subequal; nutlets 1–4, 0.5–1.1 × 0.3–1.0 mm, rounded at apex, glabrous, often glossy. $2n = 12, 32, 34, 60, 64, 72$. About 80 spp., mostly in moist to mesic forest or damp places, from S and SE Asia to China, a few species in South tropical Africa, Australia, Taiwan and Japan.

60. *Colebrookea* Sm.

Colebrookea Sm., Exot. Bot. 2: 11 (1806).

Shrub, usually gynodioecious, densely pubescent with simple hairs; leaves toothed, petiolate; inflorescence thyrsoid, dense and much branched; bracts and bracteoles small; flowers very small; calyx deciduous in fruit, 5-lobed, lobes subequal, strongly accrescent in fruit, becoming very long, narrowly linear and plumose; corolla indistinctly 2-lipped, 5-lobed (2/3), purplish to white, posterior lip short, flat, emarginate; stamens spreading, those of hermaphrodite flowers exserted, filament glabrous, anthers about as long as wide, thecae confluent; stigma-lobes subequal, rather long; nutlets 1(–4), connate to calyx, very small, rounded and eglandular-hairy at apex. $2n = 32$. One sp., *C. oppositifolia* Sm., in savannas and dry thickets, from N Pakistan and India to S China (Yunnan) and Thailand.

61. *Eurysolen* Prain

Eurysolen Prain, Sci. Mem. Off. Med. Dept. Gov. India 11: 43 (1898).

Subshrubs or shrubs with simple hairs; leaves petiolate, obovate-rhombic or oblong-obovate; inflorescence thyrsoid; calyx 5-lobed (3/2), lobes subequal, anterior often slightly longer; corolla weakly 2-lipped, 4-lobed (1/3), white, posterior lip short, hooded, emarginate or entire, corolla-tube saccate, swollen and ventrally abruptly widened, annulate; stamens shortly exserted, anthers about as long as broad, thecae confluent; stigma-lobes equal; nutlets ca. 1 × 0.8 mm, rounded at apex, glandular. One sp., *E. gracilis* Prain, in moist forests and mountain valleys, from NE India (Manipur) to S China (Yunnan) and Thailand, and in Indonesia.

62. *Craniotome* Rchb.

Craniotome Rchb., Icon. Bot. Exot. 1: 39 (1825).

Perennial subshrub or herb, with simple hairs; leaves ovate-cordate, toothed, petiolate; inflorescence thyrsoid with many-flowered, prominently pedunculate cymes; flowers small (3–5 mm long); calyx 5-lobed, lobes subequal, tube becoming urceolate-globose in fruit; corolla weakly 2-lipped, 4-lobed (1/3), reddish, posterior lip short, entire, shallowly hooded; stamens not exserted from corolla, thecae weakly distinct; stigma-lobes short and subequal; nutlets ca. 0.7 × 0.5 mm, rounded at apex, smooth, glossy. $2n = 20, 34$. One sp., *C. furcata* (Link) Kuntze, in mountain forests and thickets, from N Pakistan and N India to S China and Vietnam.

63. *Gomphostemma* Wall. ex Benth.

Gomphostemma Wall. ex Benth., Edward's Bot. Reg. 15: t. 1292 (1830); Prain, Ann. Roy. Bot. Gard. Calcutta 3: 227–271, pl. 75–105 (1891), rev.
Taitonia Yamamoto (1938).

Shrubs or perennial herbs, with branched and simple hairs; leaves toothed, petiolate; inflorescence terminal, or sometimes on separate short shoots at base of plant, thyrsoid or sometimes racemoid; calyx 5-lobed, lobes equal or subequal; corolla 2-lipped, 4-lobed (1/3), purplish-red, yellow or white, posterior lip short or moderately long, shallowly hooded, entire or emarginate, corolla-tube mostly strongly dilated in distal part; stamens not or only very shortly exserted, thecae weakly distinct, parallel or narrowly divergent; stigma-lobes subequal or unequal; nutlets 1–4, ± fleshy, rounded at apex. $2n = 34$. About 36 spp., mostly in moist forests and at forest margins, SW China, India to Indo-China, Malaysia and Indonesia.

64. *Chelonopsis* Miq.

Chelonopsis Miq., Ann. Mus. Bot. Lugduno-Batavi 2: 111 (1865).

Shrubs or perennial herbs, with simple hairs; leaves toothed; inflorescence lax, thyrsoid with lax cymes, pedunculate, up to 5-flowered; flowers large (15–40 mm long); calyx (4–)5-lobed, lobes equal or subequal (2/3), or unequal (3/1, 3/2), tube often becoming slightly inflated in fruit; corolla weakly 2-lipped, 4-lobed (1/3), purple, white or yellow, posterior lip short, scarcely hooded, entire or emarginate, corolla-tube strongly dilated in distal part, not annulate; stamens about as long as corolla-tube, anthers bearded on distal sides of thecae, thecae distinct; stigma-lobes subequal or unequal; nutlets flat, 6–12 × 2.5–5 mm, winged at apex, ± fragile and often easily releasing a seed, fibrous, ± ribbed at distal side. About 16 spp., in forests or thickets, Pakistan, China, Japan.

65. *Bostrychanthera* Benth.

Bostrychanthera Benth. in Benth. & Hooker, J.D., Gen. Pl. 2: 1170 (1876).

Perennial herbs with simple hairs; leaves toothed, shortly petiolate or sometimes subsessile; inflorescence lax, thyrsoid, with cymes lax, pedunculate, up to 4–11-flowered; flowers large (30–50 mm long); calyx 2-lipped, 5-lobed, lobes subequal or more often unequal (1/4) with posterior lip shorter; corolla weakly 2-lipped, 4-lobed (1/3), purple to white, posterior lip short, scarcely hooded, entire, corolla-tube strongly dilated in distal part, not annulate; stamens about as long as corolla-tube, anthers bearded on distal sides of thecae, thecae distinct; stigma-lobes subequal or unequal; nutlets 1(–3), with fleshy mesocarp, rounded at apex, smooth. Two spp., in forests, C and S China, Taiwan.

66. *Achyrospermum* Blume

Achyrospermum Blume, Bijdr.: 840 (1826); Bruce, Bull. Misc. Inform. 1936: 47–60 (1936), reg. rev.

Woody herbs or shrubs, often scandent, with simple hairs; leaves toothed, petiolate; inflorescence terminal and/or axillary, thyrsoid, dense; bracts distinct from and much smaller than the leaves, mostly broadly ovate to oblate, often caducous; calyx 5-lobed, lobes subequal or unequal (3/2); corolla weakly 2-lipped, 4-lobed (1/3), reddish to purple to white or sometimes red, posterior lip short, emarginate, scarcely hooded; stamens not or shortly exserted, thecae confluent; stigma-lobes equal or slightly unequal; nutlets with scales in apical half, mostly with sessile glands. $2n = 28, 84, 88$. Up to about 25 spp., in moist to mesic forests and by riverbanks, tropical Africa, Madagascar, Mascarenes, Seychelles, Comores, from Nepal and NE India to S China and Thailand, Indonesia and the Philippines.

67. *Leucosceptrum* Sm.

Leucosceptrum Sm., Exot. Bot. 2: 113 (1805).

Shrub or small tree, densely tomentose with branched and simple hairs; leaves large, subcoriaceous; inflorescence densely thyrsoid with cymes 3–5-flowered; bracts much smaller than leaves, broad, caducous, bracteoles caducous; calyx 5(–7)-lobed, lobes equal or subequal (3/2); corolla weakly 2-lipped, 4-lobed (1/3), white to purple, posterior lip short, scarcely hooded, emarginate, corolla-tube not annulate; stamens long-exserted, anthers wider than long, appearing transversally dehiscent, thecae confluent; stigma-lobes subequal; nutlets cylindrical-oblong, truncate, with subsessile glands. $2n = 44$. One sp., *L. canum* Sm., mostly at forest margins and in secondary forest, from Tibet, Nepal and N India to S China and Vietnam.

68. *Comanthosphace* S. Moore

Comanthosphace S. Moore, J. Bot. 15: 293 (1877); Bhatti & Ingrouille, Fontqueria 42: 1–10 (1995), rev.

Rhizomatous perennial herbs; leaves toothed, petiolate; inflorescence thyrsoid with cymes few-flowered; bracts much smaller than leaves, broadly ovate, caducous, bracteoles caducous; calyx 5-lobed (3/2), lobes subequal or anterior distinctly longer; corolla weakly 2-lipped, 4-lobed (1/3), reddish to purplish or pale yellowish, posterior lip short, scarcely hooded, emarginate or entire, corolla-tube annulate; stamens long-exserted, anthers wider than long, appearing transversally dehiscent, thecae confluent; stigma-lobes subequal to unequal; nutlets truncate or subtruncate, with sessile glands. Three or four spp., mostly secondary forest, forest margins and mountain slopes, China, Taiwan and Japan.

69. *Rostrinucula* Kudô

Rostrinucula Kudô, Mem. Fac. Sci. Taihoku Imp. Univ. 2: 304 (1929); Ying et al., The endemic genera of seed plants of China: 495–497 (1993); Bhatti & Ingrouille, Fontqueria 42: 1–10 (1995), rev.
Comanthosphace subg. *Rostrinucula* (Kudo) Bhatti & Ingrouille (1995).

Shrubs, tomentose with branched and simple hairs; leaves elongate, toothed, shortly petiolate; inflorescence thyrsoid, dense with up to 5-flowered cymes; bracts much smaller than the leaves, broadly ovate, deciduous; calyx 5-lobed (3/2), lobes subequal or anterior larger; corolla weakly 2-lipped, 4-lobed (1/3), purple to red, posterior lip short, scarcely hooded, entire; stamens exserted, anthers wider than long, appearing transversally dehiscent, thecae confluent; stigma-lobes subequal; nutlets rostrate and hooked at apex, with branched hairs and sessile glands. Two spp., mostly secondary forests and mountain slopes, S and C China.

70. *Anisomeles* R. Br.

Anisomeles R. Br., Prodr.: 503 (1810); Bhatti & Ingrouille, Fontqueria 44: 77–84 (1996), rev.
Epimeredi Adans. (1763), nom. gen. inval.

Stout perennial herbs or subshrubs, with simple hairs; leaves toothed, petiolate; inflorescence thyrsoid with 5- to many-flowered cymes; calyx 5-lobed, lobes equal or subequal (3/2); corolla strongly 2-lipped, 4-lobed (1/3), white to purple, posterior lip moderately long, slightly hooded, entire, rather narrow, anterior lip much longer than posterior; stamens with the filaments bearded, posterior pair longer than anterior, slightly longer than posterior lip of corolla, anthers mostly dimidiate due to abortion of one theca, thecae of anterior anthers ± distinct; stigma-lobes subequal; nutlets rounded at apex, smooth, glossy. $2n = 30, 32, 34, 40$. Three spp., secondary habitats, tropical Africa, S and SE Asia, NE Australia.

71. *Suzukia* Kudô

Suzukia Kudô, J. Soc. Trop. Agric. 2: 145 (1930); Ying et al., The endemic genera of seed plants of China: 504–505 (1993), reg. rev.

Creeping herbs, with simple hairs; leaves petiolate, blade orbicular or broadly ovate; inflorescence racemoid; calyx 2-lipped, 5-lobed (3/2), lobes subequal; corolla strongly 2-lipped, 4-lobed (1/3), purple to red; posterior lip moderately long, hooded, entire; stamens not exserted from corolla, thecae weakly distinct; stigma-lobes subequal or slightly unequal; nutlets rounded at apex, glabrous. $2n = 24$. Two spp., forests, Taiwan and Ryukyo Islands.

72. *Colquhounia* Wall.

Colquhounia Wall., Trans. Linn. Soc. Lond. 13: 608 (1822); Prain, J. Asiat. Soc. Bengal. Pt. 2, Nat. Hist. 62: 61–69 (1893), rev.

Shrubs or woody herbs, with simple and/or branched hairs; leaves toothed, petiolate; inflorescence thyrsoid, lax to dense and capitate, cymes 1–5-flowered; calyx 5-lobed, lobes equal or subequal; corolla strongly 2-lipped, 4-lobed (1/3), yellow to red or purple; posterior lip moderately long, hooded with upcurved margins, entire or emarginate, anterior lip subequally 3-lobed, corolla-tube strongly dilated in distal part; stamens not exserted from corolla, thecae ± confluent; stigma-lobes unequal or subequal; nutlets narrowly obovoid-oblong, winged at apex, glandular or glabrous. $2n = 34$. About six spp., mostly in thickets and on hill slopes, from Tibet, Nepal and N India to S China and Vietnam.

73. *Paralamium* Dunn

Paralamium Dunn, Notes Roy. Bot. Gard. Edinburgh 8: 168 (1913).

Perennial herb, densely pubescent with simple hairs; leaves with blade ovate-cordate; inflorescence thyrsoid, cymes up to 5-flowered; bracts small; calyx 2-lipped, strongly accrescent in fruit, 5-lobed (1/2/2), posterior lip very broad, rounded-truncate at apex, with upcurved margin in fruit, lateral lobes smaller and triangular, lower lobes about as long as posterior lobe, narrowly triangular-acuminate; corolla strongly 2-lipped, 4-lobed (1/3), purple, posterior lip moderately long, shallowly hooded, entire; stamens not exserted from corolla, thecae confluent; stigma-lobes unequal; nutlets ca. 1×1 mm, rounded at apex, pitted, glossy, glabrous. One sp., *P. gracile* Dunn, wet areas in forests, N Burma, S China (Yunnan) and N Vietnam.

74. *Thuspeinanta* T. Durand

Thuspeinanta T. Durand, Index Gen. Phan. 703 (1888).
Tapeinanthus Boiss. ex Benth. in A.DC. (1848), non Herb. (1837).

Small annual herbs with simple hairs; leaves usually narrow, toothed to entire, attenuate at base, petiolate; inflorescence racemose; bracteoles at base of pedicels which recurve in fruit; calyx 5-lobed, inflated in fruit, lobes equal or subequal (1/4), posterior lobe often pointing upwards; corolla strongly 2-lipped, 5-lobed (2/3), pink, posterior lip long, shallowly hooded; stamens included or only shortly exserted from corolla-tube, anther-thecae confluent; stigma-lobes equal or lower lobe ± broader; nutlets oblong, 2.2–3 times longer than wide, rounded at apex, smooth. Two spp., in dry areas, Syria, Israel and Saudi Arabia to Turkmenistan, Afghanistan and SW Pakistan (Baluchistan).

75. *Hypogomphia* Bunge

Hypogomphia Bunge, Bull. Acad. Imp. Sci. St. Petersburg, ser. 3, 18: 30 (1872).

Annual herbs, sometimes gynodioecious, aromatic, with simple hairs; leaves narrow; inflorescence thyrsoid, with up to 2-flowered cymes; calyx rather broadly campanulate, slightly zygomorphic, 5-lobed, lobes subequal, subtriangular; corolla strongly 2-lipped, 5-lobed (2/3), white, pink or purple, tube curved-geniculate, annulate within, posterior lip long, narrow, arcuate, widening near apex, emarginate, median lobe of anterior lip sinuate at apex; fertile (posterior) stamens 2, with strongly curved filaments, thickened at base, anterior pair reduced to small staminodes; stigma-lobes equal or slightly unequal; nutlets oblong, subacute at apex, granulate. One to three species. Deserts, low rocky slopes, lowlands, valleys of Iran to Kazakhstan.

76. *Synandra* Nutt.

Synandra Nutt., Gen. N. Am. Pl. 2: 29 (1818).

Biennial herbs, with simple hairs; leaves long-petiolate, broadly cordate-ovate; inflorescence racemoid, flowers sessile; calyx weakly zygomorphic, 4-lobed, posterior lobe absent, lateral lobes larger than anterior pair; corolla 2-lipped, 4-lobed (1/3), whitish, posterior lip shallowly hooded, entire, anterior lip with erose median lobe, tube exannulate; stamen-filaments conspicuously villous with long reddish hairs, apically inward-geniculate, thecae divaricate, not confluent, ovoid, prolonged to apical apicula, outer thecae of anterior stamens fused; disc with abaxial lobe equalling ovary; stigma-lobes unequal; nutlets obovoid-trigonal, smooth, glabrous. $2n = 18$. One sp., *S. hispidula* (Michx.) Baill., mesic woodlands, eastern United States (mostly in Appalachian region).

77. *Macbridea* Elliott ex Nutt.

Macbridea Elliott ex Nutt., Gen. N. Am. Pl. 2: 36 (1818).

Rhizomatous perennial herbs with simple hairs; upper leaves sessile, lower usually petiolate; inflorescence racemoid, of terminal and subterminal capitate glomerules; flowers sessile; calyx zygomorphic, 3-lobed (1/2), anterior lobes often emarginate (each probably representing 2 fused lobes); corolla 2-lipped, 4-lobed (1/3), white to lavender, posterior lip shallowly hooded, entire, tube exannulate; stamen-filaments villous, apically geniculate, anthers bearded on ventral surface, thecae parallel to divergent, not confluent, suture bordered with 5–10 conspicuous teeth, outer thecae of anterior stamens fused; disc with abaxial lobe surpassing ovary; stigma-lobes equal to subequal; nutlets obovoid, longitudinally ridged, glabrous. $2n = 18$. Two spp., wetlands and pine savannas in SE United States.

78. *Brazoria* Engelm. ex A. Gray

Brazoria Engelm. ex A. Gray, Boston J. Nat. Hist. 5: 255 (1845); Turner, Plant Syst. Evol. 203: 69 (1996), rev.

Annual herbs with simple hairs; upper leaves amplexicaul, lowest usually petiolate; inflorescence racemoid, ebracteolate; flowers short-pedicellate; calyx 2-lipped, 5-lobed (3/2), ± gaping at anthesis, becoming closed by anterior lip in fruit, posterior lip shallowly 3-lobed, lobes usually apiculate, anterior lip shallowly 2-lobed, each lobe with 2–6 teeth; corolla 2-lipped, 5-lobed (2/3), pale lavender to reddish purple, posterior lip shallowly hooded, 2-lobed, often erose, lobes of anterior lip often erose, tube annulate near base, with a 2-lobed hump on lower side partially blocking throat; stamen-filaments villous, apically inward-geniculate, anthers bearded on ventral surface, thecae asymmetrically divergent, not confluent, inner theca of each anther larger than outer one, its suture bordered by a single tooth; disc with abaxial lobe shorter than to equalling ovary; stigma-lobes equal; nutlets trigonous, puberulent, smooth. $2n = 28$. Two spp., sandy soils, United States (Texas).

79. *Warnockia* M.W. Turner

Warnockia M.W. Turner, Plant Syst. Evol. 203: 78 (1996).

Annual herbs, with simple hairs; leaves glabrate, upper amplexicaul, lower usually petiolate; inflorescence racemoid, ebracteolate, flowers short-pedicellate; calyx 2-lipped, 5-lobed (3/2), gaping at anthesis, becoming closed by anterior lip in fruit; corolla 2-lipped, 4-lobed (1/3), mostly reddish lavender but median lobe of anterior lip white, posterior lip shallowly hooded, entire, tube exannulate, with a low hump intruded on lower side near throat; stamen-filaments villous, apically inward-geniculate, thecae divergent, not confluent, inner theca often larger than outer one, suture bordered with 2–4(–6) small teeth; disc with abaxial lobe surpassing ovary; stigma-lobes equal to slightly unequal; nutlets trigonous, rounded at apex, glabrous, smooth or sometimes slightly verrucate towards base. $2n = 20$. One sp., *W. scutellarioides* (Engelm. & A. Gray) M.W. Turner, calcareous soils, United States (Texas, Oklahoma) and north-western Mexico.

80. *Physostegia* Benth.

Physostegia Benth., Edward's Bot. Reg. 15: sub t. 1289 (1829); Cantino, Contr. Gray Herb. 211: 1–105 (1982), rev.

Rhizomatous perennial herbs, glabrous or subglabrous below inflorescence, hairs simple; leaves mostly sessile, often amplexicaul, the lowest sometimes petiolate; inflorescence racemoid, ebracteolate, flowers subsessile; calyx actinomorphic, somewhat inflated in fruit, 5-lobed, lobes equal; corolla strongly 2-lipped, 4-lobed (1/3), white to reddish violet, posterior lip flat to slightly hooded, usually entire or emarginate, tube exannulate; stamens equal or anterior pair longer, filaments villous, thecae parallel, not confluent, suture bordered with 1–6(–10) small teeth; disc with abaxial lobe equal to twice as long as ovary; stigma-lobes equal to subequal; nutlets trigonous, glabrous, smooth or rarely verrucate. $2n = 38, 76$. Twelve species, northern Canada to northern Mexico. *Physostegia virginiana* (L.) Benth. is widely grown as an ornamental and sometimes naturalized elsewhere. With a base number of 19, *Physostegia* is probably of amphiploid origin from parents with $n = 9$ and 10 (Cantino 1985b). One parent may be *Warnockia* ($n = 10$), the genus with the closest phenetic relationship to *Physostegia* (Cantino 1982a). The other parent is presumably extinct.

81. *Melittis* L.

Melittis L., Sp. Pl. 2: 597 (1753); Klokov, Not. Syst. Leningrad 18: 183 (1957).

Perennial herb, with simple hairs; leaves ovate or rarely elongate, toothed, petiolate; inflorescence lax, thyrsoid, cymes up to 3-flowered, or sometimes racemoid; calyx rather broadly obconical, 2-lipped (1/2), posterior lip large, entire or irregularly dentate, anterior lip of 2 broadly triangular-obtuse lobes; corolla strongly 2-lipped, 4-lobed (1/3), large (25–40 mm long), white to purple with darker purple markings on anterior lip, posterior lip moderately long, entire, shallowly hooded; stamens not exserted from corolla, thecae distinct; stigma-lobes short and subequal; nutlets rounded at apex, shallowly and irregularly reticulate, eglandular hairy. $2n = (24), 30, 36$. One sp., *M. melissophyllum* L., mostly in forests, W, C, S and E Europe, Asia: N. Turkey.

82. *Alajja* Ikonn.

Alajja Ikonn., Nov. Syst. Pl. Vasc. Leningrad 8: 274 (1971).
Erianthera Benth. in Hook. (1833), non Nees (1832).
Susilkumara Bennet (1981).

Malodorous, perennial herbs with woody rootstock, lanate-tomentose or tomentose with simple hairs; leaves shortly petiolate, very broad, often scale-like at base of plant, often congested above; inflorescence thyrsoid with few-flowered cymes; bracteoles subulate, inconspicuous; calyx weakly zygomorphic, 5-lobed, straight, lobes subequal, lanceolate; corolla strongly 2-lipped, 5-lobed (2/3), large, purple, lanate, posterior lip long, hooded, emarginate, side-lobes of anterior lip emarginate, mid-lobe of anterior lip broadly obcordate-emarginate; stamens not or only shortly exserted from corolla, anthers ± hairy, thecae distinct; stigma-lobes equal; nutlets large, triquetrous, glabrous or pubescent at apex, truncate. One to three spp, high alpine scree plants of C Asia, E Afghanistan, N Pakistan, Kyrgyzstan, Tajikistan, W China (Xinjiang).

83. *Eriophyton* Benth.

Eriophyton Benth., Edward's Bot. Reg. 15: t. 1289 (1829).

Perennial herb, densely lanate with simple hairs, with thick roots or rhizome; lower leaves scale-like, upper rhombic to orbicular; inflorescence rather densely? thyrsoid, cymes ca. 3-flowered, ± hidden between large leaf-like bracts; bracteoles spiny; flowers large (21–28 mm long); calyx broadly campanulate, 5-lobed, lobes equal or subequal, ± spiny; corolla strongly 2-lipped, 4-lobed (1/3), purple or red, tube exannulate, posterior lip long and broad, hooded; stamens not exserted from corolla, filaments of posterior pair thickened at base, anthers hairy, thecae confluent to weakly distinct; stigma-lobes subequal; nutlets subtruncate at apex, smooth. One sp., *E. wallichii* Benth., alpine, stony alluvial areas, Tibet, Nepal, N India and S China (Yunnan).

84. *Loxocalyx* Hemsl.

Loxocalyx Hemsl., J. Linn. Soc. Bot. 26: 308 (1890).

Perennial herbs, often with slender rhizomes, pubescent with simple hairs; leaves ovate-acuminate, toothed, long-petiolate; inflorescence thyrsoid, with cymes up to 3-flowered, or racemoid; calyx 2-lipped, 5-lobed (3/2), lobes spiny, unequal, anterior lip longer; corolla strongly 2-lipped, 4-lobed (1/3), pink to dark purple, posterior lip long, hooded; stamens not exserted from corolla, thecae weakly distinct; stigma-lobes subequal; nutlets truncate to rounded, eglandular hairy or glabrous. About three spp., forest and stream sides, China and Japan. The only Japanese species, *L. ambiguus* (Makino) Makino, is often placed in *Lamium*.

85. *Ajugoides* Makino

Ajugoides Makino, Bot. Mag. (Tokyo) 29: 280 (1915).

Stoloniferous and/or rhizomatous perennial herbs; leaves crowded in upper part of stem, elliptic, toothed, attenuate at base with short petiole; flowering stems short; inflorescence rather short and dense, thyrsoid, with cymes up to 3-flowered, ± hidden between large leaf-like bracts; calyx 5-lobed, lobes equal or subequal; corolla strongly 2-lipped, 4-lobed (1/3), pink, posterior lip long, shallowly hooded; stamens not exserted from corolla, thecae weakly distinct; stigma-lobes slightly unequal; nutlets smooth, truncate at apex, glabrous. One sp., *A. humilis* (Miq.) Makino, mountain forests, Japan.

86. *Matsumurella* Makino

Matsumurella Makino, Bot. Mag. (Tokyo) 29: 279 (1915).

Perennial or annual herbs, at least sometimes tuberous and stoloniferous, with simple hairs; leaves toothed, petiolate; inflorescence thyrsoid, with cymes up to 3(–4)-flowered, or racemoid; calyx 5-lobed, lobes subequal, often ± spinescent; corolla strongly 2-lipped, 4–5-lobed (1/3, 2/3), purple to white, posterior lip long, hooded; stamens not exserted from corolla, thecae weakly distinct; stigma-lobes slightly unequal; nutlets truncate or subtruncate at apex, glabrous or sometimes hairy. One to perhaps five spp., forests, hill slopes and wet sites, China, Taiwan, Japan. *Matsumurella* is often included in *Galeobdolon* or in *Lamium*. The three Chinese species, described under *Galeobdolon*, *G. szechuaense* C.Y. Wu, *G. yangsoense* Sun and *G. kwangtungense* C.Y. Wu, may be more closely related to *Matsumurella* than to *Lamium* (incl. *Galeobdolon*).

87. *Lamium* L. Fig. 21A

Lamium L., Sp. Pl. 2: 579 (1753); Mennema, Leiden Bot. Ser. 11: 1–196 (1989), rev.
Orvala L. (1753).
Lamiastrum Heist. ex Fabr. (1759).
Galeobdolon Adans., Fam. (1763), nom. superfl.
Polichia Schrank (1781), non *Pollichia* Aiton.
Wiedemannia Fisch. & C.A. Mey. (1838).

Annual or rhizomatous or stoloniferous perennial herbs, pubescent with simple hairs; leaves toothed to deeply lobed, at least lower petiolate; inflorescence thyrsoid; calyx 5-lobed, lobes subequal or unequal (1/4); corolla strongly 2-lipped, 4–5-lobed, (1–2/3), pink, purple, yellow or white, posterior lip long, hooded, often ± arched, often densely hairy outside, anterior lip with lateral lobes short and broad with 1–2 teeth or triangular-acute, median lobe mostly broadly obcordate-emarginate, less often narrower and acute, tube usually abruptly widened above; stamens not or very shortly exserted from corolla, anthers hairy or glabrous, thecae weakly distinct, bearded or glabrous; stigma-lobes subequal or slightly unequal; nutlets truncate or subtruncate at apex, glabrous. $2n = (14), 18, 36$. Seventeen to ca. 30 spp., mainly disturbed ground, hill slopes and forests, temperate regions of Old World, from Europe and northern Africa to E Asia, introduced to other temperate or mountainous areas. Many authors place the divergent *L. galeobdolon* (L.) L. in a separate genus, characterised by having glabrous anthers, spiny bracts and calyx-lobes, with an acute, rather narrow median lobe of the lower lip of the corolla and with longer lateral corolla-lobes. The species is here included in *Lamium* due to lack of a legitimate genus name; the two genus names commonly applied, *Lamiastrum* and *Galeobdolon*, are both nomenclatoral synonyms of *Leonurus*. The very distinctive *L. galeobdolon* (L.) L. is perhaps better placed in a separate genus *Lamiastrum*, characterised by having a narrow, acute mid-lobe of anterior corolla-lip. Some of the species described under *Galeobdolon* are perhaps more closely related to *Matsumurella* (see under this genus). Most authors treat *Wiedemannia* as a separate genus characterised by having 1/4 unequal calyx-lobes (but see Ryding 2003). The generic affinities of *L. staintonii* Hedge and *L. nepalense* Hedge are uncertain. *L. tuberosum* Hedge seems to be close to the genus *Alajja*.

88. *Galeopsis* L.

Galeopsis L., Sp. Pl. 2: 579 (1753); Briq., Mem. Cour. Acad. Roy. Sci. Belg. 52: 1 (1893), rev.; Müntzing, Hereditas 13: 185–341 (1930), reg. rev.
Cannabinastrum Heist. ex Fabr. (1759), nom nud.?
Tetrahit Moench (1795) non Gérard.
Ladanum Kuntze (1891) non Spach.
Dalanum Dostál (1984).

Annual herbs with simple, often rigid hairs; leaves simple; inflorescence thyrsoid; bracteoles spiny; calyx 5-lobed, lobes equal or subequal, spiny; corolla strongly 2-lipped, 4-lobed (1/3), purple, yellow or white, posterior lip long, hooded, anterior lip with two conical protuberances near base; stamens not exserted from corolla, thecae distinct, apparently dehiscing by 2 valves, upper valve smaller, fimbriate; stigma-lobes subequal; nutlets rounded at apex, glabrous. $2n = 16, (30), 32$. About ten spp., mainly in open, disturbed habitats and lake shores, mainly Europe and W Asia, introduced to other temperate regions.

89. *Microtoena* Prain

Microtoena Prain, Hooker's Icon. Pl. 19: t. 1872 (1889) & Bull. Soc. Bot. France, ser. 3, 42: 417–427 (1895).

Perennial or perhaps sometimes annual herbs, pubescent with simple hairs; leaves toothed, petiolate; inflorescence thyrsoid, usually lax, cymes usually lax, long-pedunculate; bracteoles caducous; calyx 5-lobed (1/4), posterior lobe longer or subequal to laterals (3/2); corolla strongly 2-lipped, 4-lobed (1/3); yellow or sometimes white, posterior lip long, hooded, often deeply concave, often broadened or with 2 teeth at base, anterior lip with median lobe smaller to sometimes larger than side-lobes; stamens not or only shortly exserted, anthers ± long and narrow, thecae becoming confluent; stigma-lobes strongly unequal, posterior lobe minute; nutlets rounded at apex, glabrous. About 24 spp., mainly forests, hill slopes, mountain grassland, mostly from Nepal and N India to S China, 1 sp. [*M. insuavis* (Hance) Prain ex Briq.] also in Vietnam and Indonesia.

90. *Paraphlomis* (Prain) Prain

Paraphlomis (Prain) Prain, J. Asiat. Soc. Bengal. Pt. 2, Nat. Hist. 74: 721 (1908); H.W. Li, Acta Phytotax. Sin. 10: 57–74, pl. 16–17 (1965), reg. rev.
Phlomis sect. *Paraphlomis* Prain (1901).
Pogonanthera H.W. Li & X.H. Guo (1993) non Blume.
Sinopogonanthera H.W. Li (1993).

Rhizomatous herbs or subshrubs, hairs simple; leaves toothed, petiolate; inflorescence thyrsoid or rarely racemoid; calyx 5-lobed, lobes equal or subequal, mostly broadly triangular at the base and mucronate at apex, sometimes almost obsolete; corolla strongly 2-lipped, 4-lobed (1/3), yellow or white, posterior lip long, hooded, densely pubescent outside; stamens not or only shortly exserted from the corolla, thecae weakly distinct, parallel or narrowly divergent, sometimes hairy; stigma-lobes subequal or unequal; nutlets rounded or truncate, glabrous or eglandular hairy. About 20 spp., mostly forests and hill slopes, from NE India to Taiwan, Vietnam and Indonesia.

91. *Lamiophlomis* Kudo

Lamiophlomis Kudo, Mem. Fac. Sci. Taihoku Imp. Univ. 2: 210 (1929).

Perennial rhizomatous herb, with simple hairs; stem short or almost absent; leaves in basal rosette and often also opposite along the stem, mostly broad; inflorescence short and dense, thyrsoid; bracteoles spinescent; calyx-lobes equal, broad at the base and abruptly narrowed to a spinescent apex; corolla 2-lipped, 4-lobed (1/3), pink to purplish brown, posterior lip long, hooded, denticulate, densely villous inside; stamens not exserted from corolla, thecae confluent; stigma-lobes equal; nutlets rounded at apex, glabrous. One sp., *L. rotata* (Hook. f.) Kudo, moist alpine areas, from Tibet, Nepal and N India to C and S China. *Lamiophlomis* is often included in *Phlomis*.

92. *Phlomis* L.

Phlomis L., Sp. Pl. 2: 584 (1753).
Phlomoides Moench (1794).
Trambis Raf. (1837).

Perennial herbs or small shrubs, usually with woody rhizomes and/or tubers at root tips, with branched and/or simple hairs, rarely with simple hairs only; leaves simple, entire or toothed, at least lower petiolate; inflorescence thyrsoid; bracteoles present or sometimes absent, mostly long; calyx 5-lobed, lobes equal or sometimes unequal (3/2), mostly broad at base and abruptly narrowed to a narrow apex, often expanded at margins, sometimes obsolete or triangular-acuminate, mostly spinose; corolla strongly 2-lipped, 4-lobed (1/3), purple, pink or yellow, posterior lip long, hooded, often deeply concave, often bearded with simple hairs at margins; stamens not or only shortly exserted, posterior pair often with outgrowths near base, thecae confluent to sometimes weakly distinct; stigma-lobes unequal or sometimes equal; nutlets rounded or truncate at apex, glabrous or hairy at apex. $2n = 12, 14, 20, 22, 40, 42$. Up to ca. 100 spp., warm temperate, alpine and subtropical areas, from E Europe and the Mediterranean to Siberia, Korea, China, Nepal and N India. Adylov et al. (1986) believe that the predominantly Asiatic section *Phlomoides*, characterised by its rather shallowly hooded posterior lip of the corolla with a bearded margin, should be transferred to a separate genus, *Phlomoides* Moench, along with some species of *Eremostachys*.

93. *Eremostachys* Bunge

Eremostachys Bunge in Ledeb., Icon. Pl. Nov. Fl. Ross. 2: 10 (1830); Popov, Nouv. Mem. Soc. Nat. Mosc. 19: 1 (1966).
Paraeremostachys Adylov, Kamelin & Machm. (1986).

Stout perennial herbs, usually with woody rhizomes and/or tubers at root tips, with branched and/or simple hairs; leaves simple or laciniate to bipinnatisect; inflorescence thyrsoid or perhaps sometimes racemoid; bracteoles often connate at base; flowers mostly large; calyx 5-lobed, tubular-campanulate to broadly funnel-shaped, lobes equal or subequal, broad at the base and abruptly narrowed to a short spinose apex, broadly triangular, subulate or obsolete; corolla showy, strongly 2-lipped, 4-lobed (1/3), pink, yellow or white, posterior lip long, hooded, often deeply concave, bearded on margins; stamens not or only shortly exserted, posterior pair pectinate-fimbriate or with appendage near base, thecae distinct or confluent; stigma-lobes unequal; nutlets truncate or subtruncate and densely bearded at apex. $2n = 22$. Five to 60 spp., mostly dry places, temperate and subtropical areas from E Europe to Mongolia, W China and NW India. Several authors (e.g. Adylov et al. 1986) have recently transferred *Eremostachys* species to *Phlomoides* and *Paraeremostachys*.

94. *Pseudomarrubium* Popov

Pseudomarrubium Popov, Notul. Syst. Herb. Inst. Bot. Acad. Sci. USSR 8: 75 (1940).
Neustruevia Juz. (1954).

Subshrub, strongly rooted, with simple and glandular hairs; leaves bipinnate; inflorescence thyrsoid, with many-flowered cymes; bracteoles spinescent, connate in groups of 2–5; calyx actinomorphic, obconical, 5-lobed, lobes subulate, spinescent, tube 5-nerved; corolla strongly 2-lipped, 5-lobed (2/3), white, posterior lip straight, short, weakly hooded, anterior lip twice as long, deflexed, lateral lobes oblong-ovate, median lobe cordate, sinuate at apex, tube shorter than calyx-tube, annulate within; stamen-filaments pubescent, thecae confluent at apex; stigma-lobes unequal; nutlets triquetrous, truncate, pubescent at apex. One sp., *P. eremostachydioides* Popov. Silty and rocky slopes, Kazakhstan (Tian-Shan).

95. *Notochaete* Benth.

Notochaete Benth., Edward's Bot. Reg. 15: t. 1289 (1829).

Perennial herbs with thick rhizomes and branched and simple hairs; leaves broad, toothed, long-petiolate; inflorescence thyrsoid with several-flowered cymes; bracts similar to foliage leaves, bracteoles spiny, uncinate; calyx 5-lobed, tubular, lobes equal or subequal, spinose with spines mostly subterminal outside lobes, uncinate; corolla strongly 2-lipped, 4-lobed (1/3), purplish, posterior lip long, hooded, densely hairy outside, anterior lip with median lobe about as large as lateral lobes; stamens not exserted from corolla, thecae distinct or subconfluent; stigma-lobes subequal; nutlets truncate at apex, glabrous or with branched hairs at apex. $2n = 22$. Two spp., in forests and at forest margins, from Tibet, Nepal and N India to S China (Yunnan) and Burma.

96. *Lagopsis* (Bunge ex Benth.) Bunge

Lagopsis (Bunge ex Benth.) Bunge, Mem. Sav. Etr. St.-Petersbourg 6: 565 (1835).

Perennial herbs, densely white-lanate with simple hairs; leaves palmately lobed, orbicular to cordate; inflorescence thyrsoid, with 3 to many-flowered cymes; calyx slightly zygomorphic, (3/2) 5-lobed, 2 slightly longer, spreading in fruit, spiny; corolla 2-lipped, 4–5-lobed (1–2/3), white, yellow to brown-purple, posterior lip straight, entire or emarginate, anterior lip with lateral lobes short, median lobe spreading, cordate, tube exannulate, included in calyx-tube or slightly longer; stamens included in corolla-tube, filaments short, thecae weakly distinct; style included in corolla-tube, stigma-lobes subequal; nutlets trigonous, smooth, apically rounded. Four species, rocky and gravelly slopes of Russia (S Siberia), Kazakhstan, Mongolia, China, NW India (Kashmir), Japan.

97. *Panzerina* Soják

Panzerina Soják, Cás. Nar. Muz. (Prague) 150: 216 (1982); Krestovskaya, Novost. Sist. Vysch. Rast. 28: 140–144 (1991).
Panzeria Moench (1794), non *Panzera* Cothen. (1790).
Leonuroides Rauschert (1982).

Perennial herbs with woody rootstock, densely hairy with simple hairs; leaves palmately lobed; inflorescence thyrsoid, cymes many-flowered; flowers sessile with bracts and bracteoles; calyx zygomorphic, often weakly so, 5-lobed, lobes subequal or unequal (3/2), usually spinose; corolla strongly 2-lipped, 4-lobed (1/3), yellow, posterior lip long, hooded, anterior lip directed upwards, corolla-tube exannulate, exserted from calyx; stamens not exserted from corolla, thecae distinct, parallel or divaricate; stigma-lobes subequal; nutlets trigonous, rounded at apex, glabrous or glandular. $2n = 18$. Two to seven spp. Semi-deserts, S Siberia, C Asia (Mongolia and NW China).

98. *Leonurus* L. Fig. 21M

Leonurus L., Sp. Pl. 2: 584 (1753); Krestovskaja, Novost. Sist. Vysch. Rast. 26: 142–149 (1989) & 27: 139–144 (1990), rev.
Cardiaca Mill. (1754).

Perennial or annual herbs with simple hairs; leaves ± palmately lobed or deeply serrate; inflorescence thyrsoid, cymes 4- to many-flowered; flowers sessile; bracts and bracteoles present; calyx zygomorphic, often weakly so, 5-lobed, lobes subequal or unequal (3/2), spinose; corolla strongly 2-lipped, 4-lobed (1/3), pink, rarely white, tube annulate or rarely annulate, posterior lip long, hooded, anterior lip directed forwards, rarely upwards; stamens not exserted from corolla; thecae distinct, divaricate; stigma-lobes subequal; nutlets trigonous, truncate at apex, glabrous or with eglandular hairs or sessile glands at apex. $2n = 18, 20$. About 25 spp. Stony places, forests and disturbed ground, Eurasia except Arctic and desert regions, *L. japonicus* Houtt. is widely distributed in tropics as a weed, some other species introduced elsewhere.

99. *Chaiturus* Willd.

Chaiturus Willd., Fl. Berol. Prodr. 200 (1787).

Annual or biennial herb, with simple hairs; leaves toothed, petiolate; inflorescence thyrsoid with many-flowered cymes; flowers pedicellate with bracts and bracteoles; calyx actinomorphic, 5-lobed, lobes subequal; corolla scarcely exceeding calyx, strongly 2-lipped, 4-lobed (1/3), pale pink, posterior lip long, hooded, anterior lip directed upwards, equal or slightly exceeding upper lip, corolla-tube exannulate; stamens very shortly exserted from corolla-tube, thecae weakly distinct, divaricate; stigma-lobes subequal to unequal; nutlets trigonous, truncate, with eglandular hairs at apex. $2n = 24$. One species: *C. marrubiastrum* (L.) Ehrh. ex Rchb. Steppes and disturbed grounds, from C Europe to SW Siberia and W. China (Xinjiang).

100. *Stachys* L. Fig. 21G, 27

Stachys L., Sp. Pl. 2: 580 (1753); Bhattacharjee, Notes Roy. Bot. Gard. Edinburgh 38: 65–96 (1980), sect. Classification; Sebsebe Demissew, Kew Bull. 48: 327–341 (1993), reg. rev.
Betonica L. (1753).
Tetrahit Gérard (1761).

Perennial or annual herbs or rarely small sub-shrubs, with simple and/or branched hairs, often strongly aromatic; leaves usually toothed, petiolate or rarely sessile; inflorescence thyrsoid or racemoid; flowers pedicellate, rarely sessile; bracts and bracteoles present or absent; calyx actinomorphic to weakly 2-lipped, 5-lobed, lobes equal or unequal (3/2), often spinose, tube with throat often barbate; corolla strongly 2-lipped, 4-lobed (1/3), posterior lip long, hooded or sometimes almost flat, externally pubescent, anterior lip with median lobe usually much larger than lateral lobes, corolla-tube annulate or rarely exannulate; stamens not exserted from corolla, thecae ± distinct, usually divaricate, rarely subparallel to parallel; stigma-lobes subequal or slightly unequal; nutlets obovoid to oblong, rarely flattened trigonous, rounded or sometimes winged at apex. $2n = 10, 16, 18, 24, 30, 32, 34, 48, 64, 66, 68, 80, 102$. About 300 spp., various habitats, cosmopolitan, except for Australia and New Zealand. Several species are widely cultivated as ornamentals. *Stachys affinis* Bunge is cultivated for its edible, subterranean tubers. A cladistic analysis by Lindqvist and Albert (2002) suggests that *Betonica* and other *Stachys* may have different origins.

101. *Phlomidoschema* (Benth.) Vved.

Phlomidoschema (Benth.) Vved., Notul. Syst. Herb. Inst. Bot. Acad. Sci. USSR 9: 54 (1941).

Perennial herb, woody at base, with unpleasant smell, densely covered with branched hairs; leaves entire or finely serrulate, subglabrous above; inflorescence thyrsoid, with up to 3-flowered cymes; calyx actinomorphic, campanulate, 5-lobed, lobes equal, subtriangular, not spinescent, strongly accrescent in fruit; corolla weakly 2-lipped, 5-lobed (2/3), rose or white, lobes scarcely differentiated, rather short; stamens included in corolla-tube; filaments short, thecae distinct; stigma-lobes equal; nutlet usually only one per fruit, broadly obovoid, rounded at apex. One species, *P. parviflora* (Benth.) Vved. Rocky mountain slopes, NE Iran, Afghanistan, Uzbekistan, Pakistan, NW India (W Himalayas). Some authors include *Phlomidoschema* in *Stachys*.

102. *Prasium* L.

Prasium L., Sp. Pl. 2: 601 (1753).

Small shrub, usually glabrous or very sparsely pubescent with simple hairs; leaves toothed, petiolate; inflorescence lax, racemoid or rarely thyrsoid, cymes up to 2-flowered; bracteoles at base of pedicels; calyx broadly campanulate in fruit, slightly 2-lipped, 5-lobed (3/2), lobes equal or subequal, shortly spinose; corolla strongly 2-lipped, 4-lobed (1/3), white to pale cream, posterior lip long, hooded; stamens not exserted from corolla, thecae weakly distinct, divergent; stigma-lobes equal or subequal; nutlets drupaceous, rounded at apex, black, glabrous. $2n = 34, 36$. One sp., *P. majus* L., dry places, mainly coastal, Canary Islands, Iberian Peninsula, Morocco and Mediterranean region.

103. *Sideritis* L. Fig. 21G

Sideritis L., Sp. Pl. 2: 574 (1753); Pérez de Paz & Negrin, Phanerog. Monogr. 20: 1–327, plates 1–22 + 48 figs. (1992), rev. of subg.; Obón de Castro & Rivera, Phanerog. Monogr. 21: 1–640 (1994), rev. of section.
Leucophae Webb & Berthel. (1845).

Annual or perennial, sometimes rhizomatous herbs or shrubs, with simple and sometimes branched hairs, usually aromatic; leaves often narrow, usually toothed, sessile to petiolate; inflorescence thyrsoid or sometimes racemoid, often dense; bracteoles present or absent; calyx zygomorphic or almost regular, 5-lobed, lobes subequal or unequal (1/4), mostly triangular, often spinose, tube often barbate in throat; corolla 2-lipped, 4–5-lobed (1–2/3), usually yellow or white, sometimes purple, posterior lip flat or almost so, deeply emarginate to entire, tube shorter than calyx; stamens included in corolla-tube, thecae confluent; stigma-lobes unequal; nutlets rounded at apex. $2n = 20, 22, 24, 26, 28, 30, 32, 34, 36, 38, 40, 42, 44, 46, 50, 56$. About 140 spp., various habitats, Macaronesia, Mediterranean region to Russia, Tibet and W China. Mendoza-Heuer (1977) recognises two subgenera: subg. *Sideritis* and subg. *Marrubiastrum*, the latter endemic to Macaronesia.

104. *Haplostachys* (A. Gray.) W.F. Hillebr.

Haplostachys (A. Gray.) W.F. Hillebr., Fl. Hawaii Isl.: 346 (1888); Wagner et al., Manual of the Flowering Plants of Hawaii 1: 798–801 (1990), rev.
Phyllostegia Benth. sect. *Haplostachys* A. Gray

Perennial woody herbs, with simple hairs; leaves coriaceous; inflorescence racemoid; bracts mostly distinctly smaller than the leaves; bracteoles present at base of pedicels; calyx 5-lobed, lobes equal, short, throat ± closed in fruit; corolla 2-lipped, 4-lobed (1/3), white to sometimes purple, posterior lip moderately long, undulate, crisped, entire; stamens equal or posterior pair slightly longer, not exserted from corolla, thecae weakly distinct; style included in corolla-tube, with stigma-lobes shortly clavate; nutlets dry, hairy at apex. Five spp. (four of these probably extinct), in dry forests and shrubland at low elevations, Hawaii.

105. *Phyllostegia* Benth.

Phyllostegia Benth., Edward's Bot. Reg. 15: t. 1292 (1830); Wagner et al., Manual of the Flowering Plants of Hawaii 1: 810–826 (1990), reg. rev.; Wagner, Novon 9: 265–297 (1999), nomencl.

Perennial herbs, lianas or subshrubs, with simple hairs; leaves simple or sometimes irregularly pinnatisect; inflorescence thyrsoid or racemoid; calyx 5-lobed, becoming broader, often broadly campanulate in fruit, lobes equal or subequal (3/2) sometimes irregularly so, occasionally dentate-serrate; corolla 2-lipped, 4-lobed (1/3), white to purple, posterior lip moderately long, shorter than lower lip, shallowly hooded or almost flat, entire or emarginate, corolla-tube usually curved; stamens equal or posterior pair slightly longer, not or only shortly exserted, thecae weakly distinct; stigma-lobes short, ± clavate, subequal or anterior broader; nutlets spreading and exserted from calyx at maturity, obliquely attached at base, fleshy, with slightly winged margins, rounded at apex, glabrous. $2n = 64, 66$. About 34 spp., in mesic to moist forest, Hawaii, with one species (*P. tahitensis* Nadeaud) in Tahiti and one (*P. tongaensis* H. St. John) in Tonga Is.

106. *Stenogyne* Benth., nom. cons., non Cass.

Stenogyne Benth., Edward's Bot. Reg. ad. t. 1292 (1830); Weller and Sakai in Wagner et al. (eds.) Manual of the Flowering Plants of Hawaii 1: 831–843 (1990), rev.

Vines or rarely erect, perennial herbs, with simple hairs; leaves sometimes palmately lobed in young plants; inflorescence thyrsoid or racemoid, not terminal; calyx 5-lobed, becoming wider, often broadly campanulate in fruit, lobes equal or unequal (3/2), sometimes irregularly so; corolla 2-lipped, 4-lobed (1/3), red to pink, purple, white or green, posterior lip moderately long, usually longer than lower lip, shallowly hooded or almost flat, entire to 2-lobed, corolla-tube often long and falcate, often strongly dilated distally; stamens equal or posterior pair slightly longer, exserted or not, thecae weakly distinct; stigma-lobes subequal or slightly unequal; nutlets obovoid, often very large, fleshy, margins slightly winged, obliquely attached at base, rounded at apex, glabrous. $2n$ = 32, (34), 64. About 20 spp., in forests and subalpine shrubland, Hawaii. Some species apparently pollinated by honeycreepers.

107. *Chamaesphacos* Schrenk ex Fisch. & C.A. Mey.

Chamaesphacos Schrenk ex Fisch. & C.A. Mey., Enum. Pl. Schrenk. 1: 27 (1841).

Annual herb, subglabrous or very sparsely pubescent with simple hairs; leaves petiolate, spinescent-aristate; inflorescence thyrsoid, with up to 3-flowered cymes; calyx weakly 2-lipped, 5-lobed (3/2), lobes spinose; corolla 2-lipped, 5-lobed (1–2/3), purple, posterior lip long, flat or shallowly hooded, anterior lip deflexed, tube very long, straight, floccose-villous within at throat; stamens not exserted, filaments pubescent, thecae weakly distinct; stigma-lobes subequal; nutlets oblong, rounded at apex, smooth, narrowly membranous-winged. One sp., *C. ilicifolius* Schrenk. Sandy areas, Iran to Kazakhstan and W China (Xinjiang).

108. *Lagochilus* Bunge ex Benth.

Lagochilus Bunge ex Benth., Lab. Gen. et Sp. 640 (1834).
Chlainanthus Briq. in Engler & Prantl (1896).
Lagochilopsis Knorring (1966).

Shrubs, subshrubs or perennial herbs with woody rootstocks, with or without spines in leaf-axils, glabrous or sparsely hirsute with simple hairs; leaves entire or mostly palmatifid with spinose-mucronate or rounded lobes; inflorescence thyrsoid or racemoid, with up to 5-flowered cymes, bracteoles spinescent; calyx slightly zygomorphic, (4–)5-lobed, lobes subequal or unequal (3/2), posterior often longer, lobes spreading or erect, oblong to broadly ovate, often spinose-mucronate, equalling or longer than tube; corolla strongly 2-lipped, 5-lobed (2/3), white, yellowish-white, rarely pink, posterior lip long, straight, shallowly hooded, densely villous outside, 2-lobed, lobes sometimes 2-toothed, anterior lip with lateral lobes acute or emarginate, median lobe obcordate, tube densely pilose-annulate inside; stamens not or shortly exserted from corolla-tube, filaments complanate, anthers glabrous or hairy, thecae distinct; stigma-lobes equal; nutlets trigonous, obovoid, apex truncate or rounded, glabrous, pilose or glandular. $2n$ = 22, 34. About 40 species. Dry slopes, valleys, deserts of Iran to Mongolia, Russia (S. Siberia), NW China and N. Pakistan. Some medicinal, ornamental.

109. *Roylea* Wall. ex Benth.

Roylea Wall. ex Benth. in Edward's Bot. Reg. 15: t. 1289 (1829).

Shrub with simple hairs; leaves petiolate, blade ovate, deeply crenate; inflorescence lax, thyrsoid, cymes up to 6-flowered; calyx tubular, 5-lobed, lobes equal or subequal, oblong-elliptic, narrowed at base; corolla strongly 2-lipped, 4-lobed (1/3), pink to white, posterior lip long, hooded, densely pubescent outside; stamens not exserted from corolla, thecae weakly distinct; stigma-lobes shortly subequal; nutlets rounded at apex, smooth. $2n$ = 34. One sp., *R. cinerea* (D. Don) Baill., NW India, Nepal.

110. *Stachyopsis* Popov & Vved.

Stachyopsis Popov & Vved., Trudy Turkest. Nauch. Obsch. 1: 120 (1923).

Perennial herbs with short simple hairs; leaves toothed; inflorescence thyrsoid with many-flowered cymes; calyx actinomorphic, obconical-campanulate, 5-lobed, lobes equal, triangular, spinescent; corolla strongly 2-lipped, 5-lobed (2/3), purplish to lilac-pink, posterior lip deeply hooded, erect, entire, densely villous, anterior lip with median lobe orbicular to obovate, reflexed, emarginate, much larger than lateral lobes, tube rather broad, slightly curved, annulate; thecae distinct; stigma-lobes subequal; nutlets oblong, triquetrous, apex obliquely truncate, glabrous.

Four spp. Grassy meadows in mountainous areas, E Afghanistan, Tajikistan, Uzbekistan, Kazakhstan, N Pakistan, W China (Xinjiang).

111. *Marrubium* L. Fig. 21B

Marrubium L., Sp. Pl. 2: 582 (1753); Seybold, Stuttg. Beitr. Naturk. A. 310 (1978), reg. rev.
Atirbesia Raf. (1840).

Annual or perennial herbs, with at least some branched hairs and often also with simple hairs; leaves usually toothed, petiolate; inflorescence thyrsoid or rarely racemoid; calyx 5–15(–30)-lobed, lobes equal or variously unequal, usually ± spiny, sometimes uncinate; corolla 2-lipped, 5-lobed (2/3), posterior lip 2-lobed, long or moderately long, flat or almost so, tube shorter than calyx; stamens included in corolla-tube, thecae confluent or subconfluent; style included in corolla-tube, stigma-lobes unequal or subequal; nutlets rounded or subtruncate at apex, glabrous or with sessile glands at apex. $2n = 20, 26, 28, 34, 54$. About 40 spp., mainly in rather dry places, Europe and N Africa to Pakistan, with *M. vulgare* L. more widespread and introduced elsewhere.

112. *Metastachydium* Airy Shaw ex C.Y. Wu & H.W. Li

Metastachydium Airy Shaw ex C.Y. Wu & H.W. Li, Fl. Reipubl. Popularis Sin. 66: 28 (1977).
Metastachys Knorring (1959), non (Benth. & Hook.) Teigh.

Perennial herb, rhizomatous, with stellate hairs; basal leaves long-petiolate, sagittate; inflorescence thyrsoid, with many-flowered cymes; bracteoles partly adnate to flowers; calyx actinomorphic, 5-lobed, lobes equal, spinescent; corolla strongly 2-lipped, 5-lobed (2/3), purple, posterior lip straight, anterior lip with lateral lobes short, oblong to ovate, median lobe reniform, emarginate, tube included in calyx-tube, annulate; stamens included in corolla-tube, filaments pubescent; stigma-lobes unequal. One species, *M. sagittatum* (Regel) C.Y. Wu & H.W. Li. Steppe slopes of Kazakhstan, Kyrgyzstan, W China (Xinjiang).

113. *Pseudoeremostachys* Popov

Pseudoeremostachys Popov, Nouv. Mem. Soc. Nat. Mosc. 19: 148 (1940).

Perennial herb with rootstock, with simple and stellate hairs; leaves ovate, petiolate, rugose above; inflorescence thyrsoid, with 8–10-flowered cymes; bracteoles subulate-spinescent, longer than calyx, long pubescent; calyx actinomorphic, 5-lobed, lobes subequal, triangular, spinescent; corolla 2-lipped, 5-lobed (2/3), dark-purple, posterior lip straight, flat or slightly hooded, sinuate or 2-lobed at apex, anterior lip with lateral lobes short, oblong, median lobe reniform, tube included in calyx-tube, annulate near throat; stamens included in corolla-tube or shortly exserted, filaments short; style included in corolla-tube, stigma-lobes unequal; nutlets oblong, triquetrous, densely pubescent at apex. One species, *P. sewerzovii* (Herd.) Popov. Rocky slopes, Kazakhstan, Kyrgyzstan (Tian-Shan).

114. *Ballota* L.

Ballota L., Sp. Pl. 2: 582 (1753); Patzak, Ann. Nat. Hist. Mus. Wien 62: 57–86 (1958), rev. of section; Ann. Nat. Hist. Mus. Wien 63: 33–81 (1959), rev. of sections.

Perennial herbs or small shrubs, with branched and/or simple hairs; leaves toothed, petiolate; inflorescence thyrsoid or racemoid; bracteoles sometimes long and spinose (sect. *Acanthoprasium*); calyx mostly campanulate, often widely so, 5–10(–20)-lobed, lobes equal or variously unequal, mostly ± spiny; corolla strongly 2-lipped (1–2/3), purple to white, upper long hooded or almost flat, entire, 2-fid or dentate; stamens not or only shortly exserted from corolla, thecae ± distinct; stigma-lobes subequal or slightly unequal; nutlets rounded or subtruncate at apex, glabrous or with sessile glands at apex. $2n = 18, 20, 22, 28$. About 30 spp., Europe, North Africa, West and Central Asia, with one sp., *B. africana* (L.) Benth., in South Africa and Namibia, and *B. nigra* L. introduced to other temperate parts of the world.

115. *Sulaimania* Hedge & Rech. f.

Sulaimania Hedge & Rech. f. in Rech. f., Fl. Iran. 150: 345 (1982).

Subshrub, glabrous or with few, simple hairs; leaves subsessile, narrow, entire, spinulose at apex; inflorescence lax, thyrsoid, cymes up to 5-flowered; bracteoles 3-partite, spinose; calyx campanulate, recurved, 7–10-lobed, lobes irregularly unequal, spreading, spinose; corolla strongly 2-lipped, 4-lobed (1/3), white, posterior lip long, hooded, densely pubescent outside; stamens not exserted from corolla, thecae weakly distinct; stigma-lobes subequal; nutlets angular, truncate at apex. One sp., *S. otostegioides* (Prain) Hedge, dry places, Pakistan (Sulaiman Ranges).

116. *Moluccella* L.

Moluccella L., Sp. Pl. 2: 587 (1753).

Annual or sometimes short-lived perennial herbs, subglabrous or very sparsely pubescent with simple hairs; leaves toothed, petiolate; inflorescence thyrsoid, cymes 3–4-flowered; bracteoles short, spinose; calyx 2-lipped, very widely campanulate, recurved, 5–8(–10)-lobed, lobes subulate-spinose or obsolete; corolla strongly 2-lipped, 4-lobed (1/3), white or pink, posterior lip long, hooded, densely pubescent outside; stamens not exserted from corolla, thecae weakly distinct; stigma-lobes equal or slightly unequal; nutlets sharply triquetrous, truncate and with sessile glands at apex. Two spp., rocks and disturbed ground, S Europe to C Asia, *M. laevis* L. often cultivated and naturalized in other warm regions.

117. *Otostegia* Benth.

Otostegia Benth., Lab. Gen. et Sp. 601 (1834); Sebald, Stuttgarter Beitr. Naturk. A 363: 1–84 (1973), reg. rev.
Chartocalyx Regel (1879).
Harmsiella Briq. (1896).

Small shrubs or subshrubs with simple hairs, often with spines in leaf-axils; leaves toothed, petiolate; inflorescence thyrsoid or racemoid; bracteoles long or short, spinose; calyx slightly zygomorphic to strongly 2-lipped, very widely funnel-shaped to flat in distal part, tubular, obconical or sometimes tumescent in basal part, 5–10(–11)-lobed, lobes spreading, broad and obscure or spine-tipped, tube annulate; corolla strongly 2-lipped, 4-lobed (1/3), white or yellow, posterior lip long, hooded, often bearded and densely pubescent outside; stamens not or only shortly exserted from corolla, thecae weakly distinct; stigma-lobes slightly unequal; nutlets 1–4, with apex truncate or rounded, with or without sessile glands. About 15 spp., in rather dry, often montane areas and semi-deserts, NE Cameroun to Saudi Arabia and Yemen, NE Egypt (Sinai), Iran, Central Asia to India.

118. *Isoleucas* O. Schwartz

Isoleucas O. Schwartz, Mitt. Inst. Allg. Bot. Hamburg 10: 223 (1939).

Subshrub, densely whitish pubescent with branched hairs; leaves petiolate, blade suborbicular; inflorescence racemoid; bracteoles minute, at base of pedicel; calyx 2-lipped, obconical, 5-lobed (3/2), lobes triangular, subacute, anterior much smaller than posterior, tube densely annulate with simple hairs; corolla strongly 2-lipped, 4-lobed (1/3), white, posterior lip long, hooded, bearded with simple hairs and densely pubescent outside; stamens not or only shortly exserted from corolla, thecae weakly distinct; stigma-lobes slightly unequal; nutlets rounded and with sessile glands at apex. One sp., *I. arabica* O. Schwartz, in semi-desert, SE Yemen.

119. *Leucas* Burm. ex R. Br.

Leucas Burm. ex R. Br., Prodr. Fl. Nov. Holl.: 504 (1810); Sebald, Stuttgarter Beitr. Naturk. A. 308: 1–42 (1978), rev. of section; 341: 1–200 (1980), reg. rev. of sections; Singh, Monograph on Indian *Leucas* R. Br. (Dronapushpi) Lamiaceae (2001), reg. rev.
Lasiocorys Benth. (1834).
Physoleucas (Benth.) Jaub. & Spach (1855).

Annual or perennial herbs or subshrubs with simple hairs; leaves usually toothed, petiolate or at least narrowed at base; inflorescence thyrsoid or rarely racemoid; calyx zygomorphic, 2-lipped, or sometimes actinomorphic, 5–10(–12)-lobed, lobes variously unequal or rarely equal, mostly spinose; corolla strongly 2-lipped, 4-lobed (1/3), white or rarely slightly reddish, posterior lip long or moderately long, hooded, bearded and densely pubescent outside with whitish hairs; stamens not or only shortly exserted from corolla, thecae confluent or weakly distinct; stigma-lobes unequal to sometimes equal; nutlets with apex truncate or rounded, with or without glands, otherwise glabrous or rarely shortly eglandular hairy. $2n =$ (20), 22, (24), (26), 28 (30). About 100 spp., mostly dry or disturbed ground, tropical to southern Africa, Arabian Peninsula, Iran to S China, Taiwan, Japan and SE Asia, some spp., especially *L. martinicensis* (Jacq.) R. Br., introduced as weeds elsewhere.

120. *Leonotis* (Pers.) R. Br.

Leonotis (Pers.) R. Br., Prodr. Fl. Nov. Holl.: 504 (1810); Iwarsson, Fl. Southern Africa 28, 4: 31–38 (1985), reg. rev.
Phlomis L. sect. *Leonotis* Pers. (1807).
Hemisodon Raf. (1837).

Tall herbs, shrubs or rarely small trees, with simple hairs; leaves simple, usually toothed; inflorescence thyrsoid, cymes many-flowered, rarely long-pedunculate; bracteoles spiny; flowers large

(17–50 mm long); calyx ± zygomorphic to almost 2-lipped, 8–10-lobed, usually spinescent; corolla strongly 2-lipped, 4-lobed (1/3), posterior lip long, hooded, bearded and densely pubescent outside, with reddish, orange or cream hairs, anterior lip shorter than posterior, deflexed, soon withering; stamens not or only shortly exserted from corolla, thecae weakly distinct; stigma-lobes strongly unequal, upper lobe minute or apparently absent; nutlets oblong-trigonous, obliquely truncate and with sessile glands at apex. $2n = 24, 26, 28$. About ten spp., mostly hill slopes and rocky or disturbed ground, tropical and southern Africa, *L. nepetifolia* (L.) R. Br. introduced as a weed elsewhere in tropics. Apparently derived from within *Leucas* (Ryding 1998).

121. *Acrotome* Benth. ex Endl.

Acrotome Benth. ex Endl., Gen. Pl. 1: 627 (1838); Codd, Fl. Southern Africa 28, 4: 19–23 (1985), reg. rev.

Annual or perennial herbs or subshrubs, with simple hairs; leaves usually narrow, ± toothed, petiolate or at least narrowed at base; inflorescence thyrsoid; calyx zygomorphic or almost actinomorphic, 5–10-lobed, lobes equal or variously unequal, usually ± spinose; corolla strongly 2-lipped, 4-lobed (1/3), white; posterior lip long, slightly hooded, outside ± densely pubescent; anterior lip slightly longer than the posterior lip; stamens included in corolla-tube, thecae confluent; style included in corolla-tube, hairy, obliquely entire at apex; nutlets trigonous, truncate or subtruncate and usually with sessile glands at apex. $2n = 26$. Six spp., dry places, South tropical and southern Africa.

VII. Subfam. Nepetoideae (Dumort.) Luerss. (1882).

Shrubs and herbs, rarely trees, usually strongly aromatic with sessile glands and glandular hairs and with presence of volatile terpenoids and rosmarinic acid, iridoid glycosides and acteosides usually absent. Leaves simple, entire or lobed, sometimes compound. Inflorescence cymose, usually forming a thyrse of 1- to many-flowered cymes, pedunculate or sessile (often in verticillasters) in the axils of reduced or leaf-like bracts, bracteoles present or not. Calyx actinomorphic to strongly 2-lipped, annulate or not, usually 5-lobed, or less by fusion. Corolla zygomorphic, sometimes weakly so, often strongly 2-lipped. Stamens 4 or 2, usually by abortion of posterior pair, equal or more usually didynamous, the anterior pair often longer, divergent, ascending or declinate, usually exserted from corolla-tube and often subequalling corolla-lip, sometimes long-exserted or rarely included in corolla-tube, anther 2-thecous or rarely 1 by sterilization and modification, thecae parallel to divaricate, sometimes widely separated by extension of connective, distinct to synthecous. Pollen usually hexacolpate, rarely colpus number increased. Style gynobasic, stigma-lobes usually 2, conspicuous, equal or unequal, stigma rarely capitate. Disc usually well developed, often 4-lobed, lobes equal or anterior lobe enlarged. Ovary 4-locular, deeply 4-lobed, lobes 1-ovulate, forming 4 mericarps, rarely fewer by abortion, pericarp usually dry, rarely fleshy, often mucilaginous. Seeds exalbuminous. Embryo-investing (rarely bent).

Key to the Tribes

1. Stamens declinate **3. *Ocimeae***
– Stamens divergent or ascending 2
2. Calyx usually 10-nerved, corolla usually rather weakly 2-lipped with short, ± porrect lobes, disc asymmetric with anterior lobe elongate, nutlets globose with minute abscission-scar **1. *Elsholtzieae***
– Calyx usually 11–15-nerved, corolla often distinctly 2-lipped, more rarely weakly so, disc symmetric, or if asymmetric with anterior lobe elongate, then corolla distinctly 2-lipped, nutlets various, often ovoid with areolate abscission-scar **2. *Mentheae***

VII. 1. Tribe Elsholtzieae (Burnett) Sanders & Cantino (1984).

Annual or perennial herbs or shrubs. Stems often 4-angled. Leaves opposite, usually petiolate. Inflorescence usually an often secund raceme-like thyrse or panicle of 1(–3)-flowered cymes in the axils of usually much reduced bracts, sometimes conspicuous, broad, imbricate, often venose, bracteoles usually absent. Calyx ± 10-nerved, 5-lobed (3/2), often ± actinomorphic at anthesis, usually accrescent in fruit and usually becoming more distinctly 2-lipped, corolla ± 2-lipped, 4–5-lobed (1–2/3), lobes usually with short, scarcely spreading lobes, posterior lip 1–2-lobed, anterior lip 3-lobed, corolla-tube often short. Stamens 4 or 2 by reduction, often exserted, weakly didynamous, divergent, filaments glabrous, anthers 2-thecous, thecae parallel to divergent, separate or apically confluent. Style with 2 usually equal, subulate stigma-lobes. Disc asymmetrical with anterior lobe elongate. Nutlets usually globose, sometimes reduced to 1 by abortion, smooth, reticulate or foveolate, abscission-scar very small, often circular.

1. Nutlets strongly reticulate-foveolate, annual herbs 2
- Nutlets smooth or faintly reticulate or sometimes verrucose or tuberculate towards apex, never reticulate-foveolate, usually perennial herbs or shrubs 3
2. Fertile stamens 4 **126. *Perilla***
- Fertile stamens 2, the anterior pair much reduced **125. *Mosla***
3. Flowers in a dense, cylindrical, sometimes secund, spiciform thyrse, composed of usually several-flowered cymes or verticillasters, bracts sometimes conspicuous and fused in pairs **123. *Elsholtzia***
- Flowers in a rather lax, terminal thyrse, never densely spiciform 4
4. Cymes bracteolate, 3-flowered, distinctly pedunculate, upper cymes often 1-flowered, nutlets 4 **122. *Perillula***
- Cymes ebracteolate, 1-flowered throughout, nutlet usually 1 by abortion **124. *Collinsonia***

122. *Perillula* Maxim.

Perillula Maxim., Bull. Acad. Imp. Sci. St. Petersbourg, ser. 3, 20: 463 (1875).

Perennial, shortly rhizomatous herb, ?aromatic; leaves deeply toothed; inflorescence terminal, of lax 3(–1)-flowered, bracteolate cymes, shortly pedunculate below; flowers small, long-pedicellate; calyx campanulate-infundibuliform, posterior lobes broadly acuminate, anterior lobes longer, calyx-tube annulate shortly below throat, tube slightly gibbous at base in fruit; corolla broadly campanulate, white, obscurely 2-lipped, 5-lobed, corolla-tube slightly constricted and annulate within, near base; stamens 4, didynamous, divergent, anterior pair longer and slightly exserted from corolla-tube, posterior pair included, thecae ± parallel to slightly divergent, separate, with small connective; style and stigma-lobes short; nutlets 4, broadly obovoid, smooth. A single species, *P. reptans* Maxim., woodland near streams, Japan.

123. *Elsholtzia* Willd.

Elsholtzia Willd. in Roem. & Usteri, Bot. Mag. 11: 3 (1790); Press, Bull. Br. Mus. Nat. Hist. (Bot.) 10(1): 1–89 (1982), rev.
Aphanochilus Benth. (1829).
Paulseniella Briq. (1908).

Herbs, sometimes annual, subshrubs or shrubs, sometimes aromatic; inflorescence often cylindrical or capitate, rarely secund, of 2- to several-flowered verticillasters in axils of ovate bracts, these sometimes fused in pairs and flabellate (and often venose), bracteoles present only in sect. *Aphanochilus*, inconspicuous; calyx campanulate to cylindrical, lobes subequal, or posterior sometimes shorter, throat glabrous within; corolla white, yellowish or purplish, corolla-tube cylindrical or infundibular, usually longer than calyx, annulate or glabrous within; stamens 4, usually exserted, didynamous or not, ascending divergent, thecae divergent or divaricate, usually confluent at apex; ovary glabrous; style usually exserted beyond stamens, stigma-lobes often weakly clavate, rarely distinctly so, rarely unequal with posterior lobe shorter; nutlets 4 or rarely 1 by abortion, ovoid to oblong, glabrous or sparsely hairy, tuberculate or smooth. $2n = 16, 18, 20, 32$. About 40 species, Europe, Japan, China, Korea, introduced in N America. Press (1982) recognises three sections. Sect. *Elsholtzia* with spikes usually secund, bracts at least as broad as long, sometimes fused, membranous, veined, imbricate bracteoles absent; fruiting-calyx sometimes accrescent, never inflated; stigma-lobes subulate; nutlets rugulose or nearly smooth. Sect. *Aphanochilus* (Benth.) Benth.: spikes usually cylindrical, bracts linear to ovate, not membranous or imbricate, free; bracteoles present; fruiting-calyx as before; stigma-lobes subulate; nutlets rugulose. Sect. *Platyelasma* (Briq.) Press (series *Platyelasmeae* Briq.): spikes cylindrical, bracts at least as long as broad, not membranous or imbricate, free; bracteoles absent; fruiting-calyx inflated; style lobes with clavate tips; nutlets verrucose (*E. densa* Lab. & *E. eriostachya* (Benth.) Benth.).

124. *Collinsonia* L.

Collinsonia L., Sp. Pl. 1: 28 (1753); L., Gen. Pl. ed. 5: 16 (1754); Harley, Paton & Ryding, Kew Bull. 58: 485–489 (2003), gen. syn.
Keiskea Miq. (1865).
Micheliella Briq. (1897).

Perennial herbs often with citrus-like odour; leaves rather large, opposite or subverticillate, coarsely toothed; inflorescence of 2-flowered verticillasters, raceme-like, ebracteolate; bracts usually small, persistent; calyx campanulate, lobes often subequal, lanceolate-subulate to deltoid, sometimes posterior lip shorter, median lobe rarely minute, lobes often slightly upcurved in fruit, calyx-tube obscurely (5–)10-nerved, throat hairy, often annulate; corolla white, yellowish or purplish, anterior lip with median lobe flat, entire to laciniate, corolla-tube infundibular to subcylindrical, annulate or not, usually long-exserted; fertile stamens 4, didynamous, or 2 (posterior pair reduced to staminodes), long-exserted or rarely included, thecae divergent to weakly divaricate, often slightly confluent at apex; nutlets 1(–2) by abortion, globose or subglobose, glabrous, smooth or faintly reticulate. $2n = 50, 52$. About ten species of woodland and montane forest, four spp. in Eastern North America and about six species in China and Japan, with one endemic species in Taiwan.

125. *Mosla* (Benth.) Buch.-Ham. ex Maxim.

Mosla (Benth.) Buch.-Ham. ex Maxim., Bull. Acad. Imp. Sci. St. Petersbourg, ser. 3, 20: 456 (1875).
Orthodon Benth. & Oliver (1865) non R. Br. (1819).

Annual herbs, strongly aromatic; leaves toothed, conspicuously gland-dotted on abaxial surface; inflorescence raceme-like, congested to interrupted, often secund, of 2-flowered verticillasters; bracts small, sometimes the lower leaf-like, bracteoles absent; flowers small, pedicellate; calyx campanulate, lobes subequal or posterior lip shorter, with lobes shortly triangular, acute to obtuse, posterior median lobe sometimes much shorter, anterior lobes lanceolate, calyx-tube hairy in throat, becoming distinctly 2-lipped in fruit, base of tube gibbous, upper lip slightly spreading; corolla white to pinkish purple, campanulate, weakly 2-lipped, posterior lip emarginate, anterior lip with median lobe crenate, corolla-tube included in calyx or exserted, often slightly constricted near base and annulate (? or not) within; fertile stamens 2, posterior pair erect, divergent, shorter than corolla, anterior pair staminodal, fertile thecae divaricate, connective distinct; style equalling corolla; nutlets 4, ± globose, reticulate-foveolate. $2n = 18$. About 22 species, open or shady, mesophytic habitats, China, SE Asia to Japan.

126. *Perilla* L.

Perilla L., Gen. Pl. ed. 6, 578 (1764); Li Hsi-Wen & Hedge, Lamiaceae, in Wu Zheng-yi and Raven (eds.) Flora of China 17: 241–242 (1994), reg. rev.

Robust annual herbs, strongly aromatic; leaves toothed, long-petiolate, in cultivated forms often deep purple-tinged or laciniate; inflorescence raceme-like, lax or congested, ± secund, of 2(–4)-flowered verticillasters; bracts small, often broadly ovate-acuminate; bracteoles absent; flowers small, distinctly pedicellate; calyx campanulate, lobes triangular, posterior lip shorter with median lobe shortest, tube densely hairy outside, annulate within throat, calyx strongly accrescent in fruit, with base of tube strongly distended and gibbous, calyx-lobes becoming slightly upcurved; corolla small, white to purplish, weakly 2-lipped, lobes subequal, corolla-tube very short, included in calyx, annulate at insertion of stamens; stamens 4, subequal, divergent, shorter than or equalling corolla, thecae separate, parallel, becoming slightly divergent, connective distinct; style short; nutlets 4, subglobose, glabrous, distinctly reticulate, $2n = 20, 40$. One variable species: *P. frutescens* (L.) Britton, probably native in shady areas in the Himalayan foothills, from Pakistan eastwards, but cultivated and occurring as a casual in SE Asia to China and Japan.

VII. 2. Tribe Mentheae Dumort. (1827).

Perennial or annual herbs or shrubs, usually strongly aromatic; inflorescence highly variable, cymes 1- to many-flowered, often forming a bracteate, terminal thyrse, or series of verticillasters in the axils of leaf-like or reduced bracts, bracteoles present or not. Calyx 5- to 15-nerved, corolla ± actinomorphic to usually 2-lipped; stamens 4 or 2 by reduction, usually didynamous and ascending under posterior lip of corolla (Fig. 28) or sometimes divergent; style usually 2-lobed, stigma-lobes equal or more often unequal; disc symmetrical or not, with lobes alternate to ovary-lobes; nutlets usually ovoid or oblong, rarely truncate.

1. Fertile stamens 2, staminodes present or not, usually included* — 2
– Fertile stamens 4 — 25
2. Posterior stamens fertile — **178. *Nepeta***
– Anterior stamens fertile — 3
3. Stamens ascending towards posterior corolla-lip — 10
– Stamens divergent or parallel, rarely declinate or scarcely exserted from corolla-tube* — 4
4. Staminodes exserted from corolla — **127. *Zhumeria***
– Staminodes, if present, included — 5
5. Flowers in compact, sessile verticillasters in the axils of deeply toothed or lobed, leaf-like bracts, nutlets truncate at apex, often with toothed to tuberculate apical crest — **135. *Lycopus***
– Not as above — 6
6. Leaves densely and finely reticulate-bullate to rugose, sometimes hastate, petioles often winged above — 7
– Leaves not as above, never hastate, petioles rarely winged — 8
7. Lips of fruiting-calyx connivent, throat closed. Turkey — **128. *Dorystoechas***
– Lips of fruiting-calyx not connivent, throat open — **129. *Meriandra***

* Female plants of gynodioecious species will not always adequately key out.

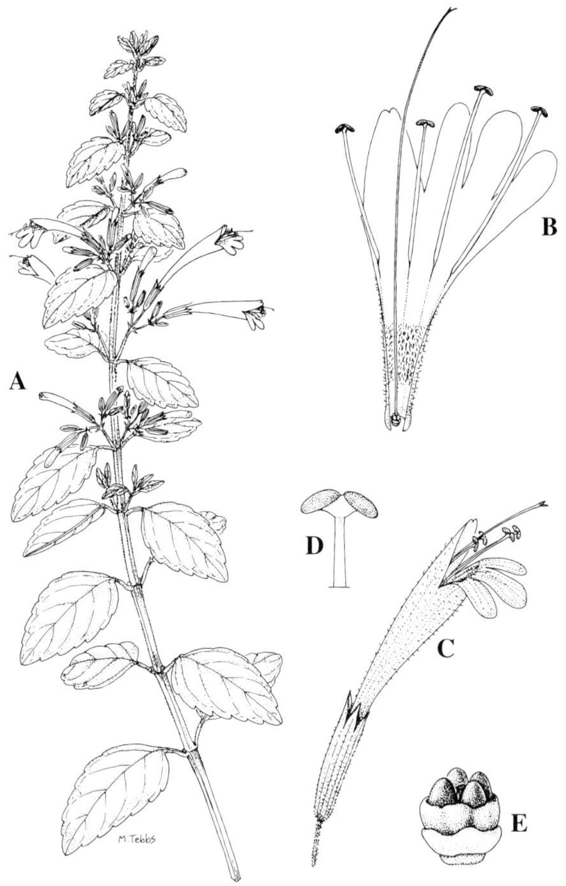

Fig. 28. Labiatae subfam. Nepetoideae tribe Mentheae. *Clinopodium multiflorum*. **A** Habit. **B** Corolla opened out. **C** Flower, side view. **D** Anther. **E** Disc and gynoecium. Drawing by M. Tebbs

8. Stamens usually long-exserted, anthers free 9
- Stamens scarcely exserted from corolla-tube, anthers connate. Texas **157. *Rhododon***
9. Posterior lip of corolla 4-lobed, stamens slightly deflexed at anthesis, posterior calyx-lip subentire to obscurely 3-lobed, tube 10-nerved. SW/C Asia **130. *Perovskia***
- Posterior lip of corolla 2-lobed or emarginate, stamens divergent, calyx distinctly 5-lobed (3/2) **166. *Cunila***
10. Anthers with 1 fertile, outer theca or rarely the inner theca subfertile and then thecae widely separated by an elongate, filamentous connective 11
- Anthers with 2 fertile thecae, connective small or extended, but thecae never widely separated 13
11. Staminal connective elongate, often articulating with filament, and separating outer, fertile theca from inner theca which is variously modified, or reduced to a small, deflexed, tooth-like appendage 12
- Staminal connective short, not articulating with filament, thecae adjacent, inner, sterile theca sometimes forming a small appendage **154. *Ziziphora***
12. Inflorescences of short, few-flowered racemes in the axils of the upper leaves. Mid-lobe of anterior corolla-lip strongly concave, stamens and styles long-exserted, strongly curved. Leaves short, cylindrical with revolute margins **132. *Rosmarinus***
- Inflorescences not as above, thyrsoid and usually terminal; flowers and leaves not as above **131. *Salvia***
13. Inflorescence a terminal thyrse of congested or interrupted verticillasters, each glomerule with an involucre of conspicuous, often distinctly veined, sometimes coloured bracts and/or bracteoles 14
- Inflorescences various, without conspicuous bracts or bracteoles, or if conspicuous, not forming an involucre 16
14. Bracts/bracteoles broadly ovate, with spinose margins, small annual herbs **172. *Acanthomintha***
- Bracts/bracteoles not spinose or absent, perennial herbs, or if annual, robust 15
15. Calyx distinctly 2-lipped **174. *Blephilia***
- Calyx ± actinomorphic **173. *Monarda***
16. Dwarf annuals with elongate, linear to spathulate bracteoles, often incurved and with margins white-ciliate, inflorescence a usually congested, terminal, spiciform or capitate thyrse. Seasonally flooded areas of western N America **158. *Pogogyne***
- Perennial herbs or shrubs, or if annuals, bracteoles inconspicuous, minute and inflorescences usually lax 17
17. Calyx-throat glabrous within, tube 5-angled, angles ciliate, leaves small, much shorter than the sharply 4-angled internodes, cymes 1–3-flowered. Perennial herbs of wet places. Eastern S America **163. *Hoehnea***
- Calyx hairy within throat, or annulate just below throat, leaves and inflorescences not as above 18
18. Posterior stamens reduced to staminodes 19
- Posterior stamens absent 22
19. Anther-thecae ± parallel, but thecae at different levels, subshrub from Florida **155. *Stachydeoma***
- Anther-thecae divergent to widely divaricate, usually at same level 20
20. Calyx-lobes connivent after anthesis, usually subequal, not annulate in calyx-throat or irregularly so, calyx-tube never saccate or gibbous, shrubs or subshrubs **160. *Poliomintha***
- Calyx-lobes erect or spreading, not connivent after anthesis, calyx usually 2-lipped with unequal lobes, distinctly annulate in calyx-throat, although sometimes reduced to a single row of hairs or annulus slightly below base of lobes and then not externally obvious, annual or perennial herbs or subshrubs 21
21. Delicate, often procumbent, perennial or annual herbs, calyx-throat sparsely annulate, tube not gibbous or saccate. Old World **150. *Clinopodium***
- Perennial or annual herbs or subshrubs, calyx-throat densely annulate, tube often gibbous or saccate at base **156. *Hedeoma***
22. Posterior lip of corolla usually longer than anterior, concave and hooded, sometimes falcate, filaments arcuate-ascending with small hairy appendages at base, corolla-tube abruptly widening above, calyx-lobes sometimes irregularly aristate **163. *Glechon***
- Posterior lip of corolla shorter than or equalling anterior, ± flat, not hooded or falcate, stamen-filaments without hairy appendages 23
23. Inflorescence a terminal, interrupted or congested spiciform thyrse, bracts inconspicuous, ± subulate, stems often virgate, internodes elongate, acutely 4-angled, leaves linear to linear-lanceolate, sometimes reduced **164. *Rhabdocaulon***
- Inflorescence of axillary cymes, from the axils of leaf-like bracts, sometimes diminishing above and cymes crowded towards stem-apex, but never spiciform, stems various, scarcely virgate 24

24. Cymes 3–6-flowered, stems acutely 4-angled, calyx-tube spreading hairy with simple hairs, posterior corolla-lip longer than anterior **165. *Eriothymus***
- Cymes 1–3-flowered, stems 4-angled but not acutely so, calyx-tube glabrous or appressed hairy with simple or minute dendroid hairs, anterior corolla-lip longer than posterior **161. *Hesperozygis***
25. Posterior pair of stamens distinctly longer than anterior (**subtribe Nepetinae**) 26
- Posterior pair of stamens shorter than or equalling anterior 37
26. Anther-thecae divergent at 180° 27
- Anther-thecae parallel or divergent at 90° 31
27. Calyx oblique-gibbous at base, nutlets with curved areole, globose, nectar-disc horned **179. *Drepanocaryum***
- Calyx not gibbous at base, nutlets elliptic to oblong, areole not curved, nectar-disc lobed 28
28. Calyx annulate within, lobes broadly ovate, becoming patent, accrescent in fruit, calyx-tube funnel-shaped **181. *Hymenocrater***
- Calyx glabrous within, lobes linear to elliptic, not markedly accrescent in fruit 29
29. Sinuses between lobes of calyx with swollen folds at base, bracteoles aristately toothed 30
- Sinuses between lobes of calyx without swollen folds at base, bracteoles entire or dentate, rarely aristately toothed **178. *Nepeta***
30. Pedicels flattened, posterior lip of corolla with 2 longitudinal folds within **187. *Lallemantia***
- Pedicels not markedly flattened, posterior lip of corolla smooth within **186. *Dracocephalum***
31. At least the upper leaves trifoliolate, pinnate or bipinnate 32
- All leaves simple 33
32. Leaves trifoliolate (very rarely 5-foliolate), calyx-lobes 3-nerved, spinose, stamens all parallel, arcuate-ascending under upper lip of corolla, nutlets glandular-pubescent **188. *Cedronella***
- Leaves pinnate or bipinnate, rarely entire, calyx-lobes acute to acuminate, not as above, anterior pair of stamens not arcuate-ascending, nutlets smooth **189. *Schizonepeta***
33. Stems densely leafy at apex, flowers in axils of upper leaves, calyx 2-lipped; high alpine scree plants **182. *Marmoritis***
- Stems leafy throughout, flowers in axils of middle and upper leaves or forming a terminal inflorescence 34
34. Anther-thecae divaricate at 90°, plants stoloniferous, flowers 1–2.5 cm long, in axils of middle and upper leaves **185. *Glechoma***
- Anther-thecae parallel, plants not stoloniferous, if stoloniferous, then flowers 3–5 cm long 35
35. Plants stoloniferous, corolla abruptly expanded in distal part of tube, anther-surface papillate **184. *Meehania***
- Plants not stoloniferous, corolla gradually expanded in distal part of tube, anther-surface smooth 36
36. Corolla resupinate, calyx annulate within, nutlets glabrous **180. *Lophanthus***
- Corolla not resupinate, calyx glabrous within, nutlets pubescent **183. *Agastache***
37. Anthers with spur-like apical appendage **169. *Dicerandra***
- Anthers without appendage 38
38. Filament minutely denticulate or bifurcate with subulate or shortly conical appendage near apex 39
- Filament entire at apex, without appendage 41
39. Annual herb with pectinate-pinnatifid, ciliate bracts, stigma-lobes 4 **176. *Cleonia***
- Annual or perennial herbs, without bracts as above, stigma-lobes 2 40
40. Posterior corolla-lip galeate, appendage of stamen-filament subulate or conical, perennial or annual herbs, leaves not in a basal rosette **175. *Prunella***
- Posterior corolla-lip ± flat, stamen-filament minutely denticulate, perennial herb with leaves in a basal rosette **177. *Horminum***
41. Corolla resupinate, corolla-tube twisted 42
- Corolla not resupinate 43
42. Corolla up to 3 mm long, stamens included or only slightly exserted, slender annual **146. *Micromeria***
- Corolla >6 mm long, stamens long-exserted, suffrutescent perennial **151. *Cyclotrichium***
43. Bracteoles conspicuous, broad, often scarious and spinose-toothed, at base of cymes **172. *Acanthomintha***
- Bracteoles or bracts sometimes conspicuous, but not as above 44
44. Inflorescence a compact, terminal thyrse, with a conspicuous involucre of broad, often coloured or scarious bracts at base, corolla-lobes narrow, linear, often longer than corolla-tube **170. *Monardella***
- Bracts and bracteoles not forming a distinct, conspicuous involucre, corolla-lobes not as above 45
45. Corolla-tube distinctly upcurved in middle and ampliate above 46
- Corolla-tube straight or weakly curved 47
46. Calyx-throat densely hairy, anther-thecae parallel, leaves ericoid, entire, narrow, subsessile **168. *Conradina***
- Calyx-throat subglabrous, anther-thecae divergent, leaves broad, toothed, petiolate **190. *Melissa***
47. Inflorescences of usually elongate to subglobose spicules, with imbricate, sometimes coloured, usually entire bracts, very different from leaves, subtending, and usually obscuring the usually 1-flowered cymes of subsessile flowers, calyx 1- or 2-lipped to almost regular, with lobes never spreading or setaceous **140. *Origanum***
- Inflorescences not as above, usually distinctly cymose and often forming either a terminal thyrse, dense or often interrupted below, of many- to few-flowered verticillasters, or these all subtended by leaf-like bracts, flowers usually distinctly pedicellate, with calyx-lobes often spreading 48
48. Calyx distinctly 2-lipped 49
- Calyx ± actinomorphic or weakly 2-lipped 56
49. Dwarf annuals with elongate, linear to spathulate bracteoles, often incurved and with margins white-ciliate, inflorescence a usually congested, terminal, spiciform or capitate thyrse. Seasonally flooded areas of western N America **158. *Pogogyne***
- Annuals or perennials, sometimes shrubby, otherwise not as above 50
50. Calyx-tube strongly curved, often gibbous at base **150. *Clinopodium***
- Calyx-tube straight, or at most only lightly curved, never gibbous 51
51. Leaves short, ericoid, glabrous above and strongly replicate, obscuring abaxial surface, flowers solitary, anterior lip of corolla almost as long as corolla-tube. Low shrubs from Colombia **152. *Obtegomeria***
- Leaves not ericoid nor strongly replicate, or if appearing so, flowers not solitary or anterior corolla-lip much shorter than tube 52
52. Calyx 15-nerved, mid-lobe of posterior calyx-lip broad and slightly decurrent at base, posterior corolla-lip with 4 short, rounded lobes, stamens long-exserted **191. *Heterolamium***

- Calyx 10–13-nerved, rarely 20–22-nerved, posterior calyx-lip not decurrent at base, posterior corolla-lip 2-lobed or 1-lobed and often emarginate, stamens exserted beyond posterior corolla-lip or not 53
53. Corolla subequally 4-lobed, posterior lobe sometimes emarginate, herbs of damp places **137. *Mentha***
- Corolla distinctly 2-lipped and 5-lobed 54
54. Calyx-tube with two dorsal or lateral, ciliate flanges, leaves conduplicate when young **138. *Thymbra***
- Calyx-tube without two dorsal or lateral, ciliate flanges, leaves not conduplicate 55
55. Flowers rather shortly pedicellate, in verticillasters forming a congested, terminal capitate or oblong thyrse, lower verticillasters sometimes distant, leaves small, usually entire, elliptic to linear, sometimes with margins recurved, and/or leaf-base long-ciliate **139. *Thymus***
- Flowers often long-pedicellate, in lax verticillasters or shortly pedunculate cymes, sometimes forming a lax, terminal thyrse. Leaves various, often toothed, not long-ciliate towards base **150. *Clinopodium***
56. Cauliflorous shrub with lax cymes below the leaves, hairs simple and dendroid, calyx-tube tinged blue, broadly campanulate, accrescent in fruit. Mexico and Guatemala **134. *Chaunostoma***
- Shrubs, subshrubs or herbs, not cauliflorous, otherwise not as above 57
57. Small shrub with small, glandular-punctate, sessile, canaliculate leaves, long-ciliate towards base, flowers minute, calyx urceolate with broadly deltate lobes, strongly accrescent, globose-urceolate and inflated in fruit. Algeria **145. *Saccocalyx***
- Calyx not globose-urceolate, or if calyx accrescent and inflated in fruit, leaves not as above and never canaliculate 58
58. Calyx 15-nerved, nerves prominent, with small, thickened fold at sinus between narrowly acuminate to aristate lobes, stamens long-exserted **136. *Hyssopus***
- Calyx variously, often less than 15-nerved, without thickened fold at sinus, lobes various 59
59. Calyx-throat glabrous or rarely calyx-teeth shortly tomentulose within, never annulate 60
- Calyx-throat annulate or barbate within 70
60. Flowers arranged in pedunculate, compound cymes in the axils of, and shorter than the leaf-like bracts, corolla annulate within, stamens included, disc orange. Robinson Crusoe Archipelago (Chile) **148. *Cuminia***
- Flowers variously arranged but cymes not clearly pedunculate or, if so, not shorter than the bracts subtending them, disc not orange 61
61. Low, ericoid shrubs with short, sessile, tightly revolute leaves and sessile flowers forming an ovoid-spiciform thyrse, anther-thecae of posterior stamens asymmetrically attached. SE United States **167. *Piloblephis***
- Not ericoid, or if so, flowers not as above, anther-thecae equal 62
62. Corolla ± actinomorphic, if weakly 2-lipped, anthers elongate to oblong 63
- Corolla distinctly 2-lipped, lobes markedly unequal, anthers shortly ellipsoid 64
63. Corolla small, subequally 4-lobed, posterior lobe sometimes emarginate, calyx-tube not strongly accrescent in fruit, stamens long-exserted in male-fertile flowers; nutlet abscission-scar areolate, herbs or rarely subshrubs of damp places **137. *Mentha***
- Corolla small to large, weakly 2-lipped, shortly 5-lobed, calyx-tube often strongly accrescent in fruit and becoming reticulate-veined, stamens exserted or included, nutlet abscission-scar minute, shrubs or herbs **133. *Lepechinia***
64. Leaves with isolateral palisade parenchyma, glabrous, elliptic to linear-oblong, distinctly gland-dotted on abaxial surface. Asia **144. *Gontscharovia***
- Leaves with dorsiventral palisade parenchyma, leaves various 65
65. Densely canescent subshrub with short, dendroid hairs and small, elliptic, glandular-punctate leaves with crenulate margins, flowers hidden among the leaf-like bracts in sessile, 3-flowered cymes **159. *Neoeplingia***
- Not as above 66
66. Bracteoles narrow, forming an involucre at base of cymose clusters, these often capitate, usually in corymbose panicles, rarely sessile in axils of leaf-like bracts **171. *Pycnanthemum***
- Bracteoles not forming an involucre, cymes not forming corymbose panicles 67
67. Leaf-margins with distinct marginal vein, usually revolute, entire. Old World **146. *Micromeria*****
- Leaf-margins without a marginal vein, revolute or flat, entire or toothed 68
68. Corolla over 15–35 mm long, distinctly 2-lipped with long, spreading lobes, calyx-lobes spine-tipped, spreading, anther-thecae elongate, widely divaricate, curved. California **150. *Clinopodium*****
- Corolla usually much smaller, if over 15 mm, less distinctly 2-lipped with short, porrect lobes, calyx-lobes short, not spine-tipped nor spreading 69
69. Calyx narrowly cylindrical, often widest below the throat, lobes usually narrowly triangular to subulate, corolla rather small, at least the posterior lip spreading. Old World **146. *Micromeria*****
- Calyx widening gradually towards mouth, sometimes shortly tubular-campanulate with shortly deltate to narrowly triangular lobes, corolla small, or if large, then calyx long-cylindrical with short teeth and corolla-tube elongate, curved with short, porrect lobes, sometimes brightly coloured. New World **150. *Clinopodium*****
70. Leaves conduplicate at least when young 71
- Leaves flat or with margin recurved 72
71. Leaves linear to oblanceolate, strongly gland-dotted, young leaves often in axillary fascicles, veins on abaxial surface obscure, calyx-tube short. Old World **143. *Satureja***
- Leaves broadly to narrowly ovate, not in axillary fascicles, veins evident on abaxial surface, calyx-tube elongate. South America **150. *Clinopodium*****
72. Leaves with a prominent marginal vein, margin often recurved, entire **146. *Micromeria*****
- Leaves without a prominent marginal vein, margin various 73
73. Leaves small, orbicular to elliptic, entire, glandular-punctate, stems white-pilose, axillary buds globose, conspicuously white-tomentose, flowers very small **141. *Zataria***
- Leaves and stems not as above, axillary buds not conspicuous, flowers larger 74
74. Calyx prominently 5-angled, intercostal nerves inconspicuous **142. *Pentapleura***
- Calyx 10–15-nerved, not angled 75
75. Stamens ascending under posterior corolla-lip, sometimes exserted 80

** The delimitation of *Micromeria* and *Clinopodium* is still unclear, and the characters here used are provisional.

- Stamens divergent, included or exserted 76
76. Herbs or rarely shrubs, usually of damp places, corolla subequally 4-lobed, posterior lobe sometimes emarginate, flowers in few to many-flowered verticillasters in the axils of leaf-like bracts **137.** ***Mentha***
- Herbs or shrubs, corolla 5-lobed, 2-lipped, sometimes weakly so, flowers not arranged in axillary verticillasters 77
77. Mat-forming perennial herbs with procumbent stems with flowers in 1- to 2-flowered cymes, calyx-lobes curved, long-subulate, plumose. Temperate S America **153.** ***Kurzamra***
- Erect shrubs or robust herbs, calyx-lobes subulate or triangular, rarely setaceous and plumose 78
78. Leaves 3 mm wide or less, corolla distinctly 2-lipped **139.** ***Thymus***
- Leaves over 5 mm wide, corolla weakly 2-lipped 79
79. Calyx-tube prominently 10-nerved, flowers in corymbiform or globose cymes. Macaronesia **147.** ***Bystropogon***
- Calyx-tube (12–)13-nerved, cymes often congested into a terminal spiciform thyrse. S America **149.** ***Minthostachys***
80. Leaf-margins with distinct marginal vein, usually revolute, entire. Old World **146.** ***Micromeria*****
- Leaf-margins without a marginal vein, revolute or flat, entire or toothed 81
81. Corolla over 15–35 mm long, distinctly 2-lipped with long, spreading lobes, calyx-lobes spine-tipped, spreading, anther-thecae elongate, widely divaricate, curved. California **150.** ***Clinopodium*****
- Corolla usually much smaller, if over 15 mm, less distinctly 2-lipped with short, porrect lobes, calyx-lobes short, not spine-tipped nor spreading 82
82. Calyx narrowly cylindrical, often widest below the throat, lobes usually narrowly triangular to subulate, corollas rather small, at least the posterior lip spreading. Old World **146.** ***Micromeria*****
- Calyx widening gradually towards mouth, sometimes shortly tubular-campanulate with shortly deltate to narrowly triangular lobes, corollas small, or if large, then calyx long-cylindrical with short teeth and corolla-tube elongate, curved with short, porrect lobes, sometimes brightly coloured. New World **150.** ***Clinopodium*****

VII.2.A. Subtribe Salviinae (Dumort.) Endl. (1838).

Perennial herbs or shrubs. Dendroid hairs sometimes present. Leaves often hastate or truncate with winged petiole, lamina often finely rugose, bullate. Flowers often with posterior stamens represented by staminodes or absent, thecae of anterior stamens distinct or reduced to 1, usually elongate, rarely ellipsoid, usually divergent or divaricate, connective sometimes elongate, stigma-lobes flattened, short. Nutlets with usually minute abscission-scar, innermost layer of pericarp usually with crystalline inclusions.

127. *Zhumeria* Rech. f. & Wendelbo

Zhumeria Rech. f. & Wendelbo, Nytt. Mag. Bot. 14: 39–43 (1967); Bokhari & Hedge, Iranian J. Bot. 1: 1–10 (1976), rev.

Shrub, strongly aromatic, white-canescent with glandular and eglandular simple and dendroid hairs; leaves small, simple, margin undulate; flowers shortly pedicellate, borne singly in axils of leaf-like bracts, bracteoles 2, inconspicuous; calyx 2-lipped, 5-lobed (3/2), lips subequal, posterior with lobes shorter, subequal, tube ± cylindrical at first, strongly accrescent, broadly campanulate in fruit, membranous, lobes spreading; corolla large, indistinctly 2-lipped, 5-lobed (2/3), blue-violet, throat and tube white; lips spreading, anterior lip with lobes rounded, tube cylindrical, included in calyx-tube, annulate within; anterior stamen pair fertile, exserted, posterior pair staminodal, shortly exserted, thecae 2, elongate, curved, slightly unequal, connective inconspicuous; style exserted, stigma-lobes flattened; disc not seen; nutlets obovoid, weakly subtrigonous, smooth, mucilaginous when wet. $2n = 40, 42$. One species, *Z. majdae* Rech. f. & Wendelbo. Open mountain slopes, S Iran (Saharo-Sindian region).

128. *Dorystoechas* Boiss. & Heldr.

Dorystoechas Boiss. & Heldr. ex Benth., as *Dorystaechas*, in DC. Prodr. 11: 261 (1848); Hedge, in Davis (ed.) Flora of Turkey 7: 461–462 (1982), rev.

Shrub, apparently gynodioecious, strongly aromatic, with simple hairs; leaves simple, hastate, rugose, crenulate, densely hairy beneath; inflorescence terminal, a compact, spiciform thyrse of many-flowered, sessile verticillasters; bracts lanceolate, caducous; bracteoles absent; flowers very shortly pedicellate; calyx ±3-lobed, 2-lipped (1/2), posterior lip sometimes obscurely 3-lobed, anterior lip shorter with triangular lobes, tube cylindrical, 8–11-veined, lips connivent in fruit; corolla small, 2-lipped, 5-lobed (2/3), white, lips spreading, anterior lip longer, with median lobe larger than laterals, tube short, exannulate; stamens 2, posterior pair absent, anterior pair weakly exserted, thecae 2, elongate, curved, slightly divergent; stigma-lobes subequal, slightly flattened; disc fleshy, reddish orange, with anterior lobe longer; nutlets elongate, triquetrous, apiculate, smooth, pale brown, abscission-scar small, elongate, not mucilaginous. One species, *D. hastata* Boiss. & Heldr. ex Benth. Montane, among rocks, SW Turkey.

129. *Meriandra* Benth.

Meriandra Benth., in Edwards's Bot. Reg. 15: t. 1282 (1829); Bokhari & Hedge, Notes Roy. Bot. Gard. Edinburgh 31 (1): 53–67 (1971), rev.

Shrub, gynodioecious (possibly dioecious in *M. strobilifera* Benth.), strongly aromatic, with simple and dendroid hairs (dendroid hairs glandular in *M. strobilifera*); leaves simple, hastate to rounded, rugose, margins serrulate to subentire; inflorescence terminal and axillary, a compact and 4-angled or interrupted spiciform thyrse of many-flowered, sessile verticillasters; bracts ovate-acuminate, caducous or persistent; bracteoles absent; calyx 2-lipped, ±3-lobed (1/2), posterior lip broader, subentire or obscurely 3-lobed, lobes connivent, anterior lip shorter, tube cylindrical, 9–11-veined, calyx erect or spreading in fruit, with lips open; corolla small, 2-lipped, 4–5-lobed (1–2/3), white or pale lilac, anterior lip usually longer than posterior, tube straight, annulate or exannulate; fertile stamens 2, posterior pair absent or staminodal (in *M. bengalensis* (Koenig ex Roxb.) Benth.), anterior pair exserted, except in female flowers, thecae 2, elongate, curved, divergent and often slightly unequal, connective inconspicuous; stigma-lobes unequal or subequal, flattened; disc 4-lobed; nutlets elongate, oblong-apiculate or obovoid-elliptic smooth, mucilaginous or not. Two species, *M. strobilifera* in N India, *M. bengalensis* in Arabia, Yemen, Ethiopia and cultivated in Indian subcontinent.

130. *Perovskia* Kar.

Perovskia Kar., Bull. Soc. Nat. Mosc. 15. t. 1 (1841).
Perowskia Benth. in DC. (1848).

Shrubs, strongly aromatic, often viscid with dendroid and simple hairs; leaves entire to laciniate or bipinnatisect; inflorescence terminal, often axillary, forming a paniculate thyrse of distant, 2- to 6-flowered verticillasters; bracts caducous, bracteoles inconspicuous; flowers ± sessile, sometimes dimorphic due to heterostyly; calyx indistinctly 5-lobed, 2-lipped (3/2), lips subequal, anterior with lobes short, deltoid, tube cylindrical, 10-nerved, often densely hairy; corolla 2-lipped, 5-lobed (4/1) violet-blue to white, posterior lip spreading to reflexed, anterior lip shorter, entire, acute, tube cylindrical, annulate within; stamens 2 with 2 staminodes, anterior stamens fertile, ± included in long-styled, or long-exserted in short-styled plants, declinate, thecae 2, elongate, parallel, connective small, staminodes borne on posterior lip of corolla; style included or exserted, declinate, stigma-lobes flattened; disc weakly 4-lobed, anterior lobe longest; nutlets elongate, rounded at apex, smooth, abscission-scar small, elongate, not mucilaginous. $2n = 20$. About seven species, SW Asia: Iran, Afghanistan, Turkmenistan, Uzbekistan, Kazakhstan, Kyrgyzstan, Pakistan, Kashmir, N India to Tibet.

131. *Salvia* L.

Salvia L., Sp. Pl. 1: 23 (1753); Benth., Lab. Gen. et Sp.: 190 (1833); Epling, Ann. Missouri Bot. Gard. 25, 1: 95–188 (1938), reg. rev.; Epling, Rep. Spec. Nov. Beih. 110: 1–383 (1938–1939), reg. rev.; Hedge, Notes Roy. Bot. Gard. Edinburgh 33: 1–121 (1974), reg. rev.; Hedge in Davis (ed.) Flora of Turkey 7: 400–461 (1982), reg. rev.
Schraderia Medik. (1791), nom. rejec.
Audibertia Benth. (1832) non *Audibertia* Benth. (1829).
Salviastrum Scheele (1849).
Polakia Stapf (1885).
Ramona Greene (1892).
Pycnosphace Rydb. (1918).
Arischrada Pobed. (1972).

Annual or perennial herbs or shrubs, often aromatic, often viscid with hairs simple, rarely dendroid; leaves simple, toothed, lobed, pinnatifid or pinnatisect, rarely spinulose; inflorescence terminal and/or axillary, often a paniculate, spiciform thyrse of distant or congested 1- to many-flowered verticillasters; bracts inconspicuous or larger and sometimes coloured, caducous or persistent, rarely spinulose, bracteoles present or not; calyx 3–5-lobed, 2-lipped (1 or 3/2), lobes rarely spinescent, tube cylindrical to campanulate, 9–15-nerved, often accrescent; corolla strongly 2-lipped, 4–5-lobed (1–2/3), white to yellow, or purple to bright red or blue, posterior lip often longer, strongly falcate, to almost straight or flattened, entire or emarginate, anterior lip often with broader median lobe, tube sometimes invaginate or ventricose below, or with paired "papillae" within, annulate or not; fertile stamens 2, with posterior pair vestigial or absent, anterior pair included to long-exserted, filaments usually short, thecae usually 1 by abortion, sometimes 2, connective elongate, often longer than filament, usually articulated at junction of filament, sometimes toothed, posterior arm ascending under corolla-lip, sometimes curved, bearing fertile theca, anterior arm bearing reduced subfertile theca, or this replaced by dolabriform or flattened callus, these sometimes connate; stigma-lobes usually unequal; disc 4-lobed, anterior lobe often larger; nutlets trigonous, ovoid or suborbicular, abscission-scar small, mucilaginous or not. $2n = 12, 14, 16, 18, 14, 20, 22,$

24, 26, 28, 30, 32, 34, 36, 38, 42, 44, 46, 48, 60, 66, 84, 86, 240. At least 900 species, open to forest and montane habitats, almost cosmopolitan with main centres of diversity in SW Asia and Central and South America. Especially the Mediterranean *S. officinalis* L. widely used as potherb, also many ornamentals, such as *S. microphylla* Kunth and *S. splendens* Sellow & Nees from the New World.

The infrageneric classification proposed, especially by Briquet (1897), largely based on staminal structure, is outdated, although the neotropical subgenera *Calosphace* Benth. and *Audibertia* (Benth.) Epling appear to be monophyletic. Pobedimova (1972) recognised the segregate genus *Arischrada* Pobed. for species with two fertile anther-thecae, but the separation seems unjustified. Hedge (1974, 1982) has informally recognised various smaller groups of closely related species.

132. *Rosmarinus* L.

Rosmarinus L., Sp. Pl. 1: 23 (1753).

Shrubs, sometimes gynodioecious, strongly aromatic, with simple and dendroid hairs; leaves linear to lanceolate, revolute, entire, sessile; inflorescences as short, 5–15-flowered racemes in axils of upper leaves; bracts inconspicuous, bracteoles absent; flowers pedicellate; calyx 2-lipped, 3 or 5-lobed (1–3/2), campanulate to infundibuliform, widening above, posterior lip curved upwards, triangular, tube 11–12-nerved, accrescent; corolla strongly 2-lipped, 5-lobed (2/3), violet to white, posterior lip with lobes spreading, anterior lip with lateral lobes, shorter, median lobe rounded, concave, toothed, tube short, exannulate; stamens 2, posterior pair absent or rudimentary, anterior pair long-exserted, ascending, much shorter in male-sterile flowers, anthers elongate, with single fertile theca separated by an elongate, downward-curving connective from the sterile theca represented by small, tooth-like appendage, connective not articulating with filament and appearing continuous; style curved, exserted, stigma-lobes short, unequal; disc 4-lobed, anterior lobe largest; nutlets ovoid, smooth, abscission-scar large with central pore, not mucilaginous, pseudostrophiolate. $2n = 24$. Three species, with natural hybrids, in the Mediterranean region and Caucasus. *Rosmarinus officinalis* L. has been widely cultivated since antiquity as an ornamental. The genus is scarcely separable from *Salvia*.

133. *Lepechinia* Willd.

Lepechinia Willd, Hort. Berol 1: 21, t. 21 (1803); Epling, Brittonia 6: 352–364 (1948); Hart, Ph.D. Diss., Harvard, Cambridge, Mass., U.S.A. (1983), rev.
Sphacele Benth. (1829).

Perennial herbs to shrubs, rarely gynodioecious, or dioecious (sect. *Parviflorae* Epling; see Hart 1985); often aromatic; leaves entire to toothed, often rugose; inflorescence terminal and often axillary, a paniculate thyrse of lax, (1–)3- to many-flowered cymes, or spiciform of ± congested, usually sessile, 6- to many-flowered verticillasters; bracts usually small, bracteoles inconspicuous; flowers large to very small; calyx ± actinomorphic to weakly 2-lipped, 5-lobed (3/2), lobes deltoid to subulate or aristate, tube campanulate, 10–15-nerved, often accrescent and inflated, lobes spreading to connivent, throat exannulate; corolla weakly 2-lipped, 5-lobed (2/3), white, yellowish, purplish to red or blue, lobes often subequal, or anterior lobe larger, tube cylindrical to campanulate, sometimes shorter than calyx, usually annulate within; stamens 4, didynamous, included to long-exserted, anterior pair usually longer, ascending, thecae 2, divergent to divaricate, elongate to oblong, sometimes confluent; stigma-lobes short, subequal; disc 4-lobed, nutlets ovoid, smooth, brown or black, abscission-scar minute, not mucilaginous. $2n = 32, 66$. About 40 species, often montane, Andes from Argentina and Chile northwards to Mexico and U.S.A. (California), with outlying species in Brazil and Hawaii, the Chilean *L. chamaedryoides* (Balbis) Epling naturalized in Réunion Is. Some species used medicinally or as ornamentals. Epling (1948) recognised eight sections, based mainly on inflorescence structure and floral morphology. Hart (1983) recognised two: sect. *Lepechinia* and sect. *Parviflorae*, considering the former to be polyphyletic.

134. *Chaunostoma* Donn. Sm.

Chaunostoma Donn. Sm., Bot. Gaz. 20: 9, t. 3 (1895).

Shrub, cauliflorous, densely hairy, with dendroid and simple hairs; leaves bullate, serrate, petioles unwinged; inflorescences axillary, borne on bare stems below leaves, forming short, racemoid thyrses, cymes single-flowered; bracts caducous, bracteoles inconspicuous; flowers medium-sized, pedicellate; calyx actinomorphic to weakly 2-lipped, 5-lobed, bright blue, lobes subequal, deltoid, tube broadly campanulate, membranous, accrescent; corolla 2-lipped, 5-lobed (4/1), ?red, posterior lip with 2 median lobes, separated by short sinus, lateral lobes rounded, almost obsolete, anterior lip rounded, tube short, exannulate; stamens 4, didynamous, ascending, long-exserted, filaments curved, thecae 2, strongly divergent, confluent, elongate; stigma-lobes short, subequal; disc fleshy, scarcely lobed; nutlets ovoid, smooth, black, with minute basal scar. One species: the rarely collected *C. mecistandrum* Donn. Sm. is known only from S Mexico (Chiapas) and Guatemala.

VII.2.B. Subtribe Menthinae (Dumort.) Endl. (1838).

Perennial or annual herbs or shrubs. Hairs simple or rarely branched. Leaves various, rarely hastate, petiole unwinged, lamina rarely finely rugose. Flowers with calyx-tube 5- to 15-nerved; stamens 4, anterior pair usually longer than posterior, or stamens divergent, subequal, or stamens 2 with posterior pair reduced to staminodes or absent; stamens usually 2-thecous, thecae usually ellipsoid, distinct to confluent, parallel to divaricate, connective often conspicuous; nectary-disc entire or often 4-lobed, with lobes usually subequal. Nutlets often areolate.

135. *Lycopus* L.

Lycopus L., Sp. Pl. 1: 21 (1753); L., Gen. Pl. ed. 5: 12 (1754); Henderson, Am. Midl. Nat. 68: 95–138 (1962), rev.

Perennial herbs, with horizontal rhizomes or runners, sometimes tuberiferous, weakly or not aromatic, with simple hairs; leaves opposite, petiolate or sessile, ovate to lanceolate or linear, variously toothed or pinnatifid; inflorescence of tight, axillary verticillasters; flowers sessile to subsessile; calyx actinomorphic to weakly zygomorphic, not accrescent, tubular-campanulate to campanulate, 4–5-nerved, 4–5-lobed, lobes ± equal, ovate to lanceolate-subulate or aristate, throat glabrous; corolla white, usually weakly zygomorphic, occasionally ± actinomorphic or 2-lipped, lobes usually 4, posterior and sometimes anterior lobes larger than lateral ones, all entire or posterior lobe bifid, tube straight; stamens 2 (posterior pair reduced to staminodes or absent), exserted or included, filaments glabrous, thecae parallel to divergent, separate at dehiscence; style often expanded near base to form flange covering top of ovary, stigma-lobes equal or unequal; disc symmetrical; nutlets obovoid-tetrahedral, often with conspicuously toothed to tuberculate apical crest, glabrous, inner and upper faces often densely glandular. $2n = 22$. About 14 species of wet habitats, mostly in the north-temperate zone (one species in Australia).

136. *Hyssopus* L.

Hyssopus L., Sp. Pl. 2: 569 (1753).

Perennial subshrub, aromatic; leaves simple, entire, usually glabrous; inflorescence of 6–many-flowered verticillasters merging to form a terminal, usually ± secund, spiciform thyrse; bracts leaf-like; flowers very shortly pedicellate, bracteolate; calyx 2-lipped (3/2), ± tubular, 15-nerved, lobes equal, deltoid, acuminate or aristate, ciliate, with thickened fold at each sinus; corolla 2-lipped, 5-lobed (2/3), violet to whitish, posterior lip deeply emarginate, concave, patent, anterior lip with median lobe ± suborbicular, emarginate, curved downwards, lateral lobes rounded and smaller, tube cylindrical; stamens 4, inserted in middle of tube, curved, exserted, thecae divergent, confluent, connective ± triangular; stigma-lobes equal, ± acicular; nutlets ± ovoid, smooth to papillose, puberulent, with mucilaginous cells. $2n = 12$. Two species in Europe, NW Africa, Central and SW Asia, including China.

137. *Mentha* L.

Mentha L., Sp. Pl. 2: 576 (1753); Harley, Flora Europaea 3: 183–186 (1972), reg. rev.; Harley, in P.H. Davis (ed.) Flora of Turkey 7: 384–394 (1982), reg. rev.
Pulegium Mill. (1753).
Preslia Opiz (1824).

Perennial, rarely annual herbs, subshrubs or shrubs, usually gynodioecious, strongly aromatic; stems often rhizomatous; inflorescence of few- to many-flowered axillary cymes, pedunculate or not, often forming either dense verticillasters along stem, in axils of leaf-like bracts, or congested into terminal spiciform thyrses in axils of reduced bracts, bracteoles inconspicuous, rarely broader or digitately lobed; calyx ± actinomorphic, 5-(rarely 4-)lobed, lobes subequal, triangular to subulate or acicular, or slightly 2-lipped, (3/2), anterior lobes often longer, tube cylindrical to infundibuliform, 10–15-nerved, throat hairy or glabrous; corolla ± actinomorphic, 4-lobed, lilac, pink or white, lobes spreading, subequal, or posterior lobe slightly broader, often emarginate, corolla-tube ± cylindrical or rarely gibbous on anterior side, usually included within calyx-tube; stamens 4, subequal, exserted, divergent (included in male-sterile plants), anthers ellipsoidal, thecae parallel, distinct; stigma-lobes subequal, shortly spreading; disc ± symmetrical; nutlets ovoid, smooth to rugose, pale to dark brown, not mucilaginous; $2n = 18, 20, 24, 36, 40, 42, 48, 54, 60, 66, 72, 78, 84, 90, 96, 108, 120, 132$. About 20 species in damp, open habitats, often by water, especially Mediterranean and Australasian (one species in New Zealand), but extending through Europe, Asia and Africa to the Cape, and with one species in North America. Some species cultivated since antiquity as potherbs and now introduced worldwide. Hybrids are of frequent occurrence in sect. *Mentha* (Harley and Brighton 1977). These can persist and spread widely by vegetative means. *Mentha spicata* L. and its hybrids, including *M.* × *piperita* L. and, in the subtropics, *M. canadensis* L., are of commercial importance. No modern infrageneric classification has yet been published, while previous systems such as those by Briquet (1897) are unsatisfactory.

138. *Thymbra* L.

Thymbra L., Sp. Pl. 2: 569 (1753); Morales, An. Jard. Bot Madrid 44, 2: 349–380 (1987), rev.
Coridothymus Rchb. f. (1857).

Subshrubs, aromatic; leaves entire, usually conduplicate, young leaves in axillary fascicles; inflorescence a condensed, terminal, spiciform or capituliform thyrse, composed of congested verticillasters; bracts not leaf-like, ovate-lanceolate or lanceolate, sometimes coloured, bracteoles elliptic to lanceolate; calyx 2-lipped, 5-lobed (3/2), cylindrical, with two ± conspicuous lateral keels, scarcely accrescent, 12–23-nerved, posterior lobes short, triangular, anterior lobes lanceolate, upcurved, throat bearded; corolla 2-lipped, 4-lobed (1/3), cream to purple, posterior lobe ± rounded, straight, emarginate, anterior lip with lobes suborbicular, median lobe downcurved; stamens 4, inserted in middle of tube, filaments curved, exserted, thecae divaricate, confluent; stigma-lobes equal to subequal; disc entire; nutlets ovoid, ± smooth. $2n = 30$. Four species, Mediterranean region, Anatolia and N Iraq.

139. *Thymus* L.

Thymus L., Sp. Pl. 2: 590 (1753); Jalas & Kaleva, Feddes Rep. 81: 93–106 (1970); Jalas, Bot J. Linn. Soc. 64: 247–271 (1971), reg. rev.; Morales, Ruizia 3: 1–324 (1986), reg. rev.

Subshrubs or shrubs, often herbaceous above, usually gynodioecious, aromatic; stems usually ± quadrangular, hairy all round, on two opposite sides or only on angles; leaves small, usually entire, frequently revolute; inflorescence of verticillasters forming a terminal, condensed, often spiciform or interrupted thyrse; bracts leaf-like or not, sometimes coloured, bracteoles inconspicuous; calyx 2-lipped, 5-lobed (3/2), sometimes nearly actinomorphic, ± campanulate or cylindrical, 10-nerved, lobes triangular or setaceous (in *T. mastichina* L.), rarely posterior lip entire, anterior lip upcurved or spreading, throat bearded; corolla 2-lipped, or rarely nearly actinomorphic, 4-lobed (1/3), white, cream, pink or violet, posterior lip ± rounded, emarginate, straight, anterior rectangular to suborbicular, rounded, spreading, tube ± cylindrical, sometimes very long (to 20 mm); stamens 4, exserted or not, thecae parallel, distinct; stigma-lobes ± equal; nutlets ovoid, smooth. $2n = 24, 26, 28, 30, 32, 42, 48, 50, 52, 54, 56, 58, 60, 84, 90$. About 220 species, open areas, Europe, Mediterranean region and Macaronesia, Asia, N Africa and mountains of Ethiopia, S Greenland. Aromatic, ornamental and medicinal plants used as flavouring, source of essential oils and honey. Many hybrids have been recorded. Sectional classification after Jalas (1971): *Micantes* Velen., *Mastichina* (Mill.) Benth., *Piperella* Willk., *Pseudothymbra* Benth., *Thymus*, *Teucrioides* Jalas, *Hyphodromi* (A. Kerner) Halácsy and *Serpyllum* (Mill.) Benth.

140. *Origanum* L.

Origanum L., Sp. Pl. 2: 588 (1753); Ietswaart, Leiden Bot. Ser. 4: 1–153 (1980), rev.
Majorana Mill. (1754).
Amaracus Hill (1756).
Dictamnus Hill. (1756).

Subshrubs or perennial herbs, sometimes gynodioecious, aromatic, branched hairs in *O. dictamnus* L.; leaves rather small, entire or toothed, hairy or glabrous, rarely glaucous; inflorescence branched, sometimes paniculate to corymbose, spicate or spiciform, spikes elongate to subglobose, erect or nodding, often pedunculate, sometimes lax or few-flowered, or compact with bracts imbricate, often showy or brightly coloured; bracteoles absent; flowers subsessile, single, or rarely in 2–8-flowered cymes in axils of all but lowest bracts; calyx actinomorphic and 5-lobed to strongly zygomorphic and 5- to 1-lobed: either 2-lipped, 3/2 to 1/2, lobes broadly triangular to lingulate, sometimes reduced, or calyx 1-lipped with posterior lip entire, rounded, bract-like with calyx-tube split almost to base, calyx-tube, when present, cylindrical to campanulate, 10–13-nerved, pilose in throat; corolla 2-lipped, usually 5-lobed (2/3), white, pink or purple, lobes subequal, tube straight or curved, sometimes ventrally saccate, short or to 3 cm in *O. amanum* Post; stamens 4, sometimes reduced or absent, usually unequal, ascending to divergent, anterior longer, usually exserted, posterior exserted to included, thecae divergent to divaricate, distinct; stigma-lobes subequal; disc ± symmetrical; nutlets ovoid to apiculate, smooth. $2n = 30$ (32). About 40 species, especially Mediterranean, extending through temperate Eurasia and N Africa, from Azores to Taiwan. Cultivated widely as culinary and medicinal herbs, especially *O. vulgare* L., *O. onites* L. and *O. majorana* L., some ornamental. Many hybrids recorded. Ietswaart (1980) recognises 10 sections.

141. *Zataria* Boiss.

Zataria Boiss., Diagn. Pl. Or. Nov. ser. 1, 5: 18 (1884).

Subshrub, apparently gynodioecious, aromatic, with whitish hairs; stems densely white-pilose; leaves small, densely gland-dotted, subentire, ± orbicular to elliptic, often with densely white-hairy, rounded buds in axils; inflorescence of few-flowered verticillasters congested to form short, spiciform thyrses, usually shortly pedunculate in axils of leaf-like bracts, reduced towards stem-apex; bracteoles inconspicuous; flowers very small, subsessile; calyx ovoid-cylindrical, actinomorphic, equally 5-lobed, lobes short, broadly triangular, hairy within, tube 5-nerved, intercostal nerves ± obsolete, 5-angled, glands on calyx and bracteoles turning reddish orange; corolla weakly 2-lipped, 5-lobed (2/3), white, lips spreading, lobes subequal, corolla-tube scarcely exserted from calyx, exannulate but hairy within; stamens in hermaphrodite flowers slightly exserted from corolla-tube, thecae slightly divergent, distinct; disc ± equal, with anterior side slightly swollen; stigma-lobes short, flattened, subequal; nutlets ovoid, smooth, often only 1 developing. One species, *Z. multiflora* Boiss. Open, rocky areas, often near seasonal streams, S Iran, E Afghanistan, Pakistan.

142. *Pentapleura* Hand.-Mazz.

Pentapleura Hand.-Mazz., Öst. Bot. Zeitschr. 63: 225 (1913); Rechinger, Flora Iranica 150: 553 (1982), reg. rev.; Davis, Flora of Turkey 7: 313 (1982), reg. rev.

Small, fastigiate shrub, aromatic; leaves simple, entire, subsessile; inflorescence of 2–8-flowered axillary cymes, shortly pedunculate or sessile, forming dense verticillasters on upper part of stem, in axils of leaf-like bracts, often somewhat congested; bracteoles inconspicuous; flowers subsessile; calyx long-cylindrical, actinomorphic, equally 5-lobed, lobes subulate, tube 10-nerved, 5-angled, 5 main nerves along crest of narrow wings, alternating with 5 weaker nerves in lower half of membranous intercostal furrows, throat hairy within; corolla ± actinomorphic to weakly 2-lipped, 5-lobed (2/3), pink, lobes subequal, corolla-tube cylindrical, very narrow, slightly longer than calyx, exannulate; stamens 4, subequal, included in corolla, filaments straight, anterior slightly longer, anthers ellipsoidal, thecae ± parallel, distinct; stigma-lobes subequal; nutlets oblong, glabrous. One sp., *P. subulifera* Hand.-Mazz., dry limestone slopes, 400–1000 m, N Iraq and SE Turkey.

143. *Satureja* L.

Satureja L., Sp. Pl. 2: 567 (1753); Doroszenko, Doct. Thesis, Edinburgh University: 175–205 (1985), rev.
Euhesperida Brullo & Furnari (1979).

Perennial or annual herbs or subshrubs, aromatic; leaves entire or scarcely toothed, usually conduplicate, glandular-punctate, young leaves in axillary fascicles; inflorescence of 2–many-flowered axillary verticillasters or cymes; bracts usually leaf-like, bracteoles inconspicuous; calyx 2-lipped or almost actinomorphic, campanulate, 10–13-nerved, rarely appearing 5-nerved, 5-lobed (3/2), posterior lobes straight to upcurved, anterior lobes longer, calyx-throat bearded; corolla 2-lipped, 4-lobed, white, lilac, pink or pale-purple; posterior lobe concave, ± rounded, emarginate, anterior lip with median lobe oblong, spreading, tube ± cylindrical, straight; stamens 4, scarcely didynamous or anterior pair longer, filaments exserted, curved; thecae ± divaricate or parallel, ellipsoidal, confluent; disc subequal; stigma-lobes subequal; nutlets ovoid, sometimes elongate or apiculate, smooth or muricate, sometimes hairy. $2n = 30$. About 38 species in the Mediterranean Region, N Africa: Morocco and Libya, Caucasus and W Asia: Saudi Arabia, Iran and Iraq. Aromatic and medicinal plants used as flavouring, source of essential oils and ornamental. Divided into two sections: *Satureja* and *Salzmannia* G. López, following López (1981).

144. *Gontscharovia* Boriss.

Gontscharovia Boriss., Bot. Mater. Gerb. Bot. Inst. Komarova Akad. Nauk. SSSR 15: 321–324 (1953).

Low subshrub, aromatic; leaves elliptic to linear-oblong, flat, entire, gland-dotted especially on abaxial surface, subsessile, palisade layer isolateral; inflorescence of verticillasters, cymes shortly pedunculate, 1–6-flowered, in axils of reduced bracts and forming a lax, terminal, spiciform thyrse; bracteoles inconspicuous; calyx ± actinomorphic, 5-lobed, lobes equal, triangular-lanceolate, calyx-tube cylindrical-campanulate, straight, 15-nerved, glabrous in throat; corolla pale pink, 2-lipped, 5-lobed (2/3), posterior lip straight, flat, anterior lip spreading, 3-lobed, equalling or slightly shorter than posterior, corolla-tube exannulate; stamens 4, didynamous, filaments straight, rather short, anterior pair longer and slightly exserted from tube, ascending under posterior lip of corolla, thecae slightly divergent, separate; stigma-lobes equal, subulate, nutlets oblong, apiculate, glabrous. One species, *G. popovii* (Fedsch. & Gontsch.) Boriss. Pamir-Alai: Uzbekistan, Tajikistan, Kazakhstan, Afghanistan, Pakistan (Chitral), India: Kashmir. Dry limestone slopes.

145. *Saccocalyx* Coss. & Durieu

Saccocalyx Coss. & Durieu, Ann. Sci. Nat. Bot. Ser. 3, 20: 80 (1853), nom. cons.

Small shrub, strongly aromatic; older stems often decumbent and rooting at base, with longitudinally fissured bark; leaves small, remote, ± spathulate, sessile, entire, canaliculate, conspicuously gland-dotted, margin ciliate with long, white hairs in lower half, cauline leaves with axillary fascicles of smaller leaves; inflorescence of verticillasters, cymes sessile, 1–3-flowered in axils of leaf-like bracts; bracteoles inconspicuous; flowers minute; calyx urceolate, actinomorphic, 5-lobed, lobes broadly deltate, slightly spreading, calyx-tube ca. 13-nerved with inconspicuous nerves, densely spreading-hairy outside, annulate at throat within, tube strongly accrescent in fruit, becoming globose-urceolate, inflated; corolla purplish-pink, ± actinomorphic, very small, 5-lobed, lobes subequal, erect, corolla-tube shortly campanulate-cylindrical, glabrous within; stamens 4, subequal, ± included within corolla, filaments glabrous, very short, thecae weakly divergent, separate, connective small; stigma-lobes slightly unequal, posterior lobe shorter; disc with elongate anterior lobe; nutlets oblong, apiculate, smooth, mucilaginous when wetted. One species, *S. satureioides* Coss. & Durieu, Algeria. Sandy margins of salt lakes, in desert.

146. *Micromeria* Benth.

Micromeria Benth., Edwards's Bot. Reg. 15: t. 1282 (1829); Perez de Paz, Monografías Secc. 4, Ciéncias Naturales 16: 1–306 (1978), reg. rev.; Morales, Bot. Complutensis 18: 157–168 (1993), reg. rev.

Perennial or annual (*M. cymuligera* Boiss. & Haussk.) herbs, subshrubs or shrubs, often aromatic; leaves entire or toothed, linear to ovate, flat or revolute; inflorescence in axillary verticillasters, 2–20-flowered, or sometimes flowers solitary in leaf-axils, bracts ± similar to leaves, bracteoles often present; flowers (possibly resupinate in *M. cymuligera*) pedicellate or not, pedicels sometimes very long; calyx actinomorphic to weakly 2-lipped, campanulate or cylindrical, scarcely accrescent, 13–15-nerved, 5-lobed, lobes ± equal or (3/2), anterior often longer, sometimes curved, spreading, throat bearded or not; corolla usually small, white to yellowish, mauve, pink or purple, 2-lipped, 4-lobed (1/3), posterior lip emarginate, ± straight, anterior lip spreading, tube cylindrical, sometimes hairy in throat; stamens 4, didynamous, included or exserted from tube, filaments straight or connivent, thecae divaricate, ellipsoidal, confluent; stigma-lobes subequal, subulate; nutlets ovoid, sometimes apiculate, ± smooth, sometimes hairy, mucilage cells present. $2n = 20, 22, 30, 50, 60$. About 70 species, Europe, Mediterranean region, Macaronesia, N Africa, Asia, mountains of E and S Africa, and W tropical Africa, Madagascar. Aromatic and medicinal plants used as flavouring and source of essential oils. Of the six sections recognised by Morales (1993), only four are included in *Micromeria* as delimited here: *Pseudomelissa* Benth., *Micromeria*, *Pineolentia* P. Pérez, *Cymularia* Boiss. The supposedly resupinate nature of the flower of *M. cymuligera* has been questioned.

147. *Bystropogon* L'Hér.

Bystropogon L'Hér., Sert. Angl. 19, t. 22, 23 (1788), nom. cons.; La-Serna, Revisión del género *Bystropogon*, J. Cramer, Vaduz (1984), rev.

Shrubs, strongly aromatic, frequently gynodioecious; leaves petiolate, crenate, dentate to subentire; inflorescence in compound, terminal or axillary, dichasial, pedunculate, few to many-flowered, often corymbiform cymes, lax or congested, in axils of leaf-like bracts, often forming spherical glomerules, often with solitary, terminal flower at lowest cyme-branch; bracteoles inconspicuous, lowest sometimes leaf-like; calyx cylindrical-campanulate, ± actinomorphic, equally 5-lobed, lobes triangular-subulate, sometimes plumose, tube 10-nerved, often long-pilose, throat hairy; corolla white, pink or lilac, weakly 2-lipped, 5-lobed (2/3), posterior lip emarginate to 2-fid, anterior lip with median lobe broader, tube included in calyx; stamens 4, shorter than corolla, didynamous, anterior pair longer, spreading, filaments glabrous, thecae parallel or divergent, distinct, connective slightly dilated; stigma-lobes unequal to subequal, long-subulate, anterior lobe broader; disc symmetrical, reddish-orange; nutlets oblong-ovoid, smooth or finely tuberculate. $2n$ = 42, 48. Seven species, endemic to Canary Is. and Madeira. Mountain slopes and clearings. Hybrids have been reported. La Serna (1984) recognised two sections. Sect. *Bystropogon*, flowers in dichasial cymes, calyces with lobes narrow, subulate, plumose, nutlets smooth, 4 spp. Canary Is. Sect. *Canariense* La-Serna, flowers in congested glomerules, calyces with lobes short, triangular-acuminate, not plumose, nutlets somewhat tuberculate, 3 spp. Madeira and Canary Is.

148. *Cuminia* Colla

Cuminia Colla, Mem. Acad. Sci. Torino 38, 6: 139 (1835); Harley, Kew Mag. 3,4: 151–156 (1986), rev.
Skottsbergiella Epling (1937), non Petrak (1927).
Johowia Epling & Looser (1937).

Small tree or shrub, aromatic, gynodioecious or gynomonoecious; leaves petiolate, sharply and irregularly serrate; inflorescence of compound, dichasial, pedunculate cymes in axils of, and shorter than, large, leaf-like bracts; bracteoles inconspicuous; calyx tubular-campanulate, rather fleshy, actinomorphic, 5-lobed, lobes triangular-lanceolate, ciliate, often slightly spreading, calyx-tube obscurely 12–14-nerved, exannulate, becoming distended at base to accommodate nutlets, throat glabrous; corolla white to lavender or dark purple, ± actinomorphic, 5-lobed, lobes rounded, posterior pair erect, tube hairy outside, longer than calyx, annulate within; stamens 4, subequal, included in corolla-tube, filaments short, thecae divergent, distinct; stigma-lobes slightly unequal, posterior shorter; disc symmetrical, fleshy; nutlets ovoid-trigonous, fleshy with thickened angles, rather large. $2n$ = 46, 44. One species, *C. eriantha* (Benth.) Benth., in lower montane forest, 200–450 m, Chile: Masatierra Is., Robinson Crusoe (Juan Fernandez) Archipelago.

149. *Minthostachys* (Benth.) Spach

Minthostachys (Benth.) Spach, Hist. Veg. Phan. 9: 164 (1840); Epling, Feddes Rep. Spec. Nov. Beih. 85: 162–168 (1936), rev.
Bystropogon sect. *Minthostachys* Benth. (1834).

Weak-stemmed, often semi-scandent shrubs, aromatic, gynodioecious; leaves shallowly toothed to subentire; inflorescence of compound, congested, sometimes 2–3, often pedunculate cymes, in axils of leaf-like bracts, these becoming reduced towards stem-apex, upper cymes often forming a terminal, spiciform thyrse; bracteoles inconspicuous; calyx cylindrical, ± actinomorphic, 5-lobed, lobes subulate to narrowly triangular, calyx-tube often externally long-pilose, 13(–12)-nerved, throat hairy; corolla white, or pale lilac, 2-lipped, 5-lobed (2/3), anterior lip with median lobe larger, corolla-tube hairy within, longer to shorter than calyx; stamens 4, didynamous, shorter than corolla, slightly exserted from tube, thecae slightly divergent, distinct, connective slightly dilated; stigma-lobes long, unequal, anterior longer; disc symmetrical, reddish orange; nutlets ovoid, finely reticulate. $2n$ = 46. Probably less than twelve spp., montane, Andes from Venezuela to Argentina. Species limits are very poorly understood.

150. *Clinopodium* L. Fig. 28

Clinopodium L., Sp. Pl. 2: 587 (1753); L., Gen. Pl. ed. 5: 256 (1754); Cantino & Wagstaff, Brittonia 50: 63–70 (1998), reg. consp.; Harley & Granda Paucar, Kew Bull. 55: 917–927 (2000), reg. consp.
Calamintha Mill. (1754).
Acinos Mill. (1754).
Gardoquia Ruiz et Pav. (1794).
Xenopoma Willd. (1811).
Antonina Vved. (1961).

Perennial, rarely annual herbs and shrubs, aromatic, with simple or branched hairs; leaves petiolate to subsessile, simple, variously toothed or entire, margin sometimes revolute; inflorescence of axillary cymes, these sometimes reduced to solitary flowers or merging to form a terminal, paniculiform to spiciform, verticillastrate, or subcapitate thyrse; calyx actinomorphic to 2-lipped, not accrescent, usually cylindrical to tubular-campanulate, straight to sigmoid, sometimes gibbous, usually 13- (11-, 15-)nerved, 5-lobed, lobes often upcurved, posterior 3 often partially fused to form a lip, anterior 2 usually free above tube, throat hairy or not; corolla white to blue, lavender, red, or orange (rarely yellow), 2-lipped (2/3), posterior lip usually notched, tube straight or curved; stamens usually 4 (rarely the posterior pair reduced to staminodes), didynamous, included or exserted, filaments usually glabrous, thecae parallel to divaricate, separate at dehiscence; stigma-lobes equal or unequal; disc usually symmetrical; nutlets ellipsoid to ovoid, obloid, obovoid, or subglobose, often somewhat trigonal, smooth or minutely sculptured, glabrous or puberulent. $2n$ = 10–72 (most frequently 18 or 20). About 100 spp., mostly in the New World (both temperate and tropical) and temperate Eurasia, but a few in Africa, tropical Asia and Indomalaysia. As delimited here, *Clinopodium* includes most of the native New World species of *Satureja* sensu Epling and Játiva (1964, 1966), see Cantino and Wagstaff (1998).

151. ***Cyclotrichium*** (Boiss.) Manden. & Scheng.

Cyclotrichium (Boiss.) Manden. & Scheng., Not. Syst. Leningrad 15: 332–337 (1953).
Calamintha sect. *Cyclotrichium* Boiss. (1879).

Perennial subshrubs, hairs simple or dendroid; leaves entire or toothed; inflorescence of axillary cymes, sometimes very shortly pedunculate, forming terminal thyrse of 6–many-flowered, usually remote verticillasters; bracts, at least the lowest, leaf-like, upper small, inconspicuous, bracteoles ± inconspicuous, smaller than to equalling calyx; calyx weakly 2-lipped, 5-lobed (3/2), lobes triangular-subulate, posterior lobes usually shorter than anterior, calyx-tube cylindrical, straight or shortly curved, 13-nerved, throat hairy; corolla resupinate, 2-lipped, 4-lobed (3/1), white or lilac, posterior (lower) lip entire or emarginate, anterior (upper) lip 3-lobed, slightly deflexed, median lobe larger, corolla-tube twisted, annulate near middle; stamens 4, ± equal, long-exserted, anthers ellipsoid, thecae 2, parallel, distinct with short connective; stigma-lobes unequal, ± subulate, posterior lobe longer; disc weakly 4-lobed; nutlets ovoid, smooth, glabrous. About eight species in SW Asia.

152. ***Obtegomeria*** P.D. Cantino & Doroszenko

Obtegomeria P.D. Cantino & Doroszenko, Novon 8: 1–3 (1998).

Low, ericoid shrub with simple hairs; stems transversely ridged at nodes; leaves subsessile, stiffly ascending, linear-oblong, entire, replicate (abaxial surface hidden); flowers axillary, solitary, with 2 lanceolate bracteoles near base of pedicel; calyx strongly zygomorphic, not accrescent, tubular-campanulate, downcurved, 11–13-nerved, 5-lobed (3/2), lobes usually slightly upcurved, posterior lobes deltoid-cuspidate, fused to form a lip, anterior lobes lanceolate-subulate, free above tube, throat densely villous; corolla usually blue to pink, 2-lipped, posterior lip deeply bifid, anterior lip 0.8× to as long as tube, 3-lobed, median lobe largest, notched, tube ± straight; stamens 4, didynamous, included, filaments glabrous, thecae divaricate, separate at dehiscence; stigma-lobes unequal; disc symmetrical; nutlets ellipsoid, smooth, glabrous. One species, *O. caerulescens* (Bentham) Doroszenko & P.D. Cantino, N Colombia.

153. ***Kurzamra*** Kuntze

Kurzamra Kuntze, Rev. Gen.: 520 (1891).
Soliera Clos (1849) non J.G. Agardh (1842).

Dwarf, mat-forming perennial herb, aromatic; stems procumbent, often rhizomatous, subterranean at base, with reduced scale-leaves; leaves, small, entire, densely hairy; inflorescence of 1(–2)-flowered cymes, sessile, subtended by leaf-like bracts and crowded towards shoot-apex; bracteoles paired, inconspicuous; calyx ± actinomorphic, 5-lobed, lobes equal, triangular, with apex long-subulate, plumose, curved, more than 2× as long as tube, spreading in fruit, calyx-tube shortly cylindrical-infundibuliform, 15-nerved, externally patent-hairy, throat annulate within; corolla lilac, 2-lipped, 5-lobed (2/3), posterior lip shallowly lobed, shorter than anterior lip, corolla-tube long-exserted from calyx, narrow at base, gradually widening towards throat, straight, glabrous within, stamens 4, didynamous, exserted from corolla-tube, divergent, anterior pair longer, thecae weakly divergent, separate, connective small; stigma-lobes flattened, slightly unequal, posterior shorter; disc ± symmetrical; nutlets oblong, mid-brown, smooth, not mucilaginous. One species, *K. pulchella* (Clos) Kuntze, Chile: Atacama and Coquimbo and adjacent Argentina.

154. ***Ziziphora*** L.

Ziziphora L., Sp. Pl. 1: 21 (1753); Hedge, Notes Roy. Bot. Gard. Edinburgh 23: 209–221 (1961); Rechinger f., Fl. Iranica 150: 480 (1982), reg. rev.
Faldermannia Trautv. (1839).

Annual or perennial herbs or subshrubs, aromatic; leaves ± entire, subsessile to shortly petiolate; inflorescence of 2–12-flowered verticillasters, of sessile or pedunculate cymes in axils of leaf-like bracts, forming a terminal, spiciform or capituliform thyrse; bracts patent or imbricate, ± conspicuous, bracteoles inconspicuous; flowers shortly pedicellate, pedicels sometimes flattened; calyx ± actinomorphic, 5-lobed (±3/2), lobes triangular, posterior lobes shorter than straight or curved anterior lobes, calyx-tube cylindrical, sometimes curved, gibbous below, 13-nerved, accrescent, throat bearded, closed; corolla 2-lipped, ±4-lobed (1/3), white, rose-red or purple, posterior lip rounded, sometimes emarginate, straight, anterior lobes ± rounded, corolla-tube cylindrical, short, sometimes scarcely exserted from calyx, exannulate; stamens 2 (posterior pair staminodal or absent), anterior pair ascending under posterior corolla-lip, exserted from tube, anthers often cohering at margins, with fertile theca ellipsoidal, divaricate, becoming confluent, or often dimidiate, lower theca sterile and forming a small appendage; stigma-lobes unequal; disc symmetrical; nutlets ovoid to trigonous, smooth or granulate. $2n = 12, 16, 18, 24, 32, 36$. About 20 species in S Europe, NW Africa and Asia to Himalayas and Altai mountains. Open, usually xeric habitats. Aromatic and medicinal plants used as flavouring and as source of essential oils.

155. *Stachydeoma* Small

Stachydeoma Small, Fl. S. E. United States 1040, 1337 (1903).

Low shrubs or subshrubs, aromatic, conspicuously glandular and setose; leaves opposite, subsessile, ovate to elliptical, entire or few-toothed, usually revolute; inflorescence a terminal thyrse or raceme; calyx zygomorphic, not accrescent, campanulate, 13-nerved, glandular-punctate, 5-lobed (3/2), setose, posterior lobes triangular, largely fused to form a shallowly lobed lip, anterior lobes subulate, slightly upcurved, free above tube, throat hairy; corolla purplish, 2-lipped (1/3), posterior lip entire, anterior lip with median lobe notched, tube straight; fertile stamens 2 (posterior pair reduced to staminodes), included to slightly exserted, filaments glabrous, thecae nearly parallel, asymmetrical (one theca attached higher than other), separate at dehiscence; stigma-lobes unequal; disc symmetrical; nutlets subglobose, smooth, glabrous. $2n = 18$. One species, *Stachydeoma graveolens* (A. Gray) Small, Florida. Differs from *Hedeoma* in chromosome number and asymmetrical anthers.

156. *Hedeoma* Pers.

Hedeoma Pers., Syn. Pl. 131 (1807); Irving, Sida 8(3): 218–295 (1980), rev.
Pseudocunila Brade (1944).

Annual or perennial herbs or subshrubs, strongly aromatic; leaves entire or toothed; inflorescence of few- to many-flowered, pedunculate to subsessile cymes, or flowers rarely solitary, in axils of often leaf-like bracts, upper cymes often congested in a terminal spiciform thyrse; bracteoles inconspicuous; flowers sometimes cleistogamous; calyx ± actinomorphic, subequally 5-lobed, or 2-lipped (3/2) with lobes unequal, posterior somewhat connate, deltoid to subulate, anterior lobes usually longer, tube 13-nerved, cylindrical to infundibular, often gibbous at base, throat densely hairy within; corolla white, lilac, pink or orange-red, 2-lipped (2/3), posterior lip sometimes subentire, concave to flat, erect or spreading, median anterior lobe usually larger, often emarginate, tube short to elongate, widening upwards, annulate or not within; stamens 2, posterior pair staminodal, anterior pair ascending under posterior corolla-lip, included to exserted, thecae ± parallel to widely divaricate, distinct, connective well developed; stigma-lobes unequal, ± subulate, posterior very small; disc symmetrical; nutlets ovoid to oblong, usually 1.5 mm long or less, areolate to foveate, sometimes glaucous, often mucilaginous. $2n = 34, 36, 44, 72, 76, 144$. About 42 species, from Canada southwards to Mexico, Central and Southern S America, from Peru and Bolivia to S Brazil, Uruguay and Argentina. Open montane or submontane, sometimes xeric habitats. Irving (1980) recognises four subgenera, based mainly on habit and reproductive characters.

157. *Rhododon* Epling

Rhododon Epling, Rep. Spec. Nov. Regn. Veg. Beih. 115: 14 (1939).

Annual herbs; leaves subsessile, entire with margins long-ciliate; inflorescence of verticillasters, cymes 1–4-flowered, in axils of leaf-like bracts, forming terminal spiciform, congested or interrupted thyrse; bracteoles long-ciliate, curved, purple-tinged; calyx often purple-tinged, 2-lipped, 5-lobed (3/2), lobes elongate, long-ciliate, anterior narrower with deeper sinus, tube 15-nerved, conspicuously ribbed, cylindrical, weakly gibbous, slightly curved, annulate at throat; corolla lilac to purple, 2-lipped, 5-lobed (2/3), posterior lip flat, anterior longer and broader, tube narrowly cylindrical, slightly gibbous towards base, annulate above gynoecium; stamens 2, posterior pair absent, anterior slightly exserted, filaments prolonged into small tooth above articulation with anther, thecae divaricate at 180°, extending parallel to corolla-tube axis, distinct but anthers connate at point of attachment, connective apparent; style hairy, stigma-lobes subequal, flattened, triangular; disc symmetrical; nutlets ellipsoid-obovate, smooth, pale brown, glabrous, weakly mucilaginous. $2n = 26$. One species, *R. ciliatus* (Benth.) Epling, or possibly two, confined to Texas, U.S.A. Open, sandy areas.

158. *Pogogyne* Benth.

Pogogyne Benth., Lab. Gen. et Sp.: 414 (1834); Howell, Proc. Acad. Sci. 22(3): 105–128 (1931), rev.; Jokerst, Aliso 13(2): 347–352 (1992), key.

Annual herbs, aromatic; leaves narrow, linear to spathulate, entire or toothed; inflorescence of (1–)3–6-flowered sessile, rarely shortly pedunculate cymes in axils of usually narrow bracts (lower leaf-like), congested into dense, terminal, spike- or head-like thyrse; bracteoles elongate, linear, curved, often conspicuously ciliate; calyx 2-lipped, 5-lobed (3/2), lobes narrowly triangular, posterior usually shorter, tube 15-nerved, narrowly infundibuliform, throat glabrous; corolla lavender to purple, rarely white, or yellowish, 2-lipped, 4-lobed (1/3), lobes short, posterior lip flat, entire, tube elongate, narrowly infundibuliform to campanulate; stamens 4, posterior pair shorter and, in some species, staminodal, anterior pair ascending under posterior lip of corolla, included to slightly exserted, thecae ± parallel, distinct, often slightly unequal, connective apparent; style usually hairy above, stigma-lobes equal to unequal; disc symmetrical; nutlets oblong-obovate with rounded, usually hairy apex. $2n = 38$. Seven species, mostly in U.S.A., California, extending to Oregon and Mexico (Baja California). Open, usually seasonally flooded, areas. Howell (1931) recognised two subgenera: subg. *Pogogyne*, with 4 fertile stamens and style densely hairy with ± equal stigma-lobes, and subg. *Hedyomoides* Gray, with 2 fertile stamens, and style sparsely hairy with unequal stigma-lobes.

159. *Neoeplingia* Ramamoorthy, Hiriart & Medrano

Neoeplingia Ramamoorthy, Hiriart & Medrano, Bol. Soc. Bot. Mexico 43: 61–65 (1982).

Subshrub, densely canescent with branched hairs; leaves toothed, glandular-punctate; inflorescence of 3-flowered sessile cymes in axils of leaf-like bracts; bracteoles linear, inconspicuous; calyx ± actinomorphic, subequally 5-lobed, lobes broadly triangular, tube 5-nerved and reticulately veined between nerves, broadly infundibuliform, glabrous within, except for hairy calyx-lobes; corolla blue, 2-lipped, 5-lobed (2/3), lobes rounded, posterior lip emarginate, flat, equalling or slightly shorter than 3-lobed anterior lip, median lobe of anterior lip larger than laterals, tube straight, widening above and scarcely exceeding calyx, annulate near mid-tube; stamens 4, usually all fertile, didynamous, posterior pair slightly shorter, anterior slightly exserted from corolla-tube, thecae slightly divergent, distinct; stigma-lobes subulate, flattened, subequal or anterior slightly longer; disc symmetrical. One species, *N. leucophylloides* Ramamoorthy, Hiriart & Medrano, Mexico, Hidalgo. Open, calcareous hillside.

160. *Poliomintha* A. Gray

Poliomintha A. Gray, Proc. Am. Acad. 8: 295 (1870); Irving, Sida 5: 8–22 (1972), rev.
Hedeoma sect. *Poliomintha* Briq. (1897).

Shrubs or subshrubs, aromatic; stems numerous, glabrate to densely canescent, with simple or rarely dendroid hairs; leaves entire or toothed, glandular-punctate; inflorescence of 3–7-flowered, pedunculate to subsessile cymes, or flowers rarely solitary, in axils of leaf-like bracts; bracteoles inconspicuous; calyx ± actinomorphic and subequally 5-lobed, or 2-lipped (3/2) with lobes unequal, posterior free or connate, triangular, anterior usually narrower and longer, lobes connivent after anthesis, tube 13–15-nerved, cylindrical, straight, not saccate, throat irregularly or distinctly annulate, or lobes canescent within; corolla lilac or orange to red, 10–35 mm long, 2-lipped (2/3), posterior lip emarginate, hooded or flat, anterior lip spreading, median lobe often larger, tube short to elongate, straight, ± annulate within; fertile stamens 2, posterior pair staminodal, anterior pair ascending under posterior corolla-lip, included to ± exserted, thecae widely divaricate, distinct, connective well developed; stigma-lobes unequal, ± subulate, anterior lobe longer; disc symmetrical; nutlets oblong, over 1.5 mm long, smooth, not mucilaginous. $2n = 36$. Seven spp, SW United States, Northern Mexico, Haiti. Open, submontane or lowland, often xeric habitats. Irving (1972) recognises two sections: sect *Poliomintha*, with tomentose, linear-oblong leaves, calyx usually 15-nerved, 3–4 mm long and annulate within; and sect. *Saturejoides* Irving, with canescent to glabrate, ovate to elliptic leaves, calyx 13-nerved, 5–12 mm long, exannulate.

161. *Hesperozygis* Epling

Hesperozygis Epling, Rep. Spec. Nov. Beih. 85: 132 (1936), reg. rev.; Epling & Stewart, Rep. Spec. Nov. Beih. 115: 12–14 (1939), rev.

Shrubs or subshrubs with simple or rarely branched hairs (*H. marifolia* (Schau.) Epling); leaves gland-dotted, entire or sometimes toothed; inflorescences of 1–3-flowered cymes, sessile or axillary in axils of usually leaf-like bracts; calyx weakly 2-lipped, 5-lobed (3/2), anterior lip longer than posterior, with deeper sinus between lobes, lobes deltoid-lanceolate or acuminate, not connivent, tube 13–15-nerved, annulate within shortly below lobes; corolla bluish purple, lilac, pink or rarely white, 2-lipped, 5-lobed (2/3), posterior lip weakly bilobed, erect, shorter to slightly longer than anterior lip, tube usually curved and widening broadly above (except in *H. marifolia*); stamens 2 (posterior pair usually absent), filaments slightly arcuate, ascending under posterior lip, thecae widely divaricate at 180°, distinct, connective strongly developed; style glabrous or rarely hairy, stigma-lobes unequal, posterior shorter; disc ± symmetrical; nutlets obovoid-oblong, dark brown, finely verrucose, slightly mucilaginous. $2n = 44$. About eight species, southern Brazil and one in Mexico.

162. *Glechon* Spreng.

Glechon Spreng., Syst. Veg. 4(2): 222, 227 (1827); Epling, Rep. Spec. Nov. Beih. 85: 142–145 (1936), rev.; Epling & Stewart, Rep. Spec. Nov. Beih. 115: 1–49 (1939), rev.

Small shrubs or subshrubs, often geoxylic, aromatic; leaves elliptic to ovate-spathulate, subentire; inflorescence of verticillasters, with sessile cymes 1–3-flowered in axils of leaf-like bracts, often crowded towards shoot apices; bracteoles inconspicuous; calyx weakly 2-lipped to ± actinomorphic, 5-lobed (3/2), lobes subequal, often spreading, broadly to narrowly triangular, often irregularly aristate, anterior lobes sometimes longer than posterior, tube 13–15-nerved, infundibular-cylindrical, throat annulate; corolla white or pink, 2-lipped, 4–5-lobed (1–2/3), posterior lip entire or often emarginate, hooded, often weakly falcate, longer than or subequal to anterior lip, tube often shorter than limb, abruptly widening above, exserted from calyx-tube and usually strongly saccate on anterior side just below throat, usually hairy within near insertion of stamens; stamens 2, posterior pair absent, filaments strongly arcuate-ascending under posterior corolla-lip, often with small hairy appendage at base and obscurely toothed at apex, thecae divergent to widely divaricate, distinct; connective well developed; disc symmetrical; style glabrous, stigma-lobes unequal, posterior shorter. Nutlets ovoid-oblong, smooth, mucilaginous. About six or seven species, southern Brazil, Paraguay, Uruguay and N Argentina. Open, damp areas. Epling (1936) recognised three sections, based mainly on vegetative characters.

163. *Hoehnea* Epling

Hoehnea Epling, Rep. Spec. Nov. Beih. 115: 8 (1939); Epling, Rep. Spec. Nov. Beih. 85: 130–132 (1936), rev. as *Keithia*.
Keithia Benth. (1834), non Spreng. (1822).

Perennial herbs; stems weak, internodes elongate, strongly 4-angled with ciliate angles; leaves small, ovate-deltate, entire or weakly toothed, ciliate, ± sessile; inflorescence of verticillasters, with 1–3-flowered sessile cymes in axils of leaf-like bracts; bracteoles inconspicuous, ciliate; calyx weakly 2-lipped, 5-lobed (3/2), lobes narrowly triangular, curved and slightly spreading, anterior lip often slightly shorter, tube cylindrical, 5-angled, 12–13-nerved, with 5 prominent, ciliate nerves, secondary nerves inconspicuous, tube glabrous within; corolla deep pink to lilac or white, 2-lipped, 5-lobed (2/3), posterior lip shorter than anterior, flat, bilobed to emarginate, median anterior lobe much longer than laterals, tube cylindrical below, widening above, longer than limb; stamens 2 (posterior pair absent), filaments glabrous, weakly curved and ascending under posterior corolla-lip; thecae widely separated and divaricate, connectives elongate; disc symmetrical, weakly 4-lobed; style glabrous, stigma-lobes unequal, posterior shorter; nutlets ovoid, brown, smooth, not mucilaginous when wetted. About four species, S Brazil, Paraguay. Marshy ground.

164. *Rhabdocaulon* (Benth.) Epling

Rhabdocaulon (Benth.) Epling, Rep. Spec. Nov. Beih. 85: 134–137 (1936), rev.; Rep. Spec. Nov. Beih. 115: 8 (1939), rev.
Keithia sect. *Rhabdocaulon* Benth. (1834).

Subshrubs or perennial herbs, stems erect, virgate, internodes elongate, strongly 4-angled; leaves narrowly linear to linear-lanceolate, distant and sometimes reduced; inflorescences of few-flowered cymes, sessile or shortly pedunculate in axils of inconspicuous bracts, forming terminal, spiciform, interrupted or congested thyrse; bracteoles inconspicuous; calyx elongate, narrowly cylindrical, rarely shorter, narrowly infundibuliform, weakly 2-lipped, 5-lobed (3/2), lobes usually subequal, shorter than tube (except in *R. strictus*), triangular to subulate, tube 13–14-nerved, nerves prominent, throat densely annulate within; corolla white, cream, lavender or red, 2-lipped, 5-lobed (2/3), lips ± subequal, or posterior lip longer, both usually shorter than tube, erect or anterior lip sometimes spreading, tube usually long-exserted from calyx, slightly curved, widening gradually above, glabrous to slightly hairy towards base within; stamens 2 (posterior pair absent), filaments slightly arcuate, ascending, rarely slightly exserted from corolla, thecae equal, divaricate, or unequal, divergent, distinct, connective well developed; disc symmetrical, ±4-lobed; style glabrous, stigma-lobes unequal, posterior shorter; nutlets ovoid-oblong, smooth, not mucilaginous when wetted. Seven species, central and southern Brazil, Paraguay and N Argentina. Savannahs and other open areas.

165. ***Eriothymus*** (Benth.) Schmidt

Eriothymus (Benth.) Schmidt, in Martius, Fl. Brasil. 8: 171, t. 32, f. 2 (1858).
Keithia sect. *Eriothymus* Benth. (1834).
Hedeoma sect. *Eriothymus* (Benth.) Briq. (1897).

Shrub with slender stems, divaricately branched, acutely quadrangular, puberulent; leaves small, coriaceous, dark green, glabrous, gland-dotted, subentire; inflorescence of 3–6-flowered, shortly pedunculate cymes, in axils of leaf-like bracts, ± reduced towards stem-apex; bracteoles inconspicuous; calyx very weakly 2-lipped, 5-lobed (3/2), posterior lip slightly longer than anterior, lobes deltate-acuminate, subequal or posterior lobes shorter than anterior, ciliate, sparsely hairy within, tube cylindrical, 13–15-nerved, externally densely villous with pale blue hairs (?), mid-tube annulate within; corolla 2-lipped, 5-lobed (2/3), posterior lip 2-lobed, slightly shorter than anterior, flat, tube cylindrical, exserted from calyx, widening above, exannulate and glabrous within; stamens 2, posterior pair absent, filaments directed under posterior corolla-lip, thecae divaricate, distinct; style glabrous, stigma-lobes slightly unequal. One species, *E. rubiaceus* (Benth.) Schmidt, southern Brazil: Minas Gerais. Montane. Only known from type collection and possibly extinct.

166. ***Cunila*** D. Royen ex L.

Cunila D. Royen ex L., Syst. Nat., ed. 10, 2: 1359 (1759), nom. cons.; Epling, Rep. Spec. Nov. Beih. 85: 138–142 (1936), rev.

Shrubs, subshrubs, and rhizomatous perennial herbs, strongly aromatic, with simple hairs; leaves sessile to short-petiolate, simple, toothed (usually finely) or entire; inflorescence of terminal and axillary cymes (occasionally reduced to solitary flowers), a terminal paniculiform or spiciform thyrse, or involucrate terminal and subterminal glomerules; calyx actinomorphic or zygomorphic, occasionally 2-lipped (3/2), not accrescent, cylindrical to tubular-campanulate, usually 10–13-nerved, 5-lobed, lobes deltoid to subulate, throat usually densely hairy; corolla lavender to white, weakly zygomorphic to 2-lipped (2/3), posterior lip usually emarginate, anterior 3 lobes occasionally forming a lip but more often fully separate above tube, tube straight to slightly downcurved; fertile stamens 2 (posterior pair reduced to staminodes or absent), exserted, filaments glabrous, thecae parallel to divaricate, separate at dehiscence; stigma-lobes equal to subequal; disc ± symmetrical; nutlets ellipsoid-trigonal, smooth, glabrous. $2n = 28$. About 15 species with a disjunct distribution, occurring from eastern North America to Panama and in southern South America (SE Brazil, NE Argentina, Uruguay, Paraguay).

167. ***Piloblephis*** Raf.

Piloblephis Raf., New Fl. 3: 52 (1838).
Pycnothymus (Benth.) Small (1903).

Low ericoid shrubs, aromatic, with simple hairs; leaves sessile, linear-lanceolate, entire, tightly revolute; inflorescence a terminal, ovoid-spiciform thyrse; flowers sessile; calyx ± actinomorphic, not accrescent, tubular-campanulate to campanulate, obscurely 5- or 15-nerved, 5-lobed, lobes subequal, lanceolate, as long as or longer than tube, throat glabrous; corolla lavender, 2-lipped, posterior lip entire or emarginate, anterior lip 3-lobed, lobes subequal, entire, corolla-tube straight; stamens 4, didynamous, all exserted or posterior pair included, filaments glabrous, thecae separate at dehiscence, those of anterior pair usually parallel, those of posterior pair often asymmetrically attached (one theca parallel with filament, the other divergent); stigma-lobes equal to subequal; disc symmetrical; nutlets broadly ellipsoid to ovoid, smooth, glabrous. One species, *Piloblephis rigida* (Bentham) Raf., dry pinelands and oak scrub in the SE United States (Florida and Georgia).

168. *Conradina* A. Gray

Conradina A. Gray, Proc. Am. Acad. Arts 8: 294 (1870); Gray, A monograph of the genus *Conradina*, Ph.D. Thesis, Vanderbilt University (1965), rev.

Low ericoid shrubs, strongly aromatic, with simple hairs; leaves sessile to subsessile, linear to spathulate, entire, revolute; inflorescence of paired, short-pedunculate, axillary cymes (forming verticillasters), sometimes reduced to solitary flowers; calyx strongly zygomorphic, not accrescent, tubular-campanulate, 13-nerved, 2-lipped, 5-lobed (3/2), posterior lip shallowly lobed, upcurved, lobes deltoid, anterior lip deeply lobed (or the lobes distinct above the tube, thus not forming an anterior lip), lobes straight to slightly upcurved, lanceolate to lanceolate-subulate, throat densely hairy; corolla white to deep purple, 2-lipped, posterior lip entire or emarginate, anterior lip 3-lobed, median lobe emarginate and sometimes erose, tube upturned in middle (usually sharply so), throat partially blocked by prominent hump on anterior surface; stamens 4, didynamous, included to slightly exserted, filaments glabrous, thecae ± parallel, separate at dehiscence, variously pubescent near base and apex, connective rather broad; stigma-lobes usually equal to subequal; disc symmetrical; nutlets broadly ovoid to obovoid, smooth or minutely reticulate, glabrous. $2n = 24$. Six species of sandy soils in the southeastern U.S., most of them rare or narrowly endemic.

169. *Dicerandra* Benth.

Dicerandra Benth., Edward's Bot. Reg. 15: sub t. 1300 (1830); Huck, Phanerog. Monogr. 19: 1–343 (1987), rev.
Ceranthera Elliott (1821), non P. Beauv. (1808).

Low shrubs, subshrubs, and annual herbs, strongly aromatic; leaves opposite, sessile or subsessile, linear to narrowly oblanceolate, oblong or lanceolate, usually entire; inflorescence of axillary cymes, merging to form a terminal thyrse (or rarely the cymes reduced to solitary flowers); calyx 2-lipped, not accrescent, tubular-campanulate, 13-nerved, posterior lip broadly rounded, truncate or minutely 2–3-toothed, anterior lip triangular, 2-toothed, both lips somewhat upcurved, throat hairy; corolla white to reddish purple, 2-lipped, of two distinct types: (a) tube infundibular, ± anteriorly geniculate, with 2-lobed palate in throat, posterior lip emarginate, nearly flat; or (b) tube subcylindrical, straight to gently downcurved, without palate, posterior lip entire, galeate; stamens 4, didynamous, declinate and exserted in type "a" corolla, or parallel and included under posterior lip in type "b" corolla, filaments glabrous, thecae broadly ovoid with apical appendage, divergent, separate at dehiscence, suture short (almost porose), connective ± deltoid; style hirtellous, stigma-lobes equal or unequal; disc ± symmetrical; nutlets ovoid, smooth, glabrous. $2n = 32, 48$. Eight to nine species of oak and pine-oak woodlands in the SE United States. Two sections are recognised: stamens exserted, declinate; upper lip of corolla flat: sect. *Dicerandra*; stamens included under galeate upper lip: sect. *Lecontea* Huck.

170. *Monardella* Benth.

Monardella Benth., Lab. Gen. et Sp.: 331 (1834); Epling, Ann. Missouri Bot. Gard. 12: 1–106 (1925), rev.; Jokerst, in Hickman (ed.) The Jepson manual of higher plants of California: 718–722 (1993), reg. rev.
Madronella Greene (1906).

Annual or perennial herbs, sometimes rhizomatous, usually aromatic, gland-dotted; leaves entire or toothed; inflorescence thyrsoid, forming a short, terminal capitulum, surrounded by an involucre of bracts, outer pair often leaf-like, inner usually broad, scarious and often conspicuously veined or coloured, bracteoles ?absent; calyx ± actinomorphic, 5-lobed (3/2), lobes subequal, triangular, tube cylindrical, longer than lobes, 10–15-nerved, sometimes accrescent, glabrous within throat; corolla purple, pink or white, 2-lipped, 5-lobed (2/3), lobes narrow, posterior lip erect, distinctly 2-lobed, anterior lip recurved, 3-lobed almost to base, 2/3 to 1/5 length of tube, tube straight; stamens 4, divergent, exserted, posterior pair usually shorter than anterior pair, filaments hairy or glabrous, thecae parallel to divergent or divaricate and then often confluent towards apex; disc symmetrical; nutlets oblong-ovoid, smooth. $2n = 42$. About 30 species. Western United States to northern Mexico. Rocky or sandy areas, desert scrubs, sagebrush, chaparral, woodland or montane forests. Many species have restricted ranges with specialised soil conditions. Epling (1925) divided the genus into two subgenera: subg. *Monardella* (as subg. *Pycnanthae*) and subg. *Macranthae* Briq., based on relative length of corolla lobes and tube and on calyx length.

171. ***Pycnanthemum*** Michx.

Pycnanthemum Michx., Fl. Bor.-Am. 2: 7 (1803), nom. cons; Grant & Epling, Univ. Calif. Publ. Bot. 20: 195–240 (1943), rev.
Koellia Moench (1794).

Rhizomatous perennial herbs, strongly aromatic, with simple hairs; leaves sessile or short-petiolate, ovate to lanceolate, elliptical or linear, variously toothed or entire; inflorescence of verticillasters, or capitate, or corymbiform, composed of compact or loose, terminal and upper-axillary glomerules (paired cymes); calyx actinomorphic to zygomorphic, not accrescent, cylindrical to tubular-campanulate, 10–13-nerved, 5-lobed, lobes triangular to ovate, lanceolate or aristate, all similar or anterior 2 more deeply cleft than posterior 3, throat glabrous; corolla usually white to lavender with purple spots, 2-lipped, posterior lip entire or emarginate, anterior lip 3-lobed, tube ± straight; stamens 4, didynamous or subequal, exserted or included, filaments glabrous, thecae parallel, separate at dehiscence; stigma-lobes unequal to subequal (rarely equal); disc symmetrical; nutlets ovoid to obloid, ± trigonal, smooth or rugose, glabrous or pubescent. $2n = 36, 38, 40, 72, 76, 78, 80$, ca. $108, 120$ ($x = 18, 19, 20$). Seventeen to 21 species, SE Canada, E and central United States, one disjunct species in California; centre of diversity in SE U.S. Most species occur in disturbed sites, naturally open habitats or forest edges.

172. ***Acanthomintha*** (A. Gray) Benth. & Hook. f.

Acanthomintha (A. Gray) Benth. & Hook. f., Gen. Pl. 2, 2: 1192 (1876); Gray, Syn. Fl. N. Am. 2, 1: 365 (1878), rev.; Jokerst, Madroño 38: 278–286 (1991), rev.

Annual herbs, aromatic; leaves petiolate, coarsely toothed to subentire; inflorescence of verticillasters with cymes 3–5-flowered, in the axils of leaf-like bracts, toothed and often apically spinose; in a terminal, congested or interrupted spiciform thyrse; bracteoles conspicuous, 2 per cyme, broadly ovate to rotund, spinose-dentate, subsessile, green or straw-coloured, equalling or overtopping calyx; calyx 2-lipped, 5-lobed (3/2), lobes subequal, triangular-lanceolate, spinose-tipped, posterior lip longer than anterior with shallower sinus between teeth, tube 13-nerved, cylindrical-campanulate, exannulate within; corolla white or pink, 2-lipped, 4–5-lobed (1–2/3), posterior lip concave, erect, anterior lip usually longer, broad, spreading, median lobe longer than laterals, often emarginate, tube narrowly cylindrical, equalling to long-exserted from calyx; stamens 4, didynamous with posterior pair shorter, or 2 and then posterior pair staminodal, anterior pair ascending, included under posterior corolla-lip, filaments glabrous, thecae divergent, distinct, sometimes unequal, glabrous or hairy, connective apparent; disc symmetrical; style hairy to glabrous, stigma-lobes unequal, subulate, posterior shorter; nutlets obovoid-oblong, smooth, mucilaginous. $2n = 19$. Four species, U.S.A. (California), one extending into Mexico (Baja California). Open areas in rocky or clay soils.

173. ***Monarda*** L.

Monarda L., Sp. Pl. 1: 22 (1753); Gen. Pl. ed. 5: 14 (1754); Scora, Univ. Calif. Publ. Bot. 41: 1–59 (1967), rev.

Rhizomatous perennial herbs or taprooted annuals or biennials (rarely low shrubs), aromatic, with simple hairs; leaves petiolate to subsessile, ovate to lanceolate, elliptic or linear, serrate to denticulate (rarely entire); inflorescence terminal or in verticillasters, of involucrate glomerules; bracts sometimes brightly coloured; flowers subsessile, ebracteolate; calyx actinomorphic, not accrescent, cylindrical, 13–15-nerved, 5-lobed, lobes lanceolate to subulate or aristate, often setose, throat usually hairy; corolla white to purple, red, or yellow, 2-lipped, posterior lip narrow, concave, entire or bifid, sometimes apically bearded, anterior lip 3-lobed, median lobe longest, tube straight or curved, corolla in bud falcate; fertile stamens 2 (posterior pair reduced to staminodes or absent), included or slightly exserted, filaments usually pubescent, anthers usually lightly fused (coherent but easily separated), thecae divaricate, separate at dehiscence; stigma-lobes usually unequal; disc symmetrical; nutlets ovoid to obloid, obscurely trigonal, smooth, glabrous. $2n = 18, 22, 36$ (rarely $24, 32, 34$). About 20 species, S Canada to S Mexico; some widely grown as ornamentals. Two subgenera are recognised (McClintock and Epling 1942; Scora 1967). Glomerules usually solitary (rarely two), terminal; upper lip of corolla straight to only slightly arcuate; stamens exserted; rhizomatous perennial herbs: subgenus *Monarda*. Glomerules several, in verticillasters; upper lip of corolla strongly arcuate; stamens usually included; taprooted annuals, biennials, short-lived perennials, or low shrubs: subgenus *Cheilyctis* (Raf.) E.M. McClint. and Epling.

174. *Blephilia* Raf.

Blephilia Raf., Am. Monthly Mag. & Crit. Rev. 4: 190 (1819); Simmers & Kral, Rhodora 94: 1–14 (1992), new sp.

Rhizomatous or stoloniferous perennial herbs, weakly to strongly aromatic; leaves petiolate, ovate to lanceolate or elliptic, serrate to serrulate; inflorescence a verticillastrate thyrse of terminal and subterminal glomerules, each subtended by an involucre of ovate to lanceolate bracts; flowers subsessile, ebracteolate; calyx zygomorphic, not accrescent, cylindrical, 13-nerved, 5-lobed (3/2), lobes lanceolate-attenuate to subulate, setose, posterior lobes partially fused to form long upper lip, anterior lobes shorter, free above tube, throat sparsely hairy; corolla purple to white, 2-lipped (1/3), posterior lip entire, anterior lip 3-lobed, median lobe narrower than other 2, tube curved, corolla in bud falcate; fertile stamens 2 (posterior pair reduced to staminodes), exserted, filaments glabrous, thecae divaricate, separate at dehiscence; stigma-lobes unequal; disc symmetrical; nutlets ellipsoid to ovoid, obscurely trigonal, smooth, glabrous. Three species, eastern North America.

175. *Prunella* L.

Prunella L., Sp. Pl. 2: 600 (1753).
Brunella Mill. (1754).
Prunellopsis Kudo (1920).

Perennial herbs, rhizomatous, not aromatic; leaves entire, serrate, laciniate to pinnatifid; inflorescence of usually 6-flowered verticillasters, in a terminal, condensed, spiciform thyrse, pedunculate or not; bracts imbricate, broadly ovate-acuminate, bracteoles present or not; calyx 2-lipped, tubular-campanulate, accrescent, 10-nerved, sometimes with 2 lateral keels, 5-lobed (3/2), posterior lip ± truncate, lobes short, mucronate, anterior lip with lobes triangular, upcurved, throat not barbate, ± closed in fruit; corolla 2-lipped, 4-lobed (1/3), purple, blue, cream or white, posterior lip hooded, concave, ± rounded, anterior lip with median lobe concave, usually dentate, lateral lobes obliquely truncate, tube ± infundibular, annulate within, contracted at throat; stamens 4, didynamous, exserted from tube, ascending under posterior lip, filaments curved, bifurcating at apex to form a subulate appendage, thecae 2, divaricate, distinct, ± parallel to filament; stigma-lobes shortly subulate; disc unlobed; nutlets ovoid, apiculate or rounded, ± angulate, mucilaginous. $2n = 28, 30, 32$. About seven species in Europe, NW Africa and Asia. *Prunella vulgaris* L. widely naturalized in temperate and montane regions of N and S America, Africa and Australasia.

176. *Cleonia* L.

Cleonia L., Sp. Pl. 2 ed. 2: 837 (1763).

Annual herb, weakly aromatic; leaves simple, crenate to dentate or pinnately lobed; inflorescence of 6-flowered verticillasters, merging to form a terminal, subcapitate thyrse; bracts conspicuous, ciliate, pectinate-pinnatifid, segments aristate; calyx 2-lipped (3/2), campanulate, 10-nerved; posterior lobes deltoid, anterior lobes longer, all straight; corolla 2-lipped (1/3), violet-purple to whitish, posterior lobe emarginate, vertical, anterior lip with median lobe whitish, ± suborbicular, emarginate, horizontal, lateral lobes suborbicular, truncate, tube ± infundibular; stamens 4, didynamous, posterior shorter, anterior pair exserted from tube, filaments glabrous, thecae ellipsoidal, confluent, connective appendaged; stigma-lobes 4, short, subequal; nutlets ovoid, smooth. $2n = 20$. One species: *C. lusitanica* L., frequently on gypsiferous soils, Iberian peninsula, NW Africa.

177. *Horminum* L.

Horminum L., Sp. Pl. 2: 596 (1753).

Perennial herb with basal rosettes of leaves, rhizomatous; leaves simple, crenate, long-petiolate; inflorescence of 2–6-flowered verticillasters forming a lax, secund, spiciform thyrse, scapose, or sometimes a few pairs of leaves present on flowering stem; bracts not conspicuous, ovate, acute, as long as or shorter than calyx; ebracteolate; calyx 2-lipped (3/2), campanulate, 13-nerved, posterior lobes deltoid, upcurved, anterior lobes long-deltoid, straight, throat glabrous; corolla 2-lipped, 4-lobed (1/3), violet, posterior lip semi-orbicular, emarginate, anterior lip with lobes ± rounded, truncate, reflexed, tube ± cylindrical, hairy in throat; stamens 4, didynamous, posterior pair shorter, inserted in middle of tube, exserted, anthers cohering in pairs, thecae confluent; stigma-lobes shortly subulate; nutlets ovoid, ± papillate. $2n = 12$. One species: *H. pyrenaicum* L., calcicole in mountains of Europe, Alps, Pyrenees, Cantabrian mountains.

VII.2.c. Subtribe Nepetinae Coss. & Germ. (1845).

Calyx-tube 15-nerved. Corolla strongly 2-lipped (2/3), often bluish. Stamens 4 (rarely anterior pair absent in some species of *Nepeta*), posterior pair longer than anterior, both usually ascending under posterior corolla-lip, sometimes exserted. Stigma-lobes subulate. Nectary disc 4-lobed or 4-horned. Nutlets elliptic to oblong, rarely curved and globose, abscission-scar usually with well-defined areole.

178. *Nepeta* L.

Nepeta L., Sp. Pl. 2: 570 (1753); Budantsev, Bot. Zhurn. 78, 1: 91–105 (1993), rev.
Kudrjaschevia Pojark (1953).
Pitardia Batt. ex Pit. (1918).

Subshrubs or annual or perennial herbs, aromatic, sometimes gynodioecious; leaves simple; inflorescences thyrsoid, cymes lax to congested, pedunculate or sessile, axillary, distant, or crowded into spike-like or ovoid heads; bracts leaf-like or reduced, sometimes longer than calyx, bracteoles inconspicuous; calyx tubular to campanulate, actinomorphic to strongly 2-lipped, 5-lobed (3/2), or rarely with lips entire, lobes subequal to unequal, throat straight to strongly oblique; corolla 2-lipped, 5-lobed (2/3), blue, violet, pink, yellow or white, posterior lip straight or curved, anterior lip with median lobe concave or ± flat, undulate or entire, tube included in or exserted from calyx, straight or curved, gradually or ± abruptly dilating upwards; stamens of hermaphrodite flowers 4, rarely 2 with anterior pair absent, usually shorter than corolla, filaments parallel, thecae divergent at 180° (stamens of male-sterile flowers rudimentary); stigma-lobes subequal; disc 4-lobed; nutlets ellipsoid to obovoid, rounded to acuminate, smooth to tuberculate, glabrous or hairy at apex, mucilaginous or not. $2n = 14$, 16, 18, 30, 32, 34, 36, 54. Over 200 species. Mainly in mountains, semi-deserts and steppes of Eurasia, extending into North Africa. Some cultivated for medicinal use or ornament. For a synopsis of infrageneric classification, see Budantsev (1993).

179. *Drepanocaryum* Pojark.

Drepanocaryum Pojark., Fl. U.R.S.S. 20: 516 (1954); Budantsev, Bot. Zhurn. 77, 12: 118–128 (1992), rev.

Annual herbs, not aromatic; leaves simple, petiolate; inflorescence thyrsoid, cymes axillary, long-pedunculate, distant; bracts similar to the leaves, bracteoles subulate to acerose, inconspicuous; calyx markedly oblique-gibbous at base, weakly 2-lipped, 5-lobed (3/2), lobes subequal or anterior lobes slightly longer, triangular, throat ± straight; corolla slightly exceeding calyx, 2-lipped, 5-lobed (2/3), blue-lilac, anterior lip with median lobe concave, margin undulate, tube gradually widening above; stamen-filaments parallel, thecae glabrous, divergent at 180°; stigma-lobes subequal; disc 4-horned; nutlets globose, tuberculate, mucilaginous, areole large, curved. One species, *D. sewerzowii* (Regel) Pojark. Mountains, W and C Asia.

180. *Lophanthus* Adans.

Lophanthus Adans., Fam. Pl. 2: 194 (1763); Budantsev, Bot. Zhurn. 77, 9: 69–77 (1992), rev.

Perennial herbs or subshrubs, aromatic; leaves simple; inflorescence thyrsoid, terminal, spike-like or in ovoid heads or paniculate, cymes often long-pedunculate or shortly pedunculate in axils of leaf-like bracts, bracteoles subulate, inconspicuous; flowers resupinate; calyx tubular or tubular-campanulate, indistinctly 2-lipped, 5-lobed, (3/2), lobes unequal, triangular, throat open, calyx-tube annulate within; corolla strongly 2-lipped, 5-lobed, (2/3), blue-violet or pink, anterior lip 3-lobed, median lobe concave or flat, margin undulate or entire, lateral lobes deltoid to ovate, posterior lip 2-lobed, tube gradually widening above, exserted or included in calyx; stamens exserted or included, filaments parallel, thecae parallel or divergent at acute angle; stigma-lobes subequal; disc 4-lobed; nutlets trigonous to ellipsoid, smooth or tuberculate. $2n = 16$. Circa 20 species. Mountains of Afghanistan, C Asia, Mongolia and China. Some species used medicinally.

181. *Hymenocrater* Fisch. & C.A. Mey.

Hymenocrater Fisch. & C.A. Mey., Index Sem. Hort. Bot. Petropol. 2: 39 (1835); Budantsev, Bot. Zhurn. 77, 12: 118–128 (1992), rev.

Shrubs, subshrubs or perennial herbs, aromatic; leaves simple, strongly toothed; inflorescence thyrsoid, spike-like or interrupted, cymes long- to short-pedunculate or sessile; bracts leaf-like, oblong-elliptic to ± ovate, bracteoles inconspicuous; flowers sometimes resupinate; calyx tubular, campanulate, straight or curved, annulate within, actinomorphic or indistinctly 2-lipped, 5-lobed (3/2), lobes subequal, broadly ovate, becoming patent, accrescent in fruit, membranous-scarious, often pale green, rose or purplish, tube funnel-shaped, distinctly ribbed; corolla 2-lipped (2/3), blue-violet, posterior lip shortly 2-lobed, anterior lip with median lobe concave or flat, margins undulate or entire, tube elongate, straight, narrow, gradually widening above; stamens exserted or included, filaments parallel, thecae divergent at 180°; stigma-lobes slightly unequal; disc 4-lobed; nutlets ovoid or elliptic, smooth or tuberculate. $2n = 18$. About ten species, Iran and Afghanistan. Some medicinal, ornamental.

182. *Marmoritis* Benth.

Marmoritis Benth., Hooker's Bot. Misc. 3: 377 (1833); Budantsev, Bot. Zhurn. 77, 12: 118–128 (1992), rev.
Phyllophyton Kudo (1929).
Pseudolophanthus Levin (1941).

Perennial herbs, aromatic; flowering stems erect or ascending, densely leafy above and bare below with small scale-leaves and long rhizomes; leaves simple, sessile or subsessile, densely imbricate, ± orbicular to reniform; inflorescence thyrsoid, terminal, cymes few-flowered, sessile, partially obscured in axils of upper leaf-like bracts; bracteoles entire, linear, shorter than calyces; flower resupinate or not; calyx tubular or campanulate, curved or ± straight, indistinctly 2-lipped, 5-lobed (3/2), lobes ovate to linear, posterior lobes longer, calyx-tube annulate or glabrous within; corolla strongly 2-lipped, 5-lobed (2/3), white or purplish, posterior lip short, straight, bifid, anterior lip with median lobe oblong or orbicular, lateral lobes oblong-ovate, tube gradually dilated towards apex, exserted; stamens included or exserted, filaments parallel, thecae parallel or divergent at 90°; stigma-lobes subequal; nutlets oblong, smooth, glabrous. Four or five species. High alpine screes of the Himalaya and China.

183. *Agastache* J. Clayton ex Gronov.

Agastache J. Clayton ex Gronov., Fl. Virgin.: 88 (1762); Lint & Epling, Am. Midl. Nat. 33: 207–230 (1945), rev.; Sanders, Syst. Bot. Mon. 15: 1–92 (1987), rev. section; Budantsev, Bot. Zhurn. 78, 2: 106–115 (1993), rev.
Dekinia M. Martens & Galeotti (1844).

Perennial herbs, aromatic; leaves simple, dentate; inflorescence thyrsoid, terminal, spike-like or panicle-like, cymes pedunculate or sessile; bracts reduced, lower often leaf-like, bracteoles subulate; calyx tubular or campanulate, straight, indistinctly 2-lipped, 5-lobed (3/2), lobes equal, throat open; corolla strongly 2-lipped, 5-lobed (2/3), pink, blue to violet or white, posterior lip straight, anterior lip with median lobe widest, spreading, entire or undulate, lateral lobes straight, tube gradually dilating to throat; stamens much exserted or included, filaments parallel or posterior pair antrorsely bent, thecae parallel; stigma-lobes equal, spreading; disc 4-lobed; nutlets ellipsoid to obovoid, smooth with apex hairy, areole oblong or orbicular. $2n = 18$. About 22 species. Mountains or deserts of U.S.A, southern Canada and Mexico, one species in W Asia. Sectional treatment after Sanders (1987). Sect. *Agastache*: stamens long-exserted, filaments of posterior stamens antrorsely bent, and crossing anterior pair. Seven species in U.S.A. and Canada, one species in Asia. Sect. *Brittonastrum* (Briq.) Lint & Epling: stamens included or shortly exserted under posterior lip of corolla, filaments of stamens all parallel. 14 species. SW United States and Mexico.

184. *Meehania* Britton

Meehania Britton, Bull. Torrey Bot. Club. 21: 32 (1894); Budantsev, Bot. Zhurn. 77, 12: 118–128 (1992), rev.

Annual or perennial, stoloniferous herbs, not aromatic; leaves simple, dentate; inflorescence thyrsoid, terminal, cymes pedunculate or sessile, few-flowered or sometimes solitary, lax, in axils of leaf-like bracts, bracteoles subulate, inconspicuous; flowers usually large; calyx tubular to campanulate, accrescent in fruit, 2-lipped, 5-lobed (3/2), lobes triangular to lanceolate, acute, those of anterior lip shorter; corolla strongly 2-lipped, 5-lobed (2/3), blue-violet to purplish, posterior lip straight, apex emarginate or 2-lobed, anterior lip with median lobe widely obovate-ovoid, concave, entire or slightly undulate, tube long-exserted, widely and abruptly dilated at throat; stamens shorter than corolla, filaments parallel, thecae parallel, papillose; stigma-lobes subequal; disc 4-lobed; nutlets orbicular to elliptic, hairy, areole orbicular. $2n = 18$. About six species. Temperate or subtropical eastern Asia, one species in N America.

185. *Glechoma* L.

Glechoma L., Sp. Pl. 2: 578 (1753); Budantsev, Bot. Zhurn. 77, 12: 118–128 (1992), rev.
Meehaniopsis Kudô (1929).

Stoloniferous perennial herbs, not aromatic, often gynodioecious; leaves simple; inflorescence of distant, few-flowered cymes, shortly pedunculate to sessile, in axils of leaf-like bracts, bracteoles subulate, inconspicuous; calyx tubular to campanulate, slightly curved near throat, indistinctly 2-lipped, 5-lobed (3/2), lobes equal, broadly triangular to linear; corolla strongly 2-lipped (2/3), blue-violet, lilac or pink, posterior lip straight, 2-lobed, anterior lip with median lobe flat or concave, flabellate or reniform, margin undulate or entire, tube longer than calyx, gradually or abruptly dilated towards throat; stamens of hermaphrodite flowers exserted from tube, included under lip of corolla, inserted near throat, filaments parallel, thecae divergent at 90°; nutlets oblong-ovoid, smooth, areole basal, orbicular or elliptic. $2n = 18, 24, 36$. Four to eight species. Mostly in grasslands, forest margins and stream sides of temperate Eurasia; cultivated in N and S America.

186. *Dracocephalum* L.

Dracocephalum L., Sp. Pl. 2: 594 (1753); Budantsev, Bot. Zhurn. 72, 2: 260–267 (1987), rev.
Moldavica Fabr. (1759).

Herbs, annual or perennial, or subshrubs, aromatic; leaves entire, dentate or pinnatisect; inflorescence thyrsoid, terminal, usually condensed, spike-like or capitate, cymes sessile or peduncles very short; bracts entire or lobed, usually aristate, sometimes coloured or leaf-like, bracteoles often similar to bracts but smaller, or subulate, often aristately toothed; calyx tubular to campanulate, straight or slightly curved, 2-lipped, 5-lobed (3/2), lobes often aristate, upper lip with lobes equal or median lobe clearly larger than laterals, lower lip with lobes shorter, sinuses between lobes with swollen folds at base, throat glabrous within; corolla 2-lipped, 5-lobed (2/3), blue to violet, pink or white, posterior lip straight or slightly concave, subentire or bifid, anterior lip with median lobe largest, tube gradually or abruptly widening from base; stamens included under posterior corolla-lip, rarely exserted, parallel, thecae pilose or glabrous, divergent at 180°; stigma-lobes subequal; disc 4-lobed; nutlets ovoid-oblong, weakly trigonous, apically truncate-rounded or slightly flattened, smooth, mucilaginous or not; areole arcuate. $2n = 10, 12, 14, 18, 20, 36, 72$. About 70 species. Usually montane, Eurasia, N Africa (one species), N America (one species). Some species medicinal, ornamental or bee plants. Three subgenera after Budantsev (1987): *Dracocephalum* (stamens included, anthers glabrous – seven sections), *Fedtschenkiella* (Kudr.) Schischk. (stamens exserted, anthers glabrous), *Ruyschiana* (Mill.) Briq. (stamens included, anthers pubescent).

187. *Lallemantia* Fisch. & C.A. Mey.

Lallemantia Fisch. & C.A. Mey., Index Sem. Hort. Bot. Petropol. 6: 52 (1839).

Annual or biennial herbs; leaves simple; inflorescence thyrsoid, terminal, spike-like or oblong, often interrupted, cymes few-flowered, sessile or rarely on very short peduncles in axils of leaf-like bracts, bracts ovate to linear, weakly toothed to entire, rarely aristate above, bracteoles ovate to rotund or sometimes linear, aristate-toothed; pedicels flattened, erect; calyx tubular, straight or curved, 2-lipped, 5-lobed (3/2), lobes unequal, triangular, aristate, posterior lip with median lobe broadest, sinus between calyx-lobes with a thickened fold or not, throat closed by connivent lips in fruit; corolla strongly 2-lipped (2/3), posterior lip with 2 longitudinal folds inside, anterior lip with lateral lobes directed downwards, tube widening above; stamens included under corolla-lip, filaments parallel, thecae glabrous, divergent at 180°; stigma-lobes subequal; disc 4-lobed; nutlets oblong, trigonous, smooth, mucilaginous. $2n = 14$. Five species. Mediterranean, SW and C Asia. Some medicinal, seed oils.

188. *Cedronella* Moench

Cedronella Moench, Meth. Pl.: 411 (1794); Budantsev, Bot. Zhurn. 78, 2: 106–115 (1993), rev.

Perennial herbs, aromatic; leaves 3(–5)-foliolate, dentate; inflorescence thyrsoid, terminal, spike-like or ovoid, cymes shortly pedunculate to sessile; bracts usually small, linear, coloured, bracteoles inconspicuous; calyx tubular, slightly curved and dilated towards throat, actinomorphic, campanulate in fruit, 5-lobed, lobes equal, narrowly triangular, acute, ± spine-tipped, throat open, slightly oblique; corolla strongly 2-lipped, 5-lobed (2/3), pinkish, posterior lip ± flattened, emarginate, anterior lip with median lobe larger, obovate, margin undulate, concave, tube exserted from calyx, abruptly dilated towards throat; stamens attached near throat, included under posterior lip of corolla, filaments parallel, thecae distinct, parallel; stigma-lobes subequal; disc 4-lobed; nutlets elliptic, smooth, glandular-pubescent, areole orbicular. $2n = 10$. One species, *Cedronella canariensis* (L.) Webb & Berth., Canary Is. and Madeira.

189. *Schizonepeta* (Benth.) Briq.

Schizonepeta (Benth.) Briq. in Engler & Prantl, Nat. Pflanzenfam. IV, 3a: 235 (1896); Budantsev, Bot. Zhurn. 78, 2: 106–115 (1993), rev.

Annual or perennial herbs, aromatic; leaves entire to bipinnate; inflorescence thyrsoid, cymes sessile, terminal, crowded into spike-like, usually unbranched heads; bracts, except sometimes the lowest, reduced, sometimes broadly ovate, coloured, or subulate, inconspicuous, bracteoles subulate, inconspicuous; calyx straight or weakly oblique, actinomorphic or 2-lipped, 5-lobed, lobes ovate-oblong to lanceolate, acute or attenuate into spinulose awns, sinuses between anterior teeth with swollen folds at base; corolla strongly 2-lipped, 5-lobed (2/3), white or blue-violet, posterior lip straight, anterior lip with median lobe flat or scarcely concave, entire to crenate, tube included or scarcely exserted from calyx, gradually widening above; stamens with anterior pair antrorsely bent, posterior pair arcuate-ascending under upper lip, thecae divergent at 90° or at an acute angle; stigma-lobes subequal; disc 4-lobed; nutlets oblong to elliptic, smooth, mucilaginous or not. $2n = 12$. Three species. Steppes of southern Siberia, Mongolia and China.

VII.2.D. Tribe Mentheae: Incertae Sedis

190. *Melissa* L.

Melissa L., Sp. Pl. 2: 592 (1753).

Perennial herbs, ± rhizomatous, aromatic; leaves simple, toothed; inflorescence of axillary verticillasters, widely spaced, with 2–16 flowers per verticillaster; bracts leaf-like, bracteoles small, patent; flowers with curved pedicels; calyx 2-lipped, 5-lobed (3/2), ± campanulate, lobes upcurved, ciliate, posterior lip with lobes very short, lower lip with lobes long-triangular, tube 13-nerved; corolla 2-lipped, hairy, white, cream or reddish, posterior lip shorter, with lobe rounded and emarginate, anterior lip with median lobe semi-orbicular, deflexed, lateral lobes very small, ± triangular, corolla-tube dilated at throat; stamens 4, didynamous, exserted from tube, anterior pair longer, filaments ± straight, thecae divergent, confluent at base; stigma-lobes subulate, subequal; nutlets ovoid, smooth, mucilaginous. $2n = 32, 34, 64$. Four species, usually in forests, Europe, N Africa, Macaronesia and Asia. Plants used as a source of essential oils, both medicinally and as flavouring.

191. *Heterolamium* C.Y. Wu

Heterolamium C.Y. Wu, Acta Phytotax. Sin. 10: 254 (1965).

Slender perennial herb; leaves cordate, toothed, long-petiolate; inflorescence secund, of 1–3-flowered cymes, shortly pedunculate to sessile, in axils of reduced bracts, forming an elongate, terminal thyrse; bracteoles inconspicuous; calyx often purple-tinged, cylindrical to campanulate, strongly 2-lipped, 5-lobed, (3/2), posterior lip with median lobe broadly ovate, rounded at apex, slightly spreading and weakly decurrent on tube at base, lateral lobes shortly triangular-acuminate, anterior lip longer, with sinus between lobes shorter, tube slightly gibbous at base, 15-nerved, annulate within, below throat; corolla white, strongly 2-lipped, 5-lobed (4/1), posterior lip of 4 short, rounded, subequal lobes, anterior lip longer, with lobe slightly concave, rounded, tube narrowly cylindrical, straight, exserted from calyx, and abruptly expanded above, exannulate; stamens 4, didynamous, long-exserted, ascending under upper lip, later becoming slightly divergent, posterior pair longer at early anthesis, anterior pair apparently extending later, anthers 2-thecous, thecae elliptic, widely divaricate, confluent at apex; style exserted, stigma-lobes subequal, subulate, spreading; disc 4-lobed, anterior lobe elongate; nutlets oblong-ovoid, smooth to finely striate, with small, elliptic areole. One species, *H. debile* (Hemsl.) C.Y. Wu, with several varieties, China.

VII.3. Tribe Ocimeae Dumort. (1829).

Aromatic herbs, shrubs or rarely trees. Stamens declinate to porrect (Fig. 29), anthers synthecous, dorsifixed. Disc with lobes alternate to or opposite ovary-lobes.

1. Inflorescence congested, spiciform, terminal at the end of a long axis (but see *Hyptis*). Posterior calyx-lobe often appendaged. Stamens included in corolla-tube. Disc-lobes opposite ovary-lobes **a. Lavandulinae: 192. *Lavandula***
– Inflorescence thyrsoid. Posterior lobe sometimes larger than others, sometimes decurrent on tube, but not appendaged. Stamens exserted from tube (except *Endoste-*

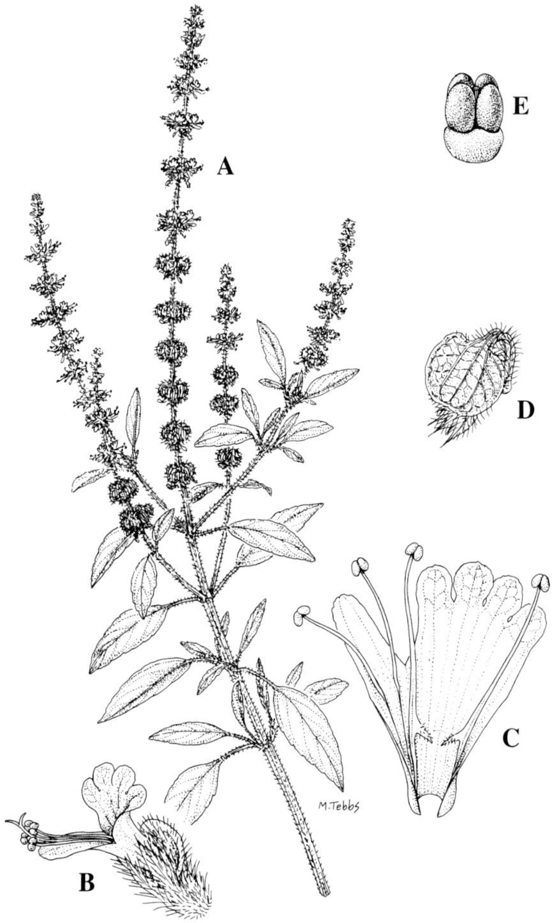

Fig. 29. Labiatae subfam. Nepetoideae tribe Ocimeae. *Ocimum americanum*. **A** Habit. **B** Flower, side view. **C** Corolla opened out. **D** Calxy. **E** Gynoecium and disc. Drawing by M. Tebbs

mon, *Fuerstia* and a few species in other genera). Disc-lobes usually alternate with ovary-lobes 2

2. Anterior lip of corolla thickened at base and reflexing at anthesis to release the stamens explosively **c. Hyptidinae** (excl. *Asterohyptis*): 29
- Anterior lip of corolla not thickened at base, very rarely reflexing at anthesis to release stamens explosively (only in *Aeollanthus* p.p.) 3
3. Anterior and posterior pairs of stamens adjacent at attachment to corolla on the anterior side of the corolla-throat **e. Plectranthinae** (excl. *Tetradenia*): 6
- Anterior and posterior pairs of stamens separated by a clear gap at point of insertion in corolla-tube, anterior pair usually attached at corolla-throat, posterior pair attached at corolla midpoint, or nearer the corolla base, if near the throat, then more regularly spaced not all together on the anterior side of the throat 4
4. Calyx with a single-lobed posterior lip and a usually 4-lobed anterior lip (rarely 3- or 7-lobed), if equally 5-lobed, then cymes sessile, condensed and lacking bracteoles **d. Ociminae** (plus *Tetradenia*): 15
- Calyx with a three-lobed posterior lip and a two-lobed anterior lip, if equally 5-lobed then cymes pedunculate and open with bracteoles at cyme nodes 5
5. Nutlets with a prominent areole at the abscission-scar; flowers gynodioecious, sessile **203. *Asterohyptis***
- Nutlets without a prominent areole at the abscission-scar; flowers hermaphroditic, pedicellate **b. Hanceolinae:** 27
6. Lateral lobes of the corolla deflexed towards the anterior lobe **220. *Thorncroftia***
- Lateral lobes of the corolla ascending towards the median lobes of the posterior lip or horizontal and held between the anterior lobe and median lobes of the posterior lip 7
7. Posterior lip of corolla not truncate, diverging from anterior lip of the corolla, with lateral lobes usually ascending and close to median lobes of posterior lip 10
- Posterior lip of corolla truncate, lateral lobes usually horizontal, held between the anterior lobe and median lobes of the posterior lip 8
8. Calyx unequally lobed, 1/4 **216. *Dauphinea***
- Calyx with 5 subequal lobes 9
9. Calyx-lobes deltoid becoming expanded and membranous in fruit; nutlets glandular, not winged **217. *Capitanopsis***
- Calyx-lobes lanceolate, not expanded and membranous in fruit; nutlets not glandular, apically winged **218. *Madlabium***
10. Calyx basally circumscissile; tube distally dorsiventrally flattened **224. *Aeollanthus***
- Calyx very rarely circumscissile; tube not dorsiventrally flattened 11
11. Calyx 3-lobed (1/2), expanded and membranous in fruit **225. *Alvesia***
- Calyx 5-lobed (1/4 or actinomorphic), rarely expanded and membranous in fruit 12
12. Calyx-teeth all equal, spinescent, cymes single-flowered **226. *Pycnostachys***
- Calyx-teeth equal or not, lanceolate to ovate 13
13. Calyx-throat closed or partially closed by deflexed upper lip, inflorescence condensed 14
- Calyx-throat usually open, sometimes closed by hairs, upper lip horizontal or curving upwards, inflorescence condensed or lax **219. *Plectranthus***
14. Stamens free, Asia **222. *Anisochilus***
- Stamens fused, Africa **223. *Leocus***
15. Anterior stamens fused 16
- Anterior stamens separate 17
16. Posterior lip of the calyx with margins decurrent on tube **205. *Hemizygia***
- Posterior lip of the calyx not decurrent on tube **206. *Syncolostemon***
17. Flowers sessile **212. *Catoferia***
- Flowers with distinct pedicels 18
18. Style bifid with stigma-lobes diverging 24
- Style capitate-clavate with stigma-lobes connivent or parallel 19
19. Posterior stamens sterile 20
- All stamens fertile 21
20. Calyx orange, swollen and fleshy in fruit, teeth ± equal **209. *Hoslundia***
- Calyx green or brown, thin, not fleshy in fruit, posterior tooth broader than others **208. *Fuerstia***
21. Calyx equally 5-toothed, actinomorphic **211. *Benguellia***
- Calyx with a broad posterior lip and a 4-lobed anterior lip 22
22. Flowers ca. 3 mm long. Disc equally 4-lobed **210. *Basilicum***
- Flowers more than 3 mm long, rarely 3 mm. Disc with anterior lobe larger 23
23. Stamens all attached to corolla at throat, rarely exserted **213. *Endostemon***

- Anterior stamens attached to corolla at throat, posterior attached around midpoint of corolla, exserted from corolla-tube **207. *Orthosiphon***
24. Bracts basally pale, rarely uniformly green; posterior stamens basally swollen or with a flattened, blade-like basal appendage; nutlets minutely striate or reticulate 25
- Bracts uniformly green, rarely uniformly coloured; posterior stamens not basally swollen, if appendiculate then appendage knee-shaped; nutlets various, not striate 26
25. Fruiting-calyx flattened in cross section, anterior and posterior lips easily split apart in fruit, lateral lobes in same plane as posterior lobe **214. *Haumaniastrum***
- Fruiting-calyx rounded in cross section, anterior and posterior lips not splitting apart in fruit, lateral lobes either midway between anterior and posterior lobes or nearer the posterior lobe, but not in the same plane **215. *Platostoma***
26. Lateral lobes of corolla held between the anterior lip and median lobes of the posterior lip, plants often dioecious **221. *Tetradenia***
- Lateral lobes of corolla ascending, held close to the median lobes of the posterior lip **204. *Ocimum***
27. Posterior stamens attached to corolla at midpoint or below midpoint of the corolla-tube **193. *Isodon***
- Posterior stamens attached near the corolla-throat 28
28. Anterior lobe of corolla deflexed, tube clearly dilating at midpoint; cymes usually pedunculate **195. *Hanceola***
- Anterior lobe of corolla pointing forwards, tube gradually dilating towards throat, cymes sessile **194. *Siphocranion***
29. Cymes usually 1-flowered, flowers with often inconspicuous, paired bracteoles at base of calyx. If cymes not as described, then flowering stems waxy with often inflated internodes 30
- Cymes many-flowered, flowers often congested or modified into bracteate heads, or rarely in axillary fascicles. Paired bracteoles not present at base of calyx. Flowering stems never waxy or inflated 31
30. Calyx strongly two-lipped; posterior lip rounded or with lobes connate. Corolla-tube strongly contracted near base **197. *Eriope***
- Calyx ± actinomorphic, not strongly 2-lipped. Corolla-tube cylindrical, not contracted near base **198. *Hypenia***
31. Flowers in lax, often few-flowered heads. Calyx-lobes triangular, often reflexing in fruit. Nutlets cymbiform with an involute, laciniate margin on inner face **199. *Marsypianthes***
- Flowers arranged variously. Calyx-lobes not reflexing in fruit. Nutlets ovoid or flattened, not as above 32
32. Calyx-lobes with a conspicuous expanded and flattened appendage at apex **201. *Peltodon***
- Calyx-lobes without appendage at apex, or absent 33
33. Flowers in spherical heads, dropping as a unit in fruit. Corolla deep purple. Calyx with 5–10 unequal spines **202. *Rhaphiodon***
- Flowers not as above, if in heads not dropping as a unit in fruit. Calyx-lobes 5, not spinose, or sometimes subulate and rigid at apex 34
34. Flowers arranged variously, often in pedunculate or sessile, congested cymes or involucrate capitula in terminal panicles, or in spikes or axillary; flowers sometimes in lax cymes, rarely fasciculate on long pedicels in the axils of the leaves. If stylopodium present, then flowers in involucrate capitula **200. *Hyptis***
- Flowers in usually lax cymes. Stylopodium present **196. *Hyptidendron***

VII.3.A. Subtribe Lavandulinae Endl. (1838).

Inflorescence spiciform, condensed at the end of a long axis. Cymes 3–9-flowered. Bracts present, bracteoles present, minute, or absent. Flowers with posterior lip of calyx sometimes larger or rarely modified into an appendage, calyx-tube 13-, 15-, or rarely 8-nerved, stamens declinate to porrect, usually didynamous, included within tube, anthers synthecous, stigma-lobes 2 or stigma unlobed, capitate. Disc usually 4-lobed, lobes borne opposite nutlets. Nutlets with areole present.

192. *Lavandula* L.

Lavandula L., Sp. Pl. 2: 572 (1753); Chaytor, J. Linn. Soc., Bot. 51: 153–204 (1937); Upson, Doct. Thesis, Reading University (1997), rev.
Stoechas Mill. (1754).
Fabricia Adans. (1763).
Styphonia Medik. (1791).
Chaetostachys Benth. (1831).
Sabaudia Buscal. & Muschl. (1913).
Isinia Rech. f. (1952).

Shrubs, subshrubs or short lived herbs, usually strongly aromatic, glabrous or indumentum variable, sometimes stellate; leaves entire or dissected; inflorescence a congested terminal spiciform thyrse, occasionally with coma of coloured bracts, usually distinctly pedunculate, simple or branched, verticillaster either a 3–9-flowered cincinnus with minute bracteoles or cyme 1-flowered, usually ebracteolate; bracts persistent, opposite or spirally arranged; flowers sessile or pedicellate with calyx persistent, ovoid-cylindrical, actinomorphic to 2-lipped, 5-lobed (3/2), lobes ± equal or posterior lip larger or modified into an appendage, tube 8-, 13- or 15-nerved, nerves to anterior lobes reaching and sometimes anastomosing at apex; corolla weakly or strongly 2-lipped, 5-lobed (2/3), blue-violet to purple or white, rarely blackish purple or yellowish, lobes spreading, variable in size, corolla-tube either just exceeding or up to 3× longer than calyx; stamens 4, declinate, usually didynamous, included within tube, anterior pair longer; filaments glabrous, anthers reniform, confluent; stigma-lobes, if present, often complanate or stigma capitate; nutlets variable in shape, colour and size, with either small basal abscission-scar or lateral scar (areole) 0.25 to 0.75× length of nutlet, usually mucilaginous. $2n$ = 18, 24, 30, 36, 42, 54. Circa 36 species from Macaronesia, Mediterranean basin, N and NE Africa, SW Asia, Arabian Peninsula, C and S India. A number of species, principally *L. angustifolia* Mill. and *L.*× *intermedia* Emeric ex Loisel., are commercially important for their essential oils, used in the perfumery and fragrance industry, aromotherapy and as ornamentals. A revised infrageneric classification of three subgenera and seven sections has been proposed (Upson 1997).

VII.3.B. Subtribe Hanceolinae (C.Y. Wu) A.J. Paton, Ryding & Harley, in Harley, Paton & Ryding, Kew Bull. 58: 485–489 (2003).

Tribe Hanceoleae C.Y. Wu in Acta Phytotax. Sin. 8: 58 (1959); Wu & Li, Flora Reip. Pop. Sin. 66: 396 (1977).

Cymes pedunculate, rarely sessile, bracteolate, rarely ebracteolate (*Siphocranion*). Bracts persistent. Calyx 2-lipped with a 3 lobed posterior lip and 2-lobed anterior lip, often actinomorphic and regularly 5-lobed. Corolla 2-lipped, anterior lip flat, concave to cucullate. Stamens declinate, not exceeding anterior lip; anterior stamens attached to corolla at throat; posterior stamens attached just below throat or near corolla base. Disc 4-lobed, lobes alternating with nutlets. Nutlets with small inconspicuous areole.

193. *Isodon* (Benth.) Schrader ex Spach

Isodon (Benth.) Schrader ex Spach, Hist. Nat. Vég. 9: 162 (1840); Li, J. Arnold Arbor. 69: 289–400 (1988), rev.; Paton & Ryding, Kew Bull. 53: 723–731 (1998).
Plectranthus L'Hér. sect. *Isodon* Schrad. ex Benth. (1832).
Plectranthus L'Hér. sect. *Melissoides* Schrad. ex Benth. (1832).
Plectranthus L'Hér. sect. *Pyramidatum* Schrad. ex Benth. (1832).
Plectranthus L'Hér. sect. *Amethystoides* Schrad. ex Benth. (1832).
Rabdosia (Blume) Hassk. (1842).
Amethystanthus Nakai (1934).
Homalocheilos J.K. Morton (1962).
Skapanthus C.Y. Wu & H.W. Li (1975).
Dielsia Kudô (1929).

Shrubs or herbs; leaves simple, opposite, rarely ternate; inflorescence thyrsoid with cymes pedunculate, bracteate, bracteolate, 3- to several-flowered; bracts persistent, not coloured, smaller than leaves; bracteoles small, usually linear; calyx campanulate, funnel-shaped, or tubular and sometimes slightly constricted or rarely curved at throat, actinomorphic or 2-lipped, 5-lobed (3/2 or regular), lobes lanceolate to deltoid, posterior never decurrent, throat open, glabrous; corolla strongly 2-lipped, 5-lobed (4/1), white, blue, pink or purplish, posterior lip ascending or reflexed, 4-lobed with lobes equal or rarely median more shallowly lobed, anterior lip horizontal, cucullate, corolla-tube straight or geniculate, saccate to spurred at base, parallel-sided distally; stamens 4, held within, or more rarely exceeding the anterior lip, posterior pair attached near base of corolla or between base and midpoint, usually with hairs, anterior pair attached between midpoint of corolla and throat, often with hairs; stigma-lobes diverging; disc 4-lobed with anterior lobe slightly larger, lobes alternating with nutlets; nutlets ovoid, rarely conical with a flattened base, glabrous, sometimes with hairs or glands, mucilaginous or not, areole small, inconspicuous. $2n = 24, 42$. Around 100 species, tropical Asia with two species in Africa.

194. *Siphocranion* Kudô

Siphocranion Kudô, Mem. Fac. Sci. Taihoku Imp. Univ. 2: 53 (1929); Li & Hedge in Wu & Raven (eds.) Flora of China 17: 268–291, reg. rev.; Paton & Ryding, Kew Bull. 53: 723–731 (1998).

Perennial herbs; leaves simple; inflorescence thyrsoid with cymes sessile, bracteate, bracteolate or ebracteolate, usually 1-, rarely 3-flowered; bracts persistent, not coloured, smaller than leaves; calyx funnel-shaped, 2-lipped, 5-lobed (3/2), lobes lanceolate to deltoid, posterior never decurrent, throat open, glabrous; corolla strongly 2-lipped, 5- or 4-lobed (4/1 or 3/1), purplish, posterior lip ascending or forward-pointing, 3- or 4-lobed with median lobe emarginate or median lobes smaller than laterals, anterior lip horizontal, cucullate, corolla-tube straight or curved, sometimes slightly inflated at base, parallel-sided and dilating slightly at throat; stamens 4, held within tube or at base of the anterior lip, posterior pair attached just below corolla-throat, glabrous, short, anterior pair attached at throat or just below, glabrous, very short; stigma-lobes diverging; disc 4-lobed with anterior lobe slightly larger, lobes alternating with nutlets; nutlets ovoid, with glands, not mucilaginous, areole inconspicuous. Two species in China and Indochina, in mixed and evergreen forest.

195. *Hanceola* Kudô

Hanceola Kudô, Mem Fac. Sci. Taihoku Imp. Univ. 2: 54 (1929); Paton & Ryding, Kew Bull. 53: 723–731 (1998).
Hancea Hemsl. (1890), nom illegit.

Herbs; leaves simple; inflorescence thyrsoid with cymes usually pedunculate, bracteate, usually bracteolate, 1–7-flowered; bracts persistent, not coloured, smaller than leaves; calyx campanulate or funnel-shaped, actinomorphic or two-lipped, 5-lobed (3/2, 1/4 or regularly 5-toothed), lobes lanceolate to deltoid, posterior sometimes decurrent, throat open, glabrous; corolla 2-lipped, 5-lobed (4/1), usually purplish, sometimes white or yellow, posterior lip spreading or forward-pointing, lateral lobes spreading or forward-pointing, anterior lip deflexed, concave or cucullate, corolla-tube curved, dilating from midpoint towards throat; stamens 4, held over anterior lip or sometimes exserted beyond it, posterior pair attached just below corolla-throat, papillate or glabrous, anterior pair attached at throat, glabrous or papillate; stigma-lobes diverging; disc 4-lobed with anterior lobe larger, lobes alternating with nutlets; nutlets ovoid, with glands, not mucilaginous, areole inconspicuous. About eight species, endemic to China, in evergreen and mixed forests.

VII.3.C. Subtribe Hyptidinae Endlicher (1838).

Cymes thyrsoid, pedunculate or not, bracteolate. Bracts persistent or deciduous, not coloured. Calyx actinomorphic or weakly 2-lipped, 5-lobed. Corolla strongly 2-lipped, rarely weakly so; anterior lip of corolla strongly concave, usually compressed, thickened at base and reflexing at anthesis to explosively release stamens which at first are held within anterior lip of corolla (excl. *Asterohyptis*). Stamens 4, attached near corolla-throat, or posterior pair attached near corolla base and anterior pair attached near throat. Style sometimes jointed above ovary, with persistent base (stylopodium). Stigma-lobes equal, connivent. Disc usually 4-lobed, lobes usually alternating with nutlets, rarely opposite (see *Hyptidendron*). Nutlet with expanded areole.

196. *Hyptidendron* Harley

Hyptidendron Harley, Bot. J. Linn. Soc. 98: 90 (1988), rev.
Hyptis Jacq. sect. *Buddleioides* Benth. (1833).
Hyptis Jacq. sect. *Umbellaria* Benth. (1833).

Trees, shrubs or subshrubs, often aromatic, hairs often branched; leaves simple, toothed; inflorescence of terminal or axillary thyrsoid, bracteate panicles, or axillary cymes; cymes (1–)3–many-flowered; bracts often leaf-like, sometimes conspicuous with white or purplish hairs, bracteoles inconspicuous; flowers shortly pedicellate; calyx tubular to infundibuliform, straight, actinomorphic to weakly zygomorphic, 5-lobed, lobes equal or slightly unequal with posterior lobe longer, erect or spreading; corolla strongly 2-lipped, 5-lobed (2/3), bluish-purple to lilac or rarely white, anterior lip with median lobe elongate, lateral lobes spreading, corolla-tube cylindrical to infundibuliform (sect. *Umbellaria*), straight; stamen filaments hairy; style with stylopodium overtopping ovary; disc weakly 4-lobed, lobes usually alternating with ovary-lobes, but opposite in *H. asperrima* (Spreng.) Harley (Rudall 1981b); nutlets ovoid to slightly flattened, sometimes winged (sect. *Hyptidendron*), weakly mucilaginous. $2n$ = ca. 64. About 19 species in savannas and margins of montane forest, mainly central, eastern and north-eastern Brazil, extending to Bolivia, with *H. arborea* (Benth.) Harley occurring in the Guayana Highlands: Venezuela, Guyana and northern Brazil, and disjunctly in the mountains of Colombia (Antioquia) and the Peru/Bolivia border. Two sections have been recognised (Harley 1988b): sect. *Hyptidendron* (five species) with corolla-tube cylindrical, 6–11 mm long, nutlets narrowly winged; and sect. *Umbellaria* (Benth.) Harley (c. 14 species), corolla-tube < 7 mm long, ± infundibular, nutlets unwinged. *H. arborea* (sect. *Hyptidendron*) can attain a height of 20 m, although usually less.

197. *Eriope* Bonpl. ex Benth.

Eriope Bonpl. ex Benth., Labiatarum Genera et Species: 142 (1833); Benth. in DC., Prodr. 12:140 (1848), rev.; Epling, Feddes Rep. Beih. 85: 188–195 (1936), rev.; Harley, Hook. Ic. Pl. 38, 3: 1–107 (1976), rev.
Eriopidion Harley (1976).

Trees, shrubs, subshrubs or herbs, often geoxylic, usually aromatic; stems hairy, or in some species virgate, with setose hairs below and upper internodes glabrous, pruinose, fistulose, swollen; leaves simple; inflorescence thyrsoid, racemoid, with cymes 1-flowered, pedunculate; bracts inconspicuous, usually caducous; flowers ± sessile, but with minute, persistent, paired bracteoles at base of calyx and appearing pedicellate; calyx turbinate to campanulate, rarely tubular, weakly 2-lipped, 5-lobed (3/2), lobes subequal with posterior usually longer and connate, often ± obsolete in fruit, or rarely with posterior lobe larger and hygroscopic in fruit, throat sometimes closed by dense white hairs; corolla strongly 2-lipped, 5-lobed (2/3), violet or pink, anterior lip with median lobe with apex acuminate, margin fimbriate or toothed, lateral lobes directed forwards, corolla-tube widening above, usually constricted near base, dorsally gibbous; stamen filaments hairy; style with stylopodium overtopping ovary, rarely absent, disc weakly 4-lobed, anterior lobe larger; nutlets slightly flattened to ovoid, rarely narrowly winged or triquetrous, usually mucilaginous. $2n$ = 20 (?40). Circa 40 species, mostly in savannas and mountains of C and E Brazil, but extending into neighbouring countries.

198. *Hypenia* (Mart. ex Benth.) Harley

Hypenia (Mart. ex Benth.) Harley, Bot. J. Linn. Soc. 98: 91 (1988); Atkinson, Ph.D. Thesis, University of St. Andrews (1998), rev. as *Eriope*.
Hyptis Jacq. sect. *Hypenia* Mart. ex Benth. (1833).

Shrubs or subshrubs, often aromatic; stems usually virgate, erect, upper internodes often glabrous, pruinose, fistulose and ± inflated, nodes and lower stems often with long, rigid setose hairs; leaves simple, toothed or lobed; inflorescence thyrsoid, lax or congested, cymes usually 1(–3)-flowered, rarely to 6-flowered, pedunculate; bracts inconspicuous, caducous or persistent; flowers ± sessile, except in *H. vitifolia* (Pohl ex Benth.) Harley, but appearing pedicellate, with usually minute, persistent bracteoles at base of calyx, flowers rarely resupinate; calyx campanulate to tubular, actinomorphic, 5-lobed or ± two-lipped (3/2), lobes subequal or posterior connate at base, deltate, rarely long-subulate at apex, throat hairy or not; corolla strongly 2-lipped, 5-lobed (2/3), lilac, blue, white or bright red to orange or pink, anterior lip with apex of median lobe rounded to acute, margin toothed, lateral lobes spreading or directed forwards, tube often elongate, cylindrical, rarely infundibuliform, often ± constricted at base; stamens exserted from tube, filaments pubescent; style jointed, but stylopodium not overtopping ovary, disc weakly lobed, anterior lobe larger; nutlets ovoid, weakly mucilaginous. $2n$ = 12, 20, 22, 28. About 23 species in seasonal to semi-arid savannas in southern and north-eastern Brazil, extending into eastern Paraguay and Bolivia.

199. *Marsypianthes* Mart. ex Benth.

Marsypianthes Mart. ex Benth., Labiatarum Genera et Species: 53, 64 (1833); Epling, Rep. Spec. Nov. Beih. 85: 184–186 (1936), rev.
Marsypianthus Benth. (sphalm., 1833).

Subshrubs or perennial herbs, often geoxylic, aromatic and glandular-viscid; leaves simple, toothed; inflorescence thyrsoid, of axillary cymes, pedunculate or not; cymes 1–3- to many-flowered and then in spherical heads; bracts and bracteoles present, narrowly elliptic-lanceolate to linear, in many-flowered cymes often forming a lax involucre; flowers shortly pedicellate; calyx infundibuliform, straight, actinomorphic, 5-lobed, often purple-tinged and sometimes conspicuously so, lobes equal, broadly to narrowly deltoid, apex acute to subulate, erect or connivent at first and often becoming spreading to reflexed in fruit; corolla strongly 2-lipped, 5-lobed (2/3), violet-blue, lilac or rarely cream, anterior lip with median lobe often much shorter than others, lateral lobes directed forwards; corolla-tube cylindrical, straight; stamen-filaments hairy; style jointed, deciduous above, its persistent base fused to carpels along their inner face; disc well developed, almost encircling ovary; nutlets cymbiform, dorsally smooth, convex, ventral surface fused to style base, on separation concave with involute, laciniate margin, dorsal surface weakly mucilaginous. $2n$ = 30. Five to six species, savannas and coastal dunes, Brazil, Paraguay and Argentina, with the weedy *M. chamaedrys* (Vahl) Kuntze from Mexico southwards and widespread in the neotropics.

200. *Hyptis* Jacq.

Hyptis Jacq., Collectanea. 1: 101, 103 (1786) [1787], nom. cons.; Epling, Rev. Mus. La Plata, Bot. 7: 1–497 (1949), rev.
Mesosphaerum P. Browne (1756).

Shrubs, subshrubs, often geoxylic, or annual to perennial herbs, strongly aromatic to odourless; leaves simple to pinnatifid or subdigitate; inflorescence thyrsoid, cymes lax or congested, pedunculate to sessile, often forming involucrate capitula, these sometimes in paniculate synflorescences, or cymes congested into spiciform or head-like thyrses, or sessile or subsessile in axils of leaf-like bracts; bracts and bracteoles usually inconspicuous to leaf-like and sometimes coloured; flowers ± pedicellate to sessile; calyx tubular to infundibuliform, actinomorphic to zygomorphic, rarely 2-lipped, 5-lobed, rarely sinuses appendaged and appearing 10 lobed, lobes sometimes obsolete, lobes usually equal, deltoid to subulate or filiform, rarely uncinate, erect to spreading or connivent in fruit, tube sometimes curved or sigmoid with deflexed throat, especially in fruit, often accrescent, sometimes annulate within; corolla strongly 2-lipped, 5-lobed (2/3), white or cream to lilac, pink or bluish purple, rarely red, posterior lip erect, anterior lip apiculate or rounded, rarely explosive pollination mechanism non-functional with lip cochleariform, corolla-tube ± infundibuliform to tubular, sometimes long and slender; stamens exserted from tube, filaments hairy to glabrous; style with stylopodium absent or present, disc weakly 4-lobed; nutlets 4 (or 1–2), flattened to ovoid, apex rounded to apiculate or toothed or comose, paired aerenchymatous appendages rarely present on dorsal surface of nutlet, mucilaginous or not. $2n$ = 16, 28, 30, 32, 40–60, 64, 96. About 280 species, mostly tropical and subtropical savannas, sometimes in humid areas, almost entirely New World, from southern U.S.A. to Caribbean and S to Argentina and Peru, a few species extending into Old World. About 26 sections currently recognised (Harley, ined), based partly on those of Epling (1949).

201. *Peltodon* Pohl

Peltodon Pohl, Pl. Bras. 1: 66 (1827); Epling, Rep. Spec. Nov. Beih. 85: 195–196 (1936), rev.

Subshrubs or herbs, often strongly geoxylic, not aromatic; leaves simple, toothed, often bullate; inflorescence a solitary, many-flowered, axillary or terminal, long-pedunculate capitulum, bracts and bracteoles persistent, involucrate, ovate, green or coloured; flowers sessile; calyx tubular, straight, actinomorphic, becoming campanulate in fruit, 5-lobed, lobes subulate, rigid, erect with apices spreading and minutely foliaceous, often coloured; corolla strongly 2-lipped, 5-lobed (2/3), cream to yellow and/or vinaceous, lobes of posterior lip triangular, anterior lip with median lobe elongate, lateral lobes directed downwards, corolla-tube narrow, widening above; stamens exserted from tube, filaments hairy; style without stylopodium; disc weakly developed; nutlets rotund, weakly trigonous, slightly mucilaginous. $2n$ = 30, 32. Six species, savannas of southern Brazil and Paraguay, extending to eastern Bolivia and northern Argentina.

202. *Rhaphiodon* Schauer

Rhaphiodon Schauer, Flora 27: 345 (1844); Epling, Rep. Spec. Nov. Beih. 85: 186–188 (1936), rev.
Zapania Nees & Mart. (1823) non Lam. (1791).

Perennial herb, strongly aromatic; stems prostrate from central taproot, rooting at nodes; leaves simple, toothed; inflorescence a solitary spherical capitulum, many-flowered, axillary, long-pedunculate, peduncle erect to ascending; capitulum with involucre of bracts and bracteoles, these long-subulate, becoming rigid, pungent; flowers sessile; calyx shortly infundibuliform, weakly zygomorphic, irregularly 5–10-lobed, lobes long-subulate, in fruit spreading and pungent, throat constricted and closed by dense white hairs, tube becoming inflated; corolla strongly 2-lipped, 5-lobed (2/3), deep purple, lobes of posterior lip rounded, anterior lip with median lobe short, directed forwards, lateral lobes longer, equalling and opposing posterior lip, corolla-tube cylindrical, widening slightly above; stamens exserted from tube, anthers revealing conspicuous, white pollen; style with stylopodium not extending above ovary, stigma ± unlobed, subcapitate to truncate; disc usually 4-lobed with anterior lobe largest or with 2 large lobes, overtopping ovary; ovules 2(–3); fruiting capitulum spinose, dispersed as a unit at maturity; nutlets large, usually 1–2 developing per flower, slightly flattened, glossy, mucilage absent. $2n$ = 26. One species, *R. echinus* (Nees & Mart.) Schauer, in dry sandy places subject to seasonal flooding, NE Brazil.

203. *Asterohyptis* Epling

Asterohyptis Epling, Bull. Torrey Bot. Club 60: 17 (1932), rev.

Shrubs, weakly aromatic, often gynodioecious; leaves simple; inflorescence thyrsoid, spike-like and paniculately branched with cymes shortly pedunculate or sessile, often 2 per leaf-axil, 6–12-flowered; bracts usually persistent, inconspicuous, often subulate and rigid, bracteoles similar but smaller; calyx tubular, straight, actinomorphic, 5-lobed, lobes ± equal, subulate, rigid, erect or more often widely spreading and spinose-tipped, throat open; corolla weakly 2-lipped, 5-lobed with lobes subequal, corolla-tube straight, hairy within; stamens with at least anterior pair weakly exserted from tube, filaments short, anterior pair longer; female plants with stamens reduced; style with stylopodium absent, stigma-lobes short; disc 4-lobed, anterior lobe larger; nutlets ovoid, mucilaginous. $2n$ = 30. Three species, in savannah, Mexico southwards to Costa Rica.

VII.3.D. Subtribe Ociminae (Dumort.) Schmidt in Martius (1858).

Subtribe *Moschosminae* Endlicher (1838).

Cymes sessile, ebracteolate. Bracts usually persistent, sometimes coloured. Calyx usually with posterior lip 1-lobed and anterior lip 4-lobed, sometimes actinomorphic and regularly 5-lobed.

Corolla usually strongly 2-lipped, rarely more regular; anterior lip flattened to concave. Stamens declinate, sometimes porrect, usually exserted, rarely included within corolla-tube; anterior pair attached at corolla-throat, posterior pair usually attached near corolla base or around corolla midpoint, rarely near throat on posterior side of the corolla. Disc usually 4-lobed, lobes alternating with nutlets. Nutlet with small, inconspicuous areole.

204. *Ocimum* L. Fig. 21C, L, Q, 29

Ocimum L., Sp. Pl. 2: 597 (1753); Gen. Pl. ed. 5: 259 (1754); Paton, Harley & Harley in Holm & Hiltunen (eds.) *Ocimum.* Medicinal and aromatic plants - industrial profiles. Harwood Academic, Amsterdam: 1–38 (1999), rev.
Becium Lindl. (1842).
Erythrochlamys Gürke (1894).
Nautochilus Brem. (1933).
Orthosiphon sect. *Serrati* Ashby (1938).
Orthosiphon subg. *Nautochilus* (Bremek.) L.E. Codd (1964).

Shrubs or herbs, often geoxylic, usually aromatic; leaves simple; inflorescence thyrsoid with cymes sessile, usually 3-flowered; bracts very rarely coloured, persistent or, if caducous, bract-scar developing into auxiliary nectary; calyx infundibular to tubular, straight, curving upwards or rarely bent downwards, strongly 2-lipped, 5-lobed (1/4), lobes unequal, posterior rounded, decurrent, curving upwards, lateral teeth deltoid, sometimes with a prominent shoulder or truncate, anterior lobes deltoid to subulate, sometimes closing the often densely hairy calyx-throat; corolla strongly 2-lipped, 5-lobed (4/1), white or pink, posterior lip ascending, equally 4-lobed or with median exceeding lateral lobes, entire or fimbriate, anterior lip ± horizontal, flat or slightly concave, corolla-tube straight or curved towards base, usually dorsally gibbous at midpoint, often dilating towards throat; stamens 4, posterior pair attached near corolla base, rarely sterile, exserted, filaments usually pubescent at base, often appendiculate, anterior pair attached at corolla-throat; stigma-lobes equal, subulate; disc equally 4-lobed; nutlets spherical to ellipsoid, rarely apically pubescent, usually mucilaginous. $2n = 24–76, 88$. Sixty-five species, pantropical, mostly in Africa, in drier areas, grassland and open woodland. Several species, especially *O. basilicum* L. and *O. tenuiflorum* L., cultivated as potherbs, for medicinal use and ritual. Some species are used as indicators of heavy metals. Subgenus *Nautochilus* (Bremek.) Paton may require recognition at generic rank.

205. *Hemizygia* (Benth.) Briq.

Hemizygia (Benth.) Briq., Engler and Prantl, Nat. Pflanzenfam. IV, 3a: 368 (1897); Ashby, J. Bot. 73: 312, 343 (1935); Codd, Bothalia 12: 1 (1976), rev.
Ocimum sect. *Hemizygia* Benth in DC. (1848).
Orthosiphon sensu Baker (1900) pro parte.
Orthosiphon sensu N.E. Br. (1910) pro parte.
Bouetia A. Chev. (1917).

Shrubs or herbs, often geoxylic, usually aromatic; leaves simple; inflorescence thyrsoid with cymes sessile, 1–3-flowered; bracts usually caducous, brightly coloured at apex; calyx shortly tubular, straight or curving slightly upwards, strongly 2-lipped, 5-lobed (1/4), lobes unequal, posterior rounded, decurrent, accrescent, curving upwards, lateral lobes deltoid, anterior subulate, throat open; corolla strongly 2-lipped, 5-lobed (4/1), white, pink or purple, posterior lip ascending with median lobes smaller and exceeding lateral lobes, usually truncate, anterior lip horizontal or slightly deflexed, slightly concave, tube straight or slightly curved, dilating slightly towards throat; stamens 4, exserted, or very rarely posterior pair included, these attached just below midpoint, ciliate or glabrous, inappendiculate, anterior pair attached at corolla-throat, connate; style bifid to clavate, stigma-lobes equal, linear or more rounded; disc 4-lobed with anterior lobe larger; nutlets ovoid, smooth with prominent dorsal vein, mucilaginous or not. $2n = 28$. Thirty species mainly in southern Africa, a few in tropical Africa, one in Madagascar and one in India. Open grassy areas or forest margins often in rocky or sandy areas.

206. *Syncolostemon* E. Mey. ex Benth. in E. Mey.

Syncolostemon E. Mey. ex Benth. in E. Mey., Comm. Pl. Afr. Austr.: 230 (1837); Codd, Bothalia 12: 21–27 (1976), rev.; Paton, Kew Bull. 53: 483–485 (1998).

Shrubs or herbs, often geoxylic, usually aromatic; stems branched; leaves simple; inflorescence thyrsoid with cymes sessile, 1–3-flowered; bracts caducous, coloured, but small and inconspicuous; calyx tubular, straight or curving slightly upwards, 5-lobed, either radially symmetrical with lobes ± equal, lanceolate or strongly 2-lipped (1/4), lobes unequal, posterior rounded, not decurrent, nor strongly curving upwards, lateral lobes deltoid, anterior subulate, throat open; corolla strongly 2-lipped, 5-lobed (4/1), white, pink or purple, posterior lip ascending with median lobes smaller and exceeding lateral lobes, usually truncate, anterior lip horizontal or slightly deflexed, slightly concave, corolla-tube straight or slightly curved, dilating slightly towards throat; stamens 4, exserted, posterior pair attached just below midpoint, ciliate or glabrous, inappendiculate, anterior attached at corolla-throat, connate; style bifid to clavate with stigma-lobes equal, linear or more rounded; disc 4-lobed with anterior lobe larger, nutlets ovoid, smooth with a prominent dorsal vein, mucilaginous or not. Ten species in South Africa and Zimbabwe, grassland or forest margins, often in rocky areas. *Syncolostemon* is very closely related to *Hemizygia*, but differs consistently in the lack of a decurrent posterior calyx-lip. The genera may need to be merged.

207. *Orthosiphon* Benth.

Orthosiphon Benth. in Edward's Bot. Reg. sub t. 1300 (1830); Ashby, J. Bot. 76: 1–10, 39–48 (1938), reg. rev.; Codd, Bothalia 8: 146–162 (1964), reg. rev.; Keng in Fl. Malesiana 8, 3: 379–382 (1978); Paton & Hedge in Hedge, Clement, Paton & Phillipson, Flore de Madagascar et des Comores 175: 117–135 (1998), reg. rev.
Clerodendranthus Kudô (1929).

Shrubs or herbs, often geoxylic, sometimes annual, aromatic or not; leaves simple, opposite, rarely ternate or quadrate; inflorescence thyrsoid with cymes sessile, usually 3-, more rarely 1-flowered; bracts persistent, not coloured, often inconspicuous; calyx shortly tubular, straight or curving upwards, strongly 2-lipped, 5-lobed (1/4), lobes unequal, posterior rounded, decurrent, curving upwards, lateral teeth deltoid, usually with a prominent shoulder, anterior lobes subulate, throat open, glabrous; corolla strongly 2-lipped, 5-lobed (4/1), white, pink or bluish, rarely red, posterior lip ascending, 4-lobed with median lobes exceeding lateral lobes, anterior lip horizontal, concave; corolla-tube straight or slightly curved, parallel-sided, dilating towards throat, usually glabrous, rarely annulate or with pubescent swellings at base of posterior stamens; stamens 4 usually held over lower lip, rarely much exserted, posterior pair usually attached above midpoint of corolla-tube, very rarely at base, usually glabrous, inappendiculate, anterior pair attached at corolla-throat; style usually capitate-clavate, stigma-lobes equal, rounded, rarely parallel-sided, connivent; disc 4-lobed with anterior lobe larger; nutlets ± ovoid, glabrous, mucilaginous or not. $2n = 22$–28, 48. Around forty species in tropical and southern Africa, Madagascar and tropical or subtropical Asia, in grassland, woodland, or forest margins.

208. *Fuerstia* T.C.E. Fries

Fuerstia T.C.E. Fries, Acta Univ. Lund Nova Ser. Avd. 2, 25, 17: 3 (1929); Taylor, J. Bot. 70: 269–274 (1932), rev.; Paton, Kew Bull. 48: 129–137 (1993), rev.

Subshrubs or herbs, often geoxylic, not aromatic; leaves simple, opposite, ternate or quadrate; inflorescence thyrsoid with cymes sessile, 1-flowered; bracts persistent, not coloured, often inconspicuous; calyx shortly tubular, straight or curving upwards, strongly 2-lipped, 5-lobed (1/4), lobes unequal, posterior rounded, decurrent, curving upwards, lateral lobes deltoid, usually with a prominent shoulder, anterior lobes subulate; throat open, glabrous; corolla strongly two-lipped, 5-lobed (4/1), white or pink, posterior lip ascending, 4-lobed with median lobes exceeding lateral lobes, anterior lip horizontal, concave; corolla-tube straight or slightly curved, parallel-sided, dilating towards throat; stamens 4, posterior pair sterile, held within tube, attached above midpoint of corolla-tube, pubescent, inappendiculate, anterior pair held over anterior lip, attached at corolla-throat, glabrous; style capitate-clavate, stigma-lobes equal, rounded, connivent; disc 4-lobed with anterior lobe larger; nutlets ± ovoid, glabrous, mucilaginous. Eight species in tropical Africa, in grassland or woodland.

209. *Hoslundia* Vahl

Hoslundia Vahl, Enum. Pl. 1: 212 (1804).

Shrubs, aromatic; leaves simple, opposite or sometimes ternate; inflorescence sometimes ternately or quadrately branched, thyrsoid with cymes sessile, 1-flowered; bracts persistent, not coloured, often inconspicuous; calyx ± regular, tubular, straight, becoming orange, globular and fleshy in fruit, 5-lobed, ± actinomorphic, lobes linear to deltoid, posterior sometimes slightly broader, not decurrent, throat closed in fruit, glabrous; corolla strongly 2-lipped, 5-lobed (4/1), white or pink, posterior lip ascending, 4-lobed with median lobes exceeding lateral lobes, anterior lip horizontal, slightly concave, corolla-tube straight, parallel-sided, dilating towards throat; stamens 4, posterior pair sterile, held within tube, attached above midpoint of tube, pubescent, inappendiculate, anterior pair held over anterior lip or just exceeding it, attached at corolla-throat, glabrous; style capitate-clavate, stigma-lobes equal, rounded, connivent; disc 4-lobed with anterior lobe larger; nutlets ± ovoid, glabrous, mucilaginous. $2n = 24$. Monotypic, *H. opposita* Vahl found in tropical Africa, open areas in grassland, woodland or forest, often on sandy or rocky soil. Sometimes used medicinally.

210. *Basilicum* Moench

Basilicum Moench, Suppl. Meth. Pl.: 143 (1802).

Annual herb, aromatic; stems branched; leaves simple; inflorescence thyrsoid with cymes sessile, usually 3-, more rarely fewer-flowered, flowers usually secund; bracts persistent, not coloured, inconspicuous; calyx shortly tubular, straight or curving upwards, strongly 2-lipped, 5-lobed (1/4), lobes unequal, posterior rounded, decurrent, curving upwards, lateral lobes deltoid, with a prominent shoulder, anterior lobes subulate, throat open, glabrous; corolla very small, 2-lipped, 5-lobed (4/1), white, purple or bluish, posterior lip ascending, 4-lobed with median lobes exceeding lateral lobes, anterior lip horizontal, concave, corolla-tube straight, parallel-sided, dilating towards throat; stamens 4 held over lower lip, posterior pair attached above midpoint of corolla-tube, glabrous, inappendiculate, anterior pair attached at corolla-throat; style capitate-clavate, stigma-lobes equal, rounded, connivent; disc equally 4-lobed; nutlets ± ovoid, glabrous, mucilaginous. $2n = 28, 30$. One species, *B. polystachyon* (L.) Moench, widespread in the Old World tropics, watersides, seasonally flooded areas.

211. *Benguellia* G. Tayl.

Benguellia G. Tayl. in J. Bot. 69 (suppl. 2): 156 (1931).

Herbs, ?aromatic; leaves simple; inflorescence thyrsoid with cymes sessile, 3-flowered; bracts persistent, not coloured, inconspicuous; calyx radially symmetrical, tubular-campanulate, straight, 5-lobed, lobes equal, deltoid; throat open, glabrous; corolla strongly 2-lipped, 5-lobed (4/1), violet, posterior lip ascending, with 2 median lobes exceeding lateral lobes, anterior lip horizontal, concave, corolla-tube straight, parallel-sided, dilating at throat; stamens 4 held over lower lip, posterior pair attached above midpoint of corolla-tube, glabrous, inappendiculate, anterior pair attached at corolla-throat; style capitate-clavate, stigma-lobes equal, rounded, connivent; disc 4-lobed with anterior lobe larger; nutlets ± ovoid, glabrous, perhaps not mucilaginous. One species, *B. lanceolata* (Gürke) G. Tayl., in marshes, Angola.

212. *Catoferia* (Benth.) Benth.

Catoferia (sphalm.: *Catopheria*) (Benth.) Benth. in Bentham & Hooker, Gen. Pl. 2: 1173 (1876); Ramamoorthy, Kew Bull 41: 299–305 (1986), rev.
Orthosiphon sect. *Catoferia* Benth. (1848).

Shrubs or herbs, woody at base or sometimes geoxylic, aromatic; leaves simple; inflorescence thyrsoid with cymes sessile, 3-flowered with pedicels fused to the inflorescence axis and calyces sessile, bracts persistent, often coloured; calyx laterally compressed at base and often constricted above nutlets, dilating distally, curved, strongly 2-lipped, 5- or 4-lobed (1/4 or 1/3), lobes unequal, posterior rounded, sometimes hooded, not decurrent, curving upwards, anterior lip either with 4 small lanceolate teeth or two rounded lateral teeth and an inconspicuous median entire or emarginate tooth, lateral lobes nearer median than posterior lobe, throat sometimes annulate, closed by connivent anterior and posterior lips; corolla strongly 2-lipped, 5-lobed (4/1), green, white or pink, posterior lip ascending, 4-lobed with median lobes exceeding lateral lobes, anterior lip horizontal, concave, corolla-tube straight or slightly curved, parallel-sided, dilating towards throat; stamens 4, much exserted, posterior pair attached above midpoint of corolla-tube near throat, glabrous, inappendiculate, anterior pair attached at corolla-throat; style capitate, stigma-lobes equal, rounded, spreading; disc 4-lobed with anterior lobe larger; nutlets ± ovoid, flattened, glabrous, mucilaginous or not. $2n = 54$. Four species in Mexico and Central America.

213. *Endostemon* N.E. Br. Fig. 21J, P

Endostemon N.E. Br. in Thistleton-Dyer, Fl. Cap. 5, 1: 295 (1910); Ashby in J. Bot. 74: 121–132 (1936), rev.; Paton, Harley, Harley & Weeks, Kew Bull. 49: 673–716 (1994), rev.
Orthosiphon sect. *Diffusi* Briq. (1897).
Pseudocimum Bremek. (1933).
Puntia Hedge (1983).

Shrubs or herbs, sometimes geoxylic, aromatic; leaves simple; inflorescence thyrsoid with cymes sessile, 1–7-flowered; bracts persistent, smaller than leaves, often inconspicuous, not coloured; calyx tubular, tubular-campanulate or spherical, straight, actinomorphic with deltoid to lanceolate lobes in one species, otherwise strongly 2-lipped, 5-lobed (1/4), lobes unequal, posterior rounded (rarely much expanded, membranous and developing coiled side-lobes), decurrent, curving upwards; lateral teeth lanceolate or deltoid, sometimes linear with a truncate or rounded apex, anterior lobes lanceolate to linear, teeth sometimes closing throat, throat sometimes densely hairy, rarely fruiting-calyx deciduous as a dispersal unit; corolla 2-lipped or very weakly so, 4–5-lobed, white, pink or blue, posterior lip forward-pointing or spreading, 3-lobed with median lobe emarginate, rarely equally 4-lobed, anterior lip horizontal or more rarely spreading, tube straight or slightly curved, dilating slightly towards throat; stamens 4, included or shortly exserted, porrect; posterior and anterior pairs regularly spaced and attached near corolla-throat, short, pilose or glabrous; style forming a small shield above nutlets, capitate-clavate at apex, stigma-lobes rounded, equal or with the posterior lobe longer, parallel; disc equally 4-lobed; nutlets obovoid to oblong-cylindrical, sometimes truncate and pubescent at apex, strongly or weakly mucilaginous. $2n = 36$. Nineteen species in tropical and southern Africa, one in India and Madagascar. Woodland or grassland.

214. *Haumaniastrum* P.A. Duvign. & Plancke

Haumaniastrum P.A. Duvign. & Plancke, Biol. Jahrb. 27: 222 (1959); idem, Trav. & Publ. Univ. Libre Lab. Bot. Syst. 33: 222 (1959); Paton, Kew Bull. 52: 293–378 (1997), rev.
Acrocephalus Benth. pro parte quoad African and Madagascan species (1830).

Shrubs or herbs, sometimes geoxylic, not or faintly aromatic; leaves simple, opposite, more rarely ternate or quadrate, regularly spaced and also sometimes in rosettes either at ground level or at apex of a short woody trunk; inflorescence thyrsoid, much condensed, forming solitary or paniculately arranged heads with cymes sessile, 3-flowered; bracts persistent, often basally coloured; calyx dorso-ventrally flattened, curving upwards, 2-lipped, 2- to 5-lobed (1–3/1–2), posterior lip entire with rounded or acute apex, 3-angled or 3-toothed, hooded or flat, anterior lip entire and rounded or truncate, emarginate or 2-toothed, lips connivent in fruit; corolla 2-lipped, usually 5-lobed (4/1), white, pink or blue, posterior lip ascending, (3–)4-lobed, median lobes usually exceeding the lateral lobes, anterior lip horizontal, entire, slightly concave, tube straight, dorsally gibbous at midpoint, dilating towards throat; stamens 4, declinate, exserted, posterior pair attached near midpoint of corolla, basally pubescent and swollen, anterior attached at corolla-throat; stigma-lobes equal, linear; disc 4-lobed with anterior lobe larger; nutlets ovoid, minutely striate, weakly mucilaginous when wet. $2n = 28, 56$. Thirty-five species in tropical Africa, grassland, open woodland, some species on mineral rich soil, others in wet areas.

215. *Platostoma* P. Beauv.

Platostoma P. Beauv., Fl. Oware 2: 61, t. 95 (1818); Paton, Kew Bull. 52: 257–292 (1997), rev.
Mesona Blume, Bijdr. (1826).
Acrocephalus Benth. (1829).
Geniosporum Wall. ex Benth (1830).
Nosema Prain (1904).
Limniboza R.E. Fr. (1916).
Ceratanthus F. Muell ex G. Taylor (1936).
Octomeron Robyns (1943).

Herbs, sometimes geoxylic, not or faintly aromatic; leaves simple; inflorescence thyrsoid, lax or more rarely condensed with cymes sessile, usually >3-flowered, more rarely 3-flowered; bracts persistent, often basally coloured; calyx tubular, campanulate or urceolate, curving upwards or straight, 2-lipped, 2- to 8-lobed (1–3/1–5), lobes unequal, posterior lip entire with rounded or acute apex or 3-lobed, with posterior lobe decurrent or not, anterior lip entire and rounded, emarginate, 2-toothed or 5-toothed, if entire then closing calyx-throat in fruit; corolla 2-lipped, 5- or 4-lobed (3–4/1), white, pink or blue, posterior lip ascending, 3–4-lobed, anterior lip horizontal, entire, slightly concave, corolla-tube straight, dorsally gibbous at midpoint or spurred near base, dilating towards throat; stamens 4, declinate or spreading, exserted, posterior pair attached around midpoint of corolla, basally pubescent and swollen, sometimes appendiculate, anterior pair attached at corolla-throat or near midpoint; stigma-lobes equal, linear; disc 4-lobed with anterior lobe larger; nutlets ovoid, minutely striate, usually weakly mucilaginous when wet. $2n = 14, 18, 28, 30, 42, 48$. Forty-five species in tropical Africa and Asia, also in Australia (Queensland), grassland, open woodland, often in wet areas.

VII.3.E. Subtribe Plectranthinae Endlicher (1838).

Cymes sessile or pedunculate, ebracteolate, rarely with minute persistent or deciduous bracteoles (*Capitanopsis*, rare in *Plectranthus*). Bracts persistent or deciduous, rarely coloured. Calyx usually with a single-lobed posterior lip and a 4-lobed anterior lip, sometimes actinomorphic and regularly 5-lobed. Corolla strongly 2-lipped, rarely actinomorphic (*Tetradenia*), anterior lip concave to cucullate, more rarely flat. Stamens declinate, rarely spreading (*Tetradenia*), exserted, anterior and posterior attached adjacently at corolla-throat. Disc usually 4-lobed, lobes alternating with nutlets. Nutlets with small inconspicuous areole.

216. *Dauphinea* Hedge

Dauphinea Hedge, Notes Roy. Bot. Gard. Edinburgh 41: 119 (1983); Hedge in Hedge, Clement, Paton & Phillipson, Flore de Madagascar et des Comores 175: 239–240 (1998), rev.

Shrubs, scarcely aromatic; leaves simple; inflorescence thyrsoid with cymes sessile, 3-flowered; bracts caducous, inconspicuous; calyx funnel-shaped, straight, strongly 2-lipped, 5-lobed (1/4), lobes unequal, posterior lanceolate, not decurrent, curving upwards, lateral lobes lanceolate with a small shoulder, closer to anterior lobes than posterior, anterior lobes lanceolate, slightly fused at base, throat open, glabrous; corolla strongly 2-lipped, 5-lobed (4/1), purple, posterior lip very short, truncate, 4-lobed with median lobes ascending, lateral lobes held between median lobes and anterior lip, anterior lip deflexed, slightly concave, corolla-tube basally geniculate, gradually dilating towards throat, glabrous within but with a basal, glabrous annulate ridge; stamens 4, free, held just beyond throat, posterior pair attached just below anterior, anterior pair attached at throat; stigma-lobes spreading, rounded; disc 4-lobed with anterior lobe larger; nutlets not seen. One species, *D. brevilabra* Hedge, in Madagascar, in forest at sea level.

217. *Capitanopsis* S. Moore

Capitanopsis S. Moore, J. Bot. 54: 249 (1916); Hedge in Hedge, Clement, Paton & Phillipson, Flore de Madagascar et des Comores 175: 253–260 (1998), rev.

Shrubs, ?aromatic; leaves simple; inflorescence thyrsoid with cymes pedunculate or shortly so, usually ebracteolate, rarely with minute bracteoles, 3-flowered; bracts caducous, not coloured, inconspicuous, calyx funnel-shaped, straight, ± actinomorphic or weakly zygomorphic, 5-lobed (1/4), lobes ± equal in flower, deltoid, often becoming membranous, reticulately nerved and much expanded in fruit with the posterior lobe broader than others, throat open, glabrous, fruiting-calyx shed as a unit; corolla strongly 2-lipped, 5-lobed (4/1), pale lilac to violet, posterior lip short, truncate, 4-lobed with median lobes ascending, lateral lobes held between median lobes and anterior lip, anterior lip deflexed or horizontal, slightly concave or flat, corolla-tube basally geniculate and saccate, parallel-sided or dilating slightly distally, glabrous within; stamens 4, free, held just beyond throat, attached near throat; stigma-lobes subulate or flattened; disc 4-lobed with anterior lobe larger; nutlets ovoid pubescent with sessile glands. Three species in Madagascar, in dry woodland or amongst rocks.

218. *Madlabium* Hedge

Madlabium Hedge in Hedge, Clement, Paton & Phillipson, Flore de Madagascar et des Comores 175: 260–263 (1998).

Shrubs, ?aromatic; leaves simple, opposite, falling before flowering, inflorescence thyrsoid with cymes sessile, bracteate, ebracteolate, 1-flowered, bracts soon deciduous, not coloured, inconspicuous; calyx ± actinomorphic, funnel-shaped, straight, calyx-teeth 5, lobes ± equal in flower, lanceolate; throat open, glabrous; corolla strongly two-lipped, 5-lobed (4/1), purple, posterior lip 4-lobed with median lobes ascending, lateral lobes held between the median lobes and the anterior lip, anterior lip deflexed or horizontal, slightly concave or flat; tube basally geniculate and saccate, dilating slightly distally, glabrous within; stamens 4, held just beyond throat, attached near throat; stigma with short subulate or flattened lobes; disc 4-lobed with anterior lobe larger, nutlets oblong, trigonous, apically winged. One species, *M. magenteum* Hedge, endemic to Madagascar, in shade in dry woodland.

219. *Plectranthus* L'Hér.

Plectranthus L'Hér., Stirp. Nov. Fasc. 4: t. 41, 42 (1788); Morton, J. Linn. Soc., Bot. 58: 231–283 (1962), reg. rev.; Codd, Bothalia 11: 371–442 (1975), reg. rev.; Keng in van Steenis (ed.) Fl. Malesiana 8, 3: 382–393 (1978), reg. rev.; Forster, Austrobaileya 4: 159–186 (1994), reg. rev.; Hedge in Hedge, Clement, Paton & Phillipson, Flore de Madagascar et des Comores 175: 137–235 (1998), reg. rev.
Germanea Lam. (1788).
Coleus Lour. (1790).
Solenostemon Schumach. (1827).
Englerastrum Briq. (1894).
Neomuellera Briq. (1894).
Capitanya Schweinf. ex Gürke (1895). (*Capitanya* Schweinf. ex Penzig (1893) nom nud.).
Burnatastrum Briq. in Engler & Prantl (1897).
Symphostemon Hiern (1900).
Isodictyophorus Briq. (1917).
Holostylon Robyns & Lebrun (1929).
Perrierastrum Guillaumin (1930).
Ascocarydion G. Tayl. (1931).
Neohyptis Morton (1963).
Rabdosiella Codd (1984).
Calchas P.V. Heath (1997).

Perennial or annual, sometimes succulent, subshrubs or herbs or geoxylic herbs, usually aromatic; leaves simple, rarely bipinnate, sometimes succulent, opposite, rarely ternate, very rarely scattered; inflorescence thyrsoid, condensed or lax, with cymes sessile or pedunculate, bracteate, very rarely bracteolate, 1- to many-flowered; bracts caducous or persistent, sometimes coloured at apex; pedicel very rarely jointed with distal part falling with calyx; calyx funnel-shaped or shortly tubular, often ventrally gibbous, straight or slightly or strongly curved, strongly 2-lipped or actinomorphic, 5-lobed (1/4 or isomorphic), lobes equal or unequal; posterior lip lanceolate to obovate, sometimes decurrent, lateral lobes rounded to lanceolate, anterior lobes lanceolate to subulate, sometimes fused for part of their length, rarely with a deep split between them, throat open, glabrous or bearded, very rarely circumscissile at base; corolla strongly two-lipped, 5-lobed (4/1), white, blue, purple or yellow, posterior lip long or short, ascending, reflexed or truncate, 4-lobed, median lobes exceeding lateral, anterior lip horizontal or sometimes finally deflexed, cucullate or concave, corolla-tube narrow, straight, variously curved or sigmoid, parallel-sided, gibbous or spurred at base, dilating distally and parallel-sided or narrowing towards throat, very rarely annulate within; stamens 4, very rarely 2, fused or free at base, held within anterior lip or exserted; stigma-lobes subulate or linear; anthers usually ± orbicular, but rarely with outer edges of locules fused in lower half to form a pouch which opens apically; disc 4-lobed with anterior lobe larger, sometimes exceeding ovary; style bifid or rarely emarginate; nutlets, ovoid, glabrous, slightly flattened or not, mucilaginous or not. $2n$ = 24–48(–84). Three-hundred species in Old World tropics. Used medicinally and a few species as tuber crops and for hedging. *P. amboinicus* (Lour.) Spreng. is widely used as a potherb. Treatment here is based on a synthetic approach, using similarity of several characters. This broad generic concept is in accordance with Keng (1978) and Hedge (1998). As a result, many traditionally recognised genera are here placed in synonymy, as they had been separated on the basis of single characters which obscure relationships and would leave *Plectranthus* s.s. as a ragbag of species only held together by not possessing characters of the segregate genera. This enlarged *Plectranthus* may still eventually prove paraphyletic due to recognition of other genera within the Plectranthinae.

220. *Thorncroftia* N.E. Br.

Thorncroftia N.E. Br., Kew. Bull. Misc. Inf. 1912: 281 (1912); Codd, Bothalia 7: 429 (1961), rev.

Semi-succulent subshrubs or herbs, aromatic; leaves simple; inflorescence thyrsoid, condensed or lax, with cymes sessile, 1-flowered; bracts persistent, sometimes caducous, leaf-like or smaller; calyx funnel-shaped or shortly tubular, straight, sometimes slightly curved, 2-lipped or weakly so, 5-lobed (1/4), lobes unequal, posterior lip lanceolate to elliptic, decurrent or not, lateral lobes lanceolate sometimes with a slight shoulder, anterior lobes lanceolate, slightly fused at base, throat open, glabrous; corolla strongly 2-lipped, 5- or 4-lobed (4–3/1), white to lilac, posterior lip with median lobes erect, emarginate or two-lobed, lateral lobes linear, spreading, deflexed, close to anterior lip, anterior lip deflexed, cucullate; tube narrow, cylindrical, parallel-sided or sometimes saccate at base and parallel-sided distally, straight; stamens 4, free, exserted; stigma-lobes subulate or linear; disc 4-lobed with anterior lobe larger, sometimes exceeding ovaries; nutlets, ovoid, glabrous, mucilaginous or not. $2n = 28$. Four species in rock outcrops, southern Africa.

221. *Tetradenia* Benth.

Tetradenia Benth. in Bot. Reg. sub t. 1300 (1830); Codd, Bothalia 14: 177–183 (1983), reg. rev.; Phillipson, Flore de Madagascar et des Comores 175: 70–94 (1998), reg. rev.
Iboza N.E. Br. (1910).

Fleshy, dioecious, rarely hermaphrodite, aromatic shrubs; leaves simple; inflorescence thyrsoid, paniculately branched with cymes sessile, 2–5-flowered; bracts persistent, inconspicuous, not coloured; calyx shortly tubular, straight, strongly 2-lipped, 5-lobed (1/4), lobes unequal, posterior ovate, not decurrent; anterior lanceolate with a sinus between the median pair giving the appearance of two emarginate lateral lobes, throat open, glabrous; corolla 2-lipped, 5-lobed (4/1), white or pink, posterior lip with the median lobes ascending, shallow, lateral lobes spreading, held between the median lobes and anterior lip, anterior lip horizontal or deflexed, concave, corolla-tube straight or sometimes curved, funnel-shaped or sometimes tubular and dilating distally, often bearded within at base; male flowers with 4 stamens, porrect, exserted, posterior attached near throat or at midpoint of corolla-tube, anterior attached at throat; stigma-lobes short, equal, connivent, often non-functional; female flowers lacking stamens;

style with 2 stigma-lobes; disc equally 4-lobed or with anterior lobe larger; nutlets oblong, glabrous, not mucilaginous. $2n = 42$. Fifteen to 20 species in Madagascar and S and tropical Africa.

222. *Anisochilus* Wall. ex Benth.

Anisochilus Wall. ex Benth. in Edward's Bot. Reg. 15: sub t. 1300 (1830); Bentham in De Candolle, Prodr. 12: 80–82 (1848); Hooker, Fl. Br. India 4: 627–630 (1885), reg. rev.; Doan in Lecompte, Fl. Gen. Indo-Chine 4: 941–943 (1936), reg. rev.
Stiptanthus (Benth.) Briq. (1897).

Annual or perennial, often succulent, aromatic subshrubs or herbs; leaves simple; inflorescence thyrsoid, often condensed, elongate or capitulate, with cymes sessile, usually 5-flowered, much condensed; bracts caducous, sometimes coloured at apex; calyx tubular, ventrally gibbous, laterally compressed basally, often dorso-ventrally compressed at throat, straight or curved slightly upwards, sometimes apically deflexed, strongly 2-lipped, 5-lobed (1/4 or rarely apparently 5/0), lobes unequal, posterior ovate or lanceolate, not decurrent, deflexed at throat, sometimes curving upwards apically, anterior lip 4-lobed with small rounded or lanceolate teeth or truncate, sometimes split between median lobes, the sinus sometimes very deep, so that posterior lip appears 5-lobed, throat closed or partially closed by the deflexed lips; corolla strongly 2-lipped, 5-lobed (4/1), white, blue or purplish, posterior lip ascending with median lobes exceeding lateral, more rarely truncate with lobes inconspicuous, anterior lip horizontal or slightly deflexed, concave or cucullate, corolla-tube narrow, parallel-sided at base, curved and dilating distally towards throat; stamens 4, free, held over or within lower lip, sometimes shortly exserted; stigma-lobes subulate, long and sometimes reflexed; disc 4-lobed with anterior lobe much larger, exceeding ovaries; nutlets ovoid, glabrous, slightly flattened, mucilaginous. $2n = 28–34, 52$. About 15 to 20 species, mainly in India and Indochina, usually in rocky areas.

223. *Leocus* A. Chev.

Leocus A. Chev. J. Bot. (Morot) II, 2: 125 (1909); Morton, J. Linn. Soc., Bot. 58: 251 (1963), reg. rev.

Herbs or shrubs, aromatic; leaves simple; inflorescence thyrsoid, often condensed, elongate or capitulate, with cymes sessile, usually 5–10-flowered, condensed; bracts caducous or persistent, sometimes coloured at apex or membranous; calyx tubular, ventrally gibbous, slightly compressed at throat, straight or curved slightly upwards, sometimes apically deflexed, strongly 2-lipped, 5-lobed (1/4), unequal, posterior lip oblong or lanceolate, not decurrent, deflexed at throat, sometimes curving upwards apically, anterior lip 4-lobed with lobes lanceolate, and a split between the median lobes, throat open or partially closed by the deflexed lips; corolla strongly 2-lipped, 5-lobed (4/1), blue or purplish, posterior lip ascending with median lobes exceeding lateral, anterior lip horizontal, cucullate; tube narrow, parallel-sided at base, curved or straight at the base, curved and dilating distally towards throat; stamens 4, fused at base, held over or within lower lip; stigma-lobes linear, often parallel; disc 4-lobed with anterior lobe much larger, exceeding ovaries; nutlets, ovoid, glabrous, flattened, mucilaginous. Five to six species in W and S tropical Africa.

224. *Aeollanthus* Mart. ex Spreng.

Aeollanthus Mart. ex Spreng. Syst. Veg. 2: 750 (1825); Ryding, Symb. Bot. Ups. 26, 1 (1986), rev.
Icomum Hua (1897).
Bovonia Chiov. (1922).

Usually succulent subshrubs or herbs, often geoxylic, aromatic; leaves simple, opposite, rarely ternate or sometimes scattered; inflorescence thyrsoid, with cymes variously arranged, sessile, 1-flowered, appearing like paniculately arranged, spike-like racemes, bracts usually inconspicuous, sometimes with an abaxial glandular structure; calyx tubular at base, distally dorsiventrally flattened, ± straight, 2-lipped, 5- or 2-lobed (3/2 or 1/1), lobes unequal, posterior 3-toothed or truncate, apex sometimes folded over anterior lip, anterior 2-toothed or truncate, lips pressed together, circumscissile near base in fruit; corolla strongly 2-lipped, 5-lobed (4/1), white, pink or blue, posterior lip erect or ascending, 4-lobed with median lobes exceeding lateral, horizontal and concave or sometimes hooded and becoming reflexed to explosively release the stamens, corolla-tube curved at midpoint or towards base, parallel-sided below, dilating towards throat, sometimes dorsally gibbous; stamens 4, free, held over or within anterior lip of corolla, glabrous; stigma-lobes equal, subulate; disc 4-lobed with anterior lobe larger; nutlets elliptic or ovoid, mucilaginous. $2n = 34$ (36). Just over 40 species in tropical Africa. Usually on rocks or thin soil.

225. *Alvesia* Welw.

Alvesia Welw., Trans. Linn. Soc. 27: 55 (1869), nom. conserv.; Matthew, Kew Bull. 25: 407–411, rev.; ibid. 26: 564–565 (1972).
Plectranthastrum T.C.E. Fries (1924).

Shrubs or herbs, not aromatic; leaves simple; inflorescence thyrsoid with cymes sessile, 2–7-flowered; bracts persistent, leaf-like but smaller, sometimes with a conspicuous abaxial glandular structure; calyx cylindrical or constricted at throat distally, becoming inflated, membranous, truncate at base, 2-lipped, 3-lobed (1/2), lobes unequal, posterior rounded, smaller than anterior, not decurrent; lateral lobes absent, anterior lobes large, rounded, throat closed by converging lobes, fruiting-calyx shed as a unit; corolla strongly 2-lipped, 5-lobed (4/1), white to purple, posterior lip reflexed, 4-lobed with median lobes exceeding lateral lobes, anterior lip horizontal, cucullate, corolla-tube basally geniculate and with small spur, parallel-sided distally, glabrous within but with a basal annulate ridge formed from the posterior staminal filament traces; stamens 4, held within anterior corolla-lip, fused or free at base; stigma-lobes linear; disc 4-lobed with anterior lobe larger; nutlets ovoid, glabrous, slightly flattened, weakly mucilaginous. Three species in S tropical Africa, in open woodland or in rocky or damp grassland.

226. *Pycnostachys*

Pycnostachys Hook., Exot. Fl. 3: t. 202 (1826); Bruce, Kew Bull. Misc. Inf. 1939: 563–592 (1939), rev.

Shrubs or herbs, annual or perennial, sometimes fleshy, weakly aromatic; leaves simple, opposite, rarely ternate or quadrate; inflorescence thyrsoid, usually condensed, capitulate, more rarely elongate, with cymes sessile, 1-flowered; bracts persistent, sometimes coloured; calyx tubular, ventrally gibbous, often laterally compressed basally, sometimes dorso-ventrally compressed at throat or regularly constricted when immature, curved upwards at base, sometimes apically deflexed, zygomorphic, regular at mouth, 5-lobed, lobes isomorphic, lanceolate, spinescent, diverging, posterior not decurrent, throat bearded or not, open at maturity or sometimes partially closed by the deflexed lips, when immature by appendages extending into sinuses between calyx-lobes, these rarely lacking; corolla strongly 2-lipped, 5-lobed (4/1), blue or purplish, posterior lip ascending with median lobes exceeding lateral, anterior lip horizontal, cucullate, corolla-tube narrow, parallel-sided at base, curved upwards or straight at base, curved and dilating towards throat distally; stamens 4, fused at base, usually held within lower lip, rarely exserted, contiguous at attachment to corolla-throat; stigma-lobes subulate or short and rounded; disc 4-lobed with anterior lobe much larger, exceeding ovaries; nutlets, ovoid, glabrous with faint dorsal vein, slightly flattened, mucilaginous. $2n$ = 34, 68. Around 40 species in Africa, Madagascar.

Labiatae: Genera Incertae Sedis

227. *Tectona* L.f.

Tectona L.f., Suppl. Pl. 20: 151 (1781).

Large trees; leaves simple, opposite, rarely alternate; inflorescence a terminal and/or axillary, dichotomous, cymose panicle; flower yellowish to white; calyx campanulate, 5–7-lobed, lobes unequal, greatly increasing during fruiting (inflated in some species); corolla infundibular, actinomorphic, 5–7-lobed, lobes subequal, corolla-tube short; stamens 5–6, epipetalous, shortly exserted, equal in length; anthers ovate or elliptic-oblong, dorsifixed, thecae parallel, opening by longitudinal slits; pollen 3-colpate; style terminal, with stigma-lobes shortly bifid, with branches subequal; ovary 2-carpellate, 4-locular, each locule 1-ovulate; ovules lateral, hemianatropous, fruit drupaceous, enclosed in enlarged persistent calyx, with thin subfleshy exocarp, a hard 4-celled endocarp with small central cavity between cells; seeds exalbuminous. $2n$ = 36. Four species, India and SE Asia.

228. *Callicarpa* L.

Callicarpa L., Sp. Pl. 1: 111 (1753); Chang, Acta Phytotax. Sin. 1, 3: 269–312 (1951), rev.; Munir, J. Adelaide Bot. Gard. 6: 5–39 (1982), reg. rev.
Geunsia Blume, Cat. Gew. Buitenz. 11, 48 (1823).

Small trees or shrubs, rarely lianas; leaves simple, decussate or in whorls of three; inflorescence axillary, cymose, flowers white, blue, violet, mauve to purple; calyx tubular or shortly campanulate, 4-lobed, not accrescent in fruit; corolla 4- (rarely 5-)lobed, tubular to infundibular or hypocratiform, actinomorphic; stamens 4, rarely 5–7, epipetalous, exserted, anthers oblong or elliptic, dorsifixed, thecae parallel, opening by longitudinal slits or by terminal pores; style terminal, usually exserted, with stigma peltate or capitate; ovary 4-locular (rarely 5-locular), fruit drupaceous, ± globose with fleshy exocarp, and hard endocarp separating into 4 stony pyrenes, each containing a single seed; seed exalbuminous. $2n = 16$ or 18. About 140 species, temperate, subtropical and tropical Asia and America, tropical Australia, and some Pacific islands. Chang (1951) has proposed two sections based on stamen structure. Sect. *Callicarpa*: anthers ovoid, dehiscing by longitudinal slits, and sect. *Verticirima* H.-T. Chang: anthers oblong, dehiscing by apical pores.

229. *Hymenopyramis* Wall. ex Griffith

Hymenopyramis Wall. ex Griffith, Calcutta J. Nat. Hist. 3: 365 (1842).

Climbing shrubs or small trees; leaves simple, decussate, opposite; inflorescence axillary and/or terminal, cymose, a dense to lax thyrse, flower white, scented; calyx 4-lobed, accrescent in fruit; corolla infundibular, zygomorphic, 4-lobed, lobes subequal; stamens 4, epipetalous, subequal, exserted, anthers ovate, thecae parallel, opening by longitudinal slits; pollen 3-colpate; style terminal, with stigma 2-lobed; ovary 2-locular, locules 2-ovuled, ovules lateral; fruit dry, schizocarpic, globose, enclosed within inflated bladder-like calyx, translucent with reticulate veins; seed exalbuminous. Six species, India to China and SE Asia.

230. *Petraeovitex* Oliv.

Petraeovitex Oliv., Hooker's Icon. Pl. 15: t. 1420 (1883); Munir, Gard. Bull. Straits Settlem. 21: 215–257 (1965), rev.

Lianas; leaves opposite, ternately to biternately compound, leaflets ovate to obovate, entire (rarely few-toothed); inflorescence of axillary cymes, often forming lax paniculiform thyrses; calyx actinomorphic, accrescent, 5-lobed, lobes greatly elongate and wing-like in fruit; corolla white to yellow, nearly actinomorphic or 1-lipped and then with only anterior lip present, subequally 5-lobed; stamens 4, equal to subequal, included or exserted, filaments ± straight, thecae parallel to slightly divergent, separate at dehiscence; pollen tricolpate, psilate, tectate-perforate; ovary unlobed; style with stigma-lobes equal or subequal; disc absent; fruit dry, indehiscent, 2-chambered, 1–2-seeded (the other ovules aborting), turbinate, apex conical, rounded or truncate-depressed, surface usually longitudinally striate, pubescent; endosperm absent. Eight species in forests of Malaysia, Indonesia, Philippines, New Guinea, and Melanesia.

231. *Peronema* Jack

Peronema Jack, Malayan Misc. 2(7): 46 (1822).

Large tree, non-aromatic; leaves opposite, petiolate, pinnately compound, leaflets 11–19, alternate to subopposite, sessile to subsessile, elliptic to lanceolate, entire; inflorescence cymose, forming a large, terminal panicle; flowers small; calyx actinomorphic, not accrescent, 5-lobed; corolla whitish, zygomorphic, limb 5-lobed, anterior lobe largest, other 4 similar; fertile stamens 2 (posterior pair reduced to staminodes), included to slightly exserted, thecae parallel, separate at dehiscence; pollen tricolpate, tectate-perforate, suprarugulose; ovary unlobed; style with stigma-lobes equal or unequal; disc poorly developed to absent; fruit dry, globose, villous, 4-seeded, remaining intact or splitting irregularly into two or four mericarps, abscission-scar as long as mericarp; endosperm absent. One species, *P. canescens* Jack, in Thailand to Malaysia and W Indonesia.

232. *Garrettia* H.R. Fletcher

Garrettia H.R. Fletcher, Bull. Misc. Inf. 1937: 71 (1937).

Shrub; leaves opposite, petiolate, simple (ovate) or trifoliolate (leaflets ovate), serrulate to subentire; cymes axillary, lax, long-pedunculate, with secund, monochasial branches, grading into a terminal paniculiform thyrse; calyx actinomorphic, accrescent, campanulate in flower, cup-shaped and essentially truncate in fruit, 5-lobed, lobes triangular, very short; corolla white with yellowish posterior lip, 2-lipped, 5-lobed (2/3), anterior lip with median lobe largest; stamens 4, didynamous (anterior pair longer), slightly exserted, thecae divaricate, confluent at dehiscence; pollen tricolpate, tectate-perforate, suprareticulate, columellae simple; ovary unlobed; stigma-lobes short, unequal; disc absent; fruit globose, indehiscent, breaking with pressure into 4 nutlets, these reticulately ridged, glabrous, glandular-punctate, abscission-scar lateral, as long as nutlet. One species, *G. siamensis* H.R. Fletcher, forests in China (Yunnan), Thailand, and Indonesia.

233. *Cymaria* Benth.

Cymaria Benth., Edward's Bot. Reg. 15: t. 1292 (1830).

Shrubs and subshrubs; leaves opposite, petiolate, usually ovate to rhombic, crenate to crenulate or repand; cymes axillary, lax, long-pedunculate, with secund, monochasial branches, grading into a terminal paniculiform thyrse; calyx actinomorphic, accrescent, campanulate in flower, broadly campanulate to urceolate or subglobose in fruit, 5-lobed, lobes triangular; corolla white to yellowish, 2-lipped (2/3), posterior lip emarginate to entire, anterior lip with median lobe largest; stamens 4, didynamous (anterior pair longer), included, thecae divaricate, confluent at dehiscence; pollen tricolpate, tectate-perforate, suprareticulate, columellae simple; ovary shallowly 4-lobed; stigma-lobes short, subequal; disc absent; nutlets obovoid, reticulately ridged, pubescent, abscission-scar lateral, 0.4–0.6× the nutlet. Two to three species of thickets and forest edges in SE Asia and Malesia.

234. *Acrymia* Prain

Acrymia Prain, Bull. Misc. Inf. 1908: 114 (1908).

Subshrub; stem prostrate, distally leafy; leaves opposite, petiolate, elliptical, crenate; cymes axillary, lax, long-pedunculate, with secund, monochasial branches; calyx ± actinomorphic, accrescent, campanulate in flower, urceolate in fruit, 5-lobed, lobes triangular; corolla white to yellowish, 2-lipped, 5-lobed (2/5), posterior lip deeply 2-lobed, anterior lip with median lobe much larger than lateral ones; stamens 4, didynamous (anterior pair longer), exserted, thecae divaricate, confluent at dehiscence; pollen tricolpate, tectate-perforate, suprareticulate, columellae simple to sparsely branched; ovary shallowly 4-lobed; stigma-lobes unequal; disc absent; nutlets obovoid, reticulately ridged, pubescent, abscission-scar lateral, 0.4–0.5× nutlet; endosperm present; embryo spathulate. One species, *Acrymia ajugiflora* Prain, Malaya (Selangor), forest understorey; endangered.

235. *Holocheila* (Kudô) S. Chow

Holocheila (Kudô) S. Chow, Acta Bot. Sin. 10: 250 (1962); Li Xi-wen and Hedge in Wu Zheng-yi and Raven (eds.) Flora of China 17: 63 (1994).
Teucrium sect. *Holocheila* Kudo (1929).

Weak-stemmed perennial herb; leaves opposite, long-petiolate, cordate to suborbicular, crenate; cymes axillary, lax, long-pedunculate; calyx weakly 2-lipped, 5-lobed (3/2), obliquely attached to pedicel, not accrescent, campanulate, posterior lip larger than anterior, lobes triangular to triangular-ovate; corolla pink, 2-lipped (1/1), both lips unlobed, posterior lip very short, rounded, anterior lip spathulate, concave, tube cylindrical; stamens 4, didynamous (anterior pair longer), exserted, filaments inserted on upper lip at top of corolla-tube, straight, anthers apparently 1-thecal, laterally attached; pollen tricolpate, tectate-perforate, suprareticulate, columellae simple; ovary shallowly 4-lobed; stigma-lobes unequal; disc well developed, symmetrical, with a slender vertical column rising between the ovary-lobes (to which the lobes are laterally attached); nutlets subglobose, obscurely trigonal, finely alveolate, glabrous, lustrous, abscission-scar lateral, 0.3× nutlet. One species, *H. longipedunculata* S. Chow, China (Yunnan).

236. *Ombrocharis* Hand.-Mazz.

Ombrocharis Hand.-Mazz., Symb. Sin. 7: 925 (1936); Li Xi-wen and Hedge in Wu Zheng-yi and Raven (eds.) Flora of China 17: 177 (1994), reg. rev.

Perennial, glandular herb, shortly rhizomatous with fine, fibrous roots and small woody tubers; stem simple, 4-angled, erect, sweetly aromatic when dry; leaves opposite, ovate-oblong, toothed, petiolate; inflorescence forming a lax, terminal raceme-like thyrse of subsessile 3-flowered cymes in the axils of short, lanceolate bracts, bracteoles inconspicuous, minute; flowers fairly long-pedicellate; calyx campanulate, 2-lipped, 5-lobed (3/2), posterior lip 3-lobed, lobes mucronate, anterior lip longer, 2-lobed, calyx-tube 11-nerved, annulate, slightly accrescent in fruit; corolla very pale violet, 2-lipped, 5-lobed (2/3), posterior lip deeply 2-lobed, weakly hooded, sparsely hairy externally, anterior lip spreading, 3-lobed, lobes subequal, rounded, corolla-tube short and broad, straight; stamens 4, didynamous, included in corolla-tube, anterior pair slightly longer, inserted in middle of corolla-tube, anthers 2-thecous, thecae rounded, parallel, separate; style shorter than corolla, stigma-lobes short, subequal; nutlets 4, ovoid, smooth, compressed. One species, *O. dulcis* Hand.-Mazz., China.

Doubtful Genera

Adelosa Blume (Viticoideae)
Pseudochamaesphacos Parsa
(? = *Chamaesphacos*)

Rejected Genera

Chingyungia T.M. Ai (as Verbenaceae aff. *Caryopteris* = *Melampyrum* Scrophulariaceae)

Selected Bibliography

Abu-Asab, M.S. 1990. Phylogenetic implications of pollen morphology in subfamily Lamioideae (Labiatae) and related taxa. Ph.D. Thesis. Ohio University, U.S.A.

Abu-Asab, M.S., Cantino, P.D. 1987. Phylogenetic implications of leaf anatomy in subtribe Melittidinae (Labiatae) and related taxa. J. Arnold Arbor. 68: 1–34.

Abu-Asab, M.S., Cantino, P.D. 1993. Phylogenetic implications of pollen morphology in tribe Ajugeae (Labiatae). Syst. Bot. 18: 100–122.

Abu-Asab, M.S. Cantino, P.D. 1994. Systematic implications of pollen morphology in subfamilies Lamioideae and Pogostemonoideae (Labiatae). Ann. Missouri Bot. Gard. 81: 653–686.

Adylov, T. Kamelin, R., Makhmedov, A. 1986. Notae de Lamiaceis. Novosti Sist. Vyssh. Rast. 23: 110–114.

Aitzetmüller, K., Tsevegüren, N. 1998. Phlomic acid in Lamioideae seed oils. Lamiales Newslett. issue 6: 13–16.

Atkinson, R. 1998. A taxonomic revision of *Hypenia* (Mart. ex Benth.) Harley (Labiatae). Ph.D. Thesis. University of St. Andrews, Scotland, 293 pp.

Barrett, S.C.H., Wilken, D.H., Cole, W.W. 2000. Heterostyly in the Lamiaceae: the case of *Salvia brandegeei*. Plant Syst. Evol. 223: 211–219.

Başer, K.H.C. 2000. Labiatae, in Index to Turkish plant chemical contents. In: Güner, A., Özhatay, N., Ekim, T., Başer, K.H.C. (eds.) Flora of Turkey, vol. 11: 531–551.

Bentham, G. 1832–1836. Labiatarum Genera et Species. London: Ridgeway & Sons, 783 pp.

Bentham, G. 1870. Labiatae. In: Flora Australiensis 5. London: Reeve and Co., pp. 70–137.

Bentham, G. 1876. Verbenaceae and Labiatae. In: Bentham, G., Hooker, J.D. Genera Plantarum, vol. 2. London: Reeve and Co., pp. 1131–1223.

Billings, F.H. 1909. The nutrition of the embryo sac and embryo in certain Labiatae. Kansas Univ. Sci. Bull. 5: 67–83.

Blunden, G., Yang, M.-H., Yuan, Z.-X., Smith, B.E., Patel, A., Cegarra, J.A., Máthé, I. Jr., Janisak, G. 1996. Betaine distribution in the Labiatae. Biochem. Syst. Ecol. 24: 71–81.

Bokhari, M.H., Hedge, I.C. 1976. *Zhumeria* (Labiatae): anatomy, taxonomy and affinities. Iranian J. Bot. 1: 1–10.

Boumann, F., Meeuse, A.D.J. 1992. Dispersal in Labiatae. In: Harley, R.M., Reynolds, T. (eds.) Advances in Labiate science. R.B.G. Kew, U.K., pp. 193–202.

Bremer, K., Backlund, A., Sennblad, B., Swenson, U., Andreasen, K., Hjertson, M., Lundberg, J., Backlund, M., Bremer, B. 2001. A phylogenetic analysis of 100+ genera and 50+ families of euasterids based on morphological and molecular data with notes on possible higher level morphological synapomorphies. Plant Syst. Evol. 229: 137–169.

Briquet, J. 1891. Les Labiées des Alpes Maritimes 1: 179–184.

Briquet, J. 1895. Verbenaceae. In: Engler & Prantl, Die natürlichen Pflanzenfamilien IV, 3a. Leipzig: W. Engelmann, pp. 132–182.

Briquet, J. 1895–1897. Labiatae. In: Engler & Prantl, Die natürlichen Pflanzenfamilien IV, 3a. Leipzig: W. Engelmann, pp. 183–375.

Budantsev, A.G. 1987. The system of the genus *Dracocephalum* (Lamiaceae). Bot. Zhurn. 72, 2: 260–267.

Budantsev, A.G. 1993. A synopsis of the genus *Nepeta* (Lamiaceae). Bot. Zhurn. 78, 1: 93–107.

Budantsev, A.L., Lobova, T.A. 1997. Fruit morphology, anatomy and taxonomy of tribe Nepeteae (Labiatae). Edinburgh J. Bot. 54, 2: 183–216.

Burkhart, A. 1939. El mecanismo floral de la Labiada *Hyptis mutabilis* (Rich.) Briq. y su convergencia hacia la flor papilionoidea. Darwiniana 3: 425–427.

Byragi Reddy, T., Subba Reddi, C. 1994. Pollination ecology of *Vitex negundo* (Verbenaceae). Proc. Indian Natl. Sci. Acad. B 60 (1): 57–66.

Cantino, P.D. 1982a. A monograph of the genus *Physostegia* (Labiatae). Contr. Gray Herb. 211: 1–105.

Cantino, P.D. 1982b. Affinities of the Lamiales: a cladistic analysis. Syst. Bot. 7, 3: 237–248.

Cantino, P.D. 1985a. Facultative autogamy in *Synandra hispidula* (Labiatae). Castanea 50, 2: 105–111.

Cantino, P.D. 1985b. Chromosome studies in subtribe Melittidinae (Labiatae) and systematic implications. Syst. Bot. 10, 1: 1–6.

Cantino, P.D. 1990. The phylogenetic significance of stomata and trichomes in the Labiatae and Verbenaceae. J. Arnold Arb. 71: 323–370.

Cantino, P.D. 1992a. Evidence for a polyphyletic origin of the Labiatae. Ann. Missouri Bot. Gard. 79: 361–379.

Cantino, P.D. 1992b. Toward a phylogenetic classification of the Labiatae. In: Harley, R.M., Reynolds, T. (eds.) Advances in Labiate science. R.B.G. Kew, pp. 27–37.

Cantino, P.D., Sanders, R.W. 1986. Subfamilial classification of Labiatae. Syst. Bot. 11: 163–185.

Cantino, P.D., Wagstaff, S.J. 1998. A reexamination of North American *Satureja* s.l. (Lamiaceae) in light of molecular evidence. Brittonia 50, 1: 63–70.

Cantino, P.D., Harley, R.M., Wagstaff, S.J. 1992. Genera of Labiatae: status and classification. In: Harley, R.M., Reynolds, T. (eds.) Advances in Labiate science. Kew: Royal Botanic Gardens, pp. 511–522.

Cantino, P.D., Wagstaff, S.J., Olmstead, R.G. 1999. *Caryopteris* (Lamiaceae) and the conflict between phylogenetic and pragmatic considerations in botanical nomenclature. Syst. Bot. 23, 3: 369–386.

Carlquist, S. 1992. Wood anatomy of Lamiaceae. A survey with comments on vascular and vasicentric tracheids. Aliso 13, 2: 309–338.

Carlson, E.M., Stuart, B.C. 1936. Development of spores and gametophytes in certain New World species of *Salvia*. New Phytol. 35: 68–91.

Chambers, H.L. 1961. Chromosome numbers and breeding systems in *Pycnanthemum* (Labiatae). Brittonia 13: 116–128.

Chang, H.-T. 1951. A review of the Chinese species of *Callicarpa*. Acta Phytotax. Sin. 1: 269–312.

Chávez, R.P. 1974. Observaciones en el polen de plantas con probable polinización quiropterófila. An. Esc. Nac. Cien. Biol. Mex. 21: 115–143.

Chen Shou-liang, Gilbert, M.G. 1994. Verbenaceae. In: Wu, Zheng-yi, Raven, P.H. (eds.) Flora of China 17: 1–48.

Codd, L.E. 1985. Lamiaceae. In: Leistner, O.A. (ed.) Flora of southern Africa, vol. 28, 247 pp.

Cole, M.D. 1992. The significance of the terpenoids in the Labiatae. In: Harley, R.M., Reynolds, T. (eds.) Advances in Labiate science. Kew: Royal Botanic Gardens, pp. 315–324.

Collinson, M.E., Boulter, M.C., Holmes, P.L. 1993. Magnoliophyta, 45. In: Benton, M.J. (ed.) The fossil record 2. London: Chapman & Hall.

Conn, B.J. 1986. A taxonomic revision of *Prostanthera* Labill. sect. *Klanderia* (F. Muell.) Benth. (Labiatae). J. Adelaide Bot. Gard. 6: 207–348.

Cousins, D.J. (ed.) 1994. Medicinal, essential oil, culinary herb and pesticidal plants of the Labiatae 1973–1993. Wallingford, U.K.: CAB International.

Cronquist, A. 1981. An integrated system of classification of flowering plants. New York: Columbia University Press, 1262 pp.

Cruden, R.W., Hermanutz, L., Shuttleworth, J. 1984. The pollination biology and breeding system of *Monarda fistulosa* (Labiatae). Oecologia 64: 104–110.

Davis, G.L. 1966. Systematic embryology of the angiosperms. London: Wiley, 528 pp.

Davis, P.H. 1982. *Pentapleura*. In: Davis, P.H. (ed.) Flora of Turkey and the East Aegean Islands, 7. Edinburgh: Edinburgh University Press, p. 313.

De Kok, R. 1997. The biology and systematics of *Oxera*, *Faradaya* and *Hosea* (Labiatae). Ph.D. Thesis. University of Oxford, 213 pp. + appendices.

De Kok, R.P.J., Mabberley, D.J. 1999a. A synopsis of *Oxera* Labill. (Labiatae). Kew Bull. 54: 265–300.

De Kok, R.P.J., Mabberley, D.J. 1999b. Generic and intrageneric delimitation of *Oxera* Labill. (Labiatae). Kew Bull. 54: 257–264.

Demissew, S., Harley, M.M. 1992. Trichome, seed surface and pollen characters in *Stachys* (Labiatae) in Tropical Africa. In: Harley, R.M., Reynolds, T. (eds.) Advances in Labiate science. Kew: Royal Botanic Gardens, pp. 149–166.

Doroszenko, A. 1985. Taxonomic studies on the *Satureja* complex (Labiatae). Ph.D. Thesis. Edinburgh University, 538 pp.

Emboden, W.A. 1964. Pollen morphology of the genus *Salvia* sect. *Audibertia*. Pollen Spores 6: 527–536.

Endress, P.K. 1994. Diversity and evolutionary biology of tropical flowers. Cambridge University Press, England, 511 pp.

Epling, C. 1925. *Monardella*. Ann. Missouri Bot. Gard. 12: 1–106.

Epling, C. 1935–1937. Synopsis of the South American Labiatae. Feddes Rep. Spec. Nov. Beih. 85: 1–341.

Epling, C. 1948. A synopsis of the tribe Lepechinieae (Labiatae). Brittonia 6: 352–364.

Epling, C. 1949. Revisión del género *Hyptis* (Labiatae). Rev. Mus. La Plata (Bot.) 7: 153–497.

Epling, C., Játiva, C. 1964. Revisión del género *Satureia* en el América del Sur. Brittonia 16: 393–416.

Epling, C., Stewart, W.S. 1939. A revision of *Hedeoma* with a review of allied genera. Rep. Spec. Nov. Regn. Veg. Beih. 115: 1–49.

Erdtman, G. 1945. Pollen morphology and plant taxonomy. IV. Labiatae, Verbenaceae and Avicenniaceae. Svensk Bot. Tidskr. 39: 279–285.

Fernandes, R.B. 1985. Notes sur les Verbenaceae. Identification des espèces d'*Holmskioldia* africaines et malgaches. Garcia de Orta Ser. Bot. Lisboa 7: 33–46.

Frei, E. 1955. Die Innervierung der floralen Nektarien dikotyler Pflanzenfamilien. Ber. Schweiz. Bot. Ges. 65: 60–114.

Garcia-Peña, M.R. 1989. A new species of *Cunila* (Lamiaceae) from southwestern Mexico. Kew Bull. 44: 727–730.

Gill, L.S. 1981. Chromosomal evolution and incidence of polyploidy in the Canadian Labiatae. Rev. Cytol. Biol. Vég. Bot. 4: 331–339.

Gill, L.S. 1984. The incidence of polyploidy in the Western Himalayan Labiatae. Rev. Cytol. Biol. Vég. Bot. 7, 1: 5–16.

Gill, L.S., Chinnappa, C.C. 1982. Pollen morphology of the west-Himalayan Labiatae. Bangladesh J. Bot. 11: 107–122.

Grayer, R.J., Kite, G.C., Goldstone, F.J., Bryan, S.E., Paton, A., Putievsky, E. 1996. Infraspecific taxonomy and essential oil chemotypes in sweet basil, *Ocimum basilicum*. Phytochemistry 43: 1033–1039.

Grayer, R.J., Veitch, N.C., Kite, G.C., Price, A.M., Kokubun, T. 2001. Distribution of 8-oxygenated leaf-surface flavones in *Ocimum*. Phytochemistry 56: 559–567.

Guignard, M.L. 1893. Recherche sur le développement de la graine et en particulier du tégument séminal: Labiées. J. Bot. 7: 241–250.

Harley, M.M. 1992. The potential value of pollen morphology as an additional taxonomic character in subtribe Ociminae (Ocimeae: Nepetoideae: Labiatae). In: Harley, R.M., Reynolds, T. (eds.) Advances in Labiate science. Kew: Royal Botanic Gardens, pp. 125–138.

Harley, M.M., Wilde, V. in prep. Hexacolpate Lamiaceae pollen from the Eocene of Messel.

Harley, R.M. 1971. An explosive pollination mechanism in *Eriope crassipes*, a Brazilian Labiate. J. Linn. Soc. (Biol.) 3: 183–186.

Harley, R.M. 1976. A review of *Eriope* and *Eriopidion* (Labiatae). Hookers Ic. Pl. 8, 3: 1–107.

Harley, R.M. 1986a. Notes on New World Labiatae X. Observations on a hybrid population of *Hyptis cruciformis* and *H. pachyphylla* in Brazil. Kew Bull. 41: 1007–1015.

Harley, R.M. 1986b. *Cuminia eriantha* (Benth.) Benth. Kew Mag. 3, 4: 151–156.

Harley, R.M. 1988a. Evolution and distribution of *Eriope* (Labiatae) and its relatives in Brazil. In: Vanzolini, P.E., Heyer, W.R. (eds.) Proc Worksh Neotropical Distribution Patterns. Rio de Janeiro: Acad. Bras. de Ciências, pp. 71–120.

Harley, R.M. 1988b. Revision of generic limits in *Hyptis* Jacq. (*Labiatae*) and its allies J. Linn. Soc. (Bot.) 98: 87–95.

Harley, R.M. 1991. The greasy pole syndrome. In: Huxley, C.R., Cutler, D. (eds.) Ant-plant interactions. Oxford University Press, pp. 430–433.

Harley, R.M., Brighton, C.A. 1977. Chromosome numbers in the genus *Mentha* L. Bot. J. Linn. Soc. 74: 71–96.

Harley, R.M., Heywood, C.A. 1992. Chromosome numbers in tropical Labiatae. In: Harley, R.M., Reynolds, T. (eds.) Advances in Labiate science. Kew: Royal Botanic Gardens, pp. 211–246.

Harley, R.M., Paucar Granda, A. 2000. List of species of tropical American *Clinopodium* (Labiatae) with new combinations. Kew Bull. 54: 917–927.

Harley, R.M., Paton, A.J., Ryding, O. 2003. New synonymy and taxonomic changes in the Labiatae. Kew Bull. 58: 485–489.

Hart, J.A. 1983. Systematic and evolutionary studies in the genus *Lepechinia* (Lamiaceae). Ph.D. Thesis. Harvard, Cambridge, Mass., U.S.A.

Hart, J.A. 1985. Evolution of dioecism in *Lepechinia* Willd. Sect. *Parviflorae* (Lamiaceae). Syst. Bot. 10: 147–154.

Hedge, I.C. 1961. *Ziziphora*. Notes Roy. Bot. Gard. Edinburgh 23: 209–221.

Hedge, I.C. 1970. Observations on the mucilage of *Salvia* fruits. Notes Roy. Bot. Gard. Edinburgh 30, 1: 79–95.

Hedge, I.C. 1972. The pollination mechanism of *Aeollanthus njassae*. Notes Roy. Bot. Gard. Edinburgh. 32, 1: 45–48.

Hedge, I.C. 1974. A revision of *Salvia* in Africa including Madagascar and the Canary Islands. Notes Roy. Bot. Gard. Edinburgh 33, 1: 1–121.

Hedge, I.C. 1982. *Salvia*. In: Davis, P.H. (ed.) Flora of Turkey 7: 400–461.

Hedge, I.C. 1990. Labiatae. In: Ali, S.I., Nasir, Y.J. (eds.) Flora of Pakistan, vol. 192, pp. 310.

Hedge, I.C. 1992. A global survey of the biogeography of the Labiatae. In: Harley, R.M., Reynolds, T. (eds.) Advances in Labiate science. Kew: Royal Botanic Gardens, pp. 7–17.

Hedge, I.C. 1998. Labiatae. In: Hedge, I.C., Clement, R.A., Paton, A.J., Phillipson, P.B. Labiatae. In: Morat, P. (ed.) Flore de Madagascar et des Comores, Fam. 175, pp. 293.
Hegnauer, R. 1966. Chemotaxonomie der Pflanzen, vol. 4. Basel: Birkhäuser.
Hegnauer, R. 1989. Chemotaxonomie der Pflanzen, vol. 8. Basel: Birkhäuser.
Hegnauer, R. 1990. Chemotaxonomie der Pflanzen, vol. 9. Basel: Birkhäuser.
Heinrich, M. 1992. Economic botany of American Labiatae. In: Harley, R.M., Reynolds, T. (eds.) Advances in Labiate science. Kew: Royal Botanic Gardens, pp. 475–488.
Heisey, R.M., Delwiche, C.C. 1984. Phytotoxic volatiles from *Trichostema lanceolatum* (Labiatae). Am. J. Bot. 71: 821–828.
Hickey, L.J. 1973. Classification of the architecture of dicotyledonous leaves. Am. J. Bot. 60: 17–33.
Hillson, C.J. 1959. Comparative studies of floral morphology of Labiatae. Am. J. Bot. 46: 451–459.
Howe, G.F. 1986. A note on the trigger pollination mechanism in the Camphor weed (*Trichostema lanceolatum*) as related to pollinator weight and behaviour. Bull. South Calif. Acad. Sci. 85 (3): 177–179.
Howell, J.T. 1931. The genus *Pogogyne*: Proc. Calif. Acad. Sci. 4th ser. 20: 105–128.
Huck, R.B. 1987. Systematics and evolution of *Dicerandra* (Labiatae). Phanerog. Monogr. 19: 1–343.
Huck, R.B. 1992. Overview of pollination biology in the Labiatae. In: Harley, R.M., Reynolds, T. (eds.) Advances in Labiate science. Kew: Royal Botanic Gardens, pp. 167–181.
Ietswaart, J.H. 1980. A taxonomic revision of the genus *Origanum* (Labiatae). Leiden Bot. Ser. 4: 1–153.
Ingrouille, M., Bhatti, G.R. 1998. Infrageneric relationships within *Pogostemon* Desf. (Labiatae). Bot. J. Linn. Soc. 128, 2: 159–183.
Irving, R.S. 1972. A revision of the genus *Poliomintha* (Labiatae). Sida 5, 1: 8–22.
Irving, R.S. 1976. Chromosome numbers in *Hedeoma* (Labiatae) and related genera. Syst. Bot. 1: 46–56.
Irving, R.S. 1980. The systematics of *Hedeoma* (Labiatae). Sida 8, 3: 218–295.
Jalas, J. 1971. Notes on *Thymus* (Labiatae) in Europe. Bot. J. Linn. Soc. 64: 199–235, 247–271.
Jamzad, Z., Harley, M.M., Ingrouille, M., Simmonds, M., Jalil, A. 2000. Pollen exine and nutlet surface morphology of the annual species of *Nepeta* L. (Lamiaceae) in Iran. In: Harley, M.M., Morton, C.M., Blackmore, S. (eds.) Pollen and spores: morphology and biology. Kew: Royal Botanic Gardens, pp. 385–397.
Junell, S. 1934. Zur Gynäceummorphologie und Systematik der Verbenaceen und Labiaten. Symb. Bot. Upsal. 4: 1–219.
Junell, S. 1937. Der Samenentwicklung bei einigen Labiaten. Svensk. Bot. Tidskr. 31: 67–110.
Kar, R.K. 1996. On the Indian origin of *Ocimum* (Lamiaceae): a palynological approach. Palaeobotanist 43, 3: 45–50.
Kästner, A. 1978–1989. Beiträge zur Wuchsformenanalyse und systematischen Gliederung von *Teucrium* L. (in 6 parts). Flora 167: 485–514 (1978); 168: 431–467 (1979); 171: 466–519 (1981); 176: 73–93 (1985); 178: 111–138 (1986); 183: 189–224 (1989).
Kästner, A. 1989. Übersicht zur systematischen Gliederung der Gattung *Teucrium* L. Biocosme Mésogéen, Nice 6: 63–77.
Kawakubo, N. 1998. Evolution of cryptic dioecy in *Callicarpa* (Verbenaceae) on the Bonin Islands. In: Stuessy, T.F., Ono, M. (eds.) Evolution and speciation of island plants. Cambridge University Press, U.K., pp. 155–168.
Keng, H. 1978. Labiatae. Flora Malesiana Ser. I, 8, 3: 301–394.
Knuth, P. 1909. Labiatae, transl. Ainsworth Davis, J.R. In: Handbook of flower pollination, vol. 3. Oxford University Press, England, pp. 243–320.
Kooiman, P. 1972. The occurrence of iridoid glycosides in the Labiatae. Acta Bot. Neerl. 21: 417–427.
Koorders, S.H. 1897. Ueber die Blüthenknospen-Hydathoden einiger tropischen Pflanzen. Leiden: Sijfert Hendrik, 129 pp.
Lam, H.J. 1919. The Verbenaceae of the Malayan Archipelago. Groningen: M. de Waal, 370 pp.
La-Serna, I. 1984. Revisión del género *Bystropogon*. Vaduz: J. Cramer, 380 pp.
Lawrence, B.M. 1992. Chemical components of Labiatae oils and their exploitation. In: Harley, R.M., Reynolds, T. (eds.) Advances in Labiate science. Kew: Royal Botanic Gardens, pp. 399–436.
Li Xi-wen, Hedge, I.C. 1994. Lamiaceae. In: Wu, Zheng-yi, Raven, P.H. (eds.) Flora of China 17: 50–299.
Lindley, J. 1829. Editorial note: In: Bentham, G., Edwards's Bot. Register 15, t. 1282.
Lindqvist, C., Albert, V.A. 2002. Origin of the Hawaiian endemic mints within North American *Stachys* (Lamiaceae). Am. J. Bot. 89: 1709–1724.
Loew, E. 1894. Blütenbiologische Floristik des mittleren und nördlichen Europa sowie Grönlands. Systematische Zusammenstellung des in den letzten zehn Jahren veröffenlichten Beobachtungsmaterials. Leipzig: W. Engelmann.
López González, G. 1981. Conspectus saturejarum. An. Jard. Bot. Madrid 38, 2: 361–415.
Lüönd, B., Lüönd, R. 1981. Insect dispersal of pollen and fruits in *Ajuga*. Candollea 36: 167–179.
Mabberley, D.J. 1997. The plant book. Cambridge: Cambridge University Press.
Mai, D.H. 1985. Entwicklung der Wasser- und Sumpfpflanzen-Gesellschaften Europas von der Kreide bis ins Quartär. Flora 176: 449–511.
Manning, J.C. 1996. Diversity of endothecial patterns in the angiosperms. In: D'Arcy, W.G., Keating, R.C. (eds.) The anther. Cambridge University Press, England, pp. 136–158.
Marin, P.D., Sajdl, V., Kapor, S., Tatic, B., Petkovic, B. 1991. Fatty acids of the Saturejoideae, Ajugoideae and Scutellarioideae (Lamiaceae). Phytochemistry 30: 2979–2982.
Marrero, A. 1992. Chromosomal evolutionary trends in the genus *Sideritis* subgenus *Marrubiastrum*. In: Harley, R.M., Reynolds, T. (eds.) Advances in Labiate science. Kew: Royal Botanic Gardens, pp. 247–256.
Martin, A.C. 1946. The comparative internal morphology of seeds. Am. Midl. Nat. 36: 513–660.
Mattern, G., Vogel, S. 1994. Lamiaceen-Blüten duften mit dem Kelch – Prüfung einer Hypothese II. Olfactorischer und gaschromatographischer Vergleich des Laub- und Kelchduftes. Beitr. Biol. Pflanzen 68: 203–248.
McClintock, E., Epling, C. 1942. A review of the genus *Monarda* (Labiatae). Univ. Calif. Publ. Bot. 20: 147–194.
Melchior, H. 1964. A. Engler's Syllabus der Pflanzenfamilien, vol. 2. Berlin: Borntraeger, pp. 666.
Mendoza-Heuer, I. 1977. Datos comparativos acerca de especies mediterráneas y macaronésicas del género *Sideritis* L. (Lamiaceae). Bot Macaronés. IV Ci. 3: 61–71.
Metcalfe, C.R., Chalk, L. 1950. Anatomy of the Dicotyledons, vol. 2. Oxford University Press, pp. 1030–1053.
Michener, C.D., Grimaldi, D.A. 1988. A *Trigona* from Late Cretaceous amber of New Jersey (Hymenoptera: Apidae: Meliponinae). Am. Mus. Nov. 2917: 1–10.

Morales, R. 1987. El género *Thymbra* L. (Labiatae). An. Jard. Bot. Madrid 44, 2: 349–380.
Morton, J.K. 1962. Cytotaxonomic studies in the West African Labiatae. J. Linn. Soc. (Bot.) 58: 231–283.
Morton, J.K. 1973. A cytological study of British Labiatae. Watsonia 9: 239–246.
Mullan, D.P. 1933. Observations on the biology and physiological anatomy of some Indian halophytes. J. Ind. Bot. Soc. 12: 165–182.
Muller, J. 1981. Fossil pollen records of extant angiosperms. Bot. Rev. 47: 1–142.
Munir, A.A. 1991. A taxonomic revision of the genus *Oncinocalyx* F. Muell. (Verbenaceae). J. Adelaide Bot. Gard. 14: 77–84.
Nabli, M.A. 1976. Etude ultrastructurale comparée de l'exine chez quelques genres de Labiatae. In: Ferguson, I.K., Muller, J. (eds.) The evolutionary significance of the exine. London: Academic Press. Linnean Society Symposium Series no. 1, pp. 499–525.
Navarro, T., El Oualidi, J. 1997. Synopsis of the genus *Teucrium* L. (Lamiaceae) in Morocco. Acta Bot. Malac. 22: 187–203.
Olmstead, R.G., dePamphilis, C.W., Wolfe, A.D., Young, N.D., Elisons, W.J., Reeves, P.A. 2001. Disintegration of the Scrophulariaceae. Am. J. Bot. 88, 2: 348–361.
Owens, S.J., Ubera-Jiménez, J.L. 1992. Breeding systems in Labiatae. In: Harley, R.M., Reynolds, T. (eds.) Advances in Labiate science. Kew: Royal Botanic Gardens, pp. 257–280.
Packham, J.R. 1983. Biological flora of the British Isles. 155. *Lamiastrum galeobdolon* (L.) Ehrend. & Polatschek. J. Ecol. 71: 975–997.
Paton, A.J. 1990. A global taxonomic investigation of *Scutellaria* (Labiatae). Kew Bull. 45: 399–450.
Paton, A.J. 1992. The adaptive significance of calyx and nutlet morphology in *Scutellaria*. In: Harley, R.M., Reynolds, T. (eds.) Advances in Labiate science. Kew: Royal Botanic Gardens, pp. 203–210.
Paton, A.J. 1998. New records and new combinations in *Hemizygia* and *Syncolostemon* (Labiatae). Kew Bull. 53, 2: 483–485.
Paton, A.J. Harley, M.M., Harley, R.M., Weeks, S. 1994. A revision of *Endostemon* (Labiatae). Kew Bull. 49: 673–716.
Pedersen, J.A. 2000. Distribution and taxonomic implications of some phenolics in the family Lamiaceae determined by ESR spectroscopy. Biochem. Syst. Ecol. 28: 229–253.
Petanidou, T. 1996. Labiatae: a key family for wild bees and the pollination ecology in Mediterranean phryganic communities. In: Harley, R.M., Paton, A.J. (eds.) Lamiales Newslett. 4: 4–6.
Petanidou, T., Vokou, D. 1993. Pollination ecology of Labiatae in a phryganic (eastern Mediterranean) ecosystem. Am. J. Bot. 80, 8: 892–899.
Pieper, W. 1928. Vorarbeiten zu einer Revision der afrikanischen Vitex-Arten mit Berucksichtigung der übrigen. Engler's Bot. Jahrb. 62, Beibl. 141: 1–91.
Pijl, L. van der 1982. Principles of dispersal in higher plants. Berlin Heidelberg New York: Springer, 214 pp.
Pobedimova, E. 1972. Taxa novae Florae URSS, 2. Novit. Syst. Pl. Vasc. (Leningrad) 9: 243–248.
Post, T., Kuntze, O. 1904. Lexicon generum phanerogamarum. *Vitex*. Stuttgart: Deutsche Verlags-Anstalt, p. 589.
Potgieter, C.J., Edwards, T.J., Miller, R.M., van Staden, J. 1999. Pollination of seven *Plectranthus* spp. (Lamiaceae) in southern Natal, South Africa. Plant Syst. Evol. 218, 1–2: 99–112.
Pozhidaev, A.E. 1989. Exine structure in pollen grains of the Lamiaceae family. Bot. Zhurnal (Moscow & Leningrad) 74: 1410–1422 & 3 pls.
Press, J.R. 1982. Taxonomic studies in the Labiatae tribe Pogostemoneae. Bull. Br. Mus. Nat. Hist. (Bot.) 10, 1: 1–89.
Primack, R.B., Duke, N.C., Tomlinson, P.B. 1981. Floral morphology in relation to pollination ecology in five Queensland coastal plants. Austrobaileya 1, 4: 346–355.
Raj, B. 1974. Pollen morphology of Indian Labiatae. 1. Tribes Ocimoideae and Satureineae. J. Palynol. 10: 89–105.
Raj, B. 1983. A contribution to the pollen morphology of Verbenaceae. Rev. Palaeobot. Palynol. 39: 343–422.
Rao, V.S. 1952. The floral anatomy of some Verbenaceae with special reference to the gynoecium. J. Ind. Bot. 31: 297–315.
Reid, E.M., Chandler, M.E.J. 1926. The Bembridge flora (catalogue of Cainozoic plants in the Department of Geology, vol. 1). London: British Museum Natural History.
Risch, C. 1956. Die Pollenkörner der Labiaten. Willdenowia 1: 617–641.
Rivera Nuñez, D., Obón de Castro, C. 1992. The ethnobotany of Old World Labiatae. In: Harley, R.M., Reynolds, T. (eds.) Advances in Labiate science. Kew: Royal Botanic Gardens, pp. 455–473.
Rodríguez-Hahn, L., Esquivel, B., Cárdenas, J., Ramamoorthy, T.P. 1992. The distribution of diterpenoids in *Salvia*. In: Harley, R.M., Reynolds, T. (eds.) Advances in Labiate science. Kew: Royal Botanic Gardens, pp. 335–347.
Rudall, P.J. 1979. Leaf and twig anatomy of *Eriope*, a xeromorphic genus of Labiatae. Bot. J. Linn. Soc. 78: 157–180.
Rudall, P.J. 1980. Leaf anatomy of the subtribe Hyptidinae (Labiatae). Bot. J. Linn. Soc. 80: 319–340.
Rudall, P.J. 1981a. Wood anatomy in the Hyptidinae (Labiatae). Kew Bull. 35: 735–741.
Rudall, P.J. 1981b. Flower anatomy of the subtribe Hyptidinae (Labiatae). Bot. J. Linn. Soc. 83: 251–262.
Rudall, P.J. 1985. Perforated ray cells in *Hyptis hagei* – a new record for Labiatae. IAWA Bull. n.s. 6, 2: 161–162.
Rudall, P.J. 1986. Leaf anatomy of *Hyptis* sect. *Pachyphyllae* (Labiatae) and related species. Kew Bull. 41: 1017–1025.
Rudall, P.J., Bateman, R.M. (2003). Evolutionary changes in flowers: evidence from naturally-occurring angiosperm terata. Trends Plant Sci. 8: 76–82.
Rudall, P.J., Clark, L. 1992. The megagametophyte in Labiatae. In: Harley, R.M., Reynolds, T. (eds.) Advances in Labiate science. Kew: Royal Botanic Gardens, pp. 65–85.
Ryding, O. 1986. The genus *Aeollanthus* s. lat. (Labiatae). Symb. Bot. Upsal. 26, 1: 1–152.
Ryding, O. 1992a. Pericarp structure and phylogeny within Lamiaceae subfamily Nepetoideae tribe Ocimeae. Nord. J. Bot. 12: 273–298.
Ryding, O. 1992b. The distribution and origin of myxocarpy in Lamiaceae. In: Harley, R.M., Reynolds, T. (eds.) Advances in Labiate science. Kew: Royal Botanic Gardens, pp. 85–96.
Ryding, O. 1993. Pericarp structure of *Leucas* and related genera (Lamiaceae subfam. Lamioideae). Nord. J. Bot. 13: 637–646.
Ryding, O. 1994a. Pericarp structure and phylogeny of Lamiaceae subfamily Pogostemonoideae. Nord. J. Bot. 14: 59–63.
Ryding, O. 1994b. Pericarp structure in the subtribe Melittidinae (Lamiaceae-Lamioideae) and its systematic implications. Bot. Jahrb. Syst. 115: 547–555.
Ryding, O. 1994c. Pericarp structure in the tribe Prasieae (Lamiaceae-Lamioideae) and its systematic implications. Bot. Jahrb. Syst. 116: 391–399.
Ryding, O. 1995. Pericarp structure and phylogeny of the Lamiaceae-Verbenaceae-complex. Plant Syst. Evol. 198: 101–141.
Ryding, O. 1998. Phylogeny of the *Leucas* Group (Lamiaceae). Syst. Bot. 23, 2: 235–247.

Ryding, O. 2002. Myxocarpy in the Nepetoideae (Lamiaceae) with notes on myxodiaspory in general. Syst. Geogr. Plants 71: 503–514.

Ryding, O. 2003. Phylogeny of the genus *Lamium* and reconsideration of the genus *Wiedemannia* (Lamiaceae). Bot. Jahrb. Syst. 124: 325–335.

Sanders, R.W. 1987. Taxonomy of *Agastache* section *Brittonastrum* (Lamiaceae - Nepeteae). Syst. Bot. Monogr. 15, p. 92.

Schultze, W., Zänglein, A., Hose, S., Kubeczka, K.H., Czygan, F.C. 1992. Volatiles in flowers of Balm (*Melissa officinalis* L.). In: Harley, R.M., Reynolds, T. (eds.) Advances in Labiate science. Kew: Royal Botanic Gardens, pp. 357–366.

Scora, R.W. 1967. Interspecific relationships in the genus *Monarda* (Labiatae). Univ. Calif. Publ. Bot. 41: 1–59.

Shi-Liang Zhou, Kai-Yu Pan, De-Yuan Hong. 1997. Pollen and nutlet morphology in *Mosla* (Labiatae) and their systematic value. Israel J. Plant Sci. 45: 343–350.

Shiu-ying Hu. 1974. A contribution to our knowledge of *Leonurus* L. I-mu-ts'ao, the Chinese motherword. J. Chin. Univ. Hong Kong 2, 2: 365–387.

Simmonds M.S.J., Blaney, W.M. 1992. Labiate-insect interactions: effects of labiate-derived substances on insect behaviour. In: Harley, R.M., Reynolds, T. (eds.) Advances in Labiate science. Kew: Royal Botanic Gardens, pp. 375–392.

Simmonds, M.S.J., Blaney, W.M., Esquivel, B., Rodriguez-Hahn, L. 1996. Effect of clerodane-type diterpenoids isolated from *Salvia* spp. on the feeding behaviour of *Spodoptera littoralis*. Pest. Sci. 47: 17–23.

Singh, T.P. 1995. Alterations in the basic chromosome number as a means of speciation in Labiatae. Feddes Rep. 106: 39–47.

Stauffer, J. 1937. The floral anatomy of the Labiatae. Abstract of thesis. Ithaca, New York: Cornell University, pp. 332–324.

Steane D.A., Mabberley, D.J. 1998. *Rotheca* (Lamiaceae) revived. Novon 8: 204–206.

Steane, D.A., Scotland, R.W., Mabberley, D.J., Wagstaff, S.J., Reeves, P.A., Olmstead, R.G. 1997. Phylogenetic relationships of *Clerodendrum* s. l. (Lamiaceae) inferred from chloroplast DNA. Syst. Bot. 22, 2: 229–243.

Steane, D.A., Scotland, R.W., Mabberley, Olmstead, R.G. 1999. Molecular systematics of *Clerodendrum* (Lamiaceae): ITS sequences and total evidence. Am. J. Bot. 86: 98–107.

Stenzel, E., Heni, J., Rimpler, H., Vogellehner, D. 1988. Phenetic relationships in *Clerodendrum* (Verbenaceae) and some phylogenetic considerations. Plant Syst. Evol. 159: 257–271.

Subba Reddi, C., Rama Das, K., Raju, A.J.S. Atluri, J.B. 1996. Sexual system and pollination ecology of *Gmelina asiatica* L. (Verbenaceae). J. Palynol. 32: 41–50.

Suddee, S. 2001. A taxonomic revision of Tribe *Ocimeae* Dumort. (Labiatae) in continental South East Asia. Doctoral Thesis. Dublin: Trinity College, 408 pp.

Tanaka, H. 1973. Moth pollination of *Clerodendron trichotomum*. J. Jap. Bot. 48, 7: 209–214.

Tangmitcharven, S., Owens, J.N. 1997. Floral biology, pollination, pistil receptivity and pollen tube growth of teak (*Tectona grandis* L. f.). Ann. Bot. (U.K.) 79, 3: 227–241.

Taylor, D.W. 1990. Palaeobiogeographic relationships of Angiosperms from the Cretaceous and Early Tertiary of the North American area. Bot. Rev. 56: 279–416.

Thorne, R.F. 1992. Classification and geography of the flowering plants. Bot. Rev. 58, 3: 225–348.

Tomás-Barberán, F.A., Gil, M.I. 1992. Chemistry and natural distribution of flavonoids in the Labiatae. In: Harley, R.M., Reynolds, T. (eds.) Advances in Labiate science. Kew: Royal Botanic Gardens, pp. 299–305.

Trudel, M.C.G., Morton, J.K. 1992. Pollen morphology and taxonomy in North American Labiatae. Can. J. Bot. 70: 975–995.

Turner, M.W. 1996. Systematic study of the genus *Brazoria* (Lamiaceae), and *Warnockia* (Lamiaceae), a new genus from Texas. Plant Syst. Evol. 203: 65–82.

Ubera-Jiménez, J.L., Hidalgo-Fernández, P.J. 1992. Temporal gynodioecy in *Rosmarinus officinalis*. In: Harley, R.M., Reynolds, T. (eds.) Advances in Labiate science. Kew: Royal Botanic Gardens, pp. 281–289.

Upson, T.M. 1997. Systematics of the genus *Lavandula* (Lamiaceae). Ph.D. Thesis. Reading University, 446 pp.

Upson, T.M., Grayer, R.J., Greenham, J.R., Williams, C.A., Al-Ghamdi, F., Fen-Hui Chen. 2000. Leaf flavonoids as systematic characters in the genera *Lavandula* and *Sabaudia*. Biochem. Syst. Ecol. 28: 991–1007.

Van Campo, E. 1976. La flore sporopollenique du gisement Miocène terminal de Venta del Moro (Espagne). Thesis. Montpellier University.

Van Tieghem 1907. Structure du pistil et du fruit des Labiées, des Boragacées et des familles voisines. Ann. Sci. Nat. (Sér. 9) 5, 6: 321–350.

Verdcourt, B. 1992. Verbenaceae. In: Polhill, R.M. (ed.) Flora of tropical East Africa. Rotterdam: Balkema, pp. 1–155.

Verdcourt, B., Trump, E.C. 1969. Common poisonous plants of East Africa. London: Collins, 245 pp.

Vestri Alvarenga, S.A., Gastmans, J.P., Rodrigues, G. do V., Moreno, P.R.H., Emerenciano, V. de P. 2001. A computer-assisted approach for chemotaxonomic studies - diterpenes in Lamiaceae. Phytochemistry 56: 583–595.

Vij, S.P., Kashyap, S.K. 1975. Pollen grain studies in some Labiatae. J. Palynol. 11: 29–42.

Vollesen, K. 1975. A taxonomic revision of the genera *Tinnea* and *Renschia* (Lamiaceae - Ajugoideae). Bot. Tidskr. 70, 1: 1–63.

Vos, W.T. 1995. A systematic study of *Leonotis* (Pers.) R.Br. (Lamiaceae) in southern Africa. Ph.D. Thesis. Pietermaritzburg: University of Natal.

Wagner, S. 1914. Contribution à l'étude anatomique du fruit des labiées. Thesis. Paris: Université de Paris.

Wagstaff, S.J., Olmstead, R.G., Cantino, P.D. 1995. Parsimony analysis of cpDNA restriction site variation in subfamily Nepetoideae (Labiatae). Am. J. Bot. 82: 886–892.

Wagstaff, S.J., Hickerson, L., Spangler, R., Reeves, P.A., Olmstead, R.G. 1998. Phylogeny and character evolution in Labiatae s.l. inferred from cpDNA sequences. Plant Syst. Evol. 209: 265–274.

Webber, J.M., Ball, P.W. 1980. Introgression in Canadian populations of *Lycopus americanus* Muhl. and *L. europaeus* L. (Labiatae). Rhodora 82: 281–304.

Weberling, F. 1992. Morphology of flowers and inflorescences, transl. Pankhurst, R.J. Cambridge University Press, England, pp. 201–313.

Whitten, W.M. 1981. Pollination ecology of *Monarda didyma*, *M. clinopodia*, and hybrids (Lamiaceae) in the southern Appalachian mountains. Am. J. Bot. 68: 435–442.

Wojciechowska, B. 1966. Morphology and anatomy of fruits and seeds in the family Labiatae with particular respect to medicinal species (in Polish). Monogr. Bot. 21: 1–244.

Wunderlich, R. 1967. Ein Vorschlag zu einer natürlichen Gliederung der Labiaten auf Grund der Pollenkörner, der Samenentwicklung und des reifen Samens. Oesterr. Bot. Z. 114: 383–483.

Yetter, T.C. 1989. Systematic and phylogenetic relationships in the *virginianum* species complex of the genus *Pycnanthemum* (Labiatae). Doct. Diss. Oxford, Ohio: Miami University.

Zohary, M. 1937. Die verbreitungsökologischen Verhaltnisse der Pflanzen Palästinas. Beih. Bot. Centralbl. 56a: 1–155.

Authorship

(Genera indicated by number)

Harley, R.M.: *Introduction, but see also Grayer and Harley, M.M., *Symphorematoideae: 1–3, Nepetoideae: *Elsholtzieae: 122–126, Mentheae: *Salviineae: 127–131, 132 (joint), 133–134, *Menthinae: 137, 140–145, 147–149, 151, 153–154, 156–165, 170, 172, Mentheae: inc. sed.: 191. *Hyptidinae 196–203, Labiatae inc. sed. 236. **Atkins and Kok**: *Viticoideae: 4–13. Labiatae inc. sed.: 227–229. **Budantsev**: Lamioideae: 75, 82, 94, 96, 101, 107–108, 110, 112–113, 117, *Nepetinae: 178–189. **Cantino**: *Ajugoideae: 14–37, *Scutellarioideae: 54–57, 58 (joint), Lamioideae 76–80, Menthinae: 135, 150, 152, 155, 166–169, 171, 173–174, Labiatae inc. sed.: 230–235. **Conn**: *Prostantheroideae: 38–53. **Grayer**: Introduction: chemistry. **Harley, M.M**: Introduction: pollen (incl. plate). **Krestovskaja**: Lamioideae: 97–100. **Morales**: Lamioideae: 103, Salviineae: 132 (joint), Menthinae: 136, 138–139, 143, 146, 154, 175–177, Mentheae: inc. sed.: 190. **Paton**: Scutellarioideae: 58 (joint), *Hanceolinae: 193–195, *Ociminae: 204–215, *Plectranthinae: 216–226. **Ryding**: *Lamioideae: 59–74, 81, 83–93, 95, 104–106, 109, 111, 114–116, 118–121. **Upson**: *Lavandulinae: 192.

---

* Principal author of section indicated.

Lentibulariaceae

E. Fischer, W. Barthlott, R. Seine and I. Theisen

Lentibulariaceae Rich. in Poit. & Turpin, Fl. Paris 1: 26 (1808).
Pinguiculaceae Dumort. (1829).
Utricularieae Endl. (1836–1840).

Herbaceous terrestrial, epiphytic or aquatic carnivorous plants. Roots frequently absent. Leaves rosulate or scattered on stolons, entire or divided, sometimes heterophyllous. Inflorescence terminal or lateral, racemose, simple, paracladia sometimes present, often reduced to a single flower, bracteate, prophylls present or absent, sometimes connate. Flowers zygomorphic. Calyx 2–4- or 5-partite. Corolla sympetalous, 2-lipped, usually spurred, upper lip entire or 2-lobed, rarely with more lobes, lower lip often with distinct gibbous palate, entire or 2–5-lobed. Stamens 2, in the abaxial half of the flower, alternipetalous, anthers bithecous. Ovary superior, unilocular, with 2 fused carpels and central placenta. Fruit usually a capsule. Seeds without endosperm.

Lentibulariaceae, comprising three genera and about 320 species, represent the most species-rich family of carnivorous plants and have a cosmopolitan distribution.

Vegetative Morphology. Roots only present in *Pinguicula*, primary root ephemeral, quickly replaced by adventitious roots. *Genlisea* and *Utricularia* are rootless, the former is reported to develop root hairs in the seedling stage (Goebel 1893). The high plasticity of shoots/leaves in *Utricularia* has been the source of ongoing debate (Sattler and Rutishauser 1990). For the purpose of the present account, the practical approach of Taylor (1989) has been taken, regarding all leaf-like organs as leaves. Vegetative shoots are commonly short and unbranched, elongate and branching shoots are present in *Utricularia*. Tubers are formed in some species of *Utricularia*. Leaves are linear to broadly ovate, and the leaf margin is entire in *Pinguicula* and *Genlisea*. Leaves are massive and fleshy in *Pinguicula*, but commonly thin in *Genlisea* and *Utricularia*. The leaves of *Utricularia* are entire or divided, mostly petiolate and linear to obovate, reniform or even peltate. Leaf dimorphism is common in all three genera. In *Pinguicula* and temperate aquatic *Utricularia*, heterophyllous leaves are associated with hibernation. Leaves constitute the traps of *Pinguicula* and *Genlisea* (Reut 1993).

Recent research indicates that there is morphological evidence for an evolutionary line from *Pinguicula* leaves to the traps of *Genlisea* and further to the traps of *Utricularia*. Uniseriate trichomes, often glandular and bearing multicellular apices, are present on most surfaces of the plants. These trichomes are very important for the carnivorous function of the species.

Vegetative Anatomy. Adventitious roots of *Pinguicula* are heptarchous to octarchous, with a two- to three-layered pericycle below the endodermis. In the shoot of *Pinguicula*, vascular bundles are arranged in a ring. In *Genlisea* and *Utricularia*, vascular bundles are absent, and phloem and xylem strands are distributed independently. Mechanical tissue is very much reduced in aquatic species of *Utricularia*. The leaves bear sessile or stalked capitate glands and, occasionally, uniseriate hairs. Secretory cells in the stalked glands of *Pinguicula* have a perforate cuticula. Stomata are present on both the upper and lower epidermis of the leaves. Crystals and crystalloids have been recorded from the epidermis in *Pinguicula* and *Utricularia*. Vascular bundles are present in the leaves of all three genera. The mesophyll is composed of isodiametric cells in *Pinguicula*, and aquatic *Utricularia vulgaris*. Aerial leaves of *U. montana* have larger intercellular spaces in the mesophyll of the lower portion of the leaf (Metcalfe and Chalk 1950; Taylor 1989).

Inflorescence Structure. The inflorescence of Lentibulariaceae consists of a polytelic raceme, sometimes reduced to a single flower (Troll 1964). Paracladia occur only rarely in some species of *Utricularia*. In *Pinguicula*, only the main florescence is developed and its axis is strongly condensed. The pedicels, however, are distinctly prolonged and the inflorescence resembles a – sometimes uniflorous – umbel with suppressed bracts.

Floral Morphology. The calyx consists of five lobes in *Pinguicula* and *Genlisea*. In *Pinguicula*, the calyx is 2-lipped with the upper lip being 3(–7)-lobed. In *Utricularia*, the species of subg. *Polypompholyx* have a 4-lobed calyx, while in the remaining species it is always 2-lobed. The morphology of the sympetalous corolla is quite uniform and it is always bilobed, with the upper lip formed by two and the lower lip by three, fused petals (Fig. 30A). The tube varies considerably in length but is always well developed. In *Pinguicula* and *Genlisea*, the lower lip is always 3-lobed, while in *Utricularia* it varies from quite entire to deeply 2-, 4- or 5-lobed. A palate is developed in *Genlisea* and most species of *Utricularia*. Only the two abaxial stamens are present. Staminodes are generally absent. The staminal morphology is rather uniform in all three genera, with short, rather flattened filaments and 2-thecous, longitudinally dehiscent anthers, the thecae being more or less distinct to confluent. In *Utricularia guyanensis*, however, long geniculate filaments are found which conform to the inner contour of the saccate spur (Taylor 1989). The gynoecium consists of two fused carpels forming an ovoid to globose ovary with a short to long style and a usually bilabiate stigma. The ovary is always unilocular, with a free placenta which usually has a stalk-like apical extension and bears numerous ovules. In some reduced *Utricularia* species (sections *Nigrescentes* and *Oliveria*, Taylor 1989), the placenta is sessile and bears only two ovules.

Pollen Morphology. The pollen of *Pinguicula* is zonocolporate with usually five to eight furrows with discrete endoapertures (Casper 1966). In *Genlisea* subg. *Genlisea*, the pollen is 3- or 4-colporate with some syncolpism (Fromm-Trinta 1981; Taylor 1989) while that of subg. *Tayloria* is 4-colporate with a narrow equatorial belt. In *G. violacea*, it proved to be spiraperturate (Taylor 1989). In *Utricularia*, according to Huynh (1968) and Taylor (1989), two main pollen types can be distinguished. In the most primitive sections *Polypompholyx*, *Tridentura* and *Pleiochasia*, it is mostly 3-colporate with discrete endoapertures. In some closely related sections, the pollen is similar but has a larger number of furrows (up to eight in sect. *Lloydia*). In section *Utricularia*, the pollen is stephanocolporate with a larger number of furrows, and with aperturate apocolpia. In the remaining species, the pollen is either more or less intermediate between the two extremes, or rarely markedly different. However, abnormalties and variation in *Utricularia* pollen are common. For

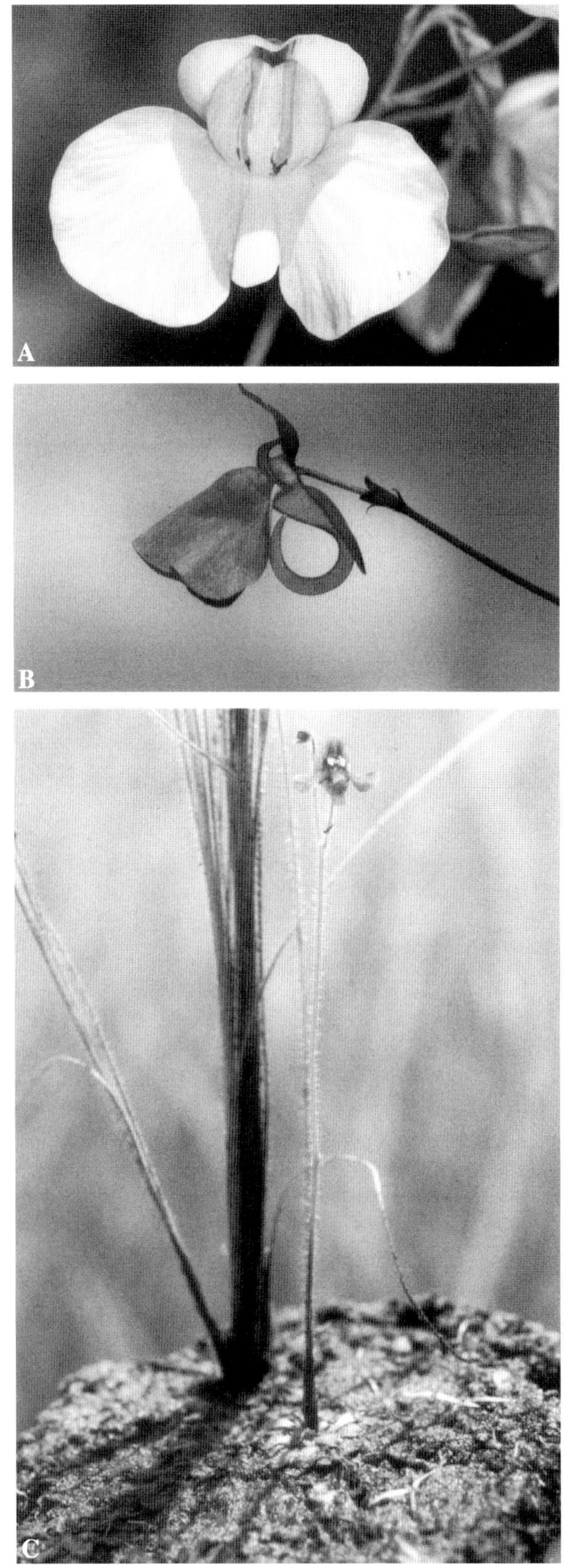

Fig. 30. Lentibulariaceae. **A** *Utricularia nelumbifolia*. **B** *Utricularia quelchii*. **C** *Genlisea stapfii*. (Photo W. Barthlott)

example, pollen grains can occur in tetrads (sections *Orchidioides* and *Utricularia*) or can be spiroaperturate monads.

Karyology. The chromosomes of Lentibulariaceae have been studied only in a few cases (Kondo 1969, 1972; Speta and Fuchs 1982; Tammaro and Pace 1987; Krahulcova and Jarolimova 1991; Mitchell et al. 1994; Zamora et al. 1996). Base numbers of x = 7, 8, 9, 10, 11, 14, 15, 21, 22 and 24 are known for *Utricularia*, and x = 8, 9, 11, 12, 16, 24 and 32 for *Pinguicula*. No chromosome counts are available for *Genlisea*. Diploid and tetraploid forms have been reported for *Utricularia juncea*.

Pollination and Reproductive Systems. In general little is known about pollination. Insects (Hymenoptera, Diptera, and Lepidoptera) are the most frequent pollinators, but also bird pollination occurs (e.g. in *Utricularia quelchii*). For some boreal taxa of *Pinguicula*, Diptera (mostly Syrphidae, rarely Anthomyriidae, Muscidae or singular *Bombus*) are reported as pollinators (Molau 1993). Self-pollination seems to be frequent and even reduced cleistogamous flowers (e.g. in *U. subulata*) occur. Some flowers are extremely long-lasting (e.g. *U. menziesii* up to 4 months).

Embryology. All three genera have anatropous, unitegmic and tenuinucellate ovules. The monocotyledonous embryo of *Pinguicula* always possesses a root pole, which is absent in *Genlisea* and *Utricularia* (Kamienski 1891). The position of the shoot pole varies: usually it is situated opposite to the micropyle, but sometimes it can be dislocated into a more lateral position.

Fruit, Seed and Dispersal. In *Pinguicula*, the capsule opens regularly with two, lateral hygroscopic valves. In *Genlisea* subg. *Genlisea*, the capsule opens with an apical lid and further ring-like circumscissions (Stopp 1958) while in subg. *Tayloria*, it opens with longitudinal slits. In *Utricularia*, some species are cleistocarpous, while in the majority of species the capsules open by a longitudinal dorsal or ventral slit, rarely with four valves or by circumscission (Taylor 1989). The seeds are usually very small (between 0.2 and 1 mm long) and lack endosperm, the testa is variously ornamented from smooth to granulose, punctate, rugose to verrucose or striate. In some *Utricularia* species, the seeds have narrow to prominent wings. Seeds of epiphytic *Utricularia* have hooks or tufts of hairs which assist wind-dispersal. Within Lentibulariaceae, dispersal mainly relies on wind and/or water. The seeds of some *Utricularia* species growing on ephemeral flush of rock outcrops seem to be adapted to stick to bird feet by either mud particles or with a mucilaginous testa.

Phytochemistry. Only few data (according to Hegnauer 1966, 1989) are available on the phytochemistry of the family. In *Utricularia* and *Pinguicula*, p-cumarid and caffeic acids have been found. In *U. vulgaris* cyanogenic glycosides have been detected. Despite earlier investigations, saponin and alkaloids are apparently missing. Flavones such as 8-hydroxyapigenin, apigenin, luteolin, 6-hydroxyluteolin and 8-hydroxyluteolin were recorded. Some iridoid glycosids could be isolated; their patterns show close resemblance to Scrophulariaceae: globularin, globulariscin, scutellariosid-II, catapol, aucubin, gardosid, mussaenodosidacid and 6-desoxycatapol. Digestive enzymes (Juniper et al. 1989) are comparatively well known and include β-glycerophosphatase, amylase, protease, leucine aminopeptidase.

Carnivory. The most striking character of Lentibulariaceae is their carnivorous habit. Three types of trap mechanism (Fig. 31) are found in the family, i.e. adhesive or flypaper-traps (*Pinguicula*), eel-traps (*Genlisea*) and suction traps (*Utricularia*); the latter two are unique in the plant kingdom. In *Pinguicula*, the attraction of prey is achieved by visual and olfactoric stimuli (Lloyd 1942; Joel et al. 1985). The leaves are covered with sessile and stalked glands. Small animals (predominantly insects, Lloyd 1942) stick to the mucilage secreted by the stalked glands. The leaves are capable of movements to enhance contact with the prey (Darwin 1875; Lloyd 1942). Digestive enzymes are secreted by the sessile glands (Heslop-Harrison and Knox 1971).

Genlisea attracts its prey, consisting of protozoa, by chemical attractants of presently unknown composition (Barthlott et al. 1998). The protozoa enter the eel-trap leaves through slits in the arms and usually accumulate in the bladder at the base of the Y-shaped part of the leaves (Reut 1993). Digestion is achieved by enzymes (Juniper et al. 1989).

Terrestrial *Utricularia* attract their protozoan prey at least partly by chemical attractants, while no chemical attraction could be demonstrated for aquatic species (Seine et al. 2002). Tactile stimuli and an abundant growth of food for potential prey on the surface of the plants have been suggested as attractants in aquatic *Utricularia* by Meyers and

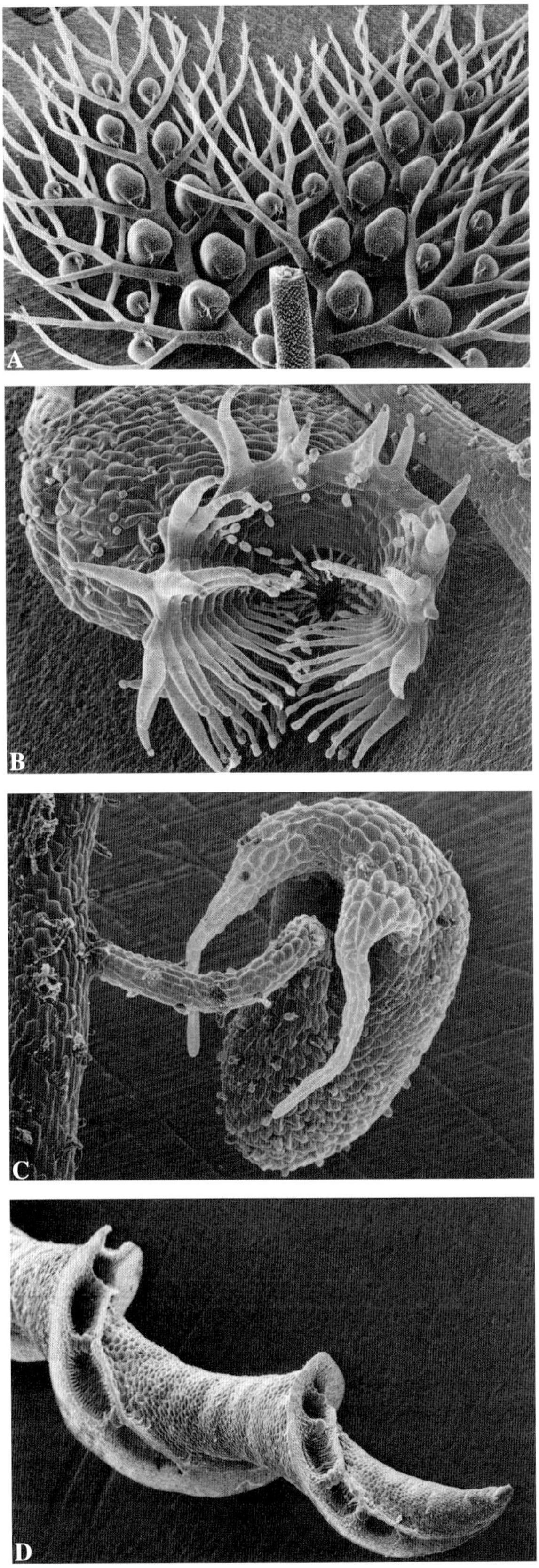

Strickler (1979). Inside the bladders, the prey is digested by enzymes secreted by bifid and quadrifid glands (Juniper et al. 1989). In contrast to a commonplace assumption, different prey strategies are found in terrestrial and aquatic *Utricularia* species.

Distribution and Habitats. Distribution of Lentibulariaceae is virtually cosmopolitan. The northern and southern extremes of distribution are held by *Pinguicula vulgaris* (73 °N on Greenland, Casper 1966) and *P. antarctica* (56 °S on Tierra del Fuego, Casper 1966). The highest altitudes are reached by some species of *Pinguicula* (*P. alpina*, 4100 m a.s.l. in the Himalayas and *P. calyptrata*, 4100 m a.s.l. in the Columbian Andes). *Pinguicula* is a holarctic element with the centre of diversity in Mexico, and extends far into Andine South America. *Genlisea* is restricted to the tropical regions of South America, Africa and Madagascar. *Utricularia* is distributed more or less worldwide, with centres of diversity in tropical regions of South America (Brazil, Venezuela, Guyana), Western Australia and India.

The family generally inhabits nutrient-poor habitats which can be grouped into four types: Open, more or less level areas with scant vegetation, such as quarzitic sands, bogs, or swamps; seepage areas on rock (Fig. 30C) or shallow soil, such as wet flush and ephemeral flush on inselbergs (Seine et al. 1995); stagnant to swiftly flowing waters, the latter inhabited by rheophytes (e.g. *Utricularia neottioides*); epiphytic habitats, such as moss, old bark, thin branches (e.g. *Pinguicula lignicola*, Casper 1987), or even the tanks of bromeliads (e.g. *U. humboldtii*).

Affinities. Flower morphology and phytochemistry (iridoid glycosids, see above) indicate a strong relationship to Scrophulariaceae. The typical corolla with palate also occurs in Scrophulariaceae s.l. as well as the reduction of endosperm (in some members of "Veronicaceae"-Gratioloideae, *Paulownia* and *Monttea*). Central placentation, at least in lower parts of ovary, can be observed in the Gratioloid genus *Limosella*, and seed structure of some *Gratiola* or *Dopatrium* closely resembles that of Lentibulariaceae. Probably subfam. Gratioloideae of "Veronicaceae" (Scrophulariaceae s.l.) has to be considered the closest

◄

Fig. 31. Lentibulariaceae. **A** *Utricularia australis*. Traps. **B** *Utricularia sandersonii*. Trap. **C** *Utricularia alpina*. Trap. **D** *Genlisea margaretae*. Trap. (Photos I. Theisen)

relative of Lentibulariaceae. This view is supported by the discovery of *Philcoxia* (Taylor et al. 2000) in quarzitic white-sand habitats in Brazil. This genus belongs to "Veronicaceae"-Gratioloideae and its bicarpellate ovary has a septum, but the root system is considerably reduced, the leaves are peltate, glandular-pubescent on their upper surface and have a circinnate vernation. The corolla resembles *Dopatrium* and some *Utricularia* of subg. *Polypompholyx*, and the flowers have two stamens. Also, Gesneriaceae show some features, e.g. seed morphology, which indicate a possible relationship. From molecular studies, Lentibulariaceae appear to be monophyletic, and *Byblis* is resolved as its closest relative within the Lamiales (Albert et al. 1992; APG 1998; Soltis et al. 2000). In the Lamiales sensu Savolainen et al. (2000), *Utricularia* comes out close to Scrophulariaceae s.l. Former speculations on a close relationship to Primulaceae, based on central placentation and the wrong record of saponins, are erroneous.

Within the family, *Pinguicula* diverges first and its species appear as a terminal clade on a long branch. *Genlisea* is sister to *Utricularia* (Müller et al. 2000). These results are confirmed by Jobson et al. (2003).

Key to the Genera

1. Corolla without palate, throat open; roots present; calyx 5-partite; peduncle without bracts; leaves all epigeic, with involute margin, upper surface covered with glands — 1. *Pinguicula*
- Corolla with palate; roots absent, but in terrestrial species root-like subterranean leaves present; calyx 5-, 4- or 2-partite; peduncle with bracts; leaves usually dimorphic, at least the hypogeic leaves with utricles or eel-trap-like, free floating plants usually with pinnate, pseudodichotomous or setose leaf segments — 2
2. Calyx 5-partite; epigeic leaves spathulate, hypogeic leaves eel-trap-like, consisting of a stalk and a tube terminating into 2 helically twisted arms; terrestrial or rarely submersed plants — 2. *Genlisea*
- Calyx 2(4)-partite; leaves extremely diverse, lamina divided into pseudodichotomous, pinnate to setose segments or entire, circular, reniform or peltate, utricles present (rarely absent in some specimens of *U. neottioides*); terrestrial or aquatic, rarely epiphytic plants — 3. *Utricularia*

1. *Pinguicula* L.

Pinguicula L., Sp. Pl.: 17 (1753); Casper, Biblioth. Bot. 127/128: 1–209 (1966), rev.

Annual or perennial herbs, terrestrial or rarely epiphytic. Stems short, erect. Leaves in a dense basal rosette, entire, ovate to elliptical or lanceolate-filiform, petiolate or sessile, upper surface with viscid glands, margin usually involute, in some species leaves dimorphic (winter buds with scales or leaves smaller than the summer leaves). True roots present. Inflorescence racemose, erect, densely congested and umbellate, with one to several (up to 10) long-pedicellate flowers. Bracts and prophylls absent. Pedicels long, usually mistaken for the inflorescence axis. Calyx 2-lipped, upper lip 3(–7)-lobed, lower lip 2-lobed, lobes ovate-lanceolate, acuminate, subequal. Corolla violet, blue, mauve, pink, white with yellow or entirely yellow, upper lip 2-lobed, lower lip 3-lobed, throat with distinct palate, spur filiform to cylindrical-conical. Stamens 2. Ovary globose, usually densely covered with short glandular hairs, style short, with unequally 2-lobed stigma. Capsule globose to ovate, opening by two valves. Seeds minute. $2n = 16, 18, 22, 24, 32, 48, 64$. About 80 species in Europe, northern Asia, Himalayas, Japan, North America, Central America (from Mexico to the Caribbean Islands) and South America (Andes from Colombia to Ecuador, Chile and Patagonia).

2. *Genlisea* A. St.-Hil. Fig. 30, 31D

Genlisea A. St.-Hil., Voy. Distr. Diam. 2: 428 (1833); Fischer et al., Nord. J. Bot. 20: 291–318 (2000), reg. rev.; Fromm-Trinta, Rodriguésia 31: 17–139 (1979), reg. rev.; Fromm-Trinta, Sellowia 36: 55–62 (1984), reg. rev.; Taylor, Carniv. Pl. Newslett. 20: 20–35 (1991), check-list.

Annual or perennial herbs. Stems short, erect. Leaves dimorphic, epigeic leaves in a dense basal rosette, spathulate, obtuse, glabrous. Subterraneous root-like leaves (rhizophylls) congested at lower part of stem, descending into the substrate, consisting of a stalk and a slender tube terminating into 2 helically twisted arms, the tube and the arms provided internally with transverse rows of stiff, inwardly directed hairs. Roots absent. Inflorescence racemose, erect, simple or sparsely branched, glabrous, glandular-pubescent or hispid, scale-like leaves few, bracts ovate-lanceolate to linear-lanceolate, acuminate, prophylls linear-lanceolate. Pedicels erect and spreading in flower and fruit or recurved in fruit. Calyx 5-lobed, divided almost to base, lobes ovate-lanceolate, acuminate, subequal, glabrous, hispid or covered with long glandular hairs. Corolla violet, blue, mauve or pink with yellowish to greenish spur or entirely yellow, lower lip and spur often covered with short glandular hairs, upper lip ovate-oblong, lower lip distinctly 3-lobed, with a palate near throat, spur clavate. Stamens 2, filaments falcate. Ovary globose, densely covered only with short glandular hairs, hispid and with glandular hairs, only hispid or glabrous, style short, with semicircular 1-lobed stigma. Capsule globose, poricidal or opening by longitudinal slits (subg. *Tayloria*). Twenty-one species, in tropical South America (11 spp.) and tropical Africa (10 spp.), one (*G. margaretae* Hutch.) also occurring in Madagascar.

3. *Utricularia* L. Fig. 30A, B, 31A–C

Utricularia L., Sp. Pl.: 18 (1753) and Gen. Pl., ed. 5: 11 (1754); Taylor, Kew Bull. Add. Ser. 14: 1–724 (1989), rev.

Annual or perennial herbs, terrestrial, epiphytic or aquatic. Vegetative parts not clearly differentiated but consisting of stems and modified leaves; sometimes horizontal, often subterraneous rhizomes, stolons or sometimes tubers present. Roots absent; root-like stems and leaves usually present in terrestrial species but absent from floating taxa. Leaves extremely diverse, rosulate at base of peduncle, or alternate, opposite or verticillate at the stolon nodes, lamina divided (in most aquatic species) into pseudodichotomous or pinnate setose to capillary segments or entire, sometimes circular to reniform or peltate, sessile to petiolate, rarely completely absent. Traps usually shortly stalked, globose to ovate vesicles (utricles), variously inserted, traps occasionally absent (*U. neottioides*). Inflorescence racemose, lax, usually with several sterile bracts on peduncle below inflorescence, in some aquatic species at the base with verticillate, spongy and inflated organs; bracts usually ovate to orbicular, prophylls present or absent. Flowers pedicellate. Calyx 2-lipped, with two or four (subg. *Polypompholyx*) equal to unequal lobes. Corolla yellow, violet, mauve or rose, rarely red, 2-lipped, upper lip erect and spreading, entire, emarginate or rarely 2-lobed, lower lip large, 5-, 3- or 2-lobed to entire, with well-developed, bipartite palate covering the throat, tube very short, spur cylindrical, obtuse, straight or incurved. Stamens 2, short and thick, filaments straight or curved, anthers bithecal, the thecae contiguous to confluent. Ovary globose to ovate, glabrous or glandular-pubescent, style short, stigma unequally 2-lipped. Capsule opening by two lateral valves, poricidal or opening irregularly, rarely indehiscent. Seeds 1 to numerous, small, often winged or with diverse testa ornamentation. $2n = 14, 16, 18, 20, 22, 24, 28, 30, 36, 42, 44, 48$. Almost cosmopolitan with about 216 species, diversity centres in tropical regions of South America (Brazil, Venezuela, Guyana, 59 spp.), Western Australia (55 spp.) and India (33 spp.).

Selected Bibliography

Albert, V.A., Williams, S.E., Chase, M.W. 1992. Carnivorous plants: phylogeny and structural evolution. Science 257: 1491–1495.

APG (Angiosperm Phylogeny Group) 1998. An ordinal classification for the families of flowering plants. Ann. Missouri Bot. Gard. 85: 531–553.

Barthlott, W., Porembski, S., Fischer, E., Gemmel, B. 1998. First protozoa-trapping plant found. Nature 392: 447.

Casper, S.J. 1966. Monographie der Gattung *Pinguicula* L. Biblioth. Bot. 127/128: 1–209.

Casper, S.J. 1987. On *Pinguicula lignicola*, an epiphytic heterophyllic member of the Lentibulariaceae in Cuba. Plant Syst. Evol. 155: 349–354.

Darwin, C. 1875. Insectivorous plants. London: John Murray.

Fromm-Trinta, E. 1981. Revisão do gênero *Genlisea* St.-Hil. no Brasil. Bol. Mus. Nac. Rio de Janeiro, Bot. 61: 1–21.

Goebel, K. 1893. Zur Biologie von *Genlisea*. Flora 77: 208–212.

Hegnauer, R. 1966. Chemotaxonomie der Pflanzen 4: 388–390.

Hegnauer, R. 1989. Chemotaxonomie der Pflanzen 8: 667–668.

Heslop-Harrison, Y., Knox, R.B. 1971. A cytochemical study of leaf-gland enzymes of insectivorous plants of the genus *Pinguicula*. Planta 96: 183–211.

Huynh, K.L. 1968. Etude de la morphologie du pollen du genre *Utricularia* L. Pollen Spores 10: 11–55.

Jobson, R.W., Playford, J., Cameron, K.M., Albert, V.A. 2003. Molecular phylogenetics of Lentibulariaceae inferred from plastid rps16 Intron and *trn*L-F DNA sequences: implications for character evolution and biogeography. Syst. Bot. 28: 157–171.

Joel, D.M., Juniper, B.E., Dafni, A. 1985. Ultraviolet patterns in the traps of carnivorous plants. New Phytol. 101: 585–593.

Juniper, B.E, Robins, R.J., Joel, D.M. 1989. The carnivorous plants. London: Academic Press.

Kamienski, F. 1891. Lentibulariaceae. In: Engler & Prantl, Die natürlichen Pflanzenfamilien IV, 3b: 108–123. Leipzig: Engelmann.

Kondo, K. 1969. Chromosome numbers of carnivorous plants. Bull. Torrey Bot. Club 96: 322–328.

Kondo, K. 1972. Comparison of variability in *Utricularia cornuta* and *Utricularia juncea*. Am. J. Bot. 59: 23–37.

Krahulcova, A., Jarolimova, V. 1991. Relationship between *Pinguicula bohemica* Krajina and *Pinguicula vulgaris* L. (Lentibulariaceae) from the karyological point of view. Preslia 63: 323–328.

Lloyd, F.E. 1942. The carnivorous plants. Mass.: Waltham.

Metcalfe, C.R., Chalk, L. 1950. Anatomy of the Dicotyledons 2: 725–1500.

Meyers, D.G., Strickler, J.R. 1979: Capture enhancement in a carnivorous aquatic plant: function of antennae and bristles in *Utricularia vulgaris*. Science 203: 1022–1025.

Mitchell, R.S., Maenza Gmelch, T.E., Barbour, J.G. 1994. *Utricularia inflata* Walt. (Lentibulariaceae), new to New York State. Bull. Torrey Bot. Club 121: 295–297.

Molau, U. 1993. Reproductive ecology of the three Nordic *Pinguicula* species (Lentibulariaceae). Nord. J. Bot. 13: 149–157.

Müller, K., Borsch, T., Legendre, L., Porembski, S., Barthlott, W. 2000. A phylogeny of Lentibulariaceae based on sequences of *mat*K and adjacent noncoding regions. Am. J. Bot. Suppl. 87: 133–134.

Reut, M.S. 1993. Trap structure of the carnivorous plant *Genlisea* (Lentibulariaceae). Bot. Helv. 103: 101–111.

Sattler, R., Rutishauser, R. 1990. Structural and dynamic descriptions of the development of *Utricularia foliosa* and *U. australis*. Can. J. Bot. 68: 1989–2003.

Savolainen, V., Fay, M.F., Albach, D.C., Backlund, A., van der Bank, M., Cameron, K.M., Johnson, S.A., Liedó, M.D., Pintaud, J.-C., Powell, M., Sheahan, M.C., Soltis, D.E., Soltis, P.S., Weston, P., Whitten, W.M., Wurdack, K.J., Chase, M.W. 2000. Phylogeny of the eudicots: a nearly complete familial analysis based on *rbc*L gene sequences. Kew Bull. 55: 257–309.

Seine, R., Porembski, S., Barthlott, W. 1995. A neglected habitat of carnivorous plants: Inselbergs. Feddes Repert. 106: 555–562.

Seine, R., Porembski, S., Balduin, M., Theisen, I., Wilbert, N., Barthlott, W. 2002. Different prey strategies of terrestrial and aquatic species in the carnivorous genus *Utricularia* (Lentibulariaceae). Bot. Jahrb. Syst. 124: 71–76.

Soltis, D.E., Soltis, P., Chase, M.W., Mort, M.E., Albach, D.C., Zanis, M., Savolainen, V., Hahn, W.H., Hoot, S.B., Fay, M.F., Axtell, M., Swensen, S.M., Prince, L.M., Kress, W.J., Nixon, K.C., Farris, J.S. 2000. Angiosperm phylogeny inferred from 18S rDNA, *rbc*L, and *atp*B sequences. Bot. J. Linn. Soc. 133: 381–461.

Speta, F., Fuchs, F. 1982. Neue *Pinguicula*-Arten (Lentibulariaceae) aus Mexiko. Stapfia 10: 111–119.

Stopp, K. 1958. Notiz über die Dehiszenzweise der Kapselfrüchte von *Genlisea hispidula* Stapf. Beitr. Biol. Pflanzen 34: 401–403.

Tammaro, F., Pace, L. 1987. Il genere *Pinguicula* L. (Lentibulariaceae) in Italia centrale ed instituzione di una nuova specie *P. fiorii* Tamm. et Pace. Inf. Bot. Ital. 19: 429–436.

Taylor, P. 1989. The genus *Utricularia* – a taxonomic monograph. Kew Bull. Add. Ser. 14: I–XI, 1–724.

Taylor, P., Souza, V.G., Giulietti, A.M., Harley, R.M. 2000. *Philcoxia*: a new genus of Scrophulariaceae with three new species from Eastern Brazil. Kew Bull. 55: 155–163.

Troll, W. 1964. Die Infloreszenzen. Typologie und Stellung im Aufbau des Vegetationskörpers. Bd. 1. Stuttgart: Gustav Fischer.

Zamora, R., Jamilena, M., Rejon, M.R., Blanca, G. 1996. Two new species of the carnivorous genus *Pinguicula* (Lentibulariaceae) from Mediterranean habitats. Plant Syst. Evol. 200: 41–60.

Martyniaceae

H.-D. Ihlenfeldt

Martyniaceae Stapf in Engler & Prantl, Nat. Pflanzenfam. IV, 3b: 265 (1895), nom. cons.

Annual, rarely perennial, erect to prostrate herbs, or rarely shrubs up to 3 m tall and with fleshy young stems (*Holoregmia*). Whole plant glandular-viscid. Leaves, simple, opposite, occasionally alternate. Flowers in terminal racemes, pedicellate, with 2 prophylls at the base of the calyx. Flowers bisexual, tetracyclic, zygomorphic. Calyx usually small, sepals 5, free or connate. Corolla gamopetalous, petals 5, yellow, purplish or white; limb spreading, lobes slightly unequal or limb distinctly bilabiate. Stamens either 4, didynamous, and 1 staminode, or stamens 2 and 3 staminodes, all inserted in the tube of the corolla and included in the tube. Annular hypogynous disk present. Ovary superior, bicarpellate, paracarpous, with 2 massive parietal T-shaped placentae. Fruits oblong, usually, especially in the upper part, curved and incompletely loculicidally deshiscent; exocarp fleshy, peeling off at maturity, endocarp woody. Upper sterile part of the fruit usually distinctly elongated (forming a "beak") and, at maturity, splitting into 2 curved short or long appendages ("horns", "claws"). Seeds 4 to many; embryo straight, endosperm very thin.

A New World family with 5 genera and about 16 species, distributed mainly in warm and arid regions from the southern USA to Argentina.

Vegetative Morphology. Most species are erect to prostrate annuals or short-lived perennials. *Proboscidea altheaefolia* has a fleshy taproot, and *Craniolaria annua* a large tuber. *Holoregmia* forms robust, much-branched shrubs with fleshy young stems. The leaves are large, opposite, occasionally also alternate in the upper part of the plant; they are petiolate, rotundate to cordate, simple or slightly lobed with entire or dentate margin.

Vegetative Anatomy. All parts of the plant are covered with glandular hairs composed of a uniserial stalk and a multicellular (8 or 16 cells), apically flattened, globular head which produces a sticky secretion rendering the whole plant sticky (Fig. 33A, B). The secretion contains reductive sugars (Inamdar 1969). Many small insects are trapped by this glue. The heads of the glandular hairs can be more or less reduced, and simple hairs also occur. The glandular hairs of the Martyniaceae differ from the mucilage-producing glandular hairs of the Pedaliaceae in which the Martyniaceae have formerly been included. All vegetative parts of the plant release an unpleasant odour.

Mechanical support of the stem is mainly due to a large amount (about 10 layers) of collenchymatous tissue and the turgidity of the parenchymatous cells of the cortex. The xylem, which forms a continuous cylinder, contains diffusely placed vessels with simple perforations and fibre tracheids. The axial parenchyma is vasicentric scanty, the rays both multiseriate and uniseriate (Carlquist 1987). Sieve-element plastids are of the Ss type (Behnke 1991).

The leaves are bifacial, and stomata are found on both sides. Most of the stomata are anomocytic or anisocytic, but also paracytic and diacytic stomata have been observed (Inamdar 1969).

Inflorescence Structure. The flowers are arranged in terminal racemes. The small rotundate to oblong bracts are membranous and ephemeral. Lateral branches from the axils of the last leaf pair below the raceme can produce additional racemes. These branches can overtop the primary raceme, giving the plant a pseudodichotomous habit. The flowers are distinctly pedicellate. At the base of the calyx there are 2 oblong prophylls which may be coloured and occasionally become somewhat fleshy.

Flower Structure and Anatomy. The 5 sepals are usually small compared to the corolla (Fig. 32A, B). They are either nearly free (*Martynia*, *Ibicella*) or laterally connate for at least 1/2 of their length, forming a tube with a ventral slit. In *Craniolaria* and especially in *Holoregmia*, the sepals are comparatively long and form a spathe-like structure, up to 55 mm long, also with a ventral slit. The

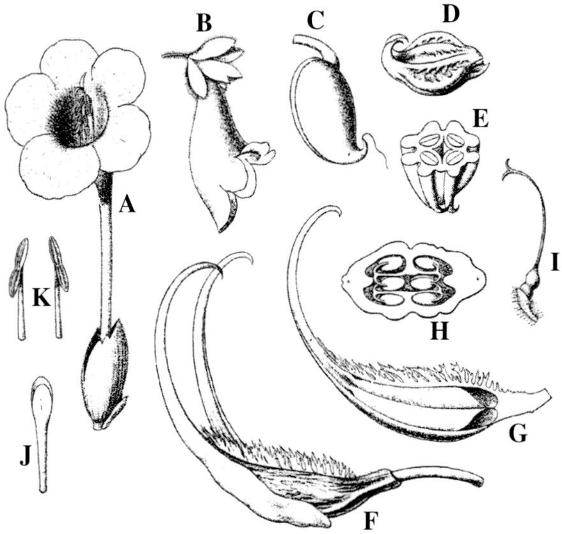

Fig. 32. Martyniaceae. **A** *Craniolaria annua*. Flower. **B** *Martynia annua*. Flower. **C** *M. annua*. Maturing fruit with excocarp still on the fruit. **D** *M. annua*. Mature fruit after the peeling off of the exocarp. **E** *M. annua*. Cross section of mature fruit. **F** *Proboscidea lousianica*. Maturing fruit with exocarp peeling off. **G** *P. lousianica*. Mature fruit, longitudinal section. **H** *P. lousianica*. Cross section of mature fruit. **I** *P. lousianica*. Ovary. **J** *P. altheaefolia*. Stigma. **K** *P. altheaefolia*. Stamens. (After Stapf 1895)

gamopetalous corolla is composed of 5 petals, aestivation is descending. The corolla (Fig. 32A, B) tube is either very long (up to 14 cm), narrow-cylindrical and abruptly dilated into a campanulate cup at the top (*Craniolaria*), or relatively short (up to 5 cm), obliquely campanulate but constricted at the base. The limb is spreading, and the lobes are nearly equal or the limb is distinctly bilabiate. Usually nectar guides such as streaks and dots, especially on the lower lobe and running down into the corolla tube, are present. The androecium consists of 5 members. Usually the median adaxial stamen is only represented by a staminode, and the 4 fertile stamens are didynamous. In *Martynia* also the 2 adaxial-lateral stamens are reduced to staminodes. All members are inserted either near the base of the tube (*Proboscidea*, *Ibicella*, *Martynia*, *Holoregmia*), or at the base of the campanulate upper portion of the tube (*Craniolaria*), and included in the tube. The thecae are completely separated and spreading at an angle of 180° (Fig. 32K); shortly before anthesis the anthers become connected into pairs. The superior ovary is composed of 2 carpels in median position and surrounded by an annular hypogynous disk (Fig. 32I). The 2 massive parietal placentae are T-shaped and divide the ovary nearly completely into 4 peripheral compartments and a central lacuna, each compartment containing 1 to numerous ovules. The style is filiform, the stigma bilobed (Fig. 32J) and presented at the mouth of the corolla, appressed to the adaxial side of the corolla.

Embryology. Only *Martynia* and *Proboscidea* have been studied in detail. The anther wall comprises the epidermis, a fibrous endothecium, 1–2 middle layers and a secretory tapetum of binucleate cells. The pollen grains, when shed, are 2-celled (Singh 1960); however, Anderson (1922) described them as 3-celled. The ovules are anatropous, pendent, unitegmic and tenuinucellate. A single cell functions directly as megaspore mother cell. After meiosis the megaspores are arranged linearly. The chalazal megaspore develops into a Polygonum-type embryo sac. The polar nuclei fuse shortly before fertilisation. The antipodes are ephemeral. Endosperm formation is cellular. The division of the primary endosperm nucleus separates a large micropylar from a small chalazal chamber. Then, a vertical division occurs in the micropylar chamber, followed by a transverse division, resulting in 2 central cells and a 2-celled micropylar haustorium. The latter subsequently becomes multicelled. The central cells produce the endosperm proper, whereas the chalazal chamber develops into a 4-celled haustorium. Embryo formation conforms to the Onagrad type-Catalpa variation.

The seed is albuminous, although the endosperm is only thin, and the embryo is straight. In *Proboscidea* the cotyledons are auricled at their base, enclosing the short, stout radicle which sometimes only slightly protrudes beyond the cotyledons (Mayberry 1947).

Pollen Morphology. Martyniaceae have a remarkable and rare pollen type. Interpretation of the pollen morphology still is a matter of debate. Principally, the exine of all genera appears to be dissected into about 20–40 platelets ("exine shields", "clypeate pollen", Halbritter and Hesse 1995; Fig. 33C) which are evenly distributed all over the surface of the globose pollen grains. Depending on species and degree of hydration, the width of the grooves between the platelets varies from 2 to 5 µm. In *Proboscidea*, *Ibicella* and *Holoregmia*, the shields are rounded-oval to elongated-angular, in *Martynia* hexagonal or pentagonal, and the exine is reticulate. In *Craniolaria* the shields are circular and consist of a raised, ring-like structure with a cavity in the centre (for details, see Bretting and Nilsson 1988). Erdtman (1952) considered pollen of *Martynia* and *Ibicella* to be rugoidate, and the regions on *Craniolaria*

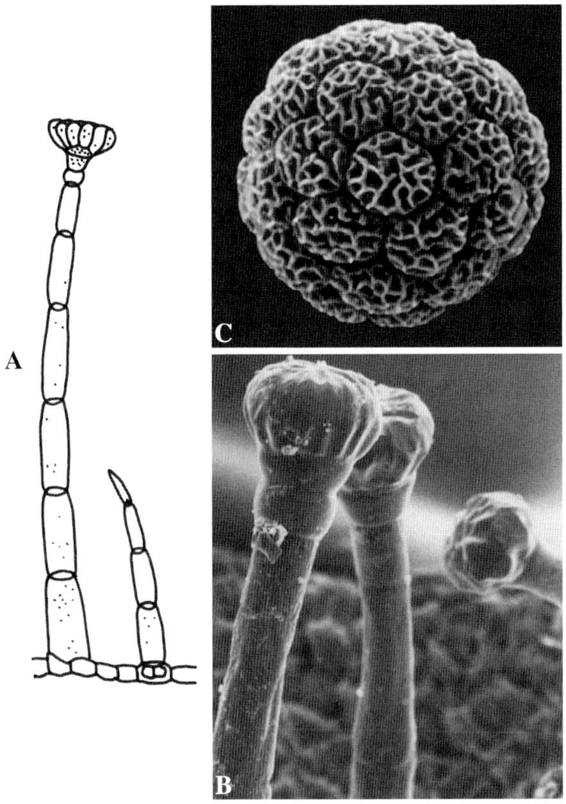

Fig. 33. Martyniaceae. **A** *Ibicella lutea*. Glandular and simple hair, ×80. **B** *Proboscidea lousianica*. Head of glandular hair, SEM, ×300. **C** *Proboscidea lousianica*. Pollen, SEM, ×625. (**A** After Abbiatti 1939; **B, C** original)

pollen encircled by rings to be aperturoid. Other authors considered the pollen of *Ibicella* as syncolpate or inaperturate. Halbritter and Hesse (1995) could demonstrate that the pollen grains of Martyniaceae should be characterised as inaperturate. Bretting and Nilsson (1988) suggested that the deviating pollen type of *Craniolaria* may be associated with a different pollination syndrome (sphingophily).

Pollen morphology of Martyniaceae vastly differs from Pedaliaceae (and Trapellaceae). Similar pollen is found in certain genera of the Bignoniaceae (e.g. *Catalpa*, Gentry and Tomb 1979). Pollen of the clypeate type, however, is also found in several families not related to Martyniaceae, for example, in Iridaceae (Halbritter and Hesse 1995).

Karyology. For *Proboscidea* and *Ibicella*, $n = 15$ has been reported. $2n = 32$ has been found in *Martynia annua*.

Pollination and Reproductive Systems. Martyniaceae exhibit two different flower syndromes: a melittophilous syndrome in *Proboscidea*, *Ibicella*, *Martynia* and *Holoregmia*, and a sphingophilous syndrome in *Craniolaria*. In the melittophilous genera the stigmas are appressed to the upper part of the corolla tube and are presented at the mouth of the tube. The anthers are positioned behind the stigma. In *Ibicella* and *Proboscidea* (probably also in *Martynia*), the lower lobe of the stigma is sensitive and moves upwards after stimulation. Sensitivity ceases as soon as compatible pollen has been deposited on the stigma.

Only the floral biology of *Proboscidea louisianica* has been investigated in detail (Thieret 1976; Phillippi and Tyrl 1979). The pollen : ovule ratio of 850 : 1 is comparatively low, probably due to the large size of the pollen grains. The main pollinators are large members of the families Apidae and Anthophoridae which utilise both pollen and nectar. Experiments under controlled conditions revealed that successful self-pollination is possible, albeit a very rare event (fruit set 4%). This is apparently due to protandry, the relative position of anthers and stigma, and sensitiveness of the latter, as artificial self-pollination resulted in good fruit set (57%). However, xenogamy, whether intrapopulational or interpopulational, resulted in better fruit set (about 80%). A (relative) prezygotic barrier to self-fertilisation is hypothesised to exist.

Fruit and Seed. Fruit structure (Fig. 32C–H), unique in Lamiales, is the only reliable character syndrome for the assignment of a taxon to Martyniaceae: a paracarpous ovary with 2 massive T-shaped placentae, a usually distinct "beak" developing from the sterile upper part of the ovary, a pericarp consisting of a woody endocarp (made up of interwoven strands of lignified fibres) and a fleshy parenchymatous exocarp, the latter peeling off in two parts at maturation of the fruit, emergences on the outer side of the endocarp, such as low longitudinal ribs, and toothed to serrate longitudinal crests and appressed spines, all hidden in the fleshy exocarp and exposed only after the peeling off of the exocarp. As to beak development during fruit formation, deshiscence (principally loculicidal), types of emergences on the mature fruit and number of seeds, there is considerable variation (for details, see generic descriptions). Features of the fruit are important characters for the delimitation of the genera. In *Holoregmia* the beak is only very short, and dehiscence of the beak at maturation renders the apex of the fruit only slightly bifurcate; after 1 year placed in the soil, the fruit may split into 2 mericarps (Harley et al. 2003). In the remaining genera, during fruit formation, the beak elongates into an arcuately-curved structure, moderately in *Craniolaria* and *Martynia*, excessively (exceeding even the length of the fruit proper) in *Proboscidea* and *Ibicella*. By deshiscence of the elongated beak, 2 apical "claws" or "horns" (rostra) are formed which make the fruits of these genera effective "trample burrs". In *Holoregmia*, *Martynia* and *Craniolaria*, the lower, only few-seeded (4–6) part of the ovary remains closed, whereas in the multiseeded fruits of *Proboscidea* and *Ibicella* dehiscence gradually proceeds towards the base of the fruit; the two halves, however, do not separate completely; the seeds are gradually released according to the advancement of dehiscence.

The structure of the seed coat apparently depends on whether the seeds are released from the fruit (*Ibicella*, *Proboscidea*) or remain included (*Martynia*, *Craniolaria*, *Holoregmia*). In *Proboscidea* (Anderson 1922), the rugose seed coat basically consists of several layers of thickened and lignified cells, heavily impregnated with tannins, which originate from the outer layers of the integument. In *Martynia* (Singh 1958), the somewhat warty seed coat consists only of the outer epidermis, whose radial and inner tangential walls become lignified, and the cuticle of the inner epidermis; all other layers become crushed during maturation. The raphe is dorsal.

As reserve material the seeds contain aleuron and oil (up to 56%, Hegnauer 1969).

Dispersal. The fruits (except for *Holoregmia*) are effective trample burrs which cling to the feet and ankles of passing animals by the aid of their curved horns. The longitudinally arranged high and sharp crests on the adaxial side of the fruits in *Proboscidea* and *Ibicella* apparently improve fixation of the fruits to the vector. The large multiseeded fruits of *Proboscidea* and *Ibicella* shed their seeds gradually during transport, whereas in *Craniolaria*, *Martynia* and *Holoregmia* the few seeds remain included in the fruit.

Phytochemistry. Martyniaceae conform to the basic chemical profile (Grayer et al. 1999) of Lamiales: iridoids, flavone glycosides and verbascoside-type compounds. A few special compounds have been described: cornoside (*Martynia*) and martynoside (*Proboscidea louisianica*).

Subdivision of and Relationships Within the Family. Based on fruit morphology, the 5 genera can be divided into 2 groups to which formal rank has not yet been assigned: *Proboscidea*/*Ibicella* (fruits with long rostra, multiseeded, dehiscence nearly complete), *Martynia*/*Craniolaria*/*Holoregmia* (fruits with short to minute rostra, few-seeded, lower part of the ovary remaining closed). Both groups contain a genus with a more northern distribution (USA, Mexico and adjacent countries, *Proboscidea* and *Martynia*), whereas the remaining genera share a southern distribution (Venezuela, Brazil to Argentina). The close relationship between *Proboscidea* and *Ibicella*, originally united in one genus, is also supported by pollen morphology and structure of the seed coat (Bretting and Nilsson 1988). The *Martynia*/*Craniolaria*/*Holoregmia* group is much less homogeneous. *Craniolaria* stands out by its sphingophilous flower syndrome and a deviating pollen morphology, *Martynia* by the possession of only 2 fertile stamens, and *Holoregmia* by its shrubby habit and the lack of distinct apical horns on the unspecialised fruit, both apparently plesiomorphic characters.

Affinities. Originally Martyniaceae (and Pedaliaceae as well, see Pedaliaceae, this volume) formed part of Bignoniaceae. Link (1829) apparently was the first to recognise *Martynia* and related genera as a distinct group to which he assigned the rank of a subordo related to, although clearly different from, Bignoniaceae. Although this view was adopted by a number of subsequent authors, most of them kept this group either included in Bignoniaceae or assigned it to Pedaliaceae (for detailed information on the history, see Ihlenfeldt 1967). The latter view apparently was substantially influenced by the occurrence of trample burrs in both Martyniaceae and Pedaliaceae, a fruit type which is quite uncommon in Lamiales. However, the structures that fix the burr to the vector are not homologous (see Pedaliaceae, this volume). Stapf (1895) was the first to point out, in some detail, the fundamental differences in morphology between Pedaliaceae and Martyniaceae, and his conclusion was that both deserve family status. However, even after Ihlenfeldt (1967) had added numerous arguments in favour of maintaining familial status for Martyniaceae, and had pointed out that Pedaliaceae and Martyniaceae by no means are as closely related as generally thought (for details, see Pedaliaceae, this volume) and, moreover, clearly distinct from Bignoniaceae, many authors kept Martyniaceae included in Pedaliaceae (e.g. Brummitt 1992). Ihlenfeldt (1967) interpreted Martyniaceae and

Pedaliaceae as two lineages, adapted to arid conditions and highly specialised in dispersal systems, which evolved in parallel from a common, remote bignoniaceous ancestor in the New and Old World respectively. This view is now supported by molecular data. In the most comprehensive study (Olmstead et al. 2001), Pedaliaceae and Martyniaceae are nested in two different major clades (which may be sister to each other), thus supporting family status for both groups. Molecular data also strongly support monophyly of the two families (Olmstead, pers. comm.); however, the position of Martyniaceae within Lamilaes remains uncertain.

Distribution and Habitats. Martyniaceae are an exclusively New World family and confined to areas of Mediterranean, subtropical and tropical climate. The distribution area ranges from the southern USA to N Argentina. *Holoregmia*, which can be considered as the most primitive extant genus, has a very restricted distribution in NE Brazil (Harley et al. 2003) whereas *Proboscidea*, native to the southern USA and northern Mexico, shows signs of recent radiation (Bretting 1983, 1985).

Most species prefer unstable or degraded, more or less arid habitats. Thus, it is not astonishing that several species have been reported to be weedy. *Martynia annua*, *Ibicella lutea* and *Proboscidea louisianica* have been transported by man to many countries with a suitable climate.

Economic Importance. Martyniaceae do not possess any essential economic importance. The fleshy roots of *Proboscidea altheaefolia* and the tuber of *Craniolaria annua* are utilised as food by local populations, as well as the young pickled fruits of *Ibicella* and *Proboscidea*. *Ibicella lutea* and *Proboscidea louisianica* are often grown as ornamentals. Mature fruits of *Proboscidea* are used in basketwork by several Indian tribes.

Key to Genera of Martyniaceae

1. Shrubs up to 3 m tall; fruits without apical horns **1. *Holoregmia***
- Annual, rarely perennial, erect to prostrate herbs, fruits with 2 apical horns 2
2. Corolla with a long narrow-cylindrical tube up to 14 cm long, apically abruptly dilated into a campanulate cup, sepals completely connate into a 4.5–5.5 cm long, spathe-like structure with a ventral slit **2. *Craniolaria***
- Corolla without a long narrow-cylindrical tube, campanulate, only constricted at the very base, sepals much shorter (up to 2.5 cm), free or, if connate, not forming a spathe-like structure 3
3. Fertile stamens 2, staminodes 3, horns of the fruit only up to 5 mm long, much shorter than the body of the fruit **3. *Martynia***
- Fertile stamens 4, 1 staminode, horns of the fruit much longer, at least equalling the length of the body of the fruit, up to 12 cm long 4
4. Sepals free, flowers pale to deep yellow, plants from Bolivia, S Brazil to Argentina **4. *Ibicella***
- Sepals connate for at least 1/2 of their length, calyx tube with a ventral slit, flowers purple, pink, lavender or cream, or yellow to orange, plants from S USA to Mexico **5. *Proboscidea***

1. *Holoregmia* Nees

Holoregmia Nees, Flora 4: 300 (1821); Harley et al., Kew Bull. 58(1): 205–212 (2003), rev.

Robust much-branched shrub, up to 3 m tall, younger stems fleshy. Sepals connate, forming a spathe-like structure with a ventral slit, only slightly shorter than the corolla tube and calyx inflated. Corolla pale ochraceous yellow, tube c. 4.5 cm long, infundibuliform-campanulate, strongly constricted to form a narrow tube c. 1 cm above the slightly bulbous base. Fertile stamens 4. Fruit smooth, with 4 shallow grooves, without distinct apical horns, apex only slightly bifurcate, no adaxial crest. The fruits may eventually split into 2 mericarps. One species, *H. viscida* Nees, Brazil, Bahia, with very restricted distribution.

This monotypic genus had been lost since its first description in 1821 and has been only recently rediscovered (Harley et al. 2003). Due to the missing description of the fruit, attribution to Martyniaceae had been doubtful.

2. *Craniolaria* L. Fig. 32A

Craniolaria L., Sp. Pl. 2: 618 (1753); Van Eseltine, New York Agric. Exp. Stat. Tech. Bull. 149: 5–10 (1929), rev.; Abbiatti, Notas Mus. La Plata Bot. 4: 464–472 (1939), reg. rev. Argentina.

Annual herbs with a fleshy tuber. Sepals completely connate, forming a spathe-like structure with a ventral slit, 4.5–5.5 cm long. Corolla white to yellowish, tube narrow-cylindrical, up to 14 cm long, apically abruptly dilated into a campanulate cup; fertile stamens 4. Fruit with horns much shorter than the body, and with a low toothed crest on the adaxial side; seeds 4–6. Three spp., Puerto Rico, Venezuela, Colombia, S Brazil to Argentina.

3. *Martynia* L. Fig. 32B–E

Martynia L., Sp. Pl. 2: 618 (1753); Stapf in Engler & Prantl, Natürl. Pflanzenfam. IV, 3b: 265–269 (1895); Van Eseltine,

New York Agric. Exp. Stat. Tech. Bull. 149: 34–39 (1929), rev.; Hevly, Taxon 18: 527–534 (1969), review.

Annual herb. Sepals free, very unequal in size. Corolla white to rose, campanulate, constricted at the base; fertile stamens 2. Fruit with horns much shorter than the body, with 4 stout ribs on both sides and a low toothed crest on the adaxial side; seeds 4, one in each of the 4 peripheral compartments. $2n = 32$. One sp., *Martynia annua* L., Mexico and West Indies, naturalised elsewhere.

4. *Ibicella* (Stapf) Van Eselt. Fig. 33A

Ibicella (Stapf) Van Eselt., New York Agric. Exp. Stat. Tech. Bull. 149: 31 (1929); l.c. 31–34, rev.; Abbiatti, Notas Mus. La Plata Bot. 4, 29: 452–463 (1939), reg. rev.
Proboscidea sect. *Ibicella* Stapf (1895).

Annual herbs. Sepals free, very unequal in size. Corolla pale to deep yellow, campanulate, constricted at the base; fertile stamens 4. Fruit with long (to 12 cm) horns, longer than the body, on the adaxial side with a serrate crest to 1 cm high, surface echinate with appressed short spines; seeds numerous. $2n = 30$. Three spp. in Bolivia, S Brazil to N Argentina, *Ibicella lutea* (Lindl.) Van Eselt. naturalised elsewhere.

5. *Proboscidea* Schmidel Fig. 32F–K, 33B, C

Proboscidea Schmidel, Ic. Pl. (ed. Keller): 49–52 (1763); Van Eseltine, New York Agric. Exp. Stat. Tech. Bull. 149: 13–31 (1929), rev.

Annual or perennial herbs, sometimes with a thickened primary root (most species of subgen. *Dissolophia* Van Eselt.). Sepals connate for at least 1/2 of their length, tube with a ventral slit. Corolla purple, pink, lavender or cream, or yellow to orange (subgen. *Dissolophia*), campanulate, constricted at the base; fertile stamens 4. Fruit with long (up to 12 cm) horns, longer than the body, on the adaxial side and occasionally also on the abaxial side with a serrate crest up to 1 cm high, surface usually not distinctly echinate. $2n = 30$. About eight spp., S USA to Mexico. *Proboscidea louisianica* (Miller) Thell. naturalised elsewhere.

Selected Bibliography

Abbiatti, D. 1939. Las Martiniáceas argentinas. Notas Mus. La Plata Bot. 4: 443–473.
Anderson, F. 1922. The development of the flowers and embryogeny of *Martynia louisiana*. Bull. Torrey Bot. Club 49: 141–157.
Behnke, H.-D. 1991. Distribution and evolution of forms and types of sieve-element plastids in the dicotyledons. Aliso 13: 167–182.
Bretting, P.K. 1983. The taxonomic relationship between *Proboscidea louisianica* and *Proboscidea fragrans* (Martyniaceae). Southw. Nat. 28: 445–449.
Bretting, P.K. 1985. Geographical intergradation in *Proboscidea parviflora* spp. *sinaloensis* (Martyniaceae). Southw. Nat. 30: 343–348.
Bretting, P.K., Nilsson, S. 1988. Pollen morphology of the Martyniaceae and its systematic implications. Syst. Bot. 13: 51–59.
Brummitt, R.K. 1992. Vascular plant families and genera. Kew: Royal Botanic Gardens.
Carlquist, S. 1987. Wood anatomy of Martyniaceae and Pedaliaceae. Aliso 11: 473–483.
Erdtman, G. 1952. Pollen morphology and plant taxonomy. Angiosperms. Stockholm: Almquist & Wiksell.
Gentry, A.H., Tomb, A.S. 1979. Taxonomic implications of Bignoniaceae palynology. Ann. Missouri Bot. Gard. 66: 756–777.
Grayer, R.J., Chase, M.W., Simmonds, S.J. 1999. A comparison between chemical and molecular characters for the determination of phylogenetic relationships among plant families: an appreciation of Hegnauer's "Chemotaxonomie der Pflanzen". Biochem. Syst. Ecol. 27: 369–393.
Halbritter, H., Hesse, M. 1995. The convergent evolution of exine shields in angiosperm pollen. Grana 34: 108–119.
Harley, R.M., Giulietti, A.M., Dos Santos, F.R. 2003. *Holoregmia* Nees, a recently rediscovered genus of *Martyniaceae* from Bahia, Brazil. Kew Bull. 58(1): 205–212.
Hegnauer, R. 1969. Chemotaxonie der Pflanzen. Vol. 5. Basel: Birkhäuser.
Ihlenfeldt, H.-D. 1967. Über die Abgrenzung und die natürliche Gliederung der Pedaliaceae R. Br. Mitt. Staatsinst. Allg. Bot. Hamburg 12: 43–128.
Inamdar, J.A. 1969. Epidermal structure and stomatal ontogeny in vegetative and floral organs of *Martynia annua* L, *Pedalium murex* L. and *Sesamum indicum* L. Flora Abt. B 158 (1968/1969): 526–537.
Link, J.H.F. 1829. Handbuch zur Erkennung der nutzbarsten und am häufigsten vorkommenden Gewächse. Vol. 1. Berlin: Haude & Spener.
Mayberry, M.E. 1947. *Martynia louisiana* Mill.: an anatomical study. Trans. Kansas Acad. Sci. 50: 164–171.
Olmstead, R.G., dePamphilis, W., Wolfe, A.D., Young, N.D., Ellison, W.J., Reeves, P.A. 2001. Disintegration of the Scrophulariaceae. Am. J. Bot. 88: 348–361.
Phillippi, A., Tyrl, R.J. 1979. The reproductive biology of *Proboscidea louisianica* (Martyniaceae). Rhodora 81: 345–361.
Singh, S.P. 1958. Morphological studies in some members of the family Pedaliaceae. Thesis. Agra University.
Singh, S.P. 1960. Morphological studies in some members of the family Pedaliaceae. Agra Univ. J. Res., Sci. 9: 218–220.
Stapf, O. 1895. Martyniaceae. In: Engler & Prantl, Die natürlichen Pflanzenfamilien IV, 3b. Leipzig: Engelmann, pp. 265–269.
Thieret, J.W. 1976. Floral biology of *Proboscidea louisianica* (Martyniaceae). Rhodora 78: 169–179.

Myoporaceae[1]

I. Theisen and E. Fischer

Myoporaceae R. Br., Prodr.: 514 (1810) nom. cons.

Small trees or shrubs with usually scattered secretory cavities. Leaves simple, usually gland-dotted, spirally arranged or rarely opposite. Flowers solitary or in axillary cymes. Calyx with five lobes, imbricate or open. Corolla sympetalous, 5(4)-lobed, sometimes 2-labiate, lobes imbricate. Stamens 4, with the posterior stamen usually lacking or reduced to a staminode, rarely stamens 5, anthers with longitudinal slits. Ovary superior, with two fused carpels and a terminal style, 2-locular, rarely 4-locular, each locule with (1)2, anatropous, unitegmic ovules pendulous from near apex, or 4–8 superposed in pairs, or the locule subdivided into 4–10 uniovulate compartments. Fruit a drupe or separating into 1-seeded drupe-like segments or a loculicidal capsule, seeds with straight embryo, endosperm scanty or lacking.

Four genera with ca. 330 species, most diverse in Australia, extending to the islands of the Pacific and Indian oceans; few species in tropical America, Hawaii, Caribbean, Japan.

Vegetative Morphology. The plants are mesophytic or xerophytic shrubs or treelets. The leaves are deciduous and usually spiral (Fig. 34A, D) or rarely opposite to whorled. They often are more or less leathery and petiolate to sessile. The lamina is entire and estipulate, non-sheathing, often gland-dotted and epulvinate, and the leaf margin is entire or crenate, serrate to dentate.

Vegetative Anatomy. Secretory cavities with resin or oil are usually present in stems and leaves (Dell 1975). A cork cambium is present and initially is surficial. Nodes are unilacunar, and internal phloem is absent. Secondary thickening starts from a conventional cambial ring. The xylem contains fibre tracheids, and vessel end-walls are simple (Carlquist and Hoekman 1986). The leaf lamina is isobilateral, and minor leaf veins lack phloem transfer cells (*Myoporum*).

[1] See also Scrophulariaceae.

Flower Structure. The inflorescence is thyrsic to racemose, usually with axillary cymes (Fig. 34D). Cymose partial florescences have numerous flowers and a distinct peduncle (*Androya*) or the peduncle is reduced and flowers are arranged in clusters (*Myoporum*, *Eremophila*). In *Eremophila* and *Bontia*, these clusters may be reduced to solitary flowers. Flowers are actinomorphic to more or less zygomorphic (Fig. 34B, E). Floral dorsiventrality involves either the androecium or the corolla and the androecium, but not the calyx. Flowers are 5–4-merous and tetracyclic. The perianth has a distinct calyx and corolla which usually are isomerous or rarely anisomerous. There are (4)5 sepals which are mostly free to nearly completely fused; when fused, the calyx is blunt-lobed or toothed. The calyx is unequal but not bilabiate, persistent (scarious), accrescent or non-accrescent, basally imbricate or open in bud, with the median sepal posterior. The 5 or rarely 4 petals are fused, imbricate, bilabiate (*Eremophila*, *Bontia*) or subrotate (*Myoporum*, *Androya*). The androecium contains exclusively fertile stamens (the upper, posterior stamen lacking) or additionally 1 posterior median staminode. The (3)4(5) stamens usually are didynamous, the anthers connivent or separate from one another, versatile, dehiscing with longitudinal slits, or dehiscing transversely, introrse, confluent at apex. The ovary is 2-locular, but sometimes shows a secondary segmentation of the locules, resulting in a 3–8-celled pistil. The style arises from a depression at the top of the ovary and in *Bontia* is stigmatic for most of its length. Stigma 1, 1–2-lobed, papillate. Placentation is axile or apical. There are (1)2 ovules per locule (from near the apex), or 4–8 per locule (superposed in pairs), or 1 pendulous ovule per locule.

Embryology. Pollen grains are 2-celled and the tapetum is glandular. Embryo-sac development follows the Polygonum type. Polar nuclei fuse prior to fertilization and there are 3 antipodal cells which do not proliferate and are persistent. The synergids are pear-shaped. Ovules are anatropous, unitegmic, and tenuinucellate. An endothelium is differentiated. Endosperm formation is cellular,

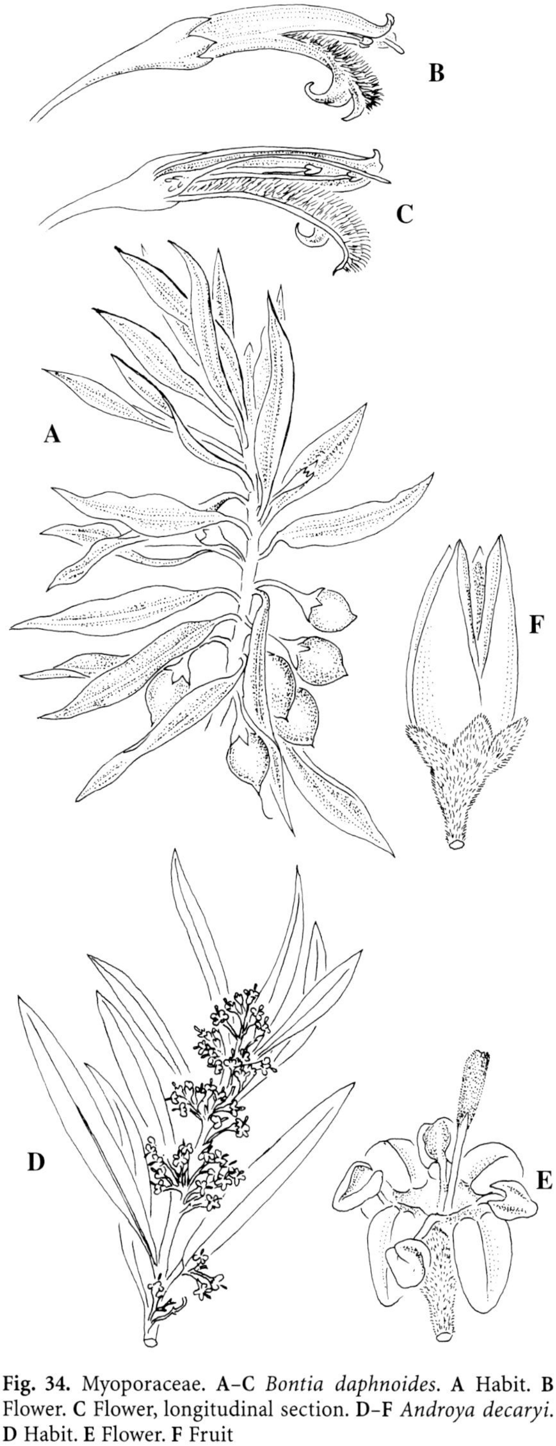

Fig. 34. Myoporaceae. **A–C** *Bontia daphnoides*. **A** Habit. **B** Flower. **C** Flower, longitudinal section. **D–F** *Androya decaryi*. **D** Habit. **E** Flower. **F** Fruit

and chalazar and micropylar endosperm haustoria present. Embryogeny is onagroid (Davis 1966).

Pollen Morphology. Pollen grains are 2–4-colporate, sometimes with two pores per furrow (Mukherjee 1974).

Karyology. Only few data on chromosomes are available and only *Eremophila* has been studied intensively (e.g. Barlow 1971). For *Myoporum*, $2n = 68$ and, for *Eremophila*, $2n = 36$ have been counted. Thus, the base numbers may be $x = 17$, 18.

Pollination and Reproductive Systems. Pollination is by insects, especially by Hymenoptera; melittophily (*Euryglossa*) has been observed in *Eremophila* (Exley 1998).

Fruit and Dispersal. The fruit is a drupe with 1-seeded pyrenes, a nut (dry, indehiscent), or a schizocarp with 2–10-seeded and drupe-like mericarps. In *Androya*, loculicdal capsules occur (Fig. 34F). Seeds are scantily or non-endospermic, usually unwinged, but in *Androya* with wings. There are 2 semi-cylindrical cotyledons. The embryo is straight and germination is phanerocotylar (Richmond and Chinnock 1994; Richmond and Ghisalberti 1994). Dispersal of fleshy fruits is by animals, mostly birds, dispersal of drupe-like segments by wind.

Phytochemistry. According to Hegnauer (1966, 1989), the Myoporaceae are not cyanogenic and do not contain phenylalanine-derived cyanogenic constituents. Alkaloids are present or absent. Iridoids and verbascosides have been detected (*Eremophila*). Proanthocyanidins, flavonols and ellagic acid are absent. Saponins/sapogenins are rarely present or absent. Skelton et al. (1997) recorded tricyclic diterpenes from *Eremophila*.

Economic Use. Some species are used for timber and some fruits are eaten locally. *Eremophila mitchelli*, Budda or Bastard Australian Sandalwood, is the source of so-called sandalwood in Queensland; *Myoporum sandwicensis* is the low-grade sandalwood of Hawaii. Also in usage is the resin of *Myoporum platycarpum*, the Australian dogwood.

Distribution and Habitats. Myoporaceae are a family of temperate to tropical regions of the southern hemisphere. The centre of diversity is Australia (especially Western Australia, *Eremophila*), but the family also extends to the islands of the Pacific (e.g. New Zealand) and Indian oceans (Mauritius, Rodriguez). *Androya* is endemic to southern Madagascar. In the northern hemisphere, a few species occur in Japan, the Hawaiian Islands, and the Carribean (*Bontia*). Typically the family can be found in dry lowland (especially *Eremophila*), but can also enter moderately wet forest (*M. sandwicense* in Hawaii). *Myoporum bontioides* is a mangrove shrub of saline habitats (Toyama and Itow 1975). Numerous species of *Eremophila* occur in semi-arid and arid regions that receive less than 250 mm annual rainfall.

Subdivision and Relationships. According to molecular systematic data (Olmstead et al. 1992; Olmstead and Reeves 1995), the Myoporaceae belong to Lamiales (Angiosperm Phylogeny Group 2003) and are part of Scrophulariaceae s.str. In a revised classification they should be included as tribe Myoporeae. Especially *Leucophyllum* (Srophulariaceae tribe Leucophylleae) shows close relationship to Myoporaceae (Young et al. 1999). This had been proposed before by several authors (e.g. Niezgoda and Tomb 1976; Karrfalt and Tomb 1983). The genera *Oftia* and *Ranopisoa*, traditionally assigned to Myoporaceae, are excluded (Dahlgren and Rao 1971) and placed in Scrophulariaceae s.str. tribe Teedieae. The Buddlejaceae also seem to be closely related and should also be included in Scrophulariaceae s.str. One of its genera, *Androya*, has recently been placed in Myoporaceae (Oxelman et al. 1999).

Key to the Genera

1. Corolla 4-lobed; fruit a capsule; southwest Madagascar — 1. ***Androya***
– Corolla 5-lobed; fruit a drupe; America, Asia, Pacific Islands, Australia, New Zealand — 2
2. Corolla 2-lipped, tube long and curved — 3
– Corolla subrotate, tube short and straight — 4. ***Myoporum***
3. Inflorescence usually thyrsic with axillary clustered cymes, rarely with solitary flowers; corolla not deeply bilobed; stamens exserted; Australia, New Zealand — 3. ***Eremophila***
– Inflorescence racemose with single flowers in leaf axils; corolla deeply bilobed; stamens subexserted to included; Caribbean to South America — 2. ***Bontia***

1. *Androya* H. Perr.

Androya H. Perr., Bull. Mus. Hist. Nat. (Paris) II, 24: 400 (1952).

Small trees. Leaves opposite, petiolate, lamina linear-lanceolate, acute at apex, finely pubescent to subglabrous. Inflorescence thyrsic with axillary cymes. Calyx campanulate, 4-lobed, lobes broadly ovate. Corolla whitish, tube short, limb subrotate, with 4 subequal petals. Stamens 4, exserted, thecae confluent. Ovary ovoid, glabrous, stigma large, tongue-shaped. Fruit a loculicidal capsule. Seeds small, winged, testa reticulate. One species, *A. decaryi* H. Perr., southern Madagascar.

2. *Bontia* L.

Bontia L., Sp. Pl. 2: 638 (1753).

Small trees or shrubs. Leaves alternate, entire, shortly petiolate, lamina linear-lanceolate, acute at apex. Inflorescence a raceme with single flowers in the axils of frondose bracts. Calyx campanulate, lobes broadly ovate. Corolla tawny-yellow with purple, campanulate, limb deeply 2-lipped, lower lip recurved, mid-lobe densely bearded, tube long and curved. Stamens 4, subexserted to included, thecae somewhat divergent. Ovary ovoid, with 4 ovules per locule, style filiform, stigmatic for most of its length. Drupe ovoid, yellow. One species, *B. daphnoides* L., Caribbean to northern South America.

3. *Eremophila* R. Br.

Eremophila R. Br., Prodr.: 518 (1810); Chinnock, J. Adelaide Bot. Gard. 1: 237–262 (1979), reg. rev.; Chinnock, J. Adelaide Bot. Gard. 2: 259–270 (1980), reg. rev.; Smith, Contrib. Queensl. Herb. 19: 1–50 (1975), reg. rev.

Small trees or shrubs. Leaves alternate, entire or dentate, shortly petiolate to sessile, lamina linear-lanceolate to broadly ovate, acute at apex. Inflorescence thyrsic with axillary clustered cymes, occasionally reduced to single flowers. Calyx campanulate, 5-lobed or rarely 4-lobed, lobes broadly ovate. Corolla whitish, pink or yellow to greenish or red, tube long, slender and curved, limb 2-lipped. Stamens 4, usually exserted, thecae somewhat divergent. Ovary ovoid, with 4 ovules per locule, style filiform, stigma bilobed. Drupe ovoid, yellow. $2n = 36$. About 210 species, Australia, especially Western Australia, one species extending to New Zealand.

4. *Myoporum* Banks & Sol. ex Forster f.

Myoporum Banks & Sol. ex Forster f., Prodr.: 44 (1786).

Small trees or shrubs. Leaves usually alternate, entire, petiolate, lamina lanceolate, acute at apex, usually dentate. Inflorescence thyrsic with axillary clustered cymes, peduncle reduced or lacking. Calyx campanulate, 5-lobed, lobes broadly ovate. Corolla usually white, subrotate, with 5 subequal petals, abaxial lobes somewhat larger. Stamens 4 or rarely 5, included to slightly exserted, thecae somewhat divergent, confluent. Ovary ovoid, bilocular or rarely 4-locular, with 1 or rarely 2 ovules per locule, style filiform, stigma capitate. Drupe ovoid, brown to blackish. $2n = 68$. Twenty-eight species, Mauritius, East Asia, eastern Malesia, New Zealand, Hawaii.

Excluded Genera

The following 2 genera belong to Scrophulariaceae s.str. and are placed in tribe Teedieae.

Oftia Adans. (1763).

Scrophulariaceae

Ranopisoa J.F. Leroy (1977).

Scrophulariaceae

Selected Bibliography

Angiosperm Phylogeny Group. 2003. An update of the Angiosperm Phylogeny Group classification of the orders and families of flowering plants: APG II. Bot. J. Linn. Soc. 141: 399–436.

Barlow, B.A. 1971. Cytogeography of the genus *Eremophila*. Austral. J. Bot. 19: 295–310.

Carlquist, S., Hoekman, D.A. 1986. Wood anatomy of Myoporaceae: ecological and systematic considerations. Aliso 11: 317–334.

Dahlgren, R., Rao, V.S. 1971. The genus *Oftia* Adans. and its systematic position. Bot. Notiser 124: 451–472.

Davis, G. 1966. Systematic embryology of the angiosperms. Myoporaceae: 182.

Dell, B. 1975. Geographical differences in leaf resin components of *Eremophila fraseri* F. Muell. (Myoporaceae). Austral. J. Bot. 23: 889–897.

Exley, E.M. 1998. New *Euryglossa* (*Euhesma*) bees (Hymenoptera: Colletidae: Euryglossinae) associated with the Australian plant genus *Eremophila* (Myoporaceae). Records West. Austral. Mus. 18: 419–425.

Hegnauer, R. 1966. Myoporaceae. Chemotaxonomie der Pflanzen 5: 132–138, 434–435.

Hegnauer, R. 1989. Myoporaceae. Chemotaxonomie der Pflanzen 8: 90–96.

Karrfalt, E.E., Tomb, A.S. 1983. Air spaces, secretory cavities and the relationship between *Leucophylleae* (Scrophulariaceae) and Myoporaceae. Syst. Bot. 8: 29–32.

Mukherjee, J. 1974. Role of palynology in the taxonomy and phylogeny of Myoporaceae. Sci. Cult. 40: 331–332.

Niezgoda, C.J., Tomb, A.S. 1976. Systematic palynology of tribe Leucophylleae (Scrophulariaceae) and selected Myoporaceae. Pollen Spores 17: 495–516.

Olmstead, R.G., Reeves, P.A. 1995. Evidence for the polyphyly of the Scrophulariaceae based on chloroplast *rbc*L and *ndh*F sequences. Ann. Missouri Bot. Gard. 82: 176–193.

Olmstead, R.G., Michaels, H.J., Scott, K.M., Palmer, J.D. 1992. Monophyly of the Asteridae and identification of its major lineages inferred from DNA sequences of *rbc*L. Ann. Missouri Bot. Gard. 79: 249–265.

Oxelman, B., Backlund, M., Bremer, B. 1999. Relationships of the Buddlejaceae investigated using parsimony jackknife and branch support analysis of *ndh*F and *rbc*L sequence data. Syst. Bot. 24: 164–182.

Richmond, G.S., Chinnock, R.J. 1994. Seed germination of the Australian desert shrub *Eremophila* (Myoporaceae). Bot. Rev. 60: 483–503.

Richmond, G.S., Ghisalberti, E.L. 1994. Seed dormancy and germination mechanisms in *Eremophila* (Myoporaceae). Austral. J. Bot. 42: 705–715.

Richmond, G.S., Ghisalberti, E.L. 1996. Population and plant growth studies of six species of *Eremophila* (Myoporaceae) from central Western Australia. J. Roy. Soc. West. Austral. 79: 175–181.

Skelton, B.W., Ghisalberti, E.L., White, A.H. 1997. A new class of tricyclic diterpenes from *Eremophila georgei* (Myoporaceae). Austral. J. Chem. 50: 705.

Toyama, S., Itow, S. 1975. The distribution and ecology of a mangrove-like plant, *Myoporum bontioides* A. Gray (Myoporaceae), in western Kyushu, Japan. Hikobia 7: 117–124.

Young, N.D., Steiner, K.E., dePamphilis, C.W. 1999. The Evolution of parasitism in Scrophulariaceae/Orobanchaceae: plastid gene sequences refute an evolutionary transition series. Ann. Missouri Bot. Gard. 86: 876–893.

Nesogenaceae

R.M. Harley

Nesogenaceae Marais, Kew Bull. 35: 798 (1981).

Annual or perennial herbs or rarely shrubs or subshrubs to 2 m, with weakly quadrangular stems, erect to prostrate and sometimes rooting at the nodes. Indumentum of unicellular or multicellular-uniseriate hairs, often gland-tipped, and with subsessile multicellular glands often present. Leaves decussate, estipulate, simple, entire to dentate or rarely deeply pinnatifid, petiolate to sessile, with base often narrowly acuminate, subglabrous, with often ciliate margins, to densely hairy. Inflorescence of small, often congested cymes, 1- to several-flowered, in axils of leaf-like bracts, or rarely congested to form a terminal thyrse, with bracts reduced, prophylls absent or rarely present, linear, inconspicuous. Flowers hypogynous, bisexual, perianth biseriate. Sepals 5, connate with 5 lobes, subequal, triangular, calyx-tube 7–10-nerved, infundibuliform, throat glabrous, calyx slightly accrescent, with lobes often spreading in fruit. Petals 5, connate, corolla usually pink or lilac, zygomorphic to weakly 2-lipped, lobes 5, imbricate, rounded, slightly unequal, posterior lobes slightly shorter than anterior, corolla-tube narrow, cylindrical, widening above. Stamens inserted in mid-tube of corolla, 4, didynamous, alternating with corolla lobes, but absent between 2 posterior lobes, scarcely exserted from tube, filaments elongate, weakly divergent, anthers dithecous, thecae divergent, distally apiculate, with small apical connective, opening by longitudinal slits. Disc absent. Gynoecium 2-carpellate, 2-locular, locules 1-ovulate, ovary weakly obovoid, often hairy at apex. Style terminal, elongate, stigma clavate to subcapitate, stigmatic tissue tuberculate. Fruit drupaceous, indehiscent, ovoid, weakly 2-lobed, pubescent around persistent style-base, pyrene 2-seeded and grooved along locular suture, imperfectly 2-locular. Seeds with testa foveolate, albuminous.

A family of one genus with about nine species in islands in southern Pacific and Indian oceans, Madagascar and E Africa: Tanzania and Kenya.

Vegetative Morphology. Species of *Nesogenes* are annual or perennial herbs, and shrubs or subshrubs in *N. tenuis* from Madagascar. Habit varies from erect to prostrate, often with adventitious roots at the nodes. The leaves lack stipules, are opposite and decussate (Fig. 35A). The lamina may be entire to dentate or, in one species, deeply pinnatifid. Indumentum typically consists of short, appressed, unicellular hairs or uniseriate gland-tipped hairs.

Vegetative Anatomy. Simple, unicellular hairs are common, and uniseriate, multicellular glandular hairs are present in most species (Cantino 1990). The presence of these, together with the absence of branched hairs, clearly distinguishes them from the genera of Chloantheae (Labiatae, formerly Verbenaceae), with which they had previously been associated. Some but not all species have conspicuous subsessile glands. Cantino (1990) reports the absence of anisocytic stomata, which are a feature of Chloantheae.

Floral and Fruit Structure. There is no nectariferous disc in the flower such as is found in most Labiatae (Marais 1981). According to Junell (1934), the ovary consists of two median carpels. The cavities of the two carpels are separated by a partition below the point of attachment of the ovules. Further up, however, the partition walls protrude on both sides and meet but do not coalesce. Marais (1981), however, claims that the ovary of *Nesogenes*, unlike that of Cyclocheilaceae, is completely 2-celled. Each locule contains a single, basal apotropous ovule (Fig. 35I, J). A hard, woody pyrene develops in the ripe fruit. This is slightly grooved at the junction of the loculi, but the two halves can only be separated with difficulty and there is no pyrene wall structure between the two seeds.

Embryology. The seeds contain endosperm (Junell 1934, and pers. obs.). This is said to be absent from the seeds of Cyclocheilaceae, with which *Nesogenes* has often been associated.

Fig. 35. Nesogenaceae. *Nesogenes decumbens*. **A** Habit. **B** Expanded flower. **C** Corolla and androecium spread out. **D** Detached stamens. **E** Gynoecium. **F** Apex of style. **G** Fruit enclosed in calyx. **H** Fruit removed from calyx. **I** Vertical section of fruit. **J** Transverse section of fruit. (After Phil. Trans. Roy. Soc. 168: t. 31 (1879), spec. vol.)

Pollen Morphology. According to Raj (1985), the pollen grains are tricolpate, subprolate to prolate, frequently perprolate, with equatorial diameter 25–33 μm and polar diameter 37–45 μm. Colpi about 28–40 μm with pointed to broad ends, margin usually thin, frequently jagged, membrane thin, densely granular and irregularly ruptured. Amb triangular to rounded-triangular. Exine tectate perforate, perforations minute, ± densely spaced and resembling a microreticulum. Sculpturing consists of units of angular, verrucoid or globular elements 5–7 in number, arranged loosely or compactly on the tectum around a perforation (perhaps best described as pilate, see below). The nexine is a thin, homogeneous layer. The sculpturing of the pollen grains strongly resembles that of Cyclocheilaceae. There is also a strong resemblance to the pollen of Orobanchaceae (sensu Olmstead et al. 2001), which has been described in detail by Minkin and Eshbaugh (1989). In this latter work, the sculpturing was described as pilate.

Dispersal. The small indehiscent fruits appear to have no specialised dispersal mechanism.

Affinities. *Nesogenes* is a small genus of closely related species, which still lacks a detailed study. This makes conclusions as to its affinities speculative. Both Bentham (1876) and Briquet (1895) considered *Nesogenes* (together with *Acharitea*, which is now considered synonymous) to be a member of Verbenaceae. Bentham placed it in tribe Chloantheae, Briquet in subfamily Chloanthoideae, tribe Achariteae. In the present treatment, *Chloanthes* and its allies are placed in family Labiatae, subfamily Prostantheroideae. Some species were originally placed in Myoporaceae (Hooker and Arnott 1832) or Scrophulariaceae (Bentham 1846), the latter corrected by Hemsley (1913). Junell's (1934) studies on gynoecial structure suggested to him that *Cyclocheilon* (Cyclocheilaceae) and *Nesogenes* are closely related, a proposal corroborated more recently by the pollen studies of Raj (1985). Marais (1981), in creating the two new families Nesogenaceae and Cyclocheilaceae, summarized the differences between the two, indicating the importance of calyx structure, which is reduced to a rim in *Cyclocheilon*, which further possesses a dehiscent, capsular fruit with non-endospermous seeds. Marais (1981) considered *Nesogenes* to be more closely related to Labiatae, while he likened Cyclocheilaceae to Acanthaceae. Recent molecular studies on both *Cyclocheilon* and *Nesogenes* suggest that they are related to Orobanchaceae (Olmstead, pers. comm. and Bremer et al. 2001). The similarity of pollen sculpturing between this latter group and both *Nesogenes* and *Cyclocheilon* (Cyclocheilaceae) is striking (see under Pollen Morphology above).

Distribution and Habitats. Good (1953) considers *Nesogenes* to be a genus of the tropical zone, having an anomalous discontinuous distribution. Many species occur on islands, especially in the Indian Ocean, the Pacific, and Madagascar, with one species extending to East Africa. The habitat for most species appears to be open savannah.

Only one genus:

Nesogenes **A. DC.** Fig. 35

Nesogenes A. DC., Prodr. 11: 703 (1847).

Characters as for family.

Selected Bibliography

Bentham, G. 1846. Scrophulariaceae. In: De Candolle, A. (ed.) Prodromus systematis naturalis regni vegetabilis 12: 597. Paris: Treuttel & Würtz.
Bentham, G. 1876. Verbenaceae and Labiatae. In: Bentham, G., Hooker, J.D. Genera Plantarum, vol. 2. London: Reeve and Co., pp. 1131–1223.
Bremer, K., Backlund, A., Sennblad, B., Swenson, U., Andreasen, K., Hjertson, M., Lundberg, J., Backlund, M., Bremer, B. 2001. A phylogenetic analysis of 100+ genera and 50+ families of euasterids based on morphological and molecular data with notes on possible higher level morphological synapomorphies. Plant Syst. Evol. 229: 137–169.
Briquet, J. 1895. Verbenaceae. In: Engler, A., Prantl, K. Die natürlichen Pflanzenfamilien, IV, 3a. Leipzig: W. Engelmann, pp. 132–182.
Cantino, P.D. 1990. The phylogenetic significance of stomata and trichomes in the Labiatae and Verbenaceae. J. Arnold Arbor. 71: 323–370.
Good, R. 1953. The geography of the flowering plants, ed. 2. London: Longmans, Green & Co., 452 pp.
Hemsley, W.B. 1913. On the genera *Radamea* Benth. and *Nesogenes* A. de Candolle. J. Linn. Soc. Bot. 41: 311–316.
Hooker, W.J., Arnott, G.A.W. 1830–1841. The botany of Captain Beechey's voyage. (*Nesogenes*: 1832, pp. 67–68) London: Henry G. Bohn, 485 pp.
Junell, S. 1934. Zur Gynäceummorphologie und Systematik der Verbenaceen und Labiaten. Symb. Bot. Upsal. 4: 1–219.
Marais, W. 1981. Nesogenaceae and Cyclocheilaceae. Kew Bull. 35: 797–812.
Marais, W. 1983. Madagascan *Nesogenes* (Nesogenaceae). Kew Bull. 38: 37–39.
Marais, W. 1984. Nesogenaceae. In: Polhill, R.M. (ed.) Flora of Tropical East Africa. Rotterdam: Balkema, pp. 3.
Minkin, J.P., Eshbaugh, W.H. 1989. Pollen morphology of the Orobanchaceae and rhinanthoid Scrophulariaceae. Grana 28: 1–18.
Olmstead, R.G., dePamphilis, C.W., Wolfe, A.D., Young, N.D., Elisons, W.J., Reeves, P.A. 2001. Disintegration of the Scrophulariaceae. Am. J. Bot. 88, 2: 348–361.
Raj, B. 1985. A contribution to the pollen morphology of Nesogenaceae and Cyclocheilaceae. Pollen Spores 27: 295–306.

Oleaceae

P.S. Green

Oleaceae Hoffmanns. & Link, Fl. Port. 1: 62 (1809), "Oleinae".
Jasminaceae Adans. (1763).
Nyctanthaceae J. Agardh (1858).

Trees, shrubs or woody climbers. Leaves opposite or rarely alternate (*Jasminum* sect. *Alternifolia*), simple, trifoliolate or imparipinnate, without stipules. Flowers regular, bisexual or unisexual (plants sometimes dioecious), sometimes with a nectar disk around base of the ovary. Calyx usually small, 4(–15)-lobed, lobes valvate, rarely calyx wanting (*Fraxinus* spp.). Corolla usually sympetalous, generally 4-lobed (to 12-lobed in *Jasminum*), imbricate, induplicate-valvate or convolute, rarely absent. Androecium of 2 stamens, rarely 4, attached to the corolla tube when sympetalous; anthers dithecal, longitudinally dehiscent, pollen grains tricolpate. Gynoecium of 2 united carpels forming a syncarpous, superior bilocular ovary, style terminal, stigma 2-lobed, dry, sometimes sessile; ovary usually with 2 ovules per loculus, sometimes 1–4, rarely numerous (*Forsythia*), anatropous or hemitropous, on axile placentas. Fruit a loculicidal or circumscissile capsule (*Menodora*), or a usually 1-seeded berry, drupe or samara; seeds with a straight, dicotyledonous embryo, with flat cotyledons and copious endosperm or with storage cotyledons and endosperm almost absent.

A world-wide family of 25 genera and perhaps 600 species.

Vegetative Morphology. Oleaceae are a woody family of trees, shrubs and climbers, although species of *Menodora* and *Dimetra* are suffruticose. In the former they may be quite small, only 4 to 6 cm tall in one variety, yet scarcely to be described as herbaceous. In *Dimetra*, and in *Jasminum kerstingii* from West Africa, their habitat – frequently burned, savannah-type grassland – ensures that they are kept suffruticose. The members of *Myxopyrum*, noted for their quadrangular stems, and most species of *Jasminum* are woody climbers, sometimes reaching high into supporting trees. The leaves are characteristically opposite (Fig. 36A), except in the nine species which constitute *Jasminum* sect. *Alternifolia* (Green 1961), and in the anomalous *J. punctulatum* from East Africa. Leaves may be simple (most members of the tribe Oleeae, and others), trifoliolate (in some species of *Jasminum*, *Menodora*, *Schrebera* and rarely in *Forsythia*, *Fraxinus* and *Syringa*), or imparipinnate in *Jasminum*, *Menodora*, *Schrebera*, *Fraxinus* and rarely in *Syringa*. In *Jasminum* there is a distinct line of abscission towards the base of the leaf rachis in the simple-leaved species, but not in the others. In *Fraxinus* and *Schrebera* there are similar lines at the base of each leaflet. Leaf margins may be entire, blunt toothed or sharp toothed, sometimes within the same genus. A number of species of *Jasminum* and *Olea paniculata* have domatia in the axils of the main nerves and midrib on the abaxial surface of the leaf. Supplementary inflorescence buds may occasionally or frequently be found in leaf axils in a number of genera.

Vegetative Anatomy. A general summary of vegetative anatomy is given by Metcalfe and Chalk (1950), but a more recent and detailed account of the wood anatomy is given by Baas et al. (1988). One feature of leaf anatomy notable and widespread in the family is the existence of idioblastic leaf sclereids. Between genera and species their presence varies widely, as does their shape and abundance. Their function, however, is debatable and their taxonomic significance uncertain. References to many of the publications dealing with this subject in relation to the Oleaceae are to be found in the bibliography provided by Rao and Bhattacharya (1979). A detailed comparative study of leaf anatomy in Malaysian species of *Chionanthus* and *Olea*, with special reference to sclereids, is given by Kiew and Ibrahim (1982). Epidermal features within the family have been reported on: stomata by Inamdar et al. (1986), trichomes by Patel and Inamdar (1974), generally for *Jasminum* by Upadhyay et al. (1991), and for representatives of several genera by Westerkamp and Demmelmeyer (1997) and Nilsson (2000), also for tertiary fossils by Sachse (2001).

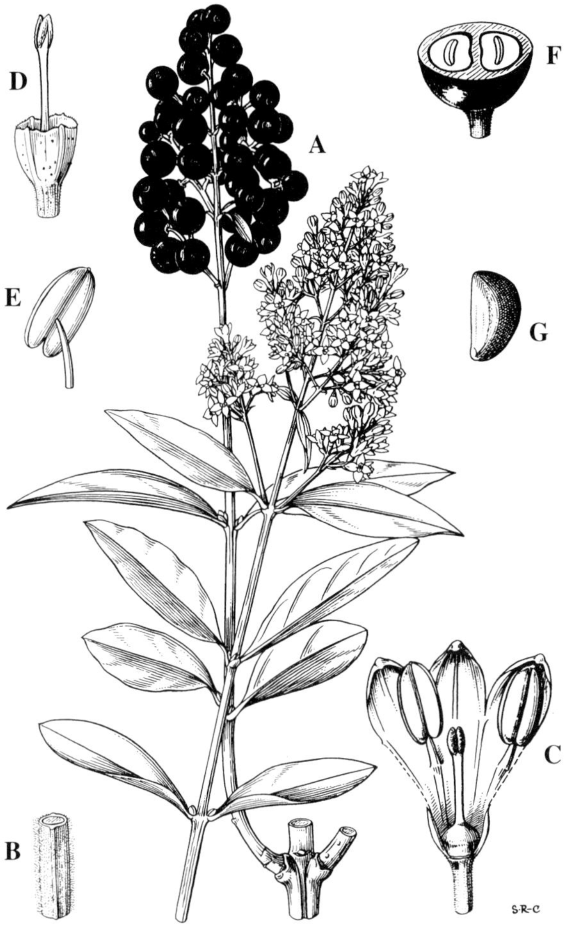

Fig. 36. Oleaceae. *Ligustrum vulgare*. **A** Flowering and fruiting shoots. **B** Cross section of shoot. **C** Opened flower. **D** Flower, corolla removed. **E** Anther. **F** Cross section of fruit. **G** Seed. (Ross-Craig 1964)

Inflorescence and Flower Structure. The inflorescence structure in the Oleaceae is basically cymose (Fig. 36A). A simple cyme, for example, in some species of *Jasminum*, is rare, but more commonly the cyme is thyrsoid and more or less paniculate. The cyme may be decussate in some species within the Oleeae, or condensed, with the inflorescence subumbellate and axillary-sessile in *Osmanthus* sect. *Osmanthus*. Oleaceae are basically sympetalous (Fig. 36C) but, in some species of *Chionanthus* and in *Osmanthus didymopetalus*, the corolla tube is extremely short, or divided to the very base with the lobes joined in pairs by the base of the staminal filament. The characteristic number of lobes to the calyx is 4 (alternating in position with those of the corolla). In *Jasminum* they may vary in number from five to nine, while the calyx may be absent in *Priogymnanthus* and in some species of *Forestiera* and *Fraxinus*. The corolla in several species of the Madagascan *Noronhia* is, to a varying degree, developed to form a corona which, in some species, surrounds the stamens and ovary. In most species of *Fraxinus* and *Nestegis* as well as in *Forestiera*, the flowers are apetalous. *Fraxinus* sect. *Ornus*, however, bears flowers with petals. In *Priogymnanthus* the corolla lobes are very early caducous. The corolla tube is strongly developed in *Syringa*, *Schrebera* and *Jasminum*. In the last two the corolla is almost consistently hypocrateriform, although in a few species of *Jasminum* it is somewhat infundibuliform. Although two stamens per flower (Fig. 36C) are characteristic for the family, four are normally borne in *Nestegis sandwicensis*, in *Osmanthus scortecinii* from Malesia, and frequently in species of *Nestegis* from New Zealand (abnormalities have occasionally led to reports of three staminate flowers). Although flowers throughout the family are characteristically hermaphrodite, in the genus (e.g. *Fraxinus*) they may be unisexual with either stamens or ovary abortive or completely absent. Flowers are heterostylous in *Forsythia*, *Nyctanthes*, *Schrebera* and at least in some species of *Jasminum* (although how widespread this condition is has not been recorded). The common flower colour is white, but yellow flowers are characteristic of *Forsythia*, *Jasminum* sect. *Alternifolia* and *Menodora*. In *Syringa* the flowers are violet to pink, in *Schrebera* suffused with pink, in *Jasminum beesianum* dull red, in some forms of *Osmanthus fragrans* orange, and in *Nyctanthes* white with orange. A notable characteristic of the Oleaceae is the flowers' sweet scent. This is true of most genera, and the fragrance is particularly prized in *Jasminum*, *Osmanthus* and *Syringa* where, because of this, they are widely cultivated.

Embryology. A general account of embryology in the Oleaceae is given by Davis (1966: 191–192). For that of *Jasminum*, see Lakshminarayana (1987), and for that of *Nyctanthes arbor-tristis*, see Kapil and Vani (1967).

Palynology. The pollen grains of the Oleaceae are tricolpate with a reticulate exine. While there is some variation in the size of the reticulum and the micromorphology of the muri, there is an overall uniformity of structure in the pollen of the whole family. As a result, and with an increasing number of taxa investigated, pollen morphology appears to be of little help with classification within the family (Nilsson 1988). Nor, because of its similarity to the pollen of some other unrelated families, e.g. some Celastraceae and Caprifoliaceae, can it be of assistance in determining the classificatory position of the family itself (Erdtman 1971). Variation in size of the grains can be attributed to heterostyly, to polyploidy or to the state of development when the pollen was shed (Kiew 1984). Within species, size appears to be related to style length, while grains from polyploid individuals can be significantly larger than those of diploid individuals.

KARYOLOGY. At least one chromosome number has been recorded for the majority of genera, several in many. The commonest diploid number is $2n = 46$, which has been recorded in 12 genera in three of the tribes in subfamily Oleoideae. In the Jasminoideae, *Menodora* has a diploid number of 22 and *Jasminum* of 26 (although tetraploidy has been recorded in several species and occasional triploidy in some much-cultivated taxa). The same base number as that of *Jasminum* has been noted for *Fontanesia* (tribe Fontanesieae). In *Forsythia* and *Abeliophyllum*, which constitute the tribe Forsythieae, $2n = 28$ is the diploid number (in *Haenianthus* $n = 20$ has been recorded, and also in one species of *Chionanthus*, under the name *Linociera lingelsheimiana*, but these counts need confirmation). Diploid numbers of 22 and 24 have been recorded for *Myxopyrum*, which genus seems to stand on its own within the family. For the sometimes uncertainly placed *Nyctanthes*, several numbers have been recorded: $2n = 22$, 36, 44 and 46.

POLLINATION. Pollination in most Oleaceae is entomophilous, and the flowers of several members of the family are noted for their sweet scent produced by osmophores situated on the corolla (Nilsson 2000). Cross-pollination is favoured and some genera are noted for being heterostylous, again, for example, in *Jasminum*. The hypocrateriform corollas of several genera, e.g. *Syringa* and *Jasminum*, provide a flat platform on which a long-tongued insect can land. In *Fraxinus* anemophily is the common condition. However, the flowers in sect. *Ornus* have a corolla, are scented and attract insects (but are not nectiferous). Those of *F. dipetala* also possess a corolla, but in species of section *Fraxinus* they are naked and also appear not to produce nectar. Pollination is therefore presumed to be carried out by wind. In the androdioecious *Phillyrea angustifolia* the relatively small, whitish-yellow flowers also appear to be anemophilous (Traveset 1994).

FRUIT AND SEED. There are three basic types of fruit in the Oleaceae: fleshy drupes or berries, capsules and samaras. Fleshy fruits are the predominant type in the tribe Oleeae, with most genera producing drupes. However, in *Ligustrum vulgare* the fruit is a one- to four-seeded berry (Fig. 36F). *Ligustrum* is also of interest because in *L. sempervirens* the fruits are fleshy and purple-black at first, but later become leathery and dehisce like a capsule, as in the closely related *Syringa* (Lawrence and Green 1993). *Jasminum*, the genus with the most species, bears didymous berries, although they may be single due to the abortion of one half. They arise by development of a separate mericarp from each carpel, similar to the nutlets in Lamiaceae. In the related *Menodora* the fruit is also didymous, although in this case a circumscissile capsule, or indehiscent in one species. Capsules are borne by members of the genera *Forsythia*, *Schrebera*, *Comoranthus* and *Syringa*. The seeds of *Schrebera* and *Comoranthus* have a large terminal wing, whereas the seeds of *Syringa* and *Forsythia* have only narrow lateral wings or none at all. *Nyctanthes* and *Dimetra* have schizocarps splitting into 2 single-seeded mericarps which may or may not split open. Samaras are the fruit-type in *Fontanesia*, *Abeliophyllum* and *Fraxinus* – the wing encircling the fruit in the first two but only developed in the upper part in *Fraxinus*. The comparative structure of the fruit in the Oleaceae has been thoroughly analysed by Rohwer (1996, 1997), both within the family and in comparison with that of related families.

DISPERSAL. The methods and agents of dispersal are reflected in the types of fruit. Fleshy drupes, which predominate in the Oleeae, or berries as in *Jasminum*, are almost certainly dispersed by birds. In the genera which produce capsules, the seeds are presumably distributed by the infructescence being shaken; this may be effected by passing animals. However, the capsular seeds of *Syringa* and *Forsythia* are often slightly winged, suggesting that in these cases the dispersal of seeds is at least facilitated by the small wings. More certainly, wind dispersal is effective in five genera: in *Comoranthus* and *Schrebera*, because of the presence of large wings on the seeds, and in *Fraxinus*, *Fontanesia* and *Abeliophyllum* which have samaras.

PHYTOCHEMISTRY. The reader is referred to Hegnauer (1969, 1990). Tannins – both hydrolysed and condensed – are absent but other phenylpropanoid derivatives are well developed. They comprise glycosides and sugar esters of phenolic acids and alcohols such as verbascoside and similar caffeic ester glycosides, and syringin and coniferin, which are accumulated in the bark, several lignans and cumarins, and flavones and flavonols. Syringin and cumarins occur vicariously in the bark of *Fraxinus* species. Iridoids and secoiridoids are prominent, including the bitter principle oleuropein; they have been used as important systematic markers by Jensen et al. (2002). Triterpenes and saponis are also present.

Relationships Within the Family. The Oleaceae comprise a rather homogeneous core group, characterized by drupaceous fruits, which most frequently has been classified as tribe Oleeae (e.g. Johnson 1957). It should be noted that the distinction of several genera within this group has been based on tradition rather than characters. The genera with other fruit types deviate more or less clearly from this core. Some of them have been placed in other tribes within subfamily Oleoideae, but most have been placed in subfamily Jasminoideae. Employing molecular techniques, Wallander and Albert (2000) have shown that the latter group is paraphyletic. Therefore, they abandoned the subfamilial rank altogether and proposed a new classification. Their tribe Oleeae (largely congruent with the subfamily Oleoideae of Taylor 1945) comprises the core group with drupaceous fruits, now called subtribe Oleinae, plus *Fraxinus* with samaras (subtribe Fraxininae), *Comoranthus* and *Schrebera* with woody capsules (subtribe Schreberinae), as well as *Ligustrum* and *Syringa* (subtribe Ligustrinae). In spite of their different fruits (drupes and berries vs. capsules), the two are closely related, linked on the one hand by *L. sempervirens*, with its dehiscent "berry", and on the other by the tree lilacs: *Syringa* subg. *Ligustrina*. The former subfamily Jasminoideae is split into four tribes with uncertain relationships one to another: *Jasminum* and *Menodora* with didymous fruits and sarcotestal seeds constitute tribe Jasmineae, *Myxopyrum* with apparently fleshy fruits plus *Nyctanthes* and *Dimetra* with dry schizocarps constitute Myxopyreae, *Forsythia* and *Abeliophyllum* Forsythieae, and Fontanesieae is monotypic. Strangely enough, the fruits of *Forsythia* and *Abeliophyllum* are extremely different from one another (many-seeded capsules vs. samaras), whereas those of *Fontanesia* agree with *Abeliophyllum* in every detail except their size. The relationships within individual genera have been studies with molecular techniques by Kim and Jansen (1998, in *Syringa*) and by Besnard et al. (2002, in *Olea*).

Affinities. The family Oleaceae has been traditionally classified in the orders Gentianales (Huber 1991 as "Contortae"), Scrophulariales (Cronquist 1981) or in an order of its own (Takhtajan 1987). Molecular data (e.g. Chase et al. 1993) consistently show that it is one of the basal clades within an expanded order Lamiales among the Euasteridae I.

Distribution and Habitats. The most widespread genus is *Fraxinus* which occurs throughout the cold and warm temperate regions of both the Old and New Worlds, reaching to the uplands in the Philippines, Java and Bali (in southeast Asia) and south to Costa Rica (in Central America). The biology of *F. excelsior* has been summarized by Wardle (1961). The large genus *Jasminum* is native only to the Old World. Where it occurs in the Americas, it is only as a cultivated plant or as one that has become naturalized. In contrast, *Menodora*, the other and vicarious member of the Jasmineae, is confined to the New World, except for three species from South Africa (one of these considered to be a variety of a species otherwise known from the USA and Mexico; Takhtajan 1987). Within the New World, it is disjunct between the south-western USA and Mexico, on the one hand (14 spp.), and southern South America on the other (6 spp.). The distribution of *Forsythia*, *Fontanesia*, *Ligustrum*, *Osmanthus* sect. *Osmanthus*, and *Syringa* illustrates an interesting and uncommon disjunction between Europe and eastern Asia. All these genera are represented by one species in Europe (except for two in *Syringa*), and otherwise occur only in eastern Asia, often represented by a number of species (Green 1972). None of them are native in the New World. *Schrebera* also has a widely disjunct distribution, with two species in India and SE Asia, one in Peru and three in tropical Africa.

Conservation. The genus *Hesperelaea* recorded from Guadalupe Island off the west coast of Mexico is almost certainly extinct, killed by goats. Watson (1876, p. 118), quoting Palmer who collected it in 1875, stated that "only three living trees were found ... no young trees seen, but many dead ones". In "The IUCN Plant Red Data Book" (1978), *Forsythia europaea* is listed under the conservation category "rare" (p. 353), and *Olea lapperinei* as "vulnerable" (p. 355).

Economic Importance. Economically the most important member of this family is the Olive, *Olea europaea* subsp. *europaea*, a native of the Mediterranean region and known from the dawn of history. Endocarp remains have been discovered, for example, from the Chalcolithic, early Bronze Age (3200 B.C.) in Jordan; see Zohary and Hopf (2000) who suggested its probable early cultivation in SW Asia. However, others have thought that it had an origin long ago in North Africa (Newberry 1937). Analysis of the relationships between and within the disjunct subspecific populations of *O. europaea* (s. lat.) have resulted in a number of publications based, for example, on an analysis of chloroplast DNA polymorphism, e.g. Vargas and Kadereit (2001), Baldoni et al. (2002) and Besnard and Bervillé (2002). *Olea europaea* is much cultivated for the oil extracted from its fruits and grown wherever a Mediterranean climate prevails, including California and parts of Australia. In some places, for example, Hawaii, S Australia and Norfolk Island, it has escaped from cultivation and has become a serious "weed" tree, spread by birds. *Jasminum sambac* is cultivated in the south of France where its flowers are gathered early in the day so that the sweetly scented essential oil can be extracted by effleurage and distillation. Flowers of *J. sambac*, *J. multiflorum* and *Osmanthus fragrans* are also used to scent tea. In forestry, several species of ash, *Fraxinus*, are grown for their resilient timber, and trees of *Chionanthus* have also been felled for their timber. The wood of several genera is noted for its strength and hardness – hence, the common names ironwood, maulwood, etc. In China *Ligustrum lucidum* is grown for culturing insects which produce a much-prized, soft wax.

Key to the Genera

1. Fruit deeply bilobed (sometimes one lobe abortive), remnant of style (if any) basal, between the two mericarps 2
- Fruit not or only slightly bilobed, remnant or scar of style apical 4
2. Corolla lobes contorted in bud; leaves scabrid-hairy; capsule apices strongly recurved **25. *Dimetra***
- Corolla lobes imbricate in bud; leaves glabrous or smooth-hairy; mericarps roundish 3
3. Fruit fleshy at maturity, very rarely dry and brittle; shrub or climber **2. *Jasminum***
- Fruit dry at maturity, circumscissile or indehiscent; small shrub or subshrub **1. *Menodora***
4. Fruit fleshy at maturity 5
- Fruit dry at maturity 19
5. Woody climber; stems distinctly quadrangular; leaves with a prominent, longitudinal vein on each side of the midrib **23. *Myxopyrum***
- Erect trees or shrubs; stems rounded or weakly angular; leaves pinnately veined 6
6. Flowers either without perianth or with only one perianth whorl (calyx or corolla) 7
- Flowers with calyx and corolla (the former sometimes rather inconspicuous) 9
7. Corolla present but very early caducous, lobes finely pilose; calyx absent. S America **22. *Priogymnanthus***
- Corolla absent, calyx absent or present 8
8. Inflorescence fasciculate or very shortly racemose-decussate; drupes dark purplish when ripe. Southern USA, C America, West Indies **21. *Forestiera***
- Inflorescence decussate; drupes purple, red or orange when ripe. New Zealand and Norfolk Island **20. *Nestegis***
9. Petals completely separate, clawed, spathulate. Mexico **15. *Hesperelaea***
- Petals united basally, at least in pairs 10
10. Corolla lobes valvate in bud 11
- Corolla lobes imbricate in bud 17
11. Corolla with scattered peltate scales on the exterior. Greater Antilles **12. *Haenianthus***
- Corolla glabrous 12
12. Corolla tube not more than 2 mm long or wanting, the lobes united in pairs by the base of the stamen filaments 13
- Corolla tube 2 mm or longer 15
13. Inflorescence with large, semi-persistent bracts enclosing the flowers. Macaronesia **16. *Picconia***
- Inflorescence bracts small, not enclosing the flowers 14
14. Corolla lobes at least 3 mm long, usually more. Widespread **13. *Chionanthus***
- Corolla lobes 1.5–2.5 mm long. Australia **14. *Notelaea***
15. Corolla with a well-developed corona embracing the stamens **17. *Noronhia***
- Corolla without such a corona 16
16. Fruit a berry or drupe, 1–4-seeded; endocarp thin, membranous to brittle; inflorescences terminal **10. *Ligustrum***
- Fruit a drupe, endocarp thick, hard; inflorescences axillary or terminal **11. *Olea***
17. Corolla somewhat chartaceous, greenish-white; endocarp thin. Mediterranean **18. *Phillyrea***
- Corolla somewhat fleshy, white; endocarp bony 18
18. Stamens 2; inflorescence fasciculate, paniculate or racemose-decussate **19. *Osmanthus***
- Stamens 4; inflorescence racemose-decussate. Hawaii **20. *Nestegis***
19. Fruit a schizocarp, splitting in the plane of the septum **24. *Nyctanthes***
- Fruit a capsule or a samara 20
20. Fruit a samara 21
- Fruit a loculicidal capsule 23
21. Trees, rarely shrubs; leaves pinnate, rarely ternate or simple; fruit elliptic to linear, winged in the upper part **6. *Fraxinus***
- Shrubs; leaves simple; fruit more or less circular with a wing all round 22
22. Corolla with a distinct tube, stamens included **3. *Abeliophyllum***
- Corolla tube extremely short to absent, stamens exserted **5. *Fontanesia***
23. Corolla with a tomentum of short, clavate hairs 24
- Corolla glabrous 25
24. Corolla only distally hairy, in the throat and on the 5–7 spreading lobes **7. *Schrebera***
- Corolla hairy throughout, lobes 4 **8. *Comoranthus***
25. Corolla tube much shorter than the lobes; flowers yellow **4. *Forsythia***
- Corolla tube more or less equal to or longer than the lobes; flowers white, reddish or lilac **9. *Syringa***

1. Tribe Jasmineae (Adans.) Lam. & DC. (1806).

Leaves opposite or alternate, simple or compound. Calyx 5–15-lobed. Corolla infundibuliform or hypocrateriform. Fruit a berry or circumscissile capsule.

1. *Menodora* Humb. & Bonpl.

Menodora Humb. & Bonpl., Pl. Aequin. 2: 98, t. 110 (1812); Steyermark, Ann. Missouri Bot. Gard. 19: 87–177 (1932), rev.; Turner, Phytologia 71: 340–356 (1991), reg. rev.

Shrubs, sometimes almost herbaceous. Leaves opposite or alternate, simple, pinnatisect or imparipinnate. Inflorescence few- to many-flowered, cymose-paniculate, terminal or axillary, flowers hermaphrodite. Calyx 5–15-lobed. Corolla infundibuliform, lobes 5 or 6, imbricate. Stamens 2. Ovary bilocular with 4, rarely 2, ovules per cell. Fruit a bilobed, membranous, circumscissile or indehiscent capsule. $2n = 22$; 23 spp., SW USA, Central America, S. America and S. Africa.

2. *Jasminum* L.

Jasminum L., Sp. Pl.: 7 (1753).

Woody climbers or rarely erect shrubs. Leaves opposite or alternate, simple, trifoliolate or imparipinnate, petiole usually articulate. Inflorescence few- to many-flowered, cymose-paniculate or subumbellate, flowers heterostylous. Calyx small, 5–10-lobed. Corolla hypocrateriform, rarely infundibuliform, white or yellow, rarely red, lobes imbricate, 5 or more. Stamens 2. Ovary bilocular with 2 axile ovules per loculus. Fruit a didymous berry (often single by abortion), dry in *J. mesnyi* and *J. nudiflorum*. $2n = (24), 26$; about 200 species throughout the Old World tropics and warm temperate regions.

2. Tribe Forsythieae H. Taylor ex L.A.S. Johnson (1957).

Leaves opposite, simple or trifoliate. Calyx 4-lobed. Corolla tube campanulate, lobes longer than the tube. Fruit a capsule or samara.

3. *Abeliophyllum* Nakai

Abeliophyllum Nakai, Bot. Mag. Tokyo 33: 1534 (1919).

Shrubs. Leaves deciduous, opposite, simple. Inflorescences axillary, racemose-decussate, flowers self-sterile. Calyx of 4 rounded lobes. Corolla white or pinkish, tube 3 mm long, lobes 4, imbricate, strap-shaped, about 1 cm long, notched at the apex. Stamens 2, included. Ovary bilocular with 1 pendulous ovule in each locule. Fruit an almost circular samara. $2n = 28$; one species, *A. distichum* Nakai, from Korea.

4. *Forsythia* Vahl

Forsythia Vahl, Enum. Pl. 1: 39 (1804).

Shrubs. Leaves deciduous, opposite, entire or trifoliolate. Inflorescences axillary, flowers bisexual, heterostylous, self-sterile. Calyx 4-lobed. Corolla yellow, tube 3–8 mm long, lobes 4, strap-shaped to elliptic, 6–20 mm long, imbricate in bud. Stamens 2, enclosed or scarcely protruding. Ovary bilocular, ovules several to numerous. Fruit a capsule, seeds slightly winged. $2n = 28$; perhaps 9 spp., one from SE Europe, the remainder from China, Korea and Japan.

3. Tribe Fontanesieae H. Taylor ex L.A.S. Johnson (1957).

Leaves opposite, simple. Calyx 4-lobed. Corolla deeply 4-lobed. Fruit a samara.

5. *Fontanesia* Labill.

Fontanesia Labill., Icon. Pl. Syr. 1: 9, t. 1 (1791).

Shrubs or small trees. Leaves deciduous, opposite, entire or minutely serrulate. Inflorescence shortly paniculate, terminal or axillary, flowers small, pedicellate, bisexual or polygamous. Calyx 4-lobed. Corolla 4-lobed, joined basally. Stamens 2, exserted. Ovary bilocular with 1 pendulous ovule per loculus. Fruit an elliptic samara, 1- or 2-seeded. $2n = 26$; one sp., *F. phillyraeoides* Labill., with a disjunct distribution from Sicily, Anatolia and Syria in the west (subsp. *phillyraeoides*) to east-central China in the east (subsp. *fortunei* (Carrière) Hegi).

4. Tribe Oleeae (Hoffmanns. & Link ex R.Br.) Dumort. (1827).

Leaves opposite, simple or pinnate. Calyx and corolla 4-lobed or absent. Fruit a drupe, capsule or samara.

4a. Subtribe Fraxininae (Vent.) Wallander & V. Albert (2000).

Leaves pinnate, rarely simple. Flowers unisexual or hermaphrodite. Calyx and/or corolla present or absent. Fruit a samara.

6. *Fraxinus* L.

Fraxinus L., Sp. Pl.: 1057 (1753); Lingelsheim in Engler, Pflanzenr. IV 243 (Heft 72): 1–65 (1920), rev.
Ornus Pers. (1805).

Trees, rarely shrubs, dioecious, hermaphrodite, androdioecious or polygamous. Leaves deciduous (rarely evergreen), imparipinnate, rarely trifoliolate or simple. Inflorescences clustered or paniculate, axillary or terminal, flowers hermaphrodite or unisexual. Calyx absent or campanulate, more or less 4-lobed. Corolla absent or 2–4 almost free petals. Staminate flowers with 2 (rarely 4) hypogynous stamens. Ovary bilocular with 2 pendulous ovules per loculus. Fruit a 1-seeded samara. $2n = 46$; about 45–50 spp. throughout the temperate regions and the subtropics of the northern hemisphere.

4b. Subtribe Schreberinae (Wight) Wallander & V. Albert (2000).

Leaves simple or pinnate. Flowers hermaphrodite. Corolla hairy. Fruit a woody capsule.

7. *Schrebera* Roxb.

Schrebera Roxb., Pl. Caromandel 2: 2, t. 101 (1799).

Trees or shrubs. Leaves opposite, simple, trifoliolate or pinnate. Inflorescences few- to many-flowered, cymose-paniculate, flowers heterostylous. Calyx campanulate, truncate or obscurely lobed. Corolla hypocrateriform, lobes 5–7, imbricate, with a patch of small, clavate hairs towards the base adaxially. Stamens 2. Ovary bilocular, with 4 pendulous ovules per loculus. Fruit a woody, clavate capsule, loculicidal, seeds winged. $2n = 46$; 6 spp., 2 in SE Asia, one in S. America and 3 in tropical Africa.

8. *Comoranthus* Knobl.

Comoranthus Knobl., Notizbl. Bot. Gart. Berlin 11: 1032 (1934); Perrier in Humbert, Fl. Madagascar 166: 68–74 (1952), rev.

Trees or shrubs. Leaves opposite, simple. Inflorescences racemose or paniculate, flowers 3 to numerous. Calyx obconic with 4 obscure teeth. Corolla somewhat infundibuliform, externally covered by a dense indumentum, lobes 4, imbricate and induplicate. Stamens 2. Ovary bilocular, with 4 pendulous ovules per loculus. Fruit a woody, clavate capsule, loculicidal. Three spp. from Madagascar and the Comoros.

4c. Subtribe Ligustrinae Koehne (1893).

Leaves simple or pinnate. Flowers hermaphrodite. Corolla glabrous. Fruit a drupe or capsule.

9. *Syringa* L.

Syringa L., Sp. Pl.: 9 (1753); McKelvey, The Lilac, a Monograph (1928); Chang & Green in Wu & Raven, Fl. China 15: 280–286 (1996), reg. rev.: Kim & Jansen, Am. J. Bot. 85: 1338–1351 (1998), rev.; Li et al., Syst. Bot. 27: 592–597 (2002), rev.

Shrubs or small trees. Leaves deciduous, opposite, simple or imparipinnate. Inflorescences paniculate, terminal or axillary, flowers hermaphrodite. Calyx small, lobes 4 or obscure. Corolla more or less cylindrical with 4 valvate, cucullate lobes. Stamens 2, Ovary bilocular with 2 pendulous ovules per locule. Fruit a loculicidal capsule with 1 or 2 narrowly winged seeds in each cell. $2n = 46$ (48); about 20 spp., 2 from Europe, the remainder from Asia, especially China.

10. *Ligustrum* L. Fig. 36

Ligustrum L., Sp. Pl.: 7 (1753).

Shrubs. Leaves evergreen or deciduous, opposite, entire. Inflorescences terminal or subterminal, cymose-paniculate, flowers hermaphrodite. Calyx campanulate, lobes 4, sometimes obscure. Corolla more or less cylindrical with 4 valvate, cucullate lobes. Stamens 2. Ovary bilocular with 2 ovules per locule. Fruit a fleshy 1–4-seeded berry or a drupe, in one species becoming leathery and eventually dehiscing. $2n = 46$ (48); about 40 spp, mainly from E and SE Asia but one species from Europe and one reaching Australia.

4d. Subtribe Oleinae Wallander & Albert (2000).

Leaves simple. Flowers hermaphrodite or unisexual. Corolla glabrous or absent. Ovary bilocular, 2 ovules per locule. Fruit a drupe.

11. *Olea* L.

Olea L., Sp. Pl.: 8 (1753); Green, Kew Bull. 57: 91–140 (2002), rev.
Tetrapilus Lour. (1790).

Shrubs or trees. Leaves evergreen, opposite, entire or toothed. Inflorescences terminal or axillary, cymose-paniculate, flowers hermaphrodite or unisexual. Calyx 4-lobed or 4-toothed. Corolla tube short, lobes 4, valvate, cucullate. Stamens 2. Ovary bilocular with 2 pendulous ovules per locule. Fruit a drupe, endocarp hard. $2n = 46$; about 32 spp. Old World tropics and warm temperate regions. Generic delimitation in need of review.

12. *Haenianthus* Griseb.

Haenianthus Griseb., Fl. Brit. W. Indies: 405 (1861); Zona in Can. J. Bot. 69: 489–493 (1991), rev.

Trees. Leaves evergreen, opposite, simple, entire. Inflorescences terminal and axillary, cymose-paniculate, flowers hermaphrodite. Calyx small, 4-lobed. Corolla tube short, lobes 4, induplicate-valvate in bud, becoming somewhat clavate. Stamens 2. Ovary bilocular, 2 ovules per locule. Fruit a drupe, endocarp thin. $2n = 40$; 2 spp., the Greater Antilles.

13. *Chionanthus* L.

Chionanthus L., Sp. Pl. 8 (1753); Green, Kew Bull. 49: 261–286 (1994), part. rev.
Linociera Sw. ex Schreb. (1791).

Trees or shrubs. Leaves evergreen or deciduous, opposite, simple. Inflorescences terminal or axillary, cymose-paniculate or rarely racemose-decussate, flowers hermaphrodite or unisexual. Calyx small, 4-lobed. Corolla 4-lobed, induplicate-valvate, joined in pairs by base of staminal filaments or with a very short tube. Stamens 2 or 4. Ovary bilocular with 2 ovules per locule. Fruit a drupe with hard endocarp. $2n = 46$; about 60 spp., worldwide in the tropics and subtropics, with 2 temperate spp.

14. *Notelaea* Vent.

Notelaea Vent., Choix Pl.: 25, t. 25 (1804); Green, J. Arnold Arbor. 49: 333–369 (1968), rev.

Shrubs or small trees. Leaves opposite, evergreen, simple, entire. Inflorescences axillary, racemose-decussate, flowers hermaphrodite. Calyx small, 4-lobed. Corolla 4-lobed, joined in pairs by the base of the staminal filaments, induplicate-valvate. Stamens 2. Ovary bilocular with 2 ovules per locule. Fruit a drupe, endocarp hard. $2n = 46$; 11 spp., Australia (Queensland, New South Wales and Tasmania).

15. *Picconia* DC.

Picconia DC., Prodr. 8: 288 (1844).

Trees or shrubs. Leaves evergreen, opposite, simple, entire or rarely serrulate. Inflorescences axillary, racemose-decussate, bracts enclosing the flowers, somewhat persistent, flowers hermaphrodite. Calyx small, 4-lobed. Corolla rotate, 4-lobed, joined in pairs by the base of the filaments, valvate in bud. Stamens 2. Ovary bilocular, with 2 ovules per locule. Fruit a drupe. $2n = 46$; 2 spp., Canary Islands, Madeira and the Azores.

16. *Hesperelaea* A. Gray

Hesperelaea A. Gray, Proc. Am. Acad. 11: 83 (1876).

Small tree. Leaves opposite, entire. Inflorescences terminal and axillary, paniculate, flowers hermaphrodite. Calyx 4-lobed, imbricate, deciduous. Corolla lobes 4, clawed spathulate, free. Stamens 4. Ovary bilocular, with 2 ovules per locule. Fruit unknown, drupaceous? One sp., *H. palmeri* A. Gray, from Guadalupe Island, Mexico. Extinct.

17. *Noronhia* Stadm. ex Thouars

Noronhia Stadm. ex Thouars, Gen. Nov. Madag.: 8 (1806); Perrier in Humbert, Fl. Madagascar 166: 15–67 (1952), rev.

Trees. Leaves opposite, simple, entire. Inflorescences axillary, racemose-decussate, paniculate or fasciculate, flowers hermaphrodite. Calyx small, 4-lobed. Corolla urceolate, campanulate or rotate, 4-lobed, valvate-induplicate, with a well-developed corona. Stamens 2. Ovary bilocular, 2 ovules per locule. Fruit a drupe, endocarp hard. About 45 spp., Madagascar and Comoro Islands.

18. *Phillyrea* L.

Phillyrea L., Sp. Pl.: 7 (1753).

Trees or shrubs. Leaves evergreen, opposite, simple, entire or dentate. Inflorescences axillary, racemose-decussate, flowers hermaphrodite or unisexual. Calyx small, more or less 4-lobed.

Corolla somewhat chartaceous, subrotate, tube short, lobes 4, imbricate in bud. Stamens 2. Ovary bilocular, 2 ovules per locule. Fruit a drupe, exocarp thin, somewhat dry, endocarp crustaceous, 1- or 2-seeded. $2n = 46$; 2 spp., the Mediterranean region.

19. *Osmanthus* Lour.

Osmanthus Lour., Fl. Cochinch.: 28 (1790); Green, Notes Roy. Bot. Gard. Edinburgh 22: 439–542 (1958), rev.; Green, J. Arnold Arbor. 44: 268–283 (1963), reg. rev.
Cartrema Raf. (1838).
Amarolea Small (1933).

Trees or shrubs. Leaves evergreen, opposite, simple, entire or dentate. Inflorescences axillary, fasciculate, racemose-decussate or cymose-paniculate, flowers hermaphrodite or unisexual. Calyx small, campanulate, 4-lobed. Corolla tube short or well developed, lobes 4, imbricate in bud. Stamens 2. Ovary bilocular, 2 ovules in each locule. Fruit a drupe, endocarp hard or crustaceous. $2n = 46$; about 32 spp., mostly Chinese, but 1 in SW Asia, 1 in SE USA and Mexico, and 1 in Malaya and Sumatra.

20. *Nestegis* Raf.

Nestegis Raf., Sylva Tellur. 10 (1838); Green, J. Arnold Arbor. 44: 377–389 (1963), rev.

Trees or shrubs. Leaves evergreen, opposite, simple, entire. Inflorescences axillary, racemose-decussate, flowers hermaphrodite or unisexual. Calyx small, cupulate, 4-lobed. Corolla absent or 4-lobed, rotate, imbricate in bud. Stamens 2 or 4. Ovary bilocular, 2 ovules per locule. Fruit a drupe. $2n = 46$; 5 spp., New Zealand and Norfolk Island, 1 sp. in Hawaii.

21. *Forestiera* Poir.

Forestiera Poir. in Lamarck, Encycl., Suppl. 2: 664 (1812).

Shrubs. Leaves deciduous or evergreen, opposite, simple, entire or dentate. Inflorescences axillary, fasciculate or shortly racemose-decussate, flowers unisexual. Calyx very small or absent. Corolla absent. Stamens 1–4, abortive in female flowers. Ovary bilocular, with 2 apical, pendulous ovules per locule, rudimentary or absent in male flowers. Fruit a drupe. $2n = 46$; 19 spp., southern USA, central America and the West Indies.

22. *Priogymnanthus* P.S. Green

Priogymnanthus P.S. Green, Kew Bull. 49: 280 (1994).

Trees. Leaves deciduous or semideciduous, opposite, simple, entire. Inflorescences axillary, racemose-decussate, flowers hermaphrodite. Calyx absent. Corolla lobes 4, thin, very early caducous. Stamens 4. Ovary bilocular. Fruit a drupe. Two spp., S. America.

5. Tribe Myxopyreae Boerl. (1899).

Shrub, small tree or woody climber. Leaves opposite, simple. Fruit a berry, capsule or schizocarp.

23. *Myxopyrum* Blume

Myxopyrum Blume, Bijdr.: 683 (1826); Kiew, Blumea 29: 499–512 (1984), reg. rev.

Woody climber, stems strongly quadrangular. Leaves opposite, simple, entire or dentate, strongly three-nerved. Inflorescences terminal or axillary, paniculate, flowers hermaphrodite. Calyx small, 4-lobed. Corolla campanulate to urceolate, lobes 4, induplicate-valvate, sometimes cucullate. Ovary bilocular, 1–3 ascending ovules per locule. Fruit a drupe, 1- to 4-seeded. $2n = 22, 24$; 4 spp., SE Asia (India to S China, Philippines and New Guinea).

24. *Nyctanthes* L.

Nyctanthes L, Sp. Pl.: 6 (1753).

Shrub or small tree. Leaves opposite, entire or dentate, scabrous. Inflorescence terminal, thyrsoid, flowers in bracteate cymes, hermaphrodite. Calyx cylindrical. Corolla hypocrateriform, lobes 4–8, convolute. Stamens 2. Ovary bilocular, one basal ovule per loculus. Fruit a schizocarp, compressed, orbiculate; seeds flattened, orbiculate. $2n = 22, 36, 44$ & 46; 2 spp. in India and mainland SE Asia.

25. *Dimetra* Kerr

Dimetra Kerr, Bull. Misc. Inf. Kew 1938: 127 (1938).

Subshrub or perennial "herb". Stems rounded 4-angled, scabrid-pilose. Leaves opposite, entire, scabrous. Inflorescence axillary, cymose-capitate, peduncle strigose, bracts linear-oblanceoate, adpressed pilose, flowers hermaphrodite. Calyx lobes 4, linear, adpressed pilose. Corolla hypocrateriform, exterior shortly pilose, lobes 4 or 5, imbricate in bud. Stamens 2. Ovary bilocular, divided to the base of the style, one basal ovule per loculus. Fruit a capsule, flattened, lobes broadly elliptic, recurved; seeds flattened. One sp., *D. craibeana* Kerr from north-eastern Thailand, in wooded grassland, subject to fires.

Selected Bibliography

Baas, P., Esser, P.M., van der Westen, M.E.T., Zandee, M. 1988. Wood anatomy of the Oleaceae. Iawa Bull. 9: 103–182.

Baldoni, L., Guerrero, C., Sossey-Aloui, K., Abbott, A.G., Angioilillo, A., Lumaret, R. 2002. Phylogenetic relationships among *Olea* species, based on nucleotide variation at a non-coding chloroplast DNA region. Pl. Biol. 4: 346–351.

Besnard, G., Bervillé, A. 2002. On chloroplast DNA variations in the olive (*Olea europaea* L.) complex: comparison of RFLP and PCR polymorphisms. Theor. Appl. Genet. 104: 1157–1162.

Besnard, G., Green, P.S., Bervillé, A. 2002. The genus *Olea*: molecular approaches of its structure and relationships to other Oleaceae. Acta Bot. Gallica 149: 49–66.

Chase, M.W. et al. 1993. Phylogenetics of seed plants: an analysis of nucleotide sequences from the plastid gene *rcb*L. Ann. Missouri Bot. Gard. 80: 528–580.

Cronquist, A. 1981. An integrated system of classification of flowering plants. New York: Columbia University Press.

Das, V.S.R., Rao, K.N. 1966. Chemotaxonomic investigation of *Nyctanthes*. Naturwissenschaften 54: 439.

Davis, G.L. 1966. Systematic embryology of the Angiosperms. New York: Wiley.

Erdtman G. 1971. Pollen morphology and plant taxonomy. New York: Hafner.

Green, P.S. 1961. Studies in the genus *Jasminum* I, Section *Alternifolia*. Notes Roy. Bot. Gard. Edinburgh 23: 355–384.

Green, P.S. 1972. *Osmanthus decorus* and disjunct Asiatic-European distributions in the Oleaceae. Kew Bull. 26: 487–490.

Green, P.S., Wickens, G.E. 1989. The *Olea europaea* complex. In: Kit Tan (ed.) The Davis & Hedge Festschrift. Edinburgh: Edinburgh University Press.

Harborne, J.B., Green, P.S. 1980. A chemotaxonomic survey of flavonoids in leaves of the Oleaceae. Bot. J. Linn. Soc. 81: 155–167.

Inamdar, J.A., Patel, R.C., Mohan, J.S.S. 1986. Structure and ontogeny of stomata in some Oleaceae. Feddes Repert. 97: 291–302.

Jensen, A.U., Franzyk, H., Wallander, E. 2002. Chemotaxonomy of the Oleaceae; iridoids as taxonomic markers. Phytochemistry 60: 213–231.

Johnson, L.A.S. 1957. A review of the family Oleaceae. Contrib. New S. Wales Natl. Herb. 2: 395–418.

Johnston, M.C. 1957. Synopsis of the United States species of *Forestiera* (Oleaceae). Southw. Nat. 2: 140–151.

Kapil, R.N., Vani, R.S. 1967. *Nyctanthes arbor-tristis* Linn.: embryology and relationships. Phytomorphology 16: 553–563.

Kiew, R. 1984. Preliminary pollen study of the Oleaceae in Malesia. Gard. Bull. Singapore 37: 225–230.

Kiew, R., Ibrahim, C.S. 1982. Comparative study of leaf anatomy of Malayan species of *Chionanthus* and *Olea* (Oleaceae) with special reference to foliar sclereids. Bot. J. Linn. Soc. 84: 79–101.

Kim, K.-J., Jansen, R.K. 1993. Phylogeny of Oleaceae based on *ndhf* sequence variation and chloroplast genome rearrangements. Abstracts XV Int. Bot. Congr., 209.

Kim, K.-J., Jansen, R.K. 1998. A chloroplast DNA phylogeny of lilacs (*Syringa*, Oleaceae): plastome groups show a strong correlation with crossing groups. Am. J. Bot. 85: 1338–1351.

Lakshminarayana, K. 1987. Embryology and systematics of *Jasminum*. Beitr. Biol. Pflanzen 61: 373–380.

Lawrence, T.J., Green, P.S. 1993. The anatomy of a dehiscent berry. Kew Bull. 48: 53–57.

Melchior, H. 1964. Engler's Syllabus der Pflanzenfamilien, ed. 12, 2: 403–405. Berlin: Borntraeger.

Metcalfe, C.R., Chalk, L. 1950. Anatomy of the Dicotyledons. Oxford: Clarendon Press.

Newberry, P.E. 1937. On some African species of the genus *Olea* and the original home of the cultivated Olive-tree. Proc. Linn. Soc. 150: 3–16.

Nilsson, S. 1988. A survey of the pollen morphology of *Olea* with particular reference to *O. europaea* sensu lat. Kew Bull. 43: 309–315.

Nilsson, S. 2000. Fragrance glands (osmophores) in the family Oleaceae. In: Nordenstam, G., El-Ghazaly, Kasas, M. (eds.) Plant systematics for the 21st Century. London: Portland Press.

Patel, R.C., Inamdar, J.A. 1974. Studies in the trichomes and nectaries of some Gentianales. In: Puri, V. et al. (eds.) Biology of land plants. Meerut: S. Prakashan.

Piechura, J.E., Fairbrothers, D.E. 1983. The use of protein-serological characters in the systematics of the family Oleaceae. Am. J. Bot. 70: 780–789.

Qin, X.-K. 1996. The use of peroxidases in the systematics of Oleaceae. Acta Bot. Yunnan. 18: 159–166.

Rao, T.A., Bhattacharya, J. 1979. A review on foliar sclereids in angiosperms. Bull. Bot. Surv. India 20: 91–99.

Rohwer, J.G. 1996. Die Frucht- und Samenstrukturen der Oleaceae. Biblioth. Bot. 148: 1–177.

Rohwer, J.G. 1997. The fruits of *Nyctanthes aculeata* (Oleaceae). Bot. Jahrb. Syst. 119: 293–299.

Ross-Craig, S. 1964. Drawings of British plants 20, pl. 23. London: Bell & Hyman.

Sachse, M. 2001. Oleaceous laurophyllous leaf fossils and pollen from the European Tertiary. Rev. Palaeobot. Palynol. 115: 213–234.

Taylor, H. 1945. Cyto-taxonomy and phylogeny of the Oleaceae. Brittonia 5: 337–367.

Thorne, R.F. 1968. Synopsis of a putative phylogenetic classification of the flowering plants. Aliso 6 (4): 57–66.

Traveset, A. 1994. Reproductive biology of *Phillyrea angustifolia* L. (Oleaceae) and effect of galling-insects on its reproductive output. Bot. J. Linn. Soc. 114: 153–166.

Upadhyay, N., Trivedi, B.S., Verma, C.L. 1991 ("1989"). Foliar epidermal studies in Indian *Jasminum*. Bull. Bot. Surv. India 31: 136–148.

Vargas, P., Kadereit, J.W. 2001. Molecular fingerprinting evidence (ISSR, INTER-Simple Repeats) for a wild status of *Olea europaea* L. (*Oleaceae*) in the Eurosiberian North of the Iberian Peninsula. Flora 196: 142–152.

Wallander, E., Albert, V.A. 2000. Phylogeny and classification of Oleaceae based on *rps*16 and *trn*L-F sequence data. Am. J. Bot. 87: 1827–1841.

Wardle, P. 1961. Biological flora of the British Isles, no.79, *Fraxinus excelsior* L. J. Ecol. 49: 739–751.
Watson, S. 1876. List of a collection of plants from Guadalupe Island, made by Dr. Edward Palmer, with his notes upon them. Proc. Am. Acad. Arts 11: 112–121.
Westerkamp, C., Demmelmeyer, H. 1997. Leaf surfaces of Central European woody plants. Berlin: Borntraeger.
Zohary, D., Hopf, M. 2000. Domestication of plants in the Old World, 3rd edn. Clarendon Press.

Pedaliaceae

H.-D. Ihlenfeldt

Pedaliaceae R. Br. Prodr. Fl. Nov. Holl.: 519 (1810), nom. cons.

Deciduous trees (up to 8 m tall) or deciduous shrubs with swollen (succulent) stem or main branches, or shrubs and half-shrubs without swollen stems, or perennial herbaceous plants, sometimes with a short swollen stem (caudex) and tuberous roots, or annuals; branches erect or ascending to procumbent, rarely with spines. All aerial parts, especially the leaves, covered with mucilage glands (at least on young parts) which produce mucilage when wetted, often with additional simple hairs. Leaves generally opposite, in large woody representatives and occasionally in the floral region of annuals alternate; leaves petiolate to subsessile, usually simple, entire to pinnatilobed, sometimes digitate, estipulate, sometimes subsucculent. Flowers usually solitary in the axils of foliage leaves, rarely in few-flowered cymes or raceme-like aggregates; pedicels usually with 2 to several nectar glands at the base. Flowers bisexual, tetracyclic, slightly to distinctly zygomorphic, hypogynous. Calyx 5-partite, small compared to the corolla tube. Corolla gamopetalous, white, yellow or variously red, tube obliquely campanulate, funnel-shaped or cylindrical, often constricted at the base, adaxially often gibbous, rarely spurred at the base; limb sub-bilabiate (anterior lobe distinctly enlarged) or with subequal lobes. Stamens 4, didynamous (fifth stamen often represented as a staminode), usually inserted near the base of the corolla and normally included in the tube; thecae 2, parallel or diverging, opening lengthwise; connective usually with an apical gland. Disc (nectary) hypogynous, fleshy, generally conspicuous, often asymmetrical. Gynoecium syncarpous, usually 2 carpels, usually equal in size, rarely distinctly different in size; the locules often completely or partially divided by false septa, each compartment containing 1 to many ovules attached to a central placenta, ovules pendent or ascendant; style filiform, exceeding the anthers; stigma usually bilobed. Fruit extremely variable, woody or coriaceous, dehiscent or indehiscent, usually provided with emergences such as horns, spines or wings. Seeds 1 to many in each compartment; testa often characteristically sculptured, sometimes forming wings; seeds containing considerable amounts of fat; endosperm very thin.

A family comprising 13 genera and about 70 species, distributed mainly in the arid regions of Africa south of the Sahara, with a few outliers in Madagascar, India, Sri Lanka, the Malayan islands and N Australia.

Vegetative Morphology. All species are monopodially branched. In general the position of the main shoots is erect (although sometimes falling over when thin branches become too long) or ascending. Definitely prostrate growth of the annual shoots is characteristic of the geophytic genera *Harpagophytum* (Fig. 37A) and *Dicerocaryum*, and of *Linariopsis prostrata*. *Sesamothamnus* is the only genus that develops distinct short shoots. The newly formed long shoots bear leaves with a much reduced and very short-lived lamina. The persisting petioles of these leaves rapidly develop into spines. Already in the same season, long-lived short shoots with subsessile leaves arise in the axils of the spines.

Most members of the Pedaliaceae are annual or perennial herbs, but several species are woody. *Sesamothamnus* and *Uncarina* exclusively contain woody species of considerable size (up to 8 m high); furthermore, in these two genera the main stems are swollen and capable of storing considerable amounts of water. In *Uncarina* the primary root develops into a succulent tuber, forming an additional underground water store. *Sesamum* sect. *Aptera*, *Ceratotheca* and *Josephinia* contain both woody species (shrubs or half-shrubs up to 2 m high) and annual species; however, the borderlines between woody plants, perennial herbs and species considered to be annuals are not distinct. Small woody plants may regularly abandon large portions of their branch system during the dry season (half-shrubs, e.g. *Ceratotheca integribracteata*), and several species of, for example, *Sesamum* and *Rogeria*, originally described as true annuals, may survive the dry season in an exceptionally favourable site and develop new shoots in the next season.

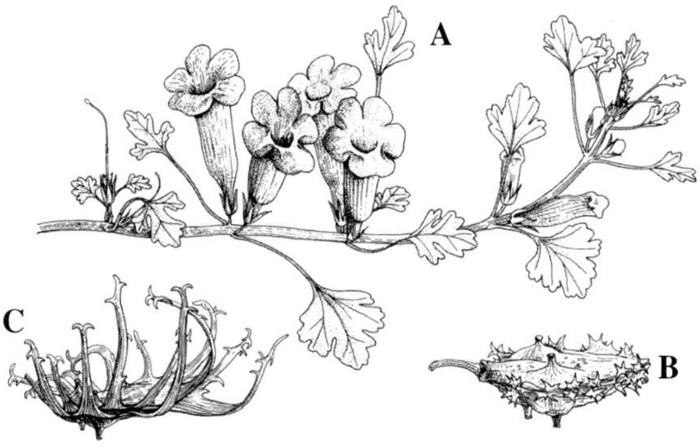

Fig. 37. Pedaliaceae. **A** *Harpagophytum procumbens*. Flowering shoot. **B** *H. zeyheri*. Fruit. **C** *H. procumbens*. Fruit. (Ihlenfeldt 1988)

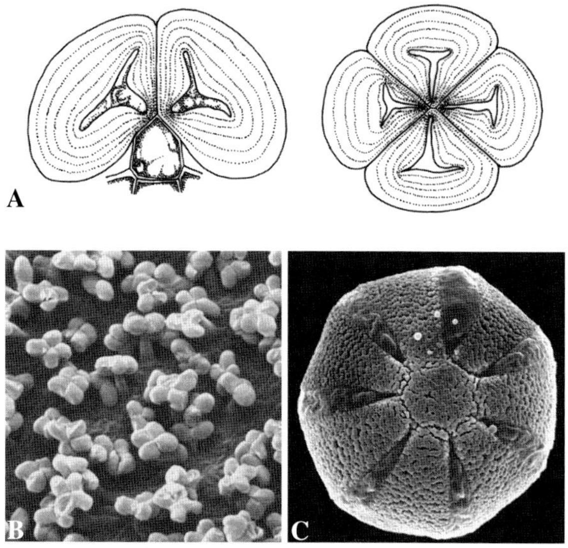

Fig. 38. Pedaliaceae. **A** *Ceratotheca triloba*. Mucilage gland with stalk cell and 4-celled head, ×600. **B** *Uncarina decaryi*. Mucilage glands on lower leaf surface. SEM ×100. **C** *Ceratotheca triloba*. Pollen, poly-zonocolpate. SEM ×650. **A** After Abels 1975; **B, C** orig.

The so-called caudex of *Pterodiscus* is a partly aerial and partly underground, succulent structure developing from a basal portion of the primary axis, the hypocotyl and the upper part of the primary root. It serves as the persisting organ of the plant, producing at its top several annual shoots which die off at the end of the season. Several genera are true geophytes. In *Harpagophytum* the primary root forms an oblong succulent tuber (up to 5 cm in diameter and up to 2 m long) with supplementary lateral tubers which develop from lateral roots. The perennial herbaceous plants of *Linariopsis* and *Dicerocaryum*, and the apparently only short-lived perennial herb *Pedaliodiscus* survive by means of a thickened, though not succulent, main root. Habit and life form are strongly functionally interconnected with dispersal (Ihlenfeldt 1967).

All woody species, whether stem succulent or not, and all perennial herbs are deciduous, even in tribe *Pedalieae*, where the leaves (except for *Uncarina*) are distinctly subsucculent. The leaves are estipulate and petiolate, although sometimes subsessile (e.g. on the short shoots of *Sesamothamnus*). There exists an extreme variation in leaf size (lamina 1–30 cm long and broad), venation and shape. The leaf blades basically are simple, the margin entire to lobed or even pinnatilobed, and only rarely is the lamina digitate (species in *Sesamum* and *Uncarina*). In several genera, the species have quite diverse lamina types (e.g. in *Uncarina*, *Sesamum*), and often the first leaves of the season and in the floral region are much smaller and less divided (e.g. in *Uncarina*, *Sesamum* and *Ceratotheca*).

Vegetative Anatomy. The most remarkable character is a unique type of glandular hair (Fig. 38A, B), not found elsewhere in the plant kingdom. The hair consists of a uniseriate, usually short stalk of one to several cells, which bears a more or less peltate head of nearly invariably four cells which is circular to quadrangular when viewed from above. In some species (*Sesamothamnus* p.pt., *Uncarina* p.pt.) the heads can be stellate, and the stalks may be much elongated (*Uncarina roeoesliana*). The four cells forming the head have extremely thickened outer cell walls (for details, see Abels 1975). Upon contact with water, the outer cell walls of these cells (except for the cuticle) dissolve completely, producing an enormous amount of mucilage. The biological function of these hairs is not yet understood. Apart from the mucilage hairs, often simple, very rarely branched uniseriate hairs occur. There are numerous transitional forms between typical mucilage hairs and simple hairs with reduced, more or less spherical heads; the latter, however, do not produce mucilage. The vegetative parts of many species have a strong, often very specific odour, the source of which is unknown.

The woody members have a vascular cylinder. Growth rings are indistinct (Carlquist 1987), and they may not represent true growth rings but rather reflect fluctuations in water supply during the season. The vascular cylinder is composed of relatively sparse vessels, solitary or in groups of 2–3, with circular pits and simple perforation plates, numerous libriform fibres (in *Uncarina* fibre-tracheids, Carlquist 1987), occasionally septate, axial parenchyma, which is vasicentric to apotracheal-banded and mostly not lignified, and numerous rays, which are uniseriate or multiseriate. In true annuals the vascular system is formed by individual bundles. Rhomboidal crystals, present in ray cells, have been described for *Rogeria* and *Uncarina*; sclereids are found in the pith and cortex of *Sesamothamnus*. The wood anatomy of Pedaliaceae is very similar to that of Martyniaceae and other related families, such as Bignoniaceae, Scrophulariaceae and Acanthaceae.

Stems of *Sesamothamnus* and *Uncarina* have concentric rings of parenchyma cells which apparently serve as a water store and enable the stems to shrink when water is withdrawn. In *Sesamothamnus* (except for *S. leistneri*) these water reserves enable the plants to flower out of season and to develop the first leaves of the season without rain; this also applies to a certain degree to *Uncarina*. The caudex plant *Pterodiscus* and the geophyte *Harpagophytum* even start to produce annual shoots without rain. A special trait of the succulent stems (including the aerial part of the caudex in *Pterodiscus*) is a translucent, papery bark over a chloroplast-containing cortex which enables the plants to perform photosynthesis even in the leafless stage.

Leaves are bifacial and amphistomatic. The mesophyll consists of 4–7, mostly very distinct layers which are made up of either palisade cells or spongy parenchyma cells. The distribution of these two basic cell types in relation to the individual layers has some taxonomic significance. In some genera (*Harpagophytum*, *Pterodiscus*, *Sesamothamnus*) all layers of the mesophyll consist of palisade cells. Usually, the density of stomata is higher on the lower surface. Most stomata are of the anomocytic or anisocytic type, but also dicytic and paracytic types occur (*S. orientale*, Karatela and Gill 1984).

Subsucculence of leaves, typical of most genera of the tribe Pedalieae, is only little reflected in leaf anatomy. None of the species has a specialised water-storing tissue. In subsucculent leaves the epidermis tends to be thicker than in non-subsucculent leaves, as do the leaves themselves. Intercellular spaces tend to be small and scarce, and at least both the first and second layers consist of densely packed palisade cells. However, there are several exceptions, e.g. the annual *Holubia*. Large hydathodes, located at the tip of leaf lobes, are found in *Uncarina*. The sieve-tube plastids of the Pedaliaceae belong to the Ss type (Behnke 1991).

Inflorescence Structure. Basically, the flowers of the Pedaliaceae are produced in small dichasia of 3–9 flowers with tiny bracts. Usually these dichasia arise in the axils of foliage leaves. However, except for the genera *Rogeria* and *Uncarina*, these dichasia are reduced to the first (central) flower, so that in the majority of genera the flowers appear solitary in the axils of leaves. The primordia of the second order of flowers usually develop into extrafloral nectaries (already identified as reduced flowers by Endlicher 1832) in the axils of minute bracts on the pedicel of the central flower (for details of the ontogeny see, e.g. Singh 1960, 1963). The number of extrafloral nectaries usually is 2, one on each side of the pedicel, although the nectaries may also occur in groups of 3–5 (e.g. *Sesamum* sect. *Sesamopteris*). The extrafloral nectaries are extensively exploited by ants, and thus it can be anticipated that the extrafloral nectaries form part of the defence system of the plants.

In *Sesamothamnus* the flowers form raceme-like aggregates on short persistent shoots. Nevertheless, in this genus, too, each single flower represents the central flower of a reduced dichasium, with the two next flowers of the dichasium developed into extrafloral nectaries. In *Rogeria* additional flower primordia of the individual dichasium (up to 5) regularly develop into flowers (and later on into fruits). Only the last flowers have nectaries in the axils of their bracts. In *Uncarina*, too, the axillary dichasia usually develop more than one flower (up to 9). In this genus, however, extrafloral nectaries are completely lacking.

Flower Structure and Anatomy. The bisexual flowers of the Pedaliaceae are to a various degree zygomorphic. The calyx consists of 5 sepals which are always fused, although the degree of fusion varies. Compared to the corolla, the calyx is noticeably small, and the lobes are usually more or less unequal in size. After anthesis the calyx usually persists until maturation of the fruit. The calyx regularly has an indumentum of both mucilage and simple hairs.

The corolla consists of 5 fused petals, and is usually large and showy. The tube is obliquely campanulate, sometimes funnel-shaped or cylindrical, often distinctly constricted at the base. The corolla may have a tubular (*Sesamothamnus* p.pt.) or sac-like (*Rogeria longiflora*, *Holubia*) adaxial spur at its base. Many flowers have a tiny protrusion in the same position, which can be interpreted as a rudimentary spur (e.g. *Pterodiscus*). The limb of the corolla, usually spreading more or less at a right angle to the tube, is distinctly subbilabiate with the anterior lobe enlarged, or the lobes are subequal. In *Sesamothamnus busseanus* the corolla lobes, except the anterior one, have long, branched fringes along their margins. The androecium consists of 4 fertile stamina and is didynamous. A fifth stamen on the adaxial side is very often represented as a staminode, with or without a rudimentary anther. Usually the androecium is inserted in the tube near the base of the corolla, in the long tubes of *Sesamothamnus* not far from the throat. Usually, the anthers are included in the tube, and more or less contiguous to its upper side. In the sphingophilous *Sesamothamnus*, however, the anthers are presented at the entrance to the tube. At the tip of the connective there is an often dark-coloured gland (except in *Linariopsis*) of unknown function. The two thecae are either ovate, spreading and pendent from the tip of the filament (tribe Pedalieae), or oblong and parallel, the filament being basifix (tribes Sesameae and Sesamothamneae). The thecae open lengthwise.

The gynoecium is superior and nearly invariably composed of 2 carpels which form a syncarpous ovary. Three and four carpels are found in certain strains of the crop plant *Sesamum orientale*, and in *Josephinia imperatricis* and *J. grandiflora*. In *Rogeria* subg. *Rogeria* the adaxial carpel is distinctly smaller and behaves different with respect to dehiscence and seed formation. The base of the gynoecium is surrounded by a mostly conspicuous, often asymmetrical nectary disc. The two locules are often divided into four compartments by false septa which can be complete or more or less incomplete. In *Linariopsis* a hole is formed in the septum during fruit formation. The uppermost, paracarpous and sterile portion of the gynoecium may develop into a distinct beak in the mature fruit, sometimes reaching nearly half the length of the fruit proper (*Uncarina*, *Rogeria*). The style is filiform, the stigma lobed, with the number of lobes corresponding to the number of carpels. The stigma usually exceeds the anthers. The style is shed shortly after the corolla has fallen off. The ovules are initiated on a central placenta, basically in 4 longitudinal rows. The number of ovules per locule ranges from numerous to few. In fruits with only 1 to very few (2–3) seeds per locule or compartment, abortive ovules can be found.

Embryology. Only *Sesamum* and *Pedalium* have been investigated in some detail. The anther wall comprises the epidermis, a one-layered fibrous endothecium (biseriate in *Sesamum laciniatum*), two or three middle layers and a secretory tapetum, which is binucleate in *S. orientale*. In *Pedalium murex* the 3–4 nuclei of tapetal cells subsequently fuse to form a single large nucleus (Singh 1958). Cell wall formation after meiosis is simultaneous, and the pollen is 3-celled when shed.

The ovule is anatropous, unitegmic and tenuinucellate. The ovules are basically pendent, although the orientation of the seeds within the mature fruit may also be more or less horizontal or even ascending. The archesporium is one- or two-celled. A single cell functions directly as megaspore mother cell. After meiosis the megaspores are arranged linearly or are T-shaped. The chalazal megaspore develops into a Polygonum-type embryo sac. The polar nuclei fuse before, or at the time of fertilisation. The antipodes are ephemeral.

Endosperm formation is cellular. The first division of the embryo sac separates a large micropylar chamber from a small chalazal chamber. Thereafter, a vertical division occurs in the micropylar chamber, followed by a transverse and then a longitudinal division in each cell of the micropylar chamber. The upper tier gives rise to a multicellular micropylar haustorium, whereas the lower tier gives rise to the endosperm proper. The chalazal chamber develops into a 4-celled haustorium. Embryo formation conforms to the Onagrad type. The seed is albuminous but the endosperm is only thin, and the embryo is straight. Both the endosperm and the cotyledons contain oil droplets which, in the oil-yielding crop plant *Sesamum orientale*, can make up 50% of the total weight of the seed.

Pollen Morphology. Pollen morphology of Pedaliaceae is very uniform with only minor variations (Straka and Ihlenfeldt 1965). Usually the pollen grains are isopolar monads, but in *Sesamothamnus* (sole genus of tribe Sesamothamneae) they are united in tetrahedral tetrads (Erdtman 1952). The medium-sized grains are prolate to sphaeroid and 5–13-zonocolpate (Fig. 38C). The number of colpi within a species and even within a single flower is rather variable. The exine consists of a thick sexine and a thin nexine, the tectum is laevigate. Minor variations, e.g. crassimarginate colpi, agree well with the subdivision of the family and the placement of the individual genera (Straka and Ihlenfeldt 1965). The fine structure of the grains of *Sesamothamnus* differs considerably from the remaining genera, which again is in accordance with the isolated position of this genus within the family.

Pollen morphology of Pedaliaceae s.str. differs strongly from that of Martyniaceae and Trapellaceae, often united with this family. There are some similarities – especially with respect to the number of colpi exceeding 3 – with pollen types found in certain genera of the Bignoniaceae, e.g. *Incarvillea* (incl. *Niedzwedzkia*) (Straka and Ihlenfeldt 1965; Gentry and Tomb 1979), and genera of the Acanthaceae, e.g. *Meyenia* (Raj 1961) and *Thomandersia* (Bremekamp 1942).

Karyology. The chromosome base number is x = 8; however, $2n = 16$ has been found only in *Pedalium murex*. Most species appear to be tetraploids with $2n = 32$, and *Sesamum radiatum* is an octoploid with $2n = 64$. In the tetraploids with $2n = 32$, the chromosomes tend to arrange in two diploid sets, which often are not located in the same plane, during mitosis (Ihlenfeldt and Grabow-Seidensticker 1979). Besides some deviating numbers recorded in the literature and apparently resulting from misidentifications (in the genus *Sesamum*, see below) or miscounts, there are only two remarkable exceptions: all species of *Sesamum* sect. *Sesamopteris* have $2n = 26$ chromosomes which, moreover, are much larger than in the remaining sections of *Sesamum*. The second exception is the crop plant *Sesamum orientale* which also possesses $2n = 26$ chromosomes, although in this species the chromosomes correspond in size both with sections *Aptera* and *Chamaesesamum* and with the second species of sect. *Sesamum*, *S. latifolium*. On account of the identical and aberrant chromosome number, the cultivated sesame has often been assigned to sect. *Sesamopteris* in the past – in spite of the lack of wings on the seeds. On the basis of morphological data, Ihlenfeldt and Grabow-Seidensticker (1979) claimed that *S. latifolium* ($2n = 32$) from NE Africa is the putative ancestor of the crop plant, and that in the process of domestication a re-arrangement of the genome took place. This hypothesis is supported by the occurrence of up to 6 B-chromosomes in *Sesamum orientale*. Tetraploid strains of *Sesamum orientale* ($2n = 52$) have been recorded and have also been created artificially (Kobayashi and Shimamura 1952).

Pollination and Reproductive System. Usually the flowers appear towards the middle of the season, rarely at the beginning of the season (*Uncarina*) or out of season (most species of *Sesamothamnus*). Anthesis usually lasts only one day, and the corolla is shed shortly after withering.

All species are insect-pollinated. Two basic flower types can be distinguished (Vogel 1954): a melittophilous type and a sphingophilous type. To the pollinating insect, the flowers offer both pollen and nectar.

Most species have the melittophilous syndrome: large showy flowers of various colour (bright yellow to numerous shades of red), limb spreading at more or less right angle, anterior lobe often enlarged; especially the latter often carries contrasting marks and lines of hairs, which may run down into the tube, and in many cases the throat is marked by a large black area (e.g. *Uncarina*). The flowers are usually presented in a more or less ascending or horizontal position (*Uncarina* spp., *Rogeria*). There are two noteworthy exceptions related to a prostrate growth form: in *Harpagophytum* the large flowers stand upright so that their limbs overtop the leaves, and in *Dicerocaryum* the pendent flowers are presented on long erect pedicels; in the latter genus the pedicels eventually bend down and place the flat trample burr on the ground. During anthesis stigma and anthers are usually found closely together in the upper half of the tube. Thus, pollination is nototribic. A special trait of many flowers is a tube that is much narrowed towards the base, with an often abrupt constriction. This may serve to exclude pollinators which are too small to come into contact with the stigma and anthers from exploitation of the nectar.

The sphingophilous syndrome is typical of *Sesamothamnus*. The white to light yellow flowers, sometimes tinged with purple, have a narrow tube up to 12 cm long, some species additionally have a narrow tubular spur. The flowers open in the evening and start to release an intensive scent. Sphingophilous flowers of more or less white colour with a large sac-like spur are found in *Rogeria longiflora* (the remaining species are melittophilous) and in *Holubia*.

Little is known about the reproductive systems. From cultivation it is known that *Pterodiscus* is self-fertile and self-pollinating, and *Uncarina* is also self-fertile but requires pollination for good fruit set. Self-pollination appears to be common in cultivated sesame, where the anthers start bursting and the stigma becomes receptive already one hour prior to anthesis (Trehan et al. 1974). In *Dicerocaryum* proteroandry combined with a change in position of anthers and stigma has been described (Vogel 1954), and in *Sesamothamnus* proterogyny has been observed.

Fruit and Seed. Compared to the small size of the family, Pedaliaceae exhibit a remarkably high number of different fruit types. Traditionally the morphology of the fruit is the most important marker for the delimitation of genera, and several generic names allude to fruit morphology.

Basically the fruit is a rigid, loculicidally (in *Uncarina* additionally septicidally) dehiscing capsule which immediately after maturation releases winged seeds. This primitive type is only found in *Sesamothamnus*; in the remaining genera the fruit has been greatly modified during the evolution of the family.

The pericarp of the fruit consists of a sclerenchymatous endocarp, made up of two crossing layers of fibres, and a parenchymatous exocarp (including the epidermis). During maturation of the fruit, the exocarp collapses and forms a thin, sheet-like dry layer which remains attached to the fruit. The mature fruits are either distinctly woody or coriaceous. Size ranges from 5 mm (*Linariopsis*) to 15 cm (*Harpagophytum procumbens*) in length and diameter.

The most outstanding feature is the occurrence of very diverse emergences on the fruits in the majority of genera (Fig. 37B, C), ranging from minute tubercles to spines, simple or armed with recurved hooks, to narrow or broad wings, pergamentaceous or rigid, sometimes also bearing recurved hooks. Several genera (e.g. *Uncarina*) have different types of emergences on the same fruit, and within a genus the set of types may differ considerably (*Rogeria*, *Pterodiscus*); for details, see generic descriptions. The emergences on the fruits play an important role in the very diverse dispersal systems of the family.

All emergences appear relatively late in the ontogeny of the fruit. Usually they become visible on the surface of the ovary at the beginning of anthesis, sometimes even later. They are initiated in, and basically formed by the endocarp, and mainly consist of fibres but do not contain vascular bundles. The persistent thin layer of the collapsed exocarp may hide small emergences and large lacunae between them (*Pterodiscus*, *Holubia*, *Pedalium*).

In the majority of genera the fruits are either tardily dehiscent (*Pterodiscus* p.pt., *Uncarina*), or more or less incompletely dehiscent (*Sesamum*, *Ceratotheca*), sometimes combined with tardy dehiscence (*Harpagophytum*), or completely indehiscent (*Linariopsis* p.pt., *Dicerocaryum*, *Pedaliodiscus*, *Pedalium*, *Holubia*, *Pterodiscus* p.pt.). In *Rogeria* subg. *Rogeria*, only the larger abaxial carpel splits at the top of the fruit to form a narrow exit, whereas the smaller adaxial carpel is indehiscent.

The number of seeds per fruit varies between two and many. As a general rule (not *Holubia*), the dehiscent fruits contain numerous seeds whereas the indehiscent fruits contain only few seeds. As another rule, fruits with extremely large appendages contain more seeds than those with smaller ones (*Harpagophytum procumbens* vs. *H. zeyheri*, Ihlenfeldt and Hartmann 1970).

The seeds are very diverse in size (2 to 10 mm in length) and shape. They are either obovate or oblong, the raphe is dorsal. The seeds are exotestal (Corner 1976), the upright cells of the outer epidermis of the testa have thick lignified walls, and may have additional annular thickenings and contain oxalate crystals. The inner layers consist of thin-walled, more or less collapsed cells and may also contain oxalate crystals. The cuticle of the inner epidermis is thick and persistent. Generally the cells of the outer epidermis of the seed coat are heavily impregnated by tannins, giving the seeds a dark brown to black colour. The testa often has a characteristic sculpturing which can provide useful characters on the species level (e.g. in the *Sesamum calycinum* complex, Grabow-Seidensticker 1988). Seeds that are not released from the fruit regularly only have a thin, light-coloured testa and lack sculptures. In *Rogeria* subg. *Rogeria* the seeds of the two unequal carpels differ in number per locule, size and testa structure, distinctly so in *R. adenophylla*.

Several genera have winged seeds, but often the wings are only rudimentary or reduced to a protruding margin. The wings develop as folds late in the ontogeny and are basically composed of only 2 layers of cells. Their walls can have complicated patterns of local thickenings. In *Sesamothamnus* the seeds have a translucent, very broad (up to 1 cm), continuous surrounding wing, in *Sesamum* sect. *Sesamopteris* three separate wings (one at the top and two lateral ones towards the base). Morphology and orientation of the wings provide useful characters at the species level, but the morphology of the wings may be influenced by the position of the seed within the capsule (*Sesamum* sect. *Sesamopteris*).

Seeds of true annuals, if released from the mature fruit, are very long-lived and often possess a pronounced dormancy; this especially applies to species of very arid habitats.

Dispersal. The family is well-known for its often very specialised and spectacular fruits, e.g. *Harpagophytum procumbens*, the Devil's claw. Despite the fragmentary status of the family, the evolution of the dispersal systems within the family can be well reconstructed (for details, see Ihlenfeldt 1967).

Basically, a change took place from a comparatively simple, anemochorous system (a rigid capsule that releases winged seeds immediately after maturation, found in *Sesamothamnus*) to more elaborate, partly very complex systems characteristic of tribes Pedalieae and Sesameae. These more advanced systems involve the fruit in the process of dispersal.

So-called wind-ballists are found in *Rogeria* (Pedalieae), and *Sesamum* and *Ceratotheca* (Sesameae): the mature capsules only open incompletely at the top and stay in an upright position on the mother plant, independent of whether the plants are perennial or annual. Seeds are gradually shed by strong movements of the mother plant. At the same time the seed wings become reduced (except in *Sesamum* sect. *Sesamopteris*), and a portion of the seeds remains completely locked in the fruit (in the basal part in *Sesamum* and *Ceratotheca*, or in the indehiscent adaxial carpel in *Rogeria* subg. *Rogeria*). These seeds are only released after the capsule has completely disintegrated which, in natural habitats, takes years. On this evolutionary level, already small emergences on the fruit with no recognisable function are found (two apical horns in *Ceratotheca* and small spines at the base of the fruit in *Rogeria*).

In tribe Pedalieae the emergences apparently evolved into 4 longitudinally arranged protrusions (wings) which form the basis for two diverging dispersal systems. In one lineage (*Pterodiscus*, *Holubia*), an anemochorous system is implemented: the wings become pergamentaceous and the exocarp forms large lacunae which reduce the specific weight of the fruit; however, the fruits are too heavy to be carried by the wind and roll on the ground. In a second lineage, epizoochorous systems evolved, either by splitting the wings into separate spines with recurved hooks (robust, large trample burrs in *Harpagophytum* and more fragile fur burrs in *Uncarina* with 8 longitudinal rows of hooked spines, both containing numerous seeds and tardily dehiscent), or by reducing the wings to four spines in the lower part of the fruit (robust, small trample burrs in *Pedaliodiscus* and *Pedalium*, both containing few seeds and indehiscent).

In tribe Sesameae only spine-like emergences occur which form the basis for the evolution of epizoochorous dispersal systems. These spines can be numerous and scattered all over the fruit (effective, small indehiscent trample burrs in *Josephinia*, spines only rudimentary in *Linariopis*), or there are only 2 spines at the top of the fruit (*Ceratotheca*, a wind-ballist, *Dicerocaryum* with a disc-like, robust indehiscent trample burr).

There are strong interdependencies between the type of burr, behaviour of the mature fruit and habit of the plant. Trample burrs are best presented on the ground. *Dicerocaryum* and *Harpagophytum* are strictly prostrate plants and the fruits are not shed. *Pedaliodiscus* and *Pedalium* are small herbs that shed their mature fruits immediately, as do the herbs and low shrubs of *Josephinia*. In contrast, the mature fur burrs of *Uncarina* are first presented high up on shrubs and trees, and only fall with considerable delay.

Phytochemistry. Pedaliaceae conform to the basic chemical profile of the Lamiales (Grayer et al. 1999): iridoids (in a very broad spectrum), flavone glycosides and verbascoside-type compounds are present. As investigations focussed on the crop plant *Sesamum orientale* and the drug plant *Harpagophytum*, little is known about the distribution of the individual compounds within the family, and therefore the usability of phytochemical data for taxonomic purposes is limited.

The oil extracted from the seeds of *Sesamum orientale* contains as main components lineolic and oleic acids and, furthermore, palmitic and stearic acids. Characteristic of sesame oil is its content of lignans (up to 2.5%, mainly sesamin and sesamolin), by which it can be discriminated from other plant oils. Higher contents of sesamin have also been found in the seeds of *Ceratotheca* (very closely related to *Sesamum* sect. *Aptera*) and in *Sesamothamnus*, whereas only traces or absence have been recorded in various genera of tribe Pedalieae (Bedigian et al. 1985). Within *Sesamum*, sesamin has also been found in the oil of several species of section *Aptera* and in *S. latifolium* (sect. *Sesamum*, the putative ancestor of the crop plant), but not in members of sect. *Sesamopteris*. This agrees well with the isolated position of sect. *Sesamopteris* within *Sesamum*, and the taxonomic position of the crop plant sesame which, in spite of the corresponding aberrant chromosome number of $2n = 26$, does not belong to sect. *Sesamopteris*. *Sesamum latifolium*, however, is reported to lack sesamolin (Bedigian et al. 1985).

Iridoids and iridoid glycosides are best studied in *Harpagophytum*. The main compound is harpagosid, accompanied by procumbid, harpagid and 8-para-coumaryl-harpagid. These compounds accumulate in the succulent main root and especially in the lateral tubers of the main root (up to 3% of the dry weight), whereas the aerial parts of the plant only contain very low concentrations.

Both species of the genus contain these iridoids; however, in *H. zeyheri* the proportion of 8-para-coumaryl-harpagid is much higher (30–60% compared to 2–6% in *H. procumbens*).

Purified mucilage isolated from the mucilage hairs of *Dicerocaryum senecioides* has a molecular weight of about 500000 Dalton, and after reduction of carboxyl groups, has been found to consist mainly of galactose, xylose, arabinose and mannose in the ratio 21:19:12:1 (Benhura and Marume 1993). Moreover, the mucilage contains 8.2% uronic acid, 2% protein and 7% ash. It behaves like a pectic material.

Subdivision of and Relationships Within the Family. The family can be subdivided into three major units, to which tribal rank has been assigned. Apart from macromorphological arguments, this grouping is also supported by pollen morphology (Straka and Ihlenfeldt 1965).

Tribe Sesamothamneae consists of *Sesamothamnus* only. This genus is very close to Bignoniaceae, so that inclusion in this family has been discussed by Bruce (1953), but this has been rejected by the same author on account of the occurrence of mucilage hairs. This tribe has a somewhat isolated position in the family by having pollen in tetrads, short shoots, and spines; noteworthy are also the sphingophilous flowers. On the other hand, it has characters that are typical of one or the other of the remaining tribes, for example, anther morphology agrees with Sesameae, whereas succulent stems and sphingophilous flowers are exclusively found in Pedalieae. Ihlenfeldt (1967, 2001) therefore claimed that this genus may be very close to, or even represent the putative ancestor of, the family.

The remaining 12 genera can be grouped into two larger units on the basis of anther morphology. In Pedalieae the more or less ovate thecae are divergent and pendent from the tip of the filament, whereas in Sesameae the thecae are oblong and not divergent, and the anthers basifix. There are further characters that can be considered as characteristic (apomorphic) of one or the other tribe, but these are always absent in one or more genera. For example, the position of the emergences on the fruit can be a good marker, but emergences are absent in *Sesamum* (Sesameae) and *Rogeria* subg. *Microrogeria* (Pedalieae), or subsucculent leaves, typical of tribe Pedalieae, are absent in *Uncarina*.

Tribe Pedalieae comprises a group of two rather isolated genera of putatively basal position, and a second group of five, generally more advanced, closely related genera, albeit very diverse in appearance and dispersal systems. The first group consists of the genera *Rogeria* and *Uncarina*. Both share small, dichasial partial inflorescences in the axils of foliage leaves (a putatively primitive character), instead of solitary flowers. *Rogeria* is a very diverse genus, the unifying characters being the distinct difference in size of the carpels, and indehiscence of the adaxial carpel (only in subg. *Rogeria*), not found elsewhere in the family. *Uncarina*, endemic to Madagascar, is much more homogeneous, but combines some primitive characters (e.g. large woody plants with succulent stems, leaves not subsucculent) with a very advanced dispersal system (highly elaborate fur burrs). Furthermore, the lack of extrafloral nectaries in this genus is unique to the family. The second group comprises a cluster of closely related genera (*Pterodiscus*, *Pedaliodiscus*, *Pedalium*, *Holubia*, *Harpagophytum*) with the paraphyletic (only kept together by the possession of a caudex) *Pterodiscus* in a basal position, from which the remaining genera may have evolved in different lineages (Ihlenfeldt 2001). The unifying character of this group is a fruit with basically four, longitudinally arranged emergences (wings), from which quite diverse and advanced fruit types evolved.

In tribe Sesameae also two groups can be distinguished. One group comprises the paraphyletic, and possibly also polyphyletic genus *Sesamum* in a basal position, and the more advanced genera *Ceratotheca* and *Dicerocaryum*. The unifying character of this group is the absence of emergences on the fruit or, when present, only two spine-like emergences at the top of the fruit in the median plane. The sole unifying character of *Sesamum* in its traditional circumscription is the absence of emergences on the fruit, undoubtedly a plesiomorphic character, and the corresponding dispersal system (wind-ballists), also considered as a relatively primitive character. *Sesamum* has been restructured by Ihlenfeldt and Grabow-Seidensticker (1979) and now comprises four sections, to which generic rank could be assigned. *Ceratotheca* has very close relationships with *Sesamum* sect. *Aptera*. *Dicerocaryum* may have evolved from *Ceratotheca*, although it also shares some characters (leaf morphology, woody fruits) with *Sesamum* sect. *Chamaesesamum*. The second group consists of *Linariopsis* and *Josephinia*. Both share small, woody burr fruits (with well-developed spines in *Josephinia* and much reduced ones in *Linariopsis*) which are indehiscent or tardily dehiscent, and small entire leaves, which indicate a relationship to *Sesamum* sect. *Chamaesesamum* from which this group may have evolved.

Linariopsis has two features unique to the family: the apical gland on the tip of the stamen is absent, and during ontogeny of the fruit a hole is formed in the septum.

Affinities. Originally Pedaliaceae (and Martyniaceae as well) formed part of Bignoniaceae (Jussieu 1789). When Brown (1810) established Pedaliaceae, he included only *Pedalium* and *Josephinia*, explicitly excluding *Sesamum*. Apparently Kunth (1818) first recognised that also *Sesamum* belongs to Brown's Pedaliaceae, but at the same time he also included in this family *Martynia* and *Craniolaria* (now Martyniaceae). Link (1829) was the first to note that *Martynia* does not fit into Pedaliaceae; however, most subsequent authors did not adopt this view. If the authors subdivided the family (e.g. into tribes Sesameae and Pedalieae or even into separate families), they always included *Martynia* and related genera in the *Pedalium* group. Apparently, this was influenced by the presence of trample burrs in both groups, a feature which is very rare in Lamiales (sensu Savolainen et al. 2000). Even after Stapf (1895) had established Martyniaceae, familial status for *Martynia* and related genera was rarely adopted (e.g. Brummitt 1992; APG 1998).

The family became even more heterogeneous when Oliver (1887) added his new genus *Trapella*, a genus of aquatic plants with burr-fruits from East Asia. However, as some characters of this genus are very unusual for Pedaliaceae, the author was uncertain whether assignment to Pedaliaceae was really justified or whether the genus should not be better placed in a separate family. Although more detailed studies (Oliver 1888; Glück 1940) revealed many additional characters not found elsewhere in Pedaliaceae, status as a separate family was only established by Honda and Sakisaka in 1930. As in Martyniaceae, familial status of *Trapella* is not yet generally adopted (e.g. APG 1998).

Against the background of often indistinct borderlines between generally adopted families in Lamiales (sensu Savolainen et al. 2000), Pedaliaceae, Martyniaceae and Trapellaceae are well-defined taxonomic units and deserve familial rank (Ihlenfeldt 1967). The main arguments are: structure of the inflorescence (small cymes in the axils of foliage leaves, usually reduced to a single flower, and flowers of higher orders reduced to extrafloral nectaries in Pedaliaceae; terminal racemes in Martyniaceae; solitary flowers in the axils of foliage leaves in Trapellaceae), structure of the gynoecium (syncarpous and superior in Pedaliaceae; paracarpous and superior in Martyniaceae; syncarpous and inferior in Trapellaceae), non-homology of the fruit appendages (emergences without vascular bundles on carpels of a superior ovary in Pedaliaceae; split upper sterile curved part of the superior ovary in Martyniaceae; emergences with vascular bundles on the hypanthium of an inferior ovary in Trapellaceae), behaviour of the exocarp in the mature fruit (collapsing and staying on the fruit in Pedaliaceae and Trapellaceae; peeling off in Martyniaceae), pollen morphology (5–12-colpate in Pedaliaceae, inaperturate (clypeate) in Martyniaceae, 3-colpate in Trapellaceae), and nature of the glandular hairs (4-celled head, producing mucilage in Pedaliaceae; multicelled head, producing a sticky substance in Martyniaceae; 4–8-celled head, not producing mucilage in Trapellaceae).

Pedaliaceae s.l. have always been placed in the vicinity of Bignoniaceae. Pedaliaceae s.str. have been interpreted as a highly specialised (mainly with respect to dispersal systems) Old World branch from a pantropical bignoniaceous stock, with Martyniaceae as a possible sister group in the New World (Ihlenfeldt 1967). Both families share a preference for unstable or degraded habitats in dry, warm climates. In this respect, the monotypic Trapellaceae basically differ from both Pedaliaceae and Martyniaceae by representing water plants from East Asia. Moreover, Trapellaceae exhibit certain features that are very unusual for Lamiales. The three families not only differ in many important characters from each other, but equally from Acanthaceae, Bignoniaceae and Scrophulariaceae s.l. as possible close relatives (see Ihlenfeldt 1967).

Distribution patterns indicate a high age of Pedaliaceae. Raven and Axelrod (1974) claimed that Pedaliaceae arose after the separation of South America from Africa, although apparently prior to the separation of India and Madagascar from Africa (about 100 m.y.b.p., Ihlenfeldt 1994).

Several phylogenetic analyses of Lamiales (sensu Savolainen et al. 2000) based on molecular data have been published recently (e.g. Chase et al. 1993; Olmstead and Reeves 1995; Olmstead et al. 2001). As to the position, affinities and the status of the three families in question, the results are not unequivocal. However, there seems to be consent that Pedaliaceae and Martyniaceae are not as closely related as thought, as they are nested in two different major clades (Olmstead et al. 2001) which may be sister to each other, indicating that the two families may have evolved in parallel from a common ancestor. This conforms well to the conclusion of Ihlenfeldt (1967), which was based on morphological data. Monophyly of Pedaliaceae (and Martyniaceae) is also strongly supported (Olmstead, pers. comm.), and both groups, therefore, deserve familial status. Trees presented by Chase et al. (1993), Olmstead and Reeves (1995) and Olmstead et al. (2001) indicate relationships between Pedaliaceae and Thunbergiaceae/Acanthaceae. Mainly on the basis of pollen morphology, affinities between these two families have already been discussed in the literature (see Ihlenfeldt 1967). Possible affinities, also discussed in the literature and again mainly based on similarities in pollen morphology, between Pedaliaceae and the heterogeneous Bignoniaceae (e.g. *Incarvillea* group, see Ihlenfeldt 1967), are still unresolved. Bignoniaceae are either nested in the clade that contains Martyniaceae (*Catalpa*, *Kigelia*, Olmstead et al. 2001) or appear as sister group (*Jacaranda*) to both Pedaliaceae and Martyniaceae (Olmstead, pers. comm.).

As to the status and position of Trapellaceae, see Trapellaceae (this Vol.).

Distribution and Habitats. Pedaliaceae basically are an Old World family centred in Africa south of the Sahara, with a few outliers in Madagascar (*Uncarina*, endemic, *Dicerocaryum*, also in E and S Africa), India and Sri Lanka (*Sesamum* sect. *Chamaesesamum*), and Malayan Islands and N Australia (*Josephinia*, also in E Africa). The crop plant *Sesamum orientale* (probably native to NE Africa) today is cultivated in many countries of the New and Old World with a suitable climate. *Pedalium murex* is found in many tropical countries all over the world as a weed in plantations; it is probably native to NE Africa and has been spread by man.

Those genera or sections of genera that are confined to Africa (including Madagascar) exhibit a distinct bi-centred distribution pattern: about one half occurs either in N Africa south of the Sahara, preferably in NE Africa (north of the equator = northern centre: *Pedalium*, *Pedaliodiscus*, *Sesamum* sect. *Sesamum*) or in southern Africa (= southern centre: *Harpagophytum*, *Holubia*, *Uncarina*, *Dicerocaryum*). The remaining genera/sections occur in both centres (*Sesamothamnus*, *Sesamum* sect. *Sesamopteris* and sect. *Aptera*, *Rogeria*, *Linariopsis*, *Ceratotheca*, *Pterodiscus*). In most members of the latter group, however, the whole range again is definitely bipartite with gaps up to 3500 km; only *Ceratotheca*, *Sesamum* sect. *Aptera* and *Pterodiscus* exhibit a distribution pattern that can be considered as continuous between the two centres. For more detailed information, see Ihlenfeldt (1994).

In those genera/sections which occur both in the northern and southern centres, the representatives of the respective centres nearly exclusively belong to different species. A remarkable exception is *Rogeria adenophylla*, which grows from the Cape Verde Islands all along the southern edge of the Sahara and is found again in a comparatively small area, 2500 km away, in N Namibia and SW Angola.

A considerable number of species possesses very small distribution areas. This especially applies to those species (Ihlenfeldt 1994) which, within a genus, are considered (relatively) primitive, e.g. *Sesamothamnus leistneri*.

All Pedaliaceae are adapted to a warm climate with at least one dry season. The most extreme example is *Sesamum abbreviatum*, an ephemeral annual exclusively at home on the sand dunes of the Namib desert.

The vast majority of species prefer open, unstable or degraded habitats; thus, it is not astonishing that several species are weedy invaders of plantations or colonisers of abandoned fields: *Pedalium murex* (also considered as a saline soil indicator), *Ceratotheca sesamoides*, *Dicerocaryum* species, and even the crop *Sesamum orientale*. Dense populations of *Harpagophytum*, native of grassland savannahs, are considered as an indicator of overgrazing. However, the woody representatives are usually found in more stable habitats such as rock crevices or among boulders. This especially applies to the two genera of large woody plants, *Sesamothamnus* and *Uncarina*, which moreover have a preference for alkaline soils; the latter genus stands out by its ability to successfully compete even in dense deciduous forests.

Economic Importance. *Sesamum orientale*, often cited under the incorrect name *Sesamum indicum* (Seegeler 1989), is an important oil-yielding plant. It is an ancient crop plant. Archaeological evidence indicates that sesame was being cultivated in Palestine and Syria already about 5000 years B.P. or even earlier (Nayar 1976; Bedigian and Harlan 1986). Today it is grown in all continents in areas with a tropical to subtropical climate. The total world-production is very difficult to assess, since most of the crop is only produced for local markets. In 1972 FAO assessed a yearly production of 1.7 million tons (Nayar 1976). Although the average yield per ha (about 300 kg, Seegeler 1983), compared to other oil crops, is rather low, *Sesamum orientale* is a much-esteemed crop plant, due to the high quality and stability against rancidity of the oil, its ability to thrive on soils poor in nutrients and to easily adapt to drought stress, and the fact that it matures in only 105 days or even less. Under favourable conditions, a yield of 2.7 tons/ha is possible. The oil is mostly employed for cooking, but it is also used as an emollient and in the manufacture of soap, margarine and cosmetics. Sesame meal is an excellent source of protein for both human and animal consumption, and is also rich in vitamin E. The antioxidants sesamin and sesamolin, extracted from the oil, are often used to stabilise antiseptics, bactericides and viricides, and they show synergistic activities with certain insecticides, such as rotenone and pyrethrin.

On a much smaller scale, also *Sesamum radiatum*, *S. alatum* and *S. angolense* (the oil of the latter species contains up to 9% sesamin), and *Ceratotheca sesamoides* are locally grown as oil plants in tropical Africa.

Harpagophytum, native to southern Africa, has long been known as a drug in traditional medicine, and it is still used by the native populations to cure, mostly as an infusion from the dried tubers, various diseases such as indigestion and skin lesions; it is also administered to women who anticipate a difficult birth (Watt and Breyer-Brandwijk 1962). During the last decade, extracts from the tubers of *Harpagophytum procumbens* have become a favourite medicine also in Europe. The aqueous extract from the dried tubers contains, beside harpagosid which is suggested to be the main agent, numerous other compounds such as flavones, acteosid and various sugars. It is mainly applied to cure rheumatic and arthritic diseases, also disorders of the digestive tract, and as a palliative (for a survey, see Wegener 1998).

Mucilage, gained from the mucilage hairs of various members of this family, in Africa, Madagascar and India, is widely used for body care, as a substitute for soap, and as a lubricant in human and veterinary medicine; it also serves to cure indigestion. In Madagascar the fruits of *Uncarina* serve to trap mice.

Key to Genera of Pedaliaceae

1. Deciduous trees or large shrubs (to 8 m high) with slightly to distinctly succulent main stems. Leaves (at least on long shoots) alternate 2
- Deciduous shrubs (to 2 m high) or dwarf shrubs, perennial or annual herbs (sometimes with a persistent ± succulent caudex). Leaves (except sometimes in the floral region) opposite 3
2. Plants with long shoots bearing spines (originating from petioles) and short shoots with fasciculate sessile leaves. Flowers in raceme-like aggregates. Fruits flattened woody capsules without any emergences. Africa **1. *Sesamothamnus***
- Plants without short shoots, spines absent. Flowers in small dichasia in the axils of foliage leaves, very rarely solitary. Fruits woody to coriaceous capsules with numerous long, apically hooked spines. Madagascar **2. *Uncarina***
3. Shrubs, dwarf shrubs or half-shrubs 4
- Herbaceous plants, sometimes with persistent succulent caudex 7
4. Capsules with 2 lateral horn-like emergences at the apex, these sometimes rather indistinct **10. *Ceratotheca***
- Fruits without 2 apical horns, sometimes with other emergences such as tubercles or spines 5
5. Capsules devoid of emergences (except tip of the fruit which may form a distinct beak, and low longitudinal ribs), sometimes slightly gibbous at the base **9. *Sesamum***
- Fruits woody, indehiscent or tardily dehiscent, with emergences such as spines, tubercles or wings 6
6. Fruits ± globose, indehiscent and covered with numerous simple spines **13. *Josephinia***
- Fruits indehiscent or tardily dehiscent, covered with numerous tiny tubercles, these sometimes rather indistinct **12. *Linariopsis***
7. Plants with a persistent succulent caudex which produces annual shoots **4. *Pterodiscus***
- Plants without a caudex, although sometimes with a succulent or non-succulent, thickened primary root, annuals or perennials 8
8. Flowers (and fruits) in small dichasia in the axils of foliage leaves. Fruits woody, with 2 lateral wings or 2 to several spines (subg. *Rogeria*), or fragile and devoid of emergences (subg. *Microrogeria*) **3. *Rogeria***
- Flowers (and fruits) always solitary in the axils of foliage leaves 9
9. Fruits coriaceous to pergamentaceous 10
- Fruits woody 12
10. Fruits pergamentaceous with 4 longitudinal broad wings, indehiscent. Flowers with a large basal sac-like spur **8. *Holubia***
- Fruits without longitudinal wings. Flowers without a spur 11
11. Capsules with two lateral horn-like emergences at the apex, these sometimes rather indistinct **10. *Ceratotheca***
- Capsules devoid of emergences (except tip of the fruit which may form a distinct beak, and low longitudinal ribs), sometimes slightly gibbous at the base **9. *Sesamum***
12. Capsules devoid of emergences, incompletely dehiscent. India and Sri Lanka **9. *Sesamum*** (sect. *Chamaesesamum*)
- Fruits with emergences such as spines and wings 13
13. Fruits large (5–15 cm long and broad) and flattened, tardily dehiscent only apically, with 4 longitudinal wings with recurved hooks at the margins, or with 4 rows of curved arm-like emergences, also bearing recurved hooks. Prostrate perennials with a tuberous succulent primary root **7. *Harpagophytum***
- Fruits small (up to 3 cm long), woody, indehiscent with 2 to numerous simple spines. Erect annuals or prostrate perennials without succulent primary root 14
14. Fruits ± globose, indehiscent and covered with numerous simple spines. Erect annuals. Malayan Islands and N Australia **13. *Josephinia***
- Fruits not globose, indehiscent, with only 2–4 simple spines. Africa and Madagascar (*Pedalium* today also in other continents as a weed) 15

15. Fruits longitudinally compressed (± disc-like), with 2 erect conical spines. Prostrate perennials. E and S Africa and Madagascar **11. *Dicerocaryum***
- Fruits not longitudinally compressed, contracted into an obconical base, with 4 lateral spines. Prostrate to ascending annuals or short-lived perennials 16
16. Fruits laterally compressed, upper part with 4 longitudinal narrow wings and 4 rows of minute emergences on the flattened faces. Kenya and Tanzania **5. *Pedaliodiscus***
- Fruits not laterally compressed, upper part 4-angled, without distinct wings, faces rugose to tuberculate. NE Africa and weed in many tropical countries **6. *Pedalium***

I. Tribe Sesamothamneae Ihlenf. (1967).

Plants with distinct long shoots and short shoots; long shoots with alternate petiolate leaves with much reduced and short-lived lamina, petioles persistent and developing into spines; short shoots in the axils of the spines, leaves on short shoots fasciculate and almost sessile; leaves not subsucculent. Flowers in few-flowered raceme-like inflorescences. Anthers basifix, the oblong thecae not divergent; pollen in tetrads. Fruits without any emergences.

1. *Sesamothamnus* Welw.

Sesamothamnus Welw., Trans. Linn. Soc. 27: 49, t. 19 (1869); Ihlenfeldt in Eggli, U. (ed.) Illustrated Handbook of Succulent Plants, Dicotyledons: 355–356 (2002), review.
Sigmatosiphon Engl. (1894).

Trees (up to 5 m tall) or shrubs; trunks and main branches swollen at the base. Leaves deciduous, entire, usually obovate. Flowers large, white, cream, pinkish or yellow; sometimes tinged with purple, corolla tube long cylindrical or narrowly funnel-shaped, curved or straight, usually with a conspicuous spur at the base; lobes subequal, entire or rarely fringed. Fruit a woody, laterally compressed capsule. Seeds numerous, obovate and compressed, with a broad translucent surrounding wing. Six spp., NE and southern Africa.

II. Tribe Pedalieae Meisn. (1840).

Plants without distinct short shoots and without spines; leaves subsucculent (except *Uncarina*). Flowers either in small contracted dichasia or solitary in the axils of foliage leaves. The ± ovate thecae divergent and pendent from the tip of the filament; pollen in monads; fruits (except *Rogeria* subg. *Microrogeria*) with emergences such as spines or wings.

2. *Uncarina* (Baill.) Stapf Fig. 38B

Uncarina (Baill.) Stapf in Engler & Prantl, Natürl. Pflanzenfamilien IV, 3b: 261 (1895); Humbert, Fl. Madagascar, Pedaliacées (1971), rev.; Ihlenfeldt in Eggli, U. (ed.) Illustrated Handbook of Succulent Plants, Dicotyledons: 356–360 (2002), review.
Harpagophytum DC. ex Meisn. sect. *Uncarina* Baill. (1887).

Trees (up to 8 m tall), small bottle trees or shrubs with subterraneous tuber; leaves usually large (up to 30 cm long and broad), deciduous, very variable in shape, often polymorphic, first leaves of the season usually much smaller and simplified, entire or dissected into lobes, not subsucculent. Flowers large, in small contracted dichasia (up to 9 flowers) without extrafloral nectaries, very rarely solitary, yellow, white or purple, corolla tube cylindrical, lobes subequal. Fruits large (up to 12 cm diameter), woody to coriaceous tardily dehiscent capsules, with long spines bearing 3–4 recurved hooks at the top, often with additional simple spines and occasionally wings. Seeds numerous, obovate, flattened, often with a surrounding wing. About 13 spp., all endemic to the arid regions of N, W and S Madagascar.

3. *Rogeria* J. Gay ex Del.

Rogeria J. Gay ex Del., Cent. Pl. Afr.: 78 (1826); Ihlenfeldt, Mitt. Bot. München 4: 605–608 (1967), review; Ihlenfeldt, Mitt. Staatsinst. Allg. Bot. Hamburg 12: 69–73, 112–113 (1967), review.

Short-lived deciduous perennial or annual herbs, erect or ± rosulate (subg. *Microrogeria* Ihlenf.), leaves sinuate to lobed. Flowers in small contracted dichasia, purple, red, white or cream, corolla tube tubular or long cylindrical (then with a distinct gibbous spur at the base), lobes subequal. Fruits either (subg. *Rogeria*) woody and persisting on the mother plant for many years, carpels distinctly different in size with only the larger abaxial carpel opening at the beak-like top of the fruit, and short spines or wings, or (subg. *Microrogeria*) fruit pergamentaceous and fragile, carpels only slightly different in size, both carpels completely dehiscent, and without any emergences. Seeds numerous. $2n = 32$. Four spp., southern edge of the Sahara and SW Africa (Angola, Namibia, NW Republic of South Africa).

4. *Pterodiscus* Hooker

Pterodiscus Hooker Bot. Mag. 70, t. 41417 (1844); Ihlenfeldt, Mitt. Inst. Allg. Bot. Hamburg 28/29: 5–21 (2001), key; Ihlenfeldt in Eggli, U. (ed.) Illustrated Handbook of Succulent Plants, Dicotyledons: 351–355 (2002), review.
Pedaliophyton Engl. (1902).

Perennial herbs with a caudex as persisting organ, with annual, usually erect and unbranched shoots, up to 30 cm long; leaves linear to broadly oblong, entire to pinnatifid. Flowers solitary, yellow, brilliant orange, red or purple, corolla tube funnel-shaped or narrowly cylindrical and constricted in the lower part, lobes subequal or anterior lobe enlarged. Fruits large (up to 30 mm diameter), laterally compressed, either woody or coriaceous, with 4 spreading longitudinal, either woody or pergamentaceous wings, the latter usually very broad, although occasionally rather narrow, curled and appressed to the body, indehiscent or tardily dehiscent. Seeds either 6–20, then obovate and testa rugose (in dehiscent fruits), or mostly 2, then oblong and testa smooth (in indehiscent fruits). About 13 spp., NE, CW and southern Africa.

A very inhomogeneous and apparently paraphyletic genus (Ihlenfeldt 2001).

5. *Pedaliodiscus* Ihlenf.

Pedaliodiscus Ihlenf., Ber. Deutsch. Bot. Ges. 81: 149 (1968), et Taxon 18: 347 (1969).

Short-lived ascending perennial herb with thickened, although not succulent primary root, leaves oblong. Flowers solitary, cream, tubular, lobes subequal. Fruits small (up to 15 mm long), laterally compressed, contracted into an obconical base, woody, indehiscent, upper part with 4 longitudinal rigid narrow wings, and in addition 4 distinct spines at the base of the upper part and 4 rows of minute emergences on the flattened faces. Seeds 4–6, oblong. One sp., *P. macrocarpus* Ihlenf., E Africa (Kenya, Tanzania).

6. *Pedalium* Royen ex L.

Pedalium Royen ex L., Syst. Nat. ed. 10: 1123 (1759).
Cacatali Adans. (1763).

Ascending annual herb, leaves oblong to elliptic, entire or coarsely dentate. Flowers solitary, yellow, corolla tube tubular to narrowly funnel-shaped, lobes subequal. Fruits small (1–2 cm long), woody, subpyramidal, abruptly contracted into an obconical base, indehiscent, upper part 4-angled, with a spreading spine at the base of each angle, faces of the upper part rugose or tuberculate. Seeds 2–4, oblong. $2n = 16$. One variable sp., *P. murex* L., probably native to NE Africa, but today a widespread weed in many tropical countries and considered a saline soil indicator.

7. *Harpagophytum* DC. ex Meisn. Fig. 37

Harpagophytum DC. ex Meisn., Pl. Vasc. Gen. 1: 298 and 2: 206 (1840); Ihlenfeldt & Hartmann, Mitt. Staatsinst. Allg. Bot. Hamburg 13: 15–69 (1970), monogr.

Perennial herbs with several prostrate, usually unbranched annual shoots from a succulent tuberous primary root, additional tubers often present on lateral roots. Leaves very variable in shape and often polymorphic, entire to deeply lobed. Flowers solitary, purple, pink or yellow, corolla tube tubular, constricted at the base, lobes subequal. Fruits large (up to 15 cm diameter), laterally compressed woody capsules, tardily dehiscent only at the top, either with 4 longitudinal rigid wings with recurved hooks along the margins, or with 4 rows of long and curved arm-like emergences, also bearing recurved hooks; each of the flattened faces of the fruit with two additional, solid obtuse protuberances. Seeds numerous, obovate. Two spp., southern Africa.

8. *Holubia* Oliv.

Holubia Oliv., Hooker, Ic. Pl. t. 1475 (1884); Ihlenfeldt in Fl. Zamb. 8, 3: 89 (1988), review.

Erect annual herb with spreading branches, leaves circular to ovate, slightly lobed or sinuate. Flowers solitary, yellow-green, corolla tube cylindrical in the middle, funnel-shaped at the mouth and with a large basal sac-like spur, lobes subequal. Fruits coriaceous, indehiscent, ± rectangular in cross section, with a broad pergamentaceous wing on each of the four edges, ± circular in lateral view. Seeds numerous, obovate. One sp., *H. saccata* Oliver, E southern Africa.

III. Tribe Sesameae (Endl.) Meisn. (1840).

Plants without distinct short shoots and without spines; leaves not subsucculent. Flowers always solitary in the axils of foliage leaves. Anthers ± basifix, the oblong thecae not divergent; pollen in monads. Fruits (except *Sesamum*) with emergences such as apical horns, terminal spines or numerous tubercles or simple spines.

9. *Sesamum* L.

Sesamum L., Sp. Pl. ed. 1: 634 (1753); Ihlenfeldt & Grabow-Seidensticker, Proc. IX Plen. Meet. AETFAT Las Palmas: 53–60 (1979), sections, origin of cultivated sesame.
Sesamopteris DC. ex Meisn. (1840).
Simsimum Bernh. (1842).
Gangila Bernh. (1842).

Deciduous shrubs, half-shrubs, short-lived perennials or annuals, erect or procumbent to ascending. Leaves very variable in shape, often polymorphic (upper and lower leaves often much smaller and simplified). Flowers white, pink or purple, corolla tube tubular and constricted at the base, funnel-shaped or campanulate, lobes of the corolla tube subequal or anterior lobe enlarged. Fruits coriaceous or rarely woody (sect. *Chamaesesamum*) capsules devoid of any emergences, often incompletely dehiscent. Seeds numerous. $2n = 26, 32, 64$. About 19 species, all native to the arid parts of Africa S of the Sahara, except sect. *Chamaesesamum* (India, Sri Lanka).

A very inhomogeneous and apparently paraphyletic, possibly also polyphyletic genus. *Sesamum orientale* L. (sect. *Sesamum*) is an important crop plant, cultivated worldwide for oil which is extracted from the seed.

Four sections can be distinguished (Ihlenfeldt and Grabow-Seidensticker 1979): sect. *Aptera* Seidenst., sect. *Chamaesesamum* Benth., sect. *Sesamopteris* Endl., sect. *Sesamum*.

10. *Ceratotheca* Endl. (1832). Fig. 38A, C

Ceratotheca Endl., Linnaea 7: 5, t. 1–2 (1832); Abels, Mém. Soc. Brot. 25: 1–358 (1975), monogr.
Sporledera Bernh. (1842) non Hampe (1837).

Small deciduous shrubs or half-shrubs, short-lived perennials or annuals, erect or semi-prostrate. Shape of leaves very variable, circular, ovate, cordate, reniform, lanceolate or trilobed, often polymorphic (upper and lower leaves often much smaller and simplified). Flowers white, pink or lilac, corolla tube funnel-shaped, sub-bilabiate, the anterior lobe enlarged. Fruits laterally compressed coriaceous capsules, loculicidally dehiscent, with two lateral horns (sometimes rather inconspicuous) at the apex. Seeds numerous, compressed, obovate. $2n = 32$. Five species, E and southern Africa.

The genus is very closely related to *Sesamum* sect. *Aptera*. *Ceratotheca sesamoides* Endl. is occasionally grown for its seeds and can be weedy, and *C. triloba* (Bernh.) Hook. f. is grown as an ornamental.

11. *Dicerocaryum* Bojer

Dicerocaryum Bojer, Ann. Sci. Nat. II, 4: 268–269, t. 10 (1835); Abels, Mem. Soc. Brot. 25: 1–358 (1975), monogr.
Pretrea J. Gay (1840).

Perennial herbs with several prostrate annual shoots from a non-succulent thickened primary root, annual shoots often with numerous lateral branchlets. Leaves very variable in shape, narrowly to broadly ovate, pinnipartite to serrate. Flowers on long erect pedicels, white, pink, violet or occasionally yellow; corolla tube campanulate, limb sub-bilabiate, anterior lobe enlarged. Fruits longitudinally much compressed (± disc-like), woody, indehiscent, with two erect conical spines from near the raised centre of the disc (= top of the fruit). Seeds 4–10, obovate to oblanceolate. Three spp., Madagascar, E Africa to southern Africa.

Species of *Dicerocaryum* can be weedy.

12. *Linariopsis* Welw.

Linariopsis Welw., Trans. Linn. Soc. 27: 53 (1869).

Small (10–40 cm tall) prostrate to erect perennial herbs or half-shrubs with thickened persisting non-succulent primary root. Leaves lanceolate, entire. Flowers mauve, corolla tube tubular, constricted at the base, anterior lobe enlarged; gland at the tip of the filament absent. Fruits small (6–10 mm long), woody, obovate with a terminal mucro, indehiscent or tardily dehiscent only at the top, covered with numerous tiny tubercles, these sometimes very indistinct; septum with a large hole. Seeds 2, oblong. Two or three spp., CW Africa and Angola.

13. *Josephinia* Vent.

Josephinia Vent., Jard. Malm.: 67, t. 67 (1804).
Pretreothamnus Engl. (1905).

Low shrubs or half-shrubs, or erect annuals. Leaves obovate or trilobed, usually dentate. Flowers mauve, pale pink or cream, corolla tube campanulate, limb sub-bilabiate, anterior lobe enlarged; carpels occasionally 3–4. Fruits small (1–2 cm long), woody, subglobose, indehiscent, armed with numerous spines. Seeds either 8–12, then obovate and testa rugose, or 4, then oblong and testa smooth. About 4 spp., E Africa (Kenya, Somalia), Malayan Islands, N Australia.

Selected Bibliography

Abels, J. 1975. Die Gattungen *Ceratotheca* Endl. und *Dicerocaryum* Boj. (Monographien der afrikanischen Pedaliaceae III–IV). Mem. Soc. Brot. 25: 1–358.

APG 1998. An ordinal classification for the families of flowering plants. Ann. Missouri Bot. Gard. 85: 531–553.

Bedigian, D., Harlan, J.R. 1986. Evidence of cultivation of sesame in the ancient world. Econ. Bot. 40: 137–154.

Bedigian, D., Seigler, D.S., Harlan, J.R. 1985. Sesamin, sesamol and the origin of sesame. Biochem. Syst. Ecol. 13: 133–139.

Behnke, H.-D. 1991. Distribution and evolution of forms and types of sieve-element plastids in the dicotyledons. Aliso 13: 167–182.

Benhura, M.A.N., Marume, M. 1993. The mucilaginous polysaccharide material isolated from *ruredzo* (*Dicerocaryum zanguebarium*). Food Chem. 46: 7–11.

Bremekamp, C.E.B. 1942. The position of the genus *Thomandersia* Baill. Recueil Trav. Bot. Néerl. 39: 166–175.

Brown, R. 1810. Prodromus Florae Novae Hollandiae et Insulae van-Diemen. London: Johnson.

Bruce, E.A. 1953. Notes on African Pedaliaceae. Kew Bull. 1953: 417–429.

Brummitt, R.K. 1992. Vascular plant families and genera. Kew: Royal Botanic Gardens.

Carlquist, S. 1987. Wood anatomy of Martyniaceae and Pedaliaceae. Aliso 11: 473–483.

Chase, M.W., Soltis, D.E., Olmstead, R.G., Morgan, D., Les, D.H., Mishler, B.D., Duvall, M.R., Price, R.A., Hills, H.G., Qiu, Yin-Long, Kron, K.A., Rettig, J.H., Conti, E., Palmer, J.D., Manhart, J.R., Sytsma, K.J., Michaels, H.J., Kress, W.J., Karol, K.G., Clark, W.D., Hedrén, M., Gaut, B.S., Jansen, R.K., Kim, Ki-Joong, Wimpee, C.F., Smith, J.F., Furnier, G.R., Strauss, S.H., Xiang, Qiu-Yun, Plunkett, G.M., Soltis, P.S., Swensen, S.M., Williams, S.E., Gadek, P.A., Quinn, C.J., Eguiarte, L.E., Golenberg, E., Learn, G.H. Jr., Graham, S.W., Barrett, S.C.H., Dayanandan, S., Albert, V.A. 1993. Phylogenetics of seed plants: an analysis of nucleotide sequences from the plastid gene *rbc*L. Ann. Missouri Bot. Gard. 80: 528–580.

Corner, E.J.H. 1976. The seeds of Dicotyledons, vol. 1. Cambridge: Cambridge University Press.

Endlicher, S. 1832. *Ceratotheca*, eine neue Pflanzengattung aus der Ordnung Sesameen. Linnaea 7: 1–42.

Erdtman, G. 1952. Pollen morphology and plant taxonomy. Angiosperms. Stockholm: Almquist & Wiksell.

Gentry, A.H., Tomb, A.S. 1979. Taxonomic implications of Bignoniaceae palynology. Ann. Missouri Bot. Gard. 66: 756–777.

Glück, H. 1940. Die Gattung *Trapella*. Bot. Jahrb. Syst. 71: 267–336.

Grabow-Seidensticker, H. 1988. Der *Sesamum calycinum*-Komplex (Pedaliaceae R.Br.). Mitt. Inst. Allg. Bot. Hamburg 22: 217–241.

Grayer, R.J., Chase, M.W., Simmonds, S.J. 1999. A comparison between chemical and molecular characters for the determination of phylogenetic relationships among plant families: an appreciation of Hegnauer's "Chemotaxonomie der Pflanzen". Biochem. Syst. Ecol. 27: 369–393.

Honda, M., Sakisaka, M. 1930. Daiko Nippon Shokubutsu Bunruigaku [Syst. Pl. Japon.].

Ihlenfeldt, H.-D. 1967. Über die Abgrenzung und die natürliche Gliederung der Pedaliaceae R.Br. Mitt. Staatsinst. Allg. Bot. Hamburg 12: 43–128.

Ihlenfeldt, H.-D. 1988. Pedaliaceae. In: Launert, E. (ed.) Flora Zambesiaca 8, 3: 86–113. London: Flora Zambesiaca Managing Committee, British Museum Natural History.

Ihlenfeldt, H.-D. 1994. Phytogeography of Pedaliaceae R.Br. In: Proc. 13th Plen. Meet. AETFAT Malawi, vol. 2, pp. 1063–1075.

Ihlenfeldt, H.-D. 2001. Fruits in *Pterodiscus* Hook. and a key to the species. Mitt. Inst. Allg. Bot. Hamburg 28/29: 5–21.

Ihlenfeldt, H.-D., Grabow-Seidensticker, U. 1979. The genus *Sesamum* L. and the origin of the cultivated Sesame. In: Proc. 9th Plen. Meet. AETFAT Las Palmas 1978, pp. 53–60.

Ihlenfeldt, H.-D., Hartmann, H. 1970. Die Gattung *Harpagophytum* (Burch.) DC. ex Meissn. (Monographien der afrikanischen Pedaliaceae II). Mitt. Staatsinst. Allg. Bot. Hamburg 13: 15–69.

Jussieu, A.L. de 1789. Genera Plantarum. Paris: Herissant et Barrois.

Karatela, Y.Y., Gill, L.S. 1984. Phytodermology and ontogeny of stomata in some Pedaliaceae. J. Econ. Tax. Bot. 5: 237–240.

Kobayashi, T., Shimamura, T. 1952. Morphological and cytological studies on induced polyploidy in *Sesamum indicum* L. Jap. J. Genet. 27: 157–171.

Kunth, C. 1818. Révision de la famille des Bignoniacées. J. Phys. Chim. Hist. Nat. Arts 86: 444–454.

Link, J.H.F. 1829. Handbuch zur Erkennung der nutzbarsten und am häufigsten vorkommenden Gewächse, vol. 1. Berlin: Haude & Spener.

Nayar, N.M. 1976. Sesame. In: Simmonds, N.W. (ed.) Evolution of crop plants. London: Longman.

Oliver, D. 1887. *Trapella sinensis*, Oliv. Hooker's Icon. Pl. 16: t. 1595.

Oliver, F.W. 1888. On the structure, development, and affinities of *Trapella*, Oliv., a new genus of Pedalineae. Ann. Bot. 2: 75–15.

Olmstead, R.G., Reeves, P.A. 1995. Evidence for the polyphyly of the Scrophulariaceae based on chloroplast *rbc*L and *ndh*F sequences. Ann. Missouri. Bot. Gard. 82: 176–193.

Olmstead, R.G., dePamphilis, W., Wolfe, A.D., Young, N.D., Ellison, W.J., Reeves, P.A. 2001. Disintegration of the Scrophulariaceae. Am. J. Bot. 88: 348–361.

Raj, G. 1961. Pollen morphological studies in the Acanthaceae. Grana Palynol. 3: 3–108.

Raven, P.H., Axelrod, D. 1974. Angiosperm biogeography and past continental movements. Ann. Missouri. Bot. Gard. 61: 539–673.

Savolainen, V., Fay, M.F., Albach, D.C., Backlund, A., van der Bank, M., Cameron, K.M., Johnson, S.A., Lledó, M.D., Pintaud, J.-C., Powell, M., Sheahan, M.C., Soltis, D.E., Soltis, P.S., Weston, P., Whitten, W.M., Wurdack, K.J., Chase, M.W. 2000. Phylogeny of the eudicots: a nearly complete familial analysis based on *rbc*L gene sequences. Kew Bull. 55: 257–309.

Seegeler, C.J.P. 1983. Oil plants in Ethiopia, their taxonomy and agricultural significance. Wageningen: PUDOC.

Seegeler, C.J.P. 1989. *Sesamum orientale* (Pedaliaceae): Sesame's correct name. Taxon 38: 656–658.

Singh, S.P. 1958. Morphological studies in some members of the family Pedaliaceae. Thesis. Agra University.

Singh, S.P. 1960. Morphological studies in some members of the Family Pedaliaceae. Agra Univ. J. Res., Sci. 9(2): 218–220.

Singh, S.P. 1963. Morphological studies in some members of the family Pedaliaceae. II. *Pedalium murex* L. Agra Univ. J. Res., Sci 12(1): 143–161.

Stapf, O. 1895. Pedaliaceae. In: Engler & Prantl, Die natürlichen Pflanzenfamilien IV, 3b. Leipzig: Engelmann, pp. 253–265.

Straka, H., Ihlenfeldt, H.-D. 1965. Pollenmorphologie und Systematik der Pedaliaceae R.Br. Beitr. Biol. Pflanzen 41: 175–207.
Trehan, K.B., Chand, H., Mehta, S.K., Baijal, S.K. 1974. Studies on floral biology in sesame (*Sesamum indicum* L.). Oilseeds J. 4: 7–11.
Vogel, S. 1954. Blütenbiologische Typen als Elemente der Sippengliederung. Bot. Stud. 1. Jena: Fischer.
Watt, J.M., Breyer-Brandwijk, M.G. 1962. The medical and poisonous plants of southern and eastern Africa. Edinburgh: Livingstone.
Wegener, T. 1998. Die Teufelskralle (*Harpagophytum procumbens* DC.) in der Therapie rheumatischer Erkrankungen. Zeitsch. Phytotherapie 19: 284–294.

Phrymaceae[1]

P.D. Cantino

Phrymaceae Schauer in A.DC., Prodr. 11: 520 (1847), nom. cons.

Perennial herbs from short, vertical rhizome, non-aromatic, with simple hairs; mature root system wholly adventitious; stem erect, usually unbranched, basally terete, quadrangular upward, usually swollen above lower nodes. Leaves opposite, petiolate, with pinnate venation. Inflorescence racemose, terminal and axillary. Flowers perfect, hypogynous, short-pedicellate, with 2 prophylls, becoming sharply reflexed soon after anthesis; calyx synsepalous, zygomorphic, accrescent, 5-nerved, 5-lobed, 3 posterior lobes subulate-uncinate, indurated in fruit, 2 anterior lobes triangular, calyx shed with fruit enclosed; corolla white and lavender, 2-lipped, posterior lip upturned, usually emarginate, anterior lip much longer, ± horizontal, 3-lobed; stamens 4, inserted on corolla, didynamous (anterior pair longer), included, filaments glabrous, anther-thecae parallel to divergent; pollen tricolpate; gynoecium syncarpous, 2-carpellate, pseudomonomerous, ovary 1-locular, 1-ovular, ovule subbasal, hemitropous, style terminal, glabrous, persistent after corolla abscission but not to fruit maturity, stigma lobes 2, unequal. Fruit an achene, pericarp smooth, glabrous; seed coat vestigial, endosperm scanty, cotyledons convolute.

The only species occurs disjunctly in mesic forests of eastern North America and eastern Asia.

Vegetative Morphology and Anatomy. The detailed descriptions by Holm (1913) and Whipple (1972) are the source of the following summary, unless otherwise indicated. Germination is hypogeous, the cotyledons remaining inside the persistent calyx. The small primary root is short-lived, soon replaced by a diffuse, adventitious root system. The roots are not mycorrhizal. The vertical, subterranean portion of the stem bears both the adventitious roots and the overwintering buds. Two buds form each fall, one of which develops into a new shoot the following spring, the other remaining dormant. The stem in mature plants is usually swollen above the lower nodes. The swellings may facilitate the bending of the stem, helping it maintain a vertical orientation (Whipple 1972). The primary vascular tissue in the stem is arranged in six collateral bundles, except at the base of the stem where the vascular cylinder is continuous. An interfascicular cambium produces a small amount of secondary vascular tissue and is bordered internally by a region of sclerenchyma. The pith is hollow in the lower aerial internodes. The nodes are unilacunar, with a 3-parted leaf trace.

The leaves are opposite, petiolate, and mostly ovate, with margins coarsely and often doubly dentate to crenate. The mesophyll is bifacial, with a palisade layer one cell thick and a spongy layer two to four cells thick. Collenchyma is limited to a few hypodermal layers above and below the major veins. The leaves are hypostomatic. Stomata are mostly anomocytic, with some anisocytic stomata and rarely a few diacytic ones present (Cantino 1990, and unpubl. data).

Stems and leaves bear both glandular and uniseriate, non-glandular trichomes. Glandular trichomes consist of a two-celled cap borne on a short foot cell ("subsessile") or a uniseriate stalk (Holm 1913; Cantino 1990, and unpubl. data). The stalked glandular hairs occur on the stem, particularly in the inflorescence. Leaf glands are all subsessile.

Floral Morphology and Anatomy. The flowers are arranged in racemes (Fig. 39A), each flower subtended by a bract, with a pair of prophylls at the base of the pedicel. The flowers are more or less horizontal at anthesis but become sharply reflexed 24–48 hours after corolla abscission (Thieret 1972) and remain so until fruit maturity (Fig. 39F); hence, the common name "lopseed." Floral vascularisation and organogenesis were described by Whipple (1972) and Ramana et al. (1983).

The flowers (Fig. 39B) are complete and hypogynous with a pentamerous perianth. The synsepalous calyx is markedly zygomorphic but not

[1] See also Scrophulariaceae.

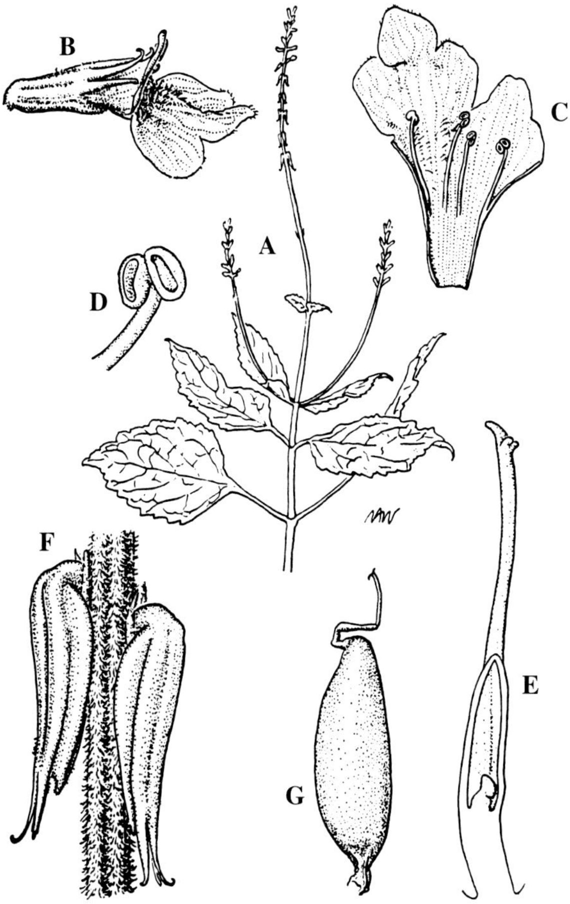

Fig. 39. Phrymaceae. *Phryma leptostachya*. A Upper part of plant with flowers and young fruit. B Flower. C Corolla opened. D Anther. E Vertical section of gynoecium. F Nearly mature fruiting calyx. G Fruit. (Thieret 1972)

bilabiate because all lobes arise at the same level on the tube. The calyx (Fig. 39B) is cylindrical in flower, 5-nerved, the two anterior lobes triangular, the three posterior lobes much longer and subulate-uncinate. The fruiting calyx is lanceolate and sulcate, with indurated posterior lobes (Fig. 39F). The white and lavender corolla (Fig. 39B, C) is sympetalous and bilabiate. The posterior lip is upturned and usually emarginate. The anterior lip is much longer, more or less horizontal, 3-lobed, the middle lobe convex. The corolla tube is straight and cylindrical. The four stamens are inserted on the corolla, didynamous (the anterior pair longer), and included within the corolla (Fig. 39C). The filaments are glabrous. The anthers (Fig. 39D) are tetrasporangiate and dithecal, and the thecae parallel to divergent. The pistil (Fig. 39E) is pseudomonomerous, i.e. bicarpellate but with a unilocular ovary due to suppression of the abaxial carpel (Whipple 1972; Ramana et al. 1983). The single ovule is subbasal and hemitropous. It begins its development at the base of the carpel but shifts to a lateral position due to the more rapid growth of the abaxial side of the ovary (Cooper 1941). The style is terminal, glabrous, and persistent after corolla abscission but not until fruit maturity. The two stigma lobes are unequal.

Embryology. The mature ovule is epitropous, unitegmic, and tenuinucellar (Cooper 1941; Thieret 1972; Whipple 1972). It is initially orthotropous (Cooper 1941) but becomes hemitropous when mature (Whipple 1972). Embryo sac development is of the Polygonum type, resulting in the usual seven-celled megagametophyte (Cooper 1941). See Cooper (1941) for additional information on ovule development and fertilisation. At maturity, the anther wall consists of an inner endothecium with thickened bands and an outer epidermis (Whipple 1972). The pollen is 2-celled when shed (Whipple 1972).

Pollen Morphology. Pollen grains are tricolpate, suboblate to oblate spheroidal, with a circular amb, tectate-perforate to microreticulate exine, and simple columellae (Chadwell et al. 1992).

Karyology. The chromosome number is variably reported as $2n = 28$ (Sugiura 1936; Cooper 1941; Löve and Löve 1982; Probatova et al. 1991) and $2n = 14$ (Vasudevan 1975).

Pollination. The flowers are protandrous (Thieret 1972). They open in the early morning and are withered by early afternoon, with fertilisation occurring 6–8 hours after pollination (Cooper 1941). According to Robertson (1892), *Phryma* is pollinated by small bees; Thieret's (1972) observations suggest that the flowers are infrequently visited.

Fruit and Seed Structure and Dispersal. The fruit (Fig. 39G) is an achene enclosed in the accrescent calyx. The seed coat, reduced to a cuticular layer, is closely appressed to the thin pericarp (Whipple 1972). Endosperm is ab initio cellular, with a 4-celled chalazal haustorium that extends into the funiculus, forming a bridge between the developing endosperm and the conducting cells of the ovule (Cooper 1941; Thieret 1972). At seed maturity, the endosperm consists of only two layers of cells surrounding the embryo (Cooper 1941; Whipple 1972), which has convolute cotyledons (Thieret 1972).

The dispersal unit is the accrescent calyx (Fig. 39F) containing a single achene. Dispersal is presumably epizoochorous (Holm 1913; Ridley 1930), the hooked upper calyx-lobes serving to attach the fruit to clothing and animal fur.

Phytochemistry and Economic Importance. Ground root of *Phryma* is used to kill houseflies in Japan (Taniguchi and Oshima 1972a; Tseng 1973), and extracts of all parts of the plant are toxic to mosquito larvae (Patterson et al. 1975). Four lignans extracted from the roots of *Phryma* (phrymarolin-I and -II, haedoxan A, and leptostachyol acetate) are either insecticidal or synergistic with commercial insecticides (Taniguchi and Oshima 1972a, b; Tseng 1973; Taniguchi et al. 1989). Otherwise, little is known about the phytochemistry of Phrymaceae (Hegnauer 1969). The only economic importance of Phrymaceae is as a possible source of insecticides.

Affinities. *Phryma* has been widely regarded as related to Verbenaceae and has often been submerged within that family (e.g. Bentham and Hooker 1876; Cronquist 1981), but it differs from Verbenaceae in having a pseudomonomerous gynoecium that develops into an achene. Other authors (e.g., Benson 1957) have placed the family in the Scrophulariales, as an order separate from Lamiales.

Whipple (1972) and Ramana et al. (1983) proposed that *Phryma* be given tribal status in Verbenaceae. In vegetative and inflorescence morphology, *Phryma* resembles various Verbenaceae, particularly *Stachytarpheta* of tribe Lantaneae (Whipple 1972). There are also similarities in floral vascularisation and gynoecial morphology between *Phryma* and Lantaneae (Whipple 1972; Ramana et al. 1983). In both taxa, the abaxial carpel is suppressed, and it has been hypothesised (Whipple 1972; Ramana et al. 1983) that the unilocular ovary of *Phryma* evolved by the further suppression of one of the two locules in the remaining carpel. However, Chadwell et al. (1992) demonstrated that the pollen morphology of *Phryma* is inconsistent with a close relationship to Lantaneae; they suggested that *Phryma* is closely related to, but not a member of, Verbenoideae (= Verbenaceae, as circumscribed in this volume).

A cladistic analysis of *rbc*L sequence data (Wagstaff and Olmstead 1997) found *Phryma* to be part of a large polytomy, which also includes representatives of many other families of Lamiales and Scrophulariales. *Phryma* did not emerge as sister group to Verbenaceae; in the strict consensus tree, Verbenaceae are more closely related to several other families than to Phrymaceae. There is recent molecular evidence (Beardsley and Olmstead 2002) that the sister group of *Phryma* lies within *Mimulus* (traditionally placed in Scrophulariaceae); thus, *Mimulus* is paraphyletic. These authors suggest that Phrymaceae be expanded to include *Mimulus* and eight related genera (*Berendtiella*, *Elacholoma*, *Glossostigma*, *Hemichaena*, *Lancea*, *Leucocarpus*, *Mazus*, and *Peplidium*).

Distribution. The single species of Phrymaceae occurs in mesic forests, with a disjunct distribution in eastern North America (New Brunswick to Manitoba, south to northern Florida and eastern Texas) and eastern Asia (Himalayas to Manchuria, southeastern Siberia, Japan, and Vietnam) (Thieret 1972). The Asian and American plants are usually treated as varieties. They are only marginally different in morphology and have the same chromosome number. They show similar variation in the shape, size, and pubescence of the leaves (Hara 1962), and in pollen morphology (Chadwell et al. 1992). At anthesis, the upper calyx-lobes of most Asian plants are ≤2 mm long, while those of most American plants are >2 mm long.

Although their morphological divergence has been minimal, the degree of genetic divergence of the Asian and North American varieties of *Phryma*, as measured by allozyme genetic identity and DNA (ITS) sequence divergence, led Lee et al. (1996) to estimate that the varieties have been disjunct for 12.3 to 25 million years.

Only one genus:

Phryma L. Fig. 39

Phryma L., Sp. Pl. 2: 601 (1753); Gen. Pl. ed. 5: 262 (1754).

Characters as for family.
One species, *Phryma leptostachya* L.

Selected Bibliography

Beardsley, P.M., Olmstead, R.G. 2002. Redefining Phrymaceae: the placement of *Mimulus*, tribe Mimuleae, and *Phryma*. Am. J. Bot. 89: 1093–1102.

Benson, L. 1957. Plant classification. Boston: D.C. Heath & Co.

Bentham, G., Hooker, J.D. 1876. Genera Plantarum, vol. 2. London: Reeve & Co.

Cantino, P.D. 1990. The phylogenetic significance of stomata and trichomes in the Labiatae and Verbenaceae. J. Arnold Arbor. 71: 323–370.

Chadwell, T.B., Wagstaff, S.J., Cantino, P.D. 1992. Pollen morphology of *Phryma* and some putative relatives. Syst. Bot. 17: 210–219.

Cooper, D.C. 1941. Macrosporogenesis and the development of the seed of *Phryma leptostachya*. Am. J. Bot. 28: 755–761.

Cronquist, A. 1981. The evolution and classification of flowering plants. Boston: Houghton Mifflin Co.

Hara, H. 1962. Racial differences in widespread species with special reference to those common to Japan and North America. Am. J. Bot. 49: 647–652.

Hegnauer, R. 1969. Chemotaxonomie der Pflanzen. Vol. 5. Basel: Birkhäuser.

Holm, T. 1913. *Phryma leptostachya* L., a morphological study. Bot. Gaz. 56: 306–318.

Lee, N.S., Sang, T., Crawford, D.J., Yeau, S.H., Kim, S.C. 1996. Molecular divergence between disjunct taxa in eastern Asia and eastern North America. Am. J. Bot. 83: 1373–1378.

Löve, A., Löve, D. 1982. IOPB chromosome number reports LXXV. Taxon 31: 344–360.

Patterson, B.D., Wahba Khalil, S.K., Schermeister, L.J., Quraishi, M.S. 1975. Plant-insect interactions. I. Biological and phytochemical evaluation of selected plants. Lloydia 38: 391–403.

Probatova, N.S., Sokolovskaya, A.P., Rudyka, E.G. 1991. Chromosome numbers in some species of vascular plants from the Soviet far east and other regions of the USSR. Bot. Zhurn. 76: 1174–1178.

Ramana, R.V., Prakasa Rao, P.S., Narayana, L.L. 1983. A contribution to the floral anatomy of *Phryma leptostachya* L. Curr. Sci. 52: 922–924.

Ridley, H.N. 1930. The dispersal of plants throughout the world. Ashford: Reeve & Co.

Robertson, C. 1892. Flowers and insects. VII. Bot. Gaz. 17: 65–71.

Sugiura, T. 1936. Studies on the chromosome numbers in higher plants with special reference to cytokinesis I. Cytologia 7: 544–595.

Taniguchi, E., Oshima, Y. 1972a. Phrymarolin-I, a novel lignan from *Phryma leptostachya* L. Agric. Biol. Chem. 36: 1013–1025.

Taniguchi, E., Oshima, Y. 1972b. New gmelinol-type lignan, leptostachyol acetate. Tetrahendron Lett. 8: 653–656.

Taniguchi, E., Imamura, K., Ishibashi, F., Matsui, T., Nishio, A. 1989. Structure of the novel insecticidal sesquilignan, Haedoxan A. Agric. Biol. Chem. 53: 631–643.

Thieret, J.W. 1972. The Phrymaceae in the southeastern United States. J. Arnold Arbor. 53: 226–233.

Tseng, A.L. 1973. A phytochemical study of *Phryma leptostachya* L. and *Echinocystis lobata* Michx. M.Sc. Thesis. Fargo: North Dakota State University.

Vasudevan, K.N. 1975. Contribution to the cytotaxonomy and cytogeography of the flora of the western Himalayas (with an attempt to compare it with the flora of the Alps). Part II. Ber. Schweiz. Bot. Ges. 85: 210–252.

Wagstaff, S.J., Olmstead, R.G. 1997. Phylogeny of Labiatae and Verbenaceae inferred from *rbc*L sequences. Syst. Bot. 22: 165–179.

Whipple, H.L. 1972. Structure and systematics of *Phryma leptostachya* L. J. Elisha Mitchell Sci. Soc. 88: 1–17.

Plantaginaceae[1]

Andrea E. Schwarzbach

Plantaginaceae Juss., Gen. Pl.: 89 (1789), nom. cons. ("Plantagines").

Herbs or occasionally small shrubs, terrestrial (to aquatic). Leaves mostly all basal and alternate, rarely cauline and alternate or opposite, simple or rarely pinnately nerved and lobed, often sheathing at the base, hairs present in the leaf axils; blade phyllodial and more or less parallel-veined, usually entire, linear, lanceolate, oblanceolate, ovate, or orbicular, sometimes succulent. Flowers in spikes or heads, rarely solitary, bracteate or ebracteate, small, mostly actinomorphic and perfect, usually protogynous and wind-pollinated; sepals (3)4(5), imbricate, gamosepalous or partially so, calyx lobes unequal in *Plantago* subg. *Bougueria*; petals 4(5), gamopetalous; stamens (1)4, adnate to the corolla, opposite the calyx lobes, filaments mostly long; anthers versatile, introrse, dehiscing via longitudinal slits. Gynoecium bicarpellate, syncarpous, style 1, stigma usually 2-lobed, often plumose; ovary superior, mostly bilocular; placentation axile with 2–50 ovules or basal with one ovule (*Plantago* subg. *Bougueria* and subg. *Littorella*); ovules unitegmic, tenuinucellate. Fruit a capsule or nut, capsules circumscissile; seeds straight or rarely curved, concave, mostly small, endosperm abundant.

The family is cosmopolitan with one genus and about 270 species.

Vegetative Morphology and Anatomy. Most species form a basal rosette with leafless inflorescence stalks (Fig. 40A); others are erect herbs with a multi-branched stem (Kuiper and Bos 1992). Trichomes, common on leaves, scapes and floral parts, have been studied intensively in the Plantaginaceae (e.g., Rahn 1992, 1996; Andrzejewska-Golec and Swietoslawski 1993). Trichomes are non-glandular or glandular with a head of one or a few cells. Except for the head, the hairs are mostly uniseriate, but vary in the number, length and mode of connection of the component cells, as well as in the thickness of the wall and the width of the lumen (Rahn 1992).

Flower, Seed and Fruit Dispersal. Flowers are arranged in spikes (Fig. 40A), rarely heads or solitary; they are small and mostly scentless and without nectar. Often long filaments raise the anthers well above the floral tube (Fig. 40B). The anthers are mostly large with abundant pollen, except in a few cleistogamous species. Rahn (1992) described an extension of the anther connective as a character common to all Plantaginaceae. The anther extension is small in some subgenera (subg. *Litorella* and subg. *Bougueria*) but constitutes a large, flap-like structure in others. Some insect-pollinated species have colored filaments and are fragrant. Placental growth in the ovary can separate seeds completely from each other in some species with small seed number. Individual plants with a large numbers of seeds per capsule (Fig. 40C) can produce large quantities of seeds (Casper 1975; 40 000 seeds/plant/growing season in *P. major*). Van der Pijl (1982) lists endozoochory (seeds eaten by birds) and epizoochory (mucilaginous seeds sticking to animal vectors) as the main dispersal strategies.

Pollen Morphology, Pollination, Reproductive Systems. Pollen characters do not vary substantially within the family. All pollen is spheroidal and panto-polyporate (Dietrich 1969) with 4–15 apertures (Rahn 1992). Most *Plantago* species are wind-pollinated. However, insect-pollination is found as well. Pollinators are syrphid and muscid flies, as well as certain bees and beetles (Leerefeld et al. 1976; Sharma et al. 1993). There is a great deal of variation in breeding system among the species. Outcrossing rates vary from 0 to 100% (Wolff et al. 1988). Flowers of many species partially avoid selfing by protogyny. However, the younger, upper flowers in the inflorescence may get pollinated by the more mature, pollen-producing flowers in the lower part of the inflorescence.

Karyology. Chromosome numbers vary extensively in the family. The basic chromosome numbers in the family are x = 6, 5 or 4, but many

[1] See also Scrophulariaceae.

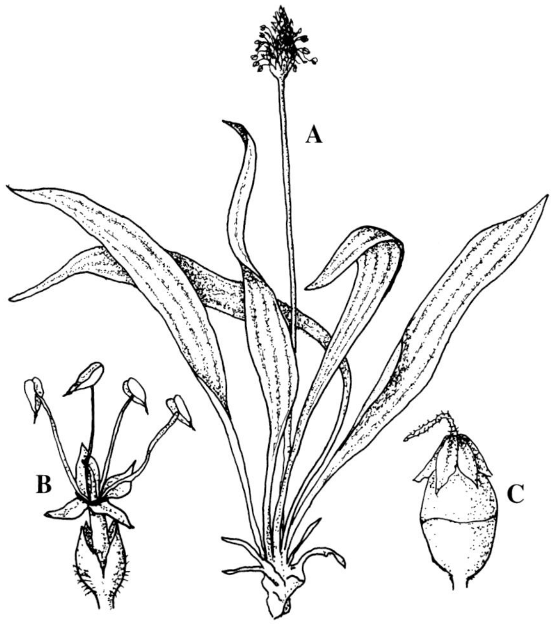

Fig. 40. Plantaginaceae. *Plantago lanceolata*. **A** Habit. **B** Individual flower. **C** Mature fruit, dried corolla parts remain, calyx removed.

polyploids are found with up to 16x in some individuals of *P. altissima* (Dietrich 1980).

Phytochemistry. *Plantago* produces a wide variety of chemical compounds. Since leaves and seeds of some *Plantago* species are widely used for medicinal purposes, their chemistry is well studied. The main groups of components are aucubin and related iridoid glucosides (Taskova et al. 2002), alkaloids, terpenes and saponins. The mucilaginous seeds contain large quantities of xylose and galacturonic acid (Hegnauer 1969).

Affinities. Plantaginaceae have been placed either in their own order or regarded as part of Scrophulariaceae. Recently, Judd et al. (2002) presented a new circumscription of Plantaginaceae containing many genera (113) of the former Scrophulariaceae, such as *Penstemon*, *Veronica*, and *Anthirrinum*. Olmstead et al. (2001) listed *Plantago* in a similar clade which they named Veronicaceae.

Distribution and Habitat. Members of the family show a cosmopolitan distribution in aquatic (*P.* subg. *Littorella*), high Andean (*P.* subg. *Bougueria*), coastal (e.g., *P. coronopus*, *P. maritima*), and hot desert habitats. Several species such as *P. major* and *P. lanceolata* are weedy with a worldwide distribution.

Paleobotany. Seeds that resemble those of modern *Plantago* have been described from the Pliocene of Poland (Szafer 1954). Pollen attributed to *Plantago* was reported from the Middle to Late Miocene and more recent deposits (summarized in Muller 1981).

Economic Importance. Laxatives are extracted from the seeds of *Plantago psyllium* ("Psyllium") and *P. ovata* ("isphagul" seeds). Psyllium, a colloid mucilage, is contained in the seed coat. The seed coats are removed, dried and ground to a fine powder. The powder absorbs water in the intestinal tract, producing a bulky mass that is unaffected by bacteria as it moves through the intestines. It provides relief from constipation and chronic diarrhea. Chemical compounds in the leaves of *Plantago major* are used in traditional medicine for treatment of a wide variety of ailments, such as a wound healing remedy, and to combat urinary tract infections and tumors (reviewed in Samuelsen 2000).

Only one genus:

1. *Plantago* L. Fig. 40

Plantago L., Sp.Pl.:112 (1753).
Littorella P.J. Bergius (1768).
Psyllium Juss. (1789).
Asterogeum Gray (1821).
Bougueria Decne. (1836).

Description as for the family. Rahn (1996) has included the previously separate genera *Littorella* P.J. Bergius (Dietrich 1971) and *Bougueria* Decne. (Dietrich 1970) in *Plantago* as *P.* subg. *Littorella* (P.J. Bergius) Rahn and *P.* subg. *Bougueria* (Decne.) Rahn. This decision was based on a phylogenetic study of the family using morphological, embryological, and chemical data. More recently, preliminary DNA sequence data were presented supporting these changes (Rønsted et al. 2002). However, subg. *Litorella* is a monophyletic sister group to the rest of *Plantago* and some authors prefer the separation into two different genera, *Plantago* and *Littorella* (Hoggard et al. 2003). Four additional subgenera are recognized by Rahn (1996) – *P.* subg. *Coronopus* (Lam. & D.C.) Rahn, *P.* subg. *Plantago*, *P.* subg. *Albicans* Rahn and *P.* subg. *Psyllium* (Juss.) Harms and Reiche. Rønsted et al. (2002) identified subg. *Albicans* as paraphyletic and suggested to submerge subg. *Albicans* into subg. *Psyllium.*

Selected Bibliography

Andrzejewska-Golec, E., Swietoslawski, J. 1993. Hair anatomy in *Plantago* subg. *Psyllium* (Plantaginaceae). Plant Syst. Evol. 184: 113–123.

Casper, S.J. 1975. Plantaginaceae. In: Hegi, G. Illustrierte Flora von Mitteleuropa, 2. Aufl., vol. VI/1, pp. 559–608.

Dietrich, H. 1969. Über die Aussagekraft pollenmorphologischer Fakten bei stenopalynen Verwandtschaftsbereichen am Beispiel der Plantaginaceae. Feddes Repert. 79: 347–353.

Dietrich, H. 1970. *Bougueria nubicola* Decne. - eine interessante Plantaginaceen-Gattung. Wiss. Zeitschr. Friedrich-Schiller-Univ. Jena, Math.-Nat. R. 19: 293–305.

Dietrich, H. 1971. Blütenmorphologische und palynologische Untersuchungen an *Littorella*. Feddes Repert. 82: 155–165.

Dietrich, H. 1980. Cytologische Untersuchungen innerhalb der Familie der Plantaginaceae III. Cytotaxonomische Ergebnisse. Wiss. Zeitschr. Friedrich-Schiller-Univ. Jena, Math.-Nat. R. 29: 559–589.

Hegnauer, R. 1969. Chemotaxonomie der Pflanzen. Vol. 5. Basel: Birkhäuser, pp. 330–337.

Hoggard, R.K., Kores, P.J., Molvray, M., Hoggard, G.D., Broughton, D.A. 2003. Molecular systematics and biogeography of the amphibious genus *Littorella* (Plantaginaceae). Am. J. Bot. 90: 429–435.

Judd, W.S., Campbell, C.S., Kellogg, E.A., Stevens, P.F., Donoghue, M.J. 2002. Plant systematics. A phylogenetic approach, 2nd edn. Sunderland, Mass.: Sinauer.

Kuiper, P.J.C., Bos, M. (eds.) 1992. *Plantago*: a multidisciplinary study. Berlin Heidelberg New York: Springer.

Leerefeld, H., Meeuse, A.D.J., Stelleman, P. 1976. Anthecological relations between reputedly anemophilous flowers and syrphid flies. II. *Plantago media* L. Acta Bot. Neerl. 25: 205–211.

Muller, J. 1981. Fossil pollen record of extant angiosperms. Bot. Rev. 47: 1–142.

Olmstead, R.G., DePamphilis, C.W., Wolfe, A.D., Young, N.D., Elisons, W.J., Reeves, P.A. 2001. Disintegration of the Scrophulariaceae. Am. J. Bot. 88: 348–361.

Rahn, K. 1992. Trichomes within Plantaginaceae. Nord. J. Bot. 12: 3–12.

Rahn, K. 1996. A phylogenetic study of the Plantaginaceae. Bot. J. Linn. Soc. 120: 145–198.

Rønsted, N., Chase, M.W., Albach, D.C., Bello, M.A. 2002. Phylogenetic relationships within *Plantago* (Plantaginaceae): evidence from nuclear ribosomal ITS and plastid *trn*L-F sequence data. Bot. J. Linn. Soc. 139: 323–338.

Samuelsen, A.B. 2000. The traditional uses, chemical constituents and biological activities of *Plantago major* L. A review. J. Ethnopharmacol. 71: 1–21.

Sharma, N., Koul, P., Koul, A.K. 1993. Pollination biology of some species of genus *Plantago* L. Bot. J. Linn Soc. 111: 129–138.

Szafer, W. 1954. Pliocene flora from the vicinity of Czorsztyn (West Carpathians) and its relationship to the Pleistocene. Inst. Geol. Prace, XI. Warsaw: Wydawnictwa Geologiczne.

Taskova, R., Evstatieva, L., Handjieva, N., Popov, S. 2002. Iridoid patterns of genus *Plantago* L. and their systematic significance. Z. Naturforsch. 57: 42–50.

Van der Pijl, L. 1982. Principles of dispersal in higher plants. Berlin Heidelberg New York: Springer.

Wolff, K., Friso, B., van Damme, J.M.M. 1988. Outcrossing rates and male sterility in natural populations of *Plantago coronopus*. Theor. Appl. Gen. 76: 190–196.

Plocospermataceae

LENA STRUWE and S.R. JENSEN

Plocospermataceae Hutch., Fam. Fl. Pl., ed. 3: 469 (1973).

Terrestrial dioecious shrubs or small trees, much branched, partially pubescent. Leaves (sub)opposite, simple, ovate-oblong, obtuse, short-petiolate, without stipules. Inflorescence subtended by two leaves, axillary, in 1–7-flowered congested racemes or dichasia, often reduced to only 1–2 flowers. Flowers bisexual, but functionally unisexual, hypogynous, 5(6)-merous; corolla sympetalous, imbricate. Stamens 5(6), inserted in middle of corolla tube alternating with corolla lobes; filaments free; anthers free, basifixed, longitudinally dehiscent, orientation introrse to extrorse, 2- or 4-celled; functionally female flowers with stamens but without viable pollen. Ovary 1-celled, stipitate, with nectariferous disk at base of ovary in female flowers; placentas 2, parietal; ovules 2–4, in two basal-parietal pairs, one on each side; in functionally male flowers the ovary is smaller and there is no nectary disc, style, or stigma; style elongate, deciduous, twice dichotomously divided into four slender branches; stigma apical. Capsule bivalved, with 1–4 seeds. Seeds with tuft of hairs (coma) on chalazal end.

This monotypic family occurs in northern Meso-America (Costa Rica, Guatemala, and Mexico).

VEGETATIVE MORPHOLOGY AND ANATOMY. The wood has numerous, crowded vessels, solitary or in radial rows of 4–12, simple perforations, alternate intervascular pitting, and many small and elliptical pits (D'Arcy and Keating 1973). There are numerous (mostly) uniseriate and heterocellular rays. The axial parenchyma is scarce, lignified, and includes some druses and crystals (not seen in other parts of the wood). Intraxylary (internal) phloem is lacking. The fiber-tracheids are thick-walled, often septate, and have circular-bordered pitting. The leaf nodes are unilacunar.

The coriaceous leaves are entire to slightly dentate. The leaf anatomy was investigated by D'Arcy and Keating (1973), who reported the absence of sclereids and secretory cells in the mesophyll. Indumentum on the leaf surfaces occurs as either long, unbranched, narrow hairs, or as shorter, unicellular, pyramid-shaped, and sharp-pointed trichomes. Glandular hairs are also present on the leaf lamina, but colleters (multicellular, finger-shaped glands) are absent from this plant. The stomata are anomocytic and a thick cuticle is present. Druses and extraxylary fibers are common along the veins in the leaves, and venation is pinnate with a strongly developed midrib.

INFLORESCENCE, FLORAL STRUCTURE AND FLOWER ANATOMY. The flowers are 5-merous, positioned 1–7 together on very short peduncles and pedicels in the leaf axils (Fig. 41A), and are functionally unisexual (Chiang and Frame 1987). The campanulate calyx has simple hairs on the outside and subequal, imbricate lobes with ciliate margins (D'Arcy and Keating 1973). The blue-violet to purple corolla is broadly funnel-shaped to campanulate, slightly zygomorphic, has non-glandular hairs on the outer surface and is glabrous within. The obtuse corolla lobes are shorter than the corolla tube, and the anthers are inserted halfway down the length of the corolla tube, alternating with the corolla lobes. The staminal filaments are of about equal length, and are tipped with deeply sagittate, introrse to extrorse, and versatile anthers.

Plocosperma has bisexual flowers that are functionally unisexual and occur on male and female plants, respectively. The presence of two types of flowers in *Plocosperma* was not noted until rather recently. Previously plants with male flowers were known as the genus *Lithophytum*, and the name *Plocosperma* was used only for plants with female flowers. Chiang and Frame (1987) showed that *Lithophytum* is identical with *Plocosperma* and only differs in having functionally male flowers. *Lithophytum* was described by Brandegee (1911) and it was initially (and hesitantly) assigned to Solanaceae. It was transferred to Verbenaceae by D'Arcy and Keating (1973; in its own tribe), and later to Hydrophyllaceae (Cronquist 1981).

The floral anatomy of functionally male flowers (lacking nectary disk, style and stigma) was investigated by D'Arcy and Keating (1973). The flowers have a vascular system consisting of ten calyx traces (calyx laterals are fused) originating from a circular stele. This is followed by five petal traces, then five stamen traces, and finally six vascular traces to the gynoecium. The unilocular ovary was interpreted as a reduced 4-carpellate structure by D'Arcy and Keating (1973), based on the number of gynoecial traces. Cronquist (1981) disagreed with this interpretation, and suggested two carpels as the ancestral state. The number of placentas is also 2.

Functionally female flowers have a characteristic style that is dichotomously divided into four style branches (Fig. 41B; Endress et al. 1996). In the ovary, the 2–4 ovules are subapical-parietal and pendulous, or sub-basal and erect (Leeuwenberg 1967). A nectariferous disk is present at the base of the stipitate, 1-celled ovary. Stamens are present but the pollen is not viable.

Pollen Morphology. The pollen is shed as tricolporate monads, which have very long colpi and a reticulate exine (Punt 1980). Fresh pollen grains are viscid (D'Arcy and Keating 1973).

Pollination and Reproductive System. No reports on possible pollinators of *Plocosperma* have been published, but the plants are almost certainly animal-pollinated. Supporting this is their dioecy, the presence of a nectary in the female flower, the viscid pollen, and the well-developed, open corolla.

Fruit, Seed and Dispersal. The thin, linear, and long capsule has usually only one seed (rarely up to four) with a straight embryo and fleshy endosperm. The long and narrow seeds have a sessile tuft of multicellular hairs at the apex (Fig. 41C) – hence, the name *Plocosperma*). The tufted seeds of *Plocosperma* are presumably wind- or animal-dispersed, but no field observations have been made.

Phytochemistry. Only a single report on the chemistry of *Plocosperma* has appeared (Jensen 1992). This reported the presence of the quinol glycoside cornoside as well as two caffeoyl phenylethanoid glycosides (CPGs), namely lugrandoside and echinacoside belonging to the verbascoside group. CPGs have a distribution limited to Lamiales (Jensen 1992), where they occur in most taxa, although with a few systematic exceptions (Lamiaceae: Nepetoideae and Verbenaceae: *Vitex*; Pedersen 2000). CPGs have rarely

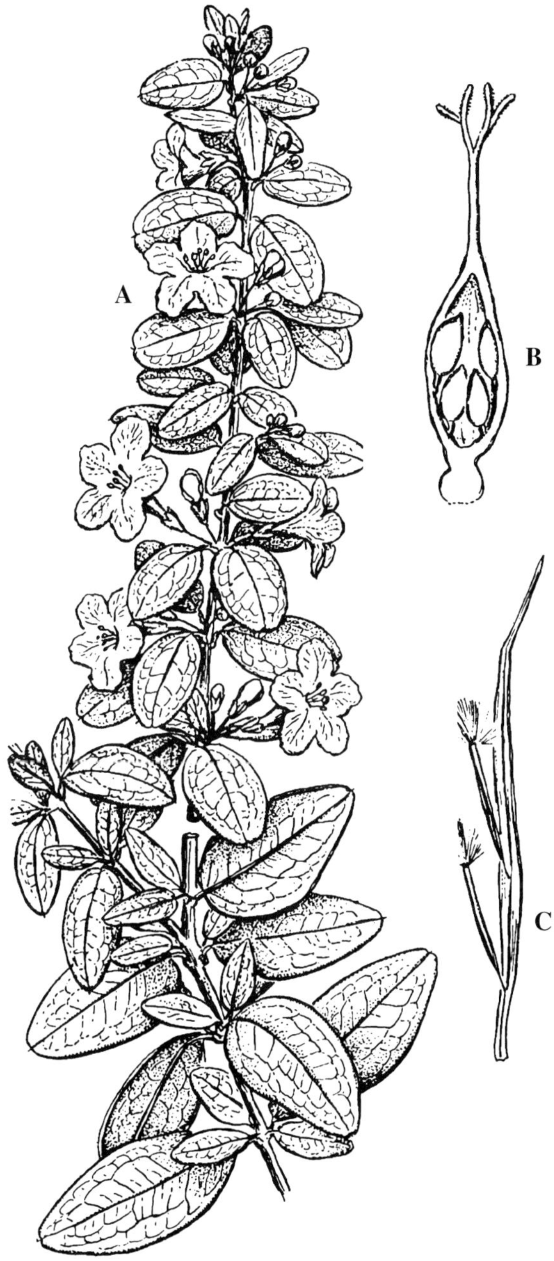

Fig. 41. Plocospermataceae. *Plocosperma buxifolium*. **A** Flowering branch with male flowers. **B** Pistil, opened, with four ovules and twice-branched styles. **C** Opened fruit with two seeds, each with a tuft of hairs at apex. (After Leeuwenberg 1967)

been reported from other orders, and have never been found in Gentianales. No iridoids have been isolated from *Plocosperma* (Jensen 1992).

Distribution and Habitats. *Plocosperma* is found in (dry) secondary forests and scrub habitats in Costa Rica, Guatemala, and Mexico at altitudes of 200–700 m.

Affinities. *Plocosperma* has been a difficult species to classify since its discovery. Initially, it was placed in tribe Gelsemieae of Loganiaceae by Bentham (1876), based on the common presence of imbricate corolla aestivation and twice dichotomously divided stigmas. The tuft of hairs (coma) on the seeds is superficially similar to many Apocynaceae, and caused Hutchinson (1973) to classify the genus as a new family in the Apocynales. Later, *Plocosperma* was included in Apocynaceae, despite the lack of latex (Cronquist 1981). However, Apocynaceae have comas with fragile, unicellular hairs which easily abscise, whereas the coma of *Plocosperma* is permanent and has robust, multicellular hairs (Endress et al. 1996). Leeuwenberg (1967, 1980) kept *Plocosperma* as tribe Plocospermeae in Loganiaceae.

Recent molecular-based and phytochemical analyses of *Plocosperma* have shown that Plocospermataceae is supported as being one of the most basally placed families in Lamiales, and is not closely linked to Gentianales or Solanales (Struwe et al. 1994; Endress et al. 1996; Oxelman et al. 1999).

The presence of CPGs and cornoside in *Plocosperma* supports a position in Lamiales. Cornoside is usually present instead of iridoids, as is also seen in Gesneriaceae, and in the phylogenetically more basally positioned Plocospermataceae and Tetrachondraceae (Jensen 1996, 2000; Wagstaff, this volume). Altogether, the chemical data agree with molecular evidence and strongly suggest a position within the Lamiales.

Also morphological and anatomical data support the move of *Plocosperma* from Loganiaceae (Gentianales) to the Lamiales. Wood anatomy of Plocospermataceae shows good correlation with Verbenaceae, but not with Solanaceae (D'Arcy and Keating 1973). If the ovary is derived from a 4-carpellate condition (see discussion above), this would also support a close relationship to Verbenaceae (D'Arcy and Keating 1973). Reticulate, tricolporate pollen grains and unilacunar nodes are also characteristic of both Verbenaceae and Plocospermataceae. Punt (1980) also noted that the pollen features of Plocosperma do not suggest a close relationship to any Loganiaceae.

One genus:

Plocosperma **Benth. in Hooker** Fig. 41

Plocosperma Benth. in Hooker, Icon. 12: 82, t. 1195 (1876); Leeuwenberg, Acta Bot. Neerl. 16: 56–61 (1967), rev.
Lithophytum Brandegee (1911).

See family description.

The genus is considered to include only one species, *Plocosperma buxifolium* Benth (Leeuwenberg 1967). Three different species have earlier been identified based on variation in hairiness and size of leaves, fruits, and capsules. *Plocosperma* is found in (dry) secondary forests and scrub habitats in Costa Rica, Guatemala, and Mexico at altitudes of 200–700 m.

Selected Bibliography

Bentham, G. 1876. *Plocosperma*. In: Bentham, G., Hooker, J.D. Genera Plantarum 2: 789.

Brandegee, T.S. 1911. *Lithophytum*. In: Plantae Mexicanae Purpusianae, III, pp. 188–189). Univ. Calif. Publ. Bot. 4(11): 177–194.

Chiang, F., Frame, D. 1987. The identity of *Lithophytum* (Loganiaceae, Plocospermeae). Brittonia 39: 260–262.

Cronquist, A. 1981. An integrated system of classification of flowering plants. New York: Columbia University Press.

D'Arcy, W.G., Keating, R.C. 1973. The affinities of *Lithophytum*: a transfer from Solanaceae to Verbenaceae. Brittonia 25: 213–225.

Endress, M.E., Sennblad, B., Nilsson, S., Civeyrel, L., Chase N.W., Huysmans, S., Grafstrom, E., Bremer, B. 1996. A phylogenetic analysis of Apocynaceae s.str. and some related taxa in Gentianales: a multidisciplinary approach. Opera Bot. Belg. 7: 59–102.

Hutchinson, J. 1973. The families of flowering plants, ed. 3. Dicotyledons. London: Oxford University Press.

Jensen, S.R. 1992. Systematic implications of the distribution of iridoids and other chemical compounds in the Loganiaceae and other families of the Asteridae. Ann. Missouri Bot. Gard. 79: 284–302.

Jensen, S.R. 1996. Phenylethanoid glycosides in *Sanango racemosa* and in the family Gesneriaceae. Phytochemistry 43: 777–783.

Jensen, S.R. 2000. Chemical relationships of *Polypremum procumbens*, *Tetrachondra hamiltonii* and *Peltanthera floribunda*. Biochem. Syst. Ecol. 28: 45–51.

Leeuwenberg, A.J.M. 1967. Notes on American Loganiaceae I. Revision of *Plocosperma* Benth. Acta Bot. Neerl. 16: 56–61.

Leeuwenberg, A.J.M. 1980. In: Engler and Prantl, Die natürlichen Pflanzenfamilien, ed. 2, 28b I. Berlin: Duncker and Humblot.

Oxelman, B., Backlund, M., Bremer, B. 1999. Relationships of the Buddlejaceae s.l. investigated using parsimony jackknife and branch support analysis of chloroplast *ndh*F and *rbc*L sequence data. Syst. Bot. 24: 164–182.

Pedersen, J.A. 2000. Distribution and taxonomic implications of some phenolics in the family Lamiaceae determined by ESR spectroscopy. Biochem. Syst. Ecol. 28: 229–253.

Punt, W. 1980. Pollen morphology. In: Engler and Prantl, Die natürlichen Pflanzenfamilien, ed. 2, 28b I. Berlin: Duncker and Humblot, pp. 162–191.

Struwe, L., Albert, V.A., Bremer, B. 1994 (1995). Cladistics and family level classification of the Gentianales. Cladistics 10: 175–206.

Scrophulariaceae[1]

E. Fischer

Scrophulariaceae Juss., Gen. Pl.: 117 (1789), nom. cons.
Veronicaceae Durande (1782).
Rhinanthaceae Vent. (1799).
Orobanchaceae Vent. (1799), nom. cons.
Antirrhinaceae Pers. (1807), nom. cons. prop.
Digitalidaceae Martinov (1820).
Gratiolaceae Martinov (1820).
Selaginaceae Choisy (1823), nom. cons.
Paulowniaceae Nakai (1949).
Oftiaceae Takht. & Reveal (1993).
Schlegeliaceae (A.H. Gentry) Reveal (1996).
Calceolariaceae (G. Don) Raf. ex Olmstead (2001).

Plants annual, biennial or perennial, herbaceous or suffrutescent, rarely climbers, shrubs or trees. Stems erect, ascending or prostrate. Leaves opposite or alternate, entire, rarely divided or pinnatifid to pinnate, leaf-bases often fused. Inflorescence a raceme or thyrse, without terminal flower. Flowers zygomorphic, prophylls present or absent. Calyx with sepals free to base or almost fused, actinomorphic or zygomorphic, often tetramerous by reduction of the median sepal. Corolla sympetalous, zygomorphic, rarely actinomorphic, the 2 upper (adaxial) petals often fused, forming an upper lip opposite to the lower (abaxial) lip with 3 lobes, throat of corolla often closed by a palate, abaxial base sometimes calcarate or at least gibbose to saccate. Stamens usually inserted on corolla-tube, sometimes 5, usually 4, the fifth (median) stamen reduced to staminode or absent, sometimes also the anterior (abaxial) stamens reduced to staminodes or lacking, anthers with two well-developed thecae or with 1 fertile theca and a smaller, reduced, sterile theca, the latter sometimes lacking. Pollen usually tricolporate or tricolpate. Ovary superior, of 2 fused carpels, sometimes one carpel reduced. Fruit a capsule with septicidal, loculicidal or poricidal dehiscence, rarely indehiscent (berry or drupe). Seeds numerous to few (4–2), with smooth, reticulate or variously ornate surface (alveolations, furrows, wings), endosperm present, embryo straight.

In the traditional circumscription, the cosmopolitan family comprises 306 genera and about 5850 species, mainly in the holarctic and the tropical mountains.

Vegetative Morphology. Woody phanerophytes are rare. Trees occur in *Paulownia* (Fig. 49) or *Shiyunghuya*; shrubs occur in *Halleria* (up to 6 m tall), *Radamaea* (Fig. 51) or *Graderia* and *Bartsia* (Fig. 52; up to 3 m tall). Few genera contain woody climbers, e.g. *Wightia* which is also a tree strangler. Chamaephytes are found in many genera, e.g. *Veronica*, *Parahebe*, *Antirrhinum*. They produce their innovation buds on shoots above ground level. Geophytes are rare and found only in few genera. Examples are *Scrophularia nodosa* or *Linaria vulgaris* with subterranean rhizomes. Hemicryptophytes and annuals or biennials are the most frequent growth forms in Scrophulariaceae. Annual therophytes can be found in most seasonal climates (e.g. *Kickxia*, *Veronica*). Biennial plants occur in *Verbascum* and *Digitalis*, in which the rosette survives the first vegetation period. True hydrophytes occur in, e.g. *Limnophila*, *Hydrotriche* and *Limosella*.

The majority of roots usually belongs to the main root system and the prolonged main root can hold a dominant position (e.g. as a depository root, *Pedicularis*, *Verbascum*). In some cases, the main root dies early during germination (*Limosella*) and the root system is entirely composed of adventitious roots produced by hypocotyl and stem. When rhizomes or stolons are present, adventitious roots can form the entire root system. All haustoria of hemi- and holoparasites probably are roots with special function (see below). These haustoria are usually situated on lateral or adventitious roots and then called secondary haustoria. Primary haustoria originate as terminal swellings of primary roots (see below).

Most Scrophulariaceae of subfamily Rhinanthoideae ("Orobanchaceae") are hemiparasites on the roots of other flowering plants. They are photosynthetic, but obtain water with diluted nutrients from their host. For water uptake, the parasite has to have a higher transpirational tension than the host. This is achieved by the guttation of water through trichome hydathodes on the lower sur-

[1] See also Buddlejaceae, Callitrichaceae, Globulariaceae, Hippuridaceae, Myoporaceae, Phrymaceae, Plantaginaceae.

faces of the leaves. In the holoparasitic *Lathraea* and *Tozzia*, which are holoparasitic early in development, these peltate hydathodes are the only structures responsible for guttation. All hemiparasites have haustoria to the xylem of the host only, and so does the holoparasitic *Lathraea*. The early (March to April) flowering *Lathraea* obtains organic compounds from the xylem of its hosts (e.g. *Corylus* or *Alnus*) when these transport such substances early in season. Holoparasites such as *Hyobanche*, some species of *Striga* and *Alectra* as well as *Orobanche* and related genera all send haustoria into xylem and phloem. *Alectra* and *Striga* are interesting models for the evolution of parasitism by containing almost autotrophic to holoparasitic species.

The parasitic Scrophulariaceae ("Orobanchaceae") probably represent a monophyletic group (Young et al. 1999). The annual species of genera like *Euphrasia*, *Odontites* or *Parentucellia* have only weakly differentiated secondary haustoria. Some species can germinate and even flower without host-connection. *Rhinanthus*, *Pedicularis* and *Bartsia* contain mainly robust, annual to perennial or even shrubby species. They also are root parasites with secondary haustoria. Within the root system also wart haustoria occur (see Vegetative Anatomy). *Tozzia* in its early development and *Lathraea* throughout its life cycle live as holoparasites with achlorophyllous scale leaves. In both genera condensed rhizomes with scale complexes occur. *Tozzia* and *Lathraea* also have only secondary haustoria. In tropical genera like *Striga* or *Alectra*, the evolution from rather unspecialised hemiparasites to highly specialised holoparasites can be observed. *Striga angustifolia* is a hemiparasite that develops only secondary haustoria. In all other *Striga* species, primary haustoria occur. The primary root emerging from the minute seeds swells terminally after contact with a host root and develops into a primary haustorium. Transport within the haustorium is mostly apoplastic along a xylem bridge. For *Striga gesneroides* as an achlorophyllous holoparasite, a symplastic transport through plasmodesma between host and parasite cells has been shown. The South African genus *Hyobanche* is an achlorophyllous holoparasite without roots, at least when adult. In *Hyobanche glabrata*, a small primary haustorium is present; in *H. sanguinea* only leaf haustoria occur (Kuijt et al. 1978). The evolution of parasitism has been thought to have started from annual hemiparasites with secondary haustoria on their roots, over hemiparasites with wart haustoria to holoparasites with primary haustoria. However, such linear evolution was not supported by analysis of plastid gene sequences (Young et al. 1999). Holoparasitism has evolved independently in different taxa such as *Striga*, *Alectra*, *Harveya*, *Hyobanche* and the traditional Orobanchaceae.

The haustoria of Rhinanthoideae-Rhinantheae can be divided into four different types (Weber and Weberling 1975). The *Tozzia alpina*-type always lacks root hairs and produces no folds on its cortex. For example, it occurs in *Tozzia*, a few *Pedicularis* species (*P. recutita*, *P. palustris*) and some *Melampyrum* species (*M. arvense*, *M. cristatum*). The *Pedicularis sylvatica*-type also lacks root hairs, but can be distinguished by the presence of usually large cortex folds. Within *Pedicularis*, this is the most frequent type. The haustoria of *Bartsia alpina* and some *Odontites* (e.g. *O. litoralis*) also belong here. The haustoria of the *Melampyrum pratense*-type have numerous root hairs while cortex folds are completely lacking. This type is recorded in some *Melampyrum* species and most species of *Euphrasia* and *Odontites*. The *Rhinanthus minor*-type usually has numerous root hairs and well-developed cortex folds. This type has been observed mainly in *Rhinanthus*, but also in some *Pedicularis* species (*P. tuberosa*, *P. rostratospicata* subsp. *helvetica*). However, some intermediate haustoria types may occur and there can be a change of haustorium type during the lifetime of a plant. In *Lathraea squamaria*, the haustoria of young stages resemble the *Melampyrum pratense*-type while in older plants they approach the *Tozzia alpina*-type.

The stem is, besides the different length and structure of hypotagma and florescence, mainly responsible for growth forms. Condensed internodes, which result in rosulate habit, can be found in *Limosella* or *Wulfenia*; creeping stems occur in *Cymbalaria* and various *Veronica* species. Accessory shoots are quite common and usually occur in leaf axils between flower or paracladium and bract (phylloscopic). They may occur either in florescence or hypotagma and they may be purely reproductive (as cymes or accessory flowers) or paracladia. Accessory flowers are found in various number and structure in the inflorescence, e.g. of *Penstemon*, *Verbascum* and *Calceolaria* (Weber 1973). Vegetative accessory shoots may occur either in florescence or hypotagma and can be stolons (e.g. in *Limosella*). Hypocotyledonous shoots, which originate from the epidermis, mainly occur in Antirrhineae. In many species of *Linaria*, the main stem remains rudimentary and the hypocotyledonous shoots exclusively represent the flowering stems. Some perennial species of *Linaria*, e.g. *L. vulgaris*, also have endogenous root shoots, which are responsible for perennial growth. In hemicryptophytes, innovation usually takes place from a rhizome, which may be orthotropous (*Scrophularia vernalis*) or plagiotropous (*Gratiola officinalis*). The rhizome usually shows sympodial growth.

Leaf position is mainly opposite (decussate), but can also be entirely alternate or may change in the inflorescence from opposite to alternate. Usually, frondose leaves of the hypotagma are replaced in the inflorescence by squamate, bracteate hypsophylls. When the hypotagma is reduced, the hypsophylls have to be of larger size for photosynthesis, which results in a frondose florescence (e.g. *Limosella*, *Gratiola*). In frondobracteate inflorescences, all intergradations between frondose and bracteate leaves are found. Some cataphylls on subterranean stems remain unfolded. In case of revolute vernation, this results in ascidiate squamate leaves of *Lathraea* with its numerous cavities (Weber 1975). Leaves of Scrophulariaceae are extremely variable in shape and venation of the lamina (entire to pinnatifid, pinnate to palmate). Sometimes, the shape and function of leaves can change in different stem areas due to environmental influence, e.g. in most species of *Limnophila*, which have dissected submersed leaves and entire aerial leaves. The leafbase may be sericaceous and dilated, resulting in a decurrent leaf-base or gamophyllous leaves, or sometimes in pseudostipules (e.g. in *Limosella*). The petiole is twining in some genera of Antirrhineae (*Maurandya*, *Maurandella*, *Epixiphium*, *Lophospermum*). Extrafloral nectaries occur in *Melampyrum* on the upper leaves close to the inflorescence and on bracts.

Vegetative Anatomy. The endodermis cells of roots are frequently divided in anticlinal direction ("supporting walls"), e.g. in adventitious roots of *Veronica alpina*. In some genera, e.g. *Alectra*, *Campylanthus* and *Craterostigma*, the roots contain an orange-yellow colour. In *Craterostigma*, this orange pigment is located in the intercellular spaces of the root cortex (Ward and Dale 1898; Fischer 1992). The anatomy of Rhinanthoid haustoria seems to be quite uniform. They consist mainly of three major parts, the vascular core, the haustorial core and the intrusive part ("endophyte"). Besides the secondary haustoria which are in contact with host roots by a xylem bridge, there exist metahaustoria which lack this contact (Weber 1976a, 1976b). Such metahaustoria have been observed in *Rhinanthus*, *Pedicularis*, *Melampyrum*, *Bartsia alpina* and *Lathraea squamaria*. They are usually larger than normal haustoria and may develop from haustoria with interrupted host-root contact or when haustoria formation was stimulated by the host root but no contact is established. In metahaustoria the endophyte is lacking and either the vascular or the haustorial cores are massively enlarged. Another type of haustoria is represented by the wart haustoria (Weber and Weberling 1975; Weber 1976a, 1976b). These are minute haustoria, usually only 0.005–0.1 mm in diameter, which regularly occur in the upper part of the main root of annual species of *Melampyrum*, *Odontites* and *Rhinanthus* as well as on robust lateral roots and adventitious roots of perennial genera such as *Pedicularis*. The wart haustoria never have root hairs, and vascular and haustorial cores are also lacking. The entire haustorium is made up of normal parenchymatous cells. Epidermal cells of wart haustoria, however, can grow like ordinary root hairs and attach to host roots. They represent the first stage in the formation of an intrusive organ and may be considered as strongly simplified haustoria, which are derived by inhibition of the development of normal haustoria.

In transverse sections of the stem, the central vascular cylinder is characterized by the lack of primary medullary rays (except *Euphrasia*, *Halleria*). Also, wood parenchyma is only sparsely developed or lacking. Cork is rarely developed, e.g. in *Leucophyllum*. The primary cortex usually has very little collenchyma. The wood vessels and tracheids are narrow and an inner (intraxylar) phloem is lacking. Ring pores occur occasionally. Fibres most commonly have simple pits and thin walls, and rarely are thick-walled with bordered pits.

The leaves of *Craterostigma* contain club-shaped bodies resembling cystoliths. Hydathodes are present in the teeth of the leaf margins of *Digitalis* or *Veronica*. Secretory elements, e.g. large tanniniferous idioblasts, are found in the mesophyll of *Scrophularia deserti*. In *Verbascum* and *Scrophularia*, subepidermal idioblasts in stems and leaves are present that show a blue fluorescence (Ziegler 1955). An overview of the occurrence of idioblasts and internal foliar secretory structures is provided by Lersten and Curtis (2001). Single-celled idioblasts are probably confined to *Scrophularia* and *Verbascum*. In *Graderia scabra* and *Radamaea montana*, multicellular "nodules", i.e. chains or clusters in the middle three strata of mesophyll, occur. They contain dense fibres or granules (Lersten and Curtis 2001). In *Capraria biflora*, secretory cavities in the mid-mesophyll has been observed. Oil cavities are also present in *Leucophyllum*. A conspicuously enlarged bundle sheath, indicative of a C_4 Kranz syndrome, was seen in species of *Anticharis*. In *Picria* and two species of *Lindernia*, only the lateral bundle sheath cells were enlarged. They joined with enlarged mesophyll cells of the same stratum to form an anatomically distinct, horizontal paraveinal mesophyll (PVM) layer (Lersten and Curtis 2001).

The leaves usually are bifacial. The hairs are quite diverse, even within one species, and represent different types (Raman 1987): unicellular conical or cylindrical to brussels-sprout-like hairs, uniseriate hairs, glandular hairs with unicellular to multicellular glandular heads and multicellular branched hairs (*Verbascum*, *Paulownia*). The heads of glandular hairs are composed of 1–60 cells (100–250 cells in *Melampyrum*) and are nearly exclusively divided vertically, in some genera of "Orobanchaceae" also horizontally. Glandular hairs that are somewhat sunk into the epidermis of both the upper and lower leaf surface function as extrafloral nectaries in *Melampyrum*. The African genus *Crepidorhopalon* bears hairs on the lower lip of the corolla, which are clavate and situated on a multicellular socle like the stinging hairs of *Urtica*. The cuticular folds on corolla trichomes have been studied by Barthlott (1980). An overview of corolla hairs is provided by Raman (1989a, 1989b, 1989c, 1989d, 1990). Bolliger (1985) showed the importance of glandular hairs for the systematics of *Odontites*. Intranuclear inclusions (protein crystals) are widely distributed and their structure may be of systematic importance (Speta and Greilhuber 1970; Speta 1972). Bigazzi (1993) recorded five types of nuclear inclusions determined by their different ultrastructure: amorphous, lamellar, fibrillar, tubular and crystalline inclusions. In Gratioleae, mainly fibrillar inclusions occurred, in Scrophulariaceae-Verbasceae mostly amorphous and rarely lamellar inclusions were found; and in Scrophularieae mainly lamellar (most species of *Scrophularia*) and rarely crystalline (*Halleria*, now placed in Stilbaceae) and amorphous inclusions were recorded. The Manuleae (incl. Selagineae) and "Calceolariaceae" had exclusively lamellar inclusions, as did the Hemimerideae, with exception of *Angelonia* (now placed in Gratioleae) which had lamellar and fibrillar inclusions. In Digitalideae, amorphous and crystalline inclusions were found, whereas *Rehmannia*, with lamellar inclusions and placed by Bigazzi (1993) into this tribe, should be excluded. In Veroniceae, with the exception of the doubtful *Campylanthus*, exclusively amorphous inclusions were observed; in "Orobanchaceae", only lamellar inclusions were found.

Inflorescence Structure. The inflorescence of Scrophulariaceae (Troll 1964; Hartl 1965–1974; Weberling and Troll 1998) is composed of a polytelic florescence and the hypotagma. Terminal flowers occur only exceptionally as peloric flowers, e.g. in *Linaria*, *Antirrhinum* or *Digitalis*. The primary lateral axes of the florescences end with a flower and two prophylls, which may be reduced or missing. In this case, the florescence is a raceme. If the axils of prophylls produce shoots with a terminal flower, the florescence is a thyrse, with partial florescences of different structure (e.g. dichasial or monochasial in *Scrophularia*). Below the florescence, the hypotagma is situated, which is characterized by paracladia, which reiterate the structure of main shoots. That part of hypotagma which produces paracladia is called the paracladial or enrichment zone. In a zone close to the base of the main stem, the development of paracladia usually is inhibited. This inhibition zone separates the basal, vegetative part of the hypotagma from its reproductive part. The upper, reproductive zone of the hypotagma with paracladia and the florescence is called the synflorescence. The relationship between hypotagma and florescence varies through the inhibition of either one of these regions. Thus, the shoot of *Limosella* spp. consists only of the florescence, while the shoots of *Wulfenia* spp., many *Veronica* species (e.g. *V. beccabunga*), *Craterostigma* spp. (Fig. 46) or some *Pedicularis* (e.g. *P. rostratocapitata*) represent only the hypotagma. All flowering shoots of these taxa are paracladia and a main florescence is lacking. This type is classified as a truncate florescence, and the main stem can proliferate (Hamann 1958; Hartl 1965–1974; Fischer 1992). *Hebe* and some of its segregates also are characterized by such prolific synflorescence. When in premature proliferation the main florescence is suppressed, a proliferation of the florescence or a retarded proliferation are found, e.g. in *Veronica filiformis* or *Bacopa monnieri*. In these cases the florescence apex becomes vegetative after anthesis and continues growth.

In some "Veronicaceae" (*Penstemon*), "Scrophulariaceae" (*Verbascum*, *Charadrophila*, Weber 1989) and "Calceolariaceae" (*Calceolaria*, Andersson and Molau 1980), accessory flowers occur. These accessory flowers arise in an outer median position at a dichasial or monochasial branching, in addition to normally positioned flowers. In *Calceolaria*, the accessory flowers lack a subtending bract. These flowers are interpreted as axillary branches from the "axil" of a suppressed prophyll (Weber 1973), by suppression of the pre-prophyllar internode and the prophyll itself. In *Sutera*, accessory shoots can arise in the florescence, in addition to the 2- or 3-flowered partial florescences. Due to their strongly vegetative character, these obscure the thyrsic nature of the inflorescence. In *Hiernia angolensis*, accessory shoots arise in leaf axils below the axillary flower (Weberling and Troll 1998). In *Calceolaria darwinii* and related species, the synflorescence has only two flowers or is even 1-flowered.

Floral Morphology. The calyx can be extremely variable. Typically, as in *Gratiola*, the sepals show a quincuncial aestivation, i.e. they are arranged spirally. Further types are ascending aestivation (in most Antirrhineae), descending aestivation (*Verbascum*, *Antirrhinum*) and valvate aestivation (*Calceolaria*, *Mimulus*, *Lathraea*). The calyx lobes are united to various degrees, i.e. from almost completely free to almost completely fused. Even within one genus (e.g. *Crepidorhopalon*, Fischer 1992), all intergradations are found. The median sepal may be reduced and, in *Veronica*, this reduction can vary from the presence of a minute fifth sepal to complete abortion. In some cases, the median vein of a sepal can be winged or angled, e.g. in *Mimulus* or *Torenia*, and this was considered to be a character of systematic value (e.g. Engler 1897). However, winged to unwinged sepals can occur within one and the same genus, e.g. *Craterostigma* (Fischer 1992). In *Torenia*, the calyx shows a special dehiscence for seed dispersal (Fischer 1992). Postfloral accrescence of the calyx can be found in various genera (e.g. *Torenia*, *Mimulus*, *Rhinanthus*).

The petals usually have an imbricate aestivation with two relatively constant patterns: ascending (rhinantoid) and descending-cochlear (antirrhinoid). The corolla aestivation has traditionally been used to divide Scrophulariaceae into two subfamilies: Scrophularioideae (= Antirrhinoideae) and Rhinanthoideae (Wettstein 1891; Thieret 1967). In Scrophularioideae the lower lip is internal in bud (antirrhinoid aestivation) whereas in Rhinanthoideae, the lower lip is external in bud (rhinanthoid aestivation). This distinction is artificial because antirrhinoid and rhinantoid aestivation may vary even within one genus (e.g. *Lindenbergia*, Brühl 1920; Hartl 1955; Hjertson 1995). Some of the groups accepted in this account, however, show a more or less dominant aestivation type, i.e. the "Antirrhinoideae" and "Scrophulariaceae" mostly have antirrhinoid buds, while most of "Veronicaceae" and "Orobanchaceae" mainly have rhinantoid aestivation. The usual number of petals is five and there are very few deviations from this pattern. In *Sibthorpia* the number of petals can be increased to eight (Hedberg 1955). Rarely four or even no petals (in *Besseya*, Hufford 1992) occur. The adaxial petals of some genera, e.g. *Veronica* (Fig. 48), are fused, resulting in an apparently 4-petalous corolla. Usually, only differences in proportions can be observed. The corolla-tube varies considerably in length due to different pollination syndromes (see below). The tube in fact consists of a corolla-stamen complex, as the stamens are fused with the corolla and participate in tube formation. A tendency of corolla-tube reduction is apparent in Digitalidoideae-Veroniceae. The long-tubed genera like *Digitalis*, *Wulfenia*, *Paederota* and *Pseudolysimachion* are probably much more primitive than the more advanced, short-tubed genera like *Veronica*. An extreme reduction of the tube is found in the North American *Synthyris ranunculina*, where secondary choripetaly occurs. The upper lip usually is formed by two fused petals, but in *Nemesia* it consists of four petals, with the fifth petal forming the lower lip. *Collinsia* has a "reversed lip flower" with a narrow lower lip, and thus resembles the flag-flowers of Fabaceae (Endress 1996). In some of the three-lobed lower lips, the mid-lobe can be reduced or aborted (*Lagotis*). Dorsiventrality is often intensified by local increase of surface area, resulting in either protrusions or introversions. Protrusions usually occur at the abaxial side of the corolla as gibbosities, pouches (e.g. *Antirrhinum*; Fig. 45) or long spurs (*Linaria*, *Nemesia*). Two adjacent spurs are found in *Angelonia* and *Diascia*. A protrusion of the entire lower lip is the "shoe" of *Calceolaria* (Fig. 49). Closure of the corolla by upward arching of the palate on the lower lip of *Antirrhinum* ("masked flowers") is quite common and occurs mainly in Antirrhineae (Fig. 45; Sutton 1988), some Gratioleae (Fig. 46; e.g. *Stemodiopsis*, Fischer 1997a) and "Orobanchaceae" (e.g. *Melampyrum*); it is absent in "Scrophulariaceae" and Digitalieae. The usual orientation of the flower is modified by resupination by 180° in *Alonsoa* and *Nemesia*. In *Pedicularis* (Fig. 53) the lower lip is not positioned in the median symmetry axis. After anthesis, the corolla, together with the stamina, usually is caducous. In most "Orobanchaceae" there is a preformed circumscission area in the corolla-tube, and the collar-like base of the tube remains on the flower.

Number and position of stamens depend mainly on flower symmetry. Generally stamens alternate with corolla lobes. In groups where monosymmetry is only weakly expressed, all five stamens are present (*Verbascum*, some *Bacopa* species, *Sibthorpia*, *Capraria*). The systematic distribution of 5-staminate flowers suggests that this condition is apomorphic within the Lamiales and is thus a reversal, because the 5-staminate condition is probably plesiomorphic at the more general level of Euasterids/Asterids (Endress 1996). Monosymmetry of flowers results in reductions in the androecium. Usually, the fifth median stamen is smaller (in some *Verbascum* species, e.g. *V. blattaria*) or may be present in the form of a small or large staminode (e.g. *Scrophularia*, *Penstemon*, *Russelia*, *Ameroglossum*) that may be larger than the stamens and can have a function in pollination biology (*Scrophularia*). In most cases there are four lateral stamens, which form two symmetrical pairs. They often are didynamic, i.e. the abaxial pair is longer than the adaxial pair or rarely vice versa. In a number of genera only two stamens are fertile. The adaxial pair is present, e.g. in *Calceolaria*, *Gratiola*, *Dopatrium*, many species of *Lindernia*, *Veronica* and relatives, and *Wulfenia*, and the abaxial pair in *Anticharis* or *Micranthemum*. In some cases, the sterile pair is reduced to large staminodes, which have a function in pollination (e.g. in *Lindernia* as false anthers, Fischer 1992; Fig. 46), or minute staminodes (e.g. *Dopatrium*) or it is completely aborted (e.g. *Philcoxia*, *Veronica*, *Wulfenia*). The almost polysymmetric flowers of *Stemotria* have three stamens (Molau 1988), the median and two lateral ones. Due to the didynamical structure, various torsions or geniculations of the filaments can be observed which often result in the convergence or connection of the anthers towards the median plane. In the Lindernieae, the abaxial filaments are conspicuously geniculate, zigzag-shaped or spurred (Fig. 46; *Craterostigma*-type and *Crepidorhopalon*-type of geniculation, Fischer 1992). Mostly those geniculations with a knob- or club-shaped outgrowth are covered with blue to yellow glandular hairs mimicking an anther with pollen, or are otherwise optically attractive (Magin et al. 1989; Fischer 1992). A similar filament structure also occurs in the Hemimerideae (*Diascia*, Steiner 1992). In the Lindernieae (*Torenia*, Armstrong 1992, *Craterostigma*, *Crepidorhopalon*, *Lindernia*, Fischer 1992), in Hemimerideae (*Diascia*, Steiner 1992) and in some Antirrhineae (*Kickxia*), postgenital fusion or connection of anthers into tandems, forming so-called anther-symplexes, are present. Another type of synorganization can be found in *Euphrasia*, where the anthers are connected to pairs by means of hairs (Hartl 1965–1974). Anthers are usually sagittate and spreading during anthesis, bearing two equal thecae. In some groups, e.g. most of the "Scrophulariaceae" (e.g. *Verbascum*, *Scrophularia*, Manuleae), the thecae are fused near the apex and the insertion at the connective (synthecous; Fig. 44). In some Gratioleae, the anthers are unilocular (monothecous) by reduction. Especially in "Orobanchaceae", the reduction of one theca is an important systematic feature (Hofmann and Fischer, 2003, in press). In *Agalinis* or *Seymeria*, both thecae are equal. In *Pseudosopubia*, *Sopubia* and *Ghikaea* (Fig. 50), the thecae are unequal and the reduced theca is always sterile and clearly differs in shape from that of the normally developed thecae. In *Pseudosopubia*, one anther pair is monothecous (Fig. 50). In *Buchnera*, *Striga* or *Rhamphicarpa*, one theca is completely aborted and the anther thus monothecous. Reduction in theca size also can be observed in some Lindernieae (e.g. *Torenia sylvicola*, Fischer 1992). A character of the thecae that occurs only in very few families is a pollen sac placentoid, a protrusion of parenchymatic tissue into each pollen sac from the connective side. It occurs in most families of Lamiales and in Solanaceae and some Convolvulaceae (Hartl 1963). In Scrophulariaceae, this placentoid is reduced in size in some Antirrhineae and *Veronica*, while in the Lindernieae, it is lacking completely (Fischer 1992).

The gynoecium is syncarpous and the two carpels are congenitally united up to the stigmatic region. It usually consists of a stigma, a simple style and ovary, and only rarely, e.g. in *Erinus*, is the style suppressed and the stigma situated directly on the ovary. The cushion-like stigma of *Erinus* is called a commissural stigma, as it is situated on the commissures with transversal orientation. The stigma can be clavate to bilobed and the lobes are mostly oriented along the median plane. Sensitive stigma lobes occur in *Mimulus*, *Glossostigma*, *Rehmannia* and most Lindernieae (e.g. *Lindernia*, *Craterostigma*, *Torenia*, *Crepidorhopalon*, Fischer 1992). They fold together after touching of the inner, receptive surface and unfold again after a certain length of time. The function of this movement may be to indicate to a pollinator that the flower has been visited shortly before and that nectar is not available yet (Endress 1996). The narrow style intergrades at its base with the ovary (*Digitalis*, *Verbascum*) or it is situated in a more or less deep, transversal furrow (Antirrhineae, *Veronica*). Sometimes, the protruding margins of this furrow fuse with the base of the style. This fused part can be recognized in transverse sections as a wall with a distinct style channel and is called an apical septum (Hartl 1962). Most commonly the ovary of Scrophulariaceae consists of two zones, a basal syncarpous (synascidiate) zone and an upper paracarpous zone, where the septum shows a median fissure, which is absent in the septum of the syncarpous zone. This fissure can be cryptic by postgenital fusions (*Gratiola*) or distinctly widened in *Lathraea* (Hartl 1956a). The proportions of both zones can vary considerably. The two extreme forms of ovary occurring in Scrophulariaceae are bilocular ovaries with a long syncarpous (synascidiate) region and a single axile and protruding placenta in each locule (most taxa), and unilocular ovaries with a reduced syncarpous region and parietal placentae with two (to four) protruding placentae on each side (mainly in *Orobanche* and related genera). Occasionally, as in *Phacellanthus*, *Platypholis* and *Xylanche*, the ovary consists of three instead of two fused carpels and is unilocular with three to six parietal placentae. A third form is a free central placenta, which occurs in Lentibulariaceae and in *Limosella*, where a septum is present in the basal part of the ovary but reduced to abortion in the upper part (Hartl 1956b). When the placenta bears many ovules, it is usually large and protrudes cushion-like into the locule or can even be revolute at the margins (*Mimulus* subg. *Diplacus*). When only few ovules are present, the placentae are only weakly developed (*Melampyrum*, *Tozzia*). Reduction in ovule number can be observed in several groups. In Manuleae, reduction to four ovules can be observed in *Tetraselago* (Junell 1961), and several genera (e.g. *Agathelpis*, *Globulariopsis*, Hilliard 1999) have uniovular locules. The epidermis of the placenta consists either entirely (*Digitalis*, *Erinus*, *Calceolaria*, *Linaria*, *Scrophularia*) or only partly of tracheidal cells. In the latter case, the tracheidal cells are concentrated nest-like at the base of the funiculi (*Gratiola*, *Penstemon*). Sometimes, the inner face of the locule (*Torenia*, Fischer 1992, *Otacanthus*) or the placenta (*Nemesia*) can bear hairs. Other variations are the reduction of the adaxial (*Antirrhinum*) or abaxial (*Chaenorrhinum*) locules. In *Calceolaria*, partially epigynous flowers may occur.

EMBRYOLOGY. The pollen sac has a secretory tapetum with usually endopolyploid cells, which is thickened at the placentoid.

The ovules are tenuinucellate, unitegmic and predominantly anatropous to hemitropous (most Gratioleae). In some taxa, hemitropy is caused by postfloral growth of the ovule between the hilum and the micropyle, and the mature seed looks atropous (*Melampyrum*; cf. Figs. 42, 43). In other groups, a dorsal growth of the ovule causes campylotropous-like, curved seeds (*Rhinanthus*). The integument usually consists of 6–7 intermediate layers, more than 8 layers (up to 20) are typical for Antirrhineae, less than 5 (usually 3–4) for the Gratioleae. The inner epidermis of the integument is developed as a glandular tissue (endothelium) which histolyses all or most intermediate layers. This mobilized cell substance, including the deposited starch, is transported by the polar endosperm haustoria (chalazal and micropylar haustoria) to the endosperm and embryo. Size, arrangement and structure of endothelial cells are variable. The cells are either large, transversally elongated, arranged in 6–8 longitudinal rows and thickened only towards the endosperm (Gratioleae), or they are small, isodiametric, irregularly arranged and with equally and weakly thickened walls (Antirrhineae, Digitalideae; Fig. 42). In the former case, the endothelial cells tend to protrude into the endosperm. When a regular protrusion of all endothelial cells occurs, the endosperm becomes longitudinally furrowed (e.g. *Calceolaria*); when the protrusion is restricted to the central part of the cell, the endosperm surface becomes alveolated (bothrospermous, e.g. *Torenia*, *Lindernia*; Fig. 42). When protruding cells (bothroblasts) alternate with non-protruding cells (abothroblasts), the endosperm surface also becomes alveolated (Scrophularioideae), and the aveoles are not round but transversally elongated in transverse section. The abothroblasts originate through an unequal division of endotheliar cells. The protrusion of all cells of the endothelium into the endosperm results in the *Torenia*-type of alveolation (Hartl 1959; Fig. 42), the protrusion only of the bothroblasts in the *Scrophularia*-type of alveolation (Hartl 1959, in Scrophulariaceae s.str.). By fusion of endosperm alveoles, longitudinally furrowed seeds arise in both types. These are called aulacospermous (Hartl 1959; Fischer 1992) and occur in different Manuleae, in *Rhabdotosperma* (Hartl 1977) and in, e.g. *Crepidorhopalon* (Fischer 1992). In some Antirrhineae and Digitalideae, the endosperm surface is sculptured independently of the endothelium (e.g. in *Linaria* and *Veronica*).

The endosperm provides valuable data for the systematics of Scrophulariaceae (Glisic 1936–1937). Thirteen different types of endosperm

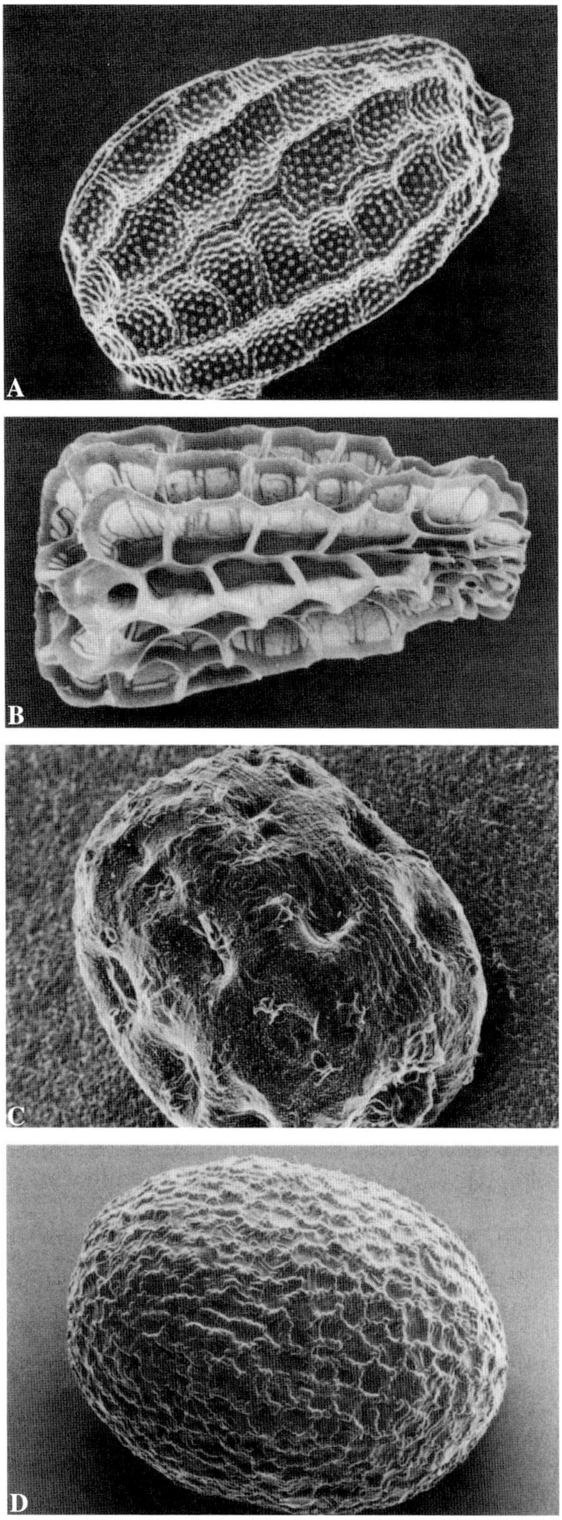

Fig. 42. Scrophulariaceae. Seeds. **A** *Achetaria ocymoides*. SEM × 216. **B** *Angelonia salicariaefolia*. SEM × 63. **C** *Lindernia crustacea*. SEM × 150. **D** *Hemiphragma heterophyllum*. SEM × 133. (Photo E. Fischer)

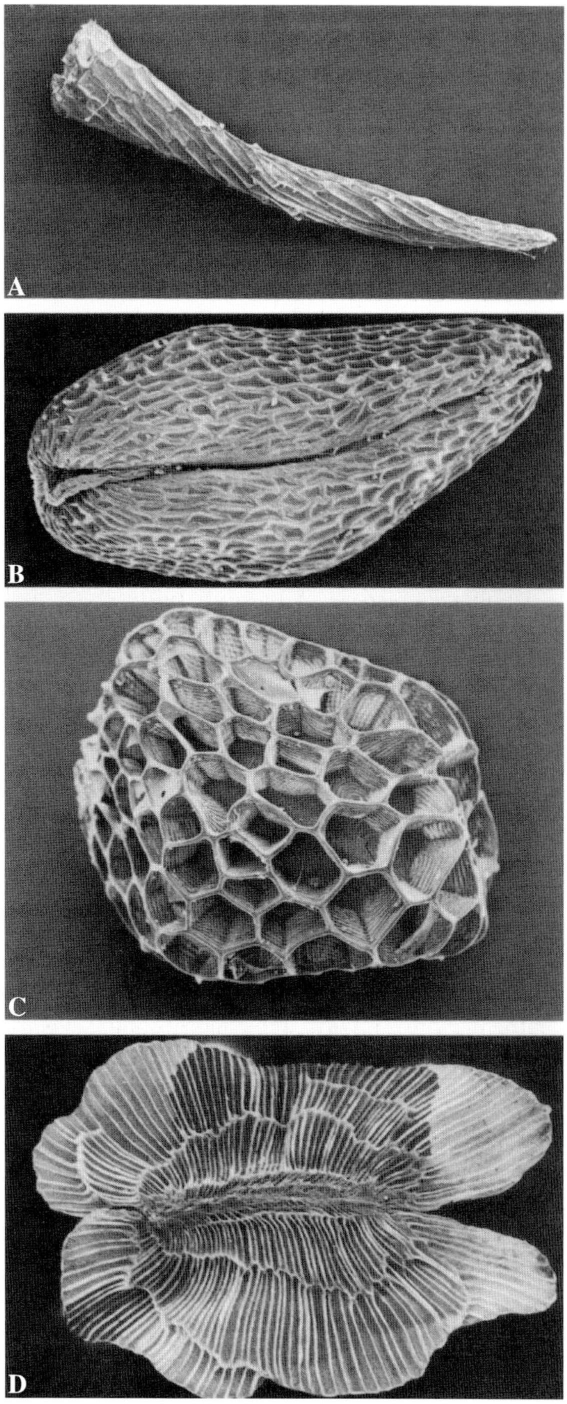

Fig. 43. Scrophulariaceae. Seeds. **A** *Alectra fruticosa*. SEM × 32. **B** *Pedicularis palustris*. SEM × 38. **C** *Cistanche phelypaea*. SEM × 93. **D** *Paulownia tomentosa*. SEM × 30. (Photo E. Fischer)

formation could be distinguished on the basis of endosperm formation and formation of the micropylar haustorium (e.g. Glisic 1936–1937; Krishna Iyengar 1939, 1940; Arekal 1963, Kapoor et al. 1975; Hakki 1977). The embryo in all cases studied is that of the Crucifer-type.

POLLEN MORPHOLOGY. The Scrophulariaceae contain several different pollen types. Pollen may occur either in tetrads or in monads. Grains united in tetrahedral or decussate tetrads and rarely in dyads are known from *Anticharis* (Varghese 1968). The individual grains are 3-zonocolporate, 3-zonocolpate, spiraperturate or 2-syncolpate. The pollen is mostly tricolpate like in *Veronica*, *Antirrhinum*, *Anarrhinum*, *Digitalis*, *Erinus*, *Paederota*, *Pseudolysimachion*, *Pedicularis sceptrumcarolinum*, *Bartsia*, *Odontites*, *Euphrasia*, *Rhinanthus*, *Melampyrum*, *Tozzia* or *Lathraea*. Tricolpate-syncolpate pollen occurs in *Pedicularis oederi* and *P. verticillata*. Dicolpate-syncolpate pollen (in bilateral pollen grains) occurs in most species of *Pedicularis* (*P. palustris*-type). Tricolporate pollen is found in *Scrophularia*, *Verbascum*, *Misopates* and *Mimulus luteus*.

General surveys of pollen morphology in Scrophulariaceae are presented by Erdtman (1952) and Varghese (1968). In Antirrhineae, pollen grains of the Old and New World genera have been studied (Elisens 1986; Bigazzi and Tardelli 1990). They are 3-colporate, typically isopolar, radially symmetrical, with colpi fusiform to narrowly oblong and colpus membrane smooth or granular. Three types could be distinguished among Old World species, i.e. type I lacking granules on the colpus membrane, with a fusiform colpus membrane and pointed ends (*Anarrhinum*, *Kickxia*); type II with fusiform colpi and a colpus membrane covered with minute granules (*Acanthorrhinum*, *Chaenorrhinum*, *Cymbalaria*, some species of *Linaria*); and type III with granulose processes on the colpus membrane and oblong colpi with rounded ends (*Antirrhinum*, *Misopates*, *Schweinfurthia* and most species of *Linaria*).

In Scrophulariaceae s.str., new data are available for Manuleae (incl. Selagineae) (Argue 1993). In Manuleae, three groups could be distinguished. In the majority of genera (*Glekia*, *Glumicalyx*, *Melanospermum*, *Phyllopodium*, *Polycarena*, *Strobilopsis*, *Tetraselago*, *Trieena*, *Zaluzianskya*, *Sutera*, *Jamesbrittenia*, *Manulea*, *Manuleopsis*, *Camptoloma*, *Antherothamnus* [both now excluded from the tribe] and part of *Lyperia*), pollen grains are single, isopolar, radially symmetrical and tricolporate with colpi length long to medium. In some *Lyperia* species, grains are isopolar, radially symmetrical, suboblate to prolate spheroidal, and convex-triangular-aperturate to 4-colporate. In two further *Lyperia* species (*L. tristis*, *L. lychnoidea*) and in remaining "Selagineae", the grains are 3- to 4-colporate with short colpi.

In Gratioleae in the traditional sense, data are available for many genera. In *Mimulus* (Argue 1980), five major pollen types were found, i.e. type I synaperturate, usually more or less spiraperturate, exine perforate to microreticulate with supratectal processes; type II tricolporate with microreticulate exine and supratectal processes present or absent; type III tricolporate, colpus membrane more or less psilate, exine with supratectal processes; type IV tricolpate with colpus covered with spinulose granules (operculate), exine microreticulate with supratectal processes; and type V 5–7-stephanocolpate, exine microreticulate or perforate with supratectal processes, spines or spinules. In *Deinostema*, *Gratiola* and *Sophronanthe*, the pollen grains are isopolar to anisopolar and radially symmetrical with three to four, equally spaced meridional colpi (Argue 1990). They are tectate-perforate, lack a baculate sexine 3 and have a reduced to intermediate sexine 1. In *Amphianthus*, *Artanema*, *Glossostigma* and *Peplidium* (Argue 1986), the pollen grains are isopolar and radially symmetrical with equally spaced meridional colpi and a columellate exine. The colpi vary from 3 to 6 in number, and the colpus membrane can be variously granular, verrucate, rugulate or spinulose. The exine varies from semitectate and microreticulate or reticulate to tectate and sparsely perforate. The Veroniceae have been studied in detail by Hong (1984). In *Veronica*, four different pollen types could be distinguished, which differ in shape of colpus and structure of colpus membrane.

In Rhinanthoideae ("Orobanchaceae"), the pollen in tribe Pedicularieae is uniform in the fundamental characters: radially symmetrical, isopolar, spherical, 3- to 4-colpate (in *Bornmuellerantha*, Bolliger 1990) and perforate-tectate, and retipilate sculpturing is most widespread (e.g. *Bartsia*, *Parentucellia*, *Euphrasia*). Reticulate sculpturing is confined to *Odontites*, while a complex retirugulate pattern was found in the monotypic genus *Bornmuellerantha*. Within the retipilate and reticulate major types, eight different minor types based on exine surface morphology could be distinguished, which correlate with taxonomic subdivisions of *Odontites* and related genera.

KARYOLOGY. The genera of tribe Antirrhineae (Sutton 1988) have numbers of $x = 6, 7, 8, 9, 13$ and 15. Higher chromosome counts for *Acanthorrhinum* ($2n = 32$), *Chaenorhinum* ($2n = 42$), *Cym-*

balaria ($2n = 28, 42, 56$) or *Linaria* ($2n = 24$) are the result of recent polyploidy. In Scrophulariaceae s.str, *Scrophularia* has basic numbers of $x = 7, 9, 10$, 12 and 13. In the closely related *Verbascum*, $2n = 32$ and 48 are apparently the most frequent, but also $2n = 30, 34, 36, 40, 50$ as well as triploid numbers may occur. Usual haploid numbers are $n = 17, 20, 23, 24, 25, 26$. The chromosomes of Manuleae have been studied by Jong in Hilliard (1994). Base numbers are $x = 6, 7, 8, 9$. In tribe Alonsoeae, a base number of $x = 14$ has been recorded in *Alonsoa* (Steiner 1996). In Hemimerideae, base numbers are $x = 7, 9$. Also in *Diascia*, $2n = 18$ has been counted in most species, but also tetraploid ($2n = 36$) and hexaploid numbers ($2n = 54$) have been found (Steiner 1996). In "Veronicaceae", the base chromosome number of *Gratiola* and *Lindenbergia* is $x = 16$ (Hjertson 1995). Chromosome counts in *Limosella* showed a base number of $x = 10$. The prevailing chromosome number in *Digitalis* (base number $x = 14$) is $2n = 56$, and probably originated by autotetraploidy. Some species, e.g. *D. lutea*, have $2n = 112$. The base chromosome number of *Pseudolysimachion* is $x = 17$ and there are diploid ($2n = 34$) and tetraploid ($2n = 68$) taxa. In *Veronica*, very different chromosome numbers have been counted ($2n = 14, 16, 18, 28, 32, 36, 42, 48, 52, 54, 56, 64$); most frequently, however, $2n = 14$, 16 and 18. The base numbers $x = 7, 8, 9$ also occur in most other Veroniceae (e.g. *Wulfenia*, *Wulfeniopsis*, *Paederota*, *Campylanthus*; see Hjertson 1997). In *Hebe* and *Derwentia*, the base chromosome number is $x = 20$ whereas in *Heliohebe*, *Chionohebe* and *Parahebe*, a base number of $x = 21$ was found (Wagstaff and Garnock-Jones 1998).

In "Orobanchaceae", the base chromosome numbers are $x = 6, 8$ in *Pedicularis*, $x = 11$ in *Rhinanthus* and *Euphrasia* where diploid and tetraploid taxa are known (Ehrendorfer and Vitek 1984; Vitek 1985a, 1985b, 1986), and $x = 10, 12$ in *Odontites*. Starting with the base number $x = 12$, which also occurs in *Bornmuellerantha*, *Macrosyringion*, *Bartsia* and *Parentucellia*, a descending series of chromosome numbers probably has evolved: 12–10–9 (Bolliger 1996). Only *Odontitella*, with $2n = 26$ ($x = 13$), holds an isolated position. The base number $x = 10$ is found in the *Odontites luteus* group (*O. luteus*: $2n = 20$, *O. lanceolatus*: $2n = 20$). Within the *O. vernus* complex, Witsch (1932) and Snogerup (1977) have found $2n = 20$ (*O. vernus*: $2n = 40$, *O. vulgaris*: $2n = 20$), while Greilhuber (1971) und Snogerup (1977) have counted also $2n = 18$ in *O. vulgaris* and *O. litoralis*. The base number $x = 9$ is also found in *Melampyrum* and *Lathraea*, while $x = 10$ has been recorded in *Tozzia*.

In *Rhinanthus*, a unique feature has been described (Hambler 1953, 1954). Within the chromosome number of $2n = 22$, two distinctly different chromosome types occur, i.e. 14 large heterochromatic chromosomes (l-chromosomes) and eight minute, euchromatic k-chromosomes.

Pollination and Reproductive Systems. The genera of "traditional" Scrophulariaceae accepted here are all zoophilous and they are mainly pollinated by insects. Anemophily occurs, e.g. in *Plantago* (see Plantaginaceae) which is assigned to be closely related to *Veronica* or *Aragoa* in most recent molecular studies (see below). Five main pollination syndromes can be discerned: melittophily, psychophily, sphingophily, ornithophily and autogamy.

Pollination by butterflies (psychophily) is found in several Manuleae (e.g. *Sutera*, *Zaluzianskya*, Vogel 1954) and *Erinus*. These species have long corolla-tubes, a regular limb and the corolla colour varies from rose to creamy white or orange. Sphingophily, i.e. pollination by hawkmoths, occurs only in rather few genera. In Manuleae, several species of *Zaluzianskya*, *Jamesbrittenia* and *Lyperia* are nocturnal and their narrowly cylindrical corolla-tube attains a length of up to 50 mm. Usually, the corolla colour is pure white to cream or bright yellow. In the Malagasy genus *Leucosalpa*, all species are sphingophilous, nocturnal and smell intensively of jasmine. One extreme case is *Leucosalpa grandiflora* with a narrow corolla-tube of 14 cm length. Also in *Rhamphicarpa fistulosa* (Fig. 50), sphingophily is found. Phalaenophily (moth-pollination) may occur in some *Manulea* species (*M. cheiranthus*, Vogel 1954), where the corolla-tube is rather short (up to 12 mm). Ornithophily occurs in North (e.g. *Penstemon*, *Mimulus cardinalis*, *Castilleja*), Central and South America (e.g. *Lophospermum*, *Russelia*, *Ameroglossum*, *Bartsia*; Fig. 44), the Canary Islands and Madeira (*Isoplexis*), Africa and Madagascar (e.g. *Halleria*, *Graderia*, *Phygelius*, *Hyobanche*) and tropical Asia (e.g. *Wightia*). Most of the bird-pollinated Scrophulariaceae can be recognized by the red or yellow colour of their corolla, a reflexed or reduced lower lip and a rich production of nectar.

The majority of entomophilous species are pollinated by bees or bumblebees (melittophilous). The deposition of pollen can be sternotribous or nototribous, dependent on the position of anthers. Bee-pollinated flowers that are borne on remarkably long and thin pedicels are called micromelittophilous (Vogel 1954). Micromelittophily occurs in some Manuleae (*Jamesbrittenia*) and in *Kickxia*.

The pedicel of *K. cirrhosa* is only about 60 μm in diameter. The numerous variations of melittophily cannot be described in detail here. However, some will be shortly mentioned. The variation in corolla form in *Pedicularis* is apparently the result of adaptive responses to pollinators of the genus *Bombus* (Macior 1983). Attractants for bees are odour and colour. Sometimes, also the intensively coloured inflorescence bracts, e.g. in *Melampyrum* species, enhance the showy flowers by their contrast. The corolla often displays coloured nectar guides which also can reflect or absorb UV-light. In several cases anthers which are hidden within the corolla are imitated by yellow-coloured corolla protrusions or intensively zigzag-shaped yellow-coloured filaments which expose the curved and yellow base towards the mouth. These "false anthers" occur in various genera, e.g. *Craterostigma* (Magin et al. 1989) or *Melampyrum* and *Antirrhinum*. Usually, the floral reward is either pollen (e.g. in *Verbascum*) or nectar (in most other genera). Nectar usually is produced by a disc between ovary and flower base. Often the disc is not clearly distinguished from the ovary, or it is present as a ring-like structure at the ovary base. In some groups, it is developed as a protrusion on the abaxial side of the ovary (in Rhinanthoideae) or as a ring, which is functional only on its abaxial side (*Chelone*).

In some genera (e.g. *Penstemon*, *Verbascum*) the disc is non-functional and either pollen is offered as a floral reward or nectar is produced outside the disc. Dense glandular hairs also may produce nectar, and these trichome nectaries are arranged at the dilated base of the filaments (*Penstemon*, *Craterostigma*) or staminodes (*Collinisia*) or on the corolla itself (*Calceolaria*). The genus *Calceolaria* is an example of the oil flower syndrome (Vogel 1988). Here, fatty oils are produced, which are collected by special oil-bees. The oil-producing trichomes are situated on the outside of the corolla which is recurved towards the inner face of the "shoe". The oil-flower syndrome occurs within the "Calceolariaceae" (most species of *Calceolaria*), *Angelonia* (Vogel and Machado 1991), *Monopera*, *Basistemon*, *Diascia*, *Bowkeria* and *Monttea* (Simpson et al. 1990; Sérsic and Cocucci 1999). The bees collect the oil from the trichomes usually with their anterior legs and transfer it to special hair cushions on their posterior legs. The South African *Diascia* represents a rather spectacular example (Vogel 1984; Steiner and Whitehead 1990). Here the oil-producing trichomes are situated within the two spurs. It was only recently noted that bees of the genus *Rediviva* with extremely prolonged anterior legs act as pollinators (Vogel and Michener 1985). Steiner (1990) could show that a translucent "window" at the base of the upper corolla lip is UV-absorptive and contrasts with the surrounding corolla which strongly reflects UV-light. It can thus probably be used for orientation by oil-collecting bees.

In some genera, e.g. *Digitalis*, the anthers open early in anthesis, when the stigma is not yet susceptible. In autogamous species of *Euphrasia* (e.g. *E. minima*, *E. frigida*), an extreme proterandry is found and the anthers open already in minute flower buds. Self-pollination thus occurs before the opening of the corolla (see below). Proterogyny, however, is much more frequent. Self-pollination can provide a reproductive assurance when pollinators are scarce. Long-tubed flowers of *Ourisia* self spontaneously when pollinators are unavailable (Arroyo and Peñazola 1990). Delayed selfing occurs in *Collinsia verna* when there is a limitation of pollinators (Kalisz et al. 1999). The flowers of *Euphrasia willkommii* are visited by only few insects belonging to generalist taxa, i.e. thrips and ants (Gómez 2002). However, reproduction is not pollen-limited in this species. Hand-pollinations demonstrated that *E. willkommii* is capable of selfing, reproductive success being similar in autogamous and allogamous crossings. Under natural conditions, the plant relies predominantly on selfing, as seed production is similar in the presence or absence of pollinators (Gómez 2002). Autogamy is probably an ecological mechanism to ensure successful reproduction in a harsh environment. Usually, selfing is prevented during the first days of anthesis by a pronounced herkogamy, when the stigma projects beyond the corolla tip. The corolla later elongates, elevating the adnate stamens, whereas the style does not increase in length. Thus, anthers and stigma come closer to each other, leading to self-pollination. It is not known whether autogamy in *Euphrasia willkommii* is promoted by delayed self-pollination or by other mechanisms, such as incomplete dichogamy or close proximity of anthers and stigmas during dehiscence (Gómez 2002). Apomixis occurs in polyploids of *Cistanche* (Pazy 1998).

Seeds. Seed sculpturing is influenced by the surface of the endosperm (Bachmann 1880; Falcão Ichaso 1978; Canne 1980; Sutton 1988), the testa epidermis, and sometimes also by subepidermal cell layers. The shape and arrangement of the testa epidermis cause the development of a strong relief in most Antirrhineae. Wing-formation, which can be transversal (*Nemesia*) or median (*Linaria*, *Rhinanthus*), can be caused by local growth of integumental intermediate layers (*Linaria*; Fig. 43). In *Melampyrum* the seed epidermis, which is rich in fat and albumen, splits off the seed at maturity (see below). The endosperm is few-layered in Gratioleae (one-layered in *Gratiola*) and multilayered in most other groups. Besides the hemicellulose of the cell walls, aleuron and fatty oil serve as storage compounds. Starch occurs only occasionally and only in very small quantities (*Melampyrum*). Exceptionally, endosperm is lacking (*Monttea*). The embryo also contains fat oil in its cells. It is usually straight or nearly straight even when the seed becomes campylotropous (see above). The size of mature seeds frequently is negatively correlated with seed number per fruit. The four-seeded fruit of *Melampyrum* contains the largest seeds (6 mm long, 15 mg) in Scrophulariaceae, while the smallest seeds are found in the many-seeded Lindernieae (0.4 mm long in *Lindernia procumbens*). The importance of seed morphology for the systematics of the family has often been emphasized (e.g. Heil 1927; Musselman and Mann 1976; Elisens and Tomb 1983; Fischer 1992; Hilliard 1994).

Fruit and Dispersal. The most frequent fruit type in Scrophulariaceae is a capsule which can open transversally by splitting of the septum and its separation from the placenta which then is positioned between two moveable valves. This dehiscence is called septicidal-septifragous (e.g. in *Scrophularia*). Additional dehiscence along the median (loculicidal dehiscence) can be observed in *Gratiola*. Exclusively loculicidal fruits occur in Rhinanthoideae. A poricidal dehiscence is found with some variations within Antirrhineae (Sutton 1988). The pores originate along preformed dehiscing lines where the surrounding tissue recurves with teeth (*Antirrhinum*), or by circumscission with an operculum (*Kickxia*, Sutton 1988). In *Torenia*, a special case of poricidy occurs. The capsule opens adjacent to the septum on each side producing longitudinal pores. Later, the septa separate from the placenta and finally, after the capsule is emptied, the septum splits. At the time of the poricidal dehiscence of the capsule, the calyx opens along longitudinal slits. Only few genera have indehiscent fruits. A rare fruit type among Scrophulariaceae is a drupe which occurs in *Tozzia* and the Madagascan genus *Leucosalpa*. Berries are found in *Radamaea* and *Teedia*.

The fruit wall consists of some layers of sclerenchymatous cells. Drying out or (rarely) wetting causes tensions which finally induce dehiscence. The outer, parenchymatous layers of the fruit wall are either completely dry at maturity and the fruit thus is a dry capsule, or the parenchymatous layers remain alive in the so-called fleshy capsules of some Rhinanthoideae.

Seed dispersal is usually effected by wind shaking the mature capsules that are open only when dry. Several genera have winged seeds which support dispersal by wind. However, in genera such as *Bartsia* or *Rhinanthus* (Molau et al. 1989), the winged seeds are apparently not effectively wind-dispersed because ca. 90% of them only reach distances of up to 1 m from the mother plant. Also very small seeds, as in *Mimulus* or *Lindernia* (see above), are wind-dispersed. Dispersal by water occurs in genera like *Veronica*, in which the capsule opens after being wetted. Some *Veronica* species also show myxospermy and the testa is mucilagineous when wet. These seeds often are dispersed together with mud particles by water birds (e.g. in *Limosella*). Ateleochorous seeds occur in species where the seeds risk to reach unfavourable habitats where they can not establish (e.g. *Cymbalaria* spp., *Stemodiopsis* spp.).

Myrmechory occurs in *Veronica hederifolia*, *Melampyrum*, *Pedicularis sylvatica* and others. Ants are attracted by nutritional bodies which assure that the seed itself is not predated. In *Melampyrum* the seed epidermis is rich in fat and albumen and splits off the seed at maturity. The seed also resembles an ant larva in shape, weight and colour (first white, later nigrescent). Most *Melampyrum* species have additional attractants for ants, such as extrafloral nectaries (glandular hairs) on the upper leaves and bracts. Once attracted by nectar, the ants creep into the opened capsule and take away the seeds. Autochory is remarkably rare in Scrophulariaceae. Ballistic fruits occur in *Lathraea clandestina* where the fleshy layers of the capsule (see above) are responsible for a turgor mechanism of its ballistic fruit.

In *Cymbalaria* and the African genus *Stemodiopsis*, the capsules are negatively phototropous and geocarpous and thus move their fruits into rock fissures.

Phytochemistry. Characteristic substances of the Scrophulariaceae (Hegnauer 1973, 1990) are iridoids, which are derived from bicyclic monoterpenes. These substances can easily polymerise, which causes the typical black colour of dry material of many taxa. Mainly C9-iridoid glycosides, especially aucubin and catapol and their derivatives, are widely distributed within the family. The iridoid glycosides may deter polyphagous herbivores due to their bitter taste, but some insects harvest them for protection against predation by birds. Pyrrolizidine alkaloids, otherwise known only from Asteraceae, Boraginaceae and Fabaceae, have been found in *Castilleja* and *Melampyrum* (Roeder and Borauel 1992).

Starch as a reserve substance is of only rare occurrence and seeds generally lack starch. As in other asterids, starch is mainly replaced by oligosaccharids (stachyose, verbascose, saccharose -galactosids, also saccharose). The scanty investigations suggest that hexites are mainly deposited in Rhinanthoideae, while stachyose is mainly found in Scrophularioideae.

Real tannins are apparently lacking. They are replaced by tannin-like derivatives of aromatic acids (e.g. benzoic acid, caffeic acid and protocatechu acid); they mostly occur as esters or depsides (verbascosid-like esterglucosides). Among the flavonoid constituents of leaves, flavones are more frequent than flavonols. Saponins are also common in the family and they generally have a triperpene structure. In *Digitalis* and *Campylanthus*, the rare steroid saponins could be isolated. *Digitalis* differs from the majority of Scrophulariaceae by lacking iridoids and containing *Digitalis* glycosides which have medical importance as a cardiac stimulant.

Fossil Records. The only reliable fossil records of a Scrophulariaceae ("Paulowniaceae") are leaves of *Paulownia* (Smiley 1961) from the Tertiary of the Ellensburg Formation (transition Miocene/Pliocene) which closely resemble the recent *P. tomentosa*. They have been described as *Paulownia columbiana*.

Distribution and Habitats. The distribution of Scrophulariaceae is almost cosmopolitan, but areas of highest species-richness are located in the holarctic region and tropical mountains. Lowland tropics are comparatively species-poor and only a few genera are restricted to these regions. Centres of diversity are often mountain ranges, e.g. the Alps and the Central Asian mountains for *Pedicularis*, the Alps and the Australian and New Zealand mountains for *Euphrasia*, the Balkans for *Melampyrum*, the Rocky Mountains for *Penstemon*, *Mimulus* and *Castilleja*, and the Andes for *Calceolaria* and *Bartsia*. Diversity centres around the Mediterranean are known for *Verbascum* and *Scrophularia* (southwest Asia) as well as for *Antirrhinum* and *Linaria* (western Mediterranean). Southern Africa is a centre of diversity for Hemimerideae (*Diascia*) and Manuleae (incl. Selagineae).

Scrophulariaceae are generally found in open, grassy habitats and are most frequently distributed in alpine lawns, savannah vegetation or steppe areas. Only a few taxa are found in Northern Hemisphere forests (e.g. *Lathraea*, *Conopholis*, *Epifagus*). Tropical rainforests also are rarely inhabited by Scrophulariaceae and then mostly in disturbed areas such as along roads and paths, in clearings, plantations or around settlements (e.g. *Lindernia*, *Torenia*, *Scoparia*). Most species of the tropical African *Harveya* and few of *Striga* (e.g. *S. hallaei*) are primary rainforest taxa.

A specialised habitat for Scrophulariaceae are granitic inselbergs and ferricretes. Here, they can be found in rock pools, on exposed rock surfaces and on rock debris. Besides annual species (e.g. *Dopatrium*, Fischer 1997b; *Lindernia conferta*, Seine et al. 1995), poikilohydric Scrophulariaceae of Lindernieae are found in such habitats in Africa and Madagascar (Fischer 1992, 1995). Desiccation tolerance can be found in all species of *Craterostigma* (Fig. 47; Gaff 1977; Fischer 1992), several *Lindernia* species (e.g. *L. pulchella*, *L. monroi*, Seine et al. 1995) and in *Chamaegigas* (Hickel 1967; Schiller et al. 1999). *Craterostigma plantagineum* can tolerate up to 96% loss of its relative water content and recover within several hours from such extreme dehydration. *Chamaegigas intrepidus* is restricted to shallow and temporarily water-filled rock pools on granitic inselbergs in Namibia (Hickel 1967), and the rosulate leaves can survive long desiccation periods. Often the plants pass through 15–20 cycles of rehydration and dehydration in a single season. Recent studies in *Craterostigma* (Bartels et al. 1990; Bernacchia et al. 1996; Ingram and Bartels 1996) and *Chamaegigas* (Schiller et al. 1997, 1998; Heilmeier et al. 2000) have shown that abscisic acid (ABA), a stress hormone, plays an important role in desiccation tolerance. During a drying cycle, ABA increased both in roots and submerged leaves in linear relation with decreasing osmotic potential. *Chamaegigas intrepidus* also showed uptake of glycine and serine from the pool water. The most abundant organic N compound there was urea (Schiller et al. 1998). Because urease activity was high even in dry sediments, catabolism of urea to $NH4^+$ may be the major pathway for providing N for *Chamaegigas* in this stressful environment.

Subdivision and Relationships. Conventional angiosperm systems (e.g. Cronquist 1981) recognized Lamiales (= Scrophulariales) to contain 12 to 15 families. The Scrophulariaceae in the circumscription of Bentham (1846, 1876) and Wettstein (1891) were suggested to be related to Acanthaceae with *Nelsonia* or to Bignoniaceae with *Schlegelia*. Closely related families like Orobanchaceae, Lentibulariaceae, Globulariaceae or Plantaginaceae were often considered to be derived from Scrophulariaceae, holding a central position in Lamiales. Only Hallier (1903) included the latter families into Scrophulariaceae. However, most authors followed Wettstein in the circumscription and division of Scrophulariaceae. He divided the family into the three subfamilies Pseudosolaneae, Antirrhinoideae and Rhinanthoideae. This concept has been only slightly modified by later authors, by uniting Pseudosolaneae and Antirrhinoideae (Thieret 1967) or including Orobanchaceae as a further subfamily (Takhtajan 1987).

Within this subdivision, the position of several genera has been discussed controversially. *Paulownia* had been placed in Scrophulariaceae by Bentham (1876) and Wettstein (1891), but in Bignoniaceae by Takhtajan (1980) and Cronquist (1981). Armstrong (1985) argued for a position in Scrophulariaceae and this view has been followed by Takhtajan (1987) and Thorne (1992). Also *Schlegelia* and *Gibsoniothamnus* were placed in Bignoniaceae (Bentham 1876; Gentry 1980; Cronquist 1981), and again Armstrong (1985) argued for a placement of *Schlegelia* in Scrophulariaceae. Subsequently, Thorne (1992) established tribe Schlegelieae in Scrophulariaceae. Another group of doubtful relationships are the Selagineae which had been maintained as a separate family by Bentham (1876), but included into Scrophulariaceae by Wettstein (1891) and most recent authors (e.g. Thorne 1992; Hilliard 1994, 1999). Cronquist (1981), however, placed Selagineae into Globulariaceae. The latter family was reduced to a tribe of Scrophulariaceae by Barringer (1993). The small aquatic families Callitrichaceae and Hippuridaceae were also considered to be closely related to Scrophulariaceae.

For several decades, it has been obvious from morphology that, in their traditional circumscription, Scrophulariaceae do not represent a monophyletic group (e.g. Hartl 1965–1974; Fischer 1992). This view more recently was supported by studies of the chloroplast genes *rbc*L and *ndh*F (Olmstead and Reeves 1995). Thus, Scrophulariaceae are polyphyletic and should be divided into several distinct families. Olmstead and Reeves (1995) showed that at least two clades are strongly supported: "Scroph I" with *Verbascum*, *Celsia* (= *Verbascum*), *Scrophularia* and *Selago* (but also containing *Buddleja* and *Nicodemia*) and "Scroph II" with *Antirrhinum*, *Digitalis*, *Veronica*, *Plantago*, *Hippuris* and *Callitriche*.

The close relationship of Buddlejaceae s.str. (comprising *Buddleja*, *Nicodemia*, *Emorya*, *Chilianthus* and *Gomphostigma*) to Scrophulariaceae s.str. was confirmed by Oxelman et al. (1999). The Scrophulariales sensu Thorne are well supported. Within the order, a Gesneriad clade, a Stilbacean clade, two "Scroph" clades, Acanthaceae, Bignoniaceae, Lamiaceae, and Verbenaceae occur as distinct lineages. Young et al. (1999) could show that all parasitic Scrophulariaceae are probably monophyletic and that parasitism has originated only once. However, there has been a multiple, independent origin of holoparasitism. Besides the parasites, the two clades of Olmstead and Reeves (1995) are again supported. The genera *Scrophularia*, *Verbascum* and *Selago* come out as closely related to each other, as do *Digitalis*, *Hemiphragma* and *Veronica*. The latter clade also contains *Antirrhinum*, *Chelone*, *Callitriche* and *Hippuris*, and the former clade *Hemimeris*, *Leucophyllum* and *Myoporum* (Young et al. 1999). *Mimulus* and *Paulownia* represent clades closer to parasitic figworts than to the remaining Scroph-clades, and the positions of *Calceolaria* and *Gratiola* are unresolved.

In a recent paper (Olmstead et al. 2001), the Scrophulariaceae were disintegrated into several distinct families: Scrophulariaceae s.str. with Hemimerideae, Aptosimeae, Verbasceae, Manuleae, Selagineae but also *Buddleja* and *Myoporum*, Veronicaceae including part of Gratioleae, Antirrhineae, Hemimerideae, Digitalideae but also *Callitriche*, *Hippuris*, *Globularia* and *Plantago*, Calceolariaceae, Paulowniaceae, Schlegeliaceae, and the parasitic Orobanchaceae. Recent studies on different tribes of Scrophulariaceae (e.g. Gebrehewit et al. 2000; Albach and Chase 2001; Beardsley and Olmstead 2002; Bello et al. 2002; Kornhall et al. 2001) now show a fairly complete picture of families included in the traditional Scrophulariaceae. The Scrophulariaceae s.str. comprise tribes Verbasceae, Scrophularieae, Alonsoeae, Hemimerideae (excluding *Angelonia* and *Collinsia*, Olmstead et al. 2001), Russelieae, Teedieae, Leucophylleae, Freylinieae and Manuleae. The Selagineae have to be included in tribe Manuleae (Kornhall et al. 2001) while *Antherothamnus* and *Camptoloma* remain in Scrophulariaceae s.str. but have to be excluded from the Manuleae. The Buddlejaceae and Myoporaceae should also be included in Scrophulariaceae s.str. The Schlegeliaceae, Calceolariaceae and Paulowniaceae should each be recognized at familial level.

A more problematical group is the "Veronicaceae" sensu Olmstead et al. (2001). It contains most of the non-parasitic Scrophulariaceae, but also the small families Callitrichaceae, Globulariaceae, Hippuridaceae and Plantaginaceae (Olmstead et al. 2001). The oldest name for the group, Veronicaceae, probably has to be replaced by the conserved name Plantaginaceae. There is moderately strong support for a clade containing *Antirrhinum, Callitriche, Hippuris, Globularia, Digitalis, Hemiphragma, Plantago* and *Veronica*. The genus *Veronica* has been shown to be paraphyletic (Albach and Chase 2001). The genus *Aragoa* is a sister group to *Plantago* (Bello et al. 2002). The Cheloneae, comprising *Chelone, Tetranema* and *Collinsia*, also form a moderately well-supported group. The tribe Gratioleae forms a strongly supported group and, together with *Angelonia*, is a clade that is sister group to the rest of Veroniceae. The position of *Lindenbergia*, however, is questionable. In several studies (Young et al. 1999; Olmstead et al. 2001) it forms a clade together with the parasitic Scrophulariaceae. This is not supported by morphological data which show a close resemblance to tribe Stemodieae (Hjertson 1995). The aestivation can vary from rhinantoid to antirrhinoid even on a single plant (Hartl 1955), and can thus not be used for systematic placement.

The genus *Mimulus* (tribe Mimuleae) as well as *Glossostigma, Peplidium, Leucocarpus, Berendtiella, Hemichaena, Mazus* and *Lancea* can be recognized at the rank of a family as Phrymaceae (Beardsley and Olmstead 2002). The genus *Mimulus* is not monophyletic and the above-mentioned genera are derived from within *Mimulus*.

The parasitic Scrophulariaceae are a well-supported clade and form a monophyletic group (Young et al. 1999; Olmstead et al. 2001). This concept, originally proposed by Bellini (1907), was supported by morphological and anatomical data (Kuijt 1969; Hartl 1965–1974; Armstrong and Douglas 1989). Only *Lindenbergia* is excluded in this treatment.

In this treatment no families are formally segregated from the traditional Scrophulariaceae. More studies are needed in this respect, especially concerning the rather broadly circumscribed Antirrhinaceae sensu Young et al. (1999) and Veronicaceae sensu Olmstead et al. (2001). In the present paper, the families to be recognized are listed and their genera are arranged according to the presently available morphological and molecular data. The keys and descriptions include *Cyrtandromoea, Rehmannia* and *Titanotrichum* (usually placed in Gesneriaceae, see treatment in this volume), and *Halleria, Bowkeria, Anastrabe* and *Ixianthes* which should be placed in Stilbaceae (see treatment of Stilbaceae in this volume). Not included are the genera traditionally assigned to Myoporaceae and Buddlejaceae which should be included in Scrophulariaceae sensu stricto, and Plantaginaceae, Globulariaceae, Hippuridaceae and Callitrichaceae which should be included in "Veronicaceae/Plantaginaceae" (see the respective treatments in this volume).

Conspectus of "Families" and Tribes

"Scrophulariaceae"

1. Tribe Verbasceae Dum.
   Genera 1–2
2. Tribe Scrophularieae.
   Genera 3–6
3. Tribe Alonsoeae Barringer.
   Genera 7–8
4. Tribe Hemimerideae Benth.
   Genera 9–14
5. Tribe Russelieae Pennell.
   Genera 15–17
6. Tribe Teedieae Benth.
   Genera 18–20
7. Tribe Freylinieae Barringer.
   Genera 21–22
8. Tribe Leucophylleae Miers.
   Genera 23–24
9. Tribe Aptosimeae (Benth.) Benth.
   Genera 25–27
10. Tribe Manuleae Benth. & Hook. f.
    Genera 28–52

Genus of uncertain placement

Genus 53

"Schlegeliaceae"

Genera 54–57

"Calceolariaceae"

Genera 58–60

"Paulowniaceae"

1. Tribe Paulownieae Pennell.
   Genera 61–62
2. Tribe "Wightieae".
   Genera 63–64

"Veronicaceae"

I. Subfam. Antirrhinoideae (Speta) D. Hartl & I. Betsche.

1. Tribe Cheloneae Benth.
   Genera 65–74

2. Tribe Antirrhineae Chav.
   Genera 75–102

II. Subfam. Gratioloideae (Benth.) Luerss.

1. Tribe Angelonieae Benth.
   Genera 103–104
2. Tribe Gratioleae Benth.
   Genera 105–120
3. Tribe Limoselleae Dum.
   Genus 121
4. Tribe Stemodieae Benth.
   Genera 122–136
5. Tribe Lindernieae Rchb.
   Genera 137–150
6. Tribe Melospermeae Rossow.
   Genera 151–152

III. Subfam. Digitalidoideae (Dum.) Luerss.

1. Tribe Digitalideae (Dum.) Dum.
   Genera 153–154
2. Tribe Veroniceae Bartling.
   Genera 155–176
3. Tribe Sibthorpieae Benth.
   Genera 177–178
4. Tribe Hemiphragmeae Rouy.
   Genus 179
5. Tribe Aragoeae (G. Don) D.Y. Hong & S. Nilsson.
   Genus 180

Genera of uncertain placement
Genera 181–182

"Phrymaceae"

1. Tribe Microcarpeae Miq.
   Genera 183–193
2. Tribe Leucocarpeae Rouy.
   Genera 194–195
3. Tribe Mimuleae Dum.
   Genera 196–201

"Orobanchaceae"

1. Tribe Gerardieae Benth.
   Genera 202–213
2. Tribe Orobancheae Lam. & DC.
   Genera 214–228
3. Tribe Escobedieae Benth.
   Genera 229–236
4. Tribe "Micrargerieae".
   Genera 237–240
5. Tribe Buchnereae Benth.
   Genera 241–258
6. Tribe "Buttonieae".
   Genera 259–263
7. Tribe "Xylocalyceae".
   Genus 264
8. Tribe Cymbarieae D. Don.
   Genera 265–270
9. Tribe "Castillejeae".
   Genera 271–277
10. Tribe Rhinantheae Lam. & DC.
    Genera 278–297

Genera of uncertain placement in "Veronicaceae" or Gesneriaceae

Tribe Rehmannieae Rouy.
Genera 298–301

"Stilbaceae"

Tribe Bowkerieae Barringer.
Genera 302–306

Key to Genera of Scrophulariaceae

1. Holoparasitic herbs with unilocular ovary and 2–6 parietal placentae 2
- Autotrophic, hemiparasitic or rarely holoparasitic herbs, shrubs or trees; ovary bilocular; placentae central 16
2. Prophylls present 3
- Prophylls absent 10
3. Anthers unequal, often one theca reduced; ovary with 2 placentae, East Asia **219. *Christisonia***
- Anthers equal; ovary with 2–4 placentae 4
4. Calyx tubular, split on one side, irregularly 4–5-dentate; lower lip of corolla shorter than upper lip, eastern North America **296. *Conopholis***
- Calyx tubular to campanulate, not split on one side; upper lip of corolla equalling lower lip or smaller 5
5. Calyx cupulate, truncate or shortly 1–5-dentate; leaves densely imbricate; northern and eastern Russia and western North America **218. *Boschniaka***
- Calyx not cupulate or truncate, 4–5-lobed or dentate; leaves lax to dense 6
6. Flowers dimorphic, lower flowers small, cleistogamous, pistillate; upper flowers chasmogamous; eastern North America south to Mexico **228. *Epifagus***
- Flowers monomorphic, all flowers chasmogamous 7
7. Upper lip of corolla galeate; placentae 2 8
- Upper lip of corolla spreading; placentae 2–4 9
8. Corolla with hairs only at margin of lobes; stigma peltate **222. *Gleadovia***
- Corolla woolly internally and externally; stigma clavate **223. *Mannagettea***
9. Plants with thick, fleshy stem; calyx tubular, 4–5-lobed, lobes obtuse to rounded; corolla cream to yellow or violet **220. *Cistanche***
- Plants not fleshy; calyx campanulate with acute teeth or lobes; corolla yellowish-brownish to blue or purple **224. *Orobanche***
10. Calyx spathe-like; anthers monothecous; pedicel much longer than stem **217. *Aeginetia***
- Calyx not spathe-like; anthers bithecous; pedicel short to long, occasionally longer than stem 11
11. Calyx tubular to campanulate, 5-dentate or 5-lobed; ovary with 2 carpels and 2–4 placentae 12
- Calyx with 2–3 free segments, bipartite or cupulate to truncate; ovary with 3 carpels and 3–6 placentae 14
12. Pedicel long, usually longer than stem; inflorescence with 1–2 flowers, corolla crimson to red, placentae 4 **221. *Phelypaea***
- Pedicel short to long; inflorescence usually a many-flowered raceme, rarely 1-flowered (*Orobanche uniflora*); corolla brownish-yellow to purple; placentae 2–4 13
13. Calyx campanulate, corolla brownish-yellow to purple, stigma capitate or lobed, placentae 2–4, plants mainly of Northern Hemisphere **224. *Orobanche***
- Calyx tubular, 5-dentate or 5-lobed, corolla yellow to purple, stigma bilobed or peltate, placentae 2, subtropical and tropical Asia **219. *Christisonia***
14. Upper lip much longer than lower lip; stigma peltate; calyx bipartite or cupulate to truncate 15
- Upper lip equalling lower lip; stigma clavate; calyx with 2–3 free segments **225. *Phacellanthus***

15. Calyx bipartite, lateral parts 2-dentate; placentae 4–6 **226. *Platypholis***
- Calyx cupulate, truncate or 3-lobed; placentae 3 **227. *Xylanche***

16 (1). Corolla with distinct abaxial spur (or 2 spurs) or at least gibbous abaxially 17
- Corolla without abaxial spur or gibbosity 56

17. Palate on lower lip of corolla absent; capsule opening loculicidal or septicidal 18
- Palate on lower lip of corolla usually present; capsule usually opening by pores 27

18. Shrubs or herbs with yellow corolla and 2 gibbous spurs or pouches; South Africa 19
- Herbs or subshrubs with white to pink, purple or blue-violet corolla, 1–2 gibbous spurs or pouches present 21

19. Inflorescence cymose; leaves 3–4-verticillate; calyx 3-lobed; prophylls present; stamens 2, staminodes 2–3 **306. *Ixianthes***
- Inflorescence racemose; leaves opposite or alternate; calyx deeply 5-lobed; prophylls absent; stamens 2–4, staminodes absent 20

20. Dwarf shrub; leaves alternate, rarely opposite; corolla sulphur-yellow, stamens 4 **14. *Colpias***
- Herb; leaves opposite; corolla bright yellow; stamens 2 **11. *Hemimeris***

21. Corolla with 2 gibbous spurs or pouches 22
- Corolla with 1 gibbous spur or pouch 24

22. Shrubs or trees; inflorescence cymose; South America **9. *Basistemon***
- Herbs; inflorescence racemose; South America or South Africa 23

23. Corolla blue to violet; leaves opposite, alternate above; stigma clavate; South America **103. *Angelonia***
- Corolla pink to purple; leaves opposite, sometimes rosulate; stigma clavate or capitate; South Africa **10. *Diascia***

24. Flowers resupinated; upper lip 4-lobed, lower lip emarginate; South and tropical Africa **13. *Nemesia***
- Flowers not resupinated; upper lip 2-lobed, lower lip 3-lobed 25

25. Corolla with distinct spur; inflorescence racemose 26
- Corolla with abaxial gibbosity on tube; inflorescence cymose, flowers 1–3 in axillary cymes; western North America **73. *Collinsia***

26. Corolla white; stigma bilobed; erect or creeping herbs of tropical Africa and Madagascar **12. *Diclis***
- Corolla blue to violet; stigma clavate; erect herbs of South America **104. *Monopera***

27. Fertile stamens 4; anthers marginally coherent and forming a ring-like structure 28
- Fertile stamens 2 or 4; anthers free or marginally coherent in pairs, not forming a ring-like structure 30

28. Cauline leaves usually deeply palmatisect, more rarely undivided; petioles not twining; capsule with each pore opening by a single, ligulate valve **86. *Anarrhinum***
- Cauline leaves often hastate or sagittate at base but not palmatisect; petioles often twining; capsule with each pore opening by an operculum or several valves or teeth 29

29. Anthers ciliate; capsule dehiscence operculate **84. *Kickxia***
- Anthers glabrous; capsule dehiscence valvate **85. *Nanorrhinum***

30. Abaxial base of corolla-tube with distinct long spur 31
- Abaxial base of corolla gibbous or sometimes saccate 37

31. Leaves lobed or dentate, palmately veined, long-petiolate **83. *Cymbalaria***
- Leaves entire, pinnately veined, sessile or shortly petiolate 32

32. Calyx with adaxial lobe much shorter than 4 abaxial lobes, usually filiform 33
- Calyx with adaxial lobe more or less equalling to greatly exceeding the 4 abaxial lobes, not filiform 34

33. Spur conical, straight, much shorter than rest of corolla **77. *Holzneria***
- Spur cylindrical or narrowly conical, often arcuate, greatly exceeding rest of corolla **87. *Linaria***

34. Leaves sessile; spur cylindrical or narrowly conical, sometimes linear or filiform, often arcuate 35
- Leaves petiolate, at least below; spur conical or fusiform, straight 36

35. Lower lip of corolla greatly exceeding upper lip; seeds with 4–7 longitudinal ridges **88. *Nuttallanthus***
- Lower lip of corolla equalling or weakly exceeding upper lip; seeds with up to 3 longitudinal ridges or with encircling wing **87. *Linaria***

36. Capsule 4-angled towards apex, loculi equal, each loculus opening by a laterally elongate, entire pore **78. *Albraunia***
- Capsule terete towards apex, loculi unequal or equal, each loculus opening by a dentate or valvate pore **76. *Chaenorrhinum***

37. Leaves linear, oblong, lanceolate, ovoid, elliptic or orbicular, pinnately veined or obscurely veined except for midrib, margin entire, base never hastate or sagittate 38
- Leaves reniform, orbicular, deltate or 5-angled, palmately veined, margin dentate or crenate, more rarely entire except for hastate or sagittate basal lobes 49

38. Seeds conspicuously dorsiventrally asymmetrical, more or less cyathiform, the ventral face with encircling wing or ridge 39
- Seeds usually radially symmetrical, if dorsiventrally asymmetrical then not cyathiform, the ventral face cristate, cristate-alate or alveolate but not with encircling wing or ridge 42

39. Corolla-tube held more or less vertically, lips much longer than tube; fertile stamens 2 **93. *Mohavea***
- Corolla-tube more or less horizontally oriented, lips shorter than or more or less equalling tube; fertile stamens 4 40

40. Calyx lobes usually exceeding corolla-tube; capsule with unequal loculi **81. *Misopates***
- Calyx lobes shorter than corolla-tube; capsule with equal loculi 41

41. Shrubs; leaves sessile; pedicels erect or erect-patent in fruit **80. *Acanthorrhinum***
- Herbs; leaves petiolate; pedicels recurved in fruit **92. *Pseudorontium***

42. Pedicels greatly exceeding corolla, usually cirrhous **91. *Neogaerrhinum***
- Pedicels shorter than or rarely equalling corolla, very rarely cirrhous 43

43. Capsule with more or less equal loculi 44
- Capsule with unequal loculi 46

44. Herbs; leaves sessile, the cauline leaves mostly alternate; corolla pale blue, tube short **88. *Nuttallanthus***
- Shrubs; leaves petiolate, cauline leaves mostly opposite or verticillate; corolla red or rarely white, tube elongate 45

45. Capsule wall thin, loculi opening by irregular ruptures; seeds more or less radially symmetrical, longitudinally cristate **101. *Galvezia***
- Capsule wall thickened, loculi opening by several teeth; seeds bilaterally symmetrical, usually with conspicuous dorsiventral asymmetry, ventral face alveolate or longitudinally cristate **102. *Gambelia***
46. Calyx with adaxial lobe usually conspicuously exceeding 4 abaxial lobes, exceeding corolla-tube; capsule with adaxial loculus few-seeded, indehiscent **79. *Schweinfurthia***
- Calyx with adaxial lobe equalling or more rarely exceeding 4 abaxial lobes, usually shorter than corolla-tube; capsule with adaxial loculus many-seeded, dehiscent 47
47. Dwarf shrubs or perennial, suffrutescent herbs; bracts or floral leaves not subtending additional branches; style erect; capsule with thickened walls; seeds radially symmetrical **75. *Antirrhinum***
- Annual or perennial herbs; bracts or floral leaves often subtending additional branches, the branches usually cirrhous; style obliquely inserted; capsule with thin walls; seeds more or less dorsiventrally asymmetrical 48
48. Leaves with midrib terminating in prominent gland, projecting from abaxial face of leaf; corolla with poorly developed palate, not occluding throat **90. *Howeliella***
- Leaves with midrib terminating in minute or more or less conspicuous gland but gland not projecting from abaxial face of leaf; corolla with prominent palate, more or less occluding throat **89. *Sairocarpus***
49. Upper leaves opposite; lower lip of corolla with conspicuously bigibbous palate **82. *Asarina***
- Upper leaves alternate; lower lip of corolla with inconspicuous or simple palate 50
50. Seeds with 2 longitudinal wings or with encircling, emarginate wing 51
- Seeds without wings or with 4 wing-like ridges 53
51. Calyx lobes gibbous or carinate, indurate in fruit; capsule with each loculus opening by broad, solitary valve **96. *Epixiphium***
- Calyx lobes not gibbous, carinate or indurate in fruit; capsule with each loculus opening by irregular ruptures 52
52. Flowers pendent; calyx divided to less than middle, somewhat inflated, membranous; corolla subequally 5-lobed or scarcely 2-lipped **100. *Rhodochiton***
- Flowers more or less horizontally oriented; calyx divided to more than middle, not inflated or membranous; corolla 2-lipped **99. *Lophospermum***
53. Corolla with conspicuous palate; capsule with unequal loculi 54
- Corolla with inconspicuous palate or palate lacking; capsule with more or less equal loculi 55
54. Leaves spinulose-dentate; calyx lobes spinulose-dentate; corolla yellow; seeds foveate **98. *Holmgrenanthe***
- Leaves entire except for basal lobes; calyx lobes entire; corolla bluish-violet; seeds tuberculate **95. *Maurandella***
55. Stems scandent; leaves deltate or 5-lobed, margin otherwise entire; petioles twining **94. *Maurandya***
- Stems trailing or ascending; leaves reniform, orbicular or cordate, margin dentate or crenate; petioles not twining **97. *Mabrya***
56 (16). Inflorescence cymose, usually composed of main axis with several to many lateral cymes 57
- Inflorescence racemose to spicate, often ramified, but ramifications with same structure as main axis 106
57. Trees or woody climbers 58
- Shrubs, subshrubs or herbs 66
58. Leaves coriaceous with spiny margin; fruit a berry; endemic tree of Cuba up to 10 m tall **56. *Synapsis***
- Leaves neither coriaceous nor spiny; fruit a capsule 59
59. Corolla blue to violet; temperate East Asia 60
- Corolla white, yellow, pink or orange; tropical Asia, South Africa or Central America 61
60. Leaves isophyllous, corolla large (5–10 cm long), stigma capitate **61. *Paulownia***
- Leaves anisophyllous, corolla small (less than 3 cm long), stigma bilobed **62. *Shiuyinghua***
61. Hemiepiphytic trees, vines or pseudolianas, often tree stranglers; corolla incurved; seeds winged; Southeast Asia 62
- Small trees, never vines or lianas; corolla not incurved; seeds not winged; South Africa and Central America 63
62. Calyx 5-lobed; corolla brownish-yellow; mainly vines **63. *Brandisia***
- Calyx 3–5-lobed or cupulate; corolla pink to orange; mainly hemiepiphytic trees or stranglers **64. *Wightia***
63. Corolla white to yellow, large (up to 8 cm long), lower lip not pouch-like; stigma clavate or bilobed; Central America and Southeast Asia 64
- Corolla white or yellow with crimson, lower lip pouch-like; stigma clavate; South Africa 65
64. Stigma capitate; corolla white to yellow; plants glabrous to sparsely pubescent; Central America **23. *Eremogeton***
- Stigma bilobed; corolla white; plants densely tomentose to villose; Borneo **304. *Brookea***
65. Corolla yellow, inside with crimson **302. *Anastrabe***
- Corolla white **303. *Bowkeria***
66. Shrubs or subshrubs 67
- Herbs 92
67. Anthers synthecous 68
- Anthers with 2 separate, equal thecae, often confluent at apex, but never synthecous 76
68. Corolla actinomorphic, rotate to subrotate; stamens 4–5 69
- Corolla zygomorphic; stamens 4 or 2, sometimes a 5th median staminode present 70
69. Corolla large, with short tube, yellow to violet; stamens 4–5, filaments densely woolly; fruit a capsule with numerous seeds; Europe, Mediterranean and Asia **1. *Verbascum***
- Corolla small, with long tube, yellow; stamens 4, filaments never woolly; fruit a capsule with 1 seed; South Madagascar **19. *Ranopisoa***
70. Corolla greenish-brown to dark brown or red, small, tube ventricose; staminode present at base of upper lip **5. *Scrophularia***
- Corolla white to yellow or blue to red, small to large, but tube never ventricose, usually cylindrical or campanulate; staminode present or absent 71
71. Corolla-tube funnel-shaped; large staminode present near base of tube on adaxial side **6. *Antherothamnus***
- Corolla-tube campanulate, cylindrical or funnel-shaped; staminode absent 72
72. Large shrubs, corolla-tube campanulate or funnel-shaped 73
- Small shrubs or shrublets; corolla-tube cylindrical or funnel-shaped 74
73. Corolla red, large (up to 80 mm); stigma capitate; shrubs of South Madagascar **28. *Barthlottia***
- Corolla white, small (up to 15 mm); stigma bifid; shrubs of Southwest Africa **29. *Manuleopsis***

74. Seeds aulacospermous; plants densely glandular-pubescent; leaves alternate **53. *Camptoloma***
- Seeds bothrospermous; plants usually not glandular-pubescent; leaves opposite, sometimes rosulate 75
75. Corolla-tube funnel-shaped to cylindrical; stamens 4; stigma bifid **32. *Sutera***
- Corolla-tube narrowly funnel-shaped; stamens 4 or 2; stigma lingulate **33. *Manulea***
76. Fruit indehiscent, a drupe or berry 77
- Fruit a dehiscent capsule 83
77. Stamens 5; corolla bright red, trumpet-shaped, subactinomorphic; seeds with stout ridges; fruit with fleshy exocarp and coriaceous endocarp; epiphytic shrub of South Africa **17. *Dermatobotrys***
- Stamens 4, sometimes a staminode present; corolla variously coloured and shaped; fruit a berry or drupe 78
78. Staminode present; corolla red, subactinomorphic, with narrow straight tube; epiphytic or hemiepiphytic shrubs of Central America 79
- Staminodes absent; corolla red, white, blue, lilac or yellow, tube curved or straight; terrestrial shrubs or herbs 80
79. Calyx 5-lobed; stigma capitate; seeds compressed, angular **54. *Gibsoniothamnus***
- Calyx 3–4-lobed; stigma bilobed or trilobed; seeds compressed, winged **55. *Schlegelia***
80. Corolla white, subrotate; seeds bothrospermous; fruit a black drupe; South Africa **18. *Oftia***
- Corolla yellow, blue to lilac or red; seeds aulacospermous or not alveolated; fruit a berry or rarely a drupe 81
81. Calyx membranaceous, cupulate, with 5 setaceous teeth; corolla yellow, 2-lipped; fruit a white berry; Central America **195. *Leucocarpus***
- Calyx not membranaceous, deeply 5-lobed or 3–5-lobed; corolla blue or red, subrotate; fruit a dark to blackish drupe or berry; Africa and Madagascar 82
82. Shrubs with blue to lilac, subrotate corolla, tube narrow; seeds aulacospermous; fruit a drupe or berry, South Africa **20. *Teedia***
- Shrubs or trees with red, campanulate corolla, tube inflated; seeds compressed, narrowly winged; fruit a black berry, African and Madagascan high mountains **305. *Halleria***
83. Leaves alternate; corolla dark purple, throat barbate; Central America **24. *Leucophyllum***
- Leaves opposite, sometimes alternate above; corolla variously coloured, throat usually glabrous 84
84. Corolla bright red 85
- Corolla white, yellow, purple or blue 87
85. Corolla campanulate, tube funnel-shaped; staminode absent; stigma bilobed; South Africa **21. *Phygelius***
- Corolla campanulate, tube cylindrical or funnel-shaped; staminode usually present, rarely absent; stigma capitate; Central and South America 86
86. Lower lip of corolla only apically weakly divided, of undivided, 1-lobed appearance; staminode nearly as long as filaments, all stamens exserted **16. *Ameroglossum***
- Lower lip of corolla distinctly 3-lobed; staminode small and filiform or lacking, stamens all included **15. *Russelia***
87. Corolla white, yellow or orange 88
- Corolla purple to blue 91
88. Stamens and style long-exserted; corolla ampliate, yellow to orange; Central America **194. *Hemichaena***
- Stamens and style usually included; corolla cylindrical, not ampliate, white to yellow 89
89. Stamens 4; stigma clavate; corolla white or yellow, lower lip pouch-like; South Africa 90
- Stamens 2; stigma capitate to bilobed; corolla white to yellow, lower lip not pouch-like; Central America **23. *Eremogeton***
90. Corolla yellow, inside with crimson **302. *Anastrabe***
- Corolla white **303. *Bowkeria***
91. Corolla violet-purple, tube elongate; stamens 4, staminode present; stigma bilobed; Central America **72. *Uroskinnera***
- Corolla purple to blue, tube funnel-shaped; stamens 4, staminode absent; stigma globose; South Africa **22. *Freylinia***
92. Corolla actinomorphic, rotate to subrotate, yellow to violet; stamens 4 to 5, filaments bearded **1. *Verbascum***
- Corolla zygomorphic, with distinct upper and lower lip; stamens 4, sometimes staminode present 93
93. Corolla with small hooded upper lip and saccate, shoe-like lower lip or lower lip with involute margin; South America 94
- Corolla tubular-campanulate, distinctly 2-lipped 96
94. Stamens 3, upper lip of corolla bilobed, lobes flat with involute margin, lower lip spreading, patent, orbicular and entire, with weakly involute margin, not saccate **60. *Stemotria***
- Stamens 2, upper lip of corolla entire or weakly emarginate, hooded, lower lip saccate or not 95
95. Lower lip of corolla saccate **58. *Calceolaria***
- Lower lip of corolla not saccate, entire, spreading, with weakly involute margin **59. *Jovellana***
96. Staminode present 97
- Staminode absent 103
97. Anthers synthecous, filaments lanose, bearded or glabrous; plants rosulate or not 98
- Anthers not synthecous, filaments usually glabrous; plants rosulate 102
98. Corolla inconspicuous, greenish-brown, dark brown or red, tube ventricose; staminode inserted at base of upper lip **5. *Scrophularia***
- Corolla conspicuous, showy, white, pink, red or blue, tube campanulate; staminode inserted near base of corolla-throat 99
99. Fertile filament bases glabrous **70. *Penstemon***
- Fertile filament bases densely hairy 100
100. Anthers glabrous; leaves palmately veined; corolla white to red **68. *Keckiella***
- Anthers woolly or glabrous; leaves pinnately veined; corolla pink to maroon 101
101. Anthers glabrous; Kamtschatka and North Japan **69. *Pennellianthus***
- Anthers woolly; western North America **66. *Nothochelone***
102. Inflorescence congested; corolla yellow; staminode subterete to bilobed; Himalayas **4. *Oreosolen***
- Inflorescence lax; corolla blue or white to mauve; staminode never bilobed, sometimes missing; Central America **71. *Tetranema***
103. Plants not rosulate, stem elongate 104
- Plants rosulate 105
104. Calyx deeply 5-lobed; corolla 2-lipped, orange-red; seeds alveolated, aulacospermous; South America and South Africa **7. *Alonsoa***
- Calyx truncate at apex, with 5 short teeth; corolla subrotate, rose to violet; seeds not alveolated; Southeast Asia **298. *Cyrtandromoea***

105. Leaves palmately veined; inflorescence congested; corolla-tube narrowly cylindrical; stigma wide and flattened; Central Asia **3. *Nathaliella***
- Leaves pinnately veined; inflorescence lax; corolla-tube campanulate; stigma capitate; Central America **71. *Tetranema***

106 (56). Large canopy trees; calyx subtruncate to irregularly 2–3-lobed; fruit globose, indehiscent, fleshy **57. *Exarata***
- Herbs or small shrubs; calyx usually 5-lobed; fruit usually a capsule 107

107. Staminode present 108
- Staminode absent 110

108. Seeds alveolated, aulacospermous; plants rosulate; staminode present or absent; South Africa **8. *Charadrophila***
- Seeds winged; plants rosulate or not; staminode always present; North America 109

109. Plants rosulate; corolla pale lavender; alpine meadows of Rocky Mountains **67. *Chionophila***
- Plants not rosulate; corolla red to rose or purple; eastern North America **65. *Chelone***

110. Anthers synthecous (unilocular); corolla large and conspicuous or small 111
- Anthers bithecous (bilocular) or monothecous, sometimes confluent at apex, but never synthecous; if unilocular, then plants small, with minute corolla (3–7 mm) 138

111. Corolla actinomorphic, rotate to subrotate, yellow to violet; stamens 4–5, filaments densely woolly 112
- Corolla zygomorphic, variously coloured; stamens 4, filaments never woolly 113

112. Seeds aulacospermous; stigma dilated to disciform; tropical Africa and Arabia **2. *Rhabdotosperma***
- Seeds bothrospermous; stigma capitate; Europe, Mediterranean and Asia **1. *Verbascum***

113. Fruit a capsule with numerous to few (–4) seeds 114
- Fruit with (1)2 cocci (nutlets), separating at maturity 129

114. Adaxial filaments not decurrent in corolla-tube 115
- Adaxial filaments decurrent in corolla-tube, often to base 118

115. Bract adnate to half of calyx tube **36. *Glekia***
- Bract not or scarcely adnate to calyx tube (at most adnate to part or all of the pedicel and sometimes extreme base of calyx) 116

116. Adaxial stamens included, inserted at middle of tube or higher, abaxial stamens either included or anthers just visible **33. *Manulea***
- One or both pairs of stamens exserted at anthesis (but both pairs deeply included in 2 species of *Sutera*, inserted near base of tube) 117

117. Inside of corolla-tube either glabrous or with longitudinal bands of clavate hairs, these sometimes coalescing, sometimes extending from throat to bases of corolla lobes; seeds with transversally elongated pits (bothrospermous) **32. *Sutera***
- Clavate hairs confined to orange/yellow patch at base of upper lip; seeds sinuously wrinkled in longitudinal bands, without pits **37. *Trieenea***

118. Stigma very short (up to 2 mm, but often much shorter), often minutely bifid, stigmatic surface ± terminal, either jelly-like or composed of minute papillae, these either terminal or terminal and extending very slightly downwards 119
- Stigma lingulate (tip occasionally minutely bifid), often ± equalling or longer than style, flattened, with 2 marginal bands of stigmatic papillae 121

119. Leaves opposite, bases decurrent forming ridges or narrow wings along stem; stigma lingulate; seeds black **31. *Lyperia***
- Leaves either alternate at least in upper part of stem or, if opposite, not decurrent; stigma capitate to minutely bifid; seeds not black 120

120. Corolla-tube scarcely expanded at apex; abaxial anthers shortly exserted, seeds with 6–8 longitudinal furrows (aulacospermous) **53. *Camptoloma***
- Corolla-tube abruptly expanded at apex, at most tips of abaxial anthers visible; seeds without furrows, testa reticulate, in 2 species seeds pitted (bothrospermous) **30. *Jamesbrittenia***

121. Bract either adnate to pedicel (or part of pedicel) only or to pedicel and base of calyx tube (never more than lower third) 122
- Bract adnate at least to half of calyx tube, rarely less 125

122. Inflorescence capitate, nodding in flower, erect in fruit **41. *Glumicalyx***
- Inflorescence either capitate or extended, always erect 123

123. Floral bracts about as long as broad, sharply differentiated from leaves **42. *Strobilopsis***
- Floral bracts longer than broad, often leaf-like 124

124. Corolla with an orange to yellow patch at base of upper lip extending into tube; seeds pallid, greenish-blue or amber-coloured, sinuously wrinkled in vertical bands **37. *Trieenea***
- Corolla patch orange to yellow, confined to upper lip; seeds black, coarsely reticulate **34. *Melanospermum***

125. Ovules 2 in each locule, one pointing upwards, one downwards; inflorescence corymb-like **43. *Tetraselago***
- Ovules at least 4 in each locule and never diametrically opposed; inflorescence a spike or raceme, sometimes condensed into a head, sometimes with numerous paracladia 126

126. Calyx distinctly 2-lipped, lower lip ±2-toothed, upper lip 3-toothed, strongly 5-ribbed, plicate in flower (expanded in fruit); stamens 4 or 2, staminodes absent **39. *Zaluzianskya***
- Calyx distinctly 2-lipped or not, rarely plicate and then 2 stamens, 2 staminodes, lobing various, often deep, sometimes to base, stamens 4 or 2, staminodes present 127

127. Hairs on stems (and elsewhere) always patent, gland-tipped; testa translucent, loosely enveloping seed, seeds 3-winged or 3-angled **35. *Polycarena***
- Hairs on stems either all eglandular or mixed with glandular hairs, some eglandular hairs always present, particularly near base of stem and there nearly always ± retrorse; testa ± opaque, tightly investing seed, seeds never winged 128

128. Calyx mostly lobed to middle or base; upper lip of corolla 2-lobed, lower lip 3-lobed **38. *Phyllopodium***
- Calyx lobes much shorter than tube; one corolla lobe isolated, the other 4 closely associated **40. *Reyemia***

129. Leaves opposite, sometimes alternate above, hairs with sculptured or unsculptured walls; annuals or perennials 130
- Leaves alternate, hairs with sculptured walls; perennials 135

130. Hairs with sculptured walls 131
- Hairs thin-walled, not sculptured 132

131. Annual or perennial plants; bract adnate to calyx; calyx lobes 3–5; corolla lobes 4–5, upper lip bearded; fruit with 2 cocci **47. *Cromidon***
- Perennial plants; bracts free; calyx lobes 5; corolla lobes 5, upper lip glabrous; fruit with one coccus **44. *Globulariopsis***

132. Calyx lobes 5; corolla with 4–5 lobes, upper lip 4-lobed 133
- Calyx lobes 1–2; corolla with 4 lobes; annual or perennial plants 134

133. Corolla 5-lobed, orange patch on upper lip present; cocci with soft walls, compressed **49. *Pseudoselago***
- Corolla 4–5-lobed, orange patch on upper lip absent; cocci hard-walled, fusiform **50. *Chenopodiopsis***

134. Calyx with one lobe; bracts adnate to calyx **52. *Hebenstretia***
- Calyx with two lobes; bracts free **51. *Dischisma***

135. Bracts adnate to calyx 136
- Bracts free 137

136. Fruit with 2 cocci **48. *Selago***
- Fruit with 1 coccus **45. *Microdon***

137. Stamens 4; fruit with 2 cocci **48. *Selago***
- Stamens 4 or 2; fruit with 1 coccus **46. *Gosela***

138 (110). Upper lip of corolla galeate 139
- Upper lip of corolla variously arranged and shaped, but never galeate 168

139. Anthers distinctly mucronate at base 140
- Anthers oblong or rounded at base 156

140. Prophylls present 141
- Prophylls absent 143

141. Stigma bifid; corolla yellowish or brownish-violet, with longitudinal palate; southern China **288. *Pterygiella***
- Stigma capitate; corolla variously coloured, longitudinal palate absent 142

142. Leaves linear, the upper leaves trifid with linear segments; corolla with biconvex palate; seeds few **269. *Bungea***
- Leaves linear-lanceolate, entire; corolla with longitudinal palate; seeds numerous **270. *Monochasma***

143. Holoparasitic herbs without green leaves **297. *Lathraea***
- Hemiparasitic herbs with green leaves 144

144. Calyx 4-lobed or 4-dentate; seeds few to numerous 145
- Calyx 5-dentate or 5-lobed; seeds numerous, small 146

145. Annual herbs; calyx 4-dentate; corolla-tube cylindrical; seeds few (1–4), large (up to 7 mm long and 3 mm wide), Southeast Europe and Asia to North America **292. *Melampyrum***
- Suffrutescent herbs or dwarf shrubs; calyx 4-lobed; corolla-tube ventricose; seeds numerous, small; Central and South America **213. *Lamourouxia***

146. Galea bilobed, lobes recurved; filaments glabrous **286. *Euphrasia***
- Galea entire, never bilobed; filaments papillose or hairy, rarely glabrous (in *Odontites* p.p.) 147

147. Lower lip equal or shorter than galea, tube equal to galea 148
- Lower lip and tube longer than galea 149

148. Epicotyl extremely shortened; stigma distinctly bilobed; middle lobe of lower lip entire **282. *Odontitella***
- Epicotyl normally developed; stigma entire, clavate-capitate; middle lobe of lower lip emarginate **284. *Odontites***

149. Leaves entire 150
- Leaves pinnatifid to pinnatisect, rarely entire 155

150. Inflorescence bracteate; bracts distinctly modified **285. *Nothobartsia***
- Inflorescence frondose; bracts resembling stem-leaves 151

151. Stigma distinctly bilobed **281. *Macrosyringion***
- Stigma entire, clavate-capitate 152

152. Pedicels much longer than calyx; theca base long-cuspidate **287. *Omphalotrix***
- Pedicels much shorter than calyx; theca base shortly mucronate 153

153. Lower lip of corolla with longitudinal gibbae; seed coat reticulate; middle lobe of lower corolla lip emarginate, Mediterranean, naturalized in Central Europe and South America **290. *Parentucellia***
- Lower lip of corolla without longitudinal gibbae; seed coat longitudinally ribbed or striate; mid-lobe of lower corolla lip not emarginate 154

154. Leaves lanceolate to linear; with special type of stalked glands (Rameauanus-type), gland-head few-celled (4–8, rarely 16), longer than wide; pollen exine coarsely retipilate **283. *Bartsiella***
- Leaves ovoid to elliptic; stalked glands with multicellular head or, if few-celled, wider than long; pollen exine finely retipilate **279. *Bartsia***

155. Calyx 5-lobed, lobes entire; upper lip not rostrate **289. *Phteirospermum***
- Calyx 5-lobed to (2–)5-dentate, lobes dentate, serrate or pinnatisect; upper lip always distinctly rostrate, often asymmetrical **291. *Pedicularis***

156 (139). Anthers unequally attached, sometimes monothecous 157
- Anthers equal, parallel, equally attached, rounded or oblong 162

157. Calyx deeply cleft, the lateral lobes entire or bilobed; lower lip of corolla very small, reduced to 3 small rudimentary teeth 158
- Calyx subequally 4-lobed or of a single, leaf-like lobe; lower lip of corolla equalling upper lip or shorter and with 3 saccate plane lobes, not rudimentary 159

158. Calyx 4-cleft, with a deeper anterior cleft, lobes entire; leaf venation always palmate, leaves entire; lower lip of corolla with 3 teeth-like recurved lobes; Mexico **275. *Gentrya***
- Calyx cleft anteriorly or along both margins, lobes entire or shortly bilobed; leaf venation palmate or pinnate, leaves entire, incised or pinnatifid; lower lip of corolla with 3 concave teeth-like lobes; western North America, few species in eastern North America, Eurasia, Central and South America **271. *Castilleja***

159. Calyx of a single sericaceous lobe or 2 distinct lobes; corolla flattened dorsiventrally and expanded laterally, lower corolla lip equalling upper lip **276. *Cordylanthus***
- Calyx tubular, 4-lobed or 4-dentate, laterally compressed; corolla not flattened dorsiventrally, lower lip either equalling upper lip or much shorter 160

160. Anthers monothecous; corolla sulphur-yellow to white or rose, throat abruptly indented, forming a fold, lips nearly equal **273. *Triphysaria***
- Anthers bithecal, corolla different 161

161. Lower lip of corolla very small, 3-saccate, corolla yellow, crimson or purplish, upper lip strongly galeate **272. *Orthocarpus***
- Lower lip of corolla equalling upper lip, with 3 nearly plane, spreading lobes, corolla white with pinkish tips, upper lip only weakly galeate **274. *Clevelandia***

162. Upper part of corolla strongly inflated, anthers distinctly exserted; Baja California **277. *Ophiocephalus***
- Upper part of corolla not strongly inflated, anthers usually included or only weakly exserted 163

163. Calyx 4-dentate, laterally compressed, inflated in fruit 164
- Calyx 5-dentate or 5-lobed, not inflated in fruit 165
164. Upper lip of corolla not rostrate; seeds with marginal wing **293. *Rhinanthus***
- Upper lip of corolla prominently rostrate; seeds not winged **294. *Rhynchocorys***
165. Leaves with palmate venation; lower lip of corolla without palate 166
- Leaves with pinnate venation; lower lip of corolla with longitudinal palate 167
166. Leaves alternate; calyx tubular, 2-lipped; corolla yellow or purplish; filaments hairy; eastern North America **265. *Schwalbea***
- Leaves opposite; calyx deeply 5-lobed; corolla yellow; filaments glabrous; Europe and Central Asia **266. *Cymbaria***
167. Calyx tubular, subequally 5-lobed nearly to middle; leaves entire or pinnatifid, eastern Asia **267. *Siphonostegia***
- Calyx 2-lipped, deeply lobed, lobes of upper lip 3 times longer than teeth of lower lip; leaves entire; eastern Mediterranean **268. *Lesquereuxia***
168 (138). Anthers with equal (rarely subequal) thecae, usually rounded at base, often confluent at apex, rarely monothecous 169
- Anthers with (equal) subequal to unequal thecae, usually mucronate to apiculate at base, never confluent at apex, if unequal, then thecae sometimes oblong and rounded, often monothecous 307
169. Leaves distinctly peltate, vernation circinnate; flowers with 2 stamens, anthers monothecous, unilocular, rounded **120. *Philcoxia***
- Leaves never distinctly peltate, vernation never circinnate; flowers with 2 or 4 stamens, anthers bithecous or rarely monothecous 170
170. Leaves alternate 171
- Leaves opposite 195
171. Plants with scale-like, reduced leaves; plant parasitic, South Central Africa **240. *Micrargeriella***
- Plants with normal-sized, variously shaped leaves 172
172. Plants rosulate 173
- Plants with leafy stem, sometimes also rosulate leaves present at base 184
173. Corolla large, only weakly 2-lipped to subrotate, cylindrical or inflated, sometimes globose; stigma bilobed 174
- Corolla small, distinctly 2-lipped; stigma usually capitate 177
174. Prophylls absent 175
- Prophylls present 176
175. Corolla inflated, usually constricted towards base, white, yellow, purple or red; testa reticulate; Central and southern Europe **153. *Digitalis***
- Corolla cylindrical, pink or white, usually not constricted towards base; testa striate-reticulate; eastern Asia **299. *Rehmannia***
176. Corolla white; calyx campanulate, 5-lobed, fused to apex **301. *Triaenophora***
- Corolla yellow with brown spots; calyx campanulate, 5-lobed almost to base **300. *Titanotrichum***
177. Stamens 4; seed coat thick, alveolate and transparent; leaves basal and rosulate 178
- Stamens 2; seed coat thick and alveolate or smooth; leaves rosulate or cauline 179
178. Corolla 5-lobed and subrotate; stamens equal, exserted, many times longer than corolla **160. *Picrorhiza***
- Corolla 4-lobed and 2-lipped; stamens weakly didynamous, shorter than corolla **161. *Neopicrorhiza***
179. Corolla 5-lobed, distinctly 2-lipped and cylindrical; stigma ±2-lobed; stamens inserted immediately below juncture of lips; leaves all basal and rosulate 180
- Corolla 4-lobed, only weakly 2-lipped to rotate; stigma marginate or capitate; stamens inserted at base of corolla-tube or at throat of corolla; leaves mostly cauline, rarely basal 181
180. Fruit berry-like and indehiscent; testa alveolate, stigma lobes spreading **159. *Kashmiria***
- Fruit a capsule; testa reticulate; stigma lobes erect or weakly spreading **157. *Wulfenia***
181. Calyx 2- or 3-lobed to dentate; fruit indehiscent, drupe-like, with 2 cocci at maturity; seeds 1 per locule **176. *Lagotis***
- Calyx 4–5-partite; fruit a capsule; seeds usually numerous 182
182. Calyx 5-partite, lobes equal or unequal; capsule turgid, tapered and marginate at apex; testa reticulate; corolla cylindrical, nearly actinomorphic; stigma very small, not thicker than style **158. *Wulfeniopsis***
- Calyx 4-partite or 5-partite with very small upper lobe; capsule ± laterally compressed, ±2-lobed at apex; testa usually smooth or somewhat rugose; corolla rotate or somewhat cylindrical; stigma capitate to clavate and distinctly thicker than style 183
183. Corolla blue or violet, upper lip plane, a little longer than lower lobes which are divided to the tube and do not form a definite lower lip; stem with few bract-like leaves below inflorescence; leaves cordate-reniform and double-crenate or dentate or deeply cleft or pinnatifid **164. *Synthyris***
- Corolla violet-purple, yellow or white, cleft nearly or completely to base between the lips, in some species rudimentary or even lacking, upper lip concave and rounded, much longer than weakly lobed lower lip; stems with several or many bract-like leaves below inflorescence; leaves cordate-ovoid to oblong, simply crenate or crenate-serrate **165. *Besseya***
184. Corolla subrotate, white, with 5 plane lobes; stamens 4–5, subequal; Central and South America **115. *Capraria***
- Corolla subrotate or zygomorphic, yellow, red, pink or purple, rarely white, never with plane lobes; stamens 4–8 (*Sibthorpia*), 4 or 2; Europe, Asia, Africa, Noth America, only *Sibthorpia* in South America 185
185. Creeping herbs; calyx 4–8–lobed, corolla rotate, yellow or white, (4)5–9(10)-lobed, tube short; stamens 4–8; seeds few, testa reticulate or smooth 186
- Erect to prostrate herbs or shrubs; calyx deeply 5-lobed, corolla zygomorphic, red, purple or pink; stamens 2–4; seeds numerous, testa reticulate to rugose 187
186. Leaves suborbicular to reniform, entire or shallowly lobed; corolla yellow with (4)5–9(10) lobes; stamens 4–8; stigma capitate; Europe, South America and Africa **177. *Sibthorpia***
- Leaves pinnately parted; corolla white with (4)5(–8) lobes; stamens 4; stigma bifid, eastern Asia **178. *Ellisiophyllum***
187. Prophylls present 188
- Prophylls absent 190

188. Annual herbs; flowers distinctly pedicellate; corolla pale mauve to blue or lilac; stamens 2–4; Africa and Asia **25. *Anticharis***
- Perennial herbs or shrubs (rarely annuals); flowers sessile or shortly pedicellate; corolla blue or purple, campanulate; stamens 4, sometimes the adaxial pair sterile; Africa 189

189. Adaxial stamens smaller, usually sterile and reduced **26. *Aptosimum***
- Adaxial stamens as long as abaxial stamens, fertile **27. *Peliostomum***

190. Stamens 2 191
- Stamens 4 193

191. Shrubs; leaves spathulate to linear-lanceolate, often succulent; corolla-tube elongate and narrow; Canary and Cape Verde islands, Northeast Africa, Yemen and Pakistan **182. *Campylanthus***
- Herbs; leaves lanceolate, never succulent; corolla-tube short, sometimes urceolate 192

192. Stigma very small, capitellate, style and stamens mostly strongly exserted; corolla subrotate, with straight and cylindrical tube; roots woolly-hairy **162. *Veronicastrum***
- Stigma rod-shaped, style and stamens included; corolla 2-lipped, upper lip 3-lobed, lower lip reflexed, tube urceolate; roots glabrous **163. *Scrofella***

193. Corolla not inflated, with narrow elongate tube and spreading limb; small subshrubs; plants of alpine habitats in Europe (Alps, Pyrenees) and North Africa (Atlas) **155. *Erinus***
- Corolla cylindrical to inflated-globose, with broad tube usually constricted towards base; large herbs or shrubs 194

194. Annual, biennial or perennial herbs or subshrubs; corolla white, yellow, purple or red; stigma bilobed; Central and southern Europe **153. *Digitalis***
- Subshrubs or shrubs; corolla red; stigma capitate; Madeira and Canary Islands **154. *Isoplexis***

195 (170). Adaxial corolla lobes or upper lip covering lateral lobes in bud (antirrhinoid aestivation, sometimes varying within one plant in *Lindenbergia*) 196
- Adaxial corolla lobes or upper lip covered by one of the lateral lobes in bud (rhinanthoid aestivation) 264

196. Abaxial stamens usually curved or twisted or Z-like with geniculate or clavate appendages, if reduced, then staminodes clavate, curved or with appendages 197
- Abaxial stamens straight or curved, but without appendages, if reduced, then staminodes filiform 219

197. Corolla minute (usually less than 2–3 mm long), usually inconspicuous; anthers often confluent at apex and sometimes fused (unilocular); interior of corolla lacking glandular hairs 198
- Corolla distinctly larger, usually showy; anthers neither confluent at apex nor fused; interior of corolla with glandular hairs 203

198. Plants erect, often submerged; anthers with weakly unequal thecae; capsule septicidal and loculicidal, with 4 valves; India **192. *Bythophyton***
- Plants prostrate to ascending, rarely erect; anthers with equal thecae, sometimes becoming unilocular; capsule septicidal or loculicidal, with 2 valves, or opening irregularly 199

199. Upper lip of corolla lacking, lobes of lower lip rounded, entire; calyx 4-lobed to middle, except for lower side, where the cleft extends to base; anthers with two separate thecae **184. *Hemianthus***
- Upper lip of corolla present, if very small and reduced, then lobes of lower lip irregularly incised-dentate; calyx 3–5-partite; anthers synthecous-unilocular 200

200. Upper lip of corolla very short and reduced, lower lip irregularly incised-dentate; plants ascending to erect, Cuba **188. *Encopella***
- Upper lip of corolla distinct, entire or 2-lobed, lower lip 3-lobed or corolla nearly subrotate, lobes entire; plants prostrate 201

201. Corolla with subequal lobes; anthers unilocular **187. *Peplidium***
- Corolla with distinct lips; anthers with separate thecae 202

202. Minute filiform staminodes present; leaves linear-lanceolate; Indian Ocean islands **191. *Bryodes***
- Minute filiform staminode absent; leaves orbicular-rounded; Central America and Caribbean **183. *Micranthemum***

203. Seeds with alveolated endosperm, the seed surface with hollows or longitudinal furrows (bothrospermous or aulacospermous), endosperm star-shaped in transverse section 204
- Seeds without alveolated endosperm, seed surface ± smooth or only weakly furrowed, endosperm weakly polygonal or undulated in transverse section 218

204. Seeds with longitudinal furrows (aulacospermous); all 4 stamens fertile 205
- Seeds with rounded pits (bothrospermous); sometimes 2 stamens reduced to staminodes 209

205. Plants procumbent, creeping; leaf venation pinnate; flowers in very lax racemes; calyx deeply divided, with 5 linear-lanceolate lobes or anisophyllous with 4 lobes; Asia 206
- Plants erect or ascending (except *Crepidorhopalon scaettae*, a creeper); leaf venation palmate; flowers generally in dense racemes or in capitules, only occasionally lax racemes; calyx with long tube and 5 short teeth, if deeply divided, then corolla less than 1 cm long; Africa and Madagascar 207

206. Calyx with 5 equal, linear-lanceolate lobes; corolla large, 2.5–4 cm long; abaxial filaments with small, spur-like appendage **140. *Schizotorenia***
- Calyx with 4 lobes, the upper outer sepal and the lower outer sepal cordate, the lateral inner pair filiform; corolla less than 1 cm long; abaxial filaments without appendage **138. *Picria***

207. Lower lip of corolla with yellow club-shaped hairs on a pluricellular socle; leaves linear-lanceolate to ovoid, smooth; filaments of the abaxial stamens with a basal spur-like appendage, tip of spur rounded and with glandular hairs; plants mostly annuals, only 2 species perennial **148. *Crepidorhopalon***
- Lower lip of corolla lacking yellow club-shaped hairs with pluricellular socle; leaves largely ovoid, leathery, somewhat shining; filament of abaxial stamens only weakly curved with a small basal knee-like appendage present or absent; perennial plants with large subterraneous woody rhizomes on heavy-metal soil 208

208. Small plants of almost rosette-like habit; stem 1–20 cm tall; leaves not connate at base; inflorescence a dense capitulum; filaments of the abaxial stamens curved, with a basal, knee-like, swollen glandular-pubescent appendage **149. *Hartliella***

- Large plants, stem more than 20 cm (up to 60 cm) tall; leaves broadly connate at base; inflorescence a dense raceme; filaments of the abaxial stamens ± straight, without basal appendage **150. *Bampsia***

209. Plants with a basal, decussate leaf rosette and lacking the terminal main inflorescence; flowering shoots all axillary 210
- Plants without basal leaf rosette; the stem with decussate leaves and long internodes; terminal main inflorescence and paracladia generally well developed 211

210. Leaves large, broadly ovoid to lanceolate, acuminate; abaxial stamens perfect, with a Z-shaped filament; roots of vivid red or orange colour **147. *Craterostigma***
- Leaves small, ovoid to lanceolate, obtuse; abaxial stamina usually reduced to staminodes (except some Madagascan taxa with anther rudiments), roots whitish **146. *Lindernia* (sect. *Scapoideae*)**

211. Leaf venation pinnate 212
- Leaf venation palmate 217

212. The 2 abaxial stamens reduced to staminodes 213
- All 4 stamens perfect 214

213. Leaves serrate or pinnatifid with aristate teeth, generally only the midvein clearly visible; corolla with long (c. 15 mm), weakly curved tube; staminodes mammiform or obtuse at apex **143. *Scolophyllum***
- Leaves entire or denticulate, rarely serrate, generally lateral veins clearly visible; corolla with short tube (less than 10 mm, often less than 5 mm); staminodes clavate **146. *Lindernia* subg. *Bonnaya***

214. Plants large; stem generally more than 50 cm tall; corolla more than 2 cm long **137. *Artanema***
- Plants much smaller, stem generally not more than 30 cm tall, if stem decumbent, occasionally up to 50 cm long; corolla generally <2 cm long (but see *Torenia*) 215

215. Calyx suborbicular, with 3 broad wings between the ridges, 3-dentate at apex, auriculate at base **141. *Legazpia***
- Calyx tubular, winged or unwinged, if winged, than with 5 wings, 5-lobed 216

216. Dehiscence of capsule septicidal; ovary without apical hairs or hairs within the locule **146. *Lindernia* subgen. *Tittmannia***
- Dehiscence of capsule poricidal, opening on both sides of the septum, at the same time as dehiscence of calyx; ovary with apical and internal hairs **142. *Torenia***

217. Inflorescence capitate, enclosed by closely overlapping large involucral leaves with a semiorbicular base and a long acuminate apex, differing from the linear-lanceolate stem-leaves **139. *Pierranthus***
- Inflorescence a lax raceme, rarely capitate (*L. crassifolia*), but then the involucral leaves not differing from stem-leaves **146. *Lindernia* subg. *Vandellia*, subg. *Pentacme***

218. Aquatic plants with basal rosette, with 2-lobed complex leaves, originating by connation of two adjacent leaves of one whorl; Namibia **145. *Chamaegigas***
- Plants of humid places or aquatic plants, if with basal rosette leaves simple, not 2-lobed **146. *Lindernia* subg. *Lindernia***

219 (196). Fertile stamens 2, abaxial or adaxial stamens reduced to staminodes or lacking 220
- Fertile stamens 4 235

220. Prophylls present 221
- Prophylls absent 225

221. Annual, submerged rosulate plants of rock pools; leaves petiolate, floating; corolla subrotate; cleistogamous flowers present; south-eastern North America **105. *Amphianthus***
- Annual or perennial terrestrial or aquatic plants; leaves not rosulate, usually sessile; corolla subrotate or 2-lipped; cleistogamous flowers absent (rarely present in *Limnophila*) 222

222. Plants small, prostrate; corolla white, subrotate; connective with 2 short arms; Argentina **110. *Braunblanquetia***
- Plants large, erect; corolla 2-lipped, white, yellow or blue to purple, with distinct upper and lower lip; thecae separate, only weakly stipitate, with rounded connective 223

223. Leaves dimorphic, emerged leaves linear-lanceolate to ovoid, entire to palmately divided, submerged leaves usually capillary-multifid; corolla blue to mauve or violet, tubular to infundibuliform; seeds with smooth testa; tropical Africa and Asia **119. *Limnophila***
- Leaves monomorphic, entire; corolla white to yellow, salverform or campanulate; seeds with striate and reticulate testa; Europe, North America and extratropical Asia 224

224. Corolla salverform, tube very narrow; capsule loculicidal; eastern North America **111. *Sophronanthe***
- Corolla campanulate, tube cylindrical; capsule loculicidal or septicidal **112. *Gratiola***

225. Water plants, growing in small ponds or pools, at least submerged at base; cleistogamous flowers present or all flowers cleistogamous 226
- Terrestrial plants, often in wetland, but usually not submerged; all flowers chasmogamous or rarely cleistogamous (*Deinostema, Psammetes*) 229

226. Anthers with fused thecae; all flowers cleistogamous; stem with 4 broad wings; leaves amplexicaulous; capsule septicidal; Namibia **190. *Dintera***
- Anthers with 2 separate thecae, connective with 2 short arms; chasmogamous flowers present; stem rounded or quadrangular, but never winged; leaves dimorphic, linear-lanceolate to capillary-multifid, never amplexicaulous; capsule loculicidal or septicidal and loculicidal 227

227. Leaves dimorphic, emerged leaves linear-lanceolate to ovoid, entire to palmately divided, submerged leaves usually capillary-multifid; abaxial stamens reduced to staminodes; stigma bifid; seeds with smooth testa; tropical Africa and Asia **119. *Limnophila***
- Leaves monomorphic, entire, submerged leaves usually linear or lanceolate; adaxial stamens reduced to staminodes; stigma peltate; seeds with reticulate testa 228

228. Corolla with narrow, cylindrical tube; basal leaves usually rosulate; anthers with pilose thecae; tropical Africa and Asia **117. *Dopatrium***
- Corolla with narrow tube, dilated at throat; leaves not rosulate but verticillate, filiform; anthers with glabrous thecae; Madagascar **118. *Hydrotriche***

229. Perennial, suffrutescent herbs; calyx 5-partite, dorsal sepal broad, lateral sepals narrow; corolla blue to violet; Brazil 230
- Annual herbs, small; calyx 3–5-lobed or -dentate, lobes equal; corolla pale purple, bluish or white; Africa, Indian Ocean islands, Asia and Australia 231

230. Corolla large, lips nearly equal and undivided, lower lip with swollen palate; thecae equal, fused **124. *Otacanthus***
- Corolla small, upper lip entire, lower lip 3-lobed, palate absent; thecae separate, unequal, one theca reduced **125. *Achetaria***

231. Plants erect; cleistogamous flowers present or all flowers cleistogamous 232
- Plants erect, prostrate or creeping; cleistogamous flowers absent 233

232. Chasmogamous flowers present; corolla purple to pale purple, longer than calyx; anthers with separate, pilose thecae; stigma bilobed; eastern Asia **116. *Deinostema***
- All flowers cleistogamous; corolla whitish, as long as calyx; anthers with fused thecae; stigma spathulate; Africa and Madagascar **189. *Psammetes***

233. Plants erect or ascending; corolla slender, filiform, nearly truncate and unlobed at the top; style with two long filiform stigma lobes; Australia **188. *Elacholoma***
- Plants prostrate, creeping; corolla tubular, with distinct lobes, stigma bilobed, but lobes not filiform, or stigma ligular; Africa, Asia and Australia 234

234. Leaves cauline or rosulate; stem creeping; calyx 3–4-dentate; staminodes present; stigma bilobed; seeds smooth **193. *Glossostigma***
- Leaves cauline, never rosulate; calyx 5-lobed; staminodes absent; stigma ligular; seeds obscurely reticulate **185. *Microcarpaea***

235 (219). Leaves with inconspicuous veination (often only midvein visible) or distinctly palmately veined, usually linear-ovoid to pinnatisect 236
- Leaves pinnately veined and ovoid-lanceolate to orbicular or deeply pinnatifid to pinnatisect 243

236. Corolla 2-lipped; thecae separated by a globose connective; anthers pubescent **129. *Darcya***
- Corolla 2-lipped (in *Dodartia*) or usually subrotate; connective never globose; anthers glabrous 237

237. Corolla distinctly 2-lipped; leaves reduced, scale-like; perennial shrub-like herbs; extratropical Asia **196. *Dodartia***
- Corolla subrotate; leaves not reduced; annual or perennial herbs of wet ground; tropical America, Africa and Asia 238

238. Calyx 3–4-dentate; plants creeping; leaves cauline or rosulate; testa smooth **193. *Glossostigma***
- Calyx deeply 5-lobed; plants prostrate to ascending or erect; leaves cauline; testa reticulate 239

239. Usually rosulate herbs, rarely with cauline leaves; corolla subrotate, 5-lobed, small; style short, incurved **121. *Limosella***
- Herbs with cauline leaves; corolla 2-lipped or subrotate (in *Bacopa* p.p.); style straight, filiform 240

240. Prophylls present; stamens of different size 241
- Prophylls absent; stamens of equal size **106. *Bacopa***

241. Prophylls at base of pedicel; capsule septicidal; corolla violet **107. *Maeviella***
- Prophylls at base of calyx; capsule septicidal or mostly loculicidal; corolla whitish to bluish 242

242. Upper lip of corolla smaller than lower lip, corolla dorsally concave; style articulate and enlarged at base; seeds mammillate **108. *Boelckea***
- Upper lip of corolla larger than lower lip or subequal, corolla not concave; style not articulate at base; seeds reticulate **106. *Bacopa* sect. *Chaetodiscus***

243. Calyx distinctly winged, 5-dentate, accrescent in fruit; corolla with distinct palate on lower lip 244
- Calyx not winged, not accrescent in fruit; corolla with palate on lower lip absent 245

244. Annual to perennial herbs or shrubs; capsule loculicidal; western North America, few species in South America, Africa, Madagascar or Asia **200. *Mimulus***
- Perennial herbs; capsule septicidal; South China **201. *Mimulicalyx***

245. At least the lower leaves deeply pinnatifid to pinnatisect 246
- Leaves entire or variously dentate to serrate, never pinnatifid or pinnatisect 250

246. Aquatic submerged herbs; leaves verticillate; corolla violet with yellow throat; capsule loculicidal; South America **109. *Benjaminia***
- Terrestrial herbs, never submerged; leaves decussate; corolla blue, lavender, white or yellow; capsule septicidal 247

247. Stigma capitate; capsule linear or ovoid; lower leaves pinnatifid to pinnatisect, upper leaves entire; testa smooth or with spirally arranged longitudinal ridges; western North and Central America 248
- Stigma bilobed; capsule ovoid; all leaves similar; testa usually with straight, never spirally arranged longitudinal ridges 249

248. Capsule linear; seeds numerous; western North to Central America **132. *Schistophragma***
- Capsule ovoid; seeds few (1–2 per locule), large, testa smooth; western North America **74. *Tonella***

249. Calyx deeply 5-lobed, lobes entire, linear; corolla pale lavender; eastern North America **134. *Leucospora***
- Calyx deeply 5-lobed, obscured by multisect lobes, unequal; corolla yellow; Brazil **135. *Schizosepala***

250. Corolla subrotate, plane, 4-lobed **114. *Scoparia***
- Corolla bilobed, upper lip 2-lobed, lower lip 3-lobed 251

251. Fruit indehiscent 252
- Fruit a capsule 254

252. Perennial herbs; leaves rosulate; corolla 2-lipped, bright mauve, lower lip with distinct palate; Himalayas **197. *Lancea***
- Perennial herbs or shrubs; leaves cauline, never rosulate; corolla subrotate, pink or blue to violet, palate absent 253

253. Perennial erect shrubs (up to 3 m tall); calyx 5-dentate; leaves monomorphic; fruit an indehiscent capsule; Chile, Argentina **151. *Monttea***
- Perennial creeping herbs; calyx deeply 5-lobed; leaves dimorphic, leaves on main stem orbicular to reniform, leaves on axillary branches linear-subulate; fruit a red berry; western Himalayas to China **179. *Hemiphragma***

254. Seeds few (6–8); shrubs of Chile **152. *Melosperma***
- Seeds numerous 255

255. Plants rosulate 256
- Plants only with cauline leaves 257

256. Plants with stolons; prophylls absent; corolla with palate on lower lip; seeds with reticulate testa, longitudinal ridges absent; East Asia **198. *Mazus***
- Plants without stolons; prophylls present; corolla lacking palate on lower lip; seeds aulacospermous; South Africa **8. *Charadrophila***

257. Thecae separated by the stipitate, 2-armed connective 258
- Thecae separate, but connective rounded, not stipitate 261

258. Capsule septicidal; seeds reticulate, without longitudinal ridges; corolla white to yellow; North, Central and South America **113. *Mecardonia***
- Capsule loculicidal and septicidal; seeds with longitudinal ridges; corolla white, yellow or blue 259

259. Corolla yellow, palate present on lower lip; tropical Africa, Arabia and Asia **122. *Lindenbergia***
- Corolla white to blue; palate on lower lip absent 260

260. Annual or perennial herbs; corolla white to blue, usually small (less than 1 cm long); connective arms distinct; pantropical **128. *Stemodia***
- Perennial herbs or subshrubs; corolla blue, usually large (more than 2–3 cm long); connective arms smaller, indistinct; Australia **131. *Morgania***

261. Prophylls present; corolla white or blue 262
- Prophylls absent; corolla yellow to purplish or white 263
262. Perennial prostrate to ascending subshrubs; corolla white, palate on lower lip present; filaments twisted or curved; fruit reflexed; tropical Africa **136. *Stemodiopsis***
- Annual or perennial erect herbs; corolla blue, palate absent; filaments straight; fruit erect; South America **133. *Conobea***
263. Prostrate to ascending herbs; calyx deeply 5-lobed; corolla white; anthers with equal thecae; seeds with longitudinal ridges; Central America **130. *Cheilophyllum***
- Erect herbs; calyx 5-dentate, lobes much shorter than tube; corolla yellow with purplish spots; anthers of abaxial stamens with smaller and reduced thecae; seeds reticulate; western North America **199. *Mimetanthe***
264 (195). Fertile stamens 2, staminodes present or absent 265
- Fertile stamens 4 277
265. Perennial hemiparasitic prostrate herbs; leaves entire or bipinnatifid; corolla yellow, tube long, funnelform; 2 staminodes present; Mexico **211. *Silviella***
- Annual or perennial herbs or shrubs; leaves entire; corolla usually blue to violet, rarely yellow (in *Paederota*, but then with erect stem), tube short; staminodes always absent 266
266. Perennial or annual herbs, rarely with woody base; leaves basal and rosulate or sparsely on stem; capsule valves thin, not compressed, herbaceous, often dehiscing first loculicidally 267
- Shrubs, small trees, rarely subshrubs, only occasionally perennial creeping herbs; leaves mostly cauline, dense, leaving a ring (pulvinus) at the node after falling off; capsule often dorsiventrally compressed, dehiscing first septicidally, capsule valves thick, woody or crustaceous 272
267. Corolla-tube about as long as lobes, limb subrotate or 2-lipped; capsule not strongly compressed; seeds numerous 268
- Corolla rotate, tube much shorter than lobes, rarely cylindrical; capsule strongly compressed, if not, containing only a few seeds 270
268. Inflorescence a dense terminal spike-like raceme; corolla usually distinctly tubular **168. *Pseudolysimachion***
- Inflorescence a terminal raceme, sometimes secund; corolla 2-lipped or subrotate 269
269. Flowers shortly pedicellate (1–2 mm long); corolla 2-lipped, yellow or blue; Alps **166. *Paederota***
- Flowers distinctly pedicellate (10–15 mm); corolla subrotate, yellow; Caucasus **167. *Paederotella***
270. Seeds more than 2 in a locule, flattened and smooth or navicular and sometime weakly rugose; perennial or annual plants **169. *Veronica* subg. *Veronica***
- Seeds 1–2 in a locule, cup-shaped or flattened on both sides and distinctly rugose; annual plants 271
271. Seeds cup-shaped, calyx lobes not coherent and entire **169. *Veronica* subg. *Cochlidiosperma***
- Seeds flattened on both sides and strongly rugose, calyx lobes coherent in pairs and dentate **169. *Veronica* subg. *Oligospermum***
272. Flowers solitary (i.e. in frondose racemes); corolla 5-lobed, 2-lipped, cylindrical **175. *Detzneria***
- Flowers usually in (bracteate) racemes, less frequently solitary; corolla 4- or 5(–6)-lobed, actinomorphic 273
273. Corolla 5(–6)-lobed; calyx 5(–6)-partite; capsule laterally compressed and 2-lobed; perennial or creeping subshrubs; flowers solitary (i.e. in frondose racemes) **173. *Chionohebe***
- Corolla 4-lobed; calyx 4-partite; capsule laterally or dorsiventrally compressed, tapered at apex or rarely 2-lobed; erect shrubs or subshrubs; flowers in (bracteate) racemes 274
274. Capsule usually weakly dorsiventrally compressed with the septum parallel to the plane of compression, occasionally laterally compressed, apex always tapered 275
- Capsule laterally compressed with the septum at right angles to the plane of compression, apex tapered, obtuse or notched 276
275. Inflorescence lateral, main axis proliferating, only lateral racemes present; stigma clavate **170. *Hebe***
- Inflorescence terminal, compound, with terminal main inflorescence and lateral racemes; stigma bifid or 4-lobed **171. *Heliohebe***
276. Stigma capitate; seeds numerous, New Guinea, New Zealand, Australia (one species) **172. *Parahebe***
- Stigma clavate; seeds usually few (6–12 per locule), Southeast Australia **174. *Derwentia***
277 (264). At least lower leaves pinnately divided; stamens with lanose filaments 278
- All leaves entire; stamens with glabrous or lanose filaments, sometimes with villose thecae 281
278. Corolla purple or rarely white, subrotate, tube inflated distally; calyx 4-lobed, lobes longer than tube **206. *Tomanthera***
- Corolla yellow, subrotate, tube not inflated distally; calyx cup-shaped, with 5 narrow lobes, lobes shorter than tube 279
279. Calyx with unequal lobes, adaxial lobe shorter and narrower than 4 abaxial lobes; stigma bilobed; seeds angulate **210. *Dasistoma***
- Calyx with equal to subequal lobes; stigma globose to capitate; seeds winged, rarely wingless 280
280. Prophylls absent; corolla with short tube; stamens exserted **208. *Seymeria***
- Prophylls present; corolla with long tube; stamens included **209. *Seymeriopsis***
281. Leaves petiolate, reniform to orbicular, leaf margin dentate or irregularly dentate-crenate 282
- Leaves sessile, linear-lanceolate, leaf margin entire 283
282. Erect subshrubs, densely villose; leaves reniform; corolla 2-lipped, white with purple stripes; capsule septicidal; South Spain **156. *Lafuentea***
- Procumbent to ascending caespitose herbs, subglabrous; leaves ovoid to orbicular; corolla subrotate, white with yellow throat or red to orange; capsule loculicidal; South America, New Zealand **181. *Ourisia***
283. Autotrophic shrubs; leaves verticillate, dense; corolla subrotate with very short tube; seeds few, 4 per locule, with distinct wings; South America **180. *Aragoa***
- Hemiparasitic, annual or perennial herbs or shrubs; leaves opposite, lax; corolla subrotate or zygomorphic with distinct, long tube; seeds few (5–6 per locule) to numerous, winged or unwinged 284
284. Anthers with rounded thecae 285
- Anthers with mucronate thecae 290
285. Calyx subequally 4-lobed; corolla yellow, zygomorphic, with distinct upper and lower lip; seeds few, 5–6 per locule, with longitudinal ridges; eastern Mediterranean **280. *Bornmuellerantha***
- Calyx 5-lobed; corolla white, rose or bright red, subrotate, campanulate, lobes subequal; seeds numerous, angulate or obovoid, reticulate 286

286. Prophylls present; anthers and thecae glabrous; Africa, Madagascar and India 287
- Prophylls absent; anthers with either pubescent filaments or thecae; southern Africa or North and South America 288
287. Annual herbs; calyx with 5 distinct lobes; corolla campanulate; stamens included; Africa, Madagascar and India **239. *Micrargeria***
- Shrubs; calyx cup-shaped, with obscure teeth; corolla with short tube and expanded, subrotate limb; stamens exserted; Madagascar **259. *Rhaphispermum***
288. Stigma capitate-globose; southern Africa **238. *Gerardiina***
- Stigma clavate; North or South America 289
289. Shrubs; corolla-tube incurved, dilated near throat; stamens with glabrous filaments, thecae villose; Brazil **212. *Esterhazya***
- Annual or perennial herbs; corolla-tube not or only weakly incurved; stamens with pubescent filaments, thecae glabrous; North to South America **202. *Agalinis***
290. Calyx 4-lobed to spathaceous 291
- Calyx equally to subequally 5-lobed (rarely unequally 3–5-lobed) 294
291. Calyx subequally 4-lobed; seeds not fusiform 292
- Calyx spathaceous; seeds fusiform 293
292. Corolla subrotate, white to pink; thecae bearded; herbs or shrubs of tropical African mountains **278. *Hedbergia***
- Corolla weakly 2-lipped, red; thecae glabrous; shrubs of Brazil **232. *Magdalenea***
293. Leaves pinnately veined, shortly petiolate; calyx strongly inflated in fruit; corolla orange-red, infundibuliform, subrotate; shrubs to shrubby vines **233. *Velloziella***
- Leaves palmately veined, sessile; calyx not strongly inflated in fruit; corolla 2-lipped, red, upper lip helmet-like; suffrutescent herbs or dwarf shrubs **234. *Nothochilus***
294. Fruit a drupe, falling down with calyx at maturity; calyx unequally 3–5-dentate; corolla yellow; rhizome with large scales; Alps and Carpathians **295. *Tozzia***
- Fruit a berry or capsule remaining on the plant; calyx 5-lobed; corolla white or cream to red or yellow; rhizome, if present, without large scales 295
295. Fruit a berry; corolla with long slender tube (up to 8–9 cm long), lianas or shrubs from Madagascar **260. *Radamaea***
- Fruit a capsule; corolla either with long slender or short tube (2–14 cm) or campanulate; prostrate or erect shrubs or herbs 296
296. Prostrate shrub from Southwest Madagascar **260. *Radamaea***
- Erect herbs, rarely shrubs, never prostrate 297
297. Corolla large (up to 14 cm long), white, salverform, subrotate; stigma bilobed; Central and South America **236. *Escobedia***
- Corolla distinctly smaller, variously coloured, never salverform but tubular or campanulate, subrotate to 2-lipped; stigma clavate to capitate 298
298. Prophylls present; seeds fusiform 299
- Prophylls absent; seeds different 302
299. Calyx not inflated in fruit; corolla yellow to red; style curved, horseshoe-like; pantropical herbs, rarely shrubs **229. *Alectra***
- Calyx distinctly inflated in fruit; corolla whitish-cream to yellow or red; style straight or only weakly curved 300
300. Leaves pinnately veined; corolla red; shrubs of Brazil **235. *Physocalyx***
- Leaves palmately veined; corolla whitish-cream to yellow; suffrutescent herbs of South America, Africa and Madagascar 301
301. Corolla yellow, tubular; inflorescence proliferating after anthesis; woody rhizome present; Madagascar **230. *Pseudomelasma***
- Corolla whitish-cream to yellow; inflorescence not proliferating after anthesis; rhizome absent; South America to Africa **231. *Melasma***
302. At least lower leaves pinnatifid 303
- Leaves all entire 306
303. Leaves alternate **244. *Petitmenginia***
- Leaves all opposite 304
304. Corolla rose, subrotate; thecae and filaments glabrous; seeds 1–2 per locule, large, testa rugulose; Central to East Asia **237. *Leptorhabdos***
- Corolla yellow or orange; at least base of filaments lanose; seeds numerous, winged or wingless; North America 305
305. Corolla orange, tubular; stamens long-exserted; stigma clavate; seeds winged **207. *Macranthera***
- Corolla yellow, funnelform or campanulate; stamens included; stigma capitate; seeds winged or wingless **205. *Aureolaria***
306. Corolla yellow, subrotate, tube inflated; seeds winged **204. *Brachystigma***
- Corolla white, rose, pink or red, subrotate to weakly 2-lipped, tube not inflated; seeds angled but never winged **202. *Agalinis***
307 (168). Calyx with unequal lobes, at least 1 sepal distinctly larger than the others 308
- Calyx with equal to subequal lobes 310
308. Outer 3 calyx lobes large, ovoid to cordate, inner 2 lobes small, linear-lanceolate; prophylls present; inflorescence lax or capitate; 1 theca of all anthers or only of the 2 abaxial anthers sterile and reduced; Southeast Asia **123. *Adenosma***
- 1 calyx lobe (usually adaxial lobe) broad, the 4 remaining lobes small and narrow; prophylls absent; inflorescence lax; all anthers with either 2 unequal thecae or monothecous; South America (Brazil) 309
309. All anthers with 2 parallel thecae, one theca sterile and reduced; plants prostrate **126. *Tetraulacium***
- Abaxial pair of stamens monothecous; adaxial pair with one theca sterile and reduced; plants erect **127. *Dizygostemon***
310. Anthers with one theca (in *Tetraspidium* a second sterile theca present) 310
- Anthers with 2 unequal thecae, one fertile and one sterile, reduced to a small appendage 322
311. Holoparasitic; leaves squamiform, lacking chlorophyll 312
- Hemiparasitic or autotroph; with green leaves 315
312. Corolla with ± long tube and expanded limb; plants not fleshy; with lax scale-like leaves 313
- Corolla tubular, limb not expanded, fleshy plants with numerous dense scale-like leaves **214. *Hyobanche***
313. Corolla with narrow tube and expanded 2-lipped limb, orifice of tube small, less than 1 mm in diameter **253. *Striga* p.p. (*S. gesnerioides*)**
- Corolla with ± large, broad tube, limb not 2-lipped 314
314. Corolla with small limb compared to the tube; blue to violet, anthers red with one large theca, the reduced one minute and hardly visible; Madagascar **252. *Tetraspidium***
- Corolla with large limb, white; anthers whitish, strictly monothecous; stigma clavate **216. *Paraharveya***

315. All corolla lobes free 316
- The 2 upper corolla lobes fused 318
316. Corolla-tube straight or only weakly curved, narrow 317
- Corolla-tube curved at right angle near throat and distinctly widened there **255. *Parastriga***
317. Plants with short corolla-tube, corolla blue or violet; filaments glabrous; leaves generally lanceolate, entire **251. *Buchnera***
- Plants with long corolla-tube, corolla whitish; filaments bearded; leaves filiform to pinnatisect with filiform segments **258. *Rhamphicarpa***
318. Corolla-tube ± straight, rarely bent or curved; fruit a capsule or berry 319
- Corolla-tube always abruptly curved; fruit a capsule 320
319. Small shrubs or subshrubs from Namibia and Angola; stamens exserted; anthers with small appendages; fruit a capsule **247. *Hiernia***
- Herbs from tropical Africa; stamens included; anthers without appendages; fruit a capsule or berry **257. *Cycnium***
320. Stigma 2-lobed **253. *Striga***
- Stigma clavate 321
321. Annual erect plants; corolla with ± short tube; filaments hairy; Southeast Asia **254. *Pseudostriga***
- Perennial prostrate herbs; corolla with long tube; filaments glabrous; East Africa **256. *Cycniopsis***
322 (310). Holoparasitic plants lacking green leaves 323
- Hemiparasitic or non-parasitic plants, always with green leaves 324
323. Corolla large, limb large and spreading, tube ± straight; sterile theca clearly visible **215. *Harveya***
- Corolla small, limb not spreading, tube curved; sterile theca minute and easily overlooked **252. *Tetraspidium***
324. Woody shrubs or climbers 325
- Annual or perennial herbs 329
325. Corolla white, with long slender tube (up to 14 cm), with an intensive odour of jasmine; leaves glaucous, with epicuticular waxes; fruit a drupe with sclerified endocarp and fleshy exo- and mesocarp; Madagascar **261. *Leucosalpa***
- Corolla rose, lilac, purple red, mauve or violet, tube campanulate to cylindrical or funnel-shaped, not slender; leaves greenish, never glaucous; fruit a capsule or indehiscent and sclerified; Africa 326
326. Shrub with calyx becoming woody and sclerified after anthesis, surrounding oblong capsule with a pronounced beak; Somalia and Socotra **264. *Xylocalyx***
- Shrubs or climbers with membranaceous calyx; capsule globose or longitudinally flattened 327
327. Shrubs with longitudinally flattened capsule; Socotra and southern Africa **248. *Graderia***
- Climbers with globose capsule 328
328. Calyx campanulate, inflated in fruit; leaves pinnatisect; East and southern Africa **262. *Buttonia***
- Calyx not inflated in fruit; leaves triangular with long teeth, never pinnatisect; Central Africa and Sao Tomé **263. *Thunbergianthus***
329. Rosulate herbs; fertile stamens 2, one pair reduced to filiform staminodes, anthers with 2 unequal mucronate thecae or monothecous; Australia **144. *Hemiarrhena***
- Shrubs or herbs with cauline leaves; fertile stamens 4, staminodes absent, anthers different; Central America, Africa, Madagascar, Asia 330
330. All anthers with one fertile theca, 2 without sterile theca and the other 2 with a sterile, appendage-like theca 331
- All anthers with 2 thecae, one sterile and one fertile 332
331. Connective not prolonged, sterile thecae filiform **249. *Ghikaea***
- Connective filiformly prolonged, with 2 unequal branches, the shorter one curved and with a small globose appendage representing the sterile theca **246. *Pseudosopubia***
332. Corolla with long, funnel-shaped tube, gradually widened at throat 333
- Corolla with short or long, not funnel-shaped tube, not distinctly widened at throat or with a gibbous, distinct widening 334
333. Calyx 5-partite **242. *Parasopubia***
- Calyx spathaceous, split down on lower side **250. *Centranthera***
334. Leaves all linear, entire; corolla-tube weakly incurved; Central America **203. *Anisantherina***
- Leaves linear or ovoid, but usually trifid to pinnately divided, rarely entire; corolla-tube straight, if incurved, then abruptly curved at right angle; Africa, Madagascar and Asia 335
335. Corolla with long tube, abruptly curved at right angle, with gibbous widening below throat, palate present; stamens included, with pilose abaxial filaments **245. *Sieversandreas***
- Corolla with short, ± straight tube, not distinctly widened at throat, palate absent; stamens exserted, with glabrous filaments 336
336. Calyx woolly tomentose on inner surface of lobes; thecae of anthers glabrous **241. *Sopubia***
- Calyx glabrous on inner surface of lobes; fertile theca with long-haired margin of stomium **243. *Baumia***

"Scrophulariaceae"

Scrophulariaceae Juss. (1789) nom. cons.
Verbascaceae Raf. (1821).
Verbasceae Dum. (1829).
Verbasceae Benth. in DC. (1846).

Subshrubs, perennial or biennial herbs. Inflorescence thyrsic or racemose. Corolla partly with abaxial protrusions and adaxial gibbosities or spurs, sometimes with 2 gibbous spurs abaxially, without palate, tube short, petal aestivation descending or rarely ascending (*Teedia*, *Oftia*). Stamens 5–4, filaments villose or glabrous, anthers mostly synthecous or thecae separate and confluent at apex, the median stamen reduced to a scale-like staminode or lacking. Ovary bilocular. Capsule with septicidal or septicidal and loculicidal dehiscence, fruit rarely a berry or drupe, or fruit separating into 2 cocci or 1 coccus (tribe Manuleae p.p.). Intermediate layers of ovule integument 3–9(10), endosperm of mature seed furrowed longitudinally, alveolated by regular alterations between bothroblasts and abothroblasts, only bothroblasts protruding into endosperm, walls of endotheliar cells thickened only towards endosperm, or endosperm of mature seed smooth or with longitudinal furrows, endotheliar cells arranged longitudinally, epidermal cells of testa usually heterogeneous, producing various ornamentation and wings. Seeds few to numerous, alveolated (bothrospermous or aulacospermous) or with smooth endosperm. Fifty-three genera with 1473 species, cosmopolitan. The Buddlejaceae and Myoporaceae should be included as tribes Buddlejeae and Myoporeae.

1. Tribe Verbasceae Dumort. (1829).

Subshrubs, perennial to biennial herbs, glandular-pubescent to densely villose, hairs usually dendroid. Inflorescence thyrsic with accessory flowers or racemose, stamens 5–4, filaments villose, anthers with confluent thecae (synthecous), stigma capitate, capsule septicidal, seeds bothrospermous or aulacospermous.

1. *Verbascum* L.

Verbascum L., Sp. Pl. 1: 177 (1753); Murbeck, Lunds Univ. Arsskrift N.F. Avd. 2, 22: 1–237 (1925), rev.; Murbeck, Lunds Univ. Arsskrift N.F. Avd. 2, 29: 1–630 (1933), rev.
Celsia L. (1753).
Staurophragma Fisch. & C. Mey. (1843).

Annual, biennial to perennial herbs or rarely small shrubs. Stems erect, rounded. Leaves cauline and in a basal rosette, alternate, linear-lanceolate to ovoid or orbicular, acuminate or obtuse, entire or pinnatifid to pinnatisect and with dentate to serrate or crenate margin. Inflorescence with axillary cymes and accessory flowers or racemose. Flowers distinctly pedicellate. Calyx deeply 5-lobed, lobes equal. Corolla yellow, rarely violet or purple, subrotate. Stamens 4–5, equal, exserted, sometimes the median stamen reduced to staminode. Ovary ovoid to globose. Seeds numerous, endosperm alveolated (bothrospermous). $2n = 18$, 26, 30, 32, 34, 36, 44, 45, 48, 64, 66. About 360 species in Europe and Asia (diversity centre in Turkey and Iran) to Northeast Africa (Ethiopia, Kenya).

2. *Rhabdotosperma* Hartl Fig. 44D, E

Rhabdotosperma Hartl, Beitr. Biol. Pflanzen 53: 57 (1977).

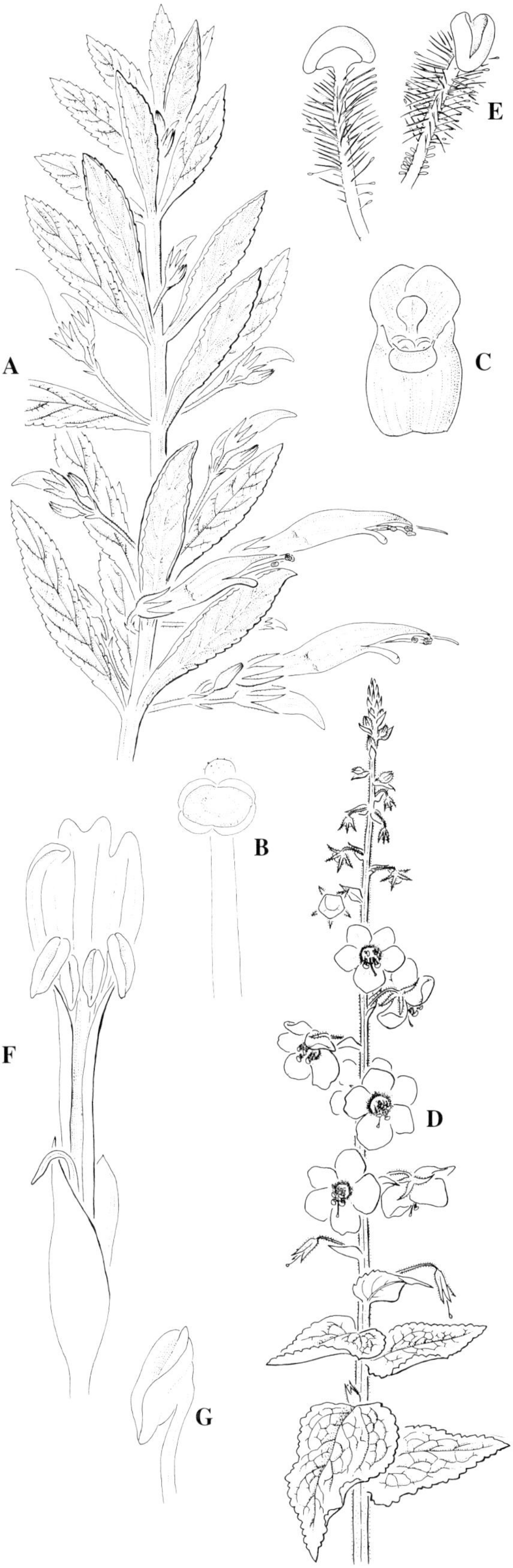

Fig. 44. Scrophulariaceae. **A, B** *Ameroglossum pernambucense*. **A** Habit. **B** Stamen. **C** *Scrophularia auriculata*. Corolla. **D, E** *Rhabdotosperma brevipedicellata*. **D** Habit. **E** Stamens. **F, G** *Hebenstretia angolensis*. **F** Flower. **G** Stamen (**A, B** after Fischer, Vogel & Lopes 1999, redrawn, **D–G** after Fischer 1999, redrawn)

Small shrubs or subshrubs. Stems erect, rounded, woody at base. Leaves cauline and in a basal rosette, alternate, sessile to petiolate, ovoid or orbicular, acuminate or obtuse, entire or pinnatisect and with dentate to serrate or crenate margin. Inflorescence racemose, never with accessory flowers, prophylls absent. Flowers shortly to distinctly pedicellate. Calyx deeply 5-lobed, lobes equal. Corolla yellow, subrotate. Stamens 4, in *R. schimperi* (Skan) Hartl 5, equal, exserted. Ovary pyriform to globose. Seeds numerous, endosperm alveolated (aulacospermous). Seven species, mountain regions of tropical Africa (Cameroon to Ethiopia, E Congo and Tanzania), Arabia.

2. Tribe Scrophularieae Dumort. (1827). Fig. 44C

Perennial to annual herbs, rosulate or with elongate stem, inflorescence cymose, anthers synthecous, stigma capitate to flattened or minutely bifid (*Antherothamnus*), capsule septicidal, seeds aveolated (bothrospermous) or not alveolated.

2a. Subtribe "Oreosoleninae"

Perennial rosulate herbs with woody rhizome, pilose or glabrous, leaves opposite, inflorescence cymose, congested, bracteate.

3. *Nathaliella* B. Fedtsch.

Nathaliella B. Fedtsch., Bot. Zurn. SSSR 17: 327 (1932).

Leaves sessile to subpetiolate, lanceolate or broadly ovoid to orbicular, obtuse, with dentate margin. Flowers sessile to shortly pedicellate. Calyx deeply 5-lobed. Corolla pink-violet, limb indistinctly 2-lipped, tube narrowly cylindrical. Stamens 4, included, staminode absent, filaments attached at base of corolla-tube, anthers divergent. Ovary ovoid. Stigma wide and flattened. Seeds numerous. One species, *N. alaica* B. Fedtsch. Central Asia, Kirghizia (Pamir-Alai).

4. *Oreosolen* Hook. f.

Oreosolen Hook. f., Fl. Brit. India 4: 318 (1884).

Leaves opposite, sessile to subpetiolate, lanceolate or broadly ovoid to orbicular, obtuse, with dentate to serrate margin. Flowers sessile to shortly pedicellate. Calyx deeply 5-lobed. Corolla yellow, limb distinctly 2-lipped, tube narrowly cylindrical. Stamens 4, exserted, the median stamen reduced to a subterete or 2-lobed staminode inserted at base of upper lip, filaments attached at throat of corolla. Ovary ovoid. Stigma capitate. Seeds numerous. Four species, Himalayas, Nepal to Bhutan, Tibet.

2b. Subtribe "Scrophulariinae"

Inflorescence terminal, cymose, congested or capitate to usually lax, forming a thyrse with axillary cymes, frondose or bracteate, prophylls present. Stamens 4, exserted, the median stamen reduced to a scale-like or filiform (*Antherothamnus*) staminode near throat of corolla.

5. *Scrophularia* L. Fig. 44C

Scrophularia L., Sp. Pl. 2: 619 (1753); Stiefelhagen, Bot. Jahrb. Syst. 44: 406–496 (1910), rev.; Dalgaard, Opera Bot. 51: 1–64 (1979), reg. rev.
Tuerckheimocharis Urban (1912).
Mosheovia Eig (1938).

Annual, biennial or perennial to suffrutescent herbs or dwarf shrubs, glabrous, glandular-pubescent to densely villose. Stems ascending to erect, quadrangular, sometimes winged. Leaves cauline, opposite, sometimes becoming alternate above, subsessile to usually distinctly petiolate, lanceolate, ovoid to orbicular, acuminate or obtuse, entire or pinnatifid to pinnatisect, with dentate to serrate margin. Flowers shortly to distinctly pedicellate. Calyx 5-lobed to middle, lobes equal. Corolla greenish-brown to dark brown or violet-purple to red with yellow, limb 2-lipped, lobes orbicular, tube ventricose, inflated. Ovary ovoid to globose. Stigma capitate to weakly bilobed. Seeds numerous, endosperm alveolated (bothrospermous). $2n = 26, 36, 42, 44, 52, 58, 60, 68, 74, 78$. About 200 species, Europe and Asia (diversity centre in Iran and Afghanistan), North to tropical America (Caribbean).

6. *Antherothamnus* N.E. Br.

Antherothamnus N.E. Br., Hooker's Icon. Pl. t. 3007 (1915); Hilliard, The Manuleae: 75–77 (1994), rev.

Shrub, foetid, glandular-pubescent. Stems erect, rigid, subspinescent with age. Leaves cauline, alternate, fasciculate, subsessile to petiolate, obovoid-cuneate, obtuse, somewhat succulent, with entire to dentate margin, base of petiole persistent.

Flowers distinctly pedicellate. Calyx deeply 5-lobed, campanulate, obscurely 2-lipped. Corolla white with yellow throat, limb 2-lipped, lobes spreading, suborbicular, tube funnel-shaped, corolla with clavate hairs inside. Stamens 4, filaments inserted near base of tube, abaxial stamens shortly exserted, adaxial stamens included. Ovary deltate. One species, *A. pearsonii* N.E. Br. in South Africa.

3. Tribe Alonsoeae Barringer (1993).

Perennial herbs, rosulate or with elongate stem, leaves opposite, inflorescence thyrsic with accessory flowers or racemose, thecae confluent at apex, capsule with septicidal or septicidal and loculicidal dehiscence (*Charadrophila*), seeds aulacospermous.

7. *Alonsoa* Ruiz & Pav.

Alonsoa Ruiz & Pav., Syst. Veg. Fl. Peruv. Chil.: 150 (1798).
Schistanthe Kunze (1841).

Perennial herbs, glabrous except for glandular-pubescent inflorescence. Stems ascending to erect, quadrangular. Leaves cauline, sessile or subsessile to shortly petiolate, linear-lanceolate to ovoid, acuminate, with dentate or serrate margin. Inflorescence racemose, bracteate, prophylls absent. Flowers distinctly pedicellate, resupinate by twisting of pedicel. Calyx deeply 5-lobed, lobes narrow and equal. Corolla orange to salmon-red, limb 2-lipped, explanate-rotate, tube almost obsolete, limb 5-lobed, adaxial lobe largest. Stamens 4, exserted, filaments short. Ovary ovoid, upcurved in fruit. Stigma capitate. $2n = 28, 56$. Eleven species, tropical South America, South Africa.

8. *Charadrophila* Marloth

Charadrophila Marloth, Bot. Jahrb. Syst. 26: 358 (1899); Weber, Bot. Jahrb. Syst. 111: 87–119 (1989), rev.

Perennial rosulate herb, pubescent. Stems erect, rounded, with short internodes. Leaves rosulate, petiolate, broad-elliptic, somewhat wedge-shaped, obtuse, with coarsely crenate margin. Inflorescence thyrsic, with accessory flowers, bracteate, prophylls present. Flowers distinctly pedicellate. Calyx 5-lobed to middle, becoming distinctly 2-lipped postflorally, lobes narrow and subequal. Corolla blue with white throat, limb 2-lipped, lobes obovoid, tube campanulate, broad, short, somewhat pouched on ventral side. Stamens 4, included, filaments short, inserted at base of tube, median fifth stamen sometimes present as staminode. Ovary ovoid, hairy outside. Stigma bilobed. One species, *C. capensis* Marloth, Southwestern Cape.

4. Tribe Hemimerideae Benth. (1835).

Herbs, rarely shrubs, inflorescence with axillary cymes of 1–5 flowers (*Basistemon*) or racemose, prophylls present (*Basistemon*) or absent, corolla-tube bisaccate, with 2 gibbous spurs abaxially, anthers with two separate to confluent equal thecae, stigma capitate to weakly bilobed, capsule loculicidal (*Basistemon*), loculicidal and septicidal or septicidal (*Diascia, Hemimeris, Diclis, Nemesia*).

9. *Basistemon* Turcz.

Basistemon Turcz., Bull. Soc. Nat. Moscou: 214 (1863); Barringer, Syst. Bot. 10: 125–133 (1985), rev.
Desdemona S. Moore (1895).
Hassleropsis Chodat (1904).
Saccanthus Herzog (1916).

Shrubs or small trees, glabrous or young branches minutely pubescent. Stems erect. Leaves cauline, opposite, sessile to subsessile, lanceolate to ovoid, acuminate, with entire or minutely serrulate margin. Flowers distinctly pedicellate. Calyx deeply 5-lobed, lobes lanceolate, subequal. Corolla light violet, limb 2-lipped, lobes spreading, tube very short, with 2 gibbous spurs abaxially. Stamens 4. Ovary conical or globose. Seeds few to numerous, endosperm absent. Eight species in tropical South America.

10. *Diascia* Link & Otto

Diascia Link & Otto, Icon. Pl. Select. 7 (1820); Hilliard & Burtt, J. S. Afr. Bot. 50: 269–340 (1984), reg. rev.
Chamaecrypta Schltr. & Diels (1942).

Annual or perennial herbs, glabrous or pubescent. Stems ascending or erect. Leaves cauline or rosulate, opposite, sometimes alternate above, sessile to petiolate, lanceolate to ovoid, acuminate, entire or pinnatifid, with entire to dentate margin. Flowers distinctly pedicellate. Calyx deeply 5-lobed, lobes lanceolate, subequal. Corolla pink to purple, limb 2-lipped, with 2 gibbous spurs, pouches or pits abaxially, lobes spreading, tube very short or obsolete. Stamens 4, included, rarely the abaxial anthers aborted. Ovary ovoid. Seeds numerous, weakly curved. $2n = 18, 36, 54$. Fifty species, South Africa.

11. *Hemimeris* L. f.

Hemimeris L. f., Suppl. Carp.: 186 (1807).

Annual herbs, glabrous or pubescent. Stems ascending or erect. Leaves cauline, opposite or the bracts subfasciculate, sessile to petiolate, lanceolate to ovoid, acuminate, entire or subpinnatifid, with entire to dentate margin. Flowers distinctly pedicellate. Calyx deeply 5-lobed, lobes lanceolate, somewhat unequal. Corolla yellow, limb 2-lipped, with 2 gibbous spurs or pouches abaxially, sometimes with 2 tooth-like appendages clasping the filaments near the corolla throat, lobes spreading, tube very short or obsolete. Stamens 2, included. Ovary ovoid. Seeds numerous, angular-ovoid or subglobose, reticulate, with or without membranous wing. $2n = 14$. Four species, South Africa.

12. *Diclis* Benth.

Diclis Benth. in Hook., Companion Bot. Mag. 2: 23 (1836).

Annual or perennial herbs, pubescent. Stems creeping, prostrate or decumbent, rarely ascending or erect. Leaves cauline, opposite, sometimes alternate above, petiolate, ovoid to elliptic or orbicular, acuminate, entire, with dentate to serrate margin. Flowers distinctly pedicellate. Calyx deeply 5-lobed, lobes lanceolate, subequal. Corolla white to cream, limb 2-lipped, lobes spreading, tube short, at base with a dependent spur. Stamens 4. Ovary elliptic, glandular-pubescent. Seeds numerous, subglobose, with reticulate membranous testa. $2n = 14$. Ten species in tropical and South Africa, Madagascar.

13. *Nemesia* Vent.

Nemesia Vent., Jard. Malm.: 41 (1804).

Annual or perennial herbs or subshrubs, glabrous or sparsely pubescent. Stems ascending or erect, quadrangular. Leaves cauline, opposite, sessile to petiolate, lanceolate to ovoid, acuminate, often cordate at base, with entire to dentate margin. Flowers usually pedicellate. Calyx deeply 5-lobed, lobes lanceolate, subequal. Corolla white, pale pink to mauve, limb 2-lipped, upper lip 4-lobed, lobes lanceolate to oblong, lower lip emarginate, with distinct palate, tube short and with abaxial sac or spur. Stamens 4, included. Ovary elliptic. Seeds numerous, cylindrical-ellipsoid, with entire or interrupted membranous wing. $2n = 18$. Sixty-five species, in south tropical and South Africa.

14. *Colpias* E. Mey. ex Benth.

Colpias E. Mey. ex Benth. in Hook., Companion Bot. Mag. 2: 53 (1836).

Dwarf shrub, densely pilose. Stems procumbent to ascending or erect, quadrangular, winged. Leaves cauline, alternate or scarcely opposite, petiolate, ovoid or orbicular-deltate, acuminate, subcordate at base, entire or incised-palmatilobed, with entire to dentate margin. Flowers distinctly pedicellate. Calyx deeply 5-lobed, lobes subequal, lanceolate. Corolla sulphur-yellow, limb 2-lipped, tube short, with 2 gibbous spurs or pouches abaxially. Stamens 4, included. Ovary ovoid. Seeds numerous, black, terete, ribbed. One species, *C. mollis* E. Mey. ex Benth., in South Africa.

5. Tribe Russelieae Pennell (1920).

Perennial herbs or terrestrial to epiphytic shrubs, inflorescence thyrsic with axillary cymes and sometimes accessory flowers (*Russelia*), prophylls present, anthers with 2 separate thecae, confluent at apex, staminode long (*Ameroglossum*), short or lacking, stigma capitate to minutely bilobed (*Dermatobotrys*), fruit a septicidal capsule or rarely a berry (*Dermatobotrys*), seeds with smooth endosperm.

15. *Russelia* Jacq.

Russelia Jacq., Enum. Pl. Carib. 6: 25 (1760); Carlson, Fieldiana Bot. 29: 231–292 (1957), rev.

Perennial suffrutescent herbs or dwarf shrubs, glabrous except for glandular-pubescent inflorescence or pubescent to densely villose. Stems erect or often recurved or pendent, quadrangular or terete. Leaves cauline, opposite or verticillate, sessile or petiolate, linear-lanceolate to ovoid, acuminate, with entire or dentate margin, sometimes reduced to small scales. Flowers usually distinctly pedicellate. Calyx deeply 5-lobed. Corolla bright red, cylindrical, limb 2-lipped, lobes rounded, tube cylindrical. Stamens 4, included. Ovary ovoid to globose, with long hyaline inner hairs. Seeds numerous, ovoid, testa exalate. $2n = 20$. Fifty-two species, Central America, Cuba.

16. *Ameroglossum* Eb. Fischer, S. Vogel & Lopez Fig. 44A, B

Ameroglossum Eb. Fischer, S. Vogel & Lopez, Feddes Repert. 110: 529 (1999).

Perennial shrub, densely pubescent, with glandular-pubescent inflorescence. Stems erect, quadrangular, woody at base. Leaves cauline, opposite, decussate or 3-verticillate, sessile to minutely petiolate, lanceolate, acuminate, with dentate margin, lower surface densely villose. Flowers pedicellate. Calyx deeply subequally 5-lobed. Corolla scarlet except the yellow lower lip, cylindrical, limb 2-lipped, upper lip 2-lobed, lower lip appearing undivided and tongue-like, minutely 3-lobed, inside glandular, with incurved rims. Stamens 4, exserted. Ovary conical, with a massive basal disc; style long exserted. Seeds numerous, ovoid, testa reticulate. One species, *A. pernambucense* Eb. Fischer, S. Vogel & Lopez in Brazil.

17. *Dermatobotrys* Bolus

Dermatobotrys Bolus, Hooker's Icon. Pl. t. 1940 (1890).

Epiphytic shrub, arising from thick woody rhizome, glabrous. Stems erect, ± quadrangular, woody. Leaves cauline, opposite, petiolate, oblong-ovoid, acuminate, with dentate margin, fleshy. Flowers shortly pedicellate. Calyx deeply 5-lobed, lobes equal. Corolla bright red, trumpet-shaped, subactinomorphic, limb 5-lobed, spreading, lobes ovoid-oblong, tube long, somewhat incurved, funnel-shaped. Stamens 5, equal, inserted at corolla throat, exserted, filaments very short. Ovary ovoid-conical. Seeds numerous, testa with short, disconnected ridges. One species, *D. saundersii* Bolus, in South Africa.

6. Tribe Teedieae Benth. (1835).

Oftiaceae Takht. & Reveal (1993).

Shrubs, inflorescence racemose (*Oftia*) or with axillary cymes (*Ranopisoa*, *Teedia*), prophylls absent (*Ranopisoa*, *Teedia*) or present, corolla subrotate, anthers with two separate, equal or confluent (*Ranopisoa*) thecae, stigma oblique and weakly bipartite or capitate (*Ranopisoa*), fruit a black drupe (*Oftia*) or berry (*Teedia*), seeds few (1–6) to numerous (*Teedia*), endosperm bothrospermous (*Oftia*) or aulacospermous (*Teedia*).

18. *Oftia* Adans.

Oftia Adans., Fam. II: 199 (1763).

Perennial suffrutescent herbs or shrubs, glabrous or glandular-pubescent to densely villose. Stems decumbent to ascending or erect, rounded. Leaves cauline, opposite, sometimes alternate above, sessile to subsessile, lanceolate to ovoid or orbicular, acuminate, with dentate to serrate, revolute margin. Flowers subsessile to distinctly pedicellate. Calyx deeply 5-lobed, lobes equal. Corolla white, subrotate, tube narrowly cylindrical, lobes rectangular to oblong and rounded. Stamens 4, included. Ovary ovoid to subglobose. Seeds 4–6. $2n = 38$. Three species in South Africa.

19. *Ranopisoa* J.F. Leroy

Ranopisoa J.-F. Leroy, Adansonia II, 17(2): 116 (1977).

Shrub, pubescent. Stems erect, rounded. Leaves cauline, alternate, petiolate, lanceolate to ovoid, obtuse, with entire margin. Flowers distinctly pedicellate. Calyx deeply 5-lobed, lobes equal. Corolla yellow, subrotate, tube narrowly cylindrical, lobes rectangular to oblong and rounded. Stamens 4, included. Ovary ovoid to subglobose, with 2 loculi, each with 2 ovules. Fruit a capsule with 1 seed (?). One species, *R. rakotosonii* (Capuron) J.F. Leroy in South Madagascar.

20. *Teedia* Rudolphi

Teedia Rudolphi, J. Bot. (Schrader) 1799(2): 288 (1800).

Perennial suffrutescent herbs or shrubs, glabrous or glandular-pubescent. Stems erect, quadrangular, woody at base. Leaves cauline, opposite, petiolate, ovoid to broadly elliptic, acuminate, with dentate to serrate margin. Flowers distinctly pedicellate. Calyx deeply 5-lobed, lobes equal. Corolla blue to lilac or mauve, subrotate, tube narrowly cylindrical, lobes rectangular to oblong and rounded. Stamens 4, included. Ovary ovoid to subglobose. Fruit a berry (in *T. lucida*) or drupe with 2 loculi. $2n = 38$. Four species in South Africa.

7. Tribe Freylinieae Barringer (1993).

Perennial herbs or shrubs, inflorescence with axillary cymes, bracteate (*Phygelius*) or frondose (*Freylinia*), anthers with two separate equal and parallel thecae. Capsule septicidal (*Phygelius*) or septicidal and loculicidal (*Freylinia*).

21. *Phygelius* E. Meyer ex Benth.

Phygelius E. Meyer ex Benth. in Hook., Companion Bot. Mag. 2: 53 (1836).

Perennial suffrutescent herbs or shrubs, glabrous except for glandular-pubescent inflorescence. Stems erect, quadrangular, very narrowly 4-winged. Leaves cauline, opposite, petiolate, ovoid to ovoid-lanceolate, obtuse or scarcely acuminate, with dentate to serrate margin. Flowers distinctly pedicellate. Calyx deeply 5-lobed, lobes equal. Corolla red, campanulate, weakly 2-lipped, tube long, narrowly funnel-shaped or trumpet-shaped, lobes spreading, rectangular to oblong and rounded. Stamens 4, included to shortly exserted. Ovary ovoid to subglobose. Stigma weakly 2-lobed. Seeds numerous, subglobose, with reticulate testa. Two species, *P. capensis* Benth. and *P. aequalis* Harv. ex Hiern in South Africa.

22. *Freylinia* Colla

Freylinia Colla, Hortus Ripul.: 56 (1824).

Shrubs or subshrubs, glabrous or young stems glandular-viscid. Stems erect, quadrangular. Leaves cauline, opposite, verticillate or alternate, sessile to shortly petiolate, obovoid to broadly lanceolate, with entire to dentate margin. Flowers shortly pedicellate. Calyx deeply 5-lobed, lobes subequal. Corolla purple to lilac or pale blue, 2-lipped, tube funnel-shaped or subcylindrical, straight, lobes rectangular to oblong and rounded. Stamens 4, included, occasionally a fifth median staminode present. Ovary ovoid. Stigma globose and capitate. Seeds numerous, with marginal wing. Four species in southern Africa, one species (*F. tropica* S. Moore) north to Tanzania.

8. Tribe Leucophylleae Miers (1850).

Shrubs or herbs, inflorescence racemose, prophylls absent.

23. *Eremogeton* Standl. & L.O. Williams

Eremogeton Standl. & L.O. Williams, Ceiba 3: 172 (1953).
Ghiesbreghtia A. Gray (1873).

Suffrutescent herb to dwarf shrub or small tree, glabrous except for densely pale tomentose young branches and leaves. Stems erect, rounded. Leaves cauline, subopposite to alternate, shortly petiolate, elliptic to ovoid or narrowly obovoid, acuminate to obtuse, with coarsely serrulate to crenate margin. Flowers distinctly pedicellate. Calyx deeply 5-lobed, lobes equal. Corolla white to yellow, large, limb 2-lipped, lobes spreading, tube broad, tubular-campanulate. Stamens 2, exserted. Ovary ovoid. Stigma capitate or weakly bilobed. Seeds numerous, narrowly oblong, sometimes angulate. One species, *E. grandiflorus* (A. Gray) Standl. & L.O. Williams, Central America.

24. *Leucophyllum* Bonpl.

Leucophyllum Bonpl., in Humb. & Bonpl., Pl. Aequin. 2: 95, t. 109 (1812).
Faxonanthus Greenm. (1902).

Small shrubs, densely pale tomentose with stellate hairs. Stems erect, rounded. Leaves cauline, alternate, subsessile to shortly petiolate, linear, linear-lanceolate or oblanceolate to spathulate, obtuse, with entire margin. Flowers distinctly pedicellate. Calyx deeply 5-lobed, lobes equal. Corolla dark purple, small, limb 2-lipped, lobes spreading, rounded, throat barbate, tube funnel-shaped-campanulate. Stamens 4, exserted. Ovary ovoid. Stigma capitate or weakly bilobed. Seeds numerous, narrowly oblong, testa reticulate. Twelve species, south-western North America to Mexico.

9. Tribe Aptosimeae (Benth.) Benth. (1876).

Perennial suffrutescent herbs or shrubs, rarely annuals, leaves alternate, inflorescence racemose or rarely cymose (*Aptosimum* p.p.), prophylls present, stamens usually 4, rarely 2–4 (*Anticharis*), included, anthers transverse, ciliate to glabrous, with confluent thecae (synthecous), stigma capitate or clavate to weakly bilobed, capsule loculicidal and septicidal, compressed at apex, endothelium of ovule with isodiametric, cube-shaped cells with equally thickened walls.

25. *Anticharis* Endl.

Anticharis Endl., Gen.: 682 (1839).
Gerardiopsis Engl. (1895).

Annual or perennial suffrutescent herbs or dwarf shrubs, glandular-pubescent. Stems erect, rounded. Leaves cauline, alternate, sessile, linear-lanceolate to ovoid, acuminate, with entire margin. Flowers pedicellate. Calyx deeply 5-lobed, campanulate, lobes linear to lanceolate. Corolla pale mauve to blue or lilac, limb weakly or distinctly 2-lipped, lobes equal, flat, rounded, spreading, tube much longer than lobes, funnel-shaped, narrow at base, dilated into throat. Ovary ovoid, subacuminate. Seeds numerous, ovoid or cylindrical, sometimes rugose. Fourteen species, Cape Verde Islands, dry regions of Africa, Arabia to India.

26. ***Aptosimum*** Burchell ex Benth.

Aptosimum Burchell ex Benth., Bot. Reg. 22: sub t. 1882 (1836), nom. cons.
Chilostigma Hochst. (1841).

Perennial suffrutescent herbs or dwarf shrubs, rarely annual, glabrous, glandular-pubescent or densely villose. Stems prostate to ascending or erect, densely tufted, rounded. Leaves cauline, alternate, sessile, linear-lanceolate, acuminate, sometimes spinose or rigid-apiculate, with entire and sometimes thickened margin. Flowers sessile to shortly pedicellate. Calyx deeply 5-lobed, campanulate, lobes linear to lanceolate. Corolla blue or purple, limb 2-lipped, upper lip 2-lobed, lower lip 3-lobed, lobes subequal, flat, rounded, tube much longer than lobes, narrow at base, widening into long throat. Ovary obcordate. Seeds numerous, compressed-globose, sometimes rugose. Twenty species, from Yemen and northern tropical Africa (1 species) to South Africa.

27. ***Peliostomum*** E. Mey. ex Benth.

Peliostomum E. Mey. ex Benth. in Lindl., Edwards's Bot. Reg. 22: t. 1882 (1836).

Perennial suffrutescent herbs or dwarf shrubs, glabrous or often glandular-viscid. Stems ascending or erect, rounded. Leaves cauline, alternate, sessile, linear-lanceolate to ovoid-elliptic, acuminate, with entire margin. Flowers shortly pedicellate. Calyx deeply 5-lobed, campanulate, lobes linear to lanceolate. Corolla dark blue or purple, limb 2-lipped, lobes subequal, flat, rounded, tube much longer than lobes, narrow at base, widening into long throat. Ovary ovoid-conical. Seeds numerous, rugose. Seven species, South Africa.

10. Tribe Manuleae Benth. & Hook. f. (1876). Fig. 44F, G

Selaginaceae Choisy (1823), nom. cons.
Hebenstretiaceae Horan. (1834).
Selagineae Wettst. (1891).

Dwarf and often ericaceous shrubs or herbs, leaves alternate or opposite, solitary or crowded in fascicles, inflorescence with axillary cymes (*Barthlottia*, *Manuleopsis*, *Sutera*, *Manulea*) or racemose, lax or congested to capitate (*Melanospermum*, *Glekia* to *Phyllopodium*, *Glumicalyx* to *Globulariopsis*), prophylls present (*Barthlottia* to *Jamesbrittenia*, *Sutera*, *Manulea*) or absent, anthers synthecous or with separate thecae (*Dischisma*, *Hebenstretia*), staminode present (*Barthlottia*) or absent, stigma capitate (*Barthlottia*), capitate to minutely bifid (*Manuleopsis*, *Jamesbrittenia*) or lingulate (*Lyperia* to *Tetraselago*), ovary 1-locular (*Globulariopsis*) or 2-locular, each locule with one ovule (*Globulariopsis* to *Hebenstretia*) or multiovular, capsule usually with septicidal dehiscence (*Barthlottia* to *Tetraselago*) or fruit separating into 2 one-seeded, hard-walled or soft (*Cromidon*, *Pseudoselago*) cocci or with 1 coccus by abortion (*Globulariopsis*, *Microdon*, *Gosela*), seeds numerous to few (*Glumicalyx* to *Hebenstretia*), bothrospermous (*Jamesbrittenia* to *Manulea*), or not alveolated.

28. ***Barthlottia*** Eb. Fischer

Barthlottia Eb. Fischer, Bull. Mus. Natl. Hist. Paris IV, 18, sect. B, Adansonia: 352 (1996).

Shrub, glabrous except for glandular-pubescent inflorescence. Stems erect, quadrangular. Leaves cauline, opposite, petiolate, elliptic-lanceolate, acuminate, with entire to dentate margin. Flowers pedicellate. Calyx deeply 5-lobed, lobes subequal, obovoid. Corolla red to purple, limb weakly 2-lipped, lobes rounded, ovoid, tube campanulate, dilated above. Stamens 4, inserted at base of corolla-tube, included. Ovary obconical. One species, *B. madagascariensis* Eb. Fischer in South Madagascar.

29. ***Manuleopsis*** Thell. ex Schinz

Manuleopsis Thell. ex Schinz, Vierteljahrschr. Naturf. Ges. Zürich 60: 405 (1915); Hilliard, The Manuleae: 78–79 (1994), rev.
Freyliniopsis Engl. (1922).

Shrub, glandular-pubescent. Stems erect, brittle, ribbed. Leaves cauline, opposite, petiolate, lanceolate, acuminate, with entire to serrate margin, base of petiole persistent. Flowers distinctly pedicellate. Calyx deeply 5-lobed, campanulate, obscurely 2-lipped. Corolla white, limb 2-lipped, lobes spreading, suborbicular, tube cylindrical, throat laterally compressed, corolla with clavate hairs inside. Stamens 4, adaxial stamens inserted near base of tube, abaxial stamens inserted near throat, shortly exserted. Ovary deltate. One species, *M. dinteri* Thell., Namibia.

30. ***Jamesbrittenia*** Kuntze

Jamesbrittenia Kuntze, Rev. Gen. 46 (1891); Hilliard, The Manuleae: 84–212 (1994), rev.

Annual, perennial or suffrutescent herbs or shrubs, glandular-pubescent. Stems ascending to erect, rounded. Leaves cauline, opposite, often alternate above, sometimes crowded, subsessile to petiolate, linear-lanceolate to ovoid, acuminate, entire or pinnatifid to pinnatisect, with entire margin. Flowers distinctly pedicellate. Calyx deeply 5-lobed, lobes elliptic, lanceolate, oblong or spathulate, campanulate. Corolla creamy-white to orange or blue to violet, throat usually yellow, limb nearly regular to distinctly 2-lipped, lobes spreading, orbicular to oblong or ovoid, often entire, sometimes retuse to bilobed, tube cylindrical, abruptly expanded near apex. Stamens 4, included or abaxial pair rarely shortly exserted. Ovary deltate, usually glandular. Seeds in 2 species bothrospermous. $2n = 24$. Eighty-three species, from South Africa to Angola and Zambia, one species (*J. dissecta* Kuntze) ranging from Egypt and Sudan to India.

31. ***Lyperia*** Benth.

Lyperia Benth. in Hook., Companion Bot. Mag. 1: 377 (1836); Hilliard, The Manuleae: 212–220 (1994), rev.

Annual herbs, one species perennial, densely glandular-pubescent. Stems ascending to erect, narrowly winged. Leaves cauline, opposite, becoming alternate above, sessile to subsessile, linear-lanceolate to ovoid, acuminate, entire, with entire to dentate margin. Flowers shortly pedicellate. Calyx deeply 5-lobed almost to base, lobes lanceolate. Corolla creamy-white to pink, mauve or blue, limb almost regular, 5-lobed, lobes rotate or reflexed, tube narrowly cylindrical, abruptly dilated near throat, throat bearded inside with clavate hairs. Stamens 4 or 2, included or just visible in throat. Ovary cuneate. $2n = 16$. Six species, South Africa.

32. ***Sutera*** Roth

Sutera Roth, Bot. Bemerk.: 172 (1807); Hilliard, The Manuleae: 220–291 (1994), rev.
Chaenostoma Benth. (1835).
Sphenandra Benth. (1836).

Perennial or suffrutescent herbs or shrublets, rarely annual, glandular-pubescent. Stems prostrate, ascending or erect. Leaves cauline, opposite, sometimes alternate above, bases either ± connate or decurrent as narrow wings or ridges, subsessile to petiolate, linear-lanceolate to ovoid or orbicular, acuminate, entire or rarely pinnatifid, with entire to dentate margin. Flowers distinctly pedicellate. Calyx deeply 5-lobed, 2-lipped, rarely divided into 6–9 lobes, lobes linear to lanceolate. Corolla white to blue to violet, throat usually yellow, limb nearly regular, 5-lobed, lobes spreading, suborbicular to oblong or ovoid, entire, tube usually funnel-shaped. Stamens 4, exserted or adaxial pair included or exserted. Ovary elliptic. $2n = 12, 14$. Forty-nine species, South Africa.

33. ***Manulea*** L.

Manulea L., Syst. Nat. ed. 12, 2: 385, 419 (1767); Mant. 12, 88 (1767), nom. cons.; Hilliard, The Manuleae: 291–387 (1994), rev.
Nemia Berg (1767), nom. rejic.
Lychnidea Moench (1802).

Annual, perennial or suffrutescent herbs or shrubs, usually glandular and eglandular-pubescent. Stems ascending to erect. Leaves cauline or rosulate, opposite, bases connate, sometimes alternate, subsessile to petiolate, linear-lanceolate to ovoid, acuminate, with entire to dentate margin. Flowers subsessile to pedicellate. Calyx 5-lobed, 2-lipped, upper lip 2-lobed, lower lip 3-lobed, lobes short to long, linear, lanceolate, oblong or subspathulate. Corolla white to yellow or blue to violet, throat usually yellow, limb almost regular to distinctly 2-lipped, lobes rotate or reflexed, suborbicular to oblong or lanceolate, obtuse entire, retuse to bilobed, tube usually narrowly funnel-shaped. Stamens 4 or 2, included or abaxial pair rarely shortly exserted. Ovary elliptic or cuneate. $2n = 16, 18$. Seventy-four species, South Africa, northwards to Angola, Zimbabwe and Mozambique.

34. ***Melanospermum*** Hilliard

Melanospermum Hilliard, Notes Roy. Bot. Gard. Edinburgh 45: 482 (1989); Hilliard, The Manuleae: 387–395 (1994), rev.

Annual or perennial herbs, glandular or eglandular-pubescent with patent or retrorse hairs. Stems ascending to erect. Leaves cauline, opposite or occasionally alternate below inflorescence, subsessile, lanceolate to ovoid or elliptic, obtuse, with entire to dentate margin. Flowers subsessile. Calyx

5-lobed to middle, 2-lipped. Corolla white or mauve to blue or violet with orange patch on base of upper lip or upper lip entirely orange, persistent, limb 2-lipped, lobes entire, tube usually slender, expanded near apex. Stamens 4, all exserted or 2 in throat, 2 exserted. Ovary elliptic. Six species, South Africa to Namibia and Zimbabwe.

35. *Polycarena* Benth.

Polycarena Benth. in Hook., Companion Bot. Mag. 1: 371 (1836); Hilliard, The Manuleae: 395–416 (1994), rev.

Annual herbs, glandular-pubescent with patent hairs. Stems usually ascending to erect. Leaves cauline, opposite, alternate above, subsessile, lanceolate to ovoid or spathulate to elliptic, obtuse, with entire to dentate margin. Flowers subsessile. Calyx 5-lobed, 2-lipped, membranous, persistent. Corolla creamy white to yellow or mauve, with orange patch on base of upper lip and throat or upper lip entirely orange, persistent, limb 2-lipped, in one sp. regular, lobes entire, tube usually slender, expanded near apex. Stamens 4, sometimes reduced to 3, 2 or 1, all exserted or adaxial pair in throat, abaxial pair exserted. Ovary elliptic. $2n = 14$. Seventeen species, South Africa, Western and South-western Cape.

36. *Glekia* Hilliard

Glekia Hilliard, Notes Roy. Bot. Gard. Edinburgh 45: 482 (1989); Hilliard, The Manuleae: 416–418 (1994), rev.

Shrublet, pubescence confined to narrow ridges spanning the internode, hairs eglandular, ± retrorse. Stems ascending to erect. Leaves cauline, opposite, bases connate, sessile, lanceolate to ovoid, acuminate with finely callose-dentate margin, thick-textured. Flowers subsessile. Calyx 5-lobed, 2-lipped, membranous, persistent. Corolla white inside with an orange patch at base of upper lip, each lobe outside with blue-purple patch, throat usually yellow, limb distinctly 2-lipped, lobes subrotund, obtuse, tube usually cylindrical below, abruptly campanulate above. Stamens 4, included, abaxial pair shortly exserted. Ovary elliptic. One species, *G. krebsiana* (Benth.) Hilliard, mountains of South Africa (Cape) and Lesotho.

37. *Trieenea* Hilliard

Trieenea Hilliard, Notes Roy. Bot. Gard. Edinburgh 45: 489 (1989); Hilliard, The Manuleae: 418–429 (1994), rev.

Annual or perennial herbs or shrubs, glabrous, glandular- or eglandular-pubescent. Stems ascending to erect. Leaves cauline, opposite, bases often connate, sometimes alternate, subsessile to petiolate, usually broadly ovoid, acuminate, with deeply and coarsely dentate margin. Flowers pedicellate. Calyx 5-lobed almost to base, obscurely 2-lipped, membranous, persistent. Corolla white to pale mauve blue, with orange or yellow patch at base of upper lip, limb 2-lipped, lobes subrotund, obtuse, entire, tube funnel-shaped. Stamens 4, either all exserted or adaxial pair included. Ovary elliptic. $2n = 12$. Nine species, South Africa (Cape).

38. *Phyllopodium* Benth.

Phyllopodium Benth. in Hook., Comp. Bot. Mag. 1: 372 (1835); Hilliard, The Manuleae: 429–460 (1994), rev.

Annual or perennial herbs, eglandular-pubescent. Stems ascending to erect. Leaves cauline, opposite below, alternate above, subsessile, lanceolate, ovoid or spathulate, acuminate, with entire to dentate margin. Flowers pedicellate. Calyx 5-lobed to middle, obscurely 2-lipped, membranous, persistent. Corolla white, creamy to pale mauve or blue, with or without orange or yellow patch at base of upper lip, limb 2-lipped, lobes subrotund, obtuse, entire, tube cylindrical to narrowly funnel-shaped. Stamens 4, 2 in one species. Ovary elliptic. $2n = 12$. Twenty-six species, South Africa, Western and South-western Cape.

39. *Zaluzianskya* F.W. Schmidt

Zaluzianskya F.W. Schmidt, Neue Selt. Pflanzen: 11 (1793), nom. cons.; Hilliard, The Manuleae: 460–522 (1994), rev.
Nycterinia D. Don in Sweet (1834).

Annual or perennial herbs, usually eglandular-pubescent, hairs retrorse or patent, glandular hairs sometimes present. Stems ascending to erect. Leaves cauline or rosulate, opposite below, often subopposite or alternate above, subsessile, usually lanceolate, ovoid or spathulate, acuminate, with entire to variously dentate margin. Flowers sessile to subsessile. Calyx 5-lobed to one third of its length, 2-lipped. Corolla limb snow-white above, pale to dark red below, creamy yellow to bright yellow above, purple brown below or white with various shades of pink to mauve, limb regular to 2-lipped, lobes rotate or reflexed, either suborbicular, spathulate, elliptic or oblong, or Y-shaped, tube narrowly cylindrical. Stamens 4 or 2. Ovary elliptic. $2n = 12$. Fifty-five species, South Africa, northwards to Zimbabwe, one isolated species in high mountains of East Africa (*Z. elgonensis* Hedberg).

40. ***Reyemia*** Hilliard

Reyemia Hilliard, Edinburgh J. Bot. 49(3): 297 (1992); Hilliard, The Manuleae: 522–525 (1994), rev.

Annual herbs, glandular-pubescent above, eglandular-pubescent below. Stems erect. Leaves mainly rosulate, opposite below, alternate above, subsessile, elliptic to ovoid, acuminate, with entire to dentate margin. Flowers shortly pedicellate. Calyx shortly 5-lobed, 2-lipped. Corolla cream to yellow, palate orange, upper lip bearded with clavate hairs, resupinate by torsion of upper part of tube, limb 2-lipped, upper lip 4-lobed, lower lip 1-lobed, lobes lanceolate, tube narrowly cylindrical below. Stamens 2 by reduction of the abaxial pair, 2 small abaxial staminodes present. Ovary oblong-elliptic. Two species, *R. chasmanthiflora* Hilliard, *R. nemesioides* (Diels) Hilliard, South Africa (Cape).

41. ***Glumicalyx*** Hiern

Glumicalyx Hiern, Hooker's Icon. Pl.: t. 2769 (1903); Hilliard, The Manuleae: 526–532 (1994), rev.

Perennial herbs or dwarf shrublets, pubescent. Stems erect. Leaves cauline, opposite below, alternate above, subsessile, lanceolate or spathulate to oblong-elliptic, subacute to obtuse, with dentate margin. Inflorescence forming rounded heads, nodding in flower. Flowers subsessile. Calyx unequally 5-lobed, rarely the 2 abaxial lobes obsolete. Corolla pale cream to orange to orange-red inside, creamy or brownish outside, thick-textured, leathery when dry, limb weakly 2-lipped, lobes spreading, oblong to suborbicular, tube funnel-shaped. Stamens 4, either 2 or 4 exserted. Ovary oblong. Seeds numerous to few (14–24). $2n = 14$. Six species, eastern mountain region of South Africa.

42. ***Strobilopsis*** Hilliard & B.L. Burtt

Strobilopsis Hilliard & B.L. Burtt, Notes Roy. Bot. Gard. Edinburgh 35(2): 172 (1977); Hilliard, The Manuleae: 532–534 (1994), rev.

Annual or perennial herb, pubescent. Stems erect. Leaves cauline, opposite below, alternate above, subsessile, narrowly elliptic to linear-oblong, subacute, with entire to dentate margin. Inflorescence capitate, not nodding in flower. Flowers subsessile. Calyx unequally 5-lobed. Corolla creamy to pale mauve, upper lip with orange patch at base, limb 2-lipped, lobes spreading, oblong to elliptic, tube funnel-shaped. Stamens 4, or 3 or 2 by reduction of the abaxial pair, exserted. Ovary oblong-elliptic, each locule with up to 6 ovules. One species, *S. wrightii* Hilliard & B.L. Burtt, Lesotho and Natal Drakensberg.

43. ***Tetraselago*** Junell

Tetraselago Junell, Svensk Bot. Tidskr. 55: 190 (1961); Hilliard, The Manuleae: 534–540 (1994), rev.

Tufted perennial herbs, pubescent. Stems erect, woody. Leaves cauline, alternate, fascicled, sessile, linear-lanceolate to narrowly elliptic or ovoid, acuminate, with entire to dentate margin. Inflorescence capitate, not nodding in flower. Flowers subsessile to shortly pedicellate. Calyx 5-toothed, weakly 2-lipped. Corolla pale blue or mauve, limb weakly 2-lipped, lobes spreading, tube funnel-shaped. Stamens 4, exserted. Ovary oblong, each locule with 2 ovules. $2n = 14$. Four species, South Africa (Transvaal, Swaziland, Natal).

44. ***Globulariopsis*** Compton

Globulariopsis Compton, Trans. Roy. Soc. S. Afr. 19: 308 (1931); Hilliard, The tribe Selagineae: 16–23 (1999), rev.

Dwarf ericaceous shrubs, much branched, pubescent. Stems ascending to erect. Leaves cauline, opposite or alternate, sessile, linear, acuminate to obtuse, with entire margin. Flowers sessile. Calyx 5-lobed to middle, lobes subequal. Corolla white, tubular, limb 2-lipped, tube cylindrical below, campanulate above. Stamens 4, included. Ovary ellipsoid. Stigma clavate. Seeds 1 by reduction of one locule. Seven species, South Africa, western and south-western Cape.

45. ***Microdon*** Choisy

Microdon Choisy, Mém. Soc. Phys. Genève 2(2): 97 (1824); Hilliard, The tribe Selagineae: 11–13 (1999), rev.
Agathelpis Choisy (1824).

Dwarf ericaceous shrubs, much branched, pubescent. Stems ascending to erect, rounded. Leaves cauline, solitary or sometimes fascicled, alternate, sessile, linear-lanceolate, oblong, elliptic or suborbicular, acuminate to obtuse, with entire margin. Flowers sessile. Calyx 5-lobed, 5-ribbed, lobes subequal. Corolla white, orange patch on upper lip present or absent, limb sub2-lipped, 5-lobed, tube cylindrical. Stamens 4, exserted. Ovary cylindrical,

2-locular, with 1 abortive locule. Stigma linear. Seeds 1 by reduction of one locule. Seven species, Western to South-western Cape.

46. *Gosela* Choisy

Gosela Choisy in DC., Prodr. 12: 22 (1848); Hilliard, The tribe Selagineae: 13 (1999), rev.

Dwarf ericaceous shrub, much branched, pubescent to villose. Stems ascending to erect, rounded. Leaves cauline, solitary or fascicled, alternate, sessile, linear, acuminate to obtuse, with entire margin. Flowers sessile. Calyx 5-lobed to middle, the 2 upper lobes ± fused. Corolla white, tubular, limb 2-lipped, tube cylindrical. Stamens 4 or 2, included, adaxial stamens sometimes with sterile anthers. Ovary ellipsoid. Stigma linear. Seeds 1 by reduction of one locule, fusiform. One species, *G. eckloniana* Choisy, South Africa, south-western Cape.

47. *Cromidon* Compton

Cromidon Compton, Trans. Roy. Soc. S. Afr. 19: 308 (1931); Hilliard, Edinburgh J. Bot. 47: 319–337 (1990), rev.

Small annual or perennial herbs, glabrous or pubescent. Stems often prostrate to decumbent. Leaves cauline, solitary or fascicled, opposite, base often connate, sometimes alternate above, sessile, lanceolate or spathulate, obtuse, with entire margin. Flowers shortly pedicellate. Calyx obliquely 3- or 5-lobed, persistent. Corolla white, orange patch on upper lip often present, limb 2-lipped, 4- or 5-lobed, tube funnel-shaped. Stamens 4, exserted. Ovary elliptic-oblong. Stigma lingulate. Seeds 2, compressed, convex on one face, plane on the other, smooth. Twelve species, South Africa.

48. *Selago* L.

Selago L., Sp. Pl. 2: 629 (1753); Hilliard, The tribe Selagineae: 24–305 (1999), rev.
Walafrida E. Mey. (1838).

Dwarf ericaceous shrubs, much branched, sometimes perennial or annual herbs, glabrous or pubescent. Stems ascending to erect, rounded. Leaves cauline, solitary or fascicled, alternate, sessile, linear-lanceolate, oblong, elliptic or spathulate, acuminate to obtuse, with entire to dentate margin. Flowers sessile to shortly pedicellate. Calyx deeply 3- to 5-lobed, lobes subequal. Corolla white, tubular, limb 2-lipped, tube short and broad or long and narrow. Stamens 4, exserted. Ovary subglobose. Stigma clavate or tridentate. Seeds 2, fusiform. Circa 190 species, mainly South Africa, few species in tropical Africa, one species (*S. muralis* Benth. & Hook. f.) in Madagascar.

49. *Pseudoselago* Hilliard

Pseudoselago Hilliard, Edinburgh J. Bot. 52(3): 245 (1995).

Dwarf ericaceous shrubs, much branched, sometimes perennial or annual herbs, pubescent to villose. Stems decumbent or prostrate, ascending to erect, usually winged or ridged. Leaves cauline, solitary or fascicled, opposite, usually becoming alternate above, sessile, linear, acuminate to obtuse, with entire margin. Flowers sessile. Calyx deeply 5-lobed, lobes subequal, adaxial lobe shortest. Corolla white, with orange patch on upper lip, limb 2-lipped, tube either funnel-shaped or cylindrical, upper lip 2-lobed or 4-lobed, lower lip 3-lobed or 1-lobed. Stamens 4, abaxial stamens exserted. Ovary elliptic-oblong. Stigma lingulate. Seeds 2, compressed, convex on one face, ± plane on the other. Twenty-eight species, South Africa (south-western Cape).

50. *Chenopodiopsis* Hilliard

Chenopodiopsis Hilliard, Edinburgh J. Bot. 47: 339 (1990).

Small annual herbs, pubescent. Stems decumbent to ascending or erect. Leaves cauline, solitary, opposite, sometimes becoming alternate above, sessile, linear, to obtuse, with entire margin. Flowers sessile. Calyx 5-lobed, the adaxial lobe sometimes only very shallowly divided, persistent. Corolla white, limb 2-lipped, 4- or 5-lobed, tube cylindrical or funnel-shaped, upper lip 4-lobed, lower lip 1-lobed, minute or sometimes missing. Stamens 4 or 2, exserted. Ovary elliptic-oblong. Stigma lingulate. Seeds 2, fusiform, smooth or transversally rugose. Three species, South Africa.

51. *Dischisma* Choisy

Dischisma Choisy, Mém. Soc. Phys. Genève 2(2): 84, 93 (1823); Roessler, Mitt. Bot. Staatssamml. München 15: 1–160 (1979), rev.

Annual or perennial herbs or shrubs, pubescent. Stems ascending to erect, rounded. Leaves cauline, solitary, opposite, sometimes alternate above, sessile, linear, oblong, elliptic or spathulate, acuminate to obtuse, with entire to dentate margin. Flowers sessile to shortly pedicellate. Calyx 2-lobed, lobes lateral, linear or lanceolate-oblong. Corolla white, tube cylindrical, divided almost to middle, limb 1-lipped, lip large, flattened or concave, 4-lobed, lobes subequal to unequal. Stamens 4, exserted. Ovary elliptic. Stigma clavate. Seeds 2, fusiform. Eleven species, South Africa to Namibia.

52. Hebenstretia L. Fig. 44 F, G

Hebenstretia L., Sp. Pl. 2: 629 (1753); Roessler, Mitt. Bot. Staatssamml. München 15: 1–160 (1979), rev.; Roessler, Mitt. Bot. Staatssamml. 18: 183–185 (1982), rev.

Dwarf ericaceous shrubs, much branched, sometimes perennial or annual herbs, glabrous or pubescent. Stems ascending to erect, rounded. Leaves cauline, solitary, alternate or opposite below, sessile, linear, obtuse, with entire margin. Flowers sessile. Calyx with 1 lobe, spathaceous. Corolla white, orange patch on limb present or absent, tube slender, divided almost to base, limb 1-lipped, lip large, flattened or concave, 4-lobed, lobes subequal to unequal. Stamens 4, exserted. Ovary oblong or elliptic. Stigma linear. Seeds 2, fusiform. $2n = 14$. Twenty-five species, tropical and South Africa, one species, *H. angolensis* Rolfe, northwards to Eritrea.

Genus of uncertain placement

53. *Camptoloma* Benth.

Camptoloma Benth. in DC., Prodr. 10: 430 (1846); Hilliard, The Manuleae: 80–84 (1994), rev.

Perennial suffrutescent herbs or dwarf shrubs, densely glandular-pubescent. Stems ascending to erect, quadrangular, winged. Leaves cauline, alternate, petiolate, ovoid to elliptic or orbicular, obtuse, with dentate to serrate margin. Flowers solitary or with axillary 2–11-flowered cymes, pedicellate. Calyx deeply 5-lobed, very obscurely 2-lipped. Corolla white to yellowish or pink to lilac with purple veins, limb 2-lipped, tube cylindrical, scarcely dilated apically, throat bearded or not with clavate hairs. Stamens 4, abaxial pair inserted in throat, very shortly exserted, adaxial pair inserted at middle of tube, included. Ovary cuneate, stigma bifid. Seeds aulacospermous. $2n = 28, 56$. Three species, in Somalia and Socotra (*C. lyperiiflorum* (Vatke) Hilliard), Angola and Namibia (*C. rotundifolium* Benth.), and the Canary Islands (Gran Canaria) (*C. canariensie* (Webb. & Berth.) Hilliard).

Camptoloma was considered to be a member of Manuleae by Hilliard (1994). Kornhall et al. (2001), however, excluded it from this tribe.

"Schlegeliaceae"

Schlegelieae (A.H. Gentry) Thorne (1992).
Schlegelieae A.H. Gentry (1980).
Schlegeliaceae (A.H. Gentry) Reveal (1996).

Trees or epiphytic to terrestrial shrubs. Inflorescence thyrsic or racemose (*Exarata*), with clustered axillary cymes, frondose or bracteate, prophylls present. Corolla subactinomorphic, limb 2-lipped, tubular. Anthers with 2 separate equal divergent thecae, staminode present or absent. Fruit a berry. Seeds numerous, compressed, angular, subtriquetrous, with membranaceous wings (*Schlegelia*, *Exarata*) or unwinged. Four genera, with 25 species, Central and South America.

54. *Gibsoniothamnus* L.O. Williams

Gibsoniothamnus L.O. Williams, Fieldiana Bot. 32: 213 (1970).

Small epiphytic or rarely terrestrial shrubs, strigose to sparsely villose. Stems erect, rounded. Leaves cauline, opposite, petiolate, ovoid or elliptic to orbicular, acuminate, with subentire to dentate margin, subcoriaceous, the leaves of a pair distinctly anisophyllous. Flowers distinctly pedicellate. Calyx 5-lobed, campanulate, laterally compressed, subrotate or cup-shaped. Corolla white to pink or purplish, subactinomorphic, limb 2-lipped, tubular, 5-lobed, lobes subequal, tube straight, narrow. Stamens 4, included. Ovary ovoid. Stigma capitate. Seeds numerous, compressed, angular. Eight species, from Mexico to Panama.

55. *Schlegelia* Miq.

Schlegelia Miq., Bot. Zeit. 2: 785 (1844); A.H. Gentry, Fl. Ecuador 7, Bignoniaceae: 129–135 (1977), reg. rev.
Dermatocalyx Oerst. (1856).

Small epiphytic or terrestrial shrubs, trees or vines, glabrous or young branches somewhat pubescent. Stems erect, rounded. Leaves cauline, opposite, petiolate, ovoid or elliptic, acuminate, with subentire margin, coriaceous. Flowers distinctly pedicellate. Calyx 3–4-lobed, campanulate, laterally compressed, cup-shaped, coriaceous. Corolla white to pink, purple or red, subactinomorphic, limb 2-lipped, tubular, 5-lobed, lobes subequal, tube straight, narrow. Stamens 4, included. Ovary ovoid. Stigma 2- or 3-lobed. Seeds numerous, compressed, subtriquetrous, with membranaceous wings. Fifteen species, from Guatemala to Caribbean, Ecuador and Brazil.

56. *Synapsis* Griseb.

Synapsis Griseb., Cat. Pl. Cub.: 187 (1866).

Tree, glabrous. Stems erect, rounded. Leaves cauline, opposite, petiolate, ovoid or elliptic, acuminate, margin with 4–6 spines, sinuate between spines, coriaceous. Flowers distinctly pedicellate. Calyx 5-lobed, campanulate, cup-shaped, the lobes reduced to short teeth, coriaceous. Corolla violet to mauve, with limb 2-lipped, tubular, 5-lobed, lobes subequal, tube straight, long, narrow, expanded towards limb. Stamens 4, included. Ovary ovoid. Stigma 2-lobed. Seeds numerous, compressed. One species, *S. ilicifolia* Griseb. from eastern Cuba.

57. *Exarata* A.H. Gentry

Exarata A.H. Gentry, Syst. Bot. 17: 503 (1992).

Large canopy trees, branchlets terete, puberulous. Leaves cauline, opposite, petiolate, obovate to elliptic, cuneate at base. Flowers distinctly pedicellate. Calyx campanulate, subtruncate to irregularly 2–3-lobed. Corolla white with purple-streaked lobes, narrowly tubular-infundibuliform, tube straight, expanded towards limb. Stamens 4, included. Ovary ellipsoid to subglobose, disc absent. Stigma 2-lobed. Fruit globose, indehiscent, fleshy. Seeds thin, with distinct vestigial wings. One species, *E. chocoensis* A.H. Gentry, Colombia, Ecuador.

"Calceolariaceae"

Calceolarieae D. Don (1835).
Calceolariaceae Raf. ex Olmstead (2001).

Shrubs or herbs. Inflorescence thyrsic, with accessory flowers, paired, each pair composed of a terminal flower and the flower from its reduced prophyll. Corolla with short tube, usually with shoe-like invagination of lower lip, with trichome nectaries. Stamens 2 or 3 (*Stemotria*), thecae separate at apex. Ovary bilocular. Capsule septicidal. Intermediate layers of ovule integument 1–2, endosperm of mature seed longitudinally furrowed, furrows produced by endothelial cells (aulacospermous), epidermal cells of testa all homogenous, their lateral walls undulate. Three genera with 307 species, Central and South America, New Zealand.

58. *Calceolaria* L. Fig. 49A, p. 401

Calceolaria L., Vet. Acad. Handl. Stockholm 31: 286 (1770), nom. cons.; Molau, Fl. Neotrop. 47: 1–326 (1988), reg. rev.; Erhart, Bibl. Bot. 153: 1–283 (2000), reg. rev.

Annual or perennial herbs and shrubs, glandular-pubescent to densely villose. Stems ascending to erect, rounded. Leaves cauline or in a basal rosette, opposite, sessile or subsessile to petiolate, linear-lanceolate to ovoid-orbicular, acuminate, entire or pinnatifid to pinnatisect, with entire or dentate to crenate margin. Flowers usually distinctly pedicellate. Corolla yellow, rarely with red spots, limb 2-lipped, upper lip ± hooded, usually much smaller than lower lip, lower lip saccate, ± inflated. Stamens 2, included. Ovary ovoid to globose. Stigma weakly bilobed. $2n = 32, 36, 60, 64$. About 300 species in tropical South America, mainly the Andes.

59. *Jovellana* Ruiz & Pav.

Jovellana Ruiz & Pav., Fl. Peruv. Chil. 1: 12 (1798).

Perennial herbs and shrubs, glabrous to glandular-pubescent or densely villose. Stems prostrate or ascending to erect, rounded. Leaves cauline, opposite, subsessile to petiolate, lanceolate or ovoid, acuminate, entire or pinnatifid to pinnatisect, with dentate to crenate margin. Flowers usually distinctly pedicellate. Corolla yellow, with red or brownish-red spots, limb 2-lipped, upper lip ± hooded, entire or weakly emarginate, usually much smaller than lower lip, lower lip spreading, patent, entire with weakly involute margin, not saccate. Stamens 2, included. Ovary ovoid to globose. Stigma capitate to weakly bilobed. Six species, Chile, New Zealand.

60. *Stemotria* Wettst. & Harms ex Engl.

Stemotria Wettst. & Harms ex Engl., Pflanzenfam. Register: 462 (1899).
Porodittia G. Don ex Kraenzl. (1907).

Shrub, sparsely pubescent to densely villose. Stems erect, rounded. Leaves cauline, opposite, petiolate, lanceolate-ovoid, acuminate or obtuse, pinnatifid, with dentate margin. Flowers usually distinctly pedicellate. Corolla yellow, limb 2-lipped, upper lip 2-lobed, with involute margin, usually much smaller than lower lip, lower lip spreading, patent, orbicular and entire with weakly involute margin, not saccate. Stamens 3 (median stamen and 2 lateral stamens), included. Ovary ovoid to globose. Stigma weakly bilobed. One species, *S. triandra* (Cav.) Wettst. & Harms, in Peru.

"Paulowniaceae"

Paulownieae Pennell (1920).
Paulowniaceae Nakai (1949).

Trees, shrubs or lianas. Inflorescence usually thyrsic. Corolla 2-lipped, obliquely spreading, tube elongate, inflated. Stamens 4, anther thecae separate or confluent. Ovary bilocular. Capsule loculicidal (*Paulownia*) or loculicidal and septicidal. Endosperm smooth, multilayered to 1-layered, epidermal cells of testa prolonged transversally to a wing, endotheliar cells thickened laterally and towards endosperm. Four genera with 22 species, East to Southeast Asia.

1. Tribe Paulownieae Pennell (1920).

Trees, glabrous except for pubescent inflorescence, young stems and leaves, inflorescence thyrsic, bracteate (*Paulownia*) or frondose (*Shiuyinghua*), prophylls present.

61. *Paulownia* Siebold & Zucc. Fig. 43D, 49B

Paulownia Siebold & Zucc., Fl. Jap. 1: 25 (1835); S.Y. Hu., Quart. J. Taiwan Mus. 12: 1–54 (1959), rev.

Trees. Stems erect, with brownish-grey bark and prominent white lenticells on smooth bark of young branches. Leaves cauline, opposite, deciduous, usually appearing after the flowers, petiolate, broadly ovoid to orbicular or angular-cordate, acuminate. Flowers distinctly pedicellate. Calyx deeply 5-lobed. Corolla blue to violet or light purple, large, limb 2-lipped, obliquely spreading, tube elongate, inflated. Stamens 4, included. Ovary ovoid to globose. Stigma capitate. Seeds numerous, testa with distinct wings. $2n = 36, 40$. Six species, East Asia.

62. *Shiuyinghua* Paclt

Shiuyinghua Paclt, J. Arnold Arbor. 43: 217 (1962).

Tree. Stems erect, woody, with brownish-grey bark. Leaves cauline, opposite, deciduous, petiolate, strongly anisophyllous, broadly ovoid (cordate) in axial position and elliptically lanceolate in abaxial position, acuminate. Flowers distinctly pedicellate. Calyx deeply 5-lobed. Corolla blue to violet or light purple, small, limb 2-lipped, obliquely spreading, tube elongate, inflated. Stamens 4, included. Ovary ovoid to globose. Stigma bilobed. Fruit unknown. One species, *S. silvestrii* (Pampanini & Bonati) Paclt, Central China. The genus, known only from one specimen, is closely related to *Paulownia* and may be part of it.

2. Tribe "Wightieae"

Trees, shrubs or vines, inflorescence thyrsic with axillary cymes, bracteate, prophylls present.

63. *Brandisia* Hooker f. & Thomson

Brandisia Hooker f. & Thomson, J. Linn. Soc., Bot. 8: 11 (1865).

Shrubs or vines, densely tomentose with stellate hairs. Stems erect or scandent, rounded. Leaves cauline, opposite or subopposite, petiolate, lanceolate to ovoid, acuminate, with entire or serrate margin. Flowers distinctly pedicellate. Calyx campanulate, 5-lobed to one third, rarely 4-dentate or bilobed, lobes subequal. Corolla brownish-yellow, infundibular, dilated above, densely stellate-tomentose outside, limb 2-lipped, lobes much reflexed. Stamens 4, exserted, thecae densely long-ciliate. Ovary globose. Stigma clavate. Seeds numerous, linear, with hyaline winged testa. Thirteen species, from Burma to China.

64. *Wightia* Wall.

Wightia Wall., Pl. Asiat. Rar. 1: 71 (1830).

Hemiepiphytic, evergreen trees or pseudolianas, usually tree stranglers, tomentose with stellate hairs. Stems erect or scandent, rounded. Leaves cauline, opposite, petiolate, elliptic or ovoid-elliptic, acuminate, with entire margin. Flowers distinctly pedicellate. Calyx campanulate, 3–5-lobed to one third or cup-shaped with indistinct lobes. Corolla pink to orange, digitaliform, tube incurved near base, dilated above, densely stellate-tomentose outside, lobes much reflexed. Stamens 4, exserted, thecae glabrous. Ovary globose. Stigma clavate. Seeds numerous, linear, with hyaline winged testa. Two species, *W. speciosissima* (D.

Don) Merr., *W. borneensis* Hook. f., from Himalayas, India and Sri Lanka to Southwest China and Malesia.

"Veronicaceae"

I. Subfam. Antirrhinoideae (Speta) D. Hartl & I. Betsche (1984).

Antirrhinaceae Pers. (1807), nom. cons. prop.
Antirrhineae Chavannes (1833).

Shrubs or herbs. Inflorescence mostly thyrsic, sometimes with accessory flowers, or racemose, prophylls present or rarely absent. Corolla usually personate, with palate on lower lip, tube often with adaxial gibbosity or spur, petal aestivation descending. Stamens 4–2, anther thecae separate, but confluent in dehiscence, the adjacent pairs of anthers marginally coherent or free, median stamen reduced to a staminode, base of filaments or staminodes with trichomatic nectaries or nectary a disc. Ovary bilocular. Capsule septicidal, loculicidal or poricidal. Intermediate layers of ovule integument 5–20, endosperm of mature seed smooth or with undulate surface produced by endothelium, endotheliar cells isodiametric, cube-shaped, with completely thickened walls, testa cells homogenous or heterogeneous, their different size causing various ornamentations, wings produced by epidermal cells or remains of intermediate-layer cells. Seeds flattened with distinct wings or scrobiculate. Thirty-eight genera with 622 species, cosmopolitan.

1. Tribe Cheloneae Benth. in DC. (1846).

Chelonaceae Martynov (1820).
Collinsieae Meisn. (1840).

Inflorescence mostly thyrsic, sometimes with accessory flowers, rarely racemose (*Chelone*, *Chionophila*), prophylls present or rarely absent (*Chionophila*), flowers personate or at least with abaxial protrusion (palate) or pouch (*Collinsia*), fifth median stamen reduced to a staminode or lacking, base of filaments or staminodes often with trichomatic nectaries, stigma capitate to bilobed (*Uroskinnera*), testa cells homogenous, capsule septicidal to loculicidal (*Uroskinnera*). Seeds flattened with distinct wings, wingless (*Tonella*) or scrobiculate (*Uroskinnera*), not alveolated or bothrospermous (*Tetranema*).

65. *Chelone* L.

Chelone L., Sp. Pl. 2: 611 (1753).

Perennial herbs, glabrous or glandular-pubescent. Stems erect, quadrangular. Leaves cauline, sessile or petiolate, opposite, lanceolate-ovoid, acuminate with dentate or serrate margin. Flowers subsessile. Calyx deeply 5-lobed. Corolla red or rose-purple, throat inflated, somewhat triangular in cross section, bottom of the throat lanose inside, upper lip galeate. Stamens 4, included, filaments lanose to pubescent, staminode usually glabrous, anthers with two confluent thecae, lanose. $2n = 28$. Four species from eastern North America.

66. *Nothochelone* (A. Gray) Straw

Nothochelone (A. Gray) Straw, Brittonia 18: 85 (1966).
Chelone sect. *Nothochelone* A. Gray (1878).

Perennial herb, glabrous or glandular-pubescent. Stems erect, quadrangular. Leaves cauline, shortly petiolate, opposite, lanceolate-ovoid, base cordate or rounded, acuminate with dentate or serrate margin. Flowers pedicellate. Calyx deeply 5-lobed. Corolla pink to maroon, throat inflated, glabrous inside, lobes spreading. Stamens 4, included, filaments lanose to pubescent, anthers with two confluent, spreading thecae, lanose. Ovary ovoid with distinct nectary disc. One species, *N. nemorosa* (Douglas ex Lindl.) Straw from western North America.

67. *Chionophila* Benth.

Chionophila Benth. in DC. Prodr. 10: 331 (1846).
Penstemonopsis Rydb. (1917).

Perennial suffrutescent herbs, glabrous except for glandular-pubescent inflorescence. Stems erect, quadrangular. Leaves in a basal rosette, petiolate, opposite, linear-lanceolate, acuminate with entire margin. Flowers pedicellate. Calyx lobes equalling the tube or distinctly shorter. Corolla pale lavender, tubular. Stamens 4, included, anthers with two distally confluent thecae, glabrous. Two species, *C. jamesii* Benth. and *C. tweedyi* (Canby & Rose) Henders of alpine meadows near timberline in Rocky Mountains, North America.

68. *Keckiella* Straw

Keckiella Straw, Brittonia 19: 203 (1967).
Lepidostemon Lem. (1861).
Keckia Straw (1966).

Perennial subshrubs or shrubs, glabrous or glandular-pubescent. Stems erect, quadrangular. Leaves cauline, shortly petiolate, opposite, lanceolate-ovoid, base cordate or rounded, with dentate or serrate margin. Flowers pedicellate. Calyx deeply 5-lobed. Corolla whitish to red, upper lip hooded, lobes often reflexed. Stamens 4, included, filaments lanose to pubescent at base, anthers with two confluent, spreading thecae, glabrous. Ovary ovoid, with a distinct nectary disc. $2n = 16$. Seven species from western North America.

69. ***Pennellianthus*** Crosswh.

Pennellianthus Crosswh., Am. Midl. Nat. 83: 362 (1970).
Leiostemon Raf. (1825).

Perennial herb, glabrous or glandular-pubescent. Stems erect, quadrangular. Leaves cauline, shortly petiolate, opposite, lanceolate-ovoid, base cordate or rounded, with dentate or serrate margin. Flowers pedicellate. Calyx deeply 5-lobed. Corolla pink to maroon, strongly 2-lipped, throat inflated, glabrous inside, lobes spreading. Stamens 4, included, filaments lanose to pubescent, anthers with two confluent, spreading thecae, lanose. One species, *P. frutescens* (Lamb.) Crosswh. from North Japan and Kamchatka.

70. ***Penstemon*** Schmidel

Penstemon Schmidel, Icon. Pl. ed. Keller: 2 (1763); Krautter, Publ. Univ. Pennsylvania 3, 2: 93–206 (1908), rev.

Perennial herbs or shrubs, glabrous or glandular-pubescent. Stems erect, quadrangular. Leaves cauline, shortly petiolate to sessile, opposite, linear-filiform to lanceolate-ovoid, with entire dentate or serrate margin. Flowers pedicellate. Calyx deeply 5-lobed. Corolla blue, purple, red or white, strongly 2-lipped, throat inflated, glabrous inside, lobes spreading. Stamens 4, included, filaments glabrous to pubescent, anthers with two distinct or confluent, spreading thecae, glabrous or lanose. $2n = 16, 32$. About 250 species in North America from Canada to Mexico.

71. ***Tetranema*** Benth. ex Lindl.

Tetranema Benth. ex Lindl., Edwards's Bot. Reg. 29: 52 (1843), nom. cons; Pennell, Proc. Acad. Nat. Sci. Philad. 77: 269–272 (1925), rev.

Perennial rosulate, sometimes suffrutescent herbs, sparsely pubescent. Stems erect, rounded, with short internodes. Leaves opposite, narrowed to a sessile or weakly clasping base, obovoid, obtuse, with entire to crenate or dentate margin. Flowers distinctly pedicellate. Calyx 5-lobed to base, lobes narrow and subequal. Corolla white to mauve or blue, campanulate, limb 2-lipped, lobes elliptic-ovoid, tube campanulate, broad, short. Stamens 4, included, filaments short, inserted at base of tube. Ovary ovoid. Seeds numerous, endosperm alveolated (bothrospermous). Eight species, Mexico to Guatemala, Nicaragua and Costa Rica.

72. ***Uroskinnera*** Lindl.

Uroskinnera Lindl., Gard. Chron. 1857: 36 (1857); Schultes, Bot. Mus. Leafl. Harvard Univ. 9: 65–83 (1941), rev.

Shrubs, roughly pilose to densely villose. Stems erect, rounded. Leaves cauline, opposite, petiolate, ovoid to elliptic-oblong, pinnately veined and with dentate to serrate margin. Flowers shortly pedicellate. Calyx deeply 5-lobed, campanulate. Corolla violet-purple, limb 2-lipped, lobes spreading, tube elongate. Stamens 4, included, anthers with two divergent thecae confluent at apex. Ovary ovoid. Three species in Central America.

73. ***Collinsia*** Nutt.

Collinsia Nutt., J. Acad. Nat. Sci. Philad. 1: 190. t. 9 (1817).

Annual herbs, glabrous or pubescent. Stems erect. Leaves cauline, opposite, sessile to subsessile, linear-lanceolate to ovoid, acuminate, often cordate at base, with entire to dentate margin. Flowers sessile to shortly pedicellate. Calyx deeply 5-lobed, lobes lanceolate, subequal. Corolla blue to violet, limb 2-lipped, central lobe of lower lip forms a pouch enclosing style and stamens, tube flattened and gibbous abaxially. Stamens 4, included. Ovary conical or globose. Seeds numerous to few (sometimes 1 per locule), large, winged or wingless. $2n = 14$. About 20 species in western North America.

74. ***Tonella*** Nutt. ex A. Gray

Tonella Nutt. ex A. Gray, Proc. Am. Acad. Arts 7: 378 (1868).

Annual herbs, glabrous except for glandular-pubescent inflorescence. Stems erect. Leaves cauline, opposite, verticillate in inflorescence, sessile to petiolate, linear-lanceolate to ovoid, acuminate, lower leaves distinctly pinnatisect, nearly trifoliolate, with entire to dentate margin.

Flowers distinctly pedicellate. Calyx deeply 5-lobed, lobes lanceolate, subequal. Corolla blue and white, limb weakly 2-lipped, nearly subrotate, lobes spreading, oblong to rounded. Stamens 4, exserted. Ovary conical or globose. Seeds 1–2 per locule, large, wingless. Two species, *T. floribunda* A. Gray, *T. tenella* (Benth.) Heller, western North America to Mexico.

2. Tribe Antirrhineae Chav. (1833). Fig. 45

Linariaceae Martynov (1820).

Inflorescence racemose, frondose or bracteate (*Antirrhinum*, *Acanthorrhinum*, *Anarrhinum* to *Sairocarpus*, *Gambelia*), prophylls lacking, corolla usually personate, with palate on lower lip, corolla-tube often with adaxial gibbosity or spur, petal aestivation descending, stamens 4 or 2 (*Mohavea*), included or somewhat exserted (*Rhodochiton* to *Gambelia*), the median stamen reduced to a minute staminode, anthers with two separate spreading thecae, the adjacent pairs of anthers marginally coherent (*Antirrhinum* to *Epixiphium*) or free (*Mabrya* to *Gambelia*), nectary a disc, ovary with apical septum, stigma capitate to weakly bilobed (*Antirrhinum*, *Linaria*), epidermal cells of testa heterogeneous, their different size causing various ornamentations, wings produced by epidermal cells or remains of intermediate-layer cells, capsule dehiscence various, mainly poricidal.

75. *Antirrhinum* L.

Antirrhinum L., Sp. Pl. 2: 612 (1753); Sutton, Revision Tribe Antirrhineae: 67–97 (1988), rev.

Perennial suffrutescent herbs or dwarf shrubs, glabrous or glandular-pubescent. Stems ascending to erect, sometimes twining. Leaves cauline, opposite, alternate above, sessile to petiolate, linear-lanceolate to ovoid or orbicular, with entire margin. Flowers distinctly pedicellate. Calyx deeply 5-lobed. Corolla white or cream, yellow, pink to purplish with yellow palate, limb 2-lipped, with broad tube, gibbous or saccate abaxially, palate occluding throat. Ovary ovoid to suborbicular. Capsule poricidal. Seeds numerous, testa reticulate-alveolate or cristate. $2n = 16$. Twenty species from western Mediterranean.

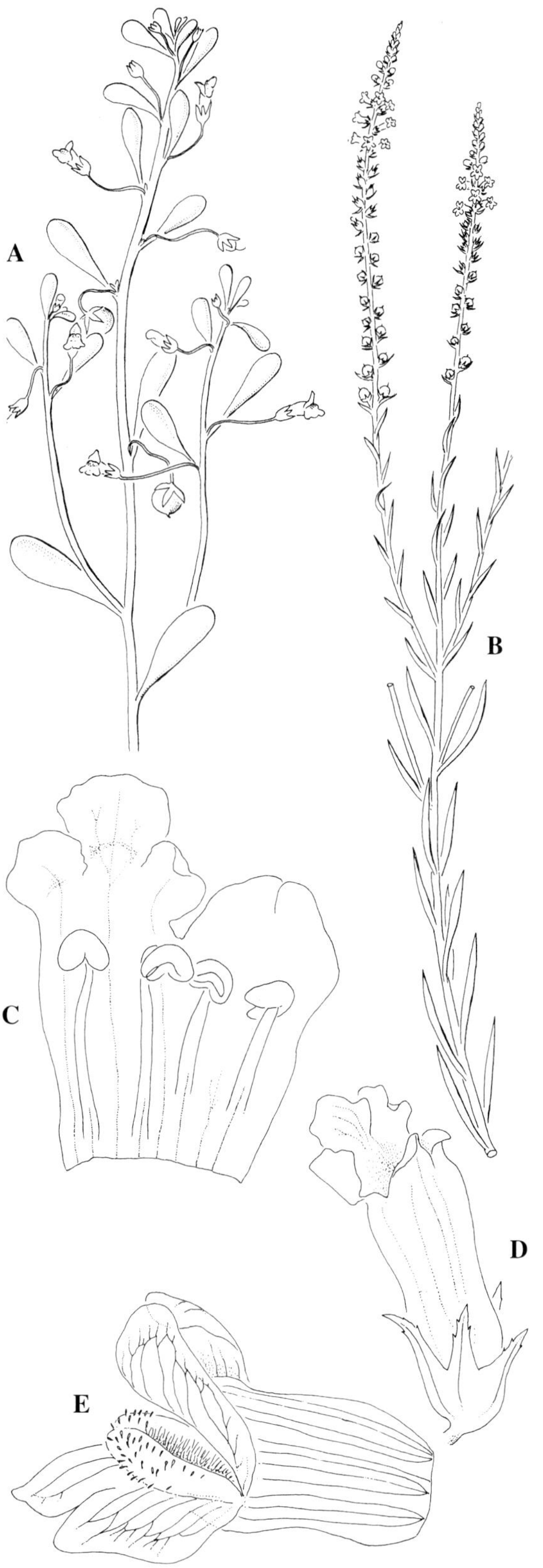

Fig. 45. Scrophulariaceae. **A, E** *Schweinfurthia pterosperma*. **A** Habit. **E** Corolla. **B–D** *Anarrhinum forskahlei*. **B** Habit. **C** Corolla, dissected. **D** Flower

76. *Chaenorrhinum* (DC.) Rchb.

Chaenorrhinum (DC.) Rchb., Consp.: 123 (1828); Sutton, Revision Tribe Antirrhineae: 97–129 (1988), rev.
Linaria sect. *Microrrhinum* Endl. (1839).
Microrrhinum (Endl.) Fourr. (1869).
Hueblia Speta (1982).

Annual or suffrutescent herbs, glabrous or glandular-pubescent. Stems ascending to erect. Leaves cauline, opposite, alternate above, sessile to petiolate, linear-lanceolate to ovoid, with entire margin. Flowers distinctly pedicellate. Calyx deeply 5-lobed. Corolla white, pink, bluish-lilac, usually with yellow palate, limb 2-lipped, calcarate abaxially at base, palate not occluding throat. Ovary ovoid to globose. Capsule poricidal. Seeds numerous, testa cristate or rarely reticulate, with longitudinal ridges. $2n = 14, 42$. Twenty-one species in Europe and Mediterranean to Iran and Afghanistan.

77. *Holzneria* Speta

Holzneria Speta, Bot. Jahrb. Syst. 103: 16 (1982); Sutton, Revision Tribe Antirrhineae: 130–131 (1988), rev.

Annual herbs, glabrous to glandular-pubescent. Stems erect. Leaves cauline, opposite, alternate above, petiolate, linear-lanceolate to ovoid and spathulate, obtuse with entire margin. Flowers distinctly pedicellate. Calyx deeply 5-lobed, lobes conspicuously unequal, adaxial lobe much shorter than abaxial lobes. Corolla white, tinged with pink, limb 2-lipped, calcarate abaxially at base, palate not occluding throat, spur inconspicuous, conical. Ovary ovoid to globose. Seeds numerous, testa reticulate-alveolate, with longitudinal ridges. Two species, *H. spicata* (Korovin) Speta and *H. microcentron* (Bornm.) Speta in Southwest Asia.

78. *Albraunia* Speta

Albraunia Speta, Bot. Jahrb. Syst. 103: 32 (1983); Sutton, Revision Tribe Antirrhineae: 131–135 (1988), rev.

Annual herbs, glabrous to glandular-pubescent. Stems erect. Leaves cauline, alternate, petiolate below to subsessile above, oblong-lanceolate to spathulate, with entire margin. Flowers distinctly pedicellate. Calyx deeply 5-lobed, lobes subequal, adaxial lobe somewhat shorter than abaxial lobes. Corolla bluish-purple with white tube and yellow palate, limb 2-lipped, tube calcarate abaxially at base, palate occluding throat, spur conical. Ovary oblong-ovoid to globose. Seeds numerous, testa irregularly reticulate or cristate, usually with long hair-like multicellular tubercles. Three species in Southwest Asia.

79. *Schweinfurthia* A. Braun Fig. 45A, E

Schweinfurthia A. Braun, Sitzungsber. Ges. Naturf. Freunde Berlin 1866: 24 (1866); Sutton, Revision Tribe Antirrhineae: 135–141 (1988), rev.

Annual or suffrutescent herbs, glabrous or glandular-pubescent. Stems prostrate to erect, sometimes woody at base and spinescent. Leaves cauline, opposite, alternate above, subsessile to petiolate, linear-lanceolate to ovoid, obtuse with entire margin. Flowers distinctly pedicellate. Calyx deeply 5-lobed, lobes unequal, adaxial lobe exceeding 4 lateral lobes. Corolla whitish with purple veins and yellow palate, limb 2-lipped, tube ecalcarate, gibbous abaxially at base, palate occluding throat. Ovary globose, loculi very unequal. Seeds numerous, testa cristate or tuberculate, with longitudinal ridges. Six species, from Northeast Africa (Somalia, Ethiopia, Kenya) to Yemen, Arabia and India.

80. *Acanthorrhinum* Rothm.

Acanthorrhinum Rothm., Feddes Repert. Spec. Nov. Regni Veg. 52: 31 (1943); Sutton, Revision Tribe Antirrhineae: 143–145 (1988), rev.

Perennial dwarf shrub, glabrous or glandular-pubescent. Stems erect, spinescent. Leaves cauline, alternate, sessile, linear, obtuse and with entire margin, caducous. Flowers distinctly pedicellate. Calyx deeply 5-lobed, lobes subequal. Corolla white tinged with mauve, palate yellow, limb 2-lipped, with broad tube, ecalcarate, inconspicuously gibbous abaxially, palate occluding throat. Ovary ovoid to subglobose, loculi subequal. Seeds numerous, testa reticulate-alveolate or cristate. $2n = 32$. One species, *A. ramosissimum* (Cosson & Durieu) Rothm. from Northwest Africa.

81. *Misopates* Raf.

Misopates Raf., Aut. Bot.: 158 (1840); Sutton, Revision Tribe Antirrhineae: 145–155 (1988), rev.

Annual herbs, glabrous or pilose to glandular-pubescent. Stems ascending to erect. Leaves cauline, opposite, alternate above, sessile to shortly petiolate, linear-lanceolate, obtuse and with entire margin. Flowers distinctly pedicellate. Calyx deeply 5-lobed, lobes unequal, adaxial lobe longest. Corolla dull pink to whitish or cream, limb 2-lipped, with broad cylindrical tube, gibbous or saccate abaxially, palate occluding throat. Ovary ovoid to suborbicular, loculi unequal. Seeds numerous, testa reticulate-alveolate or cristate, with a narrow wing. $2n = 16$. Seven species, Europe, Mediterranean to Cape Verde Islands, Ethiopia and Northwest India.

82. *Asarina* Mill.

Asarina Mill., Gard. Dict. ed. 7 (1757); Sutton, Revision Tribe Antirrhineae: 155–158 (1988), rev.

Perennial suffrutescent herb, densely pubescent. Stems prostrate to trailing or ascending. Leaves cauline, opposite, petiolate, reniform to ovoid-cordate or orbicular, entire or lobed, acuminate and with crenate-dentate margin. Flowers distinctly pedicellate. Calyx deeply 5-lobed, lobes ± equal. Corolla white with pale yellow lips and deep yellow palate, limb 2-lipped, with broad tube, dorsiventrally compressed, subgibbous abaxially, palate occluding throat. Ovary ovoid to subglobose, loculi equal. Seeds numerous, testa longitudinally cristate. One species, *A. procumbens* (L.) Mill. in Southwest Europe.

83. *Cymbalaria* Hill

Cymbalaria Hill, Brit. Herbal.: 113 (1756); Sutton, Revision Tribe Antirrhineae: 158–169 (1988), rev.

Short-lived perennial herbs, sometimes behaving as annuals, glabrous to puberulent or villose. Stems procumbent or decumbent to trailing. Leaves cauline, opposite or alternate, petiolate, reniform to ovoid-cordate or orbicular, lobed or rarely entire, acuminate and with entire margin. Flowers distinctly pedicellate. Calyx deeply 5-lobed, lobes subequal to unequal with adaxial lobe exceeding abaxial lobes. Corolla lilac to violet with yellow or reddish palate, limb 2-lipped, with broad tube, calcarate or rarely gibbous abaxially, palate occluding throat, spur conical or cylindical. Ovary ± globose, loculi equal. Seeds numerous, testa tuberculate, cristate, alveolate or smooth. $2n = 14, 28, 42, 56$. Nine species in Europe, Central and East Asia.

84. *Kickxia* Dumort.

Kickxia Dumort., Fl. Belg. 35 (1827); Sutton, Revision Tribe Antirrhineae: 169–249 (1988), part. rev.; Gebrehiwet Medhanie, Acta Univ. Upsal.: 19–25 (1999), rev.; Nord. J. Bot. 20: 655–689 (1999), rev.
Elatinoides (Chav.) Wettst. (1891).

Dwarf shrubs, perennial or annual herbs. Stems prostrate to ascending. Leaves mostly alternate, simple, linear-lanceolate to suborbicular, entire, dentate or lobed, hastate or sagittate, subsessile to petiolate, petioles occasionally twining. Flowers distinctly pedicellate, pedicels not twining or reflexed. Calyx deeply 5-lobed, lobes ± equal. Corolla white to yellow, usually with purple lower lip, limb 2-lipped, with cylindrical to campanulate tube, calcarate or rarely saccate abaxially, palate occluding throat, spur conical or cylindrical. Anthers forming ring-like structure. Ovary ovoid to subglobose, loculi equal. Capsule dehiscence operculate. Seeds numerous, reniform to ellipsoid, testa reticulate-alveolate or tuberculate. $2n = 18, 36$. Nine species from Europe, West African Islands (Canary Islands to Cape Verde Islands), North and Northeast Africa (south to Kenya), South Central and Southwest Asia.

85. *Nanorrhinum* Betsche

Nanorrhinum Betsche, Courier Forschungsinst. Senckenberg 71: 131 (1984); Sutton, Revision Tribe Antirrhineae: 169–249 (1988), part. rev.; Gebrehiwet Medhanie, Acta Univ. Upsal.: 25–60 (1999), rev.; Nord. J. Bot. 20: 655–689 (1999), rev.
Pogonorrhinum Betsche (1984).

Dwarf shrubs, perennial or annual herbs, glabrous or pubescent, stems prostrate to erect or climbing by means of twining petioles. Leaves mostly alternate, simple or sometimes deeply divided, linear to suborbicular, entire, dentate or lobed, hastate or sagittate. Flowers pedicellate, pedicels not twining or reflexed. Calyx deeply 5-lobed, lobes ± equal. Corolla white to yellow, usually with purple lower lip, limb 2-lipped, with cylindrical to campanulate tube, calcarate or rarely saccate abaxially, palate partially occluding throat, glabrous to lanate internally, spur conical or cylindrical. Anthers forming ring-like structure. Capsule dehiscence valvate. Seeds oblong-ovoid, ellipsoid, or rarely reniform, tuberculate, tubercles rounded to columnar. Ten species from Canary Islands, Cape Verde Islands, North Africa, subtropical to tropical Northeast Africa (south to Kenya), Arabia, South Central and Southwest Asia to India and Burma in the east. Sutton (1988) accepted 37 species, while Ghebrehiwet (1999, 2000) reduced the number to ten.

86. *Anarrhinum* Desf. Fig. 45B–D

Anarrhinum Desf., Fl. Atl. 2: 51 (1798), nom. cons.; Sutton, Revision Tribe Antirrhineae: 249–260 (1988), rev.

Biennial or perennial herbs or dwarf shrubs, glabrous or glandular-pubescent to hirsute. Stems erect. Leaves heteromorphic, alternate, basal leaves entire, dentate or lobed, petiolate, often forming a rosette, cauline leaves usually deeply palmatisect, sometimes entire. Flowers subsessile or pedicellate. Calyx deeply 5-lobed, lobes equal. Corolla white or creamy white to dull blue or violet, limb 2-lipped, with cylindrical tube, calcarate or gibbous abaxially, palate poorly developed. Anthers forming ring-like structure. Ovary subglobose, loculi equal. Seeds numerous, testa tuberculate. $2n = 18$. Eight species, mainly western Mediterranean (Spain, Portugal), one species (*A. bellidifolium* (L.) Willd.) to Germany (Mosel valley), one species (*A. forskaohlii* (Gmelin) Cufodontis) to Ethiopia.

87. *Linaria* Mill.

Linaria Mill., Gard. Dict. Abr. ed. 4 (1754); Sutton, Revision Tribe Antirrhineae: 260–455 (1988), rev.
Trimerocalyx (Murb.) Murb. (1940).

Annual, biennial or perennial herbs, glabrous or glandular-pubescent to villose. Stems ascending to erect, the hypocotylary secondary stems often abundant and dominant. Leaves cauline, opposite or verticillate below, alternate above, sessile to petiolate, linear-lanceolate to ovoid. Flowers sessile or pedicellate. Calyx deeply 5-lobed, the adaxial lobe usually longest, rarely shortest. Corolla white, yellow, mauve to blue or lilac, palate usually yellow, limb 2-lipped, with broad tube, calcarate abaxially, palate occluding throat, spur narrowly conical, linear or filiform. Ovary ovoid to globose, loculi unequal or equal. Seeds numerous, testa reticulate-alveolate, cristate, tuberculate or smooth. $2n = 12, 24$. About 150 species, from North America to Europe and Asia.

88. *Nutallanthus* D.A. Sutton

Nutallanthus D.A. Sutton, Revision Tribe Antirrhineae: 455 (1988), rev.

Annual or biennial herbs, glabrous or glandular-pubescent. Sterile stems prostrate or decumbent, fertile stems erect. Leaves cauline, opposite or verticillate on sterile stems, alternate on fertile stems, sessile, linear-lanceolate to ovoid. Flowers pedicellate. Calyx deeply 5-lobed, lobes equal. Corolla bluish or pale lilac to white, palate usually white, limb very unequally 2-lipped, with narrow tube, very narrowly calcarate, palate scarcely occluding throat. Ovary oblong-ovoid to subglobose, loculi equal. Seeds numerous, with 4–7 longitudinal ridges, testa tuberculate or smooth. $2n = 12$. Four species in North and South America.

89. *Sairocarpus* D.A. Sutton

Sairocarpus D.A. Sutton, Revision Tribe Antirrhineae: 461 (1988), rev.

Annual or perennial herbs, glabrous or glandular-pubescent. Stems ascending to erect, often with flexuous or cirrhous twining lateral branches in inflorescence. Leaves cauline, opposite below, alternate above, sessile to petiolate, linear-lanceolate to ovoid or cordate-orbicular. Flowers distinctly pedicellate. Calyx deeply 5-lobed, lobes subequal or unequal. Corolla bluish-violet to reddish with usually whitish throat and tube, limb unequally 2-lipped, with cylindrical tube, gibbous or saccate abaxially, palate prominent, ± occluding throat. Ovary ovoid to oblong, loculi equal or subequal. Seeds numerous, testa tuberculate, reticulate-alveolate or cristate. $2n = 30, 32$. Thirteen species in Southwest North America and Mexico.

90. *Howeliella* Rothm.

Howeliella Rothm., Leafl. W. Bot. 7: 115 (1954); Sutton, Revision Tribe Antirrhineae: 477–478 (1988), rev.

Annual herb, densely glandular-pubescent. Stems ascending to erect, branches not cirrhous. Leaves cauline, opposite below, alternate above, subsessile to petiolate, ovoid to oblong-elliptic. Flowers distinctly pedicellate. Calyx deeply 5-lobed, lobes very unequal, the adaxial lobe greatly exceeding the 4 abaxial lobes. Corolla dull white with pinkish upper lip and yellowish palate, limb unequally 2-lipped, with cylindrical tube, gibbous or saccate abaxially, palate poorly developed, not occluding throat. Ovary obliquely ovoid, loculi somewhat unequal. Seeds numerous, testa irregularly reticulate-alveolate or partially cristate. $2n = 16$. One species, *H. ovata* (Eastw.) Rothm., in California.

91. *Neogaerrhinum* Rothm.

Neogaerrhinum Rothm., Feddes Repert. Spec. Nov. Regni Veg. 52: 31 (1943); Sutton, Revision Tribe Antirrhineae: 479–481 (1988), rev.

Annual herbs, glabrous except for flowers or densely villose. Stems erect or scrambling, often climbing by means of twining pedicels. Leaves cauline, opposite below, alternate above, subsessile to petiolate, ovoid or lanceolate-ovoid. Flowers distinctly pedicellate, pedicels elongate, cirrhous. Calyx deeply 5-lobed, lobes ± equal. Corolla purple or yellow, limb unequally 2-lipped, with cylindrical tube, gibbous or subentire abaxially, palate poorly developed, not occluding throat. Ovary globose, loculi equal or subequal. Seeds numerous, testa tuberculate or cristate. Two species, *N. strictum* (Hook. & Arn.) Rothm. and *N. filipes* (A. Gray) Rothm. in Southwest North America.

92. *Pseudorontium* (A. Gray) Rothm.

Pseudorontium (A. Gray) Rothm., Feddes Repert. Spec. Nov. Regni Veg. 52: 33 (1943); Sutton, Revision Tribe Antirrhineae: 481–482 (1988), rev.

Annual herb, glandular-pubescent to villose. Stems erect, branches not cirrhous. Leaves cauline, opposite below, alternate above, petiolate, ovoid or rhombic to suborbicular. Flowers distinctly pedicellate, pedicels strongly recurved in fruit, not cirrhous. Calyx deeply 5-lobed, lobes ± equal. Corolla purple with yellow palate, limb subequally 2-lipped, with cylindrical tube, scarcely gibbous abaxially, palate ± occluding throat. Ovary globose, loculi equal. Seeds numerous, cyathiform, with longitudinal ridges and inrolled, scarcely emarginate wing. One species, *P. cyathiferum* (Benth.) Rothm. in Southwest North America and Mexico.

93. *Mohavea* A. Gray

Mohavea A. Gray, Rep. Explor. Railroad Pacific 4: 122 (1856); Sutton, Revision Tribe Antirrhineae: 482–483 (1988), rev.

Annual herbs, glandular-pubescent to villose. Stems erect, branches not cirrhous. Leaves cauline, opposite below, alternate above, petiolate, linear to lanceolate. Flowers distinctly pedicellate, pedicels not cirrhous. Calyx deeply 5-lobed, lobes subequal, adaxial lobe longest. Corolla yellow, limb subequally 2-lipped, with cylindrical tube, somewhat compressed dorsiventrally, saccate abaxially, palate partially occluding throat. Fertile stamens 2. Ovary subglobose, loculi subequal. Seeds numerous, cyathiform, with longitudinal ridges and inrolled, scarcely emarginate wing. $2n = 26$. Two species, *M. confertiflora* (DC.) Heller and *M. breviflora* Coville in Southwest North America and Mexico.

94. *Maurandya* Ortega

Maurandya Ortega, Hort. Matrit. Descr. 1(2): 21 (1797); Sutton, Revision Tribe Antirrhineae: 484–487 (1988), rev.

Perennial suffrutescent herbs, glabrous except for flowers. Stems climbing by means of twining petioles and pedicels. Leaves cauline, alternate, petiolate, deltate or rarely 5-lobed. Flowers distinctly pedicellate, pedicels twining, becoming deflexed distally in fruit. Calyx deeply 5-lobed, lobes unequal, adaxial lobe exceeding the 2 median lobes. Corolla blue to violet, limb unequally 2-lipped, with cylindrical tube, somewhat compressed dorsiventrally with 2 longitudinal abaxial folds, subgibbous abaxially at base, without basal palate. Ovary oblong-globose to oblong-ovoid, loculi subequal. Seeds numerous, radially symmetrical, tuberculate with irregular ridges. $2n = 24$. Two species, *M. scandens* (Cav.) Pers. and *M. barclaiana* Lindley from Mexico to Guatemala.

95. *Maurandella* (A. Gray) Rothm.

Maurandella (A. Gray) Rothm., Feddes Repert. Spec. Nov. Regni Veg. 52: 26 (1943); Sutton, Revision Tribe Antirrhineae: 487–488 (1988), rev.

Perennial suffrutescent herb, glabrous. Stems climbing by means of twining petioles. Leaves cauline, alternate, petiolate, deltate or rarely 5-lobed. Flowers distinctly pedicellate, pedicels not twining, becoming deflexed in fruit. Calyx deeply 5-lobed, lobes subequal, 2 abaxial lobes scarcely exceeding the adaxial lobes. Corolla dark blue to violet, the palate spotted with yellow or white, limb unequally 2-lipped, with cylindrical tube, somewhat compressed dorsiventrally, subgibbous abaxially at base, with well-developed basal palate. Ovary oblong-globose to oblong-ovoid, loculi unequal. Seeds numerous, somewhat dorsiventrally asymmetrical, tuberculate with irregular ridges. $2n = 24$. One species, *M. antirrhiniflora* (Willd.) Rothm. from Southwest North America to Mexico.

96. *Epixiphium* (A. Gray) Munz

Epixiphium (A. Gray) Munz, Proc. Calif. Acad. Sci. ser 4, 15(12): 380 (1926); Sutton, Revision Tribe Antirrhineae: 488–490 (1988), rev.

Annual herb with taproots, glabrous. Stems climbing by means of twining petioles. Leaves cauline, alternate, petiolate, deltate-cordate to obscurely 5-lobed, base hastate or sagittate. Flowers distinctly pedicellate, pedicels not twining. Calyx deeply 5-lobed, lobes subequal. Corolla pale bluish-violet, limb subequally 2-lipped, with broad tube, somewhat dorsiventrally compressed, gibbous abaxially at base, palate partially occluding throat. Ovary broadly conical-ovoid, somewhat laterally compressed, loculi subequal. Seeds numerous, strongly dorsiventrally compressed, with encircling, emarginate wing, testa tuberculate. $2n = 24$. One species, *E. wislizenii* (A. Gray) Munz from Southwest North America to Mexico.

97. *Mabrya* Elisens

Mabrya Elisens, Syst. Bot. Monogr. 5: 57 (1985); Sutton, Revision Tribe Antirrhineae: 490–495 (1988), rev.

Perennial suffrutescent herbs with fibrous roots, glandular-viscid to villose. Stems erect, ascending or pendent, often forming mats. Leaves cauline, alternate, petiolate, orbicular to reniform, crenate, dentate or somewhat palmatisect to subentire. Flowers distinctly pedicellate, pedicels becoming recurved in fruit. Calyx deeply 5-lobed, infundibuliform or urceolate, lobes subequal. Corolla yellow to purplish-pink or red, limb subequally 2-lipped, with narrowly infundibuliform tube, somewhat compressed dorsiventrally, with 2 prominent longitudinal abaxial folds, not gibbous abaxially at base, without basal palate. Ovary subglobose to broadly ovoid, loculi ± equal. Seeds numerous, radially symmetrical, tuberculate or longitudinally cristate. $2n = 24$. Five species, Southwest North America and Mexico.

98. *Holmgrenanthe* Elisens

Holmgrenanthe Elisens, Syst. Bot. Monogr. 5: 54 (1985); Sutton, Revision Tribe Antirrhineae: 495–497 (1988), rev.

Perennial suffrutescent herb, glandular-villose. Stems forming clumps or mats. Leaves cauline, alternate, petiolate, orbicular, spinulose-dentate, petioles not twining. Flowers distinctly pedicellate, pedicels not twining, becoming recurved in fruit. Calyx deeply 5-lobed, lobes subequal, with adaxial lobe scarcely exceeding the abaxial lobes, lobes conspicuously spinulose-dentate. Corolla pale yellow, limb subequally 2-lipped, with broad tube, somewhat compressed dorsiventrally, with 2 longitudinal abaxial folds, ± entire abaxially at base, with poorly developed basal palate. Ovary globose, loculi unequal. Seeds numerous, radially symmetrical, foveate, tuberculate. One species, *H. petrophila* (Coville & C.V. Morton) Elisens, Southwest North America.

99. *Lophospermum* D. Don

Lophospermum D. Don, Trans. Linn. Soc. Lond. 15(2): 351 (1827); Sutton, Revision Tribe Antirrhineae: 497–505 (1988), rev.

Perennial suffrutescent herbs, glabrous to glandular- or eglandular-pubescent. Stems climbing by means of twining petioles or pedicels, woody at base. Leaves cauline, alternate, petiolate, deltate, cordate, suborbicular or reniform, margin dentate or crenate to subentire. Flowers distinctly pedicellate, pedicels often cirrhous, not recurved in fruit. Calyx narrowly infundibuliform or urceolate, deeply 5-lobed, lobes subequal. Corolla purplish-pink or reddish-purple, tube sometimes white, limb unequally 2-lipped, with broad tube, somewhat compressed dorsiventrally, not gibbous abaxially at base, without basal palate. Ovary globose to subglobose, loculi ± equal. Seeds numerous, somewhat dorsiventrally symmetrical, with 2 longitudinal wings, tuberculate with irregular ridges. $2n = 24$. Six species, Mexico to Guatemala.

100. *Rhodochiton* Zucc. ex Otto & Dietrich

Rhodochiton Zucc. ex Otto & Dietrich, Verh. Ver. Beförd. Gart. Königl. Preuss. Staaten 10(26): 153 (1833); Sutton, Revision Tribe Antirrhineae: 505–509 (1988), rev.

Perennial suffrutescent herbs, glandular-pubescent to glabrescent. Stems climbing by means of twining petioles. Leaves cauline, alternate, petiolate, cordate to deltate, irregularly mucronate-dentate or subentire with minute teeth. Flowers distinctly pedicellate, pedicels pendent, not cirrhous. Calyx shallowly 5-lobed, broadly infundibuliform or campanulate, lobes equal, entire. Corolla blackish-purple or deep pinkish-purple, subactinomorphic, limb subequally 2-lipped, with broad terete or somewhat 5-sided tube, with 4 shallow longitudinal folds, entire abaxially at base, without basal palate. Ovary oblong-globose, loculi equal. Seeds numerous, somewhat dorsiventrally subsymmetrical, tuberculate with 2 longitudinal wings. $2n = 24$. Three species in Mexico, among them *R. atrosanguineum* (Zucc.) Rothm., a twiner with sensitive petioles.

101. *Galvezia* Dombey ex Juss.

Galvezia Dombey ex Juss., Gen. Pl.: 119 (1789); Sutton, Revision Tribe Antirrhineae: 510–514 (1988), rev.

Shrubs, glabrous to glandular-pubescent. Stems erect or ascending, branches rarely cirrhous. Leaves cauline, opposite or the uppermost alternate, entire. Flowers pedicellate, pedicels sinuate or deflexed in fruit. Calyx deeply 5-lobed, lobes subequal. Corolla white to dark red with whitish throat, limb subequally 2-lipped, with elongate subcylindrical tube, saccate abaxially, palate poorly developed, not occluding throat. Ovary ovoid to subglobose, loculi ± equal. Seeds numerous, longitudinally cristate, testa tuberculate. Four species in Ecuador and Peru.

102. *Gambelia* Nutt.

Gambelia Nutt., Proc. Acad. Nat. Sci. Philad. 4(1): 7 (1848); Sutton, Revision Tribe Antirrhineae: 514–517 (1988), rev.

Spreading or juncoid shrubs, glabrous except for glandular-pubescent inflorescence and corolla. Stems erect or ascending, the branches rarely cirrhous. Leaves cauline, 3-verticillate, opposite or rarely the uppermost alternate, entire. Flowers pedicellate, pedicels sinuate or deflexed in fruit. Calyx deeply 5-lobed, lobes subequal. Corolla red to scarlet, limb subequally 2-lipped, with subcylindrical tube, saccate abaxially, palate partially occluding throat. Ovary ovoid to subglobose, loculi subequal, abaxial locule larger than adaxial. Seeds numerous, bilaterally symmetrical, with longitudinal ridges, testa alveolate or cristate. $2n = 26$. Four species, Southwest North America to Mexico.

II. Subfam. Gratioloideae (Benth.) Luerss. (1882).

Gratiolaceae Martynov (1820).
Gratioleae Benth. (1835).

Annual or perennial herbs to small shrubs. Inflorescence usually racemose or sometimes with axillary 2(–3)-flowered cymes, terminal or truncate with axillary racemes. Corolla subrotate to bilabiate, often with abaxial protrusion on lower lip (palate), petal aestivation usually descending (not *Lindenbergia*, *Hemiarrhena*). Stamens 2–4, rarely 5, the abaxial pair often reduced to staminodes or lacking, filaments of the abaxial pair geniculate, Z-shaped or with spur-like appendages, anthers with two separate parallel pilose or glabrous thecae or anthers monothecous. Ovary usually bilocular. Capsule loculicidal, loculicidal and septicidal or septicidal, rarely fruit indehiscent. Intermediate layers of ovule integument 1–3 (rarely 3–10 in tribe Angelonieae), endosperm of mature seed smooth, furrowed or alveolated, alveolation by protrusion of all endotheliar cells, epidermal cells of testa homogenous. Seeds with longitudinal ridges, testa cells with hook-like wall thickenings, alveolated (bothrospermous, aulacospermous) or not alveolated. Fifty genera with 546 species, cosmopolitan. The monotypic Callitrichaceae and Hippuridaceae should be included here.

1. Tribe Angelonieae Benth. (1846).

Herbs, inflorescence racemose, prophylls present (*Monopera*) or absent, corolla-tube bisaccate, with 2 or 1 gibbous spurs abaxially, anthers with two separate to confluent equal thecae, stigma capitate to weakly bilobed, capsule loculicidal and septicidal.

103. *Angelonia* Bonpl. Fig. 42B

Angelonia Bonpl. in Humb. & Bonpl., Pl. Aequin. 2: 92, t. 108 (1812).

Annual or perennial herbs or shrubs, glabrous, pilose or viscid-pubescent. Stems ascending or erect. Leaves cauline, opposite or alternate above, sessile to subsessile, linear-lanceolate to ovoid, acuminate, with dentate margin. Flowers distinctly pedicellate. Calyx deeply 5-lobed, lobes lanceolate, subequal. Corolla blue to violet, limb 2-lipped, lobes spreading, orbicular, tube very short, with 2 gibbous spurs abaxially. Stamens 4, included. Ovary conical or globose. Seeds numerous, testa hyaline-membranous, reticulate-foveolate, endosperm absent. $2n = 20$. Twenty-five species in tropical America, some species naturalized in Old World tropics.

104. *Monopera* Barringer

Monopera Barringer, Brittonia 35: 111 (1983).

Annual or perennial herbs, glabrous except for glandular-pubescent pedicel and calyx. Stems erect. Leaves cauline, opposite, sessile, linear-lanceolate to ovoid, obtuse, with entire margin. Flowers distinctly pedicellate. Calyx deeply 5-lobed, lobes lanceolate, subequal. Corolla blue to violet, limb 2-lipped, lobes spreading, linear-lanceolate, tube very short, with 1 gibbous spur. Stamens 4, included. Ovary conical or globose. Seeds numerous, testa hyaline-membranous, reticulate-echinate, endosperm absent. Two species, *M. micrantha* (Benth.) Barringer, *M. perennis* (Chodat & Hassler) Barringer, in South America (Brazil, Paraguay).

2. Tribe Gratioleae Benth. (1835).

Terrestrial to aquatic herbs, endosperm not alveolated.

2a. Subtribe "Gratiolinae"

Herbs, often aquatic, leaves opposite, inflorescence racemose, frondose, corolla subrotate to 2-lipped, stamens 2–4, anthers with 2 parallel contiguous or separate thecae, stigma capitate (*Amphianthus*) or bilobed, capsule loculicidal, loculicidal and septicidal (*Boelckea*, *Braunblanquetia*, *Gratiola*) or septicidal (*Amphianthus*, *Maeviella*, *Mecardonia*).

105. *Amphianthus* Torr.

Amphianthus Torr., Ann. Lyc. New York 4: 82 (1837).

Annual herb, glabrous or sparsely glandular-pubescent. Leaves in a basal rosette, opposite, petiolate, ovoid, obtuse, with entire margin. Inflorescence with one fertile flower and a pair of floating leaves on water surface, and cleistogamous flowers submersed at level of basal rosette, prophylls present. Flowers sessile. Calyx 5-partite, the lobes ovoid, subequal. Corolla of chasmogamous flowers white, subrotate. Stamens 2, included. Ovary obcordate. Capsule weakly flattened. Seeds numerous, testa striate, black or dark brown. One species, *A. pusillus* Torr., in Georgia and South Carolina, growing in rock pools on granitic rocks.

106. *Bacopa* Aubl.

Bacopa Aubl., Hist. Pl. Guiane 1: 128, t. 48 (1775), nom. cons.
Moniera Juss. (1756).
Herpestis Gaertn. f. (1805).
Geochorda Cham. & Schltdl. (1828).
Ildefonsia Gardner (1842).
Allocalyx Cordemoy (1895).
Hydrotrida Small (1913).
Monocardia Penn. (1920).
Ancistrostylis T. Yamaz. (1980).
Sinobacopa D.Y. Hong (1987).

Annual or perennial herbs, glabrous, rarely glandular-pubescent or villose. Stems prostrate to ascending or erect, rounded or weakly quadrangular. Leaves cauline, opposite, sessile to subsessile, linear-lanceolate, ovoid to orbicular or deltate, acuminate to obtuse, with entire or variously dentate to crenate margin. Flowers sessile to pedicellate. Calyx 5-lobed almost to base, upper lobe largest, ovoid, lateral two lobes usually smallest, linear to lanceolate. Corolla white to bluish-violet, subrotate, 2-lipped, tubular or campanulate, subequally 5-lobed. Stamens 4, included. Ovary ellipsoid to globose or ovoid, glabrous. Seeds numerous, testa coarsely reticulate, with longitudinal ridges. $2n = 68, 80$. About 60 species, pantropical.

107. *Maeviella* Rossow

Maeviella Rossow, Arq. Jard. Bot. Rio de Janeiro 27: 172 (1985).

Annual herb, glandular-pubescent. Stems erect, weakly quadrangular. Leaves cauline, opposite, sessile, linear-lanceolate, acuminate, with entire or weakly serrate margin. Flowers distinctly pedicellate. Calyx 5-lobed almost to the base, dorsal lobe cochleariform, much larger than the lanceolate ventral lobes, lateral two lobes smallest. Corolla violet, 2-lipped. Stamens 4, included. Ovary ellipsoid, glabrous. Seeds numerous, ellipsoid, testa coarsely reticulate. One species, *M. cochlearia* (Huber) Rossow, South America.

108. *Boelckea* Rossow

Boelckea Rossow, Parodiana 7(1–2): 15–24 (1992).

Annual herb, pubescent. Stems erect, weakly quadrangular. Leaves cauline, opposite, sessile, ovoid to ovoid-lanceolate, acuminate, upper side pubescent, lower side glandular-pubescent, with entire margin. Flowers sessile to subsessile. Calyx 5-lobed almost to base, with unequal sepals, upper lobe largest, ovoid to broadly ovoid, lateral two lobes usually smallest, linear to lanceolate. Corolla white with violet marks, dorsally concave, lobes obovoid. Stamens 4, included. Ovary ellipsoid, glabrous. Seeds numerous, testa coarsely reticulate. One species, *B. beckii* Rossow, Bolivia.

109. *Benjaminia* Mart. ex Benj.

Benjaminia Mart. ex Benj. in Mart., Fl. Brasil. 10: 255 (1847).

Annual aquatic herb, submerged, generally only inflorescence emerged, glabrous. Stems erect, cylindrical. Leaves verticillate, sessile, pinnatifid with filiform segments, acuminate, with entire margin. Flowers distinctly pedicellate, pendant to reflexed in maturity. Calyx 5-lobed almost to base, lobes triangular obtuse. Corolla violet with yellow mark at throat, 2-lipped, tube long, often dilated at throat, lobes spreading, rounded. Stamens 4, included, filaments twisted. Ovary conical-ovoid. Seeds numerous, testa reticulate. One species, *B. reflexa* (Benth.) D'Arcy, from Venezuela to Brazil.

110. *Braunblanquetia* Eskuche

Braunblanquetia Eskuche, Bol. Soc. Argent. Bot. 15(4): 357 (1974).

Annual herb, glabrous. Stems prostrate, rooting at the nodes, rounded. Leaves cauline, opposite, sessile, subcarnose, oblong-ovoid, attenuate, base decurrent and semiamplexicaulous, with entire margin. Flowers sessile. Calyx campanulate, 5-partite to middle, lobes triangular-obtuse. Corolla white, subrotate with cylindrical tube, 2-lipped, lobes rounded, subequal. Stamens 2, anthers with two separate thecae held apart by a connective with 2 short arms, staminodes 2, abaxial, short and filiform. Ovary ellipsoid. Seeds numerous, ovoid-oblong, testa reticulate. One species, *B. litoralis* Eskuche, Argentina (Patagonia).

111. *Sophronanthe* Benth.

Sophronanthe Benth. in Lindl., Introd. Nat. Syst. ed. 2: 445 (1836).

Perennial herb, pubescent. Stems ascending or erect, quadrangular. Leaves cauline, opposite, sessile, linear, rigid, with entire revolute margin. Flowers distinctly pedicellate. Calyx 5-lobed almost to base, lobes narrow, subequal. Corolla white, 2-lipped, salverform, tube very narrow, subulate, lobes rounded. Stamens 2, included, adaxial, abaxial pair reduced to filiform staminodes or lacking, connective not rounded. Ovary ovoid to ellipsoid. Seeds numerous, testa striate and transversally reticulate. One species, *S. hispida* Benth., south-eastern North America.

112. *Gratiola* L.

Gratiola L., Sp. Pl. 1: 17 (1753).
Fonkia Phil. (1865).
Tragiola Small & Pennell (1933).

Annual or perennial herbs, glabrous or viscid-pubescent. Stems prostrate to ascending or erect, quadrangular. Leaves cauline, opposite, sessile, linear-lanceolate to ovoid, acuminate, with dentate margin. Flowers distinctly pedicellate. Calyx 5-lobed almost to base, lobes narrow, subequal. Corolla white or yellow, 2-lipped, tube cylindrical, lobes rounded. Stamens 2, included, adaxial, abaxial pair reduced to filiform staminodes or lacking, connective rounded. Ovary ovoid to ellipsoid. Seeds numerous, testa striate and transversally reticulate. $2n = 16, 32$. About 20 species, North America, Europe, Asia, tropical mountains, lacking in Africa.

113. *Mecardonia* Ruiz & Pav.

Mecardonia Ruiz & Pav., Prodr.: 95 (1794); Pennell, Proc. Acad. Nat. Sci. Philad. 98: 83 (1946), reg. rev.
Pagesia Raf. (1817).

Perennial herbs, glabrous. Stems prostrate to ascending, quadrangular. Leaves cauline, opposite, shortly petiolate, obovoid, obtuse to rounded apically, with crenate, weakly revolute margin. Flowers distinctly pedicellate. Calyx 5-lobed almost to base, the outer 3 sepals distinctly broader than the inner 2. Corolla white or yellow, 2-lipped, tube cylindrical. Stamens 4, included, anthers with two separate thecae, held apart by a connective with 2 short arms. Ovary ellipsoid. Seeds numerous, cylindrical, testa reticulate. $2n = 22, 44$. Fifteen species in temperate and tropical America.

114. *Scoparia* L.

Scoparia L., Sp. Pl. 1: 116 (1753).

Annual or perennial herbs or subshrubs, pubescent or glabrous. Stems erect, terete. Leaves cauline, opposite or verticillate, sessile, lanceolate to oblanceolate, acuminate, with entire to dentate or serrate margin. Inflorescence terminal, racemose or with axillary clusters of 2 (rarely more) flowers produced by lateral buds. Flowers distinctly pedicellate. Calyx 4–5-lobed, lobes narrowly lanceolate to ovoid, acuminate, subequal. Corolla white, campanulate, subrotate, the 4 lobes plane, obtuse, lobes much longer than the tube, throat pubescent inside. Stamens 4, exserted. Ovary ovoid, emarginate at apex. Seeds numerous, testa reticulate. $2n = 20, 40$. 20 species, mainly in tropical America, one species (*S. dulcis* L.) pantropical.

2B. Subtribe "Caprariinae"

Caprarieae Barringer (1993).

Herbs, leaves alternate, inflorescence terminal, racemose or with axillary clusters of 2 (rarely more) flowers produced by lateral buds, lax, frondose, prophylls absent. Stamens 4–5, anthers with two equal, divergent confluent thecae, capsule loculicidal.

115. *Capraria* L.

Capraria L., Sp. Pl. 2: 628 (1753).

Annual or perennial herbs or subshrubs, pilose-pubescent throughout, rarely glabrous. Stems erect, much branched, terete. Leaves cauline, sessile or shortly petiolate, lanceolate to oblanceolate, acuminate, with dentate or serrate margin. Flowers distinctly pedicellate. Calyx 5-lobed, lobes narrowly lanceolate, acuminate, subequal. Corolla white, campanulate, subrotate, the 5 lobes plane, lobes about as long as the tube. Stamens 4–5, exserted. Ovary ovoid, emarginate at apex. Stigma bilobed. Seeds numerous, testa reticulate. $2n = 60$. Four species in tropical America.

2c. Subtribe "Dopatriinae"

Aquatic herbs, inflorescence racemose, prophylls absent, corolla 2-lipped, stamens 2, the abaxial pair usually reduced to staminodes or lacking, anthers with two separate, parallel, pilose, or glabrous (*Limnophila*) thecae, held apart by a connective with 2 short arms, or anthers monothecous (*Philcoxia*), capsule loculicidal (*Deinostema*, *Hydrotriche*) or loculicidal to septicidal (*Dopatrium*, *Limnophila*, *Philcoxia*), seeds reticulate or smooth (*Limnophila*).

116. *Deinostema* T. Yamaz.

Deinostema T. Yamaz., J. Jap. Bot. 28: 131 (1953).

Annual herbs, glabrous or glandular-pubescent. Stems erect, weakly quadrangular. Leaves cauline, opposite, sessile, linear-lanceolate to ovoid, acuminate with entire margin. Inflorescence with chasmogamous and cleistogamous flowers. Chasmogamous flowers distinctly pedicellate, cleistogamous flowers sessile, often fasciculate in leaf axils. Calyx campanulate, 5-lobed almost to base, lobes linear-lanceolate, acuminate, subequal. Corolla purple to pale purple, 2-lipped, lobes rounded. Stamens 2, included, adaxial, abaxial pair reduced to filiform staminodes. Ovary ovoid-globose. Stigma bilobed. Two species, *D. adenocaulum* (Maxim.) T. Yamaz., *D. violaceum* (Maxim.) T. Yamaz., East Asia.

117. *Dopatrium* Buch.-Ham. ex Benth.

Dopatrium Buch.-Ham. ex Benth. in Lindl., Edwards's Bot. Reg. ad t. 1770(4) (1835); Fischer, Nord. J. Bot. 17: 527–557 (1996), rev.

Annual herbs, glabrous or rarely glandular-pubescent. Stems erect, rounded. Leaves opposite, sessile, basal leaves in a dense rosette, linear-lanceolate to ovoid, acuminate to weakly obtuse, entire, cauline leaves much reduced in size. Some species with additional sessile cleistogamous flowers. Flowers distinctly pedicellate. Calyx 5-lobed almost to middle, lobes narrow, obtuse. Corolla rose to violet or yellow, 2-lipped, all lobes spreading, rounded. Stamens 2, included, adaxial pair reduced to minute filiform staminodes or lacking. Ovary ovoid to globose. Stigma weakly bilobed, somewhat peltate. $2n = 14, 28$. Twelve species, tropical Africa, India and Sri Lanka, Laos to Vietnam, one species (*D. junceum* (Roxb.) Buch.-Ham. ex Benth.) to southern Russia. Most species confined to rock pools.

118. *Hydrotriche* Zucc.

Hydrotriche Zucc., Abh. Math.-Phys. Cl. Königl. Bayer. Akad. Wiss. 1: 308 (1832); Raynal-Roques, Adansonia II, 19: 145–173 (1979), rev.

Annual or perennial submerged aquatics, generally only inflorescence emerged, glabrous. Stems cylindrical. Leaves verticillate, sessile, filiform to linear-lanceolate, acuminate, with entire margin. Chasmogamous flowers distinctly pedicellate, pendant at maturity, cleistogamous flowers sessile or shortly pedicellate. Calyx 5-lobed, lobes free almost to one third of their length, lobes triangular, obtuse. Corolla violet with yellow mark at throat, 2-lipped, all lobes spreading, rounded. Stamens 2, included, adaxial pair reduced to minute clavate staminodes. Ovary conical-ovoid. Stigma weakly 4-lobed, somewhat peltate. Four species, Madagascar, aquatics in ponds, rock pools and rice-fields.

119. *Limnophila* R. Br.

Limnophila R. Br., Prodr.: 442 (1810), nom. cons.; Philcox, Kew Bull. 24: 101–170 (1970), rev.; Yamazaki, J. Fac. Sci. Univ. Tokyo 13: 575–625 (1985), reg. rev.
Ambulia Lam. (1783).
Stemodiacra Kuntze (1891).
Bonnayodes Blatt. & Hallb. (1921).

Annual or perennial aquatic to rarely terrestrial herbs, glabrous or pubescent. Stems prostrate, creeping or erect, quadrangular. Leaves dimorphic, when submerged laciniate, pinnatifid to capillary-multifid, aerial emerged leaves opposite to verticillate, sessile or petiolate, linear-lanceolate to ovoid, with entire to serrate or pinnately divided margin. Some species with sessile cleistogamous flowers. Chasmogamous flowers sessile to pedicellate. Calyx tubular, 5-lobed, lobes subequal or adaxial lobes enlarged. Corolla tubular or infundibuliform, 2-lipped. Stamens 4, included, abaxial pair sometimes reduced to staminodes. Ovary ovoid to ellipsoid. Stigma bilobed. $2n = 30$, 34, 68. Thirty-six species, tropical Africa and Asia.

120. *Philcoxia* P. Taylor & V.C. Souza

Philcoxia P. Taylor & V.C. Souza, Kew Bull. 55: 159 (2000).

Annual or perennial herbs, glandular-pubescent. Stems erect, rounded. Leaves with circinnate vernation, basal leaves petiolate, in a dense rosette, peltate, lamina orbicular to reniform, densely glandular-pubescent on upper surface, glabrous below, with entire margin, cauline leaves lanceolate, much reduced in size, distant. Flowers distinctly pedicellate. Calyx 5-lobed almost to middle, lobes narrow, obtuse. Corolla rose to violet, 2-lipped, all lobes spreading, rounded. Stamens 2, included, adaxial pair lacking. Ovary ovoid to globose. Stigma clavate, at dilated apex of style. Three species from Brazil.

3. Tribe Limoselleae Dumort. (1827).

Limosellaceae J. Agardh (1858).
Limoselleae Rouy (1909).

Usually rosulate herbs, inflorescence racemose, frondose, usually comprising the whole rosette, prophylls absent, corolla subrotate, ovary unilocular, weakly bilocular at base, capsule loculicidal.

121. *Limosella* L.

Limosella L., Sp. Pl. 2: 631 (1753); Glück, Bot. Jahrb. Syst. 66: 488–566 (1934), rev.

Annual or perennial herbs, glabrous. Stems short, reduced, prostrate to ascending, weakly quadrangular, slender creeping stolons arising from leaf axils. Leaves rosulate or rarely cauline, alternate, sessile or petiolate, lamina subulate-filiform, linear, spathulate or oblong-ovoid, obtuse with entire margin. Flowers distinctly pedicellate. Calyx campanulate, (4–)5-lobed, the lobes usually shorter than or equal to the calyx tube. Corolla white or whitish with blue to mauve spots, subrotate, limb 5-lobed. Stamens 4, exserted. Ovary globose to ovoid. Stigma bilobed. Seeds numerous, testa reticulate-striate. $2n = 20$, 40, 60. Fifteen species, temperate Europe, Asia, North America, tropical mountains, 10 species in Africa.

4. Tribe Stemodieae Benth. (1876)

Annual or perennial herbs to small shrubs, inflorescence racemose or rarely cymose (*Stemodia* p.p.), prophylls absent or present (*Morgania*, *Conobea*, *Stemodiopsis*), corolla strongly 2-lipped, palate often present, stamens usually 4, rarely 2, anthers with 2 separate thecae held apart by a stipitate connective, stigma bilobed or capitate (*Schistophragma*), capsule usually ovoid or elliptic, rarely linear (*Schistophragma*), loculicidal and septicidal or only septicidal (*Otacanthus*, *Schistophragma*, *Leucospora* to *Stemodiopsis*), seeds with longitudinal ridges, testa cells with hook-like wall thickenings.

122. *Lindenbergia* Lehm.

Lindenbergia Lehm., Sem. Horto Bot. Hamburg 1829: 6, 8 (1829); Hjertson, Bot. J. Linn. Soc. 119: 265–321 (1995), rev.
Brachycoris Schrader (1830).
Omania S. Moore (1901).

Annual or perennial herbs, pubescent to densely villose. Stems prostrate to ascending or erect, terete. Leaves cauline, opposite, subsessile or petiolate, lanceolate to ovoid or orbicular, acuminate with dentate to serrate margin. Flowers sessile to shortly pedicellate. Calyx campanulate, 5-lobed, lobes divided to middle. Corolla yellow, strongly 2-lipped, mouth of corolla with a palate. Stamens 4, included, one theca sometimes smaller and reduced. Ovary ovoid to globose. Seeds numerous, testa with longitudinal ridges, endosperm smooth or ribbed. $2n = 28$, 30, 32. Twelve species, tropical Northeast Africa, Arabia, tropical Asia. *Lindenbergia* is placed by Young et al. (1999) and Olmstead et al. (2001) in "Orobanchaceae" based on molecular data. This is not supported by morphology, except for the variable rhinantoid aestivation.

123. *Adenosma* R. Br.

Adenosma R. Br., Prodr.: 442 (1810).
Pterostigma Benth. (1846).

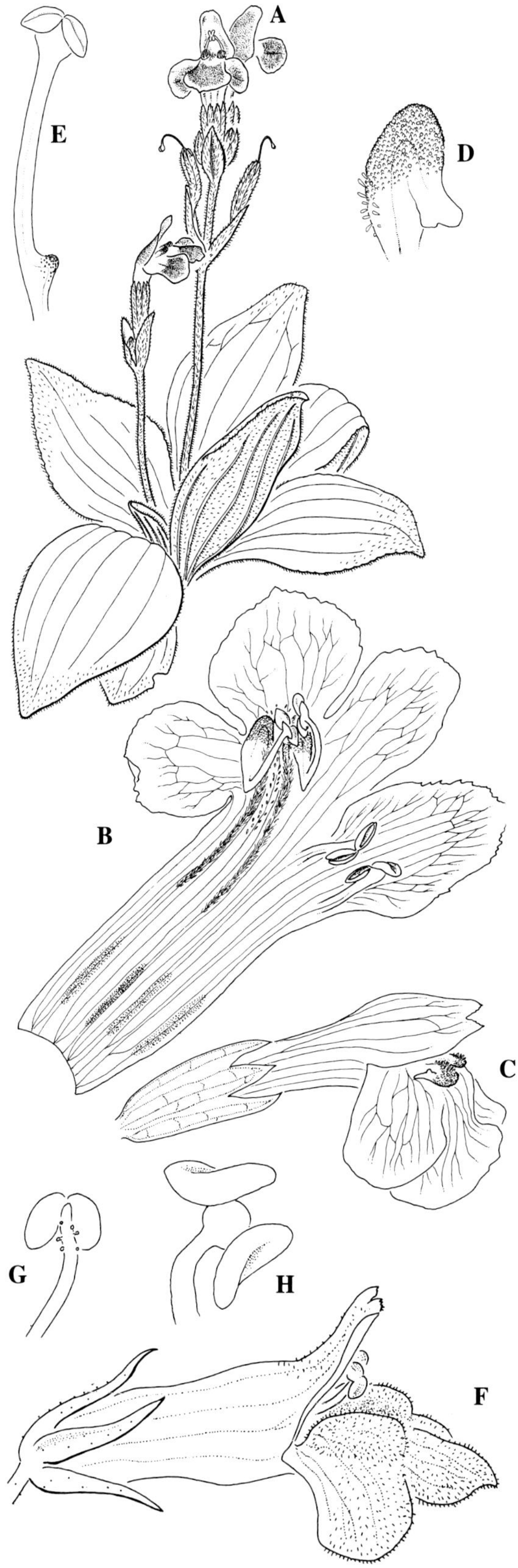

Annual or perennial herbs, glandular-pubescent to densely villose. Stems prostrate to ascending or erect, quadrangular or rounded. Leaves cauline, opposite, sessile and semiamplexicaulous to subsessile or petiolate, lanceolate, ovoid to orbicular, acuminate, with serrate margin. Flowers sessile to shortly pedicellate. Calyx irregularly 5-lobed, outer 3 lobes large, widely lanceolate to cordate-ovoid, inner 2 lobes small, linear-lanceolate. Corolla pale purple, white, mauve to blue or violet, strongly 2-lipped. Stamens 4, included, one theca of all anthers or of the 2 abaxial anthers sterile and reduced. Ovary ovoid, conical to elliptic. Seeds numerous, testa reticulate. Fifteen species, China, Indomalesia, Australia.

124. *Otacanthus* Lindl.

Otacanthus Lindl., Fl. Serres Jard. Eur. 15: 55 (1862); Ronse Decraene, Doctoraatsproefschrift: 228 Leuven (1993), rev.
Tetraplacus Radlk. (1885).

Perennial suffrutescent herbs or shrubs, glandular-pubescent. Stems ascending to erect, quadrangular. Leaves cauline or in a basal rosette, opposite, sessile or petiolate, lanceolate to obovoid, leaf-base connate-amplexicaulous, with entire to dentate-serrate margin. Flowers subsessile to distinctly pedicellate. Calyx deeply 5-lobed, the adaxial sepal ovoid and larger than the other 4 lanceolate sepals. Corolla blue to violet, whitish near throat, on lower lip hypocrateriform, strongly 2-lipped, lips nearly equal and undivided, lower lip with a swollen palate, tube cylindrical, long, slender and curved. Stamens 2, included, adaxial pair reduced to long staminodes. Ovary ovoid. Seeds numerous, truncate to ovoid, testa reticulate to minutely pitted. Six species, Brazil, one (*O. caeruleus* Lindl.) widely naturalized (Madagascar, Mascarenes).

125. *Achetaria* Cham. & Schltdl.

Achetaria Cham. & Schltdl., Linnaea 2: 566 (1827).
Beyrichia Cham. & Schltdl. (1828).

Perennial to suffrutescent herbs, glandular-pubescent to densely pilose. Stems erect, quadrangular. Leaves cauline, opposite, petiolate, ovoid,

Fig. 46. Scrophulariaceae. **A, B** *Craterostigma plantagineum*. **A** Habit. **B** Corolla dissected. **C, D** *Lindernia philcoxii*. **C** Flower. **D** Staminode. **F, G** *Stemodiopsis ruandensis*. **F** Flower. **G** Stamen. **H** *Stemodia serrata*. Stamen. (A–D After Fischer 1992, redrawn, **F, G** After Fischer 1997a, redrawn, **H** After Fischer 1999, redrawn)

acuminate, with serrate to crenate margin, often revolute. Flowers sessile. Calyx deeply lobed, campanulate, with 4 narrow and 1 broad sepal. Corolla blue to violet, strongly 2-lipped. Stamens 2, included, the adaxial pair reduced to staminodes. Ovary ovoid to globose. Seeds numerous, testa reticulate. Five species, Brazil, Caribbean.

126. *Tetraulacium* Turcz.

Tetraulacium Turcz., Bull. Soc. Imp. Nat. Moscou 16: 53 (1843).

Annual herb, densely pilose. Stems prostrate, quadrangular. Leaves cauline, opposite or verticillate, sessile, lanceolate-ovoid, acuminate, with irregularly serrate margin. Flowers distinctly pedicellate. Calyx 5-lobed, campanulate, sepals unequal, 4 narrow and 1 broad. Corolla bluish, strongly 2-lipped. Stamens 4, included, one theca reduced. Ovary globose, with 4-winged style. Seeds numerous. One species, *T. veronicaeforme* Turcz., Brazil.

127. *Dizygostemon* (Benth.) Radlk. ex Wettst.

Dizygostemon (Benth.) Radlk. ex Wettst. in Engler & Prantl, Nat. Pflanzenfam. IV, 3b: 74 (1891).

Annual herbs, pubescent to densely pilose. Stems erect, quadrangular. Leaves cauline, opposite, shortly petiolate, linear-lanceolate, acuminate, with serrate, revolute margin. Flowers shortly pedicellate. Calyx deeply 5-lobed, campanulate, adaxial sepal larger. Corolla blue to violet, strongly 2-lipped. Stamens 4, included, abaxial pair with monothecous anthers, anthers of adaxial pair with two separate thecae, one theca sterile and reduced. Ovary ovoid. Seeds numerous, testa reticulate. Two species, *D. floribundum* (Benth.) Radlk. and *D. angustifolium* Giul., Brazil.

128. *Stemodia* L. Fig. 46H

Stemodia L., Syst. Nat. ed. 10: 1091, 1118, 1374 (1794), nom. cons.; Minod, Contr. Étude genre *Stemodia*, Thèse Univ. de Genève: 5–103 (1918), rev.; Turner & Cowan, Phytologia 74: 61–103 (1993), reg. rev.
Poarium Desvaux (1825).
Choadophyton Minod (1918).
Valeria Minod (1918).
Verena Minod (1918).
Lendneria Minod (1918).
Anamaria V.C. Souza (2001).

Annual or perennial herbs, glabrous or glandular-pubescent to densely villose. Stems prostrate to ascending or erect, quadrangular. Leaves cauline, opposite or verticillate, sessile to subsessile or petiolate, lanceolate, ovoid or orbicular, acuminate, with subentire to variously dentate or serrate margin. Flowers pedicellate. Calyx deeply 5-lobed, campanulate, lobes narrow, equal or subequal. Corolla white to bluish, strongly 2-lipped, tube cylindrical. Stamens 4, included. Ovary ovoid to globose. Seeds numerous, testa with longitudinal ridges. $2n = 28$. Fifty-six species, tropical America, Asia, Africa.

129. *Darcya* B.L. Turner & Cowan

Darcya B.L. Turner & Cowan, Phytologia 74: 268 (1993).
Stemodia L., p.p.

Suffruticose perennial herbs, glabrous or glandular-pubescent. Stems ascending or erect, quadrangular. Leaves cauline, opposite, petiolate, ovoid or triangular-ovoid, with serrulate margin. Flowers pedicellate. Calyx deeply 5-lobed, campanulate, lobes narrow, equal or subequal. Corolla white to blue, strongly 2-lipped, tube cylindrical. Stamens 4, included. Ovary ovoid to globose. Seeds numerous, testa alveolate-reticulate. Three species, Costa Rica and Panama to Colombia.

130. *Cheilophyllum* Pennell ex Britton

Cheilophyllum Pennell ex Britton, Mem. Torrey Bot. Club 16: 103 (1920).

Annual herbs, glandular-pubescent. Stems prostrate to ascending, quadrangular, winged. Leaves cauline, opposite, sessile, ovoid, acuminate, with dentate margin. Flowers pedicellate. Calyx deeply 5-lobed, campanulate. Corolla white, strongly 2-lipped, tube cylindrical. Stamens 4, included. Ovary ovoid. Seeds numerous, testa with longitudinal ridges. Eight species, Caribbean.

131. *Morgania* R. Br.

Morgania R. Br., Prodr.: 441 (1810).

Perennial herbs or subshrubs, glabrous or glandular-pubescent. Stems erect, quadrangular. Leaves cauline, opposite or rarely verticillate, sessile, linear-lanceolate, acuminate, with entire to dentate margin. Flowers sessile or pedicellate. Calyx deeply 5-lobed, campanulate. Corolla blue, strongly 2-lipped. Stamens 4, included, filaments of the abaxial pair geniculate. Ovary ovoid. Seeds numerous, testa with longitudinal ridges. Four species, Australia.

132. *Schistophragma* Benth. ex Endl.

Schistophragma Benth. ex Endl., Gen.: 679 (1839).

Annual herbs, glabrous. Stems ascending to erect, quadrangular, winged. Leaves cauline, opposite, sessile, linear, acuminate, simple or pinnatisect, segments usually linear. Flowers distinctly pedicellate. Calyx deeply 5-lobed, the lobes linear. Corolla violet, tubular, strongly 2-lipped. Stamens 4, included. Ovary linear. Capsule linear. Seeds numerous, testa with spirally arranged longitudinal ridges. $2n = 40$. Two species, western North America to Central America.

133. *Conobea* Aubl.

Conobea Aubl., Hist. Pl. Guiane: 639 (1775).

Annual or perennial herbs, glabrous or pubescent. Stems erect, quadrangular. Leaves cauline, opposite, sessile or petiolate, linear-lanceolate or ovoid, acuminate, with dentate margin. Flowers distinctly pedicellate. Calyx deeply 5-lobed. Corolla blue, strongly 2-lipped. Stamens 4, included. Ovary ovoid or globose. Seeds numerous, testa with longitudinal ridges. Seven species, tropical America.

134. *Leucospora* Nutt.

Leucospora Nutt., J. Acad. Philad. 7: 87 (1834).

Annual herb, pubescent throughout. Stems prostrate at base, erect, quadrangular. Leaves cauline, opposite, petiolate, triangular-ovoid in outline, deeply pinnatifid with 3–7 linear-oblong segments, with entire margin. Flowers pedicellate. Calyx deeply 5-lobed, lobes linear. Corolla pale lavender, tubular, strongly 2-lipped. Stamens 4, included. Ovary ovoid. Seeds numerous, testa with longitudinal ridges. One species, *L. multifida* (Michx.) Nutt., eastern North America.

135. *Schizosepala* G.M. Barroso

Schizosepala G.M. Barroso, Arquiv. Jard. Bot. Rio de Janeiro 14: 259 (1956).

Annual (?) herb, glandular-pubescent. Stems erect. Leaves cauline, opposite, petiolate, lanceolate in general outline, deeply pinnatifid. Flowers pedicellate. Calyx deeply 5-lobed, number of lobes obscured by multisect lobes with unequal segments, campanulate. Corolla yellow, strongly 2-lipped, lobes with crisp, irregularly incised margins. Stamens 4, included. Ovary ovoid. One species, *S. glandulosa* G.M. Barroso, Matto Grosso, Brazil.

136. *Stemodiopsis* Engl. Fig. 46F, G

Stemodiopsis Engl., Annuario Reale Ist. Bot. Roma 7: 25 (1897), Bot. Jahrb. Syst. 23: 497 (1897); Eb. Fischer, Bot. Jahrb. Syst. 119: 305–326 (1997), rev.

Perennial subshrubs, glabrous or densely pubescent. Stems prostrate to ascending-erect, often pending, terete. Leaves cauline, opposite, sessile to petiolate, linear-filiform, lanceolate to ovoid, acuminate to obtuse, with entire, dentate or crenate margin. Flowers distinctly pedicellate, pedicel curved and reflexed in fruit. Calyx 5-lobed, lobes longer than tube, subequal. Corolla white, strongly 2-lipped, with distinct palate, tube cylindrical. Stamens 4, included, abaxial longer filaments curved or twisted. Ovary conical, elongate, attenuate into slender curved style. Seeds numerous, testa with longitudinal ridges, endosperm smooth or ribbed. Six species, tropical Africa, confined to fissures in rock outcrops.

5. Tribe Lindernieae Rchb. (1831).

Linderniinae Benth. in DC. (1846).
Vandellieae Benth. (1876).
Torenieae Rouy (1909).

Herbs to suffrutices, some taxa poikilohydric, inflorescence racemose, terminal or truncate with axillary racemes (*Chamaegigas*, *Lindernia* p.p., *Craterostigma*), prophylls absent, corolla strongly 2-lipped, with glandular hairs on inside, stamens 4, filaments of the abaxial pair geniculate, Z-shaped or with spur-like appendages or 2, the abaxial pair reduced to filiform or geniculate staminodes (*Picria*, *Scolophyllum*, *Hemiarrhena*, *Lindernia* p.p., *Bampsia* p.p.), anthers touching in pairs, with two equal and spreading thecae, unequal mucronate thecae or monothecous anthers in *Torenia* p.p. and *Hemiarrhena*, stigma bilobed, capsule septicidal, seeds with alveolated (bothrospermous, aulacospermous) endosperm or not alveolated.

137. *Artanema* D. Don

Artanema D. Don in Sweet, Brit. Fl. Gard. II, 3: t. 234 (1834), nom cons.
Ourisianthus Bonati (1924).

Perennial herbs, glabrous. Stems erect, quadrangular, ribbed. Leaves cauline, opposite, sessile, semi-amplexicaulous, lanceolate or oblong-lanceolate, acuminate, with distantly minutely serrate margin. Flowers distinctly pedicellate. Calyx 5-lobed campanulate. Corolla purple to violet, strongly 2-lipped, tube long, cylindrical. Stamens 4, didynamous, exserted, filaments of the abaxial pair geniculate, each with double geniculate projections, connective of lower theca forming long spur. Ovary globose. Seeds bothrospermous. $2n = 40$. Four species, tropical Africa (*A. longifolium* (L.) Vatke), Asia.

138. *Picria* Lour.

Picria Lour., Fl. Cochinch.: 359, 392 (1790).
Caranga Juss. (1804).
Curanga Juss. (1807).

Annual herb, densely minutely pubescent. Stems prostrate to ascending, quadrangular. Leaves cauline, opposite, petiolate, ovoid, chartaceous, acuminate, with crenate margin. Flowers pedicellate. Calyx flat, of 4 separated sepals, arranged in two series, upper outer sepal largest, cordate, entire, lower outer sepal 2-lobed, weakly smaller than the upper, lateral inner pair smallest, filiform. Corolla red-brown, tubular, strongly 2-lipped. Stamens 2, included, abaxial pair reduced to staminodes, each with a rudimentary anther. Ovary deltate-ovoid. Seeds aulacospermous. One species, *P. fel-terrae* Lour., Indomalesia.

139. *Pierranthus* Bonati

Pierranthus Bonati, Bull. Soc. Bot. Genève II, 4: 254 (1912).

Annual herb, sparsely hirsute. Stems erect, quadrangular. Leaves cauline, opposite, sessile, linear-filiform, acuminate, with entire margin. Inflorescence capitate, bracts rounded-rhombic with distinct long apex. Flowers sessile. Calyx tubular, deeply 5-lobed, lobes thinly membranous. Corolla yellow, strongly 2-lipped, tube cylindrical. Stamens 4, exserted, filaments of abaxial pair geniculate. Ovary obliquely ellipsoid. Seeds bothrospermous. One species, *P. capitatus* (Bonati) Bonati, Southeast Asia.

140. *Schizotorenia* T. Yamaz.

Schizotorenia T. Yamaz., J. Jap. Bot. 53: 101 (1978).

Perennial herbs, glabrous except for few scattered hairs. Stems prostrate, quadrangular, rooting at the nodes. Leaves cauline, opposite, petiolate, lamina deltate-ovoid, acuminate, with serrate margin. Flowers distinctly pedicellate. Calyx deeply 5-lobed almost to base, lobes linear-lanceolate, cylindrical. Corolla dark purple, large, strongly 2-lipped, tube cylindrical. Stamens 4, exserted, filaments of abaxial pair geniculate, with spur-like appendages. Ovary obliquely oblong, glabrous. Seeds aulacospermous. Two species, *S. atropurpurea* (Ridley) T. Yamaz., *S. finetiana* (Bonati) T. Yamaz., Southeast Asia.

141. *Legazpia* Blanco

Legazpia Blanco, Fl. Filip. ed. 2: 338 (1845).

Perennial herb, glabrous except for pubescent calyx. Stems prostrate, quadrangular, rooting at the nodes. Leaves cauline, opposite, petiolate, lamina ovoid or orbicular-ovoid, acuminate, with serrate margin. Inflorescence with lateral paracladia. Flowers distinctly pedicellate. Calyx suborbicular, with 3 broad wings, 3-toothed at apex, shortly auriculate at base. Corolla yellow, strongly 2-lipped, tube cylindrical. Stamens 4, filaments of abaxial pair geniculate, with clavate appendages. Ovary oblong, asymmetrical, glabrous. Seeds bothrospermous. One species, *L. polygonoides* (Benth.) T. Yamaz., East Asia, Micronesia, New Guinea.

142. *Torenia* L.

Torenia L., Sp. Pl. 2: 619 (1753); Yamazaki, J. Fac. Sci. Univ. Tokyo 13: 575–625 (1985), reg. rev.; Fischer, Trop. Subtrop. Pflanzenw. 81: 332–348 (1992), reg. rev.

Annual or perennial herbs, usually glabrous, rarely pubescent. Stems prostrate to ascending or erect, quadrangular. Leaves cauline, opposite, subsessile to petiolate, lamina lanceolate-ovoid to orbicular, acuminate or obtuse, with dentate to crenate margin. Flowers distinctly pedicellate. Calyx 5-lobed, fused to middle or above, sepals with distinct wings on midvein, accrescent. Corolla white to cream with blue spots, yellow with mauve spots, purple or violet to blue, strongly 2-lipped, tube cylindrical. Stamens 4, exserted, filaments of abaxial pair geniculate, with clavate appendages or Z-shaped, thecae rounded to weakly mucronate, abaxial anthers sometimes with two very unequal, mucronate thecae or monothecous. Ovary oblong-ovoid, asymmetrical, apex pilose, locules with hairs inside. Capsule poricidal. Seeds bothrospermous. $2n = 16, 18, 34$. Fourty species, Southeast Asia, Africa, Madagascar, one species (*T. thouarsii* (Cham. & Schtdl.) Kunze) introduced in tropical America.

143. *Scolophyllum* T. Yamaz.

Scolophyllum T. Yamaz., J. Jap. Bot. 53: 98 (1978).

Perennial herbs, glabrous. Stems ascending or erect, quadrangular. Leaves cauline, opposite, sessile, simple or pinnatifid, lanceolate, acuminate with aristate teeth on margin. Flowers distinctly pedicellate. Calyx 5-lobed nearly to middle, lobes linear-lanceolate, setose. Corolla pink to mauve, strongly 2-lipped, tube slender and cylindrical. Stamens 2, abaxial pair reduced to shortly clavate staminodes, adaxial anthers with two subequal to unequal thecae. Ovary obliquely cylindrical, asymmetrical. Seeds bothrospermous. Two species, *S. ilicifolium* (Bonati) T. Yamaz., *S. spinifidum* (Kerr ex Barnett) T. Yamaz., Southeast Asia.

144. *Hemiarrhena* Benth.

Hemiarrhena Benth., Fl. Austral. 4: 518 (1868).

Perennial (?) herb, glabrous. Stems erect, weakly quadrangular. Leaves in a basal rosette, opposite, sessile or subsessile, ovoid or broadly oblong, obtuse, with entire margin. Flowers pedicellate. Calyx 5-lobed nearly to base, lobes acuminate, membranaceous. Corolla blue to violet, strongly 2-lipped, tube slender and cylindrical. Stamens 2, abaxial pair reduced to filiform staminodes, anthers with two unequal mucronate thecae or monothecous. Ovary ovoid. Seeds bothrospermous. One species, *H. plantaginea* Benth. (= *Lindernia plantaginea* (Benth.) F. Mueller), tropical Northwest Australia.

145. *Chamaegigas* Dinter

Chamaegigas Dinter, Ber. Tätigk. naturw. Ges. Isis 1913–1915: 38 (1916); Feddes Repert. Beih. 3: 123, 168 (1928); Fischer, Trop. Subtrop. Pflanzenw. 81: 328–332 (1992), rev.

Perennial herb, glabrous except for some glandular hairs. Stems ascending, condensed, forming a plagiotropous rhizome. Leaves in a basal rosette, opposite, sessile, lanceolate, 2-lobed with two unequal apices by connation of the two leaves of one node. Inflorescence lateral, with axillary racemes, with two pairs of bracts at nearly the same level, each pair with one flower floating on the water surface. Flowers subsessile to shortly pedicellate. Calyx 5-lobed, divided to above middle, campanulate. Corolla white with blue spots on lower lip, strongly 2-lipped, tube short, cylindrical. Stamens 4, exserted, filaments of the abaxial pair geniculate. Ovary globose. Seeds oblong, testa reticulate, longitudinally ribbed, endosperm smooth. One species, *C. intrepidus* Dinter, Namibia, growing as a poikilohydric plant in seasonally water-filled rock pools.

146. *Lindernia* All. Fig. 42C, 46C, D

Lindernia All., Melanges Philos. Math. Soc. Roy. Turin 3: fig. 5 (1766); Philcox, Kew Bull. 22: 1–72 (1968), reg. rev.; Yamazaki, J. Fac. Sci. Univ. Tokyo 13: 1–64 (1981), reg. rev.; Fischer, Trop. Subtrop. Pflanzenw. 81: 214–328 (1992), reg. rev.
Vandellia L. (1767).
Bonnaya Link & Otto (1820).
Ilysanthes Raf. (1820).
Tittmannia Rchb. (1824).
Strigina Engl. (1897).
Trichotaenia T. Yamaz. (1953).

Annual or perennial herbs, subshrubs or dwarf shrubs. Stems prostrate to ascending or erect, quadrangular. Leaves cauline or in a basal rosette, opposite, sessile to subsessile or petiolate, linear-lanceolate, spathulate, ovoid or orbicular, acuminate or obtuse. Inflorescence terminal or lateral with axillary racemes. Flowers subsessile to pedicellate, in some species sessile cleistogamous flowers present. Calyx 5-lobed, campanulate, deeply divided or only weakly dentate, in some species winged. Corolla white to cream with blue, rarely entirely blue, violet, mauve or purple, strongly 2-lipped, tube cylindrical. Stamens 4, exserted, or 2, the abaxial pair then reduced to geniculate or filiform staminodes, in some species staminodes with rudimentary anthers. Ovary ovoid to globose. Seeds bothrospermous or with smooth endosperm. $2n = 14, 16, 18, 24, 28, 42$. About 100 species, Europe (2), tropical Africa (40), Madagascar (17) Southeast Asia (40), North America (5), Central and South America (7). Several African species, mainly on rocks and inselbergs, survive the dry seasons as poikilohydrics.

147. *Craterostigma* Hochst. Fig. 46A, B, 47

Craterostigma Hochst., Flora 24, 2. Teil: 668 (1841); Fischer, Trop. Subtrop. Pflanzenw. 81: 85–126 (1992), rev.

Perennial herbs with subterranean orange-red to yellow rhizome and roots, usually pilose.

Fig. 47. Scrophulariaceae. **A, B** *Craterostigma plantagineum*. **A** Plant wet. **B** Plant dried. **C** *Craterostigma pumilum*. (**A, B** Photo E. Fischer, **C** Photo W. Barthlott)

Stems ascending, congested, quadrangular. Leaves in a basal rosette, opposite, subsessile, linear-lanceolate to broadly ovoid, acuminate, with entire to dentate or crenate margin. Inflorescence lateral, with axillary racemes. Flowers distinctly pedicellate. Calyx 5-lobed, campanulate, divided to middle or above, wingless or winged (*C. alatum* Hepper). Corolla white to cream with blue, rarely entirely purple, strongly 2-lipped, tube cylindrical. Stamens 4, exserted, filaments of abaxial pair geniculate, Z-shaped. Ovary ovoid. Seeds bothrospermous. $2n = 56$. Nine species, tropical and South Africa, Arabia, Yemen, India, surviving the dry seasons as poikilohydrics.

148. *Crepidorhopalon* Eb. Fischer

Crepidorhopalon Eb. Fischer, Feddes Repert. 100: 443 (1989); Fischer, Trop. Subtrop. Pflanzenw. 81: 126–203 (1992), rev.

Annual herbs or perennial subshrubs, glabrous, rarely pubescent. Stems prostrate to ascending or erect, quadrangular. Leaves cauline or in a basal rosette, opposite, sessile, linear-filiform, lanceolate to orbicular, acuminate to obtuse, with entire to dentate margin. Flowers subsessile or pedicellate. Calyx 5-lobed, campanulate, deeply divided with almost free lobes to only weakly dentate, cupulate, sepals winged or wingless. Corolla blue to violet or yellowish-whitish, always with yellow mark on lower lip formed by club-shaped hairs on a multicellular base, strongly 2-lipped, tube cylindrical. Stamens 4, exserted, filaments of the abaxial pair geniculate, with clavate appendages. Ovary globose to oblong-ovoid. Seeds aulacospermous. Twenty-eight species, tropical and South Africa, Madagascar (1), several species on copper or cobalt soil.

149. *Hartliella* Eb. Fischer

Hartliella Eb. Fischer, Trop. Subtrop. Pflanzenwelt 82: 204 (1992).

Perennial suffrutescent herbs with woody rhizome, glabrous. Stems erect, quadrangular, congested, woody at base. Leaves cauline or rosulate, opposite, sessile, lanceolate, ovoid or broadly orbicular, leathery and shiny, acuminate with entire margin. Inflorescence congested to densely capitate. Flowers sessile to subsessile. Calyx 5-lobed, divided almost to base, lobes acuminate. Corolla blue to violet, strongly 2-lipped, tube long, cylindrical. Stamens 4, exserted, filaments of the abaxial pair geniculate, with short clavate appendages. Ovary oblong-ovoid. Seeds aulacospermous. Four species, Central Africa (Congo, Katanga), growing on heavy-metal soil.

150. *Bampsia* Lisowski & Mielcarek

Bampsia Lisowski & Mielcarek, Bull. Jard. Bot. Natl. Belg. 53: 377 (1983).

Perennial suffrutescent herbs with woody rhizome, glabrous. Stems erect, quadrangular, wings produced by decurrent leaf-base. Leaves cauline, opposite, sessile, connate at base, leathery, oblanceolate, obovoid or elliptic-orbicular, acuminate, with entire margin. Inflorescence congested to capitate. Flowers shortly pedicellate. Calyx infundibuliform, 5-lobed, lobes divided to above middle. Corolla pale bluish to yellowish, strongly 2-lipped, tube straight, cylindrical. Stamens 2 (*B. lawalreana* Lisowski & Mielcarek), adaxial, abaxial stamens absent or 4 (*B. symoensiana* Lisowski & Mielcarek), included, filaments straight. Ovary obovoid. Seeds aulacospermous. Two species in Congo (Katanga).

6. Tribe Melospermeae Rossow (1985).

Shrubs, inflorescence racemose or sometimes with axillary 2(3)-flowered cymes (*Monttea*), stamens 4, staminode present, anthers with two equal rounded thecae, stigma capitate, fruit an indehiscent to subindehiscent (*Monttea*) or loculicidal and septicidal capsule.

151. *Monttea* Gay

Monttea Gay, Hist. Chile Bot. 4: 416 (1849); Rossow, Parodiana 3: 365–396 (1985), rev.

Perennial shrubs, glabrous or pubescent on young stems. Stems ascending to erect, much branched, terete. Leaves cauline, opposite, rarely alternate, sessile or petiolate, subulate-terete, linear-lanceolate to broadly ovoid-orbicular, acuminate, with entire margin. Flowers subsessile to shortly pedicellate. Calyx campanulate, 5-dentate, tube much longer than the lobes, accrescent and fleshy in fruit. Corolla blue to violet, subrotate, 5-lobed, lobes subequal, rounded, oblong, with ciliate margin, pubescent, tube campanulate, narrow at base, widened near throat. Ovary ovoid. Seeds few, 1(–2) per locule, large, testa finely reticulate. Three species, Chile, Argentina.

152. *Melosperma* Benth.

Melosperma Benth. in DC., Prodr. 10: 374 (1846); Rossow, Parodiana 3: 365–396 (1985), rev.

Perennial xerophytic shrub or subshrub, glabrous to glandular-pubescent. Stems prostrate, terete. Leaves cauline, opposite, shortly petiolate, ovoid, acuminate, with entire margin. Flowers pedicellate. Calyx 5-dentate, oblique, with 10 prominent veins, lobes triangular-ovoid. Corolla blue to violet, rarely white, obscurely 2-lipped, tube longer than lobes, weakly curved. Ovary ovoid. Seeds few, 6–8(12), large, with convex dorsal keel, testa finely reticulate. One species, *M. andicola* Benth., Chile, Argentina.

III. Subfam. Digitalidoideae (Dum.) Luerss. (1882).

Veronicaceae Durande (1782).
Digitalidaceae Martynov (1820).
Digitalideae (Dum.) Dum. (1829).

Herbs or shrubs. Inflorescence racemose, often spike-like, terminal or lateral with axillary racemes. Corolla tubular-bilabiate to nearly actinomorphic, tube often short and reduced, without abaxial protrusions, petal aestivation ascending. Stamens 2–4(–8), anthers sagittate with two separate equal and rounded thecae, confluent at apex. Ovary bilocular. Capsule septicidal and loculicidal or fruit a berry. Intermediate layers of ovule integument 2–6(–8), endosperm of mature seed smooth or with undulate surface produced by endothelium, testa cells of mature seed homogenous. Seeds numerous to few, 4 per locule, convex dorsally, flattened ventrally, testa membranaceous, often with distinct wings. Thirty genera with 523 species, cosmopolitan. The Plantaginaceae and Globulariaceae should be included here.

1. Tribe Digitalideae (Dum.) Dum. (1829).

Rosulate herbs to shrubs, inflorescence racemose, lax, often secund, prophylls absent, corolla tubular to campanulate, stamens 4, anthers sagittate with two separate equal and rounded thecae, ovary with basal nectary disc, capsule septicidal.

153. *Digitalis* L.

Digitalis L., Sp. Pl. 2: 621 (1753); Werner, Bot. Jahrb. Syst. 79: 218–254 (1960), rev.; Werner, Feddes Repert. 70: 109–135 (1965), rev.

Annual, biennial or perennial to suffrutescent herbs, glabrous or glandular-pubescent to densely villose. Stems erect, rounded. Leaves usually in a basal rosette, alternate, subsessile to petiolate, linear-lanceolate to ovoid, acuminate, with entire or dentate to serrate or crenate margin. Flowers shortly pedicellate. Calyx deeply 5-lobed, campanulate. Corolla tubular, cylindrical or inflated-globose, usually constricted towards base, purple-violet, whitish or yellowish with brownish veins, limb 2-lipped. Stamens 4, included. Ovary ovoid conical. Stigma bilobed. Seeds numerous, testa reticulate. $2n = 56, 112$. About 19 species, Mediterranean, Europe to Central Asia.

154. *Isoplexis* (Lindl.) Loudon

Isoplexis (Lindl.) Loudon, Encycl. Pl.: 528 (1829); Werner, Feddes Repert. 70: 109–135 (1965), rev.

Shrubs, glabrous or glandular-pubescent to densely villose. Stems erect, rounded. Leaves usually congregated at the end of branches, alternate, subsessile, attenuate at base, lanceolate to ovoid-oblong, acuminate, with serrate or serrulate margin. Flowers subsessile. Calyx deeply 5-lobed, campanulate. Corolla cylindrical to campanulate or inflated-globose, usually constricted towards base, orange-yellow with purple veins, limb 2-lipped. Stamens 4, weakly exserted. Ovary ovoid to conical. Stigma capitate. Seeds numerous, testa reticulate. $2n = 56$. Four species, Canary Islands and Madeira.

2. Tribe Veroniceae Duby (1820).

Erinaceae Duvau ex Pfeiff. (1873).

Herbs or shrubs, inflorescence sometimes truncate, corolla tubular, 2-lipped to almost actinomorphic, tube often short and reduced, stamens usually 2, rarely 4, anthers with two separate equal thecae, these confluent at apex, capsule usually loculicidal and septicidal.

2a. Subtribe "Wulfeniinae"

Inflorescence racemose, terminal (*Erinus*, *Lafuentea*) or lateral, truncate, with axillary racemes, prophylls absent, stamens 4 or 2, the abaxial stamens aborted (*Wulfenia* to *Kashmiria*), capsule septicidal and loculicidal or fruit berry-like and indehiscent (*Kashmiria*).

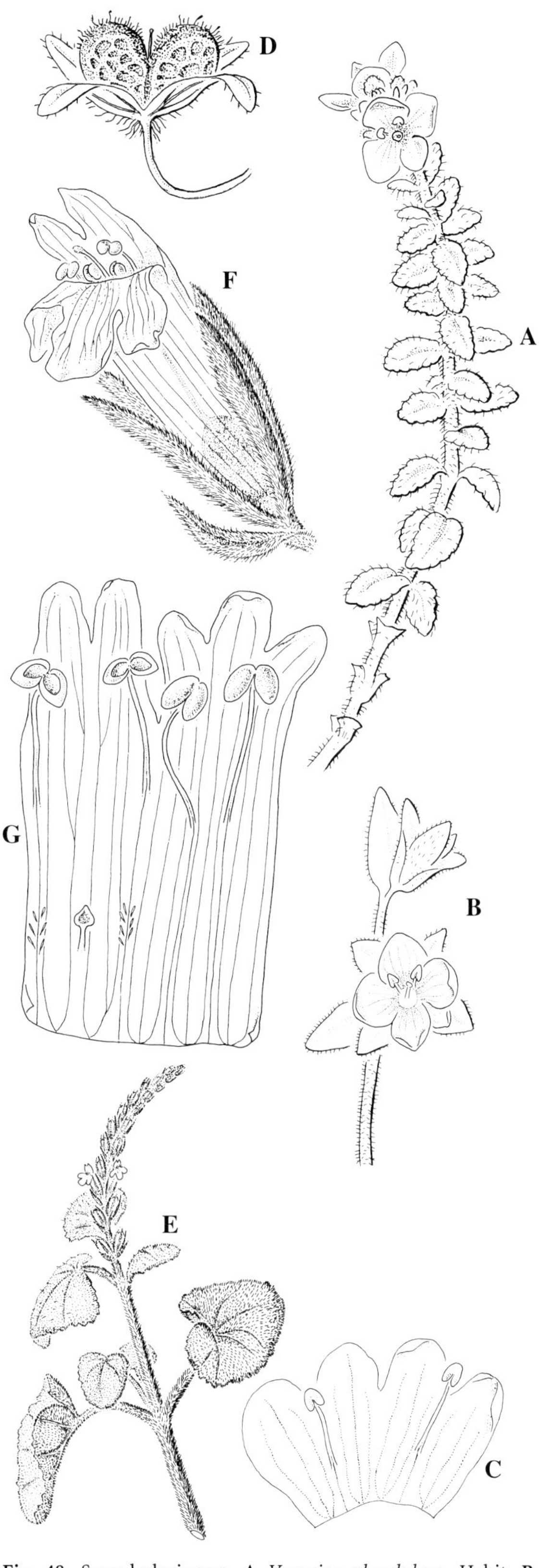

Fig. 48. Scrophulariaceae. **A** *Veronica glandulosa*. Habit. **B** *Veronica abyssinica*. Inflorescence. **C** *Veronica* spec. Corolla dissected. **D** *Veronica opaca*. Fruit. **E–G** *Lafuentea rotundifolia*. **E** Habit. **F** Flower. **G** Corolla, dissected. (**A, B** After Fischer 1999, redrawn)

155. *Erinus* L.

Erinus L., Sp. Pl. 2: 630 (1753).

Perennial suffrutescent herbs, glandular-pubescent to villose. Stems ascending to erect, rounded, forming lax cushions. Leaves cauline, alternate, subsessile, lanceolate to cuneate, with coarsely dentate margin. Flowers shortly pedicellate. Calyx deeply 5-lobed, campanulate, lobes linear-oblong. Corolla bright purple (rarely white), limb 2-lipped, lobes subequal, tube narrow. Stamens 4, included. Ovary ovoid with narrow apex, style lacking. Stigma sessile, capitate. Seeds numerous, testa reticulate. Two species, *E. alpinus* L., Pyrénées and Alps, *E. thiabaudii* Jahand. & Maire, Morocco, sometimes considered as one species, *E. alpinus*, with two subspecies.

156. *Lafuentea* Lag. Fig. 48F, G

Lafuentea Lag., Gen. Species Pl. Nov.: 19 (1816).

Subshrub, densely greyish-pubescent to hirsute, glandular. Stems erect, rounded. Leaves cauline, opposite, petiolate, reniform to orbicular-deltate, with irregularly crenate-dentate margin. Flowers shortly pedicellate. Calyx deeply 5-lobed, campanulate, lobes linear-lanceolate. Corolla white with purple stripes, limb 2-lipped. Stamens 4, exserted. Ovary ovoid, emarginate. Stigma capitate. Seeds numerous, testa reticulate, with small capillary appendages. One species, *L. rotundifolia* Lag., South Spain.

157. *Wulfenia* Jacq.

Wulfenia Jacq., Misc. Austriaca Bot. 2: 60 (1781); Lakusic, Bot. Jahrb. Syst. 99: 443–461 (1978), rev.

Perennial herbs, glabrous or glandular-pubescent. Stems erect to ascending, short, proliferating. Leaves in a basal rosette, alternate, subsessile, narrowed near base, ovoid to orbicular, acuminate, with crenate margin. Flowers shortly pedicellate. Calyx deeply 5-lobed, campanulate. Corolla tubular, blue to violet-blue, limb 2-lipped, tube long, cylindrical. Stamens 2, included. Ovary conical to ovoid. Stigma capitate. Seeds numerous, testa reticulate. $2n = 18$. Four species, Alps, Albania and former Yugoslavia, and southern Turkey and Syria.

158. *Wulfeniopsis* D.Y. Hong

Wulfeniopsis D.Y. Hong, Acta Phytotax. Sin. 18(1): 51 (1980).

Perennial herbs, glabrous or weakly pubescent. Stems erect to ascending, short, proliferating. Leaves in a basal rosette, alternate, subsessile, narrowed near base, oblong-ovoid, with coarsely and irregularly lobed or dentate margin. Flowers shortly pedicellate. Calyx deeply 5-lobed, campanulate. Corolla nearly actinomorphic, tubular, blue to purple, limb with 4 equal lobes, tube long, cylindrical. Stamens 2, included. Ovary conical to ovoid. Stigma capitate. Seeds numerous, testa reticulate. $2n = 16$. Two species, *W. amherstiana* (Benth.) D.Y. Hong, *W. nepalensis* (Benth.) D.Y. Hong, Afghanistan, western Himalayas.

159. *Kashmiria* D.Y. Hong

Kashmiria D.Y. Hong, Bot. Not. 133: 565 (1980).
Falconeria Hook. f. (1883).

Perennial herb, ± pubescent. Stems erect to ascending, short, proliferating. Leaves in a basal rosette, alternate, petiolate, petiole broad, distally winged, lamina obovoid to oblong, with irregularly and deeply double-crenate margin. Flowers shortly pedicellate. Calyx deeply 5-lobed, campanulate. Corolla tubular, blue, limb 2-lipped, tube short, cylindrical. Stamens 2, included. Ovary conical to ovoid. Stigma bilobed. Seeds numerous, testa alveolate. $2n = 16$. One species, *K. himalaica* (Hook. f.) D.Y. Hong, Himalayas.

160. *Picrorhiza* Royle ex Benth.

Picrorhiza Royle ex Benth., Scroph. Ind.: 47 (1835).

Perennial herb, glabrous or weakly pubescent. Stems ascending, short, proliferating. Leaves in a basal rosette, alternate, subsessile, narrowed near base, spathulate to narrow elliptic, acute, with sharply serrate margin. Flowers shortly pedicellate. Calyx deeply 5-lobed, campanulate, lobes lanceolate, ciliate. Corolla almost actinomorphic, shorter than sepals (4–5 mm long), tubular, pale blue to blue, lobes subequal, ciliate within, tube as long as the lobes, cylindrical. Stamens 4, exserted, many times longer than corolla. Ovary conical to ovoid-attenuate. Stigma capitate. Seeds numerous, testa alveolate and transparent. $2n = 34$. One species, *P. kurrooa* Royle ex Benth., western Himalayas.

161. *Neopicrorhiza* D.Y. Hong

Neopicrorhiza D.Y. Hong, Op. Bot. 75: 56 (1984).

Perennial herb, brown-pubescent. Stems ascending, short, proliferating. Leaves in a basal rosette, alternate, subsessile, narrowed near base, spathulate to narrowly elliptic, acute, with sharply serrate margin. Flowers shortly pedicellate. Calyx deeply 5-lobed, campanulate, lobes unequal, adaxial lobe smallest, lanceolate, ciliate. Corolla longer than sepals, tubular, dark blue to purple, limb 2-lipped, lobes of upper lip longest, tube shorter than the lobes, cylindrical. Stamens 4, exserted, shorter than corolla. Ovary conical to ovoid-attenuate. Stigma capitate. Seeds numerous, testa alveolate and transparent. One species, *N. scrophulariiflora* (Pennell) D.Y. Hong, Himalayas.

2B. Subtribe "Veroniciinae"

Inflorescence racemose, often spike-like, terminal or lateral with axillary racemes (*Heliohebe* p.p., *Veronica* p.p., *Hebe*, *Parahebe*, *Derwentia*, *Lagotis*), sometimes compound with main florescence and lateral racemes (*Heliohebe*), bracteate, prophylls absent or present (*Chionohebe*, *Detzneria*), stamens 2, stigma capitate to clavate or 2–4-lobed (*Heliohebe*), capsule loculicidal and septicidal (*Veronicastrum*, *Scrofella*, *Paederota*, *Veronica* to *Detzneria*) or loculicidal, rarely indehiscent (*Lagotis*).

162. *Veronicastrum* Heist. ex Fabr.

Veronicastrum Heist. ex Fabr., Enum.: 111 (1759).
Calorhabdos Benth. (1835).
Botryopleuron Hemsl. (1900).

Perennial herbs to subshrubs, glabrous or pubescent. Stems procumbent to ascending or erect, rounded. Leaves cauline, alternate or verticillate, subsessile to shortly petiolate, linear-lanceolate to ovoid, acuminate, with dentate to serrate margin. Flowers sessile to shortly pedicellate. Calyx deeply 5-lobed, campanulate, lobes acuminate. Corolla subrotate, tubular, blue-purple, with 4 subequal lobes, throat with a ring of hairs. Stamens 2, exserted. Ovary ovoid to conical. Seeds numerous, testa reticulate. $2n = 34$. Five or six species, Northeast North America, eastern Himalayas to Taiwan.

163. *Scrofella* Maxim.

Scrofella Maxim., Bull. Acad. Imp. Sci. Saint-Pétersbourg 32: 511 (1888).

Perennial herb, glabrous except for glandular-pubescent peduncle, calyx and bracts. Stems erect, rounded. Leaves cauline, alternate, sessile, oblong-lanceolate, obtuse, margin entire. Flowers very shortly pedicellate. Calyx deeply 5-lobed, adaxial lobe minute and tooth-like, remaining lobes linear-lanceolate, acute. Corolla white, urceolate, limb 2-lipped, lower lip liguliform, conspicuously reflexed. Stamens 2, included. Ovary ovoid. Stigma rod-shaped. Seeds numerous, testa alveolate. One species, *S. chinensis* Maxim., Northwest China (Gansu, Qinghai, Sichuan).

164. *Synthyris* Benth.

Synthyris Benth. in DC., Prodr. 10: 454 (1846); Pennell, Proc. Natl. Acad. Sci. 85: 77–106 (1933), rev.

Perennial herbs, glabrous or pubescent. Stems erect, rounded. Leaves in a basal rosette, alternate, petiolate, lamina cordate-reniform, obtuse, entire or deeply cleft or pinnatifid, with double-dentate or crenate margin. Flowers shortly to distinctly pedicellate. Calyx deeply 4-lobed. Corolla subrotate to weakly zygomorphic, blue or violet, unequally 4-lobed, lower lobes divided to tube, not forming a lower lip. Stamens 2, exserted. Ovary ovoid to orbicular, somewhat compressed and emarginate at apex. Seeds numerous to 2 per locule, flat, ovoid or orbicular, testa reticulate. $2n = 24$. About 15 species, western North America.

165. *Besseya* Rydb.

Besseya Rydb., Bull. Torrey Bot. Club 30: 279 (1903); Pennell, Proc. Natl. Acad. Sci. 85: 77–106 (1933), rev.

Perennial herbs, glabrous or pubescent. Stems erect, rounded. Leaves in a basal rosette, alternate, petiolate, lamina cordate-ovoid, obtuse, entire, with crenate or crenate-serrate margin. Flowers shortly pedicellate. Calyx deeply 4-lobed, lobes subequal. Corolla violet-purple, yellow or white, 2-lipped, cleft almost or completely to base between the lips, upper lip much larger than weakly lobed lower lip, in some species corolla rudimentary or even lacking. Stamens 2, exserted, inserted at the clefts between corolla lips, or, if corolla lacking, on outer side of nectary disc. Ovary nearly orbicular, only weakly compressed and emarginate at apex. Seeds numerous, flat, ovoid or orbicular, testa reticulate. $2n = 24$. Nine species, western North America.

166. *Paederota* L.

Paederota L., Opera Varia: 200 (1758).

Perennial herbs, subglabrous to crispate-pubescent or hirsute. Stems ascending to erect. Leaves cauline, opposite, alternate in inflorescence, shortly petiolate, ovoid, with dentate margin. Flowers very shortly (1–2 mm) pedicellate, pedicels erect in fruit. Calyx deeply 5-lobed, campanulate, the adaxial lobe smaller than the others. Corolla yellow or blue, limb 2-lipped, lobes narrow, as long as the tube. Stamens 2, exserted. Ovary ovoid, not compressed. Seeds numerous, testa reticulate. $2n = 36$. Two species, *P. bonarota* (L.) L., *P. lutea* Scop., Alps and Balkan.

167. *Paederotella* (E. Wulff.) Kemul-Nath.

Paederotella (E. Wulff.) Kemul-Nath., Not. Syst. Geogr. Inst. Bot. Tphilis, Fasc. 17: 21 (1953).

Perennial herbs, glabrous to glandular-pubescent or curly-hirsute. Stems ascending to erect. Leaves cauline, opposite, alternate in inflorescence, shortly petiolate, ovoid to oblong-ovoid, with remotely obtuse-dentate margin. Flowers very distinctly pedicellate, pedicels curved in fruit. Calyx deeply 5-lobed, campanulate, the adaxial lobe smaller than the others. Corolla yellow, limb ± rotate, the 5 lobes obtuse, erect, free almost to base, tube much shorter than lobes. Stamens 2, exserted. Ovary ovoid, not compressed. Seeds numerous, testa reticulate. Three species, Southwest Asia, Caucasus.

168. *Pseudolysimachion* Opiz

Pseudolysimachion Opiz, Seznam rostlin Kveteny Ceske: 80 (1852).

Perennial, sometimes suffrutescent herbs, glabrous or glandular-pubescent to densely villose. Stems ascending or erect. Leaves cauline, opposite or verticillate, shortly petiolate, linear-lanceolate to ovoid, with entire to dentate margin. Inflorescence many-flowered (40–200). Flowers sessile to shortly pedicellate. Calyx 4-lobed, divided to base. Corolla subrotate, bluish to violet, tube slightly longer than broad, limb shorter than tube, with 4 ± unequal lobes. Stamens 2, exserted. Ovary ovoid, scarcely compressed laterally. Seeds numerous, planoconvex. $2n = 34, 68$. About 15 species in temperate Europe and Asia.

169. *Veronica* L. Fig. 48A–D

Veronica L., Sp. Pl. 1: 9 (1753); Fischer, Flora of Turkey 6: 689–753 (1978), reg. rev.; Fischer, Fl. Iranica: 52–165 (1981); Fischer, Fl. China 18: 65–80 (1998).
Diplophyllum Lehm. (1814).
Cochlidospermum Rchb. (1826).
Odicardis Raf. (1838).
Oligospermum D.Y. Hong (1984).

Annual or perennial, sometimes suffrutescent herbs, glabrous or glandular-pubescent to densely villose. Stems procumbent to ascending or erect. Leaves cauline or in a basal rosette, opposite, usually alternate above, sessile to petiolate, linear-lanceolate to ovoid or orbicular, entire or variously pinnatifid to pinnatisect, with entire to dentate margin. Inflorescence terminal or lateral. Flowers distinctly pedicellate. Calyx 4–5-lobed, divided to base, median calyx lobe very small or (usually) absent. Corolla rotate, bluish, purple, reddish or white, limb longer than tube, with 4 unequal lobes. Stamens 2, exserted. Ovary ovoid, often laterally compressed. Seeds numerous to 1–2 per locule (subgenera *Cochlidiosperma* and *Oligospermum*), often flattened or excavate (cup-shaped), often with a distinct enlarged elevation on chalazal (ventral) surface. $2n = 14, 16, 18, 32, 34, 36, 48, 52, 54, 64$. About 250 species in the Northern Hemisphere and the tropical mountains, several species as neophytic weeds.

170. *Hebe* Comm. ex Juss.

Hebe Comm. ex Juss., Gen.: 105 (1789).
Leonohebe Heads (1987).

Shrubs, glabrous or young shoots pubescent, leaf margin often ciliate. Stems erect. Leaves cauline, opposite, sessile or shortly petiolate, leaf-base connate, linear-lanceolate to ovoid or oblong, entire. Flowers shortly pedicellate. Calyx 4-lobed, divided to base. Corolla rotate, bluish, purple, reddish or white, limb longer than tube, with 4 unequal lobes. Stamens 2, exserted. Ovary ovoid, often dorsiventrally compressed with the septum parallel to the plane of compression, occasionally laterally compressed, tapered at apex or rarely 2-lobed. Seeds numerous, often flattened or excavate. $2n = 40, 42$. About 70 species, Australasia.

171. *Heliohebe* Garnock-Jones

Heliohebe Garnock-Jones, New Zeal. J. Bot 31: 323 (1993).

Shrubs or subshrubs, glabrous or pubescent, leaf margin often ciliate. Stems ascending to erect. Leaves cauline, opposite, sessile or shortly petiolate, leaf-base connate, linear-lanceolate to ovoid, oblong or spathulate, entire, with crenate to serrate margin. Flowers shortly pedicellate, protogynous, usually with stigma emerging from bud. Calyx 4–5-lobed, lower pair sometimes united, upper lobe shorter or absent. Corolla rotate, violet, mauve, bluish, pink or white, limb longer than tube, with 4 subequal lobes. Stamens 2, exserted. Ovary ovoid, dorsiventrally compressed, occasionally laterally compressed, tapered at apex or rarely 2-lobed. Seeds (1–)3–6 per locule, fusiform to irregular, compressed, smooth, or rugulose on abaxial side, winged at ends or at proximal end and narrowly winged around edges. $2n = 42$. Five species from New Zealand.

172. *Parahebe* W.R.B. Oliver

Parahebe W.R.B. Oliver, Rec. Domin. Mus. 1: 229 (1944).

Shrubs or suffrutescent herbs, pilose. Stems erect or ascending to procumbent. Leaves cauline, opposite, petiolate, leaf-base not connate, linear-lanceolate to ovoid or oblong, entire, with crenate-dentate margin. Flowers shortly pedicellate. Calyx 4-lobed, divided to base. Corolla rotate, lavender, pinkish or white, shortly tubular, with 4 unequal widely spreading lobes. Stamens 2, exserted. Ovary ovoid, laterally compressed, tapered, obtuse or notched at apex. Seeds numerous, often flattened, not reticulate. $2n = 42$. Thirty species, New Guinea, New Zealand, Australia (1).

173. *Chionohebe* W.R. Briggs & Ehrend.

Chionohebe W.R. Briggs & Ehrend., Contr. Herb. Austral. 25: 1 (1976).
Pygmaea Hook. f. (1864).

Low or cushion-like shrub, glabrous or young shoots pubescent. Stems prostrate to ascending. Leaves cauline, opposite, sessile or shortly petiolate, imbricate, coriaceous, leaf-base connate, ovoid or oblong, entire, with entire margin. Flowers shortly pedicellate. Calyx 5(–6)-lobed, divided to base. Corolla rotate, bluish, purple or white, limb longer than tube, with 5(–6) lobes. Stamens 2, exserted. Ovary ovoid, laterally compressed and 2-lobed. Seeds numerous, often flattened. $2n = 42$. One species in Tasmania, six species in New Zealand.

174. *Derwentia* Raf.

Derwentia Raf., Fl. Tellur. 4: 55 (1836).

Semi-woody shrubs or herbs, glabrous or pubescent. Stems erect or ascending. Leaves cauline, opposite, leaf-base not connate, sessile or shortly and broadly petiolate, entire or pinnatisect, linear-lanceolate to ovoid, with entire to dentate margin. Flowers shortly pedicellate. Calyx 4-lobed, divided to base. Corolla rotate, lavender, deep blue or rarely pink, limb longer than tube, with 4 unequal spreading lobes. Stamens 2, exserted. Ovary ovoid, weakly or distinctly laterally compressed, tapered at apex or rarely 2-lobed. Seeds few (6–12 per locule) to numerous, flattened adaxially, rugose abaxially, testa not reticulate. $2n = 38, 40$. Eight species, Southeast Australia.

175. *Detzneria* Schlecht. ex Diels

Detzneria Schlecht. ex Diels, Bot. Jahrb. Syst. 62: 490 (1929).

Shrub, young branches with golden bristly hairs, older branches greyish floccose. Stems erect, rounded, leafy in upper parts, with leaf scars only in lower parts. Leaves cauline, opposite, sessile, linear-lanceolate, acuminate, with entire margin, glabrous. Flowers sessile. Calyx deeply 5-lobed, campanulate. Corolla tubiform, dark purplish blue, bright dark blue, violet, reddish mauve, sometimes white at base, 5-lobed, never opening completely, 2-lipped, lobes unequal. Stamens 2, included. Ovary subglobose. Seeds numerous, weakly flattened, testa smooth, indistinctly reticulate. $2n = 48$. One species, *D. tubata* Diels, in alpine regions of New Guinea.

176. *Lagotis* Gaertn.

Lagotis Gaertn., Novi Comment. Acad. Sci. Imp. Petrop. 14(1): 533 (1770).
Kokonoria Keng & K. H. Keng (1946).

Perennial rosulate herbs with creeping rhizome, glabrous. Stems ascending to erect. Leaves in a basal rosette, alternate, linear-lanceolate (then sessile and only attenuate at base) to ovoid or orbicular (then with distinct petiole), with dentate to crenate margin. Flowers sessile to shortly pedicellate. Calyx split abaxially, sometimes 2- or 3-lobed or -toothed. Corolla blue, purple or white, limb 2-lipped, tube cylindrical. Stamens 2, exserted. Ovary ovoid. Fruit indehiscent, drupe-like, with 2 cocci at maturity. Seeds 1 per locule. $2n = 22$. Twenty species, Eastern Europe (2), North and Central Asia to Caucasus, Himalayas and West China.

3. Tribe Sibthorpieae Benth. in DC. (1846).

Sibthorpiaceae D. Don (1835).
Ellisiophyllaceae Honda (1930).

Procumbent herbs, inflorescence racemose, frondose, prophylls present, corolla rotate with (4)5–9(10) lobes, stamens 4(–8), anthers sagittate, thecae contiguous or distinct, capsule compressed, loculicidal.

177. *Sibthorpia* L.

Sibthorpia L., Sp. Pl. 2: 631 (1753); Hedberg, Bot. Not. 108: 161–183 (1955), rev.

Perennial herbs, pubescent to densely villose. Stems procumbent, creeping, rooting at the nodes. Leaves cauline, alternate, petiolate, suborbicular-reniform, crenate to incised. Flowers distinctly pedicellate. Calyx deeply (4–)5(–8)-lobed, lobes weakly unequal. Corolla rotate, yellow, tube short, lobes (4)5–9(10), entire, subequal, spreading. Stamens 4–8, included. Ovary ovoid. Stigma capitate. Seeds few, testa reticulate or smooth. $2n = 20$. Five species, tropical America, Azores, Madeira, Europe, African mountains.

178. *Ellisiophyllum* Maxim.

Ellisiophyllum Maxim., Bull. Acad. Imp. Sci. Saint-Pétersbourg III, 16: 223 (1871).

Perennial herb, densely villose. Stems procumbent, creeping, rooting at the nodes. Leaves cauline, alternate, petiolate, pinnately divided, segments coarsely obtuse-dentate. Flowers distinctly pedicellate, pedicel coiled in fruit. Calyx (4–)5(–8)-lobed, lobes weakly unequal, divided to about middle. Corolla rotate, white with yellow centre, tube short, lobes (4–)5(–8), entire, subequal, spreading. Stamens 4, exserted. Ovary ovoid. Stigma bifid. Seeds few, orbicular, convex on back, testa reticulate or smooth, puberulent, viscid when moistened. One species, *E. pinnatum* (Benth.) Makino, from India to Japan and Taiwan to eastern New Guinea.

4. Tribe Hemiphragmeae Rouy (1909).

Prostrate herbs, inflorescence racemose, with axillary 1-flowered racemes, prophylls absent, stamens 4, included, anthers with two separate subconfluent thecae, sagittate, fruit a fleshy red berry.

179. *Hemiphragma* Wall. Fig. 42D

Hemiphragma Wall., Trans. Linn. Soc. Lond. 13: 611 (1822).

Perennial herb, pubescent to densely villose. Stems prostrate, quadrangular. Leaves cauline, opposite, dimorphic, leaves on main stem petiolate, orbicular to reniform, cordate at base, with coarsely dentate margin, leaves on axillary branches small, linear-subulate, numerous and densely tufted. Flowers sessile. Calyx deeply 5-lobed, lobes linear. Corolla pink, subrotate, weakly 2-lipped, Ovary globose. Stigma clavate, acute. Seeds numerous, testa smooth. One species, *H. heterophyllum* Wall., western Himalayas to Assam, China.

5. Tribe Aragoeae (G. Don) D.Y. Hong & S. Nilsson (1993).

Aragoaceae D. Don (1835).

Shrubs, leaves cauline, verticillate, very dense. Inflorescence racemose, frondose, prophylls absent. Corolla subrotate, white, tube very short, limb 4-lobed, lobes spreading. Anthers with two separate equal thecae, confluent at apex. Capsule loculicidal and septicidal. Seeds few, 4 per locule, convex dorsally, flattened ventrally, testa membranaceous, with distinct wings.

180. *Aragoa* Kunth

Aragoa Kunth in Humb., Bonpl. & Kunth, Nova Gen. Sp.: 120 (1819).

Shrubs, glabrous. Stems erect, rounded, woody. Leaves cauline, verticillate, very dense, sessile, linear-lanceolate, acuminate, with entire, sometimes ciliate margin. Flowers sessile to very shortly pedicellate. Calyx deeply 4–5-lobed, campanulate. Corolla subrotate, white, tube very short, limb 4-lobed, lobes spreading. Stamens 4, exserted. Ovary ovoid. Stigma clavate to capitate. Five species, South America (Andes).

Genera of uncertain placement in Digitalioideae

181. *Ourisia* Comm. ex Juss.

Ourisia Comm. ex Juss., Gen. Pl.: 100 (1789).

Perennial caespitose herbs, glabrous or pubescent. Stems procumbent to ascending, rounded. Leaves cauline, opposite, petiolate, lanceolate to ovoid or orbicular, with dentate to crenate margin. Inflorescence racemose, frondose or bracteate, prophylls present or absent. Flowers distinctly pedicellate. Calyx deeply 5-lobed, campanulate. Corolla subrotate, white with yellow throat or red or orange, tubular-campanulate, limb weakly 2-lipped. Stamens 4, included or exserted. Ovary ovoid to globose. Stigma capitate. Capsule loculicidal. Seeds numerous, testa reticulate. $2n = 32, 48$. Thirty-six species, 30 species in South America (Andes), six species in New Zealand.

182. *Campylanthus* Roth

Campylanthus Roth, Novae Pl. Species: 4 (1821); Hjertson, Edinburgh J. Bot. 60: 131–174 (2003), rev.
Chamaeacanthus Chiov. (1929).

Shrubs, glabrous, except for glandular-pubescent inflorescence. Roots reddish. Stems ascending to erect, rounded, woody. Leaves cauline, alternate, sessile to subsessile, linear-lanceolate to spathulate, succulent, acuminate to obtuse, with entire margin. Inflorescence racemose, often secund, bracteate, prophylls present. Flowers shortly pedicellate. Calyx deeply 5-lobed or 5-dentate, campanulate. Corolla subrotate, purple to pink or whitish, limb weakly 2-lipped, tube incurved, slender, elongate. Stamens 2, included, anthers with two separate equal thecae confluent at apex. Ovary ovoid, compressed. Stigma capitate. Capsule septicidal and loculicidal. Seeds numerous, orbicular, compressed, testa sometimes with distinct wing, reticulate. $2n = 14$. Twelve species, Canary Islands, Cape Verde Islands, Northeast Africa (Somalia), Socotra, Yemen, Pakistan.

Fig. 49. Scrophulariaceae. A *Calceolaria darwinii*. B *Paulownia fargesii*. (Photo W. Barthlott)

"Phrymaceae"[1]

Small aquatics to suffrutescent herbs. Inflorescence usually racemose, rarely with axillary cymes. Corolla often with abaxial protrusion on lower lip (palate). Stamens 2–4, the abaxial pair usually reduced to staminodes or lacking, filaments often with clavate geniculations, anthers with two separate thecae or synthecous. Ovary usually bilocular. Capsule with loculicidal

[1] See Phrymaceae, this Volume.

or septicidal to irregular dehiscence, or fruit a berry. Intermediate layers of ovule integument 1–3, endosperm of mature seed smooth or furrowed, epidermal cells of testa homogenous. Seeds often with anchor-like wall thickenings in testa cells. Nineteen genera with 234 species, cosmopolitan. The genus *Phryma* should be included here.

1. Tribe Microcarpeae Miq. (1857).

Small aquatic herbs, inflorescence racemose, prophylls absent, stamens usually 2, representing the abaxial pair, rarely 4 (*Glossostigma* p.p.), filaments with clavate geniculations at base (*Micranthemum*, *Hemianthus*, *Peplidium*, *Encopella*, *Bryodes*) or geniculations lacking, anthers with two separate thecae (*Micranthemum*, *Hemianthus*, *Psammetes*) or synthecous, unilocular, capsule with loculicidal (*Microcarpaea*, *Elacholoma* to *Encopella*, *Glossostigma*) or septicidal to irregular (*Hemianthus*) dehiscence.

183. *Micranthemum* Michx.

Micranthemum Michx., Fl. Bor.-Am. 1: 10 (1803), nom. cons.
Amphiolanthus Griseb. (1866).
Hemisiphonia Urban (1909).

Annual, rarely perennial small herbs, glabrous. Stems prostrate, rooting at the nodes, weakly quadrangular. Leaves cauline, opposite, sessile, usually ovoid or orbicular, obtuse with entire margin. Flowers sessile to subsessile. Calyx regularly 4-partite to 5-partite, divided nearly to base. Corolla whitish, 2-lipped, with very short tube. Stamens 2, exserted, filaments with clavate geniculations covered with glands. Ovary ovoid to globose. Stigma bilobed. Seeds numerous to few, testa rugose, with 4 longitudinal ridges. Seventeen species, North to South America, Caribbean.

184. *Hemianthus* Nutt.

Hemianthus Nutt., J. Acad. Philad. 1: 119, t. 6 f. 2 (1817).

Annual to rarely perennial small herbs, glabrous. Stems prostrate, rooting at the nodes, weakly quadrangular. Leaves cauline, opposite, sessile, usually ovoid or orbicular, obtuse with entire margin. Flowers sessile to subsessile. Calyx 5-partite, divided to middle except below, where the cleft extends to the base. Corolla whitish, unilabiate, with very short tube, upper lip lacking, lower lip 3-lobed, median lobe distinctly exceeding the others. Stamens 2, exserted, filaments with clavate geniculations. Ovary ovoid to globose. Stigma bilobed. Seeds numerous to few, testa rugose, with 4 longitudinal ridges. Three to four species, Central America.

185. *Microcarpaea* R. Br.

Microcarpaea R. Br., Prodr.: 435 (1810).

Annual small herb, glabrous. Stems prostrate, rooting at the nodes, weakly quadrangular. Leaves cauline, opposite, sessile, linear-oblong or linear-oblanceolate, obtuse at apex, attenuate at base, with entire margin. Flowers sessile. Calyx tubular-campanulate, 5-ribbed, 5-lobed, divided nearly to a third of length, lobes deltate-ovoid, acute, ciliate on margin. Corolla pink, tubular, 2-lipped, with very short tube. Stamens 2, included. Ovary ovoid to globose. Stigma inflexed, ligular. Seeds numerous, testa obscurely reticulate. One species, *M. minima* (Koen.) Merr., tropical Asia.

186. *Elacholoma* F. Muell. & Tate

Elacholoma F. Muell. & Tate, Vict. Nat. 12: 14 (1895).

Annual small herb, glabrous. Stems erect. Leaves cauline, opposite, sessile, linear-oblong or linear-oblanceolate. Flowers sessile to very shortly pedicellate. Calyx inflated, 5-ribbed, 5-lobed, divided nearly to a third of length, lobes narrowly ovoid. Corolla-tube slender, filiform, widened, but almost unlobed at the top. Stamens 2, exserted. Ovary ovoid to globose. Stigma bilobed. Seeds numerous, testa obscurely reticulate. One species, *E. hornii* F. Muell. & Tate, Central Australia and Queensland.

187. *Peplidium* Delile

Peplidium Delile, Fl. Egypte: 50, 148 (1813).

Annual small herbs, glabrous. Stems prostrate, rooting at the nodes, weakly quadrangular. Leaves cauline, opposite, sessile, usually ovoid or orbicular, obtuse with entire margin. Flowers sessile to subsessile. Calyx regularly 5-partite, divided nearly to middle. Corolla whitish, 2-lipped to subrotate, with very short tube, lobes subequal. Stamens 2, exserted. Ovary ovoid to globose. Stigma bilobed. Seeds numerous to few, testa rugose, with 4 longitudinal ridges. $2n = 20$. Nine species, tropical Africa, Asia, Australia.

188. *Encopella* Pennell

Encopella Pennell, Mem. Torrey Bot. Club 16: 106 (1920).
Encopa Griseb. (1866).

Annual small aquatic herb, glabrous. Stems ascending to erect, rounded to weakly quadrangular. Leaves cauline, opposite, the upper sometimes 3–4 verticillate, sessile, linear-lanceolate to filiform, acuminate, with entire margin. Flowers distinctly pedicellate. Calyx usually 4-partite, sometimes 3- to 5-partite, lobes much longer than calyx tube, acuminate, filiform. Corolla white, 2-lipped, with very short campanulate tube, upper lip very small and reduced, much shorter than lower lip, lobes oblong-obovoid, irregularly incised-dentate. Stamens 2, exserted, adaxial pair reduced to clavate staminodes. Ovary globose, unilocular. Stigma bilobed. Seeds numerous to few, testa reticulate. One species, *E. tenuifolia* (Griseb.) Pennell, Cuba.

189. *Psammetes* Hepper

Psammetes Hepper, Hooker's Icon. Pl. t. 3582 (1962).

Annual small herb, glabrous. Stems erect, quadrangular, narrowly 4-winged. Leaves cauline, opposite, sessile, linear-lanceolate, amplexicaulous, acuminate with minutely dentate margin. Inflorescence congested. Flowers subsessile or shortly pedicellate, cleistogamous. Calyx 5-partite, divided nearly to base, lobes linear-filiform, minutely dentate. Corolla whitish, as long as calyx, 2-lipped, with inflated tube, lobes rounded. Stamens 2, included. Ovary ovoid, septum complete. Stigma spathulate, 1-lobed. Seeds numerous to few, testa reticulate. One species, *P. madagascariensis* (Bonati) Eb. Fischer & Hepper, Nigeria, Madagascar.

190. *Dintera* Stapf

Dintera Stapf in Stapf & Schinz, Mém. Herb. Boissier 20: 28 (1900).

Annual small submerged herb, glabrous. Stems erect, quadrangular, with 4 broad wings in upper part. Leaves cauline, opposite, sessile, ovoid, acuminate, amplexicaulous, with entire margin. Flowers sessile, cleistogamous. Calyx 5-partite, divided nearly to base, lobes linear, obtuse, much longer than calyx tube. Corolla white, 2-lipped, with very short tube. Stamens 2, included, anthers with two unequal thecae. Ovary ovoid, septum reduced. Stigma weakly bilobed. Seeds numerous to few, testa foveolate. One species, *D. pterocaulis* Stapf, Southwest Africa.

191. *Bryodes* Benth.

Bryodes Benth. in DC., Prodr. 10: 433 (1846).

Perennial small herb, glabrous. Stems prostrate, rooting at the nodes, weakly quadrangular. Leaves cauline, opposite, sessile, usually ovoid or orbicular, obtuse with entire margin. Flowers sessile to subsessile. Calyx regularly 4-partite to 5-partite, divided nearly to base. Corolla whitish, 2-lipped, with very short tube. Stamens 2, exserted, adaxial pair reduced to filiform staminodes. Ovary ovoid to globose. Stigma bilobed. Seeds numerous to few, testa rugose, with 4 longitudinal ridges. One species, *B. micrantha* Benth., Madagascar, Mascarenes.

192. *Bythophyton* Hooker f.

Bythophyton Hooker f., Fl. Brit. India 4: 286 (1884).

Annual submerged herb, glabrous. Stems erect, wholly submerged, terete. Leaves cauline, opposite, sessile, subulate-lanceolate, acuminate with entire margin. Flowers shortly pedicellate, cleistogamous. Calyx regularly 4-partite, lobes much longer than tube. Corolla white, minute, obscurely 2-lipped, with very short tube. Stamens 2, included, anthers with two parallel unequal thecae. Ovary ovoid, upper part with free placenta, lower part with septum. Stigma subcapitate. Seeds numerous, narrowly oblong, testa reticulate. One species, *B. indicum* (Hook. f. & Thomson) Hook. f., Indomalesia.

193. *Glossostigma* Wight & Arn. ex Arn.

Glossostigma Wight & Arn. ex Arn., Nova Acta Phys.-Med. Acad. Caes. Leop.-Carol. Nat. Cur. 18: 355 (1836), nom. cons.

Annual small herbs, glabrous. Stems short, reduced, weakly quadrangular, slender creeping stolons arising from leaf axils. Leaves cauline or rosulate, opposite, sessile, linear or spathulate, obtuse with entire margin. Flowers distinctly pedicellate. Calyx campanulate, 3–4-lobed, lobes much shorter than calyx tube. Corolla whitish to bluish, tubular, weakly 2-lipped, with very short tube. Stamens 4 or 2, if 2 only the abaxial stamens developed, exserted, adaxial pair reduced to filiform staminodes. Ovary globose. Stigma bilobed. Seeds numerous, testa smooth. $2n = 10$,

60. Five species, Africa, India, Australia, New Zealand.

2. Tribe Leucocarpeae Rouy (1909).

Suffrutescent herbs, inflorescence laxly cymose, with axillary cymes, prophylls small, anther thecae divergent, stigma bilobed, fruit a loculicidal capsule or white berry (*Leucocarpus*).

194. *Hemichaena* Benth.

Hemichaena Benth., Pl. Hartweg.: 78 (1841); Thieret, Fieldiana Bot. 34: 89–99 (1972), rev.
Berendtia A. Gray (1868).
Berendtiella Wettst. & Harms (1899).

Perennial suffrutescent herbs or shrubs, glabrous to viscid-villose. Stems erect, quadrangular. Leaves cauline, opposite, sessile to shortly petiolate, lanceolate to ovoid, acuminate, with dentate margin. Flowers large and showy, shortly pedicellate. Calyx campanulate or tubular-campanulate, 5-costate, 5-lobed. Corolla subrotate, 2-lipped, the tube exceeding the calyx, lobes rounded. Stamens 4, exserted. Ovary ovoid. Seeds numerous, linear-oblong, testa reticulate to reticulate-striate. Five species, Central America.

195. *Leucocarpus* D. Don

Leucocarpus D. Don in R. Sweet, Brit. Fl. Gard. II, 2: t. 124 (1831).

Perennial herb, glabrous to somewhat glandular-pubescent. Stems erect, distinctly quadrangular, narrowly 4-winged. Leaves cauline, opposite, sessile, linear-lanceolate, cordate-amplexicaulous at base, acuminate, with serrate margin. Flowers shortly pedicellate. Calyx membranaceous, tubular-campanulate, conspicuously 5-costate, with 5 setaceous teeth. Corolla yellow, zygomorphic, lips spreading, tube elongate. Stamens 4, included. Ovary ovoid to globose. Seeds numerous, subglobose, testa reticulate. One species, *L. perfoliatus* Humb., Bonpl. & Kunth., tropical America.

3. Tribe Mimuleae Dum. (1829).

Annual to perennial herbs, inflorescence racemose, prophylls absent, corolla strongly 2-lipped, palate present, anthers with two equal thecae confluent at apex, stigma bilobed, fruit a loculicidal capsule or indehiscent, seeds usually with anchor-like wall thickenings in testa cells.

196. *Dodartia* L.

Dodartia L., Sp. Pl. 2: 633 (1753).

Perennial suffrutescent herb with woody rhizome, glabrous. Stems erect, terete. Leaves cauline, opposite, alternate above, sessile, oblong-lanceolate to linear, squamiform, very reduced, acuminate, with remotely denticulate to entire margin. Flowers shortly pedicellate. Calyx 5-lobed, lobes much shorter than calyx tube, campanulate. Corolla deep violet to dark purple, strongly 2-lipped, lower lip with hairy palate, tube long cylindrical. Stamens 4, exserted. Ovary globose, obtuse. Capsule loculicidal or subindehiscent. Seeds numerous, testa smooth. One species, *D. orientalis* L., southern Russia, western Asia.

197. *Lancea* Hooker f. & Thomson

Lancea Hooker f. & Thomson, Hooker's J. Bot. Kew Gard. Misc. 9: 244 (1857).

Perennial herbs, glabrous. Stems erect to prostrate, if stoloniferous, quadrangular. Leaves in a basal rosette, opposite, sessile, oblong-ovoid to spathulate, acuminate, with entire margin, narrowed at base. Flowers subsessile. Calyx deeply 5-lobed. Corolla bright mauve, strongly 2-lipped, tube cylindrical, narrow, dilated towards throat. Stamens 4. Fruit indehiscent. Two species, Tibet, China.

198. *Mazus* Lour.

Mazus Lour., Fl. Cochinch.: 385 (1790).

Annual or perennial herbs, glabrous or pubescent. Stems erect, ascending to prostrate, often forming stolons, quadrangular. Leaves cauline or in a basal rosette, opposite, alternate above, sessile to petiolate, lanceolate-ovoid, acuminate, with entire to dentate or crenate margin. Flowers distinctly pedicellate. Calyx 5-lobed, fused to middle, campanulate. Corolla blue to violet, tubular, strongly 2-lipped, tube shorter than calyx. Stamens 4, included. Ovary ovoid. Capsule loculicidal. Seeds numerous, ellipsoid, testa reticulate. $2n = 40$. Thirty species, Asia, Australia.

199. *Mimetanthe* Greene

Mimetanthe Greene, Bull. Calif. Acad. 1, 4: 181 (1885).

Annual herb, densely pilose (with spreading hairs) and glandular-pubescent. Stems erect, terete to weakly quadrangular. Leaves cauline, opposite, sessile to subpetiolate, oblanceolate to oblong or broadly linear, acuminate, with entire, dentate serrate or crenate margin. Flowers distinctly pedicellate. Calyx 5-lobed, lobes much shorter than tube, 5-sulcate, but not angled or winged. Corolla yellow or purplish, weakly 2-lipped, palate absent. Stamens 4, included, anthers of the abaxial pair much smaller, reduced or lacking. Ovary ovoid, glandular-pubescent. Capsule loculicidal. Seeds numerous, testa reticulate. One species, *M. pilosa* Greene, North America.

200. *Mimulus* L.

Mimulus L., Sp. Pl. 2: 634 (1753); Grant, Ann. Missouri Bot. Gard. 11: 99–374 (1924), rev.
Diplacus Nutt. (1838).
Eunanus Benth. (1846).

Annual or perennial herbs or subshrubs (*M. orbicularis* Benth. a floating aquatic), glabrous to pubescent. Stems prostrate to ascending or erect, quadrangular to winged. Leaves cauline or in a basal rosette, opposite, sessile or petiolate, linear-lanceolate, ovoid or broadly orbicular, acuminate or obtuse, with entire or serrate margin. Flowers distinctly pedicellate. Calyx tubular, 5-dentate, lobes much shorter than the calyx tube, usually winged on midveins of lobes, accrescent in fruit. Corolla white with purple or bluish marks, blue or yellow to orange or red, strongly 2-lipped, tube cylindrical. Stamens 4, included. Ovary oblong to ovoid-oblong. Capsule loculicidal. Seeds numerous, testa with distinct wall thickenings, reticulate. $2n$ = 28, 44, 46, 54, 60, 64. About 150 species, western North America to Mexico, tropical Africa (1), Madagascar (1), Asia (2).

201. *Mimulicalyx* P.C. Tsoong

Mimulicalyx, P.C. Tsoong, Fl. Reip. Pop. Sin. 67(2): 400 (1979).

Perennial herbs, glabrous. Stems erect to ascending, quadrangular. Leaves cauline or in a basal rosette, opposite, sessile, lanceolate, obtuse, with entire margin. Flowers distinctly pedicellate. Calyx 5-dentate, lobes much shorter than calyx tube, winged upon midvein of each sepal, accrescent in fruit. Corolla cylindrical, subrotate to weakly 2-lipped. Stamens 4, included. Ovary ovoid. Capsule septicidal. Two species, *M. paludigenus* P.C. Tsoong, *M. rosulatus* P.C. Tsoong, China.

"Orobanchaceae"

Pedicularidaceae Juss. (1789).
Rhinanthaceae Vent. (1799).
Orobanchaceae Vent. (1799), nom. cons.
Rhinantheae Lam. & DC. (1806).
Pediculariđeae Duby (1828).
Rhinanthoideae Benth. (1846).

Hemiparasitic to holoparasitic herbs, shrubs or vines. Inflorescence racemose, prophylls usually lacking. Corolla strongly 2-lipped, upper lip galeate or plane, often with abaxial protrusions (palate) on lower lip, petal aestivation ascending. Stamens (2–)4, anther thecae separate, with equal or rarely unequal thecae, acuminate at apex or rounded to mucronate, sterile theca reduced, or anthers monothecous. Ovary bilocular to usually unilocular with 2 fused carpels and 2–4 free placentae or with 3 fused carpels and 6 free placentae. Capsule loculicidal, bivalved or trivalved, septicidal or a berry or drupe. Intermediate layers of ovule integument 2–4(5) in tribes Geradieae to Xylocalyceae, 5–10 in tribes Cymbarieae to Rhinantheae, endosperm of mature seed smooth, endotheliar cells only thickened towards endosperm, peripheral cells of intermediate layer usually preserved in tribes Cymbarieae to Rhinantheae, epidermal cells of testa homo- or heterogeneous. Ninety-six genera with 2059 species, cosmopolitan.

1. Tribe Gerardieae Benth. (1835).

Eugerardieae Benth. (1876).

Hemiparasitic herbs, inflorescence racemose, prophylls absent or present (*Anisantherina*), anthers with two separate equal thecae, acuminate at apex or rounded to mucronate (*Lamourouxia*), rarely weakly unequal (*Anisantherina*), ovary bilocular, stigma clavate to capitate or bilobed (*Dasistoma*), capsule loculicidal.

202. *Agalinis* Raf.

Agalinis Raf., New Fl. Am. 2: 61 (1837); Pennell, Proc. Acad. Nat. Sci. Philad. 81: 111–242 (1929), reg. rev.
Gerardia L. (1753).

Annual or perennial herbs, glabrous or scabrid. Stems erect, terete. Leaves cauline, opposite, in one species alternate, sessile, linear-lanceolate to filiform and subulate, entire or rarely lobed at base, acuminate. Flowers distinctly pedicellate. Calyx campanulate, 5-dentate or 5-lobed. Corolla white, rose-purple, pink or bright red, campanulate to funnelform, subrotate to weakly 2-lipped, tube weakly inflated distally. Stamens 4, included, filaments usually pubescent. Ovary oblong to globose. Seeds numerous, usually angled, testa reticulate. $2n = 13, 26, 28, 30$. Fourty-five species from North America to South America (Bolivia, Argentina).

203. *Anisantherina* Pennell ex Britton

Anisantherina Pennell ex Britton, Mem. Torrey Bot. Club 16: 106 (1920); Pennell, Proc. Acad. Nat. Sci. Philad. 81: 242–244 (1929), rev.

Annual herb, hirsute-pubescent. Stems erect, terete. Leaves cauline, opposite, sessile, linear, acuminate, margin strongly revolute. Flowers distinctly pedicellate. Calyx campanulate, 5-lobed. Corolla pink, campanulate, subrotate, 5-lobed, the lobes shorter than tube, tube weakly incurved, distally inflated. Stamens 4, included. Ovary globose. Seeds numerous, oblong-linear, testa reticulate. One species, *A. hispidula* (Mart.) Pennell, from Mexico to Panama, Cuba and Brazil.

204. *Brachystigma* Pennell

Brachystigma Pennell, Proc. Acad. Sci. Philad. 80: 432 (1928).

Perennial herb with woody rootstock, retrorse-pubescent. Stems erect, terete, simple or branched. Leaves cauline, opposite or ternate, sessile, linear, acuminate, entire or sometimes with minute lateral lobes, margin revolute. Flowers distinctly pedicellate. Calyx campanulate, 5-lobed, lobes triangular-cuspidate. Corolla yellow, campanulate, subrotate, 5-lobed, lobes shorter than tube, spreading, tube distally strongly inflated. Stamens 4, included, filaments lanose. Ovary ovoid. Seeds numerous, ovoid, testa with reticulations, these partly forming wings. One species, *B. wrightii* (A. Gray) Pennell, Arizona and northern Mexico.

205. *Aureolaria* Raf.

Aureolaria Raf., New Fl. Am. 2: 58 (1837); Pennell, Proc. Acad. Nat. Sci. Philad. 80: 372–427 (1928), rev.
Dasystoma Raf. (1819) p.p.

Annual or perennial herbs, glabrous, glandular- or eglandular-pubescent. Stems erect, terete. Leaves cauline, opposite to subopposite, whorled or alternate, sessile or petiolate, 1–2 pinnatifid, oblong, lanceolate, usually broadest at base, the upper leaves entire with entire or dentate to serrate margin. Flowers pedicellate. Calyx campanulate or turbinate. Corolla yellow, subrotate to weakly zygomorphic, funnelform or campanulate, limb spreading, tube villose or pubescent within. Stamens 4, included, at least the two longer filaments ± lanose. Ovary oblong. Seeds numerous, reticulate, winged or wingless. $2n = 24, 28$. Ten species, eastern North America, one species in Mexico.

206. *Tomanthera* Raf.

Tomanthera Raf., New Fl. Am. 2: 65 (1837); Pennell, Proc. Acad. Nat. Sci. Philad. 80: 439–446 (1928), rev.
Otophylla Benth. (1846).

Annual herbs, restrorse-hispid. Stems erect, terete. Leaves cauline, opposite, sessile, lanceolate to ovoid, acuminate obtuse, entire or bipinnatifid, the upper usually auriculate at base. Flowers sessile to shortly pedicellate. Calyx 5-lobed. Corolla purple or rarely white, subrotate, lobes much shorter than the tube, spreading, tube broadly campanulate, inflated distally. Stamens 4, included, filament lanose distally or glabrous. Ovary ovoid to obovoid. Stigma clavate or linear. Seeds numerous, ellipsoid-ovoid to triangular, testa reticulate, wingless. Two species, *T. auriculata* (Michx.) Raf. and *T. densiflora* (Benth.) Pennell in central North America.

207. *Macranthera* Nutt. ex Benth.

Macranthera Nutt. ex Benth., Edwards's Bot. Reg. 21: ad t. 1770; Companion Bot. Mag. 1: 174 (1835); Pennell, Proc. Acad. Nat. Sci. Philad. 80: 436–439 (1928), rev.
Conradia Nutt. (1834).

Annual or biennial herb, retrorse-puberulent. Stems erect, stout, terete. Leaves cauline, opposite, petiolate, lanceolate to ovoid, acuminate, pinnatifid. Flowers distinctly pedicellate. Calyx with 5 linear lobes. Corolla orange throughout, subrotate, lobes much shorter than the tube, lower lip distinctly spreading, densely pubescent inside with bearded hairs at base of filaments, tube tubular, straight, not inflated distally. Stamens 4, exserted, lanose. Ovary broadly ovoid. Seeds numerous, angular, testa with distinct wings, reticulate. $2n = 26$. One species, *M.*

flammea (Bartram) Pennell in south-eastern North America.

208. *Seymeria* Pursh

Seymeria Pursh, Fl. Am. Sept. 2: 736 (1813), nom. cons.; Pennell, Proc. Acad. Nat. Sci. Philad. 77: 335–373 (1926), rev.
Afzelia Gmel. (1796).
Brachygyne Small (1903).

Annual or perennial herbs, glandular-pubescent. Stems erect, terete, profusely branched. Leaves cauline, opposite, sessile, pinnate or bipinnate or divided into linear to filiform segments. Flowers pedicellate, prophylls absent. Calyx with cup-shaped tube, the tube shorter than the 5 narrow linear lobes. Corolla yellow, often marked purple, zygomorphic, tube shorter than lips, lips widely spreading. Stamens 4, exserted, filaments pubescent. Ovary ovoid. Seeds numerous, winged or wingless. $2n = 26$. Twenty-five species, south-eastern North America to Caribbean. *Seymeria cassioides* (Gmel.) S.F. Blake is a hemiparasite on pine roots.

209. *Seymeriopsis* N.N. Tsvelev

Seymeriopsis N.N. Tsvelev, Bot. Zhurn. 72(12): 1662 (1987).

Annual herb, glandular-pubescent. Stems erect, terete, profusely branched. Leaves cauline, opposite, sessile, pinnately divided into linear to filiform segments. Flowers pedicellate, prophylls present. Calyx with cup-shaped tube, the tube shorter than the 5 narrow linear lobes. Corolla yellow, often marked purple, zygomorphic, tube longer than lips, lips widely spreading. Stamens 4, included, filaments pubescent. Ovary ovoid. Seeds numerous, winged. One species, *S. bissei* N.N. Tsvelev, in Cuba.

210. *Dasistoma* Raf.

Dasistoma Raf., J. Phys. 89: 99 (1819); Pennell, Proc. Acad. Nat. Sci. Philad. 80: 427–431 (1928), rev.

Annual herb, retrorse-pubescent. Stems erect, terete. Leaves cauline, opposite, petiolate, the lowest broadly ovoid, deeply pinnatifid to bipinnatifid, segments serrate or dentate, upper leaves progressively reduced to lanceolate and entire. Flowers shortly pedicellate. Calyx weakly zygomorphic, tube cup-shaped, the 4 lower lobes oblong, obtuse, more or less equalling the tube, the upper median lobe shorter and narrower. Corolla yellow, tube narrowly campanulate, more or less equalling or longer than the lobes, densely pubescent within, limb 5-lobed, lobes all separate to the apex of the tube, subequal, widely spreading. Stamens 4, included or weakly exserted, filaments villose. Ovary ovoid to subglobose. Seeds numerous, angulate, testa epidermis papery, reticulate. One species, *D. macrophylla* (Nutt.) Raf., in south-eastern North America.

211. *Silviella* Pennell

Silviella Pennell, Proc. Acad. Sci. Philad. 80: 432 (1928).
Silvia Benth. (1846) non *Silvia* Allem. (1845).

Perennial herbs, pubescent. Stems prostrate, terete. Leaves cauline, opposite, petiolate, ovoid, acuminate, entire or bipinnatifid. Flowers pedicellate. Calyx 5-lobed, lobes entire or pinnatifid. Corolla yellow, subrotate, 5-lobed, lobes spreading, tube funnelform, weakly inflated distally. Stamens 2, included, abaxial pair reduced to staminodes. Ovary ovoid. Seeds numerous, ovoid, wingless, testa finely reticulate. Two species, *S. serpyllifolia* (Humb., Bonpl. & Kunth) Pennell and *S. prostrata* (Humb., Bonpl. & Kunth) Pennell in Mexico.

212. *Esterhazya* Mikan

Esterhazya Mikan, Delect. Fl. Faun. Brasil.: t. 8 (1821).

Perennial shrubs. Stems erect. Leaves cauline, opposite, sessile or petiolate, linear-lanceolate, acuminate with entire margin. Flowers distinctly pedicellate. Calyx campanulate or subtubular, 5-dentate. Corolla rose to bright red, 5-lobed, lobes subequal, orbicular, limb expanded, tube incurved, dilated above. Stamens 4, exserted, thecae densely villose. Ovary ovoid to globose. Seeds numerous, angulate, testa smooth. Four species, from Bolivia to Brazil.

213. *Lamourouxia* Kunth

Lamourouxia Kunth in Humb., Bonpl. & Kunth, Nova Gen. Species 2: ed. fol. 269; ed. qu. 335 (1818), nom. cons.

Perennial suffrutescent herbs or dwarf shrubs, pubescent, sometimes viscid. Stems ascending to erect, quadrangular. Leaves cauline, opposite, sessile to shortly petiolate, linear-lanceolate, ovoid, sometimes cordate at base, acuminate, entire to bipinnatifid, with entire to dentate margin. Flowers distinctly pedicellate. Calyx campanulate, 4-lobed, sepals subequal, divided to or above middle. Corolla usually orange-red, strongly 2-lipped, upper lip galeate, tube ventricose. Stamens 4, included, all fertile or the adaxial pair reduced to staminodes. Ovary ovoid to globose. Seeds numerous, testa reticulate. Twenty-six species from Mexico to Peru.

2. Tribe Orobancheae Lam. & DC. (1806).

Phelypaeaceae Horan. (1834).
Hyobancheae Benth. (1876).
Aeginetiaceae Livera (1927).

Holoparasitic herbs (except some hemiparasites in *Harveya*), inflorescence racemose, usually bracteate, anthers with two equal or unequal thecae, the sterile theca reduced (*Hyobanche*, *Harveya*, *Christisonia*) or monothecous (*Paraharveya*, *Aeginetia*), thecae usually mucronate, ovary bilocular (*Hyobanche* to *Paraharveya*) to usually unilocular with 2 fused carpels and 2–4 free placentae or with 3 fused carpels and 6 free placentae (*Phacellanthus* to *Xylanche*), capsule loculicidal (*Hyobanche* to *Paraharveya*), bivalved or trivalved (*Phacellanthus* to *Xylanche*).

214. *Hyobanche* L.

Hyobanche L., Mant. 2: 155 (1771).

Perennial holoparasitic herbs, densely pubescent. Stems short, erect, fleshy or sometimes weakly woody. Leaves squamate, sessile, broadly ovoid, obtuse, with entire margin. Inflorescence with red violet or purple bracts. Flowers sessile. Calyx pubescent. Corolla purple red, strongly 2-lipped, tubular, limb not expanded, upper lip galeate, tube weakly curved. Stamens 4, exserted. Ovary ovoid. Stigma capitate. Fruit a fleshy capsule. Seven species, South Africa.

215. *Harveya* Hook.

Harveya Hook., Icon. Pl.: t. 118 (1837).

Perennial holoparasitic to hemiparasitic herbs, densely glandular-pubescent to glabrous. Stems short, erect, fleshy or sometimes weakly woody. Leaves squamate, sessile, broadly ovoid, obtuse, with entire margin. Flowers usually large, sessile or pedicellate. Calyx tubular or tubular-campanulate, 5-dentate or 5-lobed to below middle. Corolla white, cream, purple-red with yellow or lilac marks, tube usually narrow at base, often much divided above and towards throat, incurved or weakly incurved, lobes 5, erect-spreading, subequal. Stamens 4, usually included, sometimes exserted. Ovary ovoid. Stigma clavate. Seeds numerous, oblong in outline, truncate. Forty species, in tropical and South Africa, Madagascar.

216. *Paraharveya* Eb. Fischer & Siedentop

Paraharveya Eb. Fischer & Siedentop, Syst. Geogr. Plants (2004, in press).

Perennial holoparasitic herb. Stems short, erect. Leaves squamate, sessile, broadly ovoid, obtuse, with entire margin. Flowers usually large, sessile or pedicellate. Calyx tubular or tubular-campanulate, 5-dentate. Corolla white, tube usually narrow at base, often much dilated above and towards throat, incurved or weakly incurved, lobes 5, erect-spreading, subequal. Stamens 4, included. Ovary ovoid, with curved style. Stigma oblong, tongue-shaped. Seeds numerous, oblong in outline, truncate. One species, *P. alba* (Hepper) Eb. Fischer & Siedentop in Central and East Africa.

217. *Aeginetia* L.

Aeginetia L., Sp. Pl. 2: 632 (1753); Beck-Mannagetta, Pflanzenreich 261: 16–21 (1930), rev.

Perennial holoparasitic herbs, glabrous. Stems very short, erect, fleshy. Leaves squamate, sessile, broadly ovoid, obtuse, with entire margin. Flowers usually large, pedicel usually many times longer than stem. Calyx spathe-like, split abaxially nearly to base, acute or obtuse, usually entire, sometimes obtusely keeled on back. Corolla pale rose-purple, tube usually broad and elongate, often much dilated above and towards throat, incurved, lobes 5, erect to spreading, subequal. Stamens 4, included. Ovary ovoid, with curved included style. Stigma peltate. Seeds numerous, scobiform, oblong, truncate. Three or four species in Indomalaysia and East Asia. *Aeginetia indica* L. is a pest of sugarcane.

218. *Boschniaka* C.A. Mey. ex Bong.

Boschniaka C.A. Mey. ex Bong, Veg. Sitka: 41, Mém. Acad. St.-Pétersbourg sér. 6, 2: 157 (1833); Beck-Mannagetta, Pflanzenreich 261: 304–306 (1930), rev.
Orobanche Sect. *Kopsiopsis* Beck (1890).
Kopsiopsis (Beck) Beck (1930).

Perennial holoparasitic cone-like herbs, glabrous. Stems short, erect, fleshy, with a large tuber-like base where attached to the host. Leaves closely imbricated, red or yellowish, squamate, sessile, broadly ovoid, obtuse, with entire margin. Flowers usually small, very shortly pedicellate or sessile. Calyx cupulate, truncate or 1–5-dentate. Corolla pale to dark yellowish-brown, tube incurved, dilated at base, lobes 5, erect to spreading, unequal. Stamens 4, included or exserted. Ovary ovoid. Stigma peltate or distinctly (2–)3–4-lobed. Seeds numerous, scobiform, ellipsoid or globose, testa alveolate. Four species in northern and arctic Russia and western North America and China and Japan, usually as parasites of forest trees (e.g. *Alnus* spp.).

219. ***Christisonia*** Gardner

Christisonia Gardner, Calcutta J. Nat. Hist. 8: 153 (1847); Beck-Mannagetta, Pflanzenreich 261: 308–317 (1930), rev.
Oligopholis Wight (1850).
Campbellia Wight (1850).
Legocia Livera (1927).

Perennial holoparasitic herbs, glabrous. Stems very short or elongate, erect, fleshy. Leaves squamate, sessile, broadly ovoid, obtuse, with entire margin. Flowers usually large, pedicel in some species many times longer than stem. Calyx tubular, 5-dentate or 4–5-lobed, lobes often unequal. Corolla yellowish to purple, tubular-campanulate or infundibuliform, tube usually elongate, often dilated towards throat, incurved or straight, limb erect to spreading, weakly bilobed. Stamens 4, included. Ovary ovoid. Stigma weakly bilobed or peltate. Seeds numerous, scobiform, subglobose. Seventeen species in Southwest China and Indomalesia. Root parasites on Acanthaceae and bamboos.

220. ***Cistanche*** Hoffmanns. & Link Fig. 43C, 51A

Cistanche Hoffmanns. & Link, Fl. Port. I: 318, t. 63 (1809); Beck-Mannagetta, Pflanzenreich 261: 26–41 (1930), rev.

Perennial holoparasitic herbs, glabrous. Stems short, erect, very fleshy, often dilated at base. Leaves squamate, sessile, broadly ovoid, acuminate, with entire margin. Flowers usually large, shortly pedicellate to sessile. Calyx 4- to 5-lobed, lobes obtuse, rounded, equal or adaxial two lobes narrower. Corolla cream, yellow or violet, funnel-shaped, tube more or less curved or bent, often much dilated above and towards throat, incurved, limb erect to spreading. Stamens 4, exserted. Ovary ovoid. Stigma globular. Seeds numerous, scobiform, globose. $2n$ = 20, 40. Ten species, Mediterranean (Europe 2), North and West Africa, Ethiopia to western India and north-western China, eaten as asparagus by Tuareg.

221. ***Phelypaea*** L.

Phelypaea L., Opera Var.: 237 (1758); Beck-Mannagetta, Pflanzenreich 261: 41–44 (1930), rev.
Phelipaea Tourn. ex Desf. (1807).

Perennial holoparasitic herbs, glabrous. Stems short, erect. Leaves squamate, somewhat amplexicaulous, sessile, broadly ovoid, acuminate to obtuse, with entire margin. Inflorescence usually only 1-(rarely 2-)flowered. Flowers usually large, long-pedicellate. Calyx tubular, 5-lobed, more or less 2-lipped, lobes acute. Corolla red to crimson, tube curved, dilated towards base, limb erect to spreading, distinctly 2-lipped. Stamens 4, included. Ovary ovoid. Stigma capitate. Seeds numerous, scobiform, reticulate. $2n$ = 24. Four species from the eastern Mediterranean to Iran and Armenia (Europe 2), parasites on *Psephellus* spp. and *Centaurea* spp.

222. ***Gleadovia*** Gamble & Prain

Gleadovia Gamble & Prain, J. Asiat. Soc. Bengal 69, 2: 488 (1901); Beck-Mannagetta, Pflanzenreich 261: 306–308 (1930), rev.

Perennial holoparasitic herbs, glabrous. Stems short, erect, very fleshy, often dilated at base. Leaves squamate, sessile, broadly ovoid, acuminate or obtuse, with entire margin. Flowers usually large, shortly pedicellate to sessile. Calyx 5-lobed, lobes obtuse, rounded, equal. Corolla pale purple, tube ± straight, limb bilobed, margin pubescent, upper lip helmet-like, much larger than lower lip, lower lip 3-lobed, lobes acuminate. Stamens 4, included. Ovary ovoid. Stigma peltate. Seeds numerous, scobiform, ellipsoid. Six species, from western Himalayas to western China.

223. ***Mannagettaea*** Harry Sm.

Mannagettaea Harry Sm., Acta Hort. Gothoburg 8: 135 (1933).

Perennial holoparasitic herbs, glabrous except for calyx and corolla. Stems short, erect, very fleshy, often dilated at base. Leaves squamate, sessile, broadly ovoid, acuminate, with entire margin. Flowers usually large, shortly pedicellate to sessile. Calyx 5-lobed, lobes acuminate, woolly at margin, equal. Corolla purplish-violet, woolly-pilose inside and partly outside, tube more or less straight, limb bilobed, upper lip helmet-like, larger than lower lip, lower lip 3-lobed, lobes acuminate. Stamens 4, included. Ovary ovoid. Stigma clavate. Three species, from eastern Siberia to western China.

224. *Orobanche* L.

Orobanche L., Sp. Pl. 2: 632 (1753); Beck-Mannagetta, Pflanzenreich 261: 44–304 (1930), rev.
Thalesia Raf. (1818).
Necranthus Gilli (1968).

Annual or perennial holoparasitic herbs, glabrous or variously pubescent. Stems very short or long, erect. Leaves squamate, sessile, broadly ovoid, obtuse or acuminate, with entire margin. Flowers sessile or pedicellate, sometimes pedicel many times longer than stem. Calyx campanulate, 3–5-lobed or divided to base into 2 bifid or bilobed segments. Corolla yellowish-brownish to white, blue, purplish or violet, tube usually curved, much dilated above and towards throat, limb erect-spreading, distinctly bilobed. Stamens 4, included to weakly exserted. Ovary ovoid to ovoid-globose. Stigma capitate or lobed. Seeds numerous, scobiform, subglobose. $2n = 24, 38, 48, 72, 96$. About 150 species, mainly in northern temperate regions, root parasites on various dicotyledonous families.

225. *Phacellanthus* Siebold & Zucc.

Phacellanthus Siebold & Zucc., Abh. Akad. München 4, 3: 141 (1846).
Tienmuia Hu (1939).

Perennial holoparasitic herb, glabrous. Stems short, erect, fleshy. Leaves closely imbricate, red or yellowish, squamate, sessile, broadly ovoid, obtuse, with entire, membranous margin. Flowers usually large, very shortly pedicellate or sessile. Calyx of 2 (or 3) separate, sometimes much-reduced sepals. Corolla pale yellowish, tubular, tube rather narrow, straight, bilobed, limb erect to spreading. Stamens 4, included. Ovary ovoid. Stigma clavate. One species, *P. tubiflorus* Siebold & Zucc. in East Asia and Japan.

226. *Platypholis* Maxim.

Platypholis Maxim., Bull. Acad. St.-Pétersbourg 30: 88 (1886).

Perennial holoparasitic herb, glabrous except for calyx and corolla. Stems short, erect, fleshy, with tuberous rhizome. Leaves laxly squamate, sessile, broadly oblong, obtuse, with entire margin. Flowers small, very shortly pedicellate or sessile. Calyx bipartite, lateral parts ovoid-oblong, bidentate or bilobed, teeth acute, glandular-pubescent. Corolla inflated, pilose, broadly ellipsoid, bilobed, upper lip erect-retuse, longer than lower lip, lower lip 3-lobed, lobes acute, middle lobe shortest. Stamens 4, long-exserted. Ovary ovoid. Stigma peltate, indistinctly lobed. One species, *P. boninsimae* Maxim., Bonin Sima Island (Japan).

227. *Xylanche* Beck

Xylanche Beck in Engler & Prantl, Natürl. Pflanzenfam. IV, 3b: 132 (1893).

Perennial holoparasitic herbs, glabrous except for corolla. Stems short, erect, fleshy, with tuberous rhizome. Leaves densely squamate, sessile, broadly oblong, obtuse or acute, with entire margin. Flowers small, very shortly pedicellate or sessile. Calyx short, cupulate, truncate or 3–5-lobed, lateral lobes unequal, fused with abaxial lobes. Corolla reddish-brown, inflated above middle, margin pilose, broadly ellipsoid, bilobed, upper lip erect-retuse, much longer than lower lip, lower lip inconspicuous, 3-lobed, lobes very small and rounded. Stamens 4, long-exserted. Ovary ovoid. Stigma peltate, indistinctly lobed. Seeds numerous, scobiform, complanate or subglobose. Two species, *X. himalaica* (Hook. f. & Thoms.) Beck and *X. kawakamii* (Hayata) Beck from Himalayas to Taiwan as parasites on *Rhododendron* spp.

228. *Epifagus* Nutt.

Epifagus Nutt., Gen. Am. 2: 60 (1818); Beck-Mannagetta, Pflanzenreich 261: 23–25 (1930), rev.
Leptamnium Raf. (1818).
Mylanche Wallr. (1825).

Perennial holoparasitic herb, pale brown or yellowish brown, glabrous. Stems erect, rounded. Leaves cauline, alternate, sessile, lax, reduced to scales, ovoid or triangular, acuminate. Flowers shortly pedicellate, dimorphic, the lower flowers small, cleistogamous, pistillate, the upper flowers chasmogamous, hermaphroditic, often sterile. Lower cleistogamous flowers with cup-shaped calyx, the 5 short lobes triangular-subulate, corolla calyptriform, not opening, stamens lacking. Upper chasmogamous flowers with similar, but larger calyx. Corolla tubular, zygomorphic, strongly 2-lipped, white with two brown-purple stripes, upper lip concave, almost erect, sometimes galeate, lobes acuminate. Stamens 4, included. Ovary ovoid to globose.

Stigma capitate. Seeds numerous, small, subglobose, angled, testa reticulate. One species, *E. virginianus* (L.) Bart., eastern Canada to eastern U.S.A., parasite on trees (especially *Fagus*).

3. Tribe Escobedieae Benth. (1846).

Hemiparasitic herbs, inflorescence racemose, usually frondose, prophylls present or absent (*Magdalenea*), anthers with two equal apiculate to mucronate thecae, capsule loculicidal.

229. *Alectra* Thunb. Fig. 43A, 50C, D

Alectra Thunb., Nova Gen. Pl.: 81 (1784); Melchior, Notizbl. Bot. Gart. Berlin 15: 423–446 (1941), rev.; Philcox, Fl. Zamb. 8, 2: 85–93 (1990), reg. rev.; Fischer, Bull. Mus. Natl. Hist. Nat., Paris IV, 18, sect. B, Adansonia: 45–65 (1996), reg. rev.
Pseudorobanche Rouy (1909).

Annual herbs or perennial subshrubs, one species (*A. fruticosa* Eb. Fischer) a shrub. Stems hispid with retrorse hairs. Leaves sessile, subsessile to shortly petiolate, opposite, alternate in inflorescence, linear-lanceolate, linear-ovoid to broadly or narrowly lanceolate, subentire or crenate to coarsely toothed, acute, cuneate, rounded to cordate at base. Flowers sessile or shortly pedicellate. Calyx campanulate, glabrous to ciliate on veins and margins of lobes, not inflated in fruit. Corolla yellow to dark orange with reddish-purple venation or red, Ovary ovoid to globose, style horseshoe-shaped, recurved. Stigma clavate. Seeds numerous, fusiform, linear or clavate, slender, truncate, outer testa transparent, reticulate. Fourty species, tropical America, Africa, Madagascar and Asia.

230. *Pseudomelasma* Eb. Fischer

Pseudomelasma Eb. Fischer, Bull. Mus. Natl. Hist. Nat., Paris IV, 18, sect. B, Adansonia: 61 (1996).

Perennial suffrutex with subterranean woody rhizome. Stem and leaves densely tomentose. Leaves opposite ovoid-lanceolate, subentire to weakly crenate, obtuse. Inflorescence a reduced, umbel-like raceme with small scaly bracts, proliferating after anthesis. Flowers distinctly pedicellate, pedicel tomentose, elongating in fruit. Calyx 5-lobed, inflated in fruit, lobes triangular, acute. Corolla yellow, tubular, with oblong and obtuse lobes. Ovary ovoid, style straight, pilose. Stigma clavate, ± straight, pilose. Seeds numerous, ovoid, not fusiform, outer testa not transparent. One species, *P. pedicularioides* (Baker) Eb. Fischer in Central Madagascar.

231. *Melasma* P.J. Bergius

Melasma P.J. Bergius, Descript. Pl. Cap.: 162, t. 3 (1767); Melchior, Notizbl. Bot. Gart. Berlin 15: 119–127 (1940), rev.
Hatschbachia Smith (1953).

Perennial suffrutescent herbs, scabrid or hispid. Stems erect, terete. Leaves cauline, opposite, occasionally alternate above, sessile or subsessile, ovoid to orbicular, acuminate with entire or coarsely dentate margin. Flowers distinctly pedicellate. Calyx campanulate, 5-lobed, 10-veined, inflated in fruit, lobes valvate, subequal. Corolla whitish-cream to yellow or red, tubular-campanulate, 5-lobed, lobes spreading. Stamens 4, included. Ovary ovoid to globose, with straight style. Stigma clavate, flattened. Seeds numerous, fusiform, truncate, outer testa transparent, reticulate. Twenty species, tropical Africa and America.

232. *Magdalenaea* Brade

Magdalenaea Brade, Arq. Inst. Biol. Veg. 1: 235 (1935).

Perennial shrub, glabrous. Stems erect. Leaves cauline, opposite, sessile, lanceolate, acuminate. Flowers distinctly pedicellate. Calyx campanulate, divided to above middle, 4-lobed, lobes acute. Corolla subrotate, weakly 2-lipped, with oblong lobes, margin of corolla lobes truncate, tube straight. Stamens 4, included. Ovary ovoid, with straight filiform style. Stigma capitate. One species, *M. limae* Brade, south-eastern Brazil.

233. *Velloziella* Baill.

Velloziella Baill., Bull. Mens. Soc. Linn. Paris 1: 715 (1887).

Perennial herbs to shrubby vines, glabrous or sparsely pubescent. Stems erect, flexuose, climbing, terete. Leaves cauline, opposite, shortly petiolate, lanceolate to ovoid, acuminate with sinuate-dentate margin. Flowers distinctly pedicellate, pedicel abruptly curved at apex. Calyx spathaceous, ellipsoid-inflated, with entire margin, acuminate. Corolla orange-red, infundibuliform, curved, subrotate with 5 subequal, weakly emarginate lobes. Stamens 4, included. Ovary ovoid, with straight filiform style. Stigma capitate, weakly 4-fid. Seeds numerous, fusiform, truncate, outer testa transparent, reticulate. Two species, *V. dracocephaloides* (Vell.) Baill., *V. westermannii* Dusen, Brazil.

234. ***Nothochilus*** Radlk.

Nothochilus Radlk., Sitzungsber. Math.-Phys. Cl. Königl. Bayer. Akad. Wiss. München 19: 216 (1889).

Perennial suffrutescent herb or dwarf shrub, pubescent. Stems erect, terete. Leaves cauline, opposite, sessile, lanceolate-ovoid, acuminate with dentate margin. Flowers shortly pedicellate. Calyx spathaceous, dentate, angled with prominent veins. Corolla red, strongly 2-lipped, upper lip forming a large galea, lower lip small, 3-lobed. Stamens 4, included. Ovary ovoid, with straight filiform style. Stigma capitate. Seeds numerous, fusiform, truncate, outer testa transparent, reticulate. One species, *N. coccineus* Radlk., Brazil.

235. ***Physocalyx*** Pohl

Physocalyx Pohl, Flora 10: 147; Pl. Brasil. Ic. 1: 63, t. 53 (1827).

Perennial suffrutescent shrubs, pubescent. Stems erect, terete. Leaves cauline, opposite, sessile to subsessile, broadly ovoid to orbicular, acuminate with entire margin, coriaceous. Flowers distinctly pedicellate. Calyx campanulate, 5-dentate, teeth acuminate, accrescent in fruit. Corolla red, densely glandular-pubescent outside, strongly 2-lipped. Stamens 4, included. Ovary ovoid, with curved filiform style. Stigma clavate. Two species, *P. aurantiacus* Pohl, *P. major* Mart., Brazil.

236. ***Escobedia*** Ruiz & Pav.

Escobedia Ruiz & Pav., Prodr.: 91 (1794); Pennell, Proc. Acad. Nat. Sci. Philad. 83: 411–426 (1931), rev.; Thieret, Ceiba 8: 92–101 (1961), rev.

Perennial herbs, glabrous or pubescent, roots yellow. Stems erect, simple, terete. Leaves cauline, opposite, sessile, thick, linear to linear-lanceolate, acuminate, with entire or obscurely dentate margin. Flowers distinctly pedicellate, very large. Calyx funnelform or tubular, lobes usually much shorter than tube, with conspicuous veins. Corolla white, salverform, 5-lobed, lobes orbicular, rounded, tube long (up to 12 cm) and slender. Stamens 4, included. Ovary ellipsoid to obovoid, with long curved style. Stigma bilobed. Seeds numerous, clavate. About 15 species, tropical America.

4. Tribe "Micrargerieae"

Hemiparasitic herbs, inflorescence racemose, prophylls present (*Micrargeria*, *Micrargeriella*) or absent, anthers with two separate equal to subequal thecae, capsule septicidal or loculicidal (*Leptorhabdos*).

237. ***Leptorhabdos*** Schrenk

Leptorhabdos Schrenk, in Fisch. & E. Mey., Enum. 1: 23 (1841).
Dargeria Decne. (1844).

Annual herb, glabrous or glandular-pubescent. Stems erect, terete. Leaves cauline, opposite, sessile, pinnatifid with narrow linear segments, the upper leaves often entire, acuminate. Flowers shortly pedicellate. Calyx shortly campanulate-tubular, 5-dentate, glandular-pubescent, lobes triangular-lanceolate. Corolla rose, 5-lobed, lobes 2-lobed, limb spreading, tube short. Stamens 4, included. Ovary ellipsoid to globose. Stigma capitate, obtuse. Seeds 1–2 per locule, large, testa rugulose. $2n = 14$. One species, *L. parviflora* (Benth.) Benth., Caucasus and Iran to Central Asia and Himalayas.

238. ***Gerardiina*** Engl.

Gerardiina Engl., Bot. Jahrb. Syst. 23: 507 (1897); Mielcarek, Fragm. Flor. Geobot. 41: 81–85 (1996), rev.

Perennial herbs, minutely hispid to subglabrous. Stems erect, simple, subquadrangular, sulcate. Leaves cauline, opposite, sessile, decurrent at base, linear-lanceolate, obtuse or rarely subacute with entire margin. Flowers distinctly pedicellate. Calyx campanulate, 5-lobed, lobes subequal, broadly ovoid-deltate. Corolla pink to purple or blue, occasionally white, campanulate, oblique, subrotate, lobes rounded, tube narrow at base, dilated above. Stamens 4, included. Ovary ovoid, glabrous. Stigma globose-capitate. Seeds numerous, narrowly cylindrical. Two species, *G. angolensis* Engl., *G. kundelungensis* Mielcarek, tropical Southeast Africa.

239. ***Micrargeria*** Benth.

Micrageria Benth. in DC., Prodr. 10: 509 (1846).

Annual herbs, minutely pubescent to scabrid. Stems erect, weakly quadrangular to terete. Leaves cauline, opposite to alternate, sessile, linear-filiform, acuminate, entire or trifid. Flowers pedicellate. Calyx campanulate, 5-lobed, lobes ovoid to lanceolate, acute or obtuse. Corolla pinkish-white, lilac to reddish-purple, campanulate, subrotate, limb subequally 5-lobed, lobes entire, tube dilated above, sometimes incurved. Stamens 4, included. Ovary ovoid to globose. Stigma clavate. Seeds numerous, obovoid. $2n = 32$. Four to five species, tropical Africa, Madagascar, India.

240. *Micrargeriella* R.E. Fries

Micragеriella R.E. Fries, Wiss. Ergebn. Schwed. Rhodesia-Kongo-Exped.: 290 (1916).

Perennial herb, glabrous to sparsely pubescent. Stems erect, terete. Leaves cauline, alternate, scale-like, entire, subulate, acuminate, scabrid. Flowers distinctly pedicellate. Calyx campanulate, shortly 5-lobed. Corolla white to very pale pink, 5-lobed, strongly 2-lipped, tube dilated in throat, weakly incurved. Stamens 4, included. Ovary globose. Stigma globose-capitate. Seeds numerous, testa reticulate. One species, *M. aphylla* R.E. Fries, Congo, Zambia.

5. Tribe Buchnereae Benth. (1835).

Buchneraceae (Benth.) Lilja (1870).

Hemiparasitic to holoparasitic herbs, anthers with two unequal thecae or monothecous.

5a. Subtribe "Sopubiinae"

Inflorescence racemose, prophylls present or absent (*Hiernia*, *Ghikaea*), anthers with unequal thecae, one theca usually reduced and different in shape (but see *Pseudosopubia* to *Ghikaea*), capsule loculicidal, loculicidal and septicidal (*Petitmenginia*), rarely septicidal (*Pseudosopubia*, *Ghikaea*).

241. *Sopubia* Buch.-Ham. ex D. Don

Sopubia Buch.-Ham. ex D. Don, Prodr. Fl. Nepal.: 88 (1825); Hansen, Kew Bull. 30: 543–558 (1975), reg. rev.; Hofmann & Fischer, Adansonia III, 20: 299–312 (1998), reg. rev.

Annual or mostly perennial herbs or subshrubs, glabrous, scabrid, woolly or tomentose. Stems erect to ascending, terete below, quadrangular above, with longitudinal rows of hairs separated by

Fig. 50. Scrophulariaceae. **A**, **B** *Ghikaea speciosa*. **A** Habit. **B** Stamens. **C**, **D** *Alectra sessiliflora*. **C** Stamen. **D** Pistil. **E**, **F** *Rhamphicarpa fistulosa*. **E** Habit. **F** Stamen

glabrous sections. Leaves cauline, opposite or verticillate, sometimes alternate above, sessile, lower leaves trifid, upper leaves simple, with linear-lanceolate to filiform segments. Flowers pedicellate. Calyx campanulate, 5-lobed, lobes triangular, inner margins of lobes densely pilose. Corolla pink, rose-violet to white, cream or yellow, subrotate, with 5 subequal, usually entire lobes, tube short. Stamens 4, exserted, anthers all coherent at anthesis. Ovary ovoid to globose. Stigma tongue-shaped. Seeds numerous, testa reticulate. Forty species, mostly in Africa, Madagascar, one species from the Himalayas to Indochina, Indomalaysia, New Guinea and Australia.

242. *Parasopubia* Hofmann & Eb. Fischer

Parasopubia Hofmann & Eb. Fischer, Bot. Jahrb. Syst. (2004, in press).

Annual or perennial herbs or subshrubs, glabrous, scabrid, woolly or tomentose. Stems erect. Leaves cauline, opposite or verticillate, sometimes alternate above, sessile, lower leaves trifid with linear-lanceolate to filiform segments, upper leaves simple. Flowers sessile or pedicellate. Calyx campanulate, 5-lobed, inner margins of lobes glabrous. Corolla pink to rose-violet, funnel-shaped, tube campanulate, long. Stamens 4, included. Ovary ovoid to globose. Stigma capitate. Seeds numerous, testa reticulate. Two species, *P. delphinifolia* (L.) Hoffmann & Eb. Fischer, *P. bonatii* Hoffmann & Eb. Fischer in Southeast Asia.

243. *Baumia* Engl. & Gilg (1903).

Baumia Engl. & Gilg in Warb., Kunene-Samb.-Exped. Baum: 365, t. 9 (1903).

Annual herb, greyish-pilose. Stems erect, terete. Leaves cauline, opposite, sessile, entire or pinnatifid with linear-lanceolate to filiform segments. Flowers sessile. Calyx campanulate, 5-lobed, lobes triangular, inner margins of lobes glabrous. Corolla pink to pinkish-white, subrotate, with 5 subequal, entire lobes, tube short. Stamens 4, exserted, anthers not coherent at anthesis. Ovary ovoid to globose, winged at apex, with horseshoe-shaped, recurved style. Stigma clavate. Seeds numerous, testa reticulate. One species, *B. angolensis* Engl. & Gilg in Angola.

244. *Petitmenginia* Bonati

Petitmenginia Bonati, Notul. Syst. (Paris) 1: 335 (1911).

Annual (?) herbs. Stems erect, stiff, with densely appressed hairs. Leaves cauline, alternate, sessile, pinnatifid with three linear segments or entire, linear-lanceolate, margin revolute. Flowers sessile. Calyx campanulate, 5-angular, with 5 acuminate lobes, scabrid, revolute at margin. Corolla rose-violet, tubular, subrotate with 5 subequal lobes or weakly 2-lipped. Stamens 4, included. Ovary subglobose. Stigma clavate, inconspicuous. Seeds numerous, oblong, testa reticulate. Two species, *P. comosa* Bonati, *P. stricta* (Benth.) Hofmann & Eb. Fischer, southern China, Southeast Asia.

245. *Sieversandreas* Eb. Fischer

Sieversandreas Eb. Fischer, Bull. Mus. Natl. Hist. Nat., Paris IV, 18, sect. B, Adansonia: 213 (1996).

Perennial hemiparasitic shrub. Stems prostrate to ascending, densely pubescent. Leaves opposite, linear-lanceolate, pinnatifid with one or two pairs of broad and short acuminate segments. Flowers subsessile. Calyx pilose, tube 3 times as long as lobes, accrescent in fruit. Corolla white, palate yellow with red-purple margin, tube more than twice as long as lobes, slender, with gibbous widening at right angle below throat. Stamens 4, included. Ovary ovoid. Stigma capitate. Seeds numerous, ovoid with reticulate testa. One species, *S. madagascarianus* Eb. Fischer in southern Madagascar.

246. *Pseudosopubia* Engl.

Pseudosopubia Engl., Ann. Ist. Bot. Roma 7: 28 (1897) et in Bot. Jahrb. Syst. 23: 511 (1897).

Perennial suffrutescent herbs or shrubs, pilose. Stems ascending to erect, rarely procumbent, rounded to quadrangular. Leaves cauline, opposite, sessile or subsessile, linear-lanceolate to broadly ovoid, acuminate, with entire margin. Flowers distinctly pedicellate. Calyx campanulate, 5-lobed, divided to above middle, lobes triangular to ovoid. Corolla rose to mauve or lilac, strongly 2-lipped, lobes obovoid, tube narrow at base and dilated towards throat. Stamens 4, included, adaxial stamens with short filaments, anthers monothecous, abaxial stamens with shorter filaments and long filiform connective, one branch of the connective long with a fertile theca, the other short, with a sterile, rounded to globose theca. Ovary globose. Stigma capitate. Seeds numerous, obliquely compressed, truncate. Five species, tropical Northeast Africa (Somalia, Ethiopia, Kenya) and southern Africa (Angola).

247. *Hiernia* S. Moore

Hiernia S. Moore, J. Bot. 18: 196, t. 211 (1880); Engl. & Gilg in Warb., Kunene-Samb.-Exped. Baum: 363, t. 8 (1903), rev.

Perennial subshrub, pubescent. Stems erect, branches horizontally arranged, terete. Leaves cauline, opposite or weakly alternate, linear-lanceolate. Flowers sessile to shortly pedicellate. Calyx campanulate, with 5 linear lobes, divided to above middle. Corolla bluish-violet, tube as long as calyx, weakly incurved, dilated above, limb 5-lobed, expanded, lobes orbicular. Stamens 4, exserted, anthers with two thecae. Ovary ovoid to subglobose. Stigma capitate. One species, *H. angolensis* S. Moore, Angola, Namibia.

248. *Graderia* Benth.

Graderia Benth. in DC., Prodr. 10: 521 (1846).

Perennial suffrutescent herbs or subshrubs with woody rhizome, pilose. Stems ascending to erect, rarely procumbent, obtusely quadrangular. Leaves cauline, opposite or alternate above, sessile or subsessile, linear-lanceolate, acuminate, pinnatisect or entire, with entire or dentate margin. Flowers pedicellate. Calyx 5-lobed, campanulate. Corolla purple, mauve or pink, tubular, 5-lobed, lobes almost entire, limb spreading. Stamens 4, included, anthers with two separate divergent thecae, one theca usually slightly smaller and reduced. Ovary globose, compressed. Stigma clavate. Seeds numerous, obovoid-cylindrical. Five species, Africa and Socotra.

249. *Ghikaea* Schweinf. & Volkens Fig. 50A, B

Ghikaea Schweinf. & Volkens in Ghika, Cinq Mois Pays Somalis: 214 (1898).

Perennial shrub or climber, pilose. Stems ascending to erect. Leaves cauline, opposite, alternate above, shortly petiolate, spathulate-ovoid, obtuse, with entire margin. Flowers pedicellate, pedicel shortly fused with stem. Calyx 5-lobed, campanulate, lobes shorter than tube, triangular. Corolla rose to pink or purple, subrotate, 5-lobed, with subequal lobes, tube narrow at base and dilated towards throat. Stamens 4, included, adaxial stamens with bithecal anthers, thecae unequal, fertile theca mucronate at base, sterile theca filiform, abaxial stamens monothecous. Ovary ovoid. Stigma capitate. Seeds numerous, oblong, with reticulate testa. One species, *G. speciosa* (Rendle) Schweinf. & Volk. ex Pilger, Somalia, Ethiopia, Kenya.

250. *Centranthera* R. Br.

Centranthera R. Br., Prodr.: 438 (1810).

Annual herbs, pubescent. Stems ascending to erect, terete. Leaves cauline, opposite to alternate above, sessile, linear-lanceolate, acuminate, with entire to distantly serrate margin. Flowers sessile to subsessile. Calyx spathaceous, obliquely acute, split along abaxial side, shortly 3-lobed at adaxial apex. Corolla yellowish, strongly 2-lipped, lobes elliptic or orbicular, tube curved, dilated upwards. Stamens 4, included, thecae subequal or one theca smaller and sterile. Ovary ovoid. Stigma biflabellate. Seeds numerous, oblong, testa with spirally arranged narrow reticulation. $2n = 30$. Five or six species, China to Australia.

5B. Subtribe "Buchneriinae"

Inflorescence racemose, prophylls present or absent (*Parastriga*, *Cycnium* p.p.), anthers monothecous by abortion of the second theca, capsule septicidal, rarely loculicidal (*Pseudostriga*) or fruit a berry (*Cycnium adonense*).

251. *Buchnera* L.

Buchnera L., Sp. Pl. 2: 630 (1753); Philcox, Fl. Zamb. 8, 2: 96–127 (1990), reg. rev.; Mielcarek, Fragm. Flor. Geobot. 41: 96–161 (1996), reg. rev.
Stellularia Benth. (1880).
Benthamistella Kuntze (1891).
Eylesia S. Moore (1908).

Annual or perennial rigid herbs, usually scabrid. Stems procumbent, ascending or erect, quadrangular or rounded. Leaves cauline or in a basal rosette, opposite, subopposite or alternate above, sessile or shortly petiolate, filiform to linear-lanceolate, ovoid or orbicular, acuminate or obtuse with entire to regularly or coarsely dentate margin. Flowers sessile or subsessile. Calyx tubular, 10-veined or -ribbed, 4–5-dentate. Corolla blue to violet or purple, rose to scarlet, yellow to white, limb subequally 5-lobed, tube cylindrical, straight, weakly curved to arcuate. Stamens 4, included. Ovary ovoid, with short style. Stigma clavate to bilobed. Seeds numerous, testa reticulate. $2n = 28, 40$. About 100 species, pantropical, mainly in tropical Africa.

252. ***Tetraspidium*** Baker

Tetraspidium Baker, J. Linn. Soc., Bot. 20: 215 (1884).

Perennial suffrutescent herb, densely villose. Stems erect, rounded. Leaves cauline, opposite, alternate above, sessile, linear-lanceolate to ovoid, acuminate, with entire margin. Flowers pedicellate. Calyx 5-lobed, campanulate. Corolla 2-lipped, tube campanulate, pilose with glandular and eglandular hairs outside. Stamens 4, included. Ovary ovoid. Stigma clavate. Seeds numerous. One species, *T. laxiflorum* Baker, in Central Madagascar.

253. ***Striga*** Lour.

Striga Lour., Fl. Cochinch.: 22 (1790); Mohamed, Musselman & Riches, Ann. Missouri Bot. Gard. 88: 60–103 (2001), reg. rev.

Annual, rarely perennial herbs, hirsute or scabrous. Root system greatly condensed, adventitious roots arising from subterranean scales. Stems erect, quadrangular, often ridged. Leaves opposite or nearly so, sessile or subsessile, in most species reduced to small linear-lanceolate scales, with entire margin. Flowers sessile or subsessile. Calyx tubular, 5-lobed or 5- (rarely 4-)toothed. Corolla white, rose, lilac, red or cream to orange, 2-lipped, strongly 2-lipped, tube narrow, curved at right angle, throat of tube small. Stamens 4, included. Ovary tubular. Stigma clavate. Seeds numerous, minute (scobiform seeds), testa with prominent encircling ridges. $2n = 24, 36, 40, 80$. Thirty-three species, Africa (27 spp.), Asia, Australia.

254. ***Pseudostriga*** Bonati

Pseudostriga Bonati, Notul. Syst. (Paris) 1: 338 (1911).

Annual herb, pubescent. Stems erect, quadrangular, winged. Leaves opposite, sessile, oblong-lanceolate to linear-lanceolate, acuminate with entire margin. Flowers sessile or subsessile. Calyx spathe-like, upper sepals fused. Corolla rose, 2-lipped, tube curved at middle. Stamens 4, included. Ovary oblong. Stigma clavate, obtuse. Seeds numerous, oblong or cylindrical, testa with longitudinal ridges. One species, *P. cambodiana* Bonati, Southeast Asia.

255. ***Parastriga*** Mildbr.

Parastriga Mildbr., J. Arnold Arbor. 11: 52 (1930).

Annual herb, glabrous. Stems erect, quadrangular. Leaves opposite, sessile, broadly ovoid, acuminate with entire margin. Flowers subsessile. Calyx campanulate, 5-lobed, sepals divided to less than 1/4 of their length. Corolla white to rose, 5-lobed, zygomorphic, tube bent at right angle and widening below throat. Stamens 4, included. Ovary ovoid. Stigma tongue-shaped, dilated, 2-lobed at apex. Seeds numerous, fusiform to elliptic, testa reticulate. One species, *P. alectroides* Mildbr., tropical Central and East Africa.

256. ***Cycniopsis*** Engl.

Cycniopsis Engl., Bot. Jahrb. Syst. 36: 233 (1905); Backlund, Hunde & Langström, Nord. J. Bot. 13: 185–194 (1993), rev.; Nord. J. Bot. 14: 64 (1994), rev.

Perennial herbs, glabrous or pubescent. Stems procumbent. Leaves opposite, sessile, lanceolate to ovoid, obovoid or elliptic, acuminate to obtuse with attenuate base. Inflorescence proliferating with vegetative shoots. Flowers subsessile. Calyx tubular, straight or weakly curved, 5-lobed. Corolla white, yellow or pink to mauve-lilac, tube curved and inflated near throat, limb 4-lobed, lobes otuse and emarginate. Stamens 4, included. Ovary ovoid globose. Stigma clavate. Seeds numerous, black, ± cylindrical with prominently reticulate testa. Two species, *C. humifusa* (Forrsk.) Engl., *C. humilis* (Hochst. ex Benth.) Backlund, Hunde & Langström, Northeast Africa (Ethiopia, Uganda, Kenya), Yemen.

257. ***Cycnium*** E. Mey. ex Benth.

Cycnium E. Mey. ex Benth. in Hook., Companion Bot. Mag. 1: 368 (1836); Hansen, Dansk Bot. Arkiv 32: 7–72 (1978), rev.

Annual or perennial herbs or dwarf shrubs with tuberous or elongated rhizomes, pubescent. Stems procumbent, ascending to erect. Leaves opposite, occasionally alternate or ternate, sessile, subsessile to petiolate, entire, lanceolate-ovoid and acuminate or pinnatisect to bipinnatisect, with entire to dentate or serrate margin. Flowers sessile to distinctly pedicellate. Calyx 4–5-lobed, lobes equal to subequal. Corolla 5-lobed, salverform, white, rose-lilac or cream, 2-lipped, zygomorphic, tube curved and slender. Stamens 4, included. Ovary variable in shape, compressed or not. Stigma capitate. Seeds numerous, oblong-ovoid. $2n = 38, 48$. Fifteen species, tropical Africa.

258. *Rhamphicarpa* Benth. Fig. 50E, F

Rhamphicarpa Benth. in Hook., Companion Bot. Mag. 1: 368 (1836); Hansen, Bot. Tidsskr. 70: 103–125 (1975), rev.

Annual herbs, glabrous. Stems erect, quadrangular or rounded, slender. Leaves opposite or nearly so, sometimes alternate above, sessile, entire, filiform or pinnatisect with filiform segments. Flowers pedicellate. Calyx 5-lobed. Corolla white, 5-lobed, subrotate, with long, curved tube. Plants flowering at night. Stamens 4, included. Ovary ovoid or globose. Stigma bilobed. Capsule ± distinctly rostrate, vertically compressed. Seeds numerous, cylindrical or obovoid. Six species, tropical Africa, South Africa, Madagascar, Turkey, Russia, India, tropical Australia.

6. Tribe "Buttonieae"

Shrubs or vines, inflorescence racemose, rarely in *Radamaea* with 3-flowered axillary cymes, prophylls present, anthers bithecal with equal (*Rhaphispermum*) rounded to subequal thecae (*Radamaea*) or one theca reduced and sterile (*Leucosalpa* to *Thunbergianthus*), fruit a septicidal capsule, drupe or berry.

259. *Rhaphispermum* Benth.

Rhaphispermum Benth. in DC., Prodr. 10: 509 (1846).

Perennial shrub, glabrous. Stems erect, rounded. Leaves cauline, opposite, sessile, lanceolate, acuminate with entire margin. Flowers distinctly pedicellate. Calyx campanulate, obscurely 5-dentate, about as long as broad. Corolla white, subrotate, 5-lobed, lobes orbicular, with crisped margin, limb broadly expanded, tube short. Stamens 4, exserted. Ovary globose. Stigma globose-capitate. Capsule septicidal, compressed, with truncate apex. Seeds numerous, testa reticulate. One species, *R. gerardioides* Benth., Central Madagascar.

260. *Radamaea* Benth. Fig. 51B

Radamaea Benth. in DC., Prodr. 10: 509 (1846).

Fig. 51. Scrophulariaceae. **A** *Cistanche phelypaea*. **B** *Radamaea perrieri*. (**A** Photo W. Barthlott, **B** Photo I. Theisen)

Perennial shrubs or vines, densely pilose. Stems ascending to erect, rarely prostrate, weakly quadrangular or rounded. Leaves cauline, opposite, sessile to shortly petiolate, linear-lanceolate to ovoid or orbicular, acuminate with entire margin. Flowers pedicellate. Calyx campanulate, 5-lobed, hispid or glandular-pubescent, divided to middle. Corolla white with violet marks or cream, 5-lobed, lobes subequal, orbicular, limb expanded, tube narrow. Stamens 4, included. Ovary globose. Stigma weakly bilobed. Fruit a black berry. Seeds up to 15, large, ridged, oblong, testa with reticulate thickenings. Five species, rain and dry forests of Madagascar.

261. ***Leucosalpa*** Scott-Elliot

Leucosalpa Scott-Elliot, J. Linn. Soc., Bot. 29: 35 (1891); Humbert, Boissiera 7: 281–292 (1943), rev.

Perennial shrubs or vines, glabrous. Stems ascending to erect, quadrangular. Leaves opposite, sessile or petiolate, lanceolate to ovoid, acuminate, with entire margin, weakly succulent. Flowers distinctly pedicellate. Calyx deeply or weakly 5-lobed, campanulate. Corolla subrotate, white, smelling of jasmine, salverform, tube up to 14 cm long. Stamens 4. Ovary ovoid, with two unequal locules. Stigma capitate. Fruit an ovoid drupe with woody endocarp and fleshy, at maturity hardened exo- and mesocarp. Seeds 2–4 per locule, testa with reticulate thickenings. Three species, dry forests of Madagascar.

262. ***Buttonia*** McKen ex Benth.

Buttonia McKen ex Benth., Hooker's Icon. Pl. 11: 63 (1871).

Perennial climbing shrubs, glabrous or shortly pubescent. Stems ascending, quadrangular. Leaves opposite, petiolate, pinnatisect, lower lobes elliptic, obtuse, central lobes broadly lanceolate to rhombic, with variously dentate or laciniate margin. Flowers distinctly pedicellate, pedicels reflexed in fruit. Calyx 5-lobed, campanulate to urceolate, inflated in fruit. Corolla carmine, rose-lilac to violet, throat darker, fragrant, 5-lobed, lobes rounded, subequal. Stamens 4, included, sterile theca aborted on lower stamens. Ovary ovoid, bilocular with ± equal locules. Stigma clavate. Seeds numerous, filiform, testa reticulate. Two species, *B. hildebrandtii* Engl., *B. natalensis* McKen, tropical East and South Africa.

263. ***Thunbergianthus*** Engl.

Thunbergianthus Engl., Bot. Jahrb. Syst. 23: 509 (1897).

Perennial climbing shrubs, glabrous. Stems ascending. Leaves opposite, petiolate, petiole with small spine-like hairs, lamina triangular to ovoid, cordate at base, acuminate, with dentate margin. Flowers distinctly pedicellate. Calyx campanulate, the 2 abaxial sepals fused to a 2-lobed lobe. Corolla rose to clear lilac, subrotate with 5 rounded lobes. Stamens 4, included. Ovary ovoid. Stigma globose to clavate. Seeds numerous, linear. Two species, *T. quintasii* Engl., São Tomé, *T. ruwenzoriensis* Good, tropical Central and East Africa (Congo, Rwanda, Burundi, Uganda).

7. Tribe "Xylocalyceae"

Shrubs, inflorescence racemose, prophylls present, corolla subrotate, anthers with two separate parallel or divergent thecae, dehiscing by apical pores. Fruiting calyx woody, completely hardened and enclosing the fruit.

264. ***Xylocalyx*** Balf. f.

Xylocalyx Balf. f., Proc. Roy. Soc. Edinburgh 12: 84 (1883); Carter, Kew Bull. 16: 147–152 (1962), rev.; Thulin, Nord. J. Bot. 7: 267–269 (1987), reg. rev.

Perennial dwarf shrubs, glabrous. Stems erect, quadrangular. Leaves subsessile, opposite or subopposite, narrowly elliptic, acute to subacute, margin entire. Calyx 5-lobed, campanulate. Corolla pale blue, subrotate, funnel-shaped, 5-lobed, lobes subequal. Stamens 4, included. Ovary with filiform style. Stigma capitate. Capsule with a beak at apex. Seeds numerous, testa finely pitted. Four species, Somalia, Socotra.

8. Tribe Cymbarieae D. Don (1835).

Herbs, inflorescence racemose, prophylls present, calyx tubular, weakly dorsiventral, corolla strongly two-lipped, upper lip galeate, lower lip usually with longitudinal palate, anthers with two separate equal and rounded or mucronate (*Bungea*) thecae, stigma capitate, capsule loculicidal.

265. ***Schwalbea*** L.

Schwalbea L., Sp. Pl. 2: 606 (1753).

Perennial herb, minutely pubescent and glandular. Stems erect, terete. Leaves cauline, alternate, sessile, lanceolate with entire margin. Flowers shortly pedicellate. Calyx tubular, 2-lipped, divided to above middle, lobes unequal, dorsal lobe smallest. Corolla yellow or purplish, strongly 2-lipped, tube cylindrical. Stamens 4, included. Ovary ovoid oblong. Seeds numerous, testa reticulate. $2n = 36$. One species, *S. americana* L., eastern North America.

266. *Cymbaria* L.

Cymbaria L., Sp. Pl. 2: 618 (1753).
Cymbochasma Endl. (1839).

Perennial suffrutescent herbs, pubescent to densely greyish villose. Stems erect, terete. Leaves cauline, opposite, sessile, linear-lanceolate, acuminate, with entire margin. Flowers shortly pedicellate. Calyx deeply 5-lobed. Corolla yellow, strongly 2-lipped, tube cylindrical, longer than lips. Stamens 4, somewhat exserted. Ovary ovoid, with straight filiform style. Seeds few, elliptical, flat, testa with distinct wings. Four species, Europe, Central and East Asia.

267. *Siphonostegia* Benth.

Siphonostegia Benth. in Hook. & Arn., Bot. Beech. Voy.: 203 (1835).

Annual or perennial herbs, pubescent to densely villose. Stems erect, terete. Leaves cauline, opposite, alternate above, petiolate, acuminate, entire or pinnatifid with linear-lanceolate segments, with entire or acutely dentate margin. Flowers sessile or subsessile. Calyx tubular, subequally 5-lobed nearly to middle, 10-ribbed. Corolla yellowish, strongly 2-lipped. Stamens 4, included. Ovary narrowly oblong. Seeds numerous. Two species, *S. chinensis* Benth., *S. laeta* S. Moore, East Asia.

268. *Lesquereuxia* Boiss.

Lesquereuxia Boiss., Diagn. I, 12: 44 (1853).
Siphonostegia Benth. (1835) p.p.

Perennial herb, glandular-pubescent, viscid. Stems erect, terete. Leaves cauline, opposite, sessile, oblong-elliptic, acuminate, with entire margin. Flowers distinctly pedicellate. Calyx tubular, 2-lipped, deeply lobed, lobes of upper lip 3 times longer than teeth of lower lip, oblong-ovoid. Corolla purplish with yellow lower lip, strongly 2-lipped. Stamens 4, included, Ovary oblong-linear. Seeds numerous, flattened, curved, irregularly obpyriform, horned at apex. One species, *L. syriaca* Boiss. & Reut., eastern Mediterranean.

269. *Bungea* C.A. Mey.

Bungea C.A. Mey., Verzeichniss Pfl. Caucasus: 108 (1831).

Perennial suffrutescent herbs, densely pilose. Stems ascending to erect, terete, tufted. Leaves cauline, opposite, subsessile to petiolate, lower leaves entire, linear-lanceolate, acuminate, upper leaves trifid, with linear segments. Flowers sessile. Calyx shortly tubular with 4 narrow, linear lobes. Corolla strongly 2-lipped. Stamens 4, included. Ovary ovoid. Seeds few, large. Two species, Southwest and Central Asia and China.

270. *Monochasma* Maxim. ex Franch. & Sav.

Monochasma Maxim. ex Franch. & Savat., Enum. Pl. Jap. 2: 458 (1878).

Annual or biennial herbs, densely villose. Stems ascending to erect, terete, tufted. Leaves cauline, opposite, sessile, linear-lanceolate, acuminate, with entire to dentate margin. Flowers shortly pedicellate. Calyx tubular, 4-lobed to middle, accrescent in fruit. Corolla pale rose-purple, strongly 2-lipped. Stamens 4, included. Ovary ovoid. Seeds numerous. Four species, East Asia.

9. Tribe "Castillejeae"

Herbs, inflorescence racemose, prophylls absent, calyx tubular, laterally compressed, cleft anteriorly or laterally, often spathe-like, corolla strongly two-lipped, upper lip galeate or plane (*Clevelandia*), anthers with two equal oblong thecae or only one theca (*Orthocarpus*, *Triphysaria*), capsule loculicidal.

271. *Castilleja* Mutis ex. L.f.

Castilleja Mutis ex. L.f., Suppl. 47: 293 (1782); Eastwood, Proc. Am. Acad. 44: 563–591 (1909), rev.
Orthocarpus Nutt. p.p.

Annual or perennial herbs, glabrous or pubescent. Stems erect. Leaves cauline, alternate, the lowest opposite, sessile, linear-lanceolate, acuminate, entire or incised, rarely pinnatifid. Inflorescence with leaf-like yellow, red or white bracts. Flowers sessile to shortly pedicellate. Calyx tubular, laterally compressed, often dilated at base, cleft anteriorly or laterally, the lobes entire or shortly 2-lobed. Corolla red or purple to yellow and white, strongly 2-lipped, lobes of lower lip concave, spreading, usually less than half as long as galea. Stamens 4, included. Ovary ovoid or oblong. Seeds numerous, testa reticulate. $2n = 24, 46, 48, 72, 96, 120$. About 200 species, three species in eastern North America, five species in Eurasia, eight species in Central America, five species in the Andes, the rest in western North America.

272. *Orthocarpus* Nutt. Fig. 53A

Orthocarpus Nutt., Gen. 2: 56 (1818).

Annual herbs, glabrous or pubescent. Stems erect or procumbent. Leaves cauline, alternate, the lowest opposite, sessile, linear-lanceolate, acuminate, usually entire at base and incised to pinnatifid above. Inflorescence with leaf-like yellow, red or white bracts. Flowers sessile to shortly pedicellate. Calyx tubular, laterally compressed, often dilated at base, 4-lobed or toothed. Corolla yellow, red or purple, strongly 2-lipped, lower lip very small, lobes concave, spreading, usually less than half as long as galea. Stamens 4, included. Ovary ovoid or oblong. Seeds numerous to few, testa reticulate. $2n = 22, 24, 28, 30, 40, 48, 96$. Nine species, western North America.

273. *Triphysaria* Fisch. & C. Mey.

Triphysaria Fisch. & C. Mey., Ind. Sem. Hort. Petrop. 2: 52 (1835).

Annual herbs, glabrous or pubescent. Stems erect to decumbent. Leaves cauline, alternate, the lowest opposite, sessile, linear, acuminate, entire, upper leaves finely divided. Flowers sessile to shortly pedicellate. Calyx tubular, ± equally 4-lobed. Corolla sulphur-yellow or white to rose, strongly 2-lipped, lips nearly equal. Stamens 4, included. Ovary ovoid or oblong. Seeds numerous, testa reticulate. Five species from western North America.

274. *Clevelandia* Greene

Clevelandia Greene, Bull. Calif. Acad. Sci. 1 (4): 182 (1885).

Annual herb, pubescent. Stems erect. Leaves cauline, alternate, sessile, linear-lanceolate, acuminate, pinnatifid. Flowers sessile to shortly pedicellate. Calyx tubular, campanulate, 4-lobed, lobes equalling tube, linear-filiform. Corolla white with pink tinge, strongly 2-lipped, with narrow tube, upper lip as long as lower lip, lobes plane. Stamens 4, included. Ovary ovoid or oblong. Seeds numerous, testa minutely reticulate-foveolate. $2n = 24$. One species, *C. beldingii* (Greene) Greene, California.

275. *Gentrya* Breedlove & Heckard

Gentrya Breedlove & Heckard, Brittonia 22: 20 (1970).

Annual herb, pubescent. Stems erect. Leaves cauline, alternate, the lowest opposite, sessile, linear-lanceolate, acuminate, entire. Flowers shortly pedicellate. Calyx tubular, laterally compressed, 4-cleft with a deeper anterior cleft, lobes entire. Corolla strongly 2-lipped, tube narrow and longer than calyx, lower lip very small, lobes teeth-like, recurved. Stamens 4, included. Ovary ovoid or oblong. Seeds numerous, testa reticulate. $2n = 24$. One species, *G. racemosa* Breedlove & Heckard, Mexico.

276. *Cordylanthus* Nutt. ex Benth.

Cordylanthus Nutt. ex Benth. in DC., Prodr. 10: 597 (1846), nom. cons.
Adenostegia Benth. (1846).
Dicranostegia (A. Gray) Pennell (1947).

Annual herbs, pubescent. Stems erect to ascending or procumbent. Leaves cauline, alternate, the lowest opposite, sessile, linear-lanceolate, acuminate, entire or pinnatifid, up to 11-lobed. Inflorescence with outer bracts leaf-like, inner bracts calyx-like. Flowers sessile to shortly pedicellate. Calyx divided to base abaxially, spathe-like, apex generally entire or shallowly notched. Corolla strongly 2-lipped, lower lip ± equalling upper lip, mid-lobe generally weakly recurved. Stamens 4, included. Ovary ovoid or oblong. Seeds 10–20, testa reticulate. $2n = 26, 32$. Eighteen species, western North America.

277. *Ophiocephalus* Wiggins

Ophiocephalus Wiggins, Contr. Dudley Herb. 1: 175 (1933).

Annual herb, pubescent. Stems erect. Leaves cauline, alternate, sessile, linear to subulate-linear, acuminate, entire. Flowers sessile to shortly pedicellate. Calyx tubular, 4-cleft with a deeper anterior cleft, the lobes entire. Corolla 2-lipped, tube narrowly cylindrical and longer than calyx, widened into a strongly inflated upper part, lips short and incurved. Stamens 4, exserted. Ovary ovoid or oblong. Seeds numerous, testa reticulate. $2n = 24$. One species, *O. angustifolius* Wiggins, Baja California.

10. Tribe Rhinantheae Lam. & DC. (1806)

Pedicularideae Duby (1828).
Euphrasieae Benth. (1846).
Euphrasiaceae Martynov (1820).
Melampyraceae Rich. ex Hook. & Lindl. (1821).

Herbs to small shrubs, inflorescence racemose, prophylls absent or present (*Conopholis*), corolla usually strongly two-lipped, upper lip galeate or expanded (*Hedbergia*, *Bornmuellerantha*, *Tozzia*), longitudinal palate present or absent (*Hedbergia*, *Bornmuellerantha*, *Macrosyringion*, *Euphrasia*, *Omphalotrix*, *Rhynchocorys* to *Lathraea*), anthers with two equal mucronate or rounded (*Bornmuellerantha*) thecae, bearded or glabrous (*Bornmuellerantha*, *Omphalotrix*, *Parentucellia* p.p., *Pedicularis*, *Melampyrum*, *Rhynchocorys*), stigma capitate to clavate or bilobed (*Macrosyringion*, *Odontitella*, *Pterygiella*), capsule loculicidal, septicidal (*Omphalotrix*) or rarely fruit indehiscent (*Tozzia*).

278. *Hedbergia* Molau Fig. 52D–F

Hedbergia Molau, Nord. J. Bot. 8: 194 (1988).

Perennial suffrutescent herb, densely pubescent. Stems erect, rounded. Leaves cauline, opposite, sessile, elliptic to lanceolate, with revolute, serrate or crenate margin. Flowers shortly pedicellate. Calyx subequally 4-lobed, divided to above middle, weakly 2-lipped, campanulate. Corolla subrotate, white to pink or magenta, limb 5-lobed, lobes subequal, short, rounded at apex, tube campanulate, curved, shorter than limb and calyx. Stamens 4, included. Ovary narrowly elliptic to oblong, not emarginate. Seeds numerous, testa reticulate, with longitudinal wings. One species, *H. abyssinica* (Benth.) Molau, mountains of tropical Africa (Cameroon, Central and East Africa, south to Malawi and Zambia).

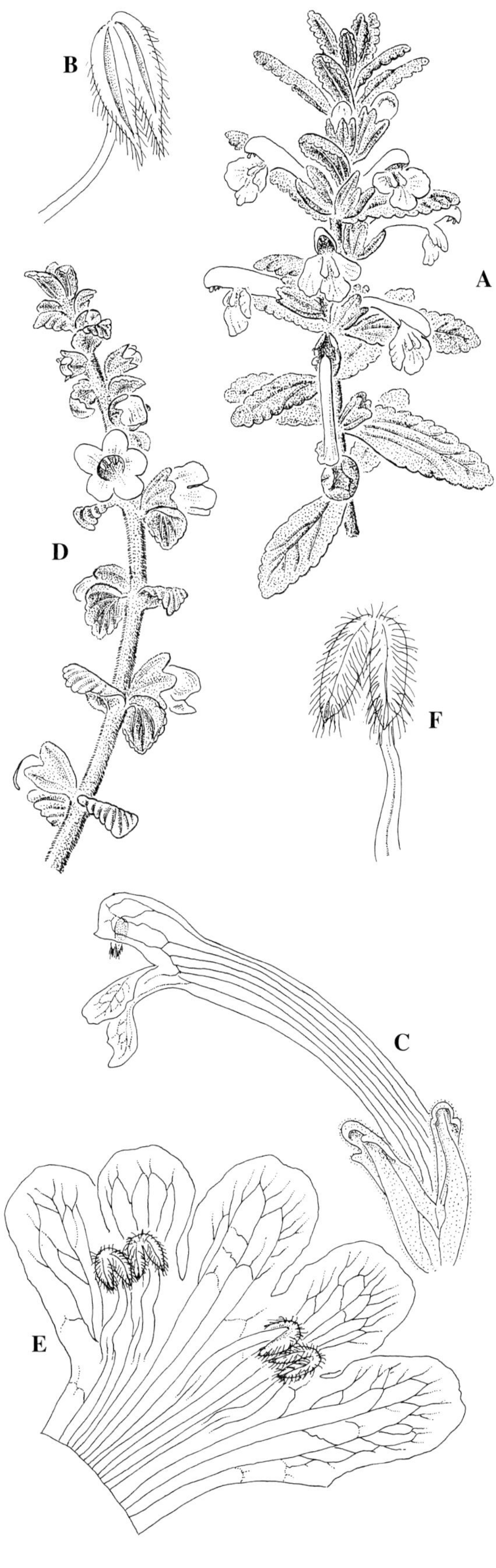

Fig. 52. Scrophulariaceae. **A, B** *Bartsia longiflora*. **A** Habit. **B** Stamen. **C** *Bartsia recurva*, flower. **D–F** *Hedbergia abyssinica*. **D** Habit. **E** Corolla, dissected. **F** Stamen

279. *Bartsia* L. Fig. 52A–C

Bartsia L., Sp. Pl. 2: 602 (1753), nom. cons.; Molau, Opera Bot. 102: 1–99 (1990), rev.
Bellardia All. (1785).

Annual or perennial herbs or subshrubs, densely pubescent and often glandular. Stems prostrate to ascending or erect, rounded. Leaves cauline, opposite, sessile, herbaceous or subcoriaceous, linear to ovoid or elliptical, with revolute, crenate, rarely dentate, serrate or entire margin. Flowers shortly pedicellate. Calyx subequally 4-lobed, divided usually to above middle, weakly 2-lipped, the median clefts usually deeper than lateral ones, campanulate. Corolla yellow, white, red, purple to dark brownish-violet, 2-lipped, tubular, tube longer than calyx and galea. Stamens 4, mostly included. Ovary narrowly elliptic to oblong or subglobose, not emarginate. Seeds numerous, testa reticulate, with longitudinal wings, the wings finely cross-striate. $2n = 24$. Fifty species in South and Central America, three species in tropical Africa, Europe and North America one species (*B. alpina* L.), mainly high mountain elements (except *B. trixago* L.).

280. *Bornmuellerantha* Rothm.

Bornmuellerantha Rothm., Mitt. Thüring. Bot. Ver. N.F. 1: 226 (1943); Bolliger, Willdenowia 26: 61–64 (1996), rev.

Annual herb, hirsute with stout hairs. Stems erect, rounded. Leaves cauline, opposite, sessile, linear-lanceolate, acuminate, with entire margin. Flowers shortly pedicellate. Calyx subequally 4-lobed, divided to above middle, weakly 2-lipped, campanulate. Corolla yellow, weakly 2-lipped, tube as long as calyx. Stamens 4, included. Ovary ovoid to obovoid, not or only weakly emarginate. Seeds 5–6 per locule, up to 1.6 mm long, testa reticulate, with longitudinal ridges. $2n = 24$. One species, *B. aucheri* (Boiss.) Rothm. from Turkey to Iran.

281. *Macrosyringion* Rothm.

Macrosyringion Rothm., Mitt. Thüring. Bot. Ver. N.F. 1: 227 (1943); Bolliger, Willdenowia 26: 64–71 (1996), rev.

Annual herbs, pubescent and shortly glandular. Stems erect, rounded. Leaves cauline, opposite, sessile, linear, acuminate to obtuse, with entire margin. Flowers shortly pedicellate. Calyx subequally 4-lobed, divided to middle, weakly 2-lipped, campanulate. Corolla yellow, 2-lipped, tube 2–4.5 times as long as calyx, slender. Stamens 4, included. Ovary ovoid to oblong, not emarginate. Seeds 8–20 per locule, up to 1.6 mm long, testa reticulate, with longitudinal ridges. $2n = 24$. Two species, *M. longiflorum* (Vahl) Rothm. in eastern Spain, *M. glutinosum* (Bieb.) Rothm. in Greece and Turkey.

282. *Odontitella* Rothm.

Odontitella Rothm., Mitt. Thüring. Bot. Ver. N.F. 1: 226 (1943); Bolliger, Willdenowia 26: 72–76 (1996), rev.

Annual herb, hispid. Stems erect, rounded. Leaves cauline, opposite, sessile, linear, obtuse, with entire margin. Flowers shortly pedicellate. Calyx subequally 4-lobed, divided to above middle, weakly 2-lipped, campanulate. Corolla yellow, becoming orange or reddish brown, 2-lipped, tube shorter than calyx. Stamens 4, included. Ovary narrowly elliptic to oblong, not emarginate. Seeds 8–15 per locule, up to 1.7 mm long, testa reticulate, with longitudinal ridges. $2n = 26$. One species, *O. virgata* (Link) Rothm. in Portugal and western Spain.

283. *Bartsiella* Bolliger

Bartsiella Bolliger, Willdenowia 26: 76 (1996).

Annual herb, hispid with stiff hairs and glandular hairs. Stems prostrate to ascending or erect, rounded. Leaves cauline, opposite, sessile, linear-lanceolate, obtuse, with entire margin. Flowers shortly pedicellate. Calyx subequally 4-lobed, divided to above middle, weakly 2-lipped, campanulate. Corolla brownish-violet, 2-lipped, tube about as long as calyx. Stamens 4, partially included. Ovary narrowly elliptic to oblong, emarginate. Seeds few, 5–6 per locule, up to 5.3 mm long, testa reticulate, narrowly winged. One species, *B. rameauana* (Emb.) Bolliger, in the Atlas Mountains, Morocco.

284. *Odontites* Ludw.

Odontites Ludw., Inst. Regn. Veg. ed. 2: 120 (1757); Bolliger, Willdenowia 26: 79–161 (1996), rev.
Orthantella Rauschert (1983).

Annual or rarely short-lived perennial herbs, hispid or glandular-viscid. Stems erect, rounded. Leaves cauline, opposite, sessile, linear-lanceolate, acuminate to obtuse, with entire to dentate or crenate margin. Flowers shortly pedicellate. Calyx subequally 4-lobed, fused to above middle, weakly 2-lipped, campanulate. Corolla yellow, pink, purple, purplish-brownish or brownish-violet, 2-lipped, tube shorter or weakly longer than calyx. Stamens 4, included. Ovary narrowly elliptic to oblong, emarginate. Seeds 2–20 per locule, testa reticulate, with longitudinal ridges. $2n = 20, 24, 40$. Thirty-two species, Madeira, West and South Europe, North Africa to West Asia (Himalayas).

285. *Nothobartsia* Bolliger & Molau

Nothobartsia Bolliger & Molau, Plant Syst. Evol. 179: 60 (1992).

Perennial herbs, densely pubescent. Stems erect, rounded. Leaves cauline, opposite, sessile, ovoid to elliptical, with revolute, dentate or crenate margin. Flowers shortly pedicellate. Calyx subequally 4-lobed, divided to above middle, weakly 2-lipped, campanulate. Corolla yellow with red gibbae or purple-violet with dark violet gibbae, 2-lipped, tube ca. 1.3 times longer than calyx. Stamens 4, mostly included. Ovary narrowly elliptic to oblong, not emarginate. Seeds 12–19 per locule, testa reticulate, with longitudinal narrow wings. Two species, *N. aspera* (Brot.) Bolliger & Molau, *N. spicata* (Ramond) Bolliger & Molau, Southwest Spain, Portugal to Morocco and Pyrenees.

286. *Euphrasia* L.

Euphrasia L., Sp. Pl. 2: 604 (1753); Ehrendorfer & Vitek, Plant Syst. Evol. 144: 25–44 (1984), reg. rev.; Sell & Yeo, Bot. J. Linn. Soc. 63: 189–233 (1970), reg. rev.; Vitek, Plant Syst. Evol. 148: 215–237 (1985), reg. rev.; Vitek, Plant Syst. Evol. 149: 1–18 (1985), reg. rev.; Vitek, Plant Syst. Evol. 151: 241–269 (1986), reg. rev.; Yeo, Bot. J. Linn. Soc. 77: 223–334 (1978), reg. rev.; Barker, J. Adelaide Bot. Gard. 5: 1–304 (1982), reg. rev.

Annual or perennial herbs or subshrubs, glabrous to densely pubescent, often glandular. Stems erect, rounded, sometimes woody at base. Leaves cauline, opposite, sometimes alternate above, sessile, linear-lanceolate, ovoid to elliptical, with revolute, dentate or crenate margin. Flowers shortly pedicellate. Calyx subequally 4-toothed, divided to more than 2/3 of its length, weakly 2-lipped, campanulate. Corolla yellow, white with yellow and blue or blue to lilac marks, with distinct blackish-violet streaks, 2-lipped, tube longer or shorter than calyx, cylindrical. Stamens 4, mostly included. Ovary narrowly elliptic to oblong, obtuse to emarginate. Seeds numerous, testa reticulate, with longitudinal furrows. $2n = 22, 44$. About 350 species, Europe, Azores, North America, North and Central Asia, New Guinea, New Zealand, Australia.

287. *Omphalotrix* Maxim.

Omphalotrix Maxim., Mém. Acad. Imp. Sci. Saint-Pétersbourg VI, Sci. Math., Seconde Pt. Sci. Nat. 9: 208 (1859).

Annual herb, pubescent with simple, white, curved hairs mostly directed downwards, leaves glabrous. Stems erect, rounded. Leaves cauline, opposite, sessile, oblong-lanceolate, obtuse at apex, scabrous at margin, serrate-crenate. Flowers distinctly pedicellate. Calyx campanulate-tubular, 4-lobed, divided almost to up to half into two bidentate lips. Corolla white, strongly 2-lipped, tube cylindrical. Stamens 4, included. Ovary ovoid-oblong. Seeds few, 6–10 per locule, $1\,mm \times 0.6\,mm$, testa with longitudinal furrows. One species, *O. longipes* Maxim., Northeast Asia (south-eastern Russia, China, Korea, Mongolia).

288. *Pterygiella* Oliv.

Pterygiella Oliv., Hooker's Icon. Pl. 25: pl. 2463 (1896).
Xizangia D.Y. Hong (1986).

Annual herbs, hispid. Stems erect, quadrangular, narrowly winged or sometimes cylindrical and wingless. Leaves cauline, opposite, sessile, linear-lanceolate or widely ovoid, acuminate to obtuse, with entire or remotely serrate margin. Flowers shortly pedicellate. Calyx subequally 5-lobed, lobes triangular, divided to above middle, weakly 2-lipped, campanulate. Corolla yellow or brownish-violet, 2-lipped, tube funnel-shaped, shorter or slightly longer than calyx. Stamens 4, included. Ovary ovoid, densely hirsute. Seeds numerous, oblong or reniform, testa reticulate, with longitudinal ridges. Four species, southern China.

289. *Phteirospermum* Bunge ex Fisch. & C. Mey.

Phteirospermum Bunge ex Fisch. & C. Mey., Index Sem. Hortus Bot. Petrop. 1: 35 (1835).
Emmenospermum C. B. Clarke ex Hook. f. (1883).
Centrantheropsis Bonati (1914).

Annual or perennial herbs, glandular-pubescent. Stems erect, terete. Leaves cauline, opposite, alternate above, petiolate, obtuse at apex, pinnatifid to pinnatisect with ovoid-oblong segments, with entire or acutely dentate margin. Flowers sessile or subsessile. Calyx campanulate, subequally 5-lobed almost to middle. Corolla yellow, strongly 2-lipped. Stamens 4, included. Ovary narrowly oblong. Seeds numerous, oblong, with reticulate testa. Four species, East Asia.

290. *Parentucellia* Viv.

Parentucellia Viv., Fl. Lybicae Spec.: 31 (1824).
Eufragia Griseb. (1844).
Pseudobartsia D.Y. Hong (1979).

Annual herbs, glandular-viscid. Stems erect, terete. Leaves cauline, opposite, alternate above, sessile, acuminate, entire or pinnatifid with linear-lanceolate segments, with entire or dentate to serrate margin. Flowers subsessile to shortly pedicellate. Calyx tubular, subequally 4(–5)-lobed nearly to middle. Corolla yellow, white or reddish, strongly 2-lipped. Stamens 4, included. Ovary narrowly oblong. Seeds numerous, smooth or reticulate. $2n = 24$. Two species, *P. viscosa* (L.) Caruel and *P. latifolia* (L.) Caruel, Western Europe, Mediterranean to Central Asia and Iran. *Parentucellia latifolia* has been introduced to South America (e.g. Chile) and Asia (China).

291. *Pedicularis* L. Fig. 43B, 53B

Pedicularis L., Sp. Pl. 2: 607 (1753); Li, Proc. Acad. Nat. Sci. Philad. 100: 205–378 (1948), reg. rev.; Li, Proc. Acad. Nat. Sci. Philad. 101: 1–214 (1949), reg. rev.; Mayer in Tutin et al., Fl. Europ. 3: 269–276 (1972); Tsoong, Acta Phytotax. Sin. 4: 71–147 (1955), reg. rev.; Tsoong, Acta Phytotax. Sin. 5: 15–73, 239–278 (1956), reg. rev.; Tsoong, Acta Phytotax. Sin. 9: 252–274 (1961), reg. rev.

Annual, biennial or perennial, half-rosulate or rosulate herbs, glabrous to densely villose. Stems ascending or erect, rarely procumbent, rounded, sometimes winged. Leaves cauline or in a basal rosette, opposite, alternate or verticillate, sessile or petiolate, linear-lanceolate to ovoid, usually 3–4-pinnatifid or pinnatisect, rarely entire with dentate to serrate margin. Inflorescence terminal, sometimes terminal inflorescence lacking. Flowers sessile to distinctly pedicellate. Calyx deeply 5-lobed to 2–5-dentate, lobes dentate, serrate or pinnatifid, campanulate or tubular, accrescent after anthesis. Corolla strongly 2-lipped, white, greenish, yellow, red or purple, sometimes multicoloured, upper lip laterally compressed, sometimes asymmetrical, usually rostrate, lower lip spreading, tube cylindrical to obconical. Stamens 4, included. Ovary ovoid to lanceolate, compressed, rostrate. Seeds numerous, globose to oblong, straight or curved, testa reticulate. $2n = 16$, 32. About 600 species, Northern Hemisphere to Central America and China.

Fig. 53. Scrophulariaceae. **A** *Orthocarpus purpurascens*. **B** *Pedicularis sceptrum-carolinum*. (**A** Photo W. Barthlott, **B** Photo E. Fischer)

292. *Melampyrum* L.

Melampyrum L., Sp. Pl. 2: 605 (1753); Beauverd, Mém. Soc. Phys. Genève 38: 291–657 (1916), rev.; Soó, Repert. Spec. Nov. Regni Veg. 23: 159–176 (1926), reg. rev.; Soó, Repert. Spec. Nov. Regni Veg. 23: 385–397 (1927a), reg. rev.; Soó, Repert. Spec. Nov. Regni Veg. 24: 127–193 (1927b), reg. rev.

Annual herbs, glabrous or strigose. Stems erect, rounded. Leaves cauline, opposite, sometimes alternate above, sessile to rarely shortly petiolate, linear-lanceolate to ovoid, acute, with dentate to serrate margin. Inflorescence sometimes secund, with leaf-like, sometimes weakly pinnatifid red violet or purple bracts. Flowers sessile to distinctly pedicellate. Calyx 4-dentate, tubular, not 2-lipped. Corolla strongly 2-lipped, white, yellow or purple, sometimes multicoloured, upper lip laterally compressed, lower lip shorter than upper lip, tube cylindrical. Stamens 4, included. Ovary ovoid to lanceolate, compressed, rostrate. Seeds 1–4, up to 7×3 mm, globose to oblong, testa smooth, very thin, often falling off or removed by ants. $2n = 18$. Thirty-five species, Europe, Asia to Japan and Southeast Asia, North America.

293. *Rhinanthus* L.

Rhinanthus L., Sp. Pl. 2: 603 (1753); Soó & Webb in Tutin & al., Fl. Europ. 3: 276–280 (1972); Sterneck, Abh. Zool.-Bot. Ges. Österreich 1(2): 1–150 (1901), reg. rev.
Alectorolophus Hall. (1742).

Annual herbs, glabrous or strigose. Stems erect, rounded. Leaves cauline, opposite, sessile to rarely shortly petiolate, linear-lanceolate to ovoid, acute, with dentate to crenate margin. Inflorescence sometimes secund, with leaf-like bracts, often differing in shape from the foliage leaves, usually with long teeth or pinnatifid. Flowers sessile to shortly pedicellate. Calyx 4-dentate, laterally compressed, ovoid to suborbicular, glabrous or villose, accrescent after anthesis. Corolla strongly 2-lipped, yellow, upper lip laterally compressed, with 2 teeth just below the apex, lower lip weakly shorter than upper lip, tube long, cylindrical. Stamens 4, included. Ovary ovoid to lanceolate, laterally compressed. Seeds numerous, up to 6 mm long, discoid, flattened transversally, testa reticulate, very thin, sometimes falling off, usually with marginal wing. $2n = 14, 22$. About 45 species, Europe, temperate Asia and North America.

294. *Rhynchocorys* Griseb.

Rhynchocorys Griseb., Spicil. Fl. Rumel. 2: 12 (1844), nom. cons.

Annual or perennial herbs, glabrous or glandular-pubescent. Stems erect, rounded. Leaves cauline, opposite, rarely alternate below, sessile to rarely shortly petiolate, linear-lanceolate to ovoid, acute, with dentate to crenate or serrate margin. Flowers sessile to shortly pedicellate. Calyx 4-dentate, laterally compressed, ovoid to suborbicular, glabrous or villose, accrescent after anthesis. Corolla strongly 2-lipped, yellow, upper lip compressed laterally and prominently rostrate, lower lip slightly shorter than upper lip, tube cylindrical. Stamens 4, included. Ovary orbicular, compressed laterally. Seeds numerous to few, discoid, flattened, testa reticulate-striate, not winged. Six species, Southeast Europe to Iran.

295. *Tozzia* L.

Tozzia L., Sp. Pl. 2: 607 (1753).

Biennial or perennial herb with rhizome covered by fleshy, whitish, imbricate scale leaves with revolute margin, glabrous except for pubescent angles of stem. Stems ascending to erect, quadrangular. Leaves cauline, opposite, sessile, ovoid, obtuse with entire to coarsely crenate (near base) revolute margin. Flowers distinctly pedicellate. Calyx unequally 3- to 5-dentate. Corolla 2-lipped, yellow, with purple spots inside. Stamens 4, weakly exserted. Ovary ovoid, bilocular, only weakly compressed transversally. Fruit a drupe, usually detaching together with the surrounding calyx at maturity. Seeds 1, the remaining 3 ovules degenerating, globose, testa smooth, falling off at maturity. $2n = 20$. One species, *T. alpina* L., Alps, Carpathians.

296. *Conopholis* Wallr.

Conopholis Wallr., Orob. gen. diask.: 78 (1825); Beck-Mannagetta, Pflanzenreich 261: 22–23 (1930), rev.

Perennial holoparasitic herbs, pale brown or yellowish brown, with subterranean tuber, glabrous. Stems ascending to erect, rounded, fleshy. Leaves cauline, opposite, sessile, densely imbricate, reduced to achlorophyllous scales, ovoid or orbicular, acuminate. Flowers sessile. Calyx tubular, split on abaxial side, irregularly 4–5-dentate. Corolla strongly 2-lipped, pale brown or yellowish, lower lip shorter than upper lip. Stamens 4, exserted. Ovary ovoid to globose, unilocular with 4 parietal placentae. Seeds numerous, scobiform, 1 mm long, angled, testa thick, reticulate. $2n = 40$. Two species, south-eastern North America to

Mexico, *C. americana* (L.) Wallr. and *C. mexicana* A. Gray.

297. *Lathraea* L.

Lathraea L., Sp. Pl. 2: 605 (1753); Beck-Mannagetta, Pflanzenreich 261: 317–332 (1930), rev.; Heinricher, Monogr. Gattung *Lathraea*: 1–151 (1931).
Clandestina Adans. (1763).

Perennial holoparasitic herbs, entirely whitish-purplish, rhizomes with imbricate, fleshy scale-like leaves with internal chambers, glabrous except for glandular-pubescent upper stem and inflorescence. Stems ascending to erect, rounded. Leaves cauline, opposite, sessile, reduced to achlorophyllous scales, linear-lanceolate to ovoid or orbicular, acuminate. Flowers shortly to distinctly pedicellate, sometimes inflorescence axis subterranean and only flowers exposed on long pedicels above ground. Calyx 4-dentate, campanulate. Corolla strongly 2-lipped, white to purplish, lower lip shorter than upper lip. Stamens 4, included or weakly exserted. Ovary ovoid to globose, unilocular with 2 parietal placentae or bilocular. Capsule loculicidal, opening explosively in subg. *Clandestina* (Adans.) Heinricher. Seeds numerous, subglobose, testa reticulate to rugulose. 2n = 18, 36, 42. Seven species, temperate Eurasia, Europe.

Genera of close relationship to "Veronicaceae"-Gratioloideae or Gesneriaceae (included in the key)

298. *Cyrtandromoea* Zoll.

Cyrtandromoea Zoll., Syst. Verz. Ind. Arch. 3: 55, 58 (1858).

Perennial, suffrutescent herbs, glabrous to pubescent. Stems erect, quadrangular. Leaves petiolate, opposite, linear-lanceolate to ovoid, acuminate, with dentate to serrate margin. Calyx tubular, truncate at apex, with 5 short teeth, enlarged in fruit. Corolla rose to violet, subrotate, all lobes orbicular. Stamens 4. Ovary pubescent to glabrous, conical to cylindrical. Seeds numerous, ellipsoid, with reticulate testa. Eleven species, Burma, Thailand, western Malesia.

Tribe Rehmannieae Rouy (1909).

Usually rosulate herbs, inflorescence racemose, frondose, prophylls present or absent (*Rehmannia*), corolla cylindrical, anthers with two separate thecae, stigma bilobed, capsule loculicidal or loculicidal and septicidal (*Titanotrichum*), endosperm not alveolated.

299. *Rehmannia* Libosch. ex Fisch. & C. Mey.

Rehmannia Libosch. ex Fisch. & C. Mey., Index Sem. Hortus Petrop. 1: 36 (1835), nom. cons.

Perennial herbs, pubescent. Stems ascending to erect, terete. Leaves in a basal rosette and cauline, alternate, sessile or subsessile, lanceolate to ovoid, acuminate, with serrate margin. Calyx 5-lobed, campanulate, divided to middle. Corolla pink or white with darker spots, cylindrical, strongly 2-lipped. Stamens 4. Ovary ovoid. Seeds numerous, testa striate-reticulate. Nine species, eastern Asia.

300. *Titanotrichum* Soler.

Titanotrichum Soler., Ber. Deutsch. Bot. Ges. 27: 393 (1909).

Perennial herb, pubescent. Stems ascending to erect, terete. Leaves cauline, alternate, petiolate, lanceolate to ovoid, acuminate, with serrate margin. Inflorescence often with axillary bulbils. Flowers distinctly pedicellate. Calyx deeply 5-lobed almost to base, campanulate. Corolla yellow with brownish spots, cylindrical, strongly 2-lipped. Stamens 4, included. Ovary ovoid. Flowers often sterile. Seeds numerous, testa striate-reticulate. One species, *T. oldhami* Sol., Borneo, Sumatra, Taiwan, China, southern Japan.

301. *Triaenophora* (Hook.) Soler.

Triaenophora (Hook.) Soler., Ber. Deutsch. Bot. Ges. 27: 393, 399 (1909).
Spirostegia Ivanina (1955).

Perennial herbs, pubescent. Stems ascending to erect, terete. Leaves cauline or rosulate, alternate, petiolate, lanceolate to ovoid, acuminate, with serrate margin. Flowers distinctly pedicellate. Calyx shortly 5-lobed, campanulate. Corolla white with darker spots, cylindrical, strongly 2-lipped. Stamens 4, included. Ovary ovoid. Seeds numerous, testa striate-reticulate. Three species, North-east Asia, China.

"Stilbaceae"[1]

The following taxa are assigned to Stilbaceae according to recent molecular data. They are

[1] See also Stilbaceae, this Volume.

included in the key and taxonomic treatment of traditional Scrophulariaceae to facilitate their determination.

TRIBE BOWKERIEAE Barringer (1993).

Shrubs or small trees, inflorescence a thyrse with axillary cymes, sometimes 1-flowered (*Bowkeria*), corolla with pouch-shaped lower lip, anthers with two equal thecae, confluent at maturity, stigma clavate or bilobed (*Brookea*), capsule loculicidal and septicidal (*Anastrabe*) or septicidal, or fruit a berry (*Halleria*).

302. *Anastrabe* E. Mey. ex Benth.

Anastrabe E. Mey. ex Benth. in Hook., Companion Bot. Mag. 2: 54 (1836).

Shrub or small tree with numerous branches, pubescent to tomentose with stellate hairs. Stems erect, rounded. Leaves cauline, opposite or subopposite, petiolate, lanceolate or oblong, acuminate, with entire or serrate margin, leathery. Flowers distinctly pedicellate. Calyx 5-lobed to middle, campanulate, tomentose with stellate hairs. Corolla yellow, inside with crimson marks, limb 2-lipped, lobes spreading, tube shorter than lobes, campanulate. Stamens 4, included. Ovary subglobose. One species, *A. integerrima* E. Mey. ex Benth., coastal districts, from Mozambique to South Africa (Natal and Cape).

303. *Bowkeria* Harv.

Bowkeria Harv., Thes. Cap. 1: 24 (1859).

Shrubs or small trees with numerous branches, glandular-pubescent to tomentose. Stems erect, rounded. Leaves cauline, usually ternate, sessile or petiolate, lanceolate, elliptic or ovoid, acuminate, with subentire to crenate or dentate margin, leathery. Flowers distinctly pedicellate. Calyx 5-lobed, campanulate, lobes subequal, adaxial lobe broadest. Corolla white or yellow, limb 2-lipped, tube shortly cylindrical below, broadly campanulate above. Stamens 4, included, staminode sometimes present. Ovary cylindrical or conic, 2–3-locular. Stigma clavate. Five species, South Africa, frequently cultivated trees and shrubs.

304. *Brookea* Benth.

Brookea Benth. in Benth. & Hook. f., Gen. Pl. 2: 939 (1876).

Shrubs or small trees with numerous branches, densely tomentose to villose. Stems erect, rounded. Leaves cauline, opposite, petiolate, lanceolate or ovoid, acuminate, with serrate to dentate margin, leathery. Flowers subsessile or shortly pedicellate. Calyx 4–5-lobed to one third of its length, campanulate, tomentose. Corolla white, limb 2-lipped, lobes spreading, tube longer than lobes, cylindrical. Stamens 4, included. Ovary ovoid. Three species, Borneo.

305. *Halleria* L.

Halleria L., Sp. Pl. 2: 625 (1753).

Shrubs or small trees, glabrous. Stems erect, quadrangular to winged. Leaves cauline, opposite, petiolate, ovoid or elliptic to orbicular, acuminate, with subentire to dentate margin. Flowers distinctly pedicellate. Calyx 3- to 5-lobed, usually 4-lobed, subrotate or cup-shaped. Corolla red to orange, limb 2-lipped, subactinomorphic, campanulate, 4–5-lobed, tube curved or straight, inflated above to funnel-shaped. Stamens 4, exserted. Ovary ovoid. Stigma capitate to bilobed. Seeds numerous, compressed, narrowly winged. Four species, tropical and South Africa (two spp.), Madagascar (two spp.).

306. *Ixianthes* Benth.

Ixianthes Benth., Companion Bot. Mag. 2: 53 (1836).

Shrubs with numerous branches, glabrous except for leaves and flowers. Stems erect, rounded, woody. Leaves cauline, densely 3–4-verticillate, sessile, linear, acuminate, pinnately veined and with serrate margin in upper half, leathery, hispid or glabrous. Inflorescence terminal, cymose, with axillary cymes, frondose, bracteoles present, cymes often 1-flowered and thus resembling a raceme. Flowers distinctly pedicellate. Calyx 3-lobed to mid, campanulate, pilose. Corolla yellow, densely pilose-viscid outside, limb 2-lipped, upper lip 2-lobed, lower lip 3-lobed, lobes spreading, equalling tube, tube widely campanulate, longitudinally gibbous on back and with corresponding fold within, densely glandular. Stamens 2, inserted in basal part of corolla-tube, included, staminodes 2–3, filaments thickish, anthers with two separate equal thecae, reniform, confluent at maturity. Ovary ovoid, densely glandular, with straight filiform style. Stigma clavate. Fruit a capsule with loculicidal and septicidal dehiscence. Seeds numerous, oblong, testa membranous, reticulate. One species, *I. retzioides* Benth., in South Africa

(Tulbagh to Clanwilliam, Cape region), growing as rheophyte.

Genus of uncertain placement (not included in the key)

Heteranthia Nees & Mart.

Heteranthia Nees & Mart., Nova Acta Phys. Med. Acad. Caes. Leop.-Carol. Nat.-Cur. 11: 41 (1823).

Glabrous herbs. Leaves cauline, alternate, long-petiolate. Inflorescence racemose, lax, bracteose, prophylls absent. Flowers shortly pedicellate. Calyx campanulate, 5-lobed, lobes equal. Corolla small, limb 2-lipped, upper lip entire to emarginate, lower lip 3-lobed, lobes spreading, rounded, tube campanulate, short. Stamens 4, included, anthers with two equal thecae and orbicular connective. Capsule globose septicidal. Seeds numerous, testa striate. One species, *H. decipiens* Nees & Mart. from Brazil.

Bentham (1876) in Bentham and Hooker (1876) placed this genus with some doubts close to *Leucophyllum* and *Eremogeton*, which is followed by Brummitt (1992). A placement in Solanaceae seems, however, more appropriate, as Solereder (1915) has shown the similarity to *Browallia*, *Melananthia* and *Schwenkea*. Also, the anatomy of the genus resembles that of Solanacae. More material is needed to settle its systematic position.

Doubtful genus (not included in the key)

Trungboa Rauschert

Trungboa Rauschert, Taxon 31(3): 562 (1982).
Cyphocalyx Gagnepain (1950).

Shrubs, densely yellow-tomentose. Stems creeping, rounded. Leaves cauline, opposite, subsessile, elliptic to oblong, cordate, acuminate, with entire margin, subglabrous above, pilose below. Inflorescence terminal, racemose, subcapitate, prophylls absent. Flowers distinctly pedicellate. Calyx deeply 2-lobed, lobes obtuse, gibbose to galeate. Corolla up to 6 mm long, 4-lobed, semiorbicular, lobes subequal, upper lobes erect, lower lobes patent, tube short and not exceeding calyx. Stamens 4, filaments flattened, anthers with two suborbicular equal thecae. Ovary globose-depressed, glabrous. Stigma capitate or weakly bilobed. Fruit unknown. One species, *T. poilanei* (Gagnepain) Rauschert, in Annam, Vietnam.

After study of the type specimen in P, I can not assign this taxon to Scrophulariaceae or any related family. Unfortunately, the flowers are poorly preserved and fruits are lacking. Later workers (e.g. Yamazaki, Flore Cambodge, Laos et Vietnam 1988) ignore this taxon.

Excluded genera

Amalophyllon T.S. Brandegee, Univ. Calif. Publ. Bot. 6: 63 (1914).

Gesneriaceae (*Achimenes* Pers.)

Digomphia Benth., Lond. J. Bot. 5: 364 (1846).

Bignoniaceae

Dolichostemon Bonati, Bull. Soc. Bot. Genève, sér. 2, 15: 109 (1924).

Labiatae

Gynocraterium Bremek., Kew Bull. 1939: 557 (1939).

Acanthaceae

Isaloa Humbert, Bull. Soc. Bot. France 87: 316 (1937).

Acanthaceae

Jerdonia Wight, Icon. Pl. Ind. Orient. 4(2): 10 (1848).

Gesneriaceae

Namation Brand, Feddes Repert. 10: 280 (1912).

Solanaceae (*Petunia* Juss.)

Neozenkerina Mildbr., Notizbl. Bot. Gart. Berlin 7: 491 (1921).

Acanthaceae

Sanango Bunting & Duke, Ann. Missouri Bot. Gard. 48: 270 (1961).

Gomara Ruiz & Pav. (1794).
Gomaranthus S. Rauschert (1982), nom. nov.

Gesneriaceae

Selected Bibliography

Albach, D.C., Chase, M.W. 2001. Paraphyly of *Veronica* (Veroniceae, Scrophulariaceae): evidence from internal transcribed spacer (ITS) sequences of nuclear ribosomal DNA. J. Plant Res. 114: 9–18.

Andersson, L., Molau, U. 1980. The inflorescence of *Calceolaria*. Bot. Not. 133: 21–32.

Arekal, G.D. 1963. Embryological studies in Canadian representatives of the tribe Rhinantheae, Scrophulariaceae. Can. J. Bot. 41: 267–302.

Argue, C.L. 1980. Pollen morphology in the genus *Mimulus* (Scrophulariaceae) and its taxonomic significance. Am. J. Bot. 67: 68–87.

Argue, C.L. 1985. Pollen morphology in the genera *Monttea* and *Melosperma* (Scrophulariaceae). Am. J. Bot. 72: 1248–1255.

Argue, C.L. 1986. Pollen morphology of *Amphianthus*, *Artanema*, *Curanga*, *Glossostigma*, and *Peplidium* (Scrophulariaceae-Gratioleae). Am. J. Bot. 73: 1570–1576.

Argue, C.L. 1990. Pollen morphology of *Deinostema*, *Geochorda*, *Gratiola*, *Ildefonsia*, *Sophronanthe*, and *Tragiola* (Scrophulariaceae, Gratioleae, Gratiolinae). Can. J. Bot. 68: 1651–1660.

Argue, C.L. 1993. Pollen morphology in the Selagineae, Manuleae (Scrophulariaceae) and selected Globulariaceae, and its taxonomic significance. Am. J. Bot. 80: 723–733.

Armstrong, J.E. 1985. The delimitation of Bignoniaceae and Scrophulariaceae based on floral anatomy and the placement of problem genera. Am. J. Bot. 72: 755–766.

Armstrong, J.E. 1992. Lever action anthers and the forcible shedding of pollen in *Torenia*. Am. J. Bot. 79: 34–40.

Armstrong, J.E., Douglas, A.W. 1989. The ontogenetic basis for corolla aestivation in Scrophulariaceae. Bull. Torrey Bot. Club 116: 378–389.

Arroyo, M.K.T., Peñazola, A. 1990. Genetic self-incompatibility in a South American species of *Ourisia* (Scrophulariaceae). New Zeal. J. Bot. 28: 467–476.

Atsatt, P.R. 1970. Hemiparasitic flowering plants: phenotypic canalization by hosts. Nature 225: 1161–1163.

Bachmann, E.T. 1880. Die Entwicklungsgeschichte und der Bau der Samenschalen der Scrophularineen. Nova Acta Leop. 43: 1–179.

Barringer, K. 1993. Five new tribes in the Scrophulariaceae. Novon 3: 15–17.

Bartels, D., Schneider, K., Terstappen, G., Piatkowski, D., Salamini, F. 1990. Molecular cloning of abscisic acid modulated genes which are induced during desiccation of the resurrection plant *Craterostigma plantagineum*. Planta 181: 27–34.

Barthlott, W. 1980. Morphogenese und Mikromorphologie komplexer Cuticularfaltungsmuster an Blütentrichomen von *Antirrhinum* L. (Scrophulariaceae). Ber. Deutsch. Bot. Ges. 93: 379–390.

Beardsley, P.M., Olmstead, R.G. 2002. Redefining Phrymaceae: the placement of *Mimulus*, tribe Mimuleae, and *Phryma*. Am. J. Botany 89: 1093–1102.

Beck v. Mannagetta, G. 1930. Orobanchaceae. In: Engler, A. (ed.) Das Pflanzenreich 4, 261: 1–348.

Bellini, R. 1907. Criteri per una nuova classificazione delle Personatae (Scrophulariaceae et Rhinanthaceae). Ann. Bot. 6: 131–145.

Bello, M.A., Chase, M.W., Olmstead, R.G., Rønsted, N., Albach, D. 2002. The páramo endemic *Aragoa* is the sister genus of *Plantago* (Plantaginaceae; Lamiales): evidence from plastid *rbcL* and nuclear ribosomal ITS sequence data. Kew Bull. 57: 585–597.

Bentham, G. 1846. Scrophulariaceae. In: De Candolle, A. Prodromus systematis Naturalis 10: 186–586. Paris: Victor Masson.

Bentham, G. 1876. Scrophulariaceae. In: Bentham, G., Hooker, J.D. (eds.) Genera Plantarum 2: 913–980. London: Reeve & Co.

Bernacchia, G., Salamini, F., Bartels, D. 1996. Molecular characterization of the rehydration process of the resurrection plant *Craterostigma plantagineum*. Plant Physiol. 111: 1043–1050.

Bigazzi, M. 1993. A survey of the intranuclear inclusions in the Schrophulariaceae (sic) and their systematic significance. Nord. J. Bot. 13: 19–31.

Bigazzi, M., Tardelli, M. 1990. Pollen morphology and ultrastructure of the Old World Antirrhineae (Scrophulariaceae). Grana 29: 257–275.

Bolliger, M. 1985. Die Drüsenhaare der Gattung *Odontites* Ludwig (Scrophulariaceae) und ihre systematische Bedeutung. Bot. Jahrb. Syst. 107: 153–175.

Bolliger, M. 1990. The pollen morphology of *Odontites* (Scrophulariaceae) and its taxonomic significance. Plant Syst. Evol. 173: 159–178.

Bolliger, M. 1996. Monographie der Gattung *Odontites* (Scrophulariaceae) sowie der verwandten Gattungen *Macrosyringion*, *Odontitella*, *Bornmuellerantha* und *Bartsiella*. Willdenowia 26: 37–168.

Brühl, P. 1920. On the systematic position of *Lindenbergia* Lehmann. J. Dept. Sci. Univ. Calcutta, Bot. 2: 11–16.

Brummitt, R.K. 1992. Vascular plant families and genera. Kew: Royal Botanic Gardens, 804 p.

Canne, J.H. 1980. Seed surface features in *Aureolaria*, *Brachystigma*, *Tomanthera*, and certain South American *Agalinis* (Scrophulariaceae). Syst. Bot. 5: 241–252.

Cronquist, A. 1981. An integrated system of classification of flowering plants. New York: Columbia University Press.

Ehrendorfer, F., Vitek, E. 1984. Evolution alpiner Populationen von *Euphrasia* (Scrophulariaceae). Entdeckung kleinblütiger diploider Sippen. Plant Syst. Evol. 144: 25–44.

Elisens, W.J. 1986. Pollen morphology and systematic relationships among New World species in the tribe Antirrhineae (Scrophulariaceae). Am. J. Bot. 73: 1298–1311.

Elisens, W.J., Tomb, A.S. 1983. Seed morphology in New World Antirrhineae (Scrophulariaceae). Systematic and phylogenetic implications. Plant Syst. Evol. 142: 23–47.

Endress, P.K. 1996. Diversity and evolutionary biology of tropical flowers. Cambridge: Cambridge University Press, 511 p.

Engler, A. 1897. Scrophulariaceae africanae II. Bot. Jahrb. Syst. 23: 497–517.

Erdtman, G. 1952. Pollen morphology and plant taxonomy. Angiosperms. Stockholm: Almquist Wiksell.

Falcão Ichaso, C.L. 1978. Tipos de Sementes Encotradas nas Scrophulariaceae. Rodriguesia 30: 335–344.

Fischer, E. 1992. Systematik der afrikanischen Lindernieae (Scrophulariaceae). Trop. Subtrop. Pflanzenw. 82: 1–365.

Fischer, E. 1995. Revision of the Lindernieae (Scrophulariaceae) in Madagascar. 1. The genera *Lindernia* Allioni and *Crepidorhopalon* E. Fischer. Bull. Mus. Natl. Hist. Nat. Paris, 4ème sér., 17, sect. B, Adansonia: 227–257.

Fischer, E. 1997a. Revision of the genus *Stemodiopsis* Engl. (Scrophulariaceae). Bot. Jahrb. Syst. 119: 305–326.

Fischer, E. 1997b. A revision of the genus *Dopatrium* Buch.-Ham. ex Benth. (Scrophulariaceae). Nord. J. Bot. 17: 527–555.

Fischer, M.A. 1969. Einige Chromosomenzahlen aus den Gattungen *Veronica*, *Pseudolysimachion*, *Paederota*, *Wulfenia* und *Lagotis* (Scrophulariaceae-Veronicinae). Oesterr. Bot. Z. 116: 430–443.

Gaff, D. 1977. Desiccation tolerant vascular plants of Southern Africa. Oecologia 31: 95–109.

Gentry, A.H. 1980. Bignoniaceae. Flora Neotrop. Monogr. 25. New York Botanical Garden.

Ghebrehiwet, M. 1999. Phylogenetic and taxonomic studies in the tribe Antirrhineae (Scrophulariaceae) with special reference to the genera *Kickxia* nad *Nanorrhinum*. Acta Univ. Ups. Comprehensive Summaries of Uppsala Dissertations from the Faculty of Science and Technology 489.

Ghebrehiwet, M. 2000. Taxonomy, phylogeny and biogeography of *Kickxia* and *Nanorrhinum* (Scrophulariaceae). Nord. J. Bot. 20: 655–689.

Ghebrehiwet, M., Bremer, B., Thulin, M. 2000. Phylogeny of the tribe Antirrhineae (Scrophulariaceae) based on morphological and *ndh*F sequence data. Plant Syst. Evol. 220: 223–239.

Glisic, L.M. 1936–1937. Ein Versuch der Verwertung der Endospermmerkmale für typologische und phylogenetische Zwecke innerhalb der Scrophulariaceen. Bull. Inst. Jard. Bot. Univ. Belgrade 4: 42–73.

Glück, H. 1934. *Limosella*-Studien. Beiträge zur Systematik, Morphologie und Biologie der Gattung *Limosella*. Bot. Jahrb. Syst. 66: 488–566.

Gómez, J.M. 2002. Self-pollination in *Euphrasia willkommii* Freyn (Scrophulariaceae), an endemic species from the alpine of the Sierra Nevada (Spain). Plant Syst. Evol. 232: 63–71.

Greilhuber, J. 1971. Chromosomenzahlen von *Verbascum austriacum*, *V. lanatum* und einigen Rhinanthoideen (Scrophulariaceae). Mitt. Bot. Arbeitsgem. Oberösterr. Landesmus. Linz 3: 31–35.

Guédès, M. 1974. Le gynécée de *Paulownia* et *Schlegelia* et le problème de la délimination des Scrofulariacées et Bignoniacées. Compt. Rend. Acad. Sci., Paris 278: 2629–2632.

Hakki, M.I. 1977. Embryologische und morphologische Untersuchungen an Pflanzen aus Südafrika. I. Über die Embryologie, Morphologie und systematische Zugehörigkeit von *Dermatobotrys saundersii* Bolus. Bot. Jahrb. Syst. 98: 93–119.

Hallier, H. 1903. Ueber die Abgrenzung und Verwandtschaft der einzelnen Sippen bei den Scrophularineen. Bull. Herb. Boissier, 2ème sér. 3: 181–207.

Hamann, U. 1958. Morphologische und histogenetische Studien an *Veronica*-Arten. 1. Zur Morphologie der Blütenstände im Hinblick auf einen allgemeinen Infloreszenzbegriff. Bot. Jahrb. Syst. 78: 69–118.

Hambler, D.J. 1953. Prochromosomes and supernumary chromosomes in *Rhinanthus minor* Ehrh. Nature 172: 629–630.

Hambler, D.J. 1954. Cytology of the Scrophulariaceae and Orobanchaceae. Nature 174: 838.

Hartl, D. 1955. Das Vorkommen rhinanthoider Knospendeckung bei *Lindenbergia* Lehm., einer Gattung der Scrophulariaceae-Antirrhinoideae. Oesterr. Bot. Z. 102: 80–83.

Hartl, D. 1956a. Morphologische Studien am Pistill der Scrophulariaceen. Oesterr. Bot. Z. 103: 185–242.

Hartl, D. 1956b. Die Beziehungen zwischen den Plazenten der Lentibulariaceen und Scrophulariaceen nebst einem Exkurs über die Spezialisationsrichtungen der Plazentation. Beitr. Biol. Pflanzen 32: 471–490.

Hartl, D. 1959. Das alveolierte Endosperm bei Scrophulariaceen, seine Entstehung, Anatomie und taxonomische Bedeutung. Beitr. Biol. Pflanzen 35: 95–110.

Hartl, D. 1962. Die morphologische Natur und die Verbreitung des Apicalseptums. Beitr. Biol. Pflanzen 37: 241–330.

Hartl, D. 1963. Das Placentoid der Pollensäcke, ein Merkmal der Tubifloren. Ber. Deutsch. Bot. Ges. 76: 71–72.

Hartl, D. 1965–1974. Scrophulariaceae. In: Hegi, G. (ed.) Illustrierte Flora von Mitteleuropa 6(1), 2. Aufl. München: Paul Parey.

Hartl, D. 1977. *Rhabdotosperma*, eine neue, aus Gliedern von *Verbascum* L. und *Celsia* L. gebildete Gattung der Scrophulariaceen. Beitr. Biol. Pflanzen 53: 55–60.

Hedberg, O. 1955. A taxonomic revision of the genus *Sibthorpia* L. Bot. Not. 108: 161–183.

Hegnauer, R. 1973. Scrophulariaceae. Chemotaxonomie der Pflanzen 6: 343–386.

Hegnauer, R. 1990. Scrophulariaceae. Chemotaxonomie der Pflanzen 9: 527–556.

Heil, H. 1927. Vergleichend-anatomische Studien an Samen von *Chamaegigas* und verwandten Gattungen. Ber. Deutsch. Bot. Ges. 45: 555–561.

Heilmeier, H., Ratcliffe, G., Hartung, W. 2000. Urea: a nitrogen source for the aquatic resurrection plant *Chamaegigas intrepidus* Dinter. Oecologia 123: 9–14.

Hickel, B. 1967. Zur Kenntnis einer xerophilen Wasserpflanze: *Chamaegigas intrepidus* Dtr. aus Südwestafrika. Int. Rev. ges. Hydrobiol. 52: 361–400.

Hilliard, O.M. 1994. The Manuleae. A tribe of Scrophulariaceae. University of Edinburgh, 579 p.

Hilliard, O.M. 1999. The tribe Selagineae (Scrophulariaceae). Kew: Royal Botanical Gardens, 312 p.

Hjertson, M. 1995. Taxonomy, phylogeny and biogeography of *Lindenbergia* (Scrophulariaceae). Bot. J. Linn. Soc. 119: 265–321.

Hjertson, M. 1997. Systematics of *Lindenbergia* and *Campylanthus* (Scrophulariaceae). Acta Univ. Ups. Comprehensive Summaries of Uppsala Dissertations from the Faculty of Science and Technology 331: 1–46.

Hofmann, H.P., Fischer, E. 2003. Generic delimitation of *Sopubia* Buchanan-Hamilton (Scrophulariaceae) with description of *Parasopubia* gen. nov. Bot. Jahrb. Syst. (in press).

Hong, D.Y. 1984. Taxonomy and evolution of the Veroniceae (Scrophulariaceae) with special reference to palynology. Opera Bot. 75: 1–60.

Hufford, L.D. 1992. Floral structure of *Besseya* and *Synthyris* (Scrophulariaceae). Int. J. Plant Sci. 153: 217–229.

Ingram, J., Bartels, D. 1996. The molecular basis of dehydration tolerance in plants. Ann. Rev. Plant Physiol. Plant Mol. Biol. 47: 377–403.

Junell, S. 1961. Ovarian morphology and taxonomical position of Selagineae. Svensk Bot. Tidskr. 55: 168–192.

Kalisz, S., Vogler, D., Fails, B., Finer, M., Shepard, E., Herman, T., Gonzales, R. 1999. The mechanism of delayed selfing in *Collinsia verna* (Scrophulariaceae). Am. J. Bot. 88: 447–454.

Kampny, A.M. 1995. Pollination and flower diversity in Scrophulariaceae. Bot. Rev. 61: 350–366.

Kapoor, T., Parulekar, N.K., Vijayaraghavan, M.R. 1975. Contribution to the embryology of *Celsia coromandeliana* Vahl. with a discussion on its affinities with *Verbascum thapsus.* Bot. Not. 1975, 128: 338–449.

Kornhall, P., Heidari, N., Bremer, B. 2001. Selagineae and Manuleae, two tribes or one? Phylogenetic studies in Scrophulariaceae. Plant Syst. Evol. 228: 199–218.

Krishna Iyengar, C.V. 1939. Development of the embryo sac and endosperm-haustoria in some members of Scrophulariaceae II. *Isoplexis canariensis* Lindl. et *Celsia coromandeliana* Vahl. J. Indian Bot. Soc. 18: 13–20.

Krishna Iyengar, C.V. 1940. Development of embryo sac and endosperm-haustoria in some members of Scrophulari-

aceae III. *Ilysanthes hyssopioides* Benth. and *Bonnaya tenuifolia* Spreng. J. Indian Bot. Soc. 19: 5–18.
Kuijt, J. 1969. The biology of parasitic flowering plants. Berkeley: University of California Press.
Kuijt, J., Visser, J.H., Weber, H.C. 1978. Morphological observations on leaf haustoria and related organs of the South African genus *Hyobanche* (Scrophulariaceae). Can. J. Bot. 56: 2981–2986.
Lakusic, R. 1978. Die chorologisch-ökologische und morphologisch-zytologische Differenzierung der europäischen Arten der Gattung *Wulfenia* Jacq. Bot. Jahrb. Syst. 99: 443–461.
Lersten, N.R., Curtis, J.D. 2001. Idioblasts and other unusual internal foliar secretory structures in Scrophulariaceae. Plant Syst. Evol. 227: 63–73.
Macior, L.W. 1983. The pollination dynamics of sympatric species of *Pedicularis* (Scrophulariaceae). Am. J. Bot. 70 (6): 844–853.
Magin, N., Classen, R., Gack, C. 1989. The morphology of false anthers in *Craterostigma plantagineum* and *Torenia polygonioides* (Scrophulariaceae). Can. J. Bot. 67: 1931–1937.
Metcalfe, C.R., Chalk, L. 1950. Anatomy of the Dicotyledons 2: 725–1500.
Molau, U. 1988. Scrophulariaceae. Part I. Calceolarieae. Flora Neotrop. Monogr. 47: 1–326.
Molau, U., Eriksen, B., Teilmann Knudsen, J. 1989. Predispersal seed predation in *Bartsia alpina*. Oecologia 81: 181–185.
Musselman, L.J., Mann, W.F. 1976. A survey of surface characteristics of seeds of Scrophulariaceae and Orobanchaceae using scanning electron microscopy. Phytomorphology 26: 370–378.
Olmstead, R.G., Reeves, P.A. 1995. Evidence for the polyphyly of the Scrophulariaceae based on chloroplast *rbc*L and *ndh*F sequences. Ann. Missouri Bot. Gard. 82: 176–193.
Olmstead, R.G., DePamphilis, W., Wolfe, A.D., Young, N.D., Ellisons, W.J., Reeves, P.A. 2001. Disintegration of the Scrophulariaceae. Am. J. Bot. 88: 348–361.
Oxelman, B., Backlund, M., Bremer, B. 1999. Relationships of the Buddlejaceae investigated using parsimony jackknife and branch support analysis of *ndhF* and *rbcL* sequence data. Syst. Bot. 24: 164–182.
Pazy, B. 1998. Diploidization failure and apomixis in Orobanchaceae. Bot. J. Linn. Soc. 128: 99–103.
Raman, S. 1987. A code proposed for the classification of trichomes as applied to the Scrophulariaceae. Beitr. Biol. Pflanzen 62: 349–367.
Raman, S. 1989a. The trichomes on the corolla of the Scrophulariaceae. I. Tribes Aptosimeae and Scrophularieae. Beitr. Biol. Pflanzen 64: 127–140.
Raman, S. 1989b. The trichomes on the corolla of the Scrophulariaceae. II. Tribes Hemimerideae and Calceolarieae. Beitr. Biol. Pflanzen 64: 141–155.
Raman, S. 1989c. The trichomes on the corolla of the Scrophulariaceae. III. Tribe Digitalieae. Beitr. Biol. Pflanzen 64: 199–212.
Raman, S. 1989d. The trichomes on the corolla of the Scrophulariaceae. V. Tribe Antirrhineae Chavannes. Beitr. Biol. Pflanzen 64: 357–375.
Raman, S. 1990. The trichomes on the corolla of the Scrophulariaceae. 6. Tribe Pedicularieae D.Don. 1825, (Rhinantheae Benth. 1835). Beitr. Biol. Pflanzen 64: 377–390.
Roeder, E., Borauel, T. 1992. Pyrrolizidine alkaloids from *Melampyrum pratense*. Nat. Toxins 1: 35–37.
Rothmaler, W. 1943. Zur Gliederung der Antirrhineae. Feddes Repert. 52: 16–39.
Schiller, P., Heilmeier, H., Hartung, W. 1997. Abscisic acid (ABA) relations in the aquatic resurrection plant *Chamaegigas intrepidus* under naturally fluctuating environmental conditions. New Phytol. 136: 603–611.
Schiller, P., Heilmeier, H., Hartung, W. 1998. Uptake of amino acids by the aquatic resurrection plant *Chamaegigas intrepidus* and its implication for N nutrition. Oecologia 117: 63–69.
Schiller, P., Wolf, R., Hartung, W. 1999. A scanning microscopical study of hydrated and desiccated submerged leaves of the aquatic resurrection plant *Chamaegigas intrepidus*. Flora 194: 97–102.
Seine, R., Fischer, E., Barthlott, W. 1995. Notes on the Scrophulariaceae on Zimbabwean inselbergs with description of *Lindernia syncerus* sp. nov. Feddes Repert. 106: 7–12.
Sérsic, A.N., Cocucci, A.A. 1999. An unusual kind of nectary in the oil flowers of *Monttea*: its structure and function. Flora 194: 393–404.
Simpson, B.B., Neff, J.L., Dieringer, G. 1990. The production of floral oils by *Monttea* (Scrophulariaceae) and the function of tarsal pads in *Centris* bees. Plant Syst. Evol. 173: 209–222.
Smiley, C.J. 1961. A record of *Paulownia* in the tertiary of North America. Am. J. Bot. 48: 175–179.
Snogerup, B. 1977. Chromosome numbers of scandinavian *Odontites* species. Bot. Not. 130: 121–124.
Solereder, H. 1915. Über die Versetzung der Gattung *Heteranthia* von den Scrophulariaceen zu den Solanaceen. Beih. Bot. Zbl. 23: 113–117.
Speta, F. 1972. Über Eiweikörper in Zellkernen bei Scrophulariaceen; Vorkommen, Form und systematische Bindung. Oesterr. Bot. Z. 120: 117–136.
Speta, F. 1977. Proteinkörper in Zellkernen: neue Ergebnisse und deren Bedeutung für die Gefäpflanzensystematik nebst einer Literaturübersicht für die Jahre 1966–1976. Candollea 32: 133–163.
Speta, F., Greilhuber, J. 1970. Über das gleichzeitige Vorkommen von zweierlei Eiweisskörpern in den Zellkernen von *Pseudolysimachion spicatum* und einigen anderen Scrophulariaceen. Oesterr. Bot. Z. 118: 1–16.
Steiner, K. 1990. The *Diascia* (Scrophulariaceae) window: an orientation cue for oil-collecting bees. Bot. J. Linn. Soc. 102: 175–195.
Steiner, K. 1992. A new *Diascia* species from the Richtersveld, South Africa. S. Afr. J. Bot. 58: 36–38.
Steiner, K. 1996. Chromosome numbers and relationships in tribe Hemimerideae (Scrophulariaceae). Syst. Bot. 21: 63–76.
Steiner, K., Whitehead, V.B. 1990. Pollinator adaptation to oil-secreting flowers – *Rediviva* and *Diascia*. Evolution 44: 1701–1707.
Sutton, D.A. 1988. A revision of the tribe Antirrhineae. London: British Museum of Natural History, 575 p.
Takhtajan, A. 1980. Outline of the classification of flowering plants (Magnoliophyta). Bot. Rev. 46 (3): 225–359.
Takhtajan, A. 1987. Systema Magnoliophytorum (in Russian). Leningrad: Nauka.
Thieret, J.W. 1955. The seeds of *Veronica* and allied genera. Lloydia 18: 37–45.
Thieret, J.W. 1967. Supraspecific classification in the Scrophulariaceae: a review. Sida 3: 87–106.
Thorne, R.T. 1992. Classification and geography of the flowering plants. Bot. Rev. 58: 225–348.
Troll, W. 1964. Die Infloreszenzen. Typologie und Stellung im Aufbau des Vegetationskörpers. Jena: Gustav Fischer, 615 p.

Varghese, T.M. 1968. Studies in the family Schrophulariaceae (sic). II. Pollen morphology. J. Palynol. 4: 91–97.
Vitek, E. 1985a. Evolution alpiner Populationen von *Euphrasia* (Schrophulariaceae): Die mittel- bis kleinblütigen drüsenhaarigen Arten. Plant Syst. Evol. 148: 215–237.
Vitek, E. 1985b. Evolution alpiner Populationen von *Euphrasia* (Schrophulariaceae): *E. alpina* und *E. christii*. Plant Syst. Evol. 149: 1–18.
Vitek, E. 1986. Evolution alpiner Populationen von *Euphrasia* (Schrophulariaceae): Die tetraploide *E. minima*. Plant Syst. Evol. 151: 241–269.
Vogel, S. 1954. Blütenbiologische Typen als Elemente der Sippengliederung, dargestellt anhand der Flora Südafrikas. Bot. Stud. 1: 1–338.
Vogel, S. 1984. The *Diascia* flower and its bee- and oil-based symbiosis in southern Africa. Acta Bot. Neerl. 33: 509–518.
Vogel, S. 1988. Die Ölblumensymbiosen - Parallelismus und andere Aspekte ihrer Entwicklung in Raum und Zeit. Z. zool. Syst. Evol.-forsch. 26: 341–362.
Vogel, S., Machado, I.C. 1991. Pollination of four sympatric species of *Angelonia* (Scrophulariaceae) by oil-collecting bees in North East Brazil. Plant Syst. Evol. 153–178.
Vogel, S., Michener, C.D. 1985. Long bee legs and oil-producing floral spurs, and a new *Rediviva* (Hymenoptera, Melittidae; Scrophulariaceae). J. Kansas Entomol. Soc. 58: 359–364.
Wagstaff, S.J., Garnock-Jones, P.J. 1998. Evolution and biogeography of the *Hebe* complex (Scrophulariaceae) inferred from ITS sequences. New Zeal. J. Bot. 36: 425–437.
Ward, H.M., Dale, E. 1898. On *Craterostigma pumilum* Hochst., a rare plant from Somali-Land. Trans. Linn. Soc. Bot. II, 5: 343–355.
Weber, A. 1973. Die Struktur der paarblütigen Partialfloreszenzen der Gesneriaceen und bestimmter Scrophulariaceen. Beitr. Biol. Pflanzen 49: 129–160.
Weber, A. 1989. Family position and conjectural affinities of *Charadrophila capensis* Marloth. Bot. Jahrb. Syst. 111: 87–119.
Weber, H.C. 1975. Vergleichende Betrachtungen über die unterirdischen Organe von *Lathraea squamaria* L. und *Tozzia alpina* L. (Scrophulariaceae). Beitr. Biol. Pflanzen 51: 1–15
Weber, H.C. 1976a. Anatomische Studien an den Haustorien einiger parasitischer Scrophulariaceen Mitteleuropas. Ber. Deutsch. Bot. Ges. 89: 57–84.
Weber, H.C. 1976b. Studies on new types of haustoria in some Central European Rhinanthoideae (Scrophulariaceae). Plant Syst. Evol. 125: 223–232.
Weber, H.C. 1976c. Über Wirtspflanzen und Parasitismus einiger mitteleuropäischer Rhinanthoideae (Scrophulariaceae). Plant Syst. Evol. 125: 97–107.
Weber, H.C. 1980. Zur Evolution des Parasitismus bei den Scrophulariaceae und Orobanchaceae. Plant Syst. Evol. 136: 217–232.
Weber, H.C., Weberling, F. 1975. Zur Morphologie der Haustorien von Rhinanthoideen (Scroph.). Ber. Deutsch. Bot. Ges. 88: 329–345.
Weberling, F., Troll, W. 1998. Die Infloreszenzen. Typologie und Stellung im Aufbau des Vegetationskörpers, Bd. 2, 2. Jena: Gustav Fischer, 483 p.
Wettstein, R. v. 1891. Scrophulariaceae. In: Engler & Prantl, Die natürlichen Pflanzenfamilien IV, 3b. Leipzig: Engelmann, pp. 39–107.
Witsch, H. 1932. Chromosomenstudien an mitteleuropäischen Rhinantheen. Oesterr. Bot. Z. 81: 108–141.
Yamazaki, T. 1953. On the floral structure, seed development, and affinities of *Deinostema*, a new genus of Scrophulariaceae. J. Jap. Bot. 28: 129–133.
Young, N.D., Steiner, K.E., dePamphilis, C.W. 1999. The evolution of parasitism in Scrophulariaceae/Orobanchaceae: plastid gene sequences refute an evolutionary transition series. Ann. Missouri Bot. Gard. 86: 876–893.
Ziegler, A. 1955. Die blau fluoreszierenden Idioblasten der Scrophulariaceen. Sitzungsber. Österr. Akad. Wiss., Math.-Naturwiss. Kl. Abt. 1, 164: 419–485.

Stilbaceae[1]

H.P. Linder

Stilbaceae Kunth, Handb. Bot.: 393 (1831), nom. cons.
Retziaceae Bartling (1830).
Buddlejaceae K. Wilhelm (1910) p.p.

Trees, small shrubs or shrublets; shrubs single-stemmed, or multistemmed from a lignotuber, branching monopodially or sympodially, young branches often angular and puberulous. Leaves estipulate or with a stipular ring (*Nuxia*), opposite, ternate, or in dense (pseudo)whorls of 3–7, rarely alternate on some lateral branches, sessile or petiolate (*Nuxia*), blade simple, in the genera other than *Nuxia* more or less ericoid, linear, with recurved margins and one or two stomatal grooves. Inflorescences terminal or rarely on axillary branches (*Retzia*), thyrsoid or dichotomously branched, ultimate branches with solitary flowers or heads of three or more flowers, often with a single compact head. Bracts various, but often with two foliose prophylls, often also bibracteate at the inflorescence nodes. Flowers bisexual, small to large. Calyx of 4–5 sepals, usually fused and tubular basally with 4–5 free, imbricate to valvate lobes; occasionally completely free. Corolla white, creamy or red, tubular, sometimes circumscissile (*Nuxia*), with 4–5 free lobes, radially symmetrical or bilabiate; corolla tube cylindrical to funnel-shaped, much longer to shorter than the lobes, throat often with a ring of hairs; corolla lobes glabrous, or densely villous on the inner surface (*Kogelbergia*, *Thesmophora*), or sparsely hairy on the outer surface, in *Nuxia* with some glandular hairs at the base of the lobes. Stamens usually 4 or 5 (*Retzia*), alternipetalous, arising between corolla lobes or from the corolla tube, equal; filaments free, short to long, slender or stout; anthers 2-thecous, dorsifixed, thecae usually confluent at apex, parallel or often diverging below, longitudinally dehiscent. Nectary small to obsolete. Ovary superior, usually slightly compressed, with 2 locules, often becoming 1-locular by abortion or loss of septum; ovules usually 1–3 to many (*Nuxia*) in each locule, placentation axile-basal, axile, axile-apical (*Thesmophora*, *Retzia*) or axile-peltate (*Nuxia*). Style terminal on ovary, filiform, erect, stigma capitate or minutely bilobed. Fruit a dehiscent capsule. Seeds ovoid to fusiform, usually longitudinally striate, reticulate, or reticulate-striate.

A family of seven genera and 28 species, widespread in Africa, Madagascar and the Mascarene Islands.

Vegetative Morphology and Anatomy. All species in this remarkably homogenous family are woody perennials. Many of the species have a lignotuber, from which they resprout after fires. These shrubs are characterised by being multistemmed, and are often quite short plants. Plants that are killed by fire, and subsequently re-establish from seed, are single-stemmed, and frequently grow taller than individuals that resprout from lignotubers. It is common to find closely related species in the Cape flora that differ in their response to fire, with some resprouting from a lignotuberous base, while others are killed by fire (Schutte et al. 1995); this seems to be the result of rampant, convergent adaptation to the frequent fires in the heath vegetation.

The wood anatomy of Stilbaceae s.s. was described by Carlquist (1986), and of *Nuxia* by Mennega (1980). Carlquist relates the salient anatomical features of Stilbaceae s.s. (little axial parenchyma, minimal ray tissue in young stems, and the presence of fibriform vessels) to maximizing mechanical strength. He regards the family as anatomically xerophytic, comparable to southern Californian Chaparral plants. The most drought-adapted wood is found in species from the dry coastal plains; those from the mountains are less xerophytic.

Growth rings are absent or poorly demarcated, and the vessels typically are small, 20–90 μm in diameter, mostly in short to long radial multiples, rarely solitary. The vessel elements are slender, fibriform, and the perforation plates mostly simple, rarely scalariform. A small number of vasicentric tracheids are present. All other imperforate tracheary elements are libriform fibres with apparently simple pits and without septa. Axial

[1] See also Scrophulariaceae.

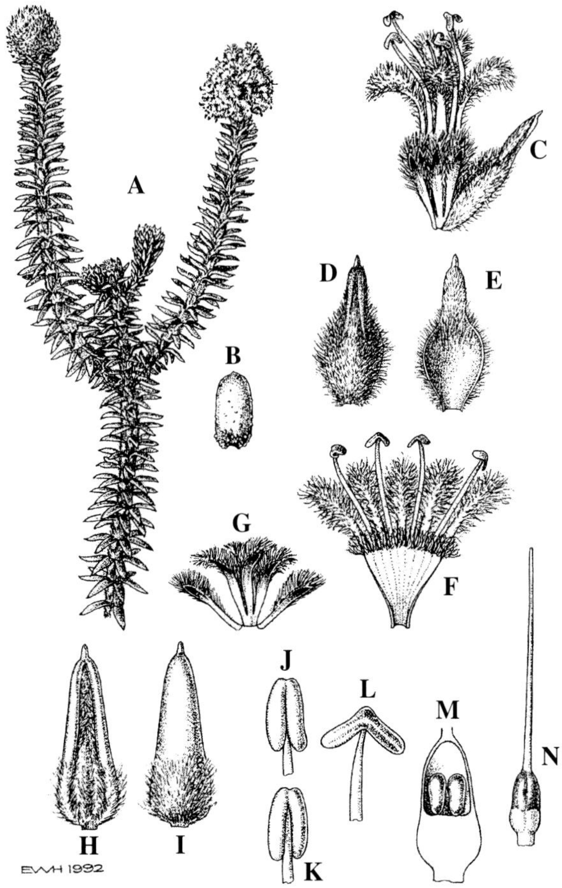

Fig. 54. Stilbaceae. *Kogelbergia verticillata*. **A** Flowering shoot. **B** Mature seed. **C** Open flower and subtending prophyll. **D** Prophyll, abaxial view. **E** Prophyll, adaxial view. **F** Opened corolla. **G** Calyx and two floral prophylls. **H** Adaxial view of leaf. **I** Adaxial view of leaf. **J, K** Pre-dehiscence anthers. **L** Anther after dehiscence. **M** Longitudinal section through ovary. **N** Gynoecium. (Rourke 2000a; with kind permission of the Editor, National Botanical Institute, Pretoria, South Africa)

parenchyma is almost absent. The rays are both uniseriate and multiseriate, with long uniseriate wings present on multiseriate rays. Uniseriate rays and the wings of multiseriate rays are composed of upright cells, while multiseriate portions are composed of procumbent cells. Crystals and silica are absent.

The leaf arrangement of Stilbaceae s.s. is somewhat unusual, with the leaves being more or less whorled. There are 3–8 leaves per whorl, and from node to node the leaves alternate. Almost invariably the leaves are also imbricate, so that the axes have a dense, cylindrical, tail-like appearance (Fig. 54A). The leaves are relatively hard and persistent, dropping from the plants after 2–4 years. On the lower leaf surfaces two stomatal grooves are visible. Such "ericoid" leaves are common in the Cape flora, and may be an adaptation either to the dry summers or to the low levels of soil nutrients. Unfortunately, the leaf anatomy has not yet been investigated; consequently, the construction and distribution of the stomata are unknown. The leaves of *Nuxia* are more "normal", with expanded, pinnately veined blades. In all except one species the leaves are petiolate, and they are mostly ternate, although opposite and quaternate leaves also occur. Alternate leaves are very rare, and restricted to lateral branches in a few species.

The indumentum of the plants consists mostly of simple hairs, although small, stalked, glandular hairs have also been reported for Stilbaceae s.s. (Metcalfe and Chalk 1950; Dahlgren and Van Wyk 1988). In *Nuxia* small glandular hairs are frequent on the calyx and corolla, often on the vegetative parts of the plants (Leeuwenberg 1975).

Inflorescence Structure. The superficial appearance of the inflorescences is highly variable, from dense capitate structures to more open structures, and this variation is possibly associated with a range of pollination types. Furthermore, in some taxa (e.g. *Retzia*) the plants are monopodial, and the stem continues to grow through the "inflorescence". In other groups (e.g. *Stilbe*, *Nuxia*), the inflorescence terminates the growth of the stem, and stem growth continues from a lateral leaf-bud below the inflorescence. This typically sympodial growth form results in a very distinctive, scarred, old inflorescence axis, surrounded by the lateral branches of the new growth in Stilbaceae s.s., while in *Nuxia* the old inflorescences appear to be shed readily.

The structure of the inflorescences has not yet been analysed in detail. Superficially it appears as if the flowers in *Stilbe* are sessile along the main inflorescence axis. In other genera, e.g. *Retzia*, the flowers are clustered on multifoliate side shoots. The densely capitate heads of *Kogelbergia* (Fig. 54A) may represent a condensed panicle (Rourke 2000a). In all species of Stilbaceae s.s. (Fig. 54C), the flowers are subtended by two large floral bracts, these presumably representing prophylls. These are often of the same size as the sepals.

The inflorescences of *Nuxia* are cymose, each node being subtended by a pair of bracts. While the lower bracts are mostly leaf-like, the upper bracts (prophylls) are scale-like and smaller than the calyx. The homologies between the bracts in *Nuxia* and those in Stilbaceae are not clear. In most species of *Nuxia* the inflorescence is a large, open panicle with single flowers at the apices of the terminal branches. However, in many species the terminal branches are short, and in some absent, resulting in an inflorescence of several globose heads.

Floral Structure. The flowers in most Stilbaceae are small, less than 10 mm long, while in *Retzia* the flowers are 55–65 mm long.

The four to five sepals are always fused basally (Fig. 54C), and the calyx lobes are either imbricate or valvate in bud. Although mostly radially symmetrical, there is a tendency in the calyx to be somewhat bilabiate. The calyx lobes are either glabrous or setose, membranous (usually) or hard (in *Stilbe*). They are probably mostly persistent after the corolla is shed. In *Nuxia* the calyx is leathery, forming a cup around the ovary, and with minute glandular hairs. The calyx encloses the ripening ovary.

The four to five corolla lobes are always fused in the lower half, with the stamens originating from the sinuses or from the tube (Fig. 54F). The corolla tube is generally straight, rarely somewhat curved. The corolla lobes can be radially or bilaterally symmetrical, with the lobes equally fused, and of unequal size. The corolla indumentum shows a wide range of variation, but the corolla is very rarely completely glabrous. Generally a ring of hairs is found at the entrance to the corolla throat (Fig. 54F). In various groups the inside of the lobes has strong and long hairs, forming a dense wool (*Kogelbergia*, *Thesmophora*; Fig. 54F). *Nuxia* has a clearly circumscissile corolla.

The androecium in *Retzia* and *Nuxia* is isomerous with the corolla lobes; thus, the full complement of five and four stamens respectively is present. In all other genera the stamens are reduced relative to the corolla lobes, with the posterior stamen lost. The stamens are alternipetalous and always inserted near the sinus of the corolla lobes or on the corolla tube, and the four or five stamens are always identical with each other. The anthers are dorsifixed, and the two thecae are confluent at the apex, their apices are divergent, and often the whole structure is U-shaped (Fig. 54J–L). The degree of exsertion of the stamens is variable, and probably linked to pollination biology.

The ovary (Fig. 54M, N) in Stilbaceae is basically bilocular, with a single basal (apical in *Thesmophora*) ovule, or with two ovules, one ascending, the other descending, in *Retzia*, while *Nuxia* has numerous ovules on a peltately expanded axile placenta. There are frequent reductions in the gynoecium, and these follow two routes. One is the loss of the septum, thus resulting in a unilocular, biovulate ovary (e.g. in *Kogelbergia*; Rourke 2000a). The other is the abortion of a locule, also resulting in a unilocular, uniovulate ovary (e.g. in *Xeroplana* = *Stilbe overbergensis* and *Stilbe gaymnopharyngia*; Rourke 1977). The style is always simple.

Embryology. *Retzia* was studied by Engell (1987), but the rest of the family is embryologically still unknown. The ovules are anatropous, unitegmic and tenuinucellate. With age the integumentary tissue becomes several-layered, developing a long micropyle. The embryo sac development is of the Polygonum type, but only two antipodals develop. Basal to the embryo sac, a hypostase develops. The nucellus and embryo sac accumulate starch grains, but these disappear after fertilization. Endosperm development could not be observed for all stages, but appears to be cellular.

Pollen Morphology. The pollen morphology of *Retzia* has been reported in several studies (Goldblatt and Keating 1976; Punt 1980), while that of Stilbaceae s.s. was described by Raj (1993), and that of *Nuxia* by Punt (1980). The grains in Stilbaceae s.s. and *Nuxia* are tricolporate, subprolate to prolate, 15–24 µm along the polar axis. *Retzia* is described as 3-colporate, spheroidal, with a polar axis of 35 µm. In both groups the colpi are long and narrow, tapering to an acute point, while the endoaperture is lalongate to lolongate. The exine is tectate-perforate, and has been studied by TEM only in *Retzia* (Goldblatt and Keating 1976), and this indicates that the tectum and the columellae are both about twice as thick as the foot layer. There is substantial variation in the tectal ornamentation in the family. The muri are variable in width; various tectal elements may be found enclosing minute openings.

Karyology. The only available karyological investigations are of *Retzia*, for which Goldblatt and Keating (1976) reported $n = 12$, and *Nuxia floribunda*, for which Gadella reported $2n = 38$ (Gadella 1980).

Pollination. *Retzia* is recorded as being bird-pollinated, and is visited by the orange-breasted sunbird, *Anthobaphes violacea* (Dahlgren et al. 1979). *Retzia* has a stiff inflorescence with the flowers and birds supported by the rigid leaves (Rebelo 1987). South African sunbirds perch while feeding on flowers, as they are not able to hover. Furthermore, the large, red and black flowers are also considered as typical adaptations for bird pollination, and in these features the genus is sharply distinct from the rest of the family.

There are no reports on the pollination biology of the rest of the family. However, it seems most unlikely that the small, white to cream-coloured flowers will also be bird-pollinated.

FRUIT AND SEED. The fruits are dehiscent capsules, the seeds (Fig. 54B) are ovoid and generally reticulate, also colliculate, or tuberculate, generally small, ca. 2 mm long (Rourke 2000a). In *Retzia* the seeds are larger, up to 6 mm long, covered with longitudinal septate crests, and with a large cylindrical embryo and starchy endosperm (Leeuwenberg and Leenhouts 1980). In *Nuxia* the embryo is straight and the endosperm fleshy (Leeuwenberg 1975).

PHYTOCHEMISTRY. Stilbaceae can be characterised by the presence of C-8 iridioid glucosides (Frederiksen et al. 1999). Typically most asterids have C-10 or C-11 iridioid glucosides, and C-8 iridioids are found only in *Nuxia*, *Thunbergia*, *Arbutus*, and *Arctostaphylos*, besides also being recorded in *Stilbe*, *Xeroplana* and *Retzia* in Stilbaceae. Of these, unedoside and stilbericoside have been recorded from all three genera (Dahlgren et al. 1979). Capensioside, holmioside, s-deoxiholmioside and thunbergioside have to date been recorded only from *Retzia* (Damtoft et al. 1993). These compounds are probably biogenetically linked, and indicate the close relationship between *Retzia* and Stilbaceae s.s., as well as the affinities between Stilbaceae and *Nuxia*.

CLASSIFICATION. Stilbaceae s.s. were generally regarded either as part of Verbenaceae or as allied to Verbenaceae, on account of their opposite (whorled) leaves, superior bicarpellate ovary, sympetalous corolla, and anisomerous androecium (5 corolla lobes, 4 stamens). However, the taxon was generally characterised by the ericoid habit, bilocular ovaries, uniovulate locules and endospermous seed. Family rank was granted by Endlicher (1836–1840), Hutchinson (1973) and Dyer (1976), subfamily rank by Melchior (1964) and Cronquist (1981), and tribal rank by Bentham and Hooker (1876). The South African literature (e.g. Pearson 1912; Phillips 1926) followed Bentham and Hooker.

Retzia was originally included in Convolvulaceae by Harvey (1838), and in Apocynaceae by Lindley (1836). A much more common assignation was either in or next to Solanaceae, based on the sympetalous flowers, superior bicarpellary ovary, and isomerous stamens. Endlicher (1836–1840) and Hutchinson (1973) placed the genus into a monotypic family, De Candolle (1852) in the Retzieae in Solanaceae, and Bentham and Hooker (1876), Wright (1904) and Phillips (1926) included *Retzia* in the tribe Cestrineae of Solanaceae. Fedde (1896) excluded the genus from the Solanaceae, on the absence of intraxylary phloem which is universal in Solanaceae, and moved it to Loganiaceae. Here it was maintained either as a separate family next to Loganiacae (e.g. Cronquist 1981), or as tribe Retzieae, next to the Buddlejeae in Loganiaceae (Leeuwenberg and Leenhouts 1980). A morphological-cladistic analysis of the Gentianales showed that *Retzia* does not belong in the Loganiaceae, but rather groups with Buddlejaceae (Struwe et al. 1994), consistent with Leeuwenberg's grouping. The broad concept of Loganiaceae, as advanced by Leeuwenberg (1975), is no longer accepted. Within this broad concept, however, *Retzia* has always been associated with Buddlejaceae which, in more recent systems, is placed in Lamiales, while Loganiaceae s.s. are in Gentianales (APG 1998).

The combination of *Retzia* and Stilbaceae seems to have been first argued by Dahlgren et al. (1979). They advanced 18 similarities between the two clades, and six differences. Many of the similarities could be considered to be plesiomorphic; possibly the most impressive are chemical. Both *Stilbe* and *Retzia* have unedoside and stilbericoside, both of which are C-8 iridoids (Damtoft et al. 1993). These are rare in plants, and are also found in *Thunbergia* and several species of *Erica*. The occurrence of several C-8 iridoids indicates the common possession of a biosynthetic pathway, which is a strong synapomorphy for these two genera, contrary to the arguments advanced by Cronquist (1981). Carlquist (1986) could not find any differences in the wood anatomy between Stilbaceae and Retziaceae, and argued on this basis that the families could not be distinguished. This was taken up taxonomically by Thorne (1992), who combined both taxa in Stilbaceae, as subfamilies Stilboideae and Retzioideae. Takhtajan (1997) maintained the two families separate but closely related, and with Buddlejaceae in the general relationship of Scrophulariceae.

Nuxia was first associated with *Buddleja* by Bentham (1836), and since has been kept in this relationship. However, some authors recognize this group of genera as Buddlejaceae (e.g. Cronquist 1981; Takhtajan 1997; APG 1998), while others include the group at various ranks within Loganiaceae (e.g. Bentham and Hooker 1876;

Leeuwenberg and Leenhouts 1980). In the African floras the latter option was generally followed, except for Leistner (2000). Recently, molecular analyses demonstrated that *Nuxia* is more closely related to Stilbaceae s.s. than to the remaining genera of Buddlejaceae, but this has not yet been translated into a formal taxonomy.

Here we follow the classification of Rourke, which includes Stilbaceae s.s. (as Stilboideae) and Retziaceae (as Retzioideae, following Thorne 1992). Molecular evidence suggests that both *Nuxia* and *Halleria* should be included in Stilbaceae. There is some morphological support for the inclusion of *Nuxia* in Stilbaceae - in particular, the leaves are in whorls of three (whorls of three or more in Stilbaceae), and the corolla tube has a ring of hairs (as is found in all species of Stilbaceae) - consequently, *Nuxia* is also included in the family. However, there seems to be no morphological support for the inclusion of *Halleria* in Stilbaceae, and this is left in Scrophulariaceae.

Affinities. The affinities of the monotypic family Retziaceae, and the oligotypic Stilbaceae, have stimulated much discussion and some research over the past decades. Cladistic analysis of nucleotide sequence data of the chloroplast-encoded gene of *rbc*L grouped together *Retzia* and *Euthystachys*, and indeed the sequences for the two species are very similar. However, the analysis failed to indicate what the sister-group of this clade is (Bremer et al. 1994). The Bremer et al. (1994) analysis demonstrated that Retziaceae and Stilbaceae could be combined as one family, as previously suggested by Carlquist (1986) and Thorne (1992), a treatment subsequently followed by Rourke (2000a, b) and Goldblatt and Manning (2000). This grouping could be confirmed by the inclusion of sequences from the *ndh*F gene in the studies of Oxelman et al. (1999) and Backlund et al. (2000). These two analyses also demonstrated that *Nuxia* occupies a sister-position to Stilbaceae s.l., and that this genus should rather be included here than in Buddlejaceae. A further expansion of the family is indicated by the analysis of Olmstead et al. (2001), which placed *Halleria* in a sister-relationship to *Retzia*, but unfortunately neither Stilbaceae s.s. nor *Nuxia* were included in this analysis. Furthermore, the statistical support for this grouping is rather weak. Consequently, the molecular data can be interpreted as indicating an expanded clade consisting of Stilbaceae s.s., Retziaceae, *Nuxia* and *Halleria*. The DNA sequence data also indicate clearly that the affinities of Stilbaceae lie with the Lamiales (Scrophulariaceae s.l., Buddlejaceae), rather than with Gentianales (Loganiaceae, Apocynaceae, Rubiaceae) or Solanaceae (Backlund et al. 2000). This is corroborated morphologically by the presence of a well-developed and persistent hypostase, which suggests a position near Scrophulariaceae and Buddleyaceae, but not in Loganiaceae (Engell 1987).

The evolutionary history of Stilbaceae s.s. has not been investigated. The clade is endemic to the Cape flora, and the simplest explanation would be a radiation within this species-rich region. However, the timing and mechanisms of the radiation are not yet known.

Distribution and Ecology. Stilbaceae, excluding *Nuxia*, are endemic to the Cape Floristic Region (CFR) as delimited by Goldblatt (1978) and, within the Cape Floristic Region, they are restricted to the mountains and plains of the Western Cape. As such they are the second-largest of the five endemic families of the CFR, the largest being Penaeaceae with 23 species (Goldblatt and Manning 2000). The concentration of species in the family in the Caledon mountains, at the interface between the all-year rainfall region along the South African south coast and the summer-dry region of the west coast, is also typical of biogeographical patterns in the Cape flora (Oliver et al. 1983). *Nuxia* has a wide distribution in tropical Africa. A single species is found in West Africa, 5 in southern Africa and 3 in East Africa, 10 occur in Madagascar (9 endemic), and the type species, *N. verticillata*, is restricted to the Mascarene Islands of Réunion and Mauritius.

Stilbaceae, excluding *Nuxia*, are heathland plants that grow exposed to sun in fire-prone habitats. In these habitats they are exposed to severe drying conditions in mid-late summer, whereas the winters are cool and damp, and the ericoid leaves could be interpreted as adaptation to this environment. Many species also show adaptations to resprouting after fire, with a lignotuber at the base. The adaptations to fire by the non-lignotuberous species have not been studied. The species in the family occupy a wide range of ecological habitats. Most species are found on well-drained, gravelly mountain slopes, with a few species found in other, more specialized habitats: coastal plains on deep sand or limestone (*Stilbe ericoides*) or rock crevices (*Thesmophora scopulosa* and *Stilbe rupestris*).

Nuxia is most commonly found in montane forest in tropical Africa and Madagascar. Some species are found in open habitats, others in forest, many in open forest or scrub, rarely in savannah.

The altitude range is from sea level in South Africa to over 3000 m at the highest margins of the montane forest. The genus is absent from tropical forest and from arid regions. It may be described as an afromontane element.

Key to the Genera

1. Leaf blades expanded; ovary with numerous seeds per locule **1.** ***Nuxia***
- Leaf blades more or less ericoid with incurved margins; ovary with 1–3 seeds per locule 2
2. Corolla tube 45–55 mm long; each locule with 2–3 ovules **7.** ***Retzia***
- Corolla tube less than 12 mm long; locules uniovulate 3
3. Corolla 4-lobed 4
- Corolla 5-lobed 5
4. Sepals free; corolla lobes glabrous **2.** ***Campylostachys***
- Sepals fused below; corolla lobes pubescent **5.** ***Thesmophora***
5. Corolla bilabiate with larger, erect posterior petals and three narrow anterior petals **6.** ***Stilbe***
- Corolla regular, lobes more or less the same length 6
6. Corolla lobes pubescent 4. *Kogelbergia*
- Corolla lobes glabrous 3. *Eustachys*

1. *Nuxia* Lam.

Nuxia Lam., Illustr. 1: 295 (1791); Commerson ex Lamarck, Tabl. Encycl. Meth., Bot. t. 71. 3 March (1791).
Lachnopylis Hochst. (1843).

Shrubs or trees, bark fissured. Leaves opposite, ternate, quaternate, rarely alternate, mostly petiolate, blade expanded. Flowers in a terminal, thyrsoid inflorescence, either solitary on terminal branches or aggregated into compact heads. Calyx campanulate to cylindrical, 4-lobed, outside of calyx with minute glandular hairs. Corolla regular, 4-lobed, circumscissile, tube cylindrical and often widened around the ovary, often with glandular hairs on the outside and a pilose ring in the throat. Stamens 4, filaments slender, anthers much exserted from the corolla. Ovary mostly appressedly pubescent, bilocular, placenta axile-peltate, ovules numerous in each locule, style much exerted. Capsule about as long or somewhat longer than the persistent calyx, seeds small and brown. Fifteen species, from South Africa to Liberia, Ethiopia, southern Arabia, Madagascar, and the Mascarenes.

2. *Campylostachys* Kunth

Campylostachys Kunth, Abh. Königl. Akad. Wissens. Berlin 1831: 207 (1832).

Lignotuberous, multistemmed shrublets. Leaves ericoid linear, sulcate, patent, in whorls of 4, acute. Inflorescence a terminal, slightly drooping, globose spike. Flowers sessile, subtended by 2 opposite, lanceolate-acute prophylls. Calyx glabrous, sticky; sepals 5, free, imbricate, narrowly lanceolate cymbiform, apices acute. Corolla regular, broadly tubular, of 4 equal, glabrous petals, tube and petals approximately same length; throat fringed with a ring of hairs. Stamens 4, anthers dorsifixed, much exserted from the corolla. Ovary ovoid-cylindric, slightly compressed, 2-locular with a single basal ovule in each locule; style straight, terete, glabrous. One species, *Campylostachys cernua* (L.f.) Kunth, and possibly one further undescribed species, both from the Western Cape of South Africa.

3. *Euthystachys* A.DC.

Euthystachys A.DC., Prodr. 12: 606 (1848).

Lignotuberous, multistemmed shrublets. Leaves ericoid, linear, in whorls of 4, initially pubescent, soon glabrous. Inflorescence a sessile ovoid spike. Flowers sessile, subtended by 2 opposite, softly pubescent prophylls. Calyx regular, sepals equal, fused into a tube in their lower third, soft-textured, externally pubescent. Corolla regular, lower half tubular-funnel form, upper half with 5 equal, spreading, ovate-acute lobes; glabrous but throat fringed with a ring of hairs. Stamens 4, slightly exserted. Ovary laterally compressed, glabrous, obovoid, 2-locular, with a basal ovule in each locule; style straight, filiform, glabrous. One species, *Euthystachys abbreviata* (E. Mey) A.DC. from the central Western Cape mountains.

4. *Kogelbergia* Rourke Fig. 54

Kogelbergia Rourke, Bothalia 30: 12 (2000), rev.
Stilbe P.J. Bergius (1767), p.p.

Single-stemmed or multistemmed, lignotuberous shrublets. Leaves sulcate, in whorls of 5–7, narrowly lanceolate-linear, erect, patent or reflexed. Inflorescence a sessile, globose or ovoid terminal spike. Flowers sessile, subtended by 2 opposite floral prophylls. Calyx regular, sepals 5, either free or fused and tubular at the base, apically silky hairy. Corolla regular, funnelform, 5-lobed, glabrous externally but inner surface of lobes densely villose to fimbriate, the indumentum terminating in the corolla-tube throat. Stamens 4, exserted; anthers dorsifixed, ovate. Ovary ovoid, bilaterally flattened, 1-locular with 2 basal, erect ovules; style straight, terete. Two species from the Western Cape.

5. *Thesmophora* Rourke

Thesmophora Rourke, Edinburgh J. Bot. 50: 89 (1993).

Dwarf, single-stemmed sprawling shrublets inhabiting rock crevices, older branches with lanate tufts between the leaf scars. Leaves in whorls of 4, very narrowly lanceolate-acute, sulcate, margins minutely serrate, glabrous. Inflorescence a subterminal spike of 20–35 flowers. Flowers sessile, subtended by 2 opposite prophylls. Calyx tubular, regular, glabrous, 5-lobed. Corolla tubular, bilabiate, 4-lobed; upper lip greatly enlarged, rounded, galeate; lower lip of 2 rounded lateral lobes and a minute acute median lobe; lobes densely setose on outer surface. Stamens 4, slightly exserted. Ovary obconic, with a lobed fleshy disc, minutely stipitate, glabrous, 2-locular with a single pendulous ovule in each locule; style filiform, glabrous. A single species, *Thesmophora scopulosa* Rourke, from the Ceres mountains.

6. *Stilbe* Berg.

Stilbe Berg., Descript. Plant. Cap. Bon Spei: 30 (1767).
Eurolobium Hochst. (1842).
Xeroplana Briq. (1896).

Single-stemmed, or lignotuberous multistemmed shrublets. Leaves sulcate, in whorls of 3–5, linear-subulate to narrowly lanceolate. Inflorescence a dense, sessile, terminal spike; old inflorescences persisting as bare, scarred axes. Flowers sessile, subtended by 2 opposite floral prophylls. Calyx tubular, usually hard, occasionally slightly abaxially curved, with 5 lobes. Corolla tubular, bilabiate with 2 erect, glabrous posterior lobes and 3 narrower, glabrous anterior lobes; interior of throat usually with dense ring of hairs, rarely glabrous. Stamens 4, slightly exserted. Ovary oblong, compressed, potentially 2-locular but abaxial locule abortive, leaving a single locule with a single, basally attached ovule; style straight, terete. Seven species from the Western Cape.

7. *Retzia* Thunb.

Retzia Thunb., Physiograph. sälskapets Handlinger 55 (1776).

Lignotuberous, multistemmed, monopodial shrubs. Leaves in whorls of 3–5, very narrowly lanceolate-linear to linear, straight, sulcate, softly pubescent, later glabrescent. Flowers on short axillary shoots, in groups of up to 3, subtended by 2 opposite foliaceous prophylls as long as the calyx. Calyx regular, sericeous, lower half tubular; upper half with 5 free, equal, narrowly lanceolate lobes. Corolla regular, tubular, very slightly curved at base, red, softly sericeous; lobes broadly lanceolate, black, densely sericeous. Stamens 5, equal, inserted between corolla lobes on short filaments. Ovary conic-ovoid, glabrous, 2-locular with 2–3 ovules in each locule, with 1 (or 2) erect, the others pendent; style glabrous, filiform, exserted; stigma bilobed. Seeds ovoid, reticulate, yellowish brown. $2n = 24$. One species, *Retzia capensis* Thunb. from the Western Cape mountains.

Selected Bibliography

APG 1998. An ordinal classification for the families of flowering plants. Ann. Missouri Bot. Gard. 85: 531–553.
Backlund, M., Oxelman, B., Bremer, B. 2000. Phylogenetic relationships within the Gentianales based on *ndhF* and *rbcL* sequences, with particular reference to the Loganiaceae. Am. J. Bot. 87: 1029–1043.
Bentham, G. 1836. Observations on some new or little known genera, XI. *Nuxia*, XII. *Buddleia*. Companion Bot. Mag. 2: 59–60.
Bentham, G., Hooker, J.D. 1876. Genera plantarum, vol. 2.2. London: Lovell Reeve.
Bremer, B., Olmstead, R.G., Struwe, L., Sweere, J.A. 1994. *rbc*L sequences support exclusion of *Retzia*, *Desfontainia*, and *Nicodemia* from the Gentiales. Plant Syst. Evol. 190: 213–230.
Carlquist, S. 1986. Wood anatomy of Stilbaceae and Retziaceae: ecological and systematic implications. Aliso 11: 299–316.
Cronquist, A. 1981. An integrated system of classification of flowering plants. New York: Columbia University Press.
Dahlgren, R.M., Van Wyk, A.E. 1988. Structures and relationships of families endemic to or centered in southern Africa. Monogr. Syst. Bot. Missouri Bot. Gard. 25: 1–94.
Dahlgren, R., Nielsen, B.J., Goldblatt, P., Rourke, J.P. 1979. Further notes on Retziaceae – its chemical contents and affinities. Ann. Missouri Bot. Gard. 66: 545–556.
Damtoft, S., Franzyk, H., Jensen, S.R., Nielsen, B.J. 1993. Iridoids and verbascosides in Retzia. Phytochemistry 34: 239–243.
De Candolle, A. 1852. Prodromus Systematis Naturalis, vol. 13. Paris: Masson.
Dyer, R.A. 1976. The genera of southern African flowering plants. Vol. 2. Gymnosperms and Monocotyledons. Pretoria: Department of Agricultural Technical Services.
Endlicher, S. 1836–1840. Genera Plantarum secundum Ordines Naturales disposita. Wien: Beck.
Engell, K. 1987. Embryology and taxonomical position of *Retzia capensis* (Retziaceae). Nord. J. Bot. 7: 117–124.
Fedde, F. 1896. Beiträge zur vergleichende Anatomie der Solanaceae. Breslau.
Frederiksen, L.B., Damtoft, S., Jensen, S.R. 1999. Biosynthesis of iridoids lacking C-10 and the chemotaxonomic implications of their distribution. Phytochemistry 52: 1409–1420.
Gadella, T.W.J. 1980. Cytology. In: Engler and Prantl, Die natürlichen Pflanzenfamilien, ed. 2, 28b I. Berlin: Duncker & Humblot.
Goldblatt, P. 1978. An analysis of the flora of Southern Africa: its characteristics, relationships, and origins. Ann. Missouri Bot. Gard. 65: 369–436.

Goldblatt, P., Keating, R.C. 1976. Chromosome cytology, pollen structure, and relationship of *Retzia capensis*. Ann. Missouri Bot. Gard. 63: 321–325.
Goldblatt, P., Manning, J. 2000. Cape plants. A conspectus of the Cape flora of South Africa. Pretoria: National Botanical Institute.
Harvey, W.H. 1838. The genera of South Africa plants. Cape Town: A.S. Robertson.
Hutchinson, J. 1973. The families of flowering plants. Oxford: Clarendon Press.
Leeuwenberg, A.J.M. 1975. The Loganiaceae of Africa. XIV. A revision of *Nuxia* Lam. Med. Landbouwhogesch. Wageningen 75: 1–80.
Leeuwenberg, A.J.M., Leenhouts, P.W. 1980. Taxonomy. In: Engler and Prantl, Die natürlichen Pflanzenfamilien, ed. 2, 28b I. Berlin: Duncker & Humblot, pp. 8–96.
Leistner, O.A. 2000. Seed plants of southern Africa: families and genera. Strelitzia. Pretoria: National Botanical Institute, pp. 1–775.
Lindley, J. 1836. A natural system of botany. London: Longman, Rees, Orme, Brown and Green.
Melchior, H. 1964. Syllabus der Pflanzenfamilien. Berlin: Borntraeger.
Mennega, A.M.W. 1980. Anatomy of the secondary xylem. In: Engler and Prantl, Die natürlichen Pflanzenfamilien, ed. 2, 28b I. Berlin: Duncker & Humblot, pp. 112–161.
Metcalfe, C.L., Chalk, L. 1950. Anatomy of the Dicotyledons. Oxford: Clarendon Press.
Oliver, E.G.H., Linder, H.P., Rourke, J.P. 1983. Geographical distribution of present-day Cape taxa and their phytogeographical significance. Bothalia 14: 427–440.
Olmstead, R.G., DePamphilis, C.W., Wolfe, A.D., Young, N.D., Elisons, W.J., Reeves, P.A. 2001. Disintegration of the Scrophulariaceae. Am. J. Bot. 88: 348–361.
Oxelman, B., Backlund, M., Bremer, B. 1999. Relationships of the Buddlejaceae s.l. investigated using parsimony jackknife and branch support analysis of chloroplast *ndhF* and *rbc*L sequence data. Syst. Bot. 24: 164–182.
Pearson, H.H.W. 1912. Verbenaceae. In: Thiselton-Dyer, W.T. (ed.) Flora Capensis, vol. 5, part 1. London: Lovell Reeve, pp. 180–226.
Phillips, E.P. 1926. The genera of South African flowering plants. Pretoria: Botanical Survey of South Africa 10.
Punt, W. 1980. Pollen. In: Engler and Prantl, Die natürlichen Pflanzenfamilien, ed. 2, 28b I. Berlin: Duncker & Humblot, pp. 162–191.
Raj, B. 1993. A contribution to the pollen morphology of Stilbaceae Kunth. Pollen Spores 25: 395–408.
Rebelo, A.G. 1987. Bird pollination in the Cape flora. In: Rebelo, A.G. (ed.) A preliminary synthesis of pollination biology in the Cape flora. Pretoria: South African National Scientific Programmes Rep., pp. 83–108.
Rourke, J.P. 1977. A revision of *Xeroplana* Briq. (Stilbaceae). J. S. Afr. Bot. 43: 1–8.
Rourke, J.P. 2000a. A review of generic concepts in the Stilbaceae. Bothalia 30: 9–15.
Rourke, J.P. 2000b. Stilbaceae. In: Leistner, O.A. (ed.) Seed plants of southern Africa: families and genera. Pretoria: National Botanical Institute, pp. 541–543.
Schutte, A.L., Vlok, J.H.J., Van Wyk, B.-E. 1995. Fire-survival strategy – a strategy of taxonomic, ecological and evolutionary importance in fynbos legumes. Plant Syst. Evol. 195: 243–259.
Struwe, L., Albert, V.A., Bremer, B. 1994. Cladistics and family level classification of the Gentianales. Cladistics 10: 175–206.
Takhtajan, A. 1997. Diversity and classification of flowering plants. New York: Columbia University Press.
Thorne, R.T. 1992. Classification and geography of the flowering plants. Bot. Rev. 58: 225–348.
Wright, C.H. 1904. Solanaceae. In: Thiselton-Dyer, W.T. (ed.) Flora Capensis, vol. 4, part 2. London: Lovell Reeve, pp. 87–121.

Tetrachondraceae

S.J. Wagstaff

Tetrachondraceae Skottsb. ex R.W. Sanders & P.D. Cantino, Taxon 33: 72 (1984).

Small perennial herbs, mat-forming to upright. Leaves opposite, subsessile, joined at the base by an interstipular sheath; lamina linear to broadly ovate. Flowers mostly solitary or in leafy cymes, actinomorphic, 4-merous, minute. Corolla white, tube subrotate to campanulate, with imbricate aestivation. Calyx campanulate, persistent, with lanceolate to ovate-obtuse lobes. Sepals and petals connate at the base. Stamens alternate with the petals, inserted on corolla; anthers introrse, dorsifixed, bilocular with two distinct thecae; dehiscence longitudinal. Pollen tricolporate. Gynoecium bicarpellate, syncarpous. Ovary partly inferior, style terminal (*Polypremum*), or ovary superior, style gynobasic (*Tetrachondra*), stigma slightly truncate to capitate. Ovules anatropous, unitegmic, and tenuinucellate. Placentation basal or the ovules on peltate placentae attached to septum. Fruit a bivalved capsule or reduced to four nutlets. Seeds with copious endosperm. Embryo straight.

A small family of 2 genera with 3 species, distributed from the SE United States to Central and South America, and New Zealand.

Vegetative Morphology. The Tetrachondraceae are small, mesophytic to hydrophytic, perennial herbs (Fig. 55A, G). *Polypremum* arises from a distinct taproot, and the branches are procumbent to ascending from a central crown, whereas *Tetrachondra* forms dense mats with adventitious roots developing at the lower leaf nodes. The vegetative axis shows sympodial branching. The branches are slightly angular or fluted. The leaves are opposite, sessile or with a short petiole. They are joined at the base by an interstipular sheath (Fig. 55B, H). The lamina is pinnately veined, and minutely ciliolate to slightly scabrid on the margin. A variety of unicellular and multicellular trichomes have been described. According to Rogers (1986), the surface of the lamina of *Polypremum* is covered with distinctive, sunken multicellular trichomes composed of a basal cell, a stalk cell and 4 or more cells comprising a broad cap. Similar trichomes were described in *Tetrachondra* and illustrated by Skottsberg (1912).

Floral Structure and Pollination. The flowers are mostly solitary in the upper leaf axils or in reduced leafy cymes. They are minute, 4-parted, and radially symmetrical. The throat is ringed with a line of transverse white hairs in *Polypremum*, whereas a ring of hairs is lacking in *Tetrachondra*. The stamens are attached to the corolla and alternate with the lobes (Fig. 55C, I). The anthers are bilocular with two distinct thecae; dehiscence is longitudinal (Fig. 55D, J). The ovary is partly inferior in *Polypremum*, with the style terminal, whereas it is superior in *Tetrachondra*, which has a gynobasic style (Fig. 55K).

The small white flowers of Tetrachondraceae are presumably visited by unspecialized pollinators such as thrips, flies, and short-tongued bees, but little is known about the pollination or their breeding systems.

Embryology. The female archesporium of *Polypremum* develops from a single hypodermal primary cell (Moore 1948). The primary archesporial cell does not cut off a cover cell, but directly functions as the megaspore mother cell. The megaspore mother cell divides, forming four megaspores arranged in a linear sequence, of which the chalazal megaspore develops into the embryo sac. The mature embryo sac is unusual in that its micropylar end extends to the surface of the ovule and is not enclosed by the integument.

Division of the endosperm nucleus of *Polypremum* precedes that of the zygote in the developing ovule (Moore 1948). Endosperm development is cellular with a cell wall being formed during the first and all subsequent mitotic divisions. The first division is always transverse, and the second divides these cells longitudinally, so that two tiers of two cells are formed. The micropylar and chalazal endosperm cells form feeble, non-aggressive haustoria; only the central portion develops into the endosperm proper.

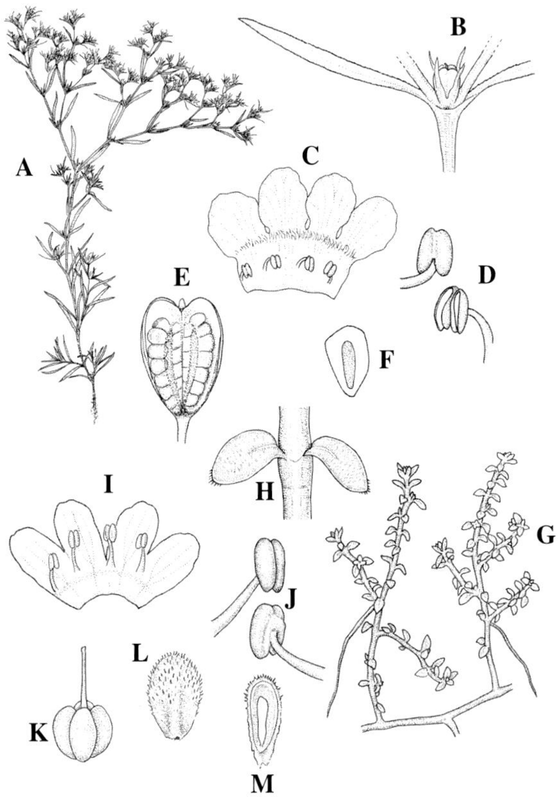

Fig. 55. Tetrachondraceae. **A–F** *Polypremum procumbens.* **A** Habit. **B** Vegetative shoot with attached leaves showing the interstipular sheath and solitary fruit. **C** Corolla opened to show the attached stamens and pubescent throat. **D** Stamens. **E** Fruit in longitudinal section with numerous ovules attached to peltate placenta. **F** Seed in longitudinal section showing the embryo. **G–M** *Tetrachondra hamiltonii.* **G** Habit. **H** Vegetative shoot with attached leaves showing the interstipular sheath. **I** Corolla opened to show the attached stamens. **J** Stamens. **K** 4-lobed fruit with gynobasic style. **L** Nutlet, adaxial view. **M** Nutlet in longitudinal view showing the seed and embryo. (Illustration by Rebecca Wagstaff)

The developing zygote of *Polypremum* is initially situated at the micropylar end of the embryo sac, and then migrates to the center of the endosperm tissue and divides. Embryo development follows the Onagrad type. A 6-celled proembryo is formed with the micropylar cell being elongate, somewhat vacuolate and functioning as a suspensor (Moore 1948).

Karyology. Little is known of the karyotype of Tetrachondraceae and a critical study is needed. Moore (1947, 1948) reported a triploid chromosome number $3n = 33$ from endosperm cells and a diploid chromosome $2n = 22$ from the root tips of *Polypremum*. However, the use of sectioned material rather than squashes makes it desirable to confirm Moore's counts. His illustrations show chromosomes that are variable in length and centromere position. The chromosome number of *Tetrachondra* is not known.

Pollen Morphology. The pollen morphology of Tetrachondraceae is described by Punt (1980) and Moar (1993). The pollen of *Polypremum* is tricolporate, with a tectate baculate structure, short bacules, and a psilate surface with minute perforations. It is suboblate or oblate in equatorial view and angular in polar view. The endoaperture is large, porous-like, lolongate, and the sexine is distinctly raised over the endoaperture. The pollen of *Tetrachondra* is similar to that of *Polypremum*. It is also tricolporate; the exine is thin with a tectate baculate structure, and a psilate surface with minute perforations. The endoaperture is indistinct. The pollen of *Tetrachondra* has a rhomboidal shape that is triangular in polar view (Moar 1993).

Fruit and Seed. The fruits of Tetrachondraceae develop from a bicarpellate ovary. The fruit of *Polypremum* is a bivalved capsule that is obovoid to ovoid, and laterally compressed (Rogers 1986). The septum forms a sunken vertical line that separates into two adaxial lobes and usually also opens along the abaxial sutures. Each locule has one peltate placenta attached to the base of the septum, with 20–30 anatropous ovules (Fig. 55E). Numerous small seeds are produced that are translucent and light yellow when dry.

In contrast to *Polypremum*, the fruit of *Tetrachondra* forms four locules by the development of a false septum (Fig. 55K–M. Each locule has a single anatropous ovule that is attached to a basal placenta (Skottsberg 1912). The mature fruit consists of four dry nutlets that are weakly attached at the base of the receptacle. The nutlets are broadly obovate and triquetrous, with a dull brown surface.

Phytochemistry. According to Jensen (1999), Tetrachondraceae are united by a unique chemical profile characterized by the presence of the reduced sugars sorbitol and cornoside along with caffeoyl – phenylethanoid glycosides, and the lack of iridoids. Sorbitol is fairly common in members of the Rosaceae, but elsewhere it commonly appears only in the Plantaginaceae. Cornoside is reported from *Cornus* (Cornaceae) and *Forsythia* (Oleaceae). Salidroside and conandroside are minor components of the reduced sugars in

Polypremum, and verbascoside is a minor component in *Tetrachondra* (Jensen 1999).

AFFINITIES. Partly because of their reduced flowers and distinctive morphology, *Tetrachondra* and *Polypremum* have been difficult to classify. *Tetrachondra* was initially included in the Crassulaceae (Hamilton 1885), but differs in its syncarpous gynoecium derived from two carpels. The gynoecium of Crassulaceae is derived from five or more carpels that are free or united only at the base. The gynoecial morphology of *Tetrachondra* resembles that of the Lamiaceae and Boraginaceae, but Wagstaff and Olmstead (1997) suggested that this reflects convergent evolution, rather than descent from a common ancestor. *Tetrachondra* can be distinguished from members of the Boraginaceae by its leaves which are strictly opposite, and its seeds with copious endosperm. While opposite leaves are characteristic of the Lamiaceae, most members of the family have a bilabiate, rather than the actinomorphic corolla characteristic of *Tetrachondra*.

Polypremum has been misplaced in the Buddlejaceae, Loganiaceae, or Rubiaceae (see discussion in Moore 1948; Rogers 1986; Oxelman et al. 1999). *Polypremum* exhibits many characteristics of the Buddlejaceae and many authors place it in this family, but it differs in regard to the absence of endosperm haustoria, early development of the proembryo, and a chromosome number of $2n = 22$ (Moore 1948). *Polypremum* can be distinguished from members of the Rubiaceae by its superior ovary and lack of distinct interpetiolar stipules. The female archesporium of *Polypremum* consists of a single cell, whereas it is composed of many cells in most members of the Rubiaceae. Endosperm formation in the Rubiaceae and Loganiaceae is also typically nuclear, while it is cellular in *Polypremum* (Moore 1948).

Phylogenetic analysis of *rbc*L and *ndh*F sequences provides strong support for a clade consisting of *Polypremum* and *Tetrachondra*, both genera now comprising the Tetrachondraceae (Oxelman et al. 1999; Wagstaff et al. 2000). The family emerges near the base of the Lamiales (sensu APG 1998), sister to a large clade that includes the Acanthaceae, Bignoniaceae, Gesneriaceae, Lamiaceae, Scrophulariaceae, Stilbaceae, and Verbenaceae, all with a range of corolla symmetry from weakly to strongly zygomorphic.

DISTRIBUTION AND HABITATS. Tetrachondraceae are widely distributed from the southeastern United States to Central and South America, and New Zealand. The genus *Polypremum* is monotypic, with the sole species *P. procumbens* found in northern South America, Central America, Mexico, West Indies, Mexico and the southeastern United States (Rogers 1986). A single disjunct population is reported from Paraguay, but it is not known whether it is the result of accidental human introduction. Outlier reports of *P. procumbens* in Hawaii, Guam, Wallis (near Samoa), and Palau, however, are probably recent introductions. It inhabits wet or more often dry habitats at both low and high elevations, usually on sandy soils (Rogers 1986).

The genus *Tetrachondra* includes two species with a disjunct distribution, *T. hamiltonii* in New Zealand and *T. patagonica* in Chile and Argentina. *Tetrachondra hamiltonii* is distributed in New Zealand from Stewart Island, Otago, and Southland, with a few specimens known from Northwest Nelson and the Ruahine Range (Johnson and Brooke 1989) where it is found along lake or river margins and damp grasslands from low elevation to alpine. At various times during the year the plants can be completely submerged, and carpet river bottoms to a considerable depth (Hamilton 1885; Oliver 1892). Two geographically isolated subspecies of *T. patagonica* are recognized in South America (Moore 1970): subsp. *patagonica* from southern Patagonia and subsp. *fuegiana* from the southern tip of Tierra del Fuego. They are typically found in wet habitats with open mineral soils or bryophyte communities above tree line from 500 to 950 m.

KEY TO THE GENERA

1. Plants mostly terrestrial, procumbent to upright with ascending branches. Leaves linear. Corolla with a ring of hairs in the throat. Stamens included. Fruit a capsule with numerous seeds **1. *Polypremum***
– Plants mostly subaquatic to aquatic, forming dense mats. Leaves broadly ovate to obovate-oblong, rather fleshy. Corolla lacking a ring of hairs in the throat. Stamens slightly exserted. Fruit consisting of four nutlets each with a single seed **2. *Tetrachondra***

1. *Polypremum* L. Fig. 55A–F

Polypremum L., Sp. Pl. 1: 111 (1753); Rogers, J. Arnold Arbor. 67: 143–185 (1986), reg. rev.; Norman, Flora Neotrop. 81: 185–186 (2000).
Symphoranthus Mitch. App. 219 (1748).
Cleyera Adans. (1763).

Small perennial (reportedly sometimes annual) herbs arising from a central taproot with numerous sympodial branches. Branches procumbent to ascending, fluted, often with 4 sharp flanges and 2 or 4 rounded ridges. Leaves linear, slightly scabrid along the margins and stem ridges. Flowers small, 4-merous, sessile, mostly solitary or in terminal leafy cymes. Calyx persistent, divided almost to the base, lobes lanceolate with hyaline margins. Corolla white, shorter than the calyx, campanulate, with a ring of hairs in the throat. Stamens 4, included, filaments about as long as the anthers. Ovary 2-celled, partly inferior. Style short, persistent with a capitate stigma. Ovules anatropous on peltate placentae that are attached to the base of the septum. Capsule laterally compressed, obovoid to ovoid, loculicidal, later septicidal. Seeds small, cuboidal, with a smooth surface, translucent and light yellow when dry. $2n = 22$. Flowering chiefly from March to October, but throughout the year in warm climates. One species, *P. procumbens* L. from northern S America to SE USA.

2. ***Tetrachondra*** Petrie ex D. Oliver Fig. 55G–M

Tetrachondra Petrie ex D. Oliver, Hook. Icon. pl. xxiii., t. 2250 (1892); Skottsberg, Bot. Jahrb. 107: 17–26 (1912); Moore, Bol. Soc. Argent. Bot. 13: 1–9 (1970), reg. rev.

Small perennial aquatic or semiaquatic herbs forming dense mats. Branches slender, angular, developing adventitious roots at the nodes. Leaves broadly ovate to obovate-oblong, margin entire, minutely ciliolate, rather fleshy, and obscurely gland-dotted. Flowers minute, 4-merous. Corolla white, longer than the calyx, subrotate-campanulate, lacking a ring of hairs in the throat. Calyx persistent; the lobes ovate-obtuse, shorter than the corolla. Stamens 4, exserted, filaments slightly longer than the anthers. Gynoecium bicarpellate, later forming 4, 1-seeded locules by the development of a false septum. Ovary superior, and deeply 4-parted with a gynobasic style. Stigma slightly truncate. Fruit of 4 erect nutlets; each ovoid-triquetrous, brown, and densely setulose to glabrous with a small basal attachment scar. Flowering chiefly in January and February. Two species in New Zealand, Chile and Argentina.

Selected Bibliography

APG (Angiosperm Phylogeny Group) 1998. An ordinal classification for the families of flowering plants. Ann. Missouri Bot. Gard. 85: 531–553.

Hamilton, W.W. 1885. Notes on the occurrence and habits of some of our New Zealand plants. Trans. Proc. New Zealand Inst. 17: 290–293.

Jensen, S.R. 1999. Chemical relationships of *Polypremum procumbens*, *Tetrachondra hamiltonii*, and *Peltanthera floribunda*. Biochem. Syst. Ecol. 28: 45–51.

Johnson, P., Brooke, P. 1989. Wetland plants in New Zealand. Wellington: DSIR Publishing.

Moar, N.T. 1993. Pollen grains of New Zealand dicotyledonous plants. Canterbury, New Zealand: Manaaki Whenua Press.

Moore, R.J. 1947. Cytotaxonomic studies in the Loganiaceae. I. Chromosome numbers and phylogeny in the Loganiaceae. Am. J. Bot. 34: 527–538.

Moore, R.J. 1948. Cytotaxonomic studies in the Loganiaceae. II. Embryology of *Polypremum procumbens* L. Am. J. Bot. 35: 404–410.

Moore, D.M. 1970. Additions to the vascular flora of Tierra del Fuego. Bol. Soc. Argent. Bot. 13: 1–9.

Oliver, D. 1892. *Tetrachondra hamiltonii*. Hooker's Icon. Pl. vol. III, part II, pl. 2250.

Oxelman, B., Backlund, M., Bremer, B. 1999. Relationships of the Buddlejaceae s.l. investigated using branch support analysis of chloroplast *ndh*F and *rbc*L sequence data. Syst. Bot. 24: 164–182.

Punt, W. 1980. Pollen morphology. In: Leeuwenberg, A.J.M. (ed.) Die natürlichen Pflanzenfamilien, Fam. Loganiaceae, Band 28bI. Berlin: Duncker & Humblot, pp. 162–191.

Rogers, G.K. 1986. The genera of Loganiaceae in the southeastern United States. J. Arnold Arbor. 67: 143–185.

Skottsberg, C. 1912. *Tetrachondra patagonica* n. sp. und die systematische Stellung der Gattung. Bot. Jahrb. Syst. Beiblatt 107: 17–26.

Wagstaff, S.J., Olmstead, R.G. 1997. Phylogeny of the Labiatae and Verbenaceae inferred from *rbc*L sequences. Syst. Bot. 22: 165–179.

Wagstaff, S.J., Martinsson, K., Swenson, U. 2000. Divergence estimates of *Tetrachondra hamiltonii* and *T. patagonica* (Tetrachondraceae) and their implications for austral biogeography. New Zealand J. Bot. 38: 587–596.

Trapellaceae

H.-D. Ihlenfeldt

Trapellaceae Honda & Sakis., Daiko Nippon Shokubutsu Bunruigaku [Syst. Pl. Japon.]: 378 (1930).

Aquatic perennial rhizomatous herbs with opposite leaves, heterophyllous, submerged leaves oblong, floating leaves deltoid-rotundate. Flowers solitary in the axils of foliage leaves, aerial flowers chasmogamous, submerged flowers cleistogamous; flowers bisexual, tetracyclic, very irregular, slightly bilabiate. Calyx small, 5-partite, persistent. Corolla gamopetalous, white or light rose, tube slightly funnel-shaped, constricted at the base. Androecium of 2 fertile stamens and 2 staminodes. Ovary bicarpellate, syncarpous, inferior, abaxial carpel abortive and empty. Fruit indehiscent, oblong, woody, vested with 3–5 long curved rigid appendages, 1-seeded; endosperm very thin.

A family with the single genus *Trapella* with one or two species, in subtropical and temperate regions of Japan, Russia (Far East), Korea and SE China.

Vegetative Morphology. The primary root of the seedling is only short-lived. The young shoot at first straggles on the mud at a water depth of 30–50 cm and develops numerous adventitious roots, predominantly at the nodes, thus forming a thread-like persistent rhizome. The leaves produced on this prostrate rhizome are oblong with remotely serrate margin. Their petiole is short and they remain submerged. Later on the tip of the primary shoot bends upwards and the elongated shoot starts floating next to the surface of the water. The leaves which are then formed are deltoid-rotundate with a crenate margin (Fig. 56A), have long petioles and float on the water surface. Occasionally leaves intermediate in form are observed. Lateral branches are scarce and arise mainly from the axils of submerged leaves.

Vegetative Anatomy. As in many water plants, the small vascular cylinder of the stems is delimited by a well-developed primary endodermis. The vascular cylinder has only few vessels with spiral or scalariform thickenings. The broad cortex is transversed longitudinally by large airspaces; in the submerged parts of the plant they are arranged in two concentric zones, in the floating parts the arrangement is less regular. At the nodes the outer cortex cells have thickened walls, and a hypodermal meristem adds additional cells to the cortex. These structures apparently serve to cope with the additional mechanical load exercised by the floating leaves.

The floating leaves possess both spongy and palisade mesophyll. In the submerged leaves the latter is missing. Both leaf types possess large hydathodes which are situated in the small incisions of the leaf margin. Stomata exclusively occur on the upper face of the floating leaves; they are anomocytic. Glandular hairs occur in large number on both sides of the leaves and on other parts as well. The hairs consist of a single stalk cell and a head of 4–8 cells. They look very similar to the mucilage-yielding glandular hairs of the Pedaliaceae, where they are considered as a constitutive apomorphic character. Oliver (1888) presumed that the heads of the glandular hairs in *Trapella* also produce mucilage when wetted. This has not been confirmed by subsequent studies and, since fully developed glandular hairs survive even on the lower faces of floating leaves (see, e.g. Fig. 60a in Oliver 1888), this assumption appears to be erroneous.

Inflorescence Structure. The flowers are solitary in the axils (Fig. 56A) of both the floating (usually chasmogamous flowers) and the submerged leaves (exclusively cleistogamous flowers). Usually only one of the two leaves of a node bears a flower. The chasmogamous flowers have long pedicels and are presented in an upright position above the water surface, whereas the cleistogamous flowers have very short pedicels and usually remain submerged.

Flower Structure and Anatomy. The persistent calyx is 5-partite and small compared to the corolla (Fig. 56B). The gamopetalous corolla is composed of 5 petals, aestivation is descending, the limb spreading and slightly bilabiate (2 upper and 3 lower lobes; Fig. 56). The limb is white or

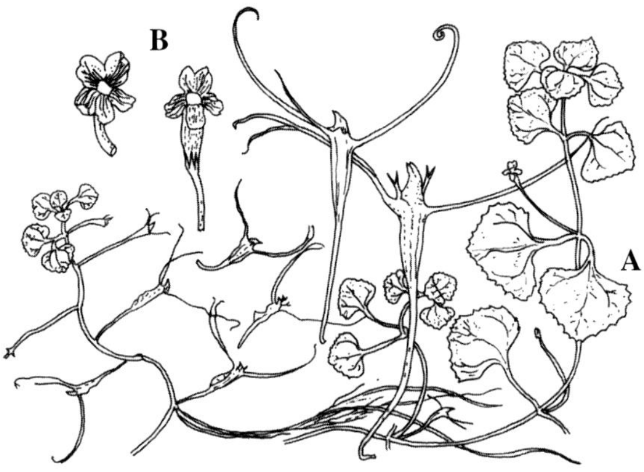

Fig. 56. Trapellaceae. *Trapella sinensis*. **A** Habit (without submerged leaves). **B** Flowers. **C** Fruits (in different stages of development). (After Wang 1990)

light rose, the tube yellowish; the tube is slightly funnel-shaped and constricted at the base. The androecium is episepalous and composed of 2 fertile stamina in adaxial-lateral position and 2 staminodes in abaxial-lateral position; no staminode in adaxial-median position is found. The androecium is inserted in the tube at about the middle of the corolla, the anthers are presented at the mouth of the tube. The filaments are filiform. The two thecae of the anthers are completely separated and are borne on a large disk-like connective, attached to the filament in a peltate manner. The anthers open lengthwise. The staminodes have much-reduced infertile anthers. The ovary is syncarpous with two carpels in median position. It is inferior, false septa are lacking. The much smaller abaxial carpel is abortive and empty. The style is filiform, the stigma a triangular to cruciform plate spreading at right angle to the style. Placentation is axile with 2 pendent ovules only in the adaxial locule. The upper ovule is sessile, whereas the lower one has a short funicle. No information is available about whether the flowers possess a nectary; at least it is very inconspicuous.

The cleistogamous flowers are tiny (2–4 mm long including pedicel); the ovary and calyx are well developed, whereas the corolla, style and stamens are much reduced; the minute thecae (0.3 mm long) contain only few pollen grains, many of them abortive, the nearly sessile stigma appears to be functional (Glück 1940).

Embryology. Detailed studies by Singh (1958, 1960) corrected several misinterpretations of an earlier study (Oliver 1888). The structure of the anther wall is still unknown. The pollen when shed is 3-cellular. The ovules are anatropous, pendent, unitegmic and tenuinucellate. Only the upper ovule of the fertile adaxial locule is functional, the lower one degenerates. The embryo sac is monosporic and of the Polygonum type. Repeated divisions of the endosperm nucleus give rise to a cellular endosperm differentiated into three portions. The chalazal portion consists of 2 much-elongated cells and functions as a chalazal haustorium. The micropylar portion consists of 4 circumaxially arranged cells and forms a micropylar haustorium which is exceptionally large and long-lived, so that it can still be observed in the mature seed; this structure has been misinterpreted as a caruncle (Glück 1940). The remaining portion of the embryo sac forms the endosperm proper.

Embryo formation conforms to the Onagrad type. The seed is albuminous although the endosperm is very thin, and the embryo is straight. The nature of the granular reserve material found both in the endosperm and the embryo is unknown.

Pollen Morphology. The pollen grains are monads, isopolar, medium-sized, suboblate to sphaeroid, 3-zonocolpate with laevigate tectum, thus representing a type which is very common in Lamiales and considered as primitive (Gentry and Tomb 1979). The sporoderm is comparatively thin. Erdtman (1952) described the pollen grains of *Trapella* as colporoid, but this was not confirmed by Straka and Ihlenfeldt (1965). Pollen morphology of *Trapella* differs vastly from that of Pedaliaceae s.str. and Martyniaceae.

Karyology. No chromosome numbers have been recorded.

Pollination and Reproductive System. The chasmogamous flowers exhibit a melittophilous syndrome. Since also the cleistogamous flowers produce fully developed fruits, at least partial self-compatibility can be assumed. However, Glück (1940) claimed fruit formation to be triggered by agamospermy.

Fruit and Seed. Both chasmogamous and cleistogamous flowers produce fruits. Those arising from chasmogamous flowers are naturally held at the level of the water or a depth of some cm, those arising from cleistogamous flowers at a depth of at least 10 cm. The indehiscent fruit contains a single seed which develops from the upper ovule in the adaxial locule. The young pericarp consists of about 15 layers of cells. Meristematic activity in the central layers gives rise to numerous fibres which are arranged in an outer layer of predominantly longitudinally elongated fibres and an inner layer of tangentially elongated fibres. Both on the outer side and the inner side, a few layers of parenchymatous cells remain. Just below the persistent calyx and alternating with the 5 sepals, the fruits bear 5 rigid filiform, apically curved appendages (Fig. 56C), up to 7 cm long, two of which may be rudimentary. Quite often the fruits have 2–3 supplementary, longitudinally arranged, narrow wings.

The spine-like appendages arise on the hypanthium. They mainly consist of fibres but also contain a vascular bundle which branches off from one of the 5 bundles that supply the corolla. Superficially these appendages somewhat resemble spine-like emergences found in certain Pedaliaceae (e.g. *Uncarina*, *Pedalium*), and these spine-like structures have often been considered as homologous. From the ontogeny of the spines of *Trapella*, however, it becomes clear that they are only analogous structures: all emergences on the fruits in Pedaliaceae arise directly on the carpels, not on a hypanthium as in *Trapella*, and they never contain vascular bundles.

The seed coat is exotestal (Singh 1960). It consists exclusively of the cells of the outer epidermis, all other cell layers having been crushed during maturation. The cells are thin-walled and do not show any thickenings. The raphe is ventral.

Dispersal. The long and curved appendages render the fruit an effective burr, and dispersal by fish and/or aquatic birds is assumed (Oliver 1888).

Phytochemistry. No data are available.

Affinities. The author of the genus *Trapella* (Oliver 1887) assigned his new genus to Pedaliaceae; apparently the burr-fruit of *Trapella* reminded him of certain burr-fruits found in Pedaliaceae. However, he was already aware of striking differences between *Trapella* and Pedaliaceae (e.g. the inferior ovary in *Trapella* and the uncommon aquatic life form), and he already considered whether this genus should not be better placed in a separate family. Oliver (1888), who published the first detailed study of this genus, added many important morphological features which do not conform to Pedaliaceae. Oliver (1888) considered possible relationships with Bignoniaceae, Gesneriaceae (Cyrtrandreae) and Scrophulariaceae (aquatic members such as *Limnophila*, *Hydrotriche*, Gratioloideae), and especially with Myoporaceae (now included in Scrophulariaceae, Savolainen et al. 2000) due to similarities in gynoecium structure, and again took into consideration the status of a separate family. In the end, however, against the background of the extreme morphological diversity found in Pedaliaceae, he decided to keep *Trapella* incorporated in Pedaliaceae, forming the new tribe Trapelleae F. Oliv. For more than four decades all subsequent authors followed this view. Familial status was only established by Honda and Sakisaka in 1930 and strongly supported by Ihlenfeldt (1967), but is still not generally accepted (e.g. Brummitt 1992; APG 1998).

There seems to be consent that Trapellaceae belong to Lamiales. However, the position within the order is still unresolved, due to a remarkable combination of very unusual and at the same time, within Lamiales, only erratically occurring characters (e.g. aquatic life form, floral diagram, inferior ovary, burr-fruit) which neither fit into any of the generally adopted families nor convincingly testify closer affinities to any of the families. Molecular data (Olmstead, pers. comm.) indicate that the family is only distantly related to Pedaliaceae (and Martyniaceae), but shows affinities to Gratioleae (Scrophulariaceae).

Distribution and Habitats. The family is at home in subtropical and temperate regions from Japan and Korea to E Russia and SE China. Many fossil records, mainly leaves and the characteristic fruits, testify that in the past the distribution area reached as far west as central Europe. The earliest record is from the Miocene of western Germany (Tralau 1965). The genus apparently has not undergone any major changes, at least in its vegetative and fruit morphology, since that time.

The natural habitats are shallow ponds. Under greenhouse conditions it is possible to grow *Trapella* as a terrestrial plant (Glück 1940). Whether terrestrial forms also occur in the field is unknown.

Only one genus:

Trapella Oliv. Fig. 56

Trapella Oliv., Hook. Icon. Pl. 16, t. 1595 (1887); Glück, Bot. Jahrb. Syst. 71 (1941): 267–336 (1940), rev.

See family description.

One or two species, native to Japan, Korea, E Russia and SE China.

Selected Bibliography

APG 1998. An ordinal classification for the families of flowering plants. Ann. Missouri Bot. Gard. 85: 531–553.

Brummitt, R.K. 1992. Vascular plant families and genera. Kew: Royal Botanic Gardens.

Erdtman, G. 1952. Pollen morphology and plant taxonomy. Angiosperms. Stockholm: Almquist & Wiksell.

Gentry, A.H., Tomb, A.S. 1979. Taxonomic implications of Bignoniaceae palynology. Ann. Missouri Bot. Gard. 66: 756–777.

Glück, H. 1940. Die Gattung *Trapella*. Bot. Jahrb. Syst. 71 (1941): 267–336.

Ihlenfeldt, H.-D. 1967. Über die Abgrenzung und die natürliche Gliederung der Pedaliaceae R.Br. Mitt. Staatsinst. Allg. Bot. Hamburg 12: 43–128.

Oliver, D. 1887. *Trapella sinensis*, Oliv. Hooker's Icon. Pl. 16: t. 1595.

Oliver, F.W. 1888. On the structure, development, and affinities of *Trapella*, Oliv., a new genus of Pedalineae. Ann. Bot. 2: 75–15.

Savolainen, V., Fay, M.F., Albach, D.C., Backlund, A., van der Bank, M., Cameron, K.M., Johnson, S.A., Lledó, M.D., Pintaud, J.-C., Powell, M., Sheahan, M.C., Soltis, D.E., Soltis, P.S., Weston, P., Whitten, W.M., Wurdack, K.J., Chase, M.W. 2000. Phylogeny of the eudicots: a nearly complete familial analysis based on *rbc*L gene sequences. Kew Bull. 55: 257–309.

Singh, S.P. 1958. Morphological studies in some members of the family Pedaliaceae. Thesis. Agra: Agra University.

Singh, S.P. 1960. Morphological studies in some members of the family Pedaliaceae. Agra Univ. J. Res., Sci. 9(2): 218–220.

Straka, H., Ihlenfeldt, H.-D. 1965. Pollenmorphologie und Systematik der Pedaliaceae R.Br. Beitr. Biol. Pflanzen 41: 175–207.

Tralau, H. 1965. New facts and new finds of fossil *Trapella* Oliver in Europe. Bot. Not. 118: 21–24.

Wang, W. 1990. Flora Reipublicae Popularis Sinicae, vol. 69. Beijing: Science Press.

Verbenaceae

S. Atkins

Verbenaceae J. St. Hilaire, Exp. Fam. Nat. vol. 1, pt. 2: 245 (1805).

Herbs, shrubs or small trees, sometimes lianas, stems terete or quadrangular with varied indumentum, unarmed, or sometimes with prickles or spines, eglandular or with resinous glands on stems, leaves, bracts, calyx or corolla; leaves opposite, usually decussate, sometimes alternate or ternate, sometimes verticillate or fascicled, sessile or petiolate, sometimes much reduced, scale-like, or spinose, simple to pinnatisect, margins entire, dentate, crenate, lobed or incised, stipules absent; inflorescence racemose, dense or lax, sometimes spicate or a compact head, axillary and/or terminal, simple or compound; bracts usually present, inconspicuous or showy; flowers hermaphrodite, rarely plants dioecious, zygomorphic or subactinomorphic, 4- or 5-merous; calyx tubular or campanulate, 4- or 5- (rarely 2-)lobed or annular or subtruncate, persistent in fruit; corolla hypocrateriform or infundibular, zygomorphic, 4- or 5-lobed, limb spreading, with lobes equal or not, tube sometimes enlarged in the middle or at throat; stamens 4(5), inserted on the corolla, alternate with the petals, usually didynamous, or reduced to 2, sometimes with 1 or more staminodes present, included, or only slightly exserted; anthers 2-locular, basifixed or dorsifixed; thecae parallel or divergent, dehiscing longitudinally, connective reduced or conspicuous and with or without glandular appendage. Ovary superior, glabrous, rarely pilose at apex, seated on a nectariferous disk, syncarpous, 2-carpellate (or 1-carpellate by abortion of the abaxial carpel, or 4-carpellate by elaboration as in *Duranta*); carpels 2-locular by the inrolling of the carpel margin; ovules 1 or 2 per locule, erect, rarely pendulous, fixed at base of locule, or at upper part of locule, anatropous, rarely hemianatropous, or subanatropous, or orthotropous; style terminal, 2-lobed or simple with stigma capitate, oblique or lateral, rarely weakly 4-lobed; fruit a dry or fleshy schizocarp separating at maturity into two or four 1-seeded mericarps, or into two 2-locular and 2-seeded mericarps, or a drupaceous fruit with one or two pyrenes with one locule and one seed, or with two or four pyrenes, each with two locules and two seeds.

A family of 34 genera with about 1200 species, mainly from the New World, with a few groups in Europe, Asia, Africa and Madagascar.

Vegetative Morphology. The family comprises herbs, subshrubs, shrubs and small trees. A few species are scandent shrubs or lianas, and a few are prostrate. *Phyla nodiflora* is a prostrate herb which roots at its nodes. The bark can be smooth or rough; some species have prickles (e.g. in *Lantana*), and some spines (e.g. in *Duranta*). Some species have woody rootstocks, and *Pitraea* has a swollen tuber. The leaves are usually herbaceous, variable in size and outline, from tiny, reduced, fleshy leaves which are adaptations to xerophytic conditions (e.g. *Acantholippia*) to large ones (leaves up to 28 cm long in *Petrea*). *Petrea* has characteristically scabrid leaves. The leaf margin is often serrate or crenate, rarely entire, occasionally lobed or pinnatisect. Venation is mostly pinnate.

Vegetative Anatomy. Growth rings are normally present in woody species. Vessels are solitary, and in irregular clusters. Perforation plates are mostly simple, or occasionally scalariform, intervascular pits are alternate, circular; vessel ray pits similar to the intervascular pits. Rays are both uniseriate and multiseriate, heterocellular, often with procumbent cells. Axial parenchyma is scarce, scanty paratracheal to vasicentric. Fibres are either septate or non-septate with simple pits on the radial and occasionally on the tangential walls. This information has been distilled from various sources, particularly ter Welle and Détienne (1988).

According to Inamdar (1969), the cells of the epidermis are polygonal, isodiametric, or elongated, and arranged irregularly, usually with thick, sinuous anticlinal walls. The surface of the cuticle has striations, radiating from the base of the hairs or stomata. Several types of trichome have been observed: (1) simple, unicellular; (2) unicellular, conical, silicified; (3) unicellular, conical, surrounded by 8–10 cells at base; (4) capitate, eglandular with long stalk and a unicellular excentric head; (5) filiform uniseriate; (6) short-stalked, capitate, glandular; (7) peltate, glandular. In addition, there are unicellular hook-shaped hairs.

The majority of stomata are anomocytic or diacytic, perigenous and with a single subsidiary cell. Abnormal stomata with a single guard cell, contiguous stomata, and stomata with unequal guard cells are common.

Extra-floral nectaries occur on the leaf surfaces, and occasionally on the calyces of *Baillonia*, *Casselia*, *Citharexylum*, *Duranta*, *Lampaya*, *Rhaphithamnus* and *Stachytarpheta*.

Inflorescence Structure. Inflorescences are axillary or terminal, always basically racemose, indeterminate, simple or compound. The inflorescence often is in the form of a spike (Fig. 57A, C), with its flowers sessile or very short pedicellate in the axils of the bracts, e.g. some species of *Verbena*, *Stachytarpheta*, *Bouchea*, etc. The head-like inflorescences of *Lippia*, *Lantana*, and *Glandularia* originate through the congestion of the flowers caused by the inflorescence axis not elongating. These inflorescences can contain many tens of flowers. In *Casselia*, the inflorescence is reduced to 2, 4 or 6 flowers, and in *Parodianthus* the partial inflorescence is a 2-flowered reduced raceme (Martínez and Múlgura, ined.).

The bracts of the inflorescence sometimes form an involucre, as in some species of *Lippia* and *Lantana*. The bracts can be small and filiform, or large and showy. For more detailed information on inflorescence structure, see Martínez et al. (1996, 1997), Múlgura de Romero et al. (1998) and Drewes and Martínez (1999).

Fig. 57. Verbenaceae. Inflorescences in tribe Verbeneae. Racemose, bracteate spikes, dense or lax, sometimes congested into a head. **A** *Junellia arequipense*. **B** *Glandularia araucana*. **C** *Tamonea curassavica*. (Martinez et al. 1996)

Flower Structure. Flowers are zygomorphic, or occasionally actinomorphic, and almost always bisexual, with functioning ovary and stamens, although dioecy occurs in some species of *Citharexylum* (Tomlinson and Fawcett 1972). The calyx (Fig. 58D) is a short or long tube, variously lobed or toothed, occasionally undulate; not ribbed except in *Petrea*, herbaceous and green (except in *Petrea* where the calyx is often the same colour as the corolla); it usually envelops the fruit (Fig. 58G), but can also form a cup at the base of the fruit, and can become woody, and may (*Petrea*) become enlarged or accrescent in fruit.

The corolla is tubular, spreading at the apex to form lobes (Figs. 58D, E, 59B). The tube can be straight or funnel-shaped; there may be constrictions or swellings at the middle or the top of tube. The 5(4) lobes can be anything from ± equal to the upper lobes slightly larger or smaller than the lower lobes. The normally 4 stamens are inserted on the corolla, alternate with the corolla lobes (Fig. 58E), arranged with the abaxial two slightly higher than the adaxial two, with the anthers lying in the lower, middle or upper parts of the tube. The filaments are short, and for the most part adhere to the side of the tube, with only a very short, free upper part. In some genera, 1 staminode is

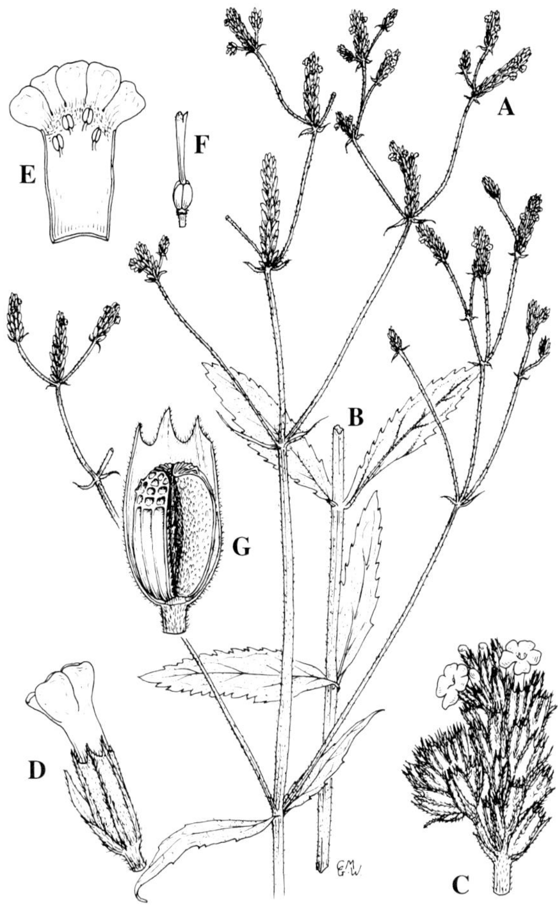

Fig. 58. Verbenaceae. *Verbena brasiliensis.* **A** Upper part of plant. **B** Two vegetative nodes of main stem. **C** Spike with subsidiary spike. **D** Flower. **E** Corolla with tube opened. **F** Gynoecium. **G** Fruit with one nutlet removed. (Yeo 1990)

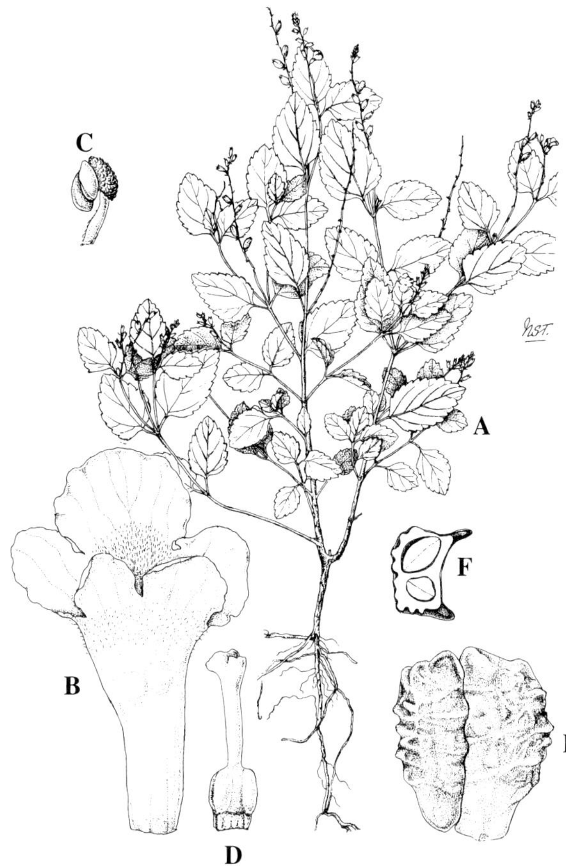

Fig. 59. Verbenaceae. *Priva boliviana.* **A** Whole plant. **B** Corolla. **C** Stamen with glandular connective. **D** Ovary and stigma. **E** Fruit. **F** Transverse section through mericarp. (Cabrera 1993)

present, representing a fifth stamen, and occasionally only 2 stamens are fertile, and the other two are reduced to staminodes. Anthers are basifixed or dorsifixed. The connective is normally reduced and contained behind the anthers, but can be dilated and visible above the anthers and with or without a glandular appendage (Fig. 59C). The anther thecae are parallel or divergent, 2-locular, opening longitudinally. The ovary (Fig. 59D) is superior, usually seated on a nectariferous disk with the style at its apex; the ovary is 2-carpellate (1-carpellate by abortion, or rarely 4-carpellate in *Duranta*), and each carpel is 1-locular or, by inrolling of the carpel margin, 2-locular; ovules 1 or 2 per locule, basal, erect, or rarely at upper part of locule and pendulous, anatropous, rarely hemi-anatropous, or subanatropous, or orthotropous; style terminal, 2-lobed or simple with stigma capitate, oblique or lateral, rarely weakly 4-lobed. The style is persistent or deciduous, sometimes leaving a short stylopodium.

Floral Anatomy. The information on floral anatomy is based on work by Rao (1952). In most taxa the vascular bundles of the different whorls arise in quick succession from the receptacular stele. The vascular supply to the calyx shows some variation. In *Lantana* there are five vascular bundles, of which one is united at the base with a corolla strand. Normally calyx and corolla midribs alternate. In *Lippia* the calyx is supplied by four vascular bundles. In *Stachytarpheta*, all the vascular bundles of the calyx are traceable to only three strands, and there is also fusion of two vascular bundles of the corolla with calyx traces. The vascular supply in *Petrea* is irregular and in *Duranta* it consists of five median and five commissural bundles. In *Citharexylum* the calyx is supplied by ten bundles, of which five are midrib bundles and the other five commissural.

The vascular supply to the corolla is more uniform. Each lobe has a midrib strand which gives out a lateral branch on either side at about the level of insertion of the stamens. In many genera these are fused at the base with the vascular traces of the calyx.

Duranta and *Citharexylum* have four vascular bundles for the functional stamens, and one for the staminode. *Petrea* has four functional stamens and a staminode, but the latter is not provided with a vascular bundle. In genera with four functional stamens, there are four vascular bundles supplying them, with no trace of the fifth strand. *Stachytarpheta* has two fertile stamens and two staminodes, each supplied by a vascular bundle.

The vascular anatomy of *Duranta* supports the view that the ovary is 4-carpellate, each having two ovules on parietal placentas. The base of the gynoecium clearly shows four dorsal strands of the carpels and four large placental bundles. The eight locules are obtained by the inwardly projecting margins of carpels and the inward projections of the ovary wall opposite the dorsal bundles. All four carpels of *Duranta* are fertile. *Citharexylum* has only two carpels. The ovary, however, although unilocular at the base and having parietal placentation, becomes 4-locular by extensions of the placentas on either side towards the dorsal bundle of each carpel. The vascular anatomy of *Lippia*, *Lantana*, *Stachytarpheta* and *Petrea* shows that, although there are two carpels present, only the posterior one is developed and ovule-bearing. The two placental bundles represent marginal traces of this carpel. The anterior carpel is represented only by its dorsal bundle. The posterior carpel becomes well-developed and fertile, with its margins incurved into the loculus.

Embryology. According to Davis (1966), anther wall formation follows the Dicotyledonous type in *Stachytarpheta*, but the Monocotyledonous type has been reported from *Phyla nodiflora*. Maheshwari (1954) showed that the outermost layer of archesporial cells divides periclinally to cut off the primary parietal layer. The latter divides once more to produce the middle layer and the tapetum. The middle layer begins to degenerate soon after the formation of the microspores. The cells of the glandular tapetum become 2–4-nucleate and Ubisch granules are present on the inner tangential walls. The tapetum shows signs of degeneration when the pollen grains are one-celled.

A hypodermal archesporial cell originates before the appearance of the integumentary initial. There is only one integument. The nucellus is thin and composed of only one layer of cells around the archesporial cell, which never cuts off a parietal cell but functions directly as the megaspore mother cell. The megaspore mother cell enlarges considerably, and subsequently undergoes meiosis whereby a linear tetrad of megaspores is formed. The chalazal megaspore functions to give rise to the embryo sac, while the other three megaspores degenerate. The embryo sac development is of the Polygonum type.

The mature embryo sac is somewhat fusiform, tapering at each end. *Stachytarpheta* and *Verbena* have a slightly enlarged micropylar region and a long, narrower chalazal region, while *Lantana* and *Lippia* are more linear, without a broader micropylar chamber or distinct chalazal epithelium (Rudall and Clark 1992). The egg cell is flask-shaped, with a large vacuole towards its micropylar end. The innermost layer of cells of the single integument forms a tapetal jacket around the embryo sac, since the nucellus degenerates. The cells of this specialised part of the integument remain undivided and uninucleate, and are rich in cytoplasm.

After fertilisation, the primary endosperm nucleus which lies near the zygote divides and the accompanying development of a thin, membranous transverse wall gives rise to a primary micropylar and a primary chalazal chamber. The nucleus of the smaller chalazal chamber then divides to form a 2-nucleate cell which functions directly as the chalazal haustorium. In most instances the cell remains bi-nucleate and elongates in the direction of the chalaza. The nucleus of the micropylar chamber divides and a longitudinal wall is formed. Subsequently, these cells divide and ultimately five tiers of cells are formed. As soon as the embryo begins to grow, the haustorial cells become functionless and degenerate. However, the chalazal haustorium persists and degenerates only when the embryo has grown considerably. At the same time the endosperm tissue grows and fills the cavity of the embryo sac. This layer of endosperm surrounding the embryo becomes rich in starch grains. Information for this section has been taken from Pal (1951) and Tatachar (1940).

Pollen Morphology. Raj (1983) presented a study of the pollen of the Verbenaceae. In general, pollen grains (Fig. 60) are radially symmetric, isopolar, 3- or rarely 4-colporate (3-colpate

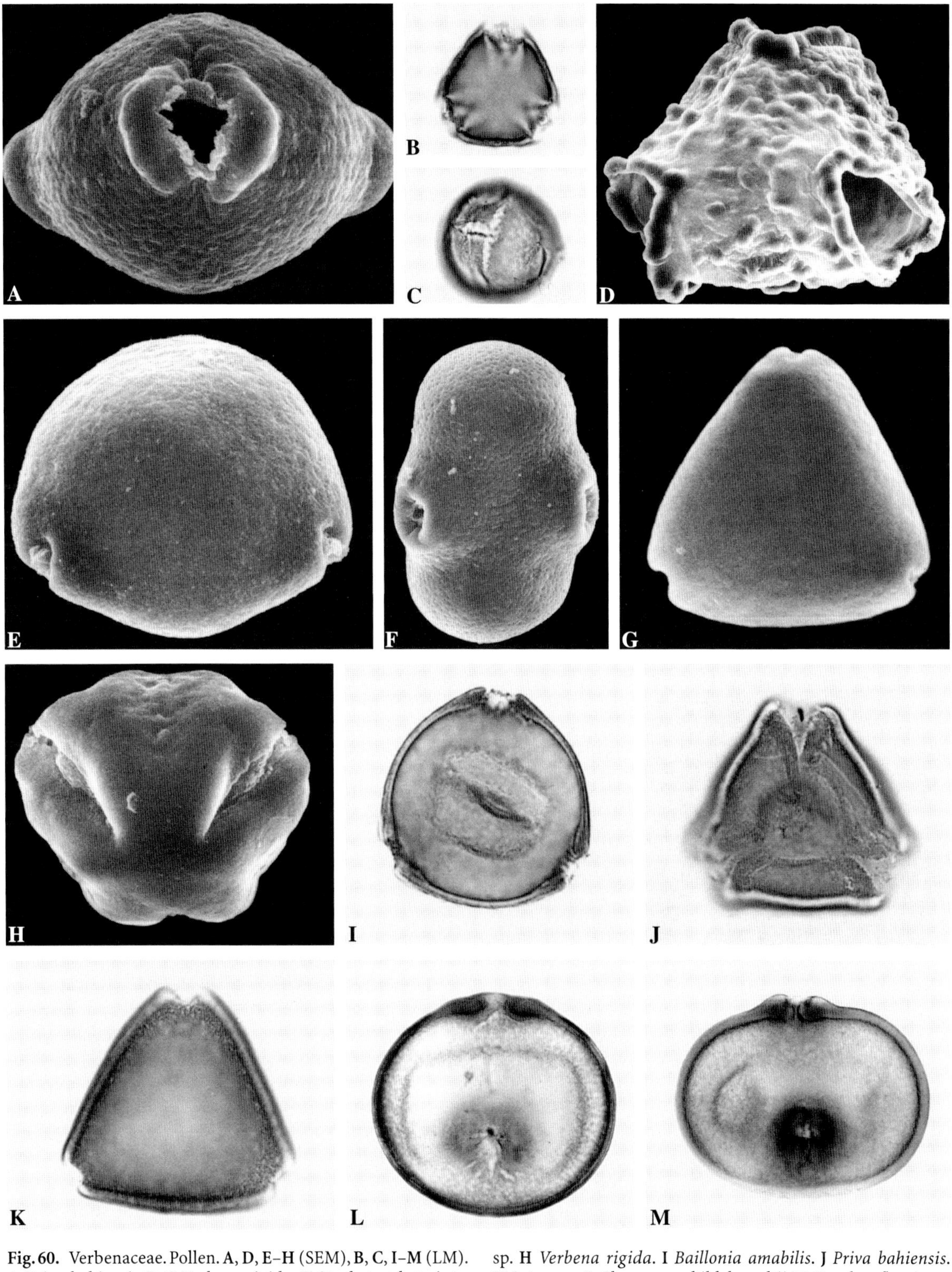

Fig. 60. Verbenaceae. Pollen. A, D, E–H (SEM), B, C, I–M (LM). A *Priva bahiensis*. B, C *Verbena rigida*. D *Stachytarpheta integrifolia*. E *Baillonia amabilis*. F *Bouchea fluminensis*. G *Petrea* sp. H *Verbena rigida*. I *Baillonia amabilis*. J *Priva bahiensis*. K *Petrea* sp. L *Chascanum hildebrandtii*. M *Bouchea fluminensis*. (Photo M.M. Harley and U. Song)

in *Petrea*). Colpi are long, narrow with tapering ends, equatorially constricted (not constricted in *Stachytarpheta*), margin thickened, membrane densely granular. Exine structure consists of a dense, finely or coarsely perforate, smooth tectum. The tectum is thin, supported by densely spaced, slender bacula. The nexine appears to form a thin homogeneous layer. Raj (1983) did not sample *Xeroaloysia*, *Xolocotzia* or *Verbenoxylum*. *Glandularia* was sampled as part of *Verbena*.

In Verbenaceae s.s., Raj (1983) identified 10 pollen types:

1. *Verbena* type, where pollen grains are 3- and occasionally 4-colporate, varying in shape from suboblate to prolate. The pollen grains have a characteristic, hexangular equatorial outline, caused by a ridge-like thickening along colpi margins. Exine structure is tectate, perforate. Size, shape and density of perforations vary in the different genera. The following genera have pollen grains of this type: *Acantholippia*, *Aloysia*, *Casselia*, *Citharexylum*, *Coelocarpum*, *Diostea*, *Dipyrena*, *Duranta*, *Hierobotana*, *Junellia*, *Lampaya*, *Lantana*, *Lippia*, *Nashia*, *Neosparton*, *Parodianthus*, *Pitraea*, *Rhaphithamnus*, *Tamonea*, *Urbania* and *Verbena*.
2. *Phyla* type, where pollen grains are 3-colporate (brevicolpate), subprolate to prolate, colpi slit-like with pointed ends, exine structure tectate-perforate, perforations sparsely spaced. Only *Phyla*.
3. *Bouchea* type, where pollen grains are 3-colporate, subprolate to prolate, pores varying in shape and size with a thickened annulus, exine structure tectate-perforate, densely spaced, bigger around pores. Only *Bouchea*.
4. *Chascanum* type, where pollen grains are 3-colporate (usually brevicolpate), prolate spheroidal to subprolate, colpi narrow, slit-like, usually equatorially constricted, occasionally branched apically and laterally. Exine structure tectate-perforate, perforations densely spaced. Only *Chascanum*.
5. *Stachytarpheta* type, where pollen grains are 3-colpate, peroblate to oblate. Outline of mesocolpi straight, alternating with wide open colpi. Exine structure tectate. Sculpture consists of sparsely or densely spaced verrucae. Only *Stachytarpheta*.
6. *Priva* type, where pollen grains are 3–4-colporate, suboblate to subprolate, colpi broad with tapering ends, margin uneven, equatorially constricted, exine structure tectate-perforate. Only *Priva*.
7. *Petrea* type, where pollen grains are 3-colpate (brevicolpate), oblate to suboblate, colpi broad with pointed ends, margin thickened. Exine structure tectate-perforate, tectum coarse, perforations densely spaced. Only *Petrea*.
8. *Recordia* type, where the pollen grains are 3-colporate, suboblate to oblate spheroidal, rarely prolate spheroidal, equatorial contour of mesocolpi slightly concave. Exine structure tectate-perforate, tectum thick, coarse with densely spaced, minute perforations. Only *Recordia*.
9. *Baillonia* type, where the pollen grains are 3(–4)-porate, suboblate. Pores ± circular to oval-shaped, with a thick annulus. A number of transparent, narrow fissures, often branched apically or laterally radiate from the margin of pores. Exine structure tectate-perforate, perforations appear in places as irregular pits. Only *Baillonia*.
10. *Rehdera* type, where pollen grains are 3-colporate (brevicolpate, occasionally brevissimicolpate), suboblate to oblate spheroidal. Exine structure tectate-perforate, perforations minute, densely spaced. Only *Rehdera*.

KARYOLOGY. There has been no overall review of the chromosomes of the Verbenaceae s.s. Generally, base numbers for genera or subgenera are x = 5, 6, 7, 8, 9, 10, 11, 12, 17, 18, 19, 20 and possibly higher. According to Sanders (2001), these diverse numbers suggest aneuploidy. His hypothesis is that x = 5 gave rise to x = 6 and 7 by ascending aneuploidy, and to x = 10 by polyploidy, which in turn gave rise to x = 8 and 9 by descending aneuploidy. Counts have been done for many of the common and cultivated species. *Verbena* has counts of $2n = 14$ and multiples of 7 up to 56. *Glandularia* has counts of $2n = 10$ or 30, and *Junellia* $2n = 20$, 40 or 60 (Botta and Brandham 1993). Extensive chromosome counts have been done on *Lantana camara* in South Africa (Spies 1984). The data show counts of $2n = 22, 33, 44, 55$ and 66. Polyploidy is common in this invasive species. Sanders (1987) was able to demonstrate that among the species of *Lantana* of the islands of the Caribbean, diploids are morphologically distinct, are elements of the native flora, and are not competitively aggressive, while polyploids are less distinct from one another, and have wide-ranging distributions. Generally in this genus, the triploids, hexaploids, and pentaploids occur in cultivated and naturalised populations.

Pollination. In all genera, the stamens are included within the corolla tube. The stigma is sometimes tucked well down inside the tube, but in some genera it is above or on the same level as the stamens. Nectar is produced in small quantities from a disk just below the ovary at the base of the corolla tube. The opening at the mouth of the corolla is often very small, sometimes only 1.5 mm at its widest point, with a further constriction in the tube below this. This suggests that the main pollinators are insects with long tongues, probing for the nectar at the bottom of the tube. The individual flowers have only a relatively small "landing area" at the point where the lobes spread at the apex of the tube but, as most inflorescences consist of tightly packed flowers, butterflies are able to land on an inflorescence and systematically take nectar from several flowers in turn. Other pollinators appear to be small Hymenoptera and Diptera, and there is evidence of at least one species (*Rhaphithamnus venustus*) being pollinated by hummingbirds (Sun et al. 1996). The majority of pollination appears to be effected by small nectar-seeking bees (Knuth 1909); in this work *Verbena officinalis* is studied, and the suggestion is made that rows of upward-pointing hairs in the throat of the corolla tube guide the insect's proboscis to brush against the anthers and then the stigma. Because of the orientation of the anthers, it does not dislodge pollen until withdrawing its proboscis, which is then wet from the nectar so that the pollen readily adheres and can be deposited onto the stigma of the next flower to be visited. Observations made by Mathur and Mohan Ram (1986) reveal evidence for the pollination of *Lantana camara* by thrips (Thysanoptera) which feed on stigmatic exudate and pollen. The flower structure does not permit the thrips to reach the nectar. Pollination is aided by the papillae at the throat of the corolla bending inwards and preventing the exit of the thrips. Each time a thrip slips back down the tube, it moves between anther and stigma and deposits a large amount of pollen on the stigma. In the same experiments it is shown that, although at certain times of the year butterflies are plentiful on the inflorescences, they are able to remove the nectar from the base of the corolla tube without effecting pollination. Similar observations have been made on *Phyla incisa* by Estes and Brown (1973).

Reproductive Systems. Almost all flowers in the Verbenaceae are hermaphroditic with functional stamens and stigma. Self-compatibility appears to be a constant feature of breeding systems in the Verbenaceae, although it is likely that there are temporal and spatial characters, e.g. protandry, which operate to encourage outbreeding as the preferred option. With obvious adaptations for insect pollination (nectar production, flower colour, and scent), and with many individual flowers in the same inflorescence, it must be supposed that pollination is likely to be geitonogamous. The relative positions of the stigma and stamens would suggest that pollination without an external agent is unlikely. Cleistogamy has been shown for *Lantana camara* but is not the general mode of pollination in this species (Mathur and Mohan Ram 1986), and autogamy has been shown to be possible in some species of *Verbena* (Cruden et al. 1990). Certain species have been known to hybridise in the wild, and this may be the result of self- and cross-compatible breeding sytems. There is no information that self-incompatibility occurs, but complete non-occurrence would be rare (Owens and Ubera-Jiménez 1992). Dioecy has developed in some species of *Lippia* (Troncoso 1961) and in *Citharexylum* (Tomlinson and Fawcett 1972), with gynodioecy well-documented in *Rhaphithamnus* (Sun et al. 1996).

Fruit and Seed. The fruit of Verbenaceae (Figs. 58G, 59E, F) is either a drupe with 1 or 2 pyrenes, which contain (1)2 locules with 1 seed in each locule, or the fruit is a dry or fleshy schizocarp, splitting at maturity into 2 or 4 mericarps, each containing 1 or 2 seeds. It has been suggested (Wagstaff and Olmstead 1997) that a schizocarp that breaks into four dry, indehiscent one-seeded mericarps is the derived state, possibly a shift from a fleshy, drupaceous fruit following colonisation of regions with a more seasonal climate, and that the dry fruit is a developmentally arrested fleshy fruit.

The outer surface of the mericarp or pyrene can be smooth or sculptured. Seeds are normally without endosperm. According to Ryding (1995), the mesocarp of some genera is made up of between 1 and 4 parenchymatous layers (the layers in *Phyla* being very slightly sclerenchymatous). The innermost cell layers of the mesocarp are much thicker than the other layers, and there are between 3 and 7 cell layers in the sclerenchyma region. *Duranta* has a smooth sclerenchyma, while *Verbena* and some *Stachytarpheta* have rough sculpturing. The endocarps of *Duranta*, *Lantana*, *Stachytarpheta* and *Verbena* are strongly sclerenchymatous. In *Stachytarpheta* and *Duranta*, the endocarp cells contain prismatic crystals. The fruit of *Verbena* separates into four elongate mericarps, each with several longitudinal ribs on its abaxial side, which are reticulately connected in the upper third. On the periphery of the fruit, the epidermal cells contain chloroplasts almost until maturity, and their outer wall is conspicuously thickened, but not lignified. In contrast, the exocarp between the mericarps consists entirely of mucilaginous cells forming interlocking papillae. Below the papillae there is a zone of parenchymous cells, four layers thick in the centre of the fruit but tapering towards the periphery. The next three mesocarp layers are continuous around the fruitlet, and on the peripheral side they border onto the epidermis. An additional parenchymous layer of slightly smaller cells with more delicate, non-impregnated walls may or may not be present. The endocarp consists of about four layers of thick-walled cells that are irregular in outline. The cells of the inner epidermis are flatter, but have similarly thickened walls; their anticlinal walls are strongly undulate. The seed coat appears as scarcely more than a thin line between the endocarp and endosperm. The fruit of *Lantana* is a two-seeded drupe, arising from a single carpel, which corresponds to the abaxial carpel in most other Verbenaceae. In contrast to *Verbena*, the exocarp of *Lantana* has a thick cuticula. The multilayered, parenchymatous mesocarp becomes fleshy at maturity. The pyrene shows bilateral symmetry. Its central part is occupied by an area of very loose tissue which is continuous from base to apex. The two seeds are on either side of the median weakness zone, slightly shifted towards an adaxial position. The endocarp cells are more or less isodiametric with relatively thin, weakly lignified walls. The seed coat consists of several layers of rather thin-walled, isodiametric cells.

DISPERSAL. There is little evidence for specialised dispersal in the Verbenaceae. The calyx normally adheres tightly to the developing mericarps, and often the calyx teeth twist together above the mature fruit, making a tight parcel. This presumably disperses from the plant as a complete unit. Once shed, the calyx would hold the fruit until, after dispersal perhaps by rain-wash or by passing animals, the calyx would eventually rot away, leaving the mericarps ready to germinate. In some species with a viscid calyx, the calyx adheres to the coats of passing animals. There is also evidence of dispersal by birds in *Lantana* and *Duranta*. These fruits are fleshy and conspicuously coloured and are adapted for the attraction of birds. The calyx of *Petrea* has long lobes which surround the fruit and may assist the fruit to be carried by wind, the long lobes acting as propeller.

PHYTOCHEMISTRY (BY R.J. GRAYER). The information given below on the comparative phytochemistry of the Verbenaceae has been largely extracted from Hegnauer (1973, 1990) and Grayer et al. (1999). To date species belonging to less than half of the genera of the Verbenaceae have been investigated for their secondary metabolites. These belong to four of the five tribes described below; species of the tribe Priveae do not yet seem to have been studied chemically. Iridoid glycosides have been reported from species of *Verbena*, *Lantana*, *Lippia*, *Stachytarpheta*, *Petrea*, *Citharexylum*, *Baillonia* and *Duranta*. They appear to be a characteristic chemical feature of the family. Essential oils are known from species of *Lantana*, *Lippia*, *Phyla*, *Aloysia* and *Acantholippia*, so they seem to be restricted to the tribe Lantaneae. Flavonoids in the Verbenaceae are usually of the flavone class and in addition they are characterised by 6-hydroxylation. These 6-hydroxylated flavones have been found as glycosides in *Verbena*, *Lantana*, *Lippia*, *Phyla*, *Stachytarpheta*, *Citharexylum*, and *Duranta*. Another class of phenolic compounds that may be characteristic constituents in the family are phenylethyl caffeoylglycosides such as verbascoside (acteoside). They have been reported from *Verbena*, *Lantana*, *Lippia* and *Phyla*, but are thought to be more widespread in the Verbenaceae. Naphthoquinones are present in the roots of some *Lantana* and *Lippia* species. Diterpenoids seem to be uncommon in the Verbenaceae s.s., although they are characteristic of the Lamiaceae, including several genera that have recently been removed from the Verbenaceae to the Lamiaceae, such as *Clerodendrum* and *Caryopteris*. Toxic triterpene esters have been found in *Lantana camara* and species of *Lippia*. Triterpene saponins have been reported from species of *Verbena*, *Duranta*, *Stachytarpheta* and *Diostea*. In *Verbena* they seem to be restricted to section *Glandularia*. The presence of nitrogen-containing compounds, such as alkaloids and cyanogenic glycosides, does not seem to be a characteristic chemical feature of the Verbenaceae, although cyanogenic glycosides are quite common in *Stachytarpheta* and *Phyla*. The oligosaccharide stachyose and related sugars are usually present as reserve carbohydrates in stems and roots, but in some tropical species such as *Bouchea ehrenbergii* they are replaced by starch.

Many secondary constituents found in the Verbenaceae have biological activity and may play a role in the ecology of the plants. For example, compounds from *Lantana camara* were shown to have allelopathic activity. This may be the reason why, in large areas of the tropics and subtropics, *L. camara* is a troublesome weed and a threat to the native flora. Some iridoids from verbenaceous species have insect-antifeedant activity and may protect their leaves from insect damage.

Subdivision and Relationships. The classification of the family is not without its problems. There have been many attempts to define it and divide it into groups. One of the earliest and virtually the only truly monographic work on the family was that of J.C. Schauer in 1847. Schauer recognised 3 tribes and 10 subtribes, embracing 42 genera. He included Avicennieae and Symphoremeae. In 1876, G. Bentham proposed the inclusion of Phrymeae, Stilbeae and Chloantheae. Until most recent times, the most widely accepted classification was that of J. Briquet in 1895. Although this is not strictly a monographic work, as it does not attempt to list, describe, and differentiate all the species in each genus, Briquet's work represents the culmination of almost 150 years of research and investigation into the group, and is a source of much, systematically arranged information. He classified the family into 7 subfamilies and 13 tribes.

His classification is based largely on gynoecial structure and supposes that most Labiatae have a deeply 4-lobed ovary with a gynobasic style, and that most Verbenaceae have an unlobed ovary with a terminal style. Examination of the variation within both families reveals a continuum, rather than a clear distinction (Cantino 1992a). Junell (1934), in his survey of gynoecium morphology, demonstrated that only subfamily Verbenoideae has ovules attached marginally and never on the lateral faces of the false septa as in the rest of Labiatae. He proposed revolutionary changes in the systematics of the group and considered that only Briquet's subfamily Verbenoideae are truly verbenaceous. The view is now taken that if the circumscription of Verbenaceae is restricted to the 34 genera (1200 spp.) of Verbenoideae, excluding tribe Monochileae, then this would make a monophyletic group. This is supported by the studies of Labiatae-Verbenaceae by Cantino (1992a, b) and has been adopted by Thorne (1992) for his phylogenetic scheme. The remaining two thirds of Verbenaceae in their traditional circumscription have been transferred to Labiatae (Cantino et al. 1992a), which now include the Caryopteridoideae, Chloanthoideae and Viticoideae. Verbenaceae are distinguished by having a 2- or 4-loculate gynoecium by the inrolling of the carpel margin, by the ovules being attached to the margins, 2 ovules in each carpel (see below), by tricolporate pollen, by thickened stigmatic tissue, and by a racemose inflorescence and salverform or funnel-shaped corollas.

Until recently it was supposed that the two families (Labiatae s.l. and Verbenaceae s.s.) are sister clades or that Labiatae are derived from Verbenaceae. However, studies of chloroplast DNA (Olmstead and Reeves 1995; Wagstaff and Olmstead 1997) reveal separate origins for Verbenaceae s.str. and Labiatae s.l. in distant parts of the Scrophulariales.

The infrafamilial classification of Verbenaceae s.str. has also been the subject of much research. Briquet's subdivisions are based on the number of locules in each carpel and the number of ovules in each locule. Junell (1934) showed that convergent abortive reduction of the abaxial carpel has occurred, making Briquet's classification unnatural. Junell transferred *Baillonia* from Lantaneae to Citharexyleae, even though *Baillonia* is 1-carpellate, because he saw no difference in the structures of the gynoecia of *Citharexylum* and *Baillonia*; and *Casselia* from Petreeae to Casselieae, again, because of the unique character of the ovule insertion.

Recently Sanders (2001) suggested that the gynoecium structures have been misunderstood. He states that the Verbenaceae schizocarps (or drupes) typically contain two 2-chambered mericarps. Each 2-chambered mericarp does not represent one carpel, but 2 adjacent half-carpels, as locule walls of two adjacent half-carpels fuse more strongly during fruit development than do those of the same carpel. He proposes that this condition is synapomorphic and primitive within Verbenaceae s.str.

Caro (1982) suggested an amendment to tribe Lantaneae sensu Briquet. He proposed the separation of the genera with dry schizocarps, with two 1-seeded mericarps (i.e. *Lippia*, *Acantholippia*, *Aloysia*, *Phyla*, *Bouchea* and *Stachytarpheta*) to form the new tribe Lippieae, from the genera with drupes with a 2-locular, 2-seeded pyrene, which would remain in tribe Lantaneae.

Sanders (2001) redefined tribe Verbeneae to include *Stachytarpheta* and *Bouchea* and related genera on the basis that, although with only one functional carpel, they are more *Verbena*-like than *Lantana*-like. He based this on putative apomorphies such as inflorescence structure, calyx and corolla shape, structure of anthers and connectives, staminode presence, style branch orientation, carpel number and fruit structure.

Martínez and Múlgura de Romero (2003) have suggested that tribe Casseliae (Schauer) Troncoso is accepted to include *Casselia*, *Tamonea* and *Parodianthus*. Schauer (1847) originally included *Casselia* and *Tamonea* in subtribe Casseliinea, based on the morphology of the inflorescences. Junell (1934) supported the affinity on the basis of ovule insertion. It has now been shown that *Parodianthus* also shares these characters, along with a thickened style which terminates in an oblique, stigmatic lobe. In all three genera the style is very short, and the stigma is held below the stamens towards the base of the corolla tube; *Tamonea* and *Casselia* share the character that the connective of the upper stamens is thickened and glandular. In *Tamonea* the fruit is drupaceous with a 4-seeded pyrene, while in *Parodianthus* and *Casselia* the fruit is a somewhat fleshy schizocarp, with 2- or 1-seeded mericarps.

In studies of *rbc*L sequences of Labiatae and Verbenaceae (Wagstaff and Olmstead 1997), *Petrea* forms a clade distinct from other genera of Verbenaceae s.str., and is shown as the sister group to Bignoniaceae.

Except for Junell's work, sampling has been rather small and incomplete. To provide a new infrafamilial classification, it would be necessary to sample more extensively and include infrageneric groups, and to look at the development of the embryo and gynoecial structures. A morphological analysis should include anther attachment and connectives, style and stigma development, reticulate fruits, inflorescence structure, calyx and corolla shape, staminode presence, style attachment, and stylopodia. In the absence of this information, this classification is based largely on the work of Junell, but has taken into account the work of Martínez and Múlgura de Romero (2003) on the genus *Parodianthus*.

The genus *Recordia* has been difficult to place. It has characters (staminode presence, anther connectives dilated, and apparent abortion of locules) which place it in Citharexyleae. Many specimens have been studied and, although flowers appear perfect, very few fruits have been found. The fruits available have been sectioned and found to be bicarpellate and 4-locular, but with 2 locules suppressed, and with no developed embryos in the locules. It is tentatively placed in Citharexyleae.

The chiefly Madagascan genus *Coelocarpum* remains unplaced. In terms of ovary and fruit characters, *Coelocarpum* is similar to genera of tribe Citharexyleae, but it lacks the 5th stamen or staminode, there is no thickening of the anther connective, and the stigma is minutely bi-lobed.

Distribution and Habitats. Verbenaceae as circumscribed here are a mainly New World family. There are only two genera without representation in the New World – *Chascanum* and *Coelocarpum*. *Chascanum*, as currently defined, is found in Madagascar, Arabia and India, but it is very closely allied to the New World *Bouchea*, and at one time the species of *Chascanum* were placed in *Bouchea*, making it a pantropical genus with a distribution similar to that of *Lantana*. This leaves only *Coelocarpum*, a genus from Madagascar Somalia and Socotra, without any close relationship to the New World. *Verbena*, which has its greatest species diversity in the New World, also has species in Europe, Asia and North Africa, while the closely allied *Glandularia* has a disjunct North-South American distribution; *Lantana* has species in Africa, as do *Lippia* and *Priva*. The main distribution area of the rest of the genera seems to be temperate and tropical America, as found in, for example, *Tamonea* which occurs from Mexico and the Caribbean to northern South America to Brazil and eastern Bolivia. Some genera are confined to the southern part of South America, i.e. *Urbania*, *Acantholippia*, *Diostea*, *Lampaya*, all of which are restricted to Argentina and Chile.

Acantholippia, *Junellia*, *Lampaya*, *Diostea*, *Neosparton*, *Dipyrena* and *Parodianthus* are typically xerophytic, and are found in desert regions or high alpine areas. Some alpine groups are found at very high altitudes in the South American Andes, e.g. *Junellia seriphioides* is found at 4000 m in Peru. Even non-xerophytic species can be found at high elevations, e.g. *Glandularia gooddingii* at 2000 m in Arizona, and *Verbena bracteata* at 2255 m in New Mexico. In Mexico, *G. teucriifolia* can be found at 3650 m. One of the highest collections is that of *Lantana haughtii* at 5180 m in Peru. At the same time, *Phyla incisa* grows at 56 m below sea level in the Salton basin in southern California, and *Verbena officinalis* grows at 396 m below sea level along the shores of the Dead Sea in Israel. Some races of *Phyla nodiflora* are alkali-resistant.

Economic Importance. Many species have medicinal properties, and infusions are made from *Lampaya*, *Verbena*, *Lippia*, and *Stachytarpheta* for a variety of complaints. Essential oils are known from species of *Lantana*, *Lippia*, *Phyla*, *Aloysia* and *Acantholippia*, and many of the essential oil plants have been used for medicinal purposes. Various species of *Aloysia* are used as a tea to aid digestion, especially *Aloysia citriodora*. *Neosparton* is used as roofing material in northern Argentina. *Phyla nodiflora* is used as a lawn, replacing grass, and sometimes it is grown for forage. The wood of *Aloysia virgata* is used to make agricultural implements. The wood of *Citharexylum* is quite hard and used for general timber. Edible tubers are provided from *Pitraea*, and hundreds of ornamentals from species of *Baillonia*, *Duranta*, *Glandularia*, *Lantana*, *Verbena*, *Citharexylum*.

Conspectus of Classification of Verbenaceae

1. Tribe Verbeneae Schauer pp. (1847) (= Euverbeneae Briq.).
   Genera 1–5
2. Lantaneae Briq. (1897).
   Genera 6–18
3. Priveae Briq. (1897).
   Genera 19–21
4. Petreeae Briq. (1897).
   Genera 22–23
5. Casselieae (Schauer) Troncoso (1974).
   Genera 24–26
6. Citharexyleae Briq. (1897).
   Genera 27–33

Incertae sedis
Genus 34

Key to the Genera

1. Calyx corolla-like, often syncolorous with corolla, 5-lobed with lobes reticulately veined and extending beyond the corolla lobes; leaves asperous 2
– Calyx not brightly coloured, lobes not extending beyond corolla lobes; leaves generally smooth; if asperous then not with accompanying characters 3
2. Lianas, scandent shrubs or trees; calyx with "inner" calyx in the form of a coronet (0.4–3.5 mm high) situated at the inner base of the calyx lobes; anther thecae introrse with enlarged connective **22. *Petrea***
– Erect shrub or small tree; calyx without "inner" calyx; anther thecae parallel without enlarged connective **23. *Xolocotzia***
3. Ovary 1-carpellate, ovules 2 4
– Ovary 2- or 4-carpellate, ovules 4 or 8 18
4. Fruit a dry or fleshy schizocarp of two 1-seeded mericarps 5
– Fruit drupaceous, with one 2-locular, 2-seeded pyrene 16
5. Fertile stamens 4 6
– Fertile stamens 2 with 2 staminodes **14. *Stachytarpheta***
6. Fruit a fleshy schizocarp with thickened pericarp and mesocarp, fissure line between the 2 mericarps usually visible 7
– Fruit a dry schizocarp with thin dry pericarp and mesocarp, readily splitting at maturity into 2 mericarps 11
7. Inflorescence a long-pedunculate, axillary, (1)2–6-flowered raceme; posterior stamens with divergent thecae and with a thick, glandular connective forming a dorsal "crest" **24. *Casselia***
– Inflorescence a sessile or short-pedunculate, terminal or axillary spike or fascicle; posterior stamens with parallel thecae and without glandular connective 8
8. Corolla campanulate, 4-lobed; plants from Cuba and the Bahamas archipelago **9. *Nashia***
– Corolla hypocrateriform, 5-lobed; plants from Brazil, Paraguay, Bolivia, Argentina and Chile 9
9. Fruiting calyx enlarged and completely surrounding fruit 10
– Fruiting calyx not enlarged, forming a short cup around the fruit **31. *Baillonia***
10. Stamens inserted in middle of corolla tube, filamentous; stigma subglobose; tall shrubs with genistoid habit **17. *Diostea***
– Stamens inserted in upper part of corolla tube just below the throat, almost sessile; stigma bi-lobed with anterior lobe thick and papillose, and reduced posterior lobe; low shrubs forming dense thickets **18. *Lampaya***
11. Perennial creeping herb, rooting at nodes; indumentum of calyx of uncinate hairs, some species with medifixed hairs on stem, leaves and bracts **11. *Phyla***
– Herbs, subshrubs or shrubs; indumentum only of simple hairs 12
12. Plants of semi-deserts, xerophytic shrubs with small (up to 1.6 mm long), fleshy leaves **13. *Acantholippia***
– Plants, herbs or shrubs, not xerophytic, without reduced leaves 13
13. Calyx 2–4-lobed or dentate; corolla 4-lobed; style short, not more than 3 times longer than ovary, stigma oblique or lateral; fruit ovoid or subrotund, rounded at apex 14
– Calyx 5-lobed or dentate; corolla 5-lobed; style long, more than 3 times longer than ovary; stigma subbilobed with one lobe much reduced; fruit linear-oblong, beaked at apex 15
14. Calyx distinctly 4-toothed; teeth acute or subulate; corolla ± actinomorphic, with lobes ± equal; inflorescence a racemose spike, erect or pendant, flowers very shortly pedicellate **12. *Aloysia***
– Calyx obscurely 2- or 4-lobed; corolla weakly zygomorphic, more or less 2-lipped; inflorescence a compact head or spike with sessile flowers **10. *Lippia***
15. Shrubs or subshrubs; fruiting calyx somewhat inflated, completely enclosing fruit; mericarps with reticulate surface; Old World **16. *Chascanum***
– Perennial herbs; fruiting calyx not inflated, shorter than fruit; mericarps with smooth surface; New World **15. *Bouchea***
16. Glabrous shrubs with cylindrical stems; leaves reduced or absent; corolla 5-lobed, ± regular **7. *Neosparton***
– Shrubs or herbs, not glabrous with stems 4-sided; leaves normal and present; corolla 4-lobed 17
17. Corolla 4-lobed, ± regular, anterior lobe emarginate; stigma capitate; infloresence a lax racemiform spike; flowers minutely pedicellate **8. *Xeroaloysia***
– Corolla 4(–5)-lobed, 2-lipped; lobes obtuse or emarginate; stigma oblique or lateral; inflorescences pedunculate heads, dense spikes or sessile clusters; flowers sessile **6. *Lantana***
18. Fruit a dry schizocarp, maturing into four 1-seeded mericarps 19
– Fruit a dry or fleshy schizocarp separating at maturity into two 2-locular, 2-seeded mericarps or fruit a fleshy drupe with 1 pyrene with four 1-seeded locules or fruit a fleshy drupe with 2 or 4 pyrenes (or less by abortion), pyrenes 2-locular and 2-seeded 23
19. Fertile stamens 2; calyx slightly inflated in fruit **5. *Hierobotana***
– Fertile stamens 4; calyx not inflated in fruit 20
20. Shrubs, sometimes forming cushions only a few centimetres high; often with deep, thick, woody rootstock; leaves reduced; sometimes with spines 21

- Perennial or annual herbs, subshrubs, not cushion-forming; without deep, woody rootstock, without spines 22
21. Calyx with long, hygroscopic hairs which completely hide the corolla; fruit globose with smooth exocarp **4. *Urbania***
- Calyx without long hygroscopic hairs; fruit oblong with reticulate exocarp **3. *Junellia***
22. Inflorescence a simple spike or panicle of spikes, often branched; style short, not more than 3 times longer than ovary **1. *Verbena***
- Inflorescence a contracted head, unbranched; style long, more than 3 times longer than the ovary **2. *Glandularia***
23. Fruit a dry or fleshy schizocarp separating at maturity into two 2-locular, 2-seeded mericarps 24
- Fruit a fleshy drupe with 1 pyrene with four 1-seeded locules or fruit a fleshy drupe with 2 or 4 pyrenes (or less by abortion), pyrenes 2-locular and 2-seeded 30
24. Perennial herbs 25
- Trees or shrubs 26
25. Plant with deep tuberous rootstock; style recurved at apex, stigma lateral, papillose **20. *Pitraea***
- Plant without deep tuberous rootstock; style 2-lobed at apex with one lobe longer and stigmatiferous, and the other lobe reduced **19. *Priva***
26. Calyx enlarged in fruit, completely surrounding the fruit **25. *Parodianthus***
- Calyx not enlarged in fruit 27
27. Fruiting calyx caducous **28. *Rehdera***
- Fruiting calyx forming a cup 28
28. Corolla lobes ± regular; 4 fertile stamens and 1 staminode; fruit obovoid, attenuate at base **29. *Verbenoxylum***
- Corolla ± 2-lipped; 4 fertile stamens and no staminode; fruit ovoid or subglobose, not attenuate at base 29
29. Glabrous, subspinescent shrubs; leaves fleshy, arranged in fascicles, alternate on young stems; stigma capitate, suboblique; Argentina **21. *Dipyrena***
- Shrubs, not glabrous or subspinescent; leaves not fleshy, usually opposite-decussate; stigma shortly 2-lobed, anterior lobe larger and stigmatiferous; Madagascar, Somalia and Socotra **34. *Coelocarpum***
30. Perennial herbs or subshrubs; posterior stamens with glandular connective, no staminode present; fruit a fleshy drupe with 1 pyrene with four 1-seeded locules; fruit often horned or woody **26. *Tamonea***
- Shrubs or small trees; posterior stamens with connective not glandular, staminode often present; fruit a fleshy drupe with 2 or 4 pyrenes (or less by abortion), pyrenes 2-locular and 2-seeded; fruit not horned or woody 31
31. Stigma capitate, obscurely 4-lobed; fruit with four 2-locular, 2-seeded pyrenes **33. *Duranta***
- Stigma subcapitate or 2-lobed; fruit with two 2-locular, 2-seeded pyrenes 32
32. Inflorescence a reduced raceme of 1–2(5) flowers; stigma 2-lobed (hooked) with one lobe stigmatiferous **30. *Rhaphithamnus***
- Inflorescence a spike, many-flowered; stigma capitate, obscurely 2-lobed 33
33. Stamens ± sessile, included; stigma short, not more than twice as long as ovary, included; fruit drupaceous with usually two 2-locular pyrenes **27. *Citharexylum***
- Stamens filamentous, posterior pair exserted; style long, more than 2 times longer than ovary, exserted; fruit dry with usually one 2-locular pyrene **32. *Recordia***

1. Tribe Verbeneae Schauer p.p. (1847).

Fruit a dry schizocarp, separating at maturity into four 1-seeded mericarps; anther thecae parallel, or only slightly divergent, often with connective dilated and glandular; style 2-lobed, with anterior lobe stigmatiferous.

1. *Verbena* L. Fig. 58, 60B, C, H

Verbena L., Sp. Pl.:18 (1753); Perry, Ann. Missouri Bot. Gard. 20: 239–358 (1933), reg. rev.; Munir, J. Adelaide Bot. Gard. 20: 21–103 (2002), reg. rev.
Stylodon Raf. (1825).
Monopyrena Speg. (1897).

Annual or perennial herbs or subshrubs, not aromatic; leaves variously dentate or lobed; inflorescence terminal, rarely axillary, a simple spike or panicle of spikes; bracts small, inconspicuous; flowers sessile; calyx tubular, persistent, 5-lobed, corolla white, pink, blue, purple, infundibular or hypocrateriform, weakly zygomorphic, 5-lobed; stamens 4, inserted near middle of tube; posterior anthers sometimes with unappendaged dorsal connective; style deciduous, not more than 3 times longer than ovary; ovary 2-carpellate, carpels 2-locular, ovules 4; fruit a dry schizocarp splitting into four 1-seeded mericarps (the fruit of *V. caroliniana* does not split). $2n = 14$. Between 200 and 250 species in tropical and temperate America, with a few in Europe, Asia and N Africa.

2. *Glandularia* J.F. Gmel. Fig. 57B

Glandularia J.F. Gmel. in L., Syst. Nat. 13, 2 (2): 920 (1796); Turner, Lundellia 2: 55–71 (1999), reg.rev.; Umber, Syst. Bot. 4: 72–102 (1979), reg.rev.
Verbena sect. *Glandularia* (J.F. Gmel.) Schauer (1847).

Herbs or subshrubs, not aromatic; creeping or semi-erect; leaves variously lobed; inflorescence of dense or lax heads, axillary and terminal, bracts minute, inconspicuous; flowers sessile, calyx tubular, persistent, 5-lobed; corolla red, mauve, yellow or white, hypocrateriform, weakly zygomorphic, 5-lobed, tube straight, or slightly curved; stamens 4, inserted in the upper middle of tube, posterior anthers often with dorsal connective, with or without glandular appendage; style persistent or deciduous, more than 3× longer than ovary; ovary 2-carpellate, carpels 2-locular, ovules 4. $2n = 10$. About 100 species in temperate and subtropical America.

3. *Junellia* Moldenke Fig. 57A

Junellia Moldenke, nom. conserv. prop., Lilloa 5: 392 (1940); Botta, Darwiniana 25: 331–353 (1984), reg. rev.; 29: 371–396 (1989), rev.
Verbena sect. *Junellia* (Moldenke) Tronc. (1974).

Shrubs, sometimes low, cushion-forming, not aromatic; sometimes spiny; leaves entire or 3-lobed, homo- or dimorphous; inflorescence a bracteate, often contracted spike, terminal or axillary, many-flowered or reduced to 1 or 2 flowers; bracts small, inconspicuous; flowers sessile, calyx tubular, persistent, 5-lobed; corolla white, cream, pink, mauve, infundibular or hypocrateriform, zygomorphic, or subzygomorphic, 5-lobed; stamens 4, inserted at middle of tube, thecae sometimes with connective visible, with or without glandular appendage; style persistent or deciduous, with a broadened base; ovary 2-carpellate, carpels 2-locular, ovules 4; fruit oblongoid with reticulate exocarp, separating into four 1-seeded mericarps. $2n = 20$. About 47 species, distributed from Peru and Bolivia to Patagonia and Chile.

4. *Urbania* Phil.

Urbania R.A. Philippi, An. Mus. Nat. Chile 2 Secc. Bot. 8: 60. Tab. II, Fig. 7 (A–K) (1891); Botta, Bol. Soc. Argent. Bot. 25 (3–4): 477–483 (1988), rev.

Perennial, low, cushion-forming subshrubs, only a few centimetres high, with subterranean stems, not aromatic; leaves entire, reduced, fleshy, densely imbricate; inflorescence a 1–4-flowered spike, terminal; bracts conspicuous; flowers subsessile; calyx tubular, persistent, deeply 5-lobed, covered with long white hygroscopic hairs almost covering the corolla; corolla infundibular, weakly zygomorphic, limb 5-lobed, tube straight; stamens 4, inserted in the upper part of the tube; style deciduous; ovary 2-carpellate, carpels 2-locular, ovules 4; fruit globose with smooth exocarp. One species, *U. pappigera* Phil., confined to the high Andean puna of Argentina and Chile.

5. *Hierobotana* Briq.

Hierobotana Briquet in Engl. & Prantl, Natürl. Pflanzenfam. IV, 3a: 148 (1897).

Woody herb, not aromatic; procumbent, leaves 3-partite, lobed; inflorescence spicate, terminal; bracts minute, inconspicuous; flowers sessile; calyx tubular, 5-lobed, lobes subequal, inflated in fruit; corolla white tinged with purple, infundibular, subactinomorphic, 5-lobed, tube straight; stamens 2, inserted near middle of tube, thecae without appendage; style persistent; ovary 2-carpellate, carpels 2-locular, ovules 4. One species, *H. inflata* (Kunth) Briq., found in Ecuador, Colombia and Peru.

2. Tribe Lantaneae (Schauer) Briq. (1897).

Ovary 1-carpellate by abortion of the adaxial carpel, carpel 2-locular, ovules 2; fruit fleshy with one 2-locular, 2-seeded pyrene, or dry or subfleshy schizocarp separating at maturity into two 1-seeded mericarps.

6. *Lantana* L.

Lantana L., Sp. Pl. 2: 626 (1753).
Camara Plum. ex L. (1763).
Riedelia Cham. (1832).

Perennial shrubs or subshrubs (sometimes forming dense thickets or scrambling shrubs to several metres), often aromatic; stems sometimes armed with prickles; leaves entire or variously dentate; inflorescence of pedunculate heads or dense, cylindrical spikes or sessile clusters, axillary; bracts linear to triangular, sometimes forming an involucre, usually conspicuous; flowers sessile; calyx tubular, persistent, truncate or shallowly lobed; corolla red, yellow, white, mauve, pink with shades variously mixed in one inflorescence, hypocrateriform, unequally 4–5-lobed, lobes obtuse or emarginate, tube straight or curved, slightly enlarged above middle; stamens 4, inserted at middle of tube, thecae parallel, connective without appendages; style short, with capitate stigma, lateral or oblique. $2n = 22, 24, 27, 33, 44$ 55 and 66. About 150 species (Verdcourt 1992) mostly in tropical and subtropical America, few in Old World tropics.

7. *Neosparton* Griseb.

Neosparton Grisebach, Plant. Lorentz.: 197 (1874); Troncoso, Darwiniana 11: 163–192 (1957), rev.

Shrubs, glabrous; not aromatic; stems cylindrical-striate, erect; leaves reduced or absent, scale-like, entire, lanceolate, soon deciduous with petiole persistent; inflorescence of short spikes or globose heads, axillary and terminal; bracts minute, inconspicuous; flowers subsessile; calyx tubular, 5-lobed, persistent; corolla white, infundibular,

± actinomorphic, 5-lobed, tube slightly enlarged at apex; stamens 4, sometimes with a staminode, inserted just above middle of tube, posterior pair shortly exserted, thecae parallel, without appendage; style deciduous, stigma capitate, terminal or oblique; fruit drupaceous, with juicy mericarp. $2n = 32$. Four species, Argentina.

8. *Xeroaloysia* Tronc.

Xeroaloysia Troncoso, Darwiniana 12: 48–57 (1960).

Low, xerophytic subshrub, not aromatic; leaves dentate; inflorescence axillary, spicate, lax; bracts filiform, inconspicuous; flowers minutely pedicellate; calyx tubular, persistent, 4-lobed, lobes ± equal; corolla white, hypocrateriform, ± actinomorphic, 4-lobed, slightly spreading, tube straight; stamens 4, inserted at upper middle part of tube, thecae parallel, without appendage; style persistent, stigma capitate; fruit drupaceous with thin, fleshy mesocarp. One species, *X. ovatifolia* (Moldenke) Tronc., mountains and chaco of NW Argentina.

9. *Nashia* Millsp.

Nashia Millspaugh, Publ. Field Columbian Mus. Bot. Ser. 2: 176 (1906).

Erect, branched shrubs, aromatic; leaves entire, sometimes revolute, verticillate or fascicled; inflorescence an axillary fascicle; bracts small, ± conspicuous; flowers sessile; calyx short, annular; corolla white or greenish, campanulate, zygomorphic, 4-lobed, lobes subequal; stamens 4, thecae parallel; style deciduous or persistent, stigma capitate; fruit subfleshy. A genus of seven species from Cuba and the Bahamas.

10. *Lippia* L.

Lippia L., Sp. Pl. 2: 633 (1753).
Zappania Scop. (1786).
Dipterocalyx Cham. (1832).
Goniostachyum (Schauer) Small (1903).

Perennial herbs, shrubs, subshrubs, sometimes with xylopodium, often aromatic; leaves entire or variously dentate; inflorescence capitate or spicate, axillary or terminal; bracts ± conspicuous, sometimes enlarged and showy (sect. *Rhodolippia*), usually imbricate, sometimes forming an involucre, or seriate; flowers sessile; calyx tubular-campanulate, obscurely 2- or 4-lobed, persistent; corolla white, creamy-yellow, pink, magenta, often yellow at throat; infundibular, weakly zygomorphic, ± 2-lipped, 4-lobed, corolla tube straight or curved; stamens 4, inserted in the middle to upper part of tube, thecae parallel, connective with or without glandular appendage; stigma oblique or lateral; fruit a dry schizocarp. Published chromosome counts are unreliable because of the confused identity of the sampling. About 200 species in tropical and subtropical America, with many species in Africa (Verdcourt 1992).

11. *Phyla* Lour.

Phyla Loureiro, Fl. Cochinchensis: 66 (1790); Munir, J. Adelaide Bot. Gard. 15(2): 109–128 (1993), reg. rev.

Procumbent, creeping, perennial herbs, often forming dense mats, not aromatic, stems trailing, rooting at nodes; indumentum of medifixed hairs (in some species) on stem; leaves entire, dentate or serrate, often verticillate; inflorescence a dense, long-pedunculate spike, axillary; bracts ± conspicuous; flowers sessile; calyx short, tubular with indumentum of uncinate hairs, 2- or 4-lobed, persistent; corolla white or lavender to purple, hypocrateriform, with short tube, irregularly 4-lobed, tube straight; stamens 4, inserted at upper part of tube, thecae parallel, without glandular connectives; style short, stigma oblique; fruit a dry schizocarp. About 15 species, mainly New World, *P. nodiflora* (L.) Greene widespread in tropics, subtropics and warmer temperate regions.

12. *Aloysia* Palau

Aloysia Palau, Parte Pract. Bot. 1: 767 (1784); Botta, Darwiniana 22: 67–108 (1979), reg. rev.
Verbena L. sect. *Aloysioides* Walp. (1845).
Lippia L. sect. *Aloysia* Schauer (1847).

Shrubs or subshrubs, usually aromatic; leaves sometimes ternate or verticillate, usually resinose-glandular, petiolate; inflorescence a racemose spike, dense or lax, erect or pendant, axillary; bracts narrow, not conspicuous; flowers minutely pedicellate; calyx tubular, distinctly 4-dentate, teeth acute or subulate, persistent, enveloping fruit; corolla cream, white, pale yellow, hypocrateriform, ± actinomorphic, limb spreading, 4-lobed, tube straight; stamens 4, inserted just above middle of tube, posterior pair sometimes slightly exserted, thecae parallel; style short, stigma lateral; fruit a dry schizocarp. $2n = 36$. About 30 species in tropical and subtropical America.

13. *Acantholippia* Griseb.

Acantholippia Grisebach, Plant. Lorentz.: 196 (1874); Botta, Darwiniana 22: 511–532 (1980), rev.; Caro, Dominguesia 3: 1–31 (1982), rev.

Low or prostrate xerophytic shrubs, very aromatic, stems often spinose; leaves sometimes alternate, reduced (up to 1.6 mm long), fleshy, margin revolute, sessile; inflorescence of few-flowered heads, elongating slightly in fruit, axillary or terminal; bracts elliptic or ovate, not conspicuous; flowers sessile; calyx tubular, 4-lobed or crenate, persistent, enveloping fruit; corolla white, lilac, throat often yellow, infundibular, limb spreading, 4- or 5-lobed, tube short; stamens 4, inserted below throat, sometimes shortly exserted, thecae parallel; style short, stigma capitate; fruit a dry schizocarp. About six species confined to the arid regions of Chile and W Argentina.

14. *Stachytarpheta* Vahl Fig. 60D

Stachytarpheta Vahl, Enum. Plant. I: 205 (1804); Munir, J. Adelaide Bot. Gard. 14: 133–168 (1992), reg. rev.
Ubochea Baill. (1892).

Woody herbs or shrubs, occasionally aromatic; leaves simple, crenate or dentate; inflorescence a terminal spike, elongate, or short; bracts conspicuous, often becoming woody and reflexed after anthesis; flowers sessile or embedded in the inflorescence axis; calyx tubular, 2–5-lobed, persistent, enveloping fruit; corolla white, blue, violet, pink, red, to almost black, hypocrateriform or infundibular, tube straight or curved, 5-lobed; fertile stamens 2 (posterior), plus 2 staminodes, inserted at middle of tube, thecae of fertile stamens divergent; style persistent, stigma capitate; fruit a dry schizocarp. About 90 species widely distributed in tropical and subtropical America, with a few weedy species naturalised in tropical and subtropical Australia, Asia, Africa.

15. *Bouchea* Cham. Fig. 60F, M

Bouchea Chamisso, Linnaea 7: 252 (1832); Grenzebach, Ann. Missouri Bot. Gard. 13: 71–100 (1926), rev.; Moldenke, Repert. Spec. Nov. Regni Veg. 48: 16–29 (1940), rev.; 49: 91–139 (1940), rev.

Perennial herbs, often woody at base, non-aromatic; leaves simple, dentate; inflorescence a terminal spike, elongate; bracts filiform; flowers sessile or very shortly pedicellate; calyx narrow-tubular, persistent, not inflated or enclosing fruit, 4- or 5-lobed; corolla deep purple, pink, mauve, blue or white, infundibular, limb spreading, 5-lobed, tube straight or slightly curved; stamens 4, inserted in the upper part of the tube, thecae parallel, or slightly divergent at base; style persistent or deciduous, stigma oblique; fruit a dry schizocarp, sometimes winged at apex. $2n = 40$. About nine species, tropical and subtropical America: from S United States, Mexico and Caribbean to Brazil, Paraguay, Bolivia and N Argentina.

16. *Chascanum* E. Mey. Fig. 60L

Chascanum E. Meyer, Comm. Pl. Afr. Austr.: 275 (1838); Moldenke, Repert. Spec. Nov. Regni Veg. 45: 113–156, 301–319 (1938), rev.
Svensonia Moldenke (1936).

Low shrubs, often woody at base; not aromatic; leaves opposite or sometimes subopposite; inflorescence a terminal spike; bracts small, not conspicuous; flowers sessile or shortly pedicellate; calyx narrowly tubular, persistent, inflated and enclosing fruit, 5-lobed; corolla white, yellowish or pink, infundibular or hypocrateriform, 5-lobed, tube straight or curved; stamens 4, inserted in upper part of tube, thecae divergent; style deciduous or persistent, stigma oblique; fruit a dry schizocarp, sometimes winged at apex. About 27 species, mainly Africa, extending to Madagascar, Arabia and India.

17. *Diostea* Miers

Diostea Miers, Trans. Linn. Soc. 27 102 (1869).

Large shrub with genistoid habit, stems cylindrical, striate, hollow; leaves small, distant, sessile or shortly petiolate; inflorescence spiciform, terminal, lax; bracts filiform, inconspicuous; flowers sessile or subsessile; calyx tubular, 5-lobed, persistent in fruit; corolla white, hypocrateriform, zygomorphic, limb 5-lobed, corolla tube widening above the middle, slightly curved; stamens 4, sometimes with a staminode, inserted at middle of tube, thecae subparallel; style deciduous, gradually thickening towards apex, stigma subglobose, papillate; fruit a schizocarp. $2n = 20$. One species, *D. juncea* (Gillies ex Hook.) Miers, Argentina and Chile.

18. *Lampaya* Phil. ex Murillo

Lampaya R.A. Philippi, An. Mus. Nac. Chile: 58, tab. II, fig. 5 (1891).

Low shrubs, much branched with prostrate stems forming thickets, not aromatic; stems with thick bark; leaves imbricate, entire; inflorescence of short, terminal, 4–5-flowered spikes; bracts scale-like, not conspicuous; flowers sessile; calyx tubular, 5-lobed, unequal; corolla violet-blue, hypocrateriform, zygomorphic, limb hardly spreading, 5-lobed, tube straight, slightly narrowed in upper part; stamens 4, inserted at top of tube, just below throat, thecae parallel; style persistent, 2-lobed, anterior lobe thick, stigmatic, posterior lobe reduced; fruit a fleshy schizocarp. Two or three species from dry altiplano of Bolivia, Chile and Argentina.

3. Tribe Priveae Briq. (1895).

Fruit a dry or fleshy schizocarp separating at maturity into two 2-locular, 2-seeded mericarps, mericarp reticulate or sculptured; posterior stamens usually with enlarged glandular connective, or with filaments glandular.

19. *Priva* Adans. Fig. 59, 60A, J

Priva Adanson, Fam. des Plantes 2: 505 (1763); Kobuski, Ann. Missouri Bot. Gard. 13: 1–34 (1926), rev.; Moldenke, Repert. Spec. Nov. Regni Veg. 41: 1–76 (1937), rev.

Perennial herbs, sometimes woody at base, not aromatic; stems erect or decumbent, leaves opposite, or sometimes subopposite, dentate; inflorescence spicate, lax, elongate, axillary or terminal; bracts small, inconspicuous; flowers sessile or subsessile; calyx tubular or tubular-campanulate, sometimes urceolate, indumentum of uncinate hairs in some species, 5-lobed, persistent; corolla white or mauve, hypocrateriform or infundibular, ± zygomorphic, limb 5-lobed; stamens 4, inserted slightly above middle of tube, posterior pair sometimes with glandular connective, thecae divergent; style persistent or deciduous, 2-lobed, anterior larger and stigmatiferous, posterior reduced; ovary 2-carpellate, carpels 2-locular, ovules 4; fruit dorsally reticulate, verrucose or spiny. $2n = 12$. About 20 species in tropical and subtropical Asia, Africa and America.

20. *Pitraea* Turcz.

Pitraea Turczaninow, Bull. Soc. Imp. Nat. Moscou 35: 328 (1862); Moldenke, Phytologia 6: 232–241 (1958), rev.; Caro, Kurtzeana 1: 271–282 (1961), reg. rev.
Castelia Cav. (1801).

Perennial herbs with long roots and a deep, tuberous rootstock, not aromatic; indumentum of uncinate hairs on rhachis, pedicels, bracts and calyx; leaves dentate; inflorescence of lax to dense terminal spikes; bracts minute; flowers shortly pedicellate; calyx tubular, 5-dentate, persistent, enveloping fruit with folded and contorted apex; corolla white or mauve-white, hypocrateriform, zygomorphic, limb spreading, 5-lobed; stamens 4, inserted in the upper middle part of the tube, normally with anterior staminode present; thecae subparallel, connective of posterior anthers thickened and glandular; style deciduous, recurved at apex, stigma lateral, papillose; ovary 2-carpellate, carpels 2-locular, ovules 4; fruit dorsally reticulate. One species, *P. cuneato-ovata* (Cav.) Caro, in temperate regions of Peru, Chile (where it is also an invasive weed), Bolivia and Argentina.

21. *Dipyrena* Hook.

Dipyrena Hooker, Bot. Misc. 1: 355 (1829).

Glabrous, subspinescent, erect shrub, with rigid stems, not aromatic; leaves small, fleshy, entire, alternate on young shoots, and in fascicles; inflorescence spiciform, dense, terminal; bracts not conspicuous; flowers shortly pedicellate; calyx short, tubular-campanulate, 5-lobed; corolla yellowish-white, infundibular, zygomorphic, limb 5-lobed; tube long, narrow, broadening at apex, curved; stamens 4, inserted in upper part of tube, filaments of posterior stamens glandular at apex, thecae divergent; style deciduous, simple, stigma capitate, suboblique; ovary 2-carpellate, carpels 2-locular, ovules 4; fruit dorsally reticulate. One species, *D. glaberrima* (Gillies ex Hook.) Hook., Argentina.

4. Tribe Petreeae Briq. (1895).

Fruit (where known) a subfleshy drupe with 2 pyrenes (or 1 by abortion), pyrenes 1-locular and 1-seeded; anther connective dilated, filaments very short, fifth stamen often present as staminode; leaves asperous; calyx lobes longer than corolla.

22. *Petrea* L. Fig. 60G, K

Petrea L., Sp. Pl.: 626 (1753); Rueda, Ann. Missouri Bot. Gard. 81: 610–652 (1993), rev.

Lianas, scandent shrubs or trees; not aromatic; leaves opposite or whorled; inflorescence a raceme, usually elongate and many-flowered, axillary or terminal; bracts small or absent; flowers pedicellate; calyx corolla-like, tubular or campanulate, blue, violet, purple or white, ribbed, usually 5-lobed, lobes longer than tube, reticulately veined; also present is an "inner" calyx in the form of a narrow coronet (0.4–3.5 mm high) situated at the inner base of the calyx lobes, apex sinuate, lobed or fimbriate, its lobes alternate with calyx lobes; corolla blue, purple, white, hypocrateriform, ± actinomorphic, limb rotate, 5-lobed, tube sometimes urceolate or infundibular; stamens 4, sometimes with staminode present, inserted above the middle of tube, thecae introrse; style persistent or deciduous, obscurely 2-lobed, stigma oblique; ovary 1-carpellate, 2-locular, ovules 2; fruit drupaceous, 2-locular, 2-seeded or by abortion the fruit 1-locular and 1-seeded, enclosed within accrescent calyx, with "inner" calyx curving inwards, completely closing calyx mouth. $2n = 34$. About 11 species, tropical America. One species naturalised in Borneo and many cultivated as ornamentals.

23. *Xolocotzia* Miranda

Xolocotzia Miranda, Bol. Soc. Bot. Mexico 29:40 (1965).

Erect shrub or small tree, not aromatic; stem terete; inflorescence shortly racemose, terminal; bracts small, deciduous; flowers shortly pedicellate; calyx tubular, corolla-like, 5-lobed, extending beyond corolla; corolla white or blue, subhypocrateriform, subactinomorphic, limb 5-lobed; stamens 4, inserted at middle of tube, slightly exserted, thecae parallel, stigma subcapitate, slightly oblique; ovary (?) 1-carpellate, carpel 2-locular, ovules 2; fruit not known. One species, *X. asperifolia* Miranda, from Mexico.

5. Tribe Casselieae (Schauer) Tronc. (1974).

Inflorescence a few-flowered, axillary (sometimes terminal) raceme; style thickened and short, terminating in an oblique stigmatic lobe; fruits fleshy; false partition fused to wall of carpel only at the top and bottom.

24. *Casselia* Nees & Mart.

Casselia Nees & Martius, Nova Actorum Acad. Caes. Leop.-Carol 11: 73 (1823); Moldenke, Fedde Rep. 39: 129–153 (1936), rev.
Timotocia Moldenke (1936).

Perennial herbs or low subshrubs, frequently with woody rootstock, not aromatic; leaves entire, dentate or serrate; inflorescence axillary, solitary, 2–6-flowered; bracts filiform, not conspicuous; flowers short-pedicellate; calyx tubular-campanulate, 5-lobed; corolla lilac or pink, infundibular, slightly zygomorphic, limb 5-lobed, tube short; stamens 4, inserted at the base of the tube, included, anthers of posterior stamens with a thick, glandular dorsal connective forming a crest, thecae divergent; style deciduous, stigma oblique, papillose; ovary 1-carpellate, anterior carpel aborted, carpel 2-locular, ovules 2; fruit a fleshy schizocarp, separating at maturity into two 1-seeded mericarps. Eleven species, Brazil, Paraguay and Bolivia.

25. *Parodianthus* Tronc.

Parodianthus Troncoso, Darwiniana 5: 31 (1941).

Shrubs, erect or decumbent, sometimes stoloniferous, with rigid stems, not aromatic; leaves entire or spiny, opposite or sometimes ternate; inflorescence in axillary 2–4-flowered racemes; bracts filiform, not conspicuous; flowers short-pedicellate; calyx tubular, 5-lobed, persistent, becoming enlarged and papery and almost completely enclosing fruit; corolla mauve, hypocrateriform, subactinomorphic, 5-lobed, lobes glabrous, tube straight, stamens 4, inserted at upper middle part of the tube, anthers with connective dilated, thecae parallel; style persistent, stigma oblique; ovary 2-carpellate, each carpel with 2 locules, ovules 4, 1 per locule, subanatropous, fixed at upper part of locule, pendulous; fruit a fleshy schizocarp with two 2-locular, 2-seeded mericarps. $2n = 32$. Two species, Argentina.

26. *Tamonea* Aubl. Fig. 57C

Tamonea Aublet, Hist. Pl. Guiane 2: 659 (1775).
Ghinia Schreber (1789).

Perennial herbs and subshrubs, not aromatic; stems erect; leaves dentate; inflorescence spiciform, terminal or axillary, bracts filiform, inconspicuous; flowers short-pedicellate; calyx tubular, 5-lobed, corolla blue, violet or white, infundibular or hypocrateriform, tube straight, unequally 5(–4)-lobed, stamens 4, inserted near middle of tube, thecae parallel, posterior pair with glandular thickening of connective; style deciduous, stigma oblique; ovary 2-carpellate, carpels 2-locular, ovules 4, fruit drupaceous with thick, sometimes woody or horned exocarp. A neotropic genus of 6 or 7 species from Mexico and the Caribbean to Brazil and E Bolivia.

6. Tribe Citharexyleae Briq. (1895).

Fleshy drupe with 2 or 4 pyrenes (or fewer by abortion), pyrenes 2-locular and 2-seeded, or a subfleshy schizocarp separating into 2-locular, 2-seeded mericarps; anther connectives dilated; 4 fertile stamens with staminode often present.

27. *Citharexylum* L.

Citharexylum L., Sp. Pl.: 625 (1753).

Trees and shrubs, unarmed or with axillary spines, not aromatic; leaves opposite or sometimes subopposite or verticillate, rarely alternate, entire or dentate; inflorescence spicate, axillary or terminal, many-flowered, elongate, erect or pendant; bracts minute, not conspicuous; flowers sessile or shortly pedicellate; calyx campanulate, truncate, entire or 5-lobed, persistent in fruit, forming a cup; corolla white or yellow, hypocrateriform, somewhat zygomorphic, limb 5-lobed, rarely 4- or 6-lobed, tube straight; stamens 4, sometimes 5, or 4 fertile and 1 staminodial, subdidynamous, usually inserted above the middle of the tube, thecae parallel; style obscurely 2-lobed, or stigma subcapitate; ovary 2-carpellate, carpels 2-locular, ovules 4; fruit drupaceous, with fleshy mesocarp. About 130 species from the southern United States and Mexico through C America and Caribbean to Argentina and Uruguay.

28. *Rehdera* Moldenke

Rehdera Moldenke, Repert. Spec. Nov. Regni Veg. 39: 48 (1935).

Trees or shrubs, not aromatic, leaves opposite, occasionally subopposite, entire; inflorescence a few (3–12)-flowered raceme with shortened axis and distinct peduncle, axillary or terminal; bracts small, inconspicuous; flowers shortly pedicellate to sessile; calyx tubular, 5-lobed, lobes densely tomentose on internal face, early deciduous in fruit, after forming a cup which then splits; corolla white, hypocrateriform, slightly zygomorphic, 5-lobed, lobes densely tomentose on internal face; stamens 4, 1 staminode, thecae subparallel; style deciduous, slightly 2-lobed, stigma subcapitate; ovary 2-carpellate, each with 2 locules, ovules 4; fruit a schizocarp, dry with thin mesocarp. Three species, C America.

29. *Verbenoxylum* Tronc.

Verbenoxylum Troncoso, Darwiniana 16: 622 (1971).

Tree or shrub, not aromatic; leaves serrate; inflorescence a solitary terminal and axillary spike; bracts minute, deciduous; flowers pedicellate; calyx campanulate, subtruncate or obscurely 5-lobed, persistent, forming a cup; corolla white, infundibular, ± actinomorphic, 5-lobed, lobes spreading, rounded, ± equal; stamens 4, 1 staminode, fixed at lower half of tube, filaments broadening at apex, glandular; anther-thecae divergent; style persistent, obscurely 2-lobed, ovary 2-carpellate, carpels 2-locular, ovules 4; fruit a fleshy schizocarp. One species, *V. reitzii* (Moldenke) Tronc., Brazil.

30. *Rhaphithamnus* Miers

Rhaphithamnus Miers, Trans. Linn. Soc. 27: 95 (1869); Moldenke, Repert. Spec. Nov. Regni Veg. 42: 62–82 (1937), rev.

Shrubs or small trees, not aromatic; often stems with long, axillary spines; leaves opposite, sometimes ternate, entire or serrate; inflorescence reduced to 2–3 flowers, axillary; bracts minute, not conspicuous; flowers shortly pedicellate; calyx urceolate or campanulate, 5-lobed, persistent, almost completely enclosing the fruit; corolla purple, infundibular, zygomorphic, 5-lobed; fertile stamens 4, inserted at lower end to middle of tube, posterior staminode sometimes present, anthers with fleshy connective, thecae divergent; style persistent or deciduous, 2-lobed; ovary 2-carpellate, carpel 2-locular, ovules 4; fruit drupaceous. $2n = 36$. Two species confined to Chile, Chiloe and Juan Fernandez islands, and SW Argentina.

31. *Baillonia* Bocq. Fig. 60E, I

Baillonia Bocquillon, Adansonia 2: 251 (1861); Moldenke, Darwiniana 5: 166 (1941), rev.

Spiny shrub, not aromatic; leaves opposite, sometimes ternate or verticillate, linear; inflorescence spiciform, axillary and terminal, lax, slightly drooping; bracts minute, inconspicuous; flowers shortly pedicellate; calyx campanulate, 5-lobed, truncate or undulate, persistent, forming a cup; corolla white, hypocrateriform, zygomorphic, limb spreading, 5-lobed, tube straight; fertile stamens 4, inserted above middle of tube, staminode often present; thecae parallel; style persistent or deciduous, obscurely 2-lobed, or stigma capitate; ovary 2-carpellate, but usually 1 by abortion, carpels 1- or 2-locular (?), ovules 2; fruit fleshy. One species, *B. amabilis* Bocq., Brazil and Paraguay.

32. *Recordia* Moldenke

Recordia Moldenke, Phytologia 1: 99 (1934).

Tree or shrub, not aromatic; leaves simple, subentire to serrate; inflorescence spicate, terminal or axillary; bracts inconspicuous; flowers pedicellate; calyx tubular-campanulate, 5-lobed, persistent, completely enclosing fruit; corolla white, strongly scented, infundibular, zygomorphic, 5-lobed, lobes ± equal; stamens 4, 1 anterior staminode, fixed at middle of tube, posterior pair of stamens exserted, anthers dorsifixed, with enlarged connective; thecae divergent; style persistent, stigma obscurely 2-lobed; ovary (?) 2-carpellate, carpel 4-locular, (sometimes only 2 developing); fruit a drupe. One species, *R. boliviana* Moldenke, from Bolivia.

33. *Duranta* L.

Duranta Linnaeus, Sp. Pl. 2: 637 (1753); Caro, Rev. Argent. Agron. 23: 1–28 (1956), reg. rev.; Sanders, Sida 10: 308–318 (1984), rev.; Munir, J. Adelaide Bot. Gard. 16: 1–16 (1995), reg. rev.

Erect or subscandent, often spinose shrubs or small trees, not aromatic; leaves opposite or sometimes verticillate, entire, serrate or dentate; inflorescence an elongate spike, lax, axillary or terminal; flowers usually pedicellate; bracts small, inconspicuous; calyx tubular or campanulate, 5-lobed, persistent, enclosing fruit; corolla blue, purple or white, hypocrateriform, ± actinomorphic, limb obliquely 5-lobed; fertile stamens 4, inserted above middle of tube, thecae parallel, posterior staminode sometimes present; style deciduous or persistent, weakly 4-lobed, stigma capitate; ovary 4-carpellate, carpels 2-locular, ovules 8; fruit a drupe, with four 2-seeded pyrenes. $2n = 16$. About 20 species from tropical and subtropical America.

Incertae sedis

34. *Coelocarpum* Balf. f.

Coelocarpum Balfour f., Proc. Roy. Soc. Edinburgh 12: 90 (1884).

Subshrubs, not aromatic; leaves dentate or subentire; inflorescence spiciform, lax, terminal, or sometimes axillary, few(2–4)-flowered; bracts minute, inconspicuous; flowers shortly pedicellate; calyx campanulate-tubular, 5-lobed; corolla yellow or white, tinged pink or mauve, infundibular, zygomorphic, 5-lobed; stamens 4, inserted in the middle of tube, thecae divergent; style deciduous, 2-lobed, anterior lobe larger, stigmatiferous; ovary 2-carpellate, carpels 2-locular, ovules 4; fruit a drupaceous schizocarp splitting into two, 2-seeded mericarps. Five species from Madagascar, Somalia and Socotra.

Selected Bibliography

Bentham, G. 1876. Verbenaceae. In: Bentham, G., Hooker, J.D. Gen. Pl. 2: 1131–1160.

Botta, S., Brandham, P.E. 1993. The taxonomic significance of chromosome number in *Junellia* (Verbenaceae). Kew Bull. 48: 143–150.

Briquet, J. 1895. Verbenaceae. In: Engler and Prantl, Die natürlichen Pflanzenfamilien IV, 3a. Leipzig: W. Engelmann, pp. 132–182.

Cabrera, A.L. 1993. Flora de la Provincia de Jujuy, vol. 9. Buenos Aires: Librart.

Cantino, P.D. 1992a. Toward a phylogenetic classification of the Labiatae. In: Harley, R.M., Reynolds, T. (eds.) Advances in Labiate science. Kew: Royal Botanic Gardens, pp. 27–37.

Cantino, P.D. 1992b. Evidence for a polyphyletic origin of the Labiatae. Ann. Missouri Bot. Gard. 79: 361–379.

Cantino, P.D., Harley, R.M., Wagstaff, S.J. 1992. Genera of Labiatae: status and classification. In: Harley, R.M., Reynolds, T. (eds.) Advances in Labiate science. Kew: Royal Botanic Gardens, pp. 511–522.

Caro, J.A. 1982. Sistematizacion del genero *Acantholippia* Grisebach (Verbenaceae) y las especies de la Flora Argentina. Dominguezia 3: 1–31.

Cruden, R.W., Baker, K.K., Cullinan, T.E., Disbrow, K.A., Douglas, K.L., Erb, J.D., Kirsten, K.J., Malik, M.L., Turner, E.A., Weier, J.A., Wilmot, S.R. 1990. The mating systems and pollination biology of three species of *Verbena* (Verbenaceae). J. Iowa Acad. Sci. 97: 178–183.

Davis, G.L. 1966. Systematic embryology of the Angiosperms. New York: Wiley.

Drewes, S.I., Martínez, S. 1999. Morfologia de la inflorescencias en Verbenaceae-Verbenoideae: 2. Tribu Petreeae. (Morphology of inflorescences in Verbenaceae-Verbenoideae: 2. Tribe Petreeae.). Darwiniana 37: 209–218.

Estes, J.R., Brown, L.S. 1973. Entomophilous, intrafloral pollinaton in *Phyla incisa*. Am. J. Bot. 60: 228–230.

Grayer, R.J., Chase, M.W., Simmonds, M.S.J. 1999. A comparison between chemical and molecular characters for the determination of phylogenetic relationships among plant families: an appreciation of Hegnauer's Chemotaxonomie der Pflanzen. Biochem. Syst. Ecol. 27: 369–393.

Hegnauer, R. 1973. Chemotaxonomie der Pflanzen, vol. 6. Basel: Birkhäuser.

Hegnauer, R. 1990. Chemotaxonomie der Pflanzen, vol. 9. Basel: Birkhäuser.

Inamdar, J.A. 1969. Epidermal structure and ontogeny of stomata in some Verbenaceae. Ann. Bot. 33: 55–66.

Junell, S. 1934. Zur Gynaceummorphologie und Systematik der Verbenaceen und Labiaten. Symb. Bot. Upsal. 4: 1–219.
Knuth, P. 1909. Handbook of flower pollination. Translated by J.R. Ainsworth Davis. Oxford: Clarendon Press.
Maheshwari, J.K. 1954. Floral morphology and embryology of *Lippia nodiflora* Rich. Phytomorphology 4: 217–30.
Martínez, S, Múlgura de Romero, M.E. 1997. Yemas axilares multiples, morfologia y tipologia de la inflorescencia en *Duranta* (Verbenaceae-Citharexyleae). (Multiple axillary buds, morphology and typology of the inflorescence in *Duranta* (Verbenaceae-Citharexyleae).) Bol. Soc. Argent. Bot. 33: 113–122.
Martínez, S., Múlgura de Romero, M.E. 2003. The taxonomic position of *Parodianthus* (Verbenaceae). A morphological survey of the gynoecium and the inflorescence. Kew Bull. (in press).
Martínez, S., Botta, S., Múlgura, M.E. 1996. Morfologia de las inflorescencias en Verbenaceae-Verbenoideae: 1. Tribu Verbeneae. (Morphology of inflorescences in Verbenaceae-Verbenoideae: 1. Tribe Verbeneae.) Darwiniana 34: 1–17.
Mathur, G., Mohan Ram, H.Y. 1986. Floral biology and pollination of *Lantana camara*. Phytomorphology 36: 79–100.
Múlgura de Romero, M.E., Martínez, S., Suyama, A. 1998. Morfologia de las inflorescencias en *Lippia* (Verbenaceae). (Morphology of inflorescences in *Lippia* (Verbenaceae.) Darwiniana 36: 1–12.
Olmstead, R.G., Reeves, P.A. 1995. Evidence for the polyphyly of the Scrophulariaceae based on chloroplast *rbc*L and *ndh*F sequences. Ann. Missouri Bot. Gard. 82: 176–193.
Owens, S.J., Ubera-Jiménez, J.L. 1992. Breeding systems in Labiatae. In: Harley, R.M., Reynolds, T. (eds.) Advances in Labiate science. Kew: Royal Botanic Gardens, pp. 257–280.
Pal, N. 1951. Studies in the embryology of some Verbenaceae. J. Ind. Bot. Soc. 30: 59–74.
Raj, B. 1983. A contribution to the pollen morphology of Verbenaceae. Rev. Palaeobot. Palynol. 39: 343–422.
Rao, V.S. 1952. The floral anatomy of some Verbenaceae with special reference to the gynoecium. J. Ind. Bot. Soc. 31: 297–315.
Rudall, P.J., Clark, L. 1992. The megagametophyte in Labiatae. In: Harley, R.M., Reynolds, T. (eds.) Advances in Labiate science. Kew: Royal Botanic Gardens, pp. 65–84.
Ryding, O. 1995. Pericarp structure and phylogeny of the Lamiaceae-Verbenaceae complex. Plant Syst. Evol. 198: 101–141.
Sanders, R.W. 1987. Taxonomic significance of chromosome observations of Caribbean species of *Lantana* (Verbenaceae). Am. J. Bot. 74: 914–920.
Sanders, R.W. 2001. The genera of Verbenaceae in the southeastern United States. Harvard Pap. Bot. 5(2): 303–358.
Schauer, J.C. 1847. Verbenaceae. In: de Candolle, A.P. Prodr. 11: 522–700.
Spies, J.J. 1984. A cytotaxonomic study of *Lantana camara* (Verbenaceae) from South Africa. S. Afr. J. Bot. 3: 231–250.
Sun, B.Y., Stuessy, T.F., Humana, A.M., Riveros, G., Crawford, D.J. 1996. Evolution of *Rhaphithamnus venustus* (Verbenaceae), a gynodioecious hummingbird-pollinated endemic of the Juan Fernandez Islands, Chile. Pacific Sci. 50: 55–65.
Tatachar, T. 1940. The development of the embryo sac and formation of haustoria in *Lantana indica* Roxb. and *Stachytarpheta indica* Vahl. J. Ind. Bot. Soc. 19: 45–52.
ter Welle, B.J.H., Détienne, P. 1988. In: Jansen-Jacobs, M.J. Verbenaceae. In: Gorts-van Rijn, A.R.A. (ed.) Flora of the Guianas. Koenigstein: Koeltz.
Thorne, R.F. 1992. Classification and geography of the flowering plants. Bot. Rev. 58: 225–348.
Tomlinson, P.B., Fawcett, P. 1972. Dioecism in *Citharexylum* (Verbenaceae). J. Arnold Arbor. 53: 386–389.
Troncoso, N.S. 1961. Sobre dioecia en especies de *Lippia*. Bol. Soc. Arg. Bot. 9: 181–185.
Verdcourt, B. 1992. Verbenaceae. In: Polhill, R.M. (ed.) Flora of tropical East Africa. Rotterdam: A.A. Balkema.
Wagstaff, S.J., Olmstead, R.G. 1997. Phylogeny of Labiatae and Verbenaceae inferred from *rbc*L sequences. Syst. Bot. 22: 165–179.
Yeo, P.F. 1990. A re-definition of *Verbena brasiliensis*. Kew Bull. 45: 101–120.

Index to Scientific Names

References to main entries in **bold-faced** print, to illustrations in *italics*.